Cancer Imaging: Lung and Breast Carcinomas

Cancer Imaging: Lung and Breast Carcinomas

Volume 1

Edited by

M. A. Hayat
Distinguished Professor
Department of Biological Sciences
Kean University
Union, New Jersey

AMSTERDAM • BOSTON • HEIDELBERG • LONDON
NEW YORK • OXFORD • PARIS • SAN DIEGO
SAN FRANCISCO • SINGAPORE • SYDNEY • TOKYO
Academic Press is an imprint of Elsevier

Elsevier Academic Press
300 Corporate Drive, Suite 400, Burlington, MA 01803, USA
525 B Street, Suite 1900, San Diego, California 92101-4495, USA
84 Theobald's Road, London WC1X 8RR, UK

This book is printed on acid-free paper. ⊗

Library of Congress Cataloging-in-Publication Data
Cancer imaging : lung and breast carcinomas / editor, M.A. Hayat.
    p. ; cm.
    Includes bibliographical references and index.
    ISBN-13: 978-0-12-370468-9 (hardcover : alk. paper)
    ISBN-10: 0-12-370468-5 (hardcover : alk. paper) 1. Lungs—Cancer—Imaging.
2. Breast—Cancer—Imaging, I. Hayat, M. A., 1940-
    [DNLM: 1. Carcinoma, Non-Small-Cell Lung—diagnosis. 2. Breast Neoplasms—diagnosis.
3. Carcinoma, Lobular—diagnosis. 4. Diagnostic Imaging—methods. 5. Lung Neoplasms—
diagnosis. WF 658 C21442008]
    RC280.L8C34 2008
    616.99′4240754—dc22   2007002969

British Library Cataloguing in Publication Data
A catalogue record for this book is available from the British Library

ISBN: 978-0-12-370468-9

For all information on all Academic Press publications
visit our Web site at www.academicpress.com

Printed in China
07 08 09 10 11   9 8 7 6 5 4 3 2 1

To

The men and women involved in the odyssey of deciphering the complexity of
cancer initiation, progression, and metastasis, its diagnosis, cure, and hopefully its prevention.

Contents of Volume 1

Part I   Instrumentation

1.1   Strategies for Imaging Biology in Cancer and Other Diseases   3

Philipp Mayer-Kuckuk and Debabrata Banerjee

1.2   Synthesis of ^{18}F-fluoromisonidazole Tracer for Positron Emission Tomography   15

Ganghua Tang

1.3   Radiation Hormesis   23

Rekha D. Jhamnani and Mannudeep K. Kalra

Part II   General Imaging Applications

2.1   Molecular Imaging in Early Therapy Monitoring   29

Susanne Klutmann and Alexander Stahl

Part III   Lung Carcinoma

Part IV Breast Carcinoma

4.30 Leiomyoma of the Breast Parenchyma: Mammographic, Sonographic, and Histopathologic Features   567

Aysin Pourbagher and M. Ali Pourbagher

4.31 Detection of Breast Cancer: Dynamic Infrared Imaging   571

Terry M. Button

4.32 Phyllodes Breast Tumors: Magnetic Resonance Imaging   581

Aimée B. Herzog and Susanne Wurdinger

Contents of Volume 2

Contributors

Abe, Yoshiyuki
Tokorozawa PET Diagnostic Imaging Clinic
Higashisumiyoshi 7-5
Tokorozawa City
359-1124 Saitama Prefecture
Japan

Alavi, Abbas
Division of Nuclear Medicine
Department of Radiology
Hospital of the University of Pennsylvania
110 Donner Building
3400 Spruce Street
Philadelphia, PA 19104

Altomari, Fiorella
Department of Radiological Sciences
University of Rome "La Sapienza"
V.le Regina Elena, 324
00161 Rome
Italy

Antoch, Gerald
Department of Diagnostic and Interventional Radiology
and Neuro-Radiology
University Hospital Essen
Hufelanstrasse 55
45122 Essen
Germany

Arai, Yasuaki
Division of Diagnostic Radiology and Nuclear Medicine
National Cancer Center Hospital
5-1-1, Tsukiji Chuo-ku
104-0045 Tokyo
Japan

Arnold, Johannes F. T.
Department of Physics
EP 5 (Biophysics)
University of Wuerzburg
Am Hubland
D-97074 Wuerzburg
Germany

Aznar, Marianne C.
Department of Radiation Physics
Copenhagen University Hospital
Blegdamsvej 9
2100 Copenhagen
Denmark

Banerjee, Debabrata
Medicine and Pharmacology
Cancer Institute of New Jersey, RWJMS/UMDNJ
195 Little Albany Road
New Brunswick, NJ 08903

Berland, Lincoln L.
Department of Radiology
UAB Hospital
619 South 19th Street
University of Alabama
Birmingham, AL 35249

Berland, Nancy W.
Department of Radiology
UAB Hospital
619 South 19th Street
University of Alabama
Birmingham, AL 35249

Bockisch, Andreas
    Department of Diagnostic and Interventional Radiology
    and Neuro-Radiology
    University Hospital Essen
    Hufelanstrasse 55
    45122 Essen
    Germany

Boers, Maarten
    Department of Clinical Epidemiology and Biostatics
    VU University Medical Center, PK 6Z 185
    P.O. Box 7057
    1007 MB Amsterdam
    The Netherlands

Boetes, Carla
    430 Department of Radiology
    University Medical Center
    Nijmegen P.O. Box 9101
    6500 HB Nijmegen
    The Netherlands

Bolan, Patrick J.
    Department of Radiology
    Center for Magnetic Resonance Research
    University of Minnesota
    2021 6th Street SE
    Minneapolis, MN 55455

Brekelmans, Cecile T. M.
    Department of Medical Oncology
    Erasmus Medical Center
    Daniel den Hoed Cancer Center
    Groene Hilledijk 301
    3075 EA Rotterdam
    The Netherlands

Bremer, Christoph
    Department of Clinical Radiology
    University Hospital Münster
    Albert-Schweitzer-Str. 33,
    D-48129 Münster
    Germany

Button, Terry M.
    Department of Radiology
    State University of New York
    Stony Brook, NY 11794

Carotenuto, Luigi
    Department of Radiological Sciences
    University of Rome "La Sapienza"
    V.le Regina Elena, 324
    00161 Rome
    Italy

Catalano, Carlo
    Department of Radiological Sciences
    University of Rome "La Sapienza"
    V.le Regina Elena, 324
    00161 Rome
    Italy

Cham, Matthew D.
    Department of Radiology
    New York Presbyterian Hospital
    Weill Cornell Medical Center
    Weill Medical College
    Cornell University
    525 East 68th Street
    New York, NY 10021

Degenhard, Andreas
    Applied Neuroinformatics Group
    Faculty of Technology
    Bielefeld University
    P.O. Box 10031
    D-33501 Bielefeld
    Germany

del Cura, José Luis
    Servicio de Radiodiagnostico
    Hospital de Basurto
    Avenida Montevideo 18
    48013 Bilbao
    Spain

Demirkan, Hatice Mirac Binnaz
    Division of Medical Oncology
    Department of Internal Medicine
    Dokuz Eylul University
    Medical School
    Inciralti
    35340 Izmir
    Turkey

Di Martino, Ercole
Ear, Nose & Throat Hospital at
Diakonissenkrankenhaus Bremen
Gröpelinger Heerstraße 406–408
D-28239 Bremen
Germany

Domann, Frederick E.
Associate Professor of Radiation Oncology
Free Radical & Radiation Biology Program
The University of Iowa
Iowa City, IA 52242

Durak, Hatice
Department of Nuclear Medicine
Dokuz Eylul University
Medical School
Inciralti
35340 Izmir
Turkey

Elmore, Joann G.
General Internal Medicine
Harborview Medical Center
325 Ninth Avenue
Box 359780
Seattle, WA 98104

Eubank, William B.
Department of Radiology
Puget Sound Veterans Affairs
Health Care System
1660 South Columbian Way
S-113-RAD
Seattle, WA 98108

Fantini, Sergio
Tufts University
Department of Biomedical Engineering
4 Colby Street
Medford, MA 02155

Fischmann, Arne
Department of Medical Radiology
University Hospital Basel
Petersgraben 4
4031 Basel
Switzerland

Flentje, Michael
Department of Physics, EP5
Uni Wuerzburg
Am Hubland
97074 Wuerzburg
Germany

Fok, S. C.
Department of Mechanical Engineering
The Petroleum Institute
Abu Dhabi
United Arab Emirates

Ganschow, Pamela S.
Rush University Medical College
Chicago, IL 60612
and
Cook County Hospital
Chicago, IL 60612

Garwood, Michael
Department of Radiology
Center for Magnetic Resonance Research
University of Minnesota
2021 6th Street SE
Minneapolis, MN 55455

Gido, Tomonori
R & D Center
Konica Minolta Medical & Graphic. Inc.
2970 Ishikawa-machi
Hachioji-shi
192-8505 Tokyo
Japan

Gilhuijs, Kenneth G. A.
Department of Radiology
Netherlands Cancer Institute
Antoni van Leeuwenhoek Hospital
Plesmanlaan 121
1066 CX Amsterdam
The Netherlands

Hemdal, Bengt Å.
Department of Radiation Physics
Malmö University Hospital
SE-205 02 Malmö
Sweden

Henschke, Claudia I.
  Department of Radiology
  New York Presbyterian Hospital
  Weill Cornell Medical Center
  Weill Medical College
  Cornell University
  525 East 68th Street
  New York, NY 10021

Hermes, Kay E.
  Clifton F. Mountain Foundation for
  Research & Education in Lung Cancer
  2028 Albans Road
  Houston, TX 77005

Herzog, Aimée B.
  Department of Gynaecologic Radiology
  Institute for Diagnostic and Interventional Radiology
  Friedrich-Schiller-University
  Bachstrasse 18
  D-07740 Jena
  Germany

Hicks, Rodney J.
  Center for Molecular Imaging
  The Peter MacCallum Cancer Center
  12 Cathedral Place
  East Melbourne, Victoria 3002
  Australia

Hoekstra, Otto S.
  Department of Nuclear Medicine
  VU University Medical Center
  De Boelelaan 1117
  1081 HV Amsterdam
  The Netherlands

Honda, Chika
  R & D Center
  Konica Minolta Medical & Graphic, Inc.
  2970 Ishikawa-machi
  Hachioji-shi
  192-8505 Tokyo
  Japan

Huang, Steve S.
  Division of Nuclear Medicine
  Department of Radiology
  Hospital of the University of Pennsylvania
  110 Donner Building
  3400 Spruce Street
  Philadelphia, PA 19104

Huang, Wei
  Department of Medical Physics
  Memorial Sloan-Kettering Cancer Center
  1275 York Avenue
  New York, NY 10021

Ishida, Jiro
  Tokorozawa PET Diagnostic
  Imaging Clinic
  Higashisumiyoshi 7-5
  Tokorozawa City
  359-1124 Saitama Prefecture
  Japan

Isomoto, Ichiro
  Department of Radiological Sciences
  Unit of Translational Medicine
  Nagasaki University Graduate School
  of Biomedical Sciences
  1-12-4 Sakamoto, Nagasaki City
  852-8523 Nagasaki
  Japan

Jakob, Peter M.
  Department of Physics, EP 5
  Uni Wuerzburg
  Am Hubland
  97074 Wuerzburg
  Germany

Jasanoff, Alan
  Department of Nuclear Science & Engineering
  & Brain & Cognitive Sciences
  Francis Bitter Magnet Laboratory
  Department of Brain & Cognitive Sciences
  Massachusetts Institute of Technology
  77 Massachusetts Avenue
  Cambridge, MA 02139

Jhamnani, Rekha D.
  Department of Radiology
  Emory University School of Medicine
  1364 Clifton Road NE
  Atlanta, GA 30322

Kalra, Mannudeep K.
  Department of Radiology
  Emory University School of Medicine
  1364 Clifton Road NE
  Atlanta, GA 30322

Kalra, Naveen
Department of Radiodiagnosis
Postgraduate Institute of Medical Education and
Research, PGIMER
Sector 12, Chandigarh
India

Kim, Tae Sung
The Department of Radiology and
Center for Imaging Science
Samsung Medical Center
Sungkyunkwan University School of Medicine
50 Ilwon-dong, Gangnam-gu
135-710 Seoul
Republic of Korea

Kimura, Tomoki
Department of Radiology
School of Medicine
Kagawa University
1750-1 Ikenobe
Miki-cho, Kita-gun
761 0793 Kagawa
Japan

Kishi, Kazuma
Toranomon Hospital
Department of Clinical Oncology
2-2-2 Toranomon Hospital
Minato-ku
105-8470 Tokyo
Japan

Klijn, Jan G. M.
Department of Medical Oncology
Erasmus Medical Center
Daniel den Hoed Cancer Center
Groene Hilledijk 301
3075 EA Rotterdam
The Netherlands

Klutmann, Susanne
Department of Nuclear Medicine
University Hospital Hamburg-Eppendorf
Martinistrasse 52
D-20246 Hamburg
Germany

Knoll, Peter
Wilhelminenspital
Department of Nuclear Medicine
Montleartstr. 37
A-1171 Vienna
Austria

Kotas, Markus
Department of Radiation Oncology
Uni Wuerzburg
Am Hubland
97074 Wuerzburg
Germany

Kriege, Mieke
Department of Medical Oncology
Erasmus Medical Center
Daniel den Hoed Cancer Center
Groene Hilledijk 301 3075 EA
Rotterdam
The Netherlands

Kurosaki, Atsuko
Toranomon Hospital
Department of Diagnostic Radiology
2-2-2 Toranomon Hospital
Minato-ku
105-8470 Tokyo
Japan

Lee, Kyung Soo
Department of Radiology
Samsung Medical Center
Sungkyunkwan University School of Medicine
50 Ilwon-dong
Kangnam-ku
135-710 Seoul
Korea

Lewin, John M.
Diversified Radiology of Colorodo, P.C.
938 Bannock Street, Suite 300
Denver, CO 80204

Machida, Kikuo
Tokorozawa PET Diagnostic Imaging Clinic
Higashisumiyoshi 7-5
Tokorozawa City
359-1124 Saitama Prefecture
Japan

Madeddu, Giuseppe
Department of Nuclear Medicine
University of Sassari
Viale San Pietro 8
07100 Sassari
Italy

Maeda, Testuo
Division of Diagnostic Radiology and Nuclear Medicine
National Cancer Center Hospital
5-1-1, Tsukiji, Chuo-ku
104-0045 Tokyo
Japan

Mann, Ritse M.
Department of Radiology (667)
Radboud University Nijmegen Medical Center
Geertgrooteplein 10
P.O. Box 9101
6500 HB Nijmegen
The Netherlands

Martin, Colin J.
Health Physics
Department of Clinical Physics and Bio-Engineering
West House (Ground Floor)
Gartnavel Royal Hospital
1055 Great Western Road
G12 OXH Glasgow
Scotland
United Kingdom

Matthews, Suzanne
Department of Diagnostic Imaging
Northern General Hospital
Herries Road
S5 7AU Sheffield
United Kingdom

Matsuo, Satoru
Shiga University of Medical Science
Department of Radiology
Tsukinowa-cho, Seta, Otsu
520-2192 Shiga
Japan

Mayer-Kuckuk, Philipp
Hospital for Special Surgery
535E. 70th Street
New York, NY 10021

Meisamy, Sina
Department of Radiology
Center for Magnetic Resonance Research
University of Minnesota
2021 6th Street SE
Minneapolis, MN 55455

Michaelson, James S.
Department of Pathology
Harvard Medical School
Division of Surgical Oncology
Yawkey Building 7th Floor
Massachusetts General Hospital
55 Fruit Street
Boston, MA 02114

Miller, Janet C.
Center for Biomarkers in Imaging
Department of Radiology
Massachusetts General Hospital
175 Cambridge Street
Mail Code CPZ 175-200
Boston, MA 02114

Milne, Eric N. C.
Department of Clinical Research
Imaging Diagnostic Systems
6531 NW 18th Court
Plantation, FL 33313

Mirzaei, Siroos
Wilhelminenspital
Department of Nuclear Medicine
Montleartstr. 37
A-1171 Vienna
Austria

Morcos, Sameh K.
Department of Diagnostic Imaging
Northern General Hospital
Herries Road
S5 7AU Sheffield
United Kingdom

Morishita, Mariko
Department of Molecular Medicine
Atomic Bomb Disease Institute
Nagasaki University Graduate School
of Biomedical Sciences
1-12-4 Sakamoto, Nagasaki City
852-8523 Nagasaki
Japan

Mostbeck, Gerhard
Otto Wagner Hospital
Department of Radiology
Sanatoriumsstrasse 2
1140 Vienna
Austria

Mountain, Clifton F.
Division of Cardiothoracic Surgery
The University of California at San Diego
1150 Silverado Dr., Suite 110
La Jolla, CA 92037

Nagata, Masayoshi
Iruma Heart Hospital
Koyata 1258-1
Iruma City
358-0026 Saitama
Japan

Nagata, Yasushi
Department of Therapeutic Radiology and Oncology
Kyoto University Graduate School of Medicine
606-8501 Kyoto
Japan

Nakamura, Masato
Department of Pathology
Tokai University School of Medicine
Namiki 3-1
Tokorozawa City
359-8513 Saitama
Japan

Nambu, Atsushi
Department of Radiology
University of Yamanashi
1110 Shimokato Tamaho-cho Nakakoma-gun
409-3898 Yamanashi
Japan

Nattkemper, Tim W.
Applied Neuroinformatics Group
Faculty of Technology
Bielefeld University
P.O. Box 10031
D-33501 Bielefeld
Germany

Ng, E.Y-K
School of Mechanical & Aerospace Engineering
College of Engineering
Nanyang Technological University
50 Nanyang Avenue
Singapore 639798

Niu, Gang
Department of Radiology
Molecular Imaging Program at Stanford
Stanford University
1201 Welch Road
Stanford, CA 94041

Ogata, Toshiro
Department of Surgery II
National Defense Medical College
Namiki 3-2
Tokorozawa 359-8513
359-1124 Saitama
Japan

Ohtsuru, Akira
Department of Molecular Medicine
Atomic Bomb Disease Institute
Nagasaki University Graduate School of Biomedical
Sciences
1-12-4 Sakamoto, Nagasaki City
852-8523 Nagasaki
Japan

OJili, Vijaynath
Department of Radiology
Emory University School of Medicine
1364 Clifton Road NE
Atlanta, GA 30322

Onishi, Hiroshi
Department of Radiation Oncology
University of Yamanashi
1110 Shimokato Tamaho-cho Nakakoma-gun
409-3898 Yamanashi
Japan

Ozeki, Yuichi
Department of Surgery II
National Defense Medical College
Namiki 3-1
Tokorozawa City
359-8513 Saitama
Japan

Padula, Simona
Department of Radiological Sciences
University of Rome "La Sapienza"
V.le Regina Elena, 324
00161 Rome
Italy

Partridge, Savannah C.
Department of Radiology
University of Washington
Seattle Cancer Care Alliance
825 Eastlake Avenue East
G3-200, P.O. Box 19023
Seattle, WA 98109

Passariello, Roberto
Department of Radiological Sciences
University of Rome "La Sapienza"
V.le Regina Elena, 324
00161 Rome
Italy

Pediconi, Federica
Department of Radiological Sciences
University of Rome "La Sapienza"
V.le Regina Elena, 324
00161 Rome
Italy

Perk, Lars R.
Section of Tumor Biology
Department of Otolaryngology/
Head and Neck Surgery
Vrije Universiteit
University Medical Center
De Boelelaan 1117
1081 HV Amsterdam
The Netherlands

Pien, Homer H.
Center for Biomarkers in Imaging
Radiology Department
Massachusetts General Hospital
149 13th Street, Bldg 36, Rm. 2301
Charlestown, MA 02129

Pourbagher, Aysin
Department of Radiology
Baskent University and Medical Research Center
Dadaloglu Mah 39 sk
Yuregir, Adana 01250
Turkey

Pourbagher, M. Ali
Department of Radiology
Baskent University and Medical Research Center
Dadaloglu Mah 39 sk
Yuregir, Adana 01250
Turkey

Pracht, Eberhard D.
Department of Physics, EP 5
Uni Wuerzburg
Am Hubland
97074 Wuerzburg
Germany

Prosch, Helmut
Otto Wagner Hospital
Department of Radiology
Sanatoriumsstrasse 2
1140 Vienna
Austria

Richardson, Julie
Moor Instruments, Ltd.
Millwey, Axminster
13 5HU Devon, Essex
United Kingdom

Sakata, Ikuko
Tokorozawa PET Diagnostic Imaging Clinic
Higashisumiyoshi 7-5
Tokorozawa City
359-1124 Saitama Prefecture
Japan

Schillaci, Orazio
Department of Biopathology and Diagnostic Imaging
University Tor Vergata
00195 Rome
Italy

Schober, Otmar
Klinik und Polikllinik für Nuklearmedizin
Universitätsklinikum Münster
Albert-Schweitzer-Str. 33
D-48149 Münster
Germany

Scholbach, Jakob
University of Leizpig
Faculty of Mathematics and Computer Science
Augustusplatz 10/11
D-04109 Leipzig
Germany

Scholbach, Thomas
"St. Georg" Municipal Children's Hospital
Delitzscher Str. 144
D-04129 Leipzig
Germany

Seirberlich, Nicole
Department of Physics
EP 5 (Biophysics)
University of Wuerzburg
Am Hubland
D-97074 Wuerzburg
Germany

Sorensen, A. Gregory
Center for Biomarkers in Imaging
Department of Radiology
Massachusetts General Hospital
Boston, MA 02114

Spanu, Angela
Department of Nuclear Medicine
University of Sassari
Viale San Pietro 8
07100 Sassari
Italy

Stahl, Alexander
Department of Nuclear Medicine
Technical University of Munich
Klinikum rechts der Isar
Ismaninger Strasse 22
81675 München
Germany

Sutton, David G.
Medical Physics Department
Ninewells Hospital and Medical School
DD1 9SY Dundee
Scotland
United Kingdom

Tamura, Katsumi
Tokorozawa PET Diagnostic Imaging Clinic
Higashisumiyoshi 7-5
Tokorozawa City
359-1124 Saitama Prefecture
Japan

Tanaka, Toyohiko
Shiga University of Medical Science
Department of Radiology
520-2192 Shiga
Japan

Tang, Ganghua
PET Center and Nuclear Medicine Department
Nang Fang Hospital
Southern Medical University
51055 Guangzhou
China

Taroni, Paola
INFM-Dipartimento di Fisica and IFN-CNR
Politecnico di Milano
Piazza Leonardo da Vinci 32
I-20133 Milano
Italy

Tateishi, Ukihide
Division of Diagnostic Radiology and Nuclear Medicine
National Cancer Center Hospital
5-1-1, Tsukiji, Chuo-ku
104-0045 Tokyo
Japan

Tudorica, Luminita A.
Department of Radiology
State University of New York
Stony Brook, NY 11794

Twellmann, Thorsten
Department of Biomedical Engineering
Eindhoven University of Technology
P. O. Box 513
NL-5600 MB Eindhoven
The Netherlands

Uyl-de Groot, Carin A.
Institute for Medical Technology Assessment iMTA
Erasmus MC (University Medical Centre Rotterdam)
P. O. Box 1738
3000 DR Rotterdam
The Netherlands

Vallone, Paolo
Department of Diagnostic and Interventional Radiology
National Cancer Institute
80131 Naples
Italy

van Dongen, Guus A. M. S.
Department of Otolaryngology/Head and Neck Surgery
Vrije University Medical Center
De Boelelaan 1117
1081 HV Amsterdam
The Netherlands

van Tinteren, Harm
Comprehensive Cancer Center Amsterdam
Plesmanlaan 125
1066 CX Amsterdam
The Netherlands

Visser, Gerard W. M.
  Nuclear Medicine & PET Research
  Vrije Universiteit University Medical Center
  De Boelelaan 1117, P. O. Box 7057
  1007 MB Amsterdam
  The Netherlands

Wall, Alexander
  Department of Clinical Radiology
  University Hospital Muenster
  Albert-Schweitzer-Str.33
  D-48129 Münster
  Germany

Ware, Robert E.
  The Center for Molecular Imaging
  The Peter MacCallum Cancer Center
  12 Cathedral Place
  East Melbourne, Victoria 3002
  Australia

Weckesser, Matthias
  Klinik und Polikllinik für Nuklearmedizin
  Universitätsklinikum Münster
  Albert-Schweitzer-Str. 33
  D-48149 Münster
  Germany

Wurdinger, Susanne
  Department of Gynaecologic Radiology
  Institute for Diagnostic and Interventional Radiology
  Friedrich-Schiller-University
  Bachstrasse 18
  D-07740 Jena
  Germany

Yamashita, Shunichi
  Department of Molecular Medicine
  Atomic Bomb Disease Institute
  Nagasaki University Graduate School of Biomedical
  Sciences
  1-12-4 Sakamoto, Nagasaki City
  852-8523 Nagasaki
  Japan

Yankelevitz, David F.
  Department of Radiology
  New York Presbyterian Hospital
  Weill Cornell Medical Center
  Weill Medical College
  Cornell University
  525 East 68th Street
  New York, NY 10021

Yip, Rowena
  Department of Radiology
  New York Presbyterian Hospital
  Weill Cornell Medical Center
  Weill Medical College
  Cornell University
  525 East 68th Street
  New York, NY 10021

Zheng, Bin
  Imaging Research Center
  Department of Radiology
  University of Pittsburgh, School of Medicine
  300 Halket Street, Suite 4200
  Pittsburgh, PA 15213

Preface

The primary objective of this series, CANCER IMAGING, is to present the readers with the most up-to-date instrumentation, general applications, as well as specific applications of imaging technology to diagnose various cancers. The present work concentrates on the application of this technology to the diagnosis of lung and breast carcinomas, the two major worldwide cancers.

In this work we discuss strategies for imaging cancer and clinical applications of this technology and explain the role of molecular imaging in early therapy monitoring. In addition, we cover the following topics: synthesis and use of different contrast agents, especially the most commonly used tracer, 18F-fluorodeoxyglucose (FDG), in conjunction with imaging modalities (e.g., magnetic resonance imaging [MRI]); the advantages and limitations of the tracers; and rational and protocol details of whole-body screening with computed tomography (CT), positron emission tomography (PET), and PET/CT. We also describe the role of imaging in drug development, gene therapy, and therapy with monoclonal antibodies; for example, we describe the advantages of preclinical immuno-PET. The advantages of hybrid modalities, such as PET/CT (e.g., in lung cancer), are also presented.

The importance of the use of imaging for clinical diagnosis is presented in detail, and the relationship between radiation dose and image quality is discussed. In addition, we present detailed methods for absorbed X-ray dose measurement in mammography to avoid the risk of radiation-induced carcinogenesis.

We also detail lung cancer screening, staging, and diagnosing, applying different imaging modalities and point out false-negative and false-positive images potentially encountered in various body parts when imaged. Imaging modalities, including high-resolution CT, thin-section CT, computer-aided diagnosis with CT, low-dose helical CT, and FDG-PET/CT, used for staging and diagnosing lung cancer (e.g., non-small cell lung carcinoma), are discussed.

Details of a large number of imaging modalities, including optical mammography, digital mammography, contrast-enhanced MRI, SPECT, color-Doppler sonography, and sonography, used for diagnosing breast cancer are presented. Other topics include use of imaging for diagnosing invasive lobular carcinoma, density of breast carcinoma, axillary lymph node status in breast cancer, small-size primary breast cancer, and microcalcification in breast lesions. The role of Doppler sonography in differentiating benign and malignant solid breast tumors is discussed, as are breast scintigraphy and magnetic resonance spectroscopy of breast cancer. Use of MRI for monitoring response to breast cancer treatment is explained. Other methods such as dynamic infrared imaging and thermography are also presented.

This work consists of 61 chapters and has been developed through the efforts of 132 authors and coauthors representing 17 countries. The high quality of each manuscript made my work as the editor an easy one. Strictly uniform style of manuscript writing has been accomplished. The results are presented in the form of both black-and-white and color images that are appropriately labeled.

I am indebted to the contributors for their promptness in accepting my suggestions, and appreciate their dedication and hard work in sharing their knowledge with the readers. Each chapter provides unique individual, practical knowledge based on the expertise of the authors. The chapters contain the most up-to-date information, and it is my hope that the book will be published expeditiously.

I am thankful to the Board of Trustees and Dr. Dawood Farahi for recognizing the importance of scholarship in an institution of higher education, and for providing resources for completing this project. I appreciate the cooperation shown by Phil Carpenter (developmental editor), and I am grateful to Betsy Mathew for her help in preparing this volume.

September 2006
M. A. Hayat

Selected Glossary

ABI: Analyzer-based imaging

Ablation: Consists of the removal of a body part or the destruction of its function.

ACF: Autocorrelation function

ACS: Autocalibrating signals

ADC: Apparent diffusion coefficient; a measure of the mean-square displacement of an ensemble of molecules within a unit of time.

Adenocarcinoma: A malignant neoplasm of epithelial cells in a glandular or glandlike pattern.

Adenoma: A benign epithelial neoplasm in which the tumor cells form glands or glandlike structures; does not infiltrate or invade adjacent tissues.

Adjuvant: Additional therapy given to enhance or extend the primary therapy's effect, as in chemotherapy's addition to a surgical regimen; a treatment added to a curative treatment to prevent recurrence of clinical cancer from microscopic residual disease.

Algorithm: A systematic process consisting of an ordered sequence of steps, each step depending on the outcome of the previous one; a step-by-step protocol management of a health-care problem.

Antibody (immunoglobulin): A protein produced by B-lymphocytes that recognizes a particular foreign antigenic determinant and facilitates clearance of that antigen; antigens can also be carbohydrates and even DNA.

APTI: Amide proton transfer imaging

ATR: Attenuated total reflection

BCT: Breast-conserving therapy

BOLD: Blood oxygen level dependent

BPAS-MR: Basic parallel anatomic scanning–magnetic resonance

Brachyradiotherapy: Radiotherapy in which the source of radiation is placed close to the surface of the body or within a body cavity.

CAD: Computer-aided detection

CADx: Computer-aided diagnosis

Cancer Chemoprevention: The prevention of cancer or treatment of identifiable precancers; histopathologic or molecular intraepithelial neoplasia.

Carcinoma: Various types of malignant neoplasm arising from epithelial cells, mainly glandular (adenocarcinoma) or squamous cell. Carcinoma is the most common cancer and displays uncontrolled cellular proliferation, anaplasia, and invasion of other tissues, spreading to distant sites by metastasis. The origin of carcinoma in both sexes is in skin, and in the prostate in men and in the breast in women. The most frequent carcinoma in both sexes is bronchogenic carcinoma.

CBCT: Cone beam computed tomography

CCA: Conical correlation analysis

CCDC: Charge-coupled device camera

CDMAM: Contrast detectability mammography phantom

CDUS: Color-Doppler ultrasonography

CECT: Contrast-enhanced computed tomography

CED: Convection-enhanced drug delivery

CE-MRA: Contrast-enhanced magnetic resonance angiography

CEUS: Contrast-enhanced ultrasound

CHARMED: Composite hindred and restricted model of diffusion

CIMS: Chemical imaging mass spectrometry provides both the chemical information of a mass spectrometer and the spatial organization of each component on a surface, including biological surfaces.

CISS: Constructive interference in a steady state

***Clinical Guidelines*:**   Statements aimed at assisting clinicians in making decisions regarding treatment for specific conditions. They are systematically developed, evidence-based, and clinically workable statements that aim to provide consistent and high-quality care for patients. From the perspective of litigation, the key question has been whether guidelines can be admitted as evidence of the standard of expected practice, or whether this should be regarded as hearsay. Guidelines may be admissible as evidence in the United States if qualified as authoritative material or as a learned treatise, although judges may objectively scrutinize the motivation and rationale behind guidelines before accepting their evidential value. The reason for this scrutiny is the inability of guidelines to address all the uncertainties inherent in clinical practice. However, clinical guidelines should form a vital part of clinical governance.

CMRI: Cardiac magnetic resonance imaging

CMT: Continuously moving table

CNB: Core needle biopsy

COX: Cyclooxygenase

CR: Computed radiography

CSI: Chemical shift imaging

CTA: Computed tomography arteriography

CTC: Computed tomography colonography

CTDI: Computed tomography dose index

CTF: Computed tomography fluoroscopy

CTHA: Computed tomography hepatic arteriography

CTLM: Computed tomography laser mammography

CTP: Computed tomography portography

CTV: Clinical target volume

CW-NMRI: Continuous-wave nuclear magnetic resonance imaging

DBT: Digital breast tomosynthesis

DBTM: Digital breast tomosynthesis mammography

DCIS: Ductal carcinoma *in situ*

DEDM: Dual-energy digital mammography

DEI: Diffraction-enhanced imaging

DEPT: Distortionless enhancement by polarization enhancement

DFM: Dipolar field microscopy

DFS: Disease-free survival

***Diagnosis*:**   The differentiation of malignant from benign disease or of a particular malignant disease from others. A tumor marker that helps in diagnosis may be helpful in identifying the most effective treatment plan.

DOP: Depth of penetration

DOT: Diffuse optical tomography

DPI: Doppler perfusion index

DSA: Digital subtraction angiography

DSC: Dynamic susceptibility contrast

DSE: Dobutamine stress echocardiography

DSR: Dynamic spatial reconstructor

DTI: Diffusion tensor imaging

DT-MRI: Diffusion tensor magnetic resonance imaging

DVH: Dose volume histogram

DWI: Diffusion-weighted imaging

DW-MRI: Diffusion-weighted magnetic resonance imaging

EBCT: Electron beam computed tomography

EBCTA: Electron beam computed tomographic angiography

EBP: Evidence-based practice

EBRT: External beam radiotherapy

EBT: Electron beam tomography

ECD: Electrochemical detector

ECGI: Electrocardiographic imaging

ECR: Equivalent cross-relaxation rate

ECRI: Equivalent cross-relaxation rate imaging

ECS: Echocontrast cystosonography

EEG/MEG: Electro- and magnetoencephalography

EFG: Electric field gradient

EGFR: Epidermal growth factor receptor

EPI: Echo-planar imaging

EPI: Echo-portal imaging

EPID: Electronic portal imaging device

EPR: Electron paramagnetic resonance

EPR: Enhanced permeation and retention

EPSI: Echo-planar spectroscopic imaging

ERUS: Endorectal ultrasound

ESFT: Ewing sarcoma family of tumors

EUS: Endoscopic ultrasonography

FA: Fractional anisotropy

FBP: Filter back projection

FDG: ^{18}F-fluoro-2-deoxy-D-glucose

FDPM: Frequency-domain photon migration

FDTD: Finite difference time domain

FFDM: Full-field digital mammography

FFT: Fast Fourier transform

FIGO: Federation of International Gynecology and Obstetrics

FISP: Fast imaging with steady precession

FLAIR: Fluid attenuation inversion recovery

FLASH: Fast low-angle shot

FLIM: Fluorescence lifetime imaging microscopy

FLIP: Functional lumen imaging probe

FLT: [^{18}F] 3′-deoxy-3′-fluorothymidine

fMRI: Functional magnetic resonance imaging

FMISO: ^{18}F-fluoromisonidazole

FMT: Fluorescence-mediated tomography

FNAB: Fine-needle aspiration biopsy

FNH: Focal nodular hyperplasia

fNIRS: Functional near-infrared spectroscopy

FOV: Field of view

FOX: Field of excitation

FPA: Focal plane array

FPI: Flat-panel imager

FPI: Fluorescent protein imaging

FPT: Fast Padè transform

FRET: Fluorescence resonance energy transfer

FSCT: Fast-scan computed tomography

FSE: Fast-spin echo

3D FSE: 3D T2-weighted fast-spin echo

FSEI: Fast-spin echo imaging

FTIR: Fourier transform infrared resonance

Gallium-68 (^{68}Ga): A positron-emitting cyclotron-independent radionuclide with a short half-life of 68 min

*Gastritis***:** The inflammation, especially mucosal, of the stomach

Gd-DTPA: Gd-diethylenetriaminepentaacetic acid

*Gene Therapy***:** A therapy in which a gene(s) or gene-transducer cells are introduced to a patient's body for a therapeutic or gene-making purpose. Gene therapy by definition is not necessarily a molecular targeting therapy, but the reason for the high expectations is that the new mechanisms of cancer cell targeting can be integrated into therapy.

GIST: Gastrointestinal stromal tumor

GRAPPA: Generalized autocalibrating partially parallel acquisition

HCC: Hepatocellular carcinoma

HCT: Helical computed tomography

HDR: High-dose rate

HGGT: High-grade glial tumor

HIPAA: Health Insurance Portability and Account Ability Act

HPLC: High-performance liquid chromatography

HRCT: High-resolution computed tomography

HRMAA: High-resolution melting amplicon analysis; used primarily to screen for mutationally activated proteins.

HT: Helical tomography

ICRP: International Commission on Radiological Protection

IGRT: Image-guided radiotherapy

*Immunotherapy***:** Involves delivering therapeutic agents conjugated to monoclonal antibodies that bind to the antigens at the surface of cancer cells. Ideal antigens for immunotherapy should be strongly and uniformly expressed on the external surface of the plasma membrane of all cancer cells. Many solid neoplasms often demonstrate regional variation in the phenotypic expression of antigens. These regional differences in the immunophenotypic profile within the same tumor are referred to as intratumoral heterogeneity. Therapeutic agents that have been used include radioisotopes, toxins, cytokines, and immunologic cells.

IMRT: Intensity-modulated radiation therapy; a special form of CFRT (conformal radiotherapy). IMRT is the delivery of radiation to the patient via fields that have nonuniform radiation fluence. However, it is fluence, not intensity, that is modulated.

IOC: Intraoperative cholangiogram

IRFSE: Inversion recovery fast-spin echo

IRSE: Inversion recovery spin echo

IVM: Intravital microscopy

LDR: Low-dose rate

LGGT: Low-grade glial tumor

LINAC: Linear accelerator

LRRT: Locoregional radiotherapy

LSI: Laser speckle imaging

LSS: Light-scattering spectroscopy

Lymph: The intracellular tissue fluid that circulates through the lymphatic vessels.

Lymphadenopathy: The enlargement of the lymph nodes.

Lymph nodes: Small secondary lymphoid organs containing populations of lymphocytes, macrophages, and dendric cells that serve as sites of filtration of foreign antigens and activation of lymphocytes.

Lymphoma: A cancer of lymphoid cells that tend to proliferate as solid tumors.

MADD: Maximum allowed dose difference

Malignant: Tumors that have the capacity to invade and alter the normal tissue.

MALT: Mucosa-associated lymphoid tissue

MBF: Myocardial blood flow

MCE: Myocardial contrast echocardiography

MCMLI: Multicolumn multiline interpolation

MCR-ALS: Multivariate curve resolution-alternating least square

MDCT: Multidetector row computed tomography

MDEFT: Modified driven equilibrium Fourier transform

Mediastinoscopy: An invasive procedure used for staging mediastinal lymph node metastases, which has a sensitivity of ~ 90%.

MEG: Magnetoencephalography

MEMRI: Manganese-enhanced magnetic resonance imaging

Metastasis: Initially tumor growth is confined to the original tissue of origin, but eventually the mass grows sufficiently large to push through the basement membrane and invade other tissues. When some cells loose adhesiveness, they are free to be picked up by lymph and carried to lymph nodes and/or may invade capillaries and enter blood circulation. If the migrant cells can escape host defenses and continue to grow in the new location, a metastasis is established. Approximately more than half of cancers have metastasized by the time of diagnosis. Usually it is the metastasis that kills the person rather than the primary (original) tumor.

Metastasis itself is a multistep process. The cancer must breakthrough any surrounding covering (capsule) and invade the neighboring (surrounding) tissue. Cancer cells must separate from the main mass and be picked up by the lymphatic or vascular circulation. The circulating cancer cells must lodge in another tissue. Cancer cells traveling through the lymphatic system must lodge in a lymph node. Cancer cells in vascular circulation must adhere to the endothelial cells and pass through the blood vessel wall into the tissue. For cancer cells to grow, they must establish a blood supply to bring oxygen and nutrients; this usually involves angiogenesis factors. All of these events must occur before host defenses can kill migrating cancer cells.

If host defenses are to be able to attack and kill malignant cells, they must be able to distinguish between cancer and normal cells. In other words, there must be immunogens on cancer cells not found on normal cells. In the case of virally induced cancer circulating cells, viral antigens are often expressed, and such cancer cells can be killed by mechanisms similar to those for virally infected tissue. Some cancers do express antigens specific for those cancers (tumor-specific antigens), and such antigens are not expressed by normal cells.

Although metastasis is known to be the principal cause of death in individuals with cancer, its molecular basis is poorly understood. To explore the molecular difference between human primary tumors and metastases, gene expression profiles of adenocarcinoma metastases of multiple tumor types have been compared with unmatched primary adenocarcinomas. A gene-expression signature that distinguished primary from metastatic adenocarcinomas was found. More importantly, it was found that a subset of primary tumors resembles metastatic tumors with respect to this gene-expression signature. The results of this study differ from most other earlier studies in that the metastatic potential of human tumors is encoded in the bulk of a primary tumor. In contrast, some earlier studies suggest that most primary tumor cells have low metastatic potential, and cells within large primary tumors rarely acquire metastatic capacity through somatic mutation. The emerging notion is that the clinical outcome of individuals with cancer can be predicted using the gene profiles of primary tumors at diagnosis.

MIP: Maximum-intensity projection

MITS: Matrix inversion tomosynthesis

***Molecular Genetics*:**   A subdivision of the science of genetics, involving how genetic information is encoded within the DNA and how the cell's biochemical processes translate the genetic information into the phenotype.

***Molecular Imaging*:**   The *in vivo* characterization and measurement of biological processes at the cellular and molecular levels. In other words, in contrast to conventional diagnostic imaging, molecular imaging probes the molecular abnormalities that are the basis of disease, including cancer, rather than image the end effects of these molecular alterations.

***Monitoring*:**   Repeated assessment if there are early relapses or other signs of disease activity or progression. If early relapse of the disease is identified, a change in patient management will be considered, which may lead to a favorable outcome for the patient.

MPR: Multiplanar reconstruction

MRA: Magnetic resonance angiography

MRA: Magnetic resonance arthrography

MRCP: Magnetic resonance cholangiopancreatography

MRDSA: Magnetic resonance digital subtraction angiography

MRE: Magnetic resonance elastography

MREIT: Magnetic resonance electrical impedance tomography

MRS: Magnetic resonance spectroscopy analyzes specific atomic nuclei and their compounds using the phenomenon of magnetic resonance and chemical shift. This method provides information on the metabolism of organs and cells, biochemical changes, and quantitative analysis of compounds in humans, with no harm to the body.

MRSI: Multivoxel magnetic resonance spectroscopic imaging

MSCT: Multislice computed tomography

MTC: Magnetization transfer contrast

MTD: Maximum tolerated dose

MTI: Microwave tomographic imaging; provides quantitative maps of tissue dielectric properties that may correlate with tissue functional information.

MVCT: Megavoltage computed tomography

NDD: Normalized dose difference

NEC: Noise-equivalent quanta describes the equivalent number of quanta or counts required by an ideal imaging system to produce the same noise characteristics as does an actual system that is degraded by noise.

NECR: Noise-equivalent counting rate

***Neoplasia*:**   Pathologic process that causes the formation and growth of an abnormal tissue.

***Neoplasm*:**   An abnormal tissue that grows by cellular proliferation faster than normal and continues to grow.

NIOI: Near-infrared optical imaging

NIR: Near-infrared

NIRL: Near-infrared light

NIRS: Near-infrared spectroscopy

NMR: Nuclear magnetic resonance

NPV: Negative predictive values

NSA: Number of signal averages

NTCP: Normal tissue complication probability

OAP: Oblique axial plane

OCT: Optical coherence tomography is based on imaging probes inserted into a body lumen or directly into a soft tissue through thin catheters. Such catheter based OCT has been developed primarily for gastrointestinal and intravascular imaging.

ODT: Optical diffusion tomography

ODT: Optical Doppler tomography

OGTT: Oral glucose tolerance test

OHR: Optimized head and neck reconstruction

OIS: Optical imaging spectroscopy

OOSCC: Oropharyngeal squamous cell carcinoma

OPET: Optical positron emission tomography

OPSI: Orthogonal polarization spectral imaging

OPT: Optical projection tomography

OSEM: Ordered subsets expectation maximization

***Palliative*:**   Reducing the severity of a disease; denotes the alleviation of symptoms without curing the underlying disease.

PAM: Photoacoustic mammoscope

***Pancreatitis*:**   Inflammation of the pancreas can be caused by alcoholism, endocrine diseases, hereditary, viral, parasitic, allergic, immunologic, pregnancy, drug effects, and abdominal injury.

PAT: Parallel acquisition technique

PAT: Photoacoustic tomography

PCA: Principal component analysis

PCM: Phase-contrast mammography

PCNA: Proliferating cell nuclear antigen

PCT: Perfusion computed tomography

PDT: Photodynamic therapy is a promising treatment for accessible tumors; localized to the tumor tissue by using photosensitive drugs, which may lead to tumor regression or even death.

PEDRI: Proton electron double-resonance imaging

Phase-contrast X-ray Imaging: Utilizes refractive index variations (phase information) in addition to conventional absorption information with conventional X-ray absorption techniques. Phase-contrast images can be recorded with a significantly lower dose than conventional images.

phMRI: Pharmacological MRI; a technique that can be used to monitor the neurophysiological effects of central nervous system–active drugs.

PI: Parallel imaging

PIT: Parallel imaging technique

PMRI: Pharmacologic magnetic resonance imaging

PPI: Partially parallel imaging

PPILS: Partially parallel imaging with localized sensitivities

PPV: Positive predictive values

Prognosis: The prediction of how well or how poorly a patient is likely to fare in terms of response to therapy, relapse, survival time, or other outcome measures.

PSI: Probabilistic similarity index

PTC: Percutaneous transhepatic cholangiogram

RCT: Randomized controlled trial

RECIST: Response evaluation criteria in solid tumors

RET: Resonance energy transfer

RF: Radiofrequency

RGCT: Respiratory-gated computed tomography scanning

ROCM: Receiver operating characteristic method

ROI: Region of interest

RP: Radical prostatectomy

SAR: Specific absorption ratio/rate

Sarcoma: A connective tissue neoplasm that is usually highly malignant; formed by proliferation of mesodermal cells.

Sarcomatoid: A neoplasm that resembles a sarcoma.

SCI: Spatial compound imaging

Scintigraphy: A diagnostic method consisting of the administration of a radionuclide having an affinity for the organ or tissue of interest, followed by photographic recording of the distribution of the radioactivity with a stationary or scanning external scintillation camera.

Screening: The application of a test to detect disease in a population of individuals who do not show any symptoms of their disease. The objective of screening is to detect disease at an early stage, when curative treatment is more effective.

SEA: Single echo acquisition

SENSE: Sensitivity encoding

SFRT: Stereotactic fractionated radiation therapy

SIMS: Secondary ion mass spectrometry

SLEPI: Spin-locked echo planar imaging

SLN: Sentinel lymph node

SLNB: Sentinel lymph node biopsy

SMART: Simultaneous modulated accelerated radiation therapy

SMASH: Simultaneous acquisition of spatial-harmonics

SMRI: Stereotactic magnetic resonance imaging

SNB: Sentinel node biopsy

SNR: Signal-to-noise ratio

Specificity: The capacity for discrimination between antigenic determinants by antibody or lymphocyte receptor.

SPECT: Single photon emission computed tomography; a cross-sectional, quantitative functional imaging modality in routine use in oncology for the initial staging of the cancer.

SPI: Single-point imaging

SPIO: Superparamagnetic iron oxide

Sporadic: A multilocal genocopy, occurring irregularly is a disease occurring only rarely without regularity; extreme variability in the expression of a gene.

SPRITE: Single-point ramped imaging with T_1 enhancement

SSCT: Slow-scan computed tomography

SS-NMR: Solid-state nuclear magnetic resonance

STAT: Signal transducers and activators of transcription

STEAM: Stimulated echo acquisition mode

STIR: Short T1 inversion recovery

STRAFI: Stray field imaging

SURLAS: Scanning ultrasound reflector linear array system

SUV: Standard uptake value

SWI: Stiffness-weighted image

SWR: Standardized whole-body reconstruction

TACE: Transarterial chemoembolization

TACT: Tuned aperture computed tomography

TCP: Tumor control probability

TDI: Tissue Doppler imaging

TEE: Transesophageal echocardiography

TEM: Transverse electromagnetic

TERUS: Tracked endorectal ultrasound

TESO: Time-efficient slice ordering

TGC: Time gain compensation

THI: Tissue harmonic imaging is a grayscale ultrasound mode that can provide images of higher quality than conventional sonography by using information from harmonics. Harmonics are generated by nonlinear wave propagation of ultrasound in tissue.

Tomography: The making of a radiographic image of a selected plane by means of reciprocal linear or curved motion of the X-ray tube and film cassette; images of all other planes are out of focus and blurred.

TPPM: Two-pulse phase-modulated

TRAIL: Two reduced acquisitions interleaved

Transmit SENSE: Transmit sensitivity encoding

TTE: Transthorasic echocardiography

Tumor Microenvironment: The interaction between epithelial tumors and their stroma, including fibroblasts, blood vessels, and extracellular matrix. This definition is extended to interactions between potential tumor cells and immediately surrounding cells of the same tissue type.

TVDT: Tumor volume doubling time

UHF: Ultrahigh magnetic field

US: Ultrasonography

USP: United Sates pharmacopoeia

VEGFR: Vascular endothelial growth factor receptor

XeCT: Xenon-enhanced computed tomography

XRF: X-ray fluoroscopy

carcinogenesis by chemical, physical, and biological agents and into inherited susceptibility to cancer.

Although anatomic imaging is in common use, functional imaging has the advantage of discerning the underlying biochemical pathways that confirm the presence or absence of pathology. In other words, molecular imaging advances our understanding of biology and medicine through noninvasive *in vivo* investigation of molecular events involved in normal and pathologic processes. In addition, the importance of molecular imaging in early diagnosis and assessment of therapy becomes apparent considering that alterations in molecular processes during treatment generally precede anatomical changes. Positron emission tomography is ideally suited to detect changes in molecular processes resulting, for example, from chemotherapy. The transition of molecular imaging from the animal-imaging environment to the clinical human environment is progressing rapidly. An effort has been made in this work to integrate molecular imaging into clinical radiology. It is not an exaggeration to say that we are observing the practice of medical imaging in an era of molecular medicine. In other words, we have entered molecular medicine from molecular imaging. Structure, function, and biochemistry have become fused in some areas of clinical practice. Spatial resolution obtained with imaging has progressed, as stated earlier, from whole body to organs, from tissues to cells, and to molecules. It is possible now to study not only biological states but also biochemical processes that precede a state or phenotype at a specific point in time.

A number of tracers (radionuclides) are in use; the following data indicate the application of different tracers in the descending order of their use: ^{18}F, ^{11}C, ^{99m}Tc, and ^{131}I. However, ^{18}F remains the dominant tracer. Only two of these tracers are summarized here; these and other tracers are discussed in detail in some other chapters in this and Volume 2 of this series.

^{18}F-fluorodeoxyglucose (FDG) is the most common tracer used in imaging for cancer diagnosis. It is a radioactively labeled glucose analog that can be localized accurately due to its emission of positrons. This tracer is essentially a radiopharmaceutical agent that provides the ability for imaging glucose metabolism that closely mimics endogenous molecules. The usefulness of FDG is due to the evidence that tumor cells have increased levels of glucose metabolism and rapid cell proliferation. The increased glucose uptake observed in malignant cells is attributed to either an increase in the transcription and translation of glucose transport proteins in tumor cells or to hexokinase activity. Because of this characteristic, FDG uptake is elevated in malignant cells compared with normal cells. As a result, neoplasms are reliably distinguishable from surrounding tissues on FDG-PET imaging. Quantification of FDG uptake is also useful to determine if the tissue or organ involved is benign or malignant. The tracer is also crucial in determining the therapeutic response and prognosis. High sensitivity and high-negative predictive value have led to its important role in cancer diagnosis. Thus, the application of this tracer in the diagnosis of lung and breast cancers is discussed extensively in this work.

In contrast to anatomic imaging modalities such as CT and MRI, FDG-PET imaging is based on metabolism and tissue perfusion. In other words, this protocol is a functional imaging modality that characterizes different tissues in the body according to glucose metabolism. It is well known, as mentioned earlier, that most malignant cells require a high glucose uptake because of increased glycolysis. This is the reason to use FDG-PET as the screening tool.

Positron emission tomography using FDG is a noninvasive imaging modality for assessment of glucose metabolism in a variety of malignancies, including adenocarcinomas. This protocol is widely used as a marker in many oncology and nuclear medicine centers throughout the world for the diagnosing and staging of cancer. The medium half-life of fluorine-18 (~ 110 min) makes it ideally suitable to accomplish these goals.

In clinical practice, FDG-PET scanning is used for staging tumors, detecting tumor recurrence, monitoring efficacy of therapy, and differentiating malignant from nonmalignant tissues. This system is more accurate than CT alone in the detection of both primary tumors and distant metastases. It is interesting to find out whether increased uptake of FDG precedes the onset of malignancy or whether it follows the initiation of cancer development. If the former is true, then tissue abnormalities associated with cancer can be determined at the premalignant dysplasia stage. This advantage allows early diagnosis of cancer and screening of high-risk groups. The question still remains as to the nature of the mechanism responsible for initiating the accumulation of FDG in premalignant abnormal cells.

It should be noted, however, that FDG is also taken up by normal tissues as well as in inflamed conditions. Nevertheless, there is an increasing uptake of FDG with time, for example, in breast malignancies, whereas such uptake in inflammatory lesions and normal breast tissues decreases with time. In any case, it is necessary to be aware of normal variants, artifacts, and other causes of false-positive results. Another PET tracer is 3-deoxy-3-^{18}F-fluorothymidine (^{18}F-FLT), which is thought to be a superior predictor, for example, of brain tumor progression and survival when compared with FDG.

FLT is a thymidine analog developed for imaging tumor proliferation with PET. To quantitatively assess images, the blood activities of FLT and its glucoronidated metabolites can be measured and its kinetics analyzed.

Dynamic measurements of FLT retention can be used to calculate metabolic rates using a limited set of specimens and correction for metabolites measured in a single specimen obtained in 1 hour.

Thymidine analogs have the advantage of being readily labeled with ^{18}F, have limited metabolism *in vivo*, and are trapped in the DNA synthetic pathways. The mechanism of trapping is via monophosphorylation to [^{18}F]-FLT-phosphate catalyzed by the cytosolic enzyme thymidine kinase 1, which being cell-cycle regulated, provides a surrogate measure of cells in the S phase of the cell cycle. Thus, changes in [^{18}F]-FLT uptake can be correlated directly with cell proliferation, although FLT is not incorporated into DNA. Although this protocol has high detection sensitivity, some limitations in specificity, such as the difficulty of distinguishing between proliferating tumor cells and inflammation, have been reported. FLT-PET can be further developed as a generic pharmacodynamic readout for early quantitative imaging of drug-induced changes in cell proliferation *in vivo*.

It is well established that early diagnosis is key to a cancer "cure." Prognosis is highly dependent on the stage of the disease. Thus, a simple and reliable screening method would be of tremendous advantage. For example, mammography and CT colonography have established a niche for imaging in cancer screening. Imaging techniques in clinical practice are used for staging tumors, detecting tumor recurrence, monitoring efficacy of therapy, and differentiating between malignant and benign tissues. Unfortunately, however, there are no or only a few associated early symptoms of some cancer types. Pancreatic and ovarian cancers and lymphoma are examples of malignancies difficult to diagnose at an early stage.

It cannot be overemphasized that careful training and thoughtful use of imaging technology undeniably enhance patient care. It is worth remembering that imaging information by itself is of limited value; it is the careful interpretation and thoughtful application of what we have seen that gives our data significance. This technology enhances physical examination, can image the disease process, and can bring to light new issues in patient management with greater clarity than patient history and conventional examination alone. Medical or surgical treatment plans can be modified according to the information extracted from imaging.

It is estimated that imaging testing will cost more than $100 billion in the year 2006 in the United States. This staggering amount of money will be spent because imaging is also a defensive medicine. Nevertheless, excessive use of imaging will exact a heavy toll on available monetary resources.

Although the question of whether exposure to medical radiation increases the incidence of cancer in the general population is controversial, it is known that repeated exposure, for example, to mammography and computed tomography screening (depending on the radiation dose) is harmful to the patient. Therefore, imaging modalities that introduce radiation should be used only when necessary. In this work, Eric Milne points out the risk of developing breast cancer after repeated thoracic and/or abdominal CTs; this subject is also discussed by James S. Michaelson, and Pamela S. Ganschow and Joann G. Elmore in their chapters in this work. In addition, the European Union has adopted a directive cautioning against occupational exposure to electromagnetic fields produced by MRI; it applies only to workers, however, not to patients.

Four major topics discussed in this work are imaging instrumentation, imaging applications, and imaging of lung and breast carcinomas. Although cancer therapy is not our main subject, the crucial role of imaging in selecting the type of therapy and its posttreatment assessment is discussed. The major emphasis in this work is on cancer imaging; however, differentiation between benign tumors and malignant tumors is also covered. Continued investment of time and expertise by researchers worldwide has contributed significantly to a greater understanding of the cancer process. In most cases, the methodologies presented were either introduced or refined by the authors and routinely used in their clinical facilities. Some of the new topics that are at an experimental stage are also included for further testing and refinement.

I
Instrumentation

1

Strategies for Imaging Biology in Cancer and Other Diseases

Philipp Mayer-Kuckuk and Debabrata Banerjee

Introduction

Therapy for many urgent health problems, including cancer and osteoporosis, has improved during the last decade, but in the majority of cases the patient's expectation of a cure is unmet. The ongoing search for better treatments in most academic and industry laboratories builds on basic research, which in turn drives drug discovery. The hope is that a more detailed understanding of the cellular biology and the characterization of its alterations during disease development will ultimately lead to the identification of highly specific and hence effective drug targets. A prime example of this effort is the development of tyrosine kinase inhibitors, such as imatinib (Gleevec), as targeted therapy for cancers, including chronic myeloid leukemia. Similarly, therapeutic antibodies for the targeted therapy of breast and colorectal cancer, such as trastuzumab (Herceptin) or beva-cizumab (Avastin), respectively, have proven to be effective. There is, however, an apparent limitation to this approach. Not all patients may present an identical target. Particularly critical is the occurrence of target variations that are often observed in advanced cancers as a result of mutations. There is thus a rationale for target validation in patients. Molecular imaging is expected to fulfill the role of a diagnostic tool for this demand and has gained significant attention in recent years. Molecular imaging has been previously defined as "the *in vivo* characterization and measurement of biologic processes at the cellular and molecular level" (Weissleder and Mahmood, 2001). In the longer term, molecular imaging is likely to further develop into a diagnostic tool capable of tissue profiling at the molecular level. For example, selection of cancer therapy would greatly benefit from a pretreatment assessment of tumor biology. This includes the knowledge of signaling and transcription pathways used by the cancer cells to escape the normal life cycle and whether the cells are likely to undergo metastasis.

Molecular imaging is not restricted to diagnostic imaging in patients. Its application to animal models, perhaps most importantly the use of imaging reporter genes in mice, has great potential for a better understanding of biological pathways under physiological conditions. One advantage of studying biology *in vivo* lies in a more efficient target identification and validation as part of drug discovery. Furthermore, pharmacokinetics, pharmacodynamics, and toxicity can be assessed in imaging reporter mice (Maggi and Ciana, 2005). In addition, molecular imaging is likely to reduce labor and cost particularly during late-stage drug discovery and may help to decide faster whether to enter clinical phase (Rudin and Weissleder, 2003).

During the last several years, many different protocols for imaging biology have been reported; both the portfolio of biological constituents that can be imaged and the imaging strategies available for particular constituents have increased. This development requires the biologist to make choices from available imaging options. Our goal is to help the reader in his or her selection of a suitable strategy for biological imaging. Emphasis is naturally placed on cancer research because molecular imaging has developed out of this field. To offer a structured and simple view on molecular imaging, we have divided current imaging strategies into two basic steps: (1) conferring imaging visibility using peptides, proteins, probes, or reporter genes, and (2) subsequent image acquisition utilizing suitable imaging modalities (all discussed in the section Imaging Strategies). In our subsequent overview of strategies for imaging biology, we give a concise yet comprehensive update on preclinical (see Preclinical Applications) as well as potential or early-stage clinical applications (Imaging Strategies for Clinical Applications) of molecular imaging for the *in vivo* assessment of biology.

This contribution largely excludes strategies that image tumor physiology, including energy state, proliferation and hypoxia, or tumorigenesis rather than a specific constituent. Furthermore, imaging strategies with a long history such as antibodies, peptides, or certain radiopharmaceuticals have only been covered partially, though attempts have been made to touch on recent developments.

Imaging Strategies

Molecular imaging has strong roots in classical diagnostic imaging. In fact, molecular imaging directly evolved from classical radiology. The tremendous diagnostic improvement provided by X-ray radiology in the twentieth century along with the invention of magnetic resonance imaging (MRI) 50 years ago prompted the search for diagnostics capable of registering information beyond anatomical structures. Mention must be made of two concepts developed for assessment of the cell, and hence tissue energy state, which mark the transition to molecular imaging. First, in 1979 fluorine-labeled 2-fluorodeoxy-D-glucose (FDG) was introduced as a metabolic marker, and the combination of the radioactive tracer ^{18}FDG and positron emission tomography (PET) became a valuable tool for detection of energy states in tissues such as brain, heart, and cancer. This method served as a template for early molecular imaging reports (Tjuvajev *et al.*, 1996; Gambhir *et al.*, 1999) that monitored accumulation of nuclide-labeled thymidine analogs in imaging reporter gene-transduced cells. Many newer imaging strategies have been derived from these pioneering studies, but they are all based on the same concept: An external imaging modality is used for detection of a labeled molecule, which is recognized selectively by a biological constituent postsystemic administration.

The second imaging concept parental to molecular imaging is magnetic resonance spectroscopy in living subjects, which developed in the early 1980s. The ability to collect information about specific endogenous molecules that contain phosphorus—for example, nucleoside triphosphates, inorganic phosphate, or phosphocreatine—by simply taking advantage of their intrinsic nuclear spin property through MR technology represents an early molecular imaging strategy. However, this approach appears to be limited to certain molecules that harbor appropriate MR detectable nuclei.

Conferring Imaging Visibility

To image a biological constituent or activity *in vivo*, it must possess a specifically detectable property. Cells and most macromolecules, however, do not exhibit intrinsic properties that would permit detection with currently available imaging modalities. For this reason molecular imaging is regularly carried out in two sequential steps. First, by taking advantage of a variety of means, imaging visibility is conferred on an *in vivo* constituent or activity. This is then followed by the actual imaging process, which employs an external imaging device that allows "seeing" the constituent or activity within the body. In this context, "seeing" refers to noninvasive detection and measurement.

To confer imaging visibility, the strategy requiring the least number of components is the use of peptides, proteins, or probes that are recognized by a specific *in vivo* target. For use in molecular imaging, peptides, proteins, or probes are modified with a detectable label. The choice of label (e.g., nucleotide or fluorophore) will determine the type of imaging modality (e.g., nuclear or optical) and vice versa. The described category also encompasses probes carrying a target-activated label; a more detailed description is provided in the next paragraph. Conferring imaging visibility through labeled peptides, proteins, or probes allows direct analysis of endogenous biology and is usually a reversible process, with the label disappearing from the organism over time. In contrast, imaging visibility conferred by reporter genes assesses biology indirectly through the generation of an additional exogenous cellular reporter protein and frequently requires genetic manipulation. In addition to the reporter gene, a second component, such as a tracer or substrate, is usually necessary for imaging the reporter gene *in vivo*. Such gene-marked cells tend to become part of the endogenous biology.

Peptides, Proteins, and Probes

Peptides, as well as proteins such as antibodies or Annexin V, offer two important advantages for conferring imaging visibility to an *in vivo* constituent or activity. First, because many peptides and proteins are derived from organisms, including humans, they often exhibit favorable *in vivo* characteristics, including pharmacokinetics. Second, the chemistry and molecular biology for manipulating them is well established. It includes the capability for isolation or synthesis at required quantities and engineering options, as exemplified in the generation of novel low-molecular-weight antibodies. Importantly, routine labeling protocols, specifically for isotope labeling, are available or can be designed readily, making nuclear imaging the traditional modality of choice.

Newer optical-imaging approaches make use of labels such as indocyanine dye Cy5.5 fluorophores or polymer-coated quantum dots. Because light absorption in biological tissues reaches a minimum ~750 nm wavelength, the emission maxima of fluorescent labels are preferentially in the near-infrared (NIR) window of the electromagnetic spectrum. Although many successful applications have been reported, these labels can reach a considerable size that might occasionally impact function. For MRI, polypeptides have been marked with MR contrast agents. They contain paramagnetic Gd^{3+} or Fe^{3+} ions formulated as organic molecule complexes or dextran-coated microcrystalline magnetite cores (particles or spheres). Finally, to confer visibility for ultrasound imaging a sensor can be modified with contrast-enhancing microbubbles (gas bubbles surrounded by a lipid or protein shell).

Probes are synthetic, nontoxic substances designed to localize to an *in vivo* target after systemic administration. Two types can be distinguished: probes modified with a fluorescent dye, which can be imaged following their *in vivo* concentration at a target site; and protease-activated probes, which are designed to gain fluorescence after proteolytic cleavage. Protease-activated probes for optical imaging are comprised typically of a backbone displaying cleavage sites recognized by particular proteases (Weissleder *et al.*, 1999). Attached via the lysine amino groups are methoxypolyether glycol side chains for enhanced biodistribution characteristics, as well as Cy5.5 fluorophores. Within the polymer, the close proximity of the fluorophores results in efficient self-quenching. The probe is activated through enzymatic cleavage, which results in the release of short peptides containing isolated and therefore fully fluorescent NIR fluorophores (excitation and emission maxima of 673 nm and 689 nm, respectively). Activation of the probe through proteases can result in a > 10-fold increased *in vitro* fluorescence intensity (Weissleder *et al.*, 1999). In recent years, proof-of-principle studies on protease-activated probes have been completed in mouse models of protease overexpressing tumors. Whether they have adequate activity, safety, and immunogenicity for use in patients remains to be determined.

Significant features of many of the peptides, proteins, and probes discussed in this paragraph include (a) availability in physiological formulation, (b) ability to be administered

intravenously, (c) pharmacokinetics for imaging applications, and (d) no toxicity at doses used for imaging. Therefore, they are either used or likely to be used in clinical applications. In contrast, the clinical application of imaging reporter genes discussed in the following paragraph ultimately requires gene transfer into human cells. This step is inevitably linked to the fate of genetic medicine in general. Unacceptable adverse side effects, including oncogenic transformation, have raised concerns regarding the clinical potential of gene transfer. Major improvement in regard to safety of gene transfer will be necessary to regain a wide acceptance of this approach. Safety measures currently under investigation include self-inactivating vectors, shielded gene expression cassettes, and suicide genes for the control of unwanted cell proliferation.

Imaging Reporter Genes

Imaging reporter genes typically code for exogenous proteins that can be detected readily and specifically in a non-invasive manner. They are commonly placed downstream of a genetic control element that regulates reporter gene expression. The ability to choose and manipulate the control element makes imaging reporter genes excellent tools for gene expression studies. Furthermore, as cDNA fragments, reporter genes are ideally suited for protein engineering. A reporter gene can be manipulated, for example, to create multimodality fusion reporter genes or even split reporter genes, as described later in this chapter. Alternatively, endogenous proteins can be fused and hence tagged with a reporter gene. In addition, the dual usage of several reporter genes in molecular imaging as well as molecular biology provides easy transition from *in vitro* to *in vivo* studies. Taken together, reporter genes represent a versatile and powerful preclinical imaging tool. We found it convenient to classify imaging reporter genes based on the imaging modality as: reporter genes for (a) optical, (b) nuclear, and (c) MRI applications.

Reporter Genes for Optical Imaging

A fairly simple reporter gene for optical imaging codes for the green fluorescent protein (GFP). To study, for example, GFP-tagged proteins in a living subject and thus facilitating optical fluorescence imaging, it is sufficient to implant the GFP fusion protein expressing cells into a living subject such as a mouse and to use an external light source and detector for fluorescence excitation and emission registration, respectively. Efficient deep tissue imaging might be restricted to nude mouse models and might need high GFP-expressing tumors and sensitive instrumentation. Nevertheless, this modality remains an important option for cancer research, and it can be implemented in laboratories with modest time and cost investments. Moreover, several animal models for GFP imaging in mice are now commercially available. In recent years, many laboratories have implemented optical bioluminescence imaging (BLI), which represents the method of choice for the majority of preclinical applications discussed in this chapter. We will describe the currently available luciferase-based BLI reporter genes in the following three paragraphs.

Luciferases represent a class of light-producing enzymes present in certain cells of bioluminescent organisms (Greer and Szalay, 2002). Depending on their origin, luciferases exhibit distinct biochemical characteristics, and several different types have been utilized as imaging reporter genes. The beetle-derived *Photinus* Firefly luciferase (FLuc) is the most widely used optical-imaging reporter gene and is well known as a reporter gene in molecular biology. A second less-well-described synthetic luciferase reporter gene has been derived from Click beetle (CBLuc). Two other luciferases utilized for imaging have been isolated from sea organisms. *Renilla* luciferase (RLuc) from sea pansy has a proven record as an imaging reporter gene, while the application of a mainly secreted marine copepod-derived *Gaussia* luciferase (GLuc) is fairly new. Of note, to optimize luciferases, for example, with respect to mammalian gene expression or cytosolic luminescence, engineered variants have been made. Specific designations for some of those manipulated enzymes are being used, including the synthetic RLuc (hRLuc) or the humanized GLuc (hGLuc).

Firefly luciferase oxidizes the benzothiazole *Firefly* D-luciferin (often simply designated luciferin) in an ATP- and Mg^{2+} cofactor-dependent bioluminescence reaction that yields AMP, PP_i, and CO_2 and emits an excess of energy as light. The emission of green-yellow light *in vitro* at room temperature (E_{max} ~ 562 nm) undergoes a red-shift *in vivo* at body temperature and peaks at 612 nm, which enhances tissue penetration of the photons detected by BLI (Zhao *et al.*, 2005). The luciferase substrate luciferin exhibits an *in vivo* profile excellent for many imaging applications. It is nontoxic, can pass the cell membrane freely, and crosses the blood-brain and placenta barriers. Recent work suggests a potential of another luciferase as imaging reporter gene. Two mutants derived from wild-type CBLuc that emit green light at 537 nm (CBGr68 and CBGr99) and one 613-nm red light emitting CBLuc mutant (CBRed) have been described (Zhao *et al.*, 2005). All catalyze a reaction similar to FLuc. A comparison of *in vivo* characteristics indicated that CBRed has a light output and transmission efficiency similar to FLuc (Zhao *et al.*, 2005).

With a size of ~ 36 kDa and 20 kDa, *Renilla* and *Gaussia* luciferase, respectively, are smaller than to the beetle-derived luciferases, which are ~ 60 kDa in size, a feature that can be of interest when designing complex, multifunctional reporter gene constructs. Both *Renilla* and *Gaussia* luciferase catalyze the production of CO_2 and light from coelenterazine and its derivatives instead of luciferin, and their enzymatic activities are cofactor independent. Although the *in vivo* characteristics of coelenterazine are

incompletely described, it is known to be less stable than luciferin and is also a substrate for p-glycoprotein. *Renilla* and *Gaussia* luciferase produce blue light with an emission peak ~ 480 nm, suggesting less efficient tissue penetration in comparison to *Firefly* luciferase. Experiments comparing the *in vivo* kinetics of *Renilla* luciferase to *Firefly* luciferase indicated a faster peak in light production (flash kinetics) for *Renilla* and a prolonged time of light emission for *Firefly* luciferase (glow kinetics) (Bhaumik and Gambhir, 2002). Despite the distinct kinetics, it has been demonstrated that both enzymes can be used for simultaneous imaging of expression of two reporter genes in a single mouse (Bhaumik and Gambhir, 2002; Tang *et al.*, 2003). Studies systematically comparing all luciferases with respect to their sensitivity as imaging reporter genes, for example, are currently not available. Two reports, however, suggest a hRLuc performance similar to Fluc, while hGLuc outperforms hRLuc (Bhaumik *et al.*, 2004a; Tannous *et al.*, 2005). Firefly luciferase, however, is often the investigator's first choice because it is robust, well established, and has not been reported to fail as an optical reporter gene in any mouse tissue.

Reporter Genes for Nuclear Imaging

The herpes simplex virus thymidine kinase (HSV TK) is the most widely used nuclear reporter gene and the only one utilized in imaging studies addressing biology. Cells transduced with HSV TK are specifically labeled with an exogenous viral kinase activity and hence only those cells readily phosphorylate and trap specific HSV TK substrates. The latter can be synthesized as radioactively labeled derivatives and administered in trace amounts to subjects, permitting nuclear imaging. Currently, thymidine and acycloguanosine analogs, such as 2′-fluoro-2′-deoxy-5-iodouracil-β-D-arabinofuranoside (FIAU) or [4-fluoro-3-(hydoxymethyl)butyl] guanine (FHBG), respectively, are the two major classes of HSV TK tracers (Gambhir et al., 2000b). Mutant HSV TK reporters have also been developed, such as the six amino acid–substituted HSV sr39TK, which utilizes acycloguanosines more efficiently (Gambhir *et al.*, 2000a).

Other nuclear reporter genes have been developed and have potential in future applications. In contrast to HSV TK, these reporters are expressed in human tissues, though in a very restricted fashion. This reduces specificity of the reporter, but opens the window for potential applications in humans. Moreover, these reporter genes do not metabolize the tracer in an enzymatic reaction but rather increase cellular tracer binding or uptake. This mode of action may restrict their ability for signal amplification. Tracers derived from the ligand 3-(2′-fluoroethyl)-spiperone (FESP) have been developed for imaging a dopamine D2 receptor (D2R) reporter gene, which has been mutated for elimination of receptor-mediated signaling (Liang *et al.*, 2001). A second less-well-studied receptor developed as imaging reporter for nuclear imaging is the somatostatin receptor (SSTR) (Zinn and Chaudhuri, 2002). Lastly, expression of the 13-pass transmembrane protein sodium iodide symporter (NIS) as reporter gene has been demonstrated to increase uptake of iodine isotopes, for example, for PET imaging in small animals (Groot-Wassink *et al.*, 2002).

Reporter Genes for Magnetic Resonance Imaging

The development of simple, specific, and sensitive reporter genes for broad applications in animal and human MRI is challenging. Only a few have been described and none has been utilized in the studies described in this chapter, but the potential of reporter gene MRI justifies a brief summary. One strategy is based on the intracellular accumulation of an iron-ion-based contrast agent that confers MR imaging visibility due to modulation of the NMR relaxation rates. Expression of an engineered transferrin receptor reporter gene has been shown to enable MR detection based on increased cellular uptake of iron-oxide particles fused to holo-transferrin (Weissleder *et al.*, 2000). A more recent study demonstrated that expression of a metalloprotein reporter from the ferritin family results in a cellular accumulation of endogenous iron that is sufficient for detection by MRI (Genove *et al.*, 2005).

Imaging Modalities

Nuclear

Nuclear modalities make use of established clinical radiology tools, such as the gamma camera, single photon emission computerized tomography (SPECT), and PET. These methods permit *in vivo* detection of free isotopes as well as compounds labeled with isotopes. Typical isotopes used in biomedical research are the gamma-emitters technetium-99m (^{99m}Tc) and indium-111 (^{111}In) or the positron emitter fluorine-18 (^{18}F). The introduction of dedicated imaging equipment has accelerated the application of nuclear modalities to small animal imaging. Advantages of nuclear imaging include sensitive deep tissue imaging, easy signal quantification in tissue samples, and with most techniques, tomographic imaging, but it also requires complex and expensive equipment, the use of radioactive procedures, and frequent in-house radiochemistry for tracer production.

Optical

Optical modalities have been developed for preclinical molecular imaging. They utilize sensitive external cameras for detection of light emission from the body. Light either stems from fluorescence or is produced in biochemical reactions. Optical-imaging methods use both visible (400 nm–700 nm) and NIR (700 nm–900 nm) light. The latter is particularly useful for molecular imaging because light absorption through biological tissues decreases significantly at wavelengths longer than ~ 600 nm reaching a minimum ~ 750 nm (Weissleder and Ntziachristos, 2003). Hence, opti-

mal tissue penetration can be achieved at this wavelength. Optical modalities offer straightforward and fast imaging at moderate costs, but deep tissue optical imaging is restricted as compared to the other imaging techniques described. Nevertheless, recent developments including near-infrared fluorescence molecular tomography (FMT) may allow for sufficient tissue penetrations for human diagnostic imaging (Weissleder and Ntziachristos, 2003).

Magnetic Resonance Imaging

Like nuclear modalities, MRI is already in use in the clinic. Typically, MR images visualize anatomical structures based on their water content via measurement of a signal generated from proton nuclei in response to excitation by radio waves matching the intrinsic frequency of the precessing proton nucleus. Similar imaging can be performed routinely in small-animal MRI. In the molecular imaging arena, the strength of MR is a tailored application for detection of specific cells or even molecules inside anatomical structures at high resolution. These advanced applications may require the basic scientist to seek assistance from an MRI scientist.

Ultrasound

Ultrasound imaging, similar to MRI, provides high-resolution, anatomical information and is widely used in the clinic. Essentially, this modality sends out pulses of ultrasound and detects the returning echos. In ultrasound imaging resolution and penetration depth are inversely related. Small-animal equipment operating at high frequencies is available and typically provides 30–100 μm resolution at 5–15 mm penetration depth.

Preclinical Applications

Reporter gene imaging is currently the modality of choice for preclinical applications. Three principal strategies may be used to introduce an imaging reporter gene into mouse tissue. Either the animal is generated transgenic for the imaging reporter gene, or the reporter gene is delivered through *in vivo* gene transfer. The third option is to modify cells with the reporter gene *ex vivo* and subsequently reintroduce them into the animal. Transplantation of imaging reporter gene-modified cells has been extensively exploited in cancer research, in particular because the proliferative capacity of cancer cells allows for effective nonviral or viral reporter gene modification of the cells as well as rapid tumor formation after host inoculation.

Gene Transcription

Most signal transduction pathways specifically activate target genes. Imaging has been shown to be very useful in the study of transcriptional activation as originally demonstrated in cancer using PET imaging (Doubrovin *et al.*, 2001). Cancer cells were modified to express a HSV TK-GFP dual-reporter gene under control of a p53-dependent enhancer element and subsequently propagated as tumors in rats. In this model, reporter gene activation following treatment with the p53-activating drug N, N′-bis(2-chloroethyl)-N-nitrosourea (BCNU) was imaged. Later, a similar strategy was developed for imaging p53-dependent activation of FLuc (Wang and El-Deiry, 2003b). To allow for detection of cell number in addition to p53 activity, the reporter was combined with a second construct constitutively expressing RLuc. In two studies, combined reporter gene imaging was used to elucidate p53 transcriptional activity during therapy-induced cell death (Wang and El-Deiry, 2003a; Wang *et al.*, 2005). Employing the HSV TK-GFP dual-reporter gene described earlier, yet under control of a TGF-β element responsive to the Smad2/3-Smad4/RUNX complex, imaging provided evidence for Smad signaling in bone metastasis from breast cancer (Kang *et al.*, 2005). Another study addressed cell-cycle progression downstream of p53 and combined a transgenic mouse expressing luciferase from a truncated human E2F 1 promoter with a genetic mouse model of platelet-derived growth factor (PDGF)-induced glioma (Uhrbom *et al.*, 2004). Light emission increased and decreased during tumor growth and treatment with PDGF receptor inhibitor, respectively. Lastly, hypoxia signaling has been imaged (Serganova *et al.*, 2004). In a series of elegant tumor hypoxia models, induction of hypoxia was imaged utilizing a hypoxia response element-controlled HSV TK-GFP reporter. Transcriptional activation in T-cells has also been studied. Nuclear factor of activated T-cells (NFAT)-dependent gene activation in human Jurkat cells modified to express HSV TK-GFP under control of an artificial NFAT-enhancer element upstream of a minimal CMV promoter was reported (Ponomarev *et al.*, 2001).

Imaging tissue-specific transcription has been extensively studied in prostate cancer models. Adams et al. made use of an artificial protein specific antigen (PSA) enhancer/promoter to control expression of the FLuc reporter gene (Adams *et al.*, 2002). Bioluminescence imaging permitted visualization of the dynamics of FLuc expression in PSA-expressing tumors following systemic administration of an adenovirus carrying the imaging reporter gene. To overcome the relatively weak activity of the PSA promoter, a two-step transcriptional activation strategy (TSTA) has been employed (Zhang *et al.*, 2002). In this single-expression cassette strategy, PSA drives expression of a strong transcriptional activator, such as a GAL4-VP-16 fusion protein, which then activates transcription of the reporter gene. This effort resulted in a series of studies utilizing different combinations of reporter gene variants and animal models not only for establishing feasibility but also for investigating androgen receptor signaling (Zhang *et al.*, 2003; Sato *et al.*, 2005). Imaging transcription in tissues other than prostate has also been described. As part of a genetic brain tumor, FLuc expression restricted to the pituitary glands by the pro-opiomelanocortin promoter has been reported (Vooijs *et al.*, 2002).

Imaging angiogenesis is of interest in many diseases. Optical imaging of transcriptional activity during angiogenesis has been reported in a transgenic mouse harboring the FLuc reporter gene under control of a vascular endothelial growth factor (VEGF) promoter (Wang *et al.*, 2006). The investigators improved the relatively weak transcriptional activity of VEGF promoter element by a TSTA amplification strategy similar to the one previously described. The model enabled longitudinal bioluminescence imaging of VEGF activity post-tumor inoculation.

A growing number of imaging-competent transgenic mouse models have been developed in academic as well as industrial laboratories, and many of them are based on luciferase bioluminescence detection. Because they might be of value to the reader's research, we will provide an incomplete list based on the promoters used: cytochrome P450, estrogen receptor, glial fibrillary acidic protein, heme oxygenase, inducible nitric oxide synthase, NFκB, osteocalcin, serum amyloid A protein, Smad, and VEGFR2.

Ribonucleic Acid Biology

In two studies, reporter gene imaging has been utilized for the study of RNA biology. The first study explored the post-transcriptional antifolate regulation of dihydrofolate reductase (DHFR) translation (Mayer-Kuckuk *et al.*, 2002). DHFR binds its cognate mRNA and inhibits its own translation. Addition of methotrexate relieves this translational inhibition by removing DHFR protein from the RNA-protein complex, thus allowing translation to resume, which results in increased levels of DHFR protein. Using DHFR linked to the HSV TK reporter gene, this mechanism was imaged using PET. In the second study, bioluminescence imaging resulted from spliceosome-mediated RNA trans-splicing (SMaRT) (Bhaumik *et al.*, 2004b). Simplified, one part of a split RLuc was expressed as a pre-mRNA containing the 5′-RLuc exon followed by an intron, while the other part was expressed as the 3′-RLuc exon coding pre-trans-splicing molecule (PTM). Utilizing a PTM-coded intron binding domain, the SMaRT reaction trans-splices the split RLuc exons. For *in vivo* imaging of SMaRT generation of RLuc, PEI polycation components were used to systemically deliver the PTM to the pre-mRNA expressing tumor.

Protein Biology

Protein–Protein Interactions

An elegant transcriptional strategy for imaging protein–protein interactions has been derived from the two-hybrid system. Similar to this well-established protein interaction assay, potentially interacting proteins A and B are fused to the GAL4 DNA-binding domain and to the transcriptional activator VP16, respectively. Upon interaction of the proteins and hence restoration of transcriptional activation, an imaging reporter gene placed under control of a minimal promoter containing a GAL4 binding site is expressed and detectable. Independent investigations demonstrate imaging of the known interactions of the ID protein with MyoD (Ray *et al.*, 2002) and of large T-antigen and p53 (Luker *et al.*, 2002, 2003b).

Protein complementation and reconstitution provide an alternative to the approach based on the two-hybrid system. Protein complementation makes use of split proteins, which regain integrity if they come into close proximity. The restored, though not covalently recoupled protein, may serve as intact enzyme. Reactivation of a split luciferase for monitoring protein–protein interactions in a living mouse is an attractive imaging strategy. To demonstrate complementation of a split luciferase protein as a result of a protein interaction, the C- and N-terminal luciferase domains were fused to the two interacting proteins MyoD and ID, respectively (Paulmurugan *et al.*, 2002). The results obtained from this model indeed suggested imaging of heterodimeric protein–protein interactions using a split optical reporter assay *in vivo*. Later, this strategy was explored for the *in vivo* study of HSV TK homodimerization (Massoud *et al.*, 2004). The protein complementation method could also be used to detect drug-mediated protein interaction as described for the rapamycin-mediated heterodimerization of the human proteins FK506-binding protein (FKBP) and rapamycin-binding domain (Luker *et al.*, 2004; Paulmurugan *et al.*, 2004). To achieve improved sensitivity of this strategy, a recent report described fusion of the two human protein-split luciferase dimers using a peptide spacer (Paulmurugan and Gambhir, 2005).

The protein reconstitution method is based on the remarkable process of protein splicing, which occurs in unicellular organisms but not in animals or humans. This self-catalyzed post-translational rearrangement yields a mature protein through the excision of an intervening polypeptide segment, known as intein, from a nonfunctional protein precursor (Paulus, 2000). To utilize this mechanism for imaging the interaction of two proteins, two triple fusion proteins were used (Paulmurugan *et al.*, 2002). One contained the N-terminal fragments of luciferase and intein plus one interacting protein. The other harbored the corresponding C-terminal fragments of luciferase and intein plus the second interacting protein partner. It was shown that following protein–protein interaction the split intein regains function and processes the intein-luciferin polypeptide sequences to yield a fused and therefore functionally active luciferase. Lastly, protein–protein interactions were imaged utilizing bioluminescence resonance energy transfer. In a recent proof-of-principle report, FKBP was fused to the N-terminus of a renilla luciferase donor protein, while the rapamycin-binding domain was coupled to the C-terminus of a mutant GFP acceptor protein (De and Gambhir, 2005). As expected, in the presence of rapamycin, heterodimerization occurred and resulted in resonance energy transfer as detected by increased light emission from the GFP acceptor protein following coelenterazine administration.

Transporters

A unique characteristic of the *Renilla* luciferase substrate coelenterazine is that it enters the cell via p-glycoprotein transport. Hence, in RLuc-modified cells coelenterazine uptake and subsequent light production are a function of p-glycoprotein activity. This allowed for imaging of enhanced P-glycoprotein activity in drug-resistant cells or the forced downregulation of p-glycoprotein following RNA interference (Pichler *et al.*, 2005). Another elegant study described bioluminescence imaging of nuclear transport of protein (Kim *et al.*, 2004). In a triple fusion protein, cytosolic androgen receptor was linked to the C-termini of both RLuc and DnaE intein. The complementary N-terminal fragments of RLuc and intein were expressed in the nucleus. Upon 5α-dihydrotestosterone-induced nuclear entry of the triple fusion protein, intein fragment interaction resulted in RLuc reconstitution due to protein splicing. This event was successfully imaged in mice.

Protein and Cell Degradation

Protein degradation in intact cells under physiological conditions is mediated mainly by the proteasomal degradation machinery. A relatively simple method that allows monitoring of proteasome activity in an indirect fashion has been described (Luker *et al.*, 2003a). The bioluminescence imaging reporter luciferase was tagged with four mutant ubiquitin units. The mutant ubiquitins are resistant to ubiquitin hydrolases, but direct the tagged luciferase for proteasomal degradation. Optical signals correlating with proteasomal activity were obtained indirectly by inhibiting proteasome activity using specific inhibitors and hence accumulating tagged luciferase in the cell.

In a notable report, cells were modified to express a p27-FLuc fusion protein under control of a constitutive viral promoter (Zhang *et al.*, 2004). As postulated, luciferase activity was shown to mimic cell-cycle-dependent p27 turnover. Pharmacological inhibition of cyclin-dependent kinase 2, which is involved in p27 degradation, was detected in tumors by bioluminescence imaging of p27-FLuc accumulation. A more recent study also made use of a luciferin fusion protein strategy for imaging IκBα state (Gross and Piwnica-Worms, 2005). A constitutive promoter drove expression of an IκBα-FLuc protein. Because induction of inhibitor of NFκB Kinases (IKK) is essential for IκBα degradation, this reporter protein indirectly monitored IKK activity. In mice, $I\kappa B\alpha$ degradation during liver inflammation and pharmacological IKK inhibition in tumors were imaged.

To directly image molecular pathways during programmed cell death, a strategy for release of functional FLuc due to caspase-3 activity has been developed (Laxman et al., 2002). Firefly luciferase activity was downregulated by means of sterical interference. The enzyme was fused C- and N-terminal to the estrogen receptor (ER) regulatory domain. Each of the two spacers between ER and FLuc contained a caspase-3 specific cleavage site. The resulting triple fusion protein was shown to release active FLuc following tumor necrosis factor-related apoptosis-inducing ligand (TRAIL) treatment. An alternative strategy based on tetrapeptide-modified DEVD-aminoluciferin has been described (Shah et al., 2005). In FLuc-modified cells, activation of caspase-3 or -7 led to the proteolytic removal of the DEVD sequence and initiation of bioluminescence. Although imaging in living mice was achieved, the animals were killed postimaging apparently due to DEVD-aminoluciferin toxicity.

Imaging Strategies for Clinical Applications

Receptors and Cell-Surface Targets

Imaging strategies that take advantage of antibodies for targeting cell-surface constituents have been investigated for many years, mostly along with the development of radioiummunotherapy. Premier antigens are Lewis Y, glycoprotein 72, and in particular carcinoembryonic antigen (CEA). In the vast majority of studies, the antibodies were radioactively labeled for detection by nuclear imaging. The development of antibody imaging has recently undergone significant improvement with the implementation of engineered monoclonal antibody (mAb) fragments (e.g., diabodies or minibodies) and site-specific radiolabeling (Wu and Senter, 2005). The lower molecular weight mAb fragments exhibit shorter circulation half-lives and increased tumor penetration. This results in maximal tumor uptake within hours and high tumor-to-blood ratio values in less than 24 hr. Because of its clinical relevance due to the introduction of targeted therapy directed against Her2/neu growth factor receptor, recent work has evaluated a variety of antibody fragments for PET imaging of Her2 (Smith-Jones et al., 2004; Olafsen et al., 2005; Robinson et al., 2005). Other advanced strategies, including pretargeting of multivalent bispecific antibodies followed by in vivo labeling using nuclide-labeled hapten, have been recently reported (Sharkey et al., 2005); against CEA this strategy provided a significantly better tumor-to-blood ratio as compared to an anti-CEA Fab' agent. Moreover, imaging other constituents

such as endoglin has been tested (Bredow et al., 2000). An alternative imaging strategy relies on MRI for detection of antibodies carrying MR contrast agent. A dilemma associated with this strategy is the relatively low sensitivity of MRI, which requires delivering as much contrast agent to the presented antigens as possible. As demonstrated for imaging endothelial integrin $\alpha_v\beta_3$, an additional strategy is to couple antibodies to MR contrast agent via biotin-avidin linkers; this strategy has increased target specificity due likely to an antibody contrast agent ratio of greater than one (Sipkins et al., 1998; Anderson et al., 2000). These relatively large immunoconjugates, however, have limited tumor penetration because of their size. To circumvent this problem, a pretargeting approach was tested (Artemov et al., 2003). For imaging, the anti-Her2 mAb was coupled in vivo to contrast agent using an avidin-biotin linker. Newer experimental strategies with respect to imaging include quantum dot-labeled antibodies for optical imaging of prostate cancer (Gao et al., 2004), as well as the use of microbubble-labeled antibodies for ultrasound detection of integrin $\alpha_v\beta_3$ in tumors (Ellegala et al., 2003).

In a relatively straightforward strategy, cell receptors can also be imaged by using isotope-labeled receptor ligands. A significant body of work in this regard has been directed toward neuroimaging, including detection of somatostatin receptor expressing neuroendocrine malignancies (Weiner and Thakur, 2005). Besides being significant for imaging these rather rare cancers, clinical imaging targeting the somatostatin receptors has been extended to other malignancies. In many cases, however, it is likely not to exceed the diagnostic value of ^{18}FDG/PET. Nevertheless, recent work in metastatic breast cancer suggests that ^{99m}Tc-depretoide might be a good predictive imaging marker for early endocrine treatment response (Van Den Bossche et al., 2006). In preclinical studies, a variety of other receptors, including cholecystokinin B receptor, epidermal growth factor receptor, gastrin-releasing peptide receptors, α-melanocyte-stimulating hormone receptor, neurotensin receptor, sigma-2 receptor, and vasoactive intestinal peptide/pituitary adenylate cyclase-activating peptide receptor, are under investigation. However, for successful clinical imaging, novel ligands have to meet stringent requirements, for instance, marked specificity and minimal physiological activity.

Integrins are mediators of cell adhesion and signaling. Participating in both tumor angiogenesis and metastasis, the integrin $\alpha_v\beta_3$ represents a desirable imaging target. An attractive imaging strategy relies on nuclide-labeled peptides that display the RGD motif. Peptides containing arginine-glycine-asparate (RGD) can be designed to recognize integrin $\alpha_v\beta_3$, with high specificity, and attempts have been made to find peptides possessing satisfactory in vivo tracer characteristics. An excellent overview listed over 20 radiolabeled $\alpha_v\beta_3$ antagonists under investigation (Haubner and Wester, 2004).

Currently, three classes of RGD peptides for imaging have received the most attention. First, Galacto-RGD peptides were shown to have favorable *in vivo* characteristics, including high tumor-to-blood and tumor-to-muscle ratios (Haubner *et al.*, 2001). Indeed, first clinical studies were initiated recently with ^{18}F-Galacto-RGD (Beer *et al.*, 2005; Haubner *et al.*, 2005). The results support the tracer's feasibility. However, large inter- and intraindividual variations in integrin presence were detected, and a need for better interpretation of the imaging results was suggested. Second, modified di- or multimeric RGD peptides can serve as Galacto-RGD alternatives (Zhang *et al.*, 2006). Third, RGD peptides modified with NIR fluorophores or MR contrast agents represent a new class of RGD peptides for optical imaging (Chen *et al.*, 2004) or MRI (Winter *et al.*, 2003), respectively.

Enzyme Activities

In an important contribution, optical imaging of tumor-associated protease activity was achieved using protease-activated probes that present a poly-L-lysine cleavage site (Weissleder *et al.*, 1999). The paradigm behind this strategy is the enzymatic, on-site generation of active fluorophore from inactive precursors. Using a lysosome lysate-based assay, it was shown that cysteine protease inhibitors, trypsin inhibitors, and trypsinlike serine protease inhibitors effectively blocked probe activation (Weissleder *et al.*, 1999). A significant feature of the probe is a good biodistribution characteristic suitable for *in vivo* use. This aspect, along with the modular design of the probe, prompted development of derivatives with different protease specificities. A cathepsin D specific imaging probe in which release of the active fluorophore required recognition and cleavage of a cathepsin D-specific 7 amino acid peptide has been developed (Tung *et al.*, 2000). Furthermore, a metalloprotease-sensitive probe with high specificity for metalloprotease 2, which in a similar fashion made use of a metalloprotease-specific cleavage site, has been described (Bremer *et al.*, 2001). The clinical implementation of the discussed strategies is likely.

Transporters

The idea of imaging folate receptor activity through increased cellular accumulation of labeled folates has been explored in a variety of studies over the past 10 years. Folate receptor is frequently overexpresssed in cancer cells, and enhanced expression of subtypes can be observed in activated macrophages and brain cells. Several reports demonstrated the feasibility of imaging cancer cell uptake of radioactively labeled folate conjugates, and early clinical studies were initiated (Siegel *et al.*, 2003; Reddy *et al.*, 2004). This approach has been extended to optical and MRI imaging utilizing near-infrared fluorophore and MR contrast agent-conjugated folate derivatives, respectively (Moon *et al.*, 2003; Choi *et al.*, 2004). As an alternate to folate, the anti-neoplastic folate analog methotrexate has been studied for imaging purposes (Ilgan *et al.*, 1998). Moreover, imaging macrophage infiltration in dysplastic intestinal adenoma was facilitated with a fluorescently labeled folate (Chen *et al.*, 2005).

The great clinical importance of multidrug resistance (MDR) in cancer therapy warrants predictive imaging methods to spare patients from unnecessary treatment. Principal drug efflux transporters such as the p-glycoprotein have a considerable broad specificity, and imaging its activity is readily facilitated by detection of lack of cellular accumulation of a p-glycoprotein-dependent tracer. The radiopharmaceutical MIBI, a p-glycoprotein substrate, was used to first demonstrate this imaging strategy (Piwnica-Worms *et al.*, 1993). Since then, nuclide-labeled 2-methoxy isobutyl isonitrile (MIBI) has entered clinical testing, and taking advantage of the same principle, many alternative radiopharmaceuticals have been explored as imaging markers for 2-methoxy isobutyl isonitrile (MDR). The latest preclinical work suggests that a variety of factors, including accessibility, hypoxia, or early apoptosis, can influence cellular MIBI accumulation (Moretti *et al.*, 2005), while current clinical studies attempt to optimize imaging protocols, for example, through dynamic imaging. Furthermore, attention is given to the question if imaging of MDR can be predictive.

Cell Death

Cell death via the apoptotic pathway is mediated through a series of highly regulated molecular events. Among them are the increase in phosphatidylserine (PS) translocation from the inner to the outer membrane leaflet as well as the caspase cascade. A noted study demonstrated imaging of apoptosis using radioactively labeled Annexin V, a protein that binds with high affinity to PS (Blankenberg *et al.*, 1998). This report stimulated further research in this area and has resulted in a very large number of similar radioligand-based strategies, as recently reviewed (Lahorte *et al.*, 2004). A first clinical study suggesting response prediction with ^{99m}Tc-Annexin V imaging was reported in 2002 (Belhocine *et al.*, 2002), while recent experimental work brought this strategy back to the mouse model for further studies in respect to therapy response (Mandl *et al.*, 2004). Similarly, fluorescent Annexin V can be used for nonradioactive apoptosis imaging, including FMT (Ntziachristos *et al.*, 2004). An alternate strategy utilized the C_2 domain of synaptotagmin I instead of Annexin V (Zhao *et al.*, 2001). The C_2 domain of synaptotagmin I was labeled with iron oxide particles for MRI of drug-induced apoptosis. Finally, a caspase-1-sensitive fluorescent imaging probe has also been described (Messerli *et al.*, 2004). Activation of the probe in tumors overexpressing caspase-1 was adequate for imaging.

In conclusion, although we have described several exciting developments in the field of molecular imaging in preclinical settings, the area of molecular imaging in the clinic is still in its infancy. Barriers that need to be overcome include development of clinically compatible optical-imaging probes and technologies with sensitivities powerful enough to overcome deep tissue challenges. Nuclear imaging using clinically applicable probes is increasing, and it is hoped that this approach will facilitate clinical investigations and have an impact not only for diagnostic purposes but also for dynamic response measurements as well pharmacokinetic and pharmacodynamic studies in clinical trial settings.

Acknowledgments

The authors are indebted to the following for introducing them to molecular imaging: Dr. Ronald G. Blasberg, Dr. Mikahil Doubrovin, Dr. Juri Gelovani, and Dr. Jason A. Koutcher. We thank Dr. Joseph R. Bertino for critically reading the manuscript. We wish to acknowledge all the important contributions we had to exclude from the references due to space limitations.

References

Adams, J.Y., Johnson, M., Sato, M., Berger, F., Gambhir, S.S., Carey, M., Iruela-Arispe, M.L., and Wu, L. 2002. Visualization of advanced human prostate cancer lesions in living mice by a targeted gene transfer vector and optical imaging. *Nat. Med. 8*:891–897.

Anderson, S.A., Rader, R.K., Westlin, W.F., Null, C., Jackson, D., Lanza, G.M., Wickline, S.A., and Kotyk, J.J. 2000. Magnetic resonance contrast enhancement of neovasculature with alpha(v)beta(3)-targeted nanoparticles. *Magn. Reson. Med. 44*:433–439.

Artemov, D., Mori, N., Ravi, R., and Bhujwalla, Z.M. 2003. Magnetic resonance molecular imaging of the HER-2/neu receptor. *Cancer Res. 63*:2723–2727.

Beer, A.J., Haubner, R., Goebel, M., Luderschmidt, S., Spilker, M.E., Wester, H.J., Weber, W.A., and Schwaiger, M. 2005. Biodistribution and pharmacokinetics of the alphavbeta3-selective tracer 18F-galacto-RGD in cancer patients. *J. Nucl. Med. 46*:1333–1341.

Belhocine, T., Steinmetz, N., Hustinx, R., Bartsch, P., Jerusalem, G., Seidel, L., Rigo, P., and Green, A. 2002. Increased uptake of the apoptosis-imaging agent (99m)Tc recombinant human Annexin V in human tumors after one course of chemotherapy as a predictor of tumor response and patient prognosis. *Clin. Cancer Res. 8*:2766–2774.

Bhaumik, S., and Gambhir, S.S. 2002. Optical imaging of *Renilla* luciferase reporter gene expression in living mice. *Proc. Natl. Acad. Sci. USA 99*:377–382.

Bhaumik, S., Lewis, X.Z., and Gambhir, S.S. 2004a. Optical imaging of *Renilla* luciferase, synthetic *Renilla* luciferase, and firefly luciferase reporter gene expression in living mice. *J. Biomed. Opt. 9*:578–586.

Bhaumik, S., Walls, Z., Puttaraju, M., Mitchell, L.G., and Gambhir, S.S. 2004b. Molecular imaging of gene expression in living subjects by spliceosome-mediated RNA trans-splicing. *Proc. Natl. Acad. Sci. USA 101*:8693–8698.

Blankenberg, F.G., Katsikis, P.D., Tait, J.F., Davis, R.E., Naumovski, L., Ohtsuki, K., Kopiwoda, S., Abrams, M.J., Darkes, M., Robbins, R.C., Maecker, H.T., and Strauss, H.W. 1998. *In vivo* detection and imaging of phosphatidylserine expression during programmed cell death. *Proc. Natl. Acad. Sci. USA 95*:6349–6354.

Bredow, S., Lewin, M., Hofmann, B., Marecos, E., and Weissleder, R. 2000. Imaging of tumour neovasculature by targeting the TGF-beta binding receptor endoglin. *Eur. J. Cancer 36*:675–681.

Bremer, C., Tung, C.H., and Weissleder, R. 2001. *In vivo* molecular target assessment of matrix metalloproteinase inhibition. *Nat. Med. 7*:743–748.

Chen, W.T., Khazaie, K., Zhang, G., Weissleder, R., and Tung, C.H. 2005. Detection of dysplastic intestinal adenomas using a fluorescent folate imaging probe. *Mol. Imaging 4*:67–74.

Chen, X., Conti, P.S., and Moats, R.A. 2004. *In vivo* near-infrared fluorescence imaging of integrin alphavbeta3 in brain tumor xenografts. *Cancer Res. 64*:8009–8014.

Choi, H., Choi, S.R., Zhou, R., Kung, H.F., and Chen, I.W. 2004. Iron oxide nanoparticles as magnetic resonance contrast agent for tumor imaging via folate receptor-targeted delivery. *Acad. Radiol. 11*:996–1004.

De, A., and Gambhir, S.S. 2005. Noninvasive imaging of protein–protein interactions from live cells and living subjects using bioluminescence resonance energy transfer. *FASEB J. 19*:2017–2019.

Doubrovin, M., Ponomarev, V., Beresten, T., Balatoni, J., Bornmann, W., Finn, R., Humm, J., Larson, S., Sadelain, M., Blasberg, R., and Gelovani Tjuvajev, J. 2001. Imaging transcriptional regulation of p53-dependent genes with positron emission tomography *in vivo*. *Proc. Natl. Acad. Sci. USA 98*:9300–9305.

Ellegala, D.B., Leong-Poi, H., Carpenter, J.E., Klibanov, A.L., Kaul, S., Shaffrey, M.E., Sklenar, J., and Lindner, J.R. 2003. Imaging tumor angiogenesis with contrast ultrasound and microbubbles targeted to alpha(v)beta3. *Circulation 108*:336–341.

Gambhir, S.S., Barrio, J.R., Phelps, M.E., Iyer, M., Namavari, M., Satyamurthy, N., Wu, L., Green, L.A., Bauer, E., MacLaren, D.C., Nguyen, K., Berk, A.J., Cherry, S.R., and Herschman, H.R. 1999. Imaging adenoviral-directed reporter gene expression in living animals with positron emission tomography. *Proc. Natl. Acad. Sci. USA 96*:2333–2338.

Gambhir, S.S., Bauer, E., Black, M.E., Liang, Q., Kokoris, M.S., Barrio, J.R., Iyer, M., Namavari, M., Phelps, M.E., and Herschman, H.R. 2000a. A mutant herpes simplex virus type 1 thymidine kinase reporter gene shows improved sensitivity for imaging reporter gene expression with positron emission tomography. *Proc. Natl. Acad. Sci. USA 97*:2785–2790.

Gambhir, S.S., Herschman, H.R., Cherry, S.R., Barrio, J.R., Satyamurthy, N., Toyokuni, T., Phelps, M.E., Larson, S.M., Balatoni, J., Finn, R., Sadelain, M., Tjuvajev, J., and Blasberg, R. 2000b. Imaging transgene expression with radionuclide imaging technologies. *Neoplasia 2*:118–138.

Gao, X., Cui, Y., Levenson, R.M., Chung, L.W., and Nie, S. 2004. *In vivo* cancer targeting and imaging with semiconductor quantum dots. *Nat. Biotechnol. 22*:969–976.

Genove, G., DeMarco, U., Xu, H., Goins, W.F., and Ahrens, E.T. 2005. A new transgene reporter for *in vivo* magnetic resonance imaging. *Nat. Med. 11*:450–454.

Greer, L.F., 3rd, and Szalay, A.A. 2002. Imaging of light emission from the expression of luciferases in living cells and organisms: a review. *Luminescence 17*:43–74.

Groot-Wassink, T., Aboagye, E.O., Glaser, M., Lemoine, N.R., and Vassaux, G. 2002. Adenovirus biodistribution and noninvasive imaging of gene expression *in vivo* by positron emission tomography using human sodium/iodide symporter as reporter gene. *Hum. Gene Ther. 13*:1723–1735.

Gross, S., and Piwnica-Worms, D. 2005. Real-time imaging of ligand-induced IKK activation in intact cells and in living mice. *Nat. Methods 2*:607–614.

Haubner, R., Weber, W.A., Beer, A.J., Vabuliene, E., Reim, D., Sarbia, M., Becker, K.F., Goebel, M., Hein, R., Wester, H.J., Kessler, H., and Schwaiger, M. 2005. Noninvasive visualization of the activated alphavbeta3 integrin in cancer patients by positron emission tomography and [18F]Galacto-RGD. *PLoS Med. 2:*e70.

Haubner, R., and Wester, H.J. 2004. Radiolabeled tracers for imaging of tumor angiogenesis and evaluation of anti-angiogenic therapies. *Curr. Pharm. Des. 10:*1439–1455.

Haubner, R., Wester, H.J., Weber, W.A., Mang, C., Ziegler, S.I., Goodman, S.L., Senekowitsch-Schmidtke, R., Kessler, H., and Schwaiger, M. 2001. Noninvasive imaging of alpha(v)beta3 integrin expression using 18F-labeled RGD-containing glycopeptide and positron emission tomography. *Cancer Res. 61:*1781–1785.

Ilgan, S., Yang, D.J., Higuchi, T., Zareneyrizi, F., Kim, E.E., and Podoloff, D.A. 1998. Imaging tumor folate receptors using 111In-DTPA-methotrexate. *Cancer Biother. Radiopharm. 13:*177–184.

Kang, Y., He, W., Tulley, S., Gupta, G.P., Serganova, I., Chen, C.R., Manova-Todorova, K., Blasberg, R., Gerald, W.L., and Massague, J. 2005. Breast cancer bone metastasis mediated by the Smad tumor suppressor pathway. *Proc. Natl. Acad. Sci. USA 102:*13909–13914.

Kim, S.B., Ozawa, T., Watanabe, S., and Umezawa, Y. 2004. High-throughput sensing and noninvasive imaging of protein nuclear transport by using reconstitution of split Renilla luciferase. *Proc. Natl. Acad. Sci. USA 101:*11542–11547.

Lahorte, C.M., Vanderheyden, J.L., Steinmetz, N., Van de Wiele, C., Dierckx, R.A., and Slegers, G. 2004. Apoptosis-detecting radioligands: current state of the art and future perspectives. *Eur. J. Nucl. Med. Mol. Imaging 31:*887–919.

Laxman, B., Hall, D.E., Bhojani, M.S., Hamstra, D.A., Chenevert, T.L., Ross, B.D., and Rehemtulla, A. 2002. Noninvasive real-time imaging of apoptosis. *Proc. Natl. Acad. Sci. USA 99:*16551–16555.

Liang, Q., Satyamurthy, N., Barrio, J.R., Toyokuni, T., Phelps, M.P., Gambhir, S.S., and Herschman, H.R. 2001. Noninvasive, quantitative imaging in living animals of a mutant dopamine D2 receptor reporter gene in which ligand binding is uncoupled from signal transduction. *Gene Ther. 8:*1490–1498.

Luker, G.D., Pica, C.M., Song, J., Luker, K.E., and Piwnica-Worms, D. 2003a. Imaging 26S proteasome activity and inhibition in living mice. *Nat. Med 9:*969–973.

Luker, G.D., Sharma, V., Pica, C.M., Dahlheimer, J.L., Li, W., Ochesky, J., Ryan, C.E., Piwnica-Worms, H., and Piwnica-Worms, D. 2002. Noninvasive imaging of protein–protein interactions in living animals. *Proc. Natl. Acad. Sci. USA 99:*6961–6966.

Luker, G.D., Sharma, V., Pica, C.M., Prior, J.L., Li, W., and Piwnica-Worms, D. 2003b. Molecular imaging of protein–protein interactions: controlled expression of p53 and large T-antigen fusion proteins *in vivo. Cancer Res. 63:*1780–1788.

Luker, K.E., Smith, M.C., Luker, G.D., Gammon, S.T., Piwnica-Worms, H., and Piwnica-Worms, D. 2004. Kinetics of regulated protein–protein interactions revealed with firefly luciferase complementation imaging in cells and living animals *Proc. Natl. Acad. Sci. USA 101:*12288–12293.

Maggi, A., and Ciana, P. 2005. Reporter mice and drug discovery and development. *Nat. Rev. Drug Discov. 4:*249–255.

Mandl, S.J., Mari, C., Edinger, M., Negrin, R.S., Tait, J.F., Contag, C.H., and Blankenberg, F.G. 2004. Multi-modality imaging identifies key times for annexin V imaging as an early predictor of therapeutic outcome. *Mol. Imaging 3:*1 8.

Massoud, T.F., Paulmurugan, R., and Gambhir, S.S. 2004. Molecular imaging of homodimeric protein–protein interactions in living subjects. *FASEB J. 18:*1105–1107.

Mayer-Kuckuk, P., Banerjee, D., Malhotra, S., Doubrovin, M., Iwamoto, M., Akhurst, T., Balatoni, J., Bornmann, W., Finn, R., Larson, S., Fong, Y., Gelovani Tjuvajev, J., Blasberg, R., and Bertino, J.R. 2002. Cells exposed to antifolates show increased levels of proteins fused to dihydrofolate reductase: a method to modulate gene expression. *Proc. Natl. Acad. Sci. USA 99:*3400–3405.

Messerli, S.M., Prabhakar, S., Tang, Y., Shah, K., Cortes, M.L., Murthy, V., Weissleder, R., Breakefield, X.O., and Tung, C.H. 2004. A novel method for imaging apoptosis using a caspase-1 near-infrared fluorescent probe. *Neoplasia 6:*95–105.

Moon, W.K., Lin, Y., O'Loughlin, T., Tang, Y., Kim, D.E., Weissleder, R., and Tung, C.H. 2003. Enhanced tumor detection using a folate receptor-targeted near-infrared fluorochrome conjugate. *Bioconjug. Chem. 14:*539–545.

Moretti, J.L., Hauet, N., Caglar, M., Rebillard, O., and Burak, Z. 2005. To use MIBI or not to use MIBI? That is the question when assessing tumour cells. *Eur. J. Nucl. Med. Mol. Imaging 32:*836–842.

Ntziachristos, V., Schellenberger, E.A., Ripoll, J., Yessayan, D., Graves, E., Bogdanov, A., Jr., Josephson, L., and Weissleder, R. 2004. Visualization of antitumor treatment by means of fluorescence molecular tomography with an annexin V-Cy5.5 conjugate. *Proc. Natl. Acad. Sci. USA 101:*12294–12299.

Olafsen, T., Kenanova, V.E., Sundaresan, G., Anderson, A.L., Crow, D., Yazaki, P.J., Li, L., Press, M.F., Gambhir, S.S., Williams, L.E., Wong, J.Y., Raubitschek, A.A., Shively, J.E., and Wu, A.M. 2005. Optimizing radiolabeled engineered anti-p185HER2 antibody fragments for *in vivo* imaging. *Cancer Res. 65:*5907–5916.

Paulmurugan, R., and Gambhir, S.S. 2005. Novel fusion protein approach for efficient high-throughput screening of small molecule-mediating protein–protein interactions in cells and living animals. *Cancer Res. 65:*7413–7420.

Paulmurugan, R., Massoud, T.F., Huang, J., and Gambhir, S.S. 2004. Molecular imaging of drug-modulated protein–protein interactions in living subjects. *Cancer Res. 64:*2113–2119.

Paulmurugan, R., Umezawa, Y., and Gambhir, S.S. 2002. Noninvasive imaging of protein–protein interactions in living subjects by using reporter protein complementation and reconstitution strategies. *Proc. Natl. Acad. Sci. USA 99:*15608–15613.

Paulus, H. 2000. Protein splicing and related forms of protein autoprocessing. *Annu. Rev. Biochem. 69:*447–496.

Pichler, A., Zelcer, N., Prior, J.L., Kuil, A.J., and Piwnica-Worms, D. 2005. *In vivo* RNA interference-mediated ablation of MDR1 P-glycoprotein. *Clin. Cancer Res. 11:*4487–4494.

Piwnica-Worms, D., Chiu, M.L., Budding, M., Kronauge, J.F., Kramer, R.A., and Croop, J.M. 1993. Functional imaging of multidrug-resistant P-glycoprotein with an organotechnetium complex. *Cancer Res. 53:*977–984.

Ponomarev, V., Doubrovin, M., Lyddane, C., Beresten, T., Balatoni, J., Bornman, W., Finn, R., Akhurst, T., Larson, S., Blasberg, R., Sadelain, M., and Tjuvajev, J.G. 2001. Imaging TCR-dependent NFAT-mediated T-cell activation with positron emission tomography *in vivo. Neoplasia 3:*480–488.

Ray, P., Pimenta, H., Paulmurugan, R., Berger, F., Phelps, M.E., Iyer, M., and Gambhir, S.S. 2002. Noninvasive quantitative imaging of protein–protein interactions in living subjects. *Proc. Natl. Acad. Sci. USA 99:*3105–3110.

Reddy, J.A., Xu, L.C., Parker, N., Vetzel, M., and Leamon, C.P. 2004. Preclinical evaluation of (99m)Tc-EC20 for imaging folate receptor-positive tumors. *J. Nucl. Med. 45:*857–866.

Robinson, M.K., Doss, M., Shaller, C., Narayanan, D., Marks, J.D., Adler, L.P., Gonzalez Trotter, D.E., and Adams, G.P. 2005. Quantitative immuno-positron emission tomography imaging of HER2-positive tumor xenografts with an iodine-124 labeled anti-HER2 diabody. *Cancer Res. 65:*1471–1478.

Rudin, M., and Weissleder, R. 2003. Molecular imaging in drug discovery and development. *Nat. Rev. Drug Discov. 2:*123–131.

Sato, M., Johnson, M., Zhang, L., Gambhir, S.S., Carey, M., and Wu, L. 2005. Functionality of androgen receptor-based gene expression imaging in hormone refractory prostate cancer. *Clin. Cancer Res. 11:*3743–3749.

Serganova, I., Doubrovin, M., Vider, J., Ponomarev, V., Soghomonyan, S., Beresten, T., Ageyeva, L., Serganov, A., Cai, S., Balatoni, J., Blasberg, R., and Gelovani, J. 2004. Molecular imaging of temporal dynamics and spatial heterogeneity of hypoxia-inducible factor-1 signal transduction activity in tumors in living mice. *Cancer Res. 64*:6101–6108.

Shah, K., Tung, C.H., Breakefield, X.O., and Weissleder, R. 2005. *In vivo* imaging of S-TRAIL-mediated tumor regression and apoptosis. *Mol. Ther. 11*:926–931.

Sharkey, R.M., Cardillo, T.M., Rossi, E.A., Chang, C.H., Karacay, H., McBride, W.J., Hansen, H.J., Horak, I.D., and Goldenberg, D.M. 2005. Signal amplification in molecular imaging by pretargeting a multivalent, bispecific antibody. *Nat. Med. 11*:1250–1255.

Siegel, B.A., Dehdashti, F., Mutch, D.G., Podoloff, D.A., Wendt, R., Sutton, G.P., Burt, R.W., Ellis, P.R., Mathias, C.J., Green, M.A., and Gershenson, D.M. 2003. Evaluation of 111In-DTPA-folate as a receptor-targeted diagnostic agent for ovarian cancer: initial clinical results. *J. Nucl. Med. 44*:700–707.

Sipkins, D.A., Cheresh, D.A., Kazemi, M.R., Nevin, L.M., Bednarski, M.D., and Li, K.C. 1998. Detection of tumor angiogenesis *in vivo* by alphaVbeta3-targeted magnetic resonance imaging. *Nat. Med. 4*:623–626.

Smith-Jones, P.M., Solit, D.B., Akhurst, T., Afroze, F., Rosen, N., and Larson, S.M. 2004. Imaging the pharmacodynamics of HER2 degradation in response to Hsp90 inhibitors. *Nat. Biotechnol. 22*:701–706.

Tang, Y., Shah, K., Messerli, S.M., Snyder, E., Breakefield, X., and Weissleder, R. 2003. *In vivo* tracking of neural progenitor cell migration to glioblastomas. *Hum. Gene Ther. 14*:1247–1254.

Tannous, B.A., Kim, D.E., Fernandez, J.L., Weissleder, R., and Breakefield, X.O. 2005. Codon-optimized Gaussia luciferase cDNA for mammalian gene expression in culture and *in vivo*. *Mol. Ther. 11*:435–443.

Tjuvajev, J.G., Finn, R., Watanabe, K., Joshi, R., Oku, T., Kennedy, J., Beattie, B., Koutcher, J., Larson, S., and Blasberg, R.G. 1996. Noninvasive imaging of herpes virus thymidine kinase gene transfer and expression: a potential method for monitoring clinical gene therapy. *Cancer Res. 56*:4087–4095.

Tung, C.H., Mahmood, U., Bredow, S., and Weissleder, R. 2000. *In vivo* imaging of proteolytic enzyme activity using a novel molecular reporter. *Cancer Res. 60*:4953–4958.

Uhrbom, L., Nerio, E., and Holland, E.C. 2004. Dissecting tumor maintenance requirements using bioluminescence imaging of cell proliferation in a mouse glioma model. *Nat. Med. 10*:1257–1260.

Van Den Bossche, B., Van Belle, S., De Winter, F., Signore, A., and Van de Wiele, C. 2006. Early prediction of endocrine therapy effect in advanced breast cancer patients using 99mTc-depreotide scintigraphy. *J. Nucl. Med. 47*:6–13.

Vooijs, M., Jonkers, J., Lyons, S., and Berns, A. 2002. Noninvasive imaging of spontaneous retinoblastoma pathway-dependent tumors in mice. *Cancer Res. 62*:1862–1867.

Wang, S., and El-Deiry, W.S. 2003a. Requirement of p53 targets in chemosensitization of colonic carcinoma to death ligand therapy. *Proc. Natl. Acad. Sci. USA 100*:15095–15100.

Wang, W., and El-Deiry, W.S. 2003b. Bioluminescent molecular imaging of endogenous and exogenous p53-mediated transcription *in vitro* and *in vivo* using an HCT116 human colon carcinoma xenograft model. *Cancer Biol. Ther. 2*:196–202.

Wang, W., Ho, W.C., Dicker, D.T., Mackinnon, C., Winkler, J.D., Marmorstein, R., and El-Deiry, W.S. 2005. Acridine derivatives activate p53 and induce tumor cell death through Bax. *Cancer Biol. Ther. 4*: 893–898.

Wang, Y., Iyer, M., Annala, A., Wu, L., Carey, M., and Gambhir, S.S. 2006. Noninvasive indirect imaging of vascular endothelial growth factor gene expression using bioluminescence imaging in living transgenic mice. *Physiol. Genomics 24*:173–180.

Weiner, R.E., and Thakur, M.L. 2005. Radiolabeled peptides in oncology: role in diagnosis and treatment. *BioDrugs 19*:145–163.

Weissleder, R., and Mahmood, U. 2001. Molecular imaging. *Radiology 219*:316–333.

Weissleder, R., Moore, A., Mahmood, U., Bhorade, R., Benveniste, H., Chiocca, E.A., and Basilion, J.P. 2000. *In vivo* magnetic resonance imaging of transgene expression. *Nat. Med. 6*:351–355.

Weissleder, R., and Ntziachristos, V. 2003. Shedding light onto live molecular targets. *Nat. Med. 9*:123–128.

Weissleder, R., Tung, C.H., Mahmood, U., and Bogdanov, A., Jr. 1999. *In vivo* imaging of tumors with protease-activated near-infrared fluorescent probes. *Nat. Biotechnol. 17*:375–378.

Winter, P.M., Caruthers, S.D., Kassner, A., Harris, T.D., Chinen, L.K., Allen, J.S., Lacy, E.K., Zhang, H., Robertson, J.D., Wickline, S.A., and Lanza, G.M. 2003. Molecular imaging of angiogenesis in nascent Vx-2 rabbit tumors using a novel alpha(nu)beta3-targeted nanoparticle and 1.5 tesla magnetic resonance imaging. *Cancer Res. 63*:5838–5843.

Wu, A.M., and Senter, P.D. 2005. Arming antibodies: prospects and challenges for immunoconjugates. *Nat. Biotechnol. 23*:1137–1146.

Zhang, G.J., Safran, M., Wei, W., Sorensen, E., Lassota, P., Zhelev, N., Neuberg, D.S., Shapiro, G., and Kaelin, W.G., Jr. 2004. Bioluminescent imaging of Cdk2 inhibition *in vivo*. *Nat. Med. 10*:643–648.

Zhang, L., Adams, J.Y., Billick, E., Ilagan, R., Iyer, M., Le, K., Smallwood, A., Gambhir, S.S., Carey, M., and Wu, L. 2002. Molecular engineering of a two-step transcription amplification (TSTA) system for transgene delivery in prostate cancer. *Mol. Ther. 5*:223–232.

Zhang, L., Johnson, M., Le, K.H., Sato, M., Ilagan, R., Iyer, M., Gambhir, S.S., Wu, L., and Carey, M. 2003. Interrogating androgen receptor function in recurrent prostate cancer. *Cancer Res. 63*:4552–4560.

Zhang, X., Xiong, Z., Wu, Y., Cai, W., Tseng, J.R., Gambhir, S.S., and Chen, X. 2006. Quantitative PET Imaging of Tumor Integrin {alpha}v{beta}3 Expression with 18F-FRGD2. *J. Nucl. Med. 47*:113–121.

Zhao, H., Doyle, T.C., Coquoz, O., Kalish, F., Rice, B.W., and Contag, C.H. 2005. Emission spectra of bioluminescent reporters and interaction with mammalian tissue determine the sensitivity of detection in vivo. *J. Biomed. Opt. 10*:41210.

Zhao, M., Beauregard, D.A., Loizou, L., Davletov, B., and Brindle, K.M. 2001. Non-invasive detection of apoptosis using magnetic resonance imaging and a targeted contrast agent. *Nat. Med. 7*:1241–1244.

Zinn, K.R., and Chaudhuri, T.R. 2002. The type 2 human somatostatin receptor as a platform for reporter gene imaging. *Eur. J. Nucl. Med. Mol. Imaging 29*:388–399.

2

Synthesis of
^{18}F-fluoromisonidazole Tracer for
Positron Emission Tomography

Ganghua Tang

Introduction

Hypoxia is known to be an important prognostic factor in human oncology (Mahy *et al.*, 2004). The presence of hypoxia within tumors can be correlated with a resistance to traditional chemotherapy and radiotherapy (Lewis *et al.*, 2001; Lewis and Welch, 2001). Detection and quantification of oxygen levels in human tumors may have a significant influence on the outcome evaluation of patients treated with radiation and on the use of hypoxia-specific chemotherapeutic agents (Gronroos *et al.*, 2001). A variety of techniques such as histological studies, oxygen concentration (pO_2) probe measurements, and scintigraphic studies, for measuring oxygen in tissues, have been developed over the years. To date, the invasive technique of using computerized polar graphic oxygen-sensitive probe measurements is regarded to be the gold standard for measuring the hypoxic fractions in tumors (Sorger *et al.*, 2003). However, the development for a noninvasive approach to detect tumor hypoxia is still a challenge for oncological molecular imaging. The most encouraging imaging techniques have been obtained using the selective uptake of radiotracers in hypoxic tumors, combined with the functional positron emission tomography (PET) or single photon emission computerized tomography (SPECT) imaging.

The use of appropriately radiolabeled 2-nitroimidazole radiosensitizers represents an attractive alternative to scintigraphic imaging of tumor hypoxia. 2-nitroimidazoles undergo an intracellular oxygen-reversible single-electron chemical reduction in hypoxic environments, and subsequently the forming reactive oxygen radicals covalently bind to micromolecules, mainly to thiol-containing proteins (Mahy *et al.*, 2004). The nitroimidazole adducts hence trapped into hypoxic cells can be detected by the use of radiolabeled tracers. During the last decade, several potential ^{123}I or ^{99m}Tc-labeled 2-nitroimidazole derivatives have been synthesized and validated in preclinical and early clinical studies with SPECT imaging.

Among noninvasive techniques, the most common method uses 1H-1-(3-[^{18}F]fluoro-2-hydroxypropyl)-2-nitroimidazole (fluorine-18 fluoromisonidazole, FMISO) with PET imaging. However, FMISO exhibits some disadvantages, such as the slow clearance kinetics resulting in delaying specific imaging for 2–3 hr postinjection, and high lipophilicity of FMISO resulting in substantially high background, which can interfere with the image's quality. More recently, other ^{18}F-labeled compounds, such as ^{18}F]fluoroetanidazole (Rasey et al., 1999); ^{18}F]fluoroery-thronitroimidazole (FETNIM) (Gronroos et al., 2001); 2-(2-nitroimidazol-l-yl)-N-(3,3,3-^{18}F]fluoropropyl)acetamide (EF3) (Mahy et al., 2004); 2-(2-nitro-1[H]-imidazol-l-yl)-N-(2,2,3,3,3-^{18}F]pentafluoropropyl)- acetamide (EF5) (Ziemer et al., 2003); ^{18}F]fluoroazomycinarabinofura-noside (FAZA) (Sorger et al., 2003); [124]I]iodo-azomycin-galactoside (IAZG) (Zanzonico et al., 2004); and [60/61/62/64]Cu] copper-diacetyl-bis(N4-methylthio semicarbazone) (Cu-ATSM) (Chapman et al., 2001; Laforest et al., 2005; Lewis et al., 2001), have been synthesized and validated in pre-clinical or early clinical studies with PET imaging. Although these potential PET tracers can overcome some limitations of FMISO, FMISO still is now the most widely used radio-tracer in clinical PET imaging. FMISO is currently being used as an *in vivo* marker to noninvasively assess hypoxia in human malignancies and in the hearts of patients with myocardial ischemia with PET imaging (Graham et al., 1997; Kamarainen et al., 2004), and plays an important role in predicting the oxygenation status in tumors during radio-therapy.

The synthetic strategies of FMISO can be divided into two main groups: (1) a nucleophilic substitution on a pro-tected precursor with subsequent removal of the protection group (Kamarainen et al., 2004; Lim and Berridge, 1993; Patt et al., 1999) or epoxide ring-opening (Jerabek et al., 1986); and (2) the production of a ^{18}F-labeled intermediate epifluorohydrin with subsequent coupling to the nitroimida-zole moiety under basic conditions (Kamarainen et al., 2004). Both approaches have their limitations. The second method and the epoxide ring-opening in the first method give low radiochemical yield (Kamarainen et al., 2004; Jerabek et al., 1986). On the whole, the most promising direct-labeling approach of FMISO seems to be the nucle-ophilic substitution of the tosylate leaving group by ^{18}F]-fluoride on the tetrahydropyranyl-protected precursor 1-(2′-nitro-1′-imidazolyl)-2-O-tetrahydropyranyl-3-O-toluenesulphonyl propanediol (NITTP), following hydroly-sis of the protecting group with high radiochemical yield (Kamarainen et al., 2004; Patt et al., 1999). Currently, the automated synthesis of FMISO by this promising method has been reported using high-performance liquid chro-matography (HPLC) purification, with the radiochemical yield of >60% in the synthesis time of ~ 60 min (Patt et al., 1999) and using Sep Paks purification with the radiochemi-cal yield of ~ 34% in the synthesis time of ~ 50 min at com-mercial ^{18}F]fluorodeoxyglucose (FDG) synthesis module (Kamarainen et al., 2004), respectively. More recently, fully automated synthesis of FMISO via HPLC purification using a modified commercial FDG synthesizer has also been reported (Oh et al., 2005). However, HPLC purification is quite cumbersome and too time consuming to be performed routinely, and Sep Paks purification gives a low radiochem-ical yield. Therefore, a fully automated synthesis of FMISO using Sep Paks purification with a high yield within short synthesis time has been carried out by our research group (Tang et al., 2005).

Methods

[18]F]fluorination of the epoxide 1-(2,3-epoxypropyl)-nitroimidazole and epoxide ring-opening leads to FMISO. (Jerabek et al., 1986)

Aqueous ^{18}F]fluoride was produced by proton bombard-ment of a target containing enriched (>95%) [18]O water. A small aliquot of the aqueous ^{18}F]fluoride activity obtained from the target was placed in a platinum crucible containing 20 μmol of tetrabutylammonium fluoride (TBAF) and dried under a stream of nitrogen at 110°C. The residue remaining was further dried by azeotropic distillation using two 100 μL aliquots of acetonitrile. The carrier-added ^{18}F-labeled TBAF was then taken up in 100 to 200 μL of an aprotic solvent N,N-dimethylacetamide (DMA), and was added to a reac-tion vial containing 1.0 to 1.3 mg of 1-(2, 3-epoxypropyl)-nitroimidazole. The vial was sealed and the reaction mixture heated for 30 min at 60°C, which was purified by HPLC using a silica gel column. Radiochemical yields were deter-mined by HPLC and TLC using silica gel eluted with chlo-roform/methanol 9/1 (v/v).

Reaction of ^{18}F-labeled intermediate epifluorohydrin with the nitroimidazole moiety under basic conditions leads to FMISO. (Kamarainen et al., 2004)

Fluorine-18 fluoromisonidazole (FMISO) was produced by displacement of the tosyl group for (2R)-(−)-glycidyl tosylate (GOTS) with ^{18}F]fluoride to afford ^{18}F]epifluo-rohydrin (FEPI), and subsequent nucleophilic ring-open-ing of FEPI with 2-nitroimidazole afforded FMISO. First, no-carrier-added (NCA) aqueous ^{18}F]fluoride was added to a mixture of Kryptofix[2.2.2] (K222) (23 mg) and K_2CO_3 (4.6 mg) in CH_3CN/water (86/14, v/v). Azeotropic distillation at 130°C under a N_2-flow was carried out by adding three 1.5 mL aliquots of CH_3CN. Second, GOTS (40 mg) in 0.6 mL dimethylsulfoxide (DMSO) was added to the dried residue. Then FEPI was distilled with CH_3CN to a borosilicate vial containing 0.6 mL of dimethylfo-mamide (DMF) kept at 0°C. CH_3CN was added in small portions of 0.1–0.2 mL during 25 min. Third, 2-nitroimi-dazole (40 mg) dissolved in KOH solution was added and reacted with FEPI kept at 120°C for 45 min. The crude FMISO was first purified with silica gel column chro-matography using 30% CH_3CN/chloroform at a flow rate of 2 mL/min. Subsequently, after the collected fractions were evaporated and the residue was dissolved in 5%

ethanol, FMISO was purified by the HPLC system consisting of a pump, an automatic sample injector, a UV absorption detector, and an analytical reversed-phase Spherisorb 10 ODS C18 column (250×4.6 mm) with 5% ethanol/water solution as mobile phase at a flow rate of 2 mL/min. The analysis of the final product and the reference compounds was carried out with the same HPLC system as mentioned earlier and with the TLC chromatography developed using 30% CH_3CN/chloroform solution.

Nucleophilic substitution on a protected precursor with subsequent removal of the protection group leads to FMISO. Two-pot automated synthesis procedure by Sep Paks separation. (Kamarainen et al., 2004)

Fluorine-18 fluoromisonidazole (FMISO) was prepared using an automated FDG synthesis module made by IBA (Ion Beam Applications, Belgium). After NCA aqueous [^{18}F]fluoride was transferred to the synthesis vessel containing Kryptofix[2.2.2] (13.5 mg) and K_2CO_3 (1.7 mg) in CH_3CN/water (8/1), the solvents were evaporated under an argon flow by adding 1 mL of CH_3CN two times. Then 1-(2'-Nitro-1'-imidazolyl)-2-O-tetrahydropyranyl-3-O-toluene-sulfonyl propanediol (NITTP) in 2 mL of CH_3CN was added, and the reaction was carried out at 100°C for 10 min. Subsequently, 10 mL of diethyl ether was added, and the product was passed through two silica Sep Paks to a second vessel (in two 5 mL portions). The ether was evaporated, and 2 mL of 1 N HCl was added to the residue for hydrolysis at 100°C for 3 min. Next, 1 mL of 2 N NaOH was added to neutralize the solution. The solution was passed through a Sep Pak, C 18, an alumina Sep Pak, and a Millipore filter connected in series to the product vial containing 1 mL of 1 N $NaHCO_3$. Finally, the column was rinsed with 4 mL of 10% ethanol. The synthesis time was ~ 50 min. The analyses of the intermediate compounds, the chemical purity, and the radiochemical purity were performed by using HPLC on the semipreparative µBondapack C 18 (7.8×300 mm, 10 µm, Waters) column with 20% CH_3CN/H_2O as mobile phase at 2 mL/min, TLC chromatography with ethyl acetate as the mobile phase, and LC-MS.

One-pot automated synthesis procedure at TracerLab FX_{F-N} synthesis module from GE Healthcare Technologies by Sep Paks and HPLC separation. (Tang et al., 2005)

Fluorine-18 fluoromisonidazole (FMISO) was synthesized in a commercially-available TracerLab FX_{F-N} synthesis module that was substantially modified and adapted to the production of the referred compound using HPLC purification and Sep Paks purification, respectively. The module was operated and monitored via a process control box connected to a personal computer. In order to allow the automated synthesis of FMISO and manual activation of any of the individual components at any time during the process,

the positions of valves were reassigned to their adequate locations in the control interface. The synthesizer was programmed by the software package in a step-by-step time-dependant sequence of events, such as valve opening or closing, helium purging, vacuum, temperature going up/down, and HPLC flow rate increase/decrease. Several live parameters records were performed during the synthesis process, including activity in the target vial, reaction vessel and product vial, temperature, and pressure in the reactor, and UV absorbance and radioactivity measurements for the semipreparative HPLC.

The automated radiochemical synthesis of FMISO was modified from the TracerLab FX_{F-N} through a two-step one-pot procedure that consisted of ^{18}F-fluorination of NITTP as the precursor molecule and subsequent hydrolysis of the tetrahydropyranyl (THP)-protected product. The individual synthesis steps involved in the preparation of FMISO can be divided into three main categories. *Category 1*, Fluorination of the precursor, including: A. [^{18}F]Fluoride trapped by QMA cartridge; B. Addition of eluent to elute [^{18}F]fluoride from QMA cartridge; C. Drying of [^{18}F]fluoride with acetonitrile; D. Addition precursor solution to reaction vessel; E. [^{18}F]Fluorination at 100°C for 10 min. *Category 2*, Hydrolysis of the protective group, including: F. Evaporation of acetonitrile; G. Addition 1 N HCl for hydrolysis; H. Hydrolysis at 100°C for 5 min. *Category 3*, Neutralization/HPLC and formulation, including: I. Neutralization with 30% sodium acetate; J. Addition of HPLC eluent to reaction vessel; K. Transfer the solution to HPLC loop; L. Elute using EtOH:H_2O=5:95, flow: 8 mL/min, detection: UV 254 nm; M. Collect the radioactive peak of FMISO (retention time: 17–19 min); N. Sterile filtration or Sep Paks purification and formulation, including: O. Neutralization with 1 N NaOH; P. Addition HPLC eluent to reaction vessel; Q. Transfer the solution to Sep Paks; R. Elute using EtOH:H_2O=5:95; S. Collect [^{18}F]FMISO eluent; T. Sterile filtration.

For HPLC purification, the first 10 steps (A–E, G–K) are carried out in the FMISO synthesis unit, and the last 2 steps (L–M) are performed within the HPLC purification unit. For Sep Paks purification, the first 11 steps (A–H, O–Q) are carried out in the FMISO synthesis unit, and the last three steps (R–T) are performed within the Sep Paks purification unit. After synthesis, cleaning procedures were carried out in between the runs. Before delivery of [^{18}F]fluoride to the module, vial 1 was filled with a mixture of 15 mg of K222, 3 mg of K_2CO_3, 1 mL of acetonitrile, and 0.5 mL of water, vial 2 was added with 1 mL of HPLC eluent or 10 mL of HPLC eluent, vial 3 was added with 1 mL of 1 N HCl, vial 4 was filled with 0.5 mL of 30% sodium acetate or 2 mL of 1 N NaOH solution, and vial 5 was added with 5 mg of the precursor NITTP dissolved in 1 mL of acetonitrile.

[^{18}F]fluoride was obtained through the nuclear reaction ^{18}O(p, n)^{18}F by irradiation of a 95% ^{18}O-enriched water target with a 16.5 MeV proton beam at the PETtrace cyclotron (GEMS). After the delivery of [^{18}F]fluoride from cyclotron, the radioactivity was collected on a QMA Sep Pak cartridge, where [^{18}F]fluoride was trapped and ^{18}O-water was collected for recycling. 1.5 mL of a K222 solution in vial 1 was eluted through the QMA Sep Pak cartridge, in which the trapped ^{18}F$^-$ was eluted into the reaction vessel. The solvent was evaporated under a stream of helium at 85°C. After complete removal of solvent, the precursor in vial 6 was added to the reaction vessel containing the dried [K/K222]$^{+18}$F$^-$ complex, and the vessel was heated at 100°C for 10 min. Then, the reaction mixture was cooled (or evaporated to dryness for Sep Pak separation) and from the resulting reaction mixture was added to 1 N HCl in vial 4. The mixture was hydrolyzed to remove the THP-protected group by heating for 5 min at 100°C. For HPLC purification, after cooling the mixture was neutralized with 30% sodium acetate and passed through an alumina Sep Pak cartridge. The eluate was collected in a glass vial. Before HPLC purification, 1 mL of HPLC eluent (H$_2$O/C$_2$H$_5$OH 95/5, v/v) in vial 3 was added to the reaction vessel, and the solution was then passed through the same alumina Sep Pak cartridge. The eluate was collected in the same glass vial. Finally, FMISO purification was carried out in the TracerLab FX$_{F-N}$ synthesis module built-in HPLC system with a semipreparative reverse-phase C18 column (10 mm × 250 mm) and C18 precolumn equipped with a UV detector and a radioactivity detector. The mobile phase used was H$_2$O/C$_2$H$_5$OH (95/5, v/v) at a flow rate of 8 mL/min. The peak corresponding to FMISO was collected and passed through 0.22 μm sterile filtration into a sterile vial to obtain the final formulation. For Sep Pak purification, after cooling the mixture was neutralized 1 N NaOH solution and passed through an alumina Sep Pak cartridge, an SCX cartridge, a C18 Sep Pak cartridge, and a Millipore filter connected in series. The Sep Paks were further eluted with HPLC eluent, and finally the eluates were combined. The radiochemical yield was expressed as the amount of radioactivity in the FMISO fraction divided by the total ^{18}F-radioactivity.

High performance liquid chromatography analysis, which was used for checking radiochemical purity and specific activity eluted with H$_2$O/C$_2$H$_5$OH (95/5, v/v) at a flow rate of 1 mL/min, was carried out on a modular HPLC system with a reverse-phase analytical C18 column (4.6 mm × 150 mm, Shimadzu Corporation, Japan). The column consisted of two LC-10ATvp pumps (Shimadzu Corporation, Japan) and variable-wavelength SPD-10ATvp UV detector (Shimadzu Corporation, Japan), an LB 508 Radioflow Detector with two-channel analyzer (EG &G, Germany), and computer (Shimadzu Corporation, Japan). The UV signal was monitored with a UV Lambda Max detector at 254 nm. A radioTLC (ethyl acetate as solvent system) was also used for checking radiochemical purity. The pyrogen test was carried out using a LAL test kit. The K222 detection test

was performed on the silica gel 60-coated plate developed with methanol/ammonium hydroxide (9/1, volume-to-volume ratio) as solvent system and iodine vapor was used for staining the spots to render them visible (Chaly and Dahl, 1989). Radiochemical stability was checked with a 10 mL solution (saline 9 mL + FMISO solution (5% ethanol) 1 mL) using a radioTLC and an analytical HPLC up to 6 hr.

One-pot automated synthesis procedure by HPLC separation at TracerLab Mx FDG synthesis module from GE Healthcare Technologies. (Oh et al., 2005)

The steps for preparing FMISO consisted of three categories. *Category 1*, Fluorination of the precursor, including: A. Addition of 5 mg K$_2$CO$_3$/300μL H$_2$O and 22 mg Kryptofix[2.2.2]/300μL CH$_3$CN to elute [^{18}F]fluoride from the QMA; B. Drying of [^{18}F]fluoride with acetonitrile; C. Addition precursor solution to the reaction vial; D. [^{18}F]fluoriation at 105°C for 360 s and 75°C for 280 s; E. Evaporation of the reaction solvent. *Category 2*, Hydrolysis of the protective group, including: F. Trapping of the reaction mixture onto tC18 cartridge; G. Washing of the tC18 cartridge to remove unreacted [^{18}F]fluoride with H$_2$O; H. Elution of reaction mixture from tC18 cartridge to reaction vial with 2 mL ethanol; I. Evaporation of ethanol; J. Addition 1 N HCl for hydrolysis to the reactor; K. Hydrolysis of THP-protective group at 105°C for 300 min. *Category 3*, Neutralization/HPLC purification and formulation, including: L. Neutralization with 2 N NaOH and citrate buffer; M. Transfer of the solution to HPLC loop; N. Purification: EtOH:H$_2$O=5:95, 8 mL/min, UV 254 nm; O. Collect the radioactive peak of FMISO by three-way valve (retention time: 9–11 min); P. Sterile filtration. The first 13 steps (A–M) were carried out in the FMISO synthesis module, and the last 3 steps (N–P) were performed within the HPLC purification unit. For method I, experiments of Categories 1 (A–E), 2 (only J and K steps), and 3 (L–P) were completed. For method II, FMISO production by adding solid-phase purification steps consisted of all steps (A–P) except Step E in Category 1.

[^{18}F]fluoride radioactivity was trapped by QMA cartridge and the ^{18}O-enriched water was removed to the ^{18}O water collection vial in the synthesis module, after delivery of [^{18}F]fluoride water from the cyclotron. The trapped [^{18}F]fluoride was eluted to the reaction vial with a Kryptofix[2.2.2]/K$_2$CO3 mixture solution. After complete removal of solvent, the precursor NITTP (5–20 mg/2 mL) in acetonitrile was added to the reaction vial. The [^{18}F]fluorination was carried out in a closed reaction vial at 95–120°C for 300–600 s and then at 75°C for 280 s. For Method I, acetonitrile was removed by evaporation at 90°C for 5 min (Step E in Category 1) followed by hydrolysis steps. For Method II, the reaction mixture was diluted with 30 mL of water after [^{18}F]fluorination. The [^{18}F]fluorination reaction mixture and unreacted precursor in the mixed solution were trapped on the tC18 cartridge. The tC18 cartridge was washed twice with 20 mL of water to remove unreacted [^{18}F]fluoride and polar organic impurities for easy HPLC purification. The reaction mixture was eluted from the tC18 cartridge with ethanol and removed to the reaction vial. Residual ethanol was removed at 85°C for 5 min.

Hydrolysis with 1 N HCl was performed in a closed reaction vial at 105°C for 300 s. Neutralization was performed with the NaOH and buffer solution after hydrolysis. Neutralized FMISO crude solution was transferred to HPLC. Water (3 mL) was sent into the reaction vial for flushing of the reaction vial and the disposable cassette. The water was sucked up and transferred to the HPLC injector, and its purification was carried out on an Econosil C18 column (Alltech, 10 μm, 10 × 250 mm) eluted with ethanol/water (5/95) at a flow rate of 5 mL/min and monitored with a radioactivity detector and a UV detector at 254 nm. Radiochemical purity and specific radioactivity were measured with an analytical radio-HPLC. A radio-TLC (EtOAc/ethanol=1:1) was also used for checking the radiochemical purity.

Results and Discussion

Like other PET tracers, the routine production of FMISO presents many challenging laboratory requirements. To assure timely delivery of PET tracers, the synthesis process must be fast, efficient, reliable, and designed to comply with the applicable regulatory guidelines. A fully automated system will not only help to achieve these goals, but also decrease the radiation burden to radiosynthesis personnel.

Fluorine-18 fluoromisonidazole (FMISO) synthesis was performed by the use of [^{18}F]fluorination of the epoxide 1-(2,3-epoxypropyl)-2-nitroimidazole and epoxide ring-opening with a low average radiochemical yield of < 1% at end-of-synthesis based on HPLC and TLC analysis in an overall synthesis time of 1 hr (Jerabek et al., 1986). The specific activity was estimated to be 2750 mCi/mmol based on the known specific activity of the ^{18}F addition reaction yield. The change of reaction conditions did not improve the yield. Several reasons may account for the low radiochemical yields obtained. The reaction can occur at either carbon atom of the epoxide ring resulting in the formation of two different ^{18}F-labeled fluorohydrins: the desired 2-hydroxy-3-fluoroisomer and the undesired 2-fluoro-3-hydroxy isomer. It is also possible that [^{18}F]fluoride can react by nucleophilic substitution of the nitro group on the imidazole; however, in fact, experiments show it is almost impossible. The most possible reason for the low radiochemical yields of FMISO is that the epoxide is unreactive toward nucleophilic attack by fluoride ion under basic conditions (Jerabek et al., 1986). Therefore, we must investigate other routes of synthesis for FMISO.

Fluorine-18 fluoromisonidazole (FMISO) was prepared via the fluorination of GOTS and nucleophilic ring-opening of FEPI with 2-nitroimidazole, with the uncorrected radiochemical yield ~ 10% and the radiochemical purity > 99% within a total synthesis time of ~ 180 min. The low yields could be mainly due to the volatility of the FEPI, the high losses on the walls of the reaction vials, and the long synthesis time. This method isn't preferred for synthesis of FMISO (Kamarainen et al., 2004).

Synthesis of FMISO was carried out by two-pot automated synthesis procedure via Sep Paks separation on an automated FDG synthesis module made by IBA. The uncorrected radiochemical yield was 21% on an average, the radiochemical purity was > 97%, and the total synthesis time, ~ 50 min. The advantage of this approach was that no HPLC was needed for the purification of FMISO and the synthesis time could be reduced to 50 min, compared to the 60 min and 180 min for the two methods described earlier. However, the most problematic issues were the high losses of activity on silica Sep Paks (mostly ^{18}F$^-$), which indicates an unsatisfactory labeling reaction, and on the walls of the reaction vials, which was mainly a labeled unhydrolyzed FMISO intermediate, consequently decreasing the synthesis yield. In addition, the synthesis of FMISO was too complicated to be performed routinely. The synthetic process required two separate Sep Pak purifications and two evaporations (Kamarainen et al., 2004). These operations were time consuming as well as inconvenient in automated control work. Therefore, this approach is still waiting to be improved.

Synthesis of FMISO was performed by a one-pot automated synthesis procedure via HPLC separation at TracerLab Mx FDG synthesis module. The optimal labeling conditions for the automated synthesis of FMISO was 10 mg of precursor in acetonitrile (2 mL) at 105°C for 360 s, followed by heating at 75°C for 280 s. Using 3.7 GBq of [^{18}F]fluoride as a starting activity with these optimized conditions, FMISO was obtained with high uncorrected radiochemical yields of (58.5±3.5)% and radiochemical purity of (98.2±1.5)% for (60.0±5.2) min using HPLC purification (Method I). When solid-phase purification steps were added (Method II), the uncorrected radiochemical yields were (54.5±2.8)% for (70.0±3.8) min. Using 5 mg of the precursor, the uncorrected radiochemical yields of (40.5±2.7)% and (42.2±4.9)% for the two methods, respectively, were obtained. Use of 20 mg of precursor resulted in a marginal increase of radiochemical yields of 60.2% and 56.2% for Method I and Method II, respectively, with a more cumbersome HPLC purification procedure due to larger amounts of organic impurities (Oh et al., 2005). The experimental results show that the synthesis time is waiting to be shortened further and HPLC purification should be avoided though the radiochemical yields for these two methods.

Fluorine-18 fluoromisonidazole (FMISO) was prepared in a two-step one-pot protocol, starting from the precursor NITTP at the TracerLab FX$_{F-N}$ synthesis module (Tang *et al.*, 2005). The corrected radiochemical yield for FMISO by HPLC purification was 60% on an average after a synthesis time of ~ 60 min, while the corrected radiochemical yield for FMISO by Sep Paks purification was 51% on an average after a synthesis time less than 40 min. The HPLC purification gave a little higher corrected radiochemical yield than Sep Paks purification, but the two purification methods gave almost similar uncorrected radiochemical yield. The HPLC purification of crude products showed only two radioactive peaks. One peak was for FMISO, and the other peak was for [^{18}F]fluoride. For HPLC purification, the extent of acetonitrile evaporation also affected the radiochemical yield. Complete acetonitrile evaporation after fluorination showed reduced radiochemical yield of about 5% compared to the case that left 0.5 mL–1.0 mL of residual acetonitrile after [^{18}F]fluorination. The typical distribution of activity by using HPLC purification is shown as follows (all the results are decay corrected to the initial activity of [^{18}F]fluoride, $n = 4$): Final product FMISO injection, (60±18)%; Sep Pak alumina cartridge, (16±4)%; Sep Pak QMA, (2±1)%; Reaction vessel, (10±8)%; HPLC column, (3±2); Activity not accounted for, (9±7)%. The typical distribution of activity by using Sep Paks purification is also shown as follows (all the results are decay corrected to the initial activity of [^{18}F]fluoride, $n = 4$): Final product FMISO injection, (51±12)%; Sep Pak C18, SCX, and alumina cartridges, (23±8)%; Sep Pak QMA, (2±1)%; Reaction vessel, (11±9)%; Activity not accounted for, (13±7)%.

The identity of the intermediates and the final product was confirmed by comparing the chromatograms with unlabeled reference materials. The radiochemical purity of final product after the two purification methods was > 97%, confirmed by TLC and HPLC. We did not find radiochemical and chemical impurities by analytical HPLC chromatogram and radioTLC. Color spot test for detection of K222 by TLC also showed no detection of K222 in final injectable FMISO solution. With higher initial radioactivity, FMISO showed good stability and > 95% radiochemical purities at 6 hr after synthesis. The purified FMISO in 10 mL of 5% ethanolic saline solution kept more than 95% radiochemical purities after 6 hr.

In our study, we described the automated one-pot synthesis device amenable for the routine clinical production of FMISO by HPLC purification and Sep Paks purification using commercial TRACERlab FX$_{F-N}$ synthesis module. The HPLC purification gave higher corrected radiochemical yield of FMISO than Sep Paks purification; however, the most well-recognized limitation of HPLC purification had difficulty in performing a fully automated synthesis of PET tracers. Also, HPLC purification was time consuming and took a longer time to get FMISO injection than Sep Paks purification. In fact, we obtained similar uncorrected radiochemical yields (~ 40%) of using these two purifications. Radiosynthesis of FMISO using Sep Paks purification was shown to be reliable and easy to perform automated control

work, and gave high yield with > 99% chemical purity and > 97% radiochemical purity. This approach was quite easy to implement into the automated process and gave good results. Hence, an automated FDG-synthesis module could not only be used to produce FMISO in high yield by Sep Paks purification with only minor modifications, but also further reduce the whole radiosynthesis time.

Patt *et al.* (1999) reported that the radiochemical yield of FMISO was dependent mainly on the amount of precursor and on the labeling reaction temperature. Based on their results, a reaction time of 10 min was chosen for the optimization of the fluorination reaction temperature. The highest corrected radiochemical yield of 86% was observed for 10 mg of precursor, and high corrected radiochemical yield of 60% was obtained for 5 mg of precursor after 10 min fluorination reaction, whereas a dramatic decrease in the labeling yield was seen when low amounts of precursor were used for the reaction. In other words, the radiochemical yield was less than 1% with 1 mg of precursor after 10 min fluorination reaction. In order to save the amount of precursor and money, we chose 5 mg of precursor after 10 min fluorination reaction for the optimization conditions. Our results were in good agreement with those reported by Patt *et al.* (1999) and Oh *et al.* (2005). In particular, the automated one-pot synthesis of FMISO by Sep Paks purification not only provided a simple and convenient procedure, but also showed short synthesis time, good reproducibility, and good radiochemical yields.

Although several invasive and noninvasive techniques have been used to detect tumor hypoxia, none has been widely accepted as a clinical method for such purposes owing to the technical limitations of these techniques. A practical noninvasive PET imaging modality such as FMISO PET would aid in treatment management decisions and observing response to traditional chemo- and radiotherapies. In addition, a routine clinical FMISO PET imaging for the detection of hypoxia would help rationalize the administration of hypoxia-specific therapies such as radiosensitizers, neutron radiation, and hypoxic cell cytotoxins, and facilitate the clinical investigation of drugs designed to increase tumor oxygenation.

In conclusion, synthesis of FMISO was performed via a one-pot, two-step synthesis procedure using the modified commercial TRACERlab FX$_{F-N}$ synthesis module. Nucleophilic fluorination of the precursor molecule 1-(2′-nitro-1′-imidazolyl)-2-O-tetrahydropyranyl-3-O-toluenesulphonylpropanediol (NITTP) using noncarrier-added [^{18}F]fluoride, followed by hydrolysis of the protecting group with 1 mol/L HCl and purification with Sep Paks instead of HPLC, gave FMISO. The overall uncorrected radiochemical yield was > 40%, the whole synthesis time was < 40 min, and the radiochemical purity was > 95%. The new automated synthesis procedure can be applied to the fully automated synthesis of FMISO at commercial FDG synthesis module.

References

Chaly, T., and Dahl, J.R. 1989. Thin layer chromatographic detection of kryptofix 2.2.2 in the routine synthesis of [¹⁸F]2-fluoro-2-deoxy-D-glucose. *Nucl. Med. Biol. 16*:385–387.

Chapman, J.D., Zanzonico, P., and Ling, C.C. 2001. On measuring hypoxia in individual tumors with radiolabeled agents. *J. Nucl. Med. 42*:1653–1655.

Graham, M.M., Peterson, L.M., Link, J.M., Evans, M.L., Rasey, J.S., Koh, W.J., Caldwell, S.H., and Krohn, K.A. 1997. Fluorine-18-fluoromisonidazole radiaton dosimetry in imaging studies. *J. Nucl. Med. 38*:1631–1636.

Gronroos, T., Eskola, O., Lehtio, K., Minn, H., Marjamaki, P., Bergman, J., Haaparanta, M., Forsback, S., and Solin, O. 2001. Pharmacokinetics of [¹⁸F]FETNIM: a potential hypoxia marker for PET. *J. Nucl. Med. 42*:1397–1404.

Jerabek, P.A., Patrick, T.B., Kilbourn, M.R., Dischino, D.D., and Welch, M.J. 1986. Synthesis and biodistribution of ¹⁸F-labeled fluoronitromisonidazoles: potential *in vivo* markers of hypoxic tissue. *Appl. Radiat. Isot. 37*:599–605.

Kamarainen, E.L., Kyllonen, T., Nihtila, O., Bjork, H., and Solin, O. 2004. Preparation of fluorine-18-labelled fluoromisonidazole using two different synthesis methods. *J. Label. Compd. Radiopharm. 47*:37–45.

Laforest, R., Dehdashti, F., Lewis, J.S., and Schwarz, S.W. 2005. Dosimetry of ⁶⁰/⁶¹/⁶²/⁶⁴Cu-ATSM: a hypoxia imaging agent for PET. *Eur. J. Nucl. Med. Mol. Imaging 32*:764–770.

Lewis, J.S., Sharp, T.L., Laforest, R., Fujibayashi, Y., and Welch, M.J. 2001. Tumor uptake of copper-diacetyl-bis(N⁴-methylthiosemicarbazone): effect of changes in tissue oxygenation. *J. Nucl. Med. 42*:655–661.

Lewis, J.S., and Welch, M.J. 2001. PET imaging of hypoxia. *Q. J. Nucl. Med. 45*:183–188.

Lim, J.L., and Berridge, M.S. 1993. An efficient radiosynthesis of [¹⁸F]fluoromisonidazole. *Appl. Radiat. Isot. 44*:1085–1091.

Mahy, P., De Bast, M., Leveque, P.H., Gillart, J., Labat, D., Marchand, J., and Gregoire, V. 2004. Preclinical validation of the hypoxia tracer 2-(2-nitroimidazol-1-yl)-N-(3,3,3-[¹⁸F]trifluoropropyl)acetamide, [¹⁸F]EF3. *Eur. J. Nucl. Med. Mol. Imaging 31*:1263–1272.

Oh, S.J., Chi, D.Y., Mosdzianowski, C., Kim, J.Y., Gil, H.S., Kang, S.H., Ryu, J.S., and Moon, D.H. 2005. Fully automated synthesis of [¹⁸F]fluoromisonidazole using a conventional [¹⁸F]FDG module. *Nucl. Med. Biol. 32*:899–905.

Patt, M., Kuntzsch, M., and Machulla, H.J. 1999. Preparation of fluoromisonidazole by nucleophilic substitution on THP-protected precursor: yield dependence on reaction parameters. *J. Radioanalyt. Nucl. Chem. 240*:925–927.

Rasey, J.S., Hofstrand, P.D., Chin, L.K., and Tewson, T.J. 1999. Characterization of [F-18]fluoroethanidazole, a new radiopharmaceuticals for detecting tumor hypoxia. *J. Nucl. Med. 40*:1072–1079.

Sorger, D., Patt, M., Kumar, P., Wiebe, L.I., Barthel, H., Seese, A., Dannenberg, C., Tannapfel, A., Kluge, R., and Sabri, O. 2003. [¹⁸F]Fluoroazomycinarabinofuranoside (¹⁸FAZA) and [¹⁸F]fluoromisonidazole (¹⁸FISO): a comparative study of their selective uptake in hypoxic cells and PET imaging in experimental rat tumors. *Nucl. Med. Biol. 30*:317–326.

Tang, G., Wang, M., Tang, X., Luo, L., and Gan, M. 2005. Fully automated one-pot synthesis of [¹⁸F]fluoromisonidazole. *Nucl. Med. Biol. 32*:553–558.

Zanzonico, P., O'Donoghue, J., Chapman, J.D., Schneider, R., Cai, S., Larson, S., Wen, B., Chen, Y., Finn, R., Ruan, S., Gerweck, L., Humn, J., and Ling, C. 2004. Iodine-124-labeled iodo-azomycin-galactoside imaging of tumor hypoxia in mice with serial microPET scanning. *Eur. J. Nucl. Med. Mol. Imaging 31*:117–128.

Ziemer, L.S., Evans, S.M., Kachur, A.V., Shuman, A.L., Card, C.A., Jenkins, W.T., Karp, J.S., Alavi, A., Dolbier Jr., W.R., and Koch, C.J. 2003. Noninvasive imaging of tumor hypoxia in rats using the 2-nitroimidazole ¹⁸F-EF5. *Eur. J. Nucl. Med. 30*:259–266.

3

Radiation Hormesis

Rekha D. Jhamnani and Mannudeep K. Kalra

Introduction

Humans are exposed to radiation on a day to day basis from natural background radiation and human-made radiation, including radiation used for diagnosis and treatment. The average background radiation exposure per capita in the United States is ~ 3.6 mSv (or 360 mrem) per year. This radiation exposure has several sources. Earth is bombarded by radiation from outside the solar system from positively charged ions from protons and iron nuclei. The interaction of this radiation with Earth's atmosphere generates secondary radiation in the form of X-rays, muons, protons, alpha particles, pions, electrons, and neutrons. In addition, the sun produces solar cosmic rays that are composed of low-energy particle radiation. Radioactive material is also found in the soil, rocks, water, air, and vegetation. Potassium, uranium, and thorium are the major sources of such terrestrial radiation. Interestingly, radon gas leaks out of uranium-containing soil and may even affect well-sealed homes.

Human-made radiation is similar in nature and effect to natural radiation. Tobacco, building materials, combustible fuels such as gas and coal, ophthalmic glass, televisions, luminous watches and dials, airport X-ray systems, smoke detectors, road construction materials, electron tubes, fluorescent lamp starters, and lantern mantles are other small sources of human-made radiation. With rising numbers of radiation-based medical procedures (such as X-rays), radiation exposure from medical sources has received tremendous attention.

Recently, there have been concerns over increasing radiation exposure from radiation-based medical procedures, such as X-ray-based radiological techniques and nuclear medicine studies. Notably, computed tomography (CT) has been singled out as the medical procedure that contributes the most radiation exposure among all medical sources of radiation. It is estimated that CT contributes to < 10% of all radiological investigations, but to two-thirds of medical radiation exposure. Based on existing scientific evidence, the National Institute of Environmental Health Sciences (NIEHS), a subsidiary of the Food and Drug Administration (FDA), has declared that X-ray radiation at low doses (similar to doses that can be associated with CT scanning and certain other radiological procedures such as angiography) is a known human carcinogen. Regardless of this declaration, some investigators believe that some controversy still remains surrounding radiation carcinogenesis and the benefits of radiation at these low levels of exposure (Calabrese and Baldwin, 2002; Feinendegen, 2005). In fact, some studies have shown that radiation hormesis, or beneficial radiation at low levels, occurs in both animals and humans (Wang *et al.*, 2005).

This chapter introduces the concept of hormesis and discusses some pertinent animal and human studies that support radiation hormesis.

Hormesis

Hormesis has been defined as an adaptive response characterized by biphasic dose responses of similar quantitative features with respect to amplitude and range of the stimulatory response that are either directly induced or the result of compensatory biological processes following an initial disruption in homeostasis. Radiation hormesis is a lesser known effect of low-dose radiation, which propounds that, on the basis of animal and human studies, toxic radiation acts like a stimulant (beneficial) in low doses. Support for hormesis comes from studies that show lower incidence of cancer in subjects exposed to low-radiation doses, and reduction in chromosomal aberrations secondary to high-dose radiation exposure following preexposure to low-dose radiation and stimulation of the immune system and tumor-killer lymphocytes by low-dose radiation.

In the early twentieth century, a significant issue was whether stimulatory or "excitatory" responses due to low-radiation dose were the result of direct stimulation or an overcompensatory response following radiation-induced injury. In the 1880s, Hugo Schulz hypothesized, in what is now referred to as the Arndt-Schulz law, that a direct stimulatory response accounted for the stimulation that occurs at low doses. The direct stimulatory response states that hormetic responses occur via direct biopositive mechanisms, which is an adaptive response that operates within normal maintenance functions that allow for metabolic excursions within the twofold range of background. However, Shields Warren, a leading expert in radiation biology and health, showed that stimulation was due to an overcompensation reparative response due to an initial disruption in homeostasis (Calabrese and Baldwin, 2002). The overcompensation stimulation hormesis specifically states that stimulation is an adaptive response due to low levels of stress or damage, which, in turn, results in enhanced fitness for some physiological systems for certain periods of time (Calabrese and Baldwin, 2002).

Mechanisms

DNA repair at the molecular level has been proposed as a possible mechanism for hormesis. According to this hypothesis, a low-radiation dose induces the production of special proteins that are involved in DNA repair processes. Low-dose radiation causes DNA double-strand breaks that activate induced resistance, which protects cells even against damage due to metabolism and thus may protect against cancer development. In addition, if cells killed by low-dose radiation are replaced by cells that have intact DNA with a smaller potential for cancer development, then the cell killing and replacement process may protect against cancer (Wolff et al., 1988).

Free radical detoxification at the molecular level has been proposed as a second possible mechanism for hormesis. Low doses of radiation cause a temporary inhibition in DNA synthesis, which gives irradiated cells a longer time to recover. The inhibition also induces the production of free

radical scavengers, which makes irradiated cells more resistant to further exposure to radiation its been added (Feinendegen et al., 1987). With increasing age, there is an accumulation of damage caused by endogenous stressors, such as reactive oxygen molecules produced during fuel utilization, and a spectrum of environmental factors, such as chemical toxins and infectious agents. Low-dose radiation lessens the damage done by these various agents, thus retarding deterioration and extending life (Kirkwood, 1977).

The third mechanism for stimulation of the immune system occurs at the cellular level. Low doses of radiation stimulate the function of the immune system. In 1909, it was shown that mice treated with low-level radiation were more resistant against bacterial disease (Russ, 1909). It has also been shown that upon receiving low doses of radiation, the body produces more of those proteins that are involved in regulating the cell cycle, cyclones, kinases, enzymes responsible for DNA repair, and genes that induce cell proliferation (Wiencke et al., 1986; Fornace, 1992).

Animal Studies

Animal studies have shown the suppressive effects of low-dose radiation on cancer induction, growth, and metastasis (Liu, 2003). A study of the response of mouse immune systems showed that low levels of radiation (0.01 to 10 Gy) stimulated intercellular reactions among lymphocytes via secretion of regulatory cytokines. Similarly, low-dose radiation also prevented DNA damage in mouse cells. It was found that 4 hr after radiation, free glutathione and superoxide dimutase levels increased and lipid peroxidation decreased. These changes caused alterations in thymidine kinase enzyme activity in mouse bone cells, which, in addition to transcriptional regulation of gamma-glutamylcysteine synthetase gene, prevented DNA damage. These effects were shown to last up to a few weeks after radiation. Low-dose radiation also showed reduction of carcinogenesis in rodent cells. A single low-dose irradiation of two-month-old mice heterozygous for the *Trp-53* gene delayed the appearance of spontaneous lymphoma and spinal osteosarcoma (Feinendegen, 2005).

Human Studies

Perhaps the most interesting human study on hormesis was Matanoski's study of the effects of nuclear radiation on British shipyard workers (1991). In the study, 28,000 nuclear shipyard workers were compared to 32,500 same-age nonexposed shipyard workers. Interestingly, the nuclear workers did not suffer harm from radiation. Furthermore, the number of deaths from cancer among them was substantially lower than that of the controls, and their death rate from all causes was 24% lower than that of the unexposed controls. However, certain study limitations existed. For example, the fact that more nonexposed workers were used than exposed workers represents a flaw in the study methodology. Also, the day-to-day habits of the nonexposed workers (such as smoking and alcohol) were not observed. Opponents of hormesis argue that the nonexposed workers could have been subjected to unhealthier day-to-day habits, thus causing them to have a higher death rate independent and unaffected by exposure to radiation (Matanoski, 1991).

Another study done on British radiologists entering the field of oncology after 1920 showed the positive effects of low-dose radiation. Interestingly, the British radiologists had much higher cancer rates than other British men; however, their deaths from causes other than cancer decreased significantly. Thus, radiation stimulated their immune system and may have canceled out additional cancer deaths. This study has limitations similar to those of the shipyard study because it, too, did not consider the day-to-day habits of both groups (i.e., other confounding risk factors such as smoking, alcohol, and diet). The radiologists could have had healthier day-to-day habits or the normal British men could have had unhealthier day-to-day habits, thus affecting the death rates for both groups (Smith and Doll, 1981).

Similarly, Lee's study done on communities in Chinshan, Taiwan, showed that citizens living near nuclear power plants benefited from the radiation that they experienced. Measurements of peripheral blood components were performed among 3602 men and women (35 years of age and above) living near two nuclear power plants in Taiwan. Positive associations were found for hemoglobin, hematocrit, platelet, white blood cell, and red blood cell counts. For example, platelet count increased greatly if exposure from the nuclear plant increased by one exposure unit. However, it has been argued that the workers exposed to radiation could have been participating in day-to-day activities that increased blood cell counts independent of radiation exposure (Lee et al., 2001).

The most important limitation of the aforementioned human studies is the *healthy worker effect*, a term coined by McMichael (1976). This term suggests that people who have an occupation (i.e., who work) are generally healthier than those who do not. According to McMichael, workers usually exhibit lower overall death rates than the general population because the severely ill and chronically disabled are ordinarily excluded from employment. Moreover, workers have better access to health care and insurance policies, thus making them healthier than the general population. Workers also participate in regular physical activity, thus making them healthier than the general population, who may or may not participate in such activities.

Controversy

Despite reports of the positive effects of low-level radiation, the linear nonthreshold theory of radiation-induced cancer persists, which states that all levels of radiation exposure can lead to cancerous growth. The linear nonthreshold theory describes radiation-induced cancer as a stochastic effect, which implies that there is no threshold for radiation and all doses of radiation are harmful. Several studies have documented these negative effects of radiation at all levels.

Radiation risks have been categorized into deterministic and stochastic effects. The deterministic effects are associated with cell death and occur when radiation dose to a particular region exceeds a threshold level. These effects are rarely seen with diagnostic X-ray-based or nuclear medicine-based imaging. The major risks to the subjects from diagnostic imaging are from stochastic effects, which include cancer in the irradiated subjects and genetic effects in the offspring of the irradiated subjects. These effects do not require the radiation dose to exceed a threshold level; however, incidence increases with rise in absorbed dose. The International Commission on Radiological Protection (ICRP) Special Task Force Report 2000 on CT radiation exposure stated that radiation doses from CT are relatively high and that these doses to tissues can often approach or exceed the levels known to increase the probability of cancer (Task Group on Control of Radiation Dose in Computed Tomography, 2000).

Pogribny et al. (2005) showed that whole-body application of 0.5 Gy X-rays leads to a decrease in histone tyimethylation in the thymus, which results in a significant decrease in global DNA methylation. In addition, irradiation at these levels also caused ~ 20% decrease in levels of methyl-binding proteins (MeCP2 and MBD2). These alterations caused genomic destabilizations that ultimately led to cancer.

Another interesting study, conducted by Per et al. (2004), attempted to determine whether low-radiation dose exposure in infancy affected cognitive function in adults. This study encompassed 3094 men who had received radiation for cutaneous haemangioma before the age of 18 months and were tested in adulthood by cognitive tests focusing on learning ability, logical reasoning, spatial recognition, and high school attendance. Results showed that the proportion of boys who attended high school decreased with increasing doses of radiation to the frontal and posterior parts of the brain.

The long-term effects of radiation doses on atomic-bomb survivors in Hiroshima and Nagasaki, from the 1950s to the 1990s, have also been studied extensively. Data show that excess risks associated with radiation began 5 to 10 years after exposure. Significant excess risks were seen for all cancers, specifically stomach, lung, liver, colon, bladder, breast, ovarian, thyroid, and skin cancers, and multiple myeloma. For doses between 0.005 and 0.2 Sv (relatively lower doses), there were 63 excess deaths due to cancer. For doses between 0.2 and 0.5 Sv (relatively higher doses), there were 76 excess cancer deaths. For doses between 0.5 and 1 Sv, there were 79 excess cancer deaths, and for doses over 1 Sv, there were 121 excess cancer deaths. Thus, the chance that given cancer death is associated with atomic-bomb radiation exposure was found to be linearly correlated with dose (Pierce *et al.*, 1996; Thompson *et al.*, 1994). The findings of these studies are supported by the Techa River study, which also reported the association of low doses of radiation with long-term carcinogenic effects (Krestinina *et al.*, 2005). In the 1950s, people living in rural villages along the Techa River received internal and external exposures due to ionizing radiation coming from the release of radioactive material from the Mayak plutonium production complex.

In conclusion, radiation at high doses leads to cell damage and cancer. However, some animals and humans studies have reported that low-dose radiation can have beneficial hormetic effects. Conversely, increased risk of cancer at low levels of radiation exposure has also been documented. Until the "controversy" between these opposing concepts—risk (carcinogenesis) versus benefits (hormesis)—is resolved, one should lean toward the side of caution and ensure that radiation exposure is as low as reasonably achievable (ALARA principle).

References

Calabrese, E.J., and Baldwin, L.A. 2002. Defining hormesis. *Hum. Exp. Toxicol. 21:*91–97.

Feinendegen, L.E. 2005. Evidence for beneficial low-level radiation effects and radiation hormesis. *Brit. J. Radiol. 78:*3–7.

Feinendegen, L.E., Muehlensiepen, H., Bond, V.P., and Sondhaus, C.A., 1987. Intracellular stimulation of biochemical control mechanisms by low-dose, low-LET irradiation. *Health phys. 52:*663–669.

Fornace, A.J. 1992. Mammalian genes induced by radiation: activation of genes associated with growth control. *Annu. Rev. Genet. 26:*507–526.

Kirkwood, T.B.L. 1977. Evolution of ageing. *Nature 270:*301–304.

Krestinina, L.Y., Preston, D.L., Ostroumova, E.V., Degteva, M.O., Ron, E., Vyushkova, O.V., Startsev, N.V., Kossenko, M.M., and Akleyev, A.V. 2005. Protracted radiation exposure and cancer mortality in the Techa River cohort. *Radiat. Res. 164:*602–611.

Lee, Y.T., Sung, F.C., Lin, R.S., Hsu, H.C., Chien, K.L., Yang, C.Y., and Chen, W.J. 2001. Peripheral blood cells among community residents living near nuclear power plants. *Sci. Total Environ. 280:*165–172.

Liu, S.Z. 2003. On Radiation hormesis expressed in the immune system. *Crit. Rev. Toxicol. 33:*431–441.

Matanoski, G., 1991. Health effects of low-level radiation in shipyard workers: final report. *EV 10095:*471–479.

McMichael, A.J. 1976. Standardized mortality ratios and the "healthy worker effect": scratching beneath the surface. *J. Occup. Med. 18:*1169–1173.

Per, H., Hans-Olov, A., Dimitrios, T., Pedersen, N., Lagiou, P., Ekbom, A., Ingvar, M., Lundell, M., and Granath, F. 2004. Effects of low doses of ionizing radiation in infancy on cognitive function in adulthood: Swedish population based cohort study. *Brit. Med. J. 328:*19.

Pierce, D.A., Shimizu, Y., Preston, D.L., Vaeth, M., and Mabuchi, K. 1996. Studies of the mortality of A-bomb survivors. Report 12, Part I. Cancer: 1950–1990. *Rad. Res. 146:*1–27.

Pogribny, I., Koturbash, I., Volodymyr, T., Hudson, D., Stevenson, S., Sedelnikova, O., Bonner, W., and Kovalchuk, O. 2005. Fractionated low-dose radiation exposure leads to accumulation of DNA damage and profound alterations in DNA and histone methylation in the murine thymus. *Mol. Cancer Res. 3:*553–561.

Russ, V.K. 1909. Consensus of the effect of X-rays on bacteria. *Hygie 56:*341–344.

Smith, P.G., and Doll, R., 1981. Mortality from all causes among British radiologists. *Br. J. Radiol. 54:*187–194.

Task Group on Control of Radiation Dose in Computed Tomography. 2000. Managing patient dose in computed tomography: a report of the International Commission on Radiological Protection. *Ann. ICRP 30:*7–45.

Thompson, D.E., Mabuchi, K., Ron, E., Soda, M., Tokunaga, M., Ochikubo, S., Sugimoto, S., Ikeda, T., Terasaki, M., Izumi, S., and Preston, D.L. 1994. Cancer incidence in atomic bomb survivors. Part II: Solid tumors, 1958–1987. *Rad. Res. 137:*17–67.

Wang, Y.Z., Evans, M.D.C., and Podgorsak, E.B. 2005. Characteristics of induced activity from medical linear accelerators. *Med. Phys. 32:*2899–2910.

Wiencke, J.K., Afzal, V., and Oliveri, G. 1986. Evidence that H-3 thymidine induced adaptive response of human lymphocytes to subsequent doses of X-rays involves the induction of a chromosomal repair mechanism. *Mutagenesis 1:*375–380.

Wolff, S., Afzal, V., Wiencke, J.K., Olivieri, G., and Michaeli, A., 1988. Human lymphocytes exposed to low doses of ionizing radiations become refractory to high doses of radiation as well as chemical mutagens that induce double strand breaks in DNA. *Int. J. Rad. Biol. 53:*39–48.

II

General Imaging Applications

1

Molecular Imaging in Early Therapy Monitoring

Susanne Klutmann and Alexander Stahl

Introduction

In the conventional diagnostic work-up, therapy response is usually assessed by clinical or radiological methods. Using conventional criteria for therapy monitoring, imaging evaluation of tumor response is based on a reduction of the tumor size, which in general is measured by radiological techniques. However, the reduction in size is only a *late* sign of a potentially effective therapy. Therefore, the tumor response is assessed weeks or even months after treatment initiation. Thus, the effectiveness of a given therapy cannot be defined in an early course.

Advances in research on novel therapeutic strategies have indicated that the application of different therapeutic agents induces different responses of the same tumor cells in humans. This observation leads to the conclusion that tumor therapy must be more or less individually adapted. The prediction of tumor response in a very early course of therapy will avoid side effects of an ineffective therapy. This situation represents a typical oncological problem since patients with nonresponding tumors are treated without any benefit for a long time. Moreover, an inaccurate assessment of therapeutic outcome delays the point in time of focusing on a potentially more effective second-line therapy. With regard to both aspects, early monitoring of cancer therapy has the enormous potential to optimize the cost of a "tailor-made therapy."

The Place of Early Therapy Monitoring in the Management of Cancer

Recommendations regarding the optimal therapeutic strategy of cancer are changing. They are based on clinical trials with typical end points as median survival or clinical (or pathological) remission. The entry criteria for the clinical trials are usually based on the TNM (tumor, nodes, and metastases)-staging of a disease. Results of numerous trials show, however, that only moderate differences between novel strategies and established reference methods are achieved. One possible interpretation is that the therapy should be more individualized. A possible contribution of imaging is the early recognition of patients who may or may not benefit from the therapeutic regime.

During the past decades, the tremendous advances made in molecular and cell biology have extended our knowledge of human pathophysiology and diseases. Advances in molecular biology have defined properties that determine the response of various tumors to a given therapy. These factors may be intrinsic to the tumor cell itself or may correlate to the microenvironment of the tumor. The development of imaging signals of tumor cells using radioactive compounds allows the visualization of tumor cell characteristics prior to and directly after an initiated therapy corresponding to occurring metabolic changes. In this context, a wide spectrum of diagnostic targets have been proposed for functional imaging of cancer cells, that is, assessing glucose metabolism by 2-[^{18}F]fluoro-2-deoxy-D-glucose (F-18-FDG), defining the proliferation rate with F-18-FLT, or imaging the apoptotic cell fraction with Tc-99-annexin. The tracer approach using radioactive-labeled compounds directed against various targets excels because of its high sensitivity. However, problems such as lack of highly selective tumor markers, biological inhomogeneity of tumor masses, and different-acting mechanisms of therapeutic agents and special features, for example, of external irradiation therapy are yet to be solved.

Since not all the aforementioned tracers are tumor-specific, detailed knowledge of normal distribution and potential pitfalls is indispensable to evaluate the use of functional imaging in monitoring early therapy. The following discussion presents an overview of the present role of the targets (with special regard to FDG) and positron emission tomography (PET) in cancer imaging. However, it is clear that the debate on the role of functional imaging for *early* therapy monitoring is far from over.

What Can be Expected from Positron Emission Tomography Imaging?

Downstaging (or upstaging) during therapy is usually based on a visual interpretation of images—that is, significant changes in selected characteristic parameters indicating therapeutic response. However, therapy monitoring requires quantitative data. Tumor uptake of FDG is usually defined using the standard uptake value (SUV) as a quantitative parameter. Under standardized circumstances, a change in SUV of > 20% can be considered significant. Unfortunately, several factors influence the SUV, and, therefore, it is indispensable that SUV measurements follow a strict protocol.

Plasma glucose levels have a significant influence on SUV since both FDG and glucose compete for glucose transport. A correction within the normal range is probably not necessary, but elevated glucose levels represent a common problem in diabetic patients. Moreover, it is important to know that the uptake of FDG is time dependent. A common protocol follows an image acquisition time from 30 to 90 min after injection. Only a short delay of 15 min may induce a change of > 15% in SUV measuring. In addition, partial-volume effects may cause an underestimation of the true accumulation of FDG in the tumor. This holds especially true for small tumor lesions. Finally, physical properties of the PET scanner (geometric resolution, noise level, and sensitivity), the definition of SUV (definition of the region of interest, slice thickness, pixel size, number of pixels considered, etc.), and reconstruction algorithms are important factors influencing SUV measurements. This list of possible factors indicates the inherent potential of the method, which is not yet sufficiently explored. Methodological differences between each particular case make it difficult to integrate experiences. Briefly, clinical decisions based on quantitative PET parameters need a considerably higher degree of reproducibility.

F-18-FDG in Therapy Monitoring

Knowledge of enhanced glucose utilization in tumors is based on the publications of Warburg. The development of PET and the synthesis and evaluation of the glucose analog FDG by Sokolov opened a new perspective on tumor imaging. Presently, FDG-PET has become a widely accepted tool for diagnosis, staging, and restaging in various tumor entities. Quantitative measurements of FDG uptake were reported by several groups (Minn *et al.*, 1995; Weber *et al.*, 1999). This observation is an important background for using FDG imaging in therapy monitoring. A large number of quantitative parameters estimating FDG metabolism by SUV and the rate of phosphorylation are accepted as having the most predictive value (Weber *et al.*, 2003).

The first report regarding the normalization of FDG uptake after chemotherapy was published by Paul (1987). In a patient with retroperitoneal sinus carcinoma who had a positive F-18-FDG scan before treatment, two PET scans were true-negative after effective treatment. Wahl (Wahl *et al.*, 1993) compared the predictive value of FDG metabolic imaging to radiographic determination of tumor size in patients with primary advanced breast cancer. In pathologically proven partial or complete remission, maximal tumor uptake of FDG decreased immediately during treatment: day 8, 78 +/− 9.2% (P < .03); day 21, 68.1 +/− 7.5% (P < .025); day 42, 60 +/− 5.1% (P < .001); day 63, 52.4 +/− 4.4% (P < .0001) of the basal values. Interestingly, the tumor diameter did not decrease significantly during this period. Moreover, no significant decrease of the SUV after three cycles of treatment was observed in nonresponding patients (Wahl *et al.*, 1993). These findings confirm the observation that metabolic response occurs early after chemotherapy, and it even precedes reduction of tumor size.

Recently, Dose Schwarz (Dose Schwarz *et al.*, 2005) reported successful prediction of therapeutic response using FDG-PET. In patients with metastatic breast cancer, a significant decrease of FDG uptake was observed in the responder group soon after the completion of the first cycle of chemotherapy. Metabolic responders identified after the first cycle of chemotherapy had an overall survival of 19.2 months, compared to the nonresponders with a survival of 8.8 months. Subsequently, several groups used FDG-PET for therapy monitoring. In general, a decrease of FDG uptake is interpreted as a good prognostic parameter of cell death. However, inflammation, stromal host response, macrophage infiltration, and activation of salvage pathways may lead to a transitional increase of FDG uptake and therefore to false-positive results.

In animal experiments Spaepen (Spaepen *et al.*, 2003) analyzed the time course of changes in FDG uptake after cyclophosphamide therapy in severe combine immune deficient mice inoculated with human B tumor cells. Initially, 1 to 3 days after therapy, SUV reduction was similar to the changes in viable cell fraction. These changes preceded the reduction in tumor weight that occurred only 8 days after therapy. Eight to 10 days after therapy, FDG uptake stabilized despite a further reduction in the viable tumor cell fraction. Fifteen days after therapy, the viable tumor fraction increased again, as accurately predicted by an increase in SUV, while the tumor weight remained unchanged and the stromal host response and the necrosis were strongly reduced. Moreover, several groups observed initial enhancement of FDG uptake in cell cultures a few hours after chemotherapy ("metabolic flare phenomenon"). Haberkorn (Haberkorn *et al.*, 1994) investigated initial changes in FDG uptake after therapy with gemcitabine in cultured rat prostate adenocarcinoma cells. They concluded that the enhanced uptake might be based on an increased number of transport proteins. They considered the redistribution of transport proteins from the intracellular pool as the most likely explanation.

Owing to limited resolution of PET imaging, only overall data can be obtained from the tumor mass. This integrated approach may, however, mimic regional differences within the tumor tissue. Although F-18-FDG is a useful tumor imaging agent, an understanding about the molecular mechanisms of its cellular uptake is necessary in using it for therapy monitoring. Granulation tissue around the tumor and macrophages infiltrating the marginal areas surrounding the necrotic area of the tumor may show a higher uptake of F-18-FDG than the viable tumor cell itself (Kubota *et al.*, 1992). These results demonstrate that one should consider not only the properties of the tumor cells but also the characteristics of the nonneoplastic cellular elements, which appear in association with growth or necrosis of the tumor cells.

Monitoring Neoadjuvant Therapy

Neoadjuvant chemotherapy is an accepted method, for example, for locally advanced breast cancer and is under investigation in other tumor types. The rationale of neoadjuvant therapeutic approaches is to eliminate micrometastases and to shrink the primary tumor in order to improve operability. Generally, nonresponders to neoadjuvant chemotherapy have a poorer outcome and show a reduced overall survival. Furthermore, response to neoadjuvant therapy may influence the time interval to subsequent tumor resection (Ansquer *et al.*, 2001). Regarding overall outcome after neoadjuvant treatment in patients with, for example,

esophageal cancer, prognosis for patients with nonresponding tumors seems to be even worse than for those patients treated by surgery alone. However, patients with insufficient response suffer from all toxic side effects of chemotherapy, and point in time of surgical treatment is delayed.

In patients with advanced-stage ovarian cancer, Avril performed FDG-PET after the first and third circles of neoadjuvant chemotherapy (Avril *et al.*, 2005). A significant correlation to PET findings and survival was observed already after the first cycle of chemotherapy. By using a threshold for decrease in SUV (from the baseline) of 20% after the first cycle, median overall survival was 38.8 months in metabolic responders compared with 23.1 months in metabolic nonresponders. At a threshold of 55% decrease after the third cycle, median overall survival was 38.9 months for metabolic responders and 19.7 months for nonresponders. Thus, FDG was shown to play a major role in the evaluation of adjuvant chemotherapy. The role of FDG-PET in monitoring therapeutic effects (i.e., predicting complete response, evaluating neoadjuvant or palliative chemotherapy) in selected tumor types will be discussed in the following section.

Therapy Monitoring in Non-Small Cell Lung Cancer (NSCLC)

Pathologically complete response (CR) is defined as being tumor cell free (Milleron *et al.*, 2005), or only a few cancer cells containing (Betticher *et al.*, 2003) histopathologic material at surgery. It requires clearance of both the primary tumor and lymph nodes. Pathologic CR after preoperative chemotherapy for stage IIIA NSCLC has been shown to be associated with excellent survival (Pisters *et al.*, 1993). In clinical routine, response evaluation in most patients is based on the CT scan. A recent study investigated the clinical value of CT and fiber-optic bronchoscopy on biopsy-based classification of therapeutic response. In this study only 7 out of 19 patients with pathologic CR were correctly identified (Milleron *et al.*, 2005).

In a retrospective analysis conducted in 76 patients operated for a stage III NSCLC, none of the eight pathologic CR were clinically correct predicted (Margaritora *et al.*, 2001). Several studies evaluating PET scan after preoperative chemotherapy or chemoradiation in NSCLC showed a high sensitivity (97–100%) but limited specificity (58–67%) for detecting residual primary tumors, and good specificity (75–99%) but poor sensitivity (50–61%) for lymph node involvement (Ryu *et al.*, 2002; Cerfolio *et al.*, 2003). One of these studies compared PET and CT scan monitoring in 34 patients treated with chemotherapy or chemoradiation therapy for NSCLC. Positron emission tomography was shown to be more specific (67% vs. 0%) for detecting residual tumor and more sensitive (50% vs. 30%) for N2 disease compared to CT scan (Cerfolio *et al.*, 2003).

In patients with locally advanced NSCLC and neoadjuvant chemotherapy, tumor clearance from the mediastinal lymph nodes is considered as a marker for favorable prognosis. There is a strong need for a reliable noninvasive alternative because restaging by CT is considered to be unreliable and mediastinoscopy is often difficult to perform.

Cerfolio (Cerfolio et al., 2004) investigated 56 patients with NSCLC and neoadjuvant therapy. PET was performed prior to and post-treatment. PET findings were validated by surgical resection. It was demonstrated that the change in maximum SUV correlates better to histopathologic response (percent nonviable cells in the resected tumor tissue) than to the change of size on CT scan.

Palliative systemic therapy is performed in patients with stage IV NSCLC. This therapy is often associated with high toxicity and life-threatening side effects. Thus, an early interruption of unnecessary therapies is an important issue. This problem was investigated by Weber et al. In patients with advanced NSCLC they compared metabolic response with best response according to Response Evaluation Criteria in Solid Tumors (RECIST). In this group of patients, early prediction of therapeutic success is of particular importance because the majority of patients with unresectable disease (stage IIIB, IV) undergo palliative therapy with platinum-based chemotherapy regimens. However, palliative chemotherapy prolongs median survival only by ~ 2 months (Breathnach et al., 2001). Weber (Weber et al., 2003) found a close correlation between best response according to RECIST criteria and metabolic response. Sensitivity and specificity for prediction of best response were 94% and 74%, respectively. Based on the metabolic response, PET was able to clearly differentiate between median survival of 252 days for responders and 151 days for nonresponders. Corresponding data for median time to progression were 163 versus only 54 days. As the investigations were performed 21 days after treatment began, the results allow an early termination of ineffective therapies and thus change to second-line alternatives.

Therapy Monitoring in Non-Hodgkin's Lymphoma (NHL)

Previous studies suggested that patients with rapid response to induction treatment are likely to have a better and durable response (Armitage et al., 1986). For this reason, it is important to distinguish between responders to standard approaches and nonresponders, who may benefit from an early change to an alternative or more experimental treatment (Spaepen et al., 2002). For the response assessment of NHL, International Workshop Criteria (IWC) are the most accepted tools. A drawback of these criteria, however, is the dominant role of CT (Cheson et al., 1999). Early response to induction chemotherapy is difficult to assess, as patients with NHL often present with residual masses of uncertain dignity. These residual masses may consist of both fibrotic tissue and viable tumor cells. However, conventional radiographic characteristics cannot differentiate between active tumor tissue and fibrosis (Canellos, 1988; Hill et al., 1993). Moreover, tumor volume reduction measured by CT is only a late sign of effective therapy (Spaepen et al., 2002; Thomas et al., 1988).

A complete remission after first-line chemotherapy is associated with a longer progression-free survival, compared with partial remission, which is associated with a poorer clinical outcome. However, differentiation between tumor and fibrosis after first-line treatment is a difficult problem, especially in those patients with a large remaining tissue mass. There is considerable evidence that FDG-PET after first-line chemotherapy is useful to identify the dignity of residual masses diagnosed by CT (Spaepen et al., 2001; Jerusalem et al., 1999; De Wit et al., 1997).

Jerusalem (Jerusalem et al., 1999) compared CT and FDG-PET in the evaluation of therapeutic response. Only 5 out of 24 patients with residual masses on CT were PET positive. In 30 patients without residual masses on CT, only one exhibited pathological glucose accumulation. In contrast, all PET-positive patients exhibited relapse. On the other hand, 5 out of 19 PET-negative patients (26%) with residual masses and 3 of 29 PET-negative patients (10%) without residual mass showed a tumor relapse (Jerusalem et al., 1999). In contrast, the results published by Naumann (Naumann et al., 2001) are somewhat discordant. They investigated 43 patients with Hodgkin's disease (HD) and 15 patients with NHL. All patients had post-therapeutic residual masses. Patients with PET-positive residual masses had a recurrence rate of 62.5% (5/8 patients), whereas patients with PET-negative residual masses had a recurrence rate of only 4% (2/50). No recurrence occurred in any of the 39 HD patients with a negative PET scan. All 4 NHL patients with a positive PET scan relapsed.

The excellent prognostic value of a negative PET scan after therapy was emphasized by Zinzani et al. (2004). They investigated 41 patients with HD and 34 patients with aggressive NHL. After treatment, 4 out of 5 (80%) patients who were PET(+)/CT(−) relapsed as compared with none out of 29 patients in the PET(−)/CT(−) subset. Among the 41 CT(+) patients, 10 out of 11 (91%) who were PET(+) relapsed as compared with none out of 30 who were PET(−). They concluded that PET negativity at restaging strongly suggests the absence of an active tumor disease. The additional role of PET imaging in evaluating therapeutic success was investigated by Juweid (Juweid et al., 2005) in a retrospective multicenter study. The authors defined criteria for the combined IWC and PET-based evaluation of therapeutic response. Fifty-four patients with aggressive NHL were included. CT and PET investigations were performed after four to eight cycles of chemotherapy. They demonstrated that the combination of IWC response criteria and PET provides a more accurate response classification compared to IWC alone. Involving PET in the classification process, the number of CR was increased. Positron emission tomography was especially helpful in identifying a subset of IWC-PR patients with a more favorable prognosis.

Spaepen (Spaepen *et al.*, 2002) reported the results of FDG-PET at midtreatment with doxorubicin-containing chemotherapy in 70 patients with aggressive NHL. Presence or absence of abnormal FDG-PET was related to progression-free survival. At midtreatment, 33 patients showed abnormal FDG uptake. None of these patients achieved a durable complete remission. A negative PET scan was obtained in 37 patients, 31 of whom remained in complete remission, with a median follow-up of 1107 days. The prognostic power of the PET scan was higher than the International Prognostic Index (The International Non-Hodgkin's Lymphoma Prognostic Factors Project, 1993). Using PET at midtreatment as a prognostic marker, Haioun *et al.* reported slightly different results. This group used an anthracycline-containing therapeutic regimen in 90 patients, supplemented with rituximab in 41% of cases. The event-free survival after 2 years was 82% in PET negatives and 42% in PET positives (Haioun *et al.*, 2005).

Therapy Monitoring in Carcinomas of the Esophagus, Esophagogastric Junction, and Stomach

Carcinomas of the upper digestive tract are heterogeneous in terms of etiology, histopathology, and epidemiology. In the upper two-thirds of the esophagus, squamous cell carcinomas (SCC) predominate, with alcohol and smoking being the most prominent risk factors. Carcinomas of the esophagogastric junction, on the other hand, are mostly adenocarcinomas (AEG), as are carcinomas of the stomach. AEGs arise from metaplastic epithelial cells at the esophagogastric junction, which have been transformed into an intestinal-type mucus layer in response to prolonged irritation from gastric juice. Known risk factors are gastroesophageal reflux disease and obesity, while alcohol and smoking seem to play no major etiological roles. In western countries, AEGs are among the carcinomas with the highest increase in incidence, having surpassed SCC of the esophagus in frequency. The overall incidence of gastric carcinomas, except for carcinomas of the gastric cardia (same etiologic factors as in AEGs), has decreased in western countries during the last several decades. This is in contrast to eastern countries, such as Korea and Japan, where gastric carcinomas are still among the most common carcinomas. The changing pattern of gastric carcinomas in western countries as well as the geographical distribution of this disease has been related to the prevalence of *Helicobacter pylori*.

General Aspects of Early Therapy Monitoring with FDG-Positron Emission Tomography

Despite these profound differences, carcinomas of the upper digestive tract share common features during cytotoxic therapy with regard to therapy monitoring with FDG-PET. In general, response rates to chemotherapy and/or radiotherapy are below 50% for all carcinoma types of the upper digestive tract as measured by histopathological examination of the surgical specimen (after termination of therapy, e.g., in a neoadjuvant setting). A common criterion for histopathological response to therapy is the presence of no or only scattered tumor cells (< 10% viable tumor cells) in the tumor bed. In a neoadjuvant (preoperative) setting, the overall survival of patients is strongly dependent on whether or not histopathological response is achieved. In carcinomas of the upper digestive tract, FDG-PET allows the prediction of histopathological response to therapy as early as two weeks after the onset of cytotoxic therapy by measuring therapy-induced changes in glucose metabolism of the tumor. The accuracy for prediction of histopathological response is > 80%. Correspondingly, metabolic changes in the tumors as measured by FDG-PET are a strong predictor of patient survival.

Also later than 14 days of therapy and after termination of therapy, tumor changes in the PET signal between baseline and follow-up are predictive of histopathological response and patient survival. However, the accuracy for response prediction does not appear to increase significantly as compared with evaluation at 14 days after onset of therapy. This observation is explained by the fact that the PET signal decreases in both, carcinomas with histopathological response and those without though at different speeds. Both groups reach a rather low tumor glucose metabolism after therapy (Fig. 1). As a consequence, late after initiation of therapy, changes in the tumor PET signal from baseline to follow-up do not further diverge between responders and nonresponders but rather are parallel to each other (Fig. 2). The decline in metabolic activity even in nonresponders becomes clear from the fact that even in nonresponders a tumor cell kill as high as 90% is observed (see response definition above). Thus, in some nonresponding tumors, the remaining viable tumor cell fraction may be too small to significantly increase the PET signal over that of responding tumors.

Absolute measurements of tumor glucose metabolism can be derived from a single PET study prior to, during, or after therapy. However, these single measurements are of no or little value for prediction of response or survival. To some extent, the lack of predictive power resides in the considerable variability of glucose metabolism in tumors of the upper digestive tract. This variability overshadows the course of the PET signal in responding tumors compared with nonresponding tumors. For this reason, measurement of relative changes between baseline and follow-up is preferred to single measurements of absolute tumor FDG uptake. Nevertheless, pretherapeutic values of tumor FDG uptake tend to be higher in responding tumors, whereas those after therapy tend to be lower compared with nonresponding tumors.

A frequently raised concern in therapy monitoring with FDG-PET is the development of inflammatory changes

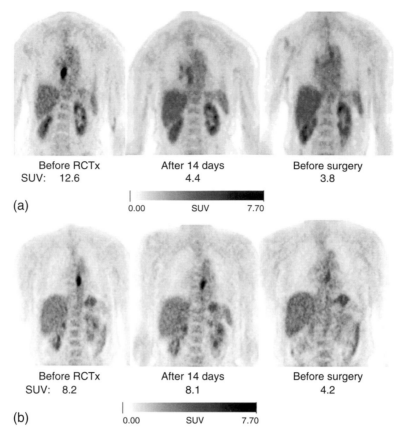

Before RCTx   After 14 days   Before surgery
SUV: 12.6   4.4   3.8

(a)
0.00   SUV   7.70

Before RCTx   After 14 days   Before surgery
SUV: 8.2   8.1   4.2

(b)
0.00   SUV   7.70

Figure 1 (a) Tumor FDG accumulation over time in a responding esophageal carcinoma. Tumor FDG uptake is already significantly reduced two weeks after onset of chemoradiotherapy and is at background levels at completion of therapy. (b) Corresponding course in a nonresponding esophageal carcinoma. Although tumor FDG uptake is low after completion of therapy, there is still marked tumor uptake at two weeks.

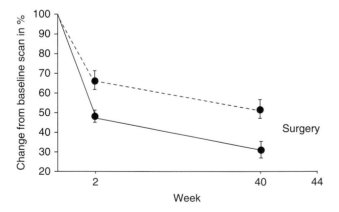

Figure 2 Time-activity course in responding (full line) and non-responding (dotted line) adenocarcinomas of the esophagogastric junction following chemotherapy. There is a pronounced difference in FDG uptake between both subgroups at two weeks after onset of therapy. This difference remains nearly unchanged until completion of therapy.

owing to cytotoxic therapy, in particular radiotherapy. Since inflamed tissues show an enhanced glucose use, these changes are thought to interfere with correct evaluation of therapy response by FDG-PET. Inflammatory changes are occasionally observed during and after cytotoxic therapy, especially in SCC of the esophagus during chemoradiotherapy. However, the findings on FDG-PET are readily discernible from tumor uptake by their extended configuration and relatively low FDG uptake (Fig. 3). Thus, in our experience inflammatory changes—at least at 14 days after onset of therapy—do not significantly hamper therapy monitoring with FDG-PET.

Specific Aspects

In SCC of the esophagus, during chemoradiotherapy, a cut-off value of > 30% for the decrease in tumor SUV from baseline to two weeks after onset of therapy has been found

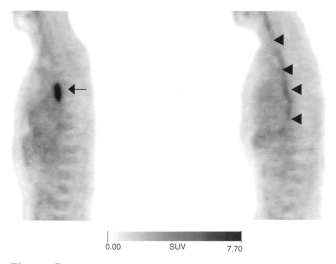

Figure 3 FDG uptake in the esophagus owing to inflammatory changes following chemoradiotherapy differs from tumor uptake by lower intensity and diffuse configuration.

to detect almost all histopathological responding tumors in a neoadjuvant therapy setting (sensitivity 93%). Specificity is ≈ 88% (Wieder *et al.*, 2004). A > 30% decrease in tumor SUV is also associated with an increased rate of complete resections (R0) of the primary tumor (100% vs. 63%) and a favorable patient survival (two-year survival rate 79% vs. 38%). Therapy monitoring is possible in almost all patients: virtually all SCCs of the esophagus show markedly enhanced glucose use and are thus suited for quantitative measurements of relative changes in the PET signal.

In AEGs, histopathological response (and favorable survival) can be predicted with a 93% sensitivity and 95% specificity (two-year survival rate 60% vs. 37%) by using a cut-off value of > 35% at two weeks after onset of therapy (Weber *et al.*, 2001). The rate of complete resections is enhanced in tumors with a > 35% decrease in tumor FDG uptake (87% vs. 41%). Only a few AEGs elude therapy monitoring because of a pretherapeutic FDG uptake that is too low for assessment of percent decrease during therapy.

Therapy response in gastric carcinomas can be predicted with a sensitivity of 77% and a specificity of 86% by using a cut-off value of > 35% (Ott *et al.*, 2003). Correspondingly, the PET result is highly predictive of survival (two-year survival rate 90% vs. 25%). Although not significant, the rate of complete resections was found to be slightly higher in primary tumors with a > 35% decrease in FDG uptake (92% vs. 74%). However, at least in western countries, a significant percentage of gastric carcinomas escape from therapy monitoring because of a too low pretherapeutic tumor FDG uptake. This percentage may be as high as 30% if PET protocols are used, which apply a rather short interval between FDG injection and image acquisition (e.g., 40 min). A longer time interval (e.g., 90 min), owing to progressive

tumor FDG uptake, may allow for inclusion of some more gastric carcinomas for therapy monitoring.

Procedural Aspects

At two weeks after onset of therapy, in carcinomas of the upper digestive tract, the cut-off values for discriminating responders from nonresponders by PET are in the range of a 30–35% decrease in tumor FDG uptake. This range is above the known intraindividual variability of tumor SUVs of ~ 20%. Thus, a measured 30% or 35% decrease reliably indicates reduction in tumor metabolism. However, as a precondition in order to allow for measurements of 30% to 35% decrease, the initial tumor FDG uptake must also exceed background levels. In practice, a twofold elevation of tumor FDG uptake over background levels is considered sufficient.

Because tumor FDG uptake relies on a multitude of factors a highly standardized protocol is indispensable in order to obtain reliable results from therapy monitoring with FDG-PET. These factors include the interval from injection to imaging (e.g. 40 min. versus 90 min.), the interval from baseline to follow-up, and the reconstruction algorithm (filtered backprojection versus iterative). If kept unchanged for baseline and follow up, these methodical aspects have no significant impact on the accuracy for response prediction. Overall, if the above demands are met, FDG-PET can be considered a robust method for early therapy monitoring in carcinomas of the upper digestive tract.

Colorectal Cancer

Adenocarcinomas of the colon or rectum are among the most common cancers in western countries. Chemotherapy is commonly used with curative intention (adjuvant and neoadjuvant protocols in combination with surgery, in conjunction with radiotherapy) or palliative intention. 5-fluorouracil (FU) in combination with folinic acid has been the mainstay of chemotherapy for decades. Response to chemotherapy is generally observed in < 50% of patients. In metastatic liver disease, using FDG-PET, prediction of tumor response to chemotherapy with 5-FU is feasible by measuring relative changes in tumor metabolism between baseline and 4–5 weeks after onset of therapy (cut-off value −15% for tumor-to-liver ratios, sensitivity 100%, specificity 90% on a lesion-by-lesion basis, 75% on a patient-by-patient basis) (Findlay *et al.*, 1996). At an earlier point in time after onset of therapy (1–2 weeks), response prediction is hampered by transient increase in tumor metabolism in some responding lesions (so-called flare phenomenon) and slightly decreasing tumor metabolism in some nonresponding lesions. Prior to chemotherapy, single PET measurements of tumor metabolism are not predictive of response. During chemotherapy, a slight (at 1–2 weeks) and a marked (at 4–5 weeks) association, respectively, has been observed between tumor metabolism and tumor response as obtained from a single PET examination.

In colorectal cancer, for the site of tumor origin, early response monitoring has not yet been studied in detail with FDG-PET. Rather, the FDG-PET result after completion of chemoradiotherapy has been correlated with tumor response and feasibility of sphincter preserving surgery. Positron emission tomography tracers other than FDG have been assessed for response prediction in colorectal cancer. The positron emitting analog [^{18}F] fluorouracil biochemically behaves in exactly the same way as the chemotherapeutic agent 5-FU. The degree of accumulation of [^{18}F] FU in liver metastases from colorectal cancer, prior to therapy, has been shown to positively correlate with survival following treatment with 5-FU. Another PET tracer, [^{11}C]methyl-L-methionine, is representative for cellular amino acid uptake, which has been shown to be enhanced in malignancies including rectal cancer (Wieder *et al.*, 2002). However, after preoperative chemoradiotherapy, amino acid uptake in rectal cancer is reduced in both responding and nonresponding tumors. This observation has been attributed to a stunning phenomenon of amino acid uptake even in nonresponding tumors following therapy. Changes in amino acid uptake are therefore not useful for differentiating responding tumors from nonresponding ones therefore becomes difficult.

In conclusion, therapy monitoring using PET is more than 15 years old. Despite continuously growing experience in numerous centers, the acceptance of this promising technology is still poor. Despite its clear potential, it is not yet routinely part of controlled investigations. However, numerous studies have shown that FDG-PET is a helpful method in early assessment of therapeutic response. Several investigations were conducted in patients with NSCLC, NHL, or carcinoma of the esophagus showing promising results with respect to individual therapy monitoring. Evaluation of therapeutic effect is currently based on the RECIST (Response Evaluation Criteria in Solid Tumors) guidelines. This guideline considers simplified evaluation of tumor size using transaxial CT images. Given the broad acceptance of the response criteria, there is a discussion about translating additional metabolic information into the evaluation process. Whether, or in what form it will be, depends on the results of clinical studies currently in progress. These results will also promote the discussion regarding the innovative technology of combined morphologic and metabolic imaging (O'Connell, 2004).

References

Ansquer, Y., Leblanc, E., Clough, K., Morice, P., Dauplat, J., Mathevet, P., Lhomme, C., Scherer, C., Tigaud, J.D., Benchaib, M., Fourme, E., Castaigne, D., Querleu, D., and Dargent, D. 2001. Neoadjuvant chemotherapy for unresectable ovarian carcinoma—a French multicenter study. *Cancer 91*:2329–2334.

Armitage, J., Weisenburger, D., Hutchins, M., Moravec, D., Dowling, M., Sorensen, S., Mailliard, J., Okerbloom, J., Johnson, P., and Howe, D. 1986. Chemotherapy for diffuse large-cell lymphoma—rapidly responding patients have more durable remissions. *J. Clin. Oncol. 4*:160–164.

Avril, N., Sassen, S., Schmalfeldt, B., Naehrig, J., Rutke, S., Weber, W.A., Werner, M., Graeff, H., Schwaiger, M., and Kuhn, W. 2005. Prediction of response to neoadjuvant chemotherapy by sequential F-18-fluorodeoxyglucose positron emission tomography in patients with advanced-stage ovarian cancer. *J. Clin. Oncol. 23*:7445–7453.

Betticher, D.C., Hsu Schmitz, S.F., Totsch, M., Hansen, E., Joss, C., von Briel, C., Schmid, R.A., Pless, M., Habicht, J., Roth, A.D., Spiliopoulos, A., Stahel, R., Weder, W., Stupp, R., Egli, F., Furrer, M., Honegger, H., Wernli, M., Cerny, T., and Ris, H.B. 2003. Mediastinal lymph node clearance after docetaxel-cisplatin neoadjuvant chemotherapy is prognostic of survival in patients with stage IIIA pN2 non-small cell lung cancer: a multicenter phase II trial. *J. Clin. Oncol. 21*:1752–1759.

Breathnach, O.S., Freidlin, B., Conley, B., Green, M.R., Johnson, D.H., Gandara, D.R., O'Connell, M., Shepherd, F.A., and Johnson, B.E. 2001. Twenty-two years of phase III trials for patients with advanced non-small-cell lung cancer: sobering results. *J. Clin. Oncol. 19*:1734–1742.

Canellos, G. 1988. Residual mass in lymphoma may not be residual disease. *J. Clin. Oncol. 6*:931–933.

Cerfolio, R., Ojha, B., Mukherjee, S., Pask, A., Bass, C., and Katholi, C. 2003. Positron emission tomography scanning with 2-fluoro-2-deoxy-d-glucose as a predictor of response of neoadjuvant treatment for non-small cell carcinoma. *J. Thoracic and Cardiovas. Surg. 125*:938–944.

Cerfolio, R.J., Bryant, A.S., Winokur, T.S., Ohja, B., Bartolucci, and A.A. 2004. Repeat FDG-PET after neoadjuvant therapy is a predictor of pathologic response in patients with non-small cell lung cancer. *Annals Thoracic Surg. 78*:1903–1909.

Cheson, B.D., Horning, S.J., Coiffier, B., Shipp, M.A., Fisher, R.I., Connors, J.M., Lister, T.A., Vose, J., Grillo-Lopez, A., Hagenbeek, A., Cabanillas, F., Klippensten, D., Hiddemann, W., Castellino, R., Harris, N.L., Armitage, J.O., Carter, W., Hoppe, R., Canellos, and G.P. 1999. Report of an international workshop to standardize response criteria for non-Hodgkin's lymphomas. *J. Clin. Oncol. 17*:1244.

De Wit, M., Bumann, D., Beyer, W., Herbst, K., Clausen, M., Hossfeld, and D.K. 1997. Whole-body positron emission tomography (PET) for diagnosis of residual mass in patients with lymphoma. *Annals of Oncol. 8*:57–60.

Dose Schwarz, J., Bader, M., Jenicke, L., Hemminger, G., Janicke, F., and Avril, N. 2005. Early prediction of response to chemotherapy in metastatic breast cancer using sequential 18F-FDG PET. *J. Nucl. Med. 46*:1144–1150.

Findlay, M., Young, H., Cunningham, D., Iveson, A., Cronin, B., Hickish, T., Pratt, B., Husband, J., Flower, M., and Ott, R. 1996. Noninvasive monitoring of tumor metabolism using fluorodeoxyglucose and positron emission tomography in colorectal cancer liver metastases: correlation with tumor response to fluorouracil. *J. Clin. Oncol. 14*:700–708.

Haberkorn, U., Morr, I., Oberdorfer, F., Bellemann, M., Blatter, J., Altmann, A., Kahn, B., and van Kaick, G. 1994. Fluorodeoxyglucose uptake *in vitro*: aspects of method and effects of treatment with gemcitabine. *J. Nucl. Med. 35*:1842–1850.

Haioun, C., Itti, E., Rahmouni, A., Brice, P., Rain, J.D., Belhadj, K., Gaulard, P., Garderet, L., Lepage, E., Reyes, F., and Meignan, M. 2005. [18F]fluoro-2-deoxy-D-glucose positron emission tomography (FDG-PET) in aggressive lymphoma: an early prognostic tool for predicting patient outcome. *Blood 106*:1376–1381.

Hill, M., Cunningham, D., MacVicar, D., Roldan, A., Husband, J., McCready, R., Mansi, J., Milan, S., and Hickish, T. 1993. Role of magnetic resonance imaging in predicting relapse in residual masses after treatment of lymphoma. *J. Clin. Oncol. 11*:2273–2278.

The International Non-Hodgkin's Lymphoma Prognostic Factors Project. 1993. A predictive model for aggressive non-Hodgkin's Lymphoma. *N. Engl. J. Med. 329*:987–994.

Jerusalem, G., Beguin, Y., Fassotte, M.F., Najjar, F., Paulus, P., Rigo, P., and Fillet, G. 1999. Whole-body positron emission tomography using 18f-fluorodeoxyglucose for post-treatment evaluation in Hodgkin's disease and non-Hodgkin's Lymphoma has higher diagnostic and prognostic value than classical computed tomography scan imaging. *Blood 94*:429–433.

Juweid, M.E., Wiseman, G.A., Vose, J.M., Ritchie, J.M., Menda, Y., Wooldridge, J.E., Mottaghy, F.M., Rohren, E.M., Blumstein, N.M., Stolpen, A., Link, B.K., Reske, S.N., Graham, M.M., and Cheson, B.D. 2005. Response assessment of aggressive non-Hodgkin's Lymphoma by integrated international workshop criteria and fluorine-18-fluorodeoxyglucose positron emission tomography. *J. Clin. Oncol. 23*:4652–4661.

Kubota, R., Yamada, S., Kubota, K., Ishiwata K, Tamahashi N., and Ido, T. 1992. Intratumoral distribution of fluorine-18-fluorodeoxyglucose *in vivo*: high accumulation in macrophages and granulation tissues studied by microautoradiography. *J. Nucl. Med. 33*:1972–1980.

Margaritora, S., Cesario, A., Galetta, D., D'Andrilli, A., Macis, G., Mantini, G., Trodella, L., and Granone, P. 2001. Ten-year experience with induction therapy in locally advanced non-small cell lung cancer (NSCLC): is clinical re-staging predictive of pathological staging? *Eur. J. Cardio-Thoracic Surg. 19*:894–898.

Milleron, B., Westeel, V., Quoix, E., Moro-Sibilot, D., Braun, D., Lebeau, B., Depierre, A., for the French Thoracic Cooperative Group. 2005. Complete response following preoperative chemotherapy for resectable non-small cell lung cancer: accuracy of clinical assessment using the French trial database. *Chest 128*:1442–1447.

Minn, H., Zasadny, K.R., Quint, L.E., and Wahl, R.L., 1995. Lung cancer—reproducibility of quantitative measurements for evaluating 2- F-18-fluoro-2-deoxy-d-glucose uptake at pet. *Radiology 196*:167–173.

Naumann, R., Vaic, A., Beuthien-Baumann, B., Bredow, J., Kropp, J., Kittner, T., Franke, W-G., and Ehninger, G. 2001. Prognostic value of positron emission tomography in the evaluation of post-treatment residual mass in patients with Hodgkin's disease and non-Hodgkin's Lymphoma. *Brit. J. Haematol. 115*:793–800.

O'Connell, M. 2004. PET-CT modification of RECIST guidelines. *J. Natl. Cancer Inst. 96*:801–802.

Ott, K., Fink, U., Becker, K., Stahl, A., Dittler, H.J., Busch, R., Stein, H., Lordick, F., Link, T., Schwaiger, M., Siewert, J.R., and Weber, W.A. 2003. Prediction of response to preoperative chemotherapy in gastric carcinoma by metabolic imaging: results of a prospective trial. *J. Clin. Oncol. 21*:4604–4610.

Paul, R. 1987. Comparison of fluorine-18-2-fluorodeoxyglucose and gallium 67 citrate imaging for detection of lymphoma. *J. Nucl. Med. 28*:288–292.

Pisters, K., Kris, M., Gralla, R., Zaman, M., Heelan, R., Martini, N. 1993. Pathologic complete response in advanced non-small cell lung cancer following preoperative chemotherapy: implications for the design of future non-small cell lung cancer combined modality trials. *J. Clin. Oncol. 11*:1757–1762.

Ryu, J.S., Choi, N.C., Fischman, A.J., Lynch, T.J., and Mathisen, D.J. 2002. FDG-PET in staging and restaging non-small cell lung cancer after

neoadjuvant chemoradiotherapy: correlation with histopathology. *Lung Cancer 35*:179–187.

Spaepen, K., Stroobants, S., Dupont, P., Bormans, G., Balzarini, J., Verhoef, G., Mortelmans, L., Vandenberghe, P., and De Wolf-Peeters, C. 2003. [18F]FDG PET monitoring of tumour response to chemotherapy: does [18F]FDG uptake correlate with the viable tumour cell fraction? *Eur. J. Nucl. Med. Molec. Imaging. 30*:682–688.

Spaepen, K., Stroobants, S., Dupont, P., Vandenberghe, P., Thomas, J., de Groot, T., Balzarini, J., De Wolf-Peeters, C., Mortelmans, L., and Verhoef, G. 2002. Early restaging positron emission tomography with 18F-fluorodeoxyglucose predicts outcome in patients with aggressive non-Hodgkin's Lymphoma. *Ann. Oncol. 13*:1356–1363.

Spaepen, K., Stroobants, S., Dupont, P., Van Steenweghen, S., Thomas, J., Vandenberghe, P., Vanuytsel, L., Bormans, G., Balzarini, J., De Wolf-Peeters, C., Mortelmans, L., and Verhoef, G. 2001. Prognostic value of positron emission tomography (PET) with fluorine-18 fluorodeoxyglucose ([18F]FDG) after first-line chemotherapy in non-Hodgkin's Lymphoma: is [18F]FDG-PET a valid alternative to conventional diagnostic methods? *J. Clin. Oncol. 19*:414–419.

Thomas, F., Cosset, J.M., Cherel, P., Renaudy, N., Carde, P., and Piekarski, J.D. 1988. Thoracic CT-scanning follow-up of residual mediastinal masses after treatment of Hodgkin's disease. *Radiotherapy and Oncology 11*:119–122.

Wahl, R., Zasadny, K., Helvie, M., Hutchins, G., Weber, B., and Cody, R. 1993. Metabolic monitoring of breast cancer chemohormonotherapy using positron emission tomography: initial evaluation. *J. Clin. Oncol. 11*:2101–2111.

Weber, W.A., Ott, K., Becker, K., Dittler, H.J., Helmberger, H., Avril, N.E., Meisetschlager, G., Busch, R., Siewert, J.R., Schwaiger, M., and Fink, U. 2001. Prediction of response to preoperative chemotherapy in adenocarcinomas of the esophagogastric junction by metabolic imaging. *J. Clin. Oncol. 19*: 3058–3065.

Weber, W.A., Petersen, V., Schmidt, B., Tyndale-Hines, L., Link, T., Peschel, C., and Schwaiger, M. 2003. Positron emission tomography in non-small cell lung cancer: prediction of response to chemotherapy by quantitative assessment of glucose use. *J. Clin. Oncol. 21*:2651–2657.

Weber, W.A., Ziegler, S.I., Thodtmann, R., Hanauske, A.R., and Schwaiger, M. 1999. Reproducibility of metabolic measurements in malignant tumors using FDG PET. *J. Nucl. Med. 40*:1771–1777.

Wieder, H.A., Brucher, B.L., Zimmermann, F., Becker, K., Lordick, F., Beer, A., Schwaiger, M., Fink, U., Siewert, J.R., Stein, H.J., and Weber, W.A. 2004. Time course of tumor metabolic activity during chemoradiotherapy of esophageal squamous cell carcinoma and response to treatment. *J. Clin. Oncol.* 900–908.

Wieder, H., Ott, K., Zimmermann, F., Nekarda, H., Stollfuss, J., Watzlowik, P., Siewert, J.R., Fink, U., Becker, K., Schwaiger, M., and Weber, W.A. 2002. PET imaging with [11C]methyl- L-methionine for therapy monitoring in patients with rectal cancer. *Eur. J. Nucl. Med. Mol. Imaging 22*: 789–796.

Zinzani, P.L., Fanti, S., Battista, G., Tani, M., Castellucci, P., Stefoni, V., Alinari, L., Farsad, M., Musuraca, G., Gabriele, A., Marchi, E., Nanni, C., Canini, R., Monetti, N., and Baccarani, M. 2004. Predictive role of positron emission tomography (PET) in the outcome of lymphoma patients. *Brit. J. Cancer 91*:850–854.

2

Positron Emission Tomography in Medicine: An Overview

Abass Alavi and Steve S. Huang

Introduction

Positron emission tomography (PET) is a noninvasive imaging modality for whole-body imaging of cellular and metabolic functions. The concept of PET was first conceived in the early 1970s. Collaboration between researchers at the University of Pennsylvania and the Brookhaven National Laboratory culminated in the first positron emission tomography of the human brain with Fluorene-18-Fluorodeoxyglucose (FDG) in 1976. Modern PET cameras differ greatly from the gamma camera used in the first PET scan. The modern coincident two-photon-detecting PET instruments provide outstanding images with superb spatial resolution and sensitivity compared to other imaging modalities. Paralleling the refinement in PET cameras has been an explosion in the development of PET tracers serving all aspects of modern medicine. With a chemically diverse repertoire of positron-emitting radionuclide (Table 1), PET chemists can devise tracers that target almost every aspect of the human biochemical milieu. This chapter briefly discusses current and upcoming applications of PET tracers in oncology, neurology, cardiology, infectious disease, and inflammatory disorders.

Positron Emission Tomography in Oncology

FDG is the most widely available and currently the dominant PET tracer in oncology. Cancer imaging with FDG-PET takes advantage of increased glucose metabolism in cancer cells first described by Otto Warburg in the 1930s. FDG is taken up by cells with glucose transporters such as GLUT-1. Once in the cytoplasm, FDG enters the first step of the glycolytic pathway and is converted to FDG-6-phosphate. Unlike glucose-6-phosphate, FDG-6-phosphate is not a substrate of phosphoglucose isomerase, the second enzyme in the glycolytic pathway. Hence, the radioactivity accumulates inside cells. Selective imaging of the cancer cells is practical because of persistent glucose uptake in many cancer cells in low-serum insulin state when most cells outside of the central nervous system consume glycogen or fatty acid instead of glucose for energy needs. This produces a high cancer to surrounding tissue contrast in a fasting patient. This "contrast resolution," based on differential cellular metabolism, makes FDG-PET unique compared with other imaging modalities such as modern X-ray computed tomography (CT) and magnetic resonance imaging (MRI.) Severity and extent of metastatic disease are

Table 1 Positron-Emitting Nuclide with Potential Medical Applications and Their Corresponding Half-lives

Radionuclide	Half-life	Radionuclide	Half-life
^{18}F	110 min	^{44}Sc	3.97 hr
^{11}C	20.4 min	^{55}Co	17.5 hr
^{15}O	2.03 min	^{61}Cu	3.4 hr
^{13}N	9.9 min	^{62}Cu	9.74 min
^{66}Ga	9.49 hr	^{64}Cu	12.7 hr
^{68}Ga	1.14 hr	^{76}Br	16.2 hr
^{86}Y	14.7 hr	^{110m}In	1.15 hr
^{124}I	100 hr	^{94m}Tc	52 min
^{82}Rb	1.25 min		

Source: Lundqvist, H., and Tolmachev, V. 2002. Targeting peptides and positron emission tomography. *Biopolymers 66*:381-392, and Takalkar A., Mavi A., Alavi A., and Araujo L. 2005, PET in cardiology, *Radiol. Clin. N. Am. 43*:107–119.

often poignantly illustrated with an FDG-PET scan in a few frames of projection images (Fig. 4).

Clinically, FDG-PET is used for (1) localization of suspected malignancies that appear as foci of high-glucose metabolism in low-insulin state, and (2) ascertaining metabolic activities of a known focus of abnormality. Encompassing these two unique roles are the tasks of initial staging of patients newly diagnosed with cancer, restaging cancer patients with suspected recurrence, assessment of recurrence, and cancer treatment monitoring. Accurate tumor staging is essential for choosing the appropriate treatment strategy for patients with cancer. For example, FDG-PET scanning enables assessment of the local extent of thoracic malignancies as well as the presence of nodal or distant metastasis. At the time of diagnosis, PET is useful in initial preoperative staging for non-small cell lung cancer

(NSCLC). Accurate tumor staging is essential for choosing the appropriate treatment strategy in patients with lung cancer. In one study, 30% of patients with lung cancer who were possible surgical candidates based on appearance of limited disease with conventional imaging techniques were found to have metastasis with FDG-PET (Mavi *et al.*, 2005). A PET scan is helpful in evaluating mediastinal-spread lung cancer and in distinguishing stage N3 from potentially resectable stage N1 and N2 disease. Positron emission tomography detects unexpected extrathoracic metastases in 10–20% of patients and changes therapeutic management in about 20% of patients with lung cancer. By staging patients more accurately, FDG-PET utilization decreases unwarranted surgery and morbidity associated with aggressive treatment when palliation should be considered.

For restaging, FDG-PET has become an invaluable tool in the evaluation of patients with suspected recurrent cancer based on increased serum tumor marker levels such as thyroglobulin, carcinoembryonic antigen, or CA-125. The role of PET for detecting the recurrence of a tumor following the initial response has become the hallmark of this technique in almost every malignancy. Anatomic imaging techniques, such as CT, suffer from the general shortcoming of morphologic criteria for initial staging. In addition, structural changes following surgery and/or radiation/chemotherapy render anatomic imaging techniques inconclusive in such settings. Restaging with FDG-PET was shown to have sensitivity consistently between 80 and 95%, specificity of 75–90%, and accuracy of 80–90% between 1993 and 2000 (Juweid and Cheson, 2006). It is likely that the performance of the most recent generation of PET scanners will be even better.

Conventional treatment monitoring using CT relies on changes in the size of the lymph nodes, which is a slow process and may not be conclusive in the early phases of

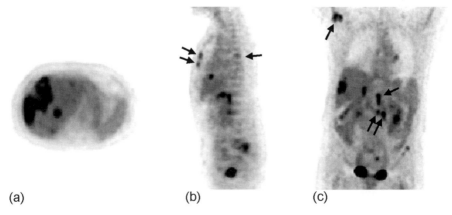

(a)   (b)   (c)

Figure 4 Cancer staging with FDG-PET. Representative cross-sectional images of a PET scan of a patient with metastatic esophageal cancer. (a) Transaxial section showing liver metastasis. (b) Sagittal section illustrating bone marrow involvement in the sternum and a vertebral body (arrows). (c) Coronal section demonstrating liver metastasis, para-aortic lymph node metastasis, and metastatic disease in the right humeral head (arrows).

favorable response. Furthermore, CT is unable to distinguish between active disease and residual scar tissue after therapy. Given that a treatment course typically lasts several months, undergoing an ineffective treatment not only wastes valuable health-care resources but also causes unnecessary morbidity for patients. Being able to determine treatment effectiveness and to redirect management as necessary can save both valuable time for patients and resources for the society at large. For example, Cachin *et al.* studied treatment monitoring clinical trial by looking at the therapeutic efficacy of high-dose chemotherapy with bone marrow transplant and concluded that response to therapy as measured by FDG-PET is the single most powerful independent predictor of survival. Eighty-three percent of the patients who underwent the high-dose chemotherapy treatment with complete response as shown by FDG-PET were alive at 12 months. By comparison, responses by Response Evaluation Criteria in Solid Tumors (RECIST) based on conventional imaging techniques were unable to predict survival (Cachin *et al.*, 2006). For Hodgkin's Lymphoma (HL), a negative FDG-PET scan post-therapy is highly indicative of long-term, disease-free survival and is particularly useful in the presence of residual CT mass (Kumar *et al.*, 2004.) Figure 5 illustrates treatment monitoring of a patient with Hodgkin's Lymphoma. For aggressive non-Hodgkin's Lymphoma (NHL), a positive FDG-PET scan is predictive of disease persistence or recurrence. For both NHL and aggressive HL, early assessment of response appears to be predictive of long-term outcome. Optimal time of FDG-PET scan during therapy needs to be determined. For indolent NHL, the high rate of false-negative FDG-PET scans raises questions regarding its clinical role in treatment monitoring.

The role of FDG-PET continues to expand as more evidence of its utility and cost-effectiveness in various areas of medical practice continues to emerge. Table 2 offers a glimpse of current applications in oncology and representative sensitivity and specificity for each application. The utility of FDG-PET, however, is limited in detecting lesions < 5 mm in diameter, and FDG-PET is not meant to be a substitute for pathologic examination for detecting micrometastasis. In fact, FDG-PET is likely to be part of the imaging work-up for selection of biopsy sites. PET-guided biopsy/resection has been heavily studied and will likely emerge as the standard of care for neoplastic diseases such as insulinoma or mesothelioma in the near future.

Positron Emission Tomography in Lung and Breast Cancer

Lung cancer and breast cancer are two major causes of cancer-related mortality in the United States. The number of deaths due to lung cancer annually is greater than that of breast, colon, and prostate cancers combined (Bunyaviroch and Coleman, 2006). Breast cancer is a leading cause of cancer mortality among women, second only to lung cancer. One in nine women is diagnosed with breast cancer in her lifetime. Diagnosis of lung cancer often starts with a solitary pulmonary nodule incidentally noted during a chest X-ray or a chest CT exam. Positron emission tomography is useful as an adjunct modality in evaluation of solitary pulmonary nodules, especially when results from conventional imaging modalities are equivocal. The sensitivity and specificity of FDG-PET for evaluating radiologically indeterminate nodules

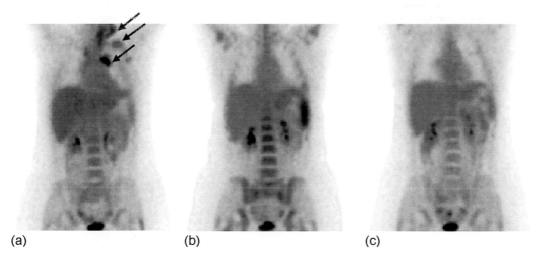

(a) (b) (c)

Figure 5 Treatment monitoring with FDG-PET. (a) Patient with Hodgkin's disease involving the neck, mediastinum, and left supraclavicular lymph nodes (arrows). (b) Post-chemotherapy scan conducted 46 days after the initial scan illustrating resolution of the malignant foci. Prominent bone marrow and spleen uptake were observed due to the recovering hematopoietic system post-chemotherapy. (c) Repeat PET scan 91 days after the initial scan showed resolution of bone marrow and splenic FDG uptake.

Table 2 Selected Indications for FDG-PET in Oncology and the Representative Sensitivities and Specificities

Clinical Indication	Sensitivity	Specificity
Head and neck tumor, restaging	95%	83%
Thyroid cancer restaging		
Follicular/papillary	75%	90%
Medullary	78%	79%
Lung cancer		
Solitary pulmonary nodule	97%	78%
Bone marrow metastasis	92%	99%
Adrenal metastasis	100%	80%
Restaging (non-small cell)	97–100%	62–100%
Colon cancer		
Diagnosis	74%	84%
Staging	82%	94%
Restaging	97%	76%
Breast cancer, restaging	97%	92%
Esophageal cancer, staging	51%	94%
Lymphoma, staging	90%	93%
Pancreatic cancer, staging	79%	86%
Cervical cancer, staging	80%	> 95%
Ovarian cancer, restaging	90%	86%
Renal cell carcinoma, staging	64%	100%

Source: Lin, E., and Alavi, A. 2005, *PET and PET/CT: A Clinical Guide* (New York: Thieme Medical Publishers).

are 92% and 90%, respectively (Mavi *et al.*, 2005). FDG-PET may not be the best modality for evaluating solitary pulmonary nodules that are highly suspicious of cancer by conventional radiography; the negative predictive value of FDG-PET is low in this scenario. Clinicians should also be aware that FDG uptakes in bronchoalveolar carcinoma and carcinoid tumors are lower compared to typical non-small cell lung carcinoma. Hilar and mediastinal staging of NSCLC is necessary to differentiate unresectable N3 disease from N1/N2 disease. Positron emission tomography with FDG has a sensitivity of 85–95% and a specificity of 81–100% for mediastinal/hilar staging (Mavi *et al.*, 2005). Some investigators do feel that one cannot completely exclude mediastinal involvement with negative PET finding. Mediastinoscopy may still be necessary. In one study, mediastinoscopy exhibited 3% false-positive rate, while PET had 11.7% false-positive rate. The use of PET/CT and next-generation time-of-flight PET, which has higher spatial resolution, may lower the false-negative rate of PET for mediastinal/hilar staging.

The sensitivity of detecting primary breast cancer with PET has been reported as being between 76 and 100% (Kumar and Alavi, 2004). Our experience indicates that PET is not sensitive enough as a single modality for early detection of breast cancer. The deficiency is mainly due to spacial resolution; hence, microscopic lesions such as those found in ductal carcinoma *in situ* cannot be detected with PET. Although PET has demonstrated good sensitivity and specificity in detecting axillary nodal metastasis, the sensitivity is

not optimal when compared with sentinal lymph node biopsy. The sensitivities of axillary lymph node metastasis detection ranges from 20–100%, with more recent studies showing lower sensitivity (Kumar and Alavi, 2004). This is largely due to improvement in pathologic diagnosis. Although PET currently has no role in primary breast cancer detection and axillary node staging, it should be used for evaluation of distant metastasis, restaging, and treatment monitoring. The sensitivity of PET in detecting distant metastasis ranges from 86–100% and specificities range from 79–97%. For detecting recurrence, FDG-PET has sensitivity ranges from 79–89%. FDG-PET has been proven useful in monitoring treatment response in breast cancer, especially for patients undergoing chemotherapy. Separating responders from nonresponders in breast cancer treatment is achievable with FDG-PET with sensitivities of 90–100% and specificities of 74–85%.

Positron Emission Tomography in Brain Imaging

FDG-PET was originally intended for tomographic imaging of the brain. Ironically, brain tumor imaging with FDG-PET has not enjoyed as much popularity as extracranial tumor imaging, owing largely to the high cortical FDG uptake regardless of patient's fasting state. Many of the current generation of cranial tumor imaging agents are labeled amino acids that have more muted uptake in the normal brain structure. Examples of such tracers are O-(2-^{18}F-fluoroethyl)-L-tyrosine (FET) and [^{11}C]-L-methionine. However, FDG-PET does play an important role in patients with suspected recurrent brain tumors and inconclusive, contrast-enhanced MRI or CT examinations. Because radiation-induced necrosis and recurrent tumors appear enhanced on these scans, metabolic imaging with FDG, which reflects disease activity at the cellular level, is invaluable in this setting. The sensitivity for distinguishing necrosis from brain tumor recurrence is ~ 86% and specificity ~ 56% (Newberg and Alavi, 2005).

One prominent use of FDG-PET for brain imaging is for early diagnosis of Alzheimer's disease. A pattern currently used for diagnosing early Alzheimer's disease (AD) is bi-parietal, bi-temporal hypometabolism. Using such glucose metabolism patterns, one can differentiate early AD from frontal-temporal dementia or Pick's disease. A typical Pick's disease patient exhibits frontal lobe hypometabolism on FDG-PET scan that is not present in early AD. The task of identifying early AD patients is becoming important now that better therapeutics are on the horizon. FDG-PET can also be used for localization of epileptic foci in patients with partial seizures. While a seizure focus may be identified as a region of increased FDG uptake if FDG is injected during seizure, such an "ictal" study is difficult to conduct. The more practical approach is to examine a patient during the "interictal" state. Contrary to the result of an "ictal" study, a seizure focus appears as a region of hypometabolism if FDG is injected during the "interictal" period. This type of study can identify seizure foci in 55–80% of epileptic patients with focal EEG abnormalities.

Other PET tracers for brain imaging include Fluorene-18 fluorodopa ([^{18}F]-FDOPA) for dopamine uptake and metabolism in the brain. FDOPA is useful for use in patients with movement disorders such as Parkinson's disease. [^{15}O]-H2O can be used for general brain perfusion studies. [^{11}C]-raclopride can be used to image the presence of D2 receptors. There are also seretonine receptor and muscarinic receptor tracers currently undergoing clinical trials.

Positron Emission Tomography in Cardiac Imaging

Major categories of tracers for cardiac imaging are those for assessment of (1) blood flow, (2) regions of viability, and (3) regions of hypoxia. Cardiac flow tracers for PET include oxygen-15-labeled water ([^{15}O]-H$_2$O), [^{13}N]-ammonium, and the generator produced rubidium-82 chloride (Takalkar *et al.*, 2005). Among these tracers, rubidium-82-chloride has been approved by the FDA. ^{15}O- and ^{13}N-based tracers require a well-coordinated cyclotron facility and stress lab due to the relatively short half-life of these tracers and the cyclotron requirement for production. ^{82}Rb also has a short half-life, but production and administration can be better coordinated with the strontium-rubidium generator near the bedside. Copper-

62-labeled pyruvaldehyde bis (N4-methylthio-semicarbazone) ([^{62}Cu]-PTSM) is also a flow tracer, but lack of a steady supply of ^{62}Cu has prevented its widespread evaluation. [^{62}Cu]-ATSM is a hypoxia tracer that shows great promise in cardiac imaging. Instead of inferring regions of ischemia based on differential cardiac perfusion between rest and stress, hypoxia tracers can directly illuminate regions of hypoxia.

Positron Emission Tomography in Infection and Inflammation

In addition to cancer cells, activated granulocytes and monocytes also have elevated glucose metabolisms that can lead to markedly increased FDG uptake. In fact, FDG is an excellent tracer for detecting infection and inflammation (Zhuang *et al.*, 2005.) FDG uptake has been reported in sarcoidosis, esophagitis, sinusitis, fungal and bacterial pneumonia, abscesses in various locations, mastitis, osteomyelitis, prosthetic infection, inflammatory bowel disease, and even in atherosclerotic plaques. Recent studies have suggested that FDG-PET may be useful in the work-up of patients with fever of unknown origin (FUO). FDG is ideal because of its sensitivity in detecting metabolically active regions that may be the source of infection or neoplasm in this clinical setting.

FDG-PET may become the study of choice in orthopedic infections. PET is particularly useful in evaluating prosthetic loosening and infection. MRI evaluation is problematic in hip and knee prosthetics due to artifacts from metal implants. The overall sensitivity of detecting lower extremity prosthetic infection with FDG-PET is reported to be greater than 90%, with specificities of 80 to 89%. FDG-PET can also evaluate acute and chronic osteomyelitis, especially in diabetic feet.

Cell Proliferation Agents

Positron emission tomography of cell proliferation is currently performed by indirectly measuring DNA synthesis. 3-deoxy-3-[^{18}F]-fluorothymidine (FLT) is a promising cell proliferation agent. FLT is a thymidine analog. However, unlike [^{3}H]-thymidine, which is a standard compound for *in vitro* proliferation study, FLT is not incorporated into DNA (Mankoff *et al.*, 2005). The biochemical fate of FLT is rather analogous to that of FDG in that it is phosphorylated upon cellular uptake and trapped within a cell without further modification. FLT has the potential to be used as a specific agent for assessing disease activity in various stages of different malignancies, particularly for determining response to therapy because cytotoxic chemotherapeutic agents affect cell division earlier and more prominently than glucose metabolism. Therefore, FLT may prove to be superior to FDG for assessing response to treatment. Also, inflammatory reactions that may confound the use of FDG will not affect the use of FLT in this setting. Ongoing clinical trials are investigating the role of ^{18}F-FLT in imaging proliferation of liver metastasis of colorectal carcinoma, and treatment monitoring of head and neck tumors.

Hypoxia Positron Emission Tomography Imaging

Hypoxia in tumor tissue appears to be an important prognostic indicator of response to either chemotherapy or radiation therapy. Therefore, detection of hypoxia in advance of such interventions may help to optimize the use and outcome of different therapeutic modalities. Also, utilizing hypoxic agents for evaluating the efficacies of emerging anti-angiogenic therapies is a promising approach. Several compounds have been synthesized for detecting hypoxia. These compounds form covalent bonds with cellular proteins under low-oxygen conditions; they are retained substantially in higher concentrations in the hypoxic tissues, which can be detected by external imaging techniques. Examples of hypoxia imaging tracers are [^{18}F]-fluoromisonidazole (FMISO), copper-62 diacetyl-bis (N4-methylthiosemicarbazone) (^{62}Cu-ATSM), 2-(2-nitroimidazol-1[H]-yl)-N-(3-[^{18}F]fluoropropyl)acetamide ([^{18}F]-EF1), and [2-(2-nitro-1[H]-imidazol-1-yl)-N-(2,2, 3,3,3-pentafluoropropyl)-acetamide] ([^{18}F]-EF5). The [^{18}F]-EF compounds have performed well in animal studies and may also prove to be effective for noninvasive imaging of tumor hypoxia (Alavi *et al.*, 2004). There are currently four clinical trials involving the use of EF5. There is one clinical trial with FMISO in conjunction with 18F-FLT for evaluating treatment response of radiation therapy. ^{62}Cu-ATSM appears to have better pharmacodynamic properties than FMISO and could be a better imaging agent for hypoxia.

Peptide and Protein Positron Emission Tomography Tracers

Most tracers with positron-emitting nuclides are small molecules < 1000 Daltons. Increasingly, polypeptides and macromolecules such as Annexin V and various antibodies with molecular weight ~ 150,000 Daltons are being evaluated for clinical use. Positron nuclides with short half-life are less practical for macromolecules. Large molecules in general require hours to days for optimal background clearance. Radionuclides with half-life less than 6 hours decay before optimal images may be acquired. ^{124}I has the advantage of ease of protein labeling and long half-life. This makes ^{124}I a practical imaging radionuclide for proteins and antibodies. ^{64}Cu is also a very attractive radionuclide for large molecules because of the 12.7 hr half-life. Synthetically, ^{64}Cu is incorporated through a chelate such as DOTA, which can be attached chemically to large proteins days or months prior to use. Other chelate-based positron-emitting radionuclides include ^{94m}Tc and ^{111m}In (Lundqvist and Tolmachev, 2002). Being able to image with antibodies can drastically increase diagnostic information, given the almost endless array of antibodies targeting known cell-surface receptors.

In conclusion, modern instrumentations of positron emission tomography allow the detection of nanograms to micrograms of tracers with increasing sensitivity and resolution. The ability to image trace amounts of chemicals in a human body with high sensitivity opens the possibility for constructing and visualizing compounds that target selected cell-surface receptors or biochemical pathways. This gives clinicians and scientific investigators multiple tools for greater understanding and assessment of pathophysiology *in vivo* in cellular context. Positron emission tomography, therefore, is an unprecedented platform that may bridge the gaps between clinical practice and latest understanding of molecular biology of diseases.

References

Alavi, A., Lakhani, P., Mavi, A., Kung, J., and Zhuang, H.M. 2004. PET: a revolution in medical imaging. *Radiol. Clin. N. Am.* 42:983–1001.

Bunyaviroch T., and Coleman, R.E. 2006. PET evaluation of lung cancer. *J. Nucl. Med.* 47:451–469.

Cachin, F., Prince, H.M., Hogg, A., Ware, R.E., and Hicks, R.J. 2006. Powerful prognostic stratification by [^{18}F]fluorodeoxyglucose positron emission tomography in patients with metastatic breast cancer treated with high-dose chemotherapy, *J. Clin. Oncol.* 24:3026–3031.

Juweid, M.E., and Cheson, B.D. 2006. Positron emission tomography and assessment of cancer therapy. *N. Engl. J. Med.* 354:496–507.

Kumar R., and Alavi, A. 2004. Fluorodeoxyglucose-PET in the management of breast cancer *Radiol. Clin. N. Am.* 42:1113–1122.

Kumar, R., Maillard, I., Schuster, S.J., and Alavi, A. 2004. Utility of Fluorodeoxyglucose-PET imaging in the management of patients with Hodgkin's and non-Hodgkin's lymphomas. *Radiol. Clin. N. Am.* 42:1083–1100.

Lundqvist, H., and Tolmachev, V. 2002. Targeting peptides and positron emission tomography. *Biopolymers* 66:381–392.

Mankoff, D.A., Shields, A.F., and Krohn, K.A. 2005. PET imaging of cellular proliferation. *Radiol. Clin. North. Am.* 43:153–167.

Mavi, A., Lakhani, P., Zhuang, H., Gupta, N.C., and Alavi, A. 2005. Fluorodeoxyglucose-PET in characterizing solitary pulmonary nodules, assessing pleural diseases, and the initial staging, restaging, therapy planning, and monitoring response of lung cancer. *Radiol. Clin. North. Am.* 43:1–21.

Newberg, A.B., and Alavi, A. 2005. The role of PET imaging in the management of patients with central nervous system disorders. *Radiol. Clin. N. Am.* 43:49–65.

Takalkar, A., Mavi, A., Alavi, A., and Araujo, L. 2005. PET in cardiology. *Radiol. Clin. N. Am.* 43:107–119.

Zhuang, H.M., Yu, J.Q., and Alavi, A. 2005. Applications of fluorodeoxyglucose-PET imaging in the detection of infection and inflammation and other benign disorders. *Radiol. Clin. N. Am.* 43:121–134.

3

Radiation Dose and Image Quality

Colin J. Martin and David G. Sutton

Introduction

Medical imaging fulfills a crucial role in the diagnosis and treatment of disease. A range of techniques are available that can provide detailed images of anatomy and physiology based on different properties of the organs and tissues. These techniques use different forms of radiation that are able to pass through tissue (Martin et al., 2003). The radiation fields are modified by the tissue, and so information on the body structure can be obtained by interrogating the radiation signals at the surface. Several imaging modalities use ionizing radiations, which can potentially damage cells within the body. In these modalities in particular, it is important that the amount of radiation used, and consequently the radiation dose to the patient, are not unnecessarily high, but on the other hand, the images produced must be adequate to fulfill the clinical need. There is therefore a delicate balance between radiation dose and image quality. This chapter describes the various techniques using ionizing radiations and considers the factors that affect the radiation dose and image quality. First, the various imaging techniques that are available using both ionizing and nonionizing radiations are described briefly and the type of information that each provides is summarized. It should be noted that techniques involving the use of any type of radiation will deposit some energy in the tissue, but the effects of the higher energy ionizing radiations are potentially more harmful.

The most widely used medical imaging techniques employ X-rays. Conventional X-ray images show the variations in tissue density and composition as a two-dimensional projection. Movement and operation for some of the body's systems can be demonstrated through the use of contrast media. Computed tomography (CT) is another X-ray technique that has developed during the last 30 years with the availability of rapid computer processing technology. This technique delivers cross-sectional images through the body, reconstructed from X-ray transmission data. It allows the imaging of more subtle differences in tissue density and has been the most significant advance in the history of X-ray imaging. The other investigative technique that uses ionizing radiation is nuclear medicine, which involves the use of radioactive materials as tracers. Radiopharmaceuticals are prepared, the uptake of which depends on the physiological processes that are occurring in organs within the body. When these radiopharmaceuticals are administered to patients, the distribution of radioactivity can be determined by forming an image of the radiation emitted. This branch of nuclear medicine, which is called radionuclide imaging, enables functional changes to be both imaged and measured.

Other imaging techniques involving the use of nonionizing radiations, such as ultrasound and radiofrequency waves, are now established components of the imaging portfolios of all major hospitals. Ultrasound uses pulses of high-frequency

45

sound. Organ boundaries and inhomogeneities within tissues are detected by means of the echoes reflected from their surfaces. The principle of the technique resembles that of radar. The interactions depend on the mechanical nature of the tissue, which involves elastic and inertial properties, that are different from those imaged by X-rays. Magnetic resonance imaging (MRI) uses radiofrequency energy and very strong magnets to produce images from the hydrogen nuclei (protons) in water, soft tissue, and fat molecules. The signals obtained from the protons are strongly influenced by their biochemical environment. This technique is a particularly sensitive technique for diagnosing diseases of the central nervous and musculoskeletal systems.

Images obtained using ionizing radiations can provide answers to numerous questions with regard to disease. The benefits in the diagnosis, treatment, and management of disease are obvious, and failure to carry out an examination that is indicated may have a significant detrimental effect on a patient's health. However, the risks associated with the radiation dose from any exposure must not be discounted. If a previous examination has been carried out, the results from this exam should be reviewed first. If there are nonionizing radiation techniques that can provide the same information, then consideration should be given to whether these should be used. Apart from radiation considerations, unwarranted investigation of any type will waste valuable resources and may increase the anxiety of the patient. In addition, some procedures are invasive and therefore carry additional risk to the patient. Therefore, imaging investigations, not simply those involving ionizing radiation, should only be used to answer specific questions that have been generated by clinical inquiry. If the outcome of a procedure will not affect the management of a patient, it is not justified and should not be performed. In order to ensure that imaging procedures are performed only if they are justified, referring clinicians should provide full information regarding the patient's relevant clinical history and indicate the clinical objectives, so that the radiologist can select the most appropriate test to meet the patient's needs.

The quality of an image and the anatomical detail that can be seen within it depend on the properties of the imaging system and the radiation used. In general, use of more radiation will improve the quality of the image within certain limits, although other factors will need to be taken into consideration because the imaging process involves a complex interplay among several different variables. More radiation will give the patient a higher radiation dose. Therefore, the aim should be to provide an image that is adequate for the clinical task with a radiation dose to the patient that is as low as possible. Staff in an imaging department should select the most suitable equipment and the appropriate level and energy of radiation for each task. For example, visualization of a catheter insertion will require far less radiation than the subsequent images of the vasculature obtained by adminis-

tration of contrast through the catheter. In all cases, the level of image quality should be sufficient to ensure that any clinical diagnostic information that could be obtained is imaged, but that the radiation dose to the patient is not significantly higher than is necessary.

Radiation Dose

Many factors affect imaging performance and radiation dose to the patient for X-ray and radionuclide techniques. This chapter discusses the important ones in order to provide the reader with an understanding of their interdependence, and describes some of the options available for reducing doses with different imaging systems. However, before embarking on this discussion, it is useful to consider what we actually mean by the terms *radiation dose* and *image quality*.

Several different quantities are used for measuring and evaluating doses to patients in diagnostic radiology and radionuclide imaging. Of these, the effective dose is the most closely related to the risk associated with radiation (ICRP, 1991). This quantity takes account of doses to radiosensitive organs in different parts of the body, and is linked to the risk of stochastic effects such as cancer induction and genetic effects. However, effective dose is an abstract rather than a measurable quantity. The factors employed to derive it are based on epidemiological evidence of risks from studies of human populations exposed to large radiation doses, in particular the Japanese survivors of the atomic bombs dropped on Hiroshima and Nagasaki. Effective dose attempts to provide a quantity, which represents the uniform dose to the whole body that would have a similar health detriment in terms of stochastic effects in the long term. As mentioned earlier, effective dose cannot be measured directly, so it must be assessed from dose quantities that can be measured. The measurable dose quantities for diagnostic radiology patients are the entrance surface dose and the dose-area product. The entrance surface dose is the dose to the skin at the point where an X-ray beam enters the body. It can be measured with small thermoluminescent crystal dosemeters placed on the skin or calculated from radiographic exposure factors, coupled with measurements of X-ray tube output. The dose-area product is the product of the dose in air (air kerma) within the X-ray beam and the beam area. It quantifies all the radiation that enters a patient, and it can be measured by using an ionization chamber fitted to the X-ray tube. Dose-area product and entrance surface dose are useful quantities: they can easily be applied to monitor, audit, and compare radiation doses from a wide variety of radiological examinations. If required, estimates of effective dose (and hence the associated risk) can be calculated from the practical dose measurement quantities through the use of coefficients derived from computer simulations of the interactions of X-rays as they pass through the various organs within the body.

Effective doses are measured in terms of a unit called the sievert. One sievert (Sv) is a very large dose, and a whole-body dose of 6 Sv is likely to be fatal. Doses for diagnostic examinations are measured in terms of the millisievert (mSv), which is a thousand times smaller. Examples of average effective doses for two common radiographic examinations are chest postero-anterior (PA), 0.02 mSv and lumbar spine antero-posterior (AP), and lateral, 1.0 mSv (Hart et al., 2002; Martin and Sutton, 2002; Martin et al., 2003; CRCPD, 2003b). More complex procedures involving fluoroscopy are associated with higher doses, examples being barium meals, which have a range of 0.7–3 mSv, and barium enemas, which have a range of 2–7 mSv (Martin, 2004). The effective doses for diagnostic investigations are of a similar order of magnitude to those that are received from the natural sources of exposure in everyday life—for example, 0.1 mSv from cosmic ray exposure during a return air flight from Europe to America and the 2–3 mSv received on average from natural background radiation in a year.

Some organs of the body are more sensitive to radiation than others, and when these more sensitive organs are irradiated, the associated risk will be higher. Thus, radiographs of the abdomen and pelvis, where there are several organs that are more sensitive to radiation, tend to have higher effective doses (0.7 mSv) than those of the chest, skull, and limbs (< 0.03 mSv). However, for nuclear medicine examinations, the radiation dose depends on the organs in which the radiopharmaceutical accumulates, which may include tissues distant from the organ of interest. In particular, as radionuclides are excreted, they will accumulate in the bladder and thereby expose organs within the pelvis. Extensive studies have been carried out on the uptake and excretion of radiopharmaceuticals by the body. The Medical Internal Radiation Dose Committee of the United States Society of Nuclear Medicine (Stabin, 1996) has developed a schema for calculating absorbed doses from administered radiopharmaceuticals. Typical effective doses from some common nuclear medicine examinations are bone scan–3 mSv, lung perfusion dashes–1 mSv, and lung ventilation–0.4 mSv.

Effective doses from CT scans are higher than those from other techniques (head 2 mSv, chest 8 mSv, and pelvis 10 mSv). These now make up approximately half of the radiation dose received collectively by all patients from X-rays in developed countries, although they form only a small percentage of the examinations carried out. The flexibility of new CT scanners and the speed with which scans can now be performed have led to a proliferation of CT examinations and have greatly increased the ease and speed at which they can be performed. However, the justification for each procedure should be given careful consideration. Screening programs should be given special attention. If hundreds of thousands of CT examinations are performed unnecessarily, as

part of a health screening program, this could potentially initiate cancers in many individuals. For example, Brenner (2004) has reported that if 50% of all current and former smokers in the U.S. population aged from 50 to 75 were to receive annual CT screening, the estimated number of lung cancers associated with the radiation from the screening based on current risk models would be ~ 36,000 which would represent a 1.8% increase over the baseline. In addition to the increased radiation risk, the high false-positive rate encountered in CT with its concomitant increase in other patient investigations has to be taken into account (Beinfeld et al., 2005). So, given that the number of serious abnormalities detected is small and the false-positive rate is high, such programs may well not provide a net health benefit and could divert valuable health-care resources from areas where they could be used more effectively.

The magnitude of the risk arising from any individual diagnostic procedure is very low even when effective doses are of the order of 10 mSv. Nevertheless, consideration should be given to whether those examinations that deliver higher doses are really required and whether an alternative technique, possibly using nonionizing radiation, could be used. In this context, the risk from a simple examination with a dose of 1 mSv or less can be regarded as negligible.

The risks from exposure to an embryo or fetus are greater than those to children or adults, so particular care is needed when considering whether an investigation of a pregnant woman should be carried out. The examination should only be performed if the risk of not making a diagnosis at that stage is greater than that from irradiating the fetus. Where the examination can be delayed without undue risk to the patient, this may be the better option, or if an acceptable technique is available using nonionizing radiation, this option may be chosen. Despite the precautions taken, radiation examinations of pregnant patients are undertaken from time to time without the patient or medical practitioner realizing that they are pregnant. The absorbed doses to the fetus are usually small, so even though the risks to the fetus are greater than those to an adult, the risks of any effect are still very low. Concerns may be raised regarding potential health effects, but the risks from any diagnostic radiation examination should never justify the termination of a pregnancy. When concerns are expressed, an assessment should be made by a qualified expert in radiation protection prior to counseling the patient.

Dose limits are not applied to medical exposures, but in order to provide a check or comparator, diagnostic reference or guidance levels have been established by various organizations in terms of measurable quantities, such as entrance surface dose or dose-area product. The levels apply to particular examinations. The mean dose in a hospital for a selection of patients of average weight should be less than the relevant diagnostic reference level. If the reference level is exceeded, this should trigger an investigation into whether further optimization is required. Levels are also recommended for the maximum entrance surface dose rates for fluoroscopic examinations with particular image intensifier sizes. In radionuclide imaging, diagnostic reference levels are specified in terms of the amount of radioactivity administered.

Image Quality

Medical image quality relates to the subjective interpretation of visual data and does not have a simple analytical definition despite several attempts to provide one (Rose, 1974; ICRU, 1996). It is the clinical information contained in the image and whether this can be interpreted by the observer that is important rather than whether the appearance of the image is pleasing to the eye. Image quality should be defined in relation to a specific medical task. The ideal set of parameters describing image quality should give a measure of the effectiveness with which an image can be used for its intended purpose. However, in order to follow this through, it would be necessary to carry out expensive clinical trials and to analyze the results in terms of receiver operator characteristics. Even then, since the interpretation and diagnosis made from an X-ray involve subjective opinions from the radiologist, results are likely to vary at different centers. The situation is further complicated because image requirements are different for every examination.

Physical properties that determine the imaging performance of an X-ray system are the resolution and the contrast reproduction of the image receptor, that is, the film/screen system or digital imaging plate on which the image is formed (Hendee and Ritenour, 2002; Dendy and Heaton, 1999). The innate resolution of a film/screen system is determined by both the thickness of the phosphor layer, with which the X-rays interact to produce the light that is recorded by the film, and the grain size of the film emulsion. The resolution for a digital detector relates to the size of each picture element or pixel. These factors determine the resolution for an ideal arrangement in which a thin object is placed next to the image receptor. Because structures within the body are being imaged, the ideal resolution will not occur in practice. X-ray images have an innate unsharpness due partly to the distance between the object and the image receptor, which results because the X-ray tube filament is not infinitesimally small (geometric unsharpness) and because the structures in the body have finite thicknesses (patient unsharpness).

The ability of the receptor to reproduce differences in the intensity of the transmitted X-rays (or contrast) within the useful range is also important in determining imaging performance. Film can reproduce intensity differences only within a relatively narrow range before the film saturates, but digital detectors are much more flexible, and the viewing conditions can be adjusted to match the recorded data. The threshold contrast level gives an indication of the smallest contrast that can be detected by an imaging system and is used as a measure of performance.

The signal-to-noise ratio relates to the visibility of an object. It depends on the object size, the X-ray attenuation or contrast, and the radiation intensity. Noise is the random variation in background density on film and analogue imaging devices, or in pixel values in digital systems. Any image produced using a limited number of photons will show a variation in background noise, even when the object is completely uniform. For example, in the CT image of a section through the abdomen (Fig. 6a), tissues such as the liver, which are relatively uniform, have a range of gray levels. Images that have been obtained using successively more X-ray photons are shown in Figures 6b, 6c, and 6d. As the dose level increases, the random background variation within the uniform tissues decreases because the signal-to-noise level increases. The signal is related to the number of photons (N) and the noise to the square root of the number of photons $\sqrt{N}$, so the signal-to-noise ratio is proportional to $\sqrt{N}$. The tissue detail that can be identified within the images in Figure 1 gradually increases from Figure 6a to Figure 6d. Choosing the level that is sufficient to provide all the clinical diagnostic information is perhaps the most important decision to be made for every radiological technique.

A methodology for assessing the basic aspects of quality for radiographic images dependent on technique and imaging performance has been set up (EC, 1996). This methodology is useful for ensuring that techniques employed within a department provide clinical images of acceptable quality and that any changes made, such as the introduction of dose-saving techniques, do not have a detrimental effect on the clinical image. More objective methods of evaluating image quality are used by standards laboratories in evaluating the performance of imaging systems. Quantities applied include the modulation transfer function, which measures the imaging performance in terms of the signal reproduction for details of different sizes, and the detective quantum efficiency and noise equivalent quanta, which relate to the visibility of details of varying size within the noise generated by the imaging system (ICRU, 1996). These should relate more directly to the visualization tasks that a radiologist will perform in making a diagnosis, but closer links between the perception tasks and the physical measurements need to be established.

Subjective assessments of clinical images are not satisfactory for ensuring that the imaging performance of equipment is maintained. Imaging performance of fluoroscopic and digital equipment will be monitored as part of a quality assurance program. A variety of test objects are employed for this purpose. Some have details with different contrast levels and are used to assess the threshold contrast when objects are just visible. Others have sets of lines of different spacing that are used to evaluate resolution. Upon installation checks are made that the image quality for visualization

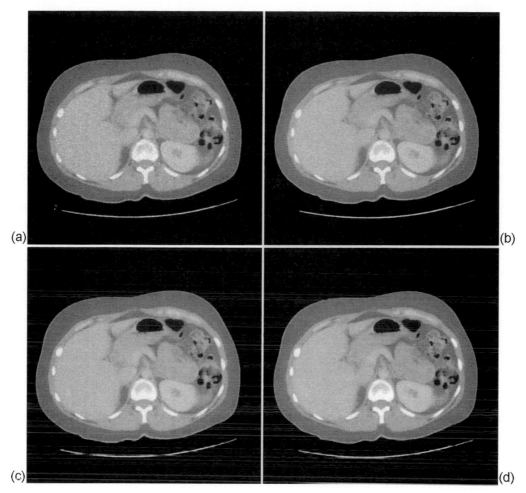

Figure 6 CT images of a slice through the abdomen showing how quantum noise varies with mA for (a) 20 mA, (b) 50 mA, (c) 100 mA, and (d) 200 mA. The radiation dose will increase proportionately with mA.

of details meets the specified requirements. Thereafter, checks on the constancy of the imaging performance will be made at intervals, together with dose performance to identify any deterioration that occurs, so that steps can be taken to restore performance (Martin *et al.*, 1999a). Criteria are set for resolution and threshold contrast that can be used to judge the acceptability of the equipment for clinical imaging. This chapter will not go further into the practical application of these methods, but will discuss the factors that affect image quality and radiation dose.

X-ray Beam Interactions

Consider first a simple X-ray image or radiograph. This is a two-dimensional projection or shadow image produced by an irradiating part of a body with X-rays and recording

the pattern of transmitted X-ray photons on film or other form of image receptor. A metal target within an X-ray tube is bombarded with high-energy electrons to produce the X-ray beam, which contains photons with a range of energies. The X-ray beam will pass through all the different tissues such as skin, solid organs, lungs, and bone within a patient. Some photons will be transmitted straight through without being affected, and others will interact with the tissues and will either be absorbed or scattered away in a different direction. Those that pass through without deviation form the useful part of the image.

The differences in X-ray transmission between tissue structures that determine the image contrast depend on the mechanisms through which X-ray photons interact with tissue, and these vary with photon energy. The important interactions in diagnostic radiology are the photoelectric effect and Compton scattering (Hendee and Ritenour, 2002; Dendy and Heaton, 1999; Martin and Sutton, 2002). In a photo-electric interaction, all the energy from the X-ray photon is absorbed. The probability of photoelectric interactions increases rapidly with the atomic number of atoms present in the tissue and with the tissue density, so photoelectric interactions produce good contrast between various structures in tissues with different elemental compositions. Compton scattering is an inelastic process in which the X-ray photon loses some of its energy and is deflected from its original path. The probability of Compton scattering depends only on tissue density, so it contributes much less to image contrast than the photoelectric effect does. Moreover, because the photons deviate from their original direction, Compton scattering creates a background of random events or noise, so that the net result is an overall blurring of the image. For diagnostic X-rays, the number of photons interacting through the photoelectric effect decreases rapidly with photon energy, while the probability of Compton scattering is largely independent of photon energy. Thus, the proportion of photons interacting via the photoelectric effect changes with the energy spectrum of the X-ray beam.

X-ray beams with different ranges of photon energies can be produced by changing the tube potential or voltage (measured in kilovolts [kV]) applied across the X-ray tube, which accelerates the electrons. X-ray beams produced with lower kVs have a greater proportion of lower energy photons, so the proportion of interactions via the photoelectric effect is higher and this gives better contrast images. However, the absorption of energy from the X-ray beam through the photoelectric effect reduces the penetration, so a higher radiation intensity has to be used and this increases the radiation dose to the tissue. A greater proportion of the photons in a higher kV/higher energy X-ray beam will penetrate through the body, and therefore this will tend to give a lower radiation dose. The increased beam penetration occurs because there are fewer photoelectric interactions and so the image contrast will be poorer. Thus, choice of tube potential for an X-ray beam is a balance between image quality and radiation dose.

Radiographic Imaging

Radiography is both the simplest and most commonly used method for forming a medical image. A radiographic image is formed by irradiating an object with a uniform beam of X-ray photons and recording the transmitted radiation on a film or imaging plate. The number of photons contributing to an image affects both the magnitude of the signal and the fluctuations in background noise. The noise is primarily "quantum mottle" resulting from the finite number of photons employed, although there are other contributions from anatomic and system structural noise. If more photons are used, the potential quality of the image that can be obtained

is higher, as the size of the signal containing useful information increases relative to the background noise. However, this also means that the radiation dose is greater. Thus, the relationship is similar to that for the CT images shown in Figure 6. The most important aspect of imaging performance is whether clinically significant tissue detail can be detected above the background noise. Choosing the appropriate equipment to obtain the most suitable image and the amount of radiation that should be used involves a delicate balance, in which there is a trade-off between higher image quality and increased radiation dose (Martin et al., 1999b).

Conventional film/screen radiography has a so-called limited dynamic range. This means that exposures must be constrained to within the range that will give perceptible differences in film blackening for the human visual system. The exposure range used for any radiograph is predetermined through the choice of the combination of film and the fluorescent screen used with it (Hendee and Ritenour, 2002; Dendy and Heaton, 1999). The film, which is sensitive to light, is sandwiched between two fluorescent screens inside a light-tight cassette. The screens are coated with a fluorescent phosphor, which is usually a rare-earth compound such as lanthenum oxy-bromide or gadolinium oxy-sulphide, which converts X-ray photons into visible light photons. One compromise between radiation dose and image quality is made by the choice of thickness of the phosphor layer. A thick screen will have a high efficiency for conversion of X-rays to light, but the image will be more blurred through spread of the light photons. Thin screens will give better resolution but will require a higher radiation exposure. Film/screen combinations are allocated speed indices, which relate to the reciprocal of the doses required to produce an image and are analogous to the speeds used in conventional photography. The higher the speed index the faster the film—that is, the less radiation is required to produce an image. However, just as in conventional photography using film, the image that is obtained when a higher speed index system is used will be noisier (more grainy). A speed index of 400 is standard, 200 might be used for fine detail, and 600 or 800 are very high-speed systems. The choice of speed index is the most important factor affecting patient dose in radiography.

The situation with so-called digital radiography is somewhat different. In conventional images produced using film-screen combinations, the interaction of each X-ray photon is recorded where it strikes the film. In digital images, the image space is divided into a large number of discreet picture elements or pixels. The energy deposited by photons interacting with each pixel in the image receptor is recorded, so that the image is a matrix of numbers. Pixel size varies between ~ 0.2 mm and 0.05 mm depending on the application. The image itself is displayed as discreet brightness or gray levels, and a gray level is assigned to every pixel with an associated exposure number within a particular range to generate the image. The total number of gray levels is defined by the bit depth of the imaging system that is the number of information bits used to record the gray level. A depth of 2 is associated with 4 levels, 3 with 8, 4 with 16, and so on. Examples of the appearance of a CT image displayed with 4, 8, 16, and 32 gray levels are shown in Figure 7. Digital radiographs are generally digitized to between 12 and 14 bits, corresponding to 4096 and 16,384 gray levels, respectively. The human eye is only able to distinguish between 35 and 64 gray levels, but storage of the data to a much greater contrast definition gives the potential for the gray level presentation to be expanded for a limited range of X-ray contrasts in order to allow lower contrast features to be visualized.

The X-rays that have passed through a patient can be converted into a digital signal in different ways. Computed radiography (CR) utilizes phosphor plates contained within cassettes, similar to those used for film. An image is stored on the phosphor plate, which is subsequently scanned by a laser to release stored energy as light, which is then con-verted to a digital signal. Direct digital radiography (DR) uses arrays of detector elements. Many of the devices incorporate a phosphor plate, usually made from caesium iodide, in front of the detector array to convert the X-ray photons into light, but others use a photodiode (selenium), which is itself sensitive to X-rays. In order to record tissue detail with sufficient resolution to avoid loss of diagnostic information, the number of pixels along each side of the image is usually between 1000 and 2000, so that each image may contain several million pixels.

Digital imaging systems generally have a wide dynamic range, enabling a range of different exposure levels to be used. The viewing conditions can then be adjusted after the exposure has taken place for optimum presentation for the human visual system. The minimum level of contrast that can be seen in an image depends on the number of X-ray photons that are detected. When a recorded digital image contains more pixels and consequently has a higher resolu-tion, then if the threshold for contrast discrimination is to be

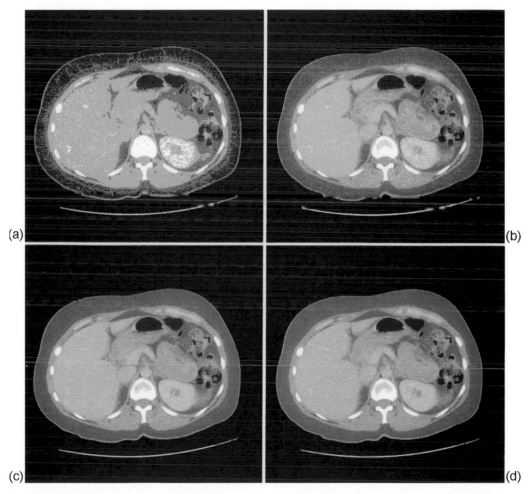

Figure 7 CT images of a slice through the abdomen displayed with (a) 4, (b) 8, (c) 16 and (d) 32 gray levels.

kept the same at that resolution, the mean number of X-ray photons detected by each pixel will need to remain the same. Thus, the amount of radiation used to produce the image and so the radiation dose to the patient will increase proportionately. Therefore, again a choice must be made regarding the level of resolution required for a digital system, which is a balance between the level for one aspect of image quality and the dose to the patient, analogous to the selection of the speed for a film/screen system. An example showing the effect of pixel size on a CT slice through the abdomen is given in Figure 8. The detail can be seen most clearly in the 512×512 image, but at the size at which it is reproduced in this book the blurring in the 256×256 image appears minimal. The number of pixels per unit distance must be at least double the spatial frequency or spacing of the smallest detail that it is required to image. For digital radiography systems made from discreet detector elements, the pixel size normally chosen is between 0.14 mm and 0.2 mm, but 0.05 mm to 0.1 mm will be used for mammography.

As already stated, digital radiography detectors have a much larger dynamic range than X-ray film. Digital images are viewed and manipulated on a monitor. Therefore, they are not constrained by the viewing requirements of images on film, so that a much higher or lower dose can be employed and the gray-level settings can be adjusted to display the data recorded with gray levels in the optimum range for the human visual system. Different image display options are used for different parts of the body and are included in display software provided by the manufacturer and tailored to the requirements of the user. The attenuations of body tissues cover a wide range, so that if an image is displayed with a gray scale that varies uniformly with a signal, it may not be possible for the eye to distinguish all the small changes within the tissues in one image. Viewing consoles can be used to display a selected range of contrasts. The contrast window can be adjusted to concentrate on high- or low-density structures, or widened or narrowed to give different contrast resolutions. This technique, known as windowing, allows more informa-

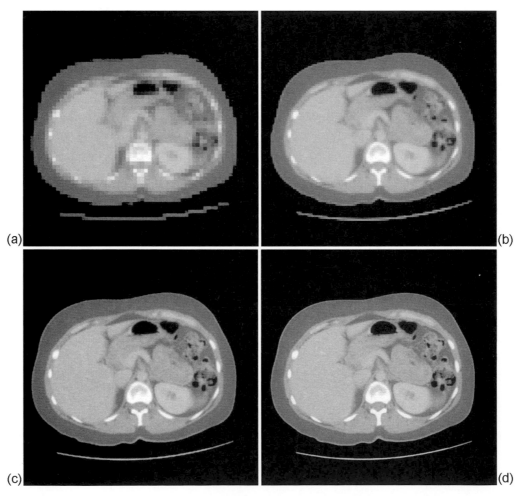

(a)   (b)   (c)   (d)

Figure 8 CT images of a slice through the abdomen acquired with (a) 64×64, (b) 128×128, (c) 256×256, and (d) 512×512 pixels. For the same level of contrast visualization and image noise, the relative doses would be (a) 1D, (b) 4D, (c) 16D, and (d) 64D.

tion to be gleaned by displaying the digital image in different ways. The image can also be magnified, and processing options such as edge enhancement can be used to improve the appearance of the image or the visibility of certain features. However, the level of image processing must not be such that relevant clinical details are obscured or imaginary features are created from fluctuations in noise. For this reason, in-house development of image-processing tools is not a trivial matter. The capability to manipulate the relationship between dose and image brightness gives digital radiography more scope for achieving a balance between dose and image quality than film. Ideally, the levels to which exposures can be reduced before background noise produced by quantum mottle becomes a problem could be set for individual applications in order to achieve optimum imaging performance.

The introduction of digital radiography systems has not led to a significant change in patient doses. Digital radiography systems tend to be marginally more efficient, and where they are replacing film/screen systems, they should allow some degree of dose reduction. However, in order to maintain the level of noise in CR images at an acceptable level, doses used tend to be similar to and sometimes greater than those for conventional 400-speed index film/screen systems. Since the digital display system adjusts the image brightness and contrast to optimize the display, there is a risk that use of higher exposure levels will go undetected unless users regularly check and monitor the system exposure indicator or other dose-related variable.

In both digital and analog radiography, in order to achieve a consistent exposure level, a device referred to as automatic exposure control is usually employed in fixed radiographic imaging facilities. This comprises a set of X-ray detectors behind the patient and in front of the image receptor, which measure the level of transmitted radiation. These detectors are used to terminate the exposure when a predetermined dose level is reached, thereby ensuring that consistent exposure levels are given to the image receptor for imaging patients of different sizes. Radiography can be performed using mobile equipment, but the quality of the image is likely to be lower because alignment of the radiographic cassette cannot be set up as accurately as with a fixed unit, the distance of the cassette from the X-ray tube will be variable, and an automatic exposure control cannot be used to terminate the exposure. In addition, the output of mobile units is lower, so the range of exposures that can be obtained is limited and longer exposure times may be required. Therefore, mobile radiography should only be used when an examination on a fixed installation is not feasible.

Fluoroscopy

Fluoroscopy is the other conventional type of X-ray image vizualization. Fluoroscopy produces images in real time, and allows organ and tissue movement to be portrayed. An image intensifier is employed to allow an image to be obtained at a lower radiation level. This incorporates a phosphor to form a light image of the X-rays that pass through the patient. The light photons are then converted into electrons, which are accelerated in order to amplify the signal. A smaller, brighter image is produced, which is captured by a video camera and displayed on the monitor. Some of the new, more complex fluoroscopy units use flat-panel detectors to record the X-ray image, instead of image intensifiers. These detectors are similar to those employed in direct or indirect digital radiography. The real-time capability of fluoroscopy allows complex structures to be studied using X-rays. It also allows key-hole surgery procedures to be viewed as they are performed with minimal trauma. Modern systems are based on digital images, and the last image frame can be displayed for study without the need for further irradiation of the patient. The stored images can also be used to provide a permanent record of the examination.

Movement of soft tissue structures, and in particular movement of fluid through vessels in the body, would be difficult to visualize with the normal range of fluids one would expect to find within the body, because the tissues and fluids transmit similar proportions of the X-ray photons. Therefore, substances containing elements with higher atomic numbers are introduced to enhance radiological contrast artificially. These substances are referred to as contrast media. Iodine contrast medium is used for visualization of blood vessels. It is normally administered through a peripheral vein and movement tracked in a series of digital images. Besides investigation of the patency of the vasculature itself, alterations in vascularity can be used for detection of metastatic growth. Another contrast medium used for investigation of the gastrointestinal tract is barium sulfate, which is administered as a suspension, either orally or via the rectum.

Conventional fluoroscopy uses an X-ray tube emitting a continuous beam of X-rays. Images are recorded and displayed at 25 or 30 frames per second, with X-rays being emitted throughout the 33 ms–40 ms length of each frame. This means that any movement of tissues that occurs during this period will be superimposed, which will result in blurring of images of moving structures. However, if the X-ray beam is pulsed, this can both improve the image quality and reduce radiation dose (Martin *et al.*, 1999b). If a pulse is applied for only 5 ms out of 40 ms, and the image is held for the remaining 35 ms, there will be less degradation of image quality for individual frames due to movement. Thus, pulsing at 25–30 frames per second will produce a clearer image than continuous fluoroscopy. Pulsed fluoroscopy can also reduce dose, but enough radiation must still be delivered to maintain an acceptable level of quantum noise. An effective way of decreasing the dose is to reduce the frequency of pulses. If the number of pulses per second is halved and every frame is repeated on the monitor, the radiation dose to the patient will be halved. An observer is unlikely to notice a difference in any image unless the structures are moving very rapidly. If the organs being viewed only move slowly, then pulse rates as low as 2 or 3 pulses per second can be used, which will reduce the dose by a factor of 10.

In complex examinations where fluoroscopy is used, a permanent record is generally required of features observed, both for review and consultation in the future. Stored digital images or hard-copy images may be taken from the image intensifier with doses that are only 10–20% of those for a radiograph. The spatial resolution of the image is less than that of a radiographic film, but the images are usually adequate for review of examinations such as barium meals and enemas. These imaging methods must be used with caution, as dose reduction is dependent on exposure factor selection by the anatomical program; if lower tube potentials are employed to improve image quality, the actual dose to the patient may increase.

Digital acquisition imaging for angiography and similar applications use larger patient doses to provide a higher level of image quality. If a sequence of vascular images of the same area is recorded when a contrast medium is administered, an early image before the iodine passed through can be used as a mask and subtracted from later images, so that the only contrast in the image is related to flow through the vessel. This technique is called digital subtraction imaging and is used in digital subtraction angiography.

Radiation Quality

Radiation quality refers to the proportions of photons with different energies within an X-ray beam. The contrast between different structures in an X-ray image results from removal of photons from the primary beam. The radiation quality influences the image quality and radiation dose through the mechanisms by which the X-ray photons of different energy interact with the tissue, as already discussed in the section on X-ray beam interactions. X-ray beams that contain more low-energy photons give better image contrast. However, a greater proportion of the photons are absorbed in the body. It is therefore necessary to use a larger radiation intensity and thus give a higher radiation dose to the patient in order to have sufficient photons transmitted through the body to form an image. The X-ray beam quality chosen for a particular radiological examination should achieve a balance between these factors. The main factors that affect the radiation quality of an X-ray beam are the tube potential and the beam filtration.

Tube Potential

The tube potential determines both the maximum photon energy and the proportion of high-energy photons. The optimum potential will depend on the part of the body being imaged, the size of the patient, the type of information required, and the response of the image receptor. Significant reductions in patient dose have been achieved by using higher tube potentials for particular radiographic projections, where the resulting images still provided the diagnostic information required for the examination despite having a lower contrast. Tube potentials used for radiographic examinations have been established through experience. A tube potential of 80 kV will produce X-ray photons with a broad range of energies up to 80 keV; this is a typical value used for radiographs of the abdomen or pelvis. A 50 kV to 60 kV X-ray beam will give better contrast, but less beam penetration and so a higher radiation dose. This dose is used for thinner or less attenuating regions, such as the arms, hands, and feet, while even lower energy X-rays produced with voltages < 30 kV are used for mammography (breast radiography). On the other hand, 90 kV to 120 kV X-rays will give better beam penetration and so a lower radiation dose; these are employed for thicker, more attenuating parts of the body, such as the lumbar spine. Standard kV ranges have been recommended for a range of radiographic examinations based on practices in different countries (EC, 1996). Selection of tube potentials lower than those recommended is likely to lead to larger radiation doses than are necessary.

Fluoroscopic units operate under automatic exposure rate control whereby the tube potential and tube current (mA) are adjusted automatically with the attenuation of the patient in order to maintain a constant dose-rate at the image intensifier. This may be achieved in different ways depending on the equipment, manufacturer, and model. The option used most frequently is for kV and mA to be raised together according to a predetermined relationship. Most units have a range of alternatives, so that the operator can choose to raise tube potential more rapidly and so employ a relatively high kV for small patients to reduce dose or raise mA and maintain a low kV to provide better image contrast if required. This is appropriate for cardiology and angiography procedures in order to visualize small vessels. The absorption characteristics of the iodine contrast medium coincide with the energy spectrum of a 65–70 kV X-ray beam, so the highest level of image quality can be achieved for tube potentials in this range. Reductions in effective dose of 20–60% can be made through changes in the kV/mA options used.

Filtration

All diagnostic X-ray beams are filtered using thin sheets of metal such as aluminium or copper. The purpose of this process is to reduce the proportion of low-energy photons in the beam, which are not transmitted through the patient and therefore contribute little to the image. In the United Kingdom, the minimum degree of filtration permissible is 2.5 mm aluminium equivalent, and in the United States the specification is defined in terms of the half-value layer that is the amount of filter material required to reduce the transmitted beam intensity by half (IPEM, 2002; CRCPD, 2001). A larger proportion of the X-ray photons will interact through the photoelectric effect in copper than in aluminium, because copper has a higher atomic number. So, copper will absorb a higher proportion of the photons with energies between 20 and 40 keV that make a significant contribution to patient entrance surface dose. The disadvantage of using copper filters is that an increased tube output is required in order to compensate for the reduction in photon fluence resulting from attenuation in the copper filter. With tube potentials of 70–80 kV, reductions of over 50% in entrance surface dose and 40% in effective dose can be achieved by using a 0.2 mm thick copper filter. Additional copper filters with thicknesses of 0.1 mm to 0.5 mm are included in high-performance fluoroscopic systems used in cardiology and interventional radiology, and filters of 0.1 mm and 0.2 mm of copper may be used for pediatric radiography and fluoroscopy.

The use of alternative filter materials for mammography involves different criteria, as here much smaller detail within a thinner volume of tissue needs to be imaged. A lower energy X-ray beam is required, and a molybdenum filter with a molybdenum anode, which allowed the higher intensity emissions of characteristic X-rays from molybdenum at 17 keV and 19 keV to be used, was the standard technique. However, alternative anode/filter combinations, such as a molybdenum anode and a rhodium filter, and a tungsten anode and a rhodium filter can give good image quality with lower absorbed doses, especially when digital detectors are used. Many mammography units allow the operator to select the filter for each patient because of the wide range in breast size within the population.

The greatest emphasis on achieving the correct balance between image quality and patient dose for mammography is required in breast screening programs. Large numbers of women undergo mammography examinations, most of whom will have normal mammograms. Only the small group of women who have breast cancer diagnosed at an earlier stage, where there is a better prognosis for successful treatment, receive the benefit. Risk-benefit analyses are carried out for screening programs, which take account of the prevalence of breast cancer in the population and the age distribution. The female breast is particularly sensitive to radiation, so mean glandular dose for the breast tissue is evaluated. More regulation is applied for mammography than for other radiology procedures in a number of countries, including the United States, with strict requirements for quality assurance on the equipment and procedures.

Scattered Radiation

Another factor that affects image quality is scattered radiation. A proportion of the radiation that is scattered from tissues within the body will reach the image receptor and so contribute to the background noise. If the noise level is too high, this can severely affect the quality of the image (Fig. 6). The level of scattered radiation in radiography and fluoroscopic examinations can be reduced through the use of an antiscatter grid. An antiscatter grid consists of thin strips of lead sandwiched between low-attenuation material such as fiber or paper. X-rays originate from a point within the X-ray tube known as the focal spot. X-ray photons that are not scattered or changed in direction as they are transmitted through the patient will pass between the lead strips with little attenuation. However, radiation scattered from within the patient will travel in different directions, and when passing toward the image receptor, it is more likely to be attenuated by the lead strips. The lead strips may be parallel, but can be angled toward the focal spot of the X-ray tube to improve transmission. In addition to removing scattered radiation, the grid also reduces the intensity of the transmitted primary beam to some extent, so that more radiation is required and the radiation dose to the patient is higher.

The characteristics of the grid and the question of whether a grid should be used at all require careful consideration. A radiographic examination of an adult abdomen performed without a grid is unlikely to show the detailed tissue structure required for diagnosis, while a similar examination for a young child is likely to be perfectly satisfactory because the amount of scattered radiation generated is much less. Grids are not used routinely for fluoroscopic examinations of children in many departments and may also be removed for parts of adult fluoroscopic examinations where a high level of image quality is not required.

For examinations where a grid is used, the choice of grid characteristics is important for optimizing imaging performance. The grid ratio between the depth of the lead strips in the direction of the X-ray beam and the width of the fiber interspace perpendicular to the beam determines the effectiveness of the grid in removing scattered radiation, but also affects the transmission of the primary beam. Typical values are between 16:1 and 4:1, with higher ratios being more effective in removing scatter but attenuating the X-ray beam more. For thicker parts of the body, such as the lumbar spine, the use of a high-grid ratio with a high-tube potential will yield satisfactory image quality; for examination of less attenuating regions with lower scatter levels, a lower grid ratio together with a lower tube potential may give a better result.

An air gap of ~ 200 mm between the patient and image receptor can be employed as an alternative to a grid in order to reduce the scatter level in chest examinations. The lung produces much less scatter because of the lower tissue density, although the heart and spine will scatter radiation. Because scatter spreads out in all directions from the patient, when the image receptor is moved away from the patient, so that there is an air gap between the patient and image receptor, a much smaller proportion of the scattered radiation will reach the image receptor. A larger focus to film distance of 3 m to 4 m is required when an air-gap technique is used in order to reduce the magnification that results from moving the image receptor away from the patient. The output from the X-ray tube must be increased to compensate for the greater distance.

Optimization of Technique in Fluoroscopy

The radiologist, radiographer, cardiologist, or other operators participate more directly in a fluoroscopic examination than in other techniques, in which the exposure parameters are determined beforehand. In fluoroscopy, the operator directs the procedures based on the images viewed in real time, while the patient is being examined. Thus, the operator's technique and his or her skill in using the equipment are important in determining both the outcome of the examination and the radiation dose to the patient. Good technique during fluoroscopy involves minimizing the X-ray "beam-on" time and the number of images recorded, as well as collimating the beam to the area of interest. The latter expedient serves to minimize the primary beam dose to the patient and to keep the amount of scattered radiation to a minimum with consequent benefit to image quality. Other general technique factors are positioning the image intensifier as close to the patient as possible and moving the X-ray tube as far away in order to keep the dose to the skin as low as possible. There will be a range of kV/mA options, which can be selected to suit the particular application; choosing the lowest dose that is suitable for the purpose is important, as selection of a higher dose alternative can increase the patient dose by several times.

Modern fluoroscopy equipment has many dose-saving facilities that can assist in the process of optimization. There may be automatic selection of thicker copper filters for examination of lighter patients and less attenuating parts of the body on certain programs. Graphical overlays may be available on the last image hold to allow collimation to be set without further exposure, and there may be fluoro-loop options that allow dynamic scenes to be stored for later review. Efficient dose management depends on the operators having instruction in the capabilities of the equipment, so that they understand the advantages and disadvantages of the features available, including the doses associated with different pulsing options, so that they can achieve fully optimized use of the equipment. X-ray equipment is continually being developed and improved. Techniques are only likely to be optimum if operators are fully conversant with all the facilities available on their equipment. Extensive training and setting up of equipment by company applications specialists when equip-

ment is installed are important in making operators aware of new facilities, thereby enabling optimum use of techniques.

Radiologist and radiographer operators in most hospitals carry out fluoroscopic examinations, such as barium studies, following protocols with agreed sequences of image projections. They will develop their technique for the various examinations as they gain more experience, selecting the imaging options that provide the information they need. Operators should consider the level of image quality required for each part of the procedure, whether much of the examination can be performed with low-dose options and whether the higher intensity or lower kV programs need only be brought in for specific components of the examination. A reduction in the number of images that need to be recorded and in the fluoroscopy time will minimize the radiation dose. It is important for operators to review their techniques periodically in order to continually improve—for example, to decide whether all the views recorded are necessary or whether alternative ones might serve the purpose of the procedure better. Medical physicists evaluate equipment performance and advise users on dose options. Surveys of patient doses will alert departments if doses are high and will also act as a general audit tool for appraising the technique. Results of such surveys should be taken into account when operators are reviewing their technique. As part of the periodic review, changes should be implemented to address issues of high dose or poor image quality.

Complex cardiology and interventional radiology procedures may involve risk of high skin doses, and a number of cases of skin damage have been reported (ICRP, 2000). Although these are not simply imaging procedures, some of the same factors already mentioned, such as minimizing beam-on time, using close collimation of the X-ray beam, employing low-dose kV/mA options, and removing the grid where possible, are involved. Problems are more likely to arise for larger patients for whom the radiation entrance dose rates will need to be higher in order to obtain the required dose rate at the image intensifier. Keeping the image intensifier close to the patient and the X-ray tube as far away as possible, and using copper filtration options that are available, are particularly important for minimizing skin dose. The latter option has limited usefulness in the case of large patients as the load required from the X-ray tube may exceed the maximum available. If there is likely to be a risk with a particular procedure, then care should be taken to change the angulation of the X-ray beam in order to avoid irradiating the patient's skin primarily in one area, thereby accumulating a high localized radiation dose.

Computed Tomography

Computed Tomography Scanners

Computed tomography (CT) is a specialized form of radiology that is becoming increasingly important with the development of rapid computer analysis techniques (Kalender, 2005). In the simplest CT scanners, a fan-shaped X-ray beam irradiates a narrow section or slice across the body, and the transmitted radiation is recorded by an array of detectors positioned on the opposite side of the patient. The transmitted radiation recorded by the detector array gives the X-ray attenuation in one dimension along a projection through the body. In order to provide the data required for a CT image, many sets of attenuation data are acquired at different angles by rotating the X-ray tube and detector array to different positions around the body. A two-dimensional image of a slice through the body can then be reconstructed from mathematical analysis of the data sets. Computer reconstruction produces an image showing the attenuation of tissues within the slice (see Figs. 6–8). The attenuation is expressed in CT numbers, which describe the attenuation coefficient of each tissue relative to that for water. The CT number in each pixel represents the mean attenuation coefficient for a volume of tissue within the slice, equal to the pixel area multiplied by the slice thickness, and this is known as a voxel. Variations in CT number and image contrast are due primarily to differences in tissue density. This is because CT images are usually obtained with high X-ray tube voltages of 120 kV or 140 kV, where Compton scattering is the predominant interaction mechanism, although 80 kV may be used for children (EC, 1999). However, optimum X-ray spectra for different CT applications have not been determined. These scanners employ a high level of filtration in order to produce an X-ray spectrum with smaller proportions of lower energy photons than are used for other forms of X-ray imaging. So, X-ray tubes with high outputs must be used. There are significant differences in the level of filtration in scanners manufactured by different companies. As a result, tube current levels (mAs) on different scanners will not give an indication of relative dose.

In CT scanners, the patient couch moves slowly through the X-ray tube/detector gantry, so that the X-ray beam follows a helical path around the patient. X-ray images of slices throughout the volume of the body scan are reconstructed from the data recorded. In modern scanners, two-dimensional arrays of detectors are used, which can record sufficient data to reconstruct 64 or more slices from one rotation of the X-ray tube. In this way, scans of large volumes of the body can be produced in short scan times, enabling a rapid patient throughput. Modern computing methods, which permit rapid manipulation of CT image data, can be used to bring out different characteristics. Reprocessing allows the information to be reformatted and displayed in different ways that may be more useful for specific applications such as representing the volume of the examination as a three-dimensional image, in which surfaces of organs and tissues are portrayed that can then be rotated to provide an appreciation of the anatomy in three dimensions. Computed tomography provides images of the organs within the body without the overlying structures seen in projection images. Because there is less noise due to anatomical structure, low-contrast objects can be seen more clearly in CT.

Radiation Dose and Image Quality

As with other techniques, there is a trade-off between patient dose and image quality (Kalender, 2005). If a smaller number of photons is employed, then the random variation in pixel values or noise may be too large to allow detection of small variations in contrast. The number of photons is determined by the X-ray tube current (mA) multiplied by the scan time (s) or mAs, and the width of the slice being imaged. The tube current will be increased for imaging thinner slices and decreased for thicker ones if noise levels are to be similar. The selection of the slice width and noise level for a particular application is a balance between achieving the necessary level of image quality and keeping the radiation dose to the minimum.

The effective doses from CT examinations of the abdomen and pelvis (10 mSv) are higher than those for most other X-ray techniques. Consideration should therefore be given to the possibility of using an alternative nonionizing radiation test such as ultrasound or MRI in preference to CT (Martin et al., 2003). This is particularly important in children and young patients with benign disease. If the technique is available at the hospital and is as accurate as CT for the particular investigation, then CT is not indicated. Even if the alternative is less accurate than CT, it may still provide sufficient information for management of some patients, without the associated radiation dose. Guidelines drawn up by professional groups of radiologists (EC, 2000, RCR, 2003) are valuable for indicating whether CT or alternative modalities should be considered. When in doubt the clinician should always discuss the clinical problem with a radiologist before requesting a CT scan.

In order to ensure that radiation doses are not higher than necessary, the scan parameters should in theory be optimized for each patient. The radiologist must therefore understand the clinical problem. The referring clinician should present the clinical history and set out clearly the question to be answered, so that the examination can be targeted at the clinical need. The key parameters that can be adjusted to reduce patient dose are the volume of the patient examined, the thickness of the exposed slices, the pitch for helical scanning, the X-ray tube current and exposure time (mAs), and the number of repeat exposures for enhancements. The relationship between these parameters is complex. Departments will normally have sets of standard protocols, but these should be adjusted to suit the clinical task. If they are followed to the letter, the likely result is unnecessarily high doses for some patients. However, in order to optimize individual patient scans, each request needs to be reviewed by a specialist CT radiologist or radiographer. There is a body of opinion that supports this option very strongly (Golding and Shrimpton, 2002).

For CT, as for radiography, the exposure determines the level of noise, and images obtained at lower radiation doses with higher noise levels may impede diagnosis. Thus, the exposure factors kV and mAs should be determined by a combination of the diameter of the patient, and so the expected attenuation, and the noise level that is acceptable for the imaging task. The dose for a given slice at a set kV will be directly proportional to mAs on a particular scanner. Much lower tube currents should be employed for imaging pediatric patients because of the lower attenuation levels. In areas of high natural contrast, such as the lung or bone, it may be possible to operate at lower dose levels and still obtain acceptable images. In principle, appropriate noise levels could be defined for each imaging task in order to ensure that the structures of interest are clearly visible above the background noise (Fig. 6). There is increasing evidence that acceptable image quality can be achieved with lower tube voltages than are currently used in CT with lower patient doses in both contrast and noncontrast studies. This better quality can be achieved because the improvement in image contrast with the lower tube voltage more than compensates for the higher attenuation for certain examinations in the 90–140 kV range. For example, Nakayama *et al.* (2005) have shown that by decreasing the tube voltage, the radiation dose in abdominal scans can be reduced by over 55%, and the amount of contrast required is also reduced. The same group (Funama *et al.*, 2005) has suggested that a reduction in tube voltage from 120 to 90 kV can lead to a 35% radiation dose reduction without sacrificing low-contrast detectability. For quantitative coronary calcium measurements, there is a consensus view on an appropriate compromise that will allow a reliable measurement, while keeping the dose at an acceptably low level. For other investigations, however, the radiologist will need to make a judgment based on experience. Considerable progress has been made in reconstruction methods to reduce noise in recent years, and the selection of the most appropriate reconstruction parameters is also important when low-contrast or soft tissue differentiation is required.

The volume of the patient exposed should only be that which is relevant to the clinical question to be addressed. The path followed by the X-ray beam around the patient also affects the exposure. The pitch of the helix relates to the distance the patient couch moves during one tube rotation. If the X-ray tube follows an extended spiral through the volume being imaged with gaps between adjacent sweeps of the X-ray beam (large pitch), the overall radiation dose to the patient will be lower than if a tight spiral with each sweep of the beam adjacent to or even overlapping the previous one (small pitch) is used.

Various equipment features are available to help to minimize radiation dose in current CT scanners. An example is the modulation of tube output in real time as the X-ray tube rotates around the patient. A lower intensity (lower mA) X-ray beam is used for the thinner antero-posterior directions and a higher one for the lateral directions to give a similar transmitted intensity, with a similar noise level, in both orientations. In addition, as the scan moves along the body, a lower tube current can be used when transmission is through less attenuating tissues such as the lung. The exposure modulation function can be chosen based on an initial "scout" view, recorded at the start of the scan, to enable the scan volume to be defined. Since the speed of data processing is increasing, the modulation function may in the future be performed in real time, based on data detected as the scan progresses.

Computed Tomography Dose Assessment

The quantity measured for assessment of CT dose is the CT dose index (CTDI), which is the integral of the dose across the volume irradiated in producing an image of a slice. This can be measured with special radiation detectors placed parallel to the scanner axis across the slice width. Weighted CTDIs derived from measurements made in the center and at the periphery of Perspex phantoms, representing either a body or head, are used in evaluation of patient doses. These are combined with the number of slices and slice thickness to derive a quantity called the dose length product, which takes account of the volume of the patient scanned and is used in assessment and comparison of CT patient exposures based on the clinical protocols used. It is difficult for the radiologist or radiographer to know how the performance of their unit compares with that of other hospitals unless surveys of patient dose are performed from time to time, and compared with those of other hospitals in the area and either local or national diagnostic reference levels based on performance and practice at other hospitals. Effective doses can be derived from the weighted CTDI and the dose length product value, but information on the scanner filters is required for this. Software is available for carrying out these calculations, which requires data to be input on the characteristics of each scanner, as well as details of the scan protocol.

Radionuclide Imaging

The other form of imaging using ionizing radiation does not use the transmission of radiation through the body, but looks at radioactivity inside it (Martin *et al.*, 2003). Radionuclide imaging involves the administration of a pharmaceutical, to which a radioactive atom or radionuclide is bound. Pharmaceuticals are chosen, which take part in particular biological processes and become localized in specific organs. The technique can be used in the study of a variety of functions including perfusion, regional blood flow, metabolic pathways, and biokinetics. Because an image of the distribution of radioactivity is formed, which provides a map of physiological function, it can also be used to assess the distribution of functioning tissue. The ideal radionuclide for imaging should be capable of binding to a range of pharmaceutical carriers to form a variety of radiopharmaceuticals. It should, if possible, only emit gamma rays suitable for detection by imaging equipment, and it should not remain active and so irradiate the patient for too long a period after the imaging task has been completed. Technetium-99m is the radionuclide used most widely for imaging. It emits 140 keV gamma rays and does not emit any particle radiations that would increase the dose to the patient. It has a physical half-life of 6 hrs. This means that the activity level in a sample falls by one-half every 6 hrs. This is long enough to allow useful information to be obtained, but not so long that the dose becomes excessive, because the activity will have fallen to one-sixteenth of the original after 24 hrs. In addition to the physical decay, the radionuclide will gradually be removed from the body tissues and excreted. These biological processes serve to further reduce the dose to the patient.

Imaging Technique

Radionuclide imaging is performed using gamma cameras that incorporate large scintillation crystals (400 mm diameter, 6–12 mm thick) to form an image (Hendee and Ritenour, 2002; Dendy and Heaton, 1999). The crystal absorbs gamma-ray photons and reemits the energy as visible light. In order to avoid radiation from throughout the body being imaged in random parts of the detector crystal, a collimator consisting of an array of holes separated by lead septa is used to limit detection to the region of the body adjacent to each part of the crystal. The choice of the diameter of the holes for the collimator to be used for a particular application represents a compromise between resolution, improved by the use of smaller holes, and sensitivity, because fewer photons will be transmitted through the smaller holes. The detector crystal is optically coupled to an array of photomultiplier tubes, which detect the light photons and amplify the signal. The position of the gamma-ray interaction in the crystal is determined from analysis of the strengths of the signals detected by each photomultiplier tube.

Radionuclide uptake is linked to functioning tissue, and so gamma camera images provide different information from other imaging techniques. The anatomical detail that can be seen is relatively poor, the resolution being 7–10 mm, but the demonstration of function is unique. Radionuclide imaging is often simply used to produce single images of the functioning parts of particular organs, such as perfusion and ventilation of the lungs, or the blood supply to the heart,

liver, or thyroid. Because the technetium-99m remains active for some time, sequential images can also be recorded to follow the movement of the radiopharmaceutical with time as it is processed by the body—for example, giving a dynamic assessment of renal function. Radionuclide imaging has also proved useful in detecting tumors, and whole-body imaging is the technique of choice for detecting and diagnozing bone metastases. Thus, radionuclide imaging can be employed either to image the functioning parts of an organ, to study the function or rate of uptake and clearance of the radiopharmaceutical from the organ, or to detect regions of rapid tissue growth. The images recorded are usually planar from a simple gamma camera image, but tomographic images may be obtained using reconstruction algorithms similar to those employed with CT scanners via the technique called single photon emission computerized tomography (SPECT). For this technique, data are recorded for 64 or 128 projections at different angles around the patient, often using a dual-headed camera.

Positron emission tomography (PET) is another form of radionuclide imaging using positron-emitting radionuclides. The positron (positively charged electron) is again released from an administered radiopharmaceutical. A positron will interact rapidly with an electron soon after it is emitted. The two particles will be annihilated with the production of two high-energy (511 keV) photons traveling in opposite directions. In a PET scanner, the patient lies at the center of a ring of detectors that are designed to record the pairs of high-energy photons, and these are used to build up an image of a transverse slice through the body. The significant advantage of PET imaging is that many of the radionuclides, which can be incorporated readily into organic molecules for imaging of tissue physiology, emit positrons. A number of these have very short half-lives (carbon-11—20 min; nitrogen-13—10 min; and oxygen 15—2 min); therefore, a cyclotron is required on-site in order to produce them. However, fluorine-18, which is also a positron emitter, has a sufficiently long half-life (110 min) to allow it to be transported to centers within a 2-hr journey time of a production facility, and fluorine-18 incorporated into a glucose molecule has proved one of the most useful PET radiopharmaceuticals. Because PET and SPECT images have few anatomical landmarks, modern scanners incorporate X-ray CT and fuse the CT and PET images. This allows localization of uptake to be seen within a detailed anatomical image. These developments are likely to be important in planning radiotherapy treatments, as the PET images relate to tissue function and offer the potential for more accurate delineation of tumor volumes within CT images (Price, 2005).

Radiation Dose and Image Quality

In radionuclide imaging, the balance between patient dose and image quality involves the choice of radiopharmaceutical and the amount of radioactivity. Radiopharmaceuticals undergo trials before approval for medical use, and guidelines are published on radionuclide activities that should be employed (ARSAC, 1998; CRCPD, 2003a). The activities will be reduced proportionately for examinations of children. Tables of scaling factors as a function of weight are available, which should be applied to the activity administered to an adult (EANM, 1990). Activities for administration may be modified by the clinical practitioner for examinations of particular patients.

Doses to patients having imaging examinations will not be trivial. In general, the effective dose to a patient from a radionuclide imaging study will be similar to that for a patient undergoing a radiographic or fluoroscopic examination (1 mSv–6 mSv), although some examinations have doses close to those for CT scans, with cardiology examinations using Thallium-201 having doses up to 18 mSv. A number of radiation protection issues are specific to the use of radionuclides. The radiation dose received by the patient will depend on where the radiopharmaceutical accumulates in the body. Radiopharmaceuticals will be taken up by organs other than those being investigated, and so parts of the patient's body distant from the tissues undergoing investigation may receive a dose of radiation following injection of the radiopharmaceutical. Some protection can be achieved for particular organs. For example, doses to the bladder and surrounding organs can be reduced by encouraging patients both to drink after a procedure is completed and to empty their bladder frequently. Administration of competing substances to block thyroid uptake is used when compounds labeled with radioiodine are employed to investigate parts of the body other than the thyroid.

The administration of a radiopharmaceutical to a pregnant patient will inevitably result in exposure of the fetus to radiation. Some radiopharmaceuticals such as radioactive iodide cross the placental boundary freely and are taken up in fetal tissues. The uptake of radionuclide in fetal tissues can be particularly high if radioiodine is administered or even remains in the mother's body during the stages when the fetal thyroid is developing, and the dose to the fetal thyroid can be extremely high. Other radiopharmaceuticals do not cross the placental boundary, but will still irradiate the fetus externally if they are taken up in nearby organs. For radiopharmaceuticals, which are rapidly eliminated via the kidneys, the maternal urinary bladder may be a major source of irradiation. In these cases frequent voiding of the bladder after administration of the radiopharmaceutical is particularly important. It may be possible in particular cases to administer a lower activity to a pregnant patient and acquire images over a longer period of time, but care must be taken to ensure that this does not compromise the diagnostic quality of the image. A radionuclide imaging examination of a pregnant woman will require particular justification because of the higher risk to the fetus, as indicated in the section on radiation dose.

When a radiopharmaceutical is administered to a mother who is breastfeeding, radioactivity may be secreted in her milk and her infant will suffer internal exposure from the radiation emitted by the ingested radioactivity. Information is available regarding the radiopharmaceuticals for which this should be considered (ARSAC, 1998; Martin and Sutton, 2002; CRCPD 2003a). If the effective dose received by the infant is likely to be too high, then either the examination might be delayed until after the mother has ceased breastfeeding or an interruption in breastfeeding may be recommended. The mother may be able to express milk before the radiopharmaceutical is administered, which can be given to the baby for the next feed. In some cases, use of an alternative radiopharmaceutical can reduce the dose to the infant.

Other problems arise because the radioactivity is present for some time after a patient leaves the radionuclide imaging department. Patients are sources of radiation after their scans are completed and will irradiate other persons with whom they come into contact after they leave the hospital. However, for most diagnostic scans, the dose levels are low, so any precautions that do need to be taken will be limited.

The quality of the image depends on the technique. The technique factors include spatial resolution, noise, and scattered radiation, which are determined by the imaging equipment and the collimator used, and are influenced by the acquisition protocol and image processing algorithm employed. Regular quality assurance is important to ensure that performance is maintained. The images are acquired over a longer period of time than X-rays, which require the patient to remain stationary. It may be necessary on occasion to immobilize or sedate children, in order to carry out a successful examination. The activity could be increased for elderly patients who may be suffering with pain and may find the position required for the examination difficult and painful to maintain.

Conclusions

The formation of images of the body involves interplay between many different factors. In order to achieve the correct balance between patient dose and image quality, it is necessary both to understand the way in which the images are formed and to know the factors that influence image quality and the patient's radiation dose. Considerable improvements have been made in imaging equipment in recent years. Many dose-saving facilities are now included, and if these are fully utilized, considerable reductions in patient dose can be achieved. It is part of the professional responsibility of radiologists and radiographers to ensure that they understand the capabilities of their equipment and use the options available in order to enable the optimum balance between image quality and radiation dose to be achieved.

The flexibility of digital imaging techniques provides more scope for optimization of exposure parameters. Despite all the effort that has been put into optimization in recent years, the dose levels used for similar examinations in different hospitals vary substantially. But dose reduction without regard for image quality could produce images that are inadequate for diagnosis. This must be avoided just as much as unnecessarily high doses must be avoided. Image quality is difficult to assess objectively on a routine basis. Consensus is required regarding what makes an image acceptable, particularly with regard to noise levels for different types of examinations. Although this is to be accomplished, the analytical potential that is now available with images in digital form and the ever more powerful computer processing may facilitate these developments.

A major area of expansion in the evolution of medical imaging in recent years has been the continued increase in the use of CT. This has a considerable influence on the collective patient dose from medical procedures. This trend is likely to continue for the foreseeable future. The possibilities that multislice CT scanners offer will further enhance the potential for diagnosis and could make it the technique of choice in some areas where conventional radiography is now used. Detailed assessments of a variety of factors must be performed in order to establish the appropriate role for CT. Important factors to be considered are the clinical benefits in diagnosis, including the reliability of the diagnoses; the potential improvements in patient management, including treatment and comfort; the health-care resources that would need to be available in terms of capital and time required to perform the examinations, review all the images taken, and carry out the necessary follow-up; and the risks from the additional radiation exposure to both patients and staff. The extent of CT's role in medical diagnosis will evolve as more evidence becomes available and different compromises are reached in different countries, depending on economic resources and national practices. Computed tomography now also plays an important part in the radiotherapy treatment of cancer by providing detailed anatomical images, which are applied directly in planning the treatment dose distribution. The combination of CT images with other techniques such as PET has the potential to provide more accurate data for delineating tumor volumes for treatment by radiotherapy.

The contribution of imaging to medicine is becoming ever more important. Wherever the path of imaging leads, there will be a need to establish how techniques should be applied and the optimum parameters for their use. It is an exciting time in the development of medical imaging, and there are still many questions to be answered and problems to be solved. The contribution of medical imaging in the diagnosis and management of cancer will continue to expand and contribute to the success of new treatment methods as they are developed.

References

Administration of Radioactive Substances Advisory Committee. 1998. Notes for guidance on the clinical administration of radiopharmaceuticals and use of sealed radioactive sources. Department of Health: London, UK.

Beinfeld, M.T., Wittenberg E., and Gazelle, G.S. 2005. Cost effectiveness of whole body screening. *Radiology* 234:415–422.

Brenner, D.J. 2004. Radiation risks potentially associated with low dose CT screening of adult smokers for lung cancer. *Radiology 231*: 440–445.

Conferences of Radiation Control Programme Directors. 2001. Diagnostic X-ray and imaging systems in the healing arts. In: *Suggested State Regulations for Control of Radiation*, Volume I (Ionizing Radiation), Part F, CRCPD dynamic document. Available: http://www.crcpd.org/SSRCRs/F_2001.pdf.

Conferences of Radiation Control Programme Directors. 2003a. Use of radionuclides in the healing arts. In: *Suggested State Regulations for Control of Radiation*, Volume I (Ionizing Radiation), Part G, CRCPD dynamic document. Available: http://www.crcpd.org/SSRCRs/gpart.pdf.

Conferences of Radiation Control Programme Directors. 2003b. Nationwide Evaluation of X-ray Trends. Twenty-five years of NEXT. Available: http://www.crcpd.org/publication.asp #NEXT.

Dendy, P.P., and Heaton, B. 1999. *Physics for Diagnostic Radiology* (2nd ed.). Philadelphia: Institute of Physics Publishing.

European Association of Nuclear Medicine. 1990. Paediatric Task Group of the EANM. A radiopharmaceutical schedule for imaging in paediatrics. *Eur. J. Nucl. Med.* 17:127–129.

European Commission. 2000. Referral guidelines for imaging, Radiation Protection 118 Luxembourg: European Commission.

European Commission. 1999. European guidelines on quality criteria for computed tomography, EUR 16262 EN. EC. Luxembourg.

European Commission. 1996. European guidelines on quality criteria for diagnostic radiographic images, EUR 16260 EN. EC. Luxembourg.

Funama, Y., Awai, K., Nakayama, Y., Kakei, K., Nagasue, N., Shimamura, M., Sato, N., Sultana, S., Morishita, M., and Yamashita, Y. 2005. Radiation dose reduction without degradation of low contrast detectability at abdominal multisection CT with a low tube voltage technique: phantom study. *Radiology* 237:905–910.

Golding, S., and Shrimpton, P. 2002. Radiation dose in CT: are we meeting the challenge? *Br. J. Radiol.* 75:1–4.

Hart, D., Hillier, M., and Wall, B.F. 2002. Doses to patients from medical X-ray examinations in the UK-2000 review. National Radiological Protection Board Report, NRPB-W14, Chilton, UK.

Hendee, W.R., and Ritenour, E.R. 2002. *Medical Imaging Physics* (4th ed.). New York: Wiley-Liss.

Institute of Physics and Engineering in Medicine. 2002. Medical and dental guidance notes: a good practice guide to implementing ionizing radiation protection legislation in the clinical environment. York, UK: Institute of Physics and Engineering in Medicine.

International Commission on Radiation Units and Measurement (ICRU). 1996. Medical imaging—the assessment of image quality. ICRU Report 54, ICRU, Bethesda, MD.

International Commission on Radiological Protection. 2000. *Avoidance of Radiation Injuries from Medical Interventional Procedures.* ICRP Publication 85. Oxford, UK: Elsevier Science.

International Commission on Radiological Protection. 1991. *1990 Recommendations of the International Commission on Radiological Protection.* ICRP Publication 60. Oxford, UK: Pergamon Press.

Kalendar, W.A. 2005. *Computed Tomography* (2nd ed.). Erlangen, Germany: Publicis Corporate Publishing.

Martin, C.J. 2004. A review of factors affecting patient doses for barium enemas and meals. *Br. J. Radiol.* 77:864–868.

Martin, C.J., Dendy, P.P., and Corbett, R.H. 2003. Medical imaging and radiation protection for medical students and clinical staff. London, UK: British Institute of Radiology.

Martin, C.J., Sharp, P.F., and Sutton, D.G. 1999a. Measurement of image quality in diagnostic radiology. *Appl. Radiat. Isot. 50*:21–38.

Martin, C.J., and Sutton, D.G. 2002. *Practical Radiation Protection in Health Care*. Oxford, UK: Oxford University Press.

Martin, C.J., Sutton, D.G., and Sharp, P.F. 1999b. Balancing patient dose and image quality. *Appl. Radiat. Isot. 50*:1–19.

Nakayama, Y., Awai, K., Funama, Y., Hatemura, H., Imuta, M., Nakaura, T., Ryu, D., Morishita, S., Sultana, S., Sato, S., and Yamashita, Y. 2005. Abdominal CT with low tube voltages: preliminary observations about radiation dose, contrast enhancement, image quality and noise. *Radiology 237*: 945–951.

Price, P.M. 2005. *PET Scanning in Radiotherapy. Brit. J. Radiol.* Special Issue, Supplement 28.

Rose, A. 1974. *Vision, Human and Electronic*. New York: Plenum Press.

Royal College of Radiologists. 2003. *Making the Best Use of a Department of Clinical Radiology* (5th ed.). London, UK: RCR.

Stabin, M.G. 1996. MIRDOSE: personal computer software for internal dose assessment in nuclear medicine. *J. Nucl. Med. 37*: 538–546.

4

Contrast Agents for Magnetic Resonance Imaging: An Overview

Alan Jasanoff

Introduction

Chemical compounds have been used to manipulate magnetic resonance image contrast since the development of magnetic resonance imaging (MRI) in the 1970s. The purpose of most of these MRI contrast agents is to delineate structures of clinical or scientific importance, beyond the extent achievable with intrinsic MRI signal alone. Selective contrast enhancement is achieved when the agent localizes preferentially to the structures of interest, or when the agent's capacity to influence contrast varies spatially. More recently, so-called smart contrast agents have been developed for measurement of physiological parameters and detection of metabolites. Several agents are FDA-approved for medical applications, including cancer imaging, and many more have been synthesized and are subjects of ongoing preclinical investigation. The goal of this chapter is to introduce the major classes of MRI contrast agents, the physical parameters that influence their efficacy, and their comparative advantages and disadvantages for *in vivo* imaging. For further discussion of specific contrast agents and applications to cancer diagnosis, the reader is directed to a number of excellent reviews (Allen and Meade, 2004; Caravan *et al.*, 1999; Mansson *et al.*, 2006; Merbach and Toth, 2001; Yu *et al.*, 2005) and to other chapters in this volume.

The differences between classes of contrast agents can best be understood with reference to the underlying physics of MRI (an introduction to MRI instrumentation and physics is presented in Part 1 of this series). In MRI scans of human or animal subjects, image contrast arises from the distribution of spin-1/2 nuclei in tissue, most prominently the protons in water and fat molecules, and from interactions of the MRI scanner with these nuclei through the phenomenon of nuclear magnetic resonance (NMR). In a typical scan, polarized nuclear spins in the sample are perturbed from equilibrium ("excited" and "prepared") by one or more radiofrequency (RF) electromagnetic pulses. The excited spins evolve in the presence of magnetic field gradients and perhaps further RF pulses, and are ultimately detected by the scanner electronics, yielding raw data that may be mathematically transformed into an image. Key aspects of this process that vary spatially and therefore influence the resulting image contrast are (a) the amount of spin polarization present after excitation and preparation, (b) the time constant with which the polarization decays before detection (the T_2 or transverse relaxation time), and (c) the time constant with which equilibrium polarization is recovered between repeated applications of the MRI pulse sequence (the T_1 or longitudinal relaxation time).

Three classes of contrast agents are discussed in this chapter. First, relaxation agents work by influencing the relaxation times T_1 and T_2. They are the most widely used contrast agents, and they will be treated here in greatest depth. A second and relatively new class of agents is based on an effect called chemical exchange-dependent saturation transfer (CEST) (Ward *et al.*, 2000). This effect involves a spin preparation technique that reduces the MRI signal detected in the presence of a CEST agent. The third class of contrast agents contains spin-1/2 nuclei other than protons, and are designed to be directly imaged by the MRI hardware. Detection of signals from these agents produces images distinct from and often in parallel to standard ^{1}H MRI images.

Relaxation Agents

Basic Principles of Relaxation Contrast

It was observed many years before the invention of MRI that paramagnetic solutes increase the rates associated with T_1 and T_2 relaxation in NMR spectroscopy experiments (Bloch *et al.*, 1946; Bloembergen *et al.*, 1948). Paramagnetic compounds are characterized by an electronic structure containing unpaired electrons; examples include many of the transition metals, lanthanides, and free radicals such as nitroxide (NO). Because observed MRI signal intensity varies systematically with T_1 and T_2, paramagnetic chemicals can function as contrast agents. The MRI methods most frequently applied in anatomical and functional imaging use variants of the spin echo and gradient echo pulse sequences, and produce MRI signal given by Eqs. [1] and [2], respectively:

$$I^{SE} = C\left[1 - 2e^{-\frac{(T_R - T_E/2)}{T_1}} + e^{-\frac{T_R}{T_1}}\right]e^{-\frac{T_E}{T_2}} \quad [1]$$

$$I^{GE} = C\left(1 - e^{-\frac{T_R}{T_1}}\right)e^{-\frac{T_E}{T_2}} \quad [2]$$

where C is a constant incorporating scan, specimen, and hardware-dependent factors, and T_R and T_E are the pulse sequence repetition time and echo time (time between excitation and readout), respectively. From these equations, it is clear that a compound that shortens T_1 will have the effect of intensifying (brightening) MRI signal (Figs. 9a and b). A compound that shortens T_2 will, on the other hand, tend to darken images, producing "negative" contrast.

The majority of contrast agents used in MRI today are paramagnetic relaxation agents for proton imaging, based on T_1 and T_2 shortening effects. Relaxation agents produce bigger absolute signal changes than other types of MRI contrast agents, although the amount of signal change due to a relaxation agent may be difficult to measure explicitly under circumstances where a subject's intrinsic image intensity distribution is unknown. Most relaxation contrast agents are capable of producing changes in both T_1 and in T_2, but conventional imaging conditions tend to generate one or the other form of contrast more effectively. The most widely used T_1 agents are organic chelates incorporating gadolinium [Gd(III)], a highly paramagnetic lanthanide; Gd-DTPA (Magnevist™) is the best known of these (Weinmann *et al.*, 1984) (Fig. 9c). T_2 agents have been based most frequently on iron and iron oxides. Several iron-containing proteins (e.g., ferritin and hemoglobin) produce endogenous T_2 contrast (Gillis and Koenig, 1987). "Superparamagnetic" iron oxide nanoparticles generate particularly potent T_2 effects, and are gaining popularity as anatomical and molecular imaging agents (Bulte and Kraitchman, 2004).

Acceleration of relaxation rates by paramagnetic contrast agents is due to their interactions with spin-1/2 nuclei in solvent that contribute directly to the image (here assumed to be water protons, e.g., in tissue). Generally speaking, both types of relaxation enhancement arise due to fluctuating magnetic fields generated by the contrast agent. At low concentrations (< 10 mM), interactions between the contrast agent molecules themselves are negligible, and solvent relaxation rates are observed to vary linearly with the concentration of contrast agent:

$$1/T_i = 1/T_{i0} + r_i[M] \quad i = 1, 2 \quad [3]$$

where $1/T_{i0}$ is the T_1 or T_2 relaxation rate in the absence of the solute, [M] is the concentration of the paramagnetic solute, and r_i is a constant of proportionality called the T_1 or T_2 relaxivity, typically given in units of mM^{-1}s^{-1}.

A number of microscopic physical phenomena combine to determine the r_1 and r_2 relaxivities of a contrast agent; these factors have been discussed elsewhere in considerable detail (Lauffer, 1987; Muller *et al.*, 2001; Toth *et al.*, 2001) and will be only briefly introduced here. Two classes of contributing solvent–solute interactions have been defined, so-called inner and outer sphere (Fig. 9c). Inner sphere interactions involve direct bonding or coordination by solvent molecules of the paramagnetic center. Relaxation in the inner sphere regime is dominated by dipole–dipole (through space) and scalar (through bond) coupling effects, both requiring close contact between the contrast agent and water molecules. Outer sphere interactions occur over larger distances and arise due to diffusion of water molecules through magnetic field gradients generated by the contrast agent. Both r_1 and r_2 can be expressed as a sum of inner and outer sphere components.

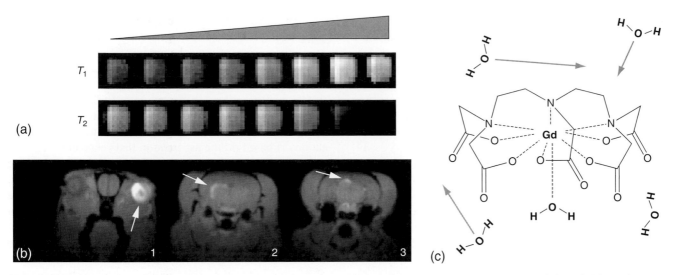

Figure 9 Enhancement of contrast by relaxation agents. (a) Increasing concentrations of an imaging agent (left to right) produce progressive brightening in T_1-weighted imaging, and progressive darkening in T_2-weighted imaging. (b) Similar effects can be observed *in vivo*. Here, a neural tract-tracing T_1 agent (Mn^{2+} ions; Pautler *et al.*, 1998) has been injected into the right eye of a rat. In coronal MRI slices through the rat's head, the injected eye (arrow, panel 1), and anatomically connected structures (arrows, panels 2 and 3), appear brighter. (c) Inner and outer sphere mechanisms contribute both to T_1 and T_2. Inner sphere water molecules participate in direct coordination (gray dotted line) of a paramagnetic metal ion, here the gadolinium center in Gd-DTPA. Outer sphere water molecules diffuse in the neighborhood of the relaxation agent (arrows), experiencing local field gradients that promote relaxation.

Determinants of Inner Sphere Relaxivity

Inner sphere relaxivity is the principal contribution to the effects of some of the most common contrast agents, particularly the T_1 agents that incorporate Gd(III). A theory of inner sphere relaxivity was developed by Solomon (1955), Bloembergen (1957), and Bloembergen and Morgan (1961), and summarized by a set of equations named for them. Inner sphere r_1 and r_2 are given by:

$$r_1^{IS}[M] = \frac{1}{T_1^{IS}} = \frac{cq}{55.5}\left(\frac{1}{T_{1M} + \tau_M}\right) \quad [4]$$

$$r_2^{IS}[M] = \frac{1}{T_2^{IS}}$$
$$= \frac{cq}{55.5}\left(\frac{1}{\tau_M}\right)\left[\frac{T_{2M}^{-1}\left(\tau_M^{-1} + T_{2M}^{-1}\right) + \Delta\omega_M^2}{\left(\tau_M^{-1} + T_{2M}^{-1}\right)^2 + \Delta\omega_M^2}\right] \quad [5]$$

where c is the molal concentration of contrast agent (~ equal to [M]), q is the number of exchangeable water coordination sites on the agent, τ_M is the time constant for exchange at those sites, T_{1M} and T_{2M} are the longitudinal and transverse relaxation time constants for bound water protons, and $\Delta\omega_M$ is the precession frequency difference between bound and free protons. The dependences of r_1^{IS} and r_2^{IS} on q and τ_M are qualitatively intuitive: the more exchange sites (higher q), and the faster the exchange (lower τ_M), the more water molecules are "relaxed" per unit time and the higher the relaxivity. Both T_1

and T_2 inner sphere relaxivities also increase as the corresponding bound water relaxation times decrease.

T_{1M} and T_{2M} are functions of the geometry, electronic structure, and dynamics of the contrast agent; tuning of these properties has been a major focus of contrast agent development. The following expressions for these quantities are valid under several assumptions:

$$\frac{1}{T_{1M}} = \frac{2}{15}\frac{\gamma_I^2\gamma_S^2\hbar^2 S(S+1)}{r^6}\left(\frac{\mu_0}{4\pi}\right)^2$$
$$\left[\frac{3\tau_{c1}}{1+\omega_I^2\tau_{c1}^2} + \frac{7\tau_{c2}}{1+\omega_S^2\tau_{c2}^2}\right] \quad [6]$$
$$+ \frac{2}{3}S(S+1)\left(\frac{A}{\hbar}\right)^2\left[\frac{\tau_{e2}}{1+\omega_S^2\tau_{e2}^2}\right]$$

$$\frac{1}{T_{2M}} = \frac{1}{15}\frac{\gamma_I^2\gamma_S^2\hbar^2 S(S+1)}{r^6}\left(\frac{\mu_0}{4\pi}\right)^2$$
$$\left[\frac{3\tau_{c1}}{1+\omega_I^2\tau_{c1}^2} + \frac{13\tau_{c2}}{1+\omega_S^2\tau_{c2}^2} + 4\tau_{c1}\right] \quad [7]$$
$$+ \frac{1}{3}S(S+1)\left(\frac{A}{\hbar}\right)^2\left[\frac{\tau_{e2}}{1+\omega_S^2\tau_{e2}^2} + \tau_{e1}\right]$$

where γ_I and γ_S are the proton and electron gyromagnetic ratios, $\hbar$ is Planck's constant divided by 2π, S is the electron spin number ($n/2$, where n is the number of unpaired electrons),

r is the distance between the paramagnetic metal and bound proton, μ_0 is the vacuum permeability constant, $A/\hbar$ is the electron-nuclear hyperfine coupling constant, and ω_I and ω_S are the proton and electron angular Larmor precession frequencies (equal to γB_0, where B_0 is the magnetic field strength). The correlation times for proton and electron spin fluctuations, τ_{ci} and τ_{ei}, are given by:

$$\frac{1}{\tau_{ci}} = \frac{1}{T_{ie}} + \frac{1}{\tau_M} + \frac{1}{\tau_R} \qquad i = 1, 2 \qquad [8]$$

$$\frac{1}{\tau_{ei}} = \frac{1}{T_{ie}} + \frac{1}{\tau_M} \qquad i = 1, 2 \qquad [9]$$

where T_{1e} and T_{2e} are the electronic longitudinal and transverse relaxation times, respectively, τ_M is again the bound water residence time, and τ_R is the correlation time for rotational motion of the contrast agent complex (for spheroid molecules approximately equal to $4\pi a^3 \eta / 3k_B T$, where a is the radius of the agent, η is the solvent viscosity, k_B is the Boltzmann constant, and T is temperature).

The first terms in Eqs. [6]–[7] arise from dipole–dipole coupling between proton and electron spins, and the second terms in each equation (modulated by $A/\hbar$) arise from scalar coupling between the two spins. For common paramagnetic lanthanide complexes, interaction with water protons is primarily through space, so the dipolar relaxation mechanism tends to dominate. Both dipolar and scalar coupling interactions are strongly field dependent, as governed by the terms in square brackets; T_{1e} and T_{2e} also vary with field. Both the dipole–dipole and scalar coupling contributions to r_1^{IS} approach zero at high magnetic fields (i.e., T_{1M} becomes infinite), whereas the corresponding contributions to r_2^{IS} approach values determined by the high field limits of τ_{c1} and τ_{e1}. The field dependence of relaxivity is referred to as nuclear magnetic relaxation dispersion (NMRD).

Equations [6]–[7] indicate that T_{1M} and T_{2M} are shorter for contrast agents with more unpaired electrons (i.e., those with higher S), implying that relaxivity is greatest for strongly paramagnetic contrast agents. Similarly, bound water relaxation times are shorter and relaxivity larger for contrast agents with a smaller coordination distance r. Equations [6]–[9], the previous discussion, and the fact that $\omega_I \ll \omega_S$, together predict that agents with large values of τ_{c1} will have the greatest maximum relaxivities. Because τ_{c1} is determined jointly by T_{1e}, τ_M, and τ_R, and short τ_M is required for high inner sphere relaxivity in Eqs. [4]–[5], r_1^{IS} and r_2^{IS} will be highest in agents with intermediate water exchange rate ($\tau_M \sim 10$ ns), and with T_{1e} and τ_R greater than τ_M. Effective inner sphere relaxation agents are designed around paramagnetic centers with high spin and long T_{1e}. These agents typically have τ_M near or greater than 10 ns, and it is the rotational correlation time τ_R that most frequently limits τ_c. For small molecule agents like Gd-DTPA, τ_R is approximately 0.1 ns at room temperature (Lauffer, 1987). Rotational motion is significantly slower for macromolecular

contrast agents, however, and greater relaxivity for these larger agents has been observed at magnetic fields below 5 T.

Determinants of Outer Sphere Relaxivity

Outer sphere relaxivity is important to T_2 contrast agents and to T_1 agents with few or no inner sphere water exchange sites. A standard physical description of outer sphere interactions is based on the assumption that relaxation is due to random diffusion of water molecules in the neighborhood of the contrast agent (Freed, 1978). This approach omits explicit consideration of so-called second sphere water molecules that make up a contrast agent's hydration shell, but are not directly coordinating the metal. Both r_1^{OS} and r_2^{OS} depend largely on the time scale for diffusional motion relative to the size of the contrast agent:

$$\tau_D = R^2/D \qquad [10]$$

where R is the distance of closest approach of water molecules to the agent and D is the sum of diffusion constants for water and the contrast agent. Outer sphere relaxivities are then given by:

$$r_1^{OS}[M] = \frac{1}{T_1^{OS}}$$
$$= \frac{32\pi}{405} \frac{\gamma_I^2 \gamma_S^2 \hbar^2 S(S+1) N_A[M]}{1000 RD} \qquad [11]$$
$$\times \left[3j(\omega_I) + 7j(\omega_S) \right]$$

$$r_2^{OS}[M] = \frac{1}{T_2^{OS}}$$
$$= \frac{32\pi}{405} \frac{\gamma_I^2 \gamma_S^2 \hbar^2 S(S+1) N_A[M]}{1000 RD} \qquad [12]$$
$$\times \left[1.5j(\omega_I) + 6.5j(\omega_S) + 2j(0) \right]$$

where N_A is Avogadro's number, [M] is the contrast agent concentration, γ_I and γ_S are again the proton and electron gyromagnetic ratios, S is the electron spin number, and $\hbar$ is the reduced Planck's constant.

Magnetic field dependence of r_1^{OS} and r_2^{OS} is determined by the so-called spectral density terms in Eqs. [11]–[12], $j(\omega_I)$ and $j(\omega_S)$, which are defined as follows:

$$j(\omega) = \text{Re} \left\{ \frac{1 + \frac{1}{4}z}{1 + z + \frac{4}{9}z^2 + \frac{1}{9}z^3} \right\}, \qquad [13]$$

$$Z = \sqrt{i\omega\tau_D + \tau_D/T_{1e}}$$

These expressions approach zero for magnetic fields where $\omega\tau_D \gg 1$. As a result, r_1^{OS} (like r_1^{IS}) becomes marginal at high field, whereas r_2^{OS} is dominated by a single field-independent component (the so-called secular contribution, proportional to $j(0)$ in Eq. [12]). For typical gadolinium chelates in water, τ_D is less than a nanosecond; these contrast agents therefore have nonzero outer sphere relaxivity at all practical imaging field strengths, but the magnitude of this contribution is still small compared with inner sphere relaxivity. Because of the R^2-dependence of τ_D, macromolecular or nanoscale contrast agents have significantly longer diffusional correlation times and may produce potent T_2-specific outer sphere effects at moderate to high magnetic fields.

Characteristics of T_1 Agents

The majority of T_1 agents in clinical or laboratory use today are small molecule complexes incorporating paramagnetic metals (Caravan et al., 1999). Gd(III) has been the favorite choice of metal because of its long electronic relaxation time and seven unpaired electrons ($S = 7/2$). Mn(II) and Fe(III) have a lower spin number ($S = 5/2$), but have also been used in effective T_1 agents (Lauffer, 1987). The free metals themselves are unsuitable contrast agents due to their toxicity and, in the case of gadolinium, their tendency to precipitate. Polydentate chelators form tight ($K_d < 10^{-20}$), kinetically stable, and soluble complexes with the metals, rendering them tolerable for in vivo use. Gadolinium is most often applied in conjunction with the chelators DTPA or tetraazacyclododecane tetraacetic acid (DOTA), or with derivatives of the two. These chelates have far fewer water exchange sites than uncomplexed (aquo) Gd^{3+} ions ($q = 1$ vs. 8–9 for free Gd), but also have significantly longer τ_R. As a result, the relaxivity of Gd-based contrast agents is similar to that of free gadolinium.

Equations [4]–[9] and consideration of the relevant correlation times imply that small molecule paramagnetic contrast agents should have comparable T_1 and T_2 relaxivities at fields below 5 T. These agents produce more formidable T_1 contrast in most contexts (including in vivo) because background T_1 relaxation times are usually significantly longer than T_2 relaxation times. The r_1 of Gd-DTPA is 4.7 mM^{-1}s^{-1} at 25°C and 0.47 T (Powell et al., 1996). Assuming this relaxivity and a background T_1 of 1 second, ~ 22 μM of this agent would be required to give a 10% change in T_1, corresponding to a 6% change in T_1-weighted spin echo MRI intensity under conditions that maximize the signal change ($T_R \sim T_1$). Higher concentration would be required to produce an equivalent change at fields above 0.47 T, where r_1 is reduced. Concentrations near 100 μM are typical of T_1 agents in clinical and biological applications.

Additional T_1 agents based on gadolinium include Gd-DOTA (Dotarem™), with r_1 of 4.7 mM^{-1}s^{-1}, and Gd-DTPA-BMA (Omniscan™) with r_1 of 4.4 mM^{-1}s^{-1} (both at 0.47 T and 25°C) (Powell et al., 1996). Like Gd-DTPA, these compounds have $q = 1$. A manganese(II)-based complex, Mn-DPDP (Teslascan™), is one of few relatively widely used contrast agents with $q = 0$. Relaxivity of Mn-DPDP is 2.8 mM^{-1}s^{-1} (at 1 T and 25°C) and arises from outer sphere

relaxation effects (Elizondo et al., 1991). Although the relaxivities of these agents are all comparable, each may be favored for reasons other than relaxivity. Gd-DTPA-BMA, for instance, is less negatively charged than Gd-DTPA and is preferred for applications where hydrophobicity promotes selective retention (Caravan et al., 1999). Mn-DPDP produces anomalously high contrast in the liver, making it a useful hepatobilary agent (Young et al., 1989, 1990). Gd-DTPA is widely applied because of its solubility in blood plasma; applications include angiography and perfusion imaging. Structures of several T_1 agents are presented in Figure 10a.

Characteristics of T_2 Agents

An important requirement for useful T_2-selective contrast agents is a large r_2/r_1 ratio, in combination with high absolute r_2. As the preceding discussion of r_2 relaxivity indicates, this condition can be achieved for agents with predominantly outer sphere relaxation mechanisms, usually along with large τ_D and hence large size. A family of contrast agents that satisfies these criteria are superparamagnetic iron oxide nanoparticles (SPIOs) (Saini and Ferrucci, 1988), which are the principal T_2 agents used in MRI today. Several SPIO agents (e.g., Feridex™) are currently in clinical use for liver imaging applications (Weinmann et al., 2003). SPIO contrast agents are colloidal suspensions, sometimes called ferrofluids, in which each particle contains one or more nanometer-scale crystals of magnetite (Fe_3O_4) or its more stable oxidation product, maghemite (γ-Fe_2O_3) (Muller et al., 2001). At relevant experimental temperatures, crystals of either material function as discrete magnetic domains with aggregate magnetic moments arising from cooperativity among the component electron spins. Fluctuations of the superparamagnetic moments induce T_1 and T_2 relaxation effects that are significantly more pronounced than those of an equivalent number of isolated paramagnetic moments.

Most SPIO contrast agents consist of an iron oxide core of 2–20 nm that is rendered soluble and biologically inert with an organic coating (Fig. 10b). Specific core and coating structures determine properties of the different SPIO agents. So-called ultrasmall SPIO agents (USPIOs) have been prepared by size fractionation of larger, heterogeneous iron oxides. A subset of these contain only a single crystal of iron oxide and are referred to as mononuclear iron oxide nanoparticles (MIONs) (Shen et al., 1993). These particles are usually coated with dextran (a polysaccharide) or polyethylene glycol, and have overall particle diameters of 20–50 nm. Yet smaller USPIOs have also been produced using organic acid-based coatings (Gupta and Gupta, 2005). USPIOs typically have strong r_1 as well as strong r_2. Most SPIO contrast agents in clinical research (e.g., ferumoxides) are larger, up to several hundred nanometers in diameter; they contain variable numbers of crystals per particle and are polydisperse compared with USPIOs. These particles

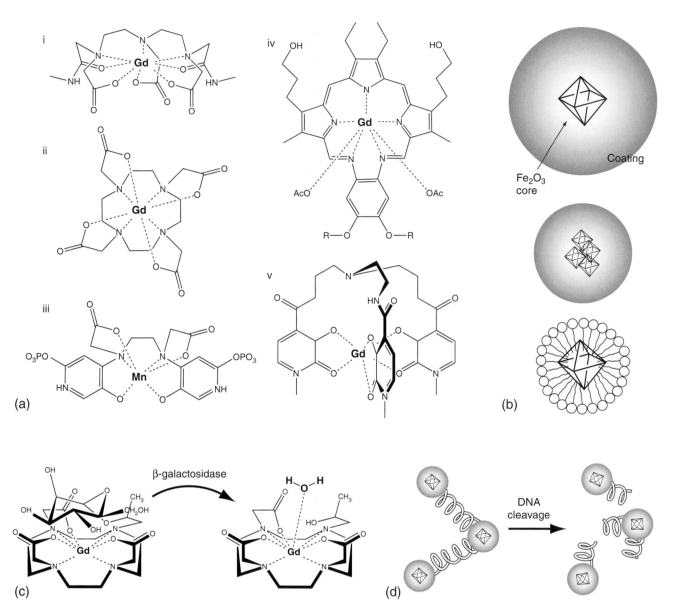

Figure 10 T_1 and T_2 contrast agent structures. (a) Structures of some representative T_1 contrast agents, with paramagnetic ions in bold: (*i*) Gd-DTPA-BMA, a neutrally charged derivative of Gd-DTPA (see Fig. 9c); (*ii*) Gd-DOTA, a stable macrocyclic complex sometimes modified at one of the acyl positions for attachment to other moieties; (*iii*) Mn-DPDP, a manganese complex with $q = 0$; (*iv*) Gd-texaphyrin, a new porphyrin-like agent for MRI or phototherapy (Sessler and Miller, 2000); (*v*) Gd-TREN-1-Me-3,2-HOPO, one of a series of high relaxivity all-oxygen liganded gadolinium complexes synthesized by Kenneth Raymond's laboratory (Raymond and Pierre, 2005). (b) Superparamagnetic iron oxides consist of an iron oxide core (labeled Fe_2O_3), typically 2–20 nm in diameter, surrounded by a coating with diameter up to several hundred nanometers. MION particles contain only a single crystal of iron oxide (top), while more conventional SPIOs may contain several particles per core (middle). Different materials are used for coating the iron oxide cores. Common choices include polysaccharides, polyethyleneglycol, and lipids or other small molecules (bottom). (c) A "smart" contrast agent synthesized by Meade and colleagues acts as a substrate for the enzyme β-galactosidase (Louie *et al.*, 2000). The agent is a hexose-caged derivative of Gd-DOTA (left). Enzymatic activity removes the sugar moiety, increasing exposure of the gadolinium center for interactions with water molecules (right). This increases r_1 of the agent. (d) A nuclease-sensitive agent prepared by Perez *et al.* (2002b) forms an aggregate due to hybridization of complementary DNA strands attached to two populations of SPIOs (left). In the presence of a specific endonuclease, the duplexes are cleaved, breaking up the aggregate and increasing T_2.

are still small enough so that thermal energy keeps them in suspension despite the gravitational and magnetic forces they experience. SPIO particles with diameters greater than a micron are sometimes used in MRI. They are noncolloidal, but produce strong T_2 and T_2^* relaxation over length scales comparable to their diameters.

The contrast changes produced by SPIOs depend primarily on their size (Gillis *et al.*, 2002). Although the SPIO coating material can also influence contrast, it is most often chosen to influence stability and localization, or to facilitate functionalization of the nanoparticles. At imaging field strengths, and for nanoparticles smaller than ~ 20 nm in diameter, the outer sphere transverse relaxivity expression (Eq. [12]) reduces to a "motional averaged" limit:

$$r_2^{SPIO}[M] = \frac{1}{T_2^{SPIO}} = \frac{16}{45} \frac{\gamma_I^2 \mu^2 V}{R^4 D} \qquad [14]$$

where R is the distance of closest approach to the particle, D is the diffusion constant of water, V is the volume fraction occupied by the particles, and μ is the magnetic moment of each particle. The relaxivity r_2^{SPIO} is often expressed in units of $mM^{-1}s^{-1}$ with respect to the concentration of iron atoms. The particle moment μ is equal to Mv, where [M] is the magnetization of the superparamagnetic core material (4.12 $\times 10^5$ A/m for maghemite at high magnetic fields) and v is the core volume. An implication of Eq. [14] is that r_2^{SPIO} increases (for fixed [Fe]) as the particle volume increases. The relaxivity reaches a limit for larger nanoparticles, defined by the condition that:

$$\Delta\omega_r \tau_D > 1, \quad \Delta\omega_r = \sqrt{\frac{4}{5}} \gamma_I \mu / R^3 \qquad [15]$$

where $\Delta\omega_r$ expresses the proton frequency shift at the particle equator and τ_D is the diffusional correlation time defined by Eq [10]. Relaxation caused by these larger particles enters a regime in which radiofrequency pulses may be used to refocus some of the transverse relaxation (i.e., T_2^* relaxation). Observed T_2 then declines as particle diameter increases, with its value dependent on the pulse sequence used for the measurement.

A conventional SPIO, ferumoxides (diameter 120–160 nm), has a measured r_1 of 40 and r_2 of 160 (mM Fe)$^{-1}$s^{-1} at 37°C and 0.47 T (Jung and Jacobs, 1995). Under the same conditions, a representative USPIO, MION-46 (diameter 20 nm), has r_1 of 17 and r_2 of 35 (mM Fe)$^{-1}$s^{-1} (Shen *et al.*, 1993). The crystalline cores of ferumoxides and MION-46 contain on the order of 50,000 and 2000 atoms of iron, respectively, so the r_2 values translate into per particle relaxivities of roughly 8 × 10^6 mM^{-1}s^{-1} for ferumoxides and 7 × 10^4 mM^{-1}s^{-1} for MION. In a T_2-weighted imaging scheme with a background T_2 of 100 ms and T_E optimized for absolute signal change, concentrations of 1 nM ferumoxides or 100 nM MION-46 particles would give rise to ~ 20% con-

trast. These are several orders of magnitude below the concentrations of small molecule agents required for comparable signal changes in T_1-weighted imaging.

Advances in the Design of Relaxation Agents

Active research is directed at improving the relaxivity of Gd-based contrast agents, developing methods to target T_1 and T_2 agents, and producing MRI-based sensors from agents that undergo relaxivity changes in response to biochemical signals. The preceding discussion indicated that significant contrast changes require relatively high concentrations of conventional T_1 agents, on the order of 25–100 μM. For a contrast agent to be useful at lower concentrations, it must have considerably higher relaxivity. Raymond and coworkers have introduced a series of hydroxypyridinone (HOPO) related compounds with over twice the relaxivity of Gd-DTPA (Raymond and Pierre, 2005) (Fig. 2a). The advantage in r_1 is due largely to their near optimal τ_M (Xu *et al.*, 1995). These complexes retain their relaxivity advantage at magnetic fields over 5 T, making them particularly significant for high-resolution scanning applications. High-relaxivity T_1 agents have also been formed by combining several Gd coordination sites into single macromolecular structures like dendrimers and fullerenes (Caravan *et al.*, 1999; Merbach and Toth, 2001). Because they have relatively long τ_R, these compounds enjoy higher relaxivity per gadolinium than Gd-DTPA. Single polyvalent gadolinium compounds also make more effective labeling reagents than conventional chelates.

The ability to label specific cells with contrast agents and identify them *in vivo* would be of great significance for diagnostic medicine and a number of fields in biology. A number of techniques have been introduced for performing this identification with T_1 and T_2 agents. Work of several laboratories has demonstrated that coupling of SPIOs or gadolinium-containing nanoparticles to protein targeting domains enables specific cell populations to be identified *in vivo* (Artemov *et al.*, 2004). Recent examples include selective labeling of transferrin receptor-expressing cells with transferrin-conjugated SPIOs (Weissleder *et al.*, 2000), detection of apoptotic cells with synaptotagmin and annexin SPIO conjugates (Schellenberger *et al.*, 2002; Zhao *et al.*, 2001), and identification of Her-2/neu receptor-expressing cells using a biotinylated antibody and streptavidin-coated nanoparticles (Artemov *et al.*, 2003). A related family of applications has involved using nanoparticle relaxation agents to label cells (lymphocytes and stem cells) *ex vivo*, and then transplanting them and tracking them *in vivo* (Bulte *et al.*, 2002). Cell targeting has also been possible using small molecule relaxation agents. Activity-dependent labeling with manganese chloride has emerged as a powerful technique in neuroimaging (Koretsky and Silva, 2004). Mn(II) is infused into animals and enters active neurons through calcium channels, where it accumulates while animals perform tasks or receive stimulation (Lin and Koretsky, 1997). Apparent selective labeling of tumors by porphyrin derivatives is a less well-understood but important phenomenon in cancer research. It was observed some time ago that Mn-TPPS and related compounds accumulate preferentially in tumors and can be visualized by T_1-weighted imaging (Patronas et al., 1986). More recently, a family of modified porphyrins (Texaphyrins) with the ability to form stable, high-relaxivity Gd complexes was introduced for combined imaging and phototherapeutic applications (Young et al., 1996) (Fig. 10a).

The development of metabolically sensitive relaxation agents has also been a major trend in the past several years. A group of T_1 agents has been synthesized by Meade and colleagues to sense the activity of enzymes and metabolites (Meade et al., 2003). In one agent, a gadolinium chelate is "caged" by a sugar monomer (Louie et al., 2000). In the presence of the enzyme β-galactosidase, the sugar is cleaved, exposing the gadolinium and increasing the effective q of the complex (Fig. 10c). The cleaved compound therefore has a higher relaxivity, and can be used to map patterns of β-galactosidase expression in transgenic animals. A related mechanism was used to generate a Gd-based calcium sensor, combining Gd-DO3A with the calcium buffering molecule bisaminophenoxy ethane tetraacetic acid (BAPTA) (Li et al., 1999). In the contrast agent's calcium-free state, a free ligand from the calcium binding site is thought to participate in chelating Gd. When calcium binds, this ligand is withdrawn, again increasing the effective q and r_1. Another mechanism for molecular sensing is derived from the τ_R dependence of T_1 agents, and has been exploited to produce protein-binding sensors. In one example, a Gal80-binding peptide was fused to Gd-DO3A; in the presence of the target Gal80, the peptide formed a bimolecular complex with significantly longer τ_R and higher relaxivity at 1.5 T (De Leon-Rodriguez et al., 2002).

A family of SPIO-based sensors has been based on T_2 relaxivity changes that accompany particle clustering (i.e., size changes) (Perez et al., 2004). In one example, Weissleder, Josephson, and colleagues developed crosslinked arrays of nanoparticles held together by strands of complementary DNA (Perez et al., 2002b). In the presence of a restriction endonuclease, the DNA was cut, breaking apart the complex and dramatically reducing T_2 (Fig. 10d). This group used similar mechanisms to sense protein–protein interactions and protease activity (Perez et al., 2002a). A non-SPIO T_2 sensor for oxygen can be formed using the iron-containing protein hemoglobin, which undergoes a change in spin state when O_2 binds (Sun et al., 2003). This property of hemoglobin is also the basis of functional imaging with endogenous blood-oxygenation-dependent contrast (Ogawa et al., 1990). Allen and Meade

(2004) have reviewed additional examples of molecular sensing by contrast agents. A general limitation of these agents, particularly for quantitative measurement applications, is that the same factors that sensitize them to specific targets may also be influenced by environmental variables (e.g., pH, viscosity, nonspecific binding) that are beyond an experimenter's means to control or measure.

Chemical Exchange-dependent Saturation Transfer Agents

The CEST Effect

Populations of protons in distinct chemical environments are associated with slightly different Larmor frequencies (chemical shifts). Polarization associated with sufficiently resolved proton pools may be manipulated using frequency-selective RF pulses and imaged using band-limited detection. Irradiation by a continuous wave or rapid train of selective RF pulses at the frequency associated with a given proton pool has the effect of "saturating" NMR signal from that pool. Polarization of the saturated pool is diminished, and a delay of several times T_1 is required for a return to equilibrium. If the saturated protons are in slow exchange with a second pool of (nonirradiated) protons, some of the saturation will be "transferred" to the second pool, resulting in diminished signal from the nonirradiated exchanging pool (Fig. 11a). Balaban and colleagues showed that this chemical exchange-dependent saturation transfer (CEST) effect, a form of what is generally known as magnetization transfer (MT), can be used as a ^{1}H MRI contrast mechanism (Ward et al., 2000). In MRI with CEST contrast, the saturated proton pool consists of labile protons bound to a dilute solute (CEST agent), and the nonirradiated pool corresponds to water protons that make up the bulk of the MRI signal. The MRI signal is reduced (darkened) in proportion to the concentration of the CEST agent and the efficiency of saturation transfer. A potential advantage of these agents over relaxation agents is that they may be "turned on and off" and calibrated in various ways, depending on whether or at what frequency the RF saturation is applied (Ward and Balaban, 2000).

The saturation transfer efficiency of a CEST agent depends on the number of proton exchange sites per agent (n), the time constant for proton exchange (τ_E), the saturation power (B_1) and duration (t_{sat}), the Larmor frequency difference ($\Delta\omega_{s\text{-}w}$) between the saturated CEST agent protons and bulk water protons, and relaxation rates for both proton pools (Woessner et al., 2005; Zhou et al., 2004). An analytical expression has been derived for the relative fraction of the bulk MRI signal remaining after saturation transfer:

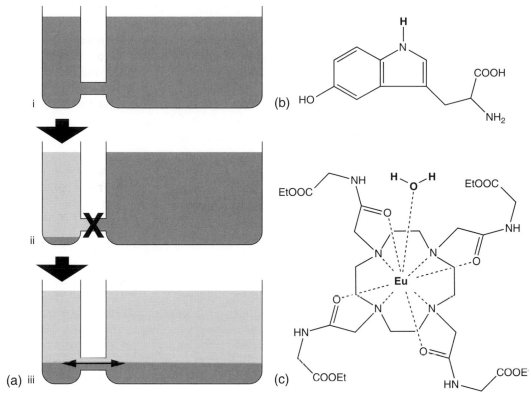

Figure 11 CEST agents. (a) The principle of chemical exchange-mediated saturation transfer relies on physical exchange between protons in two pools (schematized as the contents of left and right "containers" in this drawing; schematic adapted from Zhang *et al.* (2003a). (*i*) At equilibrium, both pools have steady-state magnetization. (*ii*) In the absence of exchange, radiofrequency saturation of protons associated with a CEST agent (left container) lowers the magnetization of the irradiated protons, but leaves the bulk proton pool (right container) unaffected. (*iii*) When the two pools are exchanging, saturation of the CEST agent proton pool lowers the magnetization of both proton pools and leads to signal darkening in MRI images. (b) 5-hydroxytryptophan was identified as an effective nonparamagnetic CEST agent by Ward *et al.* (2000). Saturating the resonance of the indole proton of this compound (bold) produces the greatest CEST contrast; the other exchangeable protons have less favorable combinations of chemical shift and exchange rate (τ_E). (c) A PARACEST compound introduced by Zhang *et al.* (2001). Water molecules coordinating a europium atom (both in bold) are in fast exchange with bulk water molecules. Saturation of the lanthanide-shifted bound water resonance produces CEST contrast.

$$\frac{M_Z^w(t_{sat})}{M_0^w} = \frac{R_1^w + (1-\alpha)\,k_{ws}}{R_1^w + k_{ws}}$$
$$+ \frac{\alpha k_{ws}}{R_1^w + k_{ws}} \exp\left\{-\left(R_1^w + k_{ws}\right)t_{sat}\right\} \quad [16]$$

Here M_Z^w and M_0^w are the water magnetization present after and before saturation, respectively, R_1^w is the water proton T_1 relaxation rate, α is the steady-state saturation fraction of the CEST pool ($\alpha = 1$ if the pool is fully saturated), and k_{ws} is the rate constant for proton transfer from the bulk water pool to the CEST pool. k_{ws} is equal to $cn/(111\tau_E)$, where c is the concentration of the (dilute) CEST agent and $111\,M$ is the concentration of water protons. It is assumed that the system is at equilibrium before irradiation, and also that steady-state saturation of the CEST pool is achieved in negligible time with respect to t_{sat} (Zhou *et al.*, 2004).

Equation [16] indicates that progressive saturation of the water proton pool is possible as t_{sat} increases, but that the fractional MRI signal decrease cannot exceed

$$\frac{\alpha k_{ws}}{R_1^w + k_{ws}} = \frac{\alpha}{1 + \left(T_1^w/\tau_E\right)(cn/111)} \quad [17]$$

This limit reflects the fact that transfer of saturation from the CEST pool competes against longitudinal relaxation for influence on the bulk water magnetization. In addition, the fractional saturation of the bulk pool cannot be greater than the fractional saturation of the CEST pool, α, which may be derived from the spin dynamics of the two-pool system (i.e., the Bloch equations) as follows (Zhou *et al.*, 2004):

$$\alpha = \frac{\gamma^2 B_1^2}{\gamma^2 B_1^2 + \left(K_2^s - k_{sw}\,k_{ws}/K_2^w\right)\left(K_1^s - K_{sw}\,k_{ws}/K_1^w\right)} \quad [18]$$

$K_i^s = R_i^s + k_{sw}$ and $K_i^w = R_i^w + k_{ws}$ for $i = 1$ or 2, where R_i^s and R_i^w denote the relaxation rates associated with the CEST and bulk proton pools, respectively, and k_{sw} is the rate of proton transfer from the CEST pool to the bulk proton pool ($k_{sw} = \tau_E^{-1}$). Equation [17] shows that α is maximized for high RF power (high B_1) when relaxation rates of the CEST pool are moderate, and when the proton transfer rates are slow compared with relaxation of the bulk water pool. Due to practical limitations on RF power deposition, full saturation of the proton pool is often not possible for high-efficiency (low τ_E) CEST agents.

In order for Eqs. [16]–[17] to be valid, the frequency separation between CEST and bulk water proton pools must be large enough to allow saturation of the exchanging pool without direct (exchange-independent) saturation of the bulk water pool: $\Delta\omega_{s-w}\tau_E \gg 1$ ($\Delta\omega_{s-w}$ is equivalent to $2\pi\gamma B_0\delta_{s-w} \times 10^{-6}$, where δ_{s-w} is the chemical shift difference between CEST agent-bound and bulk water protons). For slowly exchanging systems, $\Delta\omega_{s-w}$ need not be particularly large, but the available CEST signal changes are relatively small. For instance, CEST agents based on amine or amide proton exchanges have τ_E of approximately 10^{-3} s at pH 7, meaning that an exchange site concentration (cn) over 10 mM would be required for 10% MRI signal changes. A CEST agent capable of producing 10% signal changes for cn less than 100 μM would require a τ_E on the order of 10 μs and $\Delta\omega_{s-w}$ over 100 kHz. Because of these requirements, and the added requirement for CEST agent-bound protons to be clearly resolved from other endogenous protons *in vivo*, the most effective CEST agents are paramagnetic (PARACEST) agents based on water exchange off of lanthanide coordination sites (Zhang *et al.*, 2003a; Zhang *et al.*, 2001). PARACEST agents are complexes somewhat similar to the gadolinium-containing T_1 agents, but they incorporate "low relaxivity" lanthanides with short T_{1e} (e.g., Eu^{3+}, Dy^{3+}, or Nd^{3+}) in place of Gd^{3+}. Water protons bound to these agents can have δ_{s-w} up to 700 parts per million (ppm) and τ_E (equivalent to the inner sphere τ_M) as low as 3 μs.

CEST Agents and Applications

The first CEST agents were introduced comparatively recently, and there are no CEST contrast techniques yet in widespread use. Several agents have been demonstrated *in vitro*, however, and some may prove suitable for applications in patients or animals. Diamagnetic CEST agents were described in the original studies by Ward *et al.* (2000). These authors found that nitrogen-bound protons of amines, guanidinium groups, and nitrogenous aromatic groups are most effective at reducing M_Z/M_0 of bulk water. Of these, the most effective is the indole NH proton of 5-hydroxytryptophan (Fig. 11b), which benefits from a rela-

tively high δ_{s-w} of 5.33 ppm and produces 21% signal change at a concentration of 63 mM and pH 8. Several relatively effective diamagnetic CEST agents were also observed to be pH sensitive (proton exchange is usually acid- or base-catalyzed) and could be used for pH measurements (Ward and Balaban, 2000).

Significantly more potent CEST agents can be formed by combining hundreds of exchange sites into signal polymeric macromolecules, allowing relatively low concentrations of the macromolecules to be used as contrast agents. For example, 500 kD polylysine contains roughly 2000 amide and 4000 amine proton sites per molecule, and can produce over 40% changes in M_Z at concentrations as low as 100 μM (Goffeney *et al.*, 2001). Endogenous proteins have comparable amide proton content to polylysine, and it is estimated that the total protein amide concentration in healthy tissue ($\Sigma_i\ c_i n_i$, where i indexes amide-containing species) is roughly 3 mM. Zhou *et al.* (2003) used this fact to obtain CEST contrast based on amide proton transfer (APT) *in vivo*. By comparing CEST images formed with saturation at $+\delta_{s-w}$ and $-\delta_{s-w}$ with respect to the chemical shift of water, these authors could resolve relatively subtle signal changes associated with pH-dependent perturbations of amide exchange rates in ischemic rat brains.

In another effort to improve on the efficacy of the small molecule diamagnetic CEST agents, PARACEST agents based on europium and other lanthanides have been developed by Sherry, Aime, and others (Zhang *et al.*, 2003a). A tetraamide derivative of Eu-DOTA showed changes in M_Z/M_0 several times greater than those produced by diamagnetic compounds, due to fast exchange of water at an inner sphere coordination site (Zhang *et al.*, 2001) (Fig. 11c). In studies of a related compound, Aime *et al.* (2002a) showed that the more numerous amide protons proximal to and thus chemically shifted by the paramagnetic atom could also be used to produce CEST contrast with pH dependence. The most effective was Yb-DOTAM-Gly, which at 30 mM and 14°C, and with 4 s RF saturation pulses, produced bulk M_Z ranging linearly from 100% M_0 at pH 5.5 to 30% M_0 at pH 8.0. DOTA-related lanthanide complexes have formed the bases of metabolite sensors as well. A lactate-sensitive CEST agent was formed using an anisole derivative of DO3A (Aime *et al.*, 2002b). The chemical shift of exchangeable amide protons in this agent (δ_{w-s}) changes in response to lactate, so MRI contrast effects can be produced using saturation at the amide proton frequencies of either the lactate-bound or the lactate-free complexes. A different mechanism was used to produce a CEST agent sensitive to glucose (Zhang *et al.*, 2003b). In this agent, two phenylboronate groups promote binding of the sugar with millimolar affinity. The glucose complex has longer τ_E and produces less darkening in MRI scans acquired with CEST saturation.

Nonproton Contrast Agents

Direct Detection of Nuclei Other Than Protons

Dyes and contrast agents used in optical and X-ray imaging interact directly with the electromagnetic radiation used to image specimens. Tracers for nuclear imaging themselves give rise to the emission patterns detected in PET and SPECT. Compared with agents used in these other modalities, the MRI contrast agents reviewed so far in this chapter have relatively indirect relationships with the water proton signals that form the basis of the corresponding scans. A more straightforward way to create artificial contrast in an MRI scan might be to introduce and image directly a substance that produces an NMR signal distinct in frequency from that of water protons. This approach has been taken using contrast agents containing spin-1/2 nuclei other than ^{1}H, here collectively referred to as nonproton agents.

Although nonproton contrast agents have advantages, three factors have kept these agents from rivaling proton relaxation agents in general practice: First, the sensitivity of MRI detection is intrinsically lower for nuclei other than for protons. Sensitivity is proportional to the product of sample magnetization (M_0) and frequency of detection (ω_0):

$$M_0 \omega_0 = \left| \frac{N_I \gamma_I^2 \hbar^2 B_0 I(I+1)}{3k_B T} \right| (\gamma_I B_0) \qquad [19]$$

where γ_I is the gyromagnetic ratio of the spin-1/2 nucleus in question, N_I is the number of nuclei present, and I is 1/2. Most nuclei have γ_I well below the proton gyromagnetic ratio of 2.675×10^8 s^{-1}T^{-1}, and because of the γ_I^3 dependence in Eq. [19], these nuclei produce relatively little NMR signal. The second problem with nonproton imaging agents is that feasible concentrations of useful nuclei other than ^{1}H are relatively low *in vivo*. Even an imaging agent present at a "pharmacologically high" concentration of 1 mM is $\sim 10^5$-fold more dilute than the water molecules acted on by relaxation agents. A third difficulty with nonproton imaging agents is that their detection requires special hardware (RF coils and amplifiers); many scanners are not equipped for this.

The primary advantage of nonproton contrast agents is that they can often be detected with little or no background signal. As a result, quantitative measurements with these agents are more straightforward than with relaxation contrast agents. Exogenous agents used for nonproton imaging most commonly include ^{19}F or ^{13}C, both of which have low abundance in biological specimens. When images using contrast agents containing these nuclei (or hyperpolarized nuclei; see below) are obtained, there is a high probability that the signal at any point in the image arises primarily from the contrast agent. These scans are therefore acquired in parallel with conventional ^{1}H MRI scans, so that the distribution of the nonproton imaging agent can be registered with anatomical information present only in the proton image.

^{19}F and ^{13}C Imaging Agents

From the standpoint of sensitivity, ^{19}F is the nucleus of choice for nonproton MRI (Yu *et al.*, 2005). The ^{19}F gyromagnetic ratio is 0.94 times the proton γ, making fluorine detection 83% as sensitive. Although ^{19}F accounts for 100% of the elemental abundance, there is very little fluorine naturally present in most organisms. It is also relatively simple to adapt standard MRI equipment for ^{19}F imaging. A particular strength of fluorine NMR methods is that fluorine chemical shifts are often strongly perturbed by modifications to a fluorinated compound (e.g., cleavage, protonation, or ion binding). This provides a natural mechanism by which fluorinated agents can function as sensors. A variety of ^{19}F pH sensors has been produced this way; examples include a systematic series of fluoromethylalanines with pK_as from 5.9–8.5 and 1–2 ppm resonance shifts (Taylor and Deutsch, 1983), and several aromatic species, some membrane permeable, with ^{19}F chemical shift responses from 5–17 ppm (Yu *et al.*, 2005).

Fluorinated metal ion sensors have been derived from known ion-binding buffers; design of these agents has benefited from the "chemically conservative" nature of fluorination—substitution of H by F has minimal effect on the properties of many organic compounds. Fluorinated versions of the buffers BAPTA (Fig. 12a) and aminophenoltriacetic acid (APTRA) function as sensors for calcium and magnesium, respectively (Levy *et al.*, 1988; Smith *et al.*, 1983), and fluorinated macrocyclic compounds have been designed to measure Na$^+$ and other monovalent ions (Plenio and Diodone, 1996; Smith *et al.*, 1986). Chemical shift-responsive compounds can be used for imaging with chemical shift-selective MRI pulse sequences, or for *in vivo* spectroscopy. However, relatively few imaging studies are performed with chemical shift-responsive fluorine agents because of the difficulty of obtaining high enough concentrations for meaningful image resolution.

To achieve sensitivity comparable to proton MRI, a fluorine image must be acquired with resolution lower than the proton image by a factor of $\sim 60/[F]^{1/3}$, where [F] is the fluorine concentration (in mM) in voxels of interest. Fluorine imaging is therefore most practical where high local concentrations can be attained. This condition may be met for fluorinated agents that selectively accumulate in specific regions of interest. In one recent example, a fluorinated amyloidophilic compound was introduced and incorporated specifically into brain plaques, allowing imaging studies of a mouse model of Alzheimer's disease (Higuchi *et al.*, 2005). High ^{19}F concentrations are also frequently achieved

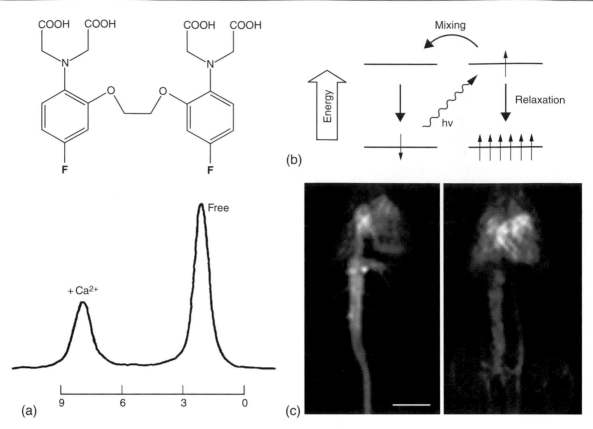

Figure 12 Nonproton imaging techniques. (a) The fluorinated agent 5F-BAPTA gives rise to a strongly calcium-dependent NMR spectrum (Smith *et al.*, 1983). The structure of 5F-BAPTA is shown above. The carboxylate groups and phenyl ether oxygen atoms participate in reversible binding of Ca^{2+} with a binding constant near 1 μM. Calcium-bound and free forms of the molecule give rise to two resonance peaks (left and right, respectively) separated by roughly 8 ppm (scale at bottom). (b) Optical pumping involves application of polarized light to energy levels of an alkali metal electron spin. Laser light ($h\nu$) is absorbed by ground state electrons in spin state 1 (bottom left), raising them to an excited state with spin state 2 (top right). Mixing and relaxation (thick arrows) redistribute the electron population among energy levels and spin states. Over time, systematic depletion of the absorbing state promotes accumulation of spin polarized electrons into the nonabsorbing ground state (bottom right). This spin polarization is transferred to ^{3}He or ^{129}Xe via molecular collisions in a gas cell, hyperpolarizing their nuclear spins. (c) Imaging with hyperpolarized contrast agents results in high-contrast images with virtually no background signal. Here, hyperpolarized ^{13}C-urea was injected into the circulatory system of a rat (Golman *et al.*, 2003). In virtually no time (left) the major arteries become visible, while after two seconds (right) the kidneys and further blood vessels begin to acquire contrast (scale bar 10 mm).

using perfluorocarbon compounds (PFCs), a family of highly fluorinated hydrophobic molecules that form droplets or emulsions in aqueous environments (Mason, 1994). In analogy to SPIO T_2 agents, PFC nanodroplets or nanoparticles have been used for cell-labeling experiments with almost no background signal and anatomically sufficient resolution (Ahrens *et al.*, 2005; Morawski *et al.*, 2004). PFCs have traditionally been applied to pO$_2$ mapping (Yu *et al.*, 2005). The ^{19}F T_1 relaxation rate in PFCs is relatively independent of variables such as pH and ion concentrations, but varies linearly as a function of oxygen tension, due to oxygen's paramagnetism and solubility in hydrophobic regions.

The spin-1/2 isotope of carbon, ^{13}C, is also sometimes incorporated into nonproton MRI contrast agents. Experiments requiring imaging of ^{13}C-labeled reagents suf-fer from two drawbacks with respect to ^{19}F MRI. First, the intrinsic sensitivity of ^{13}C magnetic resonance is only 1.6% that of protons ($\sim$ 50-fold lower than for fluorine). Second, the endogenous natural abundance of ^{13}C (1%) leads to relatively high background signal in studies using exogenous ^{13}C-labeled reagents. The sensitivity limitation can be partially overridden by using so-called indirect detection (Fitzpatrick *et al.*, 1990)—^{13}C magnetization is transferred to ^{13}C-bound protons and detected via the ^{1}H NMR signal with roughly eightfold sensitivity enhancement. Still, applications of ^{13}C imaging have generally been limited to studies of natural abundance ^{13}C (i.e., endogenous contrast) or to metabolite analogs that can be efficiently ^{13}C labeled and injected at high concentration (Shulman and Rothman, 2001). Useful features of these labeled organic molecules are that they are chemically identical to the corresponding unlabeled compounds, and that they can be distinguished using chemical shift selective methods.

Hyperpolarization Techniques

The imaging signal from some nonproton agents can be enhanced by several orders of magnitude using hyperpolarization, a process whereby the magnetization associated with the contrast agent is artificially enhanced before delivery to a subject. The polarization of a group of spin-1/2 nuclei is the difference in the populations of spin up (N^+) and spin down (N^-) state nuclei divided by the total number of nuclei. At thermal equilibrium, the observed polarization (P_{eq}) depends on the ratio between the energy splitting between spin up and down states and the thermal energy $k_B T$:

$$P_{eq} = \left| \frac{N_{eq}^+ - N_{eq}^-}{N_{eq}^+ + N_{eq}^-} \right| = \left| \tanh \left(\frac{\gamma_I \hbar B_0}{2 k_B T} \right) \right| \qquad [20]$$

At 298 K and a magnetic field of 5 T, P_{eq} would be 17 ppm for protons or 4.3 ppm for ^{13}C, in both cases a tiny fraction of the total number of nuclei. During hyperpolarization, a polarized electron spin population is generated in a paramagnetic sample. The electron magnetization is then transferred to nuclei of interest, producing a polarization much greater than P_{eq}. Once this nonequilibrium hyperpolarization (P_{hyp}) has been generated, it begins to decay back toward equilibrium with the time constant T_1. If imaging can be performed before significant T_1 decay, however, a sensitivity advantage may be obtained.

Hyperpolarization techniques are now most routinely applied to ^3He and ^{129}Xe, which are both used as volatile contrast agents for lung imaging (Moller et al., 2002). To hyperpolarize either of these noble gasses, a method called optical pumping is used (Raftery et al., 1991; Zeng et al., 1991). In a typical experiment, circularly polarized laser light of appropriate wavelength is directed at a vaporized alkali metal (usually Rb), which is mixed with the (pressurized) gas (Moller et al., 2002). Only one of the two electron spin states of the metal absorbs the light and undergoes a spin flip. The absorbing spin state is therefore systematically depopulated, yielding a high net electron polarization (Fig. 12b). Part of this polarization is imparted to ^3He or ^{129}Xe nuclei through collisions in the gas container. Nuclear polarization close to 100% is usually obtained after several hours. Pure ^3He and ^{129}Xe may be stored in hyperpolarized state, but in vivo they have T_1 of 1–10 sec, and fast low-flip angle or single-shot imaging techniques are required for three-dimensional imaging. Both ^3He and ^{129}Xe have relatively high gyromagnetic ratios (76% and 28% that of ^1H, respectively); hyperpolarization compensates for the low working concentrations of these gases to provide images with sensitivity approaching that of ^1H MRI (Albert et al.,

1994). Xenon has significantly higher water solubility than helium, making it less advantageous for lung-specific imaging, but possibly useful as a contrast agent in blood and tissue (Bifone et al., 1996; Goodson et al., 1997). Hyperpolarized ^{129}Xe has also been applied to a variety of molecular imaging applications by Alexander Pines and colleagues (Lowery et al., 2003).

Hyperpolarization of ^{13}C imaging agents has been successfully implemented by several laboratories (Mansson et al., 2006), most recently using the dynamic nuclear polarization (DNP) technique (Abragam and Goldman, 1978). In a DNP method introduced by Golman and colleagues, a ^{13}C imaging agent is brought to extremely low temperature (< 3 K) in the presence of a paramagnetic substance and at moderate to high magnetic field (Ardenkjaer-Larsen et al., 2003). Electron polarization present under these conditions is transferred to the ^{13}C nuclei using electron-nuclear double resonance methods. The frozen ^{13}C sample is then rapidly thawed and injected into a subject for imaging. Carbon polarizations up to 20–30% are attainable, a factor of 100,000 greater than thermal equilibrium levels and vastly overwhelming any background ^{13}C signal, even at low concentrations of the hyperpolarized agent (Fig. 12c). As with hyperpolarized gases, sensitivity can approach that of ^1H MRI, but T_1 relaxation limits the time period within which scanning must be performed to tens of seconds (Golman et al., 2003). To date, imaging studies using hyperpolarized ^{13}C agents have included angiography, perfusion imaging, catheter tracking, and measurements of metabolite uptake (Mansson et al., 2006). Improvements to these techniques may involve application to additional imaging agents (including ^{15}N- and ^{31}P-containing compounds), as well as enhanced hyperpolarization protocols based on higher polarizing fields and new paramagnetic polarizing substances (Bajaj et al., 2003; Hu et al., 2004).

Conclusions

The three classes of MRI contrast agents introduced in this chapter work by sharply different mechanisms, and each has advantages and disadvantages. *Relaxation agents* have been the most important in medical imaging and are still the focus of much innovative work. T_1 agents at concentrations near 100 μM, or SPIO T_2 agents at nanomolar particle levels, modulate proton MRI signals by 10% or more, making them the agents of choice in terms of raw sensitivity. New sensors based on relaxivity changes are a testament to the versatility of these agents. On the negative side, quantitative information can be difficult to obtain using relaxation agents because of environmental influences on their relaxivity and because intrinsic signal levels and relaxation rates are often poorly characterized *in vivo*. A further liability is that both T_1 and T_2 agents are based on toxic metal ions; applications in humans may therefore be limited by the potentially dire consequences of contrast agent demetallation.

CEST agents are a newer class of contrast agent, and their potential has been less thoroughly explored. Experience to date indicates that CEST agents are considerably weaker than conventional relaxation contrast agents. They also require long saturation times, which in some contexts is a possible drawback. But CEST agents are also more controllable than relaxation agents, because the frequency and extent of RF irradiation can be determined by the experimenter. This "extra dimension" to CEST imaging has facilitated quantitative measurements of pH and ligand-binding parameters.

Nonproton imaging agents also yield much lower sensitivity than relaxation agents, but may sometimes be preferred because of the absence of a significant background signal (especially in the case of fluorinated agents). Nonproton agents also have a clear advantage in terms of biocompatibility and some types of tracer experiment. They do not include metals, and isotopic ^{13}C labeling or conservative ^{19}F-^{1}H substitution can be used to form extremely close analogs of biological molecules or drugs. Hyperpolarization may extend the utility of nonproton agents, particularly as methods for polarizing solution-state molecules evolve.

Although connections between specific contrast agents and cancer imaging have not been emphasized in this chapter, the relevance to oncology of contrast enhancement techniques is self-evident, from the delineation of anatomical structures to metabolite imaging and molecular targeting. The potential of MRI contrast agents to play a role in cancer diagnosis is also expanding rapidly, with increased focus on reagents for "molecular imaging" on the one hand and increased understanding of the molecular correlates of cytopathology on the other. The noninvasiveness and relatively high spatial resolution of MRI ensure that continued exploration of contrast mechanisms and chemicals that exploit them will remain important to cancer research in the future. Several types of contrast agents used in conjunction with different imaging techniques are discussed in these two volumes.

References

Abragam, A., and Goldman, M. 1978. Principles of dynamic nuclear polarisation. *Rep. Prog. Phys. 41*:395–467.

Ahrens, E.T., Flores, R., Xu, H., and Morel, P.A. 2005. *In vivo* imaging platform for tracking immunotherapeutic cells. *Nat. Biotechnol. 23*:983–987.

Aime, S., Barge, A., Delli Castelli, D., Fedeli, F., Mortillaro, A., Nielsen, F.U., and Terreno, E. 2002a. Paramagnetic lanthanide(III) complexes as pH-sensitive chemical exchange saturation transfer (CEST) contrast agents for MRI applications. *Magn. Reson. Med. 47*:639–648.

Aime, S., Delli Castelli, D., Fedeli, F., and Terreno, E. 2002b. A paramagnetic MRI-CEST agent responsive to lactate concentration. *J. Am. Chem. Soc. 124*:9364–9365.

Albert, M.S., Cates, G.D., Driehuys, B., Happer, W., Saam, B., Springer, C.S., Jr., and Wishnia, A. 1994. Biological magnetic resonance imaging using laser-polarized ^{129}Xe. *Nature 370*:199–201.

Allen, M.J., and Meade, T.J. 2004. Magnetic resonance contrast agents for medical and molecular imaging. *Met. Ions Biol. Syst 42*:1–38.

Ardenkjaer-Larsen, J.H., Fridlund, B., Gram, A., Hansson, G., Hansson, L., Lerche, M.H., Servin, R., Thaning, M., and Golman, K. 2003. Increase in signal-to-noise ratio of > 10,000 times in liquid-state NMR. *Proc. Natl. Acad. Sci. USA 100*:10158–10163.

Artemov, D., Bhujwalla, Z.M., and Bulte, J.W. 2004. Magnetic resonance imaging of cell surface receptors using targeted contrast agents. *Curr. Pharm. Biotechnol. 5*:485–494.

Artemov, D., Mori, N., Okollie, B., and Bhujwalla, Z.M. 2003. MR molecular imaging of the Her-2/neu receptor in breast cancer cells using targeted iron oxide nanoparticles. *Magn. Reson. Med. 49*:403–408.

Bajaj, V.S., Farrar, C.T., Hornstein, M.K., Mastovsky, I., Vieregg, J., Bryant, J., Elena, B., Kreischer, K.E., Temkin, R.J., and Griffin, R.G. 2003. Dynamic nuclear polarization at 9 T using a novel 250 GHz gyrotron microwave source. *J. Magn. Reson. 160*:85–90.

Bifone, A., Song, Y.Q., Seydoux, R., Taylor, R.E., Goodson, B.M., Pietrass, T., Budinger, T.F., Navon, G., and Pines, A. 1996. NMR of laser-polarized xenon in human blood. *Proc. Natl. Acad. Sci. USA 93*:12932–12936.

Bloch, I., Hansen, W.W., and Packard, M. 1946. The nuclear induction experiment. *Phys. Rev. 70*:474.

Bloembergen, N.J. 1957. Proton relaxation times in paramagnetic solutions. *J. Chem. Phys. 27*:572–573.

Bloembergen, N.J., and Morgan, L.O. 1961. Proton relaxation times in paramagnetic solutions: Effects of electron spin relaxation. *J. Chem. Phys. 34*:842–850.

Bloembergen, N.J., Purcell, E.M., and Pound, R.V. 1948. Relaxation effects in nuclear magnetic resonance absorption. *Phys. Rev. 73*:679.

Bulte, J.W., Duncan, I.D., and Frank, J.A. 2002. *In vivo* magnetic resonance tracking of magnetically labeled cells after transplantation. *J. Cereb. Blood Flow Metab. 22*:899–907.

Bulte, J.W., and Kraitchman, D.L. 2004. Iron oxide MR contrast agents for molecular and cellular imaging. *NMR Biomed. 17*:484–499.

Caravan, P., Ellison, J.J., McMurry, T.J., and Lauffer, R.B. 1999. Gadolinium(III) chelates as MRI contrast agents: structure, dynamics, and applications. *Chem. Rev. 99*:2293–2352.

De Leon-Rodriguez, L.M., Ortiz, A., Weiner, A.L., Zhang, S., Kovacs, Z., Kodadek, T., and Sherry, A.D. 2002. Magnetic resonance imaging detects a specific peptide-protein binding event. *J. Am. Chem. Soc. 124*:3514–3515.

Elizondo, G., Fretz, C.J., Stark, D.D., Rocklage, S.M., Quay, S.C., Worah, D., Tsang, Y.M., Chen, M.C., and Ferrucci, J.T. 1991. Preclinical evaluation of MnDPDP: new paramagnetic hepatobiliary contrast agent for MR imaging. *Radiology 178*:73–78.

Fitzpatrick, S.M., Hetherington, H.P., Behar, K.L., and Shulman, R.G. 1990. The flux from glucose to glutamate in the rat brain *in vivo* as determined by ^{1}H-observed, ^{13}C-edited NMR spectroscopy. *J. Cereb. Blood Flow Metab. 10*:170–179.

Freed, J.H. 1978. Dynamic effects of pair correlation functions on spin relaxation by translational diffusion in liquids. II. Finite jumps and independent T_1 processes. *J. Chem. Phys. 68*:4034–4037.

Gillis, P., and Koenig, S.H. 1987. Transverse relaxation of solvent protons induced by magnetized spheres: application to ferritin, erythrocytes, and magnetite. *Magn. Reson. Med. 5*:323–345.

Gillis, P., Moiny, F., and Brooks, R.A. 2002. On T_2-shortening by strongly magnetized spheres: a partial refocusing model. *Magn. Reson. Med. 47*:257–263.

Goffeney, N., Bulte, J.W., Duyn, J., Bryant, L.H., Jr., and van Zijl, P.C. 2001. Sensitive NMR detection of cationic-polymer-based gene delivery systems using saturation transfer via proton exchange. *J. Am. Chem. Soc. 123*:8628–8629.

Golman, K., Ardenkjaer-Larsen, J.H., Petersson, J.S., Mansson, S., and Leunbach, I. 2003. Molecular imaging with endogenous substances. *Proc. Natl. Acad. Sci. USA* 100:10435–10439.

Goodson, B.M., Song, Y., Taylor, R.E., Schepkin, V.D., Brennan, K.M., Chingas, G.C., Budinger, T.F., Navon, G., and Pines, A. 1997. *In vivo* NMR and MRI using injection delivery of laser-polarized xenon. *Proc. Natl. Acad. Sci. USA* 94:14725–14729.

Gupta, A.K., and Gupta, M. 2005. Synthesis and surface engineering of iron oxide nanoparticles for biomedical applications. *Biomaterials* 26:3995–4021.

Higuchi, M., Iwata, N., Matsuba, Y., Sato, K., Sasamoto, K., and Saido, T.C. 2005. ^{19}F and ^{1}H MRI detection of amyloid beta plaques *in vivo*. *Nat. Neurosci.* 8:527–533.

Hu, K.N., Yu, H.H., Swager, T.M., and Griffin, R.G. 2004. Dynamic nuclear polarization with biradicals. *J. Am. Chem. Soc.* 126:10844–10845.

Jung, C.W., and Jacobs, P. 1995. Physical and chemical properties of superparamagnetic iron oxide MR contrast agents: ferumoxides, ferumoxtran, ferumoxsil. *Magn. Reson. Imaging* 13:661–674.

Koretsky, A.P., and Silva, A.C. 2004. Manganese-enhanced magnetic resonance imaging (MEMRI). *NMR Biomed.* 17:527–531.

Lauffer, R.B. 1987. Paramagnetic metal complexes as water proton relaxation agents for NMR imaging: theory and design. *Chem. Rev.* 87:901–927.

Levy, L.A., Murphy, E., Raju, B., and London, R.E. 1988. Measurement of cytosolic free magnesium ion concentration by ^{19}F NMR. *Biochemistry* 27:4041–4048.

Li, W., Fraser, S.E., and Meade, T.J. 1999. A calcium-sensitive magnetic resonance imaging contrast agent. *J. Am. Chem. Soc.* 121:1413–1414.

Lin, Y.-J., and Koretsky, A.P. 1997. Manganese ion enhances T_1-weighted MRI during brain activation: an approach to direct imaging of brain function. *Magn. Reson. Med.* 38:378–388.

Louie, A.Y., Huber, M.M., Ahrens, E.T., Rothbacher, U., Moats, R., Jacobs, R.E., Fraser, S.E., and Meade, T.J. 2000. *In vivo* visualization of gene expression using magnetic resonance imaging. *Nat. Biotechnol.* 18:321–325.

Lowery, T.J., Rubin, S.M., Ruiz, E.J., Spence, M.M., Winssinger, N., Schultz, P.G., Pines, A., and Wemmer, D.E. 2003. Applications of laser-polarized ^{129}Xe to biomolecular assays. *Magn. Reson. Imaging* 21:1235–1239.

Mansson, S., Johansson, E., Magnusson, P., Chai, C.M., Hansson, G., Petersson, J.S., Stahlberg, F., and Golman, K. 2006. ^{13}C imaging—a new diagnostic platform. *Eur. Radiol.* 16:57–67.

Mason, R.P. 1994. Non-invasive physiology: ^{19}F NMR of perfluorocarbons. *Artif. Cells Blood Substit. Immobil. Biotechnol.* 22:1141–1153.

Meade, T.J., Taylor, A.K., and Bull, S.R. 2003. New magnetic resonance contrast agents as biochemical reporters. *Curr. Opin. Neurobio.* 13:597–602.

Merbach, A.E., and Toth, E. (Eds.) 2001. *The Chemistry of Contrast Agents in Medical Magnetic Resonance Imaging*. New York: John Wiley & Sons.

Moller, H.E., Chen, X.J., Saam, B., Hagspiel, K.D., Johnson, G.A., Altes, T.A., de Lange, E.E., and Kauczor, H.U. 2002. MRI of the lungs using hyperpolarized noble gases. *Magn. Reson. Med.* 47:1029–1051.

Morawski, A.M., Winter, P.M., Yu, X., Fuhrhop, R.W., Scott, M.J., Hockett, F., Robertson, J.D., Gaffney, P.J., Lanza, G.M., and Wickline, S.A. 2004. Quantitative "magnetic resonance immunohistochemistry" with ligand-targeted ^{19}F nanoparticles. *Magn. Reson. Med.* 52: 1255–1262.

Muller, R.N., Roch, A., Colet, J.-M., Ouakssim, A., and Gillis, P. 2001. Particulate magnetic contrast agents. In: Merbach, A.E., and Toth, E., (Eds.), *The Chemistry of Contrast Agents in Medical Magnetic Resonance Imaging*. New York: John Wiley & Sons.

Ogawa, S., Lee, T.M., Kay, A.R., and Tank, D.W. 1990. Brain magnetic resonance imaging with contrast dependent on blood oxygenation. *Proc. Natl. Acad. Sci. USA* 87:9868–9872.

Patronas, N.J., Cohen, J.S., Knop, R.H., Dwyer, A.J., Colcher, D., Lundy, J., Mornex, F., Hambright, P., Sohn, M., and Myers, C.E. 1986. Metalloporphyrin contrast agents for magnetic resonance imaging of human tumors in mice. *Cancer Treat. Rep.* 70:391–395.

Pautler, R.G., Silva, A.C., and Koretsky, A.P. 1998. *In vivo* neuronal tract tracing using manganese-enhanced magnetic resonance imaging. *Magn. Reson. Med.* 40:740–748.

Perez, J.M., Josephson, L., O'Loughlin, T., Hogemann, D., and Weissleder, R. 2002a. Magnetic relaxation switches capable of sensing molecular interactions. *Nat. Biotechnol.* 20:816–820.

Perez, J.M., Josephson, L., and Weissleder, R. 2004. Use of magnetic nanoparticles as nanosensors to probe for molecular interactions. *Chembiochem.* 5:261–264.

Perez, J.M., O'Loughin, T., Simeone, F.J., Weissleder, R., and Josephson, L. 2002b. DNA-based magnetic nanoparticle assembly acts as a magnetic relaxation nanoswitch allowing screening of DNA-cleaving agents. *J. Am. Chem. Soc.* 124:2856–2857.

Plenio, H., and Diodone, R. 1996. Covalently bonded fluorine as a σ-donor for groups I and II metal ions in partially fluorinated macrocycles. *J. Am. Chem. Soc.* 118:356–367.

Powell, D.H., Ni Dhubhghaill, O.M., Pubanz, D., Helm, L., Lebedev, Y.S., Schlaepfer, W., and Merbach, A.E. 1996. Structural and dynamic parameters obtained from ^{17}O NMR, EPR, and NMRD studies of monomeric and dimeric Gd^{3+} complexes of interest in magnetic resonance imaging: an integrated and theoretically self-consistent approach. *J. Am. Chem. Soc.* 118:9333–9346.

Raftery, D., Long, H., Meersmann, T., Grandinetti, P.J., Reven, L., and Pines, A. 1991. High-field NMR of adsorbed xenon polarized by laser pumping. *Physical Review Letters* 66:584–587.

Raymond, K.N., and Pierre, V.C. 2005. Next generation, high relaxivity gadolinium MRI agents. *Bioconjug. Chem.* 16:3–8.

Saini, S., and Ferrucci, J.T. 1988. Ferrite particles as an MR contrast agent. *Radiology* 169:656.

Schellenberger, E.A., Bogdanov, A., Jr., Hogemann, D., Tait, J., Weissleder, R., and Josephson, L. 2002. Annexin V-CLIO: a nanoparticle for detecting apoptosis by MRI. *Mol. Imaging* 1:102–107.

Sessler, J.L., and Miller, R.A. 2000. Texaphyrins: new drugs with diverse clinical applications in radiation and photodynamic therapy. *Biochem. Pharmacol.* 59:733–739.

Shen, T., Weissleder, R., Papisov, M., Bogdanov, A., Jr., and Brady, T.J. 1993. Monocrystalline iron oxide nanocompounds (MION): physicochemical properties. *Magn. Reson. Med.* 29:599–604.

Shulman, R.G., and Rothman, D.L. 2001. ^{13}C NMR of intermediary metabolism: implications for systemic physiology. *Annual Rev. Physiol.* 63:15–48.

Smith, G.A., Hesketh, R.T., Metcalfe, J.C., Feeney, J., and Morris, P.G. 1983. Intracellular calcium measurements by ^{19}F NMR of fluorine-labeled chelators. *Proc. Natl. Acad. Sci. USA* 80:7178–7182.

Smith, G.A., Morris, P.G., Hesketh, T.R., and Metcalfe, J.C. 1986. Design of an indicator of intracellular-free Na$^+$ concentration using ^{19}F-NMR. *Biochim. Biophys. Acta.* 889:72–83.

Solomon, I. 1955. Relaxation processes in a system of two spins. *Phys. Rev.* 99:559–565.

Sun, P.Z., Schoening, Z.B., and Jasanoff, A. 2003. *In vivo* oxygen detection using exogenous hemoglobin as a contrast agent in magnetic resonance microscopy. *Magn. Reson. Med.* 49:609–614.

Taylor, J.S., and Deutsch, C. 1983. Fluorinated alpha-methylamino acids as ^{19}F NMR indicators of intracellular pH. *Biophys. J.* 43: 261–267.

Toth, E., Helm, L., and Merbach, A.E. 2001. Relaxivity of gadolinium(III) complexes: theory and mechanism. In: Merbach, A.E., and Toth, E., (Eds.), *The Chemistry of Contrast Agents in Medical Magnetic Resonance Imaging*. New York: John Wiley & Sons.

Ward, K.M., Aletras, A.H., and Balaban, R.S. 2000. A new class of contrast agents for MRI based on proton chemical exchange dependent saturation transfer (CEST). *J. Magn. Reson. 143*:79–87.

Ward, K.M., and Balaban, R.S. 2000. Determination of pH using water protons and chemical exchange dependent saturation transfer (CEST). *Magn. Reson. Med. 44*:799–802.

Weinmann, H.J., Brasch, R.C., Press, W.R., and Wesbey, G.E. 1984. Characteristics of gadolinium-DTPA complex: a potential NMR contrast agent. *AJR Am. J. Roentgenol. 142*:619–24.

Weinmann, H.J., Ebert, W., Misselwitz, B., and Schmitt-Willich, H. 2003. Tissue-specific MR contrast agents. *Eur. J. Radiol. 46*:33–44.

Weissleder, R., Moore, A., Mahmood, U., Bhorade, R., Benveniste, H., Chiocca, E.A., and Basilion, J.P. 2000. *In vivo* magnetic resonance imaging of transgene expression. *Nat. Med. 6*:351–355.

Woessner, D.E., Zhang, S., Merritt, M.E., and Sherry, A.D. 2005. Numerical solution of the Bloch equations provides insights into the optimum design of PARACEST agents for MRI. *Magn. Reson. Med. 53*:790–799.

Xu, J., Franklin, S.J., Whisenhunt, D.W., and Raymond, K.N. 1995. Gadolinium complex of Tris[(3-Hydroxy-1-Methyl-2-Oxo-1,2 Didehydropyridine-4-Carboxamido)Ethyl]-Amine—a new class of gadolinium magnetic resonance relaxation agents. *J. Am. Chem. Soc. 117*:7245–7246.

Young, S.W., Bradley, B., Muller, H.H., and Rubin, D.L. 1990. Detection of hepatic malignancies using Mn-DPDP (manganese dipyridoxal diphosphate) hepatobiliary MRI contrast agent. *Magn. Reson. Imaging 8*:267–276.

Young, S.W., Qing, F., Harriman, A., Sessler, J.L., Dow, W.C., Mody, T.D., Hemmi, G.W., Hao, Y., and Miller, R.A. 1996. Gadolinium(III) texaphyrin: a tumor selective radiation sensitizer that is detectable by MRI. *Proc. Natl. Acad. Sci. USA 93*:6610–6615.

Young, S.W., Simpson, B.B., Ratner, A.V., Matkin, C., and Carter, E.A. 1989. MRI measurement of hepatocyte toxicity using the new MRI contrast agent manganese dipyridoxal diphosphate, a manganese/pyridoxal 5-phosphate chelate. *Magn. Reson. Med. 10*:1–13.

Yu, J.X., Kodibagkar, V.D., Cui, W., and Mason, R.P. 2005. ^{19}F: a versatile reporter for noninvasive physiology and pharmacology using magnetic resonance. *Curr. Med. Chem. 12*:819–848.

Zeng, X., Wu, C., Zhao, M., Li, S., Li, L., Zhang, X., Liu, Z., and Liu, W. 1991. Laser-enhanced low-pressure gas NMR signal from ^{129}Xe. *Chem. Phys. Lett. 182*:538–540.

Zhang, S., Merritt, M., Woessner, D.E., Lenkinski, R.E., and Sherry, A.D. 2003a. PARACEST agents: modulating MRI contrast via water proton exchange. *Acc. Chem. Res. 36*:783–790.

Zhang, S., Trokowski, R., and Sherry, A.D. 2003b. A paramagnetic CEST agent for imaging glucose by MRI. *J. Am. Chem. Soc. 125*:15288–15289.

Zhang, S., Winter, P., Wu, K., and Sherry, A.D. 2001. A novel europium(III)-based MRI contrast agent. *J. Am. Chem. Soc. 123*:1517–1518.

Zhao, M., Beauregard, D.A., Loizou, L., Davletov, B., and Brindle, K.M. 2001. Non-invasive detection of apoptosis using magnetic resonance imaging and a targeted contrast agent. *Nat. Med. 7*:1241–1244.

Zhou, J., Payen, J.F., Wilson, D.A., Traystman, R.J., and Van Zijl, P.C. 2003. Using the amide proton signals of intracellular proteins and peptides to detect pH effects in MRI. *Nat. Med. 9*:1085–1090

Zhou, J., Wilson, D.A., Sun, P.Z., Klaus, J.A., and Van Zijl, P.C. 2004. Quantitative description of proton exchange processes between water and endogenous and exogenous agents for WEX, CEST, and APT experiments. *Magn. Reson. Med. 51*:945–952.

5

Whole-body Computed Tomography Screening

Lincoln L. Berland and Nancy W. Berland

Introduction

Few subjects in the current era of radiology have matched the intensity of controversy generated by the use of computed tomography (CT) scanning for screening asymptomatic, presumably healthy people. The most controversial of all of these screening methods is what has become known as whole-body CT screening (WBS). What is WBS, and why do some members of the medical community oppose its use? What are the consequences of performing these studies, and is the stated rationale reasonable? We will explore the reasons for the controversy, evaluating social, professional, and scientific concepts, and we will critically address the arguments of the proponents and opponents of this technique.

Although other authors deal with the ethics and medicolegal aspects of screening in this volume, some of these matters cannot be removed from a discussion of WBS, and we will address them briefly. Arguments on this emotional topic have become polarized. However, we will attempt to review them from a balanced perspective. For the reader to better appreciate our biases, we support what is described as targeted screening for specific disease states. At the University of Alabama at Birmingham, we perform screening examinations of the lungs, heart, and colon but not untargeted screening with WBS.

What Is Whole-body Computed Tomography Screening?

There is no consensus for the definition of WBS. Reports are emerging on the use of chest and abdominal CT screening. However, no one has published a description of WBS in the peer-reviewed literature as practiced in community radiology, in which most of these examinations are currently being performed. Drs. Harvey Eisenberg and Kenneth Cooper (Brant-Zawadzki, 2002) are credited with developing WBS, and Eisenberg with popularizing it through an appearance on the *Oprah Winfrey Show*. Eisenberg began performing WBS with an electron beam CT scanner, using a technique involving a 20-sec scanning acquisition from the neck to the pelvis to include coronary calcium scoring, lung cancer screening, bone densitometry, and an abdominal survey.

Presentations and web sites reflect a variety of approaches. In its broadest form, WBS includes a scan of the head, neck, chest, abdomen, and pelvis. However, WBS is practiced variably in conjunction with other targeted screening examinations. For example, a chest, abdomen, and pelvis scan may be performed with low-radiation dose to evaluate for lung nodules and any abdominal abnormalities. If a chest scan is done for lung nodules and a CT colonography is done, then the entire trunk is covered with targeted screening studies that also cover the whole body by default. However, because the use of CT colonography requires patient preps, high-resolution CT equipment, special techniques, and interpretation expertise, this is not yet widely practiced. The chest CT component of the WBS is designed primarily to detect lung cancer. If one wished to screen the chest for other diseases, a chest radiograph would likely suffice. Therefore, for the remainder of this chapter, we will consider WBS to be an untargeted examination of the abdomen.

How Is it Done? Standards, Protocols, and Informed Consent

A serious concern among critics of WBS is that no standards exist for equipment, techniques, personnel, interpretation, or quality for untargeted scans of the whole body. Unwary consumers may be told that they are being examined with the latest technology but not be informed of the breadth of quality standards that are applied to other screening tests, such as mammography. In the absence of standards, technique factors such as milliamperes, kVp, slice thickness, interval, and pitch are uncertain. Although proponents discuss low-radiation dose techniques, how low of a radiation dose is used or how widely low-dose techniques are practiced is not known. Most practitioners appear to avoid the use of intravenous contrast material because it adds risk, requires monitoring, and adds substantial time and cost. However, others observe that many findings are considerably more difficult to detect or characterize without contrast material, and that there is limited value to noncontrast examinations for the abdomen.

The use of informed consent is also variable. However, physicians usually appreciate that consent forms provide only limited legal protection. We believe that a well-written consent form can be informative, particularly regarding the risks of the examination. Unfortunately, there is currently more hype than scientific information about the risks and benefits, which should be communicated to the prospective user, even indicating that this examination is experimental. Patients might take the process of consent more seriously, compared with a promotional brochure, and might lead some of them to reflect and withdraw their request for the study. This is appropriate if the provider's reason for performing screening is to offer a service to a well-informed public rather than to build use of services as an entrepreneurial endeavor.

What Is Found on Whole-body Computed Tomography Screening?

Elsewhere in this volume, other authors discuss the purposes of and criteria for what constitutes appropriate screening tests. However, the core purpose of screening is to identify significant treatable conditions that can benefit the patient by early detection. So, we must ask specifically what conditions apply to WBS. Among the most important are renal cell carcinoma, abdominal aortic aneurysm, and ovarian carcinoma. We will consider these and other findings later.

Renal Cell Carcinoma

There are ~ 31,000 cases per year in the United States of renal cell carcinoma (RCC) and ~ 12,000 deaths, accounting for ~ 2% of new cases and deaths from cancer. By comparison, there are ~ 34,000 deaths from breast cancer, ~ 57,000 from colon cancer, ~ 160,000 from lung cancer, and more than 700,000 from cardiovascular disease every year in the United States. Therefore, if screening for RCC were equally sensitive and of equal cost per case to other beneficial screening tests, the cost-to-effectiveness ratio of screening for renal cell cancer would be much higher simply because the probability of the disease is so much lower, and thus the potential benefit to the population is so much smaller. Many more people have to be scanned to find a positive case that can lead to prolonging a life.

Approximately 61% of all RCCs are already discovered incidentally at diagnostic examinations, and, when treated, the substantial majority of small lesions are curable. However, evidence suggests that most RCCs discovered incidentally do not cause fatalities. Although there has been an increase in mortality of RCC during the past half-century, the rate of detecting RCC has increased threefold more than the increase in mortality. Population data indicate a yearly mortality rate from RCC of ~3.5/100,000 of total population, or 0.0035%. However, of the few screening studies done with ultrasound and CT, the prevalence has been a much higher 0.1–0.3%, or up to 86 times the rate of deaths. It is well known that RCC tends to be slow growing. However, that most RCC may have a benign behavior over the long term may not be fully appreciated (Rendon et al., 2000). From another perspective, RCC causes ~ 0.5% of all deaths. However, unsuspected RCC is found in 2% of autopsies. This indicates a presence of RCC many times the rate at which it kills (Swensen, 2002).

To account for the discrepancies in prevalence and mortality, one may hypothesize that a higher risk population (based on age) is being screened. However, even if one were also to hypothesize an interval occurrence of cancer at one-tenth the rate of prevalence, then still perhaps only 10–20% of renal cell cancers discovered at screening may be potentially fatal. Unfortunately, no data are available on the rate of interval incidence of RCC per year after an initial negative screening. Therefore, this information is speculative. If incidence proves to be much lower, that would imply that we are not detecting as many patients with benign-behaving RCCs as is implied by the calculations earlier.

Abdominal Aortic Aneurysm

As noted earlier, the value and cost-effectiveness of mass screening for an entity is partly related to its probability in the population and therefore its severity as a public health problem. Like RCC, abdominal aortic aneurysms (AAAs) account for a relatively small percentage of all deaths, ~ 10,000 per year. Smoking is a strong risk factor, and 1.2% of men over 65 years of age die of this condition, with a lower occurrence in women.

A study of ultrasound screening for AAA among 6058 men over age 65 showed a 21% overall mortality reduction from AAA, with a peak of 52% near the midpoint of the study. Another randomized study found a mortality rate of 3 per 10,000 person years versus 5.4 per 10,000 person years in the control group not invited for screening. The researchers concluded that screening for asymptomatic AAA could reduce the death rate by 49% (Wilmink *et al.*, 1999). Another report noted that the incidence of rupture was reduced by 55% in men in the experimental group when compared with the controls. A recent study found no death benefit from immediately treating AAA less than 5.5 cm in diameter as long as they are followed with ultrasound and treated based on growth criteria (Lederle *et al.*, 2002). Therefore, although screening may be effective in reducing deaths from ruptured abdominal aneurysms, the detection rate at screening is fairly low. In addition, many patients may undergo treatment with a major operation or aortic stent-graft procedure who would not have gone on to rupture. Perhaps the best argument against the use of CT to screen for this entity is that ultrasound is highly sensitive and specific for AAA and is less expensive.

Ovarian Carcinoma

Ovarian carcinoma has been studied as a candidate for screening because many cases present in an advanced state, and it has been hoped that earlier detection might lead to a higher survival rate. The death rate from ovarian carcinoma is similar to death rates from RCC and AAA. There are more than 14,000 deaths from ovarian cancer in the United States annually, with 24,000 newly diagnosed cases. The rate of occurrence doubles from the time a woman is 45 to 49 years old to the time she is in her sixties. Less than 15% of cases occur in women under 50 years of age.

Van Nagell *et al.* (2000) screened 57,214 women and found only 17 cancers (0.03%). The five-year survival was 87% for the screened group versus 50% for the unscreened controls. However, the cost of screening was $780,000 per stage I cancer detected, not even taking into account that ~ 10 operations were performed for every cancer detected. Another study in Japan evaluated 183,034 subjects with ultrasound as a primary screening test (Sato *et al.*, 2000). Of 22 primary tumors detected, 17 (77.3%) were classified as stage I carcinoma (0.01%), again showing the extraordinarily small yield from such screening.

The ability of CT to accurately detect ovarian carcinoma has never been evaluated. However, ultrasound is probably substantially more sensitive and specific for small ovarian lesions and is less expensive. However, the ultrasound studies cited show the extremely low yield, high risk in terms of unnecessary surgery, and high cost. Obviously, detecting this entity does not apply to men undergoing WBS, further decreasing the average statistical probability of WBS detecting significant conditions that might benefit subjects.

Other Findings on Whole-body Computed Tomography Screening

Overdiagnoses represent histologically malignant lesions or other potentially serious diseases that do not progress rapidly enough to harm the patient. Such overdiagnoses, false-positive tests, and incidental findings are the primary flaws of screening because they cause anxiety, procedures, and costs that do not lead to medical benefit. However, among these categories, perhaps incidental findings have received the least attention. Incidental findings may be defined as those that are not related to symptoms or previously detected abnormal physical findings or laboratory results. They are often benign lesions that lead to no symptoms, morbidity, or mortality if ignored. Examples include small liver cysts, granulomas and indeterminate liver nodules, adrenal nodules, and some renal lesions. Unfortunately, such findings are often reported or mistaken for significant lesions, leading to unnecessary further studies and procedures.

Liver Lesions

Two studies have evaluated the significance of liver nodules < 1.5 cm in diameter. These found a frequency of 13–17%, but no malignancy was detected in any small liver lesion in 262 patients with no primary malignancy. We informally reviewed 100 consecutive outpatient examinations, excluding patients with obvious causes for liver lesions such as polycystic kidney disease, cirrhosis, and multiple liver metastases. We observed lesions in 22%, with the increased incidence likely attributable to the use of multidetector CT scanning with thin sections. Given these studies, a small liver lesion in a screened patient will rarely represent serious disease, and when such a lesion is potentially harmful, it will rarely be curable.

Adrenal Lesions

The prevalence of adrenal nodules in patients with no endocrine complaints and no cancer history has been reported as 0.4%. In patients with prior cancers, it is ~4%. Most lesions < 4 cm in diameter are benign, < 30% grow, and < 20% develop biochemical findings. Adrenal cortical carcinoma is rare, occurring in only 4 to 12 people in 1 million and only in ~ 1 in 4000 adrenal nodules. There is less than a 50% two-year survival. Because of the aggressive nature of this lesion, early detection is unlikely to be valuable. Furthermore, there are little data suggesting any value to resecting adrenal metastases. The National Institutes of Health conducted an Adrenal Incidentaloma Consensus Conference in February 2002, concluding that these types of data "call into question the advisability of the current practice of intense clinical follow-up of this common condition [incidentally detected adrenal nodules]." Therefore, the common nature of incidental adrenal lesions and the rarity of their importance create the conditions for a high rate of follow-up studies with little benefit.

Other Miscellaneous Conditions

Many web sites, advertisements, and presentations promoting CT screening centers cite their detection of numerous diseases as evidence supporting the value of their tests. In addition to such targeted entities as lung carcinoma, colon carcinoma, and other conditions discussed earlier, carcinomas listed on these web sites and promotional materials include liver, laryngeal, bladder, sarcoma, pancreas, metastases, and lymphoma. Other nonmalignant conditions noted include renal calculi, gallstones, chronic obstructive pulmonary disease, ovarian cysts, thyroid lesions, lung infiltrates, renal lesions, adrenal lesions, ovarian/pelvic masses, hydronephrosis, congestive heart failure, pleural effusions, angiomyolipoma, undescended testes, esophageal mass, hepatitis, liver cysts, and kidney cysts. It would be difficult to argue that detecting most such entities in a preclinical phase would lead to prolonging life, even ignoring the risks of performing unnecessary tests and procedures. Many of the malignant diseases listed are not believed to be affected by early treatment. To our knowledge, no formal arguments have been made by the proponents of WBS suggesting otherwise. Therefore, the mention of such conditions may represent a marketing tactic to impress potential customers.

Risks and Costs of Positive Results

Risks of Positive Results

Although the risks of performing CT screening are minimal, the more substantial risks to subjects are the consequences of positive results. As will be discussed later, the rate of positivity may vary from under 30% to over 80%. Many people with a positive result desire further evaluation or have one specifically recommended, therefore receiving one or more of the following: diagnostic CT, follow-up low-

dose CT, MRI, ultrasound, PET, biopsy, or surgery. Each of these tests or procedures incurs a cost and risk, but the sum of effects is difficult to calculate.

Among the most feared consequences of positive or incidental findings is the cascade effect. Upon a positive result, another test may fail to resolve the question, or another questionable finding may be discovered. This process may repeat itself several times in a single patient, leading to the "million dollar work-up" or even to surgery that may cause a serious complication. One may find that some of the additional tests might not have been indicated if taken in isolation but have been performed in the context of multiple findings and pressure from the patient and families to be "on the safe side."

Although this cascade effect may be unusual, anecdotal stories are frightening. An example of this cascade effect occurred to a well-known radiologist, who published his experience. On a negative CT colonography examination, renal, hepatic, and lung masses were detected leading to additional CT scans, a PET scan, a liver biopsy, and video-aided thoracoscopy with wedge resections. He experienced excruciating postoperative pain, required 5 weeks of recuperation, and incurred over $50,000 in charges. All findings were benign. Unfortunately, screening exposes many people to the risks and costs of additional tests, whereas few benefit from early detection. Therefore, it is usually the healthy people within the screened group who pay the price of the physical risks, costs, and psychological consequences of subsequent medical procedures to save a few among them from deadly diseases.

Radiation

Radiation represents the only direct risk from noncontrast CT screening. Brant-Zawadzki (2002, p. 323) makes the following observations regarding this risk:

For a low-dose event this calculation [of mortality risk from radiation] tacitly assumes that detriment, derived by model from high-dose data, can be applied to low-dose events. There is no evidence that such is the case. Conversely there is no evidence that such is not the case. The models can be used only for a crude first order approximation of risk. Natural background radiation exposes the average American to 360 mrem per year. The effective dose delivered for whole-body (chest–abdomen–pelvis, calcium scoring) screening at our facility is 880 mrem (as measured by an independent radiation physicist). Thus, approximately twice the dose of annual background radiation is delivered.

Approximately 23% of all individuals will die of cancer (540,000 deaths a year); indeed, accepting the conservative estimates that cancer induction risk is 0.04% per rem, one can calculate that of every 100,000 people scanned, 40 will have life-threatening cancer induced by radiation during their lifetimes. On the other hand, of the same 100,000 people, 23,000 are likely to die from cancer. Assuming even a 0.005% early detection rate and resulting cure, 115 people may derive the benefit versus the potential 40 who might have cancer induced sometime in their lives.

We largely agree with these observations. A worst-case scenario of radiation exposure will probably lead to fatal malignancies in a smaller number of people than would benefit from a targeted screening test. In addition, even if the risk of exposure reaches the level estimated, the effect is substantially delayed. However, one should again appreciate that subjects with positive results are likely to receive additional radiation that they would not otherwise receive and will be exposed to other risks from medical tests and procedures. If the positive effects of screening exist, these may outweigh the risks from radiation exposure from the initial screening test itself, although different people will be harmed by the radiation than will be helped by the screening.

Costs of Positive Results

What are the cost implications of WBS? Brant-Zawadzki (2002) suggests that individuals have a right to pay for care that is not covered by third parties. However, are such examinations really performed with no cost to the third-party payer system? This question has been partly addressed by Dehn (2001):

Screening chest CT has spawned a whole new era of retail radiology, where potential patients are induced to pay cash for an as-yet unproven procedure. If this were simply the case, one could build an argument that this overdiffusion is revenue neutral to the system. Unfortunately, it is not that simple.

The NIA [National Imaging Associates, Inc.] worked to assess the extent of possible abuse. Most readers will agree that [chest CT scans that were stand-alone noncontrast scans without evidence of a conventional chest X-ray within the previous 30 days] would be suspect as truly representing screening examinations. As expected, there has been a substantial increase in the use of noncontrast chest CT scanning in the reviewed population during the 1999–2000 period. This coincides with intense public interest in the use of this technology. Inspection suggests that providers with percentages of studies greater than 30% to 40% demonstrate a variation in practice that cannot be explained by random variation. Several providers [in this study] are fraudulently eliciting payment for clinically based disease when, in fact, they are actually doing screening studies.

In addition, third-party payers usually reimburse studies generated from positive results from medical examinations. Therefore, payment from such subsequent follow-up examinations, biopsies, surgical procedures, and complications may account for the most substantial profit to health-care providers from screening.

Analyzing the Rationale of Whole-body Computed Tomography Screening

Brant-Zawadzki (2002), through his numerous presentations, interviews, and writings, has become identified as the most passionate and articulate evangelist of CT screening within radiology. Therefore, we will address his arguments systematically, with some of the following sections beginning with quotations from an article he published in the *American Journal of Roentgenology*.

Analogies to Existing Screening Practices

"Most radiologists do not think twice about doing studies requested by clinicians on patients who refer themselves for an annual physical. Individuals are already accustomed to periodic screening. Lay individuals and many physicians have been inculcated with the notion of 'an ounce of prevention'" (Brant-Zawadzki, 2002, p. 319).

This comment implies that, because self-referral for annual physicals is medically acceptable and referrals to radiographic examinations based on abnormal physicals are also medically acceptable, self-referral directly to radiographic screening (the high-tech physical) and consequent referrals for additional procedures might also be acceptable. Although it is unclear whether radiologists generally "do not think twice," about this, we certainly have "thought twice" many times when we believed that referrals from abnormal physical examinations were excessive or inappropriate. We have pondered how many relatively frivolous referrals could be curtailed. However, the following factors inhibit acting on such frustrations: (1) There are rarely standards for gauging what is medically appropriate when concern is raised by abnormal symptoms, physical findings, or laboratory results. In many environments, the confidence of the referring clinician in their diagnostic abilities strongly affects the rate of referrals. (2) There is a strong financial incentive to do most examinations requested. This incentive is not just personal but related to perceived peer pressure from the entire practice group or overt pressure from group leaders. (3) There is concern about alienating referral sources for personal and professional reasons. (4) The desire to avoid medicolegal risk strongly encourages further testing to exclude a significant disease, once a suggestive finding or laboratory result is revealed.

Considering the acceptability of the practice of self-referral for physical examinations does not address whether performing such examinations helps more patients than they harm. This practice simply represents the "facts of life" about the health-care system. This argument also does not address the question of whether the physical examination analogy applies to CT screening. This answer may lie in assessing the quantitative rather than the qualitative effects. That is, the probability of abnormal findings on WBS leading to additional procedures, multiplied by the likely higher expense and risk of these subsequent procedures (cost times risk per patient), may be much higher compared with the physical examination because CT is so much more sensitive, as proponents note. Therefore, even if one were to accept that all the referrals leading from abnormal physical examinations are appropriate, the risks and costs from WBS may be considerably higher and therefore not appropriate. This issue awaits quantitative study.

That the lay public and physicians accept the concept of "an ounce of prevention" is unquestionable. However, screening is usually applied to targeted medical conditions (breast cancer, colon cancer, cervical cancer, and so on) except for the general physical examination. Many types of screening have been discredited as ineffective or too costly. Therefore, such an untargeted screening examination under the umbrella of prevention or early detection would be unusual. Also, many question whether the WBS has the potential to help more people than to harm, as has been addressed earlier. In other words, the acceptance of screening as a concept does not justify its use in this manner.

Distrust of Authority and Self-empowerment

During the past two decades, the reputation of physicians has suffered. This deterioration started with malpractice attorneys. More recently, it has been fueled by restrictions imposed on patients by the insurance and managed-care industries and, by extension, the employers who use them. This latent American sentiment [of distrust of any professional paternalistic authority] is particularly activated when such entities betray the trust of individuals repeatedly, as has occurred with managed care. The increased standard of living and the ready access of this generation to medical information through the Internet and other mass media have produced a sense of medical self-empowerment (Brant-Zawadzki, 2002). Also, "the consumer who is made financially responsible for a product or service is better informed about its value."

We find such statements powerful and accurate. We share the opinion that medical institutions under strong political and economic forces have not adequately served patients. The ubiquitous government and insurance regulation of medicine through arcane, inequitable, and unresponsive mechanisms of payment restricts the patient from personal economic choices that could improve the quality of health care. It is startling how infrequently these mechanisms incorporate any measure of quality of the individual service into payment decisions. These forces have aggravated the alienation of health-care consumers, who rightfully want to assume more of the decision-making role in their own health.

Do these statements form a justification for WBS as one response to this problem? The limited methods available to consumers to participate in their own health-care decisions have led to a paucity of good information about the quality

or appropriateness of their options. However, there is no independent, respected body that provides comprehensive information about such services. As Brant-Zawadzki (2002) observes, many organizations simply recommend against the use of CT screening. They usually do so in general statements that are primarily intended to influence the consumer but may not be very specific, informative, or even accessible. Therefore, the provider of these services is left to provide information.

However, in particular it is the stand-alone outpatient CT screening service provider who is under strong financial incentive to encourage the use of their services, rather than offer balanced information detailing both the advantages and disadvantages of WBS. In this context, it is useful to examine an example of a testimonial based on the results of a WBS from a commercial CT screening service: "To Whom It May Concern: I thank you for saving my husband's life. He had the Virtual Physical on April 6th. Listed below are the findings: Throat—large mass at the base of the tongue. Chest—spiculated lesion on lung. Heart—blockage in three arteries. Abdomen—aortic aneurysm (5.3 cm). In six weeks he will have his final surgery and, because of the early findings, he will be fine. So again, thank you for saving his life."

The preceding statement indicates unequivocally that this individual's life was saved. However, we have no information as to whether the base of the tongue mass and the spiculated lung mass were in curable stages. Although it is plausible that there was value to detecting the coronary and aortic disease, this is not certain as the patient's wife implies. With the multisystem disease this patient experienced, the chance of prolonging his life must be questioned. However, the medically unsophisticated consumer might believe that a positive outcome is certain in this case and that finding such serious, yet curable conditions is common. Given this level of virtually unrestrained promotion, the advocate of self-empowerment should fear that when such distortions are revealed "distrust of paternalistic authority" could be generalized to this consumer movement and represent a setback rather than a respite for the frustrated consumer. Will this movement lead to breaching yet another trust? The antidotes to distrust resulting from such distortions about the screening experience may be to fully communicate both the disadvantages and advantages of screening and to obtain "informed consent" for such procedures.

Is Proof of Value Necessary?

Many physicians are beginning to think that the delay imposed by a randomized clinical trial (perhaps a decade) makes such a trial's results an unacceptable Holy Grail (Brant-Zawadzki, 2002). Much of organized radiology's opposition to patient self-referral for CT screening is based on the lack of proof or "knowledge" that it is of value for prolonging life. The same can be said for a multitude of accepted medical management procedures in which radiology partakes. We also agree with Brant-Zawadzki (2002) on this point, although this must still be placed in ethical, economic, and scientific perspectives. We believe that the following points apply.

A common-sense approach regarding the introduction of new technology has been stated in a letter to the editor of *Discover* magazine, referring to the discredited use of bone marrow transplant for breast cancer patients: "Science is always playing catch-up to clinical practice. Initial reports on a new therapy are preliminary and poorly controlled. However, we do not refuse to treat people until there is a randomized clinical trial to give a scientific basis for practice. Rather, we combine our best judgment and intuition with a patient's wishes while desperately waiting for the science to catch up and give us guidance" (p. 8). Elaborate methodology has been established for the study of technology diffusion, and many of these concepts have been accepted by insurers as justification to deny "experimental" procedures. However, we believe that their acceptance is based primarily on economic, rather than scientific, grounds, and there is considerable room for debate. Few argue that trying to prolong people's lives is a good thing until the topic of who pays for it arises.

If we were to wait for definitive evidence or even a broad consensus for the value of a test, we would, for example, still be refraining from using mammography, given the persistent controversy. However, many investigators, including ourselves, find the scientific demonstration of substantial death reduction from the use of screening mammography to be scientifically sound. The nationwide decline of breast cancer deaths is also seen as compelling supportive evidence despite well-documented disagreements. Computed tomography scanning itself would have achieved only a fraction of its current penetration into medical practice if we had been required to wait for RCT validation of applications before their widespread use. We do not believe that because reasonable studies lack the approval of specific groups, insurers should suppress the consumer's right to use them.

Some investigators have cast doubt on the ability of randomized controlled trials (RCTs) currently underway to adequately establish the scientific truth regarding diagnostic tests, specifically CT screening. For example, a current major RCT of lung cancer screening is randomizing subjects into CT screening and chest radiograph screening arms. However, this will test two screening strategies against each other rather than CT screening versus a true control group of no screening, thereby limiting the ability to detect whether screening has any effect. In addition, screening of subjects may continue for too short a period to identify the maximum benefit. We believe that these and other design components of this trial confer profound methodologic biases against

screening that will cloud the results unless the study conclusively supports CT lung screening anyway. In this context, proponents of CT screening claim that there is a substantial risk that the potential benefit of years of screening will be lost waiting for results from studies that are themselves likely to be fatally flawed. Specifically, proponents of WBS should correctly observe that no RCTs of WBS are underway or planned at all.

Given the potential value of a particular screening test, why would it not be as valid to claim that a screening test of low-risk and high probable value should be used until the value is disproven, rather than that it should not be used until its value is proven? The answer to this question depends on the risks of the testing and whether the proponents or opponents of a particular test are more likely to be correct. Can we reasonably estimate the probability of the cost-effectiveness of a test before a valid scientific study has been performed? The question of pursuing or forgoing testing before scientific proof raises profound ethical issues that we presently cannot resolve. Nevertheless, studies have suggested the plausibility of coronary artery screening, CT colonography, and lung cancer screening, but the potential risks versus benefits of WBS raise serious doubts about the plausibility of its value.

Is Whole-body Computed Tomography Screening Truly Screening?

In 1968, the classical criteria for screening were proposed: (1) the burden of disease must be sufficient, (2) the disease must be detectable in the preclinical phase, (3) there must be an effective test to help detect the disease, and (4) there must be an effective treatment. Whole-body CT scanning in self-referred individuals is not truly "screening" in the realm of medical epidemiology. Because the chest, abdomen, and pelvis are covered by the targeted screening studies, the other organs in the abdominal cavity come under scrutiny by default, thus, providing a whole body CT study. The term *screening* may therefore indeed apply to whole-body scanning even when patients self-refer without a standardized epidemiologic consensus. Obviously, guidelines for such self-referral patterns would be desirable. Currently, the entrepreneurial nature of many centers makes it difficult to exclude any individual who is willing to pay for the study (Brant-Zawadzki, 2002).

Brant-Zawadzki (2002) appropriately touches on the ambiguities of defining WBS in the context of screening. His comments appear to reflect even his own uncertainty. The only reasonably acceptable targeted CT study covering the abdomen is CT colonography. However, many screening centers do not perform CT colonography but do perform nontargeted screening studies of the abdomen. Therefore, in most cases, the abdomen is not examined by default.

However, it is likely that there would be consensus agreement that any findings on a CT colonography should be reported. As Brant-Zawadzki (2002) indicates, it would be desirable, if not essential, to establish guidelines and standards for managing such incidental findings. However, the frequent occurrence of incidental findings on targeted studies does not necessarily justify untargeted studies without any epidemiological basis.

Psychological Implications

Critics claim that asymptomatic individuals have a low prevalence of significant disease and that a large number of false-positive findings will result in excessive anxiety for patients (Brant-Zawadzki, 2002). The issue of positivity rates will be discussed later, but there has been much conjecture about the issue of the psychological effects of positive and negative findings on people undergoing screening. Some cite their concern about excessive anxiety related to positive findings and the possible inclination to abandon or ignore good health practices among people with negative scans. Proponents of screening cite their personal experiences of intense gratitude among screening subjects for negative scans and for the personal attention they receive during the screening process. Unfortunately, scientific understanding of the short-term and long-term psychological effects on screening subjects is completely lacking. Therefore, the following represent observations and speculations.

First, it should be noted that the reports of anxiety resulting from a positive test are often of equal or greater intensity in a spouse or other close relative. Thus, the negative psychological fallout affects both the patient and those close to the patient. Second, if the anxious patient or family member pressures the physician to perform additional tests, even if he or she does not readily accept their medical necessity, the psychological fallout may indirectly affect the medical outcome, including the number of procedures performed, the risks, and the costs.

The following psychological and behavioral effects are possible as a result of a positive screening result:

1. The patient may exhibit the appropriate level of concern, motivating him or her to continue with a recommended course of diagnosis and treatment. The patient may also be motivated to modify health habits to positively affect the outcome of the disease or even of other unrelated diseases (the patient taking this as a sign that it is time to change health behavior).
2. The patient may exhibit denial and fail to follow recommendations for further evaluation and treatment.
3. The patient may experience excessive anxiety, leading to distress, possibly affecting the patient's ability to function. This may also lead the patient to demand further studies and

procedures to attempt to ascertain the nature of the finding with near complete certainty.

The following may result from a negative screening examination:

1. The patient may be appropriately relieved that nothing serious was discovered and continue or even adopt healthy behaviors. The act of requesting the screening may be part of a plan to newly adopt healthy practices.
2. There may be an inappropriately intense sense of relief leading to a false sense of security and health. The patient may abandon or practice healthy behaviors less frequently because they seem less necessary.

These alternative reactions and behaviors are not mutually exclusive and are not necessarily constant. That is, a single individual may alternate between reactions. In addition, many people may have increasingly intense anxiety as a scheduled screening procedure approaches. The intensity of such anticipatory fears can be nearly disabling or at least dysfunctional. Therefore, the intensity of relief and appreciation may often reflect a release from this enhanced anxiety simply caused by the availability and imminence of the screening test.

One might speculate that individuals self-referring for screening are more likely to be among those prone to higher levels of anxiety and concern about their health, despite the absence of serious symptoms. If this is true, the screened population would be likely to have a disproportionate number of people with heightened anxiety in anticipation of screening, with aggravated anxiety regarding a positive result and with a heightened intensity of relief from a negative result. Also, the probability of a false-positive result so strongly exceeds a true-positive result in a patient who can benefit from early detection that the anxiety effects may be large. With regard to behavioral change, numerous studies of interventions for obesity and smoking have shown that it is extraordinarily difficult to affect long-term health behaviors, even with intense and focused long-term interventions. Therefore, a single screening test would seem unlikely to have strong effects on long-term health behavior, although anecdotal accounts may show positive or negative change.

Variability of Rate of Positive Results

Critics claim that asymptomatic individuals have a low prevalence of significant disease and that a large number of false-positive findings will result in needless further testing and the resulting increased risk from invasive procedures (Brant-Zawadzki, 2002). One of the pioneers of WBS, Eisenberg states: "There is not a single human being that I've examined that I haven't found some evolving pathology" (Bruce, 2001). In another article, Shope notes, "one facility claims 80% of its patients need follow-up. Other facilities put the number closer to 30%. Those differences just do not compute" (Argondezzi, 2002).

Data on positivity rates have also been cited by Brant-Zawadzki (2002), who has indicated that in a population of 1807 patients studied between January 15, 2001, and July 3, 2001, ~ 32% of screening tests were positive (with 43% of those being pulmonary nodules), and ~ 1% of screened patients were found to have cancer. Admittedly, these data and the high rate of positivity should be appreciated in light of the difference between prevalence and incidence of disease. Rates of disease detected in screening are always higher when screening first takes place (termed *prevalence*) because screening detects the summed preexisting disease. Once these same subjects return for routine interval screening examinations, one is seeing only the disease that has occurred in the time since the last examination (termed *incidence*). This is always a substantially smaller proportion.

Nevertheless, the startling variability of results among screening centers highlights lack of standards for determining positivity, although there may be other reasons for variability. Few investigators have proposed specific standards of lesion size, number, nature, and so on. The high rate of positivity also illuminates one of the core criticisms of screening. Although findings on annual physical examinations or diagnostic studies probably lead to modest numbers of additional medical procedures, CT screening converts 32% or more of asymptomatic people to patients, requiring further health-care attention. If condoned and practiced on a wide scale, this could substantially increase health-care costs with limited or, at best, uncertain benefit. Applying Brant-Zawadzki's (2002) figures of ~ 1% cancer detection rate, a minority of patients (perhaps ~ 0.3 to −0.5% of the total) would expect to have their lives "saved" (prolonged) by screening. However, more than 31% might be subjected to the risk of increased anxiety and possible additional tests, creating risk and cost.

Enhancement of Radiology's Role in Medicine

Given their knowledge of the various medical specialties with which they interact daily, their understanding of diagnostic imaging, and their knowledge of the interplay between these disciplines, radiologists may be well positioned to optimize triage for patients who need further medical care. Doing so can only enhance the image of radiologists as true physicians, whose particular expertise is diagnostic image interpretation and translation of that information into appropriate patient treatment.

Brant-Zawadzki (2002) argues that radiologists are well suited to a role in patient triage, that others wish to usurp that burgeoning role, and that becoming active in

this will enhance the position of radiologists. Although this argument may be true, if poor-quality screening practices propagate and advertising inaccurately depicts screening as unequivocally positive, the long-term effect may be negative and difficult to undo. Radiologists may become perceived as misrepresenting facts, as seeking profit at patients' expense, and as part of the paternalistic authority from which Brant-Zawadzki (2002) would like medicine to distance itself. Although properly performed screening by qualified radiologists may enhance the perception of radiology, this does not support the objective value of untargeted screening, but rather represents an advocacy position for our specialty. Although such advocacy may be warranted, the dangers to the perception of radiology are high if the means of advocacy are unregulated CT screening centers that fail to apply sound medical principles and have few incentives for high quality.

Entrepreneurial Value of Screening

As Brant-Zawadzki (2002) notes, centers dedicated to performing screening have a strong incentive to maximize revenues. How intense are these incentives to promote? "The need for promotion is unrelenting. In Los Angeles, a screening center can expect to spend $200,000 a month on TV, radio, and newspaper advertising. The monthly cost in a smaller market would be ~ $30,000." In this environment, it is difficult to apply medically principled exclusionary criteria because owners or shareholders demand profits. Unprofitable centers may be closed, affecting the jobs of the employees. Given the pressures cited earlier, it is unlikely that centers would present a balanced perspective to the potential customer. These incentives do not exist as strongly in centers that perform a variety of diagnostic studies with a small percentage of their workload from screening. Therefore, one could argue that applying medically appropriate acceptance criteria depends either on such entrepreneurial centers not existing or on regulating facilities providing screening services.

Although CT screening has been perceived as an economic windfall for those willing to undertake it, the business aspects are evolving and are not all positive. Note the following comments addressing this matter: "A whole-body screening segment [Oprah Winfrey] ran on her daytime talk show has set an entire medical subspecialty into motion. Two years after the Oprah explosion, numbers have sagged at most facilities" (Palacio, 2002). "A gold rush mentality is developing around the concept of CT screening. But, as with any gold rush, caution is in order. One of the most recent entrepreneurial trends is the use of laser surgery for vision correction. Like CT screening, it requires a fairly large capital investment, costs of the procedure are high, and it has gone through cycles of big investment, bankruptcy, and price-cutting" (Hayes, 2001).

"Radiological professional fees at some centers throughout the country have purportedly fallen to $40 for a complete total-body study. If you get a radiologist to work for a low bid, you are going to get a low bid interpretation. Additionally, the following comment pertains to a particular screening center. Compromises were made to maintain low prices. Age criteria are liberal. Electrocardiogram gating techniques that would reduce motion and misregistration artifacts are not used because of limitations encountered in a mobile environment" (Brice, 2001). These anecdotes and observations imply that competition and economic forces may limit the quality of screening in some centers. Quality may already be deteriorating in entrepreneurial practice models with marginal volume.

In conclusion, animosities among proponents and opponents of WBS have created new divisions within radiology. Proponents of whole-body CT screening claim that it represents a natural outgrowth of societal and economic trends, whereas opponents claim that screening is costly, risky, and represents brazen commercialization of radiology. We have attempted to review the rationale for such studies and to provide a balanced view of the arguments for and against whole-body CT screening. Based on these observations, we believe that, although targeted types of CT screening may be appropriate, whole-body screening cannot currently be justified based on medical, scientific, or psychological grounds. However, the debate will certainly not end here.

References

Argondezzi, T. 2002. First, do no harm. *Advance for Imaging and Oncology Administrators*, August:47–50.

Brant-Zawadzki, M. 2002. CT screening: why we do it. *Am. J. Roentgenol.* *179*:319–326.

Brice, J. 2001. At the CT screening crossroads: which way will radiologists turn? *Diagnostic Imaging*, December:42–47.

Bruce, R. 2001. A community divided. *Imaging Economics*, September: 20–25.

Dehn, T.G. 2001. Retail radiology's blowback. *Imaging Economics*, December:6–10.

Hayes, J.C. 2001. CT screening: it's entrepreneurial and not for everyone. *Diagnostic Imaging*, December:5.

Lederle, F.A., Wilson, S.E., Johnson, G.R., Reinke, D.B., Littooy, F.N., Acher, C.W., Ballard, D.J., Messina, L.M., Gordon, I.L., Chute, E.P., Krupski, W.C., Busuttil, S.J., Barone, G.W., Sparks, S., Graham, L.M., Rapp, J.H., Makaroun, M.S., Moneta, G.L., Cambria, R.A., Makhoul, R.G., Eton, D., Ansel, H.J., Freischlag, J.A., and Bandyk, D. 2002. Aneurysm detection and management veterans affairs coopeative study group. Immediate repair compared with surveillance of small abdominal aortic aneurysms. *New Eng. J. Med. 346*:1437–1444.

Meggs W.I. 2002. Letter to the editor. *Discover 8*.

Palacio, M. 2002. Standing room only. *Advance for Imaging and Oncology Administrators*, September:24–29.

Rendon, R.A., Stanietzky, N., Panzarella, T., Robinette, M., Klotz, L.H., Thurston, W., and Jewett, M.A. 2000. The natural history of small renal masses. *J. Urol. 164*:1143–1147.

Sato, S., Yokoyama, Y., Sakamoto, T., Futagami, M., and Saito, Y. 2000. Usefulness of mass screening for ovarian carcinoma using transvaginal ultrasonography. *Cancer 89*:582–588.

Swensen, S.J. 2002. CT screening for lung cancer. *Am. J. Roentgenol. 179*:833–836.

van Nagell, J.R., Jr, DePriest, P.D., Reedy, M.B., Gallion, H.H., Ueland, F.R., Pavlik, E.J., and Kryscio, R.J. 2000. The efficacy of transvaginal sonographic screening in asymptomatic women at risk for ovarian cancer. *Gynecol. Oncol. 77*:350–356.

Wilmink, T.B., Quick, C.R., Hubbard, C.S., and Day, N.E. 1999. The influence of screening on the incidence of ruptured abdominal aortic aneurysms. *J. Vas. Surg. 30*:203–208.

6

Whole-body ^{18}F-fluorodeoxyglucose Positron Emission Tomography: Is it Valuable for Health Screening?

Matthias Weckesser and Otmar Schober

Introduction

Positron emission tomography (PET) is an instrument used to noninvasively assess molecular processes in the whole body. As outlined in other chapters in this volume, the most widely available radiopharmaceutical used with PET is deoxyglucose labeled with ^{18}F (FDG). The distribution of this compound 1 hr after injection is strongly correlated with regional glucose metabolism. This distribution can be measured tomographically with PET. With the introduction of PET scanners, which are capable of assessing glucose metabolism in the whole body, this instrument is widely used for the assessment of cancer. The basis for tumor imaging with PET is the high metabolic activity of neoplastic cells as compared to surrounding normal tissue. Especially in highly malignant tumors with increased metabolic activity, this method has been shown to have a high accuracy. As

outlined in detail in the other chapters of this book, PET now has an established role in the diagnosis and staging of many malignant primaries. Furthermore, it can be used to monitor therapy and to detect recurrent tumors.

The moderate radiation exposure of PET is induced by the injection of radioactive glucose and not by the scanner itself. Thus, as in other scintigraphic methods, PET should ideally be used to assess whole-body glucose metabolism. The brain is not regularly included in whole-body imaging protocols, since a high physiological glucose metabolism of gray matter reduces the sensitivity for the detection of malignant disease. Furthermore, lower legs are not regularly included owing to the low probability of malignancy in that area.

Because PET is a whole-body imaging procedure, the idea of using it for cancer screening is inviting. Incidental findings of malignant tumors, which are unrelated to the referral diagnosis, have been observed by many nuclear medicine physicians working with FDG-PET (Fig. 13). In a series of such clinically unexpected results, Agress and Cooper (2004) showed that "incidental" does not mean "negligible" and that considerable pathology can be detected. In that study, 1550 consecutive patients with a variety of malignant tumors were referred for assessment or follow-up of malignant disease. These authors detected 58 unexpected findings. These findings were obviously unrelated to the referral diagnosis. By working with PET, metabolic hot spots can be found, which cannot be explained by the underlying malignant disease. These spots, however, are often not assessed further because they are regarded as having lower priority. Agress and Cooper (2004) succeeded in obtaining histological confirmation of 42 of these 58 unexpected findings. In 30 cases, malignant or premalignant lesions were detected. Most of those were found in the colon, where it was not possible by the intensity of glucose metabolism to differentiate malignant from premalignant lesions.

There were also cancers of the breast, larynx, ovaries, and thyroid. There where 9 benign findings, which were clinically important (cholecystitis, thyroid adenoma, prostatitis, and others). No underlying disorder was found in 3 cases. So, in ~ 3% of patients, significant disease was diagnosed, and in ~ 2%, malignant or premalignant lesions were detected. There are other reports with similar or even slightly higher rates of detection of incidental malignant disease. In a group of patients reported by Choi *et al.* (2005), the rate of malignant disease was as high as 5%. In that series, 547 patients were evaluated who were referred for conventional staging work-up and PET-CT staging work-up in the initial staging of a variety of primaries. Overall, 27 second primary malignant tumors were identified in 26 patients (4.8%). PET-CT detected 24 of 27 lesions, whereas conventional staging work-up detected only 11 of these lesions. These observations suggest the use of FDG-PET for health screening. The justification of such a concept,

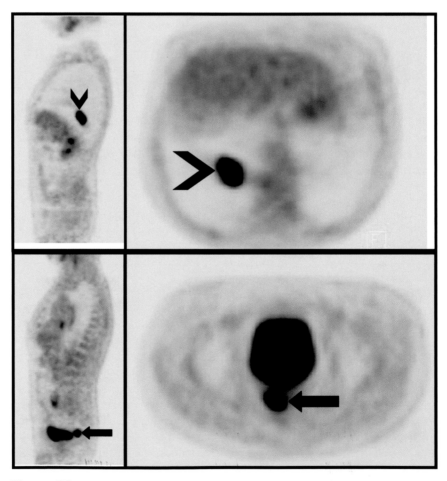

Figure 13 78-year-old male with non-small cell lung cancer referred for radiotherapy planning. Besides the hypermetabolic lung primary (upper row, arrow head), whole-body PET showed focally increased glucose metabolism behind the bladder (lower row, arrow). Rectal cancer as an independent second primary was endoscopically confirmed. Sagittal slices are shown on the left; transaxial slices are shown on the right.

however, is far from clear, and many prerequisites have to be met before it can be recommended.

Current Positron Emission Tomography Screening Programs

The largest study of cancer screening in healthy subjects has been reported from Japan (Yasuda et al., 2000; Ide and Suzuki, 2005). Nearly 40,000 healthy persons (average age: 53.6 years) were recruited for cancer screening. Positron emission tomography was only one among other screening modalities. The complementary screening modalities (ultrasound, computed tomography, magnetic resonance imaging [MRI]), however, have not been used in a systematic fashion. In the screening program, a total of 526 cancers were discovered (1.35%), which is reported to be higher than the detection rate in conventional cancer screening in Japan (0.1%). The most common cancers detected were thyroid cancer, lung cancer, colon cancer, prostate cancer, and breast cancer. The authors claimed that these cancers were detected at early stages. A total of 358 tumors were detected by PET, and 168 tumors were negative on PET. Some lesions were missed systematically, and these were cancers of prostate (negative in 82% of the cases), renal (which were negative in 72%), and bladder (which were not detected at all). In these cases, the negative results are well explained by the low metabolic activity of the primaries or by misinterpretation of the primary as urine because of renal excretion of FDG. However, there were also negative cases in tumors that are generally regarded to have high detection rates by PET (~ 10% of colon cancer and as many as 29% of lung cancer). The reason for this was small lesion size or low metabolic activity, which may be present in some histological subtypes of cancers, which are generally FDG-positive. Long-term follow-up data on the outcome of the screened population are still missing.

The Japanese screening program is actively participating in quality assurance procedures. Guidelines have been published that define the criteria of facilities appropriate for offering cancer screening. These guidelines address questions of data acquisition and analysis, and also refer to useful tests that are to be applied in addition to FDG-PET. The lower incidence of cancer as reported by the Japanese program in comparison to the data of Agress and Cooper (2004) and Choi et al. (2005) may be owing to patient selection criteria. Whereas these authors have studied patients who already had cancer, healthy volunteers were screened in the Japanese program. Thus, by predisposition the rate of second cancers is expected to be higher in patients who are referred for the evaluation of a primary. In the following section we discuss some prerequisites to screening programs as defined by the World Health Organization (WHO) and some

aspects that must be taken care of when evaluating the quality of a screening program.

Considerations on Screening Programs

The following considerations on screening programs are published on the WHO homepage (www.who.int/cancer/detection/en/):

A screening test aims to be sure that as few as possible with the disease get through undetected (high sensitivity) and as few as possible without the disease are subject to further diagnostic tests (high specificity). Given high sensitivity and specificity, the likelihood that a positive screening test will give a correct result (positive predictive value) strongly depends on the prevalence of the disease within the population. If the prevalence of the disease is very low, even the best screening test will not be an effective public health programme.

Consequently, if a diagnostic test has a sensitivity and specificity of 90% but the prevalence of a disease is as low as 1%, only 1 out of 10 positive test results is cancer related. The mathematical basis for these considerations was introduced by Bayes (1763) more than 200 years ago. It is not proven that FDG-PET has 90% accuracy in patients with asymptomatic neoplastic disease. Even in malignancies in which FDG-PET has a well-defined role, the results in early detection of disease are less favorable. For example, in colon cancer, FDG-PET regularly fails to detect small precancerous or malignant lesions, and specificity is limited by the presence of inflammatory or unspecific bowel disease (Drenth et al., 2001; Yausda et al., 2001). It is, however, the resection of small lesions that may decrease cancer-related mortality.

The definition of populations at risk is a major issue in developing screening programs. Furthermore, it has to be proved not only that cancer is diagnosed at earlier stages but that specific cancer mortality of the screened population decreases. Outcome-based prospective randomized trials on FDG-PET as a screening tool are not yet available. The potential sources of bias of such studies have been discussed by Patz et al. (2000). Three different effects have been described, which may lead to a positive evaluation of ineffective cancer screening programs.

The first effect has been termed *lead time bias*. One may assume that the time from the onset of a tumor to the death of the patient is constant and cannot be increased by therapy. If the tumor is detected earlier by screening methods, the time from tumor detection to death is longer, but the time from onset of tumor growth to death is unchanged. In a simple statistic analysis, however, survival as calculated from detection of tumor to death of the patient will be longer, if the tumor is detected earlier. But this achievement in survival time is not related to a reduction of cancer specific mortality.

The second effect has been termed *length time bias*. Tumors with a high proliferation rate are more likely to be missed by screening procedures and will more often be detected by symptoms. If the outcome of these tumors is compared to that of incidental tumors detected by screening, significant differences are expected. Again, tumors detected by screening will show better results, but overall mortality will not be reduced.

The third effect is called *overdiagnosis bias*. Some tumors may not affect survival. It has been shown for thyroid cancer, for example., that the proportion of incidentally discovered small carcinomas on postmortem studies is much higher than the incidence of the tumor. These may be indolent cancers that do not affect survival. Screening methods may result in an overdiagnosis of such cancers with an apparent increase of incidence but without any effect on mortality. Since high-glucose metabolic activity of tumors has been shown to be an indicator of poor prognosis, this type of bias may be less likely to occur. These general considerations suggest caution in accepting diagnosis of tumors at earlier stages as the only criterion of success of a screening program. Only if long-term cancer mortality is decreased, can the screening program be regarded as valid.

As we have discussed, the data available on cancer screening with PET suggest positive findings in ~ 1–2% of the screened population (Yasuda *et al.*, 2000; Chen *et al.*, 2003; Shen *et al.*, 2003; Ide and Suzuki, 2005). Thus, on an average, 50 to 100 FDG-PET scans have to be performed to identify one patient with malignancy. Furthermore, in some of these studies the authors do not state whether the patients had participated in other screening programs (e.g., patient history, fecal occult blood testing, colonoscopy and mammography, low-dose CT of the chest). So it is unclear whether or not routine diagnostic techniques would have missed the FDG-PET positive lesions. Clinical follow-up for 6–10 months is the gold standard for negative findings in these studies. This period is too short to exclude the presence of small malignant lesions (e.g., in the colon). Inclusion criteria have not been reported in detail in those studies, and results may be more favorable in well-defined populations. The definition of subjects at risk using a simple questionnaire has been studied in much detail in cardiovascular disease (e.g., Assmann *et al.*, 2002). It is, however, more difficult to identify high-risk patients when assessing cancer in general.

Cost-effectiveness is another issue in this context, which will not be addressed in detail because there are no substantial data for FDG-PET in the setting of screening. Using the data of the current screening programs, however, with detection rates of 1.5% and estimated costs of 1200 US$ for a single PET scan, the costs for detecting a single cancer are expected to be as high as 80,000 US$. Data on secondary expenses caused by false-positive findings are also not available.

¹⁸F-fluorodeoxyglucose-Positron Emission Tomography

Negative Tumors

As FDG-PET holds promise for whole-body tumor screening, the subjects who have been screened may have confidence in negative test results, which is not substantiated. It is well known that some common tumors may be negative in FDG-PET. Patients might regard further screening tests as unnecessary, and so early-stage malignancy might be missed. Prostate cancer in male subjects and breast cancer in women are examples of such tumors. A FDG-PET scan is not a substitute for the respective screening programs of these cancers, although FDG-PET may provide valuable information in restaging breast cancer. The data of the Japanese screening program, as already discussed, underline the necessity of interpreting negative PET results with caution. The combination of PET with established screening procedures should therefore be recommended. Furthermore, the definition of guidelines is mandatory, which address eligibility for participation in the program, criteria for image interpretation, and mechanisms for the treatment of abnormalities.

Radiation Protection

Although radiation exposure by FDG-PET is regarded as negligible in patients with known malignancies or high probability of malignancy, the use of FDG-PET in the general population is a different matter. The effective dose of FDG-PET can be estimated at about 10 mSv, when 370 MBq are injected (Deloir *et al.*, 1998). According to the International Commission on Radiation Protection (ICRP-report 60, 1991), the risk of radiation-induced cancer is as high as 5/10,000 for an effective dose of 10 mSv. Although these figures are estimations only, they are widely used in radiation protection. Applying these data to FDG-PET screening, we would expect one radiation-induced cancer with 2000 FDG-PET studies. If only 1 to 2% of these studies are positive, this means one additional cancer for the detection of cancer in 20–40 patients. This risk can only be accepted if a benefit for the majority of patients has clearly been demonstrated. Regulatory restrictions in many countries prohibit the use of ionizing radiation without a "legitimate indication." The use of FDG-PET in normal examinees may only be legally approved after a formal application for a scientific project. Thus, performing screening tests with exposure to ionizing radiation is a matter that needs legal clarification and may not be possible in many countries.

In conclusion, the authors are convinced that it is too early to recommend FDG-PET as a screening tool in healthy subjects. It is undoubted that FDG-PET has already incidentally detected cancer in many patients. It seems possible that by early effective treatment, patients have survived longer owing to early PET diagnosis. General considerations on sensitivity and specificity, on cost-effectiveness, and on radiation protection support skepticism about the future introduction of FDG-PET in this setting.

References

Agress, H., Jr., and Cooper, B.Z. 2004. Detection of clinically unexpected malignant and premalignant tumors with whole-body FDG PET: histopathologic comparison. *Radiology 230*:417–422.

Assmann, G., Cullen, P., and Schulte, H. 2002. Simple scoring scheme for calculating the risk of acute coronary events based on the 10-year follow-up of the prospective cardiovascular Munster (PROCAM) study. *Circulation 105*:310–315.

Bayes, T. 1763. An essay toward solving a problem in the doctrine of chances. *Philosophical Transactions of the Royal Society of London 53*:370–418.

Chen, Y.K., Kao, C.H., Liao, A.C., Shen, Y.Y., and Su, C.T. 2003. Colorectal cancer screening in asymptomatic adults: the role of FDG PET scan. *Anticancer Res. 23*:4357–4361.

Choi, J.Y., Lee, K.S., Kwon, O.J., Shim, Y.M., Baek, C.H., Park, K., Lee, K.H., and Kim, B.T. 2005. Improved detection of second primary cancer using integrated [^{18}F] fluorodeoxyglucose positron emission tomography and computed tomography for initial tumor staging. *J. Clin. Oncol. 23*:7654–7659.

Deloar, H.M., Fujiwara, T., Shidahara, M., Nakamura, T., Watabe, H., Narita, Y., Itoh, M., Miyake, M., and Watanuki, S. 1998. Estimation of absorbed dose for 2-[F-18]fluoro-2-deoxy-D-glucose using whole-body positron emission tomography and magnetic resonance imaging. *Eur. J. Nucl. Med. 25*:565–574.

Drenth, J.P., Nagengast, F.M., and Oyen, W.J. 2001. Evaluation of (pre-)malignant colonic abnormalities: endoscopic validation of FDG-PET findings. *Eur. J. Nucl. Med. 28*:1766–1769.

Ide, M., and Suzuki, Y. 2005. Is whole-body FDG-PET valuable for health screening? For. *Eur. J. Nucl. Med. Mol. Imaging. 32*:339–341.

International Commission on Radiological Protection, Report No. 60. 1991. Recommendations of the International Commission on Radiological Protection.

Patz, E.F., Jr., Goodman, P.C., and Bepler, G. 2000. Screening for lung cancer. *N. Engl. J. Med. 343*:1627–1633.

Shen, Y.Y., Su, C.T., Chen, G.J., Chen, Y.K., Liao, A.C., and Tsai, F.S. 2003. The value of ^{18}F-fluorodeoxyglucose positron emission tomography with the additional help of tumor markers in cancer screening. *Neoplasma. 50*:217–221.

World Health Organization, "Screening and Early Detection of Cancer." http://www.who.int/cancer/detection/en/.

Yasuda, S., Fujii, H., Nakahara, T., Nishiumi, N., Takahashi, W., Ide, M., and Shohtsu, A. 2001. 18F-FDG PET detection of colonic adenomas. *J. Nucl. Med. 42*:989–992.

Yasuda, S., Ide, M., Fujii, H., Nakahara, T., Mochizuki, Y., Takahashi, W., and Shohtsu, A. 2000. Application of positron emission tomography imaging to cancer screening. *Br. J. Cancer 83*:1607–1611.

World Health Organization, "Screening and Early Detection of Cancer." http://www.who.int/cancer/detection/en/.

7

Staging Solid Tumors with ^{18}F-fluorodeoxyglucose Positron Emission Tomography/ Computed Tomography

Gerald Antoch and Andreas Bockisch

Introduction

Radiological imaging procedures for tumor staging are based on morphological and functional data of viable tumor tissue. Morphological data acquired with ultrasound, computed tomography (CT), or magnetic resonance imaging (MRI) provide only limited information on tumor metabolism. This lack of functional data hampers differentiation of viable tumor tissue from adjacent structures. Functional imaging provided by positron emission tomography (PET) visualizes tumor metabolism and improves staging results over morphological imaging procedures for a variety of malignant diseases (Gambhir et al., 2001). However, a lack of anatomical landmarks on functional image sets frequently renders difficult accurate anatomical localization of an area of increased tracer uptake. Thus, it is obvious that in many cases both types of imaging complement each other.

Theoretically, separately acquired morphological and functional data sets can be fused. While software-based image fusion provides accurately co-registered images for brain studies, image co-registration inaccuracies due to respiratory motion and different patient positioning on the examination tables have limited this approach in other parts of the body (Townsend, 2001). Dual-modality PET/CT provides functional and morphological data sets in a single imaging session, which may be fused afterward using a special fusion tool (Beyer et al., 2000). Co-registration difficulties are thus minimized. Studies evaluating this imaging technique have demonstrated an increased diagnostic accuracy to that obtained from PET alone or CT alone when imaging different oncological diseases (Antoch et al., 2004b, 2003; Bar-Shalom et al., 2003; Halpern et al., 2005). In addition, fused PET/CT data sets may be of benefit compared to CT and PET viewed side by side.

This review addresses the diagnostic accuracy of [^{18}F]-2-Fluoro-2-deoxy-D-glucose (FDG)-PET/CT when assessing the tumor stage of different oncological diseases. The advantages and limitations of the combined imaging approach compared to separate PET and CT are also examined.

PET/CT Imaging Protocols for Staging Solid Tumors

Imaging protocols for PET/CT are controversial worldwide. Some authors use "low-dose" CT without intravenous contrast agents for the combined PET/CT examination, whereas others recommend a diagnostic CT with regular-dose acquisition and the use of intravenous contrast. It is important to emphasize that both PET/CT imaging protocols have been reported to be of higher diagnostic accuracy than staging alone with PET or CT (Antoch *et al.*, 2004b; Bar-Shalom *et al.*, 2003; Lardinois *et al.*, 2003). In our opinion, the question as to which PET/CT protocol should be used needs to be based on the clinical question rather than on a static acquisition scheme. Patients undergoing initial tumor staging after diagnosis of malignant disease will profit from a diagnostic CT protocol. In this setting a state-of-the-art CT will not only aid accurate localization of an area of increased tracer uptake but may also increase the accuracy of the combined imaging approach in case of PET-negative lesions (Antoch *et al.*, 2004a). Based on similar attenuation, lesions within internal organs may be missed on nonenhanced CT if they are FDG-PET negative at the same time. Setty *et al.* (2005) have shown that integration of contrast-enhanced CT into the PET/CT staging concept increases the sensitivity for detecting metastatic lesions within the liver compared to non-contrast-enhanced PET/CT. For optimal liver assessment with PET/CT, a contrast-enhanced CT component should be included. In this case the combined imaging procedure with the integration of fully diagnostic CT data obviates the need for additional imaging, thus offering a "one-stop-shopping" diagnostic work-up of malignant diseases in a single session.

There are indications, however, in which contrast enhancement does not seem to be required. In patients undergoing radiation therapy or chemotherapy after initial tumor staging, a nonenhanced low-dose CT may be sufficient for assessment of therapy response, as tumor response to treatment may be best monitored based on the functional data. The same applies to patients undergoing interventional tumor therapy (Fig. 14). A decrease in FDG uptake has been found to recede morphological response in patients undergoing therapy (Bradley *et al.*, 2004), whereas increased tracer utilization in patients undergoing therapy indicates the lack of therapy response. However, patients undergoing restaging after successful therapy may profit from full-dose CT with contrast agent application. In these cases, contrast enhancement and full-dose acquisition will aid localization of potentially newly developed metastases, local recurrences, or PET-negative lesions.

When deciding on a specific imaging protocol, it is necessary to consider radiation exposure. While radiation exposure from FDG-PET amounts to ~ 7 mSv, the radiation dose from CT may vary between ~ 3mSv (low-dose whole-body scan) and ~ 20 mSv (full-dose whole-body scan) (Brix *et al.*, 2005). Thus, radiation exposure has to be considered an issue in younger patients with potentially curable malignant diseases, and has to be taken into account when deciding on a specific protocol in this patient population.

Staging Solid Tumors with FDG-PET/CT

FDG-PET/CT has been found to be more accurate than tumor staging with CT alone or PET alone as well as staging with CT and PET viewed side by side (CT + PET) (Antoch *et al.*, 2004b; Bar-Shalom *et al.*, 2003). This higher

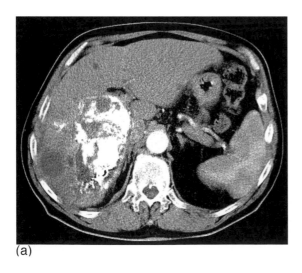

(a)

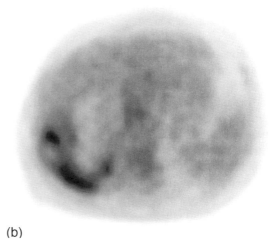

(b)

Figure 14 69-year-old male with hepatocellular carcinoma after chemoembolization. On contrast-enhanced CT (a), chemoembolization-induced tissue necrosis cannot be differentiated from potential residual tumor. FDG-PET (b) clearly

(Continued)

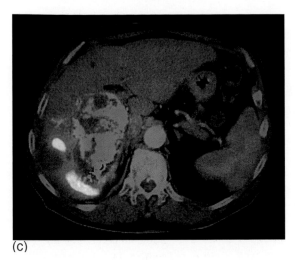

(c)

Figure 14 *continued* demonstrates areas of increased glucose utilization due to residual tumor, which can be accurately localized with PET/CT(c).

accuracy has an impact on patient management in 14–19% of patients compared to CT alone or PET alone and in 6% of patients compared to CT + PET.

T-stage

Although diagnosis of a primary lesion may be performed reliably with morphological imaging modalities, morphology alone has limitations when assessing local tumor size and a potential infiltration of the primary tumor into adjacent organs. As an example, differentiation of viable tumor tissue from adjacent pulmonary atelectasis may be difficult with CT alone (Fig. 15). This limits the accuracy of morphological imaging modalities when defining the T-stage of a malignant tumor. Positron emission tomography imaging alone, on the other hand, has been shown to be of only minor value when assessing the T-stage of different malignant tumors (Gambhir *et al.*, 2001). The limited anatomical information on FDG-PET frequently renders assessment of the potential infiltration of a tumor into adjacent organs difficult. In addition, a threshold-based image display of PET is required for characterization of the tumor size, but thresholds differ not only between tumor entities but also between different studies on a similar tumor entity. Obviously, limitations associated with a threshold-based image analysis will not be solved by FDG-PET/CT imaging that relies on the same threshold-based image display when assessing the PET component. However, fusion of function and morphology improves differentiation of viable tumor tissue from its surroundings compared to morphology alone and improves lesion localization compared to PET alone. Based on these theoretical advantages, FDG-PET/CT has

been evaluated for assessment of the T-stage of different malignant diseases.

In our population of 77 patients with different malignant diseases, FDG-PET/CT was significantly more accurate when determining the T-stage than side-by-side CT + PET, PET alone, or CT alone (Antoch *et al.*, 2004b). In this study the T-stage was accurately determined in 82% of patients with dual-modality PET/CT, in 71% of patients with CT + PET, in 64% of patients with PET alone, and in 66% of patients with CT alone. The advantage of fused PET/CT over side-by-side CT + PET was based on detection of the primary tumor in two patients, more precise evaluation of infiltration of adjacent organs in five patients, and more accurate assessment of tumor cell viability in two patients.

When discussing the accuracy of FDG-PET/CT for assessment of the T-stage in head and neck tumors, published data are rare. Branstetter *et al.* (2005) reported a higher confidence when characterizing a lesion as malignant or benign with PET/CT compared to CT alone or PET alone. In addition, a receiver operating characteristic (ROC) analysis of their data on 64 patients demonstrated PET/CT to be significantly superior to both imaging modalities alone for detection of malignancy. The sensitivity, specificity, and accuracy of lesion detection with FDG-PET/CT were 98%, 92%, and 94%, respectively. However, assessment of the T-stage was not evaluated in this study. Similarly, Syed *et al.* (2005) reported a significant increase in interobserver agreement and confidence in disease localization in head and neck tumors when using FDG-PET/CT compared to PET alone. Their study of 24 patients did not include any separate evaluation of T-stage assessment, nodal disease, or distant metastases. Unpublished data from our institution evaluating 51 patients with head and neck tumors demonstrate FDG-PET/CT to be significantly more accurate than CT alone and side-by-side image evaluation. PET/CT was able to accurately assess the T-stage in 82% of patients in this study.

Additional data are available on non-small cell lung cancer (NSCLC). Halpern *et al.* (2005) found that FDG-PET/CT accurately determined the T-stage of NSCLC in 97% of patients as compared to only 67% accurately staged patients with PET alone. The difference between PET alone and PET/CT was statistically significant in this series of 30 patients. In addition, Halpern *et al.* (2005) included software-based image fusion of PET and CT data sets in their comparison and concluded that the software-based approach failed in a substantial number of patients due to image misregistration. In cases in which image fusion was successful, staging results for NSCLC cancer were comparable to inline PET/CT. In our patient population of 27 patients (Antoch *et al.*, 2003), FDG-PET/CT was more accurate than both imaging modalities alone when assessing the T-stage. However, this difference was not statistically significant. Shim *et al.* (2005) found that FDG-PET/CT accurately assessed the T-stage of NSCLC in

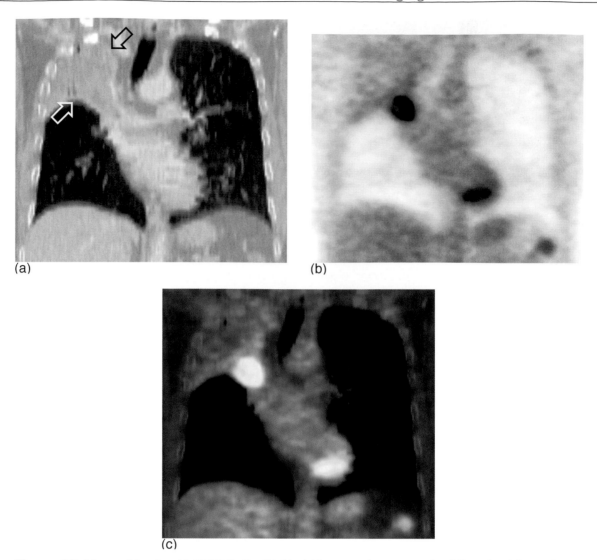

Figure 15 56-year-old male with NSCLC. On CT (a) viable tumor tissue cannot be differentiated from adjacent atelectasis (arrow). On PET alone (b), the tumor can be clearly identified, but due to the limited anatomical information, a potential infiltration into the mediastinum cannot be assessed. PET/CT (c) offers fusion of function and morphology, thus improving differentiation of viable tumor from adjacent atelectasis. Furthermore, a potential infiltration of adjacent organs by the tumor can be assessed more accurately.

86% of 106 patients as compared to only 79% with CT alone. Lardinois *et al.* (2003) reported a significantly more accurate assessment of the tumor, node, metastases (TNM) stage in their patient population of 49 NSCLC patients with PET/CT as compared to CT alone or PET alone. However, the T-stage was not assessed separately in this study.

To our knowledge, no further data on T-staging with PET/CT have been published for any other solid tumor. Thus, summarizing the available literature, we see that only few data exist on the accuracy of PET/CT when assessing the T-stage of solid tumors. The data referenced above indicate a more accurate T-staging with FDG-PET/CT in PET-positive lesions than with either imaging modality alone. However, further studies are required to

finally judge the value of combined PET/CT for assessment of the T-stage.

N-stage

Characterization of the N-status on CT images is based on lymph node size. In most areas of the body, nodes exceeding 10 mm in the smallest diameter are suspected of harboring disease (Glazer *et al.*, 1985). Limitations of this size-based node characterization system are well documented: up to 21% of nodes smaller than 10 mm are malignant, while 40% of those larger than 10 mm are benign (Deslauriers and Gregoire, 2000; Staples *et al.*, 1988) (Fig. 16). Positron emission tomography has been shown to

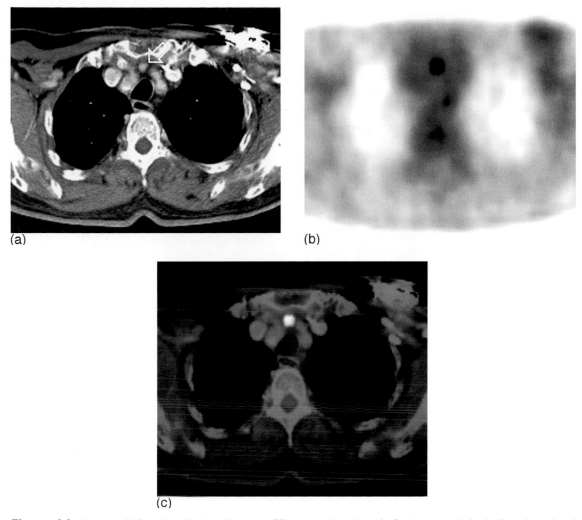

Figure 16 84-year-old female with thyroid cancer. CT shows a lymph node that is not pathologically enlarged and thus was not expected to harbor malignant disease (arrow in a). Focal FDG uptake can be detected on PET alone indicating a metastasis (b). When fusing CT and PET (c), the area of increased glucose metabolism is accurately co-registered with the lymph node. A lymph node metastasis was verified by further clinical follow-up.

be substantially more sensitive and specific than the morphological size-based evaluation (Gambhir *et al.*, 2001). Thus, integration of functional data into the node-staging concept increases the diagnostic accuracy. However, with PET alone, an area of increased tracer uptake may be difficult to localize within the body.

Combined PET/CT imaging has been found to increase the diagnostic accuracy of N-staging compared to CT or PET alone. Most data are available for NSCLC. Lardinois *et al.* (2003) reported a significantly higher diagnostic accuracy with FDG-PET/CT as compared to FDG-PET alone in 49 patients with NSCLC. The advantage of localizing an area of increased tracer uptake to a specific lymph node increased the accuracy over PET alone by improving differentiation of N1 from N2 disease. Furthermore, dif-

ferentiation of malignant nodal uptake from potentially benign lesions, such as esophageal uptake, was more easily facilitated. For FDG-PET/CT, Shim *et al.* (2005) found a sensitivity of 85%, a specificity of 84%, and an accuracy of 84% when staging N-disease in their comparison of PET/CT and CT in 106 patients. The authors conclude that FDG-PET/CT is significantly better for the staging of lung cancer than CT imaging alone. Differences between PET alone and PET/CT were not evaluated in this series. Our data of 27 patients (Antoch *et al.*, 2003) support these results with a significantly higher sensitivity, specificity, and accuracy of FDG-PET/CT as compared to CT alone. The differences between PET/CT and PET alone were not statistically significant in this study. Halpern *et al.* (2005) did not separately evaluate N-staging, but they reported a significantly higher accuracy of 83% with PET/CT as compared to only 67% with PET alone when assessing the TNM-stage in 30 patients with NSCLC. Thus, accurate anatomical correlation of an area of focally increased FDG uptake increases the diagnostic accuracy over PET alone.

Our data on 260 patients with different solid tumors (Antoch *et al.*, 2004b) demonstrated a significantly higher accuracy when determining the N-stage with FDG-PET/CT as compared to CT alone, PET alone, and CT + PET viewed side by side (Table 3). While advantages over CT alone and PET alone had to be expected to some degree, surprisingly PET/CT was also found to be superior to side-by-side CT + PET. More accurate staging results with fused PET/CT were based on localization of areas of FDG uptake to a specific lymph node that had previously not been determined to be malignant on CT + PET. Furthermore, PET/CT was able to co-register areas of FDG uptake that were considered malignant on CT + PET with organs that demonstrate physiological FDG uptake (e.g., esophagus). This led to a decrease in the number of false-positive findings.

A well-known limitation of FDG-PET imaging is false-positive findings in patients with inflammatory disease. Inflammation is characterized by increased glucose metabolism, which leads to elevated FDG uptake in these inflammatory lymph nodes. Integration of morphological data can only partially compensate this limitation. Reported positive predictive values of 80–90% for FDG-PET/CT in different tumors represent this limitation.

Summarizing the currently available literature, most studies evaluating tumor types other than NSCLC report on lesion detection, observer confidence, or interobserver variability. Initial studies evaluating the accuracy of PET/CT to assess the N-stage of a malignant disease are available. However, further studies are required to finally judge the value of FDG-PET/CT when staging a tumor for nodal metastatic spread. Obviously, the advantages of FDG-PET/CT over PET alone and CT alone are only to be expected in FDG-PET-positive tumors. In the case of a non-FDG-avid lesion, tracers other than FDG may be an option.

Otherwise, contrast-enhanced CT alone or other imaging modalities such as MRI should be employed.

M-stage

FDG-PET/CT has been found to be more accurate in staging a tumor for distant metastases compared to either imaging modality alone. Strunk *et al.* (2005) have evaluated 29 patients with colorectal cancer using FDG-PET/CT. The authors report a higher diagnostic confidence when classifying lesions as malignant or benign with higher accuracy of PET/CT as compared to PET or CT alone for M-staging. Selzner *et al.* (2004) compared contrast-enhanced CT with FDG-PET/CT for detection of liver metastases in colorectal cancer patients. They report comparable sensitivities for detection of hepatic metastases with CT and PET/CT (95% and 91%, respectively), but found PET/CT to be superior when detecting intrahepatic tumor recurrence in patients after resection of a hepatic metastasis. The diagnosis of recurrent disease in patients after operations and interventional procedures may be limited, with contrast-enhanced CT based on limitations when differentiating viable tumor tissue from scar tissue. In these cases, additional functional information will aid the correct diagnosis. A limitation of PET/CT when assessing the liver for metastatic spread may be related to PET image acquisition in shallow breathing. Small liver lesions may not be visible on PET due to a breathing-induced smearing of FDG uptake (Antoch *et al.*, 2004a) (Fig. 17). As stated earlier, Setty *et al.* (2005) have shown that integration of contrast-enhanced CT into the PET/CT staging concept increases the sensitivity for detecting metastatic lesions within the liver compared to noncontrast-enhanced PET/CT. Patients in whom small lesions are detected on CT but that are FDG-PET negative should be followed up closely for differentiation of a benign from a malignant lesion etiology. Another option to differentiate malignancy from a benign lesion may be an additional MRI of the liver.

Not only in the liver, but also in other regions of the body, focal FDG uptake may be caused by factors other than malignancy, even in patients with a known malignant tumor.

Table 3 **Sensitivities, Specificities, Positive Predictive Values (PPVs), Negative Predictive Values (NPVs), and Accuracies of the Different Imaging Procedures When Assessing the N-stage in 260 Patients with Different Malignant Diseases (Antoch *et al.*, 2004b)**

	Sensitivity	Specificity	PPV	NPV	Accuracy
PET/CT	92%	93%	88%	94%	92%
PET + CT	88%	89%	83%	92%	88%
PET	85%	88%	82%	90%	87%
CT	64%	83%	70%	79%	76%

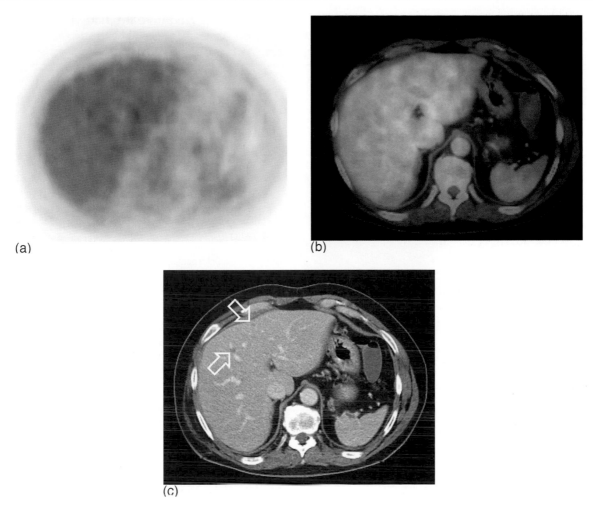

Figure 17 50-year-old male with NSCLC. PET (a) and PET/CT (b) demonstrate homogeneous tracer distribution in the liver without signs of metastases on functional imaging. The diagnosis of small liver metastases was based on the contrast-enhanced CT component of the PET/CT (c), which demonstrated small liver lesions in segment 4a later confirmed to be metastases on clinical follow-up.

Lardinois *et al.* (2005) have examined 350 patients with NSCLC for extrapulmonary lesions using FDG-PET/CT; 46% of all extrapulmonary lesions detected were not related to lung cancer and represented a wide variety of etiologies ranging from second malignancies to benign lesions. In fact, most lesions detected extrapulmonarily were found to be benign. The authors conclude that extrapulmonary lesions should be analyzed critically in patients with NSCLC to ensure accurate tumor staging.

Although sensitivities and specificities for detection of M-disease were higher for FDG-PET/CT than for PET or CT alone in our patient population (Table 4), we did not detect a statistically significant difference between PET/CT and side-by-side CT + PET. Both PET/CT and CT + PET were similarly accurate, with an accuracy of 95% and 94%, respectively. Thus, accurate image fusion

does not seem to be mandatory for assessing the M-stage. However, combining morphology and function (either fused or read side by side) was found to be significantly more accurate than either imaging modality alone. Thus, M-staging should not be based on function or morphology, but rather on a combination of the two imaging procedures.

In conclusion, FDG-PET/CT has been found to be more accurate in TNM-staging of different malignant diseases as compared to either imaging modality alone. While data on separate T-stage assessment are rare, the advantages have been documented when comparing PET/CT with separate PET or CT as well as side-by-side CT + PET for lesion detection, assessment of local nodal disease, and distant metastases. It is important to note that the advantages of FDG-PET/CT can only be expected in FDG-PET positive

Table 4 Sensitivities, Specificities, Positive Predictive Values (PPV), Negative Predictive Values (NPVs), and Accuracies of the Different Imaging Procedures When Assessing the M-stage in 260 Patients with Different Malignant Diseases (Antoch *et al.*, 2004b)

	Sensitivity	Specificity	PPV	NPV	Accuracy
PET/CT	94%	97%	97%	94%	95%
PET + CT	92%	96%	96%	93%	94%
PET	78%	99%	99%	83%	89%
CT	82%	95%	94%	85%	88%

tumors. In FDG-PET negative lesions, alternative radioactive tracers for PET/CT or other imaging modalities may be preferable to FDG-PET/CT.

References

Antoch, G., Freudenberg, L.S., Beyer, T., Bockisch, A., and Debatin, J.F. 2004a. To enhance or not to enhance? 18F-FDG and CT contrast agents in dual-modality 18F-FDG PET/CT. *J. Nucl. Med. 45, Suppl. 1*:56S–65S.

Antoch, G., Saoudi, N., Kuehl, H., Dahmen, G., Mueller, S.P., Beyer, T., Bockisch, A., Debatin, J.F., and Freudenberg, L.S. 2004b. Accuracy of whole-body dual-modality fluorine-18-2-fluoro-2-deoxy-D-glucose positron emission tomography and computed tomography (FDG-PET/CT) for tumor staging in solid tumors: comparison with CT and PET. *J. Clin. Oncol. 22*:4357–4368.

Antoch, G., Stattaus, J., Nemat, A.T., Marnitz, S., Beyer, T., Kuehl, H., Bockisch, A., Debatin, J.F., and Freudenberg, L.S. 2003. Non-small cell lung cancer: dual-modality PET/CT in preoperative staging. *Radiology 229*:526–533.

Bar-Shalom, R., Yefremov, N., Guralnik, L., Gaitini, D., Frenkel, A., Kuten, A., Altman, H., Keidar, Z., and Israel, O. 2003. Clinical performance of PET/CT in evaluation of cancer: additional value for diagnostic imaging and patient management. *J. Nucl. Med. 44*:1200–1209.

Beyer, T., Townsend, D.W., Brun, T., Kinahan, P.E., Charron, M., Roddy, R., Jerin, J., Young, J., Byars, L., and Nutt, R. 2000. A combined PET/CT scanner for clinical oncology. *J. Nucl. Med. 41*:1369–1379.

Bradley, J.D., Perez, C.A., Dehdashti, F., and Siegel, B.A. 2004. Implementing biologic target volumes in radiation treatment planning for non-small cell lung cancer. *J. Nucl. Med. 45, Suppl. 1*:96S–101S.

Branstetter, B.F., Blodgett, T.M., Zimmer, L.A., Snyderman, C.H., Johnson, J.T., Raman, S., and Meltzer, C.C. 2005. Head and neck malignancy: is PET/CT more accurate than PET or CT alone? *Radiology 235*:580–586.

Brix, G., Lechel, U., Glatting, G., Ziegler, S.I., Munzing, W., Muller, S.P., and Beyer, T. 2005. Radiation exposure of patients undergoing whole-body dual-modality 18F-FDG PET/CT examinations. *J. Nucl. Med. 46*:608–613.

Deslauriers, J., and Gregoire, J. 2000. Clinical and surgical staging of non-small cell lung cancer. *Chest 117*:96S–103S.

Gambhir, S.S., Czernin, J., Schwimmer, J., Silverman, D.H., Coleman, R.E., and Phelps, M.E. 2001. A tabulated summary of the FDG PET literature. *J. Nucl. Med. 42*:1S–93S.

Glazer, G.M., Gross, B.H., Quint, L.E., Francis, I.R., Bookstein, F.L., and Orringer, M.B. 1985. Normal mediastinal lymph nodes: number and size according to American Thoracic Society mapping. *Am. J. Roentgenol. 144*:261–265.

Halpern, B.S., Schiepers, C., Weber, W.A., Crawford, T.L., Fueger, B.J., Phelps, M.E., and Czernin, J. 2005. Presurgical staging of non-small cell lung cancer: positron emission tomography, integrated positron emission tomography/CT, and software image fusion. *Chest 128*:2289–2297.

Lardinois, D., Weder, W., Hany, T.F., Kamel, E.M., Korom, S., Seifert, B., von Schulthess, G.K., and Steinert, H.C. 2003. Staging of non-small cell lung cancer with integrated positron emission tomography and computed tomography. *N. Engl. J. Med. 348*:2500–2507.

Lardinois, D., Weder, W., Roudas, M., Von Schulthess, G.K., Tutic, M., Moch, H., Stahel, R.A., and Steinert, H.C. 2005. Etiology of solitary extrapulmonary positron emission tomography and computed tomography findings in patients with lung cancer. *J. Clin. Oncol. 23*:6846–6853.

Selzner, M., Hany, T.F., Wildbrett, P., McCormack, L., Kadry, Z., and Clavien, P.A. 2004. Does the novel PET/CT imaging modality impact on the treatment of patients with metastatic colorectal cancer of the liver? *Ann. Surg. 240*:1027–1034.

Setty, B.N., Blake, M.A., Sahani, D.V., Holalkere, N.S., and Fischman, A.J. 2005. Role of contrast-enhanced CT (CECT) in hybrid PET-CT for liver metastases in patients with colorectal cancer. *Radiology 237, Suppl. RSNA*:452.

Shim, S.S., Lee, K.S., Kim, B.T., Chung, M.J., Lee, E.J., Han, J., Choi, J.Y., Kwon, O.J., Shim, Y.M., and Kim, S. 2005. Non-small cell lung cancer: prospective comparison of integrated FDG-PET/CT and CT alone for preoperative staging. *Radiology 236*:1011–1019.

Staples, C.A., Muller, N.L., Miller, R.R., Evans, K.G., and Nelems, B. 1988. Mediastinal nodes in bronchogenic carcinoma: comparison between CT and mediastinoscopy. *Radiology 167*:367–372.

Strunk, H., Bucerius, J., Jaeger, U., Joe, A., Flacke, S., Reinhardt, M., Hortling, N., and Palmedo, H. 2005. Combined FDG-PET/CT imaging for restaging of colorectal cancer patients: impact of image fusion on staging accuracy. *Rofo. Fortschr. Geb. Rontgenstr. Neuen Bildgeb. Verfahr. 177*:1235–1241.

Syed, R., Bomanji, J.B., Nagabhushan, N., Hughes, S., Kayani, I., Groves, A., Gacinovic, S., Hydes, N., Visvikis, D., Copland, C., and Ell, P.J. 2005. Impact of combined (18)F-FDG PET/CT in head and neck tumours. *Br. J. Cancer 28*:1046–1050.

Townsend, D.W. 2001. A combined PET/CT scanner: the choices. *J. Nucl. Med. 42*:533–534.

8

Laser Doppler Perfusion Imaging: Clinical Diagnosis

E.Y-K Ng, S.C. Fok, and Julie Richardson

Introduction

The measurement of blood perfusion is important for angiogenesis studies, the results of which are useful for the development of cancer treatment. This chapter gives a brief review and discusses the applications of the laser Doppler perfusion imaging (LDPI) technique and its role in clinical diagnosis of cancer. The considerations for the successful applications of this technique along with some of the potential advancement are presented in this chapter. Through this presentation, it is hoped that both medical professionals and biomedical engineers can better understand the technique to advance future applications and improvements of LDPI systems in cancer research.

Gathering and analyzing biological and clinical data are vital processes in the prognosis, diagnosis, treatment, and understanding of many medical problems, including cancer. Doctors rely heavily on clinical tests to gather crucial data for the analysis of many medical problems so that appropriate treatments can be prescribed, while medical research scientists utilize laboratory experiments to collect relevant information so that the underlying mechanisms and the causes and effects of various diseases and treatments can be thoroughly investigated. The results of these processes have not only promoted the understanding of many illnesses

along with their early detections, but also helped develop new drugs and successful treatments of numerous medical problems.

Many tools are available for gathering biological and clinical data. Among these tools, imaging systems have become an indispensable apparatus in hospitals and medical research laboratories. The main advantage of imaging equipment is that it can provide visual representation of the measured properties or phenomena that are often imperceptible to human vision. Imaging systems used in health care include autoradiography, computed tomography (CT), magnetic resonance imaging (MRI), positron emission tomography (PET), electron microscopy, X-ray imaging, ultrasound imaging, thermography, and laser Doppler perfusion imaging (LDPI). Some imaging systems, such as CT and MRI, are often utilized to provide useful information on the structural analysis of bones and organs, whereas others such as LDPI are used mainly to assess mircovascular blood flow (Ng et al., 2003a, b).

The assessment of microvascular perfusion in tissue is important in both research and clinical practice. Many factors can influence microvascular perfusion in tissues; these factors include drugs, diseases, and medical procedures. The effects of these factors can, therefore, be indirectly assessed based on the changes in perfusion patterns. In the last two decades, applications of LDPI systems were explored in the assessment of microvascular anatomies, wound healing, assessment of burn depth, and physiological studies, including smoking effects (Bornmyr and Svensson, 1991). Port wine stains perfusion before and after argon laser treatment (Troilius and Ljunggren, 1996), skin perfusion changes after microdialysis probe insertion, and cutaneous vascular axon reflex is characterized after electrical nerve stimulation (Wardell, 1994).

This chapter discusses the role of LDPI in cancer diagnosis and presents a review of the LDPI technique and equipment along with some factors that can affect the performance. In addition, we give a brief survey of certain LDPI applications to highlight some of the considerations and factors that would facilitate the successful application of the technique in clinical diagnosis. Finally, we examine the role and potential application of LDPI in future cancer diagnosis.

Review of Laser Doppler Perfusion Imaging

Laser Doppler perfusion imaging is based on the principle that laser light undergoes a change in frequency ("Doppler shift") when it encounters a moving object. To assess microvascular perfusion, a low-intensity beam of monochromatic light generated from a laser diode is used to illuminate the exposed tissue under study. Within the stationary tissue, the red blood cells (RBCs) are the main moving particles encountered by the photons from the laser beam. The flowing RBCs will therefore impart a Doppler frequency shift on the incident light beam.

The photons from the laser beam are scattered, reflected, and absorbed in the tissue matrix. The reflected laser beam contains both the reflections from the moving RBCs and the static tissue. However, only the reflected light from the moving RBCs contain Doppler shifts. The average frequency of the Doppler shifts would be proportional to the average velocity of the RBCs, while the average amplitude of the Doppler shifts would be proportional to the mean number of RBC scatters encountered by the incident beam in the tissue volume. To measure the movement of the RBCs, a photo detector is used to retrieve a portion of the reflected light. Since this reflected light contains both shifted and nonshifted photons, it has to be compared with the original incident light wave so that the Doppler shifts can be separated. The separated Doppler shifts can be analyzed using standard signal processing techniques. The velocity distribution will give the Doppler spectrum, which is normally in the range of 30 Hz to ~ 15 kHz. The zero-order moment of the Doppler power spectrum would be related mainly to the concentration of moving RBCs. The first-order moment of the Doppler power spectrum would be related to the flow estimate.

The laser Doppler flowmetry is an early version of the LDPI system, which utilizes the laser Doppler principle in blood measurements. It uses a probe that can be attached to the tissue surface. The probe contains two main fibers: one fiber is used to direct a laser light onto a small region ($\sim < 1mm^3$) of the tissue observation site, and the other fiber is used to collect the reflected light. The instrument assigns arbitrary perfusion units to quantify the multidirectional movement of blood cells flowing through the microvasculature of the tissue. This perfusion unit is defined as the product of the concentration of the moving RBCs and their mean velocity.

The laser Doppler flowmetry has the following drawbacks:

▲ It is used mainly for continuous monitoring of perfusion at a single point.
▲ The probe attachment can give rise to problems in some applications.
▲ The movements of the fiber optics and tissue during measurement can result in artifacts that may make interpretation of the results difficult.

The laser Doppler imager is an improvement of the laser Doppler flowmetry. As single-point measurements may not give a good representation of the underlying tissue perfusion, the imager extends the measurements to cover a bigger surface area. This allows the spatial fluctuations and temporal effects over a wider tissue area to be investigated. To accomplish this, the imager replaces the probe with a scanner to record the perfusion map. This enables rapid measurement without physical contact with the tissue being studied.

Figure 18 shows a typical commercial laser Doppler imager from Moor Instruments (2005). The main components of the imager are the laser, scanner, and the detector unit. The scanner head contains lenses and mirrors mounted on stepper motors. The lenses are used mainly to focus the beam so as to minimize the aberration effects. The motor and moving mirror assembly are used to guide the beam so that a raster scan of the tissue surface area can be generated. The height and orientation of the scanner head can be adjusted. The size of the scanned area is partially dependent on these adjustments. The detector unit should be positioned above the tissue surface. During scanning, different measuring positions will give rise to different distances and incident angles to the detector unit. To avoid coherence pattern interferences that arise from these differences, the lens system can be adjusted such that the number of coherence areas on the detector surface is dependent only on the angle of the beam, which can be determined from the mirror position. With this adjustment, coherence pattern correction can be applied automatically at each position based on the known mirror position to ensure that the perfusion value is independent of the distance between the tissue and the detector.

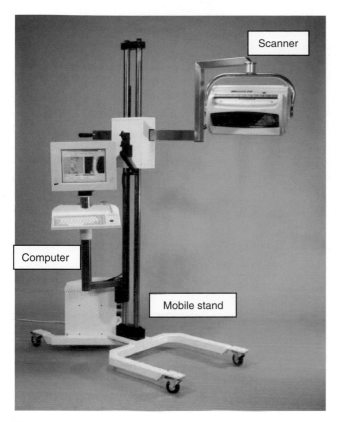

Figure 18 A typical laser Doppler imager (moorLDI2) from Moor Instruments.

Compared with the flowmetry, the imager requires a more dedicated electronics interface and post-signal processing to quantify the perfusion measurement. The simple intensity value of the reflected light can be used to generate a gray-scale photographic image. With advanced electronics and processing, the imager can also produce a two-dimensional, color-coded image of the blood movement. Figure 19 shows a typical two-dimensional color-coded image obtained from the laser Doppler imager (moorLDI burn imager with the near-infrared laser), which in this example is used to investigate the potential healing time (in days) of a burn wound. All burn images were taken with a distance between 30 and 100 cm from scan head to tissue. Over this distance range, differences are not observed in the flux values due to scaling constants in the software.

The blood flow measurements can also be integrated and averaged over the area using advanced electronic devices. The main drawback of the laser Doppler imager is that some tissue sites may not be accessible to a laser beam scanner. These sites include brain tissues, muscles, and colons. On the other hand, the laser Doppler flowmetry system can utilize probes designed specifically for invasive measurement procedures. The penetration depth of the laser light in tissue is influenced by both the tissue properties and the instrumental

factors. Tissue properties that can affect the light penetration include vascular geometry, tissue absorption, and tissue-scattering characteristics. Instrumental factors that can affect the light penetration include the laser wavelength and the configuration of the beam and detector system.

The lasers utilized in LDPI systems are either of a low intensity (< 2 mW) visible red (wavelength ~ 634 nm) or a near-infrared (~ 810 nm) laser light. The depth of penetration with the visible red laser light is typically ~ 1 mm in skin. The depth of penetration can be increased (~ 5 mm) using the near-infrared laser light. However, the penetration of the beam also depends on the skin thickness as well as on the tissue and vessel density. With increasing skin thickness, the lack of penetration could reduce the LDPI output signal in a nonlinear manner. The penetration factor can be important as the sensitivity to changes in flow will decrease in an exponential manner as a function of vascular depth. If the penetration of the beam decreases significantly owing to tissue and vessel density, the signal concentration can saturate. LDPI systems are typically capable of measuring blood flow in veins with capillary diameters of ~ 10 μm, although it is possible to measure blood flow in isolated arterioles and venules up to ~ 500 μm in diameter. However, for larger vessel diameters, the limited depth of penetration can lead to underestimation of the flow.

Laser Doppler perfusion imager systems are very sensitive to superficial blood flow, the height of scanner, or the probe position. Changes in the scanner height can result in changes in the output signals. This is mainly due to the reduction in the reflecting light signal with increasing scanner height. The recommended height for the scanning would depend on the application. When scanning multiple surfaces, it is highly recommended that a few separate scans be taken at different distances from the scanner. For the laser Doppler flowmetry, a slight change in probe position often leads to a substantial change in output signal. Hence, successful operations of the LDPI systems would require careful control of environmental effects and acclimatization of the subject to the monitoring before measurement can be taken. Before interpreting the LDPI results, confounding factors must be considered. Confounding factors are any actions, conditions, or treatments that are known to have adverse effects on the scan interpretation. These include:

▲ High reflection occurring in some areas within the region of interest.
▲ Edges of body, limb, torso, and so on, that could cause signal loss as a result of the curvature.
▲ Insufficient ambient lighting or strong direct sunlight.
▲ Tissue or subject movements.

The LDPI instruments cannot provide real-time blood perfusion in absolute units because the depth of penetration can vary between two different tissues with similar perfusion. Increased depth penetration in tissues would normally

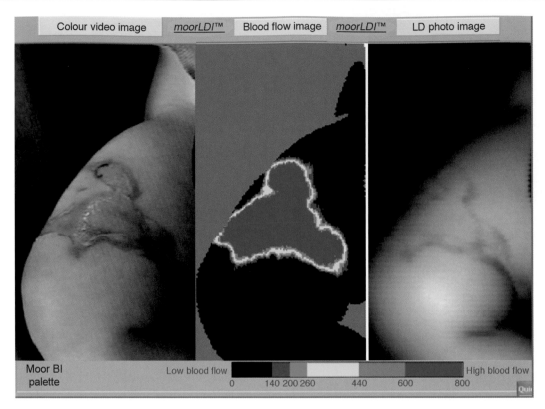

Figure 19 A moorLDI image (center) of a superficial burn wound. Red indicates high blood flow in the wound, signifying a healing potential of less than 14 days. This wound was clinically healed by post-burn day 12.

result in more Doppler-shifted photons and hence greater signal output. The varying depths of penetration also imply that the actual quantity of tissue sampled can vary from patient to patient and could change at different tissue observation sites. This issue can be further complicated by the placement and orientation of the incident beam and the photo detector. Without accurate information on the depth of penetration (and hence the tissue volume), it would not be possible to calibrate the LDPI instrument and express the measured quantity in absolute units. Accurate calibration of the LDPI instrument also requires proper technical and procedural standards, which have yet to be developed (Fullerton *et al.*, 2002). As such, although the outputs from the LDPI systems are proportional to the absolute flow in the tissue sampled, these instruments are best suited for situations requiring relative measurements based on continuous observation. The factors that have to be considered in using LDPI systems are summarized in Table 5.

Some Past and Recent LDPI Applications

Blood perfusion measurements are useful for the clinical diagnosis of medical problems that are known to affect peripheral circulation. There is a lot of reported literature on the applications of LDPI in clinical diagnosis. This section is not meant to be a comprehensive review of the literature on LDPI applications. Based only on a sample of representative works, this section merely aims to highlight some of the considerations and factors that would facilitate the successful applications of the technique in clinical diagnosis. To fully appreciate the potential applications of LDPI in cancer research, it is desirable first to examine some of the past and recent successful applications of LDPI systems.

The LDPI systems are of tremendous value in burn assessment. Clinically, it is easy to distinguish between normal and burned tissues. However, the visual evaluation of the level of burn injuries by experts is more difficult and prone to human error. The technique is less reliable and more subjective. Burn wounds affect the blood perfusion. High blood perfusion within a burn site signifies intact dermal perfusion and intact microvascular networks. For areas of third-degree burn with severely damaged microvascular networks, blood perfusion is dramatically decreased. The wavelength of the laser beam used in burn assessment is normally ~ 633 nm. This wavelength allows detection of superficial blood vessels within a depth of ~ 1–2 mm (Vongsavan and Mathews, 1993). Wavelength of ~ 830 nm

Table 5 Factors Influencing LDPI Performance

Laser Doppler Flowmetry	Laser Doppler Imager
1 Measurements are not provided in absolute units.	Measurements are not provided in absolute units.
2 Movement of the tissue can result in artifacts.	It can tolerate some movement of the tissue.
3 It is suitable for single-point measurement.	It is suitable for measurement over an area.
4 Special probes can be used for invasive measurement procedures.	Laser beam scanner may not be accessible to some tissue sites.
5 Depth of penetration could affect the output.	Depth of penetration could affect the output.
6 High tissue vessel density can have local effects on the output.	High tissue vessel density can have local effects on the output.
7 The result could be affected by the orientation and placement of the probe.	The distance between the scanner and the object is a potential source of error.
8 Control of environmental effects and acclimatization of subject are needed.	Control of environmental effects and acclimatization of subject are needed.

(near-infrared) is less prone to absorption, and backscattering from the skin and could be used to detect deeper dermal and subdermal vessels (Abbot *et al.*, 1996). Pape *et al.* (2001) used a 633 nm wavelength laser beam to assess burn depth and reported an accuracy of 97%, compared with 60–80% for established clinical methods. Vascularization and epithelialization are major factors in the healing of burn wounds. Holland *et al.* (2002) conducted an investigation using the LDPI technique using 57 children patients over a 10-month period with a 90% accuracy. They not only discovered that moderate degrees of movement of the subjects could be compensated by adjustment of the scan resolution, but also found that consistent and reliable prediction of burn wound outcome in children requires careful planning of the measurement processes as well as informed interpretations of the results. Conversant analysis of the results is necessary as large changes in skin blood flow can occur during the first 48 hr of the post-burn period. This may be prolonged due to resuscitation (currently, image interpretation had only been validated mainly for scans obtained between 2- and 5-day post-burn periods). Furthermore, confounding factors such as blisters, creams, tattoos, scars, and drugs (including inotropes or other medication that can affect peripheral blood flow) can mask the measurement results.

The accurate real-time evaluation of burn wounds based on objective measurements has great potential in "burn" patient management. The state-of-the-art treatment of thermal injury, including excision and grafting, is based mainly on clinical examination by experts, the judgment of which could be subjective. Using a blinded trial involving 23 patients with 41 representative burn wounds of indeterminate depth, Jeng *et al.* (2003) concluded that LDPI could be used to determine wounds that do not require surgery. A dedicated burn assessment palette was derived at the recent conclusion of a four-year worldwide multicenter clinical trial using the moorLDI (Spence, 2004). Figure 20 shows this palette, which uses six colors for image interpretation of burn wounds. The colors on the image correspond to potential healing times of the burn (in days). With advanced image recognition techniques, the palette could be incorporated into an automated patient management system providing tremendous benefits as unnecessary surgery could be avoided while appropriate burn wound treatments could be scheduled and adjusted according to recovery.

The potential of LDPI in assisting health-care management can also be found in the following applications:

▲ *Assessment of surgical treatments such as wound healing and in flaps*. Infections and inflammation of the wound would increase the perfusion, and the LDPI could utilize this observation for the management of wound healing. Flaps will exhibit large changes in blood perfusion during the immediate postoperative period. When the patient has settled in a more stable mode, a healthy flap should have a pulsating laser Doppler blood flow waveform due to the vasomotion and physiological responses with the cardiac cycle (Svensson *et al.*, 1992). Measurement of perfusion in flaps during the postoperative stage could provide useful data for patient

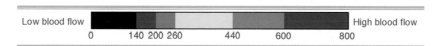

Low blood flow   0   140 200 260   440   600   800   High blood flow

Figure 20 Color palette for burn assessment based on images obtained using the moorLDI2-BI imager taken between day 2 and day 5 post-burn. (Blue: indicates a burn wound with healing potential of greater than 21 days; Green/yellow: indicates a burn wound with healing potential of 14–21 days; Pink/red: indicates a burn wound with healing potential of < 14 days.)

management in reconstructive and aesthetic surgery, a procedure that is common with some breast cancer treatments.

- ▲ *Assessment of medical problems, including Raynaud's syndrome, osteoarthritis (Ng and Tan, 2002, 2003), and epicondylithis.* For example, Ferrell *et al.* (2000) applied the technique with ultrasonography in the assessment of epicondylithis. They confirmed that LDPI can detect hyperemic areas associated with injury and inflammation with greater sensitivity than thermally-based methods. They concluded that the LDPI system would be particularly useful in the patient management as it will be able to yield information to determine the exercise schedule and its effectiveness.

- ▲ *Assessment of drug therapy and other forms of alternative medicine (Ng and Fok, 2004).* Although acupuncture has been gaining global popularity as an alternative form of treatment for many sicknesses, it is still not fully accepted in western countries due to lack of evidence of efficacy (Ng *et al.*, 2005). Litscher *et al.* (2002) studied the effects of blood perfusion following acupuncture. Their results showing the changes in blood perfusion upon acupuncture could be useful in determining of the efficacy of this alternative medicine. The influence of diabeties neuropathy on the peripheral circulation was assessed by Bornmyr *et al.* (1997). They found that diabetic disease can impair the sympathetic nervous system that controls skin blood perfusion. This finding would be useful for the assessment of drug therapy in diabetics as the microangiopathy can be detected and evaluated by perfusion methods. By measuring the blood perfusion, the effects of the drug therapy and the need for amputation could then be quantified.

The applications of LDPI in drug therapy (for example, to monitor the effects of different types of anesthetic and analgesic agents on blood perfusion) opened the door for the application of the technique in evaluating substances influencing the angiogenesis process. Angiogenesis of the cancer cell is a major factor for tumor growth. A tumor evolves when a single cell, within a group of perfectly normal cells, genetically mutates and begins to replicate itself. The reproduction of these cells can be abnormally fast. As these cells grow and mutate, a rare mutation can alter their behavior. With the growth of these abnormal cells, the tumor increases in size. *In situ* cancer occurs if the tumor remains in this state without breaking through any surrounding tissues. In some cases, the tumor can remain in this state indefinitely. In other cases, some of the tumor cells might continue to mutate and take on additional characteristics. This is the start of the angiogenesis process where circadian disturbances can be detected. The genetically altered tumor can become malignant, invade the surrounding tissue, and leak cells into the blood vessels.

Metastasis occurs when invasive tumors shed cancerous cells into the blood vessels and lymph glands. The blood vessels now act as the freeway for these cells to colonize other parts of the body, which can include vital organs such as the heart and lungs. These cells can also adhere to the inside of the blood vessel walls. In some situations, new tumor colonies can form when these cells break through the vessel walls.

The formation of new blood vessels is essential for the development of a tumor. Jain (1988) reported a pathological increase in perfusion due to the formation of new blood vessels during the development of a cancerous tumor growth, which lacks autonomic control. Microcirculation is vital for tissue nutrition. Under normal conditions, capillaries or tiny blood vessels supply nutrients and waste to and from the tissue. Capillaries do not grow rapidly under normal conditions, except in situations such as damaged tissue or menstruation. Even under these situations, autonomic control exists. Under abnormal conditions generated by tumor growth, capillaries can grow rapidly. The tumor cells release angiogenic proteins to cultivate the proliferation of these capillaries and suppress the body's normal mechanism to control angiogenic growth. These capillaries proliferate within the tumor and in turn supply both blood and more nutrients to the tumor cells. The lack of autonomic control in this "fire being fanned with oxygen" phenomenon allows for unbridled growth resulting in "circadian chaos."

The search for angiogenesis inhibitors is extremely important in the development of anticancer drugs. Angiogenesis leads to the formation of new blood vessels and requires endothelial cell proliferation, migration, differentiation, and endothelial cell survival. The importance of vascular endothelial growth factor in the development of anti-angiogenic therapies in tumor growth was highlighted by Glade-Blender *et al.* (2003). The understanding of the vascular endothelial function might offer an opportunity for the early detection and treatment of the disease. In this respect, Bosch-Marce *et al.* (2005) reinforced the notion that LDPI is one of the many techniques that can be useful in these investigations. The LDPI has been applied in research investigation of the angiogenesis process in conjunction with growth factors. The approach is based on using the perfusion parameter to measure the severity of the disease, monitor the progress of the tumor development, and determine the value of the treatment.

In skin cancer, perfusion can be used as an important parameter for discriminating between different types of tumors. Malignant skin tumors have higher perfusion than benign nevus and basal cell carcinomas. Increasing malignancy is reflected by increased tumor circulation. The possibility of differentiating between various types of tumors based on comparative investigation of the alterations in perfusion in skin cancer was discussed in Stucker *et al.* (1999). Using a laser Doppler image with a lateral resolution of 1 mm, they indicated that malignant melanomas and melanocytic nevi produce different perfusion patterns. They recommended the use of a higher resolution imager for further analysis of the perfusion patterns over a wider area, covering both the tumor and the adjacent healthy skin. Use of a high-resolution laser Doppler imager (resolution of 200 μm) in the differential diagnosis of pigmented skin tumors was subsequently reported by Stucker et al. (2002). The results of their study, involving 189 patients (22 with malignant malanomas, 39 with clinically suspicious dysplastic melanocytic nevi, 27 with basal cell carcinomas, and the remaining control group with melanocytic nevi), revealed that all malignant melanomas showed at least 1.8 times higher flow values than healthy skin. They concluded that high-resolution LDPI can be used as an additional automatic screening method for skin cancer (Hoffmann et al., 2000).

Jackson et al. (1994) have applied the LDPI technique for investigating the increase in perfusion in vulvar lesions, and Saravanamuthu et al. (2003) have presented the potential use of the laser Doppler imager in the investigation of vulvar cancer. Their findings indicated that, although the presence of moderate-to-strong expression of vascular endothelial cell growth factor is a stimulant of endothelial cell proliferation, variation in perfusion within microcirculation is complex and requires further understanding of the many contributing factors for increased blood velocity, vasodilation, recruitment of reserve vessels, and neovascularization. Furthermore, they found that calibration of LDPI can be difficult in such investigations because it can be dependent on optical properties influenced by the disease. These factors are not easily accessible for perfusion corrections. To overcome these problems, they suggested further comparative studies with normal tissues, and monitoring the changes in perfusion data under various treatments and clinical conditions.

The lifetime risks of developing breast cancer in Northern Europe and the United States are 1:8 and 1:10, respectively. The number of women affected by breast cancer for every 100,000 is 39 in Singapore, 88 (among Caucasions) in the United States, 21.2 in China, 32.2 in Hong Kong, and 19.9 in India. In Singapore 15.1 out of every 100,000 women die of breast cancer as compared to 11.1 in Hong Kong and 33 in the United States. It is known that the early detection of the tumor ensures better prognosis and higher survival rate. With tumor metabolism, it would be logical to hypothesize that there could be an increase in the breast skin blood flow. Based on this hypothesis, Seifalian et al. (1995) studied the breast skin blood flow of 101 patients (47 healthy subjects, 25 with benign tumors, and 29 with carcinomas). Their findings showed that patients with benign breast cancer have higher breast skin blood flow than symptomatic normal patients, and the highest blood perfusions were recorded in patients with breast cancer. Although the LDPI technique showed great promise for discriminating between normal and abnormal breasts, it is anticipated that deep tumors may not exhibit these differences. They suggested that further investi-

gation is warranted using improved LDPI tools with better penetration to quantitatively assess the breast conditions at different stages of treatments.

The LDPI has great potential in a variety of other medical applications. Although the technique has been applied mainly to study the blood perfusion of tissues near the skin surface, some studies on the applications of the technique do measure blood flow in internal organs. For example, Wardell et al. (2000) used the laser Doppler imager to study the heart tissue blood flow during coronary artery bypass graft surgery, while Hajivassiliou et al. (1998) assessed the blood flow distribution in the human colon during surgery. These studies suggested that LDPI instruments could also be employed in the operating theater (Liu et al., 1997). The brief survey of LDPI applications indicated that the technique can be useful in the assessment of many medical problems including cancer. The usability considerations can be broadly classified into two main categories: use of the instrument and use of images on blood perfusion.

The main factors that can affect the usability of the instrument include the instrument characteristics, performances, and procedures for its usage. Generally, instrument uncertainties can be compensated by appropriate signal conditioning circuitries, electronics, and settings. Intelligent signal processing techniques could be applied to predict and counteract the artifacts caused by tissue movements. Environmental conditions could also be standardized and controlled. However, subject-related and tissue-related factors may not be controllable. Biological and chemical characteristics of blood and tissue composition can vary from subject to subject, while tissue vessel density in a tissue can be varied at different observation sites. For the proper usage of LDPI, it is essential that factors that can be controlled should be consistent, so that their variations will not affect the LDPI results. Proper procedures need to be developed to ensure the consistency, quality, and validity of the measurements (Wardell, 1994). Considerations on the development of these procedures will depend on the application, associated confounding factors, calibration of the instruments, and quality requirements of the images. In clinical cancer research, the technique is currently confined to tumors near the skin surface. The viability of the technique for measuring deep tumors would require further improvements in resolution and penetration (which would include the design improvements to the probes). These improvements would require advancing both the hardware and software in current LDPI systems. Hardware modifications would include better probe design and configurations to access deep tumors, while software improvements would involve advanced digital signal conditioning algorithms to compensate for anticipated tissue movements and other artifacts.

The LDPI technique is still in a state of continuous development, and currently there are ongoing efforts aimed at improving its usage and performance. Some of these developments include:

▲ The development of a fully integrated optoelectronic chip for a multisource and multidetector probe for use with the laser Doppler flowmetry.
▲ Computer simulation of light transport to better understand the mechanism of scattering of photons in tissues.
▲ Development of technical and procedural standards for the calibration of LDPI systems.

The usability of the images, after eliminating all the confounding factors, would depend on understanding the underlying mechanisms affecting the blood perfusion in the application area. For example, in cancer research further comparative studies on the perfusion patterns are needed under various clinical conditions to quantify the effects of angiogenesis and treatments at different stages of tumor development. In this respect, blood perfusion may only hold one of the many keys to help unlock this complex medical problem. The next section examines the role and some of the anticipated future work needed to amalgamate LDPI into an integrated cancer diagnostic system.

Potential Integration of LDPI in Cancer Diagnosis

Cancer is not easy to detect. There are many techniques for cancer detection, and each of these has its unique advantages and disadvantages (Ng and Ng, 2006). The nature of tumors is that there are many different types. Physicians need a variety of methods to obtain the best diagnosis. Imaging techniques capable of producing pictures of areas showing specific signs, either inside the body or near the skin surface, are becoming increasingly necessary for early cancer detection. The warning symptoms are commonly based on detecting changes in physical variables such as blood flow, tissue structure, and heat generation as well as changing electrical characteristics associated with tumor growth (Ng et al., 2006). For example, presently mammography based on X-rays is the most commonly used (as gold standard) detection technique. In future, the detection of breast cancer may utilize other altered tissue characteristics such as blood perfusion, impedance, and temperature.

Studies with LDPI for the early detection of tumors near the skin surfaces have been highlighted in the previous section. To fully realize the potential of this technique as an early detection tool for these types of tumor, more comprehensive studies will have to be conducted. Blood perfusion patterns taken over a period of time for different types of tumors will have to be compared in order to determine the proper procedures and conditions for such measurements. Accuracy, sensitivity, and specificity using receiver operating characteristics (ROC) curve need to be established (Ng, 2006).

In the past few years, there has been increasing interest in developing a comprehensive screening process for breast cancer detection. Fok et al. (2003) discussed the framework for developing such a screening process. The objective of the framework is to facilitate the detection of breast cancer for the mass population. For the approach to be viable and cost-effective, it was conceptualized that future screening processes should involve several stages utilizing different, but complementing, measurement techniques. The first stage would ideally be a preliminary screening stage utilizing noninvasive techniques that are rapid, reliable, and practical for routine clinical use. At present, the first warning of cancer is still heavily dependent on the general practitioner's knowledge, perception, and experience. As a result, many people only realize that they have cancer when they are in the advanced stages of the disease. Since cancer is a consequence disease of complex interactions between genetic and social-environmental factors, it is possible to select high-risk groups based on genetic data and social habits for more frequent observations. Frequent preliminary screenings would offer an opportunity for early detection and intervention before the disease is manifested. There is great potential in utilizing LDPI as an early breast cancer warning system, as its characteristics would be ideal for part of the preliminary screening stage. A color pattern for tumor assessment similar to that shown in Figure 20 for burn assessment would be very useful not only for the doctors, but also for the development of the computer-aided early cancer warning system. Suspected patients could undertake further diagnostic tests in subsequent stages, which would involve more complicated techniques that would be more time consuming, but would allow for detailed measurements of specific target areas. Three-dimensional imaging techniques such as CT and MRI tools that could localize the tumor precisely would be vital in these stages for the surgery planning or radiation therapy.

The LDPI measurements can complement other imaging techniques such as mammography. To add value to the diagnosis, the LDPI images could also be used for the automatic matching of similar cases previously encountered by other doctors so that relevant past treatment programs could be retrieved to help establish the prescription to the current case. This would require research into the development of an intelligent management system for LDPI images. Many methodologies are available for the development of the intelligent information management system. Case-based reasoning is one of these methods (Ng et al., 2003a). It aims to solve a new problem by tailoring solutions of previously solved problems to the current problem. The framework of case-based reasoning involves four distinct stages:

1. Retrieve similar cases to the problem described.
2. Reuse the best solution (or combination of solutions) suggested by the retrieved cases.

3. Revise or adapt the best solution (or combination of solutions) to solve the current problem.

4. Retain the new solution once it has been validated.

Research into the use of LDPI images for the case indexing and matching is another anticipated area needed in an integrated cancer diagnostic system.

Another advantage of a multistage breast cancer screening process is that complementing diagnosis would give a more comprehensive picture of the medical situation based on the detection of not one but many physical variables. To realize the integration of LDPI with other measurement techniques, it is necessary first to understand the strength and weaknesses of these methods. A comprehensive review of the various medical imaging techniques for improved tumor characterization, delineation, and treatment verification can be found in the works of Forssell-Aronsson *et al.* (2002) and in many other chapters in this volume. The measurements from different techniques can be combined using sensor fusion methodologies to provide new features that cannot be discerned by individual methods. At present, correlation of multisensor measurements involving LDPI is still rarely applied. The combination of LDPI results with other measurements should be further investigated as it could offer new discovery. For example, Szili-Torok *et al.* (2002) have discussed the usefulness of combining laser Doppler flowmetry and noninvasive blood pressure monitoring for the continuous observation of cutaneous vascular resistance. In future cancer diagnosis, the fusion of images may be needed. Although the mapping of images from scintigraphy and CT or magnetic resonance imaging (MRI) can be done using external and internal markers, the fusion of images from these sources with blood perfusion and other physiological data is still a relatively unexplored domain.

The fusion of LDPI images with other images and data is not an easy task. It is anticipated that the initial effort would focus on the use of LDPI records to assist doctors in diagnosis and retrieval of records associated with similar past cases encountered. Images from other imaging techniques as well as medical data from screening, diagnostic tests, medical examination, and laboratory investigation could be gradually incorporated within an electronic patient record. This process will aid the data fusion and ultimately link to the concept of the digital hospital, the next step toward realizing the sharing of distributed medical information.

The integration of LDPI images and data could be useful for tumor characterization and investigations into the causes and effects of tumor development. These investigations could involve computer-aided tools, which could further exploit the measured data in the creation of complex mathematical models to simulate the biological phenomenon. An example is the breast model created by Ng and Sudharsan (2001) that could be used for breast cancer investigation. This model is still not fully developed because it is static and did not consider the time-changing characteristics of tumor growth. Laser Doppler assessment of microvascular blood flow could be used in conjunction with other techniques for multimodal monitoring and real-time modeling of the dynamics of the angiogenesis process. An improved dynamic model of the breast could offer better prediction of breast cancer development and treatment.

For some types of cancers the sooner they are detected, the better will be the chance of treating them effectively. Breast and skin cancers are good examples. Port wine stains have been evaluated before and after laser treatment (Troilius and Ljunggren, 1996). The tumor circulation in association with photodynamic therapy of nonmelanoma skin tumors has been followed during the treatment period (Wang *et al.*, 1997; Enejder *et al.*, 2000). A method to distinguish moles from melanomas with a high-resolution scanner (Stucker *et al.*, 1999) constitutes a promising approach for future applications in the differential diagnosis of skin tumors. In addition to simply recording the average perfusion within the tumor boundary or extension of the hyperperfused tumor area, more advanced image-processing methods should be employed to further refine the opportunity for tumor differential diagnosis offered by LDPI technology. These image-processing methods include feature extraction, analysis of boundary irregularities, and measures of intratumor blood perfusion heterogeneity.

Besides the ability to detect cancer, it would be extremely useful if the images could indicate stage of development, type of cancerous cells, and progression of the disease. LDPI images can provide a wealth of data in these categories as blood perfusion is associated with the angiogenesis process. This approach would be extremely suitable for the classification of cancerous cells and determination of their rate of progression for tumors near the skin surface. However, extended studies of the perfusion in tumors need to be performed *in vivo* and *in vitro* before color templates can be derived for easy identification of tumor types and progression. These color templates will constitute the initial steps in developing of a computerized early warning system.

Although the LDPI technique may not be suitable in developing a stand-alone noninvasive early warning tool for detecting tumors inside the body, there is great potential for the laser flowmetry to monitor the angiogenesis process in deeper tumors. For this development, the penetration limitations of the laser Doppler imager could be overcome using special probes. More research into the design of special fiber-optic probes to suit the site and size of the deep tissues would be needed. The advances in nanotechnology could be incorporated. The microvascular network sensitivity to stimuli and the influence of the probe design and probe placement on the tissue perfusion should be thoroughly investigated. Furthermore, advanced digital signal processing could be explored to compensate for the organ and probe movements.

Understanding the mechanisms within the angiogenesis process would facilitate the development of angiogenesis inhibitors. At present most of the work on the effects of drugs on blood perfusion has been conducted using animal models. For example, Adolphs *et al.* (2005) demonstrated that LDPI is a suitable method to assess the extent, intensity, and duration of skin perfusion in rats after different modes of neuraxial anesthesia. It is envisaged that the development of new drugs will increase the need for appropriate animal models not only to examine the therapeutic challenges, but also to assess the prospects, mechanisms, and risks of these treatments. To meet these challenges, a simple two-stage process could be adopted. The first stage would involve the gathering of blood perfusion data along with other relevant information in the development of angiogenesis inhibitors. The second stage would involve mapping the microvascular data and other information regarding the treatment. In the first stage, interobserver variability could be avoided during the gathering of LDPI data by taking multiple measurements during an extended period with different orientations and locations of the sensor and detector to ensure that the relative changes are accurate and free from confounding factors. Although these efforts would ensure that the data gathered are free from these errors, the process will not make it easier for the subsequent analysis of the information.

Data analysis in this domain is a challenging task as blood flow regulation may involve several mechanisms, which may vary in relative importance in the different vascular beds of the tissue. It is anticipated that the second stage could utilize some of the advanced techniques used in data mining. Research into the data mining of LDPI images and other measurement results for angiogenesis inhibitor is still relatively new. These algorithms must properly interpret the LDPI data, which may contain variable response in blood perfusion due to local anatomical factors such as heterogeneity of the microvascular bed, variable density of mast cells, and phenotype differences. There are many potential benefits in the development of data-mining algorithms utilizing LDPI images and other information for the discovery of angiogenesis inhibitors. The development of mathematical models simulating the dynamics of the angiogenesis process, together with the data mining of LDPI images and clinical information, will definitely aid and shorten the search for cancer treatments. The combination would also be extremely useful for optimal drug administration based on accurate prediction of the drug tissue distribution.

Conclusions

Many techniques are available for medical imaging. Each has advantages and disadvantages. Assessment of tumor blood flow and its pattern is significant for diagnosis and early evaluation of cancer treatment response. The applications of LDPI technique in these areas and others have been discussed in this chapter. The advancement of the LDPI instrument will open several interesting perspectives for microvascular blood flow studies in clinical settings. However, the advancement will require a lot of research to improve the performance and usage of LDPI instruments. At present, lack of standardized procedures for assessing blood perfusion may hinder the use of LDPI. This situation is likely to change once the procedures have been developed and are fool-proof.

Cancer is a consequence of complex interactions between genetic and social-environmental factors. It is envisaged that blood perfusion associated with angiogenesis only holds one of the keys to unlock part of the solution to this medical problem. For the development of a comprehensive cancer patient management system, all the keys should be utilized so that the complete solution can be determined. In this respect, the fusion of imaging techniques, physiological data, and genetic information will be necessary. Fusion of LDPI images with other measurements and clinical data would be a challenge. Data mining of the integrated information would be another challenge. By acquiring and data mining the information on blood perfusion, gene expression, biochemical subsystems, anatomical modifications, and physiological perturbations, advance computer simulations of the disease mechanisms with modeling of the effects of drug intervention could be developed to shorten the search for appropriate angiogenesis inhibitors.

Acknowledgment

The first author would like to thank Dr. Kårin Wardell, Department of Biomedical Eng., Linköping University, Sweden, for her useful discussion, sharing of views, and interests in LDPI for clinical assessment of burn injury.

References

Abbot, N.C., Ferrell, W.R., Lockhart, J.C., and Lowe, J.G. 1996. Laser Doppler perfusion imaging of skin blood flow using red and near-infrared sources. *J. Invest. Dermatol. 107*:882–886.

Adolphs, J., Schmitt, T.K., Schmidt, D.K., Mousa, S., Welte, M., Habazettl, H., and Schafer, M. 2005. Evaluation of sympathetic blockage after intrathecal and epidural lidocaine in rats by laser Doppler perfusion imaging. *Eur. Surg. Res. 37*:50–59.

Bornmyr, S., and Svensson, S. 1991. Thermography and laser Doppler flowmetry for monitoring changes in finger skin blood flow upon cigarette smoking. *Clin. Physiol. 11*:135–141.

Bornmyr, S., Svensson, S., Lilja, B., and Sundkvvvist, G., 1997. Cutaneous vasomotor responses in young type I diabetic patients. *J. Diabet. Compl. 11*:21–26.

Bosch-Marce, M., Pola, R., Wecker, A.B., Silver, M., Weber, A., Luedemann, C., Curry, C., Murayama, T., Kearney, M., Yoon, Y., Rene Malinow, M., Asahara, T., Isner, J.M., and Losordo, D.W. 2005. Hyperhomocyst(e)inemia impairs angiogenesis in a murine model of limb ischemia. *Vascul. Med. 10*:15–22.

Enejder, A.M, af Klinteberg C., Wang, I., Andersson Engels, S., Bendsoe, N., Svanberg, S., and Svanberg, K. 2000. Blood perfusion studies on basal cell carcinomas in conjunction with photodynamic therapy and cryotherapy employing laser-Doppler perfusion imaging. *Acta dermato-venereologica (Acta-Derm-Venereol)* 80:19–23.

Ferrell, W.R., Balint, P.V., and Sturrock, R.D. 2000. Novel use of laser Doppler imaging for investigating epicondylitis. *Rheumatology* 39:1214–1217.

Fok, S.C., Ng, E.Y.K., and Thimm, G.L. 2003. Developing case-based reasoning for discovery of breast cancer. *J. Mech. Med. Biol.* 3:231–245.

Forssell-Aronsson, E., Kjellen, E., Mattsson, S., and Hellstrom M., and the Swedish Cancer Society Investigation Group. 2002. Medical imaging for improved tumor characterization, delineation, and treatment verification. *Acta Oncol.* 41:604–614.

Fullerton, A., Stucker, M., Wilhelm, K.P., Wardell, K., Anderson, C., Fischer, T., Nilsso, G.E., and Serup, J. 2002. Guidelines for visualization of cutaneous blood flow by laser Doppler perfusion imaging. *Contact Dermat.* 46:129–40.

Glade-Blender, J., Kandel, J.J., and Yamashiro, D.J. 2003. VEGF blocking therapy in the treatment of cancer. *Expert. Opin. Biol. Ther.* 3:263–276.

Hajivassiliou, C.A., Greer, K., Fisher, A., and Finlay, I.G. 1998. Non-invasive measurement of colonic blood flow distribution using laser Doppler imaging. *Brit. J. Surg.* 85:52–55.

Hoffmann, M., Bormann V.C., Hoffmann, K., Altmeyer, P., and Stucker, M. 2000. Differentiation of melanocytic skin tumors by means of high resolution laser Doppler imaging. H+G *Zeitsch. Hautkrank.* (H-G-Z-HAUTKR) 490:75–78.

Holland, A.J.A., Martin, H.C.O., and Cass, D.T. 2002. Laser Doppler imaging prediction of burn wound outcome in children. *Burns* 28:11–17.

Jackson, A.E., Osborne, M.J., Seifalian, A.M., and MacLean, A.B. 1994. Assessing vulvar lesions: laser-Doppler flowmetry as a possible technique. *J. Reprod. Med.* 39:953–956.

Jain, R.K. 1988. Determinants of tumor blood flow: a review. *Cancer Res.* 48:2641–2658.

Jeng, J.C., Bridgeman, A., Shivnan, L., Thornton, P.M., Alam, H., Clarke, T.J., Jablonski, K.A., and Jordan, M.H. 2003. Laser Doppler imaging determines need for excision and grafting in advance of clinical judgment: a prospective blinded trial. *Burns* 29:665–670.

Litscher, G., Wang, L., Huber, E., and Nilsson, G. 2002. Changed skin blood perfusion in the fingertip following acupuncture needle introduction as evaluated by laser Doppler perfusion imaging. *Lasers Med. Sci.* 17:19–25.

Liu, D.L., Svanberg, K., Wang, I., Andersson Engels, S., and Svanberg, S. 1997. Laser Doppler imaging: new technique for determination of perfusion and reperfusion of splanchnic organs and tumor tissue. *Lasers Surg. Med.* 20:473–479.

Moor Instruments. http://www.moor.co.uk/ (accessed December 13, 2005).

Ng, E.Y-K., and Sudharsan, N.M. 2001. Effect of blood flow, tumour and cold stress in a female breast: a novel time-accurate computer simulation. *Int. J. Eng. Med.,* 215:393–404.

Ng, E.Y-K., and Tan, J.H. 2002. Assessment of proximal interphalangeal joints in patients with osteoarthritis via laser Doppler perfusion imaging. *J. Mech. Med. Biol.* 2:245–266.

Ng, E.Y-K., and Tan, J.H. 2003. Laser Doppler imaging of osteoarthritis in proximal interphalangeal joints. *Microvas. Res.* 65:65–68.

Ng, E.Y-K., Fok, S.C., and Goh, C.T. 2003a. Case studies of laser Doppler imaging system for clinical diagnosis applications and management. *Int. J. Eng. Med. Tech.* 27:200–206.

Ng, E.Y-K., Tan, S.L.S., and Kang, S.H. 2003b. Laser Doppler imaging of menstrual symptoms. *Int. J. Eng. Med. Tech.* 27:118–127.

Ng, E.Y-K., and Fok, S.C. 2004. Laser Doppler perfusion imaging for the study of alternative medicine. *Int. J. Comp. Appl. Tech.* 21:65–71.

Ng, E.Y-K., Wong, P.J., and Goh, C.T. 2006. Evaluation of acupuncture weight loss programme using laser Doppler perfusion imaging. *J. Mech. Med. Biol.* 6:153–173.

Ng, E.Y-K. 2006. Improved sensitivity and specificity of breast cancer thermography. *Cancer Imaging.* Elsevier Science, USA (in this volume).

Ng, E.Y-K., and Ng, W.K. 2006. Parametric optimisation of the biopotential equation for breast tumor identification using Anova and Taguchi method. *Med. and Biol. Eng. and Computing,* Springer, Netherlands 44:131–139.

Ng, E.Y-K., Rajendra Acharya, U., and Ng, W.K. 2007. Biofield potential evaluation as a new modality for early diagnosing of breast lesions: a 3D numerical model. *Int. J. of Medical Eng. and Technology,* UK (in press).

Pape, A.A., Skouras, C.A, and Byrne, P.O. 2001. An audit of the use of laser Doppler imaging in the assessment of burns of intermediate depth. *Burns* 27:233–239.

Saravanamuthu, J., Seifalian, A.M., Reid, W.M., and MaClean, A.B. 2003. A new technique to map vulva microcirculation using laser Doppler perfusion imager. *Int. J. Gynecol. Cancer* 13:812–818.

Seifalian, A.M., Chaloupka, K., and Parbhoo, S.P. 1995. Laser Doppler perfusion imaging—a new technique for measuring breast skin blood flow. *Int. J. Microcirc.* 15:125–130.

Spence, R.J. 2004. Scanning laser Doppler imaging for burn depth assessment: a preliminary report of a multicentre international trial. 27th Annual Mid-Atlantic Burn Conference, Baltimore, MD, December 5–6.

Stucker, M., Esser, M., Hoffmann, M., Memmel, U., and Hirschmuller, A. 2002. High-resolution laser Doppler perfusion imaging aids in differentiating between benign and malignant melanocytic skin tumours. *Acta Derm. Venereol.* 82:25–29.

Stucker, M., Horstmann, I., Nuchel, C., Rochling, A., Hoffmann, K., and Altmeter, P. 1999. Blood flow compared in benign melanocytic naevi, malignant melanomas and basal cell carcinomas. *Clin. Exp. Dermatol.* 24:107–111.

Svensson, H., Holmberg, I., and Svedman, P. 1992. Interpreting laser Doppler recordings from free flaps. *Scad. J. Plast. Recnstr. Surg.* 27:81–087.

Szili-Torok, T., Paprika, D., Peto, Z., Bahik, B., Bari, F., Barzo, P., and Rudas, L. 2002. Effect of auxillary brachial plexus blockage on baroreflex-induced skin vasomotor responses: assessing the effectiveness of sympathetic blockage. *Acta Anaesthesiol. Scand.* 46:815–820.

Troilius, A.M., and Ljunggren, B. 1996. Evaluation of port wine stains by laser Doppler perfusion imaging and reflectance photometry before and after pulsed dye laser treatment. *Acta Derm Venereol.* 76:291–294.

Vongsavan, N., and Mathews, B. 1993. Some aspects of the use of laser Doppler flow meters for recording tissue blood flow. *Exp. Physiol.* 78:1–14.

Wang, I., Andersson Engels S., Nilsson, G.E., Wardell, K., and Svanberg, K. 1997. Superficial blood flow following photodynamic therapy of malignant non-melanoma skin tumors measured by laser Doppler perfusion imaging. *Br. J. Dermatol.* 136:184–189.

Wardell, K. 1994. *Laser Doppler Perfusion Imaging: Methodology and Skin Applications.* Linkoping Studies in Science and Technology Dissertations no: 329. Samhall Klintland, Sweden.

Wardell, K., Hermansson, U., Nilsson, G.E., and Casimir-Ahn, H. 2000. Laser Doppler imaging of myocardial perfusion during coronary bypass surgery. *Proc. SPIE* (Optical diagnostics of biological fluids V) 3923:10–17.

9

Dynamic Sonographic Tissue Perfusion Measurement with the PixelFlux Method

Thomas Scholbach, Jakob Scholbach, and Ercole Di Martino

Introduction

This chapter introduces a novel technique (the PixelFlux technique) of color Doppler sonographic perfusion measurement, which allows the quantification of tumor tissue perfusion along with a normal ultrasound investigation of the tumor. This technique, being noninvasive and nonradiating, is patient friendly. It allows an objective calculation of perfusion signals from the tumor; it is inexpensive, needs no special ultrasound equipment, and can be made available at bedside; and frequent reevaluation during therapy is also possible. This chapter describes the first results and tries to outline future applications because of the novelty of the PixelFlux technique. The chapter also invites researchers and physicians to download the PixelFlux software from the Internet (Chameleon-Software, 2004) to try it on their own patients.

Blood flow in a certain minimum quantity and quality is an indispensable precondition of vertebrate life and for the normal functioning of a tissue. It has to meet the metabolic needs of the tissue and is therefore characteristic for a specific tissue. Moreover, perfusion is altered according to functional and structural alterations in tissues. One example for such a

state is tumor growth; others might be inflammation, aging, scarring, arterial malperfusion, and venous congestion or transplantation. Each tumor has an individual history with changing metabolic demands to the hosting organism. Rate of cell division and tumor cell energy turnover lead to a vascular network development that is characterized by a specific branching pattern of vessels and a specific perfusion intensity. Time of metastasis, capacity to enlarge, and necrosis of central parts may depend on the quality and amount of tumor perfusion. To understand these processes it is desirable to quantify tumor perfusion.

Tumor Perfusion Evaluation—State of the Art

Measurement of perfusion-related parameters of tumor homoeostasis is a recognized approach to answer questions of tumor biology (Vaupel et al., 2001). The important methods in use are measurement of tumor oxygenation, laser Doppler flowmetry in superficial tumors (Jacob et al., 2006), computed tomography (CT) (Kan et al., 2005), magnetic resonance imaging (MRI) techniques to quantify perfusion and diffusion (Weber *et al.*, 2005), scintigraphy (Wawroschek *et al.*, 2001), positron emission tomography (PET) (Bruehlmeier *et al.*, 2005), and contrast-enhanced sonography (Krix *et al.*, 2005). Some of these techniques are invasive, whereas others are restrictive in their application due to technical reasons or are expensive or demanding for the patient.

The organ most frequently investigated is the liver. Here a critical appraisal of perfusion measurement techniques is necessary. Remaining challenges are reduction of radiation with CT techniques, improvement of spatial and temporal resolution with MRI techniques, accurate quantification of tissue contrast material at MR imaging, and validation of parameters obtained from fitting enhancement curves to biokinetic models and applicable to all perfusion methods (Pandharipande *et al.*, 2005). Nevertheless, certain parameters of diverse imaging modalities (MRI and contrast-enhanced ultrasound) do correlate with each other (Kiessling *et al.*, 2003), but physical differences of contrast media (microbubbles vs. MRI contrast agents) may influence depiction of tumor vascularity, depending on the branching patterns and vessels' diameter (Galie *et al.*, 2005). So the situation at present is unclear.

Sonography allows access to many parts of the body without harming or embarrassing the patient, and it is comparatively inexpensive. Simple color Doppler sonography has the capacity to depict blood flow signals in a graduated fashion. Moreover, it is fast enough to monitor rapid changes of perfusion during the heart cycle. Many attempts have been made to use traditional sonographic parameters such as gray-scale values, and resistance index (RI) (Miyakawa *et al.*, 2005), pulsatiltity index (PI) (Okuyama *et al.*, 2004), and grade of vascularity. To classify vascularity, merely subjective scoring systems or computer-assisted algorithms have been developed. Pattern of vascularity has also been found useful to discriminate between benign and malignant tumors in lymph nodes and uterine pathologies (Alcazar *et al.*, 2003), but it was found to have no pretherapeutic predicitive value in a group of 34 early-stage breast cancers (Roubidoux *et al.*, 2005).

The perspective of sonographic and other imaging procedures with respect to prognosis of tumor response to therapy, aggressiveness, and metastatic behavior is not completely clear at present. This might be due on the one hand to tumor specifics, and on the other hand due to limitations of these techniques to reflect the important features of perfusion. Color Doppler parameters are valuable for predicting complete histological response of neoadjuvant chemotherapy in advanced breast cancer. Compared to more expensive MRI and CT or contrast-enhanced sonography, color Doppler sonography may also be an effective modality with regards to health-care expenses in less developed countries (Singh *et al.*, 2005). Conventional (i.e., single-vessel) RI and PI

may yield inconclusive results in attempts to differentiate benign from malignant tumors (Gallipoli *et al.*, 2005). Clearly, a more refined approach capable of describing momentary changes of perfusion more precisely is needed.

Novel approaches referring to both velocity and perfused area in a certain region of interest are necessary to overcome the basic restrictions and flaws of purely velocity-based perfusion estimation of only single-vessel measurements. These techniques (i.e., RI and PI) are now more than 30 years old and have definitively fallen behind the fast development of imaging techniques as offered today. Such techniques as color Doppler perfusion display and 3D blood vessel depiction, and even more the ecg-triggered 4D depiction of dynamic flow phenomena in space, offer abundant flow information still not used today. Moreover, it would be very useful to observe the changes of flow velocity and perfused area over a full heart action to achieve a dynamic appreciation of these flow phenomena. It is of foremost importance to refer all perfusion parameters not only to the vessels but also to the entire tumor. In this way, perfusion dropouts are appreciated according to their real size in relation to the tumor as a whole, and pulsatility parameters like RI and PI can be given for the whole-tissue segment under investigation. From a theoretical point of view this is the prerequisite for a more complete understanding of perfusion. Therefore, we extend present conventions for RI and PI, and calculate them as Tissue-RI (TRI) and Tissue-PI (TPI). Moreover, we extend the calculation algorithm for RI and PI from velocity measurements (as this has been in use for *single* vessels for decades) to perfused area and perfusion intensity. Such a method eventually could bring new insight into old questions and would be entitled a *dynamic* tissue perfusion measurement. It could eventually overcome limitation of a rather static apprehension of perfusion even with contrast techniques. Here the dynamic component of observation may not be precise enough to look at changes of flow during a heartbeat. Often the influx of a contrast medium is depicted over a longer period to measure only its saturation.

Dynamic Tissue Perfusion Measurement (PixelFlux)

If tumor perfusion reflects tumor behavior (growth velocity, metastatic potential, response to chemotherapy and radiotherapy, involution, necrosis), it is useful to measure tissue blood flow in a reliable, differentiated, and patient-friendly way. To achieve a reliable quantitative perfusion measurement in a tumor, it is desirable to yield data on mean flow velocity, mean perfused area, mean flow intensity, pulsatility of velocity/area/perfusion intensity, resistance of flow, spatial distribution of flow across the tissue, and quantitative distribution of perfusion intensity throughout the tissue section.

These parameters describe quantitatively the local blood flow and should be recorded in any part of the tumor in an arbitrarily chosen region of interest. To achieve this goal, we developed the method of dynamic quantitative tissue perfusion measurement (PixelFlux technique). With the PixelFlux software, such measurements can be performed in an automated fashion for all sonographically detectable tumors.

Preconditions

Every sonographic imaging depends on external as well as internal preconditions that influence the quality of the image and the image sequence of the video. To achieve comparable results, it is necessary to standardize as many conditions as possible. The most reliable data will be yielded if the ultrasound equipment and its settings are never changed; that is, one uses the same type of ultrasound machine from the same manufacturer, and the ultrasound transducer's shape and frequency also remain unchanged. Internal presetting of adjustments such as gain, depth compensation, applied frequency, imaging frame rate, spatial and time resolution, foci, type of color display in distinct hues, and others (depending on the equipment's specifications) must not be changed.

In daily routine this is not a problem. Most centers refer patients to specialized ultrasonographers or radiologists who use the same ultrasound equipment over longer periods. This allows long-term follow-ups of single patients and ensures comparability of data from larger populations. Comparability of data from different centers is not provided if differences with respect to ultrasound hardware or presetting of the equipment cannot be ruled out. It is useful, therefore, to carry out such measurements with the most sophisticated equipment available. Before the advent of quantitative tissue perfusion measurement, manufacturers aimed to develop imaging modalities focused on an excellent visual impression for the investigator. PixelFlux advances beyond a simple subjective view of the ultrasound: it extracts imaging data for a quantitative analysis. It is, therefore, crucial to standardize the imaging details of the given equipment. More sophisticated ultrasound machines analyze more precisely and produce more consistent numerical imaging data (our unpublished data). Consequently, to ensure the most reliable measurement, it is necessary to use the most sophisticated equipment.

Workflow

PixelFlux perfusion measurement is a byproduct of the conventional color Doppler sonographic investigation of the tumor. Injection of contrast-enhancing agents (CE) is possible but not necessary (see below).

Procedure

1. A relevant structure (e.g., tumor) is scanned with a suitable presetting of all available specifications of the ultrasound equipment and transducer using as high a frequency as possible to achieve a high resolution and to depict color Doppler signals as sensitively as possible.

2. An imaging plane is selected, and the transducer is held in this plane.

3. A video clip (preferably DICOM format, avi-format is also possible) with a duration of at least one full heartbeat is recorded.

4. The video clip is transferred to a PC with installed PixelFlux software.

5. The clip is opened by PixelFlux and the calibration of distances as well as color bar are carried out automatically with DICOM clips.

6. The region of interest (ROI) is cut out. Then the measurement starts. It is completed in 1–5 sec depending on the file's size and processor velocity of the PC.

7. A PACS (picture archiving and communication system) function is included in PixelFlux. It allows review of clips, ROIs, and all measurements, as well as export of measurements to a statistical software.

Output

With the PixelFlux technique the following parameters are calculated from a video sequence recording at least one full heart cycle.

1. Mean flow velocity throughout the entire region of interest (ROI).

2. Mean perfused area in relation to the ROI.

3. Area of the ROI.

4. Perfusion intensity throughout the entire ROI: perfusion intensity [cm/s] = mean perfused area [cm^2] × mean flow velocity [cm/s] /area of the ROI [cm^2].

5. Tissue Pulsatility Index (TPI) of velocity/of area/of perfusion intensity: TPI = (maximal systolic value – minimal diastolic value)/mean value—"value" may be velocity, area, or intensity.

6. Tissue Resistance Index (TRI) of velocity/of area/of perfusion intensity: TRI = (maximal systolic value – minimal diastolic value)/maximal systolic value—"value" may be velocity, area, or intensity.

7. Spatial distribution of flow across the tissue—an overlay of false colors on the ROI shows the local distribution of flow intensity.

8. Quantitative distribution of perfusion intensity throughout the tissue section: the whole range of flow intensity (resp. perfused area) over a full video sequence is divided into percentiles. Each interval's fraction of the ROI describes the distribution of perfusion intensity in numerical values.

9. Time lines of the above explained perfusion parameters of individual patients can be displayed and statistically evaluated.

Use of Contrast Enhancers

When contrast enhancers (CE) are used, an additional variable is introduced. Special attention has to be paid to keep factors such as the amount of CE, infusion or injection rate, distribution pattern, and imaging times constant. With high-end ultrasound equipment, perfusion of tumors can often be obtained without the application of CE. This reduces imponderable external influences on the measurements. The application of CE saturation of a tumor in gray-scale image sequence excludes the measurement of the pulsatility parameters inside a tumor, dependent on the heartbeat. Contrast enhancer influx is observed over a longer period of time, and the gradual rise of tumor contrast gauges perfusion intensity. If a CE is used to enhance color signals in a tumor, care should be taken to avoid blooming artifacts that are color signals massively splashed across vessel borders and thus cause errors in perfusion quantification.

Application

Perfusion measurement is desirable in many situations. Some of the numerous pathophysiologic processes accompanied by perfusion changes are outlined in this section.

After opening the video file, the first image is automatically calibrated for distances and color hues. The next step is to choose the ROI. Depending on the tissue under investigation, several options are applicable. Region of interest can be drawn free-hand, or geometric shapes can be used as a framework for definition of ROI and sub-ROIs. Their selection depends on pattern of vascularization. Often an irregular branching pattern is found in tumors. Free-hand ROIs are adequate in most cases. If the tumor has been encircled along its border, then sub-ROIs may be defined. This is helpful in any attempt to correlate perfusion to separate shells of tumor tissue from the center to periphery.

Such shells can be defined in a general manner as follows. The sub-ROI is drawn automatically according to the demands of the investigator. To achieve this and to cut out the tumor core, it is suitable to use the target mode. One can define a core region with a diameter of 50% of the whole tumor's diameter (any numerical value from 100 to 1% is selectable). This diameter is centered at the balance point of the primary ROI, and required distances are subtracted from this point to the periphery given by the primary ROI. Other possibilities are parallelograms with defined sub-ROIs as slices or parallelographic corners. These ROIs and sub-ROIs allow orientation at anatomical landmarks. For example, in

perfusion measurement of the renal cortex, it is useful to look at a complete renal segment fed by a single interlobar artery; otherwise, one would miss a subgroup of cortical vessels, and measurement bias would result.

The chosen ROI is valid for all single images of the whole video sequence. Inside the ROI every pixel is evaluated. Number and coloration of pixels change, with each image reflecting changing momentary perfusion from systole to diastole. All these changes are numerically evaluated. For each image the perfused area and the mean flow velocity are calculated. PixelFlux is able to detect the beginning and end of a complete heart action. It calculates mean perfused area and mean perfusion velocity from all images of the video from the beginning to the end of the heart cycle. From these data perfusion intensity is calculated.

PixelFlux Application in Oncology

Tumors differ with respect to their vascularity and perfusion intensity. Both vary with histology and the developmental stage of a tumor. Perfusion in central parts of a tumor may be different from perfusion in peripheral regions. Hypoxia is a relevant promoter of therapy resistance in some tumors (Vaupel and Harrison, 2004). This might trigger processes aiming at evasion of tumor cells to bring up satellites. Quantification of intratumoral oxygenation by noninvasive means or direct measurement of tumor perfusion has become pivotal to rating the consequences of new treatments such as hypoxia-mediated gene therapy, application of nonsteroidal anti-inflammatory drugs to increase tumor oxygenation (Crokart et al., 2005), or anti-angiogenic therapy (Janssen et al., 2005). Others report conflicting results with poor prognosis and response to radiation in head and neck cancers with high perfusion when measured with PET (Lehtio et al., 2004). Distribution of hypoxic tumor parts may greatly differ from individual to individual and with tumor type (Kelleher et al., 1998). It is, therefore, necessary to achieve an individual perspective of every tumor for an efficient therapy especially because tumor vasculature has become a target of tumor therapy.

Data about the state of tumor angiogenesis would be most valuable. Such numerical information could possibly be used to tailor the application of anti-angiogenic drugs and to monitor their effects. The tumor vessel network may have some peculiarities such as lacunas, caliber irregularities, arteriovenous shunts, and sudden interruption of vessel branching. These specifics should lead to a deviation from normal flow pattern in tiny tumor vessels. Here could be an application of tissue perfusion measurement techniques such as PixelFlux. It is reasonable to assume that deviations from harmonic branching patterns as a consequence of the above-mentioned peculiarities would produce deviations in flow intensity, flow distribution, pulsatility of flow, and perfused area. In fact, we were able to demonstrate such effects (Scholbach et al.,

2005a). In a series of lymph node metastases from oropharyngeal carcinoma, we found a significant difference of perfusion intensity between nodes from various N-stages.

Prostate Cancer

Another study investigated the pretherapeutic perfusion of prostate cancer (Rouviere et al., 2004). It could be demonstrated that perfusion intensity prior to high-intensity focused ultrasound (HIFU) therapy had no adverse impact on the formation of post-therapeutic necroses. Perfusion was measured here in the prostate itself and in the surrounding tissue. Perfusion intensity and perfused area did not differ with respect to uniformity of thermic tissue destruction. This way the influence of the vascularization on the effect of thermoablation of tumor tissue could be evaluated. Further studies are needed to describe tumor perfusion with respect to its influence on the abduction of thermal energy during HIFU procedures.

Tumor vascularization influences thermal conduction capacities and nutrient and oxygen supply alike. Monitoring the extent of vessel distribution and the intensity of perfusion should elucidate the individual tumor's metabolic state. One possibility to test this assumption is to correlate oxygen measurements in tumors with the perfusion measurement. In lymph node metastases from oropharyngeal carcinomas, we were able to demonstrate significant correlation of intratumoral oxygenation and sonographically determined perfusion (Scholbach et al., 2005a). In this pilot study, we could discriminate sonographically hypoxic metastases from better oxygenated ones with the dynamic perfusion measurement technique.

Head and Neck Cancer

We investigated 24 patients (44 to 78 years of age) with cervical lymph node metastases of squamous cell head and neck cancer by color duplex sonography and 17 (46 to 78 years of age) with polarography (Scholbach et al., 2005a). During defined contrast-enhancer infusion perfusion intensity, Tissue Pulsatility Index (TPI) and Tissue Resistance Index (TRI) were measured. Tumor tissue $pO2$ was measured by means of polarographic needle electrodes placed intranodally. Sonography demonstrated a significant inverse correlation between hypoxia and perfusion. The extent of hypoxic nodal fraction was weakly but significantly inversely correlated with lymph node perfusion ($r = 0.551$; $p = 0.021$). Nodes with a perfusion intensity of < 0.05 cm/sec showed significantly larger hypoxic areas ($p = 0.006$). Moreover, significant differences of TPI and TRI were demonstrated with different N-stages. In higher N-stages, lower TRI and TPI values were found. TRI values were as follows: 0.79 in N1, 0.62 in N2, and 0.47 in N3 (differences between stage N1 and N2 $p = 0.028$ and between stage N1 and N3 $p = 0.048$). A similar decline was found with TPI values: 2.07 in N1, 1.16 in N2, and 0.79 in N3 (differences between stage N1 and N2 $p = 0.028$ and between stage N1

and N3 $p = 0.030$). Figure 21 illustrates some output from dynamic perfusion measurement of a metastatic lymph node with PixelFlux. The central part shows the metastatic lymph node encircled by two lines, the outer one encompassing the outer border of the node and the smaller central circular line enclosing the central 50% of the node. Thereby, two parts of the tumor—the tumor periphery and the center—can be looked at separately. All measurements can be done separately in both zones, and results can be compared in many ways.

Evaluation of PixelFlux Results

The first step is to look at the overall perfusion intensity values as given by the columns on the left side of Figure 21. It is directly proportional to the product of the perfused area of ROI and the mean blood flow velocity over a complete heartbeat. The red column corresponds to the inferior parts of the adjacent diagram depicting the perfusion intensity of the tumor periphery. The green column reflects perfusion intensity of the tumor center. Only by a numerical description of perfusion such a clear distinction of flow becomes possible. It would be impossible to do so by simply visually evaluating moving images displaying tissue perfusion of a tumor.

The second step is to look at the spatial distribution of perfusion (the central part of Fig. 21). New possibilities of describing tumor perfusion emerge with the PixelFlux technique. Perfusion signals can be displayed in false colors signaling local perfusion intensity of the whole video sequence in one image. Thus, these colors express not only a momentary state of perfusion but the mean values over the entire video clip. Hence, additional information that is not obtainable from a simple observation of the video itself is acquired. A color spectrum from black to red is chosen with transition over white hues. Distribution of intensity is thereby overlaid as a perfusion relief and can be estimated at a first glance. This may be helpful for surgical decisions to define a border between parts of the tissue having different perfusion patterns.

While interpreting these false colors, it is necessary to keep in mind that color hues are related to the specific video and that they are not absolute. In Figure 21 the tumor center is more evenly perfused with some islands of maximal perfusion (red dots). However, the overall perfusion intensity is less than that in the periphery, as is shown by a comparison of the red and green columns. The periphery shows a more focal perfusion in the left lower quadrant where large perfused areas have only weak intensity values (gray and black).

False-color coding of perfusion parameters is but one possibility to depict numerical perfusion data. Another way to do the same is to draw a perfusion intensity distribution diagram. Such diagrams are depicted at the lower margin of the images. The x-axis describes perfusion intensity values

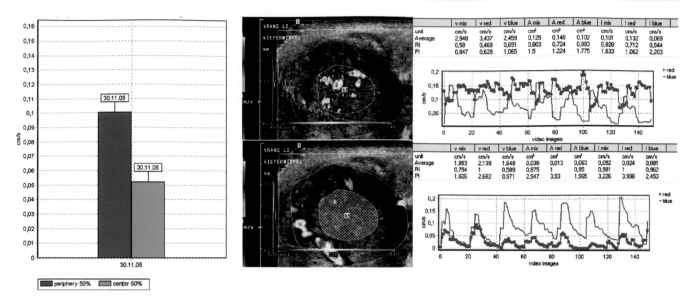

Figure 21 PixelFlux investigation of a lymph node metastasis. Left part: Perfusion intensity of lymph node periphery (red column) versus lymph node center (green column)—much stronger peripheral perfusion. Central part: Color Doppler videos (initial images) with false-color overlay displaying perfusion intensity over time. Maximum perfusion intensity is depicted as red, mean as white, and minimum as black coloration. The white curve describes distribution of perfusion intensities of the entire ROI. There are obvious differences that exist between lymph node periphery (image below) and center (image above). The lymph node center has a broader distribution with relatively strong perfusion signals, whereas in periphery perfusion intensities are almost entirely restricted to relatively low intensities (always compared to the ROI's maximum value). Right part: Corresponding time course of lymph node perfusion with numerical values in tables (see text). Red and blue lines corresponding to red and blue color pixels. It is visible at the first glance that differences between both parts of the lymph node exist.

from zero to the video's maximum value. The *y*-axis shows the frequency of occurrence of distinct intensity classes. The whole range of intensity is divided into 30 classes. Each class contains 3.33% of the whole range, and it is assigned a specific value of the *y*-axis. The curve resulting from this arrangement of perfusion data describes the perfusion intensity distribution of the investigated ROI. If the low-intensity classes prevail, the curve will show a slope from left to right. If the perfusion is more localized in areas or vessels with strong flow, then the maximum of the curve may shift to the right, or a second peak, located in the region of higher intensity values, may emerge. Both changes will shift the weighted mean value (also called expected value in statistics) to the right side of the diagram. This weighted mean value is marked on the *x*-axis and gives a fast orientation of the distribution curve's shape, when compared to the average value, which is also given. In Figure 21 perfusion intensity distribution in the center is more right-shifted than in the periphery. This tumor's central perfusion is weaker but more evenly distributed. In this way different distribution curves are generated, characterizing perfusion distribution of the tumor under investigation.

A perfusion distribution may be influenced by vascular as well as by interstitial forces. Studies to use such information have not yet been done but could bring new insight into tumor biology. The synopsis of perfusion intensity values, as given by the columns on the left side, the false-color display of perfusion intensity and its graphical description by the distribution curve, plus the numerical description of the most important flow data in tables and their course in time-line diagrams, as given in the right side of Figure 21, give a comprehensive numerical and spatial description of a tumor's perfusion. Even more numerical data may be calculated and displayed in larger tables. Actually, the PixelFlux method offers 34 numerical parameters characterizing perfusion and 15 specifying parameters.

The third step is to look at cardiac changes of blue- and red-colored pixels in Figure 21. The time course demonstrates that the perfusion changes along with the heartbeat even in the tiniest tumor vessels. The lower part of the image refers to the adjacent color Doppler image of the tumor periphery, and the upper part to the central 50% of the tumor. In both parts, blue pixels, emanating from vessels running away from the ultrasound transducer, exhibit strong pulsations. In the tumor periphery, synchronous pulsations of red-encoded flow, from vessels going toward the transducer, are visible. In contrast, red-encoded flow inside the tumor's center is faster and much less pulsating. The accompanying tables, above the time-course diagrams, present more specified data describing mean values of velocity, area, and intensity of red- and blue-encoded pixels separately. A measure of the extent of the change of these parameters is given as RI and PI values. RI and PI correspond only loosely to conventional RI and PI. We have coined the terms *Tissue-RI (TRI)* and *Tissue-PI (TPI)* instead, to mark these differences. With PixelFlux, not only velocity changes in single vessels are recorded, but we advance and record simultaneously velocity as well as area data from a whole ROI. In this way, we also take into account nonperfused parts, and so a more exact average velocity and perfused area recording becomes possible.

We extend the concept of RI and PI to the new data acquisition modality. In the tumor's center of this example, the values of TRI of the velocity of red pixels, TRI of the area of red pixels, and TRI of the intensity of red pixels are 1 each. This can be easily controlled with the red line of velocity course in the corresponding time-line diagram below the data table: red flow velocity values drop down to zero in diastole. This means that perfusion TRI of intensity will also be calculated as zero because it is directly proportional to the product of velocity and area. In these perfusion diagrams in the right part of Figure 21, only the time course of velocity is depicted. Similar curves of heartbeat-dependent changes of perfused area and perfusion intensity are not given here due to limited space.

Figure 22 shows quite impressively another phenomenon of tumor vascularization. In a single tumor, quite different perfusion patterns are found and are numerically described with the PixelFlux technique. It is a myosarcoma of the thigh extending over nearly 30% of the thigh's length. At the upper pole (upper part of the image), perfusion pulsates only weakly, whereas in central parts of the tumor highly pulsatile signals prevail. This is depicted in the lower part of the image. Perfusion intensity also differs between these tumor parts but not as strongly as pulsatility. The upper pole (green column) is not as strongly perfused as central tumor areas (red column). This could point to a prospective application of numerical descriptions of tumor perfusion: with PixelFlux the biological peculiarities not only of different tumor types but also of different parts of a tumor could become apparent. This could considerably enhance our understanding of tumor development. With PixelFlux it is possible to observe tumors in clinical settings as often as one desires because the method, being noninvasive and non-ionizing, is harmless. The oncologist can become quite familiar with the tumor's individual behavior. We can learn how tumor spreading is emerging.

Many questions may be investigated. Is low pulsatility of flow a pointer to loose intervascular tissue structures? This could eventually be a sign of fast cell division with development of intercellular matrix falling behind. Is the shift to high pulsatility a *signum mali ominis* that an imbalance between vascular bed and tissue nutrition is going to emerge? How do metabolic demands of the tissue modulate perfusion? Is hypoxia the strongest influence, or will we discover that metabolites or local factors exert an even stronger influence on perfusion changes? How would perfusion or tumor nutrition respond to internal conditions of tumor tissue along with nascence, growth, and death of tumor cells? A meticulous numerical description of perfusion could bring new insights for tumor biologists and oncologists and for

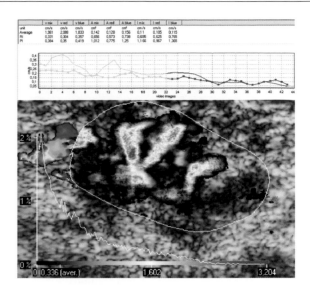

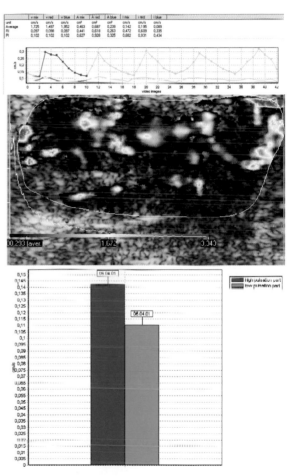

Figure 22 Quantification of perfusion in two different parts of a myosarcoma of the thigh. Upper part: Low pulsatility of flow in the tumor's periphery versus strong pulsating flow in tumor's center (below). Only slightly different intensity distribution curves with little more right-shifted curve toward stronger perfusion signals in the upper image (weighted mean value [expected value] at 10.4 resp. 8.7% [lower image] of the maximum value). The lower part of the figures again displays perfusion intensity, which is stronger in the center of the tumor.

their patients' therapy. Perfusion intensity and its local distribution could be parameters for such a description. Are there densely packed small vessels, or is there a rather wide-meshed lattice of larger vessels? Measurement of pulsation of velocity simultaneously with pulsation of vessels' diameter could describe physical forces and their impact on the tumor tissue. For example, high pulsation of flow velocity with low pulsation of vessels' diameter could point to a strong interstitial tumor. Low pulsation of flow velocity with high absolute velocity values could point to a high metabolic activity of a tumor with a strongly branched vascular tree. Separated examination of different flow direction in conjunction with the above-mentioned parameters could allow further insight.

The drop of perfusion intensity in a liver metastasis after chemotherapy is shown in Figure 23. In the right part of the image a power Doppler sonogram demonstrates the low perfused metastasis with better perfused surrounding tissue. Now we can quantify perfusion a numerical comparison of these different tissues is possible, and that the effect of therapy onto tumors and metastases is quantifiable. This is shown in the left part of Figure 23. Red bars signal a pretherapeutic situation with about one-third of the lower perfusion intensity in the metastasis. After therapy, the diverging effect onto these different tissues is evident. Perfusion of the metastasis drops to zero, whereas liver tissue on the same day shows only a moderate perfusion drop. This example points to other applications of tumor perfusion measurement. Side effects in other parts of an organ can be measured, and the effect of therapy on tumor perfusion is comparable to such side effects. This could be used in the development of new antitumor drugs. Moreover, the effect or dosage of an antitumor drug can eventually be tailored according to the actual effect on both healthy and tumor tissues.

Another application of perfusion measurement is to plan and control surgical measures. In prostate cancer, preinterventional perfusion of cancer and normal prostate tissue were measured and compared to extension of HIFU-induced necrosis (Rouviere *et al.*, 2004). In this study no correlation of perfusion intensity with necrosis quality was found. Thus, it seems that thermal energy draining by tumor blood flow is unlikely to affect the success of thermal tumor ablation.

Comprehensive numerical evaluation of a tumor's perfusion becomes feasible. PixelFlux offers the chance to have a quite individual look at each tumor in each patient. First, the measurement can be repeated at bedside without harm to the patient. A close monitoring can be achieved, with no medical or economic restrictions with regard to the imaging procedure. A wealth of numerical information is capable of describing perfusion in detail. Second, a tumor can be individually regarded as an organism with differently behaving parts at least as far as perfusion is concerned. Third, behav-

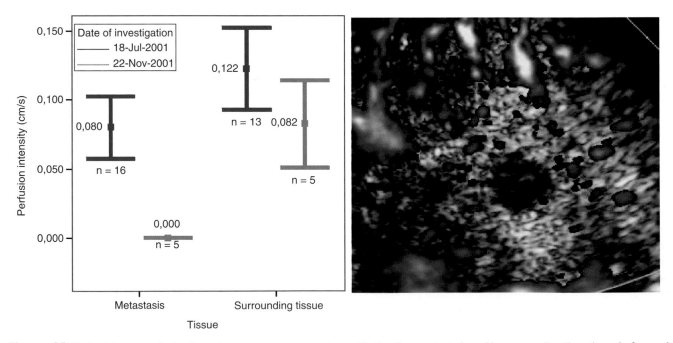

Figure 23 In the left part perfusion intensity measurements are given of both a liver metastasis and its surrounding liver tissue before and after chemotherapy. The right part shows a power Doppler image of the liver metastasis. It can be demonstrated with multiple measurements that metastasis perfusion as well as perfusion of normal liver tissue are dampened by therapy, but perfusion of the metastasis has dropped to zero while liver tissue perfusion is maintained at a level of about 75% of the initial value.

ior of the tumor as well as its environment can be monitored at the same time. Relationships of both tissues and their interactions become observable. All numerical data may be used for individual statistical analysis. One video clip contains ~ 30–50 single images with a recording time of ~ 2–3 sec. After transfer to a computer, measurement of a single clip takes ~ 5 sec. In a short period of time, many clips of a tumor can be evaluated, thus, enhancing the reliability of these measurements. The development of this technique is only at its beginning. The method is new, and its potential can only be estimated. Larger studies on various aspects of the PixelFlux technique are desirable. Next we offer some ideas on how to proceed.

Comparison of Results with Other Techniques

There is an urgent need to include the vascular pattern of tumors in consideration of their nature as well as a better interpretation of semiquantitative flow data (RI, PI) (Chammas et al., 2005; Gallipoli et al., 2005). Conventional Doppler techniques such as power Doppler (Chammas et al., 2005) or the application of contrast-enhancing media (Gallipoli et al., 2005) have led to semiquantitative scoring systems (Marret et al., 2005). Such scores suggest that peritumoral infiltration correlates with subjectively analyzed vessel density (Testa et al., 2003).

Nevertheless, the clinical value of conventional Doppler flow parameters or the application of contrast media could not be demonstrated (Blanco et al., 2003). Others (Osanai et al., 2003) found a significant correlation of the lowest measured conventional RI in breast cancer with the Nottingham Prognostic Index (NPI) (NPI = 0.2 × tumor size (cm) + histological grade (I–III) + lymph node score (1–3)). Similar results and a significantly lower RI were found in moderately or poorly differentiated tumors and advanced-stage tumors. Significantly higher peak systolic velocity (PSV) was found in moderately or poorly differentiated tumors, tumors with larger volume, and advanced-stage cervical carcinomas. No correlation was found between RI and PSV and histologic type (Alcazar et al., 2003). This finding corresponds well to our results of the dynamic flow measurement. We found significant lower TRI and TPI (Tissue-RI and -PI) in the more advanced cervical metastases of head and neck cancers (Scholbach et al., 2005a). The ambiguity of results in conventional color or power Doppler analysis points to the potential of Doppler interrogation of tumors but at the same time indicates the relevant limitations of the technique. The novel dynamic appreciation of all Doppler signals from any vessel inside a tumor, and the simultaneous automatic calculation of perfusion intensity at each point of the tumor, can be a major step forward, overwhelming some of the existing limitations. As measured with the PixelFlux technique, we found a significant inverse correlation of

intensity with intratumor oxygen saturation (Scholbach et al., 2005a). Similar results are reported from direct comparison of oxygenation with CT depiction of tumor flow (Haider et al., 2005).

Inflammation Grading

Some inflammation may accompany tumor and tumor necrosis, but it is much more frequently elicited by other stimuli. Whether differences exist between tumor-associated inflammatory hyperperfusion and other causes is unknown at present. It is reasonable to assume that both processes may influence each other. We demonstrated the use of dynamic perfusion measurements to grade inflammatory activity in chronic inflammatory bowel disease (Scholbach et al., 2004a). Inflammatory activity is measured directly via the measurement of hyperperfusion. Then each bowel segment can be classified as quiescent, chronic activated, or acute exacerbated. The same holds true for other inflammatory processes such as arthritis and thyreoiditis (our unpublished data).

Transplantation Medicine

Transplanted organs are balanced between immunosuppression with potentially devastating effects on organ function and rejection with threat of organ loss. Both processes may damage the vascular network and can be traced with a meticulous assessment of perfusion (Scholbach et al., 2004b, 2005b). Even separate levels of a vascular tree can be differentiated and evaluated independently. In kidneys a characteristic drop of perfusion intensity accompanied by an increase in overall perfusion pulsatility from central to subcapsular regions has been demonstrated (Scholbach et al., 2004b, 2005b, 2006). Loss of the tiniest vessels preceding loss of an organ's function may be recorded very early, and counteraction can be planned.

Conclusions and Outlook

Application of dynamic perfusion measurement is only at its beginning. We and others have demonstrated the feasibility, clinical interest, and direct benefit of dynamic perfusion measurements in a variety of tissues and diseases (Rouviere et al., 2004; Scholbach et al., 2004a, b, 2005a, b, 2006). Researchers and clinicians are invited to develop their own ideas and projects. The focus of this chapter is on oncology, but there are many more fields to discover: inflammation grading, function of transplanted organs, organ insufficiency, adaptation after removal of sister organs, physiologic and pathologic growth and development, microvessel diseases in diabetes, atherosclerosis, vasculitis, aging, microembolization, perfusion response to physiologic stimuli, elasticity examinations of tissues, description of complex venous networks, quantitative analysis of collateral vessels, impact of arterial stenoses on tissue perfusion, and influence of venous outflow congestion on tissues. Downloading of PixelFlux software, necessary for video analysis, is also possible at www.chameleonsoftware.de.

In oncology, a first step could be to correlate perfusion patterns with the biological properties of a tumor. The potential of metastasis could be related to a shortage of oxygen or nutrient supply inside the tumor. Local growth is dependent on the spread of tumor vessels. Observation of their growth could reveal information on the aggressiveness of the tumor. Only recently, reports have been issued on a local ultrasound-guided therapy with ultrasound contrast enhancers, so-called microbubbles (Imada *et al.*, 2005; Kessler *et al.*, 2005; Yuh *et al.*, 2005). These spherical containers can be loaded with antitumor drugs that will be transported with the bloodstream. Once they arrive at the tumor, they can be destroyed with a strong ultrasound impulse releasing their load. PixelFlux measurements might predict which tumors are eligible for such procedures and where the drug is potentially trapped inside the tiny tumor vessels. Perfusion distribution diagrams could form a comparable basis for selection of specific tumors.

The availability and affordability of a method are at least as important for their broader usage as technical and medical aspects. This is another major advantage of sonographic methods in general. In developed countries, the equipment is rather inexpensive (compared to alternative imaging equipment) and can be found in even smaller hospitals or private practices. PixelFlux is a sophisticated method to extract numerical perfusion data from conventional color Doppler images, but no new hardware or additional procedure is needed. In most circumstances, injection of microbubbles is unnecessary, making the examination itself inexpensive. This offers the possibility of more frequent follow-up studies to monitor the effect of antitumor therapies and to adapt them according to their impact on tumor perfusion. Some contraindications of other imaging modalities cease to exist, allowing inclusion of patients otherwise difficult to investigate. The next step would be a three-dimensional, heartbeat-triggered measurement of perfusion in tissue blocks instead of two-dimensional tissue slices. We made this step only recently and can now define the blood flow volume through any horizontal tumor transsectional plane in ml/s (own unpublished results). Until now, only a few but encouraging studies exist with the PixelFlux technique. There are more questions than answers, and a united effort of researchers from many centers is necessary to gain the full potential of this promising novel technique.

References

Alcazar, J.L., Castillo, G., Jurado, M., and Lopez-Garcia, G. 2003. Intratumoral blood flow in cervical cancer as assessed by transvaginal color Doppler ultrasonography: correlation with tumor characteristics. *Int. J. Gynecol. Cancer 13*:510–514.

Blanco, E.C., Pastore, A.R., Fonseca, A.M., Carvalho, F.M., Carvalho, J.P., and Pinotti, J.A. 2003. Color Doppler sonography with contrast in the differentiation of ovarian tumors. *Rev. Hosp. Clin. Fac. Med. Sao Paulo 58*:185–192.

Bruehlmeier, M., Kaser-Hotz, B., Achermann, R., Bley, C.R., Wergin, M., Schubiger, P.A., and Ametamey, S.M. 2005. Measurement of tumor hypoxia in spontaneous canine sarcomas. *Vet. Radiol. Ultrasound 46*:348–354.

Chameleon-Software. 2004. PixelFlux. www.chameleon-software.de.

Chammas, M.C., Gerhard, R., de Oliveira, I.R., Widman, A., de Barros, N., Durazzo, M., Ferraz, A., and Cerri, G.G. 2005. Thyroid nodules: evaluation with power Doppler and duplex Doppler ultrasound. *Otolaryngol. Head. Neck. Surg. 132*:874–882.

Crokart, N., Radermacher, K., Jordan, B.F., Baudelet, C., Cron, G.O., Gregoire, V., Beghein, N., Bouzin, C., Feron, O., and Gallez, B. 2005. Tumor radiosensitization by antiinflammatory drugs: evidence for a new mechanism involving the oxygen effect. *Cancer Res. 65*:7911–7916.

Galie, M., D'Onofrio, M., Montani, M., Amici, A., Calderan, L., Marzola, P., Benati, D., Merigo, F., Marchini, C., and Sbarbati, A. 2005. Tumor vessel compression hinders perfusion of ultrasonographic contrast agents. *Neoplasia 7*:528–536.

Gallipoli, A., Manganella, G., De Lutiodi di Castelguidone, E., Mastro, A., Ionna, F., Pezzullo, L., and Vallone, P. 2005. Ultrasound contrast media in the study of salivary gland tumors. *Anticancer Res. 25*:2477–2482.

Haider, M.A., Milosevic, M., Fyles, A., Sitartchouk, I., Yeung, I., Henderson, E., Lockwood, G., Lee, T.Y., and Roberts, T.P. 2005. Assessment of the tumor microenvironment in cervix cancer using dynamic contrast enhanced CT, interstitial fluid pressure and oxygen measurements. *Int. J. Radiat. Oncol. Biol. Phys. 62*:1100–1107.

Imada, T., Tatsumi, T., Mori, Y., Nishiue, T., Yoshida, M., Masaki, H., Okigaki, M., Kojima, H., Nozawa, Y., Nishiwaki, Y., Nitta, N., Iwasaka, T., and Matsubara, H. 2005. Targeted delivery of bone marrow mononuclear cells by ultrasound destruction of microbubbles induces both angiogenesis and arteriogenesis response. *Arterioscler. Thromb. Vasc. Biol. 25*:2128–2134.

Jacob, A., Varghese, B.E., and Birchall, M.B. 2006. Validation of laser Doppler fluxmetry as a method of assessing neo-angiogenesis in laryngeal tumours. *Eur. Arch. Otorhinolaryngol. 263*:444–448.

Janssen, H.L., Haustermans, K.M., Balm, A.J., and Begg, A.C. 2005. Hypoxia in head and neck cancer: how much, how important? *Head Neck 27*:622–638.

Kan, Z., Kobayashi, S., Phongkitkarun, S., and Charnsangavej, C. 2005. Functional CT quantification of tumor perfusion after transhepatic arterial embolization in a rat model. *Radiology 237*:144–150.

Kelleher, D.K., Thews, O., and Vaupel, P. 1998. Regional perfusion and oxygenation of tumors upon methylxanthine derivative administration. *Int. J. Radiat. Oncol. Biol. Phys. 42*:861–864.

Kessler, T., Bieker, R.T., Padro, C., Schwoppe, T., Persigehl, C., Bremer, M., Kreuter, W.E., Berdel, W. E., and Mesters, R.M. 2005. Inhibition of tumor growth by RGD peptide-directed delivery of truncated tissue factor to the tumor vasculature. *Clin. Cancer Res. 11*:6317–6324.

Kiessling, F., Krix, M., Heilmann, M., Vosseler, S., Lichy, M., Fink, C., Farhan, N., Kleinschmidt, K., Schad, L., Fusenig, N.E., and Delorme, S. 2003. Comparing dynamic parameters of tumor vascularization in nude mice revealed by magnetic resonance imaging and contrast-enhanced intermittent power Doppler sonography. *Invest. Radiol. 38*:516–524.

Krix, M., Plathow, C., Essig, M., Herfarth, K., Debus, J., Kauczor, H.U., and Delorme, S. 2005. Monitoring of liver metastases after stereotactic radiotherapy using low-MI contrast-enhanced ultrasound—initial results. *Eur. Radiol. 15*:677–684.

Lehtio, K., Eskola, O., Viljanen, T., Oikonen, V., Gronroos, T., Sillanmaki, L., Grenman, R., and Minn, H. 2004. Imaging perfusion and hypoxia with PET to predict radiotherapy response in head-and-neck cancer. *Int. J. Radiat. Oncol. Biol. Phys.* 59:971–982.

Marret, H., Sauget, S., Giraudeau, B., Body, G., and Tranquart, F. 2005. Power Doppler vascularity index for predicting malignancy of adnexal masses. *Ultrasound Obstet. Gynecol.* 25:508–513.

Miyakawa, M., Onoda, N., Etoh, M., Fukuda, I., Takano, K., Okamoto, T., and Obara, T. 2005. Diagnosis of thyroid follicular carcinoma by the vascular pattern and velocimetric parameters using high resolution pulsed and power Doppler ultrasonography. *Endocr. J.* 52:207–212.

Okuyama, N., Murakuni, H., and Ogata, H. 2004. The use of Doppler ultrasound in evaluation of breast cancer metastasis to axillary lymph nodes. *Oncol. Rep.* 11:389–393.

Osanai, T., Wakita, T., Gomi, N., Takenaka, S., Kakimoto, M., and Sugihara, K. 2003. Correlation among intratumoral blood flow in breast cancer, clinicopathological findings and Nottingham Prognostic Index. *Jpn. J. Cin. Oncol.* 33:14–16.

Pandharipande, P.V., Krinsky, G.A., Rusinek, H., and Lee, V.S. 2005. Perfusion imaging of the liver: current challenges and future goals. *Radiology* 234:661–673.

Roubidoux, M.A., LeCarpentier, G.L., Fowlkes, J.B., Bartz, B., Pai, D., Gordon, S.P., Schott, A.F., Johnson, T.D., and Carson, P.L. 2005. Sonographic evaluation of early-stage breast cancers that undergo neoadjuvant chemotherapy. *J. Ultrasound Med.* 24:885–895.

Rouviere, O., Curiel, L., Chapelon, J.Y., Bouvier, R., Ecochard, R., Gelet, A., and Lyonnet, D. 2004. Can color Doppler predict the uniformity of HIFU-induced prostate tissue destruction? *Prostate* 60:289–297.

Scholbach, T., Dimos, I., and Scholbach, J. 2004b. A new method of color Doppler perfusion measurement via dynamic sonographic signal quantification in renal parenchyma. *Nephron Physiol.* 96:99–104.

Scholbach, T., Girelli, E., and Scholbach, J. 2005b. Dynamic tissue perfusion measurement: a novel tool in follow-up of renal transplants. *Transplantation* 79:1711–1716.

Scholbach, T., Girelli, E., and Scholbach, J. 2006. Tissue Pulsatility Index (TPI)—a new parameter to evaluate renal transplant perfusion. *Transplantation* 81:751–755.

Scholbach, T., Herrero, I., and Scholbach, J. 2004a. Dynamic color Doppler sonography of intestinal wall in patients with Crohn disease compared with healthy subjects. *J. Pediatr. Gastroenterol. Nutr.* 39: 524–528.

Scholbach, T., Scholbach, J., Krombach, G.A., Gagel, B., Maneschi, P., and Di Martino, E. 2005a. New method of dynamic color Doppler signal quantification in metastatic lymph nodes compared to direct polarographic measurements of tissue oxygenation. *Int. J. Cancer* 114:957–962.

Singh, S., Pradhan, S., Shukla, R.C., Ansari, M.A., and Kumar, A. 2005. Color Doppler ultrasound as an objective assessment tool for chemotherapeutic response in advanced breast cancer. *Breast Cancer* 12:45–51.

Testa, A.C., Ciampelli, M., Mastromarino, C., Lopez, R., Zannoni, G., Ferrandina, G., and Scambia, G. 2003. Intratumoral color Doppler analysis in endometrial carcinoma: is it clinically useful? *Gynecol. Oncol.* 88:298–303.

Vaupel, P., and Harrison, L. 2004. Tumor hypoxia: causative factors, compensatory mechanisms, and cellular response. *Oncologist 9 Suppl* 5:4–9.

Vaupel, P., Thews, O., and Hoeckel, M. 2001. Treatment resistance of solid tumors: role of hypoxia and anemia. *Med. Oncol.* 18:243–259.

Wawroschek, F., Vogt, H., Weckermann, D., Wagner, T., Hamm, M., and Harzmann, R. 2001. Radioisotope guided pelvic lymph node dissection for prostate cancer. *J. Urol.* 166:1715–1719.

Weber, M.A., Risse, F., Giesel, F.L., Schad, L.R., Kauczor, H.U., and Essig, M. 2005. [Perfusion measurement using the T2* contrast media dynamics in neuro-oncology. Physical basics and clinical applications]. *Radiologe* 45:618–632.

Yuh, E.L., Shulman, S.G., Mehta, S.A., Xie, J., Chen, L., Frenkel, V., Bednarski, M.D., and Li, K.C. 2005. Delivery of systemic chemotherapeutic agent to tumors by using focused ultrasound: study in a murine model. *Radiology* 234:431–437.

10

Immuno-Positron Emission Tomography

Lars R. Perk, Gerard W.M. Visser, and Guus A.M.S. van Dongen

Introduction

Recent advances in molecular and cellular biology have facilitated the discovery of novel molecular targets on tumor cells (as well as on cells involved in other pathological conditions), for example, key molecules involved in cell proliferation, cell death, differentiation, metabolism, cell cell interaction, environmental interactions, and vascularization. This knowledge has boosted the design of cutting-edge pharmaceuticals, with monoclonal antibodies (MAbs) forming the most rapidly expanding category. These antibodies can be used as disease-specific contrast agents in imaging for diagnostic purposes. To date, the U.S. Food and Drug Administration (FDA) has approved five diagnostic MAbs, including four for detection of cancer. In addition, MAbs are particularly gaining momentum for use in disease-selective therapy. Presently, 17 MAbs (all intact immunoglobulins) have been approved by the FDA for therapy, most of them for systemic treatment of cancer. The yearly sales of MAbs were estimated at 5–8 billion in 2005. Hundreds of other MAbs are in clinical trials worldwide.

Introduction of immuno-PET, the combination of positron emission tomography (PET) with MAbs, is an attractive novel option to improve tumor detection because it combines the high sensitivity and resolution of a PET camera with the specificity of a MAb. For this purpose the MAb has to be labeled with a positron emitter. Immuno-PET might provide an improvement of the currently used PET tracers such as fluorine-18-labeled fluoro-2-deoxy-D-glucose (^{18}FDG), which shows increased uptake not only in tumors but also in normal tissues with high metabolic activity and other pathologic conditions such as inflammation. Apart from its diagnostic capabilities, PET also has potential for quantification of molecular interactions, which is particularly attractive when immuno-PET is used as prelude to antibody-based therapy. In an individualized therapeutic approach, immuno-PET enables the confirmation of tumor targeting and the quantification of MAb accumulation in tumor and normal tissues. Therefore, patients can be selected who have the best chance to benefit from expensive MAb therapy. In addition, immuno-PET might play a role in the selection, characterization, and optimization of novel high-potential MAb or MAb conjugate candidates for diagnosis and therapy. Introduction of hybrid scanners such as PET/CT will add a new dimension to immuno-PET, because such an approach allows simultaneous anatomical and functional imaging.

This chapter provides general information about MAb-targeted diagnosis and therapy of cancer, as well as about PET, PET systems, and quantification. The requirements for an appropriate positron emitter and characteristics of the most attractive candidate emitters for immuno-PET are discussed. An overview of preclinical and first clinical immuno-PET studies reported in literature is provided.

Diagnostic and Therapeutic Applications of Monoclonal Antibodies

A century ago, Paul Ehrlich postulated the notion that a "magic bullet" could be developed to selectively target disease. He envisioned antibodies as that magic bullet. The development of hybridoma technology for production of MAbs by Kohler and Milstein (1975) in the mid-1970s turned Ehrlich's concept into a realistic option. With hybridoma technology, an unlimited range of MAbs can be developed against any particular cellular antigen. However, the first generation of MAbs evaluated in the clinic was limited by their immunogenicity due to their murine origin. Developments in recombinant DNA technology circumvented this limitation, resulting in the production of chimeric, humanized, and complete human MAbs. Besides intact MAb molecules (molecular weight ~150 kDa), MAb fragments are also used, including F(ab')$_2$ (~ 100 kDa), F(ab') (~ 50 kDa), Fab (~ 40 kDa), and single-chain Fv (scFv, ~ 25 kDa). The scFv can be further engineered to form covalent dimers, for example, scFv$_2$ (~ 50 kDa), a diabody (~ 55 kDa), or a minibody (~ 80 kDa). These dimers all exhibit high functional affinity.

Intact MAb molecules have a long residence time in humans ranging from a few days to weeks, which results in optimal tumor-to-nontumor ratios at 2–4 days after injection. The use of MAb fragments increases blood clearance. This results in higher tumor-to-nontumor ratios at earlier time points, but the absolute tumor uptake is often lower. For example, scFv fragments are cleared from the blood within 20 hrs., resulting in high tumor-to-nontumor ratios already within a few hours after administration. These characteristics in general make intact MAbs the format of choice for therapy, while MAb fragments are considered to be more suitable for diagnosis. New strategies include the use of pretargeting approaches, which involve separating the targeting antibody from the subsequent delivery of an imaging or therapeutic agent that binds to the tumor-localized antibody (Goldenberg *et al.*, 2006).

In order to use MAbs for diagnostic purposes, MAbs have been labeled with γ-emitting radionuclides and imaged with a single photon emission computerized tomography (SPECT) camera. Until now, five radiolabeled MAbs have been approved by the FDA for diagnostic imaging, among which are four for imaging cancer (Table 6). The first radiolabeled MAb that received approval in 1992 was OncoScint. This agent consists of the indium-111-labeled (^{111}In-labeled) murine MAb satumomab (also known as B72.3) that recognizes the tumor-associated glycoprotein-72 (TAG-72) expressed on ovarian and colorectal cancer cells. In 1996, four more diagnostic antibodies labeled with ^{111}In or technetium-99m (^{99m}Tc) were approved. Three of them—CEA-scan, verluma, and Prosta Scint—are indicated for detection of several forms of cancer. These diagnostic agents are used mainly for staging disease in patients suspected of recurrent or metastatic disease, and not for therapy.

Beginning in 1997, eight MAbs have been approved for treatment of cancer (Table 7). Five of these MAbs have been approved for treatment of hematologic malignancies: rituximab, gemtuzumab ozogamicin, alemtuzumab, ibritumomab tiuxetan, and tositumomab. Three MAbs have been approved for therapy of solid tumors: trastuzumab for treatment of metastatic breast cancer, and cetuximab and bevacizumab for treatment of metastatic colorectal cancer. Most of these are unmodified MAbs and can act by mediating antibody-dependent cellular cytotoxicity (ADCC) or complement-depending cytotoxicity (CDC), by apoptosis induction, or by interfering with signal transduction pathways. To enhance its therapeutic efficacy, gemtuzumab has been armed with the supertoxic drug ozogamicin, while ibritumomab tiuxetan (Zevalin) and tositumomab (Bexxar) are radiolabeled MAbs containing the β-emitters yttrium-90 (^{90}Y) and iodine-131 (^{131}I), respectively. Through a "cross-fire" effect, radionuclides are especially attractive as warheads since, in order to be effective, not all tumor cells have to be targeted by radiolabeled MAbs. The therapeutic value of the aforementioned MAbs has been outlined in several excellent reviews (Stern and Herrmann, 2005; Adams and Weiner, 2005). Hundreds of intact MAbs and MAb fragments are currently evaluated in clinical trials.

Therapy Planning with Monoclonal Antibodies

Despite encouraging results, it needs to be stated that MAb therapy is also facing some restrictions. Because of interpatient variations in pharmacokinetics and tumor targeting, MAb therapy is not always as effective and safe as expected. In current practice, pathological analyses are often performed to confirm target expression and to select patients for MAb therapy. For example, patients with metastatic breast cancer are only eligible for therapy with the anti-HER2 MAb trastuzumab, when overexpression and gene amplification of this target protein have been confirmed on a tumor biopsy by immunohistochemistry or fluorescence *in situ* hybridization (in 20–30% of patients). However, recent data indicate that HER2 expression of the primary tumor might be different from expression of metastatic lesions, while during the course of disease upon chemotherapy and/or hormonal therapy, HER2-negative tumors can become HER2-positive, and vice versa. In addition,

Table 6 Diagnostic Radiolabeled MAbs Approved by the FDA

Table 6 Diagnostic Radiolabeled MAbs Approved by the FDA

Generic Name (Trade Name)	Target	Type	Condition
Satumomab pendetide (OncoScint)	TAG 72	^{111}In-labeled murine IgG1	Ovarian, colorectal carcinoma
Acritumomab (CEA-scan)	CEA	^{99m}Tc-labeled murine F(ab')	Colorectal carcinoma
Imciromab pentetate (Myoscint)	myosin	^{111}In-labeled murine Fab	Myocardial injury
Nofetumomab merpentan (Verluma)	EGP40	^{99m}Tc-labeled murine Fab	Small-cell lung cancer
Capromab pendetide (ProstaScint)	PSMA	^{111}In-labeled murine IgG1	Prostate cancer

myocardial HER2 expression can occur in patients shortly after anthracycline treatment and in patients with heart failure; this can be considered a contraindication for trastuzumab treatment. These data indicate that pretherapy imaging to confirm target expression and selective MAb accumulation in all tumor deposits *in vivo*, might be of help for the selection of patients with the highest chance of benefit from MAb therapy. Radioimmunoscintigraphy (RIS) can play an important role herein.

Until now, pretherapy RIS has mostly been applied as prelude to therapy with radiolabeled MAbs (radioimmunotherapy = RIT) for evaluation of tumor targeting and dosimetry. Both the Zevalin and Bexxar treatment regimens are originally preceded by an SPECT imaging procedure. In the Zevalin treatment regimen, a tracer amount of ^{111}In-ibritumomab tiuxetan is administered 7–9 days before therapy with ^{90}Y-ibritumomab tiuxetan, in order to rule out patients with unfavorable biodistribution. The Bexxar treatment schedule consists of a dosimetric dose of ^{131}I-tositumomab followed by a therapeutic dose of the same conjugate within 7–14 days. Patient-specific doses are administered on the basis of the calculated total-body residence time of dosimetric ^{131}I-tositumomab (Wahl, 2005). In case of RIT, pretherapy imaging might be particularly important because of the small therapeutic window with bone marrow being the dose-limiting organ.

Detection of the γ-ray emitting radionuclides ^{99m}Tc, ^{111}In, and ^{131}I involves planar or SPECT imaging. SPECT procedures, however, have intrinsic limitations with respect to quantification, primarily because of insufficient correction for scatter and partial absorption of γ-photons in tissue. Because of more accurate scatter and attenuation correction, PET is better qualified for tracer quantification. In addition, PET provides better spatial and temporal resolution and sensitivity than SPECT.

Immuno-PET: Imaging and Quantification

Immuno-PET is based on annihilation coincidence detection after labeling of a MAb or MAb fragment with a positron-emitting radionuclide. The emitted positron (a positively charged β-particle) will travel a distance of a few millimeters, depending on the initial positron energy and the density of the surroundings. After having lost its kinetic energy, combining with an electron leads to the so-called annihilation process, yielding two photons each with an energy of 511 keV emitted simultaneously in opposite directions. After administration of a PET conjugate to a patient, the distribution of the compound is monitored by detection of

Table 7 Therapeutic MAbs Approved for Cancer Treatment (Adams and Weiner, 2005)

FDA Approved	Generic Name (Trade Name)	Target	Type	Indication
1997	Rituximab (Rituxan)	CD20	Chimeric IgG1	Lymphoma
1998	Trastuzumab (Herceptin)	HER2/neu	Humanized IgG1	Breast cancer
2000	Gemtuzumab ozogamicin (Mylotarg)	CD33	Humanized IgG4 conjugated to calicheamicin	Acute myeloid leukemia
2001	Alemtuzumab (Campath-1H)	CD52	Humanized IgG1	Chronic lymphatic leukemia
2002	^{90}Y-Ibritumomab tiuxetan (Zevalin)	CD20	^{90}Y-radiolabeled murine IgG1	Non-Hodgkin's Lymphoma
2003	^{131}I-Tositumomab (Bexxar)	CD20	^{131}I-radiolabeled murine IgG2a	Non-Hodgkin's Lymphoma
2004	Bevacizumab (Avastin)	VEGF	Humanized IgG1	Colorectal, lung cancers
2004	Cetuximab (Erbitux)	EGFR	Chimeric IgG1	Colorectal

the annihilation photon pairs with a PET camera. A PET camera consists of a ring of detectors placed around the body of a patient. If two photons are registered by detectors on opposite sides of the body within a very short time interval (typically 5–15 ns), it is assumed that somewhere along the line between the two detectors an annihilation event has taken place. By calculating the crossing of all lines of response (LORs), the location of the radiation source can be determined.

For quantification, immuno-PET can provide reliable information when appropriate corrections are performed as described previously (Verel *et al.*, 2005). All photon pairs with energies that fall within the PET acquisition energy window (typically 350–650 keV) are called coincidences. In addition to true coincidences, the gross coincidence rate can include scatter coincidences, random coincidences, and spurious coincidences. For all latter three, however, the LOR is not representative of the annihilation location and degrades the image quality. In addition, some portion of the annihilation photons will not be detected as a result of attenuation and dead time. The probability of true coincidence detection is the product of the probabilities of each of the two annihilation photons escaping the body without interaction. For this reason, the amount of attenuation along a specific LOR is the same, independent of the event on the LOR. This fact is used in attenuation correction and requires an additional transmission scan in which the body is scanned with a rotating external radioactivity point or line source. Dead time losses result from the system's inability to process an infinite number of photons at the same time. Dead time losses can be minimized by using a system with many independent detectors, fast scintillators, and fast processing electronics. To arrive at a quantitative image, the raw PET data must be corrected for the aforementioned disturbing effects. After correction of raw PET data, the only factor influencing quantitative PET analysis of radioactivity is resolution related.

Clinical PET Imaging Systems

In 1975, the first PET scanner with thallium-doped sodium iodide (NaI[Tl]) detectors that could form true tomographic images was developed. Since then, the PET scanner has undergone several developments with respect to scintillation crystals and system design. Over the years, the hexagonal and octagonal design of the first PET systems was followed by circular designs. Today, most PET scanners contain bismuth germanate, cerium-doped gadolinium oxyorthosilicate, or cerium-doped lutetium oxyorthosilicate as scintillation material because of their high efficiency for detecting 511keV photons. Several other detector materials are under investigation. It is expected that advances in this area will further improve PET sensitivity and resolution.

To facilitate accurate interpretation of PET images and quantification, PET can be combined with CT or MRI to provide simultaneous registration of both biologic function and anatomy. In only a few years, ~ 75% of all PET sales comprise PET/CT scanners. Although combining PET with MRI is technologically more challenging because of the strong magnetic fields restricting the use of certain electronic components, the first MRI-compatible PET scanner has been developed.

Positron Emitters for Immuno-PET

For a positron emitter to be appropriate for immuno-PET, it has to fulfill several requirements. The positron emitter should have appropriate decay characteristics for optimal resolution and optimal quantitative accuracy, its production should be easy and cheap, and it should be possible to achieve easy and stable coupling to MAbs with maintenance of the antibody's *in vivo* biodistribution characteristics. It is also important that its physical half-life ($t_{1/2}$) be compatible with the time needed for an MAb or MAb fragment to achieve optimal tumor-to-nontumor ratios (typically 2–4 days for intact MAbs and 2–6 hr for MAb fragments). Although the use of shorter-lived positron emitters (with $t_{1/2}$ of hours) for immuno-PET with MAb fragments is an option, the kinetics of intact MAbs demand the use of long-lived positron emitters ($t_{1/2}$ of days) to allow imaging at later time points for obtaining maximum information.

Given these considerations, the application of positron emitters with half-life of minutes, such as the bio-isotopes oxygen-15, nitrogen-13, and carbon-11, is out of the question. Candidate positron emitters for immuno-PET include gallium-68 (^{68}Ga; $t_{1/2}$: 1.13 hr), fluorine-18 (^{18}F; $t_{1/2}$: 1.83 hr), copper-64 (^{64}Cu; $t_{1/2}$: 12.7 hr), yttrium-86 (^{86}Y; $t_{1/2}$: 14.7 hr), bromine-76 (^{76}Br; $t_{1/2}$: 16.2 hr), zirconium-89 (^{89}Zr; $t_{1/2}$: 78.4 hr), and iodine-124 (^{124}I; $t_{1/2}$: 100.3 hr) (Table 8). Most of these positron emitters have a short half-life and can only be used in combination with MAb fragments or in pretargeting approaches. In contrast, ^{89}Zr and ^{124}I are particularly suitable when used in combination with intact MAbs. The half-life of ^{89}Zr and ^{124}I also offers an advantage for logistics related to transportation and labeling of MAbs, but a potential disadvantage with respect to radiation burden to the patient, especially when coupled to compounds with long biological half-lives, such as intact MAbs. This disadvantage might be overcome by the introduction of the latest generation of high-resolution PET cameras with higher sensitivity, allowing lower doses of injected radioactivity for comparable images.

Other aspects of a positron emitter to be considered are the presence of high-energy γ-photons emitted in coincidence with positrons (^{86}Y, ^{76}Br, and ^{124}I), high ß⁺-energies (^{68}Ga, ^{76}Br, and ^{124}I) resulting in resolution loss and quantification artifacts, and the presence of radionuclidic

Table 8 Main Characteristics of Positron Emitters Used in Preclinical and Clinical RIS Studies

Positron Emitter	Production	Half-life (h)	Main ß$^+$- Energies (keV)	(%)	Intrinsic Spatial Resolution Loss (mm)
^{68}Ga	^{68}Ge/^{68}Ga-generator	1.13	1899	87.9	2.4
^{18}F	^{18}O(p,n) ^{20}Ne(d,α)	1.83	634	100.0	0.7
^{64}Cu	^{64}Ni(d,2n) ^{64}Ni(p,n)	12.7	653	17.4	0.7
^{86}Y	^{86}Sr(p,n)	14.7	1221	11.9	1.8
			1545	5.6	
^{76}Br	^{75}As(^{3}He,2n) ^{76}Se(p,n)	16.2	871	6.3	5.3
			990	5.2	
			3382	25.8	
			3941	6.0	
^{89}Zr	^{89}Y(p,n)	78.4	897	22.7	1.0
^{124}I	^{124}Te(p,n)	100.3	1535	11.8	2.3
	^{124}Te(d,2n)		2138	10.9	
	^{125}Te(p,2n)				

impurities (^{64}Cu and ^{124}I). Except for the production of ^{76}Br and ^{89}Zr as well as of ^{68}Ga, a generator-produced nuclide, all positron emitters mentioned earlier require enrichment of the target material, resulting in high costs of goods. While the positron emitters ^{68}Ga, ^{18}F, ^{64}Cu, ^{86}Y, and ^{89}Zr require indirect labeling methods using bifunctional chelates, ^{76}Br and ^{124}I can also be coupled directly to MAbs.

An attractive application is the use of immuno-PET as a scouting procedure for selecting MAb therapy candidates. This particularly holds true for RIT, to confirm tumor targeting and perform dosimetry. For this purpose, the radioimmunoconjugates used for immuno-PET and RIT should demonstrate a similar biodistribution; therefore, radionuclides (and, if required, chelates) with comparable chemical properties have to be chosen. Most commonly used beta emitters in RIT studies are copper-67 (^{67}Cu; $t_{1/2}$: 62 hr), rhenium-186 (^{186}Re; $t_{1/2}$: 89 hr), rhenium-188 (^{188}Re; $t_{1/2}$: 17 hr), lutetium-177 (^{177}Lu; $t_{1/2}$: 161 hr), ^{90}Y ($t_{1/2}$: 64 hr), and ^{131}I ($t_{1/2}$: 192 hr). Examples of PET/RIT radionuclide pairs are ^{64}Cu/^{67}Cu, ^{86}Y/^{90}Y, and ^{124}I/^{131}I. However, PET/RIT surrogate pairs also can be considered, such as ^{124}I/^{186}Re, ^{124}I/^{188}Re, ^{76}Br/^{131}I, ^{86}Y/^{177}Lu, ^{89}Zr/^{177}Lu, and ^{89}Zr/^{90}Y.

Another important consideration is whether the MAb or MAb fragment is internalized once bound to the target antigen. Degradation of ^{76}Br- and ^{124}I-labeled MAbs will result in a rapid clearance of these radionuclides from the target cells; hence, the PET image is not reflecting the actual MAb distribution. In contrast, when ^{68}Ga-, ^{18}F-, ^{64}Cu-, ^{86}Y-, and ^{89}Zr-labeled MAbs using bifunctional chelates are processed, the positron emitters will be trapped intracellular in lysosomes, so-called residualization. Therefore, this phenomenon should be taken into account when selecting a positron emitter for immuno-PET monitoring of MAb-based therapy.

Experience with Preclinical Immuno-PET

Although several positron emitters have been suggested for labeling MAbs and MAb fragments, immuno-PET research is still in its infancy. Most immuno-PET studies have focused on animal models, while clinical evaluations are rare. This is mainly due to the limited availability of suitable positron emitters, the lack of robust labeling methods, and, until recently, the relative small number of PET cameras in operation. The most widely available and applied positron emitter is ^{18}F. The success of PET as a diagnostic tool is mainly due to the use of ^{18}FDG. For coupling to MAbs, however, ^{18}F is less favorable. Major disadvantages are its short half-life and the complex, time-consuming, and specialized chemistry required for ^{18}F labeling. Although several conjugates have been evaluated in biodistribution studies in xenograft-bearing nude mice, no quantitative preclinical or clinical PET studies have been performed.

Another short-lived positron emitter is ^{68}Ga. This positron emitter is especially applied for the labeling of peptides (outside the focus of this review) and in so-called pretargeting strategies and for the labeling of small MAb fragments. Pretargeted PET with ^{68}Ga has been evaluated in mice bearing CD44v6- and MUC1-expressing tumors and in patients with breast cancer (Schumacher et al., 2001) as will be reported later. Smith-Jones et al. (2004) used micro-PET imaging with ^{68}Ga-labeled F(ab')$_2$ fragments of the anti-HER2 MAb trastuzumab to monitor HER2 expression in animal tumors during the course of treatment with 7-allylaminogeldanamycin, a Hsp90 chaperone protein inhibitor causing HER2 degradation. This is one of the few studies showing the real potential of quantitative immuno-PET imaging: a linear correlation (R = 0.972) was found between tracer uptake quantified by direct assessment of dissected tumors by γ-counter and tracer uptake estimated noninvasively by micro-PET.

Positron emitters suggested for immuno-PET with half-lives of several hours (but less than 1 day) are ^{64}Cu, ^{86}Y, and ^{76}Br. In a study performed by Wu et al. (2000), the genetically engineered anticarcinoembryonic antigen (CEA) T84.66/GS18 minibody was labeled with ^{64}Cu via the macrocyclic chelate DOTA, and micro-PET was performed in colon carcinoma bearing mice. Two to 24 hr after injection with ^{64}Cu-DOTA-minibody, tumors ranging from 27–395 mg could readily be detected. In addition, significant nonspecific uptake was seen in the livers. This will restrict the detection of hepatic lesions.

Grünberg et al. (2005) evaluated F(ab')$_2$ fragments of the internalizing anti-L1-CAM antibody chCE7 labeled with ^{64}Cu, ^{67}Cu, and ^{177}Lu for tumor diagnosis and therapy. PET images obtained 21 hr after i.v. injection of 10 MBq ^{64}Cu-chCE7F(ab')$_2$ showed clear visualization of both large tumors and small-sized peritoneal metastases with low background. Also in this study, prominent uptake of radioactivity was seen in the kidneys and liver. Radiocopper-labeled F(ab')$_2$ fragments enables the possibility of performing a PET scouting procedure with the ^{64}Cu-labeled MAb derivative prior to therapy with the ^{67}Cu-labeled counterpart. In contrast, ^{64}Cu does not seem very attractive as PET surrogate for ^{177}Lu-RIT due to differences in liver and kidney accumulation.

The positron emitter ^{86}Y has been coupled to the anti-Lewis Y humanized MAb hu3S193 and the anti-HER2 MAb trastuzumab (Herceptin), and their biodistributions were compared with those of their respective ^{111}In-labeled MAbs (Lovqvist et al., 2001; Garmestani et al., 2002). For prediction of ^{90}Y-MAb biodistribution and dosimetry in RIT studies, PET with the chemically identical positron emitter ^{86}Y coupled to MAbs might be better qualified than ^{111}In-MAb-SPECT. Although biodistribution data from both studies indicated that this might be the case, accurate quantification of ^{86}Y with PET still remains a challenge because of the prompt single γ-photons emitted in coincidence with positrons. In addition, the half-life of ^{86}Y ($t_{1/2}$: 14.7 hr) is relatively short for optimal ^{90}Y ($t_{1/2}$: 64 hr) dosimetry prediction.

More recently, Parry et al. (2005) used ^{86}Y-labeled anti-mindin/RG-1 fully human MAb 19G9 and small animal PET for visualization of prostate cancer xenografts in nude mice at different time points for up to 72 hr. At 4 hr, tumors could already be delineated in all mice ($n = 4$). Over the 72 hr imaging period, exceptionally high tumor-to-background contrast was noted in all animals. Furthermore, they used a co-registration technique for combining PET and micro-CT data to obtain PET/CT images. Unfortunately, no quantitative analysis was performed; ^{86}Y-PET images were used only for visualization.

^{76}Br has been coupled to the anti-CEA MAb 38S1 by Lovqvist et al. (1997), and biodistribution studies and PET were performed in colon carcinoma xenograft-bearing nude rats. The authors hypothesized that ^{76}Br might be an interesting alternative to ^{124}I as a PET radionuclide substitute for the iodine radioisotopes used in RIS and RIT. To this end, ^{76}Br-38S1 and ^{125}I-38S1 were prepared by direct labeling, and their biodistribution was compared in nude rats carrying human colon xenografts. Although all tumors could be readily identified by PET, deviating biodistribution was observed for ^{76}Br-38S1 and ^{125}I-38S1. These data suggest that ^{76}Br cannot be used as a PET radionuclide substitute for iodine radioisotopes.

The long-lived positron emitters ^{124}I and ^{89}Zr are particularly suitable for immuno-PET when used in combination with intact MAbs. In the early 1990s, ^{124}I-labeled MAbs already have been used for the imaging of tumors in preliminary preclinical and clinical studies. After almost a decade, interest in ^{124}I-labeled MAbs has been renewed, partly because of the improved methods for coupling of ^{124}I to MAbs (Verel et al., 2004). Lee et al. (2001) used ^{124}I to label the recombinant humanized anticolorectal cancer A33 MAb, and evaluated its biodistribution properties and PET imaging characteristics in colorectal cancer xenograft-bearing mice. Excellent tumor uptake was seen. PET detected antigen-positive tumors by 4 hr after injection, and high-resolution images were obtained 24 hr after injection. The potential of ^{124}I-labeled genetically engineered antibody fragments for tumor detection was confirmed by Sundaresan et al. (2003), who demonstrated high-resolution images and specific localization (compared to ^{18}F-FDG) of ^{124}I-labeled anti-CEA diabodies and minibodies directed to CEA-positive xenografts. In these studies, quantification of ^{124}I from the PET scans was not performed.

In a recent study, the ^{124}I-labeled anti-HER2 C6.5 diabody was used for visualization and quantification of HER2 expression with a clinical PET/CT scanner (Robinson et al., 2005). They used an indirect labeling method to circumvent loss of label in the tumors due to dehalogenation and catabolism of the labeled protein after internalization. Unfortunately, this method resulted in a decreased labeling efficiency and immunoreactivity compared to the direct labeling method. Nevertheless, high-resolution imaging of the tumors was possible. PET-based quantification of tumor uptake correlated well with necropsy-based analysis ($R^2 = 0.96$). For clinical application of this PET-based method, it will be necessary to develop reliable CT-based tumor-volume estimates.

MAbs labeled with ^{124}I may form an attractive option for imaging with PET in a scouting procedure prior to ^{131}I-radioimmunotherapy. To this end, Verel et al. (2004) evaluated the scouting performance of ^{124}I-labeled chimeric MAb (cMAb) U36 in a biodistribution experiment on coinjection with ^{131}I-labeled cMAb U36, and by PET in nude mice bearing head and neck xenografts. ^{124}I- and ^{131}I-labeled cMAb U36 showed fully congruent biodistributions, whereas selective tumor uptake was confirmed by immuno-PET imaging with visualization of all 15 tumors.

Recently, our group introduced the long-lived positron emitter ^{89}Zr as a promising residualizing radionuclide for immuno-PET. Stable coupling of ^{89}Zr to MAbs, including cMAb U36 for detection of head and neck squamous cell carcinoma (HNSCC), was accomplished using the succinylated chelate desferrioxamine B (desferal) (Verel et al., 2003a). The suitability of ^{89}Zr-labeled cMAb U36 for detection of millimeter-sized tumors (19–154 mm^3) was demonstrated in HNSCC xenograft-bearing nude mice (Fig. 24) (Verel et al., 2003b). In addition, the potential of PET to quantify ^{89}Zr-labeled cMAb U36 was proven in the same study, showing good correlation (R^2 = 0.79) between PET-image derived tumor-uptake data (noninvasive method) and uptake data derived from excised tumors (invasive method).

Studies were conducted to examine the applicability of ^{89}Zr-immuno-PET as a scouting procedure for ^{90}Y or ^{177}Lu-RIT (Verel et al., 2003b; Perk et al., 2005). The extensively internalizing anti-EGFR MAb cetuximab (Erbitux) was labeled with ^{89}Zr and with ^{90}Y as well as with ^{177}Lu. ^{89}Zr-labeled cetuximab was coinjected with either ^{90}Y- or ^{177}Lu-labeled cetuximab in xenograft-bearing nude mice, and biodistributions were assessed at 24, 48, 72, and 144 hr after injection. In both studies, ^{89}Zr-labeled MAbs on the one hand and ^{88}Y/^{177}Lu-labeled MAbs on the other hand showed similar uptake in tumor, blood, and other organs, except for sternum and thigh bone at later points in time.

Experience with Clinical Immuno-PET

As indicated earlier, clinical immuno-PET studies have been rare thus far. Nevertheless, some encouraging first results have been obtained. Pretargeted PET with ^{68}Ga has been evaluated in patients with breast cancer (Schumacher et al., 2001). Patients received anti-MUC1/anti-Ga-chelate bispecific antibody (Bs-MAb), followed 18 hr later by a blocker to remove residual circulating Bs-MAb from the blood and another 15 min later by the ^{68}Ga-labeled chelate. PET imaging was started 60–90 min after injection of the ^{68}Ga-chelate. Fourteen of 17 known lesions, averaging 25 ± 16 mm in size, were clearly visualized as foci of increased activity with PET. No false-positive readings were obtained. Detection of axillary lymph node metastases was hampered by a high activity in large blood vessels located proximal to the lymph nodes. It was concluded that PET offered better sensitivity for the detection of breast cancer at low tumor contrast than conventional RIS.

The anticolorectal carcinoma murine MAb 1A3 was labeled with ^{64}Cu and evaluated in 36 patients with suspected advanced primary or metastatic colorectal cancer (Philpott et al., 1995). After injection of the

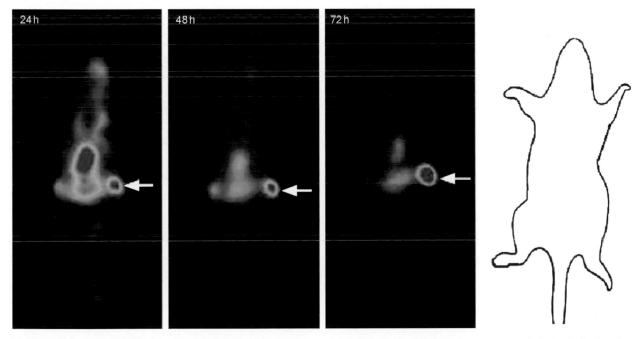

Figure 24 Immuno-PET images obtained from the same mouse bearing head and neck cancer xenografts in both flanks after injection of ^{89}Zr-labeled cMAb U36. Coronal images were acquired at 24, 48, 72 hr post-injection. Image planes are those for which only the right tumor (arrow) is clearly visible.

radioimmunoconjugate, PET was performed once or twice, 4–36 hr later. All patients underwent CT or MRI, and 18 patients were also studied with ^{18}F-FDG PET. In 29 patients, 40 of the 56 confirmed tumor sites were detected by immuno-PET (sensitivity 71%), absence of tumor was confirmed in 5 patients, and tumor status was not confirmed in 2 patients. All 17 primary and recurrent sites were clearly visualized, but only 23 of 39 metastatic sites were detected. The sensitivity of immuno-PET was best in abdomen and pelvis. In these regions, 11 new occult tumor sites were detected, including 9 small foci < 2.0 cm in diameter that were not detected by CT or MRI. In contrast, detection of metastatic disease of liver and lung was difficult because of high blood-pool activity at the early imaging time points, which were chosen because of the half-life of [64]Cu (12.7 hr). Detection of liver metastases was also hampered by nontarget background activity, caused by ^{64}Cu residualization in the liver after catabolism of the radioimmunoconjugate. In sum, for detection of small tumor foci in the abdomen or pelvis, ^{64}Cu-immuno-PET showed better sensitivity than CT or MRI and appeared as sensitivity as ^{18}F-FDG PET.

[124]I-labeled MAbs were already used for clinical immuno-PET ~ 15 years ago, but the number of patients included in these studies was small (Wilson *et al.*, 1991; Larson *et al.*, 1992). In addition, MAbs used for these studies were murine MAbs lacking the specificity of today's MAbs. Therefore, diagnostic results were far from optimal.

Because of the encouraging results obtained with [89]Zr-cMAb-U36 in animal studies, a clinical immuno-PET feasibility trial was conducted (Borjesson *et al.*, 2006; Zalutsky, 2006). The aim of this trial was to determine the diagnostic value of immuno-PET with [89]Zr-labeled cMAb U36 in patients with HNSCC, who were at high risk of having neck lymph node metastases. Twenty HNSCC patients, scheduled to undergo resection of the primary tumor and neck dissection, received 74 MBq [89]Zr-labeled cMAb U36 (10 mg) intravenously. All patients were examined by CT and/or MRI and immuno-PET prior to surgery. Six patients also underwent ^{18}F-FDG-PET. Immuno-PET scans were acquired up to 144 hr after injection. Diagnostic findings were recorded per neck side (left or right) as well as per lymph node level (six levels per side), and compared with histopathological outcome ("gold standard"). For this purpose the CT/MRI scores were combined, and the best of both scores was used for analysis. Immuno-PET detected all primary tumors ($n = 17$) as well as lymph node metastases in 18 of 25 positive levels (sensitivity 72%) and in 11 of 15 positive sides (sensitivity 73%). Representative immuno-PET images are shown in Figures 25 and 26 a–c. Interpretation of immuno-PET was correct in 112 of 121 operated levels (accuracy 93%) and in 19 of 25 operated sides (accuracy 76%). For CT/MRI, sensitivities of 60% and 73% and accuracies of 90% and 80% were found per level and side, respectively. In the six patients with seven tumor-

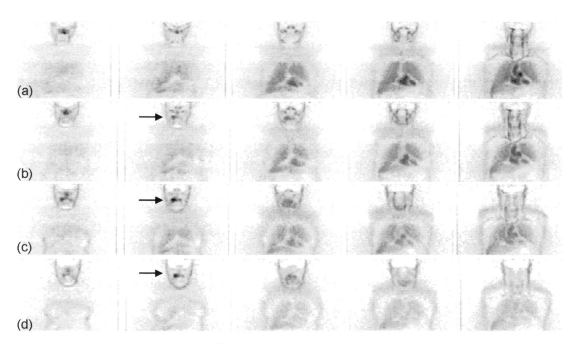

Figure 25 Immuno-PET image with [89]Zr-cMAb U36 of a head and neck cancer patient with a tumor of the right tonsil. Images were obtained within 1 hr (a), at 24 hr (b), at 72 hr (c), and at 120 hr p.i. (d). Slices from anterior (left) to posterior (right). Early images show mainly blood–pool activity with visualization of nose, heart, lungs, and liver. At later images the primary tumor is clearly visualized (arrow).

involved neck levels and sides, immuno-PET and ^{18}F-FDG-PET gave comparable diagnostic results.

The paraffin slides of the seven tumor-involved lymph node levels that had been missed with immuno-PET were reexamined by histopathological examination. The tumor-involved lymph nodes found in these levels were relatively small and contained just a small proportion of tumor tissue. Six of seven tumor-involved lymph node levels that had been missed by immuno-PET were also missed by CT and/or MRI. The smallest tumor-involved lymph node detected by immuno-PET was 5×5 mm, with 75% tumor involvement.

The authors concluded that immuno-PET with [89]Zr-cMAb U36 performed at least as good for detection of HNSCC lymph node metastases as CT/MRI. Therefore, immuno-PET with [89]Zr-cMAb U36 might also have per-spectives for the detection of distant metastases. In this study, limitations of current immuno-PET were also met. Although good resolution and sensitivity are advantages, it is fair to say that immuno-PET is hampered by the lack of anatomic structures. For example, because of the MAb's slow clearance from the blood, it might be difficult to distinguish a targeted lymph node metastasis from a cross section through a blood vessel. To deal with this problem, combining immuno-PET with anatomical imaging seems to be an interesting future approach, as demonstrated in this study by PET-CT image fusion (Fig. 26 d–f).

In conclusion, immuno-PET combines the high-resolution and quantitative aspects of PET with the high specificity and selectivity of MAbs. This makes immuno-PET an attractive imaging modality for tumor detection. In addition, immuno-PET has the potential to supersede γ-camera imaging for

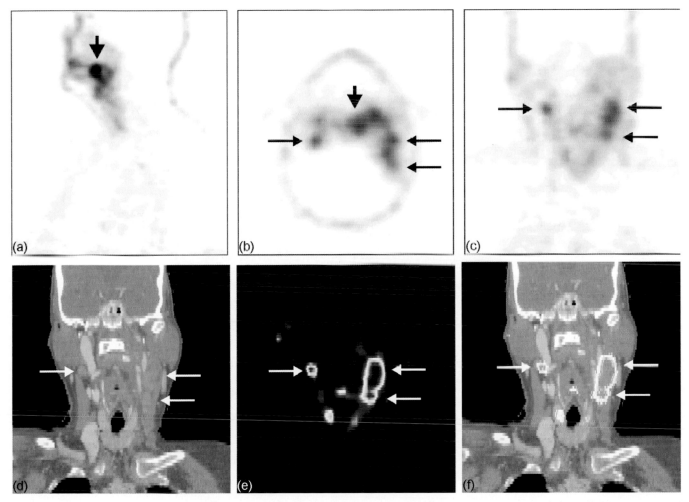

Figure 26 Immuno-PET images with [89]Zr-cMAb U36 of a head and neck cancer patient with a tumor in the left tonsil (large arrow) and lymph node metastases (small arrows) at the left (level II and III) and right (level II) side of the neck. Images were obtained 72 hr post-injection. Sagittal image (a), axial image (b), and coronal image (c). Primary tumor and lymph node metastases were clearly visualized with immuno-PET. In addition, electronic fusion (f) of CT (d) and coronal immuno-PET (e) images of the same head and neck cancer patient as above. In these slices, only the lymph node metastases are visible.

guiding MAb-based therapy because it enables the sensitive confirmation of tumor targeting and a more reliable quantification of MAb uptake in both tumor and normal tissues. Therefore, immuno-PET might be exploited as an imaging tool for the selection of high-potential MAbs for therapy as well as of patients most likely to benefit from (expensive) MAb treatment. The use of combined PET/CT scanners may add a new dimension to aforementioned applications.

Acknowledgments

This study was supported by grants from the Dutch Technology Foundation (STW, grant VBC.6120) and the European Union FP6 (LSHC-CT-2003-5032, STROMA). The publication reflects only the author's view. The European Commission is not liable for any use that may be made of the information contained. The authors thank Dr. Iris Verel and Ing. Maria J.W.D. Vosjan for their contribution to the manuscript.

References

Adams, G.P., and Weiner, L.M. 2005. Monoclonal antibody therapy of cancer. *Nat. Biotechnol.* 23:1147–1157.

Borjesson, P.K.E., Jauw, Y.W.S., Boellaard, R., De Bree, R., Comans, E.F.I., Roos, J.C., Castelijns, J.A., Vosjan, M.J.W.D., Kummer, J.A., Leemans, C.R., Lammertsma, A.A., and Van Dongen, G.A.M.S. 2006. Performance of immuno-positron emission tomography with zirconium-89-labeled chimeric monoclonal antibody U36 in the detection of lymph node metastases in head and neck cancer patients. *Clin. Cancer Res.* 12:2133–2140.

Garmestani, K., Milenic, D.E., Plascjak, P.S., and Brechbiel, M.W. 2002. A new and convenient method for purification of ^{86}Y using a Sr(II) selective resin and comparison of biodistribution of ^{86}Y and ^{111}In labeled Herceptin. *Nucl. Med. Biol.* 29:599–606.

Goldenberg, D.M., Sharkey, R.M., Paganelli, G., Barbet, J., and Chatal, J.F. 2006. Antibody pretargeting advances cancer radioimmunodetection and radioimmunotherapy. *J. Clin. Oncol.* 24:823–834.

Grünberg, J., Novak-Hofer, I., Honer, M., Zimmermann, K., Knogler, K., Blauenstein, P., Ametamey, S., Maecke, H.R., and Schubiger, P.A. 2005. *In vivo* evaluation of ^{177}Lu- and $^{67/64}$Cu-labeled recombinant fragments of antibody chCE7 for radioimmunotherapy and PET imaging of L1-CAM-positive tumors. *Clin. Cancer Res.* 11:5112–5120.

Kohler, G., and Milstein, C. 1975. Continuous cultures of fused cells secreting antibody of predefined specificity. *Nature* 256:495–497.

Larson, S.M., Pentlow, K.S., Volkow, N.D., Wolf, A.P., Finn, R.D., Lambrecht, R.M., Graham, M.C., Di Resta, G., Bendriem, B., and Daghighian, F. 1992. PET scanning of iodine-124-3F9 as an approach to tumor dosimetry during treatment planning for radioimmunotherapy in a child with neuroblastoma. *J. Nucl. Med.* 33:2020–2023.

Lee, F.T., Hall, C., Rigopoulos, A., Zweit, J., Pathmaraj, K., O'keefe, G.J., Smyth, F.E., Welt, S., Old, L.J., and Scott, A.M. 2001. Immune-PET of human colon xenograft-bearing BALB/c nude mice using I-124-CDR-grafted humanized A33 monoclonal antibody. *J. Nucl. Med.* 42:764–769.

Lovqvist, A., Humm, J.L., Sheikh, A., Finn, R.D., Koziorowski, J., Ruan, S., Pentlow, K.S., Jungbluth, A., Welt, S., Lee, F.T., Brechbiel, M.W., and Larson, S.M. 2001. PET imaging of ^{86}Y-labeled anti-Lewis Y monoclonal antibodies in a nude mouse model: comparison between ^{86}Y and ^{111}In radiolabels. *J. Nucl. Med.* 42:1281–1287.

Lovqvist, A., Sundin, A., Ahlstrom, H., Carlsson, J., and Lundqvist, H. 1997. Pharmacokinetics and experimental PET imaging of a bromine-76-labeled monoclonal anti-CEA antibody. *J. Nucl. Med.* 38:395–401.

Parry, R., Schneider, D., Hudson, D., Parkes, D., Xuan, J.A., Newton, A., Toy, P., Lin, R., Harkins, R., Alicke, B., Biroc, S., Kretschmer, P.J., Halks-Miller, M., Klocker, H., Zhu, Y., Larsen, B., Cobb, R.R., Bringmann, P., Roth, G., Lewis, J.S., Dinter, H., and Parry, G. 2005. Identification of a novel prostate tumor target, mindin/RG-1, for antibody-based radiotherapy of prostate cancer. *Cancer Res.* 65:8397–8405.

Perk, L.R., Visser, G.W.M., Vosjan, M.J.W.D., Stigter-van Walsum, M., Tijink, B.M., Leemans, C.R., and Van Dongen, G.A.M.S. 2005. ^{89}Zr as a PET surrogate radioisotope for scouting biodistribution of the therapeutic radiometals ^{90}Y and ^{177}Lu in tumor-bearing nude mice after coupling to the internalizing antibody cetuximab. *J. Nucl. Med.* 46:1898–1906.

Philpott, G.W., Schwarz, S.W., Anderson, C.J., Dehdashti, F., Connett, J.M., Zinn, K.R., Meares, C.F., Cutler, P.D., Welch, M.J., and Siegel, B.A. 1995. Radioimmuno-PET: detection of colorectal carcinoma with positron-emitting copper-64-labeled monoclonal antibody. *J. Nucl. Med.* 36:1818–1824.

Robinson, M.K., Doss, M., Shaller, C., Narayanan, D., Marks, J.D., Adler, L.P., Gonzalez Trotter, D.E., and Adams, G.P. 2005. Quantitative immuno-positron emission tomography imaging of HER2-positive tumor xenografts with an iodine-124 labeled anti-HER2 diabody. *Cancer Res.* 65:1471–1478.

Schumacher, J., Kaul, S., Klivenyi, G., Junkermann, H., Magener, A., Henze, M., Doll, J., Haberkorn, U., Amelung, F., and Bastert, G. 2001. Immunoscintigraphy with positron emission tomography: gallium-68 chelate imaging of breast cancer pretargeted with bispecific anti-MUC1/anti-Ga chelate antibodies. *Cancer Res.* 61:3712–3717.

Smith-Jones, P.M., Solit, D.B., Akhurst, T., Afroze, F., Rosen, N., and Larson, S.M. 2004. Imaging the pharmacodynamics of HER2 degradation in response to Hsp90 inhibitors. *Nat. Biotechnol.* 22:701–706.

Stern, M., and Herrmann, R. 2005. Overview of monoclonal antibodies in cancer therapy: present and promise. *Crit. Rev. Oncol. Hematol.* 54:11–29.

Sundaresan, G., Yazaki, P.J., Shively, J.E., Finn, R.D., Larson, S.M., Raubitschek, A.A., Williams, L.E., Chatziioannou, A.F., Gambhir, S.S., and Wu, A.M. 2003. I-124-labeled engineered anti-CEA minibodies and diabodies allow high-contrast, antigen-specific small-animal PET imaging of xenografts in athymic mice. *J. Nucl. Med.* 44:1962–1969.

Verel, I., Visser, G.W.M., Boellaard, R., Stigter-Van Walsum, M., Snow, G.B., and Van Dongen G.A.M.S. 2003a. Zr-89 Immuno-PET: comprehensive procedures for the production of Zr-89-labeled monoclonal antibodies. *J. Nucl. Med.* 44: 1271–1281.

Verel, I., Visser, G.W.M., Boellaard, R., Boerman, O.C., Van Eerd, J., Snow, G.B., Lammertsma, A.A., and Van Dongen G.A.M.S. 2003b. Quantitative Zr-89 immuno-PET for *in vivo* scouting of Y-90-labeled monoclonal antibodies in xenograft-bearing nude mice. *J. Nucl. Med.* 44:1663–1670.

Verel, I., Visser, G.W.M., and Van Dongen, G.A.M.S. 2005. The promise of Immuno-PET in radioimmunotherapy. *J. Nucl. Med.* 46:164S–171S.

Verel, I., Visser, G.W.M., Vosjan, M.J.W.D., Finn, R., Boellaard, R., and Van Dongen, G.A.M.S. 2004. High-quality ^{124}I-labelled monoclonal antibodies for use as PET scouting agents prior to ^{131}I-radioimmunotherapy. *Eur. J. Nucl. Med. Mol. Imaging* 31:1645–1652.

Wahl, R.L. 2005. Tositumomab and ^{131}I therapy in non-Hodgkin's lymphoma. *J. Nucl. Med.* 46:128S–140S.

Wilson, C.B., Snook, D.E., Dhokia, B., Taylor, C.V., Watson, I.A., Lammertsma, A.A., Lambrecht, R., Waxman, J., Jones, T., and Epenetos, A.A. 1991. Quantitative measurement of monoclonal antibody distribution and blood flow using positron emission tomography and ^{124}iodine in patients with breast cancer. *Int. J. Cancer* 47:344–347.

Wu, A.M., Yazaki, P.J., Tsai, S., Nguyen, K., Anderson, A.L., McCarthy, D.W., Welch, M.J., Shively, J.E., Williams, L.E., Raubitschek, A.A.,

Wong, J.Y.C., Toyokuni, T., Phelps, M.E., and Gambhir, S.S. 2000. High-resolution microPET imaging of carcinoembryonic antigen-positive xenografts by using a copper-64-labeled engineered antibody fragment. *Proc. Natl. Acad. Sci. USA. 97*:8495–8500.

Zalutsky, M.R. 2006. Potential of immuno-positron emission tomography for tumor imaging and immunotherapy planning. *Clin. Cancer Res. 12*:1958–1960.

11

Role of Imaging Biomarkers in Drug Development

Janet C. Miller, A. Gregory Sorensen, and Homer H. Pien

Introduction

Drug discovery for cancer has achieved some remarkable breakthroughs in recent years. However, the cost of development of new drugs is formidable, requiring an average of nearly 1 billion and 12 years per drug to bring to commercialization (DiMasi *et al.*, 2003). Much of this cost can be attributed to the high failure rate. It is estimated that only 1 of 5000 screened compounds actually becomes an approved drug (Pritchard *et al.*, 2003). Therefore, in order to minimize the costs of failures, the process of drug development is serial and stage-gated; it is a strategy designed to identify and eliminate unpromising compounds from further development at the earliest stage possible (Pritchard *et al.*, 2003). However, the attrition rate of compounds that reach clinical trials is over 90% (DiMasi *et al.*, 2003). This rate clearly indicates the frequent failure to meet the preclinical goals of identifying drug compounds that are most likely to have drug efficacy in humans and meet safety standards. In part, this situation is due to inadequate models of human pathophysiology (Duyk, 2003).

Another costly impediment for drug development for chronic diseases, including cancer, is the time it takes to demonstrate thoroughly and scientifically that a new drug provides clinical benefit in order to gain the required FDA approval. For cancer, the "true end point" of clinical benefit is extended survival. However, survival is not disease, drug, or patient specific. Therefore, other causes of death confound study outcome. Using survival as an end point, drug trials have to include large numbers of people, and they take years to complete. In the early 1990s, the average time for the clinical phase of new small-molecule drugs that were eventually approved was about 7 years, with about 1.5–2 years more for the approval phase (Reichert, 2003). Not only was the cost of these trials burdensome, but many people who might have benefited from more effective drugs did not have an opportunity to do so.

Recognizing this dilemma, the FDA added provisions in 1992 for accelerated drug approvals intended for serious or life-threatening diseases, provided that the drug appears to provide a benefit over available therapy (Dagher *et al.*, 2004). In 1996, FDA regulations for new cancer drugs were updated to allow accelerated approval if tumor size reduction could be demonstrated in patients with refractory disease or for disease with no effective therapy (Dagher *et al.*, 2004). Tumor size reduction was, therefore, the first imaging biomarker to be accepted as a surrogate for clinical benefit for the purpose of accelerated approval, although further studies are usually required for final approval. The first oncology drug to be granted accelerated approval came in

1995, and by January 2004, 18 oncology drugs had received accelerated approval. Of these drugs, six have since received regular approval (Johnson *et al.*, 2003).

Trends in the number of approved products and the time taken to reach approval in the mid- to late 1990s indicated that the regulatory changes had a beneficial effect. However, this trend does not seem to have been sustained into the 2000s, and the length of approval times has again increased (Reichert, 2003). The lengthy process of completing the necessary research is reflected in the costs of newly approved treatments for cancer, which can be formidable. Bevacizumab, for example, an anti-neovascular antibody recently approved for colorectal cancer, has been estimated to cost more than $40,000 for a 10-month course of treatment (Kelloff and Sigman, 2005).

Biomarkers and Surrogate Markers

Many researchers have recognized that biomarkers have considerable potential for reducing costs and development time associated with drug discovery (Frank and Hargreaves, 2003; FDA, 2004). Even small improvements in clinical trial outcomes and decision making will translate into hundreds of millions of dollars of development cost savings and faster time-to-market (Pritchard *et al.*, 2003). Biomarkers are defined as any characteristic that can be objectively measured or evaluated and that are associated with biological process, normal or pathological, or pharmacological responses to therapeutic intervention (Biomarkers Definitions Working Group, 2001). Biomarkers range from measurements of gene expression or levels of proteins in tissues to macroscopic markers of size, such as tumors. Many require biopsy or blood sampling followed by biochemical or histopathological analysis, but an increasing number of molecular, as well as macroscopic, biomarkers can be measured by imaging techniques.

The dilemma in applying biomarkers in drug discovery is that a pharmacological effect may not necessarily correspond to an effective, therapeutic response. That is, while some biomarkers may predict a clinical end point and, therefore, qualify as a surrogate end point, there are several reasons why that may not be so (Fig. 27) (Frank and Hargreaves, 2003). Indeed, poorly selected and unverified biomarkers can result in both false-negative (Fig. 27d) and false-positive results (Fig. 27e). Cardiac arrythmia is perhaps the best known example of a false surrogate end point that was used for FDA approval. Selection of cardiac arrythmia as a surrogate marker seemed logical since it is known to be associated with an increased risk of death due to cardiac complications. However, three drugs that were approved by the FDA on this basis were later withdrawn because they *increased* the likelihood of death (Fleming and DeMets, 1996). Clearly, biomarkers must be thoroughly validated before they can be accepted as surrogates of clinical response.

Although decrease in tumor size in response to therapy is the only imaging biomarker that has been accepted as a surrogate end point, and that only for accelerated approval, many other biomarkers are used in both clinical and preclinical trials and influence decision making in the drug development process. Validation studies of some of these biomarkers may result in the recognition of new surrogate end points that give early predictions of subsequent clinical outcome, thus saving millions of dollars in the drug development process.

Imaging Biomarkers

Imaging biomarkers have several advantages over those that require biopsy or blood sampling. Imaging is noninvasive and can examine the entire tumor or the entire individual and is, therefore, not prone to sampling errors, as is tissue biopsy. Information about the tumors themselves can be directly visualized, instead of indirectly as in, for example, blood analysis. Serial imaging before, during, and after therapy means that each individual can serve as his or her own control. Therefore, the number of animals required in the preclinical phases of drug development can be greatly reduced, not only because animals do not have to be sacrificed at each stage but also because interanimal variation is minimized. Similarly, imaging biomarkers have the potential to decrease the numbers required in clinical trials. Another major advantage of imaging is that it facilitates the study of orthotopic tumors, which are thought to be better models than the traditionally used subcutaneous tumors, whose principal advantage was that they could be measured with calipers.

The various techniques used in imaging are extremely versatile, including the clinically used MRI, CT, PET, SPECT, and ultrasound, as well as optical-imaging methods that are more commonly used in preclinical research. Preclinical studies are facilitated by the availability of high-resolution scanners of all clinical imaging modalities, designed for small-animal imaging (Weissleder, 2002). Because many of the same or similar imaging techniques are applicable to both preclinical research in small animals and clinical studies, imaging biomarkers are an important tool in the translation from animal models to human subjects.

The range of imaging biomarkers applicable to drug development can be classified into anatomic, physiological, and molecular. Anatomic biomarker imaging of tumor number and size has long been used for diagnosing, staging, and evaluating tumor progression as well as for monitoring response to therapy in clinical trials. Newer clinical imaging techniques have been developed for several physiological biomarkers that are commonly abnormal in tumors, including hemodynamic parameters (Miller *et al.*, 2005) and molecular diffusion rates (Ross *et al.*, 2003). Molecular

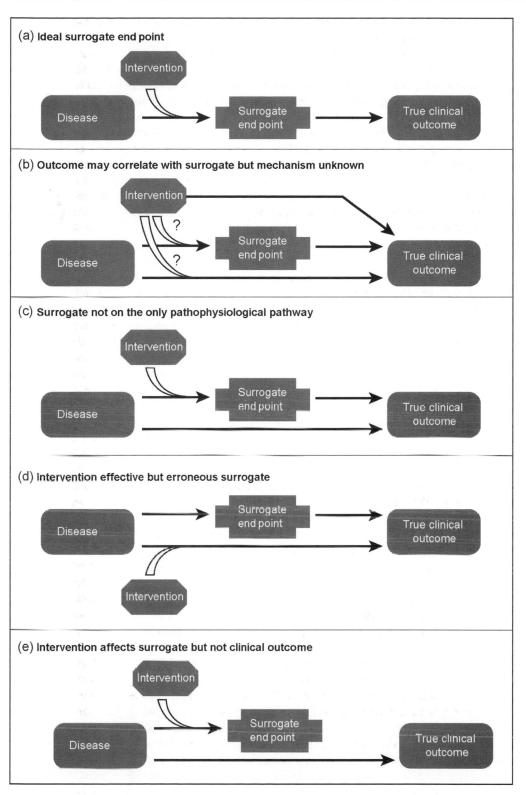

Figure 27 Classification of imaging biomarkers as surrogates of clinical end points.
Source: Adapted from Frank and Hargreaves, 2003.

imaging is used to detect characteristic biomolecules, whose concentration is often changed by disease. For example, molecular imaging agents can report on metabolic rate, DNA synthesis, enzyme activity, cell-surface receptors, or other molecular characteristics of tumors (Cook, 2003; Seddon and Workman, 2003; Shah and Weissleder, 2005). Many drugs and other therapeutic agents can be labeled and, therefore, detected by imaging techniques directly; the labeled drugs themselves become biomarkers that can be used to measure the rate of uptake, distribution, and excretion over a period of time (Fischman et al., 2002).

The concept of molecular imaging can also include acquisition of data on the distribution of natural biomolecules with magnetic resonance spectroscopy (MRS) or optical imaging. In some cases, a drug itself is detectable by imaging because of its intrinsic chemistry or, more often, by radiolabeling, which allows pharmacokinetic data collection of uptake, distribution, metabolism, and excretion of the drug (Fischman et al., 2002).

Anatomic Imaging

Decreased tumor size in response to treatment, the tumor response rate has long played a role in the approval process of new oncology drugs. Of the 14 oncology drugs granted accelerated approvals by the FDA between 1990 and 2002, 11 were based on response rate and none on survival (Johnson et al., 2003). In a more recent survey of 22 oncologic products (drugs and biologics) that received accelerated approval between 1992 and 2004, none was based on survival. Sixteen of these drugs received approval based on response rate along with other progression makers, and 9 were approved based on response rate alone (Dagher et al., 2004).

The first broadly accepted standardized criteria for assessing tumor response were developed by the World Health Organization (WHO). Under these standards, tumor size is estimated as the product of two perpendicular measurements of tumor dimensions, and tumor response is defined as a therapy-induced decrease in size of 50% or more (Miller et al., 1981). More recently, the National Cancer Institute and the European Association for Research and Treatment of Cancer made new recommendations in which tumor size is defined as a single measurement of the longest diameter of a tumor that can be accurately measured (Response Evaluation Criteria in Solid Tumors [RECIST] criteria), typically measured in MR or CT axial images (Therasse et al., 2000). Definitions of tumor response were based on the WHO criteria, with a complete response (CR) defined as the disappearance of all target lesions, a partial response as a decrease in size of 30% or more, and progressive disease (PD) as an increase in size of 20% or more or the appearance of new lesions. Stable disease was defined as tumors that did not shrink or grow sufficiently to be defined as PR or PD (Therasse et al., 2000).

Tumor response has been shown to be a highly significant ($p < .0001$) predictor of increased survival in a meta-analysis of clinical trials for colorectal cancer. However, in the same meta-analysis, the coefficient of determination between treatment effect on survival versus treatment effect on tumor response was only 0.38. This means that less than half of the treatment effects on survival could be explained by treatment effects on tumor response. This discrepancy can be explained, at least in part, by realizing that patients who survive long enough to have an opportunity to respond to treatment will, predictably, have longer overall survival (Buyse et al., 2000). These findings suggest that Phase II trials need to have larger numbers of patients to be meaningful and may explain the poor correlation ($p = 0.13$) between Phase II tumor response rates and Phase III survival demonstrated in clinical trials for small-cell lung cancer (Chen et al., 2000).

There are significant limitations in determining tumor response rates as defined by the WHO criteria or RECIST, which may also limit their usefulness as predictors of survival in the relatively small number of patients enrolled in Phase II trials. First of all, tumors can vary in shape from well-defined lobular masses to diffuse and spiculated structures. Yet the standards for tumor response rate are not based on the accuracy and reproducibility of measurements derived from imaging but from caliper or ruler measurements of wooden spheres placed on a mattress and covered with a layer of foam (Moertel and Hanley, 1976; Weber, 2005). There are no guidelines to indicate whether tumor measurements from imaging should include the entire length of the spicules or only the solid core of the tumor. Not surprisingly, the interobserver variability in the maximum diameter measurements of the same diffuse or spiculated tumor can be as much as 140% and intraobserver measurements can be as high as 37%, whereas the inter- and intraobserver variabilities for well-defined lobular tumors are about 7% and 5%, respectively (Erasmus et al., 2003). In clinical trials of, for example, non-small cell lung cancer, tumor shape may vary from one extreme to the other and the overall variability is not defined.

At times, imaging can show visible tumor shrinkage (Fig. 28), but tumor measurements do not meet either the RECIST or WHO criteria of a positive response. In such cases, tumor shrinkage can be demonstrated either by measuring tumor perimeter and calculating the volume of a single slice (Sorensen et al., 2001) or of contiguous sections of the entire tumor (Prasad et al., 2002). Discrepancies in the classification of tumor response determined by volumetric measurements and those from RECIST or WHO criteria have been observed in 26–34% of image pairs, with equal numbers being downclassified as those upclassified (Sorensen et al., 2001; Prasad et al., 2002; Tran et al., 2004). Furthermore, volumetric measurements have been found to be better than linear measurements in predicting survival after diagnosis of recurrent malignant glioma (Dempsey

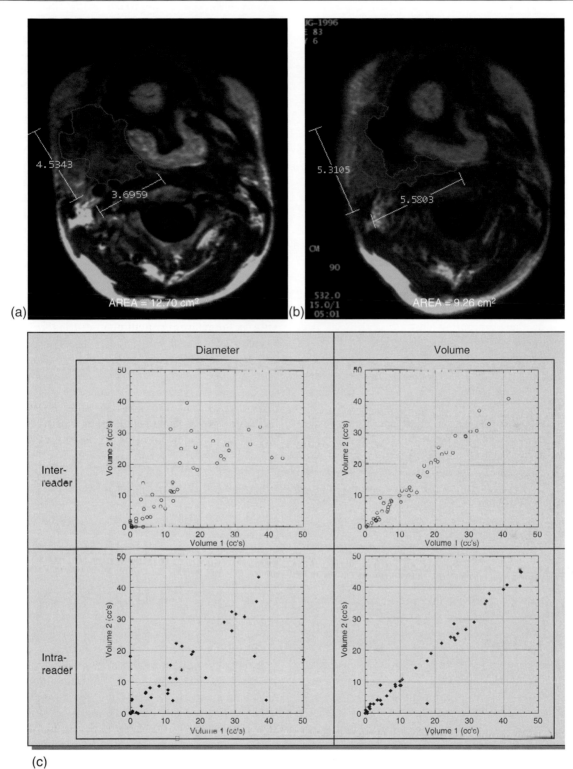

Figure 28 Accuracy of linear versus volumetric measurements of tumor size and response. (a) and (b) are images of the same patient taken 12 weeks apart. There is no difference in tumor size (outlined in red) according to RECIST or WHO criteria, but there is a clear difference in area. (c) Variability is lower for measurements of the perimeter compared to those of the diameter for both inter-reader (−0.31 ± 7.38 vs. −0.19 ± 1.76) and intra-reader measurements (0.91 ± 9.07 vs. 0.39 ± 2.51) from the same set of images ($n = 219$).

Source: From Sorensen *et al.*, 2001, Reprinted with permission from the American Society of Clinical Oncology.

et al., 2005) and for predicting disease-free survival of patients with breast cancer (Partridge *et al.*, 2005). Other advantages of employing volumetric data include decreased intra- and inter-reader variability compared to linear measurements (Fig. 28) (Sorensen *et al.*, 2001). Delineation of tumor perimeters is both rapid and precise with the use of semiautomated software.

Another potential source of error is variation in tomographic slice thickness. For example, estimates of tumor volume from CT slice thicknesses of 7.5 mm and 5 mm have been shown to be significantly different (Zhao *et al.*, 2005). These differences should not be a problem if clinical trials follow RECIST guidelines since they specify collimation of 7–7.5 mm (Therasse *et al.*, 2000). Other possible sources of variation, such as differences in CT window settings or MR instrumentation and protocols, can also affect apparent response rates. For example, in the case of lung cancer, spicules will be better visualized when lung windows rather than soft tissue windows are employed (Erasmus *et al.*, 2003). However, there are no guidelines except to say that they should be consistent throughout the study.

Another limitation is that several cancers do not result in measurable lesions, including those that result in bone cancer, leptomeningeal disease, ascites, and cystic lesions (Therasse *et al.*, 2000). Other tumors have large necrotic regions, making it inappropriate to measure the overall decrease in size when only a relatively small volume is active tumor. Finally, tumor shrinkage occurs over a substantial period of time, whereas the initial therapeutic response would be expected to occur within minutes, hours, or at most days after the start of treatment. Therefore, there is considerable interest in developing many other imaging biomarkers of tumor response.

Physiological Imaging

Tumors differ from their surrounding normal tissues in a number of ways that can be assessed by imaging. For example, abnormalities in the structure and density of tumor microvasculature result in abnormalities in blood flow, blood volume, and permeability of the vascular endothelium. Edema, necrosis, and cystic components are commonly seen in tumors, which result in altered diffusion rates. Although these abnormalities may not be tumor specific, demonstration of their normalization by imaging techniques may be useful as biomarkers of response to therapy.

Imaging Hemodynamic Responses to Therapy

The interest in hemodynamic imaging as a biomarker of therapeutic response stemmed from the development of anti-angiogenic and anti-neovascular drugs as cancer therapeutics. Anti-angiogenic drugs target specific molecules involved in new blood vessel formation, directly or indirectly (Rosen, 2000), and antivascular drugs target molecules characteristic of immature, angiogenic blood vessels (Thorpe, 2004). These targets were chosen because angiogenesis is essential for tumor growth (Folkman, 2002) but only occurs in processes such as wound healing and endometrial growth in a healthy person. Therefore, anti-angiogenic therapeutics are generally believed to pose less of a toxicity risk than traditional chemotherapeutic agents. Consequently, the optimal therapeutic dose may be less than the maximal tolerated dose, the standard for dose selection in drug development for chemotherapeutics. In addition, anti-angiogenic drugs could be expected to inhibit further growth but not necessarily reduce tumor size. Therefore, anatomic imaging may not be the most suitable method of measuring response, and alternate biomarkers of therapeutic activity were needed to assess bioactivity, determine effective doses, and halt the development of ineffective drug candidates early in clinical trials (Rehman and Jayson, 2005).

Tumors typically have high microvascular density and, therefore, high blood volume. Blood vessels are tortuous, with no organized structure or distribution, resulting in chaotic flow. In addition tumor microvasculature is immature and has abnormally high vascular endothelial permeability. Effective anti-angiogenic or antivascular therapies are expected to normalize the microvascular characteristics (Jain, 2003). Therefore, several imaging techniques for measuring blood flow, blood volume, and vascular permeability have been developed for use as imaging biomarkers for both preclinical and Phase I clinical trials of anti-angiogenic and antivascular therapies for cancer. Many of these techniques have been validated by comparison with established methods for measuring blood volume and flow, and their variance has been established by reproducibility studies (Miller *et al.*, 2005).

Dynamic contrast-enhanced MRI (DCE MRI) has been the most widely used imaging modality for measuring hemodynamic changes in clinical trials to date. MRI has several advantages, including excellent soft tissue contrast, spatial resolution (about 1.5 mm) that can demonstrate tumor heterogeneity, a high degree of signal change with contrast, low-contrast agent toxicity, and no ionizing radiation. Therefore, there are few safety restrictions on repeated scanning, although there are practical limits of scanner availability and expense. DCE MRI is versatile, allowing data collection from up to 10 slices with thicknesses of 0.5–2 mm during the first few minutes after bolus injection, as the bolus passes through the vasculature and permeates through the leaky vascular endothelium. The amount of data collected is a trade-off between obtaining information from the maximum volume of tissue while maintaining an acceptable signal-to-noise ratio.

Unlike other imaging modalities, however, MRI does not detect contrast agents directly but rather the perturbation of hydrogen atoms that surround the molecules of contrast agent, measured as T_1 recovery and T_2 or T_2^* decay. High concentrations of gadolinium chelate contrast agents, used in DCE MRI, increase the magnetic susceptibility of hydrogen atoms in their immediate vicinity, resulting in local signal loss on T_2^*-weighted images. However, at low concentrations of MRI contrast agents, T_1 enhancement is greater than the T_2^* signal loss. In tumors, gadolinium chelates rapidly diffuse out of the vasculature, resulting in an increasing level of T_1 enhancement over the first few minutes after the infusion of contrast agent.

Unfortunately, the degree of T_1 enhancement is not necessarily proportional to the concentration of contrast agent, which makes it challenging to do fully quantitative studies. Instead, changes in enhancement in T_1-weighted DCE MRI are often determined semiquantitatively from descriptive features such as maximum enhancement or the initial area under the curve (IAUC) for the first 60–90 sec after arrival of contrast (Miller et al., 2005). These semiquantitative calculations are relatively robust and simple to acquire. However, they are arbitrary and instrument dependent, which makes it difficult to compare results from different MRI scanners. Other semiquantitative parameters, such as those based on the slope of the enhancement curve, are less reproducible than IAUC (Miller et al., 2005) and have not been recommended for measuring response in clinical trials (Leach et al., 2005).

Fully quantitative kinetic parameters can be calculated using equations describing contrast agent diffusion in a physiological two compartment model: blood plasma and the extravascular extracellular space (Tofts et al., 1999). These calculations require that concentrations of contrast agent be known. This is achieved by measuring T_1 in the tissue before contrast agent administration and by selecting an imaging protocol in which the change in T_1 is both substantial and approximately proportional to contrast agent concentration. Corrections must also be made to account for vascular input, determined in each individual, in order to correct for the rapid passage of contrast agent that remains within the larger blood vessels and will otherwise distort the data from the microvasculature and surrounding tumor. The quantitative parameters derived by this method are the volume transfer constant (K^{trans}, min^{-1}) between blood plasma and the extracellular extravascular space, the extracellular extravascular space itself (v_e, ml), and the rate constant (k_{ep}, min^{-1}) between the v_e and blood plasma (Tofts et al., 1999). Calculations of quantitative parameters are computationally demanding and more subject to noise than the semiquantitative parameters. Therefore, there are advantages to both quantitative and semiquantitative analyses, and it has been recommended that both IAUC and K^{trans} be calculated when assessing anti-angiogenic therapy in early-stage clinical trials (Leach et al., 2005).

For both quantitative and semiquantitative analyses, hemodynamic parameters can be calculated from averaged MRI data or each voxel from a region of interest. The latter

has the advantage of showing the extent of tumor heterogeneity but is more susceptible to instrument noise and artifacts such as patient motion. At this time, computer software for these calculations is not yet standardized, nor is it commercially available (although several versions are freely distributed by research institutions).

PET imaging is the most accurate imaging method for quantifying regional blood volume and flow. PET can measure blood volume from steady-state imaging of the tracer, carbon monoxide ([^{11}C]-CO or [^{15}O]-CO), which remains within the vasculature because it binds rapidly and irreversibly to hemoglobin to form carboxyhemoglobin. Blood flow can be calculated from the rate of uptake of [^{15}O]-H$_2$O into the tissue, which, because water is freely diffusible, is directly proportional to blood flow. Both methods require arterial blood sampling in order to obtain the arterial concentration of tracer needed to calculate blood volume and flow.

The principal disadvantage of PET is that the positron-emitting isotopes used to measure blood flow and blood volume have extremely short half-lives (^{15}O, 2 min; ^{11}C, 20 min). Therefore, these PET studies can only be conducted in facilities that have an on-site cyclotron and chemical laboratories for the preparation of tracers. In addition, PET imaging has relatively low spatial resolution (3–4 mm in humans and 1–2 mm in small animals) and, therefore, does not reflect tumor heterogeneity. Combined PET/CT scanners, available for clinical use and for small laboratory animals, aid image interpretation because they provide anatomic information that is not available in PET images alone.

Hemodynamic imaging data measured by CT are theoretically more straightforward to calculate than those obtained from MRI because the change in X-ray attenuation is proportional to the concentration of CT contrast agent. However, the change in image intensity due to contrast attenuation is relatively small despite the relatively high concentration of contrast agent employed. CT has other advantages, including its very high spatial resolution and its lower susceptibility to imaging artifacts compared to MRI. In addition, commercially-available software can be used to calculate blood flow, blood volume, mean transit time, and vascular permeability/surface area. Unfortunately, the volume of tissue that can be scanned in dynamic-contrast CT is limited by the width of the detector, which is 2 cm in both 4- and 16-slice multidetector scanners and 4 cm for the most recent 64-slice multiple-detector scanners. There is also concern about the nephrotoxicity of CT contrast agents, especially in patients with poor renal function. This and the exposure to radiation limit the numbers of repeat scans to monitor therapy.

Other imaging modalities that have also been used to obtain information about angiogenic vasculature for drug development purposes include single photon emission computerized tomography (SPECT), ultrasound, and near-infrared optical methods (Miller *et al.*, 2005). SPECT has the advantages of wider availability than PET, and SPECT tracers are simpler to prepare locally. However, SPECT image resolution (5–6 mm) is lower than PET, and, since it is not possible to correct for attenuation, it is not fully quantitative.

Ultrasound imaging is both portable and relatively inexpensive, but the commonly used method of measuring blood flow, Doppler imaging, is not sensitive to blood vessels smaller than 100 µm and cannot, therefore, provide data about tumor microvessels (Ferrara *et al.*, 2000). Therefore, microbubble contrast agents, which can be detected in extremely small amounts, are used to measure relative blood flow. Preclinical studies using this method have compared favorably with CT perfusion data in small animals (Broumas *et al.*, 2005). However, ultrasound imaging is operator dependent, and ultrasound energy is strongly attenuated by tissue, which limits its usefulness as a quantitative method for clinical use.

Questions to be answered in preclinical and early clinical trials include whether the candidate drug has the expected bioactivity and whether that bioactivity can be achieved at a nontoxic dose. Hemodynamic imaging has been used for this purpose for a number of novel drugs in development as well as to assess the bioactivity of other more conventional treatments.

For example, PET imaging has been used to show transient tumor-specific decreases in blood volume and blood flow within 30 minutes of the administration of a single dose of combretastatin A4 phosphate (CA-4-P; OXiGENE, MA), an antitubulin drug that causes rapid collapse of neovasculature, in both preclinical and early clinical trials (Anderson *et al.*, 2003). Similarly, DCE MRI was used to measure decreases in K^{trans} and IAUC within 4 hours after the administration of a single dose of the same drug (Galbraith *et al.*, 2003). Subsequent studies of the effects of CA-4-P demonstrated a linear dose-response curve, with bioactivity well below the maximum tolerated dose in terms of both hemodynamic changes and tumor response (Stevenson *et al.*, 2003). These imaging studies played an important role in establishing a recommended dose of CA-4-P, which is now in Phase II trials in combination with conventional radiotherapy and chemotherapy (Tozer *et al.*, 2005).

Other examples in which imaging has been used to establish a dose-response relationship of anti-angiogenic or antivascular agents in preliminary clinical trials include PTK787/ZK22584 (Novartis, NJ), a small molecule inhibitor of vascular endothelial growth factor (VEGF, a promoter of angiogenesis) tyrosine kinase (Morgan *et al.*, 2003); HuMV833, an anti-VEGF antibody (Jayson *et al.*, 2002); and AG-0137376, an oral agent that inhibits VEGF tyrosine kinase (Liu *et al.*, 2005). In the case of PTK787/ZK22584, DCE MRI demonstrated that when the dose was ≤ 750 mg/day, the treatment effects on k_{ep} in colon

cancer metastases were lower after the end of the first treatment cycle (28 days) than those measured on Day 2, but the effects were sustained at doses of ≥ 1000 mg/day. These findings corresponded to an increased rate of drug elimination during the treatment cycle, which lowered the plasma AUC over time. Therefore, data from DCE MRI, combined with pharmacokinetic data, were used to establish an optimal dose of 1000–1200 mg/day for future clinical trials of this agent in patients with colorectal metastatic disease (Mross *et al.*, 2005).

The examples given above all showed positive effects on angiogenic vasculature and led to continuation of clinical trials. Hemodynamic imaging would also be useful if it were able to identify agents that are unlikely to be effective drugs, promoting their early withdrawal from clinical trials and thereby saving substantial costs. There is little published data to support this except for reports that endostatin had, at best, inconsistent effects on hemodynamic parameters measured by PET, CT, MRI, or ultrasound (Thomas *et al.*, 2003), which corresponded to the disappointing Phase I clinical results of endostatin (Twombly, 2002).

As mentioned earlier, the justification for conducting angiogenic imaging studies was largely the absence of a suitable method for observing bioactivity and determining appropriated doses for noncytotoxic drugs. The question remains whether changes in hemodynamic data reliably predict subsequent clinical response.

First of all, there is evidence that imaging techniques can be sufficiently sensitive and reproducible to obtain meaningful data. For example, PET imaging showed that the antivascular agent CA-4-P elicited an average 30% decrease in blood flow and an 11% decrease in blood volume (Anderson *et al.*, 2003). These changes are significantly greater than the coefficients of variation (CoV) of measurements of blood flow (11%) (Wells *et al.*, 2003) and blood volume (3%) (Raitakari *et al.*, 1995). Similarly, treatment effects on tumor IAUC, K^{trans}, and k_{ep} measured by DCE-MRI, have been found to be in the order of 50–60%, whereas the coefficient of variation has been found to be 12% for IAUC, 8–24% for K^{trans}, and 21% for k_{ep} (Miller *et al.*, 2005).

The next question to address is whether hemodynamic responses to therapy correspond to subsequent clinical response. At this time there is preliminary evidence that this may be so. For example, a linear correlation was found between change in k_{ep}, measured by DCE MRI on Day 2 and Day 28 of the first cycle of treatment with PTK787/ZK22584 and subsequent tumor shrinkage (Morgan *et al.*, 2003). Similarly, in a small study of patients with rectal cancer treated with bevacizumab, CT perfusion imaging 12 days after a single infusion of the drug showed an average drop in blood flow of 35%, which was followed by grade II–III tumor regression. The observed decrease in tumor profusion was also correlated with other tumor indicators, including microvessel density (via immunohistochemistry of tumor biopsies), interstitial fluid pressure (via endoscopic pressure probe), and circulating endothelial and progenitor cell levels (Willett *et al.*, 2004).

There is also evidence that a hemodynamic response can be predictive of tumor response to conventional therapy as well as anti-angiogenic and antivascular therapy. For example, a response detected by hemodynamic imaging has been shown to be predictive of subsequent clinical response for breast cancer treated with neoadjuvant therapy (Delille *et al.*, 2003), for prostate cancer treated with androgen deprivation (Padhani *et al.*, 2001), and for colorectal cancer treated with radiation (Sahani *et al.*, 2005). Finally, hemodynamic parameters have been shown to be better predictors of survival in cervical cancer than predictors such as microvessel density (Hawighorst *et al.*, 1998). The studies completed to date are not sufficient to establish hemodyamic imaging as a surrogate marker. However, because the hemodynamic response is much more rapid than tumor shrinkage, these techniques have already been applied in drug discovery as indicators of bioactivity and for dose selection.

Diffusion Imaging

Diffusion of water within tissues is restricted by cell membranes and is, therefore, faster in regions that have relatively high extracellular fluid or low cellularity. The regional distribution of apparent diffusion constant (ADC) can be assessed by diffusion-weighted MRI (DWI), a method that is commonly used clinically in stroke diagnosis to measure the extent of edema, which arises rapidly in response to ischemia. DWI pulse sequences include two counteracting magnetic gradient pulses, the first of which imparts a phase shift to hydrogen nuclei in proportion to their initial location. The second magnetic gradient pulse removes the phase shift of those nuclei, provided that they are in their original location. By timing these two gradient pulses to allow water molecules to diffuse distances comparable to the size and spacing of cells, rephasing is incomplete and the loss of signal is proportional to the rate of diffusion, from which it is possible to measure tissue ADC.

Because cancer therapeutics are intended to destroy tumor cells, it can be surmised that diffusion will be less restricted if there is a response to therapy. Alternatively, if the volume of extracellular water decreases due to loss of tumor water content or cell swelling, the rate of diffusion will be expected to decrease. Such changes will be expected to be in proportion to the effectiveness of treatment and to be apparent for a wide variety of molecular mechanisms of therapeutic action.

Preliminary evidence supports the hypothesis that DWI may be a useful means of assessing early tumor response to therapy. For example, increased ADC has been detected as early as 24 hours after the initiation of treatment in preclinical trials of a novel drug for rhabdomyosarcoma (Jordan *et al.*, 2005). In clinical trials, an increase in ADC has been observed 4 days after the start of treatment with novel chemotherapy regimens for metastatic breast cancer (Theilmann *et al.*, 2004) and 3 weeks after initiation of treatment for gliomas (Hamstra *et al.*, 2005).

These changes in ADC in response to therapy appear to correspond to subsequent tumor response, as demonstrated in both preclinical and clinical experiments (Ross *et al.*, 2003). Conversely, a subsequent reversal (decrease) in ADC several days after a single treatment has been shown to correspond to tumor regrowth (Thoeny *et al.*, 2005). In a small series of brain tumors, DWI reliably distinguished partial response (PR), stable disease (SD), and progressive disease at 3 weeks after initiation of chemotherapy, radiation, or a combination of the two treatments. In those with PR, the mean number of voxels with decreased ADC was 33%, compared to 6% in those with SD and 1% in those with PR (Moffat *et al.*, 2005).

DWI has been found to be useful for dose selection of novel drugs whose mechanism of action is not cytotoxic and, therefore, whose maximum tolerated dose is not necessarily the optimal clinical dose (Ross *et al.*, 2003). For example, in a preclinical study of glioma, a modest but significant shift in the distribution of ADC values was observed 6 days after a single dose of a novel drug, BCNU (1,3-bis(2-chlorethyl)-1-nitrosurea), an effect that was later reversed and did not lead to a subsequent tumor shrinkage. Assuming that the drug did, in fact, elicit a transient tumor response, the investigators repeated the study using a fourfold increase in dose. At the higher concentration, BCNU induced a marked and sustained drop in ADC values, which corresponded to subsequent tumor shrinkage (Ross *et al.*, 2003). In another study, DWI showed that a tumor necrosis factor-related apoptosis-inducing ligand (TRAIL) was ineffective alone seven days after initiation of treatment, although it had synergistic effects with radiation in an animal model of breast cancer, and was confirmed by subsequent histological analysis (Chinnaiyan *et al.*, 2000).

The principal disadvantage of DWI is that it is susceptible to motion artifacts. These artifacts are relatively small for brain imaging and are easier to correct compared with those of other parts of the anatomy. However, it is possible to overcome this problem in other regions. For example, image registration in the liver is particularly difficult because the position of the diaphragm is not necessarily the same during separate breath-holds and because the liver is affected by visceral movement. Nevertheless, with the aid of motion-correction algorithms, measurements of ADC in the liver can have good reproducibility, and a change in distribution of ADC values to higher values has been correlated with subsequent tumor response (Theilmann *et al.*, 2004).

Imaging studies have demonstrated the proof of principle that ADC, measured by DWI, can be a biomarker of subsequent tumor response and that it can be useful in clinical as well as preclinical phases of drug development. Because a change in ADC is likely to be independent of the therapeutic mechanism of action, DWI has the potential to be useful for the early assessment of response to a wide variety of therapies.

Molecular Imaging

Molecular imaging encompasses a wide variety of techniques that have proven to be powerful tools in cancer biology and drug development. Molecular imaging typically employs agents that interact with a specific biomolecule *in vivo* but also encompasses techniques in which endogenous biomolecules or drugs themselves can be detected and measured by imaging.

The low concentration of many molecular imaging targets demands sensitive imaging techniques. PET, which can detect tracers at concentrations as low as 10^{-12}mol/L (Fischman *et al.*, 2002), allows the detection and quantification of the regional distribution of biomarkers, such as receptors, that may be present in picomolar concentrations in tumors. The availability of suitable positron-emitting isotopes of the biologically abundant elements carbon (^{11}C), nitrogen (^{13}C), and oxygen (^{15}O), as well as elements such as fluorine (^{18}F) and iodine (^{124}I), facilitates the radiolabeling of many compounds that can act as molecular imaging agents. In comparison, SPECT is about 10-fold less sensitive and depends on isotopes such as technecium (^{99m}Tc), indium (^{111}In), and iodine (^{131}I or ^{123}I). In addition, SPECT has lower spatial resolution than PET and is not a fully quantitative technique (Miller and Thrall, 2004).

Near-infrared optical fluorescence (NIRF) imaging is another very sensitive molecular imaging technique. This method utilizes wavelengths in the range of 700–900 nm, wavelengths at which the absorption by water, lipids, and proteins as well as tissue autofluorescence is at a minimum. Therefore, the primary limitation of NIRF is light scattering, which limits the resolution of the images. These wavelengths penetrate several centimeters through tissue, which is not a limitation for *in vivo* imaging of small animals, although, in the clinical setting, NIRF imaging is only suitable for tumors that are within a few centimeters of the body surface.

Fluorescence is detected with ultrasensitive charge-coupled device (CCD) cameras. Reflectance images are relatively straightforward to obtain, but analysis of these images is not quantitative because the intensity of fluorescence is depth dependent. However, fluorescence molecular tomography (FMT), which uses advanced data processing techniques to account for the diffuse nature of photon propagation in tissue, can be fully quantitative (Ntziachristos *et al.*, 2004). Optical-imaging equipment is inexpensive and portable, the operational costs are minimal, throughput is rapid, and there are no concerns about ionizing radiation. Therefore, it is well suited to prolonged and repeated measurements of change over time, which could be used, for example, to track the bioactivity of a drug.

The most sensitive optical-imaging method utilizes agents that fluoresce after they interact with their molecular target, which are known as NIRF probes. These agents are long-circulating, high-molecular-weight, synthetic copolymers containing fluorochromes that are held together closely by peptides and are minimally fluorescent because of autoquenching. The amino acid sequence of the peptides are selected so that they are cleaved by a specific proteolytic enzyme, releasing the fluorochrome to increase fluorescence 200-fold and allowing detection in the nanomolar concentration range (Weissleder, 2002).

The backbone of NIRF probes is constructed of long-circulating high-molecular-weight, synthetic copolymers, which have been found to be nonallergenic and to have no apparent toxicity in clinical trials. NIRF probes, which attach fluorochrome molecules and short peptide sequences to the backbone, appear to be well tolerated in mice. Therefore, NIRF probes may prove to be suitable for clinical applications and become useful in the clinical arena for the detection and monitoring of human cancers such as breast, colon, and other superficial tumors. However, at this time NIRF probes are proving useful for small-animal preclinical research in target development and pharmacodynamics (Shah and Weissleder, 2005).

Bioluminescence imaging (BLI) is a useful and complementary optical-imaging technique for preclinical studies. Transgenic cells or organisms containing luciferases, which act on their substrate, luciferin, to produce light have broad emission spectra that frequently extend into the infrared. BLI can be used for real-time imaging of gene expression in cell cultures, individual cells in whole organisms, or transgenic animals. When luciferase genes are constituently active in tumor cells but not the host animal, the intensity of fluorescence is proportional to tumor burden. Since luciferases from, for example, firefly and *Renilla* have different substrates, dual labeling can increase the utility of BLI by making it possible to image two or more biological processes in a single animal simultaneously, such as the delivery of gene therapy vectors and to monitor therapy (Shah and Weissleder, 2005).

There is considerable interest in molecular imaging applications of MRI and MRS in oncology despite the limits of sensitivity, which is in the millimolar range. Techniques include MRI of agents that have a high number of paramagnetic atoms and blood oxygen–level dependent (BOLD) MRI, in which changes in the relative concentrations of deoxyhemoglobin and oxyhemoglobin are detected. Proton MRS can be used to quantify molecules such as choline and creatinine that are commonly altered in tumors, and MRS can be tuned to the naturally occurring paramagnetic atoms, ^{19}F and ^{31}P.

Molecular imaging techniques have been used both preclinically and clinically to measure characteristics that are abnormal in a wide range of cancers, such as metabolic rate, cell proliferation, or hypoxia. In addition, a number of techniques assess abnormalities that are limited to specific types of cancer, such as elevated expression of certain enzymes or receptors of genetic activity (Herschman, 2003; Seddon and Workman, 2003). PET is an especially useful modality for pharmacokinetic imaging because drugs can often be labeled with positron emitters without changing their chemical properties (Fischman *et al.*, 2002). Finally, PET has been used to track the expression of reporter genes, such as herpes simplex virus type-1 thymidine kinase, which are co-expressed with a gene of interest in transgenic animals or tumor cells (Herschman, 2003).

Glucose Metabolism and Fluorodeoxyglucose (FDG) PET

The best known and most widely used molecular imaging method is PET imaging of 2-^{18}F-fluoro-2-deoxy-D-glucose (FDG) uptake. FDG is a glucose analog that is taken up by cells as though it were glucose, aided by the action of glucose transporter proteins. Once inside the cell, FDG undergoes the first step of glycolysis, phosphorylation, after which it can neither be metabolized further nor exit through the cell membrane. Therefore, it accumulates within the cell at a rate proportional to glucose metabolism.

Uptake of FDG is substantially greater in most types of cancer compared to normal tissues, with the notable exception of prostate cancer. For a number of cancers, including non-small cell lung cancer, gastroesophageal, colorectal, breast, and head and neck cancers, as well as lymphoma and melanoma, tumor uptake of FDG has been shown to drop significantly by the end of the first cycle of chemotherapy or sooner, an effect that correlates well with subsequent tumor shrinkage and clinical outcome (Juweid and Cheson, 2006).

An elevated FDG PET uptake is not unique to cancer, but also occurs in response to processes such as inflammation, which is a possible source of confusion. Furthermore, because there is an inflammatory response to radiation therapy or surgery, FDG PET is not suitable for evaluating the response to therapy that includes radiation treatment or surgery until some months after those treatments have been completed (Juweid and Cheson, 2006). However, after inflammation has subsided, FDG PET can distinguish between radiation-induced fibrosis and residual tumor.

Although visual assessment of image contrast may be all that is necessary for cancer diagnosis and staging, quantitative or at least semiquantitative analysis of FDG PET data is necessary for the purposes of monitoring therapy and drug discovery. Quantitative measurement of the metabolic rate of glucose requires dynamic scanning to measure the rate of uptake of FDG and arterial blood sampling to measure glucose concentrations and specific activity in the blood. Patlak analysis, which assumes equilibrium between FDG in tissue

and plasma and negligible dephosphorylation, is less sensitive to noise than more fully quantitative analysis (Kelloff *et al.*, 2005). Alternatively, semiquantitative standard uptake values (SUV) can be calculated from attenuation-corrected images based on the injected dose and the estimated total volume of distribution, such as body weight or estimated surface area. Neither dynamic scanning nor arterial sampling is necessary for measuring SUV, making these measurements more practical than fully quantitative analysis. SUV measurements have been shown to be highly correlated with quantitative analysis of FDG uptake, especially when body surface area is used instead of weight. The reproducibility of FDG PET is good, with about 10% variability demonstrated in several studies (Kelloff *et al.*, 2005), which is sufficiently accurate to measure significant change due to therapy. However, when measuring SUVs after therapy, tumor shrinkage can cause partial volume errors. Therefore, the placement of ROIs is crucial, and it has been proposed that an FDG uptake be measured in a consistent circular ROI with a diameter of 1.5 cm to reduce this problem (Biersack *et al.*, 2004).

FDG PET may be particularly useful for evaluating response to therapy for tumors that cannot be accurately measured by conventional imaging, such as soft tissue sarcomas. Not only are these tumors diffuse and ill-defined but they can develop necrotic or fibrous regions in response to therapy. Therefore, tumor shrinkage may be delayed or not occur at all in responding patients. Attempts to measure tumor size of soft tissue sarcomas have not correlated well with histological assessment or survival (Stroobants *et al.*, 2003). On the other hand, FDG PET imaging in a Phase I clinical trial of imatinib (Gleevec, Novartis, Switzerland) as a treatment for gastrointestinal stromal tumors demonstrated that > 25% decrease in SUV from pretreatment levels, observed eight days after the initiation of treatment, corresponded with improved symptoms. Follow-up studies found that 92% of FDG PET responders had more than a year of progression-free survival compared to only 8% of nonresponders (Stroobants *et al.*, 2003). Similarly, in another study of therapy for soft tissue sarcoma, a ≥ 40% decrease FDG uptake observed after one to four cycles of doxorubicin-containing chemotherapy was found to be highly correlated with a significantly lower risk of disease recurrence or metastasis, and increased overall survival (Schuetze *et al.*, 2005).

FDG PET has been widely studied as a biomarker for therapeutic response in breast cancer, both for locally advanced breast cancer undergoing primary neoadjuvant therapy and cases that had previously undergone therapy. These studies have demonstrated that decreased FDG uptake in response to chemotherapy can often be observed within a week of the start of therapy, with further decreases in uptake with continued treatment. Demonstration of decreased FDG uptake has been shown to have sensitivity of 90–100% and specificity of 74–91% for predicting clinical outcome after chemotherapy for breast cancer, making FDG PET data more accurate than histopathologic criteria or tumor response in predicting clinical outcome (Biersack *et al.*, 2004). The cumulative data on the response to therapy has resulted in the approval of FDG PET as a clinical application for monitoring the response to treatment for breast cancer, which is now Medicare and Medicaid reimbursable (Juweid and Cheson, 2006).

However, there is still need for caution when using FDG PET as a biomarker of clinical response to therapy. For example, hormone therapy for breast cancer results in an initial *increase* in FDG uptake, not a decrease, and the elevated rate is sustained until the tumors become unresponsive, 3–24 months later. Because nonresponders show no change in FDG uptake, FDG PET could still predict tumor response (Biersack *et al.*, 2004). However, the biological mechanism of the hormonal effect is not understood and emphasizes the need for biomarker validation as a predictor of response.

Nevertheless, the many studies of altered uptake of FDG in response to therapy suggest that FDG PET is a valuable imaging biomarker for measuring response to novel drugs and drug combinations, not only in preclinical trials but also in the initial phases of clinical trials. In these studies, each patient can serve as his or her own control, allowing the researcher to see the effect without and with interventions with less variability. However, there is a need to establish whether semiquantitative measurements of SUVs are sufficiently accurate for this purpose or whether more quantitative analysis of dynamic scans are necessary. Second, the cut-off values that predict response need to be established. The European Organization for Research and Treatment of Cancer have proposed that a 15–25% decrease after one treatment cycle or a > 25% decrease after two cycles of therapy be defined as partial response, and a decrease to background levels be defined as complete metabolic response. These figures need to be validated by both retrospective and prospective research (Kelloff *et al.*, 2005).

Hypoxia

Tumors are commonly hypoxic because of poorly organized and regulated vascular networks, as well as ineffective delivery of oxygen to some tumor regions due to compression of blood vessels resulting from rapid tumor growth. Inadequate blood flow combined with high interstitial pressure limits the effective delivery of anticancer agents and the effectiveness of radiation therapy, which depends on the development of oxygen-free radicals. In addition, hypoxic cells are resistant to apoptosis, are more likely to migrate to less hypoxic regions (metastasize), and produce more proangiogenic factors that promote further growth. Consequently, the prognosis for successful treatment of patients with hypoxic tumors is significantly diminished using currently available therapies.

Conversely, hypoxia may be an opportunity for developing new therapies (Powis and Kirkpatrick, 2004). Several therapeutic strategies have been considered, including reducing hypoxia to make the cells more vulnerable to other treatments, selectively killing hypoxic cells, and inhibiting the responses to hypoxia, which are mediated through the hypoxia inducible factor-1α (Hif-1α), including promotion of erythropoiesis, glycolysis, and angiogenesis and inhibition of apoptosis and cell differentiation (Cook, 2003; Seddon and Workman, 2003).

Blood oxygen–level dependent (BOLD) MRI is well suited to assess changes in tumor oxygenation. This technique depends on the fact that deoxyhemoglobin is paramagnetic, while oxyhemoglobin is not. Therefore, the MR signal intensity changes when their relative concentration is modified by changes in oxygenation. BOLD MRI has been used to assess changes in oxygenation induced, for example, by carbogen (95% oxygen, 5% CO_2) breathing, which may be a useful method of determining which patients are likely to benefit from carbogen treatment to increase the effectiveness of tumor therapy (Taylor *et al.*, 2001).

In preclinical trials, BOLD MRI has been used to monitor photodynamic therapy with palladium-bacteriopheophorbide (TOOKAD). TOOKAD is an inactive prodrug that is photosensitized by near-infrared light, after which it reacts with molecular oxygen to produce cytotoxins, causing cell death. BOLD MRI was utilized as a method to determine the amount of light needed for safe but effective photodynamic therapy, in which light is delivered precisely to the tumor by fiber-optic techniques. BOLD MRI during TOOKAD photodynamic therapy of subcutaneous melanoma showed that loss of MR signal intensity in illuminated tumors was apparent within 2 minutes and reached a maximum by about 8 minutes, while unilluminated tissue showed no change. The MR signal intensity did not recover after the illumination was discontinued. These BOLD MRI effects corresponded to a drop in oxygenation measured by an optical oxygen sensor and vascular occlusion and stasis, observed with intravital microscopy. Oxygen depletion was followed by necrosis, and the mice were tumor free within a few weeks (Gross *et al.*, 2003).

Hypoxia can also be evaluated by molecular imaging agents. The most widely used of these agents, ^{18}F-fluoromisonidazole (FMISO), is a stable and robust radiopharmaceutical that, in hypoxic conditions, becomes irreversibly bound within cells through the action of intracellular reductase enzymes. Although the tumor-to-background ratio is low and PET images have low contrast, uptake of FMISO can be measured accurately and quantitatively after venous blood sampling and mathematical modeling. Similar agents with longer-lived isotopes have been developed, such as ^{64}Cu, which can be detected by PET, or ^{123}I, which can be detected by SPECT (Rajendran and Krohn, 2005).

Another fluorinated nitroimidazole, SR-4554, is under development for assessment of hypoxia using ^{19}F MRS. This agent has the obvious advantages of containing a stable isotope that does not have to be prepared on a daily basis and does not involve exposure to ionizing radiation. Preclinical studies demonstrated that SR-4554 is selectively retained in tumors and that there is good correlation between retention and tumor oxygen partial pressures, measured with an oxygen electrode. Phase I clinical trials of SR-4554 have shown that it is well tolerated and detectable in tumors. Spectroscopic peak levels and area under the curve correlate linearly with dose, and there is good reproducibility between and within patients (Seddon and Workman, 2003).

Several small molecule inhibitors of Hif-1α are under development (Powis and Kirkpatrick, 2004) and have been evaluated with imaging biomarkers. For example, the bioactivity of PX-478 (Proix Pharmaceuticals, Tucson, AZ) was examined with several imaging biomarkers in order to determine which method would be suitable to assess bioactivity in early clinical trials. In these experiments, DCE MRI demonstrated a decrease in vascular permeability 2 hours after administration of a single dose of PX-478, DWI detected changes in ADC 24–48 hours after a single dose, and MRS revealed decreased peak heights attributable to choline and other tumor markers at 12 and 24 hours after a single dose (Jordan et al., 2005). The authors concluded that each of these imaging biomarkers could be useful in evaluating bioactivity and would be employed in early clinical trials of PX-478.

Cellular Proliferation

Cellular proliferation is essential for tumor growth but only occurs at limited sites such as bone marrow and dermal tissue in healthy adults. Because DNA synthesis is essential for cellular proliferation, molecular imaging of thymidine and its analogs, incorporated exclusively into DNA (and not RNA), have been investigated as biomarkers of response to tumor therapy. Several thymidine-based agents have been investigated, including radiolabeled thymidine (2-^{11}C-thymidine, ^{11}C-*methyl*thymidine), ^{18}F-3′-fluorothymidine (FLT), and ^{18}F-1-[2′-deoxy-2′-fluoro-1b-D-arabinofurasyl]-thymidine (FMAU), all of which can be detected by PET. Unfortunately, none of these biomarkers is ideal.

Exogenous radiolabeled thymidine or its analogs are taken up into proliferating cells with the aid of nucleoside transporters, where thymidine kinase 1 (TK1) phosphorylates them to form monophosphates. However, endogenous thymidine does not arrive in proliferating cells from external sources but is made within the cells as thymidine monophosphate by methylation of deoxyuridine monophosphate with the aid of thymidylate synthase (TS). Both TK1 and TS catalyze rate-limiting steps and are upregulated in prolifer-

ating cells. Because the sources of endogenous and exogenous thymidine monophosphate depend on different enzymes, the relative expressions of TK1 and TS are potentially confounding factors that must be considered in interpreting imaging results. However, studies to date indicate that the relative use of exogenous and endogenous thymidine depends on external concentration and is predictable across species and cell types (Mankoff et al., 2005).

Radiolabeled thymidine is degraded rapidly, and, within minutes after administration, the blood pool is dominated by radiolabeled metabolites. If the radiolabel is in the 2′ position of deoxyribose, it is metabolized to ^{11}CO$_2$, which equilibrates with bicarbonate and is distributed evenly all over the body. On the other hand, if the ^{11}C is in the methyl group on the pyrimidine ring, there are a number of labeled metabolites, some of which remain trapped within the cell (Mankoff et al., 2005). As a result, analysis of ^{11}C-methythymidine PET imaging data requires blood sampling and separation of metabolites by high-pressure liquid chromatography (HPLC) (Wells et al., 2002).

Although ^{18}F-FLT and ^{18}F-FMAU are not significantly metabolized in tumor cells and contain a longer lived isotope compared to ^{11}C-thymidine, neither analog is ideal. In cell culture, proliferating cells take up thymidine twice as rapidly as FLT and fivefold faster than FMAU (Wells et al., 2004). FLT is not incorporated into DNA but remains trapped within the cells as a phosphorylated derivative, which may be a confounding factor because it is possible that TK1 activity may continue after DNA synthesis is inhibited. There are also limitations that depend on the tumor site. For example, thymidine and its analogs are taken up by proliferating cells in bone marrow, and, therefore, PET imaging of thymidine and its analogs have limited use for skeletal tumors (Shields et al., 1998). Although ^{11}C-thymidine PET can be used to assess liver tumors, neither FLT nor FMAU is suitable because they are glucuronated in the liver and, therefore, accumulate. However, slow clearance of FMAU into the bladder makes it suitable for imaging pelvic tumors (Wells et al., 2004).

Because of the complexity of thymidine uptake and degradation, simple measurements of SUV and AUC are not adequate for measuring DNA synthesis. Mankoff et al. (2005) have developed a complex compartmental method that accounts for thymidine incorporation into DNA and degradation for 2-^{11}C-thymidine, which fit data from a variety of patients. Alternatively, spectral analysis, which requires no *a priori* assumptions of the time course of the distribution of radiolabel in blood and tumor, can be used to calculate pharmacokinetic parameters. The latter method was used to demonstrate that the impulse response function (IRF$_{60\ min}$) and fractional retention of thymidine (FRT = IRF$_{60\ min}$/ IRF$_{1\ min}$) positively correlate with the proliferation rate (Ki-67 index) in biopsy samples (Wells et al., 2002). In addition, uptake of FLT PET has been found to correlate well with cell proliferation in a variety of human tumors, including lung cancer, lymphoma, and colorectal cancer (Mankoff *et al.*, 2005). Intrapatient variability of ^{11}C-thymidine PET is, with a CoV of 9% for FRT in patients with advanced gastrointestinal tumors, although there is considerable interpatient variability (Wells *et al.*, 2005).

The potential of ^{11}C thymidine PET in drug development has been demonstrated in a few Phase I clinical trials. For example, ^{11}C thymidine PET was used to establish the effective dose of AG337 (Agouron Pharmaceuticals, San Diego, CA), a TS inhibitor, in patients with advanced gastrointestinal malignancies. In previous studies of this drug, traditional pharmacokinetic and pharmacodynamic studies had shown no relationship between toxicity and drug dose following oral administration, and the relationship between dose and the plasma AUC was weak. However, ^{11}C thymidine PET imaging, one hour after commencing oral administration of AG337, showed a linear relationship between plasma AUC and an increase in FRT compared to baseline imaging data obtained a week before initiation of treatment. These results were interpreted as an increased dependence on exogenous ^{11}C thymidine because AG337 effectively inhibited endogenous production of thymidine and led to the conclusion that ^{11}C thymidine PET was a useful pharmacodynamic biomarker for clinical trials (Wells *et al.*, 2003).

The utility of FLT PET for monitoring therapy has also been examined in preclinical experiments. For example, FLT uptake was significantly decreased 48 hours after a single dose of an experimental drug, PKI-166 (Novartis AG, Switzerland), an ErbB receptor tyrosine kinase inhibitor (Waldherr *et al.*, 2005). Although these experiments indicate that FLT PET has promise for assessing response to therapy, more research on the kinetics of FLT in a variety of tumors is needed in order to establish the optimal model for analysis (Mankoff *et al.*, 2005).

Both the FLT (Waldherr *et al.*, 2005) and ^{11}C thymidine PET (Shields *et al.*, 1998; Wells *et al.*, 2004) response to therapy has been found to be significantly greater and observable earlier than responses measured by FDG PET. Therefore, it has been suggested that FLT and thymidine PET may be of use in screening patients in order to select those who are likely to benefit from novel drugs, such as ErbB receptor inhibitors, that appear to be beneficial to a subset of patients (Waldherr *et al.*, 2005). Early recognition of clinical response through imaging could affect the clinical trial design of these drugs as well as early clinical decisions on the choice of therapy.

Apoptosis

Successful cancer therapy induces tumor cells to die by apoptosis or programmed cell death, which may be initiated externally by death receptors that respond to tumor necrosis factors (TNF) or by internal mitochondrial-induced mecha-

nism. The subsequent events involve an organized cascade of increased enzyme activity, including that of several caspases. Caspases are proteolytic enzymes that cleave a variety of structural, regulatory, and DNA repair proteins and, therefore, disrupt essential activity in cells and contribute to cell death. In addition, structural changes in the cell membrane bring phosphatidyl serine to the outer leaflet and lead to fragmentation of DNA (Blankenberg *et al.*, 2003). Tumor cells are frequently resistant to apoptosis, and the promotion of this process has become important to drug development. Currently, most studies of apoptosis depend on an *in vitro* cell-staining procedure to determine the apoptotic index, terminal deoxyribosyl transferase-mediated dUTP nick ending labeling (TUNEL).

Two principal molecular approaches to imaging apoptosis are under development. The first uses a human protein, annexin V, which binds to phosphatidyl serine with very high affinity and is applicable in both preclinical and clinical drug trials (Blankenberg *et al.*, 2003). The second uses NIRF agents to measure the activity of enzymes associated with apoptosis, which are not yet approved for clinical use (Cook, 2003; Shah and Weissleder, 2005).

Phosphatidyl serine is normally confined to the inner leaflet of the phospholipid bilayer of the cell membrane and is not available to annexin V. However, soon after the initiation of apoptosis, phosphatidyl serine is translocated to the outer leaflet before loss of cell membrane integrity and cell death (Blankenberg *et al.*, 2003). If annexin V is administered during this time, apoptosis can be detected by the increased uptake of label. Biomarker imaging with annexin V can be accomplished with either PET or SPECT using radiolabeled annexin (Cook, 2003), NIRF optical imaging using annexin labeled with a fluorescent marker, Cy5.5 (Ntziachristos *et al.*, 2004), and MRI and/or NIRF using annexin conjugated with both Cy5.5 and amino cross-linked iron oxide (CLIO) nanoparticles (Schellenberger *et al.*, 2004).

In a pilot study of 15 patients, increased uptake of ^{99m}Tc-annexin V, measured by scintigraphy within three days of the start of the first course of chemotherapy, was linked to increased survival time for patients with lymphoma and lung cancer. However, in the two breast cancer patients in the study, who were treated with taxol, no increase in annexin uptake was observed, although both patients were subsequently found to respond to therapy (Belhocine *et al.*, 2002). The significance of the lack of annexin uptake is not clear. One possible explanation may be found in the timing of imaging. Animal experiments have shown that the increase in annexin uptake is biphasic and transient, with an initial peak observed between 1 and 5 hours after treatment onset and a more sustained increase between 9 and 24 hours. Therefore, the timing of imaging with annexin V appears to be critical and needs to be established before it can be used to monitor therapy effectively (Mandl *et al.*, 2004).

NIRF probes have been constructed with specific peptide sequences designed to detect and measure the activity of several caspases, including caspases 1, which is also known as interleukin 1β-converting enzyme (ICE). Caspase 1 is considered to be an initiator of apoptosis. The importance of this enzyme in cancer can be illustrated by the finding that caspase 1 expression is downregulated in the majority of human prostate cancers and that activation of caspase 1 is essential for the apoptotic response to transforming growth factor-β in human prostate cancer cells (Messerli et al., 2004).

Other Proteolytic Enzymes

Similar NIRF probes that specifically assess the activity of several other proteolytic enzymes have been found to be elevated in tumors, including several cathepsins and matrix metalloproteases (MMPs) (Shah and Weissleder, 2005). For example, cathepsin B is a lysosomal protease that is commonly overexpressed in tumors as well as the surrounding host cells, including gastric, lung, colon, brain, breast, and prostate cancers (Gondi et al., 2004). It is implicated in metastasis, and high levels of cathepsin B are predictive of poor prognoses for patient survival. Cathepsin B has been shown to initiate a proteolytic cascade involving urokinase plasminogen activator and its receptor, uPAR, which ultimately activates a latent transforming growth factor. This system has been identified as a potential therapeutic target, and the effect of antisense cathepsin B combined with antisense uPAR therapy has been explored in an animal model of glioma, demonstrating significant inhibition of tumor growth, invasion, and angiogenesis (Gondi et al., 2004).

MMPs act by breaking down proteins in the intracellular matrix, and their level of expression has been shown to be directly related to tumor stage and metastasis. A number of MMP inhibitors, such as prinomastat (AG3340), are in development as cytostatic and anti-angiogenic agents, some of which are in clinical trials. Unfortunately, assessing the bioactivity of these drugs has been indirect in these trials because only *in vitro* methods of measuring anti-MMP activity were available. However, NIRF probes activated by MMPs have been used to demonstrate *in vivo* bioactivity of prinomastat in mice within hours after the initiation of therapy (Shah and Weissleder, 2005).

Cell-Surface Receptors and Signaling Pathways

Several cell-surface receptors have been implicated in cancers, including integrins, growth factor receptors, somatostatin, and hormones. Cell-surface receptors initiate signaling pathways, such as those mediated by tyrosine kinase. Both the receptors and the enzymes involved in the signaling pathways have become important targets for drug development, which have met with some success (Krause and Van Etten, 2005). For example, Herceptin (trastazumab) targets the response to the epidermal growth factor receptor, Avastin (bevacizumab) binds to the vascular endothelial

growth receptor, and Iressa (gefitinib) inhibits some variants of the epidermal growth factor tyrosine kinase (Smith-Jones et al., 2004).

Integrins are cell-surface glycoproteins that are involved in cell-to-matrix and cell-to-cell adhesion. One of these, $\alpha_v\beta_3$ integrin, is of particular interest as a target for cancer therapy because it is highly expressed on activated endothelial cells during angiogenesis and it has been shown to play a role in the regulation of tumor growth, local invasiveness, and metastatic potential. Furthermore, inhibition of $\alpha_v\beta_3$ integrin-mediated cell–matrix interactions induces apoptosis of activated endothelial cells as well as $\alpha_v\beta_3$ integrin positive tumor cells. However, there is evidence that $\alpha_v\beta_3$ integrin is also present in higher quantities during inflammation, which may confound imaging applications (Beer et al., 2005).

There has been considerable interest in molecular imaging of $\alpha_v\beta_3$ integrin, and several molecular imaging agents have been developed that can assess $\alpha_v\beta_3$ integrin by PET, SPECT, ultrasound, and MRI (Miller et al., 2005). Most of these agents incorporate the tripeptide, arginine-glycine-aspartate (RGD), which binds to several integrins, including $\alpha_v\beta_3$ integrin. One of these, [18F]-galacto-RGD, has been found to be both specific to $\alpha_v\beta_3$ integrin and to bind with high affinity. In patients with metastatic tumors, [18F]-galacto-RGD was taken up rapidly by most but not all tumors, where it remained during a 60-minute PET scan. Because the agent is rapidly removed from the blood pool, primarily by the kidneys, [18F]-galacto-RGD PET images have high contrast with little nontumor retention except for some agent that is retained in the liver. The amount of uptake varies considerably, both within and among patients, but this may be of value when selecting patients for therapy with drugs that target $\alpha_v\beta_3$ integrin (Beer et al., 2005).

Gene Delivery and Expression

The ability to manipulate genes in experimental animals is an enormously powerful tool to improve the understanding of metabolic pathways, how they are altered by disease, and the recognition of new molecular targets for drug development. The addition of reporter genes whose expression can be assessed by radionuclide of optical imaging has added enormously to the power of these studies (Shah and Weissleder, 2005).

One gene that can act both as a reporter and a therapeutic gene for cancer is viral type-1 thymidine kinase (HSV1-TK). The expression of this gene can be monitored by PET with radiolabeled 2′-fluoro-2′-deoxy-1-b-D-arabinofuranosy-[124I]-iodouracil (FIAU), which is phosphorylated by the viral but not human thymidine kinase and, once phosphoryated, remains trapped within the cell. As a therapeutic gene, HSV1-TK acts by converting a pro-drug, ganciclovir, into a toxic compound that induces apoptosis. This treatment was successful in treating glioma in preclinical trials, but clinical trials have resulted in only occasional success. However, initial clinical trials, including an international Phase III clinical trial, were hampered by lack of information on the success of gene transfer. Only one study to date has employed FIAU PET to evaluate the efficacy of gene transfer, which was found to be successful in only one of the five patients in the study, and that was limited to the infusion site at the center of the tumor. Most importantly, signs of necrosis were observed after ganciclovir treatment in the region of increased FIAU uptake (Jacobs et al., 2003). Clearly, such monitoring of gene expression will help improve the techniques of gene transfer in humans and will have major impact on the efficient conduct of clinical trials.

An alternative optical method of tracking gene expression, bioluminescent imaging (BLI), has many advantages for preclinical phases of drug development. Tumors that grow from cell lines carrying bioluminescent proteins can be rapidly measured at any site in small animal models of cancer using low-cost optical-imaging techniques that are relatively easy to use compared to traditional imaging techniques. It can also be argued that BLI may be a better measure of tumor burden than volume because the intensity of fluorescence is proportional to the number of living bioluminescent tumor cells and is not affected by necrotic or cystic regions.

BLI has been found to be a valuable tool for the validation of therapeutic targets. For example, lymphoblastic leukemias with rearrangements of the mixed lineage leukemia (MLL) gene have a consistent and unique gene expression profile that results in the overexpression of the receptor tyrosine kinase, FLT-3, which is constitutively active in about one-third of acute myelogenous leukemias. This has led to considerable interest in FLT-3 as a therapeutic target. In experiments to validate this target, mice were injected intravenously with leukemia cells engineered to express firefly luciferase. Leukemia development was observed with BLI, which was first detected within a week in a location consistent with the femur, and progressed to involve other organs. BLI was used to demonstrate that PKC412 (Novartis Pharma AG, Switzerland), a potent small molecule inhibitor of FLT-3, prevented progression of the cancer, which was confirmed by subsequent pathological analysis, and confirmed that FLT-3 targeted therapy is a valid drug target (Armstrong et al., 2003).

Bioluminescence imaging can also be used for determining the efficacy of candidate drugs for cancers not easily visualized by other imaging techniques. For example, the efficacy of a candidate drug, AMN107, was demonstrated with BLI in mice carrying chronic myelogenous leukemia cells expressing firefly luciferase. Mice with established leukemias were treated with daily doses of AMN107 or vehicle alone. After four doses, there was a dramatic reduction in luminescence in the treated mice and a major increase in the control mice, which subsequently corresponded to a highly significant difference in survival time (Fig. 29) (Weisberg et al., 2005).

Magnetic Resonance Spectroscopy

Metabolic abnormalities due to cancer can be evaluated with both proton and ^{31}P MRS. For example, the spectral peak of choline-containing metabolites is a biomarker of membrane lipid metabolism, and an elevated level is characteristic of all types of cancers (Evelhoch et al., 2005). In brain cancers, proton MRS studies demonstrate that the spectral peak of N-acetyl aspartate (NAA), a neurotransmitter that is a biomarker of healthy neuronal tissue, is dramatically lowered in tumors as well as other neuropathologies (McKnight, 2004). Other significant spectral peaks include those of creatinine/phosphocreatinine, lipid, lactate, and myoinositol.

Both ^{31}P and proton MRS have been used to demonstrate metabolic changes response to therapy soon after the start of treatment that correlate to subsequent clinical response (Vaidya et al., 2003). For example, in vivo proton MRS showed that there was a significant drop in the spectral peak of choline, measured at 12 and 24 hours after the administration of a single dose of PX-478 (Prolx Pharmaceuticals, AZ). This drug inhibits HIF-α accumulation in response to hypoxia and inhibits growth of a variety of human tumor xenografts in mice. The authors concluded that changes in the level of choline, determined by MRS, could be an effective imaging biomarker to assess the bioactivity of PX-478 in humans (Jordan et al., 2005).

At this time, MRS has been recommended as being ready for multicenter clinical trials of therapies for breast, prostate, and brain tumors. However, imaging protocol standards must be established, and application-dependent postprocessing and analysis software must be developed in order to ensure uniform spectral processing on scanners from all vendors. Ideally, 3T scanners with quality assurance should be used to obtain high-resolution spectra and optimal analysis (Evelhoch et al., 2005).

Pharmacokinetic Imaging

Imaging with PET, SPECT, and MRS has been used to accurately measure the pharmacokinetics of a number of antineoplastic agents (Fischman et al., 2002). Pharmacokinetic imaging has the advantage of allowing repeated measurements over time to obtain data on uptake, distribution, and excretion. Unlike other pharmacokinetic methods, pharmacokinetic imaging is equally applicable in animals and humans and avoids extrapolation of animal data to humans. Therefore, in vivo human data can show, for example, whether the candidate drug reaches the target site or whether the drug accumulates in another organ in excessive amounts, giving some hints of likely drug efficacy and/or toxicity.

The availability of positron-emitting isotopes of oxygen, nitrogen, carbon, and fluorine together with the high sensitivity of PET detectors makes PET a very powerful tool for

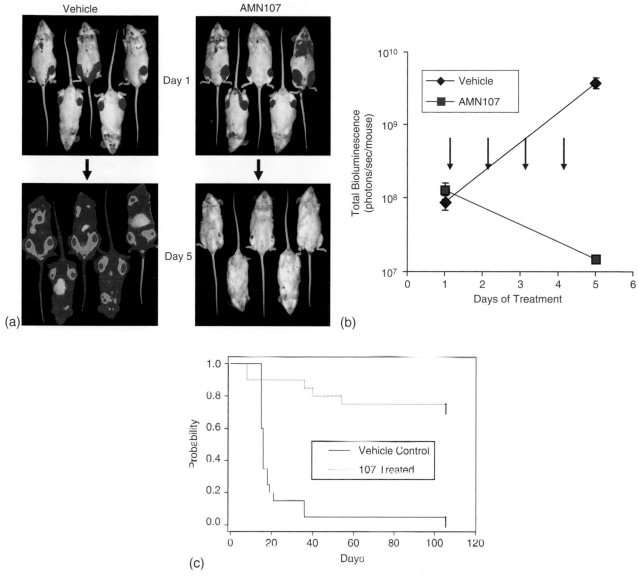

Figure 29 Bioluminescence imaging of the efficacy of AMN107 for the treatment of balb/c mince carrying human xenografts of chronic myelogenous leukemia engineered to express firefly luciferase. (a) Bioluminescence of mice treated with vehicle (left panels) and AMN107 (right panels) on Day 1 and Day 5 after the start of treatment. White represents no fluorescence, and red the most intense fluorescence and is proportional to tumor burden. (b) Quantitation of bioluminescence in vehicle and treated animals. Bars represent standard error of the mean n = 5. (c) Kaplan-Meier plot of survival of mice treated with vehicle and AMN107. *Source:* From Weisberg *et al.*, 2005, with permission.

pharmacokinetic imaging. However, the short half-lives of these isotopes limit the ability to synthesize radiolabeled drugs. This problem can be circumvented in some cases in which another radiolabeled agent that binds to the target is available, allowing displacement studies. Alternatively, MRS can detect stable isotopes of fluorine (^{19}F) and phosphorus (^{31}P), which may be useful for pharmacokinetic imaging of drugs containing these elements, although MRS can only detect much higher concentrations of drugs in the millimolar range.

The high sensitivity of PET imaging has led several groups to consider gathering pharmacokinetic data with microdoses, minute quantities of drugs labeled with PET tracers (Gupta *et al.*, 2002). However, microdosing assumes that drug distribution is not affected by drug concentration, which in at least one case has been demonstrated to be a false assumption, illustrating the importance of using clinically relevant concentrations for pharmacokinetic imaging in clinical trials (Fischman *et al.*, 2002).

The pharmacokinetics of 5-fluorouracil (5-FU) both alone and in combination with other therapies has been studied by both PET and MRS in attempts to improve the efficacy of 5-FU as a chemotherapeutic agent (Gupta et al., 2002). The therapeutic action of 5-FU depends on its conversion to 5′-fluoro-2′-deoxyuridine-5′-monophosphate, which is an inhibitor of thymidylate synthetase and, consequently, DNA synthesis and repair. Unfortunately, about 80% of 5-FU is rapidly catabolized by the enzyme dihydroxypyrimidine dehydrogenase (DPD) to form an inactive but toxic metabolite. In one strategy to improve the efficacy of 5-FU, eniuracil (Glaxo Smith Cline), which is an inhibitor of DPD, was combined with ^{18}F-5-FU. PET imaging demonstrated that the accumulation of radioactive metabolites in the liver was significantly lower when the combination of drugs was given to patients as opposed to ^{18}F-5-FU alone, and the plasma half life of the radiotracer was increased from 2.3 to more than 4 hr, indicating that eniuracil acted as predicted to block the catabolism of 5-FU.

In another example of pharmacokinetic imaging in humans, PET imaging was used to study the pharmacokinetics of a novel anti-VEGF antibody, ^{124}I-HuMV833, which had been found to have antitumor activity against a wide spectrum of human tumor xenografts. The Phase I study established tissue-specific clearance rates that varied little from patient to patient but found considerable variation in ^{124}I-HuMV833 uptake in tumors, not only varying among patients but also among tumors within individual patients. These differences could result in varied response to HuMV833, and the authors suggested that a strategy of intrapatient dose escalation and pharmacodynamic measurements would be appropriate to optimize the effectiveness of this drug (Jayson et al., 2002).

Displacement of a known receptor ligand was used to study the pharmacokinetics of aprepitant (Emend, Merck, USA), a neurokinin-1 (NK1) receptor antagonist that was under development as a treatment for emetogenic chemotherapy-induced nausea and vomiting, as well as an antidepressant. PET imaging measured the effectiveness of aprepitant in displacing a ^{18}F-labeled radioligand with known high affinity and specificity for the NK1 receptor, which is most abundant in the caudate and putamen and least abundant in the cerebellum. Analysis of the striatal/cerebellar PET signal intensity ratio demonstrated that aprepitant displaced the radioligand in a dose-dependent manner, providing data that could be used to guide dose selection in further clinical trials (Bergstrom et al., 2004).

Conclusions

Although anatomical imaging of tumor dimensions is the only imaging biomarker that has been accepted as a measure of tumor response, many other imaging biomarkers can pro-vide valuable data for drug development purposes in both preclinical and clinical trials. In preclinical stages of drug development, biomarker imaging has been used to validate targets, confirm mechanisms of action, provide prognostic indicators, assess pharmacokinetic profiles, and obtain early indicators of bioactivity of candidate drugs. Because biomarker imaging is noninvasive, data can be collected at several points in time from an entire animal or an entire tumor at any site in the body. Therefore, each animal can serve as its own control, decreasing both interanimal variability and the number of animals needed for the studies. Early indicators of bioactivity lessen the time needed for preclinical experiments. At the same time, the ability to collect data from orthotopic and/or spontaneous tumors at any site in the body facilitates the study of more realistic models of human cancers and may increase the likelihood that animal studies will translate into effective clinical therapies.

In clinical trials, pharmacokinetic and pharmacodynamic imaging can be helpful in dose selection or, if a pharmacokinetic profile is poor, may suggest toxicity and lead to early suspension of less promising drug candidates. Imaging can demonstrate whether or not the drug has its predicted bioactivity early after the onset of therapy and could, potentially, shorten the duration of clinical trials. Imaging biomarkers that assess gene expression in tumors could be used to select patients most likely to respond to a particular therapy. Prognostic accuracy and serial imaging allow each patient to serve as his or her control, increasing the statistical power of the clinical trials, while reducing the number of patients needed to provide a meaningful result.

Despite such promise, it is important to keep in mind the confounding factors that may distort conclusions derived from surrogates, as well as the degree to which biomarkers are validated for their intended purposes. Validation of new imaging biomarkers needs to be addressed in a systematic and rigorous way. Image acquisition protocols and image analysis need to be optimized and standardized. Correlations must be established between particular imaging biomarkers and the purported pathophysiological pathways, as well as between imaging biomarkers and clinical outcome. Only then can imaging biomarkers reach the standards required of a surrogate end point.

References

Anderson, H.L., Yap, J.T., Miller, M.P., Robbins, A., Jones, T., and Price, P.M. 2003. Assessment of pharmacodynamic vascular response in a phase I trial of combretastatin A4 phosphate. J. Clin. Oncol. 21:2823–2830.

Armstrong, S.A., Kung, A.L., Mabon, M.E., Silverman, L.B., Stam, R.W., Den Boer, M.L., Pieters, R., Kersey, J.H., Sallan, S.E., Fletcher, J.A., Golub, T.R., Griffin, J.D., and Korsmeyer, S.J. 2003. Inhibition of FLT3 in MLL. Validation of a therapeutic target identified by gene expression based classification. Cancer Cell 3:173–183.

Beer, A.J., Haubner, R., Goebel, M., Luderschmidt, S., Spilker, M.E., Wester, H.J., Weber, W.A, and Schwaiger, M. 2005. Biodistribution and pharmacokinetics of the alphavbeta3-selective tracer 18F-galacto-RGD in cancer patients. *J. Nucl. Med* 46:1333–1341.

Belhocine, T., Steinmetz, N., Hustinx, R., Bartsch, P., Jerusalem, G., Seidel, L., Rigo, P., and Green, A. 2002. Increased uptake of the apoptosis-imaging agent (99m)Tc recombinant human Annexin V in human tumors after one course of chemotherapy as a predictor of tumor response and patient prognosis. *Clin. Cancer Res.* 8:2766–2774.

Bergstrom, M., Hargreaves, R.J., Burns, H.D., Goldberg, M.R., Sciberras, D., Reines, S.A., Petty, K.J., Ogren, M., Antoni, G., Langstrom, B., Eskola, O., Scheinin, M., Solin, O., Majumdar, A.K., Constanzer, M.L., Battisti, W.P., Bradstreet, T.E., Gargano, C., and Hietala, J. 2004. Human positron emission tomography studies of brain neurokinin 1 receptor occupancy by aprepitant. *Biol. Psychiatry.* 55:1007–1012.

Biersack, H.J., Bender, H., and Palmedo, H. 2004. FDG-PET in monitoring therapy of breast cancer. *Eur. J. Nucl. Med. Mol. Imaging 31 Suppl. 1*:S112–117.

Biomarkers Definitions Working Group. 2001. Biomarkers and surrogate endpoints: Preferred definitions and conceptual framework. *Clin Pharm. & Ther.* 69:89–95.

Blankenberg, F., Mari, C., and Strauss, H.W. 2003. Imaging cell death in vivo. *Q. J. Nucl. Med.* 47:337–348.

Broumas, A.R., Pollard, R.E., Bloch, S.H., Wisner, E.R., Griffey, S., and Ferrara, K.W. 2005. Contrast-enhanced computed tomography and ultrasound for the evaluation of tumor blood flow. *Invest. Radiol.* 40:134–147.

Buyse, M., Thirion, P., Carlson, R.W., Burzykowski, T., Molenberghs, G., and Piedbois, P. 2000. Relation between tumour response to first-line chemotherapy and survival in advanced colorectal cancer: a meta-analysis. Meta-Analysis Group in Cancer. *Lancet* 356:373–378.

Chen, T.T., Chute, J.P., Feigal, E., Johnson, B.E., and Simon, R. 2000. A model to select chemotherapy regimens for phase III trials for extensive-stage small-cell lung cancer. *J. Natl. Cancer Inst.* 92:1601–1607.

Chinnaiyan, A.M., Prasad, U., Shankar, S., Hamstra, D.A., Shanaiah, M., Chenevert, T.L., Ross, B.D., and Rehemtulla, A. 2000. Combined effect of tumor necrosis factor-related apoptosis-inducing ligand and ionizing radiation in breast cancer therapy. *Proc. Natl. Acad. Sci. USA* 97:1754–1759.

Cook, G.J. 2003. Oncological molecular imaging: nuclear medicine techniques. *Br. J. Radiol. 76 Spec No. 2*:S152–158.

Dagher, R., Johnson, J., Williams, G., Keegan, P., and Pazdur, R. 2004. Accelerated approval of oncology products: a decade of experience. *J. Natl. Cancer Inst.* 96:1500–1509.

Delille, J.-P., Slanetz, P., Yeh, E., Halpern, E., Kopans, D., and Garrido, L. 2003. Invasive ductal breast carcinoma response to neoadjuvant chemotherapy: noninvasive monitoring with functional MR imaging—pilot study. *Radiology* 228:63–69.

Dempsey, M.F., Condon, B.R., and Hadley, D.M. 2005. Measurement of tumor "size" in recurrent malignant glioma: 1D, 2D, or 3D? *AJNR Am. J. Neuroradiol.* 26:770–776.

DiMasi, J.A., Hansen, R.W., and Grabowski, H.G. 2003. The price of innovation: new estimates of drug development costs. *J. Health Econ.* 22:151–185.

Duyk, G. 2003. Attrition and translation. *Science* 302:603–605.

Erasmus, J.J., Gladish, G.W., Broemeling, L., Sabloff, B.S., Truong, M.T., Herbst, R.S., and Munden, R.F. 2003. Interobserver and intraobserver variability in measurement of non-small cell carcinoma lung lesions: implications for assessment of tumor response. *J. Clin. Oncol.* 21:2574–2582.

Evelhoch, J., Garwood, M., Vigneron, D., Knopp, M., Sullivan, D., Menkens, A., Clarke, L., and Liu, G. 2005. Expanding the use of magnetic resonance in the assessment of tumor response to therapy: workshop report. *Cancer Res.* 65:7041–7044.

FDA. 2004. Innovation or stagnation? Challenge and opportunity on the critical path to new medical products. Rockville, MD: U.S. Food and Drug Administration.

Ferrara, K.W., Merritt, C.R., Burns, P.N., Foster, F.S., Mattrey, R.F., and Wickline, S.A. 2000. Evaluation of tumor angiogenesis with U.S.: imaging, Doppler, and contrast agents. *Acad. Radiol.* 7:824–839.

Fischman, A.J., Alpert, N.M., and Rubin, R.H. 2002. Pharmacokinetic imaging: a noninvasive method for determining drug distribution and action. *Clin. Pharmacokinet.* 41:581–602.

Fleming, T.R., and DeMets, D.L. 1996. Surrogate end points in clinical trials: are we being misled? *Ann. Intern. Med.* 125:605–613.

Folkman, J. 2002. Role of angiogenesis in tumor growth and metastasis. *Semin. Oncol.* 29:15–18.

Frank, R., and Hargreaves, R. 2003. Clinical biomarkers in drug discovery and development. *Nat. Rev. Drug Discov.* 2:566–580.

Galbraith, S.M., Maxwell, R.J., Lodge, M.A., Tozer, G.M., Wilson, J., Taylor, N.J., Stirling, J.J., Sena, L., Padhani, A.R., and Rustin, G.J. 2003. Combretastatin A4 phosphate has tumor antivascular activity in rat and man as demonstrated by dynamic magnetic resonance imaging. *J. Clin. Oncol.* 21:2831–2842.

Gondi, C.S., Lakka, S.S., Yanamandra, N., Olivero, W.C., Dinh, D.H., Gujrati, M., Tung, C.H., Weissleder, R., and Rao, J.S. 2004. Adenovirus-mediated expression of antisense urokinase plasminogen activator receptor and antisense cathepsin B inhibits tumor growth, invasion, and angiogenesis in gliomas. *Cancer Res.* 64:4069–4077.

Gross, S., Gilead, A., Scherz, A., Neeman, M., and Salomon, Y. 2003. Monitoring photodynamic therapy of solid tumors online by BOLD-contrast MRI. *Nature Medicine* 9:1327–1331.

Gupta, N., Price, P.M., and Aboagye, E.O. 2002. PET for in vivo pharmacokinetic and pharmacodynamic measurements. *Eur. J. Cancer* 38:2094–2107.

Hamstra, D.A., Chenevert, T.L., Moffat, B.A., Johnson, T.D., Meyer, C.R., Mukherji, S.K., Quint, D.J., Gebarski, S.S., Fan, X., Tsien, C.I., Lawrence, T.S., Junck, L., Rehemtulla, A., and Ross, B.D. 2005. Evaluation of the functional diffusion map as an early biomarker of time-to-progression and overall survival in high-grade glioma. *Proc. Natl. Acad. Sci. USA 102*:16759–16764.

Hawighorst, H., Knapstein, P.G., Knopp, M.V., Weikel, W., Brix, G., Zuna, I., Schonberg, S.O., Essig, M., Vaupel, P., and van Kaick, G. 1998. Uterine cervical carcinoma: comparison of standard and pharmacokinetic analysis of time-intensity curves for assessment of tumor angiogenesis and patient survival. *Cancer Res* 58:3598–3602.

Herschman, H.R. 2003. Molecular imaging: looking at problems, seeing solutions. *Science 302*:605–608.

Jacobs, A.H., Voges, J., Kracht, L.W., Dittmar, C., Winkeler, A., Thomas, A., Wienhard, K., Herholz, K., and Heiss, W.D. 2003. Imaging in gene therapy of patients with glioma. *J. Neurooncol.* 65:291–305.

Jain, R.K. 2003. Molecular regulation of vessel maturation. *Nat. Med.* 9:685–693.

Jayson, G.C., Zweit, J., Jackson, A., Mulatero, C., Julyan, P., Ranson, M., Broughton, L., Wagstaff, J., Hakannson, L., Groenewegen, G., Bailey, J., Smith, N., Hastings, D., Lawrance, J., Haroon, H., Ward, T., McGown, A.T., Tang, M., Levitt, D., Marreaud, S., Lehmann, F.F., Herold, M., and Zwierzina, H. 2002. Molecular imaging and biological evaluation of HuMV833 anti-VEGF antibody: implications for trial design of antiangiogenic antibodies. *J. Natl. Cancer Inst.* 94:1484–1493.

Johnson, J.R., Williams, G., and Pazdur, R. 2003. End points and United States Food and Drug Administration approval of oncology drugs. *J. Clin. Oncol.* 21:1404–1411.

Jordan, B.F., Black, K., Robey, I.F., Runquist, M., Powis, G., and Gillies, R.J. 2005. Metabolite changes in HT-29 xenograft tumors following HIF-1alpha inhibition with PX-478 as studied by MR spectroscopy in vivo and ex vivo. *NMR Biomed* 18:430–439.

Jordan, B.F., Runquist, M., Raghunand, N., Baker, A., Williams, R., Kirkpatrick, L., Powis, G., and Gillies, R.J. 2005. Dynamic contrast-enhanced and diffusion MRI show rapid and dramatic changes in tumor microenvironment in response to inhibition of HIF-1alpha using PX-478. *Neoplasia.* 7:475–485.

Juweid, M.E., and Cheson, B.D. 2006. Positron-emission tomography and assessment of cancer therapy. *N. Engl. J. Med. 354*:496–507.

Kelloff, G.J., Hoffman, J.M., Johnson, B., Scher, H.I., Siegel, B.A., Cheng, E.Y., Cheson, B.D., O'Shaughnessy, J., Guyton, K.Z., Mankoff, D.A., Shankar, L., Larson, S.M., Sigman, C.C., Schilsky, R.L., and Sullivan, D.C. 2005. Progress and promise of FDG-PET imaging for cancer patient management and oncologic drug development. *Clin Cancer Res. 11*:2785–2808.

Kelloff, G.J., and Sigman, C.C. 2005. New science-based endpoints to accelerate oncology drug development. *Eur. J. Cancer 41*:491–501.

Krause, D.S., and Van Etten, R.A. 2005. Tyrosine kinases as targets for cancer therapy. *N Engl. J. Med. 353*:172–187.

Leach, M.O., Brindle, K.M., Evelhoch, J.L., Griffiths, J.R., Horsman, M.R., Jackson, A., Jayson, G.C., Judson, I.R., Knopp, M.V., Maxwell, R.J., McIntyre, D., Padhani, A.R., Price, P., Rathbone, R., Rustin, G.J., Tofts, P.S., Tozer, G.M., Vennart, W., Waterton, J.C., Williams, S.R., and Workman, P. 2005. The assessment of antiangiogenic and antivascular therapies in early-stage clinical trials using magnetic resonance imaging: issues and recommendations. *Br. J. Cancer 92*:1599–1610.

Liu, G., Rugo, H.S., Wilding, G., McShane, T.M., Evelhoch, J.L., Ng., C., Jackson, E., Kelcz, F., Yeh, B.M., Lee, F.T., Jr., Charnsangavej, C., Park, J.W., Ashton, E.A., Steinfeldt, H.M., Pithavala, Y.K., Reich, S.D., and Herbst, R.S. 2005. Dynamic contrast-enhanced magnetic resonance imaging as a pharmacodynamic measure of response after acute dosing of AG-013736, an oral angiogenesis inhibitor, in patients with advanced solid tumors: results from a Phase I study. *J. Clin. Oncol. 23*:5464–5473.

Mandl, S.J., Mari, C., Edinger, M., Negrin, R.S., Tait, J.F., Contag, C.H., and Blankenberg, F.G. 2004. Multi-modality imaging identifies key times for annexin V imaging as an early predictor of therapeutic outcome. *Mol. Imaging. 3*:1–8.

Mankoff, D.A., Shields, A.F., and Krohn, K.A. 2005. PET imaging of cellular proliferation. *Radiol. Clin. North Am. 43*:153–167.

McKnight, T.R. 2004. Proton magnetic resonance spectroscopic evaluation of brain tumor metabolism. *Semin. Oncol. 31*:605–617.

Messerli, S.M., Prabhakar, S., Tang, Y., Shah, K., Cortes, M.L., Murthy, V., Weissleder, R., Breakefield, X.O., and Tung, C.H. 2004. A novel method for imaging apoptosis using a caspase-1 near-infrared fluorescent probe. *Neoplasia 6*:95–105.

Miller, A.B., Hoogstraten, B., Staquet, M., and Winkler, A. 1981. Reporting results of cancer treatment. *Cancer 47*:207–214.

Miller, J.C., Pien, H.H., Sahani, D., Sorensen, A.G., and Thrall, J.H. 2005. Imaging angiogenesis: applications and potential for drug development. *J. Natl. Cancer Inst. 97*:172–187.

Miller, J.C., and Thrall, J.H. 2004. Clinical molecular imaging. *J. Am. Col. Radiol. 1*:4–23.

Moertel, C.G., and Hanley, J.A. 1976. The effect of measuring error on the results of therapeutic trials in advanced cancer. *Cancer 38*:388–394.

Moffat, B.A., Chenevert, T.L., Lawrence, T.S., Meyer, C.R., Johnson, T.D., Dong, Q., Tsien, C., Mukherji, S., Quint, D.J., Gebarski, S.S., Robertson, P.L., Junck, L.R., Rehemtulla, A., and Ross, B.D. 2005. Functional diffusion map: a noninvasive MRI biomarker for early stratification of clinical brain tumor response. *Proc. Natl. Acad. Sci. USA 102*:5524–5529.

Morgan, B., Thomas, A.L., Drevs, J., Hennig, J., Buchert, M., Jivan, A., Horsfield, M.A., Mross, K., Ball, H.A., Lee, L., Mietlowski, W., Fuxuis, S., Unger, C., O'Byrne, K., Henry, A., Cherryman, G.R., Laurent, D., Dugan, M., Marme, D., and Steward, W.P. 2003. Dynamic contrast-enhanced magnetic resonance imaging as a biomarker for the pharmacological response of PTK787/ZK 222584, an inhibitor of the vascular endothelial growth factor receptor tyrosine kinases, in patients with advanced colorectal cancer and liver metastases: results from two Phase I studies. *J. Clin. Oncol. 21*:3955–3964.

Mross, K., Drevs, J., Muller, M., Medinger, M., Marme, D., Hennig, J., Morgan, B., Lebwohl, D., Masson, E., Ho, Y.Y., Gunther, C., Laurent,

D., and Unger, C. 2005. Phase I clinical and pharmacokinetic study of PTK/ZK, a multiple VEGF receptor inhibitor, in patients with liver metastases from solid tumours. *Eur. J. Cancer 41*:1291–1299.

Ntziachristos, V., Schellenberger, E.A., Ripoll, J., Yessayan, D., Graves, E., Bogdanov, A., Jr., Josephson, L., and Weissleder, R. 2004. Visualization of antitumor treatment by means of fluorescence molecular tomography with an annexin V-Cy5.5 conjugate. *Proc. Natl. Acad. Sci. USA 101*:12294–12299.

Padhani, A.R., MacVicar, A.D., Gapinski, C.J., Dearnaley, D.P., Parker, G.J., Suckling, J., Leach, M.O., and Husband, J.E. 2001. Effects of androgen deprivation on prostatic morphology and vascular permeability evaluated with MR imaging. *Radiology 218*:365–374.

Partridge, S.C., Gibbs, J.E., Lu, Y., Esserman, L.J., Tripathy, D., Wolverton, DS., Rugo, H.S., Hwang, E.S., Ewing, C.A., and Hylton, N.M. 2005. MRI measurements of breast tumor volume predict response to neoadjuvant chemotherapy and recurrence-free survival. *AJR Am. J. Roentgenol. 184*:1774–1781.

Powis, G., and Kirkpatrick, L. 2004. Hypoxia inducible factor-1alpha as a cancer drug target. *Mol. Cancer Ther. 3*:647–654.

Prasad, S.R., Jhaveri, K.S., Saini, S., Hahn, P.F., Halpern, E.F., and Sumner, J.E. 2002. CT tumor measurement for therapeutic response assessment: comparison of unidimensional, bidimensional, and volumetric techniques initial observations. *Radiology 225*:416–419.

Pritchard, J.F., Jurima-Romet, M., Reimer, M.L., Mortimer, E., Rolfe, B., and Cayen, M.N. 2003. Making better drugs: Decision gates in nonclinical drug development. *Nat. Rev. Drug Discov. 2*:542–553.

Raitakari, M., Knuuti, M.J., Ruotsalainen, U., Laine, H., Makea, P., Teras, M., Sipila, H., Niskanen, T., Raitakari, O.T., and Iida, H. 1995. Insulin increases blood volume in human skeletal muscle: studies using [^{15}O]CO and positron emission tomography. *Am. J. Physiol. 269*:E1000–1005.

Rajendran, J.G., and Krohn, K.A. 2005. Imaging hypoxia and angiogenesis in tumors. *Radiol. Clin. North Am. 43*:169–187.

Rehman, S., and Jayson, G.C. 2005. Molecular imaging of antiangiogenic agents. *Oncologist 10*:92–103.

Reichert, J.M. 2003. Trends in development and approval times for new therapeutics in the United States. *Nat. Rev. Drug Discov. 2*:695–702.

Rosen, L. 2000. Antiangiogenic strategies and agents in clinical trials. *Oncologist 5 Suppl. 1*:20–27.

Ross, B.D., Moffat, B.A., Lawrence, T.S., Mukherji, S.K., Gebarski, S.S., Quint, D.J., Johnson, T.D., Junck, L., Robertson, P.L., Muraszko, K.M., Dong, Q., Meyer, C.R., Bland, P.H., McConville, P., Geng, H., Rehemtulla, A., and Chenevert, T.L. 2003. Evaluation of cancer therapy using diffusion magnetic resonance imaging. *Mol. Cancer Ther. 2*:581–587.

Sahani, D.V., Kalva, S.P., Hamberg, L.M., Hahn, P.F., Willett, C.G., Saini, S., Mueller, P.R., and Lee, T.Y. 2005. Assessing tumor perfusion and treatment response in rectal cancer with multisection CT: initial observations. *Radiology 234*:785–792.

Schellenberger, E.A., Sosnovik, D., Weissleder, R., and Josephson, L. 2004. Magneto/optical annexin V, a multimodal protein. *Bioconjug. Chem. 15*:1062–1067.

Schuetze, S.M., Rubin, B.P., Vernon, C., Hawkins, D.S., Bruckner, J.D., Conrad, E.U., 3rd, and Eary, J.F. 2005. Use of positron emission tomography in localized extremity soft tissue sarcoma treated with neoadjuvant chemotherapy. *Cancer 103*:339–348.

Seddon, B.M., and Workman, P. 2003. The role of functional and molecular imaging in cancer drug discovery and development. *Br. J. Radiol. 76 Spec. No. 2*:S128–138.

Shah, K., and Weissleder, R. 2005. Molecular optical imaging: applications leading to the development of present day therapeutics. *NeuroRx. 2*:215–225.

Shields, A.F., Grierson, J.R., Dohmen, B.M., Machulla, H.J., Stayanoff, J.C., Lawhorn-Crews, J.M., Obradovich, J.E., Muzik, O., and Mangner,

T.J. 1998. Imaging proliferation in vivo with [F-18]FLT and positron emission tomography. *Nat. Med.* 4:1334–1336.

Shields, A.F., Mankoff, D.A., Link, J.M., Graham, M.M., Eary, J.F., Kozawa, S.M., Zheng, M., Lewellen, B., Lewellen, T.K., Grierson, J.R., and Krohn, K.A. 1998. Carbon-11-thymidine and FDG to measure therapy response. *J. Nucl. Med.* 39:1757–1762.

Smith-Jones, P.M., Solit, D.B., Akhurst, T., Afroze, F., Rosen, N., and Larson, S.M. 2004. Imaging the pharmacodynamics of HER2 degradation in response to Hsp90 inhibitors. *Nat. Biotechnol.* 22:701–706.

Sorensen, A.G., Patel, S., Harmath, C., Bridges, S., Synnott, J., Sievers, A., Yoon, Y.H., Lee, E.J., Yang, M.C., Lewis, R.F., Harris, G.J., Lev, M., Schaefer, P.W., Buchbinder, B.R., Barest, G., Yamada, K., Ponzo, J., Kwon, H.Y., Gemmete, J., Farkas, J., Tievsky, A.L., Ziegler, R.B., Salhus, M.R., and Weisskoff, R. 2001. Comparison of diameter and perimeter methods for tumor volume calculation. *J. Clin. Oncol.* 19:551–557.

Stevenson, J.P., Rosen, M., Sun, W., Gallagher, M., Haller, D.G., Vaughn, D., Giantonio, B., Zimmer, R., Petros, W.P., Stratford, M., Chaplin, D., Young, S.L., Schnall, M., and O'Dwyer, P.J. 2003. Phase I trial of the antivascular agent combretastatin A4 phosphate on a 5-day schedule to patients with cancer: magnetic resonance imaging evidence for altered tumor blood flow. *J. Clin. Oncol.* 21:4428–4438.

Stroobants, S., Goeminne, J., Seegers, M., Dimitrijevic, S., Dupont, P., Nuyts, J., Martens, M., van den Borne, B., Cole, P., Sciot, R., Dumez, H., Silberman, S., Mortelmans, L., and van Oosterom, A. 2003. 18FDG-Positron emission tomography for the early prediction of response in advanced soft tissue sarcoma treated with imatinib mesylate (Glivec). *Eur. J. Cancer* 39:2012–2020.

Taylor, N.J., Baddeley, H., Goodchild, K.A., Powell, M.E., Thoumine, M., Culver, L.A., Stirling, J.J., Saunders, M.I., Hoskin, P.J., Phillips, H., Padhani, A.R., and Griffiths, J.R. 2001. BOLD MRI of human tumor oxygenation during carbogen breathing. *J. Magn. Reson. Imaging* 14:156–163.

Theilmann, R.J., Borders, R., Tronard, T.P., Xia, G., Outwater, E., Ranger-Moore, J., Gillies, R.J., and Stopeck, A. 2004. Changes in water mobility measured by diffusion MRI predict response of metastatic breast cancer to chemotherapy. *Neoplasia* 6:831–837.

Therasse, P., Arbuck, S.G., Eisenhauer, E.A., Wanders, J., Kaplan, R.S., Rubinstein, L., Verweij, J., Van Glabbeke, M., van Oosterom, A.T., Christian, M.C., and Gwyther, S.G. 2000. New guidelines to evaluate the response to treatment in solid tumors. European Organization for Research and Treatment of Cancer, National Cancer Institute of the United States, National Cancer Institute of Canada. *J. Natl. Cancer. Inst.* 92:205–216.

Thoeny, H.C., De Keyzer, F., Chen, F., Ni, Y., Landuyt, W., Verbeken, E.K., Bosmans, H., Marchal, G., and Hermans, R. 2005. Diffusion-weighted MR imaging in monitoring the effect of a vascular targeting agent on rhabdomyosarcoma in rats. *Radiology* 234:756–764.

Thomas, J.P., Arzoomanian, R.Z., Alberti, D., Marnocha, R., Lee, F., Friedl, A., Tutsch, K., Dresen, A., Geiger, P., Pluda, J., Fogler, W., Schiller, J.H., and Wilding, G. 2003. Phase I pharmacokinetic and pharmacodynamic study of recombinant human endostatin in patients with advanced solid tumors. *J. Clin. Oncol.* 21:223–231.

Thorpe, P.E. 2004. Vascular targeting agents as cancer therapeutics. *Clin. Cancer Res.* 10:415–427.

Tofts, P.S., Brix, G., Buckley, D.L., Evelhoch, J.L., Henderson, E., Knopp, M.V., Larsson, H.B., Lee, T.Y., Mayr, N.A., Parker, G.J., Port, R.E., Taylor, J., and Weisskoff, R.M. 1999. Estimating kinetic parameters from dynamic contrast-enhanced T(1)-weighted MRI of a diffusable tracer: standardized quantities and symbols. *J. Magn. Reson. Imaging* 10:223–232.

Tozer, G.M., Kanthou, C., and Baguley, B.C. 2005. Disrupting tumour blood vessels. *Nat. Rev. Cancer* 5:423–435.

Tran, L.N., Brown, M.S., Goldin, J.G., Yan, X., Pais, R.C., McNitt-Gray, M.F., Gjertson, D., Rogers, S.R., and Aberle, D.R. 2004. Comparison of treatment response classifications between unidimensional, bidimensional, and volumetric measurements of metastatic lung lesions on chest computed tomography. *Acad. Radiol.* 11:1355–1360.

Twombly, R. 2002. First clinical trials of endostatin yield lukewarm results. *J. Natl. Cancer Inst.* 94:1520–1521.

Vaidya, S.J., Payne, G.S., Leach, M.O., and Pinkerton, C.R. 2003. Potential role of magnetic resonance spectroscopy in assessment of tumour response in childhood cancer. *Eur. J. Cancer.* 39:728–735.

Waldherr, C., Mellinghoff, I.K., Tran, C., Halpern, B.S., Rozengurt, N., Safaei, A., Weber, W.A., Stout, D., Satyamurthy, N., Barrio, J., Phelps, M.E., Silverman, D.H., Sawyers, C.L., and Czernin, J. 2005. Monitoring antiproliferative responses to kinase inhibitor therapy in mice with 3'-deoxy-3'-18F-fluorothymidine PET. *J. Nucl. Med.* 46:114–120.

Weber, W.A. 2005. Use of PET for monitoring cancer therapy and for predicting outcome. *J. Nucl. Med.* 46:983–495.

Weisberg, E., Manley, P.W., Breitenstein, W., Bruggen, J., Cowan-Jacob, S.W., Ray, A., Huntly, B., Fabbro, D., Fendrich, G., Hall-Meyers, E., Kung, A.L., Mestan, J., Daley, G.Q., Callahan, L., Catley, L., Cavazza, C., Azam, M., Neuberg, D., Wright, R.D., Gilliland, D.G., and Griffin, J.D. 2005. Characterization of AMN107, a selective inhibitor of native and mutant Bcr-Abl. *Cancer Cell.* 7:129–141.

Weissleder, R. 2002. Scaling down imaging: molecular mapping of cancer in mice. *Nat. Rev. Cancer* 2:11–18.

Wells, P., Aboagye, E., Gunn, R.N., Osman, S., Boddy, A.V., Taylor, G.A., Rafi, I., Hughes, A.N., Calvert, A.H., Price, P.M., and Newell, D.R. 2003. 2-[(11)C]thymidine positron emission tomography as an indicator of thymidylate synthase inhibition in patients treated with AG337. *J. Natl. Cancer Inst.* 95:675–682.

Wells, P., Gunn, R.N., Alison, M., Steel, C., Golding, M., Ranicar, A.S., Brady, F., Osman, S., Jones, T., and Price, P. 2002. Assessment of proliferation in vivo using 2-[(11)C]thymidine positron emission tomography in advanced intra-abdominal malignancies. *Cancer Res.* 62:5698–5702.

Wells, P., Gunn, R.N., Steel, C., Ranicar, A.S., Brady, F., Osman, S., Jones, T., and Price, P. 2005. 2-[11C]thymidine positron emission tomography reproducibility in humans. *Clin. Cancer Res.* 11:4341–4347.

Wells, P., Jones, T., and Price, P. 2003. Assessment of inter- and intrapatient variability in C15O2 positron emission tomography measurements of blood flow in patients with intra-abdominal cancers. *Clin. Cancer Res.* 9:6350–6356.

Wells, P., West, C., Jones, T., Harris, A., and Price, P. 2004. Measuring tumor pharmacodynamic response using PET proliferation probes: the case for 2-[(11)C]-thymidine. *Biochim. Biophys. Acta.* 1705:91–102.

Willett, C.G., Boucher, Y., Di Tomaso, E., Duda, D.G., Munn, L.L., Tong, R.T., Chung, D.C., Sahani, D.V., Kalva, S.P., Kozin, S.V., Mino, M., Cohen, K.S., Scadden, D.T., Hartford, A.C., Fischman, A.J., Clark, J.W., Ryan, D.P., Zhu, A.X., Blaszkowsky, L.S., Chen, H.X., Shellito, P.C., Lauwers, G.Y., and Jain, R.K. 2004. Direct evidence that the VEGF-specific antibody bevacizumab has antivascular effects in human rectal cancer. *Nat. Med.* 10:145–147.

Zhao, B., Schwartz, L.H., Moskowitz, C.S., Wang, L., Ginsberg, M.S., Cooper, C.A., Jiang, L., and Kalaigian, J.P. 2005. Pulmonary metastases: effect of CT section thickness on measurement—initial experience. *Radiology* 234:934–939.

III
Lung Carcinoma

1

The Role of Imaging in Lung Cancer

Clifton F. Mountain and Kay E. Hermes

Introduction

In patients with lung cancer, the stage of the disease at the time of diagnosis is generally accepted as a proxy of survival. The International System for Staging Lung Cancer (ISSLC) provides reproducible definitions for the anatomic extent of disease: the extent of the primary tumor (T), the regional lymph nodes (N), the presence or absence of distant metastasis (M), and for stage grouping TNM (Tumor/ Node/Metastases) subsets of patients. The force of mortality is reflected in the staging hierarchy—that is, from stage IV, with the poorest survival rate, to stage IA, with the highest survival rate. Accurate determination of the clinical stage of disease in patients with lung cancer is critical to selecting appropriate treatment and estimating prognosis. It is in this context that radiographic imaging contributions cannot be overemphasized. Refinements and advances in the various imaging technologies, including chest radiography (CXR), computed tomographic scans (CT), magnetic resonance imaging (MR), and positron emission tomography (PET), have improved staging accuracy. Their increased use will provide enhanced characterization of staging elements that reflect the true extent of disease prior to the start of treatment. Radiographic imaging plays an increasingly important role in appropriate restaging of patients assigned induction, multistep therapeutic programs. Complementary, radiologic noninvasive techniques and minimally invasive procedures are under investigation. The quest for patients with treatable, early lung cancer through private and public screening programs has yet to be fulfilled; however, pilot programs using improved CT scanning techniques have been undertaken.

The International System for Staging Lung Cancer

The stage of the disease at the time of diagnosis, the histological cell type of the tumor, and the patient's performance status are primary factors that determine the treatment strategy for patients with lung cancer, and thus, the ultimate potential for cure, control, or progression of the disease. For the past 20 years, the scientific community has been well served by wide application of the ISSLC. The classification was adopted in 1986 (Mountain, 1986) and, with revisions, has been in place since that time (Mountain, 1997); it has been included and endorsed in the past and present published staging manuals of the American Joint Committee on Cancer (AJCC) (2002) and the International Union Against

163

Cancer (UICC) (1997). End-result studies according to TNM factors (T—primary tumor, N—regional lymph nodes, M—distant metastasis) and the stage group provide a benchmark for (1) estimating prognosis, (2) entering patients into clinical trials, (3) comparing the effectiveness of differing treatments, and (4) evaluating new prognostic factors. Communication of new knowledge regarding the present status of treatment for specific groups of patients is facilitated by the use of a consistent, reproducible staging system that is universally understandable. It is applicable to the four major cell types of lung cancer—squamous cell carcinoma, adenocarcinoma, large cell carcinoma, and small cell carcinoma. This classification, described in the 2004 World Health Organization Classification of Lung Tumors, emphasizes the morphologic aspects of diagnosis using light microscopy. It also provides a standard nomenclature and criteria for diagnosis that can be used by pathologists worldwide (Beasley et al., 2005).

Most patients with lung cancer are not candidates for surgical treatment, and the staging definitions do not include features that could only be determined pathologically. The clinical determination of disease extent (cStage) is the basis for therapeutic management and is based on all diagnostic and evaluative information obtained before treatment is started. The clinical classification includes the results of chest roentgenograms (CXR), CT, mediastinoscopy, fine needle aspiration biopsy (FNA), diagnostic thoracoscopy, bone scanning, MR, and PET using F-18-fluorodeoxyglucose (FDG-PET). Transesophageal ultrasound (EUS) with FNA, a recent method for determining the status of regional lymph nodes, may be included; however, it is limited to examination of the posterior mediastinum (Oh et. al., 2005). The complementarity of these techniques adds to their effectiveness. Sequencing of diagnostic tests usually proceeds from the CXR and CT examinations and tests for ruling out distant metastasis to examinations for confirming the extent of the primary tumor and the status of regional lymph nodes.

In addition to clinical staging, the identical definitions are useful for classifying the extent of disease at various points in the life history of the cancer; for surgical-pathologic staging (pTNM-pStage) that is based on pathologic examination of resected specimens; or for retreatment staging (rTNM-rStage) following the initial or subsequent steps in a multistep treatment program, or any other designated point in the course of the disease. Radiologic imaging, especially FDG-PET, may prove useful in these evaluations.

Stage Groups and Survival Patterns

Stage grouping of patients according to TNM subsets results in seven stages and an additional category for those with carcinoma *in situ*. Each stage identifies a fairly precise level of disease extent whose implications for survival dura-

tion in patients with non-small cell and small cell lung cancer have been reported (Mountain et al., 1999; Mountain, 2001). The data reflect the erosion of survival expectations as the disease progresses from stage I to stage IV.

Stage IA lung cancer includes only one anatomic subset, T1 N0 M0 tumors that are 3 cm or less in greatest dimension, with no evidence of invasion proximal to a lobar bronchus. No evidence of intrapulmonary, hilar, mediastinal, or distant metastasis is found. The prognosis for these patients is significantly better than that for patients in any other stage group.

Stage IB lung cancer classifies patients with T2 tumors—that is, tumors larger than 3 cm and those of any size that either invade the visceral pleura or have associated atelectasis or obstructive pneumonitis extending to the hilar region. The proximal extent of disease must be at least 2 cm distal to the main carina, and there is no evidence of metastasis, T2 N0 M0 disease.

Stage IIA identifies patients with small primary tumors, T1, and evidence of metastasis limited to the intrapulmonary, including hilar, lymph nodes–T1 N1 M0 disease. A clinical diagnosis of stage IIA is infrequent; however, in patients undergoing surgical treatment, stage migration from other cStage categories to this pStage category is common.

Stage IIB includes two anatomic subsets of patients who have nearly identical survival rates—patients with T2 N1 M0 and T3 N0 M0 tumors. In patients with T2 tumors, metastasis is limited to the intrapulmonary, including hilar lymph nodes. No metastasis is present in those with T3 tumors—that is, with extrapulmonary extension or invasion into the (1) chest wall, including superior, sulcus tumors, (2) diaphragm, (3) mediastinal pleura or pericardium, without involving the heart, great vessels, trachea, or esophagus, (4) vertebral body, (5) a tumor in the main bronchus within 2 cm of the carina, without involving the carina, or (6) associated atelectasis or obstructive pneumonitis involving an entire lung.

Stage IIIA provides for classifying patients with (1) extrapulmonary tumor extension and evidence of metastasis to intrapulmonary, including hilar lymph nodes, T3 N1 M0 disease, and (2) T1, T2, or T3 tumors with evidence of metastasis to the ipsilateral mediastinal and subcarinal lymph nodes, N2 disease. No distant metastasis is present. The CT scan in Figure 30 illustrates a T2 N2 M0—Stage IIIA tumor presentation—a right hilar mass with right hilar adenopathy (Mountain et al., 1999).

Stage IIIB designates extensive, extrapulmonary tumor invasion (T4) and metastasis to the contralateral mediastinal and hilar lymph nodes (N3); however, no evidence of distant metastasis is present. The TNM subsets are T4 any N M0 and any T N3 M0 disease. A T4 tumor may be any size with invasion of the mediastinum, or involving heart, great vessels, trachea, esophagus, vertebral body, or carina or presence of malignant pleural or pericardial effusion, or with

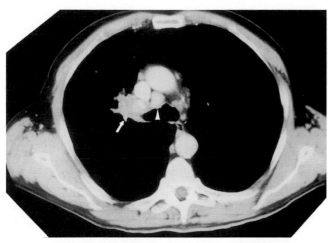

Figure 30 Computed tomographic scan of a right upper lobe bronchogenic carcinoma showing chest wall invasion (white arrow). The lymph nodes were free of disease, T3 N0 M0, stage IIB disease. *Source*: Mountain *et al.*, 1999.

satellite tumor nodules within the ipsilateral, primary tumor lobe of the lung.

Stage IV is reserved for patients with evidence of distant metastatic disease, M1, such as metastases to brain, bone, liver, adrenal gland, contralateral lung, pancreas, and other distant organs, and metastasis to distant lymph node groups such as axillary, abdominal, and inguinal. Patients with metastasis in ipsilateral nonprimary tumor lobes also are designated M1.

The Role of Imaging in Lung Cancer Staging

Appropriate use of radiologic imaging is critical to defining the clinical stage of the disease in patients with lung cancer. Advances in imaging technology have increased our ability to identify the true extent of disease. Computed tomography scanning has matured into an efficient and accurate diagnostic tool. Positron emission tomography with FDG has been cited as "the most important advance in lung cancer imaging since the introduction of CT scanning. The functional information derived from PET is complementary to the high-resolution structural imaging data from CT and MR. This combination of structural and functional imaging provides an accuracy not previously attainable in noninvasive evaluation" (MacManus and Hicks, 2003, p. 149). Increased accuracy in detecting malignant disease in solitary nodules, in lymph nodes, in the diagnosis of malignant pleural effusions, and in assessment of distant metastasis has been achieved.

In recently published guidelines for the treatment of unresectable non-small cell lung cancer, the American Society of Clinical Oncology (ASCO) recommended that chest radiography and contrast-enhanced CT that includes the liver and adrenals should be performed to stage local-

regional disease. If no evidence of distant metastasis is seen on the CT scan, FDG-PET scanning supplements the scan and is recommended (Pfister *et al.*, 2003). The complementarity of FDG-PET with CT provides a significant reduction of unnecessary thoracotomies and contributes to radiation therapy planning.

Imaging for Primary Tumor Evaluation

Chest radiography (CXR) and CT are significant for identifying the size, location, and local extent of the primary tumor, including the presence or absence of atelectasis, pleural effusion, and the status of the mediastinum. These examinations are almost universally used in the initial assessment of a lung tumor and in the detection of lymph node abnormalities. Nearly all lung cancers are initially observed on a CXR, which provides the impetus for further investigation. The CXR may provide information for T-staging—that is, estimates of the tumor size, ≤ to 3 cm, or > 3 cm; the location, central, hilar mass or peripheral; local extension; and the presence and extent of atelectasis or pneumonia. However, the CT scan more precisely identifies the anatomy of the primary tumor growth. The accuracy of tumor size and the definition of local extension involving visceral pleura or mediastinal extension are defining elements for staging T1, T2, T3, or T4 disease. The CT is more reliable than CXR for confirming chest wall invasion or invasion of the mediastinum or diaphragm. Computed tomography examination is not optimal for confirming chest wall or local mediastinal invasion, unless a chest wall mass, rib destruction, or gross encasement of mediastinal structures is evident (Ratto *et al.*, 1991). An example of a lung cancer shown on CT to be invading the chest wall, a resectable T3 N0 M0 tumor, is illustrated in Figure 31 (Mountain *et al.*, 1999). Magnetic resonance imaging is helpful in evaluating lung tumors originating in the superior sulcus. These tumors may invade unresectable structures due to the proximity of the brachial plexus, vertebral bodies, and subclavian vessels. Magnetic resonance imaging is useful to refine the anatomy of the thoracic inlet and exclude the presence of unresectable tumor. The CXR in Figure 32 shows a questionable T3–T4 tumor in the apex of the left lung. The MR examination in Figure 33 confirms T4 N2 M0, stage IIIB disease (Mountain *et al.*, 1999). Magnetic resonance imaging may be used with CT to refine the imaging of invasive characteristics, including pericardial or cardiac involvement or diaphragmatic invasion.

FDG-PET may contribute to a diagnosis of malignancy in the primary tumor, but has limited applicability for further characterization. A recent report showed that FDG-PET detected all primary lung cancers with two false-positive primary sites in a study of 97 patients under consideration for surgical resection (Saunders *et al.*, 1999). The technology has limitations in specificity, however, due

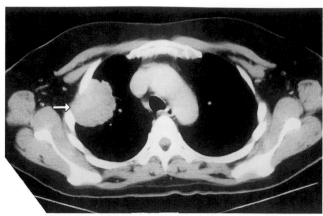

Figure 31 Computed tomographic scan of the chest shows a right hilar mass (arrow) and right paratracheal adenopathy (arrowhead). There is a suggestion of compression of the right mainstem bronchus anteriorly, T2 N2 M0, stage IIIA disease. *Source*: Mountain *et al.*, 1999.

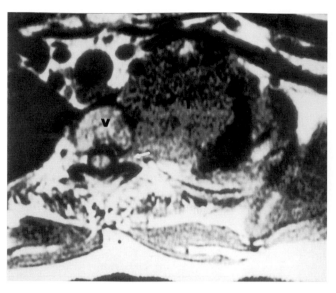

Figure 33 Magnetic resonance imaging of the chest showing tumor (T) and vertebrae (V). Arrow points to tumor extension directly into the neural foramen, T4 N0 M0, stage IIIB disease. *Source*: Mountain *et al.*, 1999.

to increased glycolytic activity in inflammatory tissue and some benign tumors. "Pure" bronchioalveolar carcinoma may not show any increase in glycolytic activity. FDG-PET has been shown to be very useful in the diagnosis of malignant pleural effusion. In a recent study of 35 patients with biopsy-proven lung cancer and radiographic findings of

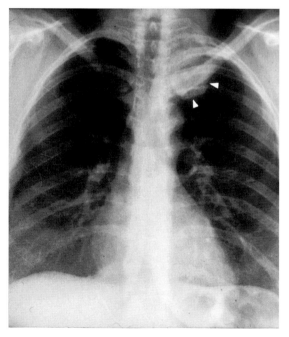

Figure 32 Posterior-anterior radiograph showing tumor in apex of left lung, questionable T3-stage IIIA or T4-stage IIIB disease. (*Source*: Mountain *et al.*, 1999) neural foramen, T4 N0 M0, stage IIIB disease. *Source*: Mountain *et al.*, 1999.

pleural effusion on CXR or CT, all patients with negative cytologic findings were read as negative for abnormal FDG uptake in pleura or pleural metastatic involvement (Gupta *et.al.*, 2002). FDG-PET imaging correctly detected the presence of malignant pleural effusion and metastatic involvement in 16 of 18 patients and excluded malignant pleural effusion or pleural metastatic involvement in 16 of 17 patients. Therefore, FDG-PET was found to have a sensitivity of 88.8%, a specificity of 94.1%, a positive predictive value of 91.4%, and an 88.8% negative predictive value for detecting malignant pleural effusion or metastatic pleural involvement. Malignant pleural effusion as a staging element reflects a T4 classification, but in patients with a negative cytology the effusion is not treated as a staging element. Thus, FDG-PET findings could alter staging and management options in patients with pleural effusion on CXR or CT, if negative cytology is not available.

If bronchioloalveolar carcinoma (BAC) presents on CXR and CT as a diffuse infiltrate, the primary tumor cannot be assessed and is designated TX. This carcinoma may be reported as the cause of a false-positive FDG-PET scan because of the low uptake in some forms of the tumor. Bilateral disease is designated M1. The carcinoma occurs as a solitary nodule or mass, and it also presents as multiple nodules within a lobe, or more than one lobe. In patients with synchronous multiple primary tumors, the patient is assigned to the highest stage. A recent report confirms the survival implications of synchronous multiple primary tumors and verifies the upstaging rule (van Rems *et al.*, 2000). Computed tomography is valuable in identifying multiple primary lung cancers and showing characteristics to differentiate them from metastasis. Although no reports are available, it is reasonable to assume that FDG-PET would show the same "hot spots" as in the primary tumor uptake, if the multiple primary tumor is of sufficient size.

Imaging for Evaluation of Regional Lymph Nodes

Existing problems of confusion and inconsistency in designating regional lymph nodes as N1, N2, or N3 have been addressed, and recommendations have been published for refined definitions for classifying regional lymph node metastasis (Mountain and Dressler, 1997). The recommendations were derived from the best features of differing nodal maps used during the past decade and a study of the literature dealing with the anatomy of the mediastinal pleura and patterns of lymph node drainage. Anatomic landmarks identify all lymph node stations within the mediastinal pleural reflection as N2 and all lymph node stations distal to the mediastinal pleural reflection and within the visceral pleura as N1. Metastasis to contralateral hilar and mediastinal lymph nodes and to ipsilateral and contralateral supraclavicular nodes are classified as N3. To identify nodes more precisely, nodes are assigned to 14 numbered stations defined by adjacent anatomic landmarks. Computed tomography correlation of anatomic definitions of stations has been published (Ko et al., 2000). Lymph node metastasis has a profound effect on prognosis; a better outcome may be anticipated when the metastatic disease is confined to the N1 stations, as opposed to those involving the N2 stations.

Size is the only radiographic criterion used for distinguishing normal from abnormal lymph nodes. The standard for defining metastasis adapted by most radiologists is a short axis lymph node diameter ≥ 1 cm on a transverse CT scan. This is thought to maximize sensitivity at the expense of specificity and to minimize false-negative results. Although CT is very accurate in demonstrating enlarged nodes, review of many studies shows that CT scan alone cannot be used as the only method to determine lymph node status because not all enlarged nodes represent metastasis and, conversely, small-normal appearing nodes may contain microscopic disease. Although metastasis in the hilar and intrapulmonary nodes usually does not preclude surgical treatment, it has implications for clinical staging end results, N0 versus N1 disease. Reported accuracy rates of MR and CT for detecting hilar node involvement are 62–88% and 68–74%, respectively (Martini et al., 1985; Glazer et al., 1985). Although the rates vary, meta-analysis of 42 studies examining the accuracy of CT for mediastinal lymph node assessment resulted in a sensitivity of 79% and a specificity of 78% (American Thoracic Society and European Respiratory Society, 1997). Others have reported sensitivity of 75% and specificity of 82%, as well as a positive-predictive value of 56% and a negative-predictive value of 83%

(Toloza et al., 2003). In a study of patients with clinical N0-1 disease by CT scan examination, 18% (68/379) of patients were found to have N2 disease at surgery (Suzuki et al., 1999). Mediastinoscopy and mediastinotomy may have an accuracy of < 80%. In a series of 859 patients undergoing CT scan and cervical mediastinoscopy or anterior mediastinotomy, 14% (103/859) had unsuspected N2 disease at thoracotomy (DeLeyn et al., 1996). A recent study, based on a decision-tree model, proposed that FDG-PET with selected mediastinoscopy could offer cost and clinical outcome benefits compared to mediastinoscopy alone in the preoperative staging of lung cancer. The authors note the high negative-predictive value of FDG-PET (93%) and conclude that, according to their model, mediastinoscopy could be reserved for patients with unresectable disease by FDG-PET. The cost savings due to fewer mediastinoscopies would not compromise patient management (Yap et al., 2005).

FDG-PET scanning is complementary to the CT findings and has become an ideal supplement owing to its higher accuracy in the detection of nodal metastasis. The use of combined CT and FDG-PET, and more recently of integrated CT-PET, has provided major improvement in detecting metastasis and thus improving staging and treatment strategy (Gilman and Aquino, 2005). A comparison of PET alone and integrated PET/CT showed that the overall staging accuracy of integrated PET/CT was significantly higher than that for PET alone. PET/CT correctly staged the T component in 97% of cases compared to 67% for PET. This difference would be largely related to the CT component (Halpern et al., 2005). Other reports confirm sensitivity of 89%, specificity of 94%, positive-predictive value of 89%, negative-predictive value of 94%, and accuracy of 93%, with integrated CT-PET for staging mediastinal lymph nodes (Antoch et al., 2003). False-positive findings related to the positive-predictive value of CT-PET are frequent enough that caution is advised in interpreting these results. In a recent study of 71 patients, the causative factors for false-positive PET scan were primarily inflammatory conditions; 7 of 10 false-positive lymph nodes showed histologically reactive lymphoid hyperplasia (Takamochi et al., 2005). Anatomic factors in centrally located tumors also are associated with false-positive staging. In patients who are otherwise operable, false-positive studies may prevent considerations for surgery; therefore, the author recommends that positive scans require biopsy confirmation. This is consistent with the recent ASCO recommendations for interpreting positive-predictive results (Pfister et al., 2003).

Imaging for Evaluation of Distant Metastasis

Two problems emerge in the attempt to understand and detect distant metastasis. First, studies of the pattern of failure after curative surgery revealed that the first site of failure was distant metastatic disease in 75% of patients who developed recurrence or metastasis within two years of treatment. This pattern was similar regardless of the pStage, histological cell type, or any other factor that could be examined. Second, many useless resections take place because of unrecognized, asymptomatic, gross metastasis. The introduction of FDG-PET is proving helpful, particularly in the area of precluding useless surgery. If initial CT scanning includes the upper abdomen, 75% of detectable abdominal metastases may be found. A problem is presented by the high incidence of benign adrenal nodules, and so FDG-PET is advocated to identify metastatic disease. The sensitivity and specificity of the examination are 100% and 80–100%, respectively (Erasmus, 1999; Marom et al., 2000; Goldsmith and Kostakoglu, 2000). Therefore, if FDG uptake in an isolated adrenal mass is normal in a patient with potentially resectable disease, surgery is advocated. An increased FDG uptake in a similar situation indicates biopsy confirmation of metastasis before surgical treatment is denied. The results of the American College of Surgeons Oncology trial reported sensitivity, specificity, positive-predictive value, and negative-predictive value of 83%, 90%, 36%, and 99%, respectively, for whole-body FGD-PET detection of M1 disease (Reed et al., 2003).

A recent evaluation of whole-body imaging with FDG-PET in 97 patients showed that the scanning altered patient management in 37% of patients, the clinical stage in 26.8%, and the nodal stage in 16.5%. It detected distant metastasis in 16.5%, but missed 7 of 10 cerebral metastases. The authors conclude that whole-body FDG-PET provides significant staging and prognostic information. Whole-body imaging detection of occult metastasis is related to the stage of disease; that is, it increases as the T and N descriptors increase, 7.5% in early-stage disease and up to 24% in advanced disease (MacManus et al., 2001). Magnetic resonance imaging has been reported useful in confirming hepatic metastasis shown on CT. Bone metastasis has been detected in 13% of a FDG-PET scanned series, with 75% being asymptomatic (Lau et al., 2000). Greater accuracy and fewer false-negative and false-positive findings are reported with FDG-PET than with radionuclide bone scans. Thus, FDG-PET may be considered a potential substitute for radionuclide scans using technetium (Peterson et al., 2003). FDG-PET has a low sensitivity for detection of brain metastasis (68%) and reportedly is least accurate in this area (Saunders et al., 1999). Brain CT or MR imaging with and without contrast material has been recommended for patients with signs or symptoms of central nervous system disease and for patients with stage III disease that are considered for aggressive local therapy, surgery, or radiation (Pfister et al., 2003; ASCO, 2003).

Restaging

A recent review of the role of FDG-PET for restaging and for assessing local recurrent disease noted the following appli-

cations currently under investigation: (1) its value as an early predictor of response to chemotherapy, (2) the potential utility following neoadjuvant therapy, and (3) the detection of tumor recurrence following potentially curative therapy (Hazelton and Coppage, 2005). The authors note that FDG-PET has a high sensitivity and reasonable specificity in confirming recurrent or residual disease, and may be useful in defining the extent of relapse. The roles of CT and FDG-PET are complementary, and further studies of this approach for restaging are warranted. It is generally accepted that sterilization of the mediastinum after induction therapy has an impact on the prognosis of patients with stage IIIA disease; accurate staging after therapy may rationally guide diverse treatment options for these patients (Bueno et al., 2000). Problems do exist with PET for restagings that are related to (1) the inability of the technique to detect micrometastases, (2) the timing of the examination, (3) the influence of radiotherapy/chemotherapy treatments, (4) the persistence of inflammatory FDG uptake, and (5) discrimination in reporting a decrease in nodal versus primary tumor standardized uptake value (Pass, 2005). Histologic confirmation of restaging is required until noninvasive methodologies and minimally invasive endoscopic techniques attain greater sensitivity and specificity (Pass, 2005).

Implications of Imaging for Lung Cancer Screening

Recent studies of CT screening for lung cancer have shown promise for detecting a high incidence of stage I lung cancers, a finding that could have potential for improving the overall cure of lung cancer. The ability to target appropriate screening populations and better tools to evaluate the tumors are being developed. With the development of CT scanning and improvement in the quality of the images, the present focus is on the use of low-dose CT scanning to screen for early lung cancer. This technique allows for low resolution of the entire thorax to be obtained with a low-radiation exposure and within a single breath hold (Ganti and Mulshine, 2005). With CT nodule enhancement and the use of FDG-PET, the number of benign biopsies has been decreased. Furthermore, ancillary findings, such as abdominal aortic aneurysms, have been diagnosed during screening that may result in an overall reduction in mortality (Hartman and Swenson, 2005). However, the major questions remain: first, whether the early detection of these lung cancers will affect the overall cure rate, and second, whether the studies will do no harm to the population screened. The main limitations of screening recently have been described as follows in a report by Jett (2005). They include: (1) a high rate of nodule detection, with over 50% of participants having at least one noncalcified nodule; (2) resulting follow-up scans, associated with increased costs; (3) cost and morbidity of biopsy or resection noncalcified nodule; and (4) a small, but difficult to quantify, risk of cancer associated with multiple follow-up CT scans.

Preliminary results of the Early Lung Cancer Action Project (ELCAP) showed that false-positive results can be kept reasonably low, are much less common on repeat screening, and may be managed with no notable excess of percutaneous or surgical biopsies when a well-defined regimen of screening is followed (Henschke *et al.*, 2005). The frequency of postiive results was low—15% for the baseline screening and 6% for subsequent cycles. The relative frequency of presurgical stage I was over 90%, and the estimated 8-year cure rate for baseline screened cancers with no evidence of metastasis is 95%, and for repeat cancers, 98%. Final results of a feasibility study for the National Lung Cancer Screening Trial established that a large-scale, randomized, controlled trial of low-dose spiral computed tomography versus chest X-ray for lung cancer screening is rational and doable for lung cancer screening (Gohagan, 2005). A recommendation for lung cancer screening for specific populations will have to await the final results of ongoing trials, the ELCAP study (Henschke *et al.*, 2006), and the National Lung Cancer Screening Trial. Research to harness the potential of new imaging modalities for routine public health application is needed to identify and manage early lung cancer.

Conclusions

In presenting the role of imaging for lung cancer staging, we recognize that in a given patient the total tumor burden cannot be precisely quantitated, and the balance between host defenses and the heterogeneity of the malignancy is not measurable. These and other complex interacting biological variables will influence the subsequent course of the disease. However, the present report supports the reality of refined clinical staging such that the true extent of disease may be reflected in end-results reports by clinical stage criteria. A significant body of literature defines selection of appropriate imaging technology for staging classification and the complementarity of various methodologies to obtain optimum results. Although the role of imaging for restaging has shown promise, further work is needed to define its specific value in this instance. Screening with low-dose CT scanning may detect stage I lung cancer in selected populations; however, application is on the drawing board. Questions regarding efficacy, safety, and patient selection remain unanswered.

References

American Joint Committee on Cancer (AJCC). 2002. Lung. In: Green, F.L., Page D.L., Fleming, I.D., Fritz, A.G., Balch, S.M., Haller, D.G., and Morrow, M. (Eds.), *AJCC Cancer Staging Manual 2002*, 6th ed., pp. 167–174. New York: Springer-Verlag.

American Thoracic Society and European Respiratory Society. 1997. Pretreatment evaluation of non-small cell lung cancer. *Am. J. Respir. Crit. Care Med. 156*:320–332.

Antoch, G., Stattaus, J., Nemat, A.T., Marnitz S., Beyer, T., Kuchi, H., Bockisch, A., Debatin, J.J., and Freundenberg, L. 2003. Non-small cancer: dual-modality PET/CT in preoperative staging. *Radiology 229*:526–533.

Beasley, M.B., Brambilla, E., and Travis, W.D. 2005. The 2004 World Health Organization classification of lung tumors. *Semin. Roentgenol. 40*:90–97.

Bueno, R., Richards, W.G., Swanson S.J., Jaklitsch, M.T., Lukanich, J.M., Mentzer, S.J., and Sugarbaker, D.J. 2000. Nodal stage after induction therapy for stage IIIA lung cancer determines patient survival. *Ann. Thorac. Surg. 70*:1836–1831.

DeLeyn, P., Schoonooghe, P., Defeffe, G., Van Raemdonck D., Coosemans, W.W., Vansteenkiste, J., and Lerut, T. 1996. Surgery for non-small cell lung cancer with unsuspected metastasis to ipsilateral mediastinal or subcarinal lymph nodes (N2 disease). *Eur. J. Cardiothoracic Surg. 10*:649–654.

Erasmus, J.J., McAdams, H.P., and Patz, E.F. 1999. Non-small cell lung cancer: FDG-PET imaging. *J. Thorac. Imaging 14*:247–256.

Erasmus, J.J., Truong, M.T., and Munden, R.F. 2005. CT, MR, and PET imaging in staging of non-small cell lung cancer. *Semin. Roentgenol. 40*:128–142.

Ganti, A.K., and Mulshine, J.L. 2005. Lung cancer screening: panacea or pipe dream? *Ann. Oncol. 16* (Suppl. 2):ii215–ii219.

Gilman, M.D., and Aquino, S.L. 2005. State-of-the-art FDG-PET imaging of lung cancer. *Semin. Rentgenol. 40*:143–153.

Glazer, G.M., Gross, B.H., Aisen, A.M., Quint, L.E., Francis, I.R., and Orringer, M.B. 1985. Imaging of the pulmonary hilum: a prospective comparative study in patients with lung cancer. *AJR 145*:245–248.

Gohagan, J.K., and The Lung Screening Research Group. 2005. Final results of the lung screening study, a randomized feasibility study of spiral CT versus chest X-ray screening for lung cancer. *Lung Cancer 47*:9–15.

Goldsmith, S.J., and Kostakoglu, L. 2000. Nuclear medicine imaging of lung cancer. *Radiol. Clin. North Am. 38*:511–524.

Gupta, N.C., Rogers, J.S., Graeber, G.M., Gregory, J.I., Mullet, D., and Atkins, M. 2002. Clinical role of F-18-fluorodeoxyglucose positron emission tomography imaging in patients with lung cancer and suspected malignant pleural effusion. *Chest 122*:1918–1924.

Halpern, B.S., Schiepers, C., Weber, W.A., Crawford T.L., Fueger, B., Phelps, M.E., and Czernin, J. 2005. Presurgical staging of non-small cell lung cancer. *Chest 128*:2289–2297.

Hartman, T.E., and Swensen, S.J. 2005. CT screening for lung cancer. *Semin. Rentgenol. 40*:193–196.

Hazelton, T.R., and Coppage, L. 2005. Imaging for lung cancer restaging. *Semin. Roentgenol. 40*:182–192.

Henschke, C.I., Shaham, D., Yankelevitz, D.F., and Altork, N.K. 2005. CT screening for lung cancer: past and ongoing studies. *Semin. Thorac. Cardiovasc. Surg. 17*:99–106.

Henschke, C.I., Shaham, D., Yankelevitz, D.E., Kramer, A., Kostis W.J., Reeves, A.P., Vasquez, M., Kotzumi, J., and Miettinen, J. 2006. CT screening for lung cancer: significance of diagnoses in its baseline cycle. *Clin. Imaging 30*:11–15.

International Union Against Cancer (UICC). 1997. Lung tumors (ICD0-162). In: Hermanek, P., Sobin, L.H., and Wittekind, C.H. (Eds.), *TNM Classification of Malignant Tumors*, 5th ed. New York: Wiley-Liss.

Jett, J.R. 2005. Limitations of screening for lung cancer with low-dose spiral computed tomography. *Clin. Cancer Res. 11(13 Pt. 2)*:4988s–4992s.

Ko, J.P., Drucker, E.A., Shepard, J.O., Mountain, C.F., Dresler C., Sabloff, B., and McCloud, T.T. 2000. CT depiction of regional nodal stations for lung cancer staging. *AJR 174*:775–782.

Lau, C.L., Harpole, D.H., and Patz, E. 2000. Staging techniques for lung cancer. *Chest Surg. Clin. N. Am. 10*:781–801.

Leuketich, J.D., Friedman, D.M., Meltzer, C.C., Belini, C.P., Townsend, D.W., Christie, N.A., and Weigel, T.L. 2001. The role of positron emission tomography in evaluating mediastinal lymph node metastases in non-small cell lung cancer. *Clin. Lung Cancer 2*:229–233.

MacManus, M.P., and Hicks, R.J. 2003. PET scanning in lung cancer: current status and future directions. *Semin. Surg. Oncol. 21*:149–155.

MacManus, M.P., Hicks, R.J., Matthews, J.P., Hogg, A., McKenzie A.F., Wirth, A., Ware, R.E., and Ball, D.L. 2001. High rate of detection of unspected distant metastases by pet in apparent stage III non-small cell lung cancer: implications for radical radiation therapy. *Int. J. Radiat. Oncol. Phys. 50*:287–293.

Marom, E.M., Erasmus, J.J., and Patz, E.F. 2000. Lung cancer and positron emission tomography with fluorodeoxyglucose. *Lung Cancer 212*:803–809.

Martini, N., Heelan, R., Wescott, J., Bains, M.D., McCormack, P., Caravelli J., Watson, R., and Zaman, M. 1985. Comparative merits of conventional, computed tomographic and magnetic resonance imaging in assessing mediastinal involvement in surgically confirmed lung cancer. *J. Thorac. Cardiovasc. Surg. 90*:639–648.

Mohammed, T-L.H., White, C.S., and Pugatch, R.D. 2005. The imaging manifestations of lung cancer. *Semin. Roentgenol. 40*:98–106.

Mountain, C.F. 1986. A new international staging system for lung cancer. *Chest 89*:225s–233s.

Mountain, C.F. 1997. Revisions in the International System for Staging Lung Cancer. *Chest 111*:1710–1717.

Mountain, C.F. 2001. Staging classification of lung cancer: a critical evaluation. *Clin. Chest Med. 23*:103–121.

Mountain, C.F., and Dressler, C.M. 1997. Regional lymph node classification for lung cancer staging. *Chest 111*:1718–1723.

Mountain, C.F., Libshitz, H.I., and Hermes, K.E. 1999. *Lung Cancer: A Handbook for Staging, Imaging and Lymph Node Classification.* Charles P. Young Company, Houston, TX.

Oh, Y.S., Early, D.S., and Azar, R.R. 2005 Clinical applications of endoscopic ultrasound to oncology. *Oncology 68*:526–537.

Pass, H. 2005. Mediastinal staging 2005: pictures, scopes, and scalpels. *Semin. Oncol. 32*:269–278.

Peterson, J.J., Kransdorf, M.J., and O'Connor, M.I. 2003. Diagnosis of occult bone metastasis: positron emission tomography. *Clin. Orthrop. 415*(Suppl.):S120–S128.

Pfister, D.G., Johnson, D.H., Azzoli, G.C., Sause, W., Smith, T.J., Baker, S. Jr., Olak, J., Stover, D., Strawn, J.R., Turrisi A.T., Sommerfield, M.R., and American Society of Clinical Oncology. 2003. American Society of Clinical Oncology treatment of unresectable non-small cell lung cancer guideline: update 2003. *J. Clin. Oncol. 22*:330–353.

Ratto, G.B., Placenza, G., Frola, C., Musanic F., Serrano, J., Giua, R., Salio, M., Jacovoni, P., and Rovida, S. 1991. Chest wall involvement by lung cancer: computed tomographic detection and results of operation. *Ann. Thorac. Surg. 51*:182–188.

Reed, C., Harpole, D.H., Posther, K., Woolson, S.L., Downey, R.J., Meyers, B.F., Heelan, R.T., MacApinlac, H.A., Jung, S.H., Silvestri, G.A., Siegel, B.A., and Rusch, V.W. 2003. Results of American College of Surgeons Oncology Group Trial Z0050: the utility of positron emission tomography in staging potentially operable non-small cell lung cancer. *J. Thorac. Cardiovasc. Surg. 126*:1943–1951.

Roberts, P.F., Follette, D.M., and von Haag, D. 2000. Factors associated with false-positive staging of lung cancer by positron emission tomography. *Ann. Thorac. Surg. 7*:1154–1159.

Saunders, C.A., Dussek, J.E., O'Dougherty, M.J., and Maisey, M.N. 1999. Evaluation of fluorine-18 fluorodeoxyglucose whole body positron emission tomography imaging in the staging of lung cancer. *Ann. Thorac. Surg. 67*:790–797.

Suzuki, K., Naga, K., Yoshida, J., Nishimura, M., Takahashi, K., and Nishiwaki, Y. 1999. Clinical predictors of N2 disease in the setting of a negative computed tomographic scan in patients with lung cancer. *J. Thorac. Surg. 117*:593–598.

Takamochi, K., Yhoshida, J., Murakami, K., Niho, N., Ishii, G., Nishimura, M., Nishiwaki, Y., Suzuki, K., and Nagai, K. 2005. Pitfalls in lymph node staging with positron emission tomography in non-small cell lung cancer patients. *Lung Cancer 47*:235–242.

Toloza, E.M., Harpole, L., and McCrory, D.C. 2003. Noninvasive staging of non-small cell lung cancer: a review of the current evidence. *Chest 123*:137S–146S.

van Rems, M.T., Zanen, P., and de La Riviere, B. 2000. Survival in synchronous vs. single lung cancer: upstaging better reflects prognosis. *Chest 118*:952–958.

Yap, K.K., Yap, K.S., Byrne, A.J., Salvatore, U., Berlangieri, A.P., Mitchell, P., Knight, S.R., Clarke, P.C., Harris, A., Tauro, A., Rowe, C.C., and Scott, A.M. 2005. Positron emission tomography with selected mediastinoscopy compared to routine mediastinoscopy offers cost and clinical outcome benefits for preoperative staging of non-small cell lung cancer. *Eur. J. Nucl. Med. Mol. Imag. 32*:1033–1040.

Yun, M., Kim, W., Alnafisi, N., Lacorte, I., Jang, S., and Alavi, A. 2001. 18F-FDG-PET in characterizing adrenal lesions detected on CT or MRI. *J. Nucl. Med. 42*:1795–1799.

2

Lung Cancer Staging: Integrated ^{18}F-fluorodeoxyglucose-Positron Emission Tomography/Computed Tomography and Computed Tomography Alone

Kyung Soo Lee

Introduction

Non-small cell lung cancer (NSCLC) accounts for 75–80% of all lung cancers and is currently the leading cause of tumor-related deaths (Tanaka *et al.*, 2000). The optimal treatment of lung cancer relies on accurate disease staging, which is based on tumor size, regional nodal involvement, and the presence of metastasis. Computed tomography (CT) has been widely used for the preoperative evaluation of tumor size and the invasion of adjacent structures. However, numerous studies have shown that CT is limited for the staging of lung cancer due to its low reliability at lymph node staging (McLoud *et al.*, 1992; Townsend, 2001). Positron emission tomography (PET) using 18-fluorine fluorodeoxyglucose (18-F FDG) has been reported to increase diagnostic accuracy in the differentiation of benign and malignant lesions and to improve nodal metastasis iden-

tification. The functional images of 18F FDG-PET are not only complementary to the images obtained using more traditional modalities, but also are more sensitive because alterations in tissue metabolism generally precede anatomic changes (Gupta *et al.*, 2001b).

The inherent limitations of PET are its failure to depict anatomic landmarks and its poor spatial resolution, which limit its ability to assess tumor size and potential infiltration of the thoracic wall, mediastinum, or other adjacent structures. The standardized uptake value (SUV) is widely used either for categorizing lesions as malignant or benign or for staging and monitoring cancer. However, many well-known factors can affect the accuracy of SUV measurement; these include patient weight, blood glucose level, uptake duration, partial-volume effect, recovery coefficient, and the type of region of interest (Paquet *et al.*, 2004). Moreover, increased glucose uptake in a benign node can be caused by either

Copyright © 2008 by Elsevier, Inc.
All rights of reproduction in any form reserved.

reactive hyperplasia or granulomatous inflammation, which may be indistinguishable from malignancy. Therefore, it is difficult to differentiate between benign lymph nodes and malignant lymph nodes using CT or 18F FDG-PET alone.

The poor spatial resolution yielded by PET due to lack of anatomic information can be overcome by combining morphologic CT and functional PET data. Although several studies that used only visual correlations between CT images and 18F FDG-PET images have demonstrated improved results for cancer staging (Townsend, 2001; Magnani et al., 1999), positional and motion-induced misregistrations limit confidence. Recently, integrated PET/CT scanners have been introduced and have produced promising initial oncologic imaging results (Goerres et al., 2002). These PET/CT findings are not simply the summation of PET and CT findings; rather, they benefit from a high level of synergism between two modalities.

This chapter will discuss the results of staging efficacy obtained by previous studies, analyze encountered problems, and hopefully provide solutions of the problems encountered and ideas of potential advancement in the staging of NSCLC using integrated PET/CT compared with CT alone.

Results Obtained by Previous Studies

T-Staging

Computed tomography is an appropriate device for determining the location and anatomic size of the primary mass and its relationship to surrounding structures such as the mediastinum, pleura, fissures, and chest wall. Moreover, CT is useful for detecting pulmonary nodules in the same lobe as the primary mass (T4 disease) or in the nonprimary lobe (M1 disease) that may not be large enough to be detected at 18F FDG-PET (currently, most PET systems have an in-plane resolution of 5 mm) (DeGrado et al., 1994). In addition, bronchioloalveolar carcinomas and carcinoid tumors may not be evaluated by 18F FDG-PET because they have a low metabolic activity (Kim et al., 1998; Chong et al., 2006).

Integrated PET/CT has been shown to be more useful than dedicated PET or CT alone in determining the T-stage of the primary tumor and in assessing the chest wall invasion (Shim et al., 2005). According to a report, although statistically not significant ($p = 0.25$), integrated PET/CT enabled staging of primary tumor (T-stage) correctly in 86% (91 of 106) of patients, whereas CT enabled in only 79% (84 of 106) of patients (Shim et al., 2005).

N-Staging

While CT and MR imaging rely on the anatomic assessment of lymph nodes, PET relies on the increased metabolic rate of neoplastic cells compared to normal cells. Mediastinal nodes containing carcinoma have been shown to have

increased uptake and accumulation of 18F FDG. Several groups have shown that PET is superior to CT in the assessment of mediastinal nodal metastases (Patz et al., 1995; Vansteenkiste et al., 1997). In one study of 99 patients, the sensitivity and specificity for the diagnosis of N2 disease were, respectively, 83% and 94% for PET compared with 63% and 73% for CT (Valk et al., 1995). In another investigation of 100 patients, mediastinal lymph nodes were staged correctly in 85% of cases with PET compared to 58% on CT (Marom et al., 1999). The authors of a meta-analysis of 14 studies published between 1990 and 1998 concluded that the sensitivity of PET for the detection of mediastinal nodal metastases is 79–84% and the specificity is 89–91% (Dwamena et al., 1999). By comparison, meta-analysis of 29 studies using CT published during the same period showed a sensitivity of 60% and a specificity of 77%.

According to an initial study (Magnani et al., 1999), integrated PET/CT results in improved sensitivity, specificity, and overall accuracy (78%, 95%, and 89%, respectively) for the detection of malignant lymph nodes, compared with visually correlated PET and CT (67%, 95%, and 86%, respectively). The greater sensitivity and accuracy of integrated PET/CT have been corroborated by several studies (Lardinois et al., 2003; Shim et al., 2005; Antoch et al., 2003). According to a report (Shim et al., 2005), for the depiction of malignant nodes, the sensitivity, specificity, and accuracy of CT on a per nodal station basis were 70% (23 of 33 nodal groups), 69% (248 of 360), and 69% (271 of 393), respectively, whereas those of PET/CT were 85% (28 of 33), 84% (302 of 360), and 84% (330 of 393) ($p = 0.249$, $p < 0.001$, $p < 0.001$, respectively) (Fig. 34). In this study, false-positive interpretations were markedly reduced (thus, improved specificity) at integrated PET/CT; 112 false-positive interpretations were rendered at CT for evaluations of 54 hilar, 16 subcarinal, 29 paratracheal, 10 subaortic, 2 pulmonary ligament, and 1 upper paratracheal nodal group, while only 58 false-positive interpretations were rendered at PET/CT for evaluations of 32 hilar, 7 subcarinal, 13 lower paratracheal, and 6 subaortic nodal groups. Ten false-negative interpretations were rendered at CT for evaluations of 4 hilar, 2 lower paratracheal, 2 subcarinal, and 1 each of prevascular and retrotracheal (group 3) and inferior pulmonary (group 9) nodal group; and 5 false-negative interpretations were rendered at PET/CT for evaluations of 1 each of paratracheal, subaortic, subcarinal, inferior pulmonary, and hilar nodal group. The authors concluded that integrated PET/CT is significantly better than stand-alone CT in lung cancer staging by providing enhanced accuracy and specificity in nodal staging.

Supraclavicular nodes are of substantial importance because the presence of supraclavicular metastases in NSCLC is associated with incurable disease. Examination of the supraclavicular lymph nodes has traditionally been performed with palpation. In cases of nonpalpable nodes, supraclavicular metastases have been assumed to be absent. However, palpation is an unreliable method for the assess-

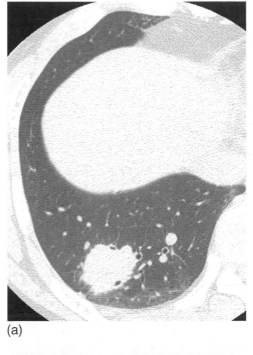

(a)

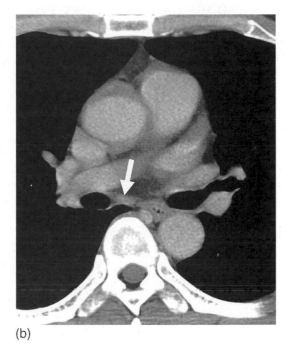

(b)

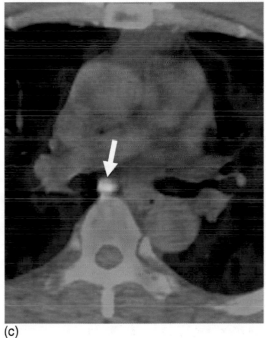

(c)

Figure 34 True-positive mediastinal lymph node metastasis at integrated PET/CT in a 46-year-old man with lung adenocarcinoma. (a) Lung-window transverse (2.5 mm section thickness, 170 mA) CT scan obtained at level of liver dome shows 28 mm nodule with lobulated margin in right lower lobe. (b) Mediastinal-window transverse unenhanced CT scan (5.0 mm section thickness, 80 mA) obtained at level of right bronchus intermedius shows 3.8 mm-sized lymph node (arrow) in short-axis diameter in subcarinal area (nodal station, 7). (c) Integrated PET/CT scan obtained at similar level to (b) demonstrate node showing markedly increased FDG uptake (maximum SUV = 7.9) (arrow) strongly suggesting malignant node, which proved to contain metastatic adenocarcinoma cells.
Source: From Shim, S.S. *et al.* 2005. Non-small cell lung cancer: prospective comparison of integrated FDG PET/CT and CT alone for preoperative staging. *Radiology 236*:1011–1019, with permission.

ment of metastases in lymph nodes in the supraclavicular regions, because the nodes with metastasis are only palpable when markedly enlarged. Currently, ultrasonography-guided fine-needle aspiration cytologic analysis is believed to be a diagnostic method of choice (van Overhagen *et al.*, 2004). Integrated PET/CT is also a promising method of supraclavicular nodal metastasis detection.

M-Staging

Conventional staging of NSCLC includes a CT scan of the chest and adrenal glands, isotope bone scan, and brain imaging reserved for selected patients with high tumor (T) and nodal (N) stage tumors. The likelihood of extrathoracic metastasis increases with conditions of higher local tumor stage, the presence of clinical or laboratory evidence of metastatic disease, and histological adenocarcinoma rather than squamous cell carcinoma. The most common sites of metastatic spread include the adrenal glands, brain bones, and liver (Silvestri *et al.*, 2003).

Whole-body PET imaging has the advantage over other imaging modalities of demonstrating not only adrenal metastases, but also other metastases that may not be apparent on CT, MR imaging, or bone scintigraphy. The advantages of integrated PET/CT for extrathoracic lung cancer staging in lung cancer patients are its abilities to detect lesions at an early stage, to accurately localize lesions, and to differentiate metastatic nodules and benign lesions, all of which aid the decision-making process in the management of these patients.

In one investigation of 100 patients with newly diagnosed pulmonary carcinoma, PET correctly indicated the M status in 40 (91%) of 44 patients with metastatic disease compared with 35 (80%) with conventional imaging. Positron emission tomography and CT correctly identified adrenal metastases of all sizes (sensitivity 100%), but PET has a specificity and positive-predictive value of 100% compared with 93% and 46% for CT. Positron emission tomography correctly identified 11 (91%) of 12 patients with bone metastases compared with 6 (50%) identified on scintigraphy; both modalities had a specificity of 92% (Valk *et al.*, 1995).

Preliminary results suggest that 18F FDG-PET imaging can characterize metabolically adrenal masses and differentiate metastatic and benign adrenal lesions (Erasmus *et al.*, 1997; Maurea *et al.*, 1999) (Fig. 35). During the past five years, 18F FDG-PET diagnostic sensitivities of 93–100%, specificities of 90–96%, and accuracies of 92–96% have been achieved (Yun *et al.*, 2001; Gupta *et al.*, 2001; Kumar

et al., 2004; Jana *et al.*, 2006). 18F FDG-PET/CT is better able to differentiate benign and malignant adrenal lesions than 18F FDG-PET alone. In one study of 175 adrenal masses in 150 patients, PET data alone (using an SUV cut-off of 3.1) yielded a sensitivity, specificity, and accuracy of 99% (67 of 68 nodules), 92% (98 of 107), and 94% (165 of 175), respectively, and combined PET/CT data yielded corresponding values of 100% (68 of 68 nodules), 98% (105 of 107), and 99% (173 of 175) (Jana *et al.*, 2006). Moreover, specificity was significantly higher for PET/CT ($P < 0.01$).

Problems and Their Solutions

It should be kept in mind that the CT component of PET/CT is obtained without intravenous contrast during shallow respiration and thus may not be adequate for playing a role as a diagnostic CT. Therefore, a diagnostic contrast-enhanced CT scan of the thorax, performed as part of the PET/CT or independently as a separate study, is still recommended for obtaining precise information concerning the anatomic relations of the primary tumor to surrounding vascular structures. Despite the fact that integrated PET/CT improved the accuracy of mediastinal nodal staging, the resolution of PET is insufficient to detect microscopic lymph node metastases (DeGrado *et al.*, 1994; Shim *et al.*, 2005; Pieterman *et al.*, 2000). If radionuclide uptake is not increased on PET, then integrated PET/CT cannot provide further information. The limited resolution also precludes the detection of subcentimeter malignant nodules in the lungs and adrenal glands. Conversely, inflammatory lymph

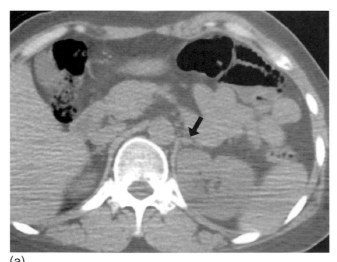
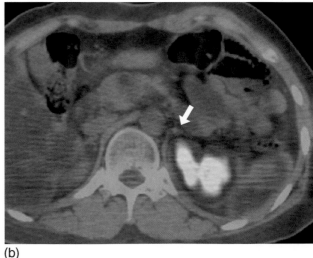

(a)   (b)

Figure 35 Adrenal nodule showing true-positive metastasis at integrated PET/CT in a 33-year-old woman with adrenal metastasis confirmed by clinical follow-up. The nodule in the left adrenal gland showed increase in size at follow-up CT. Patient also had adenocarcinoma in right lower lobe of the lung. (a) Unenhanced CT scan (5.0 mm section thickness, 80 mA) shows 11 mm-sized nodule (arrow) in left adrenal gland. (b) Integrated PET/CT demonstrates increased uptake (arrow) in left adrenal gland with maximum SUV 3.5, indicating metastatic nodule.

nodes may be present in patients with postobstructive pneumonia or chronic granulomatous infection, including tuberculosis and histoplasmosis. These nodes may appear hypermetabolic on PET images and may erroneously lead to upstaging in N-staging (Shim *et al.*, 2005; Konishi *et al.*, 2003).

Mediastinal and hilar lymph nodes showing positive uptake with calcification or higher attenuation than the surrounding great vessels usually prove to be benign, specifically in a granulomatous disease endemic area (Shim *et al.*, 2005). More than 90% of these nodes show follicular hyperplasia in the cortex and anthracotic pigmentation and macrophage infiltration with or without fibrotic micronodules formation in the medulla. These inflammatory changes of follicular hyperplasia and macrophage infiltration may have contributed to increased glucose uptake in the corresponding nodes. Therefore, nodes containing calcification or with higher attenuation than the surrounding great vessels, though showing positive uptake on PET, should be regarded as benign, especially in endemic areas of chronic granulomatous disease (Shim *et al.*, 2005).

Potential Advancements

Different 18F FDG uptake criteria may need to be used for the characterization of small subcentimeter lesions as malignant, compared to those for the characterization of large lesions. For differentiating cancer cell proliferation from infection or inflammation, other radiotracers such as 18F fluorothymidine rather than 18F FDG can be used. The 18F fluorothymidine-PET imaging is regarded as a device that allows detection of lesions that are undergoing active cell turnover (Buck *et al.*, 2003). Moreover, 11C choline (a marker of cell membrane synthesis rather than simply of glycolysis) may be a more sensitive and specific radiotracer than 18F FDG for detecting lung cancer of a low level of metabolic activity such as low-grade adenocarcinomas or bronchioloalveolar carcinoma (Tian *et al.*, 2004).

Respiratory gating can be accomplished by acquiring continuous CT images over a single respiratory cycle at each bed position and co-registering the images with the PET images according to the respiratory phase (so-called four-dimensional PET/CT). With this respiratory gating technique in PET/CT imaging, accurate localization and quantification of metabolic activity (more accurate and reproducible SUV calculations) can be improvingly allowed. The technique provides less image smearing and more accurate determination of lesion size (Pan *et al.*, 2005).

References

Antoch, G., Stattaus, J., Nemat, A.T., Marnitz, S., Beyer, T., Kuehl, H., Bockisch, A., Debatin, J.F., and Freudenberg, L.S. 2003. Non-small cell lung cancer: dual-modality PET/CT in preoperative staging. *Radiology 229*:526–533.

Buck, A.K., Halter, G., Schirrmeister, H., Kotzerke, J., Wurziger, I., Glatting, G., Mattfeldt, T., Neumaier, B., Reske, S.N., and Hetzel, M. 2003. Imaging proliferation in lung tumors with PET: 18F-FLT versus 18F-FDG. *J. Nucl. Med. 44*:1426–1431.

Chong, S., Lee, K.S., Chung, M.J., Han, J., Kwon, O.J., and Kim, T.S. 2006. Neuroendocrine tumors of the lung: clinical, pathologic, and imaging findings. *RadioGraphics 26*:41–57.

DeGrado, T.R., Turkington, T.B., Williams, J.J., Stearns, C.W., Hoffman, J.M., and Coleman, R.E. 1994. Performance characteristics of a whole-body PET scanner. *J. Nucl. Med. 35*:1398–1406.

Dwamena, B.A., Sonnad, S.S., Angobaldo, J.O., and Wahl, R.L. 1999. Metastases from non-small cell lung cancer: mediastinal staging in the 1990s—meta-analytic comparison of PET and CT. *Radiology 213*:530–536.

Erasmus, J.J., Patz, E.F. Jr., McAdams, H.P., Murray, J.G., Herndon, J., Coleman, R.E., and Goodman, P.C. 1997. Evaluation of adrenal masses in patients with bronchogenic carcinoma using 18F-fluorodeoxyglucose positron emission tomography. *AJR Am. J. Roentgenol. 168*:1357–1360.

Goerres, G.W., Kamel, E., Seifert, B., Burger, C., Buck, A., Hany, T.F., and Von Schulthess, G.K. 2002. Accuracy of image coregistration of pulmonary lesions in patients with non-small cell lung cancer using an integrated PET/CT system. *J. Nucl. Med. 43*:1469–1475.

Gupta, N.C., Graeber, G.M., Tamim, W.J., Rogers, J.S., Irisari, L., and Bishop, H.A. 2001a. Clinical utility of PET-FDG imaging in differentiation of benign from malignant adrenal masses in lung cancer. *Clin. Lung Cancer 3*:59–64.

Gupta, N.C., Tamim, W.J., Graeber, G.G., Bishop, H.A., and Hobbs, G.R. 2001b. Mediastinal lymph node sampling following positron emission tomography with fluorodeoxyglucose imaging in lung cancer staging. *Chest 120*:521–527.

Jana, S., Zhang, T., Milstein, D.M., Isasi, C.R., and Blaufox, M.D. 2006. FDG-PET and CT characterization of adrenal lesions in cancer patients. *Eur. J. Nucl. Med. Mol. Imaging 33*:29–35.

Kim, B.T., Kim, Y., Lee, K.S., Yoon, S.B., Cheon, E.M., Kwon, O.J., Rhee, C.H., Han, J., and Shin, M.H. 1998. Localized form of bronchioloalveolar carcinoma of the lung: FDG PET findings. *AJR Am. J. Roentgenol. 170*:935–939.

Konishi, J., Yamazaki, K., Tsukamoto, E., Tamaki, N., Onodera, Y., Otake, T., Morikawa, T., Kinoshita, I., Dosaka-Akita, H., and Nishimura, M. 2003. Mediastinal lymph node staging in patients with non-small cell lung cancer: analysis of false-positive FDG-PET findings. *Respiration 70*:500–506.

Kumar, R., Xiu, Y., Yu, J.Q., Takalkar, A., El-Haddad, G., Potenta, S., Kung, J., Zhuang, H., and Alavi, A. 2004. 18F FDG-PET in evaluation of adrenal lesions in patients with lung cancer. *J. Nucl. Med. 45*:2058–2062.

Lardinois, D., Weder, W., Hany, T.F., Kamel, E.M., Korom, S., Seifert, B., von Schulthess, G.K., and Steinert, H.C. 2003. Staging of non-small cell lung cancer with integrated positron-emission tomography and computed tomography. *N. Engl. J. Med. 348*:2500–2507.

Magnani, P., Carretta, A., Rizzo, G., Fazio, F., Vanzulli, A., Lucignani, G., Zannini, P., Messa, C., Landoni, C., Gilardi, M.C., and Del Maschio, A. 1999. FDG/PET and spiral CT image fusion for mediastinal lymph node assessment of non-small cell lung cancer patients. *J. Cardiovasc. Surg. 40*:741–748.

Marom, E.M., McAdams, H.P., Erasmus, J.J., Goodman, P.C., Culhane, D.K., Coleman, R.E., Herndon, J.E., and Patz, E.F., Jr. 1999. Staging non-small cell lung cancer with whole-body PET. *Radiology 212*:803–809.

Maurea, S., Mainolfi, C., Bazzicalupo, L., Panico, M.R., Imparato, C., Alfano, B., Ziviello, M., and Salvatore, M. 1999. Imaging of adrenal tumors using FDG-PET: comparison of benign and malignant lesions. *AJR Am. J. Roentgenol. 173*:25–29.

McLoud, T.C., Bourgouin, P.M., Greenberg, R.W., Kosiuk, J.P., Templeton, P.A., Shepard, J.A., Moore, E.H., Wain, J.C., Mathisen, D.J., and Grillo,

H.C. 1992. Bronchogenic carcinoma: analysis of staging in the mediastinum with CT by correlative lymph node mapping and sampling. *Radiology 182*:319–323.

Pan, T., Mawlawi, O., Nehmeh, S.A., Erdi, Y.E., Luo, D., Liu, H.H., Castillo, R., Mohan, R., Liao, Z., and Macapinlac, H.A. 2005. Attenuation correction of PET images with respiration-averaged CT images in PET/CT. *J. Nucl. Med. 46*:1481–1487.

Paquet, N., Albert, A., Foidart, J., and Hustinx, R. 2004. Within-patient variability of (18)F-FDG: standardized uptake values in normal tissues. *J. Nucl. Med. 45*:784–788.

Patz, E.F., Lowe, V.J., Goodman, P.C., and Herndon, J. 1995. Thoracic nodal staging with PET imaging with [18]FDG in patients with bronchogenic carcinoma. *Chest 108*:1617–1621.

Pieterman, R.M., van Putten, J.W., Meuzelaar, J.J., Mooyaart, E.L., Vaalburg, W., Koeter, G.H., Fidler, V., Pruim, J., and Groen, H.J. 2000. Preoperative staging of non-small cell lung cancer with positron-emission tomography. *N. Engl. J. Med. 343*:254–261.

Shim, S.S., Lee, K.S., Kim, B-T., Chung, M.J., Lee, E.J., Han, J., Choi, J.Y., Kwon, O.J., Shim, Y.M., and Kim, S. 2005. Non-small cell lung cancer: prospective comparison of integrated FDG PET/CT and CT alone for preoperative staging. *Radiology 236*:1011–1019.

Silvestri, G.A., Tanoue, L.T., Margolis, M.L., Barker, J., and Detterbeck, F. 2003. American College of Chest Physicians. The noninvasive staging of non-small cell lung cancer: the guideline. *Chest 123*: 147S–156S.

Tanaka, F., Yanagihara, K., Otake, Y., Miyahara, R., Kawano, Y., Nakagawa, T., Shoji, T., and Wada, H. 2000. Surgery for non-small cell lung cancer: postoperative survival based on the revised tumor-node-metastasis classification and its time trend. *Eur. J. Cardiothorac. Surg. 18*:147–155.

Tian, M., Zhang, H., Oriuchi, N., Higuchi, T., and Endo, K. 2004. Comparison of 11C-choline PET and FDG-PET for the differential diagnosis of malignant tumors. *Eur. J. Nucl. Med. Mol. Imaging 31*:1064–1072.

Townsend, D.W. 2001. A combined PET-CT scanner: the choices. *J. Nucl. Med. 42*:533–534.

Valk, P.E., Pounds, T.R., Hopkins, D.M., Haseman, M.K., Hofer, G.A., Greiss, H.B., Myers, R.W., and Lutrin, C.L. 1995. Staging non-small cell lung cancer by whole-body positron emission tomographic imaging. *Ann. Thorac. Surg. 60*:1573–1581.

van Overhagen, H., Brakel, K., Heijenbrok, M.W., van Kasteren, J.H., van de Moosdijk, C.N., Roldaan, A.C., van Gils, A.P., and Hansen, B.E. 2004. Metastases in supraclavicular lymph nodes in lung cancer: assessment with palpation, US, and CT. *Radiology 232*:75–80.

Vansteenkiste, J.F., Stroobants, S.G., De Leyn, P.R., Dupont, P.J., Verschakelen, J.A., Nackaerts, K.L., and Mortelmans, L.A. 1997. Mediastinal lymph node staging with FDG-PET scan in patients with potentially operable non-small cell lung cancer: a prospective analysis of 50 cases. Leuven Lung Cancer Group. *Chest 112*:1480–1486.

Yun, M., Kim, W., Alnafisi, N., Lacorte, L., Jang, S., and Alavi, A. 2001. 18F FDG-PET in characterizing adrenal lesions detected on CT or MRI. *J. Nucl. Med. 42*:1795–1799.

3

Computed Tomography Screening for Lung Cancer

Claudia I. Henschke, Rowena Yip, Matthew D. Cham, and David F. Yankelevitz

Introduction

Prior to the introduction of helical/spiral computed tomography (CT) in the early 1990s, imaging of the chest required a new breath for acquiring each image. Unless the inspiratory effort was the same for each image, potentially some of the lung was not imaged at all. The advantage of helical/spiral CT is that imaging of the chest can be reduced to < 20 sec so that all images of the lung could be obtained in a single breath.

Once helical/spiral CT was introduced, it was recognized that many more small pulmonary nodules were being detected in asymptomatic people. The CT scans on these people were being brought to our weekly Thoracic Oncology Conference at the New York–Cornell Medical Center (now New York Presbyterian–Weill Cornell Medical Center) for consultation as to the appropriate management. As no data were available on which to base these decisions, we, together with our pulmonary, oncology, and thoracic surgery colleagues, decided to organize retreats and invited others to consider these questions. As a result, it was recognized that the data needed for management decision making of these incidentally detected nodules would also provide information on the usefulness of CT screening in preventing deaths from lung cancer. Thus, we decided to simultaneously address both questions as it was thought that the advent of helical CT imaging held promise for early diagnosis of lung cancer and, thereby, for enhanced curability of this highly fatal disease.

Prior Screening Studies

Prior to designing our study, we reviewed the prior studies on screening for lung cancer. Among these studies, the most influential in determining screening policy were the four listed in this section. They were performed in the 1970s and reported in the 1980s. All four were randomized screening trials in which asymptomatic people were randomized to either the screening arm or the control arm (Fig. 36). A randomized screening trial is quite different from a randomized treatment trial (Fig. 37). In a randomized treatment trial, the disease is first determined and then the person is randomized to one of two or more treatments, whereas in a randomized screening trial the randomization is performed before the diagnosis is established. In a randomized screening trial, the null hypothesis is that the mortality rate in the screening arm is the same as that of the control arm versus the alternative hypothesis that they are different. In treatment trials, typically fatality rates are compared. The mortality rate is

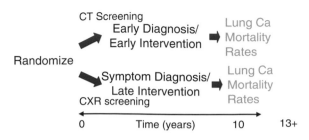

Figure 36 Schematic design of the randomized screening trial National Lung Screening Trial (NLST) showing the randomization of people at risk for lung cancer into the screening arm (CT) and control arm (CXR). Lung cancer mortality rates are determined after follow-up by ascertainment of death certificates from both arms.

defined as the number of deaths from the disease divided by the person-years of screening, while the fatality rate is defined as number of deaths from the disease divided by the total number of diagnosed cases of the disease.

Memorial Sloan-Kettering Cancer Center (MSKCC) and Johns Hopkins Medical Institution (JHMI) Studies

Both of these studies recruited male smokers (who had smoked at least one pack per day), were more than 45 years of age with an estimated survival of no less than 5 years, and had sufficient lung function to tolerate lobectomy (Melamed et al., 1984; Tockman, 1986). The design was identical in both studies. All participants were randomly assigned to two cohorts prior to baseline screening. The screening cohort had sputum cytology tests performed every 4 months and annual chest radiography, while the control cohort had only annual chest radiography. Those in whom cancer was identified in the baseline screening were excluded from further screening.

The MSKCC trial enrolled 10,040 men, and 53 lung cancers were identified as a result of the baseline screening.

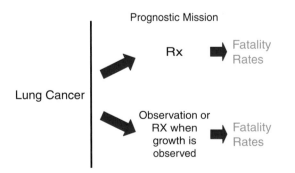

Figure 37 Schematic design of a generic treatment trial where after diagnosis of lung cancer by stage is known. The patients are randomized within stage into one of two treatment arms. Deaths are determined in each arm throughout follow-up, and case-fatality rates are calculated.

Among the 10,040 men, 4968 were randomized to the screening arm and 5072 to the control arm. The JHMI trial enrolled 10,387 men, and 79 lung cancers were identified as a result of the baseline screening. Among the 10,387 men, 5226 were randomized to the screening arm and 5161 to the control arm.

Mayo Lung Project (MLP)

Similar to the MSKCC and JHMI studies, the MLP study recruited male smokers (who smoked at least one pack per day), older than 45, who had an estimated survival of no less than 5 years, no evidence of lung cancer on initial evaluation, and sufficient lung function to tolerate lobectomy (Fontana et al., 1986). All participants had baseline screenings, and randomization was performed after baseline screening, excluding those diagnosed with lung cancer and those not eligible for surgery. The screened cohort had chest radiography and sputum cytology examination every 4 months for 6 years; the control group received a recommendation to get yearly chest radiography and sputum examination.

Of 10,933 people enrolled, 91 were diagnosed with lung cancer as a result of baseline screening. Another 1631 who did not have lung cancer were excluded after baseline screening because they either were ineligible for lobectomy or had a life expectancy < 5 years. Of the remaining 9211 men, 4618 were randomized to the screening arm and 4593 were randomized to the control arm. Subjects were then followed up for 1–5.5 years.

Czechoslovakia Study

This study recruited men who were 40–64 years of age and currently smoking. Randomization was performed after baseline screening, excluding those diagnosed with lung cancer. Those in the screening arm had chest radiography and sputum examination every 6 months for 3 years (total of six radiographs and six sputum examinations during 3 years). Those in the control arm received a single examination by the same two modalities 3 years after the prevalence screen. After that, chest radiography was performed every year in both arms for 3 years (i.e., three more times). Of the 6364 enrolled, 19 were diagnosed with lung cancer as a result of baseline screening. Of the remaining 6345 men, 3171 were randomized to the screening arm and 3174 to the control arm (Kubik and Polak, 1986).

Recommendations and Controversy Resulting from Prior Studies

Although the principal aim of these screening studies was assessment of the sputum cytology test, the results of these studies, principally those from the MLP, were also used to assess the benefit of chest radiography. None of these studies demonstrated that sputum cytology reduced the mortality rate, for the two rates were not significantly different. Thus, the null hypothesis could not be rejected. Typically, lack of rejection of the null hypothesis does not lead to acceptance of it because this depends on further considerations, including the power of the study to detect a stated difference. If the power is high, the null hypothesis may be accepted, but if the power is low, the null hypothesis should not be accepted. In this case, the power was very low, ~ 20% (Flehinger et al., 1993). Surprisingly, however, Eddy (1989) recommended accepting the null hypothesis, and as a result, the American Cancer Society (ACS, 1980), the U.S. Preventive Services Task Force (1989), and other countries worldwide with the exception of Japan made recommendations against lung cancer screening.

Although there was much debate about the interpretation of these studies and the resulting recommendation against screening, many physicians did not accept the interpretation and in fact continued chest radiographic screening of their high-risk patients (Epler, 1990). Eventually, by the late 1990s it was recognized that these studies had serious flaws and that the issue of screening for lung cancer should be reevaluated (Fontana et al., 1991; Dominioni and Strauss, 2000).

To further address the controversies arising regarding screening, a mathematical model was developed by Dr. B.J. Flehinger, chief statistician of the MSKCC study and her colleagues. Using the data from the MLP, JHMI, and MSKCC studies, they showed that had the chest radiography screening continued for some 30 years, a 13% decrease in deaths from lung cancer could have been demonstrated (Flehinger et al., 1993). It was also shown that the power for testing the hypothesis was so low that the null hypothesis should not have been accepted. A more recent analysis (Miettinen, 2000a) also showed that by focusing on the relevant deaths in the MLP from years 3 to 7, there was as much as a 40% benefit in decreasing deaths from lung cancer.

We invited Dr. Flehinger to our meetings and subsequently asked her to use this model to project the potential benefit of CT. These discussions resulted in an article showing that the probability of detecting a nodule on chest radiography was low (25%) and that if this probability could be increased to over 80%, the benefit of screening could be increased to > 80% (Flehinger et al., 1993).

The Early Lung Cancer Action Project Paradigm for Evaluation of Screening

As illustrated earlier, screening for a cancer is commonly thought of as the application of a single diagnostic test to an asymptomatic person, and testing is often assumed to reduce mortality from the cancer (Fig. 36). The diagnostic test is

viewed as an "intervention," and it is supposed to have "effectiveness" in that it should prevent the cancer's fatal outcome. These trials are very expensive and require a long time. In light of this and subsequent controversies, we decided to consider a different paradigm for evaluation of screening for cancer. In our view, this traditional viewpoint and its consequent methodology have led to much of the controversy surrounding screening, in screening not only for lung cancer but also for other cancers. The statement "after 40 years of study, mammography remains as much emotion as science" by J. Randal (2000) illustrates the controversy regarding breast cancer screening. In the controversy that erupted in 2002, Valerie Jackson (2002) pointed out that the design issues we identified (Miettinen and Henschke, 2001; Miettinen et al., 2002) explained the lack of a demonstrated benefit of screening trials.

At variance with this view, the researchers of the Early Lung Cancer Action Project (ELCAP) hold that a sharp distinction has to be made between diagnostic testing and the intervention that follows (Henschke et al., 2002a, b). A diagnostic test provides information regarding the person undergoing it, but without an associated intervention, the test has no effect on the subsequent course of health. An intervention, by contrast, is intended to improve the course of health to have effectiveness. For example, the use of chest radiography did not change the typical course of pulmonary tuberculosis; rather, the intervention with streptomycin did (Hill, 1990).

The ELCAP viewpoint defines screening as the *pursuit of early diagnosis,* which starts with an initial test and proceeds along a well-defined path (screening regimen) to diagnosis of cancer. The diagnosis that results from this pursuit is early, meaning that, at the time of diagnosis, the cancer is still in the latent, asymptomatic phase of its course, and it is also hoped that the cancer is still localized and, thereby, curable. Ultimately, the aim is to determine whether early intervention following early diagnosis provides the hope for greater curative effectiveness relative to later intervention upon the prompting of symptoms—that is, how often the pursuit of early diagnosis leads to the prevention of the cancer's fatal outcome. Thus, the real issue is the quantitative determination of the number of deaths that can be prevented.

The ideal design for evaluation of screening in preventing deaths from lung cancer separates evaluation of the diagnostic test from that of the subsequent intervention (Fig. 38). Such a study could not be done for lung cancer on ethical grounds, for it would require randomization of patients diagnosed with lung cancer to either immediate or delayed treatment and we know from long-term follow-up that the prognosis decreases dramatically when the disease is in late stage (Mountain, 1997; Inoue et al., 1998). We therefore allowed for quasi-experimental assessment of the benefit of treatment (Henschke et al., 1994, 2002b) as illustrated in Figure 39. Furthermore, we think that the focus should be on

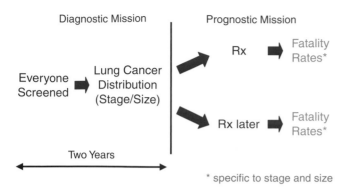

Figure 38 Ideal design for assessing the usefulness of a screening process in preventing deaths from lung cancer by early diagnosis and early treatment. The diagnostic process is assessed in terms of proportion in each stage of the disease, and the cases of lung cancer within each stage are randomized to immediate treatment or delayed treatment. Deaths are determined, and case-fatality rates are calculated.

the quantitative determination of the fatality rates and not simply on a test of hypothesis of a difference between them; that is, we should focus on the quantitative rather than the qualitative difference. This quantitative focus requires greater rigor in the specification of the screening process and its evaluation than does a hypothesis testing design. This design is not an observational or single armed study as some refer to it, but is what we call a "diagnostic-prognostic" trial with two distinctive components, the diagnostic and prognostic, with each component being evaluated separately. As for the prognostic component, alternative designs, including randomized ones, could be used for some subtypes of lung cancers depending on the questions that need to be answered.

Understanding the ultimate usefulness of CT screening requires understanding whether the various subtypes of

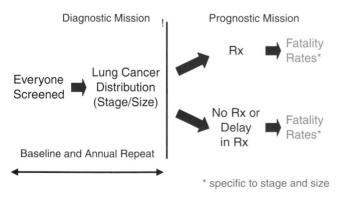

Figure 39 ELCAP design. The diagnostic process is assessed in terms of proportion in each stage of the disease and assessed as in the design shown in Figure 38. Everyone is offered early treatment, but some usually delay or refuse it. This provides quasi-experimental trial. Deaths are determined, and case-fatality rates are calculated.

diagnoses resulting from the screening regimen at issue are *genuine*. Even the management of the diagnosed cases presupposes the understanding of those matters, as in the end, the ultimate usefulness and adverse consequences of the screening depend on how these cases are actually managed. Specifically, we need to address the genuineness of the screen-diagnosed cancer, that is, whether it would lead to death if not resected, particularly in cases of stage I. Next, we need to determine how often such a genuine case is curable. In other words, when stage I (prespread detection) is achieved and early intervention is applied, we need to determine how frequently death from an otherwise fatal cancer is avoided; and on the community level, how much mortality from the cancer is reduced by such screening and intervening.

Recommendation for or against screening requires knowledge inputs as to the gain in curability of lung cancer or its corresponding reduction in case-fatality rate. Principal among these further considerations are the two that have critical bearing on the definition of the indication for screening: the person's risk for lung cancer (in the near future) and his or her life expectancy (when spared of death from lung cancer). These two inputs bear on when, if ever, to begin the screening on a given person; and insofar as it has been initiated, when to discontinue it. Thus, the decision regarding screening for lung cancer requires consideration of its benefits specific to a particular person at a particular time.

The Early Lung Cancer Action Project

Using this paradigm for evaluation of CT screening for lung cancer, ELCAP was started at two institutions in 1993 in New York City. A single cohort of 1000 high-risk subjects was recruited for experimental screening, baseline screening being performed in 1993–1998 (Henschke *et al.*, 1999) and annual repeat screening in 1994–1999 (Henschke *et al.*, 2001). The Early Lung Cancer Action Project demonstrated that 23 (85%) of the 27 diagnoses of lung cancer were of stage I, the earliest and most curable stage of lung cancer. Furthermore, 19 (83%) of the 23 stage I cases were missed on the chest radiography performed at the same time as the CT. In comparison, without CT screening, in usual care, < 15% of the cases are found to be of stage I (Henschke *et al.*, 2003a). The baseline results showed significant superiority for the CT-based screening over chest radiography screening. It was also shown that positive results could be managed with minimal use of invasive diagnostics. As expected on the annual repeat screening, the shift toward small stage IA cases has been even more pronounced, and positive test results were uncommon, < 3%.

Computed Tomography Screening in Japan

In Japan, screening through use of chest radiography was being provided on a national level to men and women 40 years of age and older. Screening sponsored by the Japanese National Cancer Center was provided to the Anti-Lung Cancer Association (ALCA), which consisted of voluntary participants. In 1993, the option to have a CT was added for those enrolled in the ALCA, presumably for similar reasons that led to the initiation of ELCAP. In 1996, they reported the results of their first 1369 screenings and showed that 14 of the 15 CT-detected cancers were of stage I (Kaneko *et al.,* 1996). They also showed the superiority of CT, as 11 of the 15 (73%) of the CT-detected cancers were missed by chest radiography.

In another study in the Nagano Prefecture, screening with CT and chest radiography was performed in a mobile CT van starting in 1996 (Sone *et al.,* 1998, 2001). Baseline screenings on 5483 men and women were followed by a total of 7303 annual repeat screenings. At baseline, 279 (5%) of the 5483 had a suspicious finding that required further work-up, and among them 23 had lung cancer. Among the 7303 repeat screenings, 309 (4%) required further work-up and 37 had cancers. Among the 60 screen-diagnosed lung cancers, 55 (88%) of them were stage I and 40 (67%) were missed on chest radiography. These two studies in Japan, like ELCAP, demonstrated the marked superiority of CT over chest radiography in detection of small, early-stage lung cancer. As both of these studies enrolled smokers and nonsmokers, the prevalence of cancer was much lower than in the ELCAP.

The New York Early Lung Cancer Action Project

Immediately after the publication of the ELCAP baseline results, an expanded study, NY-ELCAP, was started in 2000 at 12 academic medical centers in both New York City and New York State. It has resulted in a report on the updated ELCAP protocol (NY-ELCAP Investigators, 2007) and also in a report on mediastinal findings (Henschke *et al.,* 2006b). As the NY-ELCAP results are similar to those of ELCAP, its results are integrated into those of all other institutions collaborating with ELCAP.

International Conferences on Screening for Lung Cancer

The ELCAP report on baseline CT screening for lung cancer led to considerable public (Grady, 1999) and professional interest in the practice of CT-based screening for lung cancer. Suddenly, screening for lung cancer became popular with researchers initiating projects to study it, the public demanding it, and medical institutions offering it. The demand for information on screening led us to establish international conferences, inviting all institutions already performing lung screening or wishing to begin their own programs (International Conferences, 1999–2006). These conferences were an outgrowth of the already extensive role of ELCAP in helping other investigator groups at the University of Muenster in Germany (Diederich *et al.,* 2002), Hadassah Medical Center in Israel (Shaham *et al.,* 2006), the University of South Florida (Moffitt Cancer Center) and the Mayo Clinic in the United States (Swensen *et al.,* 2002), the University of Navarra (Bastarrika *et al.,* 2005) in Spain, and Hirslanden Lung Centre in Switzerland to initiate their research projects patterned after the original ELCAP.

The First International Conference on Screening for Lung Cancer in October 1999 identified the urgent need for further research as to the benefit of CT screening and how best to determine the magnitude of the benefit in preventing death from lung cancer. It was recognized that multidetector row CT scanners were rapidly replacing single-slice scanners. Consequently, slice thickness of the images decreased so that submillimeter images were being obtained in a single breath. Other advances included the use of computer monitors for the reading of the images instead of printing them on film, which provided images with much higher magnification, thereby allowing easier identification of small nodules (Henschke *et al.,* 2004a) and different types of nodules (part-solid and nonsolid) (Henschke *et al.,* 2002c). Image analytic techniques were being developed to characterize and classify pulmonary nodules on CT (Yankelevitz *et al.,* 2000), and these advances were integrated into the reading and management process. This first conference led to the general recognition that the rate of refinement of CT imaging is so rapid that today's state-of-the-art CT will become obsolete within several years.

By the Second International Conference in February 2000, it was generally agreed that CT screening presented a breakthrough opportunity in preventing death from lung cancer. Thus, there was an urgent need for rapid assessment of the magnitude of its benefit so that it could then be made available to the community at large. Critical diagnostic and intervention issues were identified, including the need for quality control and documentation of the entire screening process, starting from the initial CT to the diagnosis of lung cancer, and extending to the long-term follow-up of all diagnosed cases. The Third International Conference in October 2000 presented the new research initiatives being developed throughout the world and acknowledged the marked increase in the practice of screening. The desirability of an international consortium with a focus on pooling of data for rapid accumulation of policy-relevant information on screening for lung cancer became apparent. The subsequent conferences have provided an update on the relevant interim meetings, publications, and new trials, updates of I-ELCAP screening, and pathology protocols. They have focused on the relevant diagnostics that might impact the screening regimen or early interventions, which in turn might affect the curability of lung cancer, and they have sought to develop consensus on these topics. As of the Fourteenth Conference in April 2006, over 33,000 screenees have been enrolled.

International Early Lung Cancer Action Program

The International (I)-ELCAP was formed, drawing its members from participants in the conferences. Integral to the I-ELCAP collaboration are a shared set of principles and the use of a common protocol and management system that would allow for pooling the data from participating institutions. Its first task was to define an optimal regimen for CT screening for lung cancer, which was presented at the Fourth International Conference on Screening for Lung Cancer in April 2001 and unanimously adopted (Henschke *et al.*, 2002b). Since then it has been updated, and the latest version is publicly available on the Internet (www.IELCAP.org). All members have a shared set of principles, a common protocol, and a centralized management system for pooling the data to answer the critical questions regarding CT screening for lung cancer, including evaluation and integration of new diagnostic approaches as well as advances in alternative interventions for early lung cancers. This collaboration has grown to include 40 institutions throughout the world (www.IELCAP.org).

The ELCAP team at Weill Cornell became the Coordinating Center concerned with protocol development, quality assurance in its implementation, data flow, management, and analysis, as well as central reading, storage of images and pathology specimens, and the training of the investigators and study coordinators of the participating institutions. All participants used the Web-based management system for CT screening that has been developed by the ELCAP team. This ELCAP Management System not only guides the processes involved but also provides for documentation and data transmission as well as for central reading of images and pathology specimens.

National Cancer Institute Conferences

In 2001, the director of the National Cancer Institute organized the Lung Cancer Progress Review Group (PRG) to identify high-priority areas of research in lung cancer (Lung Cancer Progress Review Group, 2001). The resulting report made the following major recommendations: create and foster scientifically integrated, multidisciplinary, multi-

institutional research consortia organized around lung cancer; develop and expand new approaches to the biology and treatment of nicotine addiction; evaluate population-based tobacco control efforts; facilitate and accelerate the evaluation of CT to detect lung cancer at an early stage; reverse the current stage distribution at presentation and reduce mortality from lung cancer; elucidate the contributions of injury, inflammation, and infection to the genesis of lung cancer; design, implement, and study "best practices" in lung cancer management; and facilitate and encourage training programs that emphasize multidisciplinary science and clinical care. The PRG also recognized that no single approach or trial would answer all of the questions regarding lung cancer screening.

In March 2001, the National Cancer Institute and the American Cancer Society jointly sponsored the Early Lung Cancer Screening Workshop (National Cancer Institute, 2001) with the aim of bringing together experts to address issues of study design. The workshop concluded that both traditional randomized trials and nonrandomized studies were necessary to answer key questions about early lung cancer detection and that subsequent pooling of data from all these studies would be a worthwhile investment for the future because it would allow clearer interpretation of the body of evidence from all studies.

In Europe, research efforts in lung cancer had culminated in a European Union research grant being awarded to the European Union Early Lung Cancer Detection Group in 2001 (First, Second, and Third Early Lung Cancer Detection Workshop, 1999, 2001, and 2003). Stimulated by the success of this collaborative effort and the International Conferences on Screening for Lung Cancer, different trial designs were developed in Europe. Under the auspices of the EU Early Lung Cancer Detection Group and the American Cancer Society, support for the two fundamentally different study designs was expressed at multiple meetings, and it was decided to harmonize the core elements of the radiology, pathology, and biomarker protocols of the future trials with a view to future pooling of the data to gain better insight into the results.

Performance of Computed Tomography Screening for Lung Cancer

I-ELCAP's goal is to assess the effectiveness of CT screening in preventing deaths from lung cancer. It asserts that, in order to determine the effectiveness of screening, the particular *regimen* of screening as to how early diagnosis of lung cancer has to be pursued must be specified. Thus, the first task of I-ELCAP was to define a potentially *optimal* regimen, and such a regimen was adopted in 2001 (Henschke *et al.*, 2002b). A particularly notable feature of the regimen is the *difference* between the *baseline* and *repeat* screenings in the definition of a positive result and the algorithm for further work-up. In the repeat screenings, the focus is on new nodules, thus growing by definition as the prior screen is available for comparison. By contrast, such prior information is not available at baseline; therefore, it is not known how long the nodule has existed and whether it is growing or regressing. Although the fundamental nature of the screening regimen has remained stable, it is continually updated based on emerging information. Poolability of the data requires the use of the common regimen of screening, while the indications for screening (i.e., enrollment criteria) could be different and were thus left to the discretion of each participating institution.

The I-ELCAP regimen starts with the initial low-dose CT, and if the result is positive, other testing follows along a well-thought-out algorithm, which eventually leads to a (rule-in) diagnosis of lung cancer. Assessment of growth has remained a critical part of the regimen (Spratt et al., 1996; Yankelevitz et al., 1999, 2000, 2003; Kostis et al., 2003, 2004). Ultimately, the diagnosis of lung cancer derives from a biopsy of the suspicious nodule followed by the specimen's reading and interpretation. Growth and minimally invasive (i.e., percutaneous fine-needle aspiration) biopsy are important components of the regimen: they serve to limit unnecessary open surgical biopsy, and they also serve to limit resection of slow-growing cancers that would not lead to death if left untreated. Given the critical (and ultimate) role of pathology in making the diagnosis, a separate pathology protocol was developed (Vasquez et al., 2003) Following this protocol, all submitted slides were examined by a five-member panel of pulmonary experts to determine their consensus diagnosis according to the latest World Health Organization criteria (Travis et al., 2004).

The regimen provides recommendations for the work-up, but the actual decision is left to each screenee and his or her referring physician. In the I-ELCAP approach, this does not compromise the validity of the study as long as actions, results of the tests, and interventions are documented for each screenee. Adherence to the regimen, however, does affect the performance of the regimen, for it determines the frequency of unnecessary biopsy or surgery and the timeliness of the diagnosis, which ultimately determines how early (e.g., resectability, stage) the cancer is diagnosed (Henschke et al., 2004b). Thus, for adequate performance of any screening regimen, adherence to it by the screenees and their referring physicians should be stressed in physician and lay-community education.

Comparison of two regimens is provided by comparing the respective diagnostic distributions by stage and size. Two different initial diagnostic tests (e.g., CT and chest radiography) can be compared by giving both tests to all participants as illustrated by the original ELCAP of 1000 screenees (Henschke et al., 1999). Comparing different regimens of screening using other performance measures is useful. Such measures, as illustrated in the following, should not depend

on the risk indicators of the participants (e.g., age, smoking history) but on the actual performance of the regimen.

1. Proportion having a positive result of the initial CT test.

At *baseline* the result of the initial CT is positive if at least one solid or part-solid nodule 5.0 mm or more in diameter, or at least one nonsolid nodule 8.0 mm or more in diameter, is identified in lung parenchyma or in an endobronchial location when solid. When noncalcified nodules are identified but all of them are too small to imply a positive result, the result is semipositive and calls for work-up only in terms of the first annual repeat CT. On *repeat* screenings, again, the first concern with the annual repeat CT is to identify all new noncalcified nodules, but now *regardless of size*. The result of the annual repeat, low-dose CT test is positive if at least one such nodule is identified. Using these updated definitions, a positive finding occurred in < 15% of screenees on baseline and < 6% on annual repeat screening.

2. Proportion of screen-diagnosed cases.

Screen-diagnosed cases are classified as baseline or annual repeat cancers according to the screening cycle in which the nodule is first identified, regardless of when the diagnosis is actually made. A screening cycle starts with the performance of the initial test, including any diagnostic work-up, and ends before the next routinely scheduled rescreening. Any case of cancer diagnosed outside the regimen is called an *interim*-diagnosed cancer and is attributed to the cycle of screening during which it is diagnosed. The proportion of screen-diagnosed cases was > 95% in the baseline cycle and 99% in repeat cycles of screening.

3. Proportion of cases by relevant prognostic indicators.

The frequency distribution of the cases by relevant prognostic factors (e.g., stage, size, histology) are important performance measures. In terms of stage, the critical determination as to treatment depends on the clinical stage and is typically based on CT and PET results. Thus, it is the proportion of clinical stage I diagnoses that is of particular interest. We found that more than 80% of all lung cancer diagnoses, interim cases included, were of clinical stage I on baseline and annual repeat screening. Also, as expected, the median tumor size was larger at baseline than on annual repeat. There is a strong relationship of tumor stage to tumor size within stage I as 91% of those whose cancer was 15 mm or less in diameter were of stage I, 83% of those 16–24 mm, 68% of those 25–35 mm, and 55% for those 35+ mm (I-ELCAP Investigators, 2006a).

4. Proportion of cases resulting in a diagnosis of malignancy after biopsy.

Among the recommended biopsies according to the regimen of screening, 92% resulted in a diagnosis of lung cancer or other malignancy. Thus, the screening regimen turned out to be quite successful in avoiding undue invasive procedures, complications, and cost. At the same time, none of the biopsies performed outside of the regimen's recommendation resulted in a diagnosis of lung cancer.

5. Genuineness of diagnoses.

When a person with a screen-diagnosed case of lung cancer dies of some other cause before having clinical manifestations of lung cancer, the case is said to be an "overdiagnosed" case of lung cancer. The same is true of a screen-diagnosed case that is so slow-growing that, even if left untreated, it would not pose a risk for survival. While both are important topics to be addressed in the context of screening, only the latter represents overdiagnosis for us, the former being an issue of competing causes of death. We address both topics, first the proportion of screen-diagnosed cancers that are genuine, that is, leading to death if not treated, and second, the issue of competing causes of death.

Potentially detracting from the apparent benefit of CT screening is the possibility that a proportion of the screen-diagnosed cases of lung cancer are free of manifest metastases because they are growing so slowly that the cancer would not lead to death if not resected. Protection against this was built into the regimen by requiring assessment of growth prior to biopsy of nodules < 15 mm in diameter and by pathologic review by a panel of expert pulmonary pathologists. To supplement this, finding, we also determined that ~ 13% of clinical stage I cases diagnosed as a result of baseline screening had volume doubling time estimates > 400 days (Henschke *et al.*, 2006a). These slower growing baseline cancers were identified most frequently in nonsolid nodules, typically at least 10 mm in diameter. It should be noted that a cancer with a diameter of 10 mm and a volume-doubling time of 400 days would lead to death from it in ~ 10 years.

In the repeat cycles, the cancers are typically aggressive in terms of their growth since the previous screen is integral in the definition of a positive result. For example, a newly seen nodule of 3 mm means that it has grown since the prior screen when it was not visible. Assuming it had a diameter just under the visibility threshold of 2 mm on the prior screen, its slowest doubling time would be 200 days. As newly seen cancers on repeat screening are typically 3 mm or larger in diameter when first identified, these cancers are rapidly growing, aggressive, genuine cancers. Further evidence was provided by the Pathology Review Panel in its review of the pathologic specimens; all diagnoses were confirmed to be genuine lung cancers. Ultimately, evidence against overdiagnosis is provided by the untreated cases of lung cancer or those in whom the recommended biopsy and/or treatment was delayed. To date, all those cases for which treatment was delayed, progressed, and of those that were not treated, all died (NY-ELCAP Investigators, 2007).

6. Competing causes of death.

Any decision about screening for a cancer needs to consider the person's risk of dying from causes other than lung cancer. This is particularly relevant for lung cancer screening as smokers and former smokers are at higher risk of death from other, competing causes, cardiovascular diseases in particular. To shed some light on the frequency of death from a competing cause among persons who enter into CT screening for lung cancer, we determined the 5- and 10-year rates of death from causes other than lung cancer in a high-risk older cohort of 2141 smokers and former smokers who had enrolled for CT screening for lung cancer in 1993–2004 (Henschke *et al.*, 2006c); they were aged 60–75 years and had a history of 30–100 pack-years of cigarette smoking. Using Kaplan-Meier analysis, we found that the 5- and 10-year survival rates conditional on not dying from lung cancer were 96% (95% CI: 95–97%) and 91% (95% CI: 88–93%), respectively. Based on this analysis, older, high-risk smokers and former smokers seeking and receiving CT screening for lung cancer have a low 10-year risk of dying from causes other than lung cancer. Early treatment of screen-diagnosed cancer thus has a good opportunity to be life-saving.

7. Prevention of deaths from lung cancer.

The proportion of deaths prevented by CT screening can be estimated by

(proportion of clinical stage I cases) × (cure rate of genuine clinical stage I cases)

This is a conservative estimate as it assumes that all screen-diagnosed cases of higher stage die. The estimated cure rate of genuine stage I cases of lung cancer is 92%. Thus, the estimated proportion of deaths that can be prevented is 85% × 92% = 78% (I-ELCAP Investigators, 2006b). This is high when contrasted with the 5% (1–163,510/172,570) or so (American Cancer Society, 2005) in the absence of screening.

8. Indication for screening.

We performed a traditional cost-effectiveness analysis using the actual hospital costs of the original ELCAP baseline screening and subsequent work-up. It was found that CT screening is highly cost-effective, ~ $2500 per life-year saved, for smokers and former smokers 60 years and older with a history of at least 10 pack-years of smoking (Wisnivesky *et al.*, 2003). Others using actual data have found similar results (Miettinen, 2000b; Marshall *et al.*, 2001a, b; Chirikos *et al.*, 2002), except for one model-based analysis (Mahadevia *et al.*, 2003) using unrealistic assumptions (e.g., a very high rate of overdiagnosis). Instead of public policy assessment of the cost-effectiveness of screening, we think that the decision is an individual one that should be based on the benefits and risks to a specific person for a particular round of screening. Once these benefits and risks are known, the individual then has to weigh the cost of the screening (typically, $300) with its benefit of potentially not dying from lung cancer (International Early Lung Cancer Action Program Investigators 2005).

Updated Recommendations Regarding Screening

What should the physician do when faced with a person at high risk for lung cancer and with otherwise suitably long life expectancy, especially if the person asks for screening? The advisability of screening seems obvious in as much as it has been shown to provide for earlier diagnosis and treatment, and earlier treatment means a better chance to avert death from lung cancer than when the treatment is prompted by symptoms and/or signs. The physician is, however, aware of countervailing ideas that screening has not yet been demonstrated to "save lives." Furthermore, some investigators state that screening has a notable problem of "overdiagnosis," meaning that screening finds lesions that are diagnosed as cancers but are not life-threatening, which leads to unnecessary resections. Assessment of how many "lives are saved" while guarding against overdiagnosis, these same investigators say, can only be done by performing a randomized controlled trial comparing screening with no screening using a mortality end point. Other investigators, however, think that such randomized trials on screening have led to misleading answers in the past and will continue to do so. These opposing viewpoints create a serious dilemma for the physician.

To more sharply focus the debate, consider the following case. A CT for evaluation of coronary artery disease is done on a healthy 60-year-old man with a history of heavy smoking, and a suspicious nodule is identified. Biopsy is performed, and the diagnosis of stage I adenocarcinoma is made. The physician recommends early, immediate resection. No serious consideration is given to delaying the surgery until symptoms and/or signs would appear because the physician knows that resection of stage I lung cancer results in far fewer deaths from it than when it is of a higher stage. In fact, if immediate treatment were not recommended, this would even be the basis of a lawsuit.

No one seriously questions the above approach. Why then is there debate about the larger question of whether to recommend the pursuit of early diagnosis in people at high risk for lung cancer with a reasonable life expectancy? It should be understood that screening of many persons at high risk for lung cancer must be as justifiable as screening a single person at the same risk, including consideration of the person's life expectancy.

The orthodox view that a randomized screening trial is necessary to evaluate lung cancer screening has been championed by the funding of the National Lung Screening Trial (NLST), the most expensive screening study ever initiated (Aberle et al., 2002). The NLST is a randomized controlled trial comparing CT with chest radiographic screening of 50,000 current or former smokers (Fig. 36). Its designers envisioned that it would test the hypothesis that CT screening is better than chest radiography in reducing deaths from lung cancer, providing an answer ~ 10 years after the initiation of the study.

While we do not question the value of randomization for assessing alternative treatments (Fig. 37), we do question the need for randomization for assessing diagnostic tests (Henschke et al., 2003b; Miettinen et al., 2003). In treatment trials, the diagnosis of disease is established prior to the randomization, and the purpose is to compare the effectiveness of alternative treatments for the disease. For assessment of a diagnostic test, however, randomization is not necessary as the desired information is how often and how early the disease is diagnosed using that test. This requires using the diagnostic test on people with and without the disease and determining how frequently the test identifies those who really have the disease. For example, the NCI-sponsored trial evaluating digital mammography (Mammographic screening study, http://www.dmist.org) compares digital to film-screen mammography for the diagnosis of breast cancer and the NIH-sponsored PIOPED II study compares CT angiography and V-Q scanning for the diagnosis of pulmonary emboli (PIOPED II Investigators, 2006). Both studies compare competing diagnostic tests. With regard to evaluation of screening, the same evaluation of the initial diagnostic test is required. Simply because treatment follows once the disease is diagnosed does not mean that evaluation of the diagnostic test should be different.

Recognizing the flaws of the prior randomized screening trials and the new evidence emerging from CT screening, the American Cancer Society decided to update its recommendations in 2001 (Smith et al., 2001). The U.S. Preventive Task Force in 2004 (Humphrey et al., 2004) also updated its recommendations mainly due to the Japanese case-control studies (Sobue, 2000). Physicians are also recommending screening (Strauss et al., 2005). Upon review of the evidence, they changed their recommendation against screening for lung cancer to "not enough evidence for or against" it, and they suggested that persons at risk for lung cancer talk to their physician.

Problems Identified in Performing Randomized Screening Trials

Earlier we summarized the study design of the Mayo Lung Project (MLP). This study was designed to test the usefulness of sputum cytology, and yet the decisions were also reached about chest radiography. In the MLP, after 6 years of screening and between 1.5 and 5 years (median of 3 years) of follow-up of the "intervention" cohort, the cumulative number of deaths from lung cancer from the time of randomization was 122 and the aggregate person-years of screening and follow-up was ~ 38,000 (Fontana et al., 1986; Miettinen et al., 2001). The mortality rate for this cohort is the ratio of these two values, 122/38,000, which is 0.0032 per person-year or 3.2 per 1000 person-years. In the control cohort, the corresponding values were 115/38,000, or 3.0 per 1000 person-years. Because these two average mortality rates were not statistically significantly different, the "null" hypothesis that the screening does not reduce lung cancer mortality could not be rejected (Fontana *et al.*, 1986). Even after long-term follow-up of > 76,000 person-years, there was no statistically significant difference in lung cancer mortality (Marcus *et al.*, 2000). In that report, the cumulative total of 337 lung cancer deaths was identified from the time of randomization to some 20 years later, and the aggregate follow-up of the participants of this intervention cohort, while still alive, was 76,861 person-years. Thus, the mortality rate was 337/76,861 = 0.0044 per person-year, or 4.4 per 1000 person-years. In the control arm, the corresponding mortality rate was 303/76,772, or 3.9 per 1000 person-years. Again, these two rates were not statistically different.

A number of flaws were identified in this study. (1) Average mortality rates were insensitive measures. (2) There was a lack of compliance with the protocol as some 70% of those in the screening arm and more than 50% of those in the control arm underwent chest radiography screening; thus, the control cohort actually had screening, only on an irregular and less intensive basis than the screen arm. (3) Screening did not continue long enough to reach the full effect of its benefit. (4) Sample size was too small in each cohort. (5) There was a likely differential misattribution of deaths in the screened and control arm (Black *et al.*, 2002), and the misattribution of the cause of death in a small percentage of cases (2%) could have led to the opposite conclusion. These flaws clearly decreased the power of the MLP to detect the hypothesized decrease in the mortality rates; the power was estimated to be < 20% (Flehinger *et al.*, 1993). Thus, the null hypothesis should have never been accepted. Other randomized screening trials have had similar problems, and most likely, the National Lung Screening Trial will too (Kimmel *et al.*, 2004).

Consider first that mortality rates are insensitive measures of the benefit of screening. The rate in each cohort is an average rate of deaths over an arbitrary period of screening and follow-up. No detail is given as to the timing of the deaths from lung cancer with respect to the start and end of screening. Furthermore, all deaths after randomization are counted, although it is well understood that the early deaths in the screened cohort cannot be prevented by the screening because the disease was already in a late stage at the time of screening. These deaths, though found in both the screening and control cohort, typically are frequent in comparison to the deaths prevented in the screened cohort, and they have a disproportionate influence on the total deaths. Furthermore, the mortality rates are based on information abstracted from death certificates in each of the states or countries performing the study, and these rates have not been shown to be as reliable as the cause when obtained from the physician or family members (at least in the United States).

Some potential biases have been described (Black *et al.*, 2002). One is the *sticky-diagnosis bias*, where disease-specific mortality rates can be overestimated in the screening arm of a randomized controlled trial because death from another cause is falsely attributed to the screen-diagnosed cancer. This sticky-diagnosis bias has been thought to be at least partially responsible for the excess lung cancer mortality observed in the MLP screened arm. In particular, it is thought that some deaths from adenocarcinoma of other organs may have been falsely misattributed to lung cancer. The MSKCC and JHMI studies did not report all-cause mortality, thus preventing the assessment for sticky-diagnosis bias. The fourth major randomized controlled trial for lung cancer, the Czechoslovakia study, had major inconsistencies in the magnitudes of all-cause mortality and disease-specific mortality between the screened and controlled groups. Black *et al.* (2002) attributed these inconsistencies to an underreporting of deaths in the control group, further biasing the results against screening.

Another potential bias described by Black *et al.* (2002) is *slippery-linkage bias*. This bias occurs when screen-prompted interventions result in deaths that are misattributed to other causes instead of the target cancer or its screening. To the extent that screening-related deaths are misattributed to other causes, disease-specific mortality will be biased in favor of screening. When slippery-linkage bias is present, all-cause mortality is increased in the control group compared to the screened group, even when disease-specific mortality shows that the screening is beneficial.

Rather than rely solely on a single "average mortality rate," we think that greater insight is provided by identifying when the deaths from lung cancer occurred relative to the randomization. This identification can be made by calculating the yearly case-fatality rate, defined as the proportion of diagnosed cases actually having a fatal outcome. We have shown the advantage of considering the case-fatality rates in the screened and control cohorts relative to the timing of the screening on both a theoretical and a practical basis (Miettinen *et al.*, 2002). Initially, when screening starts, the ratio of the case-fatality in the screened cohort to that in the control cohort is equal to one because some of the cancers initially detected by the screening test in the screened cohort are found in people on the verge of becoming symptomatic and will presumably die at the same rate as those in the control cohort. Others in the screened cohort, however, will have cancers that are diagnosed sufficiently early to prevent their death, and additional cases will be identified in each of the successive rounds of screening. Thus, the frequency of such diagnosed cancers increases as long as screening continues (Fig. 40).

Although some of the participants, whose cancers can be cured by early diagnosis and early treatment, are found

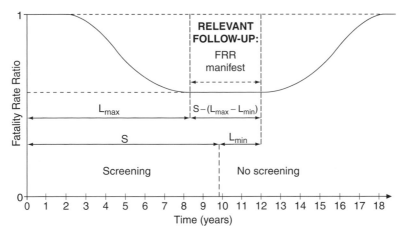

Figure 40 Follow-up experience in a randomized controlled trial comparing screening for cancer with no screening with respect to cause-specific mortality: interrelations of parameters. At any given point in the follow-up there is a particular mortality density, MD, among the screened and the not screened; for an interval of t to t + dt, with dC cases expected in it, $MD_t = dC/Pdt$, where P is the size of the population. Contrasting the screened with the not screened, there is the corresponding mortality-density ratio, MDR. This radio is depicted as a function of time since entry into the trial. The early excess mortality among the screened is not shown, since the focus is on the intended result of reduced fatality rate, FR, quantified in terms of fatality-rate ratio, FRR. MDR coincides with FRR in a particular interval of follow-up time if the duration of screening, S, exceeds the difference between the minimum, L_{max}, and minimum, L_{min}, of the time lag form early diagnosis to the death prevented by early intervention but not by late intervention (i.e. in the absence of screening). (Miettinen OS, Henschke CI, Pasmantier MW, Smith JP, Libby DM, Yankelvitz DF. Mammographic screening: No relaiable supporting evidence? *http;//www.lancet.com* Feb 2, 2002).

shortly upon the commencement of screening, the decrease in the fatality rate is only seen later when the equivalent cases in the control cohort, which are not identified early, start to die. Indeed, the purpose of screening is to provide for earlier diagnosis and consequently earlier treatment, and the fatal outcome that is prevented would typically have occurred many years in the future in the absence of screening. (This, in fact, is the basis for requiring that screening be done in people who have a suitably long life expectancy.) Initially, then, the fatality-rate ratio will be one, and if screening is effective, the ratio will decrease below one at some point in the future, depending on the disease and the screening process. As long as screening continues, the ratio will continue to decrease until a minimum value is reached. This minimum value represents the relative reduction in deaths from lung cancer due to the early diagnosis followed by early intervention. Once screening is discontinued, the ratio will start to increase again toward one at approximately the same rate it had previously decreased (Kimmel *et al.*, 2004).

We also illustrated the importance of considering the timing of deaths relative to the start of screening by reanalyzing the Malmö study in which the intervention of mammographic screening to no screening was being evaluated (Miettinen *et al.*, 2002). In that study, screening continued for ~ 10 years, but according to Gotzsche and Olsen (2000; Olsen and Gotzsche, 2001), there was no significant difference in the mortality rates of the two cohorts. We showed, however, that only after screening had continued for ~ 8–11 years after randomization, and when focusing on the relevant period of time, there was a 55% reduction in deaths in

the screened cohort demonstrated, indicating quite a significant benefit of the screening. This example demonstrated both that screening must continue long enough and that one must focus on the relevant time period in order to determine when the benefit of screening can be seen.

What happens to the ratio of the mortality or case-fatality rates when there is imperfect adherence to the study protocol? In other words, what happens when people drop out of the screening cohort or when people in the control cohort obtain screening as seen in the MLP? In these cases, the observed benefit of the screening will be much smaller than it actually is had the protocol adherence been perfect inasmuch as the deaths in the screened cohort do not decrease as much (due to dropouts in the screening program), while those in the control cohort start to decrease (due to their obtaining screening). If the screening process had a small but real benefit, this benefit might not be detected at all in the study. Recall the MLP. As shown by Flehinger and others (Flehinger and Kimmel, 1987; Gorlova *et al.*, 2001), chest radiography screening in the MLP likely had a small benefit. Even if everything in the trial was done perfectly, the mortality rate ratio would have remained close to one throughout the 6 years of screening and 3 years of follow-up. It would also have required screening for 30 years to show the small but real benefit. Since screening was done for only 6 years and protocol nonadherence was high, it could have been predicted at the outset that no mortality rate differences between the two cohorts would be found. Later reanalysis by Miettinen (Miettinen *et al.*, 2002) also showed this benefit.

Another rationale for performing randomized screening trials is the belief that lead time, length, and overdiagnosis biases can be avoided. In the case of lung cancer, as a result of the commonly accepted interpretation of the MLP, overdiagnosis bias is of greatest concern (Black, 2000). Overdiagnosis is defined as finding lung cancers that are not life-threatening, and thus leading to unnecessary resection. To avoid overdiagnosis bias, mortality rates were used: this rate only reflects deaths from lung cancer and not the total number of lung cancers found.

The idea that overdiagnosis is a significant concern in lung cancer screening was raised primarily to explain the negative results of the MLP. In it, more cases of lung cancer were reported in the "intervention" cohort of the MLP as compared with the control cohort (206 cases versus 160) (Fontana et al., 1986). Because there was no significant difference in the mortality rates, it was reasoned that this excess of 46 cases was due to overdiagnosed cases of lung cancer being found by the screening (Eddy, 1989; Black, 2000). This idea was reinforced by the extended follow-up of the MLP (Marcus et al., 2000), as the number of lung cancer deaths in the "intervention" cohort remained higher than those in the control cohort (337 vs. 303); the actual number of diagnosed cancers was not provided. The authors of this article stated that "similar mortality but better survival for individuals in the intervention arm indicates that some lesions with limited relevance may have been identified in the intervention arm" (p. 1308). Furthermore, the associated editorial stated (Black, 2000):

> In the MLP, a substantial proportion of screen-detected cases were probably pseudodisease for three reasons: 1) the mortality rate from all causes is high, about three-fold that in non-smokers (Phillips, 1996), 2) some squamous cell carcinomas detectable by sputum cytology are very small; and 3) some primary adenocarcinomas detectable by chest radiography grow very slowly (Spratt, 1996).

No further explanation of this statement was offered. The editorial did make the additional point that "overdiagnosis does not reduce disease-specific mortality endpoint [; and therefore,] disease-specific mortality is the most valid for the evaluation of screening effectiveness" (p. 1281). The associated news article (Newman, 2000), commenting on the conclusion that there was overdiagnosis in its screening arm, described it as being "totally heretical" to many advocates of lung cancer screening.

In fact, the evidence against overdiagnosed cancers found by chest radiographic screening in the MLP is compelling. Screen-detected lung cancer of stage I in the MLP, when left unresected, had a completely malignant course: < 10% of them were alive after 5 years (Flehinger et al., 1992). This suggests that, similar to the results in Japan, > 90% of the cases were genuine cancers, not overdiagnosed ones (Sobue et al., 1992).

Further detailed analysis of the growth rates of the lung cancers in the MLP and the Memorial Sloan-Kettering Cancer Project provide compelling evidence that these malignancies were aggressive (Yankelevitz et al., 2003). Instead of the estimated 50% of the screen-diagnosed cases being overdiagnosed as implied by Black (2000), we found that at most 10% of the diagnosed cancers could be considered for this category (Yankelevitz et al., 2003). In fact, all of these screen-detected stage I cancers were found on chest radiography after a prior negative chest radiograph. The median size of the cancers was nearly 2 cm, the growth rates were typical of aggressive malignancies, and when unresected, these cancers were nearly uniformly fatal (Flehinger et al., 1992). That many of these cancers could represent overdiagnosed cases strains credulity.

The continuing controversy regarding mammography shows that many randomized controlled trials comparing screening with no screening of more than a half-million women in over 40 years of study did not provide definitive answers about the benefit of screening for breast cancer (Gotzsche and Olsen, 2000). The catastrophic failure of these trials should lead experts to seriously reconsider the theoretical underpinnings of randomized controlled trials for studying screening. When the results of prior and future randomized trials comparing screening with no screening point to negative results that are counterintuitive, and these results are explained by reasons shown to be wrong, such as extensive overdiagnosis, how should this discrepancy be resolved? The most obvious recourse is to question the principles underlying the study design. Randomized trials came into prominence in medicine in the 1940s for comparing alternative treatments, particularly for tuberculosis. Since they were powerful in evaluating treatment effectiveness, it is easy to understand how the diagnostic community embraced this paradigm. However, it has become clear that this same paradigm is less useful in assessing diagnostic tests and often leads to misleading results. The time to reconsider the methodology for evaluating promising new screening tests is upon us. Innovations in diagnosis are being made rapidly, and we in the field of diagnostic radiology need to carefully consider the alternative approaches to those considered to be the "gold" standard.

References

Aberle, D.R., Black, W.C., Goldin, J.G., Patz, E., Gareen, I., and Gatsonis, C. 2002. Contemporary screening for the detection of lung cancer protocol. May 10, 2002. American College of Radiology Imaging Network (ACRIN #6654).

American Cancer Society. 1980. Cancer of the lung. *Ca—A Cancer Journal for Clinicians* 30:189–206.

American Cancer Society. Statistics 2005. Cancer Facts and Figures. Available from: http://www.cancer.org. Accessed October 11, 2005.

Bastarrika, G., Garcia-Velloso, M.J., Lozano, M.D., Montes, U., Torre, W., Spiteri, N., Campo, A., Seijo, L., Alcaide, A.B., Pueyo, J.C., Cano,

D., Vivas, I., Cosin, O., Dominguez, P., Serra, P., Richter, J.A., Monteuenga, L., and Zulueta, J.J. 2005. Early lung cancer detection with spiral computed tomography and positron emission tomography. *Am. J. Respir. Crit. Care Med. 171*:1378–1383.

Black, W.C. 2000. Overdiagnosis: an underrecognized cause of confusion and harm in cancer screening. Editorial. *J. Natl. Cancer Inst. 92*:1–6.

Black, W.C., Haggstrom, A.D., and Welch, H.G. 2002. All-cause mortality in randomized trials of cancer screening. *J. Natl. Cancer Inst. 94*:167–173.

Chirikos, T.N., Hazelton, T., Tockman, M., and Clark, R. 2002. Screening for lung cancer with CT: a preliminary cost-effectiveness analysis. *Chest 121*:1507–1514.

Diederich, S., Wormanns, D., Semik, M., Thomas, M., Lenzen, H., Roos, N., and Heindel, W. 2002. Screening for early lung cancer with low-dose spiral CT: prevalence in 817 asymptomatic smokers. *Radiology 222*:773–781.

Dominioni, L., and Strauss, G.M. 2000. Consensus Statement. International conference on prevention and early diagnosis of lung cancer, Varese, Italy. *Cancer 89*:2329–2330.

Eddy, D.M. 1989. Screening for lung cancer. *Ann. Intern. Med. 111*:232–237.

Epler, G.R. 1990. Screening for lung cancer: is it worthwhile? *Postgrad. Med. 87*:181–186.

First, Second, and Third Early Lung Cancer Detection Workshop. November 1999, 2001, and 2003. A European strategy for developing lung cancer imaging and molecular diagnosis in high risk populations. Liverpool, England.

Flehinger, B.J., and Kimmel, M. 1987. The natural history of lung cancer in a periodically screened population. *Biometrics 43*:127–144.

Flehinger, B.J., Kimmel, M., and Melamed, M.R. 1992. Survival from early lung cancer: implications for screening. *Chest 101*:13–18.

Flehinger, B.J., Kimmel, M., Polyak, T., and Melamed, M.R. 1993. Screening for lung cancer. The Mayo Lung Project revisited. *Cancer 72*:1573–1580.

Fontana, R.S., Sanderson, D.R., Woolner, L.B., Taylor, W.F., Miller, W.E., and Muhm, J.R. 1986. Lung cancer screening: the Mayo program. *J. Occup. Med. 28*:746–750.

Fontana, R.S., Sanderson, D.R., Woolner, L.B., Taylor, W.F., Miller, W.E., Muhm, J.R., Bernatz, P.E., Payne, W.S., Pairolero, P.C., and Bergstralh, E.J. 1991. Screening for lung cancer: a critique of the Mayo Lung Project. *Cancer 67 (4 Suppl.)*:1155–1164.

Gorlova, O.Y., Kimmel, M., and Henschke, C. 2001. Modeling of long-term screening for lung cancer. *Cancer 92*:1531–1540.

Gotzsche, P.C., and Olsen, O. 2000. Is screening for breast cancer with mammography justifiable? *Lancet 355*:129–134.

Grady, D. 1999. CAT scan process could cut deaths from lung cancer. Small tumors detected. *New York Times*, July 9, 1999, 1.

Henschke, C.I., Miettinen, O.S., Yankelevitz, D.F., Libby, D., and Smith, J.P. 1994. Radiographic screening for cancer: New paradigm for its scientific basis. *Clin. Imag. 18*:16–20.

Henschke, C.I., McCauley, D.I., Yankelevitz, D.F., Naidich, D.P., McGuinness, G., Miettinen, O.S., Libby, D.M., Pasmantier, M.W., Koizumi, J., Altorki, N.K., and Smith, J.P. 1999. Early Lung Cancer Action Project: overall design and findings from baseline screening. *Lancet 354*:99–105.

Henschke, C.I., Naidich, D.P., Yankelevitz, D.F., McGuinness, G., McCauley, D.I., Smith, J.P., Libby, D.M., Pasmantier, M.W., Koizumi, J., Vazquez, M., Flieder, D., Altorki, N.K., and Miettinen, O.S. 2001. Early Lung Cancer Action Project: initial findings from annual repeat screening. *Cancer 92*:153–159.

Henschke, C.I., Yankelevitz, D.F., Smith, J.P., and Miettinen, O.S. 2002a. The use of spiral CT in lung cancer screening. In: Devita V.T., Hellmam, S., and Rosenberg, S.A. (Eds.), *Progress in Oncology*. Sudbury MA: Jones and Barlett.

Henschke, C.I., Yankelevitz, D.F., Smith, J.P., and Miettinen, O.S. 2002b. Screening for lung cancer: the Early Lung Cancer Action approach. *Lung Cancer 35*:143–148.

Henschke, C.I., Yankelevitz, D.F., Mirtcheva, R., McGuinness, G., McCauley, D., and Miettinen, O.S. 2002c. CT screening for lung cancer: frequency and significance of part-solid and nonsolid nodules. *AJR 178*:1053–1057.

Henschke, C.I., Wisnivesky, J.P., Yankelevitz, D.F., and Miettinen, O.S. 2003a. Screen-diagnosed small stage I cancers of the lung: genuineness and curability. *Lung Cancer 39*:327–330.

Henschke, C.I., Yankelevitz, D.F., and Kostis, W.J. 2003b. CT Screening for lung cancer: bias, shift and controversies. In: Schoepf, U.J., (Ed.) *Multidetector-Row CT of the Thorax*. Berlin, Germany: Springer Verlag.

Henschke, C.I., Yankelevitz, D.F., Naidich, D., McCauley, D.I., McGuinness, G., Libby, D.M., Smith, J.P., Pasmantier, M.W., and Miettinen, O.S. 2004a. CT screening for lung cancer: suspiciousness of nodules at baseline according to size. *Radiology 231*:164–168.

Henschke, C.I., Yankelevitz, D.F., Smith, J.P., Libby, D., Pasmantier, M.W., McCauley, D.I., McGuinness, G., Naidich, D.P., Farooqi, A., Vazquez, M., and Miettinen, O.S. 2004b. CT screening for lung cancer: assessing a regimen's diagnostic performance. *Clin. Imag. 28*:317–321.

Henschke, C.I., Shaham, D., Yankelevitz, D.F., Kramer, A., Reeves, A.P., Vazquez, D.F., and Miettinen, O.S. 2006a. CT screening for lung cancer: significance of diagnoses in its baseline cycle. *Clin. Imag. 30*:11–15.

Henschke, C.I., Lee, I., Wu, N., Farooqi, A., Yankelevitz, D.F., Khan, A., Altorki, N.K., ELCAP, and NY-ELCAP Investigators. 2006b. CT screening for lung cancer: prevalence and incidence of mediastinal masses. *Radiology 239*:586–590.

Henschke, C.I., Yip, R., Yankelevitz, D.F., and Miettinen, O.S. 2006c. CT screening for lung cancer: competing causes of death. *Clin. Lung Cancer 7*:323–325.

Hill, A.B. 1990. Suspended judgment. Memories of the British Streptomycin Trial in Tuberculosis. The first randomized clinical trial. *Control. Clin. Trials 11*:77–79.

Humphrey, L.L., Johnson, M., and Teutsch, S. 2004. Lung cancer screening with sputum cytologic examination, chest radiography, and computed tomography: an update of the U.S. Preventive Services Task Force. *Ann. Intern. Med. 140*:738–753. Also on http://ahrg.gov/clinic/cps3dix.htm.

I-ELCAP protocol. Web site: http://www.IELCAP.org.

Inoue, K., Sato, M., Fujimura, S., Sakurada, A., Usuda, K., Kondo, T., Tanita, T., Handa, M., Saito, Y., and Sagawa, M. 1998. Prognostic assessment of 1310 patients with non-small cell lung cancer who underwent complete resection from 1980 to 1993. *J. Thorac. Cardiovasc. Surg. 116*:407–411.

International Conferences on Screening for Lung Cancer. Consensus statements of 1st to 13th Conference. Web site: http://www.IELCAP.org.

International Early Lung Cancer Action Program Investigators. 2005. CT screening for lung cancer: individualizing the benefit of screening. Scientific abstract. ASCO 2005.

International Early Lung Cancer Action Program Investigators. 2006a. CT screening for lung cancer: the relationship of disease stage to tumor size. *Arch. Intern. Med. 166*:321–325.

International Early Lung Cancer Action Program Investigators 2006b. Survival of patients with clinical stage I lung cancer detected on CT screening with computed tomography. *New Engl. J. Med. 355*: 1763–1771.

Jackson, V.P. 2002. Screening mammography: controversies and headlines. *Radiology 225*:323–326.

Kaneko, M., Eguchi, K., Ohmatsu, H., Kakinuma, R., Naruke, T., Suemasu, K., and Moriyama, N. 1996. Peripheral lung cancer: screening and detection with low-dose spiral CT versus radiography. *Radiology 201*:798–802.

Kimmel, M., Gorlova, O.Y., and Henschke, C.I. 2004. Modeling lung cancer screening. In Edler, L., and Kitsos, C., (Eds.), *Quantitative Methods for Cancer and Human Health Risk Assessment*. Hoboken, NJ: Wiley and Sons.

Kostis, W.J., Reeves, A.P., Yankelevitz, D.F., and Henschke, C.I. 2003. Three-dimensional segmentation and growth-rate estimation of small pulmonary nodules in helical CT images. *IEEE Trans. Med. Imag. 22*:1259–1274.

Kostis, W.J., Yankelevitz, D.F., Reeves, A.P., Fluture, S.C., and Henschke, C.I. 2004. Small pulmonary nodules: reproducibility of three-dimensional volumetric measurement and estimation of time to follow-up CT. *Radiology 231*:446–452.

Kubik, A., and Polak, J. 1986. Lung cancer detection: results of a randomized prospective study in Czechoslovakia. *Cancer 57*:2427–2437.

Lung Cancer Progress Review Group Report. 2001. National Cancer Institute. Chantilly, VA. August 2001. Web site: http://prg.nci.cih.gov/lung/default.html.

Mahadevia, P.J., Fleisher, L.A., Frick, K.D., Eng, J., Goodman, S.N., and Powe, N.R. 2003. Lung cancer screening with helical computed tomography in older smokers: a decision and cost-effectiveness analysis. *JAMA 289*:313–322.

Mammographic screening study. Digital mammographic imaging screening trial. Web site: http://www.dmist.org.

Marcus, P.M., Bergstralh, E.J., Fagerstrom, R.M., Williams, D.E., Fontana, R., Taylor, W.F., and Prorok, P.C. 2000. Lung cancer mortality in the Mayo Lung Project: impact of extended follow-up. *J. Natl. Cancer Inst. 92*:1308–1316.

Marshall, D., Simpson, K.N., Earle, C.C., and Chu, C. 2001a. Potential cost-effectiveness of one-time screening for lung cancer (LC) in a high risk cohort. *Lung Cancer 32*:227–236.

Marshall, D., Simpson, K.N., Earle, C.C., and Chu, C.W. 2001b. Economic decision analysis model of screening for lung cancer. *Eur. J. Cancer 37*:1759–1767.

Melamed, M.R., Flehinger, B.J., Zaman, M.B., Heelan, R.T., Perchick, W.A., and Martini, N. 1984. Screening for early lung cancer: results of the Memorial Sloan-Kettering study in New York. *Chest 86*:44–53.

Miettinen, O.S. 2000a. Screening for lung cancer. *Radiol. Clin. of North Am. 38*:479–486.

Miettinen, O.S. 2000b. Screening for lung cancer: Can it be cost-effective? *CMAJ 162*:1431–1436.

Miettinen, O.S., and Henschke, C.I. 2001. CT screening for lung cancer: coping with nihilistic recommendations. *Radiology 221*: 592–596.

Miettinen, O.S., Henschke, C.I., Pasmantier, M.W., Smith, J.P., Libby, D.M., and Yankelevitz, D.F. 2002. Mammographic screening: no reliable supporting evidence? *Lancet 359*:404–405.

Miettinen, O.S., Yankelevitz, D.F., and Henschke, C.I. 2003. Evaluation of screening for a cancer: annotated catechism of the Gold Standard creed. *J. Eval. Clin. Pract. 9*:145–150.

Mountain, C.F. 1997. Revisions in the international system for staging lung cancer. *Chest 111*:1710–1717.

National Cancer Institute. 2001. Early lung cancer screening workshop. Washington, DC. Web site: http://www3.cancer.gov/bip/dipsponsored.htm.

Newman, L. 2000. Lung project update raises issue of overdiagnosing patients. News. *Natl. Cancer Inst. 92*:1–5.

NY-ELCAP Investigators. 2007. CT screening for lung cancer: diagnoses resulting from the New York Early Lung Cancer Action Project. *Radiology. 243*:239–249.

Olsen, O., and Gotzsche, P.C. 2001. Cochrane review of screening for breast cancer with mammography. *Lancet 358*:1340–1342.

PIOPED II Investigators. 2006. Multidetector computed tomography for acute pulmonary embolism. *N. Engl. J. Med. 354*:2317–2327.

Randal, J. 2000. After 40 years, mammography remains as much emotion as science. *J. Natl. Cancer Inst. 92*:1630–1632.

Shaham, D., Breuer, R., Copel, L., Agid, R., Makori, A., Kisselgoff, D., Goitein, O., Izhar, U., Berkman, N., Heching, N., Sosna, J., Bar-Ziv, J., and Libson, E. 2006. Computed tomography screening for lung cancer: applicability of an international protocol in a single-institution environment. *Clin. Lung Cancer 7*:262–267.

Smith, R.A., von Eschenbach, A.C., Wender, R., Levin, B., Byers, T., Rothenberger, D., Brooks, D., Creasman, W., Cohen, C., Runowicz, C., Saslow, D., Cokkinides, V., and Eyre, H. 2001. American Cancer Society guidelines for the early detection of cancer: update of early detection guidelines for prostate, colorectal, and endometrial cancer; also update 2001—testing for early lung cancer detection. *CA Cancer J. Clin. 51*:38–75.

Sobue, T. 2000. A case-control study for evaluating lung cancer screening in Japan. *Cancer 89* (11 Suppl.):2392–2396.

Sobue, T., Suzuki, T., and Naruke, T. 1992. A case-control study for evaluating lung-cancer screening in Japan. Japanese Lung-Cancer-Screening Research Group. *Int. J. Cancer 50*:230–237.

Sone, S., Li, F., Yang, Z.G., Honda, T., Maruyama, Y., Takashima, S., Hasegawa, M., Kawakami, S., Kubo, K., Haniuda, M., and Yamanda, T. 2001. Results of three-year mass screening programme for lung cancer using mobile low-dose spiral computed tomography scanner. *Brit. J. Cancer 84*:25–32.

Sone, S., Takashima, S., Li., F., Yang, Z., Honda, T., Maruyama, Y., Hasegawa, M., Yamanda, T., Kubo, K., Hanamura, K., and Asakura, K. 1998. Mass screening for lung cancer with mobile spiral computed tomography scanner. *Lancet 351*:1242–1245.

Spratt, J.S., Meyer, J.S., and Spratt, J.A. 1996. Rates of growth of human neoplasms: Part II. *J. Surg. Oncol. 61*:68–83.

Strauss, G.M., Dominioni, L., Jett, J.R., Freedman, M., and Grannis, F.W. 2005. Como International Conference Position Statement. *Chest 127*:1146–1151.

Swensen, S.J., Jett, J.R., Sloan, J.A., Midthun, D.E., Hartman, T.E., Sykes, A.M., Augenbaugh, G.L., Zink, F.E., Hillman, S.L., Noetzel, G.R., Marks, R.S., Clayton, A.C., and Pairolero, P.C. 2002. Screening for lung cancer with low-dose spiral computed tomography. *Am. J. Respir. Crit. Care Med. 165*:508–513.

Tockman, M. 1986. Survival and mortality from lung cancer in a screened population. The Johns Hopkins Study. *Chest 89*:325s–326s.

Travis, W.D., Brambilla, E., Muller-Hermelink, H.K., and Harris, C.C. 2004. *World Health Organization Classification of Tumours, Pathology and Genetics of Tumours of the Lung, Pleura, Thymus and Heart*. Lyon, France: IARC Press.

U.S. Preventive Services Task Force. 1989. *Guide to Clinical Preventive Services: Screening for Lung Cancer*. Washington, DC., pp. 45–47.

Vazquez, M., Flieder, D., Travis, W., Carter, D., Yankelevitz, D., Miettinen, O.S., and Henschke, C.I. 2003. Early Lung Cancer Action Project pathology protocol. *Lung Cancer 39*:231–232. Also on Web site: http://www.IELCAP.org.

Wisnivesky, J.P., Mushlin, A., Sicherman, N., and Henschke, C.I. 2003. Cost-effectiveness of baseline low-dose CT screening for lung cancer: preliminary results. *Chest 124*:614–621.

Yankelevitz, D.F., Gupta, R., Zhao, B., and Henschke, C.I. 1999. Repeat CT scanning for evaluation of small pulmonary nodules: preliminary results. *Radiology 212*:561–566.

Yankelevitz, D.F., Kostis, W.F., Henschke, C.I., Heelan, R.T., Libby, D.M., Pasmantier, M.W., and Smith, J.P. 2003. Overdiagnosis in traditional radiographic screening for lung cancer: frequency. *Cancer 97*:1271–1275.

Yankelevitz, D.F., Reeves, A., Kostis, W., Zhao, B., and Henschke, C.I. 2000. Determination of malignancy in small pulmonary nodules based on volumetrically determined growth rates: preliminary results. *Radiology 217*:251–256.

4

Lung Cancer: Role of Multislice Computed Tomography

Suzanne Matthews and Sameh K. Morcos

Introduction

Lung cancer is the major cause of cancer death in the United Kingdom and the United States. The expected 5-year survival for all patients in whom lung cancer is diagnosed is 15%. This rises to 49% if the cancer is detected while still localized. These figures do not compare favorably with those for the other major cancers: colon 63%, breast 88%, and prostate 99%. Only 16% of lung cancer cases are detected at an early stage (American Lung Association, 2005). The majority of patients present in the later stages of the disease (stages 3 and 4). Screening programs aimed at detecting lung cancer at an earlier stage are currently being investigated. Accurate computed tomography staging is vital presurgery to prevent unnecessary thoracotomy and assists in optimal patient treatment regimes. Imaging also plays a central role in the assessment of treatment response and disease relapse.

Advances in CT technology have resulted in improved image resolution and the depiction of more structural detail. Multislice/multidetector CT, introduced in 1998, has enabled several thin contiguous slices of the body to be scanned simultaneously in less than half a second. Today it is possible to image 64 slices concurrently. This allows the whole body to be scanned in 10 sec and the whole lungs in 2–4 sec. Areas of interest can be scanned using contiguous thin collimation (1 mm) in a single breath hold, permitting high-quality volume imaging with greatly reduced movement artifact and optimal use of contrast. The acquired data allow high-quality multiplanar and three-demensional (3D) imaging, including virtual endoscopy. The ability to produce these images greatly improves the accuracy of the reporting radiologist in interpreting images.

The increase in patient radiation dose exposure is a serious consideration with this new technology. Improved sensitivity of detectors, better utilization of the X-ray beam, 3D beam modulation, and the introduction of X-ray beam-shaping filters have minimized dose increases (Toth, 2004). However, the increases in radiation doses with multidetector row CT need to be balanced against the superior imaging abilities and benefits derived from more accurate assessment of disease.

This chapter will describe the technique of using multislice CT (MSCT) in imaging patients with possible lung cancer. The advantages of MSCT in comparison to conventional CT in staging lung cancer and in monitoring patients post-treatment will be discussed. New developments provided by MSCT such as virtual bronchoscopy will be briefly presented. The chapter does not address the use of MSCT in screening for lung cancer.

Multislice Computed Tomography Technique for Diagnosis and Staging of Bronchogenic Carcinoma

Computed tomography imaging is used to investigate suspected bronchogenic carcinoma, staging confirmed tumors, and follow-up after treatment. Accurate staging is required for consideration of optimal treatment. Precise imaging protocols are dependent on the type of CT scanner available. In general, the whole chest and upper abdomen are scanned. The liver and adrenal glands are routinely assessed as these represent common, and often asymptomatic, metastatic sites for bronchogenic carcinoma.

Scanning Protocol

The patient is positioned supine with the arms raised above the head and with an 18G cannula sited in the antecubital fossa. Iodinated contrast (100 ml) is injected at a rate of 3.5 ml/sec. The whole thorax is scanned at ~ 25 sec after the start of the contrast injection so that vessels are displayed during the arterial vascular phase. Bolus tracking technology enables scanning to be synchronized with peak vascular enhancement to improve mediastinal assessment. A second scan through the upper abdomen to include the whole liver and adrenal glands is performed during the portal venous imaging phase of the liver (~ 70 sec after the start of the contrast injection). Both sequences are easily obtained during a single breath hold for most multislice CT scanners. Typical scanning parameters with 4-row multidetector CT are 2.5 mm collimation, 100 mAs, and 120 KVp. The more advanced CT scanners utilizing up to 64 rows of detectors can use ≤ 1 mm collimation.

Imaging Protocol

The scanner can be set to automatically reconstruct the images in thin (1–3 mm) axial sections for overall review. Images are typically viewed on mediastinal and lung window settings. Soft copy image review is ideally performed at a workstation. The reporter is able to reconstruct further images in any plane to help with problem solving in difficult areas.

Multislice Computed Tomography Staging of Bronchogenic Carcinoma

Lung cancer is staged using the international Tumor Node Metastases (TNM) classification first published in 1986 (Mountain, 1986). This system classifies the tumor according to size, location, and evidence of invasion into adjacent structures, such as chest wall, pleura, diaphragm, and mediastinum, and assesses the potential spread of dis-ease according to the involvement of lymph nodes or the presence of distant metastases. The TNM classification assigns the clinical stage of the tumor, which determines the feasibility of curative or palliative therapies. Accurate staging is of great importance to avoid surgery in patients who may have resectable but incurable disease or locally advanced disease that is unresectable. While CT is the primary mode of assessment of the T-stage, positron emission tomography (PET) imaging plays an increasingly important role in the assessment of the N- and M-stages. The latter scanning is particularly useful in identifying malignancy in lymph nodes that are normal in size on CT and in detecting metastases in structures that are not covered by a standard lung cancer staging CT scan.

Little research has been done specifically addressing the impact of multislice CT in the imaging of lung cancer. However, MSCT can provide high-quality multiplanar imaging, allowing accurate assessment of tumor size, shape, density, and relationship to adjacent structures. Artifacts associated with conventional CT because of partial volume effects, such as artificial enlargement of tumors of ~ 5 mm and spurious closeness of structures, are overcome by MSCT in which thin collimation (≤ 2.5 mm) is used for scanning the patient. The difficulty experienced with standard axial CT in assessing the relationship of tumor to structures lying in the same horizontal plane as the scan is overcome by MSCT. MSCT is able to depict the tumor and adjacent structures clearly in any plane, in addition to the axial plane. In this section, we discuss in detail the use of MSCT in staging lung cancer, including superior sulcus tumors. The assessment of response to treatment and the evaluation of tumor recurrence are also addressed, as is the importance of virtual bronchoscopy provided by MSCT.

T-staging

Bronchogenic carcinoma may present as a peripheral mass in the lung. However, many benign disease processes such as infections, granulomatous disease, and cryptogenic organizing pneumonia may also present as a lung mass. Furthermore, some malignant lung tumors, such as adenocarcinomas, may resemble lung infection and present as an area of consolidation containing air bronchograms with peripheral areas of ground glass opacification or bubblelike lucencies. Establishing whether the lung mass is benign or malignant with certainty requires histological examination. However, certain well-described features can be depicted clearly with MSCT, which helps to suggest whether the lesion is benign or malignant (Armstrong et al., 1995; Ko and Naidich, 2003). A malignant lesion is typically rounded in shape and is solid. Lobulation, irregular edge, or spiculation is commonly seen in malignancy and is thought to represent local tumor invasion into the surrounding lung parenchyma.

This stranding of the tumor into the lung parenchyma can produce the appearance known as corona radiata. A tumor size of ≥ 3 cm is suggestive that the lesion is malignant. A cavitating malignant tumor usually has an irregularly thickened wall, whereas a uniformly thin-walled cavity is more suggestive of a benign lesion. A mass lesion in the lung associated with hilar ± mediastinal lymphadenopathy is highly suggestive of malignancy. Invasion of chest wall, diaphragm, fissure, or mediastinum by a mass is consistent with malignancy, and detecting these features is crucially important in determining the feasibility of successful surgical resection of the lesion (details are provided later in this chapter). Eccentric calcification may be seen in a malignant tumor developing in an old tuberculous scar tissue. Lamellated, central, or popcorn patterns of calcification are seen only in benign lesions. Detection of fat within a lung mass indicates that the lesion is benign. This is seen in hamartomas and lipoid pneumonia. The depiction of all these morphological features is improved by MSCT with thin collimation (≤ 2.5 mm).

The degree of enhancement of a lung mass ≤ 3 cm following the intravenous administration of contrast can be used to differentiate benign from malignant lesions (Ko and Naidich, 2003). A rise in the density of the nodule of ≤ 15 Hounsfield Units (HU) is a good indicator that the lesion is benign and has 98% sensitivity and 58% specificity (Swensen et al., 2000). An increase in density of the mass of > 15 HU is suggestive of malignancy but can be observed with inflammatory lesions. The limitations associated with single-slice spiral CT due to respiratory motion, streak artifacts, and difficulties experienced in consistently placing a region of interest within a small nodule can be partially overcome with MSCT. The faster scanning times minimize respiratory motion artifact, and the finer sections that can be acquired permit the use of this technique on nodules < 10 mm.

The use of computer-aided diagnosis with MSCT for detecting and determining the nature of a lung mass is currently under evaluation. The technique involves the use of computer programs to detect lung nodules and analyze the detected nodule morphology against a reference data set to estimate the probability of malignancy. The rate of growth is important in differentiating benign from malignant lesions. Malignant tumors typically have a doubling time of 1 to 18 months, whereas benign lesions are stable or grow very slowly. A mass lesion can be labeled as benign if there is evidence that it has been stable with no change in size for at least 2 years. The doubling of tumor volume means an increase in tumor diameter of only 26%. This increase would be extremely difficult to evaluate with small lesions (< 10 mm) by manual measurement of the diameter. Computer-aided volumetry facilitates accurate depiction of tumor growth. Algorithms for automated 3D computer-aided segmentation and volume measurement are currently available (Wiemker et al., 2005). The accuracy of volume measurement depends on the density of the nodule, consistent decisions regarding the outer limit of the nodule, scanning slice thickness, and inherent CT spatial resolution. The latter factors are within the capabilities of MSCT, which allows fast thin-slice contiguous scanning and isotropic imaging, in turn permitting accurate volume measurement of even small lung masses.

Chest Wall Invasion

The T-stage increases depending on the presence of visceral pleural (T2) or parietal pleural/chest wall (T3) involvement. Although surgery and total disease resection are still technically possible with T3 disease, the postoperative prognosis drops with increasing the T-stage. In addition, chest wall resection has a significant impact on respiratory function and patient morbidity. Thin-section imaging (1 mm) markedly improves the diagnostic accuracy of the assessment of chest wall invasion (Uhrmeister et al., 1999). The features suggestive of chest wall invasion on CT are: contact between tumor and pleura of more than 3 cm, an obtuse angle of contact of the tumor with the chest wall, abnormalities of the extrapleural fat plane (higher density of the fat plane or circumscript thickening adjacent to the mass), identification of the tumor in the intercostal soft tissue space or deeper chest wall structures, and bone destruction or erosion (Fig. 41a). Most of these features are clearly depicted with MSCT but cannot be accurately identified using thick-slice conventional CT imaging. In particular, the extrapleural fat plane is not visualized frequently. Multislice CT improves the detection of small nodules along the pleural surface or an irregularly thickened pleural surface, which are features of pleural dissemination (Mori et al., 1998). Higashino et al. (2005) found that thin-section sagittal imaging (1.25 mm) was superior to thick- and thin-slice axial imaging for the assessment of chest wall invasion. In addition, the interobserver agreement for the assessment of invasion using thin-section multiplanar reconstructions (MPRs) was higher in comparison to standard 5 mm axial imaging. The use of 3D reconstructions allows detection of visceral pleural invasion that was not apparent on 2D CT images and in one case permitted the identification of pleural implants not seen on standard axial imaging (Kuriyama et al., 1994). The newest generation of MSCT scanners with the facility of ≥ 32 rows detectors that can scan the body at ≤ 1 mm collimation is likely to further increase CT's diagnostic accuracy in assessment of chest wall invasion. However, no data are yet available on the efficacy of these new scanners in the assessment of chest wall invasion.

Determining the presence of chest wall invasion is particularly important in the preoperative evaluation of a superior sulcus tumor because of the increased potential of early invasion related to the proximity of the lesion to the chest wall in this location. Magnetic resonance imaging has been shown to be superior to conventional CT in the depiction of superior sulcus tumor invasion of pleura, chest wall, brachial plexus,

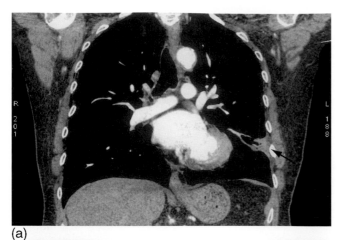

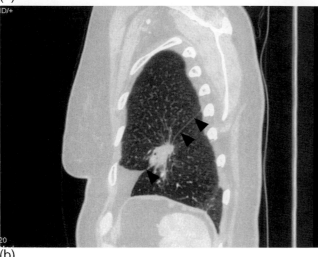

Figure 41 Staging MSCT scan of the thorax of a 64-year-old woman with bronchogenic carcinoma in the lingula of the left upper lobe. (a) A coronal image (mediastinal settings) shows early invasion of the chest wall by the bronchogenic carcinoma of the lingual (arrow). (b) A sagittal image (lung settings) shows that the bronchogenic carcinoma of the lingula has a spiculated edge consistent with malignancy and is invading the major fissure (arrowheads) to involve the left lower lobe.

subclavian vessels, and vertebral bodies. Whether modern MSCT can replace MRI in the evaluation of these tumors remains unclear. The American College of Chest Physicians and the British National Institute of Clinical Excellence recommend MRI for the assessment of superior sulcus tumors (Silvestri *et al.*, 2003; NICE Guidelines, 2005).

Invasion of Fissures and Diaphragm

The relationship of the primary tumor to the fissures is important because it determines the extent of lung resection needed. The visibility of the fissures, especially the minor, was greatly improved with thin-section coronal and sagittal (2–3 mm) images provided by MSCT. Storto *et al.* (1998)

found that multiplanar reformatted images improved the assessment of neoplastic extension across pulmonary fissures (Fig. 41b). The sensitivity in detecting fissure involvement was 100% with MSCT, whereas the sensitivity for conventional CT was only 57%. Multiplanar reformatted imaging was able to assess the involvement of the minor fissure accurately in five out of the six cases where the fissure could not be adequately identified on thick-section or thin-section high-resolution (1 mm) conventional CT. Chooi *et al.* (2005) and Higashino *et al.* (2005) have demonstrated that the assessment of fissure invasion is improved with MSCT and MPRs (coronal and sagittal images, 3 mm slice thickness) in comparison to axial imaging. Improved confidence and concordance between observers was reported with the use of MSCT and MPRs in assessing the invasion of the diaphragm, which was difficult to determine by axial imaging alone (Chooi *et al.*, 2005).

Invasion of Mediastinum

Conventional CT is limited in differentiating between tumor contiguity and tumor invasion of the mediastinum, with the sensitivity of detecting mediastinal invasion as low as 55%. Therefore, MRI, with its ability to produce images in multiple planes and good tissue characterization, was used whenever invasion of chest wall or mediastinum was suspected on conventional CT. With the advent of MSCT, the ability of CT technology to visualize the relationship between tumor and mediastinum is likely to improve (Higashino *et al.*, 2005) (Fig. 42).

N-staging

Standard lung cancer staging by CT scan includes routine assessment of the hilar and mediastinal nodes. Lymph nodes (LN) in the mediastinum are classified as abnormal by size criteria: a short axis diameter of >10 mm is the cut-off value typically used to differentiate the presence of abnormal nodes. However, differentiation between large metastatic and benign hyperplastic lymph nodes remains a problem. Computed tomography is unable to detect metastases in normal-size nodes. Clusters of normal-size lymph nodes should raise suspicion of tumor involvement and justify further investigation. The sensitivity and specificity of CT in detecting metastatic involvement are ~ 57% and 82%, respectively, if > 10 mm diameter is used as a marker of abnormal LN (Toloza *et al.*, 2003). It is improbable that MSCT can improve these values. However, Kozuka *et al.* (2003) found that 0.5 mm coronal reformatted images were slightly superior to 5 mm standard axial contrast-enhanced CT in the detection of abnormal mediastinal LN. The sensitivity and specificity of PET in the detection of mediastinal lymph node involvement by tumor have been shown to be

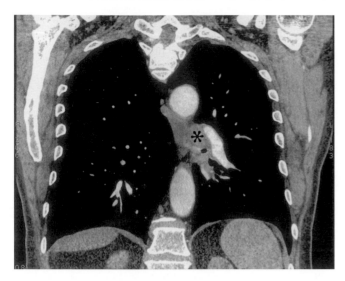

Figure 42 A coronal image (mediastinal settings) of the thorax showing a bronchogenic carcinoma in the left lower lobe (asterisk), which is lying medial to the descending interlobar pulmonary artery and invading the adjacent mediastinum.

~ 84% and 89%, respectively (Toloza *et al.*, 2003). Mediastinoscopy remains superior to imaging in the assessment of mediastinal lymph node disease, with a sensitivity and specificity of 89% and 100%, respectively (Gdeedo, 1997).

Currently, it is recommended that all patients suitable for curative surgery should have an initial PET scan. If the PET scan does not reveal abnormal mediastinal uptake or distant metastases, surgery may proceed without prior mediastinoscopy even in the presence of enlarged nodes on CT scan. A positive PET scan should be followed by histological confirmation of metastatic disease within the mediastinal lymph nodes unless there is evidence of widespread dissemination of the lung cancer (NICE Guidelines, 2005).

M-staging

Lung cancer metastases are commonly found in the lungs, pleura, liver, and adrenal glands. All these areas are readily amenable to staging with CT, and a standard CT staging scan for lung cancer should cover all these structures (NICE Guidelines, 2005; Silvestri *et al.*, 2003). The current evidence suggests that there is little benefit from routine investigation for bone, brain, or abdominal metastases in the absence of supporting clinical symptoms or signs (Toloza *et al.*, 2003). Multidetector CT imparts no major advantage over conventional CT in assessing metastases in the liver or adrenal glands apart from allowing a speedy examination. However, the facility of isotropic multiplanar image reconstruction provided by MSCT could be useful in some cases when difficulties are experienced in assessing the axial images.

Assessment of Response to Treatment and Tumor Recurrence

There is an increased risk of disease recurrence for the first 2 years after therapy and a lifelong risk of 1–2% each year of the development of a second primary lung malignancy. Therefore, a 6-month surveillance for recurrent or new disease by clinical history, physical examination, and chest X-ray or chest CT is recommended for 2 years and then annually (Colice *et al.*, 2003). It is recommended that the surveillance be continued for a maximum of 5 years (NICE Guidelines, 2005). At our institution, Sheffield Teaching Hospitals NHS Trust, a staging CT scan is recommended at 3 months after surgery, radical radiotherapy, or chemotherapy in order to establish a baseline. The CT protocol is identical to that used for the initial staging scan. Any residual disease is assessed, and the presence of mediastinal LN and distant metastases is documented. Imaging of other areas of the body is dependent on patient symptoms. A baseline CT can assist when difficulties may arise in distinguishing post-therapy scarring or radiation-induced pneumonitis from recurrent disease (Fig. 43). Positron emission tomography has proven to be helpful in detecting active disease in areas of post-treatment changes 2–3 months after completion of therapy and is likely to play an increasing role in this difficult area. There is no published data examining the utility of MSCT in the performance of follow-up CT examinations in lung cancer. However, the high quality of multiplanar imaging provided by MSCT may enable more accurate assess-

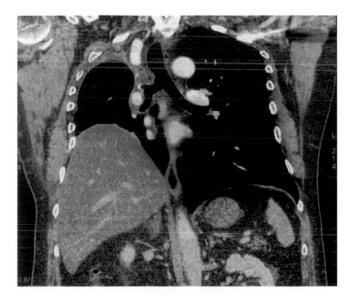

Figure 43 A follow-up MSCT scanning of the thorax of a 66-year-old man several months after right upper lobectomy. A coronal image (mediastinal settings) shows patent bronchi and no evidence of tumor recurrence. There is marked elevation of the right hemidiaphragm, suggesting injury of the phrenic nerve. Diaphragmatic paralysis was confirmed by fluoroscopy.

ment of tumor recurrence. In addition, the accuracy of monitoring change in tumor size in response to treatment is likely to be superior with MSCT and computer-aided volumetry (Wiemker *et al.*, 2005).

Virtual Bronchoscopy

Virtual bronchoscopy (VB) is a further imaging facility provided by MSCT (De Wever *et al.*, 2004). It allows noninvasive intraluminal evaluation of the airways down to the sixth- and seventh-generation bronchi and produces images very similar to those obtained by fiber-optic bronchoscopy (FOB). It is estimated that ~ 15 min are needed to view the central airways using this technique. Potential clinical uses of the technique include evaluation of bronchial stenoses/ obstruction, preoperative planning for stent insertion as the bronchi beyond the obstruction can be examined, evaluation of surgical sutures following lung resection or transplantation, and evaluation of the bronchial tree in children, including the evaluation of normal variants/malformations and the guidance of transbronchial biopsy. Virtual bronchoscopy can detect mass lesions, wall irregularity, stenoses, or obstruction, but it cannot reliably detect mucosal infiltration. The virtual bronchoscopic images should be interpreted in conjunction with the multiplanar 2D images to allow assessment of the extraluminal component of the lesion. The sensitivity of VB for the detection of obstructive lesions has been reported to be 100% and 83% for nonobstructive endobronchial lesions. However, the technique is time consuming and can produce false-positive results due to the misinterpretation of secretions or blood clots in the airways as mass lesions. In addition, tissue for pathological examination cannot be obtained via this technique. It is advisable that VB be used only in selected cases as a problem-solving tool.

Conclusions

Lung cancer is associated with a high morbidity and mortality. Optimum treatment regimes are dependent on accurate staging of the carcinoma. The introduction of MSCT, with its ability to produce high-quality multiplanar and 3D images, permits more precise prediction of the tumor T-stage. In particular, MSCT allows more accurate assessment of chest wall invasion, fissural transgression, diaphragmatic spread, and diffuse pleural disease. This technique enables the application of software computer programs that can detect lung masses, predict their nature, and analyze tumor volume. In addition, MSCT facilitates high-quality imaging even in breathless patients, as the whole chest can be scanned in only a few seconds. Pretreatment CT assessment of lung cancer should be combined with PET scanning for accurate overall staging prior to potential curative therapies. High-quality axial and multiplanar reformatted images

produced by MSCT should improve the assessment of tumor recurrence and response to treatment. Virtual bronchoscopy is another advantage of MSCT, but its clinical importance remains uncertain.

References

American Lung Association. April 2005. Lung Cancer Fact Sheet. www.lungusa.org.

Armstrong, P., Wilson, A.G., Dee, P., and Hansell, D.M. 1995. *Imaging of Diseases of the Chest*, 2nd ed. St. Louis, MO. Mosby.

Chooi, W.K., Matthews, S., Bull, M.J., and Morcos, S.K. 2005. Multislice CT in staging lung cancer: the role of multiplanar image reconstruction. *J. Comput. Assist. Tomog. 29*:257–260.

Colice, G.L., Rubins, J., and Unger, M. 2003. Follow up and surveillance of lung cancer patient following curative-intent therapy. *Chest 123*:272S–283S.

De Wever, W., Vandecaveye, V., Lanciotti, S., and Verschakelen, J.A. 2004. Multidetector CT-generated virtual bronchoscopy: an illustrated review of the potential clinical indications. *Eur. Resp. J. 23*:776–782.

Gdeedo, A., Van Schil, P., Corthouts, B., Van Mieghem, F., Van Meerbeeck, J., and Van Marck, E. 1997. Prospective evaluation of computed tomography and mediastinoscopy in mediastinal lymph node staging. *Eur. Resp. J. 10*:1547–1551.

Higashino, T., Ohno, T., Takenaka, D., Watanabe, H., Nogami, M., Ohbayashi, C., Yoshimura, M., Satouchi, M., Nishimura, Y., Fujii, M., and Sugimura, K. 2005. Thin-section multiplanar reformats from multidetector-row CT data: utility for assessment of regional tumor extent in non-small cell lung cancer. *Eur. J. Radiol. 56:*48–55.

Ko, J.P., and Naidich, D.P. 2003. Lung nodule detection and characterization with multislice CT. *Radio. Clin. N. Am. 41*:575–597.

Kozuka, T., Tomiyama, N., Johkoh, T., Honda. O., Koyama. M., Hamada, S., Nakamura, H., Yamamoto, S., and Matsumoto, T. 2003. Coronal multiplanar reconstruction view from isotropic voxel data sets obtained with mulitdetector-row CT: assessment of detection and size of medistinal and hilar lymph nodes. *Rad. Med. 21*(1):23–27.

Kuriyama, K., Tateishi, R., Kumatani, T., Kodama, K., Doi, O., Hosomi. N., Sawai. Y., Inoue, E., Kadota, Y., Narumi. Y., Fujita, M., and Kuroda, C. 1994. Pleural invasion by peripheral bronchogenic carcinoma: assessment with three-dimensional helical CT. *Radiology 191*:365–369.

Mori, K., Hirose, T., Machida, S., Yokoi, K., Tominaga, K., Moriyama, N., and Sasagawa, M. 1998. Helical computed tomography diagnosis of pleural dissemination in lung cancer: comparison of thick-section and thin-section helical computed tomography. *J. Thor. Imag. 13*: 211–218.

Mountain, C.F. 1986. A new international staging system for lung cancer. *Chest 89* (Suppl.):225S–233S.

National Institute of Clinical Excellence Guidelines. February 2005. Lung cancer: the diagnosis and treatment of lung cancer. www.nice.org.uk/ CG024NICEguideline.

Silvestri, G.A., Tanoue, L.T., Margolis, M.L., Barker, J., and Detterbeck, F. 2003. The non-invasive staging of non-small cell lung cancer: the guidelines. *Chest 123*:147S–156S.

Storto, M.L., Ciccotosto, C., Guidotti, A., Merlino, B., Patea, R.L., and Bonomo, L. 1998. Neoplastic extension across pulmonary fissures: value of spiral computed tomography and multiplanar reformations. *J. Thorac. Imag. 13*:204–210.

Swensen, S.J., Viggiano, R.W., Midthun, D.E., Muller, N.L., Sherrick, A., Yamashita, K., Naidich, D.P., Patz, E.F., Hartman, T.E., Muhm, J.R., and Weaver, A.L. 2000. Lung nodule enhancement at CT: multicenter study. *Radiology 214*:73–80.

Toloza, E.M., Harpole, L., and McCrory, D.C. 2003. Noninvasive staging of non-small cell lung cancer. *Chest 123*:137S–146S.

Toth, T.L. 2004. Scan faster. Scan longer. Scan with less dose. *Volume CT, A GE Healthcare Publication*, July, P14–18.

Uhrmeister, P., Allmann, K.H., Wertzel, H., Altehoefer, C., Laubenberger, J., Hasse, J., and Langer, M. 1999. Chest wall infiltration by lung cancer: value of thin-sectional CT with different reconstruction algorithms. *Eur. Radiol. 9*:1304–1309.

Wiemker, R., Rogalla, P., Blaffert, T., Sifri, D., Hay, O., Shah, E., Truyen, R., and Fleiter, T. 2005. Aspects of computer-aided detection (CAD) and volumetry of pulmonary nodules using multislice CT. *Brit. J. Rad. 78*:S46–56.

5

Surgically Resected Pleomorphic Lung Carcinoma: Computed Tomography

Tae Sung Kim

Introduction

The recent World Health Organization (WHO) classification of lung tumors defines pleomorphic carcinoma as "a poorly differentiated non-small cell lung cancer, namely squamous cell carcinoma, adenocarcinoma, or large cell carcinoma, containing spindle cells and/or giant cells, or a carcinoma consisting only of spindle and giant cells." In order to classify a carcinoma as pleomorphic carcinoma, at least 10% of spindle and/or giant cells should be present (Travis et al., 1999; Travis, 2002).

Several clinicopathologic investigations, including immunohistochemical studies, on pleomorphic carcinoma of the lung have been reported (Rossi et al., 2003). Recently, two radiological reports on the computed tomographic (CT) features of pleomorphic carcinoma of the lung have been published (Kim et al., 2004, 2005). In this chapter, the computed tomographic (CT) features of surgically resected pleomorphic carcinomas of the lung will be discussed to identify any specific imaging characteristics that may help in diagnosing this disease entity.

Pleomorphic Carcinoma of the Lung

Pleomorphic carcinoma of the lung is a poorly differentiated epithelial neoplasm predominantly composed of pleomorphic giant and/or spindle tumor cells (Przygodzki et al., 1996). According to the recent criteria of the WHO classification, pleomorphic carcinoma of the lung is defined as a non-small cell lung cancer (epithelial or carcinomatous component) combined with neoplastic spindle and/or giant cells (mesenchymal or sarcomatous component), or a carcinoma consisting only of spindle and giant cells (Rossi et al., 2003). Accordingly, there are various subtypes of pleomorphic carcinoma consisting of the adenocarcinoma and giant cell subtype, adenocarcinoma and spindle cell subtype, adenosquamous cell and spindle cell subtype, squamous cell carcinoma and giant cell subtype, squamous cell carcinoma and spindle cell subtype, large cell carcinoma and giant cell subtype, and large cell carcinoma and spindle cell subtype.

In a clinicopathologic and immunohistochemical study of 75 cases of pulmonary carcinomas with pleomorphic, sarcomatoid, or sarcomatous elements, Rossi *et al.* (2003) reported a strong prevalence of the tumor among male smokers (the male/female ratio was 9.7:1 and 92% of patients were smokers). According to these researchers, pleomorphic carcinoma manifested as a large, frequently peripheral, necrotic mass mainly involving upper lobes. A predilection for the upper lobes was noted in 48% of the tumors, and 33% were located in the right upper lobe. More than 70% of the tumors were located peripherally. The study results also showed that pleomorphic carcinoma of the lung had a worse prognosis than conventional non-small cell lung cancer at surgically curable stage I, justifying their separation as an independent histologic type in the WHO classification.

In a clinicopathologic study of 78 cases of pleomorphic carcinoma of the lung with various carcinomatous components, Fishback *et al.* (1994) also reported an upper lobe predilection of the tumor (65% of the tumors in the upper lobes and 47% in the right upper lobes). In their study, 60% of the tumors presented as a peripheral mass, showing chest wall invasion in 24%. On microscopic examinations, most of the tumors (91%) contained foci of necrosis. According to Kim *et al.* (2004) in a study of 10 cases of pleomorphic carcinomas of the lung, the tumors preferentially presented as a large peripheral lung mass (*n* = 9), with a central low-attenuation area and frequent invasion of the pleura (*n* = 7) and chest wall (*n* = 2). However, the histological subtypes of pleomorphic carcinoma were not specified in their results.

According to Kim *et al.* (2005) in a recent study of 30 patients (M:F = 27:3, age range 31–84 years, mean age 57 years) with surgically resected pleomorphic carcinoma of the lung, 10 central tumors and 20 peripheral tumors were identified, with diameters ranging from 1.5–10 cm (mean, 4.7 cm). The tumors also showed a predilection of the upper lobes (77%, 23/30) and particularly the right upper lobe (47%, 14/30). The marginal characteristics of the tumors were well-defined in 13 patients (43%), ill-defined in 6 (20%), lobulated in 6 (20%), and spiculated in 3 (10%). The remaining two patients revealed a central mass with distal lobar atelectasis. All tumors showed mild enhancement similar to the chest wall musculature after intravenous contrast enhancement, and 15 patients (50%) showed heterogeneous central low-attenuation areas. Tumors with a large cell component frequently showed low-attenuation areas (85%, 11 of 13), representing central necrosis on histopathologic specimens.

All nine tumors of the large cell carcinoma and giant cell subtype showed a subpleural location and a large area of low attenuation suggestive of extensive tumor necrosis at contrast-enhanced CT scans (Figs. 44–46). The attenuation values of the central low-attenuation areas of this subtype at contrast-enhanced CT scan ranged from 5–45 Hounsfield Units (HUs) (mean, 23 HU), and tiny intratumoral cavities were also noted in four patients (44%). On microscopic examination, all tumors of this subtype showed a various

degree of necrosis (20–90% in area). In this particular subtype, a poorly defined margin with peritumoral areas of ground-glass attenuation was noted in all tumors. On radiologic-pathologic correlation, these peritumoral areas of

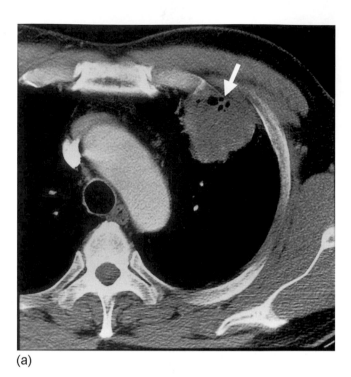

(a)

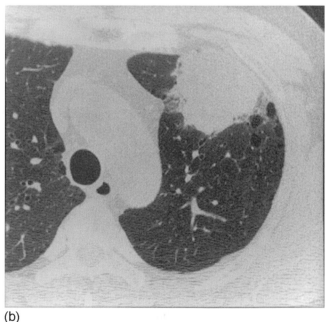

(b)

Figure 44 Pleomorphic carcinoma of the lung (large cell carcinoma and giant cell subtype) in a 57-year-old man. (a) Axial contrast-enhanced CT scan shows a peripheral low-attenuation mass with marginal irregularity in the left upper lobe. Note small intratumoral cavities (arrow). (b) Lung window image shows peritumoral halo of ground-glass attenuation.

(Continued)

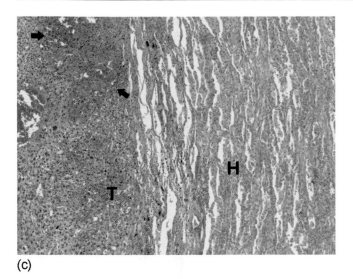

(c)

Figure 44 *continued* (c) Photomicrograph of histopathologic specimen shows pleomorphic tumor cells (T) and hemorrhagic necrosis (arrows). Peritumoral lung parenchyma (H) shows intra-alveolar macrophage aggregation and interstitial thickening due to inflammatory cell infiltration (H and E (hematoxylin and eosin), × 40).

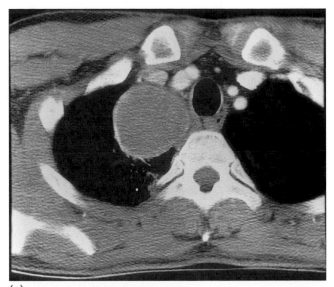

(a)

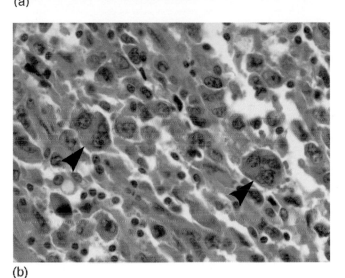

(b)

Figure 45 Pleomorphic carcinoma of the lung (large cell carcinoma and giant cell subtype) in a 31-year-old man. (a) Axial contrast-enhanced CT scan shows a peripheral low-attenuation mass in the right upper lobe. (b) Photomicrograph shows mixed composition of large cell carcinoma and pleomorphic multinucleated giant cells (arrowheads) (H and E, × 200).

ground-glass attenuation seen at CT corresponded to areas of intraalveolar macrophage collection and alveolar wall thickening with inflammatory cell infiltration and mild fibrosis. The tumor size was slightly larger (mean diameter, 5.8 cm) than that of the remaining subtypes (mean diameter, 4.3 cm). Regional invasion into the adjacent chest wall ($n = 4$) or mediastinal fat ($n = 1$) was seen or suggested in 56% of tumors (5/9) on CT scan.

Among the various subtypes included in their study, a high incidence (86%) of peripheral location of the adenocarcinoma and giant or spindle cell subtype and a strong predilection for central location (100%) of the squamous cell carcinoma and spindle cell subtype correlated well with a general predilection of adenocarcinoma for peripheral location and that of squamous cell carcinoma for central location (Silverberg, 1985). It has also been known that large cell carcinomas of the lung tend to present as a large peripheral mass with multiple necrotic foci (Yesner, 1985). Interestingly, all 13 cases (100%) of the subtypes containing the large cell carcinoma component also showed a peripheral location in their series. Consequently, the CT features of pleomorphic carcinomas of the lung appear to be dictated by the epithelial component of the tumor rather than the mesenchymal component.

Many other features similar to the previous clinicopathologic reports, including male predominance (9:1), peripheral location (67%, 20/30), frequent necrosis (50%, 15/30), and chest wall invasion (27%, 8/30), were also noted in their series. These CT features were more prominent in the large cell carcinoma and giant cell subtype when compared to the remaining subtypes: all patients were male, and all tumors showed a subpleural location, a large area of low attenua-

tion, and a high incidence of invasion into the adjacent chest wall or mediastinal fat. In postoperative pathologic tumor staging, the incidence of T3 disease was 56% (5/9) for the large cell carcinoma and giant cell subtype owing to frequent chest wall invasion, while that of the remaining subtypes was only 14% (3/21). In addition, the peritumoral area of ground-glass attenuation was characteristic of this subtype. These CT features of this specific subtype (the large cell carcinoma and giant cell subtype) are quite similar to the results of Kim *et al.* (2004) in that the tumors preferentially present as a large peripheral pulmonary mass with a central low-attenuation area. Therefore, most of the cases

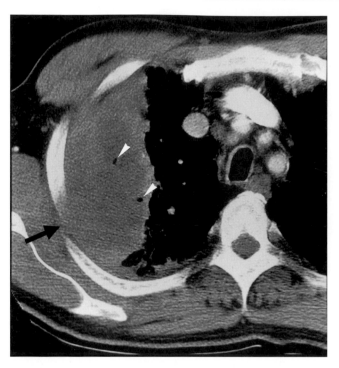

Figure 46 Pleomorphic carcinoma of the lung (large cell carcinoma and giant cell subtype) in a 60-year-old man. Axial contrast-enhanced CT scan shows a peripheral low-attenuation mass with marginal irregularity in the right upper lobe. Note small intratumoral cavities (arrowheads) and suspicious invasion into the adjacent chest wall (arrow).

included in their study seem to have represented the tumors of the large cell carcinoma and giant cell subtype.

The overall CT features of pleomorphic carcinomas of the lung are rather nonspecific and not quite different from those of conventional non-small lung cancers, such as a central or peripheral mass, marginal lobulation or spiculation, and frequent central necrosis (Fraser *et al.*, 1999). Even though the large cell carcinoma and giant cell subtype have several distinctive CT features, these results may not be directly applicable to the whole heterogeneous group of pleomorphic (giant and spindle cell) carcinoma with various epithelial components (adenocarcinoma, squamous cell carcinoma, and mixed types). However, the CT features of pleomorphic carcinomas of the lung appear to be dominated by the epithelial component of the tumor rather than the mesenchymal component. A possibility of pleomorphic carcinoma should be suggested when a subpleural necrotic tumor is seen with surrounding areas of ground-glass attenuation and regional invasion into the adjacent chest wall or

mediastinum at CT. Because of its subpleural location with an extensive area of a very low-attenuation CT value on contrast-enhanced CT scan, the large cell carcinoma and giant cell subtype of pleomorphic carcinomas can be misinterpreted as a benign cystic mass such as a mediastinal bronchogenic cyst or a neurogenic tumor with cystic degeneration (Fig. 44) or even an empyema cavity (Fig. 46).

In summary, the overall CT features of pleomorphic carcinomas of the lung appear to be dominated by the epithelial component of the tumor rather than the mesenchymal component. Among various subtypes of pleomorphic carcinomas, the large cell carcinoma and giant cell subtype showed quite consistent CT features such as a subpleural location, peritumoral areas of ground-glass attenuation, and extensive central low-attenuation areas. This subtype also shows frequent cavitation and invasion into the adjacent chest wall or mediastinal fat.

References

Fishback, N.F., Travis, W.D., Moran, C.A., Guinee, D.G., Jr., McCarthy, W.F., and Koss, M.N. 1994. Pleomorphic (spindle/giant cell) carcinoma of the lung: a clinicopathologic correlation of 78 cases. *Cancer* 73:2936–2945.

Fraser, R., Müller, N., Colman, N., and Paré, P. 1999. *Diagnosis of Disease of the Chest,* 3rd ed. Philadelphia: Saunders, pp. 1067–1250.

Kim, T.S., Han, J., Lee, K.S., Jeong, Y.J., Kwak, S.H., Byun, H.S., Chung, M.J., Kim, H., and Kwon, O.J. 2005. CT findings of surgically resected pleomorphic carcinoma of the lung in 30 patients. *AJR Am. J. Roentgenol.* 185:120–125.

Kim, T.H., Kim, S.J., Ryu, Y.H., Lee, H.J., Goo, J.M., Im, J.G., Kim, H.J., Lee, D.Y., Cho, S.H., and Choe, K.O. 2004. Pleomorphic carcinoma of lung: comparison of CT features and pathologic findings. *Radiology* 232:554–559.

Przygodzki, R.M., Koss, M.N., Moran, C.A., Langer, J.C., Swalsky, P.A., Fishback, N., Bakker, A., and Finkelstein, S.D. 1996. Pleomorphic (giant and spindle cell) carcinoma is genetically distinct from adenocarcinoma and squamous cell carcinoma by K-ras-2 and p53 analysis. *Am. J. Clin. Pathol.* 106:487–492.

Rossi, G., Cavazza, A., Sturm, N., Migaldi, M., Facciolongo, N., Longo, L., Maiorana, A., and Brambilla, E. 2003. Pulmonary carcinomas with pleomorphic, sarcomatoid, or sarcomatous elements: a clinicopathologic and immunohistochemical study of 75 cases. *Am. J. Surg. Pathol.* 27:311–324.

Silverberg, E. 1985. Cancer statistics. *CA Cancer J. Clin.* 35:19–35.

Travis, W.D. 2002. Pathology of lung cancer. *Clin. Chest Med.* 23:65–81.

Travis, W.D., Colby, T.V., Corrin, B., Shimosato, Y., and Brambilla, E. 1999. In collaboration with Sobin, L.H. and pathologists from 14 countries. *World Health Organization International Histological Classification of Tumors: Histological Typing of Lung and Pleural Tumors,* 3rd ed. Berlin, Germany: Springer-Verlag.

Yesner, R. 1985. Large cell carcinoma of the lung. *Semin. Diagn. Pathol.* 2:255–269.

6

Lung Cancer: Low-dose Helical Computed Tomography

Yoshiyuki Abe, Masato Nakamura, Yuichi Ozeki, Kikuo Machida, and Toshiro Ogata

Introduction

Lung cancer is one of the leading causes of death in developed countries, including Japan and the United States. Lung cancer is composed of small cell lung cancer (SCLC) and non-small cell lung cancer (NSCLC). The NSCLC form is usually cured with surgical operation at an early stage, and it generally shows natural or acquired resistance to anti-cancer drugs, such as platinum (Inoue et al., 2003). In contrast, SCLC (~ 20% of all lung cancers) is treated with chemotherapy and radiotherapy, even at the early stage. Small cell lung cancer has a rapid doubling time, and therefore, it is generally believed that screening is unlikely to be beneficial in patients with this cancer.

Non-small cell lung cancer is pathologically composed of adenocarcinoma, squamous cell carcinoma, large cell carcinoma, and others, and has heterogeneous biological behavior. In particular, pulmonary adenocarcinoma is composed of a variety of histopathological subgroups (acinar, papillary, bronchioloalveolar, solid with mucin, and mixed type). Some characteristics (such as mucin production in pulmonary adenocarcinoma) are known to affect the prognosis of patients with this cancer (Nishiumi et al., 2003). Paraneoplastic syndromes, such as Cushing syndrome, granulocytosis (G-CSF), and hypercalcemia, have also been reported to accompany lung cancer. Non-small cell lung cancer has a variety of histopathological features and complicated biological behaviors, and shows different case-by-case prognosis after the same treatment with surgical operation, chemotherapy, and radiation therapy.

Computed tomography (CT) is a very useful tool for the diagnosis of various cancers, including lung cancer. This scan can reveal not only the features of the primary lesion of lung cancer, but also metastatic lesions, such as mediastinal and hilar lymph nodes, liver metastasis, and brain metastasis. The prognosis of lung cancer is usually insufficient, even in cases that undergo curative operation. Some trials of the mass screening for lung cancer have been performed in various countries, especially in Japan with chest X-ray. Meanwhile, screening with conventional chest X-ray for lung cancer was reported not to contribute to a better prognosis by the Mayo project.

Some vigorous studies on lung cancer screening have been reported using low-dose helical CT (LDHCT) scanning in Japan (Kaneko et al., 1996; Sone et al., 2000). Helical CT scanning has the highest sensitivity for the detection of pulmonary nodules compared to other examinations. It also has the potential to demonstrate small, clinically unapparent, curable lung cancer, especially adenocarcinoma. The reliable detection of small pulmonary nodules is important for the early detection of lung cancer with LDHCT, because a follow-up examination can demonstrate growth as a sign of a potential malignancy. The most adequate target of lung cancer screening with LDHCT may be small adenocarcinoma of lung.

Recently, CT technology has undergone a major evolution with the introduction of multisided technology. With multislice CT (MSCT), a full lung thin slice (< 1 mm in size) scan can be performed within a single breath hold. Some pilot studies of lung cancer screening with MSCT are being undertaken to assess the detection of pulmonary nodules and the attributes required for the improved outcome of lung cancer patients. Lung cancer screening with MSCT entails a huge number of imaging studies, with possibly a true positive rate of around 1%. As mentioned in other chapters, adequate automatic nodule detection may be helpful to guide physicians to questionable structures with a computer-aided diagnosis (CAD) system. Such a system includes automatic nodule detection and a conventional viewing workstation needed to report thoracic CT examination (Abe et al., 2005). In this chapter we present our study and the Mayo Clinic project of lung cancer screening with low-dose helical CT (LDHCT) scan. Also, we discuss the usefulness and pitfalls of a mass screening for lung cancer with LDHCT scan.

Materials and Methods

Subjects. Usually, the participants are upper 50-year-old men and women given lung cancer screening with LDHCT, especially heavy smokers, including current or former smokers with informed consent (Earnest et al., 2003). Men and women younger than 40 years of age are not recommended for this lung cancer screening because of its harmful irradiation effect. The risk of breast cancer due to LDHCT is thought to increase in younger women. All participants are required to fill out questionnaires about respiratory symptoms, smoking, and their past histories.

An outline of the results of our study with LDHCT scan is as follows. Between October 2001 and January 2003, lung cancer screening with LDHCT scan was performed on 518 participants (376 men, 142 women; age range 20–85 years, mean age 57.9 years). All 518 procedures were baseline screenings, and there were no repeat screenings. The majority of participants (72.6%) were men. The frequency of current or former smokers was 174 (33.6%) and 66 (12.8%), respectively.

Interval of lung cancer screening with LDHCT. The interval of lung cancer screening with LDHCT is recommended to be once a year for heavy smokers, but the interval for nonsmokers is controversial. Previous reports showed that the detection incidence of lung cancer was higher in a baseline lung cancer screening with LDHCT than in a repeat screening. Interval lung cancer, where patients show symptoms between their annual CT screenings, is associated with poor prognosis (Diederich and Wormanns, 2004).

Computed tomography scanning. Our method for lung cancer screening with LDHCT was as follows (Abe et al., 2005). With a helical CT scanner (X vision/SR [TSX-002A], Toshiba, Japan), the scanning parameters were a 120 kilovolt peak, 50 mA, 10 mm collimation, and 2.0 pitch. The whole lung field was scanned and completed at deep inspiration during a single breath hold of ~ 15 s. The total time between entering and leaving the room was only ~ 10 min. Ten millimeter reconstructed images were stored on optical disks (650 megabytes of volume per disk). After scanning, high-resolution CT (HRCT) scanning was performed on the same day, when the lesion was strongly suspected of being lung cancer. The scanning parameters of HRCT were 120 kilovolt peak, 150 mA, 2 mm collimation, and 1.0 pitch reconstructed images at the HRCT scanning.

Lung cancer screening with multislice CT (MSCT) scan is now being studied. The Society of Thoracic Radiology in the United States recommends that the scanning parameters for a four-slice CT scan be as follows: 120–140 kilovolt peak, 20–60 mA, 1–3 mm collimation, and 3–6 pitch reconstructed images at LDHCT scanning (Aberle et al., 2001).

Checking the image and judgment. We used a CAD workstation as noted in Chapter 3.7 on computed tomography in lung cancer and computer-aided diagnosis. As for the CAD's performance, the time required to obtain images and determine diagnosis was ~ 7 min. In our study, the three respiratory physicians reached consensus according to the General Rule for the Clinical and Pathological Record of Lung Cancer (GRCPRLC), edited by the Japan Lung Cancer Society (JLCS) (2003): "a," undetermined; "b," within normal limits; "c," old inflammatory lesion; "d," suspicion of disease other than lung cancer; "e," suspicion of lung cancer. More than two respiratory physicians separately interpreted all cases according to the GRCPRLC edited by the JLCS (2003). More than two physicians interpreted separately at first and then made a final judgment with CAD assistance. When they could not reach agreement, a final conclusion on the findings was reached by consensus at the conference.

Results

An outline of our results of lung cancer screening with LDHCT scan follows. Four (0.77%) of the 518 participants were histopathologically diagnosed as having lung cancer. The size of the four lung cancer cases was 9×9 mm, 18×15 mm, 32×20 mm, and 50×45 mm in diameter. One case, with a lesion 9×9 mm in diameter, was diagnosed with pathologically bronchioloalveolar carcinoma. Besides lung cancer, other pulmonary diseases, such as active pneumonia, nontuberculous mycobacterium infection, bullae, and chronic obstructive pulmonary disease, were also diagnosed.

We showed the results of three physicians' judgments with or without CAD in Chapter 3.7. In the first half, we made judgments for 301 participants without CAD, and in the second half, for 217 participants with the CAD system. The diagnosis of each physician of the 301 cases in the first half is shown in Table 9. The judgments of "e" by three physicians were 62 (20.6%), 54 (17.9%), and 61 (20.3%), respectively. Three physicians determined 75/301 (24.9%) participants as "e" in consensus. All three physicians independently judged as "e" 46/75 (61.3%) participants, while one or two physicians judged 29 other participants as "e." Three physicians did not independently judge as "e" 14 (18.7%), 16 (21.3%), and 16 (21.3%) of the 75 participants, respectively.

Discussion

Lung cancer screening using low-dose helical CT (LDHCT) scanning is still a controversial issue (Manser et al., 2003; Bach et al., 2003a; Jett, 2005). The helical CT scan has the highest sensitivity for the detection of pulmonary nodules compared to other examinations. Nawa et al. (2002) reported that LDHCT might be a promising method for screening early lung cancer at health examinations. Most of the lung cancers detected in lung cancer screening with LDHCT are stage I (Nawa et al., 2002; Mulshine et al., 2005). Well-designed clinical trials are necessary to establish the guidelines for mass screening for lung cancer (Swensen et al., 2003; Bach et al., 2003b).

The Mayo Clinic investigated the trial of LDHCT screening for lung cancer initiated in 1999. High-risk participants more than 50 years of age and with 20 pack-years of smoking or more were listed. Low-dose helical CT with multi-slice CT (MSCT, four detectors) was used under the following conditions: 5 mm collimation with 3.75 mm reconstruction. At least one noncalcified nodule was detected in 780 of 1520 (51%) participants on the baseline CT scan, and a total of 3356 noncalcified nodules were detected in the 1118 participants from 2000 through 2003 (Swensen et al., 2005). Only 195 (6%) of all 3356 nodules were 8 mm or larger in size, while 61% of the nodules were

less than 4 mm. In the Mayo trial, a total of 68 lung cancers were detected in 66 patients.

LDHCT also has a number of limitations of screening for lung cancer. The major limitations of CT screening include (1) a high rate of nodule detection, especially with MSCT, (2) the increased cost associated with follow-up CT scans, (3) cost and morbidity of biopsy or the resection of benign nodule, and (4) the risk of cancer associated with multiple follow-up CT scans (Jett, 2005).

The helical CT scan picked up many pulmonary nodules at a range from 17–51% (Diederich et al., 2002; Swensen et al., 2005), while the sensitivity of small pulmonary nodules detected by physicians is not satisfactory. Lung cancer was detected in 0.4–2.7% cases of participants at baseline screening (Table 10). The Mayo Clinic Project picked up a total of 3356 noncalcified nodules in 1118 of 1520 participants for 4 years, and a total of 68 lung cancers were detected in 66 patients (Swensen et al., 2005). In our study with single-helical CT, 170 nodules as "e" (suspicious of lung cancer) were detected in 130 of 518 participants, and four cases were pathologically diagnosed as lung cancer.

The criteria of the decision for further examination or follow-up with CT scan is a crucial point for lung cancer screening with LDHCT. Some authors have reported that surgical operation of small cancers is not useful, because patients with smaller cancers do not have a sufficiently better prognosis than those with larger cancers in stage IA (< 3 cm) (Patz et al., 2000). A few small adenocarcinomas (< 2 cm) of the lung were reported to form metastasis of the mediastinal lymph nodes or systemic organs and to show a very poor prognosis (Nishiumi et al., 2000). The follow-up of small pulmonary nodules is important for the early detection of lung cancer with LDHCT to demonstrate growth as a sign of a potential malignancy. The General Rule for Clinical and Pathological Record of Lung Cancer (GRCPRLC) edited by the Japan Lung Cancer Society (JLCS, 2003) recommends that nodules smaller than 5 mm be followed in the next mass LDHCT screening. This suggests that too many follow-up CT scans are necessary for lung cancer screening. Further

Table 9 Judgments of 301 Participants by Each of the Three Physicians

Judgment[a]	Dr. A	Dr. B	Dr. C
b/c/d/e	82/31/126/62	91/46/110/54	99/38/103/61
% e	20.6%	17.9%	20.3%

[a]We made judgments according to the General Rule for the Clinical and Pathological Record of Lung Cancer (the Japan Lung Cancer Society): "b," within normal limit; "c," old inflammatory lesion; "d," suspicion of diseases other than lung cancer; "e," suspicion of lung cancer. The three respiratory physicians (Dr. A, Dr. B, and Dr. C) independently judged the 301 participants.

Table 10 Lung Cancer Screening with Low-dose Helical CT (LDHCT)

Author	Screened	Noncalcified Nodules (%[a])	Lung Cancer at Baseline (%[a])
Kaneko et al. (1996)	1369	588 (17%)	15 (1.1%)
Nawa et al. (2002)	7956	2099 (26%)	36 (0.4%)
Henschke et al. (2002)	1000	233 (23%)	27 (2.7%)
Pastorino et al. (2003)	1035	284 (27%)	11 (1.1%)
Swensen et al. (2005)	1520	782 (51%)	31 (2.0%)
Diederich et al. (2002)	817	350 (43%)	11 (1.3%)
Our study	518	170 (33%)	4 (0.8%)

[a] % of participants.

analysis will reveal the standard detection size besides the characteristics of small nodules for differential diagnosis of lung cancer.

Some participants with pulmonary nodules undergo thoracic surgery and/or biopsies, such as transbronchial lung biopsy (TBLB) for benign disease. The rate of invasive procedures, such as TBLB and conventional thoracotomy, for benign lesions detected by CT screening was 22–55% of participants with abnormal CT findings (Diederich and Wormanns, 2004). The GRCPRLC notes that further clinical examinations are necessary for cases with nodules > 10 mm in diameter in helical CT screening for lung cancer (JLCS, 2003). Video-associated thoracic surgery (VATS) is safely available for the diagnosis of small pulmonary nodules, including lung cancer in Japan (Okada *et al.*, 2005). The clinical management of participants with pulmonary nodules suspicious of lung cancer is a very important issue.

It is very important to assess the characteristics of small pulmonary nodules for the differential diagnosis of lung cancer. It is well known that contrast enhancement of pulmonary masses is a feasible diagnostic method in diagnosing this cancer. Angiogenesis has been reported to be controlled by complicated mechanisms involving various factors, such as vascular endothelial growth factor (VEGF) associated with human cancers. In lung cancer, our previous studies with molecular and histopathological methods have revealed that angiogenesis is related to some angiogenetic and/or angiostatic factors (Hatanaka *et al.*, 2001; Kawakami *et al.*, 2002; Nishi *et al.*, 2005).

Recent studies have also revealed that the characterization of small nodules on thin-slice sections obtained by MSCT is an important factor for the differential diagnosis of lung cancer from benign lesions (Henschke *et al.*, 2002; Li *et al.*, 2004). Tsuchiya (2005) noted the possibility of limited operation of pulmonary nodules, which can be classified into three types on thin slice by MSCT appearance: pure ground-glass opacity (GGO), GGO with solid component (mixed GGO), and solid nodules. Positron emission tomography (PET) with ^{18}F-fluorodeoxyglucose (FDG), which reflects glucose metabolism in the lesions, can also be used for performing the differential diagnosis of small pulmonary nodules detected by CT scan (Pastorino *et al.*, 2003) and for minimizing unnecessary invasive procedures for benign lesions (Bastarrika *et al.*, 2005). Positron emission tomography cannot accurately evaluate pulmonary nodules such as a tiny nodule less than 1 cm in diameter or a shadow of GGO in a conventional CT image (Nomori *et al.*, 2004).

Many pulmonary nodules, including ground-glass opacity (GGO), were picked up by the helical CT scan, and a part of GGO is thought to be pulmonary adenocarcinoma at an extremely early stage. Hasegawa *et al.* (2000) reported that 31 of 61 (50%) cancers detected in CT screening had a mean doubling time of > 340 days. Jett (2005) opined that doubling times of > 400 days would be consistent with overdiagnosis, defined as a lung cancer that would not lead to an individual's death because of a slow growth rate and competing age-related risk for death. He also noted that several cases of bronchioloalveolar carcinoma showing a GGO pattern in thin-section CT scan have minimally changed in size, and it is of questionable benefit to treat these GGO nodules with surgical operation. More data on these types of lesions are needed.

Estimation of the cost-effectiveness of CT screening for lung cancer has varied considerably, ranging from as little as $2500 per year of life saved to a high of $2.3 million (Wisnivesky *et al.*, 2003; Marshall *et al.*, 2001; Mahadevia *et al.*, 2003). The cost-based analysis by Wisnivesky *et al.* (2003) resulted in the lowest estimate for the cost-effectiveness of screening. The National Lung Cancer Screening Trials were launched in 2002 in the United States, and the results will be presented in 2009. These randomized trials, enrolling 50,000 high-risk participants from North America, compared LDHCT screening to digital chest radiography (Aberle *et al.*, 2003). Other randomized screening trials are now under way in the Netherlands and France. Health insurance systems and medical care systems for lung cancer vary between countries. We do not really have definite knowledge as to whether or not LDHCT screening will prove to be cost-effective, while we can obtain a great deal of information about lung cancer screening with LDHCT in the near future.

An area of considerable debate is the risk of cancer associated with diagnostic X-rays (Mayo *et al.*, 2003; Brenner and Elliston, 2004). A recent report from Oxford, United Kingdom, has estimated that the risk percentage of cancers attributable to diagnostic X-ray ranges from 0.6–1.8% of all cancers in most developed countries (Berrington de Gonzalez and Darby, 2004). Radiation from CT scans was reasonable for the largest number of cases in various reports, while more data are needed to obtain a sufficient result.

In conclusion, lung cancer screening using low-dose helical CT (LDHCT) scanning is still a controversial issue. It is correct that lung cancers detected at an early stage have better rates of survival in individuals. Meanwhile, lung cancer, especially NSCLC, has a variety of histopathological features and complicated biological behaviors. Heterogeneity in lung cancer is an important theme that requires deeper and wider investigations in evaluating the usefulness of mass screening with LDHCT.

References

Abe, Y., Hanai, K., Nakano, M., Ohkubo, Y., Hasizume, T., Kakizaki, T., Nakamura, M., Niki, N., Eguchi, K., Fujino, T., and Moriyama, N. 2005. A computer-aided diagnosis (CAD) system in lung cancer screening with computed tomography. *Anticancer Res.* 25:483–488.

Aberle, D.R., Gamsu, G., Henschke, C.I., Naidich, D.P., and Swensen, S.J. 2001. A consensus statement of the Society of Thoracic Radiology: screening for lung cancer with helical computed tomography. *J. Thorac. Imag. 16*:65–68.

Aberle, D.R., Black, W.C., and Goldin, J.G. 2003. Experimental design and outcome of the National Lung Cancer Trial: a multicenter randomized controlled trial of the helical CT vs. chest X-ray for lung cancer screening. *Am. J. Respir. Crit. Care Med. 167*:A736.

Bach, P.B., Kelley, M.J., Tate, R.C., and McCrory, D.C. 2003a. Screening for lung cancer: a review of the current literature. *Chest 123*:72S–82S.

Bach, P.B., Niewoehner, D.E., and Black, W.C. 2003b. American College of Chest Physicians. Screening for lung cancer: the guidelines. *Chest 123*:83S–88S.

Bastarrika, G., Garcia-Velloso, M.J., Lozano, M.D., Montes, U., Torre, W., Spiteri, N., Campo, A., Seijo, L., Alcaide, A.B., Pueyo, J., Cano, D., Vivas, I., Cosin, O., Dominguez, P., Serra, P., Richter, J.A., Montuenga, L., and Zulueta, J.J. 2005. Early lung cancer detection using spiral computed tomography and positron emission tomography. *Am. J. Respir. Crit. Care Med. 171*:1378–1383.

Berrington de Gonzalez, A., and Darby, S. 2004. Risk of cancer from diagnostic X-rays: estimates for the U.K. and 14 other countries. *Lancet 363*:345–351.

Brenner, D.J., and Elliston, C.D. 2004. Estimated radiation risks potentially associated with full-body CT screening. *Radiology 232*:735–738.

Diederich, S., and Wormanns, D. 2004. Impact of low-dose CT on lung cancer screening. *Lung Cancer 45* (Suppl. 2):S13–S19.

Diederich, S., Wormanns, D., Semik, M., Thomas, M., Lenzen, H., Roos, N., and Heindel, W. 2002. Screening for early lung cancer with low-dose spiral CT: prevalence in 817 asymptomatic smokers. *Radiology 222*:773–781.

Earnest, F., Swensen, S.J., and Zink, F.E. 2003. Respecting patient autonomy: screening at CT and informed consent. *Radiology 226*:633–634.

Hasegawa, M., Sone, S., Takashima, S., Li, F., Yang, Z.G., Maruyama, Y., and Watanabe, T. 2000. Growth rate of small lung cancers detected on mass CT screening. *Br. J. Radiol. 73*:1252–1259.

Hatanaka, H., Abe, Y., Naruke, M., Tokunaga, T., Oshika, Y., Kawakami, T., Osada, H., Nagata, J., Kamochi, J., Tsuchida, T., Kijima, H., Yamazaki, H., Inoue, H., Ueyama, Y., and Nakamura, M. 2001. Significant correlation between interleukin 10 expression and vascularization through angiopoietin/TIE2 networks in non-small cell lung cancer. *Clin. Cancer Res. 7*:1287–1292.

Henschke, C.I., Yankelevitz, D.F., Mirtcheva, R., McGuinness, G., McCauley, D., and Miettinen, O.S. ELCAP Group. 2002. CT screening for lung cancer: frequency and significance of part-solid and nonsolid nodules. *AJR Am. J. Roentgenol. 178*:1053–1057.

Inoue, Y., Tomisawa, M., Yamazaki, H., Abe, Y., Suemizu, H., Tsukamoto, H., Tomii, Y., Kawamura, M., Kijima, H., Hatanaka, H., Ueyama, Y., Nakamura, M., and Kobayashi, K. 2003. The modifier subunit of glutamate cysteine ligase (GCLM) is a molecular target for amelioration of cisplatin resistance in lung cancer. *Int. J. Oncol. 23*:1333–1339.

Japan Lung Cancer Society. 2003. General Rule for Clinical and Pathological Record of Lung Cancer *Kanehara Syuppan*, 6th ed. Tokyo, pp. 190–202.

Jett, J.R. 2005. Limitations of screening for lung cancer with low-dose spiral computed tomography. *Clin. Cancer Res. 11*(13 Pt. 2):4988s–4992s.

Kaneko, M., Eguchi, K., Ohmatsu, H., Kakinuma, R., Naruke, T., Suemasu, K., and Moriyama, N. 1996. Peripheral lung cancer: screening and detection with low-dose spiral CT versus radiology. *Radiology 201*:798–802.

Kawakami, T., Tokunaga, T., Hatanaka, H., Kijima, H., Yamazaki, H., Abe, Y., Osamura, Y., Inoue, H., Ueyama, Y., and Nakamura, M. 2002. Neuropilin 1 and neuropilin 2 co-expression is significantly correlated with increased vascularity and poor prognosis in nonsmall cell lung carcinoma. *Cancer 95*:2196–2201.

Li, F., Sone, S., Abe, H., Macmahon, H., and Doi, K. 2004. Malignant versus benign nodules at CT screening for lung cancer: comparison of thin-section CT findings. *Radiology 233*:793–798.

Mahadevia, P.J., Fleisher, L.A., Frick, K.D., Eng, J., Goodman, S.N., and Powe, N.R. 2003. Lung cancer screening with helical computed tomography in older adult smokers: a decision and cost-effectiveness analysis. *J. Am. Med. Assoc. 289*:313–322.

Manser, R.L., Irving, L.B., Byrnes, G., Abramson, M.J., Stone, C.A., and Campbell, D.A. 2003. Screening for lung cancer: a systematic review and meta-analysis of controlled trials. *Thorax 58*:784–789.

Marshall, D., Simpson, K.N., Earle, C.C., and Chu, C.W. 2001. Economic decision analysis model of screening for lung cancer. *Eur. J. Cancer 37*:1759–1767.

Mayo, J.R., Aldrich, J., Muller, N.L. Fleischner Society. 2003. Radiation exposure at chest CT: a statement of the Fleischner Society. *Radiology 228*:15–21.

Mulshine, J.L. 2005. Clinical issues in the management of early lung cancer. *Clin. Cancer Res. 11* (13 Pt. 2):4993s–4998s.

Nawa, T., Nakagawa, T., Kusano, S., Kawasaki, Y., Sugawara, Y., and Nakata, H. 2002. Lung cancer screening using low-dose spiral CT: results of baseline and 1-year follow-up studies. *Chest 122*:15–20.

Nishi, M., Abe, Y., Tomii, Y., Tsukamoto, H., Kijima, H., Yamazaki, H., Ohnishi, Y., Iwasaki, M., Inoue, H., Ueyama, Y., and Nakamura, M. 2005. Cell binding isoforms of vascular endothelial growth factor-A (VEGF189) contribute to blood flow-distant metastasis of pulmonary adenocarcinoma. *Int. J. Oncol. 26*:1517–1524.

Nishiumi, N., Abe, Y., Inoue, Y., Hatanaka, H., Inada, K., Kijima, H., Yamazaki, H., Tatematsu, M., Ueyama, Y., Iwasaki, M., Inoue, H., and Nakamura, M. 2003. Use of 11p15 mucins as prognostic factors in small adenocarcinoma of the lung. *Clin. Cancer Res. 9*:5616–5619.

Nishiumi, N., Maitani, F., Kaga, K., Iwasaki, M., Nakamura, M., Osamura, Y., and Inoue, H. 2000. Is it permissible to omit mediastinal dissection for peripheral non-small cell lung cancers with tumor diameters less than 1.5 cm? *Tokai J. Exp. Clin. Med. 25*:33–37.

Nomori, H., Watanabe, K., Ohtsuka, T., Naruke, T., Suemasu, K., and Uno, K. 2004. Evaluation of F-18 fluorodeoxyglucose (FDG) PET scanning for pulmonary nodules less than 3 cm in diameter, with special reference to the CT images. *Lung Cancer 45*:19–27.

Okada, M., Sakamoto, T., Yuki, T., Mimura, T., Miyoshi, K., and Tsubota, N. 2005. Hybrid surgical approach of video-assisted minithoracotomy for lung cancer: significance of direct visualization on quality of surgery. *Chest 128*:2696–2701.

Pastorino, U., Bellomi, M., Landoni, C., De Fiori, E., Arnaldi, P., Picchio, M., Pelosi, G., Boyle, P., and Fazio, F. 2003. Early lung-cancer detection with spiral CT and positron emission tomography in heavy smokers: 2-year results. *Lancet 362*:593–597.

Patz, E.F., Jr., Rossi, S., Harpole, D.H., Jr., Herndon, J.E., and Goodman, P.C. 2000. Correlation of tumor size and survival in patients with stage IA non-small cell lung cancer. *Chest 117*:1568–1571.

Sone, S., Li, F., Yang, Z.G., Honda, T., Maruyama, Y., Takashima, S., Hasegawa, M., Kawakami, S., Kubo, K., Haniuda, M., and Yamanda, T. 2000. Results of three-year mass screening programme for lung cancer using mobile low-dose spiral computed tomography scanner. *Br. J. Cancer 84*:25–32.

Swensen, S.J., Jett, J.R., Hartman, T.E., Midthun, D.E., Mandrekar, S.J., Hillman, S.L., Sykes, A.M., Aughenbaugh, G.L., Bungum, A.O., and Allen, K.L. 2005. CT screening for lung cancer: five year prospective experience. *Radiology 235*:259–265.

Swensen, S.J., Jett, J.R., Hartman, T.E., Midthun, D.E., Sloan, J.A., Sykes, A.M., Aughenbaugh, G.L., and Clemens, M.A. 2003. Lung cancer screening with CT: Mayo Clinic experience. *Radiology 226*:756–761.

Tsuchiya, R. 2005. Implication of the CT characteristics of subcentimeter pulmonary nodules. *Semin. Thorac. Cardiovasc. Surg. 17*:107–109.

Wisnivesky, J.P., Mushlin, A.I., Sicherman, N., and Henschke, C. 2003. The cost-effectiveness of low-dose CT screening for lung cancer: preliminary results of baseline screening. *Chest 124*:614–621.

7

Lung Cancer:
Computer-aided Diagnosis with
Computed Tomography

Yoshiyuki Abe, Katsumi Tamura, Ikuko Sakata, Jiro Ishida, Masayoshi Nagata, Masato
Nakamura, Kikuo Machida, and Toshiro Ogata

Introduction

Many advances have been made in computer-aided diagnosis (CAD) over the last two decades with regard to computer technology, software programs, and improved image-capture techniques with high resolution. Computer-aided diagnosis has become one of the major research subjects in medical imaging and diagnostic radiology (Doi, 2005). Many different types of CAD schemes are being developed for the detection and characterization of various lesions in the medical imaging of conventional projection radiography, computed tomography (CT), magnetic resonance imaging (MRI), and ultrasound. The basic concept of CAD is to provide computer output as a second opinion to assist the radiologist's image reading.

Lung cancer is one of the leading causes of a death in advanced countries. The prognosis of lung cancer is usually insufficient, even in cases that undergo curative operation. The most common methods used to detect pulmonary nodules are chest radiography and CT scan. The mayo project found that screening of lung cancer with conventional chest radiography did not contribute to a better prognosis.

Therefore, trials on lung cancer screening have been reported using low-dose helical computed tomography (LDHCT) scanning (Kaneko et al., 1996; Sone et al., 2000). Low-dose helical computed tomography scanning has the highest sensitivity for the detection of pulmonary nodules compared to other examination methods and has the potential to demonstrate small, clinically unapparent, curable lung cancer (Sone et al., 2000). The reliable detection of small pulmonary nodules is important for the early detection of lung cancer with LDHCT scanning because a follow-up examination can demonstrate growth as a sign of a potential malignancy. The sensitivity of small pulmonary nodules detected by physicians is not satisfactory (Diederich et al., 1999), especially in mass screening for lung cancer. Recently, multislice CT (MSCT) has been able to provide a full-lung, thin-slice scan within a single breath hold (Wiemker et al., 2005).

Computer-aided diagnosis schemes are being developed for the detection and characterization of various lesions in the field of the diagnosis of lung cancer, including chest radiograph (Suzuki et al., 2005) and computed tomography

(CT) (Kim *et al.*, 2005). Adequate automatic nodule detection may be helpful to guide physicians to questionable structures. These features can be integrated into the physician's work in several ways. A very convenient technique is the integration of a softcopy-viewing workstation. Utilization of automatic nodule detection in the clinical routine has become possible due to the dramatic increase in computer performance to within acceptable limits (Reeves *et al.*, 2000; Wormanns *et al.*, 2002). The Moriyama research group in Japan has developed a computer-aided diagnosis (CAD) system, which includes automatic nodule detection and a conventional viewing workstation needed to report thoracic CT examinations (Kanazawa *et al.*, 1998). In this chapter, we present this CAD system to detect pulmonary nodules performed for lung cancer screening with LDHCT scanning, and we discuss the usefulness and pitfalls of the CAD system in mass screening for lung cancer.

Materials and Methods

Subjects. All participants necessarily gave informed consent for CT screening of lung cancer and completed questionnaires about their respiratory symptoms, smoking, status, and past histories. Between October 2001 and January 2003, lung cancer screening with LDHCT scan was performed on 518 participants (376 men, 142 women; age range 20–85 years, mean age 57.9 years). All 518 procedures were baseline screenings, and no repeat screenings were made. In the first half of the period, 301 were without CAD system assistance, and in the second half 217 utilized the CAD.

Computer tomography (CT) scanning. A helical CT scanner (Xvision/SR [TSX-002A], Toshiba, Japan) was used for this study. (Abe *et al.*, 2005). The scanning parameters were 120 kilovolt peaks, 50 mA, 10 mm collimation, and 2.0 pitches. The whole lung field was scanned and completed at deep inspiration during a single breath hold of ~ 15 sec. The total time between entering and leaving the room was only ~ 10 min. Ten millimeter reconstructed images were stored on optical disks (650 megabytes of volume per disk). After scanning, high-resolution CT (HRCT) scanning was performed on the same day, when the lesion was strongly suspected of being lung cancer. The scanning parameters of HRCT were a 120 kilovolt peak, 150 mA 2 mm collimation, and 1.0 pitch reconstructed images at the HRCT scanning. Lung cancer screening with multislice CT (MSCT) scan is presented elsewhere in this volume.

Computer-aided diagnosis (CAD) workstation. The hardware of the CAD system consists of a Toshiba AS 7000 U5 workstation (Kanazawa *et al.*, 1998). A screenshot of the CAD system's user interface is shown in Figure 47. This CAD system is equipped with automatic image diagnosis and an image screening function. When the automatic image

diagnosis process is started in advance, the physician can begin lung cancer diagnosis at any time. Four consecutive slice CT images are always displayed, and the physician can easily search the chosen image. When the physician clicks the mouse button on the image, a marker is displayed at the position of the cursor. The diagnostic results from all subjects are recorded on the hard disk drive. Network transfer of CT data from the CT scanner to the CAD workstation is realized with the Digital Imaging and Communications in Medicine (DICOM) protocol. Computer-aided diagnosis software includes a detection algorithm for pulmonary nodules and a user interface.

The detection algorithm includes a complex segmentation of lung parenchyma (deletion of the CT table and soft tissue of the chest wall), followed by the detection of structures with soft tissue density within the lung parenchyma and a region analysis to evaluate the detected structures in the three-dimensional (3D) data set (Fiebich *et al.*, 1999). First, soft tissue objects within the segmented lung borders are detected using a fixed density threshold value (~ –600 Hounsfield Units [HUs]). Evaluation is performed after using a 3D region-growing algorithm. Objects with a detected volume of < 10 voxels are ignored. A spherical soft tissue density nodule of this size corresponds to a diameter

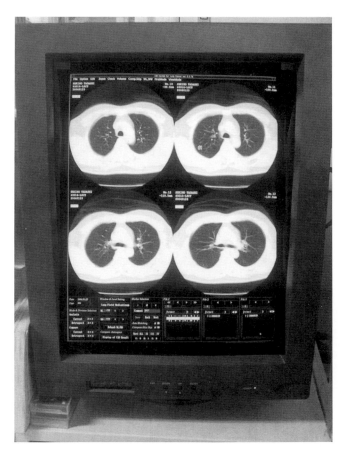

Figure 47 Screenshot of the computer-aided detection (CAD) system's user interface.

of ~ 5 mm due to the partial-volume effect and density threshold value. For the remaining objects, the distinction between probable nodules and other structures (especially vessels and scars or subsegmental atelectasis) are based on object geometry, especially on the length/width/height ratio.

Interpretation. All images were obtained at window settings appropriate for lung parenchyma (level, −600 HU; width, 1800 to 1600 HU). As for the CAD's performance, the time required to obtain images and diagnose them was ~ 7 min. The CAD cannot simultaneously provide images from a CT scan when it is running another thoracic examination. Three respiratory physicians separately interpreted all cases according to the General Rule for the Clinical and Pathological Record of Lung Cancer (GRCPRLC) edited by the Japan Lung Cancer Society (JLCS) (2003). We made judgments without CAD system assistance in the first half. In the second half, we first used the CAD system, and three physicians separately interpreted and then made a final judgment with CAD assistance (Fig. 48). When we could not reach agreement, consensus on the findings was reached at the conference.

In the first half, 75/301 (24.9%) participants were determined as "e" without CAD, while 55/217 (25.3%) participants were "e" with the CAD system in the second half (Table 11). Four participants were histopathologically diagnosed with lung cancer in all periods. The sizes of the four lung cancer cases were 9 × 9 mm, 18 × 15 mm, 32 × 20 mm, and 50 × 45 mm in diameter, respectively. Three physicians determined 75/301 (24.9%) participants as "e" in consensus without CAD in the first half. All three physicians independently judged 46/75 (61.3%) participants as "e," while one or two physicians judged 29 other participants as "e." We reevaluated the 301 participants and judged them using CAD. With CAD, three participants were added to "e" from the other judgments in the first half without CAD. Computer-aided diagnoses picked up three cases (8 mm, 5 mm, and 5 mm in diameter) that were overlooked by all three physicians.

Three physicians determined 75/301 participants as "e" in consensus without CAD. Three physicians did not independently judge as "e" 18.7%, 21.3%, and 21.3% of the 75 participants, respectively. Computer-aided diagnosis

Results

Our results of lung cancer screening with LDHCT scan are as follows. Lung cancer screening with LDHCT scan was performed on 518 participants: where 301 without CAD system assistance in the first half of the period and 217 utilized the CAD in the second half. The three respiratory physicians made judgments in consensus according to the GRCPRLC edited by JLCS (2003) ("a," undetermined; "b," within normal limit; "c," old inflammatory lesion; "d," suspicion of disease other than lung cancer; "e," suspicion of lung cancer).

Table 11 Judgments of Participants

	Without CAD	With CAD
Participant number	301	217
Judgment[a]		
b/c/d/e	73/29/124/75	56/36/70/55
% e	24.9%	25.3%

[a]We made judgments according to the General Rule for the Clinical and Pathological Record of Lung Cancer (the Japan Lung Cancer Society): "b," within normal limit; "c," old inflammatory lesion; "d," suspicion of diseases other than lung cancer; "e," suspicion of lung cancer). The three respiratory physicians judged in consensus.

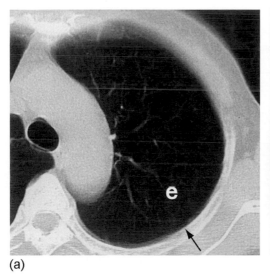

(a)

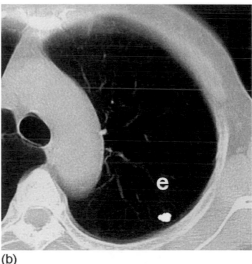

(b)

Figure 48 The detection of pulmonary nodules by computer-aided detection (CAD). (a) Original image of the location of a faint shadow (arrow). (b) Detection results—a lesion was detected by the CAD system.

could not identify 17 (22.7%) nodules of the 75 participants. Ten of the 17 nodules were diagnosed as "e" by all three physicians, while 7 other nodules were diagnosed by one or two physicians (2 nodules were diagnosed by one physician and 5 nodules by two physicians). All 17 nodules were less than 6 mm in diameter: ≤ 6 mm and > 5 mm, 8 nodules; ≤ 5 mm and > 4 mm, 4 nodules; ≤ 4 mm and > 3 mm, 5 nodules (Fig. 49). No apparent characteristics of the CT number were noted in these 17 tiny nodules. Four nodules of 17 were adjacent to the pleural lung surface, and another 4 nodules were adjacent to the vessels.

Discussion

We evaluated the CAD system during a lung cancer screening with LDHCT. In our study, three physicians determined 75/301 (24.9%) participants as "e" (suspicious of lung cancer) in consensus without CAD, while three participants were added to "e" with CAD. Three physicians did not independently judge as "e" 18.7%, 21.3%, and 21.3% of the 75 participants, while CAD could not identify 22.7% of the nodules of the 75 participants.

The CAD system is able to demonstrate some nodules overlooked by the physicians because they were located in more central areas of the lung. In this study, three cases were overlooked by all three physicians and reassessed to an "e" judgment with CAD assistance. Midthun *et al.* (2002) reported that 21% of nodules 8–20 mm in size were malignant and that lesions 8–20 mm in size require further evaluation. Some recent reports have shown the usefulness of CAD in the detection and diagnosis of pulmonary nodules

(Shah *et al.*, 2005; Peldschus *et al.*, 2005). Significant lesions are frequently missed in the routine clinical interpretation of chest CT studies, but may be detected if CAD is used as an additional reader in 100 patients (Peldschus *et al.*, 2005).

On the other hand, the CAD system has an obvious weakness in the detection of nodules adjacent to the pleural lung surface because the image segmentation algorithm recognizes the nodule as a part of the chest wall and excludes it from further image processing. The algorithm is designed to detect nodules with diameters of at least 5 mm (Kanazawa *et al.*, 1998). All 17 tiny nodules (4 adjacent to the pleural lung surface and 4 adjacent to the vessels) in our study that were detected by physicians and not identified by the CAD system were < 6 mm. Many pulmonary nodules, including ground-glass opacity (GGO), were picked up by the helical CT scan (Kaneko *et al.*, 1996). A part of GGO is thought to be pulmonary adenocarcinoma at a very early stage. Further analysis will reveal the standard detection size of small nodules.

The CAD system for analysis of pulmonary nodules has the following advantages: (1) great speed of numerical calculation, allowing precise, quantitative, and reproducible measurements, (2) an ever-increasing knowledge base to provide diagnostic information, and (3) no susceptibility to fatigue (Reeves *et al.*, 2000). Armato *et al.* (2002) reported that the CAD system correctly detected 84% of the missed cancers in a database of low-dose CT scans. Significant benefits are seen in the quantitative results that can be derived from analysis of a large number of subjects in a reproducible, efficient manner. These characteristics are very adequate for mass screenings for lung cancer. In our study,

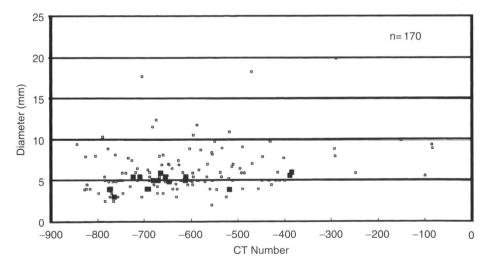

Figure 49 Characteristics of small pulmonary nodules. Computer-aided detection (CAD) could not identify 17 nodules (closed square) detected by physicians. Many other nodules (total 170 nodules, open circle) were detected both by physicians and the CAD system. All 17 nodules were 5 mm in diameter. No apparent characteristics of CT number were noted in these 17 nodules.

the "e" judgments determined by three physicians varied from 17.9% to 20.6% in the reevaluation of participants. The coincidence of all three physicians in the "e" participants was not particularly high, at 61.3%. The CAD system may provide physicians with tools to obtain more accurate diagnoses for lung cancer (Awai et al., 2004), especially in cases diagnosed by one physician.

The development of multislice (MS) CT has produced a large number of CT images that may require additional time and effort in image interpretation by radiologists (Wiemker et al., 2005; Rubin et al., 2005). Some trials are now progressing to support the automated detection of pulmonary nodules on MSCT images with morphological matching detection (Bae et al., 2005; Kim et al., 2005). Therefore, CAD is expected to assist radiologists in reducing the reading time as well as in improving the diagnostic accuracy of lung cancer screening using MSCT.

The second hopeful area in which CAD can assist the physician is the differential diagnosis between malignant and benign pulmonary nodules. Only a fraction of the pulmonary nodules are malignant carcinomas from lung cancer or metastases from cancers in other organs. Computer-aided diagnosis has the potential to improve the radiologists' diagnostic accuracy in distinguishing small benign nodules from malignant nodules on high-resolution CT (Li et al., 2004).

Some reports have compared the performance of radiologists and a CAD algorithm for pulmonary nodule detection on CT scan (Rubin et al., 2005, Marten et al., 2005; Li et al., 2005). It was reported in the diagnosis of pulmonary nodules with MSCT that CAD increased the nodule detection rates, decreased the false-positive rates, and compensated for deficient reader performance in the detection of the smallest lesions and of nodules without vascular attachment (Marten et al., 2005). Rubin et al. (2005) reported that with CAD used at a level allowing only three false-positive detections per CT scan, it was substantially higher than with conventional double reading. Marten et al. (2004) reported the evaluation of the performance of experienced versus inexperienced radiologists in comparison and in consensus with an interactive CAD system for the detection of pulmonary nodules (Marten et al., 2004). They showed that it is questionable as to whether inexperienced readers can be regarded as adequate for interpreting of pulmonary nodules in consensus with CAD, replacing an experienced radiologist.

Conclusions

The CAD system serves as a useful second opinion when physicians examine patients at lung cancer screenings with LDHCT. Computer-aided diagnosis has also been developed for the automated detection of pulmonary nodules with multislice (MS) CT scan. Computer-aided diagnosis will have a major impact on medical imaging and diagnostic radiology in the field of lung cancer in the near future.

References

Abe, Y., Hanai, K., Nakano, M., Ohkubo, Y., Hasizume, T., Kakizaki, T., Nakamura, M., Niki, N., Eguchi, K., Fujino, T., and Moriyama, N. 2005. A computer-aided diagnosis (CAD) system in lung cancer screening with computed tomography. *Anticancer Res.* 25:483–488.

Armato, S.G., 3rd, Li, F., Giger, M.L., MacMahon, H., Sone, S., and Doi, K. 2002. Lung cancer: performance of automated lung nodule detection applied to cancers missed in a CT screening program. *Radiology* 225:685–692.

Awai, K., Murao, K., Ozawa, A., Komi, M., Hayakawa, H., Hori, S., and Nishimura, Y. 2004. Pulmonary nodules at chest CT: effect of computer-aided diagnosis on radiologist detection performance. *Radiology* 230:347–352.

Bae, K.T., Kim, J.S., Na, Y.H., Kim, K.G., and Kim, J.H. 2005. Pulmonary nodules: automated detection on CT images with morphologic matching algorithm—preliminary results. *Radiology* 236:286–293.

Diederich, S., Semik, M., Lentschig, M.G., Winter, F., Scheld, H.H., Roos, N., and Bongartz, G. 1999. Helical CT of pulmonary nodules in patients with extrathoracic malignancy: CT-surgical correlation. *AJR Am. J. Roentgenol.* 172:353–360.

Doi, K. 2005. Current status and future potential of computer-aided diagnosis in medical imaging. *Br. J. Radiol.* 78 Spec. No. 1:S3–S19

Fiebich, M., Wietholt, C., and Renger, B.C. 1999. Prototype of a CAD workstation for low-dose, screening thoracic CT examination. In: Lemke, H.U., Inamura, K., and Vannier, M.W., (Eds.), *Computer-Assisted Radiology*. Amsterdam: Elsevier, pp. 99–100.

Japan Lung Cancer Society. 2003. General rule for clinical and pathological record of lung cancer. *Kanehara Syuppan*, 6th ed. Tokyo, pp. 190–202.

Kanazawa, K., Kawata, Y., Niki, N., Satoh, H., Ohmatsu, H., Kakinuma, R., Kaneko, M., Moriyama, N., and Eguchi, K. 1998. Computer-assisted diagnosis for pulmonary nodules based on helical CT images. *Comput. Med. Imag. Graph.* 22:157–167.

Kaneko, M., Eguchi, K., Ohmatsu, H., Kakinuma, R., Naruke, T., Suemasu, K., and Moriyama, N. 1996. Peripheral lung cancer: screening and detection with low-dose spiral CT versus radiology. *Radiology* 201:798–802.

Kim, J.S., Kim, J.H., Cho, G., and Bae, K.T. 2005. Automated detection of pulmonary nodules on CT images: effect of section thickness and reconstruction interval—initial results. *Radiology* 236:295–299.

Li, F., Aoyama, M., Shiraishi, J., Abe, H., Li, Q., Suzuki, K., Engelmann, R., Sone, S., Macmahon, H., and Doi, K. 2004. Radiologists' performance for differentiating benign from malignant lung nodules on high-resolution CT using computer-estimated likelihood of malignancy. *AJR Am. J. Roentgenol.* 183:1209–1215.

Li, F., Arimura, H., Suzuki, K., Shiraishi, J., Li, Q., Abe, H., Engelmann, R., Sone, S., MacMahon, H., and Doi, K. 2005. Computer-aided detection of peripheral lung cancers missed at CT: ROC analyses without and with localization. *Radiology* 237:684–690.

Marten, K., Engelke, C., Seyfarth, T., Grillhosl, A., Obenauer, S., and Rummeny, E.J. 2005. Computer-aided detection of pulmonary nodules: influence of nodule characteristics on detection performance. *Clin. Radiol.* 60:196–206.

Marten, K., Seyfarth, T., Auer, F., Wiener, E., Grillhosl, A., Obenauer, S., Rummeny, E.J., and Engelke, C. 2004. Computer-assisted detection of pulmonary nodules: performance evaluation of an expert knowledge-based detection system in consensus reading with experienced and inexperienced chest radiologists. *Eur. Radiol.* 14:1930–1938.

Midthun, D.E., Jett, J.R., and Swensen, S.J. 2002. Evaluation of nodules detected by screening for lung cancer with low dose spiral computed tomography. *Am. J. Respir. Crit. Care Med.* 165:A37.

Peldschus, K., Herzog, P., Wood, S.A., Cheema, J.I., Costello, P., and Schoepf, U.J. 2005. Computer-aided diagnosis as a second reader: spectrum of findings in CT studies of the chest interpreted as normal. *Chest 128*:1517–1523.

Reeves, A.P., and Kostis, W.J. 2000. Computer-aided diagnosis for lung cancer. *Radiol. Clin. North Am. 38*:497–509.

Rubin, G.D., Lyo, J.K., Paik, D.S., Sherbondy, A.J., Chow, L.C., Leung, A.N., Mindelzun, R., Schraedley-Desmond, P.K., Zinck, S.E., Naidich, D.P., and Napel, S. 2005. Pulmonary nodules on multi-detector row CT scans: performance comparison of radiologists and computer-aided detection. *Radiology 234*:274–283.

Shah, S.K., McNitt-Gray, M.F., De Zoysa, K.R., Sayre, J.W., Kim, H.J., Batra, P., Behrashi, A., Brown, K., Greaser, L.E., Park, J.M., Roback, D.K., Wu, C., Zaragoza, E., Goldin, J.G., Suh, R.D., Brown, M.S., and Aberle, D.R. 2005. Solitary pulmonary nodule diagnosis on CT: results of an observer study. *Acad. Radiol. 12*:496–501.

Sone, S., Li, F., Yang, Z.G., Honda, T., Maruyama, Y., Takashima, S., Hasegawa, M., Kawakami, S., Kubo, K., Haniuda, M., and Yamanda, T. 2000. Results of three-year mass screening programme for lung cancer using mobile low-dose spiral computed tomography scanner. *Br. J. Cancer 84*:25–32.

Suzuki, K., Shiraishi, J., Abe, H., MacMahon, H., and Doi, K. 2005. False-positive reduction in computer-aided diagnostic scheme for detecting nodules in chest radiographs by means of massive training artificial neural network. *Acad. Radiol. 12*:191–201.

Wiemker, R., Rogalla, P., Blaffert, T., Sifri, D., Hay, O., Shah, E., Truyen, R., and Fleiter, T. 2005. Aspects of computer-aided detection (CAD) and volumetry of pulmonary nodules using multislice CT. *Br. J. Radiol. 78* No. 1:S46–56.

Wormanns, D., Fiebich, M., Saidi, M., Diederich, S., and Heindel, W. 2002. Automatic detection of pulmonary nodules at spiral CT: clinical application of a computer-aided diagnosis system. *Eur. Radiol. 12*:1052–1057.

8

Stereotactic Radiotherapy for Non-small Cell Lung Carcinoma: Computed Tomography

Hiroshi Onishi, Atsushi Nambu, Tomoki Kimura, and Yasushi Nagata

Introduction

Non-small cell lung carcinoma (NSCLC) is one of the leading causes of mortality worldwide. Lung cancers are increasingly being detected in the early stage, due to the routine use of computed tomography (CT). For patients with stage I (T1 or 2, N0, M0) NSCLC, either a full lobar or a greater surgical resection remains the treatment of choice, with promising local control rates exceeding 80% and overall survival rates of > 50% after 5 years (Smith, 2003). However, a surgical resection is often not feasible owing to a high risk for some lung cancer patients with tobacco-related illnesses, severe cardiovascular disease, or other medical conditions. In addition, a small proportion of patients who are indicated for this surgery may also refuse surgery for personal reasons. Radiotherapy can offer a therapeutic alternative in these cases, but the outcome for conventional radiotherapy is still not satisfactory, and differences of the outcome are potentially amplified by selection bias, with local control rates ranging from 40–70% and 5-year survival rates ranging from only 5–30% (Harpole et al., 1995).

The dose of conventional radiotherapy for NSCLC has been suggested as being too low to control the condition. By increasing the accuracy of the localization of the tumor-bearing area using various imaging tools, hypofractionated or single high-dose stereotactic radiotherapy (SRT) has recently been actively investigated for the treatment of stage I NSCLC (Onishi et al., 2004). Intracranial SRT has now been performed for more than 30 years, and it has been thoroughly evaluated as a standard treatment for brain metastases. However, the translation of SRT to extracranial sites has been difficult as a result of the inherent organ motion resulting from normal body functions such as breathing and cardiac activity. In addition, the external surface anatomy does not have any structures amenable to rigid fixation to a frame.

Lax et al. (1994) reported the first experience of SRT for body tumors with the development and testing of an extracranial frame that incorporated a fiducial stereotactic coordinate system along its side panels. To decrease respiratory excursion, an abdominal press was employed, thus forcing the patient to perform relatively more chest wall breathing rather than diaphragmatic breathing. A formal verification of reproducibility study was carried out, and the target motion was thus reduced to within 0.5 cm in the axial plane and to 1.0 cm in the caudal/cephalad plane. Uematsu et al. (2001) reported a landmark study presenting the 5-year results of treatment of 50 patients with T1–2 N0, M0 NSCLC with respiratory gating, using a CT-guided frameless stereotactic radiotherapy technique. Most patients received 50–60 Gy in 5–10 fractions over a 1- to 2-week period. At 36 months, no local progression was seen in 94% of patients; the 3-year overall survival was 66%, and the 3-year cause-specific survival was 88%.

Definition of Stereotactic Radiotherapy

Stereotactic radiation therapy is an emerging treatment paradigm defined in the American Society of Therapeutic Radiology and Oncology guidelines as a "treatment method to deliver a high dose of radiation to the target with conformal dose distributions that fall off very rapidly, utilizing either a single dose or a small number of fractions with a high degree of precision within the body." It generally requires the use of multiple shaped beams or dynamic conformal arcs.

Traditionally, external beam radiotherapy has typically been administered in daily fractions of 1.8–2.0 Gy to total doses of 60–70 Gy or so over 6 or more weeks. With SRT, on the other hand, ultra-high doses in the range of 10–20 Gy per fraction are applied in a hypofractionated regimen of five or fewer fractions. Such high doses per treatment have been made feasible by recent refinements in precise, image-guided radiation treatment technology, which has thus made it possible to safely deliver very large individual doses of radiation to tumors in widely disparate extracranial locations.

The major appeal of SRT is the capacity to provide a non-invasive, highly efficient means of eradicating discrete tumor foci either at a primary or metastatic site. There are also other special situations for which SRT offers unique advantages in terms of improved spatial accuracy of dose delivery and/or greater biological potency relative to conventional techniques.

Clinical Status of Stereotactic Radiotherapy for Early-Stage Lung Carcinoma

In a short 10-year history, SRT showed a promising tumor control effect and safety for the radical treatment of stage I NSCLC or metastatic lung cancer. The local control rates of stage I non-small cell lung carcinoma with SRT has been previously reported by several authors: 94% (47/50) for 50–60 Gy in five fractions with a median follow-up of 36 months; 92% (22/24) for 60 Gy in eight fractions with a median follow-up of 24 months; 87% (30/37) for 60 Gy in three fractions with a median follow-up of 15 months; 85% for 48–60 Gy in eight fractions with a median follow-up of 17 months; 95% for 45–56.2 Gy in three fractions with a median follow-up of 10 months; and 97% (44/45) for 48 Gy in four fractions with a median follow-up of 22–30 months

(Nagata *et al.*, 2005). Onishi *et al.* (2004) recently reported the results from 14 institutions in Japan, where they summarized 245 patients, 155 with stage IA lung cancer and 90 with stage IB lung cancer. The operable and inoperable patients totaled 87 and 158, respectively, and their results showed the intercurrent death rate to be especially high in the inoperable patient group. Moreover, the 5-year survival rates of operable patients irradiated with more than BED = 100 Gy was 90% for stage IA and 84% for stage IB, and their clinical results were as good as those for surgery. As well as for primary lung carcinoma, the local control rates of SRT for metastatic lung tumor have been reported to be > 90% in cases of tumors < 3 cm in size.

Stereotactic radiotherapy seems to be a promising method for the treatment of stage I NSCLC and small metastatic lung cancer. However, a longer follow-up and more experience are still needed before any definitive conclusions can be made regarding the effects of SRT for the treatment of lung carcinoma. Several unanswered questions and currently ongoing protocols will also be reviewed. Two multi-institutional studies, RTOG0236 and JCOG0403, are presently underway.

The Significance of Computed Tomography Imaging for Stereotactic Radiotherapy

The ability to deliver a single or a few fractions of high-dose ionizing radiation with high targeting accuracy and rapid dose falloff gradients encompassing tumors within a patient provides the basis for the development of SRT. Stereotactic body radiation therapy can be applied to very localized malignant conditions in the body using minimally invasive stereotactic tumor localization and radiation delivery techniques. It also requires a high degree of precision when directing the ionizing radiation. Maneuvers to limit the movement of the target volume during treatment planning and delivery are often required to achieve the necessary precision when using this technique. The stereotactic localization of the lesion using an appropriate imaging modality, mainly CT, allows the accurate placement of one or more isocenters associated with the lesion.

Unlike conventional radiation therapy, special stereotactic equipment is employed to obtain more accurate tumor localization, planning, and treatment. The stereotactic equipment can be either frame based or frameless. Appropriate accounting of internal organ movement may be required, depending on the body site under treatment. Imaging, planning, and treatment may occur on the same day for single-fraction treatment, or the treatment could be fractionated into several sessions utilizing larger daily doses of radiation than the dose utilized during conventionally fractionated radiation therapy. Radiation delivery equipment should have a mechanical tolerance for radiation delivery of ± 2 mm. Strict protocols for quality assurance (QA) must be carefully followed. Quality assurance measures are required for the extracranial treatments given inherent organ motion, larger field apertures, and often considerably higher doses delivered. As such, SRT requires the coordination of a large and diverse team of professionals, including the radiation oncologist, medical physicist, and diagnostic radiologist.

Stereotactic body radiation therapy is an image-based treatment. All salient anatomic features of the SRT patient, both normal and abnormal, are defined with CT, positron emission tomography (PET), with or without image fusion, or any other imaging studies that may be useful in localizing the target volumes. Both high three-dimensional (3D) spatial accuracy and tissue contrast definition are very important imaging features to utilize SRT to its fullest positional accuracy. The management of patient care and treatment delivery is predicated by the ability to define the localizing target and normal tissue boundaries as well as to generate target coordinates at which the treatment beams are to be aimed. They are used for creating an anatomic patient model (virtual patient) for treatment planning, and they also contain the morphology required for the treatment plan evaluation and dose calculation.

Computed tomography and its applications are utilized during the entire process of SRT. Four aspects of the use of CT are as follows: (1) the detection, diagnosis, and staging of early-stage lung cancer; (2) treatment planning for SRT; (3) stereotactic repositioning of the 3D isocenters to the planned position during irradiation; and (4) evaluation of the treatment effect after SRT, including the differentiation between inflammatory change and the recurrent mass.

The details of imaging concerning item no. 1 (detecting, diagnosing, and staging early-stage lung cancer) are described in other chapters. The details of the three other items are presented here.

Utility of Computed Tomography for Radiotherapy Treatment Planning of Stereotactic Radiotherapy for Lung Carcinoma

Modern radiotherapy uses 3D conformal treatment planning, based primarily on CT scans, to ensure that the radiotherapy is actually delivered to the entire tumor. Computed tomography images contain information that can be used for both diagnosis and subsequent therapeutic interventions. A diagnosis needs specificity, but it is less susceptible to the accuracy of localization. For therapy, sensitivity is a fundamental reqirement. With the advance of computed image processing and its visualization tools, image guidance systems have been introduced for various radiation oncology

applications. These systems allow us to localize, target, and visualize the 3D anatomy. Using the system, one can calculate the radiation dose and optimize the trajectories for beams. For successful treatment with stereotactic radiation therapy, it is crucial to obtain accurate information about the tumor position and the potential range of movement because a concentration of beams from several different directions delivers a very high dose to the target volume, whereas there is a steep decrease in the dose at the margins of the targeted area. In particular for SRT, the imaging system should have a higher sensitivity and accuracy, while radiotherapy performed using this image guidance system is called image-guided radiotherapy (IGRT).

Definition of Target Volumes with Computed Tomography Images

The International Commission on Radiation Units and Measurements (ICRU) Report 62 refines the gross tumor volume (GTV), clinical target volume (CTV), and planning target volume (PTV) concept by introducing the definition of an internal margin to take into account variations in size, shape, and position of the CTV, and the definition of a setup margin to take into account all uncertainties in patient-beam positioning. Using the ICRU definitions, we can see that the GTV is the gross extent of the malignant growth as determined by palpation or imaging studies. The GTV, together with the surrounding volume of local subclinical involvement, constitutes the primary CTV. The ICRU Report 62 also defines the internal margin as the internal target volume (ITV). The ITV represents the movements of the CTV referenced to the patient coordinate system and is specified in relation to internal and external reference points, which preferably should be specifically related to each other through bony structures.

The outlines of the GTV were delineated on a 3D radiotherapy treatment planning system using lung CT window settings (window width 2000 Hounsfield Units [HU] and window level −700 HU, typically). A physician delineated both the solid area (tumor itself), which could be seen even using mediastinal CT window settings (window width 350 HU and window level 40 HU, typically), and the surrounding obscure area, which could be seen only under lung CT settings. The obscure area is important because it indicates either tumor microscopic invasion or respiratory tumor motion. Spiculation and pleural indentation were included within the CTV. As described earlier, Giraud et al. (2000) reported that the microscopic extension around a visible local tumor with the naked eye was different between adenocarcinoma and squamous cell carcinoma as a result of an analysis of the surgical resection specimens for which the border between the tumor and adjacent lung parenchyma were examined. According to these results, the CTV margin around GTV on

CT images must be increased to 8 mm and 6 mm for adeno-carcinoma and squamous cell carcinoma, respectively, in order to cover 95% of the microscopic extension.

Radiologic-Pathologic Correlation of Stage I Lung Carcinoma

Lung carcinoma is classified as stage I when the tumor does not have lymph node and distant metastasis and chest wall, mediastinal invasion, or extension to 2 cm proximal to the carina. Stage I lung carcinoma is further subdivided according to its maximum diameter: a tumor less than 3 cm in maximum diameter is classified as stage IA, and 3 cm or more in maximum diameter is classified as stage IB. All subtypes of lung carcinoma, including adenocarcinoma, squamous cell carcinoma, large cell carcinoma, and small cell carcinoma, thus appear as stage I disease.

In this section, first we demonstrate the usefulness of thin-section CT in the evaluation of lung carcinoma, and thereafter, we describe the thin-section CT findings of stage I lung carcinoma from the viewpoint of its attenuation, borders characteristics, and growth pattern, correlating with their pathology.

Although recognition of attenuation pattern, borders characteristics, and growth pattern is of limited value in the differentiation of histologic subtypes of lung carcinoma, we may make use of understanding these patterns not only to decide accurate GTV and CTV for SRT treatment planning, but also to distinguish the post-irradiated changes from tumor regrowth.

Usefulness of Thin-section Computed Tomography in the Evaluation of Lung Carcinoma

Although there is no definition of the term *thin-section CT* (a synonym for high-resolution CT), the generally rec-ommended criteria for thin-section CT include the thinnest available collimation, usually 1–1.5 mm, and a high-spatial frequency "sharp" algorithm for viewing the lung (Webb *et al.*, 2001). Targeted reconstruction is optional in evaluat-ing diffuse lung disease, but is strongly recommended in evaluating a pulmonary nodule as it usually appears as uni-lateral disease. A helical scan is also required in thin-section CT for a pulmonary nodule as the nodule must be wholly scanned in one breath hold.

Employing thin-section CT is strongly recommended in evaluating a pulmonary nodule because of the following rea-sons. First, thin-section CT provides more detailed informa-tion about the borders characteristics or attenuation of a nodule, which is useful in the differential diagnosis of the nodule. Second, evaluation of attenuation is also valuable in estimating a malignancy of a known lung carcinoma. It has

been shown that a lung carcinoma of either a pure GGO or GGO predominance on thin-section CT carries a better prognosis than that of a solid attenuation. As the extent of ground-glass opacity (GGO) becomes larger when using thick-section conventional CT, thin-section CT must be used to precisely evaluate the extent of GGO. Third, thin-section CT may allow us to more accurately assess the T-staging of lung carcinoma by its high-resolution images of the tumor and its surrounding structures. Finally, most recent reports regarding the CT findings of peripheral lung carcinoma or other nodular lesions employ thin-section CT as the exami-nation technique. Therefore, thin-section CT is mandatory to make an appropriate radiotherapy treatment plan and to evaluate the clinical outcomes after treatment.

Attenuation of Lung Carcinoma

The attenuation of pulmonary CT images at the lung win-dow setting is practically divided into the following five degrees: solid attenuation, ground-glass opacity (attenua-tion), the attenuation of normal lung parenchyma, the atten-uation of hyperlucent lung, and air density. Lung carcinomas usually appear as a solid, GGO, or solid-GGO mixed nodule, occasionally containing air densities. Recognizing the atten-uation of lung carcinoma is very important because it can affect management of the patient.

Solid Attenuation

Solid attenuation is defined as a high density that obscures the internal vascular structures and is equal to that of the adjacent vascular structure. Although *consolidation* would be an alternative term to express this situation, the authors prefer *solid attenuation* for expressing the attenua-tion of a pulmonary nodule because consolidation inherently indicates alveolar space filling by inflammatory cells or exudates in pneumonia. The solid areas on thin-section CT pathologically correspond to various histologies, including tumor cell nests, areas of fibrosis, lymphocyte aggregation, mucin, or areas of necrosis.

Ground-glass Opacity

Ground-glass opacity (GGO) is defined as mildly increased attenuation of the lung parenchyma, in which vas-cular structures are discernible. Pathologically, GGO corre-sponds to partial filling of the alveoli or interstitial thickening, preserving air within the lesion. If the lung pathology is beyond the spatial resolution of CT, a volume-averaging effect occurs between the pathology and surrounding air, and it appears as a hazy opacity described as GGO. Therefore, the extent of GGO is largely affected by the slice thickness employed; as the slice thickness becomes thicker, the extent of GGO becomes larger as thicker-slice CT tends to be more highly influenced by the volume-averaging effect.

In lung carcinoma, GGO pathologically corresponds to alveolar replacement growth of the tumor, which is usually seen in atypical adenomatous hyperplasia (AAH) and bronchioloalveolar carcinoma (BAC) but is also rarely observed in metastatic lung tumors (Yang et al., 2001). However, the secondary changes of lung carcinoma, such as edema, hemorrhage, poor aeration of the surrounding lung parenchyma, or spillage of mucin, also account for peritumoral GGO. The distinction between alveolar replacement growth and secondary changes are sometimes difficult. However, secondary changes usually have an ill-defined margin of the GGO, whereas alveolar replacement tumors tend to have a well-defined margin.

Lung carcinoma often consists of both GGO and solid components on thin-section CT. In such cases, the solid component corresponds to the noninvasive solid area, including alveolar collapse, cellular infiltrates or mucin within a noninvasive BAC, or areas of invasive carcinoma.

Borders Characteristics

Surface characteristics are very important not only for the differential diagnosis of a pulmonary nodule but also for estimation of the fundamental properties of a lung carcinoma. We may predict the growth pattern, the likelihood of metastasis, or the prognosis of the patient. Giraud examined 70 non-small cell carcinoma surgical resection specimens in which the border between tumor and adjacent lung parenchyma was microscopically studied (Giraud et al., 2000). The mean value of microscopic extension around the local tumor visible with the naked eye was 2.69 mm for adenocarcinoma and 1.48 mm for squamous cell carcinoma ($p = 0.01$). The usual 5 mm margin covers 80% of the ME for adenocarcinoma and 91% for squamous cell carcinoma. Taking into account 95% of the ME, a margin of 8 mm and 6 mm must be chosen for adenocarcinoma and squamous cell carcinoma, respectively. Aerogenous dissemination was the most frequent pattern observed for all groups, followed by lymphatic invasion for adenocarcinoma and interstitial extension for squamous cell carcinoma.

Spicula and Pleural Indentation

Spicula is defined as a linear opacity radiating from the tumor on thin-section CT. The pulmonary vessels involved by the tumor also appear as radiating linear opacities. The two must be strictly distinguished because each finding has a different pathological implication. Pulmonary vessels should have continuity with surrounding vessels, whereas spicula does not continue to the surrounding vessels. Spicula pathologically corresponds to fibrosis, atelectasis,

tumor invasion, or localized lymphangitic extension (Zwirewich et al., 1991). It is notable that spicula may indicate tumor invasion. Surprisingly, there are no significant criteria as to how to deal with spicula in the T-staging of lung carcinoma, although spicula are sometimes long enough to affect T-staging. Because spicula with tumor invasion is indistinguishable from that without tumor invasion, it is recommended that all spicula be regarded as tumor extension for practical purposes.

Pleural indentation, defined as a linear opacity reaching the pleural surface with an associated triangular opacity, indicates invagination of the visceral pleura. Therefore, the linear opacity seen in the pleural indentation consists of two visceral pleurae that face each other. The triangular opacity corresponds to focal small pleural effusion or elevated subparietal fat tissue secondary to pleural indentation. Typical spicula and pleural indentation should be included in the clinical target volume (CTV) when making an SRT plan.

Growth Patterns of Lung Carcinoma

Lung carcinoma can be pathologically or radiologically divided into three subtypes by its growth pattern: namely, expansive growth, invasive and contractive growth, and alveolar replacement growth. More than one of these growth patterns may coexist within a tumor.

Expansive growth indicates the expansion of tumor cells, thereby compressing the surrounding lung parenchyma. A tumor with expansive growth pathologically consists of nests of tumor cells, which often have prominent nuclei. This growth pattern can be seen in all pathological subtypes of lung carcinoma, and it is especially common in poorly differentiated tumors. Minimal ground-glass opacity may occasionally be seen around the tumor. It is more likely to represent secondary changes, such as hemorrhage or edema, rather than an alveolar replacement component in this growth pattern of tumor.

Invasive and contractive growth indicates an invasive spread of the tumor with a desmoplastic reaction. This growth pattern is most commonly seen in adenocarcinoma. This pattern is pathologically characterized by prominent fibrosis in the tumor, occasionally radiating from the tumor. On CT, there are spicula or pleural indentations at the margin of the tumor, thus suggesting contractive changes. Tumors do not always have a convex margin, and instead a linear or concaved margin may be seen. Air-containing spaces, including air bronchograms, are commonly seen. Therefore, a tumor with this type of growth pattern can be misdiagnosed as an organized pneumonia as air bronchograms and linear or concaved margin are also the features of pneumonia.

Alveolar replacement growth, also referred to as lepidic growth, indicates that tumor cells spread along the alveolar walls, thus replacing the alveolar epithelium without destroying the underlying pulmonary architecture. This pattern is only seen in bronchioloalveolar carcinoma (BAC), a subtype of adenocarcinoma, or atypical adenomatous hyperplasia (AAH), whereas expansive growth, and invasive and contractive growth are seen in virtually all types of lung carcinoma, although invasive and contractive growth is most commonly encountered in adenocarcinoma.

Limits of Computed Tomography for Evaluating Lung Tumors

CT is not always able to distinguish the limits between malignant tumors and normal tissue, particularly when atelectasis is present. (18)F-fluorodeoxyglucose positron emission tomography (FDG-PET) imaging provides functional images of the tissue glucose uptake and metabolism. FDG-PET appears to have a greater sensitivity and specificity compared with CT in non-small cell lung cancer (NSCLC). Most studies on the impact of co-registering FDG-PET with CT images obtained in the treatment position on RT volume delineation are concerned with NSCLC (Ishimori et al., 2004). Integrated hybrid PET-CT in the treatment position and co-registered images have an impact on treatment planning. A tumor with atelectasis is thus a single independent factor influencing the definition of GTV after PET. CT and FDG-PET image fusion should therefore become the standard for RT planning in NSCLC (Chapman et al., 2003). However, some difficulties remain unresolved, including the internal mobility of the GTV, unknown differing setup of patients during CT and PET-FDG, and the best correlation between the PET activity peak and tumor volume delineation.

Management of Respiratory Motion of the Target during Irradiation

Stereotactic radiation therapy for lung lesions, unlike stereotactic irradiation for intracranial lesions, entails inevitable problems, such as movement of the target because of respiratory and cardiac motion. In particular, respiration-related movement, which is often greater than 1 cm, is clinically important for lung stereotactic radiation therapy, since the diameter of lung lesions treated with this method is typically less than 4 cm.

To reduce pulmonary complications, SRT for small lung tumors can be safely performed by using a small internal margin that requires respiratory motion control. The respiratory motion control methods fall into four general categories: (1) active or voluntary breath hold, (2) suppressed respiratory motion by abdominal pressure, (3) gating, and (4) tracking or "chasing." How to take and utilize CT images for SRT treatment planning depends on the state of respiration during SRT. For SRT treatment planning, the breath hold, gating, tracking, and chasing techniques need a fast CT scan taken during breath hold, while the free or suppressed breathing technique requires a slow scan or respiration-correlated CT, which will be described later in this chapter.

Simulation Using Slow-scan Computed Tomography for Free or Suppressed Breathing Technique

Abdominal compression, if utilized, is applied to a degree that is tolerable and within the limits of tumor or diaphragm movement. The limitation of tumor and diaphragm movement should be verified by fluoroscopic examinations. CT simulation is performed in this position, and the errors added by the fusion algorithm are quantitated and included in the uncertainty shell produced by the CTV to PTV expansion.

If breath hold, gating, tracking, or chasing techniques are not used, SRT is performed under free or suppressed breathing, which thus requires four-dimensional (4D) radiotherapy treatment planning. Four-dimensional radiotherapy has been defined as the explicit inclusion of temporal changes in the anatomy during the imaging process, dose planning, and delivery of radiotherapy. Such spatial and temporal information on tumor mobility can be derived with a single CT scan procedure such as "slow"-scan CT that is performed at a rate of 4–8 seconds per section acquired (van Sörnsen de Koste et al., 2003) or respiration-correlated 4D CT scans (Mageras et al., 2004). Both techniques permit the generation of individualized internal target volumes (ITVs) to account for internal target motion for treatment planning, but 4D scans have the added advantage of permitting the retrospective selection of phases for gated radiotherapy. The commonly used slow-scan CT would demonstrate a major part of the trajectory of tumor movement due to respiration and other factors. Since free-breathing CT scans are typically acquired at the rate of one or two sections per second, the individual images might thus reflect only a limited part of the tumor motion in the respiratory cycle.

In comparison, a slow-scan CT was performed at the rate of 4–8 seconds per section acquired. The duration of scanning was programmed to be sufficiently long (i.e., longer than that of the respiratory cycle), and the trajectory of tumor movement during CT was reflected as greater or lesser attenuation because of the partial-volume-averaging effect, depending on the period for which the tumor was in the scanning range during each slow-scan CT (Takeda et al., 2005). However, slow-scan CT is suitable only for peripheral lung tumors where the contrast between tumor and normal lung tissue is high. In addition, because the pixel intensity in slow-scan CT images is proportional to the period of time that the tumor is at its extreme positions, tumor borders tend to be indistinct. Therefore, when the tumor margin is delineated on slow-scan CT images for planning of radiation therapy, thin-section CT images also should be carefully observed for reference purposes.

Respiration-correlated CT (RCCT) also can be performed with already available CT equipment and acquisition settings (Mageras *et al.*, 2004). A key feature is to use either small pitch with helical acquisition or repeated image (cine) acquisition at each couch position. RCCT can be used not only to provide three-dimensional information on intrafractional target motion but also to estimate the interfractional target variation before and during treatment. This is done by using the relationship of target and diaphragm motion in the RCCT image sets in combination with population- and patient-specific measurements of diaphragm position variation in radiographs. Underberg *et al.* (2005) revealed that significant changes in tumor volume can occur during fractionated stereotactic radiotherapy for stage I non-small cell lung cancer, which implies that repeat CT planning may be appropriate during 4D radiotherapy, thereby adding to the workload.

Three-dimensional Stereotactic Repositioning of the Isocenter during Irradiation

Precision should be validated with each treatment session by the quality control (QC) process and then be maintained throughout the entire treatment process, both during fractions and for subsequent fractions. The radiation oncologist is responsible for ensuring that the positioning and field placement are accurate for each fraction. This should include a review of the plan and direct inspection of the patient setup. In addition, treatment verification requires orthogonal X-ray to compare the bone anatomy in digitally reconstructed radiographs or by some other method, such as CT scan-based verification. Not surprisingly, a steady stream of enhanced or new methods of imaging are now being introduced to improve treatment guidance and treatment verification. High-quality megavoltage portal imaging can now be acquired with an amorphous silicon flat-panel detector. Also used to provide 3D volumetric patient data are in-room kilovoltage X-ray systems for localizing bony landmarks or implanted markers; optical camera systems that track surface reflectors; "cone beam" reconstruction from the linear accelerator portal image; a real-time tumor tracking system; and even the installation of an in-room CT.

Computed Tomography-Linear Accelerator (Linac) Unit

To improve the reproducibility of the tumor position, Uematsu *et al.* (2000) introduced a unit for radiotherapy using a common treatment couch for CT and linac so that patients do not need to be moved between CT scanning and radiotherapy. To minimize the movements of the couch and the flexures in the top plate produced by the extension toward the gantry as in the conventional scanner, this system was recently designed so that CT scanning can be performed by moving the gantry (Kuriyama *et al.*, 2003). It largely eliminates the daily differences in the target center attributable to tumor migration or setup error. The setup error using the CT-linac unit was thus confirmed to decrease to almost zero.

The center of the CT image matches the isocenter of the linac accelerator when the couch is rotated 180°. The accuracy of matching the linac isocenter with the CT image center is within 0.5 mm (Kuriyama et al., 2003). The quality assurance of the accuracy is evaluated monthly, using CT scans of a phantom center set up for the linac isocenter by rotating 180° around the couch axis.

We herein describe the breath-hold technique for SRT. All patients inhaled oxygen (2 L/min) to maintain the breath hold comfortably and as long as possible. Before irradiation, the patients underwent daily CT scans to obtain images (2 mm thick) in the vicinity of the tumor under self-breath holding. The start of CT scanning was signaled by the patient's switch. The differences in the tumor position on the CT image were calculated in three dimensions: C-C, A-P, and R-L. Next, to correct setup error, the rotational center of the CT gantry was adjusted to align with the planned isocenter of irradiation by moving the couch a distance equal to the calculated difference in tumor position. Another CT scan was performed during self–breath holding to confirm the position of the isocenter. Next, the couch was rotated 180° so that the point of the rotational center of the CT gantry matched the isocenter of the linac. Irradiation was turned on only during the patient's breath hold. Patients determined their breath-holding time and controlled the radiation beam as often as needed until the prescribed monitor units were completed.

Cone Beam Computed Tomography

Recently, there have been significant advances in 3D patient imaging in the treatment room, including use of the megavoltage imager to collect projection images during a single rotation of the gantry. These images are reconstructed using cone beam CT techniques to create a 3D representation of patient anatomy (Fig. 51). Most recently, a kilovoltage X-ray tube and an opposed amorphous silicon flat-panel imager have been mounted on a unit of linear accelerator (Létourneau *et al.*, 2005). As with the megavoltage system, this enables many projection images to be acquired during a single rotation of the gantry (Nakagawa *et al.*, 2000). Image reconstruction from the projection images is achieved using cone beam CT (CBCT) reconstruction techniques. The advantage of using kilovoltage X-rays over megavoltage in this process is the inherent increase in soft tissue contrast due to the high Z-dependence at low energies. This equipment is capable of

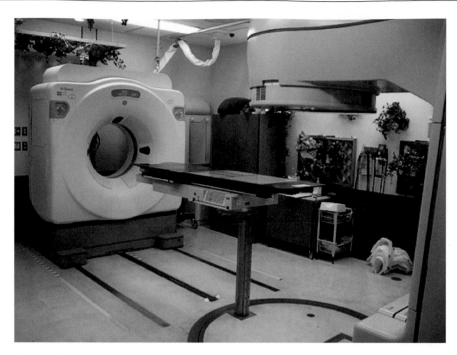

Figure 50 CT on rail-linac unit. A panoramic view of the integrated CT-linac irradiation system (a collaboration of Varian-Mitsubishi, Tokyo, Japan, and Yokogawa GE Medical, Tokyo, Japan). A linac gantry (on the right side of the figure) is placed on the opposite side of the CT gantry (on the left side of the figure). The common treatment couch is placed between these gantries. The treatment couch has two rotation axes for isocentric rotation to make noncoplanar arcs and for rotation between CT and linac. The CT gantry moved along a pair of rails on the floor and scanned the patient without sliding the table. The center of the CT image matched the isocenter of the linac accelerator when the couch was rotated 180°. The accuracy of matching the linac isocenter with the CT image center was within 0.5 mm.

performing kV radiographs, fluoroscopy, and cone beam CT acquisitions. The x-ray volumetric imaging (XVI) system allows online image guidance based on full volumetric anatomic information. Pulsed fluoroscopy can be used, for example, to track intrafraction patient variation and respiratory motion during treatment.

In contrast to other in-room CT guidance methods (tomotherapy and CT-linac unit), with on board XVI technology, the patient remains stationary in the treatment position without couch translation, which is highly desirable to minimize potential patient transfer variations. The difficulty in maintaining skin markings was less problematic because of the ability to easily obtain 3D setup verification before each treatment. The on-board imaging capability reduces the necessity of precise skin–laser positioning by providing the opportunity for image-guided repositioning. XVI acquisition, reconstruction, and patient setup analysis have to be conducted efficiently in order to minimize the probability of internal organ excursions. The capability of using the XVI system to provide real-time projection kV images to safeguard against intrafraction patient variation during treatment is a unique and highly advantageous feature. However, one limitation of the CBCT is that it needs a much longer scan

time to make a tomographic image than conventional CT scanners. Applying CBCT is not always fit to perform SRT, in particular for respiratory moving lung tumors.

Evaluation of the Treatment Effect and Differentiation between Inflammatory Change and a Recurrent Mass

Peculiarity of Radiation Injury of the Lung after Stereotactic Radiotherapy

The clinical and radiographic appearance of radiation-induced pulmonary change by hypofractionated SRT is not always considered to be similar to that induced by conventional radiotherapy because of differences in the total radiation dose, dose per fraction, dose distribution, overall treatment time, and so on. Therefore, a few reports have demonstrated that computed tomographic (CT) images associated with SRT for lung tumors are different from the CT findings after conventional radiotherapy (Kimura *et al.*, 2005). The CT appearances of pulmonary changes after conventional radiotherapy were classified by Libshitz and

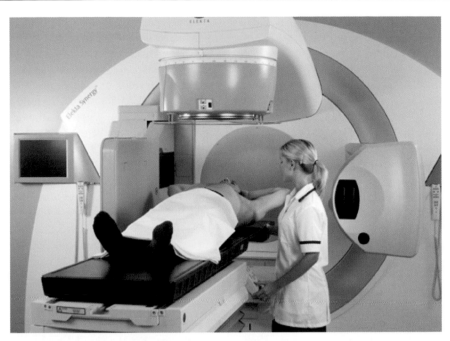

Figure 51 Cone beam CT-linac unit. A panoramic view of the integrated cone beam CT-linac system (Synergy, Elekta, Crawley, UK). This equipment is capable of performing kV radiographs, fluoroscopy, and cone beam CT acquisitions. A kilovoltage X-ray tube and opposed amorphous silicon flat-panel imager have been mounted on a treatment machine gantry. As with the megavoltage system, this enabled many projection images to be acquired during a single rotation of the gantry. Image reconstruction from the projection images is achieved using cone beam CT reconstruction techniques. The flat panel of the machine has a pixel resolution in the isocentric plane of 0.26 mm and a field-of-view (FOV) of 25.6 cm. A typical XVI imaging sequence involves the acquisition of multiple projection-radiographs (e.g., 300–500) approximately equi-angularly spaced over a continuous 360° revolution of the gantry. Projections are acquired and read out at a rate of three frames per second. Including the time for image acquisition (~1 min), an XVI data set of physical size $25.6 \times 25.6 \times 25.6$ cm³ (at a resolution of 1 mm³) can be produced within 3 min.

Shuman (1984) into four patterns as stated earlier, including a homogeneous pattern and a patchy pattern corresponding to the acute exudative phase of radiation-induced injury, a discrete pattern corresponding to the organizing or proliferative phase, and a solid pattern corresponding to the chronic fibrotic phase. High-dose SRT may cause significant large-airway damage by both mucosal injury and ultimate collapse of the airway. Furthermore, in SRT giving high daily doses, there is a clear component of bronchial injury in addition to the more typical pneumonitis seen with CRT. Along with this bronchial injury is the appearance of atelectasis of lung tissue even into areas receiving a very small dose (Timmerman and Lohr, 2004; Uno *et al.*, 2003).

Appearance Time of Radiation Injury of the Lung after Stereotactic Radiotherapy

The CT changes of the lung mostly developed 2–6 months (median 4 months) after SRT, and chronic radiation fibrosis with volume loss appeared 6–15 months (median 11 months) after SRT. Takeda *et al.* (2004) also reported, after SRT, that ground-glass opacities and dense consolidations were observed as initial lung CT findings at 3–4 months; thereafter the ground-grass opacities either disappeared or evolved into dense consolidations. The changes in images of lung injury after SRT appeared a little later than the changes observed in conventional radiotherapy.

Acute Change

Aoki *et al.* (2004) reported 31 cases after SRT (48Gy/4fr) and evaluated the patterns of radiation-induced pulmonary change by extending Libshitz's definitions while considering the dose distribution at treatment-planning CT. The results in acute change (within 6 months) were (a) homogeneously, a slightly increased density: 26%; (b) patchy consolidation: 21%; (c) discrete consolidation: 6%; and (d) solid consolidation: 0%. As a result, the patchy consolidation pattern was predominantly seen as an acute change. Takeda *et al.* (2004) reported 20 patients with 22 lesions after SRT (50Gy/5fr), and hypothesized that early or mild radiation injuries appear as ground-glass opacities (GGO);

GGO appeared on CT in 4 (18%) of 22 lesions at 3–6 months after completion of SRT. Kimura *et al.* (2006) reported 45 patients with 52 primary or metastatic lung cancers after SRT (56Gy/14fr-48Gy/4fr; median 60Gy/8fr), and classified CT appearance of acute radiation pneumonitis into five patterns based on Ikezoe's report (Ikezoe *et al.*, 1998) and their experience, as follows: (1) diffuse consolidation in 20 lesions: 38.5%; (2) patchy consolidation and GGO in 8 lesions: 15.4%; (3) diffuse GGO in 6 lesions: 11.5%; (4) patchy GGO in 1 lesion: 2.0%; and (5) no evidence of increasing density in 17 lesions: 32.6%. Diffuse or patchy consolidation was predominant as the CT appearance of acute radiation pneumonitis like other reports. On the other hand, cases that strangely had no increasing density, or only a slight change even around the tumor, accounted for 17 of 52 lesions (32.6%) at 3–6 months after SRT in this report. One reason for the unexpectedly little pulmonary change is thought to be pulmonary emphysema surrounding the tumor, which lacks normal lung interstitial tissue. Kimura *et al.* (2006) reported that 12 of 17 lesions classified as showing no evidence of increasing density were diagnosed to be pulmonary emphysema.

Pulmonary changes after SRT generally have either no clinical symptoms or only mild ones, probably because the irradiated volume of the lung is very small. However, severe radiation pneumonitis that spreads to a low-dose region and beyond the irradiation field was deemed to be produced in high-grade pulmonary fibrotic cases. Once wide-range radiation pneumonitis occurs, it may become fatal. We also need to be especially careful for patients with pulmonary fibrosis.

Similar to CRT, the first change of radiation injury produced by SRT is a diffuse haze in the irradiated region with an obscuring of the vascular outlines, and then patchy consolidations appear and make relatively sharp edges. These manifestations may gradually clear and completely disappear, thus possibly leading to fibrous change in cases of severe injury. Regarding the dose distribution of SRT, the low-dose region surrounding the high-dose region occurred owing to noncoplanar portals with many directions. It can therefore be said that the typical CT appearance of acute radiation pneumonitis in SRT is a mixture of a diffuse or patchy consolidation in high-dose regions and a diffuse or patchy GGO in low-dose regions. Ipsilateral pleural effusion hardly appears unless the tumor progresses locally.

Chronic Change

The CT images of chronic irradiated change of SRT for lung tumors are closely similar to those of conventional radiotherapy. Aoki *et al.* (2004) reported the CT appearance over 6 months after SRT of 26 cases. The results regarding late changes were: (a) homogeneously, a slightly increased density: 0%; (b) patchy consolidation: 8%; (c) discrete consolidation: 27%; and (d) solid consolidation: 65%. The conclusion was that the solid consolidation pattern, which was regarded as chronic fibrotic change was predominantly seen

as a late change. Takeda *et al.* (2004) also reported several patterns of chronic change due to severe radiation injuries after SRT. Dense consolidations appeared in 16 (73%) of 22 lesions; this consolidation did not disappear completely but persisted as solid or linear opacities. Bronchiectasis was present in 10 (45.5%) of 22 lesions and developed almost contemporaneously with dense consolidations that contained dilated or thickened bronchi. Movement of the opacity was observed in 6 (37.5%) of the 16 densely consolidated lesions.

Koenig *et al.* (2002) described the CT appearance of radiation fibrosis after 3D conformal radiation therapy, to be classified as follows: (1) a modified conventional pattern (consolidation, volume loss, and bronchiectasis similar to, but less extensive than, conventional radiation fibrosis); (2) a masslike pattern (focal consolidation limited around the tumor); or (3) a scarlike pattern (linear opacity in the region of the tumor associated with volume loss, causing mediastinal or diaphragmatic shift). Kimura *et al.* (2006) classified the CT appearance in their cases (45 patients with 52 primary or metastatic lung cancers) by Koenig's classification, as follows: (1) a modified conventional pattern in 32 lesions: 61.5%; (2) a masslike pattern in 10 lesions: 22.7%; and (3) a scarlike pattern in 9 lesions: 17.3%. The masslike pattern presents an image just like a tumor, and it is difficult to differentiate this pattern from local recurrence. It is an essential problem to determine whether or not viable cells are present in the fibrotic scar after SRT although biopsies are recommended, positron emission tomography (PET) may help to evaluate the tumor response and detect tumor recurrence. When a tumor located near the hilar bronchus is treated with SRT, high-dose SRT may cause significant bronchial injury. Chronic bronchitis or necrosis of bronchial cartilage may thus occur, which may lead to peripheral atelectasis due to bronchial stenosis (Timmerman and Lohr, 2004, Uno *et al.*, 2003).

Summary of Computed Tomography Findings of Radiation Injury of the Lung after Stereotactic Radiotherapy

1. At about 1 month after SRT, radiologic changes in normal lung tissue will not occur in many cases.

2. At about 3–6 months after SRT, diffuse or patchy consolidation in the high-dose region and diffuse or patchy GGO in the low-dose region arise as acute radiation pneumonitis. Exceptionally, these changes may hardly appear around a tumor due to pulmonary emphysema, the irradiation method (static or arc), and so on.

3. At about 6–9 months after SRT, diffuse or patchy consolidation accepted by acute radiation pneumonitis converges to solid or dense consolidation. These opacities, which usually move toward the mediastinum or hilum with shrinkage, were detected simultaneously. Cases in which

diffuse or patchy consolidation did not occur at an acute phase tended to change to a scarlike pattern at a late phase.

4. At about 1–2 years after SRT, the opacities will be stable, and if there is no recurrence, then the opacities will not change.

5. When a tumor located near the hilar bronchus is treated with SRT, chronic bronchitis or necrosis of bronchial cartilage accompanied by peripheral atelectasis may thus be produced.

Computed Tomography Evaluation of the Tumor Response and Progression

Tumor Response

Local tumor response is generally evaluated using the Response Evaluation Criteria in Solid Tumors criteria (RECIST). All cases whose tumors completely disappeared, decreased in diameter by 30% or more, and then decreased in diameter by less than 30% after radiotherapy were classified as complete response (CR), partial response (PR), and stable disease (SD), respectively. Reported as the CR and PR rate of stage I non-small cell lung carcinoma distributed 16–23% and 61–84% of total cases, respectively (Onishi et al., 2004; Nagata et al., 2005; McGarry et al., 2005). However, it is not so important to determine the tumor response according to the size because radiation pneumonitis exists after SRT on and near the high-dose irradiated area in more than two-thirds of the cases. The progression-free rate is therefore more meaningful than the response rate.

Local Recurrence

On follow-up CT examinations after SRT, it is important to precisely evaluate radiation-induced lung injury and to detect locoreginal recurrence or distant metastasis as early as possible to perform the appropriate second salvage treatment. Although the imaging findings of radiation-induced lung injury and local recurrence after conventional radiotherapy have been extensively investigated, the results of these reports could not be simply applied to SRT for the following reasons. First, SRT employs much more complex techniques, irradiating the target volume from various directions, than conventional radiotherapy. Therefore, the conformity of the opacity to the radiation portal, a most remarkable feature of radiation pneumonitis, is not identified in SRT. Second, SRT adopts a much higher biologically effective dose, the equivalent of more than 100 Gy in conventional radiotherapy, which may result in unexpected lung parenchymal change.

The detection of local recurrence is more difficult in SRT because radiation-induced lung parenchymal change is fre-

quently coexistent and does not always have typical manifestations, such as a straight border conforming to the radiation portal, as seen in that of conventional radiotherapy. In addition, radiation-induced change after SRT may assume a masslike opacity on CT after SRT, thus mimicking a recurrent tumor. There has been no report regarding CT manifestations of local recurrence after SRT. A local recurrence after SRT had relatively unique appearances on thin-section CT. The typical manifestation of local recurrence after SRT is a masslike lesion with convex borders and without air bronchograms, which continues to grow even 12 months after completion of the SRT. Ipsilateral pleural thickening with or without effusion may be observed. It should be noted that transient pleural effusion is occasionally seen in cases without local recurrence during the follow-up periods.

After SRT, the opacity including inherent tumor increases in size during the first 6 months, probably representing radiation pneumonitis. It can appear as a consolidation or mass, usually with air bronchograms. In cases without local recurrence, these changes usually subside after 6 months and then become stable as a form of consolidation or linear opacity. The borders of the mass are usually concave or linear shapes. In cases with local recurrence, however, the opacity continues to increase in size even after 6 months or to regrow after 12 months as a mass form with convex margins. In a Phase II study for T1N0M0 non-small cell lung cancer under Japan Clinical Oncology Group clinical trial number 0403, if the shadow of radiation pneumonitis increases continuously for more than 6 months on CT, it will be judged to be the local recurrence.

One may think that contrast enhancement is useful to detect local recurrence. However, the contrast-enhancement pattern is similar in both radiation-induced changes and local recurrence; they are inhomogeneously enhanced, occasionally with internal vascular enhancement (i.e., angiogram signs). Contrast enhancement is not considered to be helpful in distinguishing between radiation-induced changes and local recurrence, and so it can be omitted for purposes of detecting local recurrence or evaluating radiation-induced changes. It may be useful, however, for detecting an enlarged hilar or mediastinal lymph nodes or distant metastasis.

Usually, the shadow of radiation pneumonitis will be stable at about 1–2 years after SRT, and if there is no recurrence, then the shadow will not change. But especially in a masslike pattern, it is often difficult to make a differential diagnosis of local recurrence. Traditional CT scan-based tumor response criteria are not easily applied. Although the response evaluation by PET will attract attention from now on, it may become false-positive when radiation pneumonitis exists (Strauss et al., 1996). Residual fibrosis in a configuration that recapitulates the region of lung that received approximately 20 Gy or more can develop, and faint residual metabolic activity can sometimes be observed in this volume for many months on a PET scan. However, when 3 months or more pass after SRT, there is also a report that

PET clearly surpasses CT for clearly evaluating a residual tumor or local recurrence (Inoue *et al.*, 1995), and it is certain that PET is an effective method for the response evaluation. It is necessary to accumulate more cases with a longer follow-up for an assessment of PET.

Cases of Computed Tomography Findings after Stereotactic Radiotherapy

Representative examples of the clinical course on CT after radiotherapy are shown in Figure 52. As an acute change, a diffuse or patchy consolidation pattern with or without a contraction due to some loss of lung volume was predominantly seen (Fig. 52a, b). In some cases, no definite increase in density appeared (Fig. 52c, d). As a late change, a scarlike solid consolidation pattern was predominantly seen (Fig. 52e, f). Regarding the shape, a consolidation without recurrence has concave or linear borders, whereas a recurrent tumor has convex borders (Fig. 52g, h).

Guidelines for Quality Control of Computed Tomography Images

As has been described, imaging modalities, in particular CT, play a very important role in the treatment process of

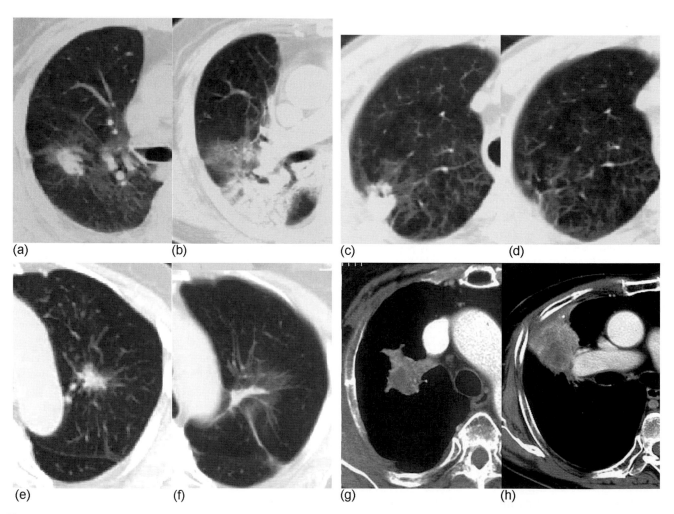

Figure 52 Representative CT findings of cases treated with SRT. (a, b): Acute phase: a pattern of patchy consolidation with peripheral GGO. A 67-year-old woman with T1N0 adenocarcinoma, 60Gy/10 fractions (a) before SBRT (b) 3 months after SRT. (A patchy consolidation with peripheral GGO. The consolidation decreased in size after 6 months and changed to a scarlike pattern.). (c, d) Acute phase: a pattern of no increasing in density. A 57-year-old man with stage IA adenocarcinoma, 60Gy/8 fractions (c) before SRT (d) 2 months after SRT. (e, f) Chronic phase: a scarlike pattern (a linear opacity in the region of the tumor associated with volume loss). An 82-year-old woman with stage IA adenocarcinoma, 60Gy/10 fractions (e) before SRT (f) 33 months after SRT (the border is concave shape). (g, h): Histologically confirmed local tumor recurrence (an inhomogeneous density mass and the border is convex shape). A 78-year-old man with stage IB squamous cell carcinoma. (g) before SRT (h) 16 months after SRT.

SRT for lung carcinoma. Therefore, general consideration should be given to the following issues concerning the quality control of images (Potters *et al.*, 2004). The targeting of lesions for SRT planning may include general radiography, CT, MRI, MRS, PET (with or without image fusion), or any other imaging studies useful in localizing the target volumes. The digital images employed for SRT must be thoroughly investigated and then corrected for any significant spatial distortions that may arise from the imaging chain. Computed tomography is the most useful, nonspatially distorted, and practical imaging modality for SRT. This modality permits the creation of the 3D anatomic patient model that is used in the treatment-planning process. Some CT considerations are as follows: partial volume averaging, pixel size, slice thickness, distance between slices, and image reformatting for the treatment planning. In some cases, target tissue and normal tissue structures may be better visualized by MRI. The considerations enumerated for CT also apply to the use of MRI. Additional caution is warranted when using MRI because of magnetic susceptibility artifacts and image distortion. As such, the use of MRI must be verified with CT images. Techniques such as combining MRI with CT images by image fusion can be used to minimize the geometrical distortions inherent in MR images.

Future Direction

On the one hand, CT technology has a potential to greatly expand high-precision radiotherapy including SRT. Topics of application of CT for radiotherapy are the development of cone beam CT quality, megavoltage CT (MVCT), and helical tomotherapy. On the other hand, CT has limitations regarding the evaluation and estimation of malignancy and the viability of tumors. As a result, CT technology is expected to progress by being combined with other imaging modalities such as PET and MRI.

Image Quality of Cone Beam Computed Tomography

The image quality of cone beam CT has not yet been satisfactorily compared to conventional CT. However, ongoing and future work using this technology aims to improve and optimize image quality and streamline the reconstruction, registration, and error-correction process so that imaging may be performed online to improve the accuracy with which treatment is delivered. The introduction of this technology can pave the way for the integration of image-guided radiotherapy into routine clinical practice, thus ensuring that complex radiation treatment can be delivered safely and precisely, ultimately leading to an improved tumor control and a reduced treatment-related toxicity.

Megavoltage Computed Tomography

An application of megavoltage CT to radiotherapy has been put to practical use (Nakagawa *et al.*, 2000). Megavoltage CT (MVCT) is performed with a treatment X-ray beam by use of a detector array that is mounted on a linear accelerator. A scan performed as half-rotational irradiation takes approximately 30 sec. The detector is small and is easily removed immediately after MVCT without disturbing radiation therapy. Megavoltage transmission data, collected by the detector array, are used for image reconstruction, thus providing patient setup verification. These data can also provide patient delivery verification and dose reconstruction, with dose reconstruction mapping the actual dose delivered. Slice-by-slice registration is possible with the alignment of the planning CT and the daily MVCT. Routine 3D setup verification is possible with the capability to obtain a daily CT with the patient in treatment position and to align with the planning CT, thus resulting in a reduction of setup errors and an improvement in the therapeutic ratio. During treatment, the beam transmitted through the patient is detected by the CT detector unit and is superimposed onto the MVCT image.

Therapists monitor the beam on a real-time computer display to verify accurate direction of the beam at the tumor. The relatively low-dose (a baseline of 3 cGy) MVCT images are of suitable quality for 3D setup verification. Verification of treatment delivery accuracy and precise, repeatable patient setup are inevitable for SRT, and they can be accomplished with MVCT. Recent advancements have further reduced the detector noise while also improving the system control to provide additional options for fractional-cGy MVCT imaging. These on-board imaging and verification capabilities will facilitate SRT, thus allowing for accurate dose delivery to the tumor and conformal avoidance of critical structures.

Helical Tomotherapy

Helical tomotherapy is an innovative means of delivering image-guided radiotherapy (IGRT) using a device that combines the features of a linear accelerator and a helical CT scanner. Some machines can generate CT images from the same megavoltage X-ray beam it uses for treatment. These MVCT images allow for the verification of the patient position prior to and potentially during radiation therapy (Forrest *et al.*, 2004). Since the unit uses the actual treatment beam as the X-ray source for image acquisition, no surrogate telemetry systems are required to register image space to treatment space. The disadvantage of using the treatment beam for imaging, however, is that the physics of radiation interactions in the megavoltage energy range may force compromises between the dose delivered and the image quality in comparison to diagnostic CT scanners. The performance of the system is therefore characterized in terms of objective measures of noise, uniformity, contrast, and spatial resolution as a function of the dose delivered by the MVCT beam. The uniformity and spatial resolutions of MVCT images are closely comparable to those of diagnostic CT images. Furthermore, the MVCT scan contrast is linear with respect to the electron density of material imaged.

MVCT images do not have the same performance characteristics as state-of-the art diagnostic CT scanners when one objectively examines noise and low-contrast resolution. These inferior results may be explained, at least partially, by the low doses delivered by our unit; the dose is 1.1 cGy in a 20 cm diameter cylindrical phantom. In spite of the poorer low-contrast resolution, these relatively low-dose MVCT scans provide sufficient contrast to delineate many soft tissue structures. Hence, these images are not only useful for verifying the patient's position at the time of therapy, but they are also sufficient for delineating many anatomic structures. In conjunction with the ability to recalculate radiotherapy doses on these images, this enables the dose guidance as well as the image guidance of radiotherapy treatments.

Imaging Supplement for Computed Tomography for Evaluating Tumor Malignancy and Extension

Computed tomography has traditionally been the standard radiographic modality for diagnosing and monitoring non-small cell lung cancer (NSCLC) after treatment. Given the limitations of CT, the utility of (18) F-fluorodeoxyglucose positron emission tomography (FDG-PET) has been investigated for the management of NSCLC, with promising findings. Its adjunctive role with CT in diagnosing and staging disease has been well established. FDG-PET also has been found to be a valuable tool for radiation treatment planning because it improves the precision of lesion definition. PET can be used to define the gross tumor more clearly (e.g., differentiation from atelectatic lung), or to determine whether sites of potential nodal involvement (that are not radiographically suspicious) should be treated at a lower dose or not treated at all. This is most often observed when atelectatic lung is present. PET imaging allowed the treatment fields to be reduced in 47% of patients with atelectatic lung in comparison to CT-based treatment planning alone in a retrospective review of 17 patients (Nestle et al., 1999). Other reports have found that PET changed treatment volume and/or fields in approximately half of the patients (Vanuytsel et al., 2000). The target volume was reduced slightly more frequently (in 24–93% of patients), but the volumes were also frequently enlarged (in 6–76% of patients), with no clear or consistent pattern among the patients. In addition, the PET data acquisition and utilization are often not obtained in the treatment position. They can thus be misleading for precise

tumor targeting, particularly when considering respiratory motion (Nehmeh et al., 2003). As a result, PET data have been primarily used clinically as a complementary imaging modality for tumor targeting and radiotherapy treatment planning.

More recently, the value of PET for determining the clinical response both during and after treatment has been explored. This review highlights the various applications of FDG-PET in the diagnosis and management of NSCLC as corroborated by clinical data, while also considering future directions. This is particularly intriguing given the data suggesting better outcomes with more aggressive radiotherapy, involving higher doses, 3D treatment planning, and also a reduced field size, such as with SRT. However, it remains to be demonstrated how much the addition of PET to radiotherapy treatment planning will affect patient outcome (i.e., local control, reduced toxicity, and survival).

In the future, further technologic improvements in PET scanners will make images easier to interpret. Although it appears to still be far in the future, new markers may offer a unique tool to further the understanding of how cancer cells grow and spread, thereby enabling us to develop more effective treatments.

In conclusion, SRT is an emerging and promising radical radiotherapeutic treatment modality for early-stage non-small cell lung cancer. Undoubtedly, much remains to be learned regarding optimal dose-fractionation schedules, side effects, patient selection, and treatment-delivery methods. Several institutions currently are developing clinical protocols, which, it is hoped, will add to our current knowledge and eventually allow proper comparisons to be made with standard conformal radiotherapy methods as well as surgical resection. CT imaging is becoming increasingly meaningful for the high-precision irradiation process of SRT for lung cancer. The advantages of CT imaging as it is utilized for SRT for NSCLC are as follows:

1. Earlier detection, more accurate diagnosis, and staging.
2. Definite treatment planning.
3. Precise three-dimensional stereotactic repositioning of the isocenter to the planned position.
4. Correct evaluation of the treatment effect after SRT, including the differentiation between inflammatory change and recurrent mass.

Furthermore, advancing CT technology, such as cone beam CT or MVCT, will be applied to the SRT treatment process. At the same time, there are limitations for CT according to its morphology regarding the evaluation and estimation of malignancy and the viability of tumors, whereby other imaging modalities like PET fusion are necessary for further precise target delineation and a correct evaluation of the effect from now on.

References

Aoki, T., Nagata, Y., Negoro, Y., Takayama, K., Mizowaki, T., Kokubo, M., Oya, N., Mitsumori, M., and Hiraoka, M. 2004. Evaluation of lung injury after three-dimensional conformal stereotactic radiation therapy for solitary lung tumors: CT appearance. *Radiology* 230:101–108.

Chapman, J.D., Bradley, J.D., Eary, J.F., Haubner, R., Larson, S.M., Michalski, J.M., Okunieff, P.G., Strauss, H.W., Ung, Y.C., and Welch, M.J. 2003. Molecular (functional) imaging for radiotherapy applications An RTOG symposium. *Int. J. Radiat. Oncol. Biol. Phys.* 55:294–301.

Forrest, L.J., Mackie, T.R, Ruchala, K., Turek, M., Kapatoes, J., Jaradat, H., Hui, S., Balog, J., Vail, D.M., and Mehta, M.P. 2004. The utility of megavoltage computed tomography images from a helical tomotherapy system for setup verification purposes. *Int. J. Radiat. Oncol. Biol. Phys.* 60:1639–1644.

Giraud, P., Antoine, M., Larrouy, A., Milleron, B., Callard, P., De Rycke, Y., Carette, M.F., Rosenwald, J.C., Cosset, J.M., Housset, M., and Touboul, E. 2000. Evaluation of microscopic tumor extension in non-small cell lung cancer for three-dimensional conformal radiotherapy planning. *Int. J. Radiat. Oncol. Biol. Phys.* 48:1015–1024.

Harpole, D.H., Jr., Herndon, J.E., Jr., Young, W.G., Wolfe, W.G., Jr., and Sabiston, D.C., Jr. 1995. Stage I nonsmall cell lung carcinoma: a multivariate analysis of treatment methods and patterns of recurrence. *Cancer* 76:787–796.

Ikezoe, J., Takashima, S., Morimoto, S., Kadowaki, K., Takeuchi, N., Yamamota, T., Nakanashi, K., Isaza, M., Arisawa, J., Ikeda, H., Masaki, N., and Kozuka, T. 1998. CT appearance of acute radiation-induced injury in the lung. *Am. J. Roentgenol.* 150: 765–770.

Inoue, T., Kim, E.E., Komaki, R., Wang, F.C., Bassa, P., Wong, W.H., Yang, D.J., Endo, K., and Podoloff, D.A. 1995. Detecting recurrence or residual lung cancer with FDG-PET. *J. Nucl. Med.* 36:788–793.

Ishimori, T., Saga, T., Nagata, Y., Nakamoto, Y., Higashi, T., Mamede, M., Mukai, T., Negoro, Y., Aoki, T., Hiraoka, M., and Konishi, J. 2004. 18F-FDG and 11C-methionine PET for evaluation of treatment response of lung cancer after stereotactic radiotherapy. *Ann. Nucl. Med.* 18:669–674.

Kimura, T., Matsuura, K., Murakami, Y., Hashimoto, Y., Kenjo, M., Kaneyasu, Y., Wadasaki, K., Hirokawa, Y., Ito, K., and Okawa, M. 2006. CT appearance of radiation injury of the lung and clinical symptoms after stereotactic body radiation therapy (SBRT) for lung cancers: are patients with pulmonary emphysema also candidates for SBRT for lung cancers? *Int. J. Radiat. Oncol. Biol. Phys.* 66:483–491.

Koenig, T.R., Munden, R.F., Erasmus, J.J., Sabloff, B.S., Gladish, G.W., Komaki, R., and Stevens, C.W. 2002. Radiation injury of the lung after three-dimensional conformal radiation therapy. *Am. J. Roentgenol.* 178:1383–1388.

Kuriyama, K., Onishi, H., Sano, N., Komiyama, T., Aikawa, Y., Tateda, Y., Araki, T., and Uematsu, M. 2003. A new irradiation unit constructed of self-moving gantry-CT and linac. *Int. J. Radiat. Oncol. Biol. Phys.* 55:428–435.

Lax, I., Blomgren, H., Naslund, I., and Svanstrom R. 1994. Stereotactic radiotherapy of malignancies in the abdomen: methodological aspects. *Acta. Oncologica.* 33:677–683.

Létourneau, D., Wong, J.W., Oldham, M., Gulam, M., Watt, L., Jaffray, D.A., Siewerdsen, J.H., and Martinez, A.A. 2005 Cone-beam-CT guided radiation therapy: technical implementation *Radiother. Oncol.* 75:279–286.

Libshitz, H.I., and Shuman, L.S. 1984. Radiation-induced pulmonary change: CT findings. *J. Comput. Assist. Tomogr.* 8:15–19.

Mageras, G.S., Pevsner, A., Yorke, E.D., Rosenzweig, K.E., Ford, E.C., Hertanto, A., Larson, S.M., Lovelock, D.M., Erdi, Y.E., Nehmeh, S.A., Humm, J.L., and Ling, C.C. 2004. Measurement of lung tumor motion using respiration-correlated CT. *Int. J. Radiat. Oncol. Biol. Phys.* 60: 933–941.

McGarry, R.C., Papiez, L., Williams, M., Whitford, T., and Timmerman, R.D. 2005. Stereotactic body radiation therapy of early-stage non-small cell lung carcinoma: phase I study. *Int. J. Radiat. Oncol. Biol. Phys.* 63:1010–1015.

Nagata, Y., Takayama, K., Matsuo, Y., Norihisa, Y., Mizowaki, T., Sakamoto, T., Sakamoto, M., Mitsumori, M., Shibuya, K., Araki, N., Yano, S., and Hiraoka, M. 2005. Clinical outcomes of a phase I/II study of 48 Gy of stereotactic body radiotherapy in 4 fractions for primary lung cancer using a stereotactic body frame. *Int. J. Radiat. Oncol. Biol. Phys.* 63:1427–1431.

Nakagawa, K., Aoki, Y., Tago, M., Terahara, A., and Ohtomo, K. 2000. Megavoltage CT-assisted stereotactic radiosurgery for thoracic tumors: original research in the treatment of thoracic neoplasms. *Int. J. Radiat. Oncol. Biol. Phys.* 48:449–457.

Nehmeh, S.A., Erdi, Y.E., Rosenzweig, K.E., Schoder, H., Larson, S.M., Squire, O.D., and Humm, J.L. 2003. Reduction of respiratory motion artifacts in PET imaging of lung cancer by respiratory correlated dynamic PET: methodology and comparison with respiratory gated PET. *J. Nucl. Med.* 44:1644–1648.

Nestle, U., Walter, K., Schmidt, S., Licht, N., Nieder, C., Motaref, B., Hellwig, D., Niewald, M., Ukena, D., Kirsch, C.M., Sybrecht, G.W., and Schnabel, K. 1999. 18F-deoxyglucose positron emission tomography (FDG-PET) for the planning of radiotherapy in lung cancer: high impact in patients with atelectasis. *Int. J. Radiat. Oncol. Biol. Phys.* 44:593–597.

Onishi, H., Araki, T., Shirato, H., Nagata, Y., Hiraoka, M., Gomi, K., Yamashita, T., Niibe, Y., Karasawa, K., Hayakawa, K., Takai, Y., Kimura, T., Hirokawa, Y., Takeda, A., Ouchi, A., Hareyama, M., Kokubo, M., Hara, R., Itami, J., and Yamada, K. 2004. Stereotactic hypofractionated high-dose irradiation for stage I nonsmall cell lung carcinoma: clinical outcomes in 245 subjects in a Japanese multi-institutional study. *Cancer.* 101:1623–1631.

Onishi, H., Kuriyama, K., Komiyama, T., Tanaka, S., Sano, N., Aikawa, Y., Tateda, Y., Araki, T., Ikenaga, S., and Uematsu, M. 2003. A new irradiation system for lung cancer combining linear accelerator, computed tomography, patient self-breath-holding, and patient-directed beam-control without respiratory monitoring devices. *Int. J. Radiat. Oncol. Biol. Phys.* 56:14–20.

Potters, L., Steinberg, M., Rose, C., Timmerman, R., Ryu, S., Hevezi, JM., Welsh, J., Mehta, M., Larson, D.A., Janjan, N.A., American Society for Therapeutic Radiology and Oncology, and American College of Radiology. 2004. American Society for Therapeutic Radiology and Oncology and American College of Radiology practice guideline for the performance of stereotactic body radiation therapy. *Int. J. Radiat. Oncol. Biol. Phys.* 60:1026–1032.

Smith, W.R. 2003. American College of Chest Physicians: treatment of stage I non-small cell lung carcinoma. *Chest.* 123:181–187.

Strauss, L.G., 1996. Fluorine-18 deoxyglucose and false-positive results: a major problem in the diagnostic of oncological patients. *Eur. J. Nucl. Med.* 23:1409–1415.

Takeda, A., Kunieda, E., Shigematsu, N., Hossain, D.M., Kawase, T., Ohashi, T., Fukada, J., Kawaguchi, O., Uematsu, M., Takeda, T., Takemasa, K., Takahashi, T., and Kubo, A. 2005. Small lung tumors: long-scan-time CT for planning of hypofractionated stereotactic radiation therapy—initial findings. *Radiology* 237:295–300.

Takeda, T., Takeda, A., Kunieda, E., Ishizaka, A., Takemasa, K., Shimada, K., Yamamoto, S., Shigematsu, N., Kawaguchi, O., Fukuda, J., Ohashi, T., Kuribayashi, S., and Kubo, A. 2004. Radiation injury after hypofractionated stereotactic radiotherapy for peripheral small lung tumors: serial changes on CT. *Am. J. Roentgenol.* 182:1123–1128.

Timmerman, R.D., and Lohr, F. 2004. Normal tissue dose constraints applied in lung stereotactic body radiation therapy. In: Kavanagh, B.D., and Timmerman, R.D. (Eds.), *Stereotactic Body Radiation Therapy.* Philadelphia: Lippincott, Williams & Wilkins, pp. 141–160.

Uematsu, M., Shioda, A., Suda, A., Fukui, T., Ozeki, Y., Hama, Y., Wong, J.R., Shoichi, A., and Kusano, S. 2001. Computed tomography-guided frameless stereotactic radiotherapy for stage I non-small cell lung cancer: a 5-year experience. *Int. J. Radiat. Oncol. Biol. Phys. 51*:666–670.

Uematsu, M., Shioda, A., Suda, A., Tahara, K., Kojima, T., Hama, Y., Kono, M., Wong, J.R., Fukui, T., and Kusano, S. 2000. Intrafractional tumor position stability during computed tomography (CT)-guided frameless stereotactic radiation therapy for lung or liver cancers with a fusion of CT and linear accelerator (FOCAL) unit. *Int. J. Radiat. Oncol. Biol. Phys. 48*:443–448.

Underberg, R.W.M., Lagerwaard, F.J., Slotman, B.J., Cuijpers, J.P., and Senan, S. 2005. Use of maximum intensity projections (MIP) for target volume generation in 4DCT scans for lung cancer. *Int. J. Radiat. Oncol. Biol. Phys. 60*:933–941.

Uno, T., Arugam T., Isobe, K., Motori, K., Kawakami, H., Uenom N., and Ito, H. 2003. Radiation bronchitis in lung cancer patient treated with stereotactic radiation therapy. *Radiat. Med. 21*:228–231.

van Sörnsen de Koste J.R., Lagerwaard, F.J., de Boer, H.C., Margriet, F.J., Nijssen-Visser, R.J., and Senan, S. 2003. Are multiple CT scans required for planning curative radiotherapy in lung tumors of the lower lobe? *Int. J. Radiat. Oncol. Biol. Phys. 55*:1394–1399.

Vanuytsel, L.J., Vansteenkiste, J.F., Stroobants, S.G., De Leyn, P.R., De Wever, W., Verbeken, E.K., Gatti, G.G., Huyskens, D.P., and Kutcher, G.J. 2000. The impact of (18)F-fluoro-2-deoxy-D-glucose positron emission tomography (FDG-PET) lymph node staging on the radiation treatment volumes in patients with non-small cell lung cancer. *Radiother. Oncol. 55*:317–324.

Webb, W.R., Müller, N.L., and Naidich, D.P. 2001. Technical aspects of high-resolution computed tomography. In: *High-resolution CT of the Lung*, 3rd ed. Philadelphia: Lippincott, Williams & Wilkins, pp. 1–47.

Yang, Z.G., Sone, S., Takashima, S., Li, F., Honda, T., Maruyama, Y., Hasegawa, M., and Kawakami, S. 2001. High-resolution CT analysis of small peripheral lung adenocarcinomas revealed on screening helical CT. *Am. J. Roentgenol. 176*:1399–1407.

Zwirewich, C.V., Vedal, S., Miller, R.R., and Muller, N.L. 1991. Solitary pulmonary nodule: high-resolution CT and radiologic-pathologic correlation. *Radiology 179*: 469–476.

9

Thin-section Computed Tomography Correlates with Clinical Outcome in Patients with Mucin-producing Adenocarcinoma of the Lung

Ukihide Tateishi, Testuo Maeda, and Yasuaki Arai

Introduction

Mucin-producing adenocarcinoma of the lung is a disorder that includes a histological category of mucinous bronchioloalveolar carcinoma and adenocarcinoma mixed subtypes (Brambilla *et al.*, 2001). Mucin-producing adenocarcinoma of the lung is composed of bronchial goblet-cell-like tall columnar cells with cytoplasmic mucin and has distinct features including genetic mutation and antigenic expression by histological examination (Tsuchiya *et al.*, 1995). The computed tomography findings of mucin-producing adenocarcinoma of the lung have been described; however, most published data refer to computed tomography (CT) findings resulting from mucinous bronchioloalveolar carcinoma (BAC), although two studies have included patients with mucin-producing adenocarcinoma of the lung in their subject groups (Dumont *et al.*, 1998; Liu *et al.*,

2000). Characteristic CT findings of mucin-producing adenocarcinoma of the lung included multiple cysts (Weisbrod *et al.*, 1992), cavitation or bubblelike lucencies (Kuhlman *et al.*, 1988), air bronchogram, interlobular bulging fissure, CT angiogram sign, and uniform low attenuation of the pulmonary consolidation (Akira *et al.*, 1999).

The development of valid outcome surrogates for mucin-producing adenocarcinoma of the lung has been recognized as an important need in lung cancer research (Barkley and Green, 1996). An essential component of the validation for prediction of patient outcome is the demonstration that thin-section CT findings correlate with true outcomes (Regnard *et al.*, 1998; Tateishi *et al.*, 2005). However, to our knowledge, there have been no published findings of large series with mucin-producing adenocarcinoma of the lung in patients who underwent surgical resection or adjuvant chemotherapy. Prognostic implications of CT findings have not been evaluated. This study was conducted to evaluate the correlation between mucin-producing adenocarcinoma of the lung as assessed by thin-section CT and a true outcome measure of morbidity. Thin-section CT was also compared with patient characteristics in terms of how well each correlated with morbidity.

Materials and Methods

Retrospective review of the pathologic records for the period between January 1996 and January 2004 identified 82 consecutive patients with mucin-producing adenocarcinoma of the lung referred to our institution for treatment. Entry criteria for this study included a diagnosis of BAC or adenocarcinoma, mixed subtype by surgical or biopsy specimen, and complete follow-up. The study population consisted of 40 men and 42 women with a mean age of 71 years (range 45–85 years). Seventy of these 82 patients underwent surgical resection, and 12 received adjuvant chemotherapy. Treatment consisted of adjuvant chemotherapy and surgical resection, which comprised wedge resection, lobectomy, or pneumonectomy in all patients. Complete resection or sampling of mediastinal or hilar lymph nodes was performed in 58 patients (71%). These included high and low ipsilateral paratracheal, subcarinal, and inferior pulmonary ligament lymph nodes and any other suspicious lymph nodes identified at surgery. The tumors were classified according to the international tumor node metastases (TNM) classification for staging lung cancer (Lababede *et al.*, 1999). Of these, 29 tumors (35%) were classified as stage Ia, 12 (15%) as stage Ib, 8 (10%) as stage IIa, 21 (26%) as stage IIIb, and 12 (15%) as stage IV. Clinical records were available for review in all patients. Clinical histories of cases to identify any underlying medical conditions, the presence or absence of symptoms, and cigarette consumption were also reviewed. The study conformed to the declaration of Helsinki, and informed written consent was obtained from each subject.

CT was performed on helical or multidetector scanners (X-Vigor, or Aquilion V-Detector, Toshiba Medical Systems, Tokyo, Japan). Helical technique in 55 patients consisted of 10 mm collimation for individual scans of the entire lung. A series of 10 mm thick images were obtained with the following scan parameters: 120 kVp, 50–100 mA/rotation, 30–40 cm of field of view (FOV), 512 × 512 matrix, and reconstruction using a standard algorithm. Additional thin-section CT images were obtained in all 55 patients using 2 mm collimation, a 20 cm FOV, 120 kVp and 200 ma per rotation, 1 sec-gantry rotation, and a high-spatial-frequency reconstruction algorithm. The remaining 27 patients were evaluated on a multidetector CT scanner using axial 2 mm × 4 modes (four images per gantry rotation), 120 kVp, 200 mA, and 0.5 sec scanning time. Thin-section CT images in these 27 patients were obtained using

2 mm sections reconstructed at 2 mm intervals using a high-spatial-frequency algorithm and retrospectively retargeted to each lung with 20 cm FOV. All patients received iodinated nonionic contrast material intravenously, and the scan delay was set at 40 sec by auto-injector (autoenhance A-50 or A-250, Nemoto Kyorindo, Tokyo, Japan). All CT examinations were performed after intravenous administration of contrast. Hard-copy images were photographed at window settings for lung (center, −600 HU [Hounsfield Unit], and width, 2000 HU) and mediastinum (center, 35 HU, and width, 400 HU). The time interval between CT and pathologic diagnosis ranged from 0–17 days.

The CT images were assessed in random order by two independent observers without reference to the clinical findings. The observers assessed the presence of areas of ground-glass attenuation, air-space consolidation, centrilobular nodules, air bronchogram, mucus plugging, bubblelike lucencies, bulging of interlobar fissure, CT halo sign, cavitation, traction bronchiectasis or bronchiolectasis, intralobular reticular opacities, interlobular septal thickening, and CT angiogram. Ground-glass attenuation was defined as an area of hazy increased parenchymal attenuation without obscuration of the underlying vascular markings. Areas of air-space consolidation were considered present when the opacity obscured the underlying vessels. Mucous plugging was defined as tubular attenuation structures resulting from mucous filling of airways. The shape of mucous plugging depends on the branching pattern of involved airways. Bubblelike lucencies were considered present when there was an enlargement unit of multiple cystic air spaces measuring 5 mm or less in diameter within the lesion surrounded by a wall of variable thickness. Bulging of interlobar fissure was defined as resulting in expansion of the lobe by the lesion. The CT halo sign was considered present when there was ground-glass attenuation surrounding the circumference of a central mass or nodule with soft tissue attenuation (Primack *et al.*, 1994). Cavitation included a circumscribed enlarged air space with a wall of variable thickness. Traction bronchiectasis or bronchiolectasis was defined as irregular bronchial dilatation within areas with parenchymal abnormality. The CT angiogram was considered to have occurred when the enhanced pulmonary vessels could be clearly identified within the lesion of low attenuation relative to the chest wall musculature. Also noted were presence of lymphadenopathy and pleural effusions. Lymphadenopathy was considered present when the short-axis diameter of the nodes was > 10 mm. The anatomic distribution was noted to be central if there was a predominance of abnormalities in the inner two-thirds of the lung, and peripheral if there was a predominance of abnormalities in the outer third of the lung. Following initial independent evaluation, the two observers reviewed all cases in which their interpretation differed and reached a final decision by consensus.

The diagnosis of mucin-producing adenocarcinoma of the lung was assigned in 70 patients (85%) by means of surgical specimens that were fixed with inflation by transpleural and transbronchial infusion of formalin. The specimens were sectioned transversely in the same plane as that of the CT. In 12 patients (15%) who received only chemotherapy, pathologic specimens were obtained by transbronchial lung biopsy. All pathologic specimens were stained with hematoxylin-eosin. The presence of mucous retention was assessed by morphologic examination with histochemical technique using periodic acid-schiff reagent. The internal characteristics of the tumors seen on thin-section CT were compared with those seen at pathologic examination of the specimens. Surgical specimens were evaluated by an expert lung pathologist for presence of intrapulmonary metastasis, thickened bronchi or bronchioles, and mucous retention. Mucinous cells were defined as those with basally located nuclei and eosinophilic to grayish cytoplasm typical of mucin-producing cells. Intracytoplasmic vacuoles of mucus in the tumor cells and mucus in the lumen of carcinomatous gland were also noted. Correlation between the CT and histologic findings was made by consensus between the radiologist, surgeon, and pathologist.

All patients were regularly followed up in our institute. Follow-up CT images were available for all patients. Disease-free survival (DFS) was calculated from the date of the surgery to the date of intrapulmonary metastasis or last contact with the patient. The mean follow-up after surgery was 34 months. After surgery, 21 of the 70 patients (30%) received chemotherapy for intrapulmonary metastases, and 10 of the 21 patients (48%) were alive with disease as of August 2004. Twenty-three patients (28%) had died by the time the present study was completed.

Univariate analysis of the thin-section CT findings was performed using the Chi square test and Fisher's exact test. The relation of thin-section CT findings and the presence or absence of intrapulmonary metastasis was tested for independent predictors using multiple logistic regression analysis, which determined odds ratio (OR) after adjusting for the other variables examined. Interobserver variation for the CT findings was quantified as the weighted kappa coefficient of agreement. Kappa values larger than zero were considered to indicate a positive correlation: value of 0–0.20, poor agreement; 0.21–0.40, fair agreement; 0.41–0.60, moderate agreement; 0.61–0.80, good agreement; 0.81–1.00, excellent agreement. Survival curves were estimated by using the Kaplan-Meier method. The univariate influence of thin-section CT findings on survival was analyzed by means of the log rank test. The Cox proportional hazard model was applied to all co-variates that had shown statistical significance ($p < 0.05$) at the univariate analysis. The Wald test was used in a backward stepwise selection procedure to identify parameters with significant independent predictive value and to estimate relative risk (RR) and 95% confidence interval (CI). All analyses were performed using SPSS statistical software (SPSS version 12.0, Chicago, IL).

Results

Twenty-nine of the 82 patients (35%) had intrapulmonary metastasis. The age at presentation was significantly higher in patients with intrapulmonary metastasis (74.1 ± 1.6 years) than those without intrapulmonary metastasis (70.0 ± 1.3 years; $p = 0.038$). A statistically significant difference was noted in the proportion of smokers between the patients with ($n = 19$) and without ($n = 16$) intrapulmonary metastasis. However, cigarette consumption was not statistically associated with the presence (9.9 ± 3.0 pack-years) or absence (12.8 ± 3.1 pack-years) of intrapulmonary metastasis. The patients with intrapulmonary metastasis were more likely to have large tumor size (9.0 ± 1.1 cm) than those without (3.2 ± 0.4 cm) intrapulmonary metastasis. There were no differences in other clinical characteristics between the two groups.

There was good interobserver variation agreement for the analysis of the thin-section CT findings (weighted kappa = 0.68–0.77). The CT findings identified frequently in mucin-producing adenocarcinoma of the lung on univariate analysis included air bronchogram ($n = 55$), areas of ground-glass attenuation ($n = 50$), areas of air-space consolidation ($n = 44$), interlobular septal thickening ($n = 43$), centrilobular nodules ($n = 29$), bubblelike lucencies ($n = 27$), intralobular reticular opacities ($n = 23$), and mucous plugging ($n = 21$). Mucous plugging, when present, was always superimposed on areas of air-space consolidation or ground-glass attenuation. Less common CT findings included bulging of interlobar fissure ($n = 18$), cavitation ($n = 13$), traction bronchiectasis or bronchiolectasis ($n = 11$), CT halo sign ($n = 10$), and CT angiogram ($n = 10$). Pleural effusion was also noted in only two patients.

Univariate analysis demonstrated that areas of ground-glass attenuation (odds ratio [OR], 2.4; $p = 0.012$), areas of air-space consolidation (or, 2.1; $p = 0.012$), centrilobular nodules (OR, 3.6; $p < 0.0001$), mucous plugging (or, 2.5; $p < 0.0001$), and CT halo sign (OR, not applicable; $p = 0.013$) were significantly associated with increased likelihood of intrapulmonary metastases. No statistically significant difference was found in the frequency of air bronchogram, bubblelike lucencies, interlobular septal thickening, bulging of interlobar fissure, cavitation, traction bronchiectasis or bronchiolectasis, intralobular reticular opacities, and CT angiogram among patients with or without intrapulmonary metastases. Multiple logistic regression analysis demonstrated that the thin-section CT findings independently associated with increased likelihood of intrapulmonary metastases were centrilobular nodules (OR, 6.6; $p < 0.05$) and mucus plugging (OR, 6.0; $p < 0.05$).

Tumors with focal type of lung involvement were well-demarcated nodular tumors growing along alveolar walls and associated with marked mucin production. Tumors with intra-pulmonary metastasis were multifocal and had poorly defined margins. Areas of ground-glass attenuation on thin-section CT correlated with the replacement growth of tumor cells or mucus. Mucus resulted in the filling of alveolar spaces in all patients. Areas of air-space consolidation corresponded to mixture of tumor cells, mucus, and decreased air content in alveolar spaces. Bubblelike lucencies on thin-section CT corresponded histologically to a mixture of mucus, goblet-type tumor cells, thickened bronchi or bronchioles, and small amounts of air in alveolar spaces. Thickened bronchi or bronchioles, which was often accompanied by localized scarring, focal alveolar collapse, and organized pneumonia, were found mainly in the distal periphery of the tumor. Intralobular reticular opacities correlated with the presence of alveoli filled with mucus and the preserved alveolar septa or underlying parenchyma. Interlobular septal thickening on thin-section CT corresponded to infiltration of the interstitium by inflammatory cells or septal edema. The sensitivity and specificity of centrilobular nodules in detecting intrapulmonary metastases were 83% (24 of 29 patients) and 83% (44 of 53 patients), respectively. Metastatic lesions often contained aerated bronchi. Nine patients (27%) with intrapulmonary metastases were minute and difficult to be identified on the corresponded CT images. The CT halo sign identified on thin-section CT corresponded to a tumor component with a bronchioalveolar or papillary growth and mucus layer excreted by surrounding tumor cells.

A statistically significant difference was found in follow-up duration after initial diagnosis between the patients with (17.5 ± 1.7 months) and without intrapulmonary metastasis (43.0 ± 2.7 months; $p < 0.0001$). Fifty-two of 82 patients (63%) had no evidence of recurrent disease. All patients without intrapulmonary metastasis and 10 of 33 patients (30%) with intrapulmonary metastasis were alive. There was a significantly greater mortality rate among patients with intrapulmonary metastasis (79%) compared to patients without one (22%; $p < 0.0001$). Univariate analysis demonstrated that the air bronchogram (OR, 2.3; $p = 0.037$), areas of ground-glass attenuation (OR, 2.2; $p = 0.019$), areas of air-space consolidation (OR, 2.3; $p = 0.005$), centrilobular nodules (OR, 5.0; $p < 0.0001$), mucous plugging (OR, 3.4; $p < 0.0001$), and bulging of interlobar fissure (OR, 1.3; $p = 0.03$) were significant predictors of poor prognosis in patients with mucin-producing adenocarcinoma of the lung. No significant predictors of poor prognosis were identified in thin-section CT findings, which included interlobular septal thickening, bubblelike lucencies, intralobular reticular opacities, cavitation, traction bronchiectasis or bronchiolectasis, CT halo sign, and CT angiogram. The 5-year DFS rates revealed as not having and having intrapulmonary metastases was 92% and 12%, respectively ($p < 0.0001$).

Age, gender, smoking history, and cigarette consumption had no significant prognostic value. The Cox proportional hazard analysis demonstrated that centrilobular nodules (relative risk [RR], 6.3; $p = 0.008$) and mucous plugging (RR, 7.2; $p = 0.002$) on CT were the independent predictors of poor prognosis. The differences of 5-year DFS rates revealed as not showing centrilobular nodules or mucous plugging were statistically significant (centrilobular nodules: 92% versus 17%: $p < 0.0001$; mucous plugging: 88% versus 0%: $p < 0.0001$).

Discussion

This study has shown that the presence of centrilobular nodules is sensitive to the presence of intrapulmonary metastasis and is predictive of poor prognosis in surgically treated patients with mucin-producing adenocarcinoma of the lung (Tateishi *et al.*, 2005). In the study, we examined the correlation between thin-section CT findings and histopathologic findings and determined the prognostic value of thin-section CT findings in treated patients with mucin-producing adenocarcinoma of the lung. We found that the presence of centrilobular nodules and mucous plugging on thin-section CT was associated with an increased likelihood of intrapulmonary metastasis. These findings were strongly predictive of poor prognosis in treated patients with mucin-producing adenocarcinoma of the lung.

The presence of thin-section CT findings as a biomarker must be closely linked to the presence of target disease. In addition, the identification of thin-section CT findings as a biomarker must be accurate, reproducible, and feasible. The biological plausibility of thin-section CT findings as an outcome surrogate is supported by the initial validation of thin-section CT findings for mucin-producing adenocarcinoma of the lung using direct comparison between thin-section CT images and macroscopic observations (Akira *et al.*, 1999; Gaeta *et al.*, 2002). Another support for biological validity is provided by using CT as the imaging modality of choice for the evaluation of findings and the fact that intrapulmonary metastases are a major source of the clinical morbidity (Barkley and Green, 1996). The link between severity of disease and thin-section CT is provided by studies showing that thin-section CT findings correlate with the extent of clinical disease. The reproducibility of thin-section CT findings in patients with mucin-producing adenocarcinoma (MPA) of the lung is supported by the previous study in which thin-section CT findings identified in the present study are similar (Tateishi *et al.*, 2005).

Our study demonstrates that thin-section CT findings correlate with true outcomes in mucin-producing adenocarcinoma of the lung. This addresses the final criterion of valid outcome surrogates. Intrapulmonary metastases are a major source of the clinical morbidity of mucin-producing adenocarcinoma of the lung. However, intrapulmonary metastases prior to surgery are considered difficult to be identified. Centrilobular nodules on thin-section CT were present in 35% of our cases. This prevalence is in concordance with that in previous studies (Aquino *et al.*, 1998; Jung *et al.*, 2001). In our study, 83% of patients with centrilobular nodules had intrapulmonary metastases. In addition, multivariate analysis demonstrated that centrilobular nodules were an independent predictor of poor prognosis. With the link between disease severity and thin-section CT findings, this study suggests that thin-section CT findings may be a useful outcome surrogate in a clinical trial.

Mucous plugging was identified in 26% of our cases on thin-section CT. This finding had a significant association with the presence of intrapulmonary metastases and survival in both univariate and multivariate analysis. Mucous plugging has been shown to increase when the intrapulmonary metastases are present. However, mucous plugging is a reversible finding that may be more related to short-term change than to long-term disease progression. Excretion of mucous plugs may be a possible cause of aerogeneous intrapulmonary metastases.

Mucin-producing adenocarcinoma of the lung has also been reported to exhibit a broad variety of CT features (Gaeta *et al.*, 2002). Previously reported thin-section CT manifestations other than centrilobular nodules and mucous plugging include areas of ground-glass attenuation, areas of air-space consolidation, and bubblelike lucencies (Aquino *et al.*, 1998; Jung *et al.*, 2001). The results of our study show that these findings correlate with the histopathologic findings. Areas of ground-glass attenuation were present in 50 of our 82 patients (61%) and correlated pathologically with lepidic growth of tumor along alveolar septa (replacement growth) or to the presence of mucus. This finding is in accordance with those in the previous studies (Aquino *et al.*, 1998; Jung *et al.*, 2001), which showed that ground-glass attenuation is one of the most common but nonspecific CT findings. Therefore, multivariate analysis showed no association between ground-glass attenuation and the presence or absence of intrapulmonary metastases or prognosis.

Areas of air-space consolidation identified in 54% of our cases correlated pathologically with a mixture of tumor cells, mucus, and decreased air content in alveolar spaces. It has been shown that in some patients, areas of ground-glass attenuation can progress to areas of air-space consolidation on sequential CT scans (Akira *et al.*, 1999). Our results suggest that areas of air-space consolidation can result from dense growth of tumor cells or by accumulation of mucus within the lesion. In the current study, significant difference was found in the occurrence of areas of air-space consolidation among the patients with or without intrapulmonary metastasis. This result is in agreement with that of a previously published study (Tateishi *et al.*, 2005). However, multivariate analysis showed that areas of air-space consolidation are not an independent predictor of intrapulmonary metastasis and prognosis. Bubblelike lucencies on thin-section CT were charterized by the presence of small focal areas of air attenuation within the lesion (Weisbrod *et al.*, 1992). The presence of bubblelike lucencies is similar to that previously described in other studies (Gaeta *et al.*, 2002). In our study, bubblelike lucencies on thin-section CT corresponded histologically to a mixture of mucus, tumor cells, thickened bronchi or bronchioles, and small amounts of air in alveolar spaces. Although bubblelike lucencies were seen in 33% of our cases, this finding was not a predictor of intrapulmonary metastasis or survival.

The presence of centrilobular nodules is strongly linked to the presence of intrapulmonary metastasis and is a good predictor of poor prognosis in surgically treated patients with mucin-producing adenocarcinoma of the lung (Tateishi *et al.*, 2005). In the present study, however, not only the presence of centrilobular nodules but also mucous plugging was significantly associated with an increased likelihood of intrapulmonary metastasis and was predictive of poor prognosis in treated patients with mucin-producing adenocarcinoma of the lung. The results of our study suggest that 95% CIs of relative risks of these findings are large for the group of treated patients. A broad range of 95% CIs likely reflects the multifocal nature of disease severity and a relatively small number of patients with intrapulmonary metastases. Furthermore, the study group consisted of a wide spectrum of disease: noninvasive, mucinous BACs and invasive adenocarcinomas. In future studies, the patient group of mucin-producing adenocarcinoma of the lung should be stratified definitely and might provide a larger number of occurrences in intrapulmonary metastases.

This study provides further support for the use of thin-section CT findings as outcome surrogates in patients with MPA of the lung. The presence of centrilobular nodules and mucous plugging are associated with greater likelihood of intrapulmonary metastases and are thus strong predictors of poor prognosis.

Acknowledgments

This work was supported in part by grants for BMS Freedom to Discovery grant and Scientific Research Expenses for Health and Welfare Programs, no. 17-12.

References

Akira, M., Atagi, S., Kawahara, M., Iuchi, K., and Johkoh, T. 1999. High-resolution CT findings of diffuse bronchioloalveolar carcinoma in 38 patients. *Am. J. Roentogenol. 173*:1623–1629.

Aquino, S.L., Chiles, C., and Halford, P. 1998. Distinction of consolidative bronchioloalveolar carcinoma from pneumonia: do CT criteria work? *Am. J. Roentogenol. 171*:359–363.

Barkley, J.E., and Green, M.R. 1996. Bronchioloalveolar carcinoma. *J. Clin. Oncol. 14*:2377–2386.

Brambilla, E., Travis, W.D., Colby, T.V., Corrin, B., and Shimosato, Y. 2001. The new World Health Organization classification of lung tumours. *Eur. Respir. J. 18*:1059–1068.

Dumont, P., Gasser, B., Rouge, C., Massard, G., and Wihlm, J.M. 1998. Bronchoalveolar carcinoma: histopathologic study of evolution in a series of 105 surgically treated patients. *Chest 113*:391–395.

Gaeta, M., Vinci, S., Minutoli, F., Minutoli, F., Mazziotti, S., Ascenti, G., Salamone, I., Lamberto, S., and Blandio, A. 2002. CT and MRI findings of mucin-containing tumors and pseudotumors of the thorax: pictorial review. *Eur. Radiol. 12*:181–189.

Jung, J.I., Kim, H., Park, S.H., Kim, H.H., Ahn, M.I., Kim, H.S., Kim, K.J., Chung, M.H., and Choi, B.G. 2001. CT differentiation of pneumonic-type bronchioloalveolar cell carcinoma and infectious pneumonia. *Br. J. Radiol. 74*:490–494.

Kuhlman, J.E., Fishman, E.K., Kuhajda, F.P., Meziane, M.M., Khouri, N.F., Zerhouni, E.A., and Siegelman, S.S. 1988. Solitary bronchioloalveolar carcinoma: CT criteria. *Radiology 167*:379–382.

Lababede, O., Meziane, M.A., and Rice, T.W. 1999. TNM staging of lung cancer: a quick reference chart. *Chest 115*:233–235.

Liu, Y.Y., Chen, Y.M., Huang, M.H., and Perng, R.P. 2000. Prognosis and recurrent patterns in bronchioloalveolar carcinoma. *Chest 118*:940–947.

Primack, S.L., Hartman, T.E., Lee, K.S., and Müller, N.L. 1994. Pulmonary nodules and the CT halo sign. *Radiology 190*:513–515.

Regnard, J.F., Santelmo, N., Romdhani, N., Charbi, N., Boucereau, J., Dulmet, E., and Levasseur, P. 1998. Bronchioloalveolar lung carcinoma: results of surgical treatment and prognostic factors. *Chest 114*:45–50.

Tateishi, U., Müller, N.L., Johkoh, T., Maeshima, A., Asamura, H., Satake, M., Kusumoto, M., and Arai, Y. 2005. Mucin-producing adenocarcinoma of the lung. Thin-section computed tomography findings in 48 patients and their effect on prognosis. *J. Comput. Assist. Tomogr. 29*:361–368.

Tsuchiya, E., Furuta, R., Wada, N., Nakagawa, K., Ishikawa, Y., Kawabuchi, B., Nakamura, Y., and Sugano, H. 1995. High k-ras mutation rates in goblet-cell-type adenocarcinomas of the lungs. *J. Cancer. Res. Clin. Oncol. 121*:577–581.

Weisbrod, G.L., Towers, M.J., Chamberlain, D.W., Herman, S.J., and Matzinger, F.R. 1992. Thin-walled cystic lesions in bronchioalveolar carcinoma. *Radiology 185*:401–405.

10

Non-small Cell Lung Carcinoma: ^{18}F-fluorodeoxyglucose-Positron Emission Tomography

Rodney J. Hicks and Robert E. Ware

Introduction

Positron emission tomography (PET) represents the most significant advance in the imaging of known or suspected lung cancer since the introduction of computed tomography (CT), and now has an established role in the diagnosis, staging, therapeutic monitoring, and restaging of this disease. Positron emission tomography allows whole-body imaging of a range of metabolic processes in tumors and normal tissues. In clinical practice, the most widely used PET tracer is ^{18}F-fluorodeoxyglucose (FDG), which, for the purpose of cancer detection, relies on the increased glycolytic metabolism that is a hallmark of many cancers. Fused PET/CT images obtained from hybrid scanners incorporating high-end PET with multidetector CT are proving to be superior to separate CT or PET images in the staging of cancer through better differentiation between abnormal and physiological uptake and better localization of the exact site of abnormalities. Although potentially a fully quantitative technique, there is an increasing acceptance among the imaging and oncological community that application of pretest probability analysis and pattern recognition are effective means of optimizing the clinical performance of this modality and thereby appropriately guiding management.

In this context, evaluation of the appropriate indications for which to perform PET, and how the information provided should be utilized for treatment selection and planning, requires a more sophisticated analysis than simple statements of sensitivity and specificity. In particular, recognition that FDG-PET represents a metabolic signature rather than a means of dichotomizing tissues into benign and malignant classifications is of fundamental importance. Previously, cut-off values of quantitative and semiquantitative measures, such as the standard uptake value (SUV), have been defined in an attempt to facilitate such categorization of lesions. However, clinical experience has taught us that many malignant conditions can have relatively low FDG uptake, whereas many benign processes can have high uptake. In the same manner as CT interpretation has long since dispensed with measuring Hounsfield Units (HU) to characterize most lesions and has become dependent on consistent thresholding of images and application of pattern recognition to optimize diagnostic precision, it is our belief that clinical PET will follow a similar course. However, in the same manner that HU values are diagnostically helpful in some specific situations, so too are measures of FDG uptake, like the SUV, valuable parameters for certain applications, particularly including comparison of serial studies in an individual patient.

Role of FDG-PET on Diagnosing Lung Cancer

With increasing utilization of CT scanning and improvements in the resolution of this modality through advances in multislice helical scanning technology, lung nodules are now more commonly identified incidentally in otherwise asymptomatic individuals. Many, and in some parts of the world most, of these nodules are not due to malignancy but rather to scarring from previous infections or to granulomatous diseases that may or may not be active. Although serial imaging by chest X-ray or CT is frequently used to exclude progressive growth due to malignancy, the opportunity to detect and treat lung cancers at the earliest possible time, and thereby to maximize the chances of cure, may be lost by an observational policy. In addition, such a strategy can generate considerable patient anxiety. Consequently, many patients undergo aggressive attempts to secure a histopathological diagnosis. Techniques utilized may include bronchoscopy with cytological examination of brushings and washings, fine-needle aspiration biopsy (FNAB), thoracoscopic and endoscopic ultrasound-guided biopsy, and thoracotomy. These procedures are to some degree invasive and have varying costs and morbidities, but all potentially add considerably to the overall health expenditure.

For the noninvasive definition of the likelihood of malignancy as the cause of a solitary pulmonary nodule (SPN), as defined by lung lesions of < 3 cm diameter without specific radiological features of either a benign or malignant nature, PET is significantly more accurate than structural imaging methods such as CT scanning (Gould et al., 2001) and has been shown to be superior to CT in almost all individual published studies. High levels of uptake of FDG are highly predictive of malignancy. However, it must be recognized that FDG is a tracer of glucose metabolism and, as such, does not only identify malignant cells. In particular, inflammatory processes can also give positive FDG-PET results. In clinical practice, it is important to integrate the information available from all possible sources prior to arriving at a hierarchical differential diagnosis. Factors that influence the likelihood of a malignant basis include the age and smoking history of the patient, any comorbidity including previous malignancy, the patient's ethnic and socioeconomic environment, the prevalence of inflammatory lung disease in the community in which they live, the results of laboratory tests, the radiological appearances, and the pattern of abnormality on PET, including the site and shape of the abnormality. Integration of such information is likely to further improve on the diagnostic performance of PET, compared to results obtained under the unrealistic conditions of blinded reading used in validation trials of PET for evaluation of SPN and particularly compared to application of a single SUV threshold.

In our own facility, we have primarily evaluated the incremental clinical impact rather than the independent diagnostic accuracy of PET. Therefore, we have used standard best-practice reporting in which PET is read with access to all clinical data that would normally be available, including the results of any previously performed CT scans. Our data confirm that FDG-PET can significantly and appropriately impact management of patients with SPN. In addition to utility in SPN, we have found that FDG-PET is also highly accurate in characterizing pulmonary mass lesions that are either unsuitable for or have failed histopathological characterization, even when they do not fulfill the classical definition of an SPN. Importantly, we have shown that in our health-care environment, FDG-PET is a cost-effective approach to evaluating SPN. Our experience is thus consistent with that in other health-care settings (Lejeune et al., 2005).

Overall, the positive predictive accuracy of FDG-PET is sufficiently high to warrant histological examination of all nodules with increased FDG uptake. In our experience, those cases that prove to be false-positive in terms of the diagnosis of malignancy are almost always due to an active disease process that warrants diagnosis and specific treatment. The most common cause of an FDG-avid but nonmalignant lesion is determined by the prevalence of various endemic lung conditions in the geographical region in which the patient lives. Globally, active tuberculosis is probably the most common cause of such a PET scan result. However, in some parts of the world, histoplasmosis and cryptococcosis can also be relatively frequent. In many of these cases, the semiquantitative measures of FDG uptake such as the SUV are within the range of values observed with malignant lesions. In the rare instance where lung cancer is strongly suspected on the basis of conventional imaging but histological confirmation is contraindicated because of severe comorbidities, a positive PET scan may be sufficient evidence to proceed to radiotherapy without histology, particularly if serology for common confounding diseases and inflammatory markers are negative.

Conversely, lesions with low FDG uptake are benign in the great majority of cases. False-negative PET scans may also occur when lesions are subject to partial-volume effects due to their small diameter or have low FDG avidity. In particular, it should be recognized that a high percentage of tumors with a diameter of < 5 mm are likely to yield false-negative results regardless of the intrinsic tumoral FDG avidity based on the count recovery characteristics of modern PET scanners (partial-volume effects) and the effects of respiratory motion. Respiratory gating of the emission data may improve detection of these small nodules. For nodules in which partial-volume effects are unlikely to account for the false-negative result, poor uptake of FDG is most commonly observed with low-grade malignancy such as carcinoid tumor and bronchioloalveolar carcinoma. Nevertheless, most such tumors still have sufficient FDG avidity to be imaged by PET (Yap et al., 2002). It is our experience that tumors with low FDG avidity have an indolent natural history and that a period of observation does not lead to adverse outcome in these cases.

Our institutional policy is to biopsy all FDG-avid lesions. The major caveat to this recommendation occurs when the pattern of uptake is highly suspicious of an inflammatory etiology. An example would be a wedge-shaped lesion with highest uptake at its base rather than its apex (the latter pattern being more typical of infective collapse beyond a malignant lesion). In this clinical setting, repeat imaging should be performed after investigation for potential infective or inflammatory etiologies and, sometimes, empirical treatment for the most likely candidate diseases. We generally do not biopsy non-FDG-avid diseases unless the *a priori* likelihood of malignancy is very high on both clinical and radiological grounds. For larger lesions without FDG avidity we recommend repeat CT at 6 months, reasoning that partial-volume effects are unlikely to account for the nonvisualization of the lesion, and that if the lesion does eventually prove to be malignant, the low-FDG avidity would generally predict an indolent growth pattern. Any lesion that grows over this observation period would be resected, whereas repeat CT in a further 12 months would be recommended for stable lesions. For lesions of small size, we recommend a repeat CT at 3 months in order to potentially identify the subgroup of lesions with rapid proliferation that are simply not visualized adequately on PET by virtue of partial-volume effects. In such cases, it is our hope that such lesions can still be identified before they become incurable. For lesions that grow, a repeat FDG-PET is performed for staging purposes on the presumption of a malignant basis. For stable lesions without significant growth, a repeat CT in 12 months is recommended.

These recommendations are somewhat at odds with the recommendations of those for lesions evaluated solely by volumetric CT (Kostis *et al.*, 2004). Because the variability in measurement of small nodules is greater than that for measurement of larger nodules, statistical considerations suggest that a longer interval should be used to follow-up small nodules than is used for larger nodules in order to increase the certainty that measured differences in size are truly indicative of growth. These considerations are based primarily on technical and statistical considerations and ignore the biological variability in the rate of growth of various lung malignancies. Accordingly, an indolent tumor of 1 cm in diameter may not grow in the 1 month recommended for repeating the CT scan, whereas aggressive tumors, such as small cell lung cancer, may grow rapidly and metastasize in the 12 months recommended for repeating CT in nodules of 2 mm diameter. Although we believe that incorporating information regarding the metabolic characteristics of the lesions can allow more rational timing of surveillance imaging, nodules of 1–2 mm are so seldom visualized even with state-of-the-art PET scanners, that we do not recommend routinely performing PET to characterize

such tiny lesions. In addition to identifying the likely biological nature of an SPN, FDG-PET scanning has the advantages of a wider survey of the thorax and body for additional sites of abnormality. As discussed in the following staging section, this incremental information may be vital to defining the most likely differential diagnosis, the most appropriate treatment, and the delivery thereof.

Preoperative PET Staging of Non-small Cell Lung Cancer

Once the diagnosis of NSCLC has been established, the key diagnostic objective is to evaluate the extent of disease because this determines the most appropriate form of treatment as well as the prognosis for survival. Although surgery provides the best chance of cure, this outcome can only be achieved with truly localized lung lesions. If regional or systemic metastasis is present, radiotherapy, chemotherapy, or surgery alone or in combination are therapeutic options, but the likelihood of cure is dramatically reduced. Multimodality therapy, including neoadjuvant chemotherapy or chemoradiation followed by surgery, is increasingly being used with curative intent in patients with advanced locoregional disease. However, aggressive locoregional therapies cannot be curative if disease exists beyond the treatment volume. Accordingly, rational treatment delivery is critically dependent on obtaining accurate staging of regional lymph nodes and excluding systemic metastasis. For this application, the evidence strongly indicates that PET is the most accurate noninvasive technique currently available.

The extent and burden of disease is becoming codified for many cancers based on characteristics of the primary tumor (T-stage), regional nodes (N-stage), and systemic metastases (M-stage). Various groupings of TNM (tumor, nodes, metastases) stage are often also combined into more conventional staging groups of I–IV, with stage I usually indicating local disease potentially curable by resection and stage IV disease representing disseminated disease incurable except by systemic therapy in highly responsive tumor types. The general strategy for staging cancer is to first characterize the primary tumor followed subsequently by evaluation for regional nodal and systemic metastatic disease. However, since lung cancer is often asymptomatic in its early stages, and patients often present with relatively advanced disease, a critical factor in broadly defining treatment options is the exclusion of systemic metastases. Accordingly, discussion of the utility of PET will commence with its role in assessing the M-stage, then for N-staging, and finally its relatively limited role in T-staging. This approach reflects our perspective that the diagnostic paradigm for staging NSCLC will change fundamentally over the next few years toward the use of PET/CT as the primary staging modality for all confirmed lung cancers, with other modalities used only to further refine information for prognostic and treatment-planning purposes pertinent to the treatment option selected on the basis of PET.

Evaluation of Distant Metastasis (M) Stage

With its ability to conveniently survey the entire body and the high contrast generally observed between tumor deposits and normal tissues, PET is capable of detecting disease in the adrenal glands, liver, and other organs, which may appear normal on CT. Positron emission tomography is also probably more accurate than radionuclide bone scan for assessing skeletal involvement. Positron emission tomography using FDG can detect unsuspected distant metastasis in 5–10% of patients with notionally curable stage I–II disease determined by conventional noninvasive imaging, potentially justifying this modality as the initial staging procedure in all but the smallest primary tumors. The rate of occult metastatic disease is even higher in those presenting with more locally advanced disease, especially those who are being considered for chemoradiation with curative intent. In a cohort of 167 patients evaluated at our facility, the rate of PET-detected metastasis increased significantly ($p = 0.016$), with increasing pre-PET stage from I (7.5%) through II (18%) to III (24%), and, in particular, was significantly higher in stage III ($p = 0.039$) than for lower stages. A similarly high rate of detection of distant metastases in apparent stage III disease was reported by others (Eschmann *et al.*, 2002). Because many patients with stage III disease are not suitable for surgical pathological staging, noninvasive imaging has the primary role in determining the patient's suitability for radiotherapy with curative intent and the planning of treatment (Hicks and MacManus, 2003).

It is clear that PET evidence of distant metastasis, even if unsupported by other evidence, is powerfully associated with subsequent progression of metastatic disease and death. In a study from the Peter MacCallum Cancer Centre of 42 patients with PET-detected distant metastasis before planned surgery ($n = 7$) or radical radiotherapy (RT)/chemoradiotherapy ($n = 35$) for NSCLC, survival was investigated as the principal end point. The influence of metastasis number and other prognostic factors was investigated using Cox regression analysis. All but 4 patients had died by the last follow-up. Median survival was 9 months overall, 12 months for 27 patients with single PET-detected metastasis, and 5 months for 15 patients with > 1 metastasis ($p = 0.009$). Eastern Cooperative Oncology Group (ECOG) performance status ($p = 0.027$) was the only other conventional prognostic factor that correlated with survival. These data provide good support for basing treatment on PET findings even in the absence of histopathological validation of all sites of suspected disease. This approach is not dissimilar to usual clinical paradigms where evidence of multifocal metastatic disease on CT or bone scan is seldom confirmed by biopsy. Given the evidence that PET scanning is more accurate than either of these modalities, it is reasonable to adopt a similar approach to such findings on PET.

In a recent study, the use of dual-modality PET/CT staging was shown to be more accurate for detection of distant metastasis in NSCLC than either PET or CT as single modalities (Lardinois *et al.*, 2003). Accurate localization of FDG-avid regions on fused PET/CT images reduces the risk of false-positive interpretations of physiologic phenomena such as uptake in bowel or metabolically active brown fat. Accordingly, previous estimates of the impact of PET on the management of lung cancer are likely to be surpassed in current clinical practice.

Evaluation of Intrathoracic Lymph Node (N) Stage

Survival in NSCLC is significantly correlated with lymph node stage and drops precipitously when mediastinal nodes contain tumor. Computed tomography has long been the standard noninvasive method for detecting intrathoracic lymph node metastasis in NSCLC. Lymph node size is usually used for characterizing nodes as being likely benign or malignant. An arbitrary cut-off point is used to distinguish positive from negative nodes. Most commonly, a short-axis diameter of 1 cm is taken to indicate malignancy. However, because reactive lymphadenopathy is a common cause of lymph node enlargement and because tumors can be present in nodes smaller than 1 cm, CT has a relatively low sensitivity and specificity for the detection of lymph nodes involved by a tumor. While increasing the diameter required for diagnosis of malignancy improves the positive-predictive value of CT, the negative-predictive value of CT becomes unacceptably low. Conversely, decreasing the size threshold improves sensitivity at the expense of a lower positive-predictive value.

The accuracy of FDG-PET in staging the intrathoracic lymph nodes has been directly estimated in numerous clinicopathologic studies. In all of these studies, including a meta-analysis (Dwamena *et al.*, 1999), PET has been shown to be more accurate than CT for staging the mediastinum. The best noninvasive results have been obtained by correlating the results of both PET and CT images, and combined PET/CT scanners are likely to become the standard noninvasive tool for staging the mediastinum. Importantly, the status of lymph node involvement as defined by PET appears to be prognostically significant. It has been reported that survival is more strongly correlated with PET stage than CT-based stage in a large group of patients, who were mostly surgical candidates (Dunagan *et al.*, 2001). We have also shown a strong association between PET stage and survival in cohorts of patients, including both surgical and radical radiotherapy candidates.

The preceding data allow reasonable confidence in selecting treatment strategies based on FDG-PET, even in the absence of histopathological verification. As with the interpretation and integration of PET data into clinical treatment planning of SPN, Bayesian (pretest probability) principles and pattern recognition should be utilized for assessing the likelihood of nodal metastasis. Non-small cell lung cancer tends to spread in a hierarchical manner to first- and second-echelon nodes. Thus, as disease spreads from the hilar nodes to more proximal nodal stations, the volume of disease tends to decrease, and, through partial-volume effects, the intensity of uptake on FDG-PET also tends to decrease. Thus, in a patient with a large right lower lobe primary and intense uptake in a right hilar node, faint uptake in the right tracheo-bronchial or subcarinal nodal stations, particularly in the absence of nodal enlargement, would be considered highly suspicious for metastatic spread. However, a similar low intensity of uptake in an enlarged contralateral mediastinal node in a patient with intense uptake in the primary and without ipsilateral nodal visualization would be considered most likely to represent reactive lymphadenopathy. While enlarged nodes with low uptake are more likely due to reactive lymphadenopathy than small nodes of similar intensity, it is important to consider the intensity of uptake in the primary lesion. The metabolic signature of the primary will usually be reflected in the metastatic sites. Thus, low uptake in the primary ought to increase the suspicion for metastatic disease in enlarged nodes even if they have low uptake. Significantly increased FDG uptake in nonenlarged mediastinal nodes, though usually indicative of malignancy, can certainly reflect granulomatous disease. In our experience, uptake due to this process has a typical pattern of bilateral hilar, subcarinal, and right paratracheal uptake of fairly uniform intensity. Unfortunately, this process can coexist with regional nodal metastasis. Accordingly, our approach is to perform "metabolically guided" biopsy of those nodal stations that would fundamentally change the therapeutic approach or prognostic grouping. For example, where confirmed metastasis in a subcarinal node would change the therapeutic choice from surgery to chemoradiation, endoscopic ultrasound-guided biopsy of any node identified in this region would be recommended.

Evaluation of Tumor (T) Stage

Because PET has limited spatial resolution and relatively poor definition of anatomical planes, FDG-PET has a limited role in T-stage definition. However, nonmalignant tissues can contribute to mass lesions related to lung cancers and thereby limit true definition of tumor size. In particular, peripheral lung collapse can make definition of tumor boundaries difficult on CT. In situations where the primary tumor abuts soft tissue of similar CT density to the primary, the FDG-PET may provide better contrast resolution between tumoral and normal tissue. Examples include pleurally based tumors that are potentially invading the chest

wall, and central tumors abutting the mediastinum. One specific example of PET findings leading to upstaging of the T-stage is the demonstration of a compensatory increase in uptake in the right vocal cord due to a left recurrent laryngeal nerve palsy from direct extension of tumor into the aorto-pulmonary window (Fig. 53). However, for practical purposes, CT and MRI remain the best methods for noninvasively determining the T-stage in most patients, and both are clearly inferior to operative and pathological evaluation. In a paradigm where PET/CT is used as the primary staging tool, detailed anatomical evaluation of the primary tumor and for small-volume nodal metastasis would be predicated on the absence of evidence of systemic or macroscopic nodal disease. The next investigation would be tailored to unanswered questions pertinent to the most likely therapeutic options. For optimal treatment planning, limited field-of-view but high-resolution diagnostic CT with contrast timed to optimally demonstrate that regional anatomy, MRI, or surgical staging might be required.

Impact of Staging FDG-PET on Patient Management

A randomized trial performed in the Netherlands evaluated the ability of PET to alter treatment in patients undergoing conventional staging prior to surgery for lung cancer. In this study, the group that was randomized to have PET had a significant reduction in the "futile thoracotomy rate" (van Tinteren et al., 2002). In addition, PET appeared to be highly cost-effective in the health-care environment in which the study was performed (Verboom et al., 2003). Although the conclusions of a further randomized trial performed in Australia (Viney et al., 2004) appeared to be at odds with the Dutch study, the cohorts were not similar, with the Australian study recruiting mainly patients with very early-stage disease (92% stage I) being planned for surgery. Review of the Australian raw data indicates that had the PET information been incorporated into decision making, ~ 25% of patients would have had their management changed. Possibly due to the local philosophies of the surgeons involved, many patients deemed to be inoperable by PET criteria still underwent thoracotomy. A decision-tree cost-effectiveness study using modeled data from the literature but incorporating actual cost data derived in the Australian health-care environment has demonstrated that FDG-PET can be cost-saving through reducing the requirement for mediastinoscopy, particularly through identification of systemic metastasis (Yap et al., 2005).

In patients being planned for radical chemoradiation, our group has demonstrated a high impact on treatment selection and delivery. In our experience, ~ 30% of patients are spared futile radical radiotherapy because of detection of unexpected distant metastasis by PET, or because of PET-detected intrathoracic disease that is too extensive for radical irradiation

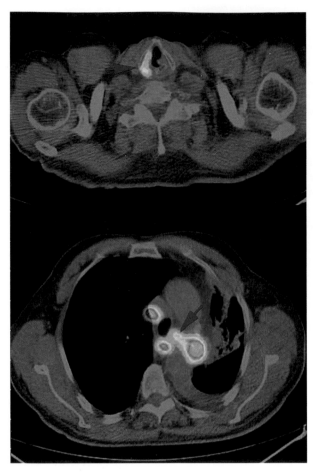

Figure 53 Increased uptake in the right vocal cord, particularly in the arytenoid muscle (horizontal arrow, upper panel), which reflects compensatory abduction of this cord in response to paralysis of the left vocal cord as a result of recurrent laryngeal nerve palsy. This finding indicates mediastinal invasion, rendering the patient inoperable. In this case the curvilinear configuration of uptake extending from the primary tumor (oblique arrow, lower panel) suggests direct mediastinal invasion rather than extranodal extension from aorto-pulmonary window metastasis. Also demonstrated are uptake in nonenlarged mediastinal nodes and the ability of PET to differentiate between tumor and adjacent collapse.

to be safely performed. We have also shown that for patients actually having radical therapies, the PET stage powerfully correlates with survival, contrasting with conventional T- and N-stage assessment in patients treated with radical radiotherapy, which is a relatively poor predictor of outcome (Ball *et al.*, 2002). The effect of PET selection on the survival of patients treated with radical radiotherapy was illustrated by a further study from our center in which two prospective cohorts were compared (MacManus *et al.*, 2001). Eighty and 77 eligible patients comprised the PET and non-PET groups, respectively. The median survival was 31 months for PET-staged patients and 16 months for non–PET-staged patients despite being well matched based on conventional

prognostic indicators including stage, ECOG status, and weight loss. This study suggests that, by using PET to exclude unsuitable patients with advanced disease and by integrating it within the radiotherapy treatment planning process, previously unattainable survival results can be achieved. These results also confirm the efficacy of radiotherapy to enhance survival in appropriately selected patients. They also warrant a cautionary note regarding comparison results of therapies in modern cohorts staged by PET with historical controls staged conventionally.

In many clinical situations, none of the available therapeutic strategies has been proved clearly superior based on available clinical trial data. More accurate evaluation of tumor extent and burden, particularly if this can be shown to more powerfully stratify prognosis, may provide a useful means of determining the true relative efficacy of different treatment paradigms. Better design and validation of therapeutic paradigms are required given the still poor outcomes of patients diagnosed with NSCLC.

Preliminary data indicate that the new combined PET/CT scanner may provide the most efficient and accurate means of integrating structural and molecular information into the treatment-planning paradigm (Ciernik *et al.*, 2003). There are significant challenges in utilizing molecular imaging data, particularly fused PET/CT data, for radiotherapy treatment planning, but these issues are being progressively addressed by a combination of phantom and clinical studies as well as refinement in the radiotherapy planning software. A particular issue is whether definition of tumor sites and the boundaries thereof should be based on qualitative or computer algorithms. We believe that by rigorous thresholding, consistent definition of gross tumor volumes can be achieved with qualitative assessment (Fig. 54). However, more sophisticated computer algorithms may be able to reduce operator dependence.

One of the major issues regarding radiotherapy planning using PET data to define tumor boundaries is whether PET or CT should take precedence when the gross tumor volumes are discordant. Although peripheral atelectasis accounts for many of these discrepancies, respiratory motion is another important factor. The acquisition of the emission data used to reconstruct PET images occurs over several minutes and therefore integrates the effect of respiratory movement. This has the effect of increasing the apparent supero-inferior size of lesions near the base of the lungs that move primarily in the coronal plane during respiration and the antero-posterior dimensions of lesions in the anterior aspect of the lungs that move mainly in the axial plane. In both circumstances, apparent activity at the edge of the lesion is slightly reduced by this movement owing to temporal undersampling, and if the lesion is relatively small, the SUV can be correspondingly reduced. Computed tomography is, however, acquired very rapidly, and multislice scanners can acquire images sufficiently fast to effectively

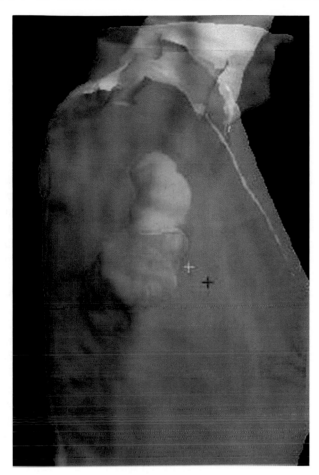

Figure 54 This three-dimensional representation of planned-treatment volumes as seen from the lateral projection demonstrates a larger treatment volume in the lungs when planned on CT (pink) and a larger mediastinal treatment volume when planned using PET (green). For radiotherapy planning, the ability to demarcate tumor from peripheral lung collapse can significantly decrease radiation treatment volumes, sparing normal lungs from potential radiation toxicity. In patients with impaired lung function, this can significantly improve the quality of life after successful treatment. Detection of disease in nonenlarged mediastinal lymph nodes can conversely extend the treatment field and thereby improve the chances of locoregional control and hopefully cure.

"freeze" respiratory motion. Alternatively, CT scan images can be acquired during breath-holding at a given phase of respiration.

As derived from instantaneous images of the relative position of organs, the location of lesions on CT planning images does not necessarily correspond to the averaged or integrated position of lesions detected by emission scanning. Respiratory movement can lead to misregistration of PET and CT lesions on fused PET and CT images whether acquired on stand-alone or combined devices. Our approach has been to assume that because the PET data represent the integrated position throughout the respiratory cycle, this should be used to plan the GTV since radiotherapy is generally delivered during normal breathing. An alternative

approach would be to perform respiratory gating of the PET, CT, and radiotherapy delivery. Efforts to develop methodology for respiratory-gated PET have been reported (Nehmeh *et al.*, 2003) and offer the potential for highly sophisticated treatment planning and delivery. Whether the resource implications of such an approach make it practical and affordable for routine clinical application remains to be seen, but for patients in whom lung function is marginal for radical therapy, these highly targeted approaches may be critical to outcome.

Role of PET in Therapeutic Response Assessment in NSCLC

The response to initial therapy may determine the further management of patients with lung cancer, given the multiple therapeutic options that are potentially available at this time. Particularly in the context of combined therapies, demonstration of therapeutic efficacy is important when considering sequential steps in the treatment process. For example, the use of neoadjuvant chemotherapy prior to surgery is probably rational only if it can be shown that this chemotherapy is successful in reducing tumor burden to the point of operability. Lack of therapeutic response and especially progression of disease during treatment should lead to abandonment of the surgical option and consideration of alternative methods for locoregional or systemic control. Similarly, new biological agents that target the epithelial growth factor receptor (EGFR) and other pathways involved in the genesis of lung cancer are becoming available. These treatments are often very expensive, and demonstration of response, or more importantly, lack thereof, may yield significant economic advantages in addition to sparing patients the unnecessary toxicity of ineffective therapies.

Three-dimensional structural imaging modalities, such as CT and MRI, have long been the most important investigations for assessment of response to nonsurgical therapies such as radiation therapy or chemotherapy. The so-called RECIST (Response Evaluation Criteria in Solid Tumors) criteria are currently applied to measurements of tumor dimensions made before and after therapy, and responses are categorized as complete response (CR), partial response (PR), no response (NR), or progressive disease. However, CT and MRI scanning have significant limitations in the assessment of tumor response in solid tumors in general and NSCLC in particular. Primary tumor boundaries may be obscured by atelectasis before or after therapy. As discussed earlier, lymph node size measured on CT is an unreliable measure of lymph node involvement by tumor. Tumors often regress gradually over several months, mandating serial measurements to assess response (Werner-Wasik *et al.*, 2001). Lesions may never regress radiologically despite having been controlled by treatment as demonstrated by long-term follow-up. Despite these well-recognized limitations, structural imaging responses have been shown to correlate with both survival and progression-free survival. While the limitations of CT in assessing therapeutic response are not as pertinent to FDG-PET and may therefore facilitate earlier assessment of response to treatment of NSCLC than structural imaging, demonstration that this assessment is accurate, and also able to stratify prognosis, is important.

Positron emission tomography results after induction therapy prior to surgery have been shown to have a significant correlation with outcome (Vansteenkiste et al., 1998; Akhurst et al., 2002). This modality also appears to be similarly useful in assessing response to chemoradiation. For example, Choi and colleagues (Choi et al., 2002) found that the FDG-PET response was strongly correlated with response to preoperative chemoradiotherapy in locally advanced NSCLC when validated by pathological examination of a tumor obtained at thoracotomy. Changes in SUV have also been shown to have prognostic significance after palliative chemotherapy for incurable NSCLC (Weber et al., 2003). In addition, our own data using a simple qualitative approach (MacManus et al., 2003) show that PET response is much more significantly correlated with survival than response measured by CT scanning.

At our facility we have developed a standardized set of PET-response categories that are defined as follows:

1. CMR (complete metabolic response)—No abnormal tumor FDG uptake; activity in the tumor absent or similar to mediastinum.
2. PMR (partial metabolic response)—Any appreciable reduction in intensity of tumor FDG uptake or tumor volume and without disease progression at other sites.
3. SMR (stable metabolic disease)—No appreciable change in intensity of tumor FDG uptake or tumor volume and without new sites of disease.
4. PMD (progressive metabolic disease)—Appreciable increase in tumor FDG uptake or volume of known tumor sites, or evidence of disease progression at other intrathoracic or distant metastatic sites.

In our NSCLC series, with a median survival after follow-up PET of 24 months, multifactor analysis including the known prognostic factors of CT response, performance status, weight loss, and stage found that only PET metabolic response was significantly associated with survival duration ($p < 0.0001$). This powerful prognostic information ought to encourage the development of investigational "response-adapted" therapeutic approaches. Despite these encouraging results, many PET specialists consider that postradiotherapy changes severely limit the utility of PET in evaluating treatment response to radiotherapy. This is not our experience. As long as some fairly simple rules, mainly based on pattern recognition, are followed, therapeutic response can still be assessed and has good prognostic value (Hicks et al., 2004).

Use of FDG-PET for Restaging Following Definitive Treatment of NSCLC

It is possible that patients with localized residual disease after definitive treatment could benefit from salvage surgery, conformal radiotherapy, or experimental chemotherapeutic regimens. There is accumulating evidence supporting the accuracy of PET scanning after radiation therapy. We have demonstrated that FDG-PET is diagnostically and prognostically superior to CT in this clinical setting by better defining both the presence and the extent of recurrent disease. Because of its ability to differentiate between viable tumor and post-treatment scarring and radiation fibrosis, metabolically targeted locoregional therapies may salvage some patients who would otherwise die of their residual disease. A recent study has confirmed the diagnostic accuracy and prognostic utility of FDG-PET in the restaging of NSCLC following surgery (Hellwig et al., 2006).

A Philosophical Perspective on the Quantitative Analysis of FDG Uptake in NSCLC

Although absolute evaluation of glucose metabolic rate of tumors has scientific appeal, fully quantitative approaches using arterial blood analysis are too invasive and complex for routine clinical use. In particular, the requirement of imaging over a very restricted field of view for a protracted time significantly compromises efficient resource utilization of an expensive piece of imaging equipment. We hold to the view that it is a better value to do whole-body scans on two patients in an hour than to image a single axial field-of-view for the same period of time in a single patient. The former approach allows detection of many more potential sites of disease and is, consequently, more likely to alter clinical management.

Whole-body survey capability is not, however, incompatible with the semiquantitative methodology that is used to calculate the standardized uptake value (SUV). Unfortunately, some practitioners have become unthinking slaves to this measure, not recognizing its limitations nor questioning its authority in guiding their diagnostic opinions. As detailed earlier, many primary lung pathologies other than cancer can have an SUV in a range that overlaps with that of malignant lesions. Similarly, malignant lesions can have relatively low FDG avidity. Methodical difficulties are also posed by the introduction of PET/CT devices due to the fact that the attenuation correction required for calculation of the SUV is generally derived from a CT-attenuation map extrapolated to reflect the tissue attenuation characteristics of a 511 keV annihilation photon. Since the CT data regarding tissue density are captured instantaneously, misregistration may serve to decrease the SUV if the lesion is attenuated as being air rather than soft tissue. The SUV will also tend to be reduced in small lesions due to respiratory blurring, and even for larger lesions it is likely to influence the apparent SUV at the extreme of the respiratory excursion. This can also influence edge finding using SUV threshold approaches in the pretreatment setting.

Use of SUV-based assessment following radiotherapy is also particularly problematic if not applied intelligently. In particular, assignment of an appropriate region of interest for analysis may be compromised by a number of factors, including the presence of the commonly observed, and sometimes intense, inflammatory reaction to radiotherapy in normal tissues within the treatment portal. These changes may have a measured value of FDG uptake in the "malignant" range but should not pose a significant problem with the visual assessment method that takes account of the distribution of normal tissue reactions. Postradiotherapy changes conform to the volume of aerated lung in the radiation treatment volume and are of a geographic rather than an anatomical distribution (Fig. 55). Furthermore, they seldom match the pattern of uptake in tumoral sites on baseline scanning. Residual disease, on the other hand, even if associated with less intense and less extensive FDG-uptake abnormalities, tends to be closely related to the position of initial tumor, allowing for anatomical distortion relating to further collapse or reexpansion of lung parenchyma, and also tends to maintain a lobular or spherical shape.

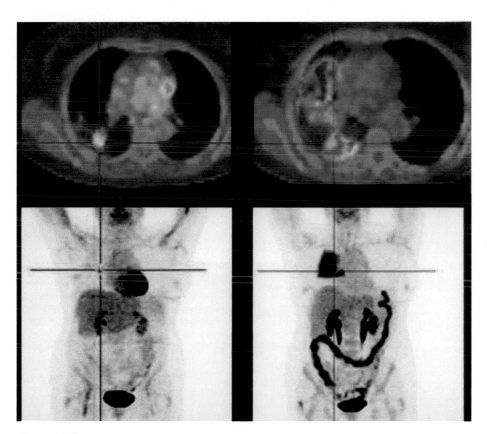

Figure 55 Despite similar intensity, the distribution of abnormality on the baseline study (left) and scans performed 8 weeks following radiotherapy (right) are quite different. The fused images (above) demonstrate a marked reduction in FDG uptake in relationship to the primary lesion but increased uptake in adjacent lung. Radiation pneumonitis tends to have a geographic distribution, often with sharply demarcated borders reflecting the radiation portal, and involves the lung beyond the primary tumor volume on baseline scanning. Where the radiation portal crosses into a different lobe, the abnormality is not confined by the fissure, as tumors usually are. The intensity of uptake can be similar to that in recurrent or residual tumor. The metabolic abnormalities associated with radiation pneumonitis are a subacute rather than an acute phenomenon. Typically, they become apparent 4–6 weeks after radiotherapy and are often quite prominent at 12 weeks post-treatment. They can persist for more than 6 months and usually progressively become associated with typical changes of radiation fibrosis on CT such as traction bronchiectasis.

Similarly, tumoral uptake tends to respect and follow natural tissue barriers such as the pleura of the oblique fissure, whereas radiation changes are not influenced by such boundaries. However, SUV can be useful for assessing the relative change in the intensity of FDG uptake in an individual lesion over time. Unless rigorous and consistent thresholding of PET images are performed, it can also "guide the eye" as to how metabolically active a lesion is.

Use of Hybrid PET-CT Images in Staging

Recognition of the complementary strengths and limitations of structural and functional imaging has underpinned the concept of correlative imaging. The traditional method for performing correlative imaging is to visually compare the qualitative appearances of the structural and functional imaging result. This cognitive integration of information by direct visual comparison is an inexpensive and useful technique that benefits from the great facility of the human brain to conceptualize three-dimensional space. Alternative methods have been developed to integrate the different data volumes into a matrix common to both. These approaches generally involve image-processing software that allows translation, scaling, and, sometimes, warping of one imaging data set to match the other. This process can be based on mutual information points—that is, structures that are visualized well within both data sets, or by providing reference fiducial markers that can be located independently on each study. These software fusion algorithms can work very well, but many are labor intensive and require particular attenuation to patient positioning to minimize the effects of posture on structural relations of organs that are deformable. Even this does not overcome the issue of structures that are independently mobile, such as the large bowel and that, therefore, may move in relationship to other structures over time.

An elegant approach to these difficulties was the development of hybrid imaging devices that allow "hardware" fusion to occur. Contemporaneous acquisition of both data sets in a known geometry, with the patient positioned identically for both studies, allows the data sets to be merged with minimal software manipulation. A combined PET-CT device was first developed through the efforts of the team at the University of Pittsburgh. Co-registration of the CT and PET images enables physiological uptake to be more confidently assigned to normal structures and pathological uptake to be both recognized and localized. As detailed earlier, use of CT-based attenuation correction can pose difficulties in calculating SUV, and misregistration related to respiratory motion can pose diagnostic challenges. Overall, however, the advantages of this technology far outweigh these limitations. Furthermore, respiratory gating of the PET data and improved reconstruction algorithms may address some of these issues in the future. At our facility we have had PET/CT since 2001, and we now use this modality routinely for staging patients planned for treatment with curative intent and increasingly for radiotherapy treatment planning.

Conclusions

PET scanning is vastly superior to conventional methods used in staging and restaging NSCLC. It provides more accurate information on the extent of disease, and it can give an early assessment of response to treatment that correlates more powerfully with survival than assessments made using other noninvasive imaging studies. Use of PET to exclude patients with incurable extensive disease from potentially toxic radical radiotherapy has the potential to significantly improve the overall results of treatment with this modality. In addition, early diagnosis of limited recurrent disease could potentially facilitate salvage therapies. Furthermore, by decreasing futile attempts at curative treatment, often scarce and expensive surgical and radiotherapy services can be more effectively used. By better defining disease extent, it may improve survival in patients receiving locoregional therapies like radiotherapy and can save incurable patients from futile and toxic treatments. PET will also potentially allow for superior selection and stratification of patients in future clinical trials and may facilitate the development of response-adapted therapies. PET-CT is a further major advance and has the potential to fundamentally change noninvasive diagnostic paradigms used in the evaluation of known or suspected NSCLC.

References

Akhurst, T., Downey, R.J., Ginsberg, M.S., Gonen, M., Bains, M., Korst, R., Ginsberg, R.J., Rusch, V.W., and Larson, S.M. 2002. An initial experience with FDG-PET in the imaging of residual disease after induction therapy for lung cancer. *Ann. Thorac. Surg. 73*:259–264; discussion 264–266.

Ball, D., Smith, J., Wirth, A., and MacManus, M. 2002. Failure of T-stage to predict survival in patients with non-small cell lung cancer treated by radiotherapy with or without concomitant chemotherapy. *Int. J. Radiat. Oncol. Biol. Phys. 54*:1007–1013.

Choi, N.C., Fischman, A.J., Niemierko, A., Ryu, J.S., Lynch, T., Wain, J., Wright, C., Fidias, P., and Mathisen, D. 2002. Dose-response relationship between probability of pathologic tumor control and glucose metabolic rate measured with FDG PET after preoperative chemoradiotherapy in locally advanced non-small cell lung cancer. *Int. J. Radiat. Oncol. Biol. Phys. 54*:1024–1035.

Ciernik, I.F., Dizendorf, E., Baumert, B.G., Reiner, B., Burger, C., Davis, J.B., Lutolf, U.M., Steinert, H.C., and Von Schulthess, G.K. 2003. Radiation treatment planning with an integrated positron emission and computer tomography (PET/CT): a feasibility study. *Int. J. Radiat. Oncol. Biol. Phys. 57*:853–863.

Dunagan, D., Chin, R., Jr., McCain, T., Case, L., Harkness, B., Oaks, T., and Haponik, E. 2001. Staging by positron emission tomography predicts survival in patients with non-small cell lung cancer. *Chest 119*:333–339.

Dwamena, B.A., Sonnad, S.S., Angobaldo, J.O., and Wahl, R.L. 1999. Metastases from non-small cell lung cancer: mediastinal staging in the 1990s—meta-analytic comparison of PET and CT. *Radiology* 213:530–536.

Eschmann, S.M., Friedel, G., Paulsen, F., Budach, W., Harer-Mouline, C., Dohmen, B.M., and Bares, R. 2002. FDG PET for staging of advanced non-small cell lung cancer prior to neoadjuvant radiochemotherapy. *Eur. J. Nucl. Med. Mol. Imaging* 29:804–808.

Gould, M.K., Maclean, C.C., Kuschner, W.G., Rydzak, C.E. and Owens, D.K. 2001. Accuracy of positron emission tomography for diagnosis of pulmonary nodules and mass lesions: a meta-analysis. *Jama* 285:914–924.

Hellwig, D., Groschel, A., Graeter, T.P., Hellwig, A.P., Nestle, U., Schafers, H.J., Sybrecht, G.W., and Kirsch, C.M. 2006. Diagnostic performance and prognostic impact of FDG-PET in suspected recurrence of surgically treated non-small cell lung cancer. *Eur. J. Nucl. Med. Mol. Imaging* 33:13–21.

Hicks, R.J., and MacManus, M.P. 2003. 18F-FDG PET in candidates for radiation therapy: is it important and how do we validate its impact? *J. Nucl. Med.* 44:30–32.

Hicks, R.J., MacManus, M.P., Matthews, J.P., Hogg, A., Binns, D., Rischin, D., Ball, D.L., and Peters, L.J. 2004. Early FDG-PET imaging after radical radiotherapy for non-small cell lung cancer: inflammatory changes in normal tissues correlate with tumor response and do not confound therapeutic response evaluation. *Int. J. Radiat. Oncol. Biol. Phys.* 60:412–418.

Kostis, W.J., Yankelevitz, D.F., Reeves, A.P., Fluture, S.C. and Henschke, C.I. 2004. Small pulmonary nodules: reproducibility of three dimensional volumetric measurement and estimation of time to follow-up CT. *Radiology* 231:446–452.

Lardinois, D., Weder, W., Hany, T.F., Kamel, E.M., Korom, S., Seifert, B., von Schulthess, G.K., and Steinert, H.C. 2003. Staging of non-small cell lung cancer with integrated positron emission tomography and computed tomography. *N. Engl. J. Med.* 348:2500–2507.

Lejeune, C., Al Zahouri, K., Woronoff-Lemsi, M.C., Arveux, P., Bernard, A., Binquet, C., and Guillemin, F. 2005. Use of a decision analysis model to assess the medicoeconomic implications of FDG-PET imaging in diagnosing a solitary pulmonary nodule. *Eur. J. Health Econ.* 6:203–214.

MacManus, M.P., Hicks, R.J., Ball, D.L., Kalff, V., Matthews, J.P., Salminen, E., Khaw, P., Wirth, A., Rischin, D., and McKenzie, A. 2001. F-18 fluorodeoxyglucose positron emission tomography staging in radical radiotherapy candidates with nonsmall cell lung carcinoma: powerful correlation with survival and high impact on treatment. *Cancer* 92:886–895.

MacManus, M.P., Hicks, R.J., Matthews, J.P., McKenzie, A., Rischin, D., Salminen, E.K., and Ball, D.L. 2003. Positron emission tomography is superior to computed tomography scanning for response-assessment after radical radiotherapy or chemoradiotherapy in patients with non-small cell lung cancer. *J. Clin. Oncol.* 21:1285–1292.

Nehmeh, S.A., Erdi, Y.E., Rosenzweig, K.E., Schoder, H., Larson, S.M., Squire, O.D., and Humm, J.L. 2003. Reduction of respiratory motion artifacts in PET imaging of lung cancer by respiratory correlated dynamic PET: methodology and comparison with respiratory gated PET. *J. Nucl. Med.* 44:1644–1648.

Vansteenkiste, J.F., Stroobants, S.G., De Leyn, P.R., Dupont, P.J., and Verbeken, E.K. 1998. Potential use of FDG-PET scan after induction chemotherapy in surgically staged IIIa-N2 non-small cell lung cancer: a prospective pilot study. The Leuven Lung Cancer Group. *Ann Oncol* 9:1193–1198.

van Tinteren, H., Hoekstra, O.S., Smit, E.F., van den Bergh, J.H., Schreurs, A.J., Stallaert, R.A., van Velthoven, P.C., Comans, E.F., Diepenhorst, F.W., Verboom, P., van Mourik, J.C., Postmus, P.E., Boers, M. and Teule, G.J. 2002. Effectiveness of positron emission tomography in the preoperative assessment of patients with suspected non-small cell lung cancer: the PLUS multicentre randomised trial. *Lancet* 359:1388–1393.

Verboom, P., van Tinteren, H., Hoekstra, O.S., Smit, E.F., van den Bergh, J.H., Schreurs, A.J., Stallaert, R.A., van Velthoven, P.C., Comans, E.F., Diepenhorst, F.W., van Mourik, J.C., Postmus, P.E., Boers, M., Grijseels, E.W., Teule, G.J. and Uyl-de Groot, C.A. 2003. Cost-effectiveness of FDG-PET in staging non-small cell lung cancer: the PLUS study. *Eur. J. Nucl. Med. Mol. Imaging* 30:1444–1449.

Viney, R.C., Boyer, M.J., King, M.T., Kenny, P.M., Pollicino, C.A., McLean, J.M., McCaughan, B.C., and Fulham, M.J. 2004. Randomized controlled trial of the role of positron emission tomography in the management of stage I and II non-small cell lung cancer. *J. Clin. Oncol.* 22:2357–2362.

Weber, W.A., Petersen, V., Schmidt, B., Tyndale-Hines, L., Link, T., Peschel, C., and Schwaiger, M. 2003. Positron emission tomography in non-small cell lung cancer: prediction of response to chemotherapy by quantitative assessment of glucose use. *J. Clin. Oncol.* 21:2651–2657.

Werner-Wasik, M., Xiao, Y., Pequignot, E., Curran, W.J., and Hauck, W. 2001. Assessment of lung cancer response after nonoperative therapy: tumor diameter, bidimensional product, and volume. A serial CT scan-based study. *Int. J. Radiat. Oncol. Biol. Phys.* 51:56–61.

Yap, C.S., Schiepers, C., Fishbein, M.C., Phelps, M.E., and Czernin, J. 2002. FDG-PET imaging in lung cancer: how sensitive is it for bronchioloalveolar carcinoma? *Eur. J. Nucl. Med. Mol. Imaging* 29:1166–1173.

Yap, K.K., Yap, K.S., Byrne, A.J., Berlangieri, S.U., Poon, A., Mitchell, P., Knight, S.R., Clarke, P.C., Harris, A., Tauro, A., Rowe, C.C., and Scott, A.M. 2005. Positron emission tomography with selected mediastinoscopy compared to routine mediastinoscopy offers cost and clinical outcome benefits for pre-operative staging of non-small cell lung cancer. *Eur. J. Nucl. Med. Mol. Imaging* 32:1033–1040.

11

Evaluating Positron Emission Tomography in Non-small Cell Lung Cancer: Moving Beyond Accuracy to Outcome

Harm van Tinteren, Otto S. Hoekstra, Carin A. Uyl-de Groot, and Maarten Boers

Introduction

The introduction of new technology in health care is often accompanied by a fundamental dilemma: rapid diffusion inspired by the perspective of considerable clinical benefit versus adequate evaluation followed by implementation in appropriate clinical situations. This dilemma was recognized shortly after the introduction of computed tomography (CT) in 1973 in the United States (Fineberg, 1978), and again with the introduction of magnetic resonance imaging (MRI) a decennium later (Steinberg, 1993). Compared to therapeutic interventions, the evaluation of a diagnostic test is particularly challenging because the results may be less directly translatable into health outcome. The value of plain radiography to diagnose bone fracture is unquestionable, and its result directly dictates patient management. However, when clinical decision making is multifactorial and test results are imperfect, improved diagnostic accuracy provided by one component of a diagnostic test sequence may not necessarily translate into meaningful therapeutic changes. Obviously, without evidence of changes in therapeutic decisions, the implementation of new tests is unlikely to have any impact on health or costs.

A hierarchical approach to the assessment of diagnostic imaging technology has been advocated by experts in health technology (Hunink and Krestin, 2002) (Fig. 56a). The framework typically assumes that evaluation of technical and image quality and diagnostic performance (sensitivity and specificity) is followed by assessment of diagnostic and therapeutic impact. Then, effectiveness on patient and societal outcomes needs to be measured, and the process is completed by defining costs and benefits of the implementation. Although this approach has been cited and discussed frequently in the literature, neither CT nor MRI was assessed accordingly before clinical adoption.

Typically, the first two levels were studied extensively, but the implementation of the technique was independent of evidence from the levels beyond accuracy. Another 20 years later, positron emission tomography (PET) was embraced with similar excitement, but it has its limitations (Balk and Lau, 2001). Although studies on the accuracy of PET have been improved by using methodological and reporting guidelines (Bossuyt *et al.*, 2003), research to establish the role of PET in changing patient outcome and cost-effectiveness failed to appear. Current constraints on health-care resources and the limited availability of the technique in our situation stimulated us to develop a program to evaluate the clinical role of ^{18}F-Fluorodeoxyglucose-PET.

The case of ^{18}F-Fluorodeoxyglucose—positron emission tomography

In 1996 the VU Medical Center (VUmc) acquired a dedicated PET scanner, at that time the second scanner in the Netherlands. The VUmc serves a population of 2.5 million people; thus a lack of scan capacity was to be anticipated. Data were urgently needed to decide for which clinical indications and, within those indications, at what point in the diagnostic work-up ^{18}FDG-PET would be most beneficial. This system is mostly suggested as an "add-on" technique, but it might also substitute for other diagnostic procedures. Our main concern was to ensure equity of access to this new technology. To avoid waiting lists, we restricted ^{18}FDG-PET to clinical research and to those applications where sufficient evidence for a clinically meaningful impact was available.

In this chapter, we first summarize the strengths and limitations of accuracy studies, exemplified by the situation in non-small cell lung cancer (NSCLC). Then, a series of coherent studies will be presented with end points beyond the level of diagnostic accuracy. This framework follows the hierarchical approach for the assessment of new technologies (Fig. 56b). The framework is not intended to be a tight straitjacket that will fit all diagnostic devices. It should merely be considered an illustration of our experience with various steps that contributed to the successful accomplishing of two randomized controlled trials that we performed with PET in NSCLC. The results of these trials were implemented in guidelines, which subsequently changed the situation of patients with suspected NSCLC in our region.

Diagnostic Accuracy of Positron Emission Tomography in Non-small Cell Lung Cancer

In NSCLC, evaluation of the extent of disease in the mediastinum is an important step in deciding about further therapy. In general, patients with preoperative identifiable N2 disease are excluded from thoracotomy. Conventional work-up of the mediastinum often includes noninvasive imaging with CT, followed by invasive exploration through mediastinoscopy. In a meta-analysis of 18 studies, including more than a thousand patients, the negative predictive value of PET was found to be 93%, which is at least as good as that of mediastinoscopy in daily clinical practice

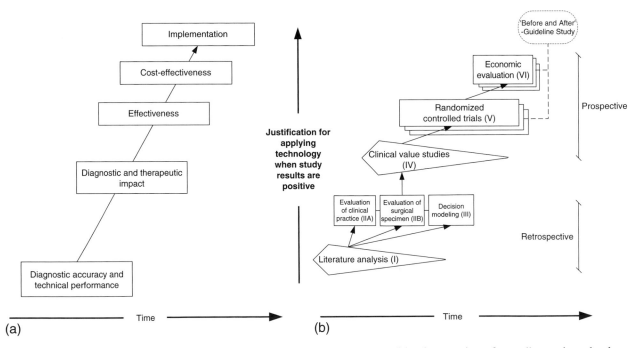

Figure 56 Theoretical hierarchical approach to development, assessment, and implementation of new diagnostic technology (a), and framework of study designs, beyond accuracy to outcome, applying the theoretical approach (b). Note: the sharp elements in (b) indicate that these processes may continue in time.

(Gould *et al.*, 2003a). However, improved diagnostic performance does not necessarily translate into meaningful changes in clinical decisions, even at the level of this relatively simple diagnostic problem. In fact, in spite of the burgeoning number of PET studies, the 2003 guidelines on NSCLC staging (Silvestri *et al.*, 2003) do not clearly recommend that mediastinoscopy can be omitted if a PET scan is negative.

Moreover, in general it will be impossible to measure impact on clinical outcomes in accuracy studies, because the design of such studies requires that the new imaging technique be assessed out of the clinical context. For example, clinically relevant PET findings (such as the suspicion of a distant metastasis in an otherwise resectable cancer) need to be confirmed because of the risk that this might be a false-positive finding (Hoekstra *et al.*, 2003). Since this is not acceptable in the context of an accuracy study on mediastinal staging, it needs to be proven that such confirmation is feasible in daily clinical practice (Pieterman *et al.*, 2000). Furthermore, in clinical practice compared with accuracy studies, it is less common to have dichotomous test results (Lardinois *et al.*, 2003), and gray areas prevail with grading of diagnostic suspicion arising from a variety of test results. This contributes to the difficulty of assessing how the technique will perform in clinical practice when it is used in combination with other diagnostic interventions. How will the clinician use the newly provided diagnostic information, and how will this usage affect the performance of other investigations and treatment?

In many ways, the level of information provided by accuracy studies can be compared to that from Phase II studies in the evaluation of treatments. These studies are a prerequisite in the process of building evidence of activity, but they are not suited to producing information on the difference between the effects of a new drug and those of other competing drugs or devices in routine practice.

The Framework

Our PET-evaluation strategy started some years before the actual installation of the PET scanner in our region. Details of these steps in this framework are discussed in the following (Fig. 55b).

Literature Analysis

When searching for evidence for determining the value of a diagnostic imaging technique, typically data on diagnostic accuracy dominate the literature. Aggregation of such data should be done in a systematic review. Standards for the design, execution, and reporting of accuracy studies and subsequent meta-analysis are now in place: in doing so we recommend applying the guidelines of the Cochrane Collaboration on

reviewing techniques (www.cochrane.org) and initiatives such as Standards for Reporting of Diagnostic Accuracy (STARD) (Bossuyt *et al.*, 2003) and the Quality Assessment of Studies of Diagnostic Accuracy included in the Systematic Reviews (QUADAS) instrument (Whiting *et al.*, 2003) for reporting diagnostic accuracy and for assessing the quality of accuracy studies, respectively. Without positive results from accuracy studies, preferably summarized in systematic reviews, higher-level efficacy studies are not warranted.

In 1996, a comprehensive report from the MDRC Technology Assessment Program was published with a systematic review of the literature on ^{18}FDG-PET as a diagnostic test for potential applications mainly in neurology, cardiology, and oncology (Flynn and Adams, 1996). Motivated by positive accuracy studies further research was suggested to define the impact of ^{18}FDG-PET on treatment decision making and on outcomes, in comparison with existing techniques.

Exploiting Clinical Data Obtained Prior to Introducing a New Test

Signaling (In)efficiency and Assessing Potential Yield in Daily Practice

To assess the potential benefit of a new test in clinical practice, it is necessary to have detailed knowledge of the situation prior to introducing that test. Local facilities, individual expertise, and diagnostic work-up practices may vary substantially even within a relatively small geographic area (Herder *et al.*, 2002). Owing to such variations, potential benefits of a new device may differ between hospitals or practices. Furthermore, the observed level and nature of the (in)efficiency provides the parameters required for sample sizes and other statistical considerations of new studies. Finally, data on the regional situation will help to interpret results from other studies and to assess the generalizability. Preferably such studies are performed on patient files and electronic registries because those reveal the actual behavior. Inclusion of different types of institutes will improve the external validity of the results and has the additional benefit that a committed network of investigators is formed for further research.

We reviewed clinical practice, yield, and costs of preoperative staging for suspected NSCLC in the medical records of all patients diagnosed between 1993 and 1995 in an academic and a large community hospital (Herder *et al.*, 2002). Crosslinking with the Dutch Cancer Registry and the Pathological Anatomical National Registry provided complete surgical, histopathological, and follow-up data. We found a high adherence to international guidelines, despite practice variation between the two hospitals. Hospitals differed in the setting of diagnostic staging (hospitalization, outpatient setting) and the extent of mediastinoscopy use. Approximately half of the operations for presumed resectable NSCLC proved futile. We linked this to limitations of the diagnostic tests undertaken at each level of the TNM [tumor, node, metastases] staging process. Together with the literature survey on [18]FDG-PET, these data clearly indicated that there was room for improvement in the preoperative diagnostic process by [18]FDG-PET.

Mining Histopathological Data

In oncology, stored surgical specimens can also provide useful information on the potential impact of a new diagnostic test. The following example illustrates this.

> When several studies claimed that [18]FDG-PET had a high accuracy and hence could qualify for lymph node staging in breast cancer and melanoma (Greco *et al.*, 2001; Rinne *et al.*, 1998), we were very skeptical as these results were counterintuitive with our experience of histopathological staging of sentinel node biopsies. We measured tumor volumes obtained in sentinel node biopsies and found that the tumor load in malignant lymph nodes was far below that which might be detected by [18]FDG-PET (Mijnhout *et al.*, 2003; Torrenga *et al.*, 2001). Subsequently, large prospective clinical studies confirmed these findings (Wagner *et al.*, 1999, 2001; Wahl *et al.*, 2004). In our *in-vitro* melanoma study (Mijnhout *et al.*, 2003) we combined physical PET principles and epidemiological data to conclude that only a select group of patients might have sufficient tumor volumes to be detected by [18]FDG-PET. Consequently, reviewing existing data sets rather than performing (expensive) prospective clinical studies led us to conclude that PET should not be used routinely for lymph node staging in breast cancer and melanoma.

Obviously, these first steps of the framework (literature analyses and assessment of clinical data prior to introduction of the technique) can be done simultaneously.

Decision Modeling

Decision analysis models the cost-effectiveness of a new diagnostic device. The model can combine results of clinical studies that cover different health-care steps. In the presence of many alternative diagnostic strategies, decision analysis can help to identify the most promising diagnostic tests or algorithms for further research (Gould *et al.*, 2003a, b). The data input is usually based on a meta-analysis of accuracy studies (Gould *et al.*, 2003b). Unfortunately, accuracy studies often fail basic quality standards (e.g., independence of test interpretation, sample size, and case selection) (Gould *et al.*, 2001; Toloza *et al.*, 2003). In addition, decision analyses studies require a large number of assumptions to make decision problems tractable (Hunink and Krestin, 2002). As a consequence, decision analysis is often of only limited value when faced with the complexity of daily clinical practice and the need to make decisions with respect to clinically meaningful outcomes. As more information is generated via clinical studies, fewer assumptions are required for decision modeling (Verboom *et al.*, 2002).

With the input of the data collected in two Dutch hospitals we considered three PET scenarios in a modeling approach: PET upfront in every patient suspected of NSCLC (1), PET after standard imaging, but prior to invasive staging (2) and PET only in patients considered operable and resectable after medical imaging and mediastinoscopy (3). From a cost perspective, the second option was considered most promising (Verboom *et al.*, 2002).

Clinical-Value Studies

Studies that determine therapeutic plans before and after the application of a new test are sometimes referred to as clinical-value studies (Freedman, 1987) or simply as before-after studies (Guyatt *et al.*, 1986). Assessments of diagnostic probabilities and provisional treatment plans are made by means of questionnaires, first without the information contributed by the imaging device and then with the information (Wittenberg *et al.*, 1980). In a third questionnaire, the physician is asked to retrospectively grade the usefulness of this additional information in diagnostic understanding and the choice of therapy.

The credibility of such studies depends on a high-quality design. Specific clinical questions should be addressed, consecutive (unselected) patients presenting with a clinical problem should be entered, and changes in diagnostic certainty and therapeutic choices should be described in sufficient detail (Guyatt *et al.*, 1986). Even with attention to these issues, limitations of the clinical value study include discrepancies between the reported intention and actual clinical behavior, expectation bias, and limited generalization. The clinical value design is most useful when the availability of the new technique is still limited; in the run-up to more complex randomized studies, every patient subjected to the technique can be included to provide relevant information. Standardized feedback also helps learning from the experience of both clinicians and diagnosticians. Furthermore, in rare diseases and indications where randomized controlled trials (RCTs) are impossible, the clinical-value study may be the highest level of evidence possible.

> Since its introduction at the VUmc in 1997 the effect of every [18]FDG-PET scan was evaluated prospectively. Clinicians completed questionnaires just before, immediately after, and several months after the scan to study diagnostic understanding and management changes. In three years more than 600 consecutive patients were included (response 95%); half of those were referrals from outside the VUmc. Diagnostic understanding increased significantly in more than 70%, and management was changed for the benefit of the patients in 40% of all cases. The added value of the scan differed by indication. A subgroup was referred to the [18]FDG-PET center because of suspected NSCLC (Herder *et al.*, 2003) with diagnostic dilemmas, such as unclear radiological findings. After PET, clinicians reported an increase in diagnostic understanding in 84% and benefi-

cial management changes in 50%, mostly canceled surgery (35%). Appreciation of [18]FDG-PET increased with time. Studies with similar designs in Australia (Kalff *et al.*, 2001) and the United States (Hillner *et al.*, 2004) also reported significant management changes (67 and 61%, respectively) due to [18]FDG-PET.

If a clinical-value study fails to show improved diagnostic understanding or therapeutic impact of an indication, it should probably be removed from the list of potential cost-effective tests, whereas promising results warrant further investigation.

Randomized Controlled Trials

The extent to which a patient may ultimately benefit from the addition of a new imaging technique (e.g., in terms of a reduction in iatrogenic toxicity or improvement in [disease-free] survival) can only be investigated through a comparison of the full implementation of the technique added to, or in (partial) substitution of, the conventional process. As only moderate benefit on patient outcome should be expected from any innovation, it is essential that both systematic and random errors are minimized. Balancing both known and unknown prognostic variables by randomly assigning patients to the new test or to the control group is the most efficient way to minimize error. Randomized controlled trials also have several qualities and benefits that arise not from the act of randomization itself but from the fact that they have many features of high-quality research (Abel and Koch, 1999): a written protocol provides transparency; the precalculation of a sample size often allows some exploration of patient and tumor characteristics that might be outcome related; and direct cost comparisons are possible in a real-life setting.

Several aspects are of particular importance for RCTs involving diagnostic imaging. In therapeutic RCTs the outcome is usually measured in terms of patient mortality and morbidity. However, diagnostic tests serve to allocate appropriate therapy to patients. A reasonable outcome measure for such studies is the extent to which appropriate therapy is applied, depending on the condition that the new test does not alter the definitions of staging and treatment for each stage that is already established. For example, there was a broad agreement among clinicians that the treatment options of NSCLC patients would be clear when the diagnostic process had been completed. In addition, because [18]FDG-PET did not produce a new stage classification, a suitable (intermediate) outcome measure could be the reduction of iatrogenic morbidity (translated into unnecessary operations) rather than survival (van Tinteren *et al.*, 2002).

For many diagnostic tests, it is likely that they will first be applied as "add-on" to conventional work-up. This is a relatively straightforward and safe approach. The moment of addition can be clearly defined in terms of logistics, as it is only necessary to have access to the technique within a reasonable timeframe. However, for some tests the challenge is to study whether it can be applied earlier in the diagnostic process and whether it may substitute other procedures. In such cases, relevant end points could shorten the work-up period, reducing morbidity by obviating invasive procedures or reducing costs.

The strategy in the control group determines the extent of the contrast and is therefore essential for interpretation of the results. Usually, the choice is between current clinical practice and state-of-the-art procedures. Current clinical practice, carefully documented as the control strategy, will provide meaningful answers for the clinical community involved (Hunink and Krestin, 2002). Our baseline study showed variances in clinical work-up between hospitals. However, through several interdisciplinary sessions in the preparation phase of the RCT, a common diagnostic work-up protocol could be agreed upon.

In 1998 nine hospitals in our region enrolled 188 patients suspected for potentially resectable NSCLC. These patients were randomly allocated to the conventional work-up approach or to the same approach with [18]FDG-PET performed just prior to mediastinoscopy or thoracotomy (van Tinteren *et al.*, 2002). The trial included about 65% of all eligible patients who were diagnosed in these institutes during that year. [18]FDG-PET-positive findings had to be confirmed by histology or ignored. In the conventional work-up group, patients were managed as in our retrospective study (Herder *et al.*, 2002). In the [18]FDG-PET add-on group the number of futile thoracotomies was reduced by 50%. The fact that objective criteria for end points and clinical consensus about management of patients after diagnosis are important, is illustrated by a recently published second RCT of [18]FDG-PET in NSCLC (Viney *et al.*, 2004). In our study, "futile thoracotomy" pertained to objective criteria (e.g., benign lung lesion, histopathological criteria, stage IIIB disease, explorative thoracotomy for any other reason or recurrent disease, or death from any cause within 1 year after randomization), whereas in the Australian trial the surgeon's decision was taken as the gold standard without validation against follow-up information (e.g., early recurrence) (van Tinteren *et al.*, 2005).

Our next RCT of [18]FDG-PET in NSCLC addressed the question of replacing conventional work-up with a [18]FDG-PET-scan, as had been simulated in the modeling study (Verhoom *et al.*, 2002). Between 1999 and 2001, 465 patients were enrolled by 23 hospitals. The study showed that application of [18]FDG-PET as the initial test had similar overall accuracy compared to traditional work-up but failed to reduce the number of tests (Herder, 2004).

The result of RCTs on a diagnostic test (sequence) should be seen in the context of patient management. In our area, there was clinical consensus that combined modality therapy (including neoadjuvant chemotherapy) should be given in case of locally advanced NSCLC. However, if an observational study on a diagnostic test identifies new prognostic subsets with unclear implication for therapy, an RCT should follow to evaluate the result of various interventions by subset rather than a trial to study the test itself.

Economic Evaluation

One advantage of adding an economic evaluation to an RCT is the possibility of concurrent data collection, using the diagnostic test as the essential contrast. Health technology assessment offers a range of techniques for the evaluation of health-care activities. The most common approach to economic evaluation in diagnostics is a cost-effectiveness analysis. In this type of analysis, outcome is expressed in natural units such as operations avoided or life years saved. Direct and indirect costs can be distinguished. Direct costs are defined as the resources related to the study intervention, for example, inpatient admission, medical procedures, surgery, pharmaceutical drugs, and laboratory tests. In economic evaluations these costs are always taken into account. The importance of the other types of costs depends on the research question and the perspective (e.g., societal or payer) (Drummond *et al.*, 1997).

Several factors are specific to the costing of diagnostic procedures. Diagnostic equipment is usually applied to many indications. For example, [18]FDG-PET is also applied to other oncological and nononcological indications. From an economic point of view the number of [18]FDG-PET scans performed for these indications should also be taken into account in calculating the cost-price of one scan. However, when the procedure is cost-effective for a certain indication,

one cannot automatically assume the total production capacity being filled up with this indication. In theory, one would want to tailor the required capacity of diagnostic equipment to the cost-effectiveness for different indications. Another dilemma emerges when the capacity to perform the procedure is limited. This may either result in waiting lists, or, for the sake of the RCT, priority can be given, resulting in unrealistically short waiting times. In such cases the indirect nonmedical costs of waiting times should also be considered in the economical evaluation.

Our RCT on add-on [18]FDG-PET provided direct data for comparison of costs in relation to diagnosis and therapy. Scenario analyses included various hospital settings, tracer accessibility and scenarios for [18]FDG-PET-scan usage (Verboom *et al.*, 2003). All scenarios proved favorable for PET. The major cost driver was the number of hospital days related to recovery from surgery.

Before and After Implementation

Studying the situation before and after implementation of the procedure, including [18]FDG-PET, closes the circle of studies evaluating the cost-effectiveness of a new diagnostic device. A full appreciation of the technique should take into account all perceptions, quality, and costs of its implementation.

Data from the Regional Cancer Center Registry, where the PLUS-study was active, indicated that the results have a substantial and lasting impact. Since the guidelines were implemented in 2000, the number of lung resections has dropped with an absolute 20% (corresponding to an estimated 50% reduction in unnecessary thoracotomies) compared to the

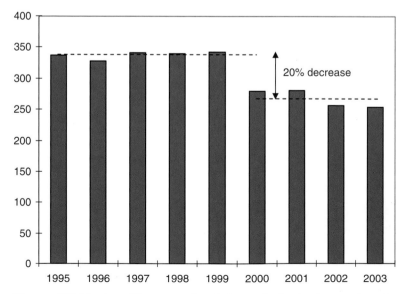

Figure 57 Number of lung resections before and after implementation of a guideline on the use of positron emission tomography in non-small cell lung cancer in 2000. The region of the Comprehensive Cancer Centre Amsterdam where the guideline was introduced serves 2.6 million inhabitants.

average over the 5 years preceding to that (Fig. 57) (Visser, 2004). A formal study is planned to investigate perceptions, quality, and costs in this region of the Netherlands.

Conclusions

Accuracy measures of diagnostic tests do not usually allow an appropriate assessment of its (cost-)effectiveness in clinical practice. Since test results are seldom black or white and are usually part of a complex work-up process, the benefit in overall patient outcome is preferably measured by comparing the diagnostic process, including the new test, concurrently with the prevailing diagnostic work-up. Obviously, a trial should be preceded by formal analyses of the residual (in)efficiency of the prevailing diagnostic work-up and of the potential diagnostic and therapeutic impact of the new test. This hierarchical approach as described by Fryback and others is generally considered the optimal way to evaluate new and costly diagnostic imaging devices (Fryback and Thornbury, 1991). However, practical examples of this stepwise process related to a particular device or indication, especially with regard to outcome levels beyond accuracy, are scarce. This chapter presents a series of coherent studies that have been performed to evaluate the use of PET for specific disease indications in cancer with clinically relevant end points beyond accuracy. These studies do not require exceptional resources.

References

Abel, U., and Koch, A. 1999. The role of randomization in clinical studies: myths and beliefs. *J. Clin. Epidemiol. 6*:487–497.

Dalk, E., and Lau, J. 2001. PET scans and technology assessment: deja vu? *JAMA 7*:936–937.

Bossuyt, P.M., Reitsma, J.B., Bruns, D.E., Gatsonis, C.A., Glasziou, P.P., Irwig, L.M., Lijmer, J.G., Moher, D., Rennie, D., and de Vet, H.C. 2003. Towards complete and accurate reporting of studies of diagnostic accuracy: the STARD Initiative. *Radiology 1*:24–28.

Drummond, M.F., Richardson, W.S., O'Brien, B.J., Levine, M., and Heyland, D. 1997. Users' guides to the medical literature. XIII. How to use an article on economic analysis of clinical practice. A. Are the results of the study valid? Evidence-Based Medicine Working Group. *JAMA 19*:1552–1557.

Fineberg, H.V. 1978. Evaluation of computed tomography: achievement and challenge. *AJR Am. J. Roentgenol. 1*:1–4.

Flynn, K., Adams, E., and Anderson, D. 1996. Positron emission tomography: descriptive analysis of experience with PET in VA and systematic reviews (FDG-PET as a diagnostic test for cancer and Alzheimer's disease). Management Decision and Research Center (MRDC) Technology Assessment Program (MTA 94-001-02), October, 1996.

Freedman, L.S. 1987. Evaluating and comparing imaging techniques: a review and classification of study designs. *Br. J. Radiol. 719*:1071–1081.

Fryback, D.G., and Thornbury, J.R. 1991. The efficacy of diagnostic imaging. *Med. Decis. Making 2*:88–94.

Gould, M.K., Kuschner, W.G., Rydzak, C.E., Maclean, C.C., Demas, A.N., Shigemitsu, H., Chan, J.K., and Owens, D.K. 2003a. Test performance of positron emission tomography and computed tomography for mediastinal staging in patients with non-small cell lung cancer: a meta-analysis. *Ann. Intern. Med. 11*:879–892.

Gould, M.K., Maclean, C.C., Kuschner, W.G., Rydzak, C.E., and Owens, D.K. 2001. Accuracy of positron emission tomography for diagnosis of pulmonary nodules and mass lesions: a meta-analysis. *JAMA 7*:914–924.

Gould, M.K., Sanders, G.D., Barnett, P.G., Rydzak, C.E., Maclean, C.C., McClellan, M.B., and Owens, D.K. 2003b. Cost-effectiveness of alternative management strategies for patients with solitary pulmonary nodules. *Ann. Intern. Med. 9*:724–735.

Greco, M., Crippa, F., Agresti, R., Seregni, E., Gerali, A., Giovanazzi, R., Micheli, A., Asero, S., Ferraris, C., Gennaro, M., Bombardieri, E. and Cascinelli, N. 2001. Axillary lymph node staging in breast cancer by 2-fluoro-2-deoxy-D-glucose-positron emission tomography: clinical evaluation and alternative management. *J. Natl. Cancer. Inst. 8*:630–635.

Guyatt, G.H., Tugwell, P.X., Feeny, D.H., Drummond, M.F., and Haynes, R.B. 1986. The role of before-after studies of therapeutic impact in the evaluation of diagnostic technologies. *J. Chronic. Dis. 4*:295–304.

Herder, G.J. 2004. Traditional versus up-front 18FDG PET staging of non-small cell lung cancer (NSCLC): a Dutch Co-operative randomized study. *J. Clin. Oncol. 14S*:7000.

Herder, G.J., van Tinteren, H., Comans, E.F., Hoekstra, O.S., Teule, G.J., Postmus, P.E., Joshi, U., and Smit, E.F. 2003. Prospective use of serial questionnaires to evaluate the therapeutic efficacy of 18F-fluorodeoxyglucose (FDG) positron emission tomography (PET) in suspected lung cancer. *Thorax 1*:47–51.

Herder, G.J., Verboom, P., Smit, E.F., van Velthoven, P.C., van den Bergh, J.H., Colder, C.D., van, M., I, van Mourik, J.C., Postmus, P.E., Teule, G.J., and Hoekstra, O.S. 2002. Practice, efficacy and cost of staging suspected non-small cell lung cancer: a retrospective study in two Dutch hospitals. *Thorax 1*:11–14.

Hillner, B.E., Tunuguntla, R., and Fratkin, M. 2004. Clinical decisions associated with positron emission tomography in a prospective cohort of patients with suspected or known cancer at one United States center. *J. Clin. Oncol. 20*:4147–4156.

Hoekstra, C.J., Stroobants, S.G., Hoekstra, O.S., Vansteenkiste, J., Biesma, B., Schramel, F.J., van Zandwijk, N., van Tinteren, H., and Smit, E.F. 2003. The value of [18F]fluoro-2-deoxy-D glucose positron emission tomography in the selection of patients with stage IIIA-N2 non-small cell lung cancer for combined modality treatment. *Lung Cancer 2*:151–157.

Hunink, M.G., and Krestin, G.P. 2002. Study design for concurrent development, assessment, and implementation of new diagnostic imaging technology. *Radiology 3*:604–614.

Kalff, V., Hicks, R.J., MacManus, M.P., Binns, D.S., McKenzie, A.F., Ware, R.E., Hogg, A., and Ball, D.L. 2001. Clinical impact of (18)F fluorodeoxyglucose positron emission tomography in patients with non-small cell lung cancer: a prospective study. *J. Clin. Oncol. 1*:111–118.

Lardinois, D., Weder, W., Hany, T.F., Kamel, E.M., Korom, S., Seifert, B., von Schulthess, G.K., and Steinert, H.C. 2003. Staging of non small-cell lung cancer with integrated positron emission tomography and computed tomography. *N. Engl. J. Med. 25*:2500–2507.

Mijnhout, G.S., Hoekstra, O.S., van Lingen, A., van Diest, P.J., Ader, H.J., Lammertsma, A.A., Pijpers, R., Meijer, S., and Teule, G.J. 2003. How morphometric analysis of metastatic load predicts the (un)usefulness of PET scanning: the case of lymph node staging in melanoma. *J. Clin. Pathol. 4*:283–286.

Pieterman, R.M., van Putten, J.W., Meuzelaar, J.J., Mooyaart, E.L., Vaalburg, W., Koeter, G.H., Fidler V., Pruim, J., and Groen, H.J. 2000. Preoperative staging of non-small cell lung cancer with positron emission tomography. *N. Engl. J. Med. 4*:254–261.

Rinne, D., Baum, R.P., Hor, G., and Kaufmann, R. 1998. Primary staging and follow-up of high risk melanoma patients with whole-body 18F-fluorodeoxyglucose positron emission tomography: results of a prospective study of 100 patients. *Cancer 9*:1664–1671.

Silvestri, G.A., Tanoue, L.T., Margolis, M.L., Barker, J., and Detterbeck, F. 2003. The noninvasive staging of non-small cell lung cancer: the guidelines. *Chest I Suppl. 123*:147S–156S.

Steinberg, E.P. 1993. Magnetic resonance coronary angiography—assessing an emerging technology. *N. Engl. J. Med. 12*:879–880.

Toloza, E.M., Harpole, L., and McCrory, D.C. 2003. Noninvasive staging of non-small cell lung cancer: a review of the current evidence. *Chest I. Suppl. 123*:137S–146S.

Torrenga, H., Licht, J., van der Hoeven, J.J., Hoekstra, O.S., Meijer, S., and van Diest, P.J. 2001. Re: Axillary lymph node staging in breast cancer by 2-fluoro-2-deoxy-D-glucose-positron emission tomography: clinical evaluation and alternative management. *J. Natl. Cancer. Inst. 21*:1659–1661.

van Tinteren, H., Hoekstra, O.S., Smit, E.F., van den Bergh, J.H., Schreurs, A.J., Stallaert, R.A., van Velthoven, P.C., Comans, E.F., Diepenhorst, F.W., Verboom, P., van Mourik, J.C., Postmus, P.E., Boers, M., and Teule, G.J. 2002. Effectiveness of positron emission tomography in the preoperative assessment of patients with suspected non-small cell lung cancer: the PLUS multicentre randomised trial. *Lancet 9315*:1388–1393.

van Tinteren, H., Smit, E.F., and Hoekstra, O.S. 2005. FDG-PET in Addition to Conventional Work-Up in Non-Small-Cell Lung Cancer. *J. Clin. Oncol. 7*:1591–1592.

Verboom, P., Herder, G.J., Hoekstra, O.S., Smit, E.F., van den Bergh, J.H., van Velthoven, P.C., and Grijseels, E.W. 2002. Staging of non-small cell lung cancer and application of FDG-PET. A cost modeling approach. *Int. J. Technol. Assess. Health Care 3*:576–585.

Verboom, P., van Tinteren, H., Hoekstra, O.S., Smit, E.F., van den Bergh, J.H., Schreurs, A.J., Stallaert, R.A., van Velthoven, P.C., Comans, E.F., Diepenhorst, F.W., van Mourik, J.C., Postmus, P.E., Boers, M.,

Grijseels, E.W., Teule, G.J., and Uyl-de Groot, C.A. 2003. Cost-effectiveness of FDG-PET in staging non-small cell lung cancer: the PLUS study. *Eur. J. Nucl. Med. Mol. Imaging 11*:1444–1449.

Viney, R.C., Boyer, M.J., King, M.T., Kenny, P.M., Pollicino, C.A., McLean, J.M., McCaughan, B.C., and Fulham, M.J. 2004. Randomized controlled trial of the role of positron emission tomography in the management of stage I and II non-small cell lung cancer. *J. Clin. Oncol. 12*:2357–2362.

Visser, O. 2004. Lung resections in the region of the Comprehensive Cancer Center Amsterdam, The Netherlands. Personal Communication.

Wagner, J.D., Schauwecker, D., Davidson, D., Coleman, J.J., III, Saxman, S., Hutchins, G., Love, C., and Hayes, J.T. 1999. Prospective study of fluorodeoxyglucose-positron emission tomography imaging of lymph node basins in melanoma patients undergoing sentinel node biopsy. *J. Clin. Oncol. 5*:1508–1515.

Wagner, J.D., Schauwecker, D.S., Davidson, D., Wenck, S., Jung, S.H., and Hutchins, G. 2001. FDG-PET sensitivity for melanoma lymph node metastases is dependent on tumor volume. *J. Surg. Oncol. 4*:237–242.

Wahl, R.L., Siegel, B.A., Coleman, R.E., and Gatsonis, C.G. 2004. Prospective multicenter study of axillary nodal staging by positron emission tomography in breast cancer: a report of the staging breast cancer with PET Study Group. *J. Clin. Oncol. 2*:277–285.

Whiting, P., Rutjes, A.W., Reitsma, J.B., Bossuyt, P.M., and Kleijnen, J. 2003. The development of QUADAS: a tool for the quality assessment of studies of diagnostic accuracy included in systematic reviews. *BMC. Med. Res. Methodol. 3*:25.

Wittenberg, J., Fineberg, H.V., Ferrucci, J.T., Jr., Simeone, J.F., Mueller, P.R., vanSonnenberg, E., and Kirkpatrick, R.H. 1980. Clinical efficacy of computed body tomography, II. *AJR Am. J. Roentgenol. 6*:1111–1120.

12

Non-small Cell Lung Cancer: False-positive Results with ^{18}F-fluorodeoxyglucose Positron Emission Tomography

Siroos Mirzaei, Helmut Prosch, Peter Knoll, and Gerhard Mostbeck

Introduction

Positron emission tomography (PET) is a molecular imaging modality that has been established as a useful tool in the management of different cancer entities. The number of clinical applications for PET and the number of PET centers is growing continuously (Hoh et al., 1997). In non-small cell lung cancer (NSCLC), PET has proven to have a higher sensitivity and specificity for the evaluation of mediastinal lymph nodes than computed tomography (CT) (Silvestri et al., 2003). Positron emission tomography has therefore been established as a routine imaging modality in the staging of NSCLC. In addition to the evaluation of mediastinal lymph node involvement, whole-body PET scanning is increasingly used to evaluate potential distant metastases in NSCLC patients. The final role of whole-body PET in the routine staging of NSCLC, however, has yet to be defined.

The most commonly used PET radiopharmaceutical in clinical practice is the glucose analog F18-fluorodeoxyglucose (^{18}F-FDG), which enters the cells by a number of membrane glucose transporter proteins. Within the cell, ^{18}F-FDG is phosphorylated to ^{18}F-FDG-6-Phosphate, which cannot be further metabolized and remains trapped within the cells. In many tumor cells, membrane glucose transporters and hexokinase tend to be overexpressed (Mueckler, 1994). In addition, glucolysis is upregulated in malignant disease (Weber, 1977). Interestingly, some tumors such as hepatocellular carcinomas may demonstrate higher glucose-6-phosphatase activity, resulting in relatively low uptake (Torizuka et al., 1995).

FDG, however, is not a tumor-specific substance, and benign lesions with increased glucose metabolism may lead to false-positive results. As is the case with other imaging modalities including radiography, ultrasound, computed tomography, magnetic resonance imaging, and conventional nuclear medicine examinations, there are normal variants, imaging artifacts, and other causes of false-positive results that have to be recognized and considered during image interpretation.

In this chapter, we aim to explain some common causes of false-positive results by FDG-PET in patients with NSCLC. We will discuss possible reasons for these false-positive findings in order to avoid the occurrence and misinterpretation of these tracer accumulations.

Thorough knowledge of potential false-positive findings is essential for correct interpretation of PET images. If a focal FDG uptake may not be unequivocally classified as benign, further imaging (CT, MRI, ultrasound) and, in many cases, tissue diagnosis may be necessary to clarify the nature of the tracer accumulation. The fusion of CT and PET images and, even more, combined CT-PET systems further increase the specificity and sensitivity of PET scans by adding anatomic information from the CT scan to the metabolic information from the PET scan (Kostakoglu *et al.*, 2004).

In the past years, a number of other PET tracers have been developed in oncologic research to assess alterations in malignant cell processes as well as changes in DNA or membrane synthesis. Although several small studies have not shown major additional clinical information using these tracers, there are some promising reports in regard of investigating potential false-positive findings in FDG-PET. Therefore, some new PET tracers will be addressed in this chapter.

Physiological High Uptake of ^{18}F-FDG in Different Tissues

Head and Central Nervous System

The brain typically shows high uptake of ^{18}F-FDG in the cortex, thalamus, and basal ganglia. Commonly, we observe symmetrical higher activity in lymphoid-tonsillar tissue compared to the background. This may become more difficult to recognize if there has been previous radiotherapy or surgery.

Neck

In the neck region there is the possibility of symmetrical physiological uptake not only in the muscles, but also in brown fat tissue. In particular, tracer accumulation in the supraclavicular area within brown fat tissue may mimic involvement of supraclavicular lymph nodes in NSCLC (Cohade *et al.*, 2003). Ultrasound evaluation of these supraclavicular lymph nodes may help to exclude lymph node involvement in such cases.

The thyroid gland has been reported to show a variable FDG uptake that may be either focal or diffuse and thus may be confounding (Shreve *et al.*, 1999). Although thyroid metastases are rare in NSCLC, focal thyroid FDG uptake should be further investigated as it may be due to a synchronous thyroid cancer (Yi *et al.*, 2005).

Chest

In the chest, it has been shown that there is greater activity in the inferior and posterior segments of the lungs com-

pared to other lung segments (Miyauchi and Wahl, 1996). The reasons for this phenomenon are not fully understood. It has been postulated that the increased blood flow in these segments as well as scatter from the heart and liver may contribute to this finding (Miyauchi and Wahl, 1996). As a consequence, sensitivity regarding lesion detection may be reduced in these lung areas.

Myocardial activity is very variable and intense because the myocardium relies on glycolitic metabolism. Problems with myocardial uptake may arise if the uptake is only regional and not homogeneous within the whole myocardium. This could then be misinterpreted as a mediastinal or pulmonary lesion. In these cases the correlation of metabolic activity with anatomic imaging is inevitable. To obtain an excellent and constant image quality in FDG-PET of the thorax, it is advisable to keep patients fasting at least 4–6 hr prior to PET. Another possible but still not validated method is to advise the patient to avoid carbohydrate-rich intake the evening before the PET investigation.

Abdomen

FDG uptake in brown adipose tissue can mimic metastasis in the suprarenal glands (Reddy and Ramaswamy, 2005). Thus, false-positive FDG-PET findings indicative for suprarenal gland metastases have been reported in patients with NSCLC. The FDG uptake in brown adipose tissue can be reduced by different premedication approaches (Gelfand *et al.*, 2005; Reddy and Ramaswamy, 2005).

Physiological uptake of FDG of varying extent is seen in the gastrointestinal system, where activity is most commonly seen in the stomach and large bowel. It has been discussed that this bowel FDG uptake might be related either to smooth muscle FDG uptake or to activity in intraluminal bowel contents (Chun *et al.*, 2003; Stahl *et al.*, 2000). Furthermore, one should be aware of attenuation differences due to physiological bowel motion in PET-CT studies, with subsequent differences in standardized uptake values (SUVs) (Nakamoto *et al.*, 2004). However, these "abnormalities" will not hamper the diagnosis in patients with lung cancer, as gastrointestinal M1 disease is rare and more attention has to be paid to metastatic involvement of the suprarenal glands. In patients with a significant nodular uptake in the colon, colonoscopy is indicated to rule out a secondary malignancy (Tatlidil *et al.*, 2002).

Urinary Tract

FDG is not totally absorbed in the renal tubules, and some urinary activity is seen in almost all patients. However, this activity may be present only in parts of the urinary tract. To avoid focal activity within the urinary tract, an iterative reconstruction algorithm should be preferred over filtered backprojection, as the latter technique could produce artifacts in regions with high FDG uptake. Furthermore, it is beneficial to hydrate the patient and apply diuretics after application of FDG, in order to accelerate the urinary excretion of the radiotracer, leading to a reduction in radiation exposure and a better image quality (Diehl *et al.*, 2004).

Breast

Glandular breast tissue often demonstrates moderate metabolic activity in premenopausal women, depending on the time within the menstrual cycle. In postmenopausal women under hormonal replacement therapy, the pattern of uptake within the breast is usually symmetric and therefore easily identified.

Skeletal Muscle

As glycolysis is the major source of energy of skeletal muscles, muscle contractions during the FDG uptake or insulin administration prior to FDG injection may cause an accumulation of the tracer in the muscle (Fig. 58) (Shreve *et al.*, 1999). If symmetrical, the tracer accumulation in skeletal muscle is easily identified as such. However, asymmetric FDG uptake, as is sometimes observed in the shoulder girdle, may be a puzzling finding (Shreve *et al.*, 1999).

Focal Uptake of FDG Due to Benign Disease

FDG is trapped not only in malignant tissue, but also in macrophages and other activated inflammatory cells (Fig. 59) (Kubota *et al.*, 1994; Yamada *et al.*, 1995). In NSCLC patients, as in other oncologic patients, this will decrease the specificity of the examination. A number of granulomatous disorders with increased uptake of FDG have

been reported. These include tuberculosis (Knopp and Bischoff, 1994), sarcoidosis (Lewis and Salama, 1994), coccidioidomycosis, and aspergillosis (Guhlmann *et al.*, 1997; Gupta *et al.*, 1999).

As already mentioned, glucose utilization is not specific for malignant tumors. Radiopharmaceuticals reflecting tissue proliferation have therefore been suggested as a means to improve the specificity of the imaging modalities. One approach to differentiate malignant disease from inflammation could be the evaluation of DNA synthesis by measuring the uptake of the thymidine analog 3'-deoxy-3'-[^{18}F]-fluorothymidine (18-F-FLT). In a study, performed by Shields *et al.*, (1998) 30 patients with solitary pulmonary nodules (SPNs) were prospectively examined with PET using 18-F FLT. The authors concluded that 18-F FLT uptake was specific in malignant lesions and may be used for the differential diagnosis of SPNs, assessment of proliferation, and estimation of prognosis (Buck *et al.*, 2002).

Another effort to differentiate malignant from inflammatory tissue has been described by applying early and delayed reading of FDG-PET images. In contrast to benign lesions, malignant lesions showed higher SUV at 3 hr than at 1 hr after tracer injection (Demura *et al.*, 2003). However, the disadvantage is the fact that another examination slot has to be used for the delayed data acquisition.

In a case study, it has been demonstrated that silicosis with anthracotic pigments in mediastinal lymph nodes may demonstrate high glucose uptake. As these lymph nodes are not metabolically active, the pathophysiological mechanism of high glucose uptake in the PET examination remained unclear (Tomita *et al.*, 2003).

As stated earlier, the overexpression of glucose transporters in human cancers, including NSCLC, is known to be responsible for the increased 18-FDG uptake delineated on PET (Higashi *et al.*, 2000). In NSCLC, PET is most useful

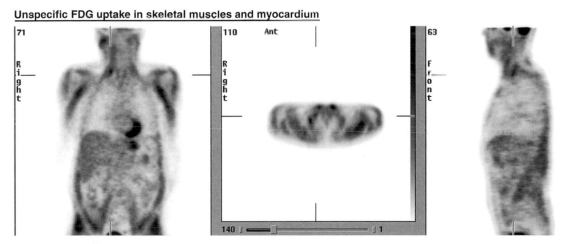

Figure 58 PET with 18-FDG: unspecific FDG uptake in the skeletal muscles and an intensive myocardial FDG uptake caused by injection of insulin prior to the PET investigation.

CT-scan and FDG-PET of a patient with suspected lung cancer

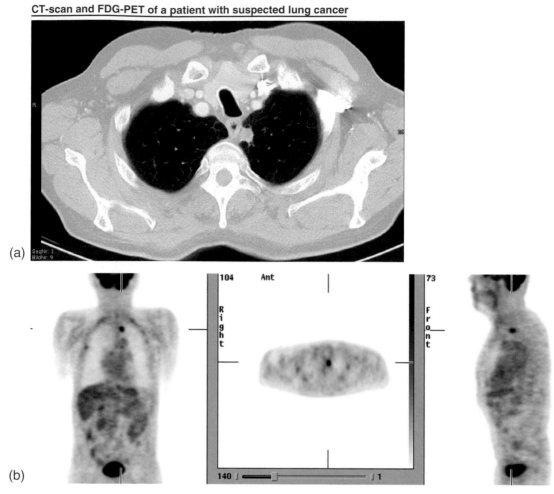

Figure 59 A 60-year-old male patient with suspected lung cancer. (a) CT scan showing a 1.2 cm pulmonary nodule in the left upper lobe. (b) FDG-PET: hypercatabolic pulmonary nodule in the left upper lobe. CT-guided biopsy revealed chronic pneumonitis.

to exclude metastatic disease, especially nodal staging. When contralateral mediastinal lymph node metastasis is suspected (N3), surgical treatment is controversial, despite the fact that surgery is the preferred treatment for NSCLC. FDG-PET has demonstrated a high sensitivity and specificity of 83–100% and 52–100% (Fischer *et al.*, 2001; Gould *et al.*, 2001; Vansteenkiste and Stroobants, 2001) and an equally high accuracy of 91% for nodal staging of NSCLC according to a recent meta-analysis including 1292 patients (Hellwig *et al.*, 2001).

However, a significant number of false-positive results in countries with a high prevalence of tuberculosis have been reported (Neuenschwander *et al.*, 2000). Pathologically, these patients present with lymphoid follicular hyperplasia, a type of chronic lymphadenopathy in patients with a history of smoking, chronic airway disease, or tuberculosis. As a consequence, false-positive results in 18-F-FDG-PET may result. It has been found that expression of glucose transporters in mediastinal lymph nodes is significantly higher in false-positive nodes than in true negative nodes, and the

grade of lymphoid follicular hyperplasia is a major pathologic factor that influences 18-F-FDG uptake in mediastinal lymph nodes (Chung *et al.*, 2004). Diffuse increased uptake of FDG may also be seen in dermatomyositis complicating malignancy, a factor that may reduce image contrast and tumor detectability (Cook, 2003).

Thymic tissue may also demonstrate enhanced uptake in children, as well as in adults following chemotherapy ("thymic rebound") (Ferdinand *et al.*, 2004; Kaste *et al.*, 2005). This thymic tissue usually presents as an inverted V-shaped activity.

In the neck region, there may be diffuse uptake within the thyroid gland in Graves' disease as well as in chronic Hashimoto's thyroiditis (Boerner *et al.*, 1998; Schmid *et al.*, 2003). As there is substantial risk of malignancy, patients with new thyroid lesions on FDG-PET scan should undergo tissue diagnosis, if the outcome of the procedure might influence outcome and management (Cohen *et al.*, 2001).

In another recent study, FDG uptake in the buttocks in patients with NSCLC has been reported (Fig. 60) (Prosch *et al.*, 2005). Sonographically, the lesions presented as poorly marginated, round hyperechogenic lesions surrounded by an unsharp hypoechogenic rim, and thus were not suspicious for metastases. Subcutaneous metastases usually present as hypoechogenic, well-demarcated nodules with a polycyclic shape within the subcutaneous fat. Using color Doppler ultrasound, these lesions demonstrate hypervascularity with multiple feeding vessels (Giovagnorio et al., 2003). Ultrasound-guided biopsy showed necrotic adipose tissue with focal scar formation and accumulation of macrophages. The authors assumed that FDG accumulation in the buttocks is highly suggestive of injection site granulomas. Interestingly, the time interval between FDG-PET demonstrating high FDG uptake and the last injection in the gluteal region was one year, indicating that metabolic activity of granulomas may last a long time. Also, benign breast diseases, including focal nodular hyperplasia and nodular adenosis, have been reported to accumulate FDG (Fig. 61) (Aznar et al., 2005; Kim et al., 2005).

High Metabolic Activity after Treatment

FDG-PET is increasingly being used to monitor chemotherapy treatment in NSCLC (Cerfolio et al., 2003). There are reports indicating that FDG-PET uptake might be a better indicator for a treatment response than use of tumor dimensions as measured on CT scans (Cerfolio et al., 2003). However, there are also reports on false-positive FDG-PET findings after chemotherapy or radiotherapy.

As stated earlier, 18-F-FDG is trapped not only in tumor cells but also in macrophages, granulation tissues, and necrosis (Kubota et al., 1994). Thus, an inflammatory reaction with invasion of macrophages and lymphocytes in a tumor-free tissue may cause false-positive findings in FDG-PET after chemotherapy (Ohtsuka et al., 2005).

Another cause of false-positive findings may be the occurrence of metaplastic and proliferative epithelial elements induced by radiochemotherapy. It has been shown that

Restaging with FDG-PET in a female patient with NSCLC

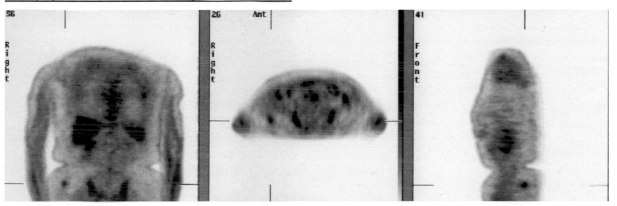

Figure 60 PET scan from a 71-year-old female patient with NSCLC. Restaging after three cycles of chemotherapy. FDG-PET scan demonstrates a hypermetabolic focus in the right buttock. Ultrasound-guided biopsy confirmed the suspicion of injection site granuloma.

FDG-PET of a patient with NSCLC after chemotherapy

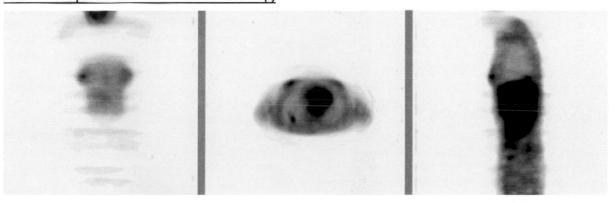

Figure 61 57-year-old female patient with a 30-month history of non-small cell lung cancer. FDG-PET scan for reevaluation after five cycles of chemotherapy demonstrating pathological tracer accumulation in the right lower dorsal lung and an additional hypermetabolic focus in the right breast (sagittal scan). Ultrasound-guided biopsy of the breast lesion revealed a hyperplastic mammary gland with signs of hyperplasia and adenosis.

radiation and chemotherapy may cause a metaplastic change of the normal bronchioalveolar cells to highly proliferative epithelial lesions during a process of fibrosis (Nishikawa *et al.*, 1995).

In order to overcome this source of error, some authors have suggested the use of other proliferation tracers (e.g., 18-F-FLT). In an animal study, it has been shown that radiolabeled pyrimidine nucleosides might be more accurate than 18-F-FDG for PET imaging of tumor response to therapy (Barthel *et al.*, 2003).

Focal Uptake of FDG Due to Artifacts

Without attenuation correction (AC), the PET images may demonstrate higher activities in superficial structures, thereby hampering identification of subcutaneous metastases (Engel *et al.*, 1996). Filtered backprojection methods (FBP) produce streak artifacts. These streaks may obscure small lesions adjacent to areas of high activity, as could be the case in central lung cancer with adjacent lymph node involvement. The vast majority of these artifacts can be avoided by using iterative algorithms for image reconstruction and attenuation correction.

Another artifact may be due to paravasal tracer injection in patients with a suspicion of lung cancer. In these patients, the SUV of the tumor has a high prognostic value (Borst *et al.*, 2005; Pottgen *et al.*, 2006), and an SUV threshold is defined to differentiate benign from malignant disease (Hara *et al.*, 2004; Hickeson *et al.*, 2002). Accordingly, it is crucial that care be taken to avoid paravasal FDG injection, which may cause not only reconstruction artifacts, but also low-count studies and inaccuracies in the measurement of the SUV. Ring artifacts may be visible in the case of misregistration between transmission and emission scans due to patient movement. This kind of artifact has become less frequent, now that the number of PET scanners that allow simultaneous acquisition of emission and transmission scans is increasing.

Focal Uptake of FDG Due to Artifacts by New PET Devices

In the early 1980s, commercial PET scanners with full-ring BGO (bismuth germinate orthosilicate) were introduced. Transmission scanning was performed using Ge-68 sources. One decade later, a new generation of commercial BGO tomographs was introduced (Mullani *et al.*, 1990). Rotating rod sources (Cs-137) were developed to replace the earlier ring sources. In this way, it became possible to perform transmission scans after tracer injection. Such examination protocols made it possible to obtain whole-body attenuation-corrected PET images. On the one hand, transmission data are necessary for attenuation correction. On the other hand, it leads to prolonged imaging time, and at the

same time the noise from the transmission scan propagates into the final attenuation-corrected images. In many clinical applications (as in intrathoracic regions) it is necessary to compare the attenuation-corrected to the nonattenuation-corrected images in order to better verify and localize small lesions.

Further improvements of combined PET-CT devices allow functional and anatomical mapping. These data are an excellent diagnostic tool in the diagnosis and staging of lung cancer and provide excellent information for the planning of radiation therapy, obtained within one examination. The X-ray attenuation map is scaled to a 511 keV transmission map (Kinahan *et al.*, 1998) with a spatial resolution of the PET, which provides a very fast transmission scan with only minimal noise and a reduced examination time of up to 15 min for whole-body studies using advanced scintillators such as GSO (gadolinium orthosilicate) or LSO (lutetium oxyorthosilicate). The scaled attenuation maps are then forward projected and used for the correction of the emission data. However, the temporal resolution of CT transmission images is significantly different from the PET emission data (few seconds versus several minutes), which could contribute to image errors.

A number of studies have demonstrated a wide range of artifacts caused by CT beam hardening, metal implants (i.e., pacemaker) or involuntary patient motion (Antoch *et al.*, 2002; Kamel *et al.*, 2003). As mentioned previously, CT artifacts may propagate into PET images in PET-CT investigations. Metal artifacts arising from dental and orthopedic implants, pacemakers, and chemotherapy ports may cause "false" high uptake, because proper attenuation-correction values for the metallic implants cannot be adequately estimated by standard CT-attenuation maps (DiFilippo and Brunken, 2005). In a recent study it has been demonstrated that segmented CT transmission maps would help to avoid such artificial focal FDG uptake by metal implants (Mirzaei *et al.*, 2005). In a similar way, contrast media-induced artifacts may occur on PET in PET-CT studies. Depending on the contrast accumulation, the resulting artifacts and biases on the attenuation-corrected PET may be significant. Furthermore, some studies have indicated an overestimation of SUV arising from elevated CT attenuation values in areas of i.v., contrast accumulation (Antoch *et al.*, 2002).

Although an improvement in diagnostic accuracy using combined PET-CT in cancer patients has been reported, a potential disadvantage of CT-based attenuation correction may arise from the significantly shorter acquisition time of CT compared to PET. This may result in co-registration errors due to respiratory movement. In the thoracic cavity, this is particularly essential for lesions near the diaphragm—that is, in the basal parts of the right lower lung lobe and in the top of the right liver lobe (Fig. 62) (Goerres *et al.*, 2002). Therefore, for lesions in this area, it is crucial to compare the attenuation-corrected with the

Champignon artifact due to diaphragmatic movement in PET-CT

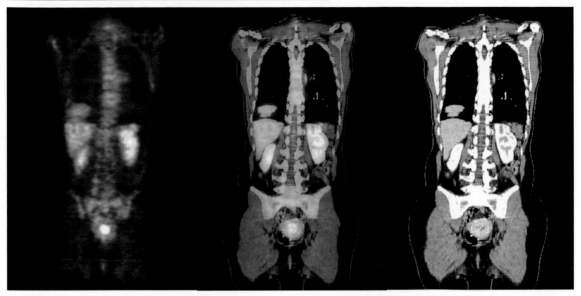

Figure 62 PET-CT with 18-FDG: coronal images demonstrate a so-called champignon artifact in the reconstructed PET (left) and fused images (middle) caused by extensive diaphragmatic movement during CT acquisition (right).

nonattenuation-corrected images, as a hepatic lesion would be falsely located in the right lung or vice versa. Respiratory gating can be used to minimize motion artifacts in PET-CT studies (Nehmeh et al., 2002). Further approaches, such as optical flow algorithms (Schafers et al., 2005), are in development. This technique estimates the change in an image over time for motion correction, thus avoiding, for example, false localization of a liver lesion in PET images into the right lower lung lobe.

Conclusions

FDG-PET has demonstrated a high sensitivity and specificity and, equally, a high accuracy of 91% for nodal staging of NSCLC (Fischer et al., 2001; Gould et al., 2001; Vansteenkiste and Stroobants, 2001). However, there is a growing number of reports on physiological processes and benign entities with hypermetabolic activity that have to be distinguished from metastatic lesions (Shreve et al., 1999). Therefore, one should keep in mind the possibility of false-positive findings due to technical and physiological reasons as well as various benign entities presenting with high glucose uptake. Accordingly, some authors have stated that FDG-PET imaging may have greater clinical value based on its excellent negative-predictive value in ruling out malignancy. Because of the limited specificity and positive-

predictive values of positive FDG-PET result, further diagnostic procedures (preferably obtaining tissue diagnosis) are inevitable in order to avoid the overstaging of malignancy (Tomita et al., 2003).

References

Antoch, G., Freudenberg, L.S., Egelhof, T., Stattaus, J., Jentzen, W., Debatin, J.F., and Bockisch, A. 2002. Focal tracer uptake: a potential artifact in contrast-enhanced dual-modality PET/CT scans. *J. Nucl. Med. 43*:1339–1342.

Aznar, D.L., Ojeda, R., Garcia, E.U., Aparici, F., Sanchez, P.A., Flores, D., Martinez, C., and Sopena, R. 2005. Focal nodular hyperplasia (FNH): a potential cause of false-positive positron emission tomography. *Clin. Nucl. Med. 30*:636–637.

Barthel, H., Cleij, M.C., Collingridge, D.R., Hutchinson, O.C., Osman, S., He, Q., Luthra, S.K., Brady, F., Price, P.M., and Aboagye, E.O. 2003. 3′-deoxy-3′-[18F]fluorothymidine as a new marker for monitoring tumor response to antiproliferative therapy in vivo with positron emission tomography. *Cancer Res. 63*:3791–3798.

Boerner, A.R., Voth, E., Theissen, P., Wienhard, K., Wagner, R., and Schicha, H. 1998. Glucose metabolism of the thyroid in Graves' disease measured by F-18-fluoro-deoxyglucose positron emission tomography. *Thyroid 8*:765–772.

Borst, G.R., Belderbos, J.S., Boellaard, R., Comans, E.F., De Jaeger, K., Lammertsma, A.A., and Lebesque, J.V. 2005. Standardised FDG uptake: a prognostic factor for inoperable non-small cell lung cancer. *Eur. J. Cancer 41*:1533–1541.

Buck, A.K., Schirrmeister, H., Hetzel, M., Von Der Heide, M., Halter, G., Glatting, G., Mattfeldt, T., Liewald, F., Reske, S.N., and Neumaier, B.

2002. 3-deoxy-3-[(18)F]fluorothymidine positron emission tomography for noninvasive assessment of proliferation in pulmonary nodules. *Cancer Res.* 62:3331–3334.

Cerfolio, R.J., Ojha, B., Mukherjee, S., Pask, A.H., Bass, C.S., and Katholi, C.R. 2003. Positron emission tomography scanning with 2-fluoro-2-deoxy-d-glucose as a predictor of response of neoadjuvant treatment for non-small cell carcinoma. *J. Thorac. Cardiovasc. Surg.* 125:938–944.

Chun, H., Kim, C.K., Krynckyi, B.R., and Machac, J. 2003. The usefulness of a repeat study for differentiating between bowel activity and local tumor recurrence on FDG PET scans. *Clin. Nucl. Med.* 28:672–673.

Chung, J.H., Cho, K.J., Lee, S.S., Baek, H.J., Park, J.H., Cheon, G.J., Choi, C.W., and Lim, S.M. 2004. Overexpression of Glut1 in lymphoid follicles correlates with false-positive (18)F-FDG PET results in lung cancer staging. *J. Nucl. Med.* 45:999–1003.

Cohade, C., Osman, M., Pannu, H.K., and Wahl, R.L. 2003. Uptake in supraclavicular area fat ("USA-Fat"): description on 18F-FDG PET/CT. *J. Nucl. Med.* 44:170–176.

Cohen, M.S., Arslan, N., Dehdashti, F., Doherty, G.M., Lairmore, T.C., Brunt, L.M., and Moley, J.F. 2001. Risk of malignancy in thyroid incidentalomas identified by fluorodeoxyglucose-positron emission tomography. *Surgery* 130:941–946.

Cook, G.J. 2003. Artifacts and Normal Variants in PET Imaging. In: Valk, D.L., and Townsend, D.W. (Eds.), *Positron Emission Tomography: Basic Science and Clinical Practice, P.E.B.* London: Springer, pp. 495–507.

Demura, Y., Tsuchida, T., Ishizaki, T., Mizuno, S., Totani, Y., Ameshima, S., Miyamori, I., Sasaki, M., and Yonekura, Y. 2003. 18F-FDG accumulation with PET for differentiation between benign and malignant lesions in the thorax. *J. Nucl. Med.* 44:540–548.

Diehl, M., Manolopoulou, M., Risse, J., Kranert, T., Menzel, C., Dobert, N., and Grunwald, F. 2004. Urinary fluorine-18 fluorodeoxyglucose excretion with and without intravenous application of furosemide. *Acta Med. Austriaca* 31:76–78.

DiFilippo, F.P., and Brunken, R.C. 2005. Do implanted pacemaker leads and ICD leads cause metal-related artifact in cardiac PET/CT? *J. Nucl. Med.* 46:436–443.

Engel, H., Steinert, H., Buck, A., Berthold, T., Huch Boni, R.A., and von Schulthess, G.K. 1996. Whole-body PET: physiological and artifactual fluorodeoxyglucose accumulations. *J. Nucl. Med.* 37:441–446.

Ferdinand, B., Gupta, P., and Kramer, E.L. 2004. Spectrum of thymic uptake at 18F-FDG PET. *Radiographics* 24:1611–1616.

Fischer, B.M., Mortensen, J., and Hojgaard, L. 2001. Positron emission tomography in the diagnosis and staging of lung cancer: a systematic, quantitative review. *Lancet Oncol.* 2:659–666.

Gelfand, M.J., O'Hara S.M., Curtwright, L.A., and Maclean, J.R. 2005. Premedication to block [(18)F]FDG uptake in the brown adipose tissue of pediatric and adolescent patients. *Pediatr. Radiol.* 35:984–990.

Giovagnorio, F., Valentini, C., and Paonessa, A. 2003. High-resolution and color Doppler sonography in the evaluation of skin metastases. *J. Ultrasound Med.* 22:1017–1022; quiz 1023–1015.

Goerres, G.W., Kamel, E., Heidelberg, T.N., Schwitter, M.R., Burger, C., and von Schulthess, G.K. 2002. PET-CT image co-registration in the thorax: influence of respiration. *Eur. J. Nucl. Med. Mol. Imag.* 29:351–360.

Gould, M.K., Maclean, C.C., Kuschner, W.G., Rydzak, C.E., and Owens, D.K. 2001. Accuracy of positron emission tomography for diagnosis of pulmonary nodules and mass lesions: a meta-analysis. *JAMA* 285:914–924.

Guhlmann, A., Storck, M., Kotzerke, J., Moog, F., Sunder-Plassmann, L., and Reske, S.N. 1997. Lymph node staging in non-small cell lung cancer: evaluation by [18F]FDG positron emission tomography (PET). *Thorax* 52:438–441.

Gupta, N.C., Graeber, G.M., Rogers, J.S., 2nd, and Bishop, H.A. 1999. Comparative efficacy of positron emission tomography with FDG and

computed tomographic scanning in preoperative staging of non-small cell lung cancer. *Ann. Surg.* 229:286–291.

Hara, M., Shiraki, N., Itoh, M., Shibamoto, Y., Iida, A., Nishio, M., and Tamaki, T. 2004. A problem in diagnosing N3 disease using FDG-PET in patients with lung cancer—high false positive rate with visual assessment. *Ann. Nucl. Med.* 18:483–488.

Hellwig, D., Ukena, D., Paulsen, F., Bamberg, M., and Kirsch, C.M. 2001. [Meta-analysis of the efficacy of positron emission tomography with F-18-fluorodeoxyglucose in lung tumors. Basis for discussion of the German Consensus Conference on PET in Oncology 2000]. *Pneumologie* 55:367–377.

Hickeson, M., Yun, M., Matthies, A., Zhuang, H., Adam, L.E., Lacorte, L., and Alavi, A. 2002. Use of a corrected standardized uptake value based on the lesion size on CT permits accurate characterization of lung nodules on FDG-PET. *Eur. J. Nucl. Med. Mol. Imag.* 29:1639–1647.

Higashi, K., Ueda, Y., Sakurai, A., Wang, X.M., Xu, L., Murakami, M., Seki, H., Oguchi, M., Taki, S., Nambu, Y., Tonami, H., Katsuda, S., and Yamamoto, I. 2000. Correlation of Glut-1 glucose transporter expression with. *Eur. J. Nucl. Med.* 27:1778–1785.

Hoh, C.K., Schiepers, C., Seltzer, M.A., Gambhir, S.S., Silverman, D.H., Czernin, J., Maddahi, J., and Phelps, M.E. 1997. PET in oncology: will it replace the other modalities? *Semin. Nucl. Med.* 27:94–106.

Kamel, E.M., Burger, C., Buck, A., von Schulthess, G.K., and Goerres, G.W. 2003. Impact of metallic dental implants on CT-based attenuation correction in a combined PET/CT scanner. *Eur. Radiol.* 13:724–728.

Kaste, S.C., Howard, S.C., McCarville, E.B., Krasin, M.J., Kogos, P.G., and Hudson, M.M. 2005. 18F-FDG-avid sites mimicking active disease in pediatric Hodgkin's. *Pediatr. Radiol.* 35:141–154.

Kim, M.J., Kim, E.K., Park, S.Y., Yun, M., and Oh, K.K. 2005. Multiple nodular adenosis concurrent with primary breast lymphoma: pitfall in PET. *Clin. Radiol.* 60:126–129.

Kinahan, P.E., Townsend, D.W., Beyer, T., and Sashin, D. 1998. Attenuation correction for a combined 3D PET/CT scanner. *Med. Phys.* 25:2046–2053.

Knopp, M.V., and Bischoff, H.G. 1994. Evaluation of pulmonary lesions with positron emission tomography. *Radiologe* 34:588–591.

Kostakoglu, L., Hardoff, R., Mirtcheva, R., and Goldsmith, S.J. 2004. PET-CT fusion imaging in differentiating physiologic from pathologic FDG uptake. *Radiographics* 24:1411–1431.

Kubota, R., Kubota, K., Yamada, S., Tada, M., Ido, T., and Tamahashi, N. 1994. Active and passive mechanisms of [fluorine-18] fluorodeoxyglucose uptake by proliferating and prenecrotic cancer cells *in vivo*: a microautoradiographic study. *J. Nucl. Med.* 35:1067–1075.

Lewis, P.J., and Salama, A. 1994. Uptake of fluorine-18-fluorodeoxyglucose in sarcoidosis. *J. Nucl. Med.* 35:1647–1649.

Mirzaei, S., Guerchaft, M., Bonnier, C., Knoll, P., Doat, M., and Braeutigam, P. 2005. Use of segmented CT transmission map to avoid metal artifacts in PET images by a PET-CT device. *BMC Nucl. Med.* 5:3.

Miyauchi, T., and Wahl, R.L. 1996. Regional 2-[18F]fluoro-2-deoxy-D-glucose uptake varies in normal lung. *Eur. J. Nucl. Med.* 23:517–523.

Mueckler, M. 1994. Facilitative glucose transporters. *Eur. J. Biochem.* 219:713–725.

Mullani, N.A., Gould, K.L., Hartz, R.K., Hitchens, R.E., Wong, W.H., Bristow, D., Adler, S., Philippe, E.A., Bendriem, B., Sanders, M., *et al.* 1990. Design and performance of POSICAM 6.5 BGO positron camera. *J. Nucl. Med.* 31:610–616.

Nakamoto, Y., Chin, B.B., Cohade, C., Osman, M., Tatsumi, M., and Wahl, R.L. 2004. PET/CT: artifacts caused by bowel motion. *Nucl. Med. Commun.* 25:221–225.

Nehmeh, S.A., Erdi, Y.E., Ling, C.C., Rosenzweig, K.E., Schoder, H., Larson, S.M., Macapinlac, H.A., Squire, O.D., and Humm, J.L. 2002. Effect of respiratory gating on quantifying PET images of lung cancer. *J. Nucl. Med.* 43:876–881.

Neuenschwander, B.E., Zwahlen, M., Kim, S.J., Engel, R.R., and Rieder, H.L. 2000. Trends in the prevalence of infection with mycobacterium tuberculosis in Korea from 1965 to 1995: an analysis of seven surveys by mixture models. *Int. J. Tuberc. Lung Dis. 4*:719–729.

Nishikawa, A., Furukawa, F., Imazawa, T., Ikezaki, S., Otoshi, T., Fukushima, S., and Takahashi, M. 1995. Cell proliferation in lung fibrosis-associated hyperplastic lesions. *Hum. Exp. Toxicol 14*:701–705.

Ohtsuka, T., Nomori, H., Watanabe, K., Naruke, T., Orikasa, H., Yamazaki, K., Suemasu, K., and Uno, K. 2005. False-positive findings on [18F]FDG-PET caused by non-neoplastic cellular elements after neoadjuvant chemoradiotherapy for non-small cell lung cancer. *Jpn. J. Clin. Oncol. 35*:271–273.

Pottgen, C., Levegrun, S., Theegarten, D., Marnitz, S., Grehl, S., Pink, R., Eberhardt, W., Stamatis, G., Gauler, T., Antoch, G., Bockisch, A., and Stuschke, M. 2006. Value of 18F-fluoro-2-deoxy-D-glucose-positron emission tomography/computed tomography in non-small cell lung cancer for prediction of pathologic response and times to relapse after neoadjuvant chemoradiotherapy. *Clin. Cancer Res. 12*:97–106.

Prosch, H., Mirzaei, S., Oschatz, E., Strasser, G., Huber, M., and Mostbeck, G. 2005. Case report: gluteal injection site granulomas: false positive finding on FDG-PET in patients with non-small cell lung cancer. *Br. J. Radiol. 78*:758–761.

Reddy, M.P., and Ramaswamy, M.R. 2005. FDG uptake in brown adipose tissue mimicking an adrenal metastasis: source of false-positive interpretation. *Clin. Nucl. Med. 30*:257–258.

Schafers, K.P., Dawood, M., Lang, N., Buther, F., Schafers, M., and Schober, O. 2005. Motion correction in PET/CT. *Nuklearmedizin 44 Suppl. 1*:S46–50.

Schmid, D.T., Kneifel, S., Stoeckli, S.J., Padberg, B.C., Merrill, G., and Goerres, G.W. 2003. Increased 18F-FDG uptake mimicking thyroid cancer in a patient with Hashimoto's thyroiditis. *Eur. Radiol. 13*:2119–2121.

Shields, A.F., Grierson, J.R., Dohmen, B.M., Machulla, H.J., Stayanoff, J.C., Lawhorn-Crews, J.M., Obradovich, J.E., Muzik, O., and Mangner, T.J. 1998. Imaging proliferation *in vivo* with [F-18]FLT and positron emission tomography. *Nat. Med. 4*:1334–1336.

Shreve, P.D., Anzai, Y., and Wahl, R.L. 1999. Pitfalls in oncologic diagnosis with FDG PET imaging: physiologic and benign variants. *Radiographics 19*:61–77; quiz 150–151.

Silvestri, G.A., Tanoue, L.T., Margolis, M.L., Barker, J., and Detterbeck, F. 2003. The noninvasive staging of non-small cell lung cancer: the guidelines. *Chest 123*:147S–156S.

Stahl, A., Weber, W.A., Avril, N., and Schwaiger, M. 2000. Effect of N-butylscopolamine on intestinal uptake of fluorine-18-fluorodeoxyglucose in PET imaging of the abdomen. *Nuklearmedizin 39*:241–245.

Tatlidil, R., Jadvar, H., Bading, J.R., and Conti, P.S. 2002. Incidental colonic fluorodeoxyglucose uptake: correlation with colonoscopic and histopathologic findings. *Radiology 224*:783–787.

Tomita, M., Ichinari, H., Tomita, Y., Mine, K., Iiboshi, H., Kisanuki, A., and Shibata, K. 2003. A case of non-small cell lung cancer with false-positive staging by positron emission tomography. *Ann. Thorac. Cardiovasc. Surg. 9*:397–400.

Torizuka, T., Tamaki, N., Inokuma, T., Magata, Y., Sasayama, S., Yonekura, Y., Tanaka, A., Yamaoka, Y., Yamamoto, K., and Konishi, J. 1995. *In vivo* assessment of glucose metabolism in hepatocellular carcinoma with FDG-PET. *J. Nucl. Med. 36*:1811–1817.

Vansteenkiste, J.F., and Stroobants, S.G. 2001. The role of positron emission tomography with 18F-fluoro-2-deoxy-D-glucose in respiratory oncology. *Eur. Respir. J. 17*:802–820.

Weber, G. 1977. Enzymology of cancer cells (second of two parts). *N. Engl. J. Med. 296*:541–551.

Yamada, S., Kubota, K., Kubota, R., Ido, T., and Tamahashi, N. 1995. High accumulation of fluorine-18-fluorodeoxyglucose in turpentine-induced inflammatory tissue. *J. Nucl. Med. 36*:1301–1306.

Yi, J.G., Marom, E.M., Munden, R.F., Truong, M.T., Macapinlac, H.A., Gladish, G.W., Sabloff, B.S., and Podoloff, D.A. 2005. Focal uptake of fluorodeoxyglucose by the thyroid in patients undergoing initial disease staging with combined PET/CT for non-small cell lung cancer. *Radiology 236*:271–275.

13

Oxygen-enhanced Proton Magnetic Resonance Imaging of the Human Lung

Eberhard D. Pracht, Johannes F.T. Arnold, Nicole Seiberlich, Markus Kotas, Michael Flentje, and Peter M. Jakob

Introduction

One of the major goals in pulmonary diagnostics is to detect pathologies in the lung using modern imaging techniques. The advantage of techniques that allow visualization of the lung as compared to other diagnostic tools, such as spirometry, is the possibility of obtaining regional rather than global information about the lung. In addition, the information provided can contain not only morphological, but also functional characteristics.

The current gold standard techniques for lung imaging are computed tomography (CT), perfusion scintigraphy, and ventilation scintigraphy. These imaging modalities allow for morphological and functional imaging, but they have the disadvantage that they employ radiating isotopes and are therefore not appropriate for all applications, such as long-term observations of lung pathologies.

In contrast, magnetic resonance imaging (MRI) is completely noninvasive and allows the monitoring of both functional and morphological information in the lung. The major drawback of proton (^{1}H) MRI of lung parenchyma is the low-proton density and the rapid signal decay due to susceptibility effects at the many air/tissue interfaces. Furthermore, respiratory motion reduces the amount of time available for acquisition and can lead to artifacts in the images.

Many techniques have been developed during the last decade to overcome these problems. These innovations recover signal from the lung and allow the achievement of morphological and functional information without artifacts induced by respiratory motion. Developments such as hardware optimization, sequence design, and investigation of different contrast agents have contributed to the improvement of lung ^{1}H MRI.

In 1992, it was demonstrated that proton lung imaging using oxygen as a contrast agent (CA) is possible (Goodrich et al., 1992). That study, showed that excised rat lung T_1 values are significantly lower in the presence of oxygen. Edelman et al. (1996) concluded that this effect could also be used to assess regional ventilation in the human lung.

Following this conclusion, researchers studied this effect on volunteers and patients with diseases such as pulmonary embolism (Nakagawa et al., 2001), emphysema (Ohno et al., 2002), and cystic fibrosis (Jakob et al., 2004). This

MR method was not only completely noninvasive, but it also improved the resolution of the images. In addition, only standard hardware for proton imaging is needed, and the contrast agent used, namely, oxygen, is readily available.

Although it was initially assumed that ventilation is the main source of signal change when oxygen is used as a contrast agent, several different mechanisms actually contribute to the signal enhancement. Perfusion and diffusion are also partially responsible for the observed signal changes (Edelman et al. 1996; Loeffler et al., 2000). In order to describe the effects of all three parameters, Jakob et al. (2004) introduced the oxygen transfer function (OTF), which characterizes the entire process of gas exchange in the lungs.

The only methods available for imaging pure regional lung ventilation were either the application of a radioactive gas (ventilation scintigraphy) or MR gas imaging. In MR gas imaging, a hyperpolarized gas such as helium (Ebert et al., 1996) or xenon (Wagshul et al., 1996) is inhaled, and the signal from the noble gas is acquired. Though promising, these methods are not appropriate for routine use because of the high cost of equipment for laser polarization and the narcotic effects of large concentrations of inhaled xenon. It has also been shown in animal studies that inhaled perfluorinated hydrocarbons can be imaged and aid in the detection of ventilation blockages (Kuethe et al., 1998). The major drawback of this method is the need for additional ^{19}F hardware equipment in order to perform such imaging.

Another proton MR imaging approach was explored by Pracht et al. (2005), who also used oxygen as the contrast agent. In contrast to previous oxygen-enhanced techniques, the susceptibility gradients between the alveoli/tissue interfaces were used as the contrast mechanism. This technique provides a more direct assessment of ventilation because the induced signal changes are dependent only on the spatial distribution of the inhaled gas.

In this chapter, the most important developments in ^{1}H MR lung imaging using oxygen as contrast agent are presented. A brief introduction to respiratory physiology is given in order to motivate functional lung imaging and provide a basic understanding of the lung function. After the introduction, the concept and theory of oxygen-enhanced imaging is presented, including the basic physics behind this phenomenon and the experimental implementation. The impact and resulting implications of oxygen-enhanced imaging are then discussed, including data from volunteers and patients. Finally, a summary of the results obtained in the last few years is given, and the potential applications for this method in routine clinical diagnostics are discussed.

Respiratory Physiology

The main function of the lung is to exchange gas between inhaled air and the blood. The oxygen contained in the inhaled air diffuses through the alveoli to the blood and is carried to body tissues. At the same time, venous blood transfers carbon dioxide from the tissues back to the lungs, where it diffuses back through the alveoli and is exhaled. This process of gas exchange in the lung parenchyma can be completely described by three parameters: alveolar ventilation (**V**), pulmonary perfusion (**Q**), and pulmonary diffusion capacity (**D$_c$**).

Alveolar ventilation (V) is defined as the volume of gas available in the alveoli per minute (units: ml/min) and determines the partial pressure of oxygen and carbon dioxide in the alveoli. In other words, alveolar ventilation is determined by multiplying the breathing frequency by the volume of air that is actually available to the alveoli and contributes to the gas exchange. During each inhalation, fresh room air is transported via convection through the airways to the alveoli. The normal alveolar ventilation for a human subject at rest is ~ 5 l/min and can increase up to 120 l/min during exertion. According to Dalton's law, the total pressure of the inhaled gas mixture is the sum of the partial pressures of the individual gases:

$$P_{total} = \sum_{gas} P_{gas}$$

At an atmospheric pressure of 760 mmHg, 78% of the total pressure of normal air is caused by nitrogen molecules and 21% by oxygen molecules. The amount of carbon dioxide is negligible (~ 0.03%). The actual alveolar oxygen pressure, $p_A O_2$, can be calculated using the alveolar gas equation:

$$p_A O_2 = p_1 O_2 - p_A CO_2 / RQ$$

where $p_1 O_2$ is the partial pressure of the inhaled oxygen and $p_A CO_2$ is the alveolar carbon dioxide pressure. The parameter RQ is the respiratory quotient, which describes the ratio between CO_2 production and O_2 consumption. Using Dalton's law for the partial pressure of the inspired oxygen, the alveolar oxygen pressure can be expressed in terms of the inhaled oxygen fraction $F_1 O_2$, which will be used in the following chapter for derivation of the oxygen transfer function (OTF):

$$p_A O_2 = F_1 O_2 (p_{atm} - p_{H_2O}) - p_A CO_2 / RQ$$

In this equation, the parameter p_{atm} denotes the atmospheric pressure and p_{H_2O} the pressure of water, which is included to describe the humidification of the air in the trachea.

After gas exchange in the alveoli (according to the alveolar gas equation), the deoxygenated air is exhaled.

Diffusion Capacity (Dc) Diffusion is the driving force for the gas exchange between the alveoli and the blood (i.e., the capillary network). The oxygen diffusion occurs due to a difference in the partial pressure of oxygen between the alveoli and the capillary network. (The same statement is valid for carbon dioxide in the opposite direction.) Under normal conditions, the oxygen partial pressure in the alveoli is 100 mmHg, whereas the oxygen partial pressure in the venous blood is only 40 mmHg. The resulting diffusion flow of oxygen, I, from the alveoli to the capillary bed is described by Fick's law:

$$I = D_c \times \Delta p o_2 = D_c \times (p_A O_2 - p_a O_2)$$

$$D_c = d_{o_2} \times \alpha_{o_2} \times \frac{A}{l}$$

where D_c is the diffusion capacity and Δp_{O2} is the difference in the oxygen partial pressure between the alveoli and the capillaries. The diffusion capacity depends on the properties of the alveoli-capillary barrier (A: area of the membrane, l: length of membrane) and the properties of the diffusing particles, which in this case are oxygen molecules (d_{O2}: diffusivity of oxygen, α_{O2}: solubility of oxygen in the blood).

The actual concentration of oxygen c_{O2} (in mmol/l) that is dissolved in the blood is given by Henry's law:

$$co_2 = \alpha o_2 \times p_a O_2$$

The concentration of oxygen that is physically dissolved in blood is therefore determined by its solubility α_{O2} in blood and the partial pressure $p_a O_2$ in the alveoli. After this diffusion from the alveoli to the capillaries, the oxygen that is physically dissolved in the blood diffuses into erythrocites where it is bound to hemoglobin. A dynamic equilibrium exists between the oxygen dissolved in the blood and the hemoglobin-bound oxygen molecules. The actual oxygen content in the blood is determined by the oxygen partial pressure in the blood, the oxygen saturation of hemoglobin, and the hemoglobin content (Lawrence, 1999).

Pulmonary Perfusion (Q), or the blood flow through the pulmonary system, ensures that the venous blood, which is coming from the body, is oxygenated in the capillary system through the diffusion processes mentioned earlier. Furthermore, pulmonary perfusion guarantees that the oxygenated blood is delivered to the left ventricle of the heart, which supplies the body with oxygen. The pulmonary blood flow is defined as the stroke volume of the heart multiplied with the heart rate, which leads at rest to an average perfusion of ~ 6 l/min.

Ventilation/Perfusion (V/Q) ratio: For efficient gas exchange, the relationship between alveolar ventilation and perfusion must be matched, which means that the respiratory quotient in the alveolar gas equation at rest is ~ 0.8 (using the average values of V and Q). This matched value is global (i.e., over the whole lung), and regional values can vary, even in healthy tissue.

V/Q mismatches can arise when:

1. The alveolar ventilation is hampered, which is the case in chronic obstructive lung diseases (COPD), asthma, emphysema, and pneumonia.
2. Perfusion deficits occur, which is the case in pulmonary embolism (PE), for example.

Another group of disorders are diffuse parenchymal lung diseases (DPLDs), also known as interstitial lung diseases. In these disorders, the diffusion barrier between the alveoli and the capillary bed (i.e., the alveolar epithelium and the pulmonary capillary endothelium) is affected, which is the case in fibrosis and emphysema. These kinds of diseases lead to poor oxygenation of the venous blood due to a blockage in the diffusion barrier even when the ventilation perfusion ratio V/Q is matched.

Theory of Oxygen-enhanced Imaging

This chapter describes the theory of oxygen-enhanced imaging in detail. The use of molecular oxygen as a contrast agent leads to an increase in the oxygen concentration in the alveoli, as was described in the Respiratory Physiology section, as well as to an increase in oxygen physically dissolved in arterial blood. An increase in the oxygen concentration leads to changes in the local magnetic field, as oxygen molecules are weakly paramagnetic. Therefore, a change in MR-relaxation parameters of the lung, namely, T_1 and T_2^*, is expected in the presence of different oxygen concentrations. After a brief review of the physics behind the particular relaxation parameters, different imaging techniques for oxygen-enhanced imaging are introduced and discussed in detail.

T_1 Relaxation in the Human Lung

Mechanisms of T_1 Relaxation in Lung Tissue

It has been shown both theoretically and experimentally that T_1 relaxation in the lung is affected by numerous factors. Zimmermann and Brittin (1957) described the T_1 relaxation in biological systems with a model of fast exchange. This model is based on the assumption that water molecules move quickly between different compartments, hence the name "fast exchange" model, leading to a measured T_1 value, which is the weighted average of the individual T_1 relaxation times in all n compartments:

$$\frac{1}{T_1} = \sum_{n}^{N} \frac{p_n}{T_{1n}}$$

where p_n denotes the fraction of protons in the nth compartment and T_{1n} is the corresponding T_1 relaxation time. In this model, the same assumptions are also valid for T_2 relaxation.

In the lung tissue, two main "compartments" exist: free water and protons bound to macromolecules.

Assuming a fast exchange between the water protons (free component, index f) and the macromolecular background (the bound component, index b), lung tissue T_1 relaxation can be approximated using the formalism above for a two-compartment system:

$$\frac{1}{T_{1, \text{native}}} = \frac{p_b}{T_{1b}} + \frac{p_f}{T_{1f}}$$

where $T_{1,\,native}$ describes the "native" T_1 relaxation of the lung tissue in the absence of oxygen (i.e., the completely deflated lung). The T_1 values of the protons in each compartment are determined mainly by the dipole–dipole interaction, leading to a T_{1f} of ~ 5 sec and a T_{1b} of ~ 1 sec at 1.5 Telsa. The T_1 value is dependent on the magnetic field strength. For the purposes of this work, all relaxation parameters are given under the assumption of a field strength of 1.5 Tesla.

According to the preceding equation, the T_1 of lung tissue depends on the ratio between free water and macromolecular protons. Although the fraction of the macromolecular background in healthy lung tissue is small compared to the water protons (Fullerton *et al.*, 1982), its effect on the native lung T_1 is significant due to the much smaller T_{1b}. Using physiologically reasonable values for the fractions of bound and free protons in healthy lung tissue, we find that the theoretical value for $T_{1,native}$ in the lung is ~ 2 sec, which is on the order of magnitude of the experimentally determined T_1 value ($T_1 \approx 1500$ ms). However, as we have already stated, the theoretical value refers to the lung in the absence of oxygen and is therefore expected to be higher than the value actually measured.

Several lung diseases lead to an increase in the fraction of protons bound to the macromolecules, that is, an increase in collagen or elastin (Starcher *et al.*, 1978; Karlinsky, 1982), which could explain the decrease in the lung T_1 relaxation time observed, for example, in pulmonary fibrosis of rat lungs (Vinitski *et al.*, 1986) or in fibrotic human lung tissue (Shioya *et al.*, 1988; Jakob *et al.*, 2004). These examples clearly show how structural and macromolecular changes in the lung can cause changes in lung T_1 and accordingly $T_{1,native}$ relaxation. Therefore, lung $T_{1,native}$, which is dependent only on lung tissue parameters, can be used as an indicator for morphological disorders in the lungs.

Influence of Oxygen and the Oxygen Transfer Function (OTF)

The T_1 relaxation of the lung is not only a function of the magnetic field strength and lung tissue parameters, but is also dependent on the concentration of oxygen physically dissolved in the blood. This phenomenon results from the paramagnetic properties of oxygen (Goodrich *et al.*, 1992) and can be used for the assessment of ventilation in the human lung (Edelman *et al.*, 1996). By administering breathing gases with different oxygen concentrations to volunteers and patients, T_1-sensitive MR imaging experiments can be performed in order to investigate lung function. Figure 63 depicts the typical results of an oxygen-enhanced (OE) imaging session.

After the acquisition of several T_1-weighted images (see Setup and Basic Imaging Techniques section) while the subject is breathing room air, the breathing gas is switched to pure oxygen and another series of images is acquired. In order to avoid wash-in and wash-out effects, which describe the transition period directly after switching the breathing gas, T_1-weighted images with the second gas are acquired at least 5 min after the switch to allow the oxygen concentration in the blood to reach an equilibrium. This process of gas exchange and image acquisition is performed several times in order to average over multiple measurements. The resulting ventilation images are obtained by subtracting the room air images from the oxygen images (Fig. 64).

The signal changes observed in healthy volunteers, when the breathing gas was switched from room air (21% oxygen) to pure oxygen (100% oxygen), were found to be homogeneous over the entire lung tissue. In contrast, the signal changes observed in patients with impaired lung tissue were inhomogeneous, indicating ventilation deficits.

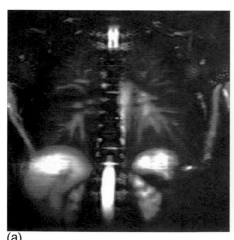 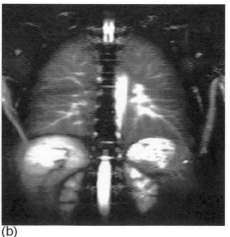

(a)   (b)

Figure 63 T_1-weighted images under room air (a) and pure oxygen (b) conditions. Both images were acquired using the same imaging parameters, and demonstrate the signal enhancement caused by oxygen.

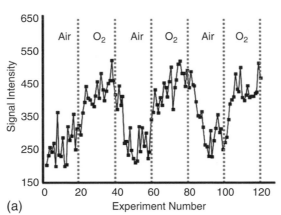

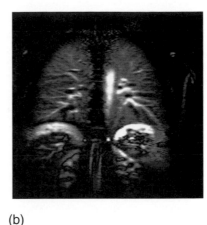

Figure 64 Time course of an OE imaging session. (a) The dotted lines indicate when the breathing gas was changed. (b) The resulting ventilation image obtained by subtracting a room air image from an oxygen image.

It was suggested, however, that these signal changes are due not only to regional ventilation, but also to diffusion and perfusion effects (Edelman *et al.*, 1996; Loeffler *et al.*, 2000). The ventilation images (i.e., the subtraction images, or the difference between room air and oxygen images) obtained with OE imaging are therefore different from those obtained with other "pure" ventilation scanning techniques, such as ventilation scintigraphy. In order to explain the origin of these signal changes due to gas transfer during OE measurements, a formalism was developed, which characterizes the complete process of gas exchange in one parameter (Jakob *et al.*, 2004). This formalism is based on both T_1 relaxation theory and respiratory physiology.

In order to calculate the actual T_1 value in the lungs, the intrinsic T_1 in the complete absence of oxygen, $T_{1,native}$ is modified by the addition of a term that describes the effect of the oxygen dissolved in the blood:

$$\frac{1}{T_1(c_{O_2})} = \frac{1}{T_{1,\,native}} + r_1 \times c_{O_2}$$

where r_1 denotes the longitudinal relaxivity of oxygen (in s^{-1} $mmol^{-1}$ L) and c_{O2} is the concentration of oxygen physically dissolved in the blood (in mmol/l).

The local oxygen concentration c_{O2} can be calculated according to Henry's law by multiplying the partial pressure of oxygen, p_aO_2, of the arterial pulmonary blood with the oxygen solubility in blood plasma. The oxygen pressure p_aO_2 in the arterial pulmonary blood depends directly on the alveolar p_AO_2 in the alveoli and on the diffusion capacity of the lung (see Respiratory Physiology section). Using these relationships, we derived an equation that describes the dependence of the relaxation time T_1 on the inspired oxygen fraction F_IO_2 (Jakob *et al.*, 2004):

$$\frac{1}{T_1(F_IO_2)} = \frac{1}{T_{1,\,native}} + OTF \times F_IO_2$$

where the oxygen transfer function (*OTF*, in units of $s^{-1}\%O_2^{-1}$) has been introduced. This value reflects the lung's capability to exchange gas, and thus is a combination of ventilation, diffusion, and perfusion effects. By measuring the actual T_1 value (see Setup and Basic Imaging Techniques section) in the presence of inhalation gases with different oxygen fractions and performing a linear fit of the results, the slope of the fit yields the OTF value, and the intercept $T_{1,native}$. With this method, both functional and morphological deficits can be detected, through the OTF and the $T_{1,native}$ values, respectively. Therefore, using this technique it is possible to get **quantitative** information about the pulmonary system, including both functional and morphological deficits. Preliminary results using the OTF approach on patients with cystic fibrosis are presented in the OE Imaging in Volunteers and Patients section.

Setup and Basic Imaging Techniques

Magnetic resonance imaging of the human lung is limited for several reasons. The primary problem is that the low-proton density in the lung provides a low signal level as compared to the noise (~ 80% of the lung volume is air). Second, the many air/tissue interfaces in the lung induce a rapid signal decay due to the short T_2^*. Finally, the image quality is often degraded by artifacts caused by breathing and motion of the heart.

Therefore, high-sensitivity imaging techniques must be applied to the lungs in order to obtain accurate lung signal and adequate image quality. Lung imaging is thus performed by using either gradient echo (GE) techniques (Haase *et al.*, 1986) with ultra-short echo times (Alsop *et al.*, 1995; Hatabu *et al.*, 1999) or turbo spin echo (TSE) techniques (Hennig *et al.*, 1986) with short inter-echo and effective echo times (Chen *et al.*, 1998). For a more detailed description of GE and TSE sequences. Both imaging techniques have a high inherent signal-to-noise ratio (SNR) and allow for short acquisition times in order to obtain these measurements in a single breath hold.

Setup Two main setups can be used to deliver oxygen to the subjects during oxygen-enhanced (OE) imaging experiments: a mask system and a mouthpiece system. The mask system is more comfortable for the volunteer or patient, because it simply lies on the face. However, because there is no seal between the mask and the skin, air from outside the mask can enter the breathing system, and the oxygen concentration delivered to the patient is slightly less than the oxygen concentration in the breathing gas. A more accurate method of delivering the breathing gas is through use of a mouthpiece in combination with a nose clamp, such that the patient can only breathe the gas supplied by the breathing system. The drawback of this method is that it can be rather uncomfortable and can increase the claustrophobic feelings that many subjects experience in the MR system. Both setups are shown in Figure 65.

A pressure regulator with which the researcher can control the oxygen concentration and breathing gas flow rate is also employed. The signal is detected using a body array, which has the advantage of increased SNR in comparison to a single coil. Headphones are supplied to the subject to allow him or her to hear breathing commands given by the researcher, who is in constant visual and auditory contact with the subject.

T_1-Weighted Oxygen-enhanced Imaging

The basic technique for oxygen-enhanced imaging is the T_1-weighted imaging scheme. This method is based on application of a global inversion pulse followed by an imaging sequence after a specific inversion time T_1. By using different inversion times, one can manipulate the T_1 contrast of the image. The optimal sequences for oxygen-enhanced imaging in terms of the SNR and image quality are TSE sequences, which are completely insensitive both to the very short T_2^* in the lung caused by field inhomogeneities and to motion artifacts due to rapid image acquisition. The most commonly used TSE sequences in OE lung imaging are the centric reordered (CR)–RARE (Rapid Acquisition with Relaxation Enhancement) and the HASTE (Half-Fourier

Acquisition Single-Shot Turbo Spin Echo) sequence, which allow acquisition of a complete image using only one excitation (Chen *et al.*, 1998). The acquisition time for such an imaging module at a typical resolution for lung imaging (~ 5 mm in plane with a slice thickness of ~ 10 mm) is in the order of 0.5 sec. With the inclusion of the inversion module and the cardiac trigger, which is used to decrease motion artifacts from the heart and artifacts due to blood flow in the pulmonary system, the time needed for a complete acquisition is ~ 2 sec. After the acquisition of the T_1-weighted image, a pause is inserted (3–5 sec) in order to wait for T_1 relaxation, and then the next image is acquired.

Once a complete T_1-weighted image series under room air (or pure oxygen) conditions is acquired, the breathing gas is switched and another series of T_1-weighted images is acquired. Because the ventilation images are actually calculated by subtracting one image from another (generally room air and pure oxygen), the inflation level of the lung for the two images has to be the same. For this reason, the complete process of image acquisition is normally performed while the subject holds his breath at a specific inflation level in order to avoid misalignment of the images. The images are typically acquired in end expiration, where the proton density is at a maximum, leading to a maximized SNR.

To obtain high-quality ventilation images, it is important to have not only a high SNR, but also a high contrast-to-noise ratio (CNR) between images made with different oxygen concentrations. If both the T_1 values of the lung in room air and in pure oxygen are known, the signal difference, and therefore the optimized inversion time TI_{opt} for the imaging session, can be calculated (Chen *et al.*, 1998):

$$TI_{opt} = \ln\left(\frac{(T_{1_{air}}/T_{1_{oxy}})}{(1/T_{1_{oxy}} - (1/T_{1_{air}})}\right)$$

Figure 66 depicts the signal dependency on T_1 for two representative T_1, as well as the signal difference at each TI. The preceding equation assumes that the T_1 values over the

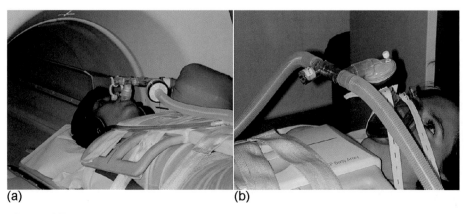

(a)   (b)

Figure 65 Setup for oxygen-enhanced measurements. (a) A volunteer breathing through the mouthpiece, and (b) breathing through the mask system.

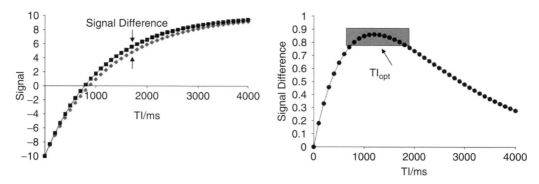

Figure 66 (Left) T_1 relaxation curves for two representative T_1 values (1300 ms, gray diamonds; 1100 ms, black squares) at different inversion times, TI. (Right) The difference between the relaxation curves for different TI times. The highest difference, and thus optimal TI time, is highlighted in the gray box.

entire lung are homogeneous. If this is not the case, and the T_1 values vary regionally, measurements made using this TI_{opt} value can produce low- or no-signal areas in those regions where the T_1 differs from the assumed value. These low-signal areas can lead to a misinterpretation of the ventilation images. This problem can be corrected by the acquisition of T_1 parameter maps that reflect the actual T_1 values, thereby signaling actual ventilation deficits through a lack of change in relaxation parameters

Quantitative Oxygen-enhanced Imaging

The quantitative oxygen-enhanced imaging approach is used to obtain actual T_1 values and is based on a T_1 mapping procedure. By acquiring a series of rapid gradient echo images after a single global inversion pulse, T_1 values can be calculated on a pixel-by-pixel basis. These T_1 parameter maps can be generally obtained in < 5 sec, thereby avoiding motion artifacts as the images are acquired in a single breath hold (Jakob *et al.*, 2001). Figure 67 shows two T_1 maps, one under room air conditions and the other using pure oxygen.

The T_1 values upon breathing pure oxygen are ~ 10% lower than those calculated upon breathing air; this change originates mainly from the larger concentration of O_2 physically dissolved in the blood in pure oxygen conditions (see OTF section).

As in T_1-weighted imaging, constant T_1 values upon the introduction of a larger oxygen concentration indicate ventilation deficits; the larger the T_1 change, the better the ventilation in the lungs. In addition, this method has the added advantage that actual T_1 values are obtained, and not just T_1 contrast. Thus, in diseased lung tissue where the T_1 values are inhomogeneous, both morphology changes (through a decreased T_1 value) and functional changes (decreased ventilation, depicted as small T_1 changes) can be evaluated.

Dynamic Oxygen-enhanced Imaging

Normally, oxygen-enhanced images are acquired while the mean pressure of oxygen in the arterial blood is at equilibrium. This avoids wash-in and wash-out effects, which can be seen directly after inhalation gas exchange when the con-

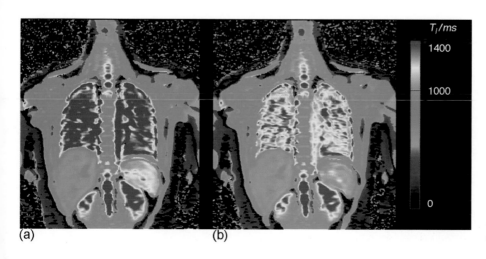

(a)   (b)

Figure 67 T_1 maps of a volunteer under room air (a) and oxygen conditions (b).

centration of oxygen in the lungs, and therefore in the blood, changes during the measurements. In dynamic oxygen-enhanced imaging, the images are acquired not only before and well after changing the breathing gas, but also during the time when the partial pressure of oxygen in the blood is not in equilibrium (i.e., directly after changing the gas).

The first dynamic OE measurement was performed using the T_1-weighted approach (Hatabu *et al.*, 2001). The time resolution of the data was 3 sec, which allowed for T_1 relaxation of the lung protons between acquisition of the images. In addition to the information available in ordinary OE imaging (i.e., lung signal in room air and pure oxygen conditions), one can also obtain the oxygen wash-in and wash-out time courses, which characterize the uptake and release of oxygen after the gases have been exchanged. These time courses can then be fit to an exponential model in order to obtain wash-in and wash-out parameters.

Arnold *et al.* (2004) performed the first completely quantitative dynamic OE measurements using the T_1 mapping procedure introduced by Jakob *et al.* (2001). As we have described, this method is especially favorable in patients with inhomogeneous T_1 distributions. To this end, a specific data analysis algorithm was developed to obtain T_1 maps with approximately the same temporal resolution as Hatabu et al. (Arnold *et al.*, 2004). Using this approach, we can determine not only room air/pure oxygen T_1 values but also wash-in and wash-out parameters (Fig. 68), which can also be correlated with physiological parameters.

T_2* Relaxation in the Human Lung

The effects of susceptibility-induced signal losses in the human lungs have been known for some time, mostly through examination of spectral linewidths from the lung, from which T_2* values can be approximated. Several different geometrical models, such as spherical shells or cubes, were

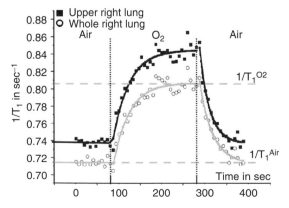

Figure 68 Time course of T_1 values in two regions of the lung during a dynamic oxygen-enhanced imaging experiment. The dotted lines indicate the gas change.

proposed to explain this dephasing behavior, and Monte Carlo simulations were performed to better understand the dynamics of the system (Case *et al.*, 1987; Durney *et al.*, 1989). However, actual parameter mapping approaches were not employed. Similarly, the dependency of the signal decay on lung inflation levels was studied (Cutillo *et al.*, 1991), but the link between T_2* and the oxygen concentration was not examined. In the OE method described in this chapter, the T_2* model used is the static dephasing approach introduced by Yablonskiy (Yablonskiy and Haacke, 1994). The advantage of this approach is that the solutions can be obtained analytically.

Mechanisms of T_2* Relaxation

When a subject is placed into a homogeneous magnetic field, susceptibility effects at interfaces between areas with different magnetic susceptibilities, such as air and tissue, lead to inhomogeneities in the magnetic field and cause an exponential signal decay in gradient echo images:

$$S(TE) = S(0) \times e^{-TE/T_2^*}$$

where TE is the echo time of the gradient echo experiment and T_2^* denotes the effective relaxation time of the sample, which can be described by a combination of two parameters, T_2 and T_2':

$$\frac{1}{T_2^*} = \frac{1}{T_2} + \frac{1}{T_2'}$$

where T_2 denotes the conventional transverse relaxation time, and T_2' reflects local inhomogeneities and characterizes the field variations ΔB across an imaging voxel. These local field variations can be divided into three categories (Yablonskiy, 1997): macroscopic, microscopic, and mesoscopic field inhomogeneities.

Macroscopic Inhomogeneities Macroscopic field inhomogeneities act over a distance larger than the voxel size and arise from field distortions such as imperfections in the magnet or the body–air interface. They generally provide no information of anatomic or physiological interest.

Microscopic Inhomogeneities Microscopic field inhomogeneities arise from magnetic field changes over distances that are comparable to the atomic or molecular size and that lead to an irreversible dephasing of the MR signal. This signal decay originates from the time-dependent dipole–dipole interaction of the spin system and is described by the relaxation parameter T_2, which characterizes dynamic field changes (in contrast to static susceptibility effects).

Mesoscopic Inhomogeneities Mesoscopic field inhomogeneities are larger than the molecular size, but they are limited to the actual voxel size. They depend on susceptibility differences within each voxel and lead to a static field shift, which results in an inhomogeneous line broadening.

Theory of Oxygen-enhanced T_2^* Imaging

T_2^* relaxation in the lung is dominated by static field inhomogeneities arising from a very short T_2' due to the many air–tissue interfaces.

T_2' Relaxation in Lung Parenchyma T_2' relaxation in the lung parenchyma is affected primarily by mesoscopic inhomogeneities (air–tissue interfaces occurring between the alveoli and capillary network) and secondarily by macroscopic inhomogeneities (the body–lung interface or the interface between lung parenchyma and large blood vessels). Because of the dominance of the mesoscopic interaction, a two-compartment model can be assumed, representing the different portions of the "homogeneous" lung parenchyma, namely, tissue and gas space. In this model, the alveoli are represented by spherical air bubbles, which are embedded in the lung tissue. The susceptibility differences between these media, $\Delta\chi$ ($\chi_{tissue} - \chi_{gas}$), induce local magnetic field changes ΔB, which also are affected by the lung inflation level. The lung inflation level, in turn, can be expressed as the ratio between the net volume of the alveoli v and the whole parenchymal volume V, or the relative volume fraction $\eta = v/V$. Applying the static dephasing approach for spherical and randomly distributed spheres, we can calculate the reciprocal of the T_2' value of the lung parenchyma (Yablonskiy and Haacke, 1994):

$$\frac{1}{T_2'} = \frac{2\pi}{3\sqrt{3}}\eta\frac{\Delta B}{3} = \frac{2\pi}{3\sqrt{3}}\eta\gamma\frac{\Delta\chi}{3}B_0$$

The susceptibility of lung tissue, which consists mainly of lung blood, can be approximated using the susceptibility of blood (50% arterial and 50% venous, yielding $\chi_{tissue} \approx -9.1$ ppm). Using this value and the susceptibility of room air, we can calculate T_2' to be in the order of 0.5 ms (dependent on the inspiration level), in contrast to T_2, which is ~ 60 ms in the lungs (Pracht *et al.*, 2005). Therefore, at a field strength of 1.5 Tesla and under room air conditions, the theoretical value of T_2^* is on the order of one millisecond and is primarily determined by T_2'.

Influence of Pure Oxygen Changing the inhalation gas from room air ($\chi = 0.35$ ppm) to pure oxygen ($\chi = 1.77$ ppm) results in an increase in the susceptibility difference between the lung tissue and the alveoli. This "modulation" of the susceptibility difference between the inhalation gases leads to a shortening of the T_2^* values. Upon application of the model and assuming an inflation level of $\eta \approx 0.7$ (at end expiration), T_2^* values of 0.9 ms for room air and 0.8 ms for oxygen are determined. It can be shown that the change in breathing gas does not appreciably

affect the T_2 values in the lung (Pracht *et al.*, 2005). For that reason, the change in T_2^* depends only on the change of T_2' of the lung:

$$\Delta\left(\frac{1}{T_2^*}\right) = \left(\frac{1}{T_2^*}\right)^{air} - \left(\frac{1}{T_2^*}\right)^{O_2} = \left(\frac{1}{T_2'}\right)^{air} - \left(\frac{1}{T_2'}\right)^{O_2}$$

Using this equation and the fact that T_2' of the lung blood does not change after gas exchange, we can show that T_2^* changes are due only to, and are directly proportional to, the susceptibility difference between the two gases (Pracht *et al.*, 2005):

$$\Delta\left(\frac{1}{T_2^*}\right) = \frac{2\pi}{9\sqrt{3}}\eta\left(\chi_{air} - \chi_{O_2}\right)B_0$$

The theoretical T_2^* change due to the change in susceptibility gradients upon the inhalation of the different gases is approximately 13% (in end expiration at 1.5 Tesla).

This equation shows that a change in the T_2^* value in the lungs can only occur when the susceptibility of the gas in the alveoli changes. The other parameters, namely, η and B_0, remain constant and thus cannot affect the T_2^* value. The only physiological parameter that describes such a change in the alveolar gas susceptibility is ventilation. Because the T_2^* is dependent only on the conditions in the alveoli, the logical conclusion is that this parameter is a direct measurement of the ventilation in the lungs. Thus, use of T_2^* as a contrast mechanism avoids the problems of the T_1 approach mentioned in the T_1 chapter (i.e., that a T_1 signal change can be linked to perfusion and diffusion in addition to ventilation).

However, the statement that T_2^* depicts pure ventilation holds true only under the idealized assumptions of the model. In reality, lung parenchyma is more complex (i.e., the alveoli are not perfect spheres surrounded by homogeneous tissue). The existence of irregularities, such as blood vessels, leads to deviations in the T_2^* values actually measured. This deviation from the theory can be seen when examining the actual change in T_2^* values. Under pure oxygen conditions, the values decrease ~ 10% as compared to values acquired under room air when averaged over the whole right/left lung. This decrease is larger in the upper right lung, up to 20%, due to the more dominant parenchyma in this region. Figure 69 shows the time course of the T_2^* values over 20 experiments in one volunteer. After 10 experiments, the breathing gas was switched from room air to pure oxygen.

Imaging Techniques Because gradient echoes are inherently modulated by T_1 and T_2^*, and both parameters are dependent on the breathing gas, it is not possible to obtain purely T_2^*-weighted images. Consequently, using T_2^*-weighted imaging techniques for ventilation scanning would lead to potential misinterpretations due to the additional T_1 weighting. In order to measure T_2^* alone,

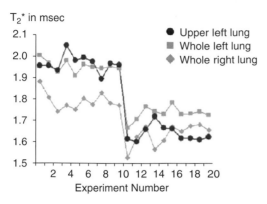

Figure 69 Time course of T_2^* values under room air (first half) and pure oxygen (second half) conditions.

a parameter mapping procedure must be employed. However, quantification of T_2^* in the human lung is rather complicated, due to the low-proton density and rapid signal decay, among other problems. To avoid these complications, images are acquired using an optimized gradient echo sequence, which is based on a FLASH sequence (Haase *et al.*, 1986). This sequence achieves minimum echo times using an asymmetric readout; the echo is read-out such that its second half is longer than the first, which allows one to use a shorter echo time. These acquisitions must be averaged several times (NS = 8) in order to achieve enough signal that the fitting routine can be applied. Due to this long acquisition time, motion artifacts tend to be stronger than, for example, in T_1 quantification. To avoid such artifacts in the T_2^* maps, a motion-insensitive acquisition scheme was developed (Pracht *et al.*, 2005). The resulting T_2^* maps offer information about the ventilation of the lungs, without significant influence from lung perfusion or diffusion.

Oxygen-enhanced Imaging in Volunteers and Patients

In the last decade, many studies have been performed in volunteers and patients using oxygen-enhanced MRI in order to improve the technique and understand how the parameters obtained correlate with pathologies. After the development of this basic method, many technical difficulties still remained, such as the optimization and acceleration of image acquisition, and the determination of the optimal oxygen flow rate. In addition, knowledge of the relationship between parameters obtained in OE imaging and actual lung function or morphology was lacking. The primary goal over the past few years was to develop a better understanding of how MR measurements could be used to provide diagnostic information in a clinical setting. This chapter details these developments.

Improvement of Imaging Technique

Oxygen Flow Rate

The first OE studies were performed without regard to the oxygen flow rate. However, the realization that different flow rates yielded different contrasts indicated that optimization of this parameter could enhance image quality. Thus, the relationship between oxygen flow rate and the changes in lung T_1 was studied in different gas environments (Mai *et al.*, 2002). At low oxygen flow rates (between 5 and 15 l/min), an approximately linear relationship between signal intensity and flow rate exists. At higher flow rates (i.e., between 15 and 25 l/min), the signal levels off due to the low-oxygen solubility in blood, and the experiment becomes uncomfortable for the patient. Thus, in order to obtain the highest signal difference, and therefore contrast, while assuring patient comfort, an oxygen flow rate of 15 l/min is generally employed.

Multislice Capability and Temporal Resolution

When global inversion pulses are applied, magnetization in the entire lung volume is inverted, and before another experiment can be performed, magnetization must return to equilibrium, which takes at least 3 sec in the lung. By using slice-selective pulses instead of global pulses for inversion in OE imaging, only the spins involved in the image are inverted, while the others remain untouched. With these pulses, it is possible to acquire many slices without waiting for full relaxation, as the acquisition can be performed in an interleaved fashion. However, in order to avoid perfusion effects, the inverted slice must be slightly thicker than the slice actually imaged. By combining this interleaved acquisition scheme with parallel imaging methods, the temporal resolution can be improved because the time needed to image each slice is reduced. With this method, it is possible to acquire dynamic oxygen-enhanced T_1-weighted images over the entire lung in approximately 10 minutes (Dietrich *et al.*, 2005).

Image Alignment

One of the major problems associated with the OE imaging approach is the fact that the final results (the "ventilation" images) are based on at least two different imaging experiments. The images from these experiments must be subtracted from each other, so it is imperative that the inhalation position is nearly identical in both images. This condition is quite difficult to fulfill because any subject motion, including lung volume changes, misaligns the images; as at least 5 min must elapse between the acquisitions to rule out in- and outflow effects, such motion effects are almost unavoidable. In order to correct for such distortions, both image acquisition and postprocessing methods were developed. An example of such an acquisition technique is the use of a respiratory trigger (implemented using a respiratory belt, for example). By using this approach, it is possible to acquire T_1-weighted images during free breathing at defined inflation levels (Dietrich *et al.*, 2005). In quantitative T_1 imaging, such free-breathing methods are not possible due to the longer acquisition time needed for measuring T_1. In order to obtain T_1 maps that can be aligned, image registration techniques after acquisition and reconstruction have been recommended (Naish *et al.*, 2005).

Studies in Patients and Correlation with Physiologic Parameters

In addition to investigations of healthy lung tissue, it has been demonstrated that oxygen-enhanced imaging is also feasible for indicating lung deficiencies in a clinical setting. Ideally, a direct correlation between OE images and lung pathologies would allow for an analysis of the effect of the disease on lung function. As an example, patients with pulmonary embolism were examined in order to determine whether the OE images of their lungs showed different MR parameter changes than the lungs of healthy individuals. The resulting oxygen transfer maps did indeed show inhomogeneous areas and a decreased oxygen transfer, indicated by a lower signal change in areas of the lung that were affected by pulmonary emphysema. The data obtained in this study was shown to be correlated to the diffusion lung capacity, which demonstrates that a decrease in oxygen transfer is directly related to the diffusion of molecular oxygen into the capillary bed.

The first quantitative assessment of the oxygen transfer function was obtained in patients with cystic fibrosis (Jakob *et al.*, 2004). The pathologic regions could be separated from healthy tissue using differences in the OTF value and the $T_{1,native}$ values. The lung $T_{1,native}$ values in these CF patients showed an inhomogeneous distribution, indicating that morphological changes had occurred in these tissue areas. In addition, the observed T_1 decrease under various hyperoxic (oxygen concentrations ranging from 21–100%) conditions depended on the actual state of the diseased lung tissue, with more damaged tissue showing smaller changes indicating reduced oxygen transfer capabilities (Fig. 70).

These results were also qualitatively compared to contrast-enhanced MR perfusion measurements. Areas with a small OTF value and a reduced lung $T_{1,native}$ showed a significant correlation with areas of reduced perfusion. This interesting result indicates that the morphological disorder hampers both gas exchange and lung perfusion.

It has also been shown that oxygen-enhanced MR imaging and contrast-enhanced perfusion MR imaging can be performed not only on CF patients, but also on patients with various lung diseases, including pulmonary embolism, lung malignancy, and bulla (Nakagawa *et al.*, 2001). The results of these studies indicate that lung perfusion can be assessed in addition to the oxygen transfer in clinical patients in a "one-stop" examination that lasts < 1 hr. Thus, by using OE imaging techniques in combination with contrast-enhanced pulmonary perfusion MRI, diseases that cause a ventilation/perfusion mismatch, such as pulmonary embolism, can be distinguished from those that lead to other abnormalities.

Other lung function parameters have also been shown to be associated with OE images. Signal intensity changes can be directly correlated to diffusing lung capacity (Mueller *et al.*, 2002) and forced expiratory volume (Ohno *et al.*, 2002). Similarly, lung function assessments performed using other modalities, such as CT, have also been shown to correspond well to images created using OE imaging in diseases such as emphysema (Ohno *et al.*, 2005).

Conclusions

Although functional oxygen-enhanced ^1H MR lung imaging techniques are not yet developed enough to be used

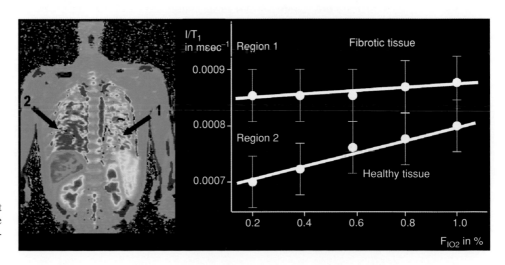

Figure 70 T_1 map of a patient with cystic fibrosis (left) and the OTF measurement in two representative regions (right).

routinely in the clinic, the preliminary results indicate that this method could be promising for future applications. Because of the fact that various parameters that indicate different aspects of the health of the lung can be examined using this technique, oxygen-enhanced proton MRI can be used for the examination of diseases that produce a variety of lung defects. The ability of OE MRI to distinguish between morphological and functional deficiencies is also of great importance when attempting to understand the mechanisms by which the disease disrupts the lung tissue. In addition, the use of OE MR gas transfer measurments in conjunction with lung perfusion measurements can provide information about the ventilation/perfusion ratio. Further examination of lung diseases must be performed in order to better understand how MR parameters correlate with lung dysfunction and how the use of MR in the clinic can be increased.

Several OE MR methods are presented in this chapter, including T_1-weighted imaging, quantitative T_1 mapping, and quantitative T_2^* mapping. Each method has its strengths and thus potential clinical uses. T_1 methods can be used to measure the ability of the lung tissue to transfer gas by measuring the OTF, as well as the lung morphology by examining the native T_1 values of the lung tissue. However, T_1 methods do not offer a representation of the pure regional ventilation, which is important for the evaluation of lung diseases such as chronic obstructive lung diseases (COPD). In contrast, the T_2^* approach, allows the ventilation of the lung to be depicted more directly, although it suffers due to the short T_2^* values in the lung. Although other MR methods, such as those that employ hyperpolarized gases, can be used to perform similar tasks, disadvantages such as high cost and difficult measurement setup make these techniques unfeasible. Thus, despite the low-proton density and large susceptibility differences that hamper lung imaging, the value of OE MR has been shown in its capability to depict various lung characteristics.

Few studies have been performed that show how these methods could be used to depict modifications in the lung caused by cancer. Even so, the topic has been the subject of much interest. As in other diseases, OE MR can provide information about the changes in morphology and functional capabilities in a cancer-stricken lung. An ideal use of this information would be to guide radiation therapy. Through detailed functional maps, radiation could be better directed at the tumor tissue, thereby avoiding damage to areas of healthy tissue. Follow-up imaging sessions could track the progress of therapy and assess the need for additional interventions. Another application of OE imaging could be to characterize the effects of different cancer types on lung tissue. With knowledge of how lung functional parameters have been affected, therapy could be tailored to the specific cancer and the particular functional damage it causes.

The probable role of oxygen-enhanced imaging is not as a replacement for today's clinical standards for lung imaging (i.e., scintigraphy, spirometry, or CT), but as a complementary technique that can offer additional information. The power of MR not only to investigate diseased tissue but also to assess the efficacy of therapy and medication makes MR useful in a variety of applications. The primary advantages of MR—namely, its noninvasive and thereby repeatable nature, cost-effectiveness, and ease of implementation—make it a prime candidate for routine use in the future.

References

Alsop, D.C., Hatabu, H., Bonnet, M., Listerud, J., and Gefter, W. 1995. Multislice, breathhold imaging of the lung with submillisecond echo times. *Magn. Reson. Med.* 33:678–682.

Arnold, J.F., Fidler, F., Wang, T., Pracht, E.D., Schmidt, M., and Jakob, P.M. 2004. Imaging lung function using rapid dynamic acquisition of T1-maps during oxygen enhancement. *MAGMA 16*:246–253.

Case, T.A., Durney, C.H., Ailion, D.C., Gutillo, A.G., and Morris, A.H. 1987. A mathematical model of diamagnetic line broadening in lung tissue and similar heterogeneous systems: calculations and measurements. *J. Magn. Reson.* 73:304–314.

Chen, Q., Jakob, P.M., Griswold, M.A., Levin, D.L., Hatabu, H., and Edelman, R.R. 1998. Oxygen enhanced MR ventilation imaging of the lung. *MAGMA 7*:153–161.

Cutillo, A.G., Ganesan, K., Ailion, D.C., Morris, A.H., Durney, C.H., Symko, S.C., and Christman, R.A. 1991. Alveolar air-tissue interface and nuclear magnetic resonance behavior of lung. *J. Appl. Physiol.* 70:2145–2154.

Dietrich, O., Losert, C., Attenberger, U., Fasol, U., Peller, M., Nikolaou, K., Reiser, M.F., and Schoenberg, S.O. 2005. Fast oxygen-enhanced multislice imaging of the lung using parallel acquisition techniques. *Magn. Reson. Med.* 53:1317–1325.

Durney, C.H., Bertolina, J., Ailion, D.C., Christman, R., Cutillo, A.G., Morris, A.H., and Hashemi, S. 1989. Calculation and interpretation of inhomogeneous line broadening in models of lungs and other heterogeneous structures. *J. Magn. Reson.* 85:574–570.

Ebert, M., Grossmann, T., Heil, W., Otten, W.E., Surkau, R., Leduc, M., Bachert, P., Knopp, M.V., Schad, L.R., and Thelen, M. 1996. Nuclear magnetic resonance imaging with hyperpolarised helium-3. *Lancet.* 347:1297–1299.

Edelman, R.R., Hatabu, H., Tadamura, E., Li, W., and Prasad, P.V. 1996. Noninvasive assessment of regional ventilation in the human lung using oxygen-enhanced magnetic resonance imaging. *Nat. Med.* 2:1236–1239.

Fullerton, G.D., Potter, J.L., and Dornbluth, N.C. 1982. NMR relaxation of protons in tissues and other macromolecular water solutions. *Magn. Reson. Imaging 1*: 209–288.

Goodrich, K.C., Hackmann, A., and Ganesan, K. 1992. Spin-lattice relaxation in excised rat lung. Book of abstracts 11th annual meeting, Society of Magnetic Resonance in Medicine, Berlin, p.1307.

Haase, A., Frahm, J., Matthei, D., and Merbold, K.D. 1986. FLASH imaging: rapid NMR imaging using low flip angle pulses. *J. Magn. Reson.* 67:258–266.

Hatabu, H., Alsop, D.C., Listerud, J., Bonnet, M., and Gefter, W.B. 1999. T2* and proton density measurement of normal human lung parenchyma using submillisecond echo time gradient echo magnetic resonance imaging. *Eur. J. Radiol.* 29:245–252.

Hatabu, H., Tadamura, E., Chen, Q., Stock, K.W., Li, W., Prasad, P.V., and Edelman, R.R. 2001. Pulmonary ventilation: dynamic MRI with inhalation of molecular oxygen. *Eur. J. Radiol.* 37:172–178.

Hennig, J., Nauerth, A., and Friedburg, H. 1986. RARE imaging: a fast imaging method for clinical MR. *Magn. Reson. Med. 3*:823–833.

Jakob, P.M., Hillenbrand, C.M., Wang, T., Schultz, G., Hahn, D., and Haase, A. 2001. Rapid quantitative lung 1H T1 mapping. *J. Magn. Reson. Imaging 14*:795–799.

Jakob, P.M., Wang, T., Schultz, G., Hebestreit, H., Hebestreit, A., and Hahn, D. 2004. Assessment of human pulmonary function using oxygen-enhanced T1 imaging in patients with cystic fibrosis. *Magn. Reson. Med. 51*:1009–1016.

Karlinsky, J.B., 1982. Glycosaminoglycans in emphysematous and fibrotic hamster lungs. *Am. Rev. Respir. Dis. 125*:85–88.

Kuethe, D.O., Caprihan, A., Fukushima, E., and Waggoner, R.A. 1998. Imaging lungs using inert fluorinated gases. *Magn. Reson. Med. 39*:85–88.

Lawrence, M., 1999. *All You Really Need to Know to Interpret Arterial Blood Gases*, 2nd ed. Philadelphia: Lippincott, Williams and Wilkins.

Loeffler R., Mueller, C.J., Peller, M., Penzkofer, H., Deimling, M., Schwaiblmair, M., Scheidler, J., and Reiser, M. 2000. Optimization and evaluation of the signal intensity change in multisection oxygen-enhanced MR lung imaging. *Magn. Reson. Med. 43*:860–866.

Mai, V.M., Liu, B., Li, W., Polzin, J., Kurucay, S., Chen, Q., and Edelman, R.R. 2002. Influence of oxygen flow rate on signal and T1 changes in oxygen-enhanced ventilation imaging. *J. Magn. Reson. Imaging 16*:37–41.

Mueller, C.J., Schwaiblmair, M., Scheidler, J., Deimling, M., Weber, J., Loeffler, R.B., and Reiser, M.F. 2002. Pulmonary diffusing capacity: assessment with oxygen-enhanced lung MR imaging preliminary findings. *Radiology 222*:499–506.

Naish, J.H., Parker, G.J., Beatty, P.C., Jackson, A., Young, S.S., Waterton, J.C., and Taylor, C.J. 2005. Improved quantitative dynamic regional oxygen-enhanced pulmonary imaging using image registration. *Magn. Reson. Med. 54*:464–469.

Nakagawa, T., Sakuma, H., Murashima, S., Ishida, N., Matsumura, K., and Takeda, K. 2001. Pulmonary ventilation-perfusion MR imaging in clinical patients. *J. Magn. Reson. Imaging. 14*:419–424.

Ohno, Y., Hatabu, H., Higashino, T., Nogami, M., Takenaka, D., Watanabe, H., and Van Cauteren, M. 2005. Oxygen-enhanced MR imaging: correlation with postsurgical lung function in patients with lung cancer. *Radiology 23*:704–711.

Ohno, Y., Hatabu, H., Takenaka, D., Van Cauteren, M., Fujii, M., and Sugimura K. 2002. Dynamic oxygen-enhanced MRI reflects diffusing capacity of the lung. *Magn. Reson. Med. 47*:1139–1144.

Pracht, E.D., Arnold, J.F., Wang, T., and Jakob, P.M. 2005. Oxygen-enhanced proton imaging of the human lung using T2*. *Magn. Reson. Med. 53*:1193–1196.

Shioya, S., Haida, M., Ono, Y., Fukuzaki, M., and Yamabayashi, H. 1988. Lung cancer: differentiation of tumor, necrosis, and atelectasis by means of T1 and T2 values measured in vitro. *Radiology 167*:105–109.

Starcher, B.C., Kuhn, C., and Overton, J.E. 1978. Increased elastin and collagen content in the lungs of hamsters receiving an intratracheal injection of bleomycin. *Am. Rev. Respir. Dis. 117*:299–305.

Vinitski, S., Pearson, M.G., Karlik, S.J., Morgan, W.K., Carey, L.S., Perkins, G., Goto, T., and Befus, D. 1986. Differentiation of parenchymal lung disorders with *in vitro* proton nuclear magnetic resonance. *Magn. Reson. Med. 3*:120–125.

Wagshul, M.E., Button, T.M., Li, H.F., Liang, Z., Springer, C.S., Zhong, K., and Wishnia, A. 1996. *In vivo* MR imaging and spectroscopy using hyperpolarized 129Xe. *Magn. Reson. Med. 36*:183–191.

Yablonskiy, D.A. 1997. Quantitation of T2′ anisotropic effects on magnetic resonance bone mineral density measurement. *Magn. Reson. Med. 37*:214–221.

Yablonskiy, D.A., and Haacke, E.M. 1994. Theory of NMR signal behavior in magnetically inhomogeneous tissues: the static dephasing regime. *Magn. Reson. Med. 32*:749–763.

Zimmermann, J.R., and Brittin, W.E. 1957. Nuclear magnetic resonance studies in multiple phase systems: lifetime of a water molecule in an adsorbing phase on silica gel. *J. Phys. Chem. 61*:1328–1333.

14

Detection of Pulmonary Gene Transfer Using Iodide-124/Positron Emission Tomography

Frederick E. Domann and Gang Niu

Introduction

Gene therapy is an evolving technique that seeks to use nucleic acids (DNA or RNA) to treat or prevent disease. There are several approaches to gene therapy, including forced expression of a therapeutic gene on the background of a mutant gene (gene addition), replacing a mutated gene that causes disease with a healthy copy of the gene (gene replacement), inactivating or "knocking out" a mutated gene that is functioning improperly, and introducing a novel gene into the body to help prevent or fight disease. Gene addition and gene replacement are currently the most commonly used approaches, and both are clinically tested methods for gene therapy. To deliver the therapeutic gene to the patient's target cells, an appropriate carrier molecule or gene delivery vehicle, often called a vector, must be used. Among the most successful vectors to date are biologically derived nanoscale virions that can be used to deliver therapeutic gene payloads consisting of cDNA expression cassettes. These include viral particles derived from the genomes of adenoviruses, retroviruses, adeno-associated viruses, and herpes simplex viruses, all of which have been genetically altered to carry a payload (for example, normal human DNA) and engineered to be replication defective. In some cases, the virion particle has been targeted specifically at certain cell types by tethering specific molecules to the surface of the particles. In other cases, elegant gene regulatory elements that are active only in the desired target cells are inserted to provide cell-type specific expression of the payload gene. Thus, efforts are being directed toward cell-type specific targeting at the levels of both payload delivery and gene expression.

Pulmonary Applications of Gene Therapy

The most widely studied lung disease targets for gene therapy thus far reported can be generally classified into two categories: (1) inherited lung diseases (such as cystic fibrosis, α-1 anitrypsin deficiency, and surfactant deficiency), and (2) acquired disorders such as asthma, lung cancers, and acute lung injury. Although pulmonary gene therapy has made significant progress, obstacles still exist. The most challenging obstacles include targeting gene expression to the appropriate cell types, inefficient gene transfer, short-lived nature of therapeutic gene expression, and host immune responses to the vector. To overcome these obstacles and engage early-phase clinical trials in gene therapy, accurate evaluation of the distribution, expression, and duration of the delivered gene is a critical goal. Various methods have been used to detect gene expression in tissue biopsy samples, including *in situ* hybridization, immunohistochemical staining, and reporter genes such as β galactosidase can be used. However, these detection methods are invasive and not suitable for routine clinical evaluation. In addition, these traditional methods to monitor the duration of gene transfer in animal models require the sacrifice of animals along a time course to achieve a temporal series of observations (Peebles *et al.*, 2004).

Ideally, the method to measure and monitor gene transfer should be noninvasive and repeatable over time. The emerging field of molecular imaging is promising to accomplish this task. In experiments based on molecular imaging, such as those with the iodine symporter, fewer animals are needed for each study. Moreover, each animal can serve as its own "control" for previous and subsequent analyses, so experimental uncertainties arising from interanimal variations are greatly reduced (Kay *et al.*, 1992). In this chapter, we will discuss the role played by molecular imaging in pulmonary gene therapy, especially positron emission tomography (PET) imaging using [^{124}]I as a radiotracer.

Gene Therapy for Inherited Lung Diseases

Cystic Fibrosis

Several diseases caused by inborn errors of metabolism might be treated by pulmonary gene delivery. Cystic fibrosis (CF) is an autosomal recessive disorder characterized by abnormal salt and water transport that leads to abnormal airway secretions, impaired mucociliary clearance, chronic bacterial infection, bronchiectasis, and premature death. Although a variety of epithelial tissues besides airways are affected in this disease (pancreatic, sweat duct, and gastrointestinal epithelia), lung disease is the major cause of morbidity and mortality in this disorder (Quinton, 1990). The CFTR gene, which codes for a protein called CF transmembrane conductance regulator (CFTR) protein, functions as both a chloride (Cl$^-$) channel and a regulator of other channels. The CF gene was cloned in 1989 and the most common mutation is a three base-pair deletion leading to deletion of phenylalanine at position 508 (ΔF508) of the gene (Rommens *et al.*, 1989). This mutation leads to abnormal intracellular processing and trafficking of the mutant CFTR, and ultimately to defective apical membrane chloride conductance in epithelial cells affected by this disorder. Directing a normal version of the CF gene to the lung, where the clinical manifestations were the most prob-

lematic, became the goal of a number of investigators. Investigations into gene transfer for CF began in 1990 when Drumm *et al.* (1990) and Rich *et al.* (1990) reported that normal Cl$^-$ transport function in CF epithelial cells was restored after introduction of a normal copy of the CFTR gene.

The first clinical trial for CF gene therapy initiated in 1993 used recombinant E1 deleted adenovirus as vector to deliver CFTR gene to human CF nasal epithelium. The treatment corrected the Cl$^-$ transport defect transiently. After treatment, the elevated basal transepithelial voltage decreased, and the normal response to a cAMP agonist was restored (Zabner *et al.*, 1993). However, the subsequent experiments performed to evaluate recombinant adenoviral-mediated delivery of CFTR to the nasal epithelium of CF patients were unable to reproducibly correct CFTR function (Knowles *et al.*, 1995). Moreover, expression analysis following the adenoviral gene transfer in other Phase I studies suggested that gene transfer was low (< 1%) with immunologic responses to the vector (Bellon *et al.*, 1997; Zuckerman *et al.*, 1999).

The first CF clinical trial using liposomes, which are less immunogenic than adenoviral vectors, was performed in 1995 and demonstrated a 20% restoration of transepithelial potential difference (TEPD) in response to low chloride at 3 days following administration of cationic pipsome/DNA complexes to the nasal epithelium (Caplen *et al.*, 1995). Despite the apparently reduced toxicity of liposome/DNA complexes compared with adenovirus trials, adverse reactions including fever, myalgias, and arthralgia still have been reported in four of eight CF patients receiving aerosolized liposome/DNA complexes administration (Ruiz *et al.*, 2001). These adverse effects are thought to be the result of unmethylated CpG nucleotides in bacterial plasmid DNA. In addition, it is also believed that the low-level correction with the utility of liposome/DNA complexes is extremely transient (Driskell and Engelhardt, 2003).

The first clinical trial with rAAV-expressing CFTR was performed in 1998 (Wagner *et al.*, 1998), and results demonstrated a dose-dependent response in the accumulation of vector genomes in sinus epithelium, with little or no immunologic consequences. A follow-up study with a second dose-escalation trial demonstrated partial correction of chloride transport abnormalities by TEPD measurement in the maxillary sinus (Wagner *et al.*, 1999). Although rAAV genomes were demonstrated to be detectable until 70–90 days postinfection with no toxicity or immune response, no vector-derived mRNA could be detected at any of the time points or vector doses by RT PCR (Aitken *et al.*, 2001). Cystic fibrosis transmembrane conductance regulator expression in the rAAV vector was driven by a weak promoter inverted terminal repeats (ITR), which likely contributed to the undetectable levels of expression in these clinical trials.

Alpha-1 Antitrypsin Deficiency

Alpha-1 antitrypsin (AAT) is a 52 kDa serum protein that is normally produced in the hepatocytes within the liver and circulates at serum levels sufficient to be detected by serum protein electrophoresis (> 20 µmol/l). This protein is the most prominent endogenous serine proteinase inhibitor (SERPIN) in humans and provides important protection against proteases such as neutrophil elastase. The deficiency of AAT is an autosomal-recessive inherited disease leading to cirrhosis of the liver and panacinar emphysema of the lung, which is the most life-threatening pathology of this disease (Song et al., 2001). This disorder is second in prevalence only to cystic fibrosis among inherited diseases of the lung, so it becomes another attractive target for pulmonary gene therapy. Airway gene transfer mediated by human AAT expressing recombinant adenovirus and DNA/liposome complexes in animal models resulted in detectable hAAT mRNA and hAAT protein in serum at least 1 week after vector administration (Rosenfeld et al., 1991; Canonico et al., 1994). Because AAT freely circulates throughout the body, nonlocal targeted gene therapy has been investigated. It was reported that direct hepatic transfer of human alpha 1-antitrypsin cDNA under the transcriptional direction of the albumin promoter-enhancer in retroviral vector led to constitutive expression of the human protein in the sera of recipients at concentrations of 30–1400 ng/ml for at least 6 months (Kay et al., 1992). Song et al. (1998) observed sustained secretion of human alpha-1-antitrypsin from murine muscle transduced with AAV vector. Using DNA/liposome complexes, Brigham et al. (2000) reported a gene therapy study for AAT correction performed in human subjects. Results showed that AAT concentrations in serial samples of nasal lavage fluid increased significantly in the transfected nostril, peaking on day 5 after transfection and returning to baseline by day 14. In a published clinical protocol, an hAAT-expressing recombinant AAV vector was tested in a population of 12 patients with AAT deficiency through intramuscular injection (Flotte et al., 2004).

Gene Therapy for Lung Cancer

According to the American Cancer Society, in 2007 there will be ~ 213,380 new cases of lung cancer in the United States and ~ 160,390 people will die of this disease. The lack of effective screening procedures means the majority of patients present in later stages in which tumors are often resistant to even multimodal therapy. During the last decade, molecular characterization and subsequent application of gene transfer methods in lung cancer have played a large role in the development of gene therapy. Different from inherited diseases, a myriad of potential causative and molecular factors are involved in the initiation, promotion, and progression of carcinogenesis. Focusing on different aspects of carcinogenesis, there are several target strategies for pulmonary cancer gene therapy such as tumor suppressor genes, angiogenesis and growth factors, and apoptosis mediators.

Tumor suppressor genes are among the most popular therapeutic targets. Mutated in > 50% of lung cancers, p53 is the classic tumor suppressor gene involved in cell-cycle regulation and apoptosis. Gene transfer of p53 in retroviral or adenoviral vectors has confirmed its therapeutic effects in animal models of human lung cancer (Zhang et al., 1994; Fujiwara et al., 1994). Following the first clinical trial for p53 gene therapy in non-small cell lung carcinoma (NSCLC) in 1996 (Roth et al., 1996), several clinical trials with p53 as therapeutic gene were reported (Schuler et al., 1998, 2001; Swisher et al., 1999, 2003; Nemunaitis et al., 2000; Kubba et al., 2000). In most cases, wild type p53 packaged in adenoviral vectors were directly injected into tumors using either bronchoscopic or CT guidance. The results from those clinical trials are promising, with tumor growth stabilization or regression observed in experimental subjects. When combined with chemotherapy, Nemunaitis et al. (2000) found that 17 out of 24 patients achieved a best clinical response of stable disease, and a mean apoptotic index increased significantly. However, Schuler et al. (2001) reported that intratumoral adenoviral p53 gene therapy provided no additional benefit in patients receiving chemotherapy for advanced NSCLC. In a Phase II study, Swisher et al. (2003) evaluated the feasibility and mechanisms of apoptosis induction after Ad-p53 gene transfer and radiation therapy in patients with NSCLC. They found that the combination was well tolerated by patients, and upregulation of the proapoptotic gene BAK was seen in tumor specimens.

Besides p53, mutations in the cyclin-dependent kinase inhibitors, including p16, p21, and p27, are frequently seen in lung carcinoma (Salgia and Skarin, 1998). The fragile histidine triad (FHIT) gene and melanoma differentiation associated gene (MDA-7) also have been shown to promote apoptosis while maintaining no toxicity to normal cells. These genes are all promising targets for pulmonary gene therapy and some of them are already in clinical trials (Ji et al., 1999; Saeki et al., 2002).

Gene Delivery Vehicles and Vectors

Based on delivery approaches, gene transfer to humans can be achieved using either ex vivo or in vivo strategies. The lung is a collapsible air-containing structure surrounded by a network of blood vessels. Due to complexity of the airway anatomy and the sparsity of progenitor cells, gene delivery to the airway epithelium will most probably require the use of in vivo gene transfer. For pulmonary cancer gene therapy, central tumors can often be reached by flexible bronchoscopy and peripheral tumors by established transthoracic radiologic-guided approaches. Inhaled delivery and endobronchial lavage also have been used in some cases for targeting epithelial cells (Gautum *et al.*, 2000; Kubba *et al.*, 2000). Compared with tumor-localized delivery, more severe regional toxicity was observed in patients treated with bronchial lavage (Kubba *et al.*, 2000). For inherited lung diseases, liquid and aerosol ventilation or inhalation are commonly used as delivery approaches because of the diffuse distribution of target cells among the whole organ.

The vectors used in gene therapy usually contain an expression cassette comprised by the cDNA of interest, and both 5′ upstream and 3′ downstream flanking regions. The 5′ region is the promoter sequence, and the 3′ flanking region usually influences the RNA splicing, and polyadenylaiton and post-translational processing. The ideal vector should possess the following characteristics: specificity to target cells over nontarget cells, ability to infect both dividing and nondividing cells, resistance to cytotoxic and humoral-mediated destruction, extended transgene expression, and low inflammatory and toxic response (Daniel and Smythe, 2003). A variety of gene transfer vectors have been utilized in human clinical trials; these can be broadly classified into viral and nonviral systems. There are also a few viral vectors available for gene transfer, including retroviruses, adenoviruses, and adeno-associated viruses.

Retroviruses

Single-stranded RNA retroviruses were the first viral vectors used in preclinical and clinical studies. Retroviruses contain two copies of a viral genetic element known as long terminal repeats, which are important for cellular integration and contain promoter elements. Their genomes also have a packaging signal and a series of viral genes—*gag*, *pol*, and *env*. A retroviral vector is constructed by deleting viral genes and inserting desired cDNAs into the remaining retrovirus genome. The main advantage of retrovirus is the ability of this vector to integrate into the host genome and thus lead to expression of the transgene in the daughter cells of the originally infected cell. However, this is also a risk factor for initiating new cancer through insertional mutagenesis. A commonly used recombinant retroviral vector is based on Maloney murine leukemia virus (MLV) that is unable to infect nondividing cells. The more recently developed lentiviral and pseudotyped lentiviral vectors exhibit not only the ability to infect nondividing cells, but also increased cell-type specificity (Coil *et al.*, 2001).

Adenoviruses

The double-stranded DNA adenovirus has been the most widely used vector in pulmonary gene therapy. Adeno-

viruses have large, complex, linear 36 kb DNA genomes and are known for their high gene transfer efficiency in multiple tissues, especially lung. The advantages of adenoviral vectors include large packaging ability, broad cellular tropism, ability to infect nondividing cells, and ease of construction and production. However, adenoviral genomes do not integrate into host genomes and therefore persist as episomes, so repetitive dosing is required for persistent transgene expression. Adenoviruses are also immunogenic, so to overcome the immunologic response generated by the vector, a third-generation adenoviral vector system, also called gutted or helper dependent adenoviral vectors, has been developed by deleting all viral genes from the recombinant vector (Zhou *et al.*, 2002).

Adeno-associated Viruses (AAV)

Adeno-associated viruses have 4.7 kb single-stranded DNA genomes. They are replication defective and require co-infection with a helper virus, such as adenovirus or herpes virus. Compared with adenoviral vectors, the AAV vector will integrate in the host genome and thus lead transgene expression in the progeny of the initially infected cell. Another attractive feature of rAAV vectors is the absence of cytotoxic T-lymphocyte responses (CTL) to the recombinant virus (Kay *et al.*, 2001). Nevertheless, humoral responses to capsid proteins remain an obstacle for repeated administration of this vector. The major limitation of rAAV vector systems has been the limited packaging capacity of the recombinant vector genome because of its small size. To address this limitation, novel techniques including *cis*-activation and *trans*-splicing have been developed to expand the packaging capacity of this vector system (Nakai *et al.*, 2000).

Nonviral Liposomal Vectors

Liposomes are synthetic lipid bilayers that form complexes with drugs or nucleic acids that facilitate entry of these products into the cell through membrane fusion or endocytosis. Compared with recombinant viral vectors, major attractions of liposome-mediated gene delivery are that they are noninfectious, generally have low toxicity, and are less immunogenic. However, low delivery efficiency and short duration of transgene expression are obstacles for their clinical application. Nevertheless they have moved forward into clinical trials.

Molecular Imaging of Pulmonary Gene Transfer

Molecular imaging is defined as the visual representation, characterization, and quantification of biological processes at the cellular and subcellular levels within intact living organisms (Shah *et al.*, 2004). Two basic elements required by molecular imaging are the molecular probes themselves and the means to measure the location and amount of these probes *in vivo* (Herschman, 2003).

Signals from light-emitting probes (fluorescence, bioluminescence, and near-infrared [NIR]–emitting molecules) are monitored by a sensitive cooled charge-coupled device (CCD) camera. Radionuclide-labeled probes are detected by PET or single photon emission computed tomography (SPECT). Probes that can change radiowave emissions are detected by magnetic resonance imaging (MRI). Optical imaging of fluorescent and bioluminescent probes has been well established in small-animal models for monitoring gene transfer due to its high sensitivity, low cost, easy operation, and short acquisition time. However, poor penetration of visible photons makes optical imaging less suitable for large animals and human beings. In addition, results from optical imaging are two-dimensional and lack tomographical information. Magnetic resonance imaging techniques can generate spectacular image resolution; however, temporal resolution is limited, and molecular probe detection is several orders of magnitude less sensitive than other imaging modalities. Although imaging of gene expression with MRI has been described, this approach is generally not suitable for lung studies because the low-proton density of the lung parenchyma is an obstacle to generate an MR image of sufficient quality (Weissleder *et al.*, 2000).

Positron emission tomography imaging has been extensively applied to the evaluation of pulmonary physiology and diagnosis of pulmonary disease (especially lung cancer). A requirement of the isotopes used in PET is that they decay by the process of positron emission. The emitted positrons are annihilated by their interaction with ambient electrons in the vicinity of the positron emission. Two photons, each of 0.511 MeV and traveling 180° apart, are emitted after each annihilation event. These photons may be counted with detectors placed on opposite sides of the body. Because the annihilation events that produce the two coincident photons must have occurred somewhere within the tissue volume between the detectors, computed tomography can be used to locate the radiation source in space. As additional detectors are added to the system, intersecting lines further establish spatial location. Thus, modern PET imagers consist of a ring of detectors. This type of electronic coincidence detection is far more efficient than that provided by the collimation by lead shielding used in conventional single photon emission detection systems because less radiation is ignored. In general, radiotracer imaging with methods such as PET has a high degree of sensitivity (the level of detection approaches 10^{-11} M of tracer) and isotropism (i.e., the ability to detect expression accurately regardless of tissue depth, unlike light-based techniques, which are largely limited to detection at the body surface). Thus, radiologic methods such as PET and SPECT are ideally suited to detect gene expression in deep organs such as the lungs.

With ^{18}F-labeled fluorodeoxyglucose (FDG), PET imaging has become widely used in diagnosing, staging, defining the treatment plan, and assessing the recurrence of pulmonary malignant diseases, especially for NSCLC (Stroobants *et al.*, 2003). Indeed, FDG-PET is more accurate for disease staging than conventional imaging in lung cancer. A study with 100 lung cancer patients reported that the accuracy of FDG-PET for staging was 83% versus 65% by chest CT and bone scintigraphy (Marom *et al.*, 1999). In addition to applications in lung cancer, PET is also powerful and quantitative in studying lung physiology and biochemistry such as peptide metabolism, blood flow, ventilation, and water content using tracers labeled with oxygen-15, carbon-11, or nitrogen-13 (Schuster, 1998).

Reporter Gene Systems

Recent integration of specific reporter genes and imaging techniques has made it possible to image endogenous molecular events. Reporter genes have been long used to study promoter/enhancer elements involved in gene expression, induction of gene expression by inducible promoters, and endogenous gene expression through the use of transgenes containing endogenous promoters fused to the reporter (Gambhir *et al.*, 1999). A common feature of all reporter vectors is a cDNA expression cassette containing a promoter and the reporter transgene(s) of interest. The promoter can be constitutive, or it can be inducible. The promoter can also be cell type specific, allowing expression of the transgene to be restricted to certain cells and organs (Blasberg and Tjuvajev, 2003). In general, reporter genes are classified as intracellular or extracellular based on the location of the gene product. Detection methods may be based on photometry, colorimetry, radiometry, fluorescence, or immunoassay. Unlike conventional reporter genes, such as chloramphenicol acetyl tranferase, LacZ/β-galactosidase, and alkaline phosphatase that require *ex vivo* analysis, molecular imaging techniques offer the possibility of monitoring the location, magnitude, and persistence of reporter gene expression *in vivo* in intact living animals or humans (Massoud and Gambhir, 2003). Most of the present reporter gene imaging is applied to monitor gene expression from genes externally transferred into cells of organ systems in living subjects. The ideal reporter gene product should have characteristics that include nonimmunogenicity, specificity, and stability and that exhibit a good correlation between gene expression and imaging signal (Gambhir *et al.*, 2000).

Several PET reporter gene systems have been investigated, each with its own advantages and disadvantages. These include cytosine deaminase (CD), herpes simplex virus type 1 thymidine kinase (HSV1-tk), dopamine 2 receptor (D2R), and sodium iodide symporter (NIS). HSV1-tk and NIS are two of the most popularly used radiological imaging reporter genes and will be discussed in the following sections. Cytosine deaminase, which deaminates cytosine to uracil, is normally expressed primarily in yeasts and bacteria. Cytosine deaminase was one of the first reporter genes to be studied for imaging reporter gene expression in mammalian cells. However, slow uptake and rapid cellular efflux of the reporter probe 6-[^{3}H]-5-fluorocytosine have been observed in human glioblastoma cells stably transfected with the *E. coli* CD reporter gene, thus limiting its further application in PET imaging (Haberkorn *et al.*, 1996).

The D2R, a 415-amino acid protein with seven transmembrane domains, is normally expressed primarily in the brain striatum and in the pituitary gland. Radiolabeled ligands of D2R, including 3-(2′-[^{18}F]-fluoroethyl)spiperone ([^{18}F]FESP) and [^{123}I]-iodobenzamine, are two examples of probes used to image D2R expression with PET and SPECT, respectively, based on receptor-ligand binding. D2R is an endogenous gene, and its expression will not invoke an immune response. The main drawbacks of D2R reporter system are perturbations of normal cell physiology and lack of signal amplification.

Herpes Simplex Virus-1 Thymidine Kinase (HSV1-TK)

Thymidine kinases are responsible for catalyzing the transfer of the γ-phosphate from adenosine triphosphate (ATP) to the 5′-terminus of thymidine to form thymidine monophosphate (dTMP). Viral TK have a different spectrum of substrate specificity than that of mammalian TK; thus, probes specific for viral tks have been developed. Two main categories of imaging probes for HSV1-tk reporter gene expression are uracil nucleoside derivatives (such as FAIU) and acycloguanosine derivatives (such as FHPG). These reporter probes are transported into cells and trapped as a result of enzymatic phosphorylation by HSV1-TK protein (Arner and Eriksson, 1995). Compared with the acycloguanosine derivatives, the uracil nucleoside derivatives probably have a higher affinity for the HSV1-TK enzyme. Comparisons of 8-[^{3}H]-ganciclovir and 2-[^{14}C]FIAU in cell culture indicate a moderate advantage of 2-[^{14}C]FIAU over 8-[^{3}H]-ganciclovir with respect to accumulation in HSV1-tk-transduced versus control cells (Tjuvajev *et al.*, 1995). Brust *et al.* (2001) compared [^{18}F]FHPG with [^{124}I]FIAU in cell culture models and in mice harboring tumors that stably expressed HSV1-tk. In HSV1-tk expressing T1115 human glioblastoma cells, the [^{18}F]FHPG uptake increased by only ~ 6-fold, whereas the [^{125}I]FIAU accumulation increased by ~ 28-fold after 120-min incubation with each radiotracer. *In vivo* studies demonstrated that the radioiodinated uracil nucleoside FIAU had a significantly higher specific accumulation than the acycloguanosine derivative [^{18}F]FHPG. These studies suggested that future clinical

research should be focused on the uridine type of HSV1-TK substrates rather than on the acylguanosine type (Brust *et al.*, 2001).

Another study comparing [^{124}I]FIAU with [^{18}F]FHBG also reported a 15-fold higher accumulation of [^{124}I]FIAU in RG2-tk+ rat xenografts (Doubrovin *et al.*, 2002). In a PET study on rats bearing multiple tumors derived from W256 rat carcinoma and RG2 rat glioma cells, a highly significant relationship between the level of [^{124}I]-FIAU accumulation and an independent measure of HSV1-tk expression was demonstrated. The results demonstrated that not only the localization but also the level of HSV1-tk expression could be obtained using radiolabeled 2′-fluoronucleoside [^{124}I]-FIAU and a clinical PET system (Tjuvajev *et al.*, 1998).

HSV1-sr39tk is a specific HSV1-tk mutant that was created by random sequence mutagenesis and selected in *E. coli* for the enhanced ability to convert the pro-drugs acyclovir and ganciclovir into cytotoxic agents (Black *et al.*, 1996). Gambhir *et al.* (2000) found that HSV1-sr39tk was a better PET reporter gene for imaging hepatic gene expression than wild-type HSV1-tk when adenoviral vectors carrying these two genes were administered to mice via tail vein injection. HSV1-sr39tk was able to utilize cycloguanosines more effectively than HSV1-TK. MicroPET imaging of mice bearing stably transfected tumors expressing HSV1-TK and HSV1-sr39tk also demonstrated greater *in vivo* imaging sensitivity of the mutant HSV1-sr39TK PET reporter gene. (Gambhir *et al.*, 2000).

Several experiments for imaging pulmonary gene transfer with PET using the HSV1-tk reporter system have been reported recently. In one study, 16 rats were studied 3 days after an intratracheal administration of 5×10^9 to 1×10^{11} viral particles of an adenoviral vector containing a fusion gene of the mutant HSV1-sr39tk and GFP. Images were obtained 1 hr after an intravenous injection of [^{18}F]FHBG. Radioactivity detection by PET imaging and gamma counting produced measurements that were tightly and linearly correlated. Positron emission tomography imaging detected thymidine kinase expression even at low viral doses that produced little to no measurable green fluorescent protein expression. Lung [^{18}F]FHBG uptake was assessed by imaging correlated with *in vitro* assays of both kinase activity and fluorescent protein. These results demonstrated that PET imaging was a sensitive and quantitative method for detecting pulmonary reporter gene expression noninvasively (Richard *et al.*, 2003a).

Subsequently, these authors quantified the magnitude and spatial distribution of transgene expression after different methods of adenoviral vector delivery (with surfactant- and saline-based vehicles) within rat lungs. They found that the average lung concentration of [^{18}F]-FHBG was significantly greater in the surfactant group than in the saline group. Lung [^{18}F]-FHBG distribution was more peripheral and more homogeneous in the surfactant group than in the saline group. Regions of increased tracer concentration in the surfactant group compared to the saline group were evenly distributed throughout the lungs (Richard *et al.*, 2003b). In another study, these authors found that pulmonary uptake of [^{18}F]-FHBG in normal rats increased as TK activity increased only at low levels of mHSV1-tk expression and then plateaued as TK activity continued to increase. Compartmental modeling failed to improve the correlation with *in vitro* assays of transgene expression. However, a linear relationship was obtained between the pulmonary uptake of [^{18}F]-FHBG and *in vitro* assays of TK activity in rats treated with ANTU (alpha-naphthylthiourea) to increase pulmonary vascular permeability. These authors concluded that in rodent lungs, [^{18}F]-FHBG uptake was a function of both transportation into tissues expressing the transgene and the level of transgene expression itself (Richard *et al.*, 2004).

Sodium Iodide Symporter

The Na$^+$/I$^-$ symporter (NIS) is an integral plasma membrane protein that mediates the active transport of iodide (I$^-$) into the thyroid follicular cells. The human NIS gene is located on chromosome 19p12–13.2 and encodes a protein of 643 amino acids, which is glycosylated and has a molecular mass of ~ 70–90 kDa. The widely accepted secondary structure model of NIS has 13 membrane-spanning domains, with the amino terminal end on the extracellular surface of the cell and the carboxyl terminal end in the cytoplasm. Shortly after its identification and cDNA cloning, several studies have described the ectopic expression of the NIS gene with the aim of achieving radioiodine-induced cell killing and tumor responses. Some of these studies were accompanied by imaging of xenograft tumors either permanently expressing NIS or transduced by intratumoral injection of adenoviruses. The results showed that xenografted tumors transduced with NIS could be imaged with gamma scintigraphy compared with untransduced control ones, using ^{123}I or ^{99m}TcO$_4^-$ as imaging agents (Mandell *et al.*, 1999; Spitzweg *et al.*, 1999, 2000, 2001a,b; Boland *et al.*, 2000; Haberkorn *et al.*, 2001; La Perle *et al.*, 2002).

Groot-Wassink *et al.* (2002) have utilized NIS in combination with ^{124}I$^-$ to monitor the biodistribution and expression of a replication incompetent adenovirus using PET imaging. They found that when injected systemically, adenovirus delivery induced gene expression essentially in the liver, adrenal glands, lungs, pancreas, and spleen. Expression of NIS in tumor xenograft models could also be detected when the virus was injected intratumorally. As a reporter gene, NIS has some potential advantages over the other reporter systems. First, NIS is a physiologically expressed protein that only rarely induces the formation of antibodies (Morris *et al.*, 1997). Second, the tracers used in combination with NIS (^{99m}Tc or iodide isotopes) are commercially available at a relatively low cost. However, the NIS reporter gene-imaging system also has disadvantages. For example, the high background signals in the thyroid and stomach will bring difficulties if the target region is near these organs. In addition, rapid washout of imaging agents will make the determination of the optimal acquisition time very important.

Marsee *et al.* (2004) developed a pulmonary metastases model in nude mice by injecting NIS expressing rat prostate tumor cells via tail vein. They found that NIS expressing tumors as small as 3 mm in diameter could be detected by SPECT with pinhole collimation using ^{125}I as imaging agent (Marsee *et al.*, 2004). Experiments by our group (Niu *et al.*, 2005) have also investigated imaging of pulmonary gene transfer with a noninvasive NIS reporter gene-imaging system. Human NIS (hNIS) expressing adenoviruses (Ad-hNIS) or empty adenoviruses (Ad-Bgl II) were instilled into the lungs of Cotton rats via nostril. The Cotton rat is a permissive host for adenovirus, and it has demonstrated extremely efficient gene transfer to the airway for CFTR gene (Rosenfeld *et al.*, 1992). After 3, 10, and 17 days post-infection, gamma camera scintigraphy with ^{99m}TcO$_4^-$ was performed to observe the distribution and duration of gene transfer. Our results showed that lungs in animals infected with Ad-hNIS were clearly visible by scintigraphy. At 17 days, lungs in Ad-hNIS-infected animals still could be visualized. We also used ^{124}I$^-$ as a PET imaging agent. The animals were infected and treated as described above. Forty-eight hours after adenoviral infection, 14.8 MBq ^{124}I$^-$ was injected intraperitoneally. One hour later, each animal was placed in the prone position into the micro-PET scanner (Philips, Mosaic). The image acquisition time was ~ 30 min. Figure 71 shows reconstructed coronal slices of the Cotton rats. On the images, selective localization of ^{124}I$^-$ to the lung regions can be clearly visualized. Additional regions of accumulation included the thyroid and salivary glands as well as the stomach, in which there is endogenous NIS gene expression and nonspecific anion transport, respectively. In contrast, although the Ad-Bgl II infected rat showed similar background localizations as in the Ad-NIS infected animal, there was no significant ^{124}I$^-$ accumulation in the lung region.

The results of the radiological image analysis showed that the distribution of adenovirus appeared to be relatively uniform across the whole lung in most animals. Given the relatively limited spatial resolution of PET imaging (several mm), it is unlikely that imaging by itself would be able to discriminate expression at different tissue levels within the lungs (e.g., airway epithelium or microvessel endothelium). A parallel fluorescence microscopy study found that gene expression was located predominantly in lung parenchyma. Airway epithelial cells are the main targets in pulmonary gene transfer. It was demonstrated that there exist apical barriers to airway epithelial cell gene transfer (Wang *et al.*,

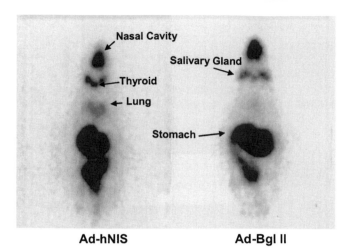

Ad-hNIS   **Ad-Bgl II**

Figure 71 hNIS pulmonary gene delivery and expression in Cotton rats was readily visualized by PET imaging. 3×10^8 PFU Ad-hNIS or Ad-Bgl II in a volume of 150 µl was administered to Cotton rats by instillation via nostril. 48 hr after infection, 14.8 MBq ^{124}I$^-$ was injected intraperitoneally. PET data were acquired for about 30 min and reconstructed to obtain tomographic images. A 0.5 mm coronal slice of each rat is shown.

2002). In addition, there will be some biological differences between rat and human airway epithelial cells when translating animal model results into clinical use. To extend our studies *in vivo* in rats to human pulmonary epithelial cells, we utilized a polarized human airway epithelial (HAE) cell layer, which was developed by immortalizing HAE by hTERT-E6/E7-transduction (Zabner et al., 2003). NuLi cell cultures retain most normal phenotypic qualities of HAE, such as the ability to form an electrically tight epithelium. Infected NuLi cells showed increasing signal intensity with increasing Ad-hNIS titer ranging from 0–40 MOI, and there was a linear relationship between signal intensity of ^{125}I$^-$ activity and its concentration (Fig. 72). These results were consistent with our previous data with human breast cancer cell line MDA-MB-435 cells, and suggest the potential value of NIS reporter system in clinical evaluation of pulmonary gene transfer to humans. The ability to noninvasively visualize gene expression tomographically by SPECT and PET in real time has significant translational implications for monitoring the success of human gene therapy.

Considerations in PET Imaging of Pulmonary Gene Transfer

Gene Transfer Barriers

The lumen of human airways is normally lined with a pseudostratified mucociliary epithelium comprised of ciliated cells and mucus-secreting goblet cells overlying basal epithelial cells. The airway surface fluid and mucous layers are natural barriers in the airway that protect against bacter-

ial and viral infections. The intense inflammatory milieu of the CF airway may also be a significant barrier (Driskell and Engelhardt, 2003). Recombinant human deoxyribonuclease (rhDNase) and gelsolin have been shown to reduce the viscosity of CF sputum by degrading or depolymerizing DNA and actin, respectively, and enhance gene transfer with adenoviral vectors in the presence of CF sputum *in vitro* (Stern et al., 1998). Other mucolytics to enhance gene transfer are under investigation.

Because of the highly polarized organization and the complexity of the apical cell surface glycocalyx, the airway epithelia multiples are much less susceptible to many types of recombinant virus infection from the apical membrane. Preferential transduction from the basolateral membrane of polarized airway epithelia has been observed for recombinant adenoviruses, AAV, and retroviruses. In addition, the low rate of endocytosis from the apical plasma membrane of polarized airway epithelia may also account for some of the observed polarity of infection. One strategy to overcome this barrier is transient disruption of tight junctions with hypotonic saline and divalent cation chelating agents such as ethylene glycol tetraacetic acid (EGTA) used in our experiments with human polarized cell layer. Ultraviolet irradiation or calcium phosphate co-precipitation also could improve receptor-independent pathways of both recombinant adenoviral- and AAV-mediated transduction (Fasbender et al., 1998). Retargeting recombinant viral vectors to apical receptors is another strategy to increase gene transfer efficiency. Retargeting has so far been achieved on adenoviruses by chemically, immunologically, or genetically modifying the adenoviral capsid coat by incorporating new receptor ligands that can target candidate receptors. Retargeting in lentiviral vectors has been accomplished through pseudotyping their surface glycoproteins to target specific cell types *in vivo*.

Iodine-124 as Imaging Agent

The half life of ^{124}I is 4.2 days, which could permit quantitative imaging over several days using PET. The decay scheme of ^{124}I is very complex, including at least 25 electron capture transitions, 6 positron transitions, and 97 gamma-ray transitions. Only ~ 23% of disintegrations result in positron emission with relatively high energy. At the same time, there are many high-energy gamma rays, some in cascade with the positron emissions (Weber et al., 1989). Theoretically, ^{124}I$^-$ is not an ideal PET imaging tracer. The high energy of the positrons from ^{124}I$^-$ will not permit high spatial resolution, particularly in a low-density tissue such as lung. Both primary and scattered photons that resulted from the high-energy gamma rays can be detected as single events within the 511 keV energy window, thus increasing the accidental coincidence count and potentially affecting quantitation. Furthermore, half of the positron emissions have a 603 keV

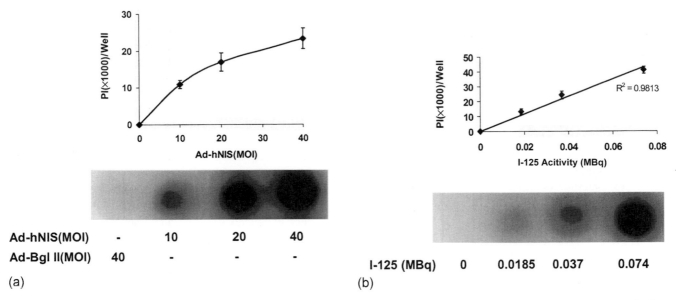

Figure 72 Efficient detection of hNIS gene transfer and expression in human airway epithelial cells. (a) NuLi cells in 24 well plates were infected with 0–40 MOI Ad-hNIS. 30 hours after infection, 0.037 MBq $^{125}I^-$ was added to each well. Washed plates were exposed on a phosphorimaging screen for quantitative analysis. (b) All cells were infected with 20 MOI Ad-hNIS, and a different concentration of $^{125}I^-$ was added. Data are expressed as the mean ± SD of three replicates. PI: pixel intensity.

gamma ray in cascade permitting detection of apparently "true" coincidences between the 603 keV gamma ray and one or both of the annihilation photons. Since the direction of these gamma rays has no correlation with the annihilation photons, the recorded line of coincidence is generally incorrect. Similarly, one can get undesirable true coincidences between cascade gamma transitions (Hoffman et al., 1981).

With different detectors and different correction schemes, Pentlow et al. (1996) imaged tumorlike phantom sources of ^{124}I surrounded by a large volume of relatively low background activity using PET. The results showed that quantitative measurements of activity within those tumorlike objects and the surrounding regions could be made. Compared with conventional PET nuclides, resolution and quantification were only slightly degraded. Another study reached a similar conclusion that quantitation of a radiopharmaceutical labeled with ^{124}I was feasible and may be improved by the development of specific corrections (Herzog et al., 2002).

Relationship between PET Signal and Reporter Gene Expression

To monitor transgene expression, PET imaging must yield not only accurate measurements of tissue radioactivity, but also measurements that are closely related to tissue levels of the gene product. In other words, a good correlation between PET-imaging signal and reporter gene expression is crucial to evaluate gene transfer precisely.

Injected vector dose is not a good standard when used to relate to radioactivity accumulation induced by reporter gene expression in vivo, although in vitro data supported the positive correlation between these two variables (Niu et al., 2004, Shin et al., 2004). Tjuvajev et al. (1999) found that the level of $[^{131}I]$FIAU-derived radioactivity accumulated in tumors increased only by a factor of two with a 10-fold increase of adenoviral vector dose from 3×10^7 to 3×10^8 pfu. Gambhir et al. (1999) demonstrated that correlation between levels of AdCMV-HSV1-tk virus injected and FGCV percent injected dose per gram of liver was poor. It has also been reported that uptake of ^{131}I-labeled FIAU in tumors in vivo was completely unrelated to the Ad-hSSTr2-TK dose (Zinn et al., 2002). This was explained by the fact that there was a big variance of the fraction of infected tumor cells per biopsy ranging between 5 and 55% (Gahèry-Sègard et al., 1997).

When transgene expression is taken as the measurement, a highly linear correlation between the PET signal and in vitro assays of transgene expression has been reported, especially at low levels of mHSV1-tk expression. With multiple subcutaneous tumors produced from wild-type RG2 cells and stably transduced RG2TK cell lines that express different levels of HSV1-tk, Tjuvajev et al. (1995) tested whether $[^{124}I]$-FIAU and PET imaging could discriminate between different levels of HSV1-tk gene expression. They found a highly significant relationship between the level of $[^{124}I]$-FIAU accumulation and an independent measure of HSV1-tk expression. Gambhir et al. (1999) also found a significant positive correlation between the percent injected dose of [^{18}F]FGCV retained per gram of liver and the levels of hepatic HSV1-tk reporter gene expression. At the same time they thought there may be a plateau of the [^{18}F]FGCV percent injected dose per gram of liver at higher levels of HSV1-tk expression.

With dual-expressing NIS and luciferase stably transfected cells, Shin *et al.* (2004) found a high correlation between radioactivity accumulation and tumor weight. Yang *et al.* (2004) evaluated the feasibility of noninvasive imaging of recombinant adenovirus-mediated NIS gene expression by $^{99m}TcO_4^-$ scintigraphy in skeletal muscle of rats. They found that the muscular NIS mRNA level quantified by real-time reverse transcription-polymerase chain reaction was positively correlated with the imaging counts.

When comparing *in vitro* measurements of tissue radioactivity with gamma counter and that *in vivo* with PET, the different denominators and low density of normal lung (typically 0.3–0.4 g/ml) should be taken into consideration. In addition, the absence of correction for photon scattering by the current generation of micro-PET devices may account for a small underestimation of lung activity measured by PET (Richard *et al.*, 2002). *In vitro* gene transfer is much less complicated than *in vivo* gene transfer. During *in vivo* gene transfer attempts, more complex physiological factors must be taken into consideration. These include transport, distribution, and degradation of imaging agents, as well as differences in blood supplies and the unique physiologies of different organs.

Resolution and Sensitivity of PET

The accuracy of PET devices to quantify tissue radioactivity is a prerequisite for the possibility that PET imaging can be used to measure transgene expression. The sensitivity of PET is relatively high in the range of 10^{11}–10^{12} mol/L and is independent of the location or depth of the reporter probe of interest. Typically, several million cells accumulating reporter probe have to be in relatively close proximity for a PET scanner to record them as a distinct entity relative to the background. Most clinical PET systems have an intrinsic resolution of ~ 6–8 mm in all spatial directions. At this resolution high-quality images can be reconstructed at a final image resolution of 8–15 mm.

In recent years, small-animal micro-PET scanners have been developed. These systems typically have a spatial resolution of ~ 8 mm^3 (2 mm × 2 mm × 2 mm) (Cherry and Gambhir, 2001), but newer generation systems in the final stages of development will have a resolution of ~ 1 mm^3 (1 mm × 1 mm × 1 mm) (Chatziioannou *et al.* 2001). Development of molecular imaging assays with PET are particularly advantageous because of the ability to validate them in cell culture and small-animal models prior to using the same reporter probe in established clinical PET centers throughout the world. The ability to perform translational research from a cell culture setting to preclinical animal models to clinical applications is one of the most unique and powerful features of PET technology.

Other issues, such as scanner sensitivity (i.e., the fraction of radioactive events actually detected by the device) and the amount of radioactivity that can be injected without causing physiologic disturbances in the system under study (a function of tracer specific activity) also affect the ability to perform imaging studies in small laboratory animals. In conclusion, the use of PET detection for pulmonary gene delivery is likely to continue to evolve with the advent of novel tracers that can more sensitively and specifically acquire functional imaging data from the normal and diseased lung. Clinical translation of basic research findings for non-invasive imaging of gene transfer to humans should remain the ultimate goal of research in this field.

Acknowledgments

The authors thank the following individuals: Ron Finn, Mark Madsen, Richard Hichwa, Michael Graham, Jim Ponto, Andrew Gaut, and Kim Krager for isotopes, image acquisition and analysis, technical support, and helpful guidance throughout these studies. This work was supported by NIH grants P01CA66081 and P20CA091709.

References

Aitken, M.L., Moss, R.B., Waltz, D.A., Dovey, M.E., Tonelli, M.R., McNamara, S.C., Gibson, R.L., Ramsey, B.W., Carter, B.J., and Reynolds, T.C. 2001. A Phase I study of aerosolized administration of tgAAVCF to cystic fibrosis subjects with mild lung disease. *Hum. Gene Ther. 12*:1907–1916.

Arner, E.S., and Eriksson, S. 1995. Mammalian deoxyribonucleoside kinases. *Pharmacol. Ther. 67*:55–86.

Bellon, G., Michel-Calemard, L., Thouvenot, D., Jagneaux, V., Poitevin, F., Malcus, C., Accart, N., Layani, M.P., Aymard, M., Bernon, H., *et al.* 1997. Aerosol administration of a recombinant adenovirus expressing CFTR to cystic fibrosis patients: a phase I clinical trial. *Hum. Gene Ther. 8*:15–25.

Black, M.E., Newcomb, T.G., Wilson, H.-M.P., and Loeb, L.A. 1996. *Proc. Natl. Acad. Sci. USA 93*:3525–3529.

Blasberg, R.G., and Tjuvajev, J.G. 2003. Molecular-genetic imaging: current and future perspectives. *J. Clin. Invest. III*:1620–1629.

Boland, A., Ricard, M., Opolon, P., Bidart, J.M., Yeh, P., Filetti, S., Schlumberger, M., and Perricaudet, M. 2000. Adenovirus-mediated transfer of the thyroid sodium/iodide symporter gene into tumors for a targeted radiotherapy. *Cancer Res. 60*:3484–3492.

Brigham, K.L., Lane, K.B., Meyrick, B., Stecenko, A.A., Strack, S., Cannon, D.R., Caudill, M., and Canonico, A.E. 2000. Transfection of nasal mucosa with a normal alpha-1 antitrypsin (AAT) gene in AAT deficient subjects: comparison with protein therapy. *Human Gene Therapy 11*:1023–1032.

Brust, P., Haubner, R., Friedrich, A., Scheunemann, M., Anton, M., Koufaki, O.N., Hauses, M., Noll, S., Noll, B., Haberkorn, U., Schackert, G., Schackert, H.K., Avril, N., and Johannsen, B. 2001. Comparison of [18F]FHPG and [124/125I]FIAU for imaging herpes simplex virus type 1 thymidine kinase gene expression. *Eur. J. Nucl. Med. 28*:721–729.

Canonico, A.E., Conary, J.T., Meyrick, B.O., and Brigham, K.L. 1994. Aerosol and intravenous transfection of human alpha 1-antitrypsin gene to lungs of rabbits. *Am. J. Respir. Cell Mol. Biol. 10*:24–29.

Caplen, N.J., Alton, E.W., Middleton, P.G., Dorin, J.R., Stevenson, B.J., Gao, X., Durham, S.R., Jeffery, P.K., Hodson, M.E., Coutelle, C., Huang, L., Porteous, D.J., Williamson, R., and Geddes, D.M. 1995. Liposome-mediated CFTR gene transfer to the nasal epithelium of patients with cystic fibrosis. *Nat. Med. 1*:39–46.

Chatziioannou, A., Tai, Y.C., Doshi, N., and Cherry, S.R 2001. Detector development for microPET II: a 1 microl resolution PET scanner for small animal imaging. *Phys. Med. Biol. 46*:2899–2910.

Chen, Z.Y., Yant, S.R., He, C.Y., Meuse, L., Shen, S., and Kay, M.A. 2001. Linear DNAs concatemerize *in vivo* and result in sustained transgene expression in mouse liver. *Mol. Ther. 3*:403–410.

Cherry, S.R., and Gambhir, S.S. 2001. Use of positron emission tomography in animal research. *ILAR J. 42*:219–232.

Coil, D.A., Strickler, J.H., Rai, S.K., and Miller, A.D. 2001. Jaagsiekte sheep retrovirus Env protein stabilizes retrovirus vectors against inactivation by lung surfactant, centrifugation, and freeze-thaw cycling. *J. Virol. 75*:8864–8867.

Crawford, D.C., Flower, M.A., Pratt, B.E., Hill, C., Zweit, J., McCready, V.R., and Harmer, C.L. 1997. Thyroid volume measurement in thyrotoxic patients: comparison between ultrasonography and iodine-124 positron emission tomography. *Eur. J. Nucl. Med. 24*:1470–1478.

Daniel, J.C., and Smythe, W.R. 2003. Gene therapy of lung cancer. *Semin. Surg. Oncol. 21*:196–204.

Dharmarajan, S., Hayama, M., Kozlowski, J., Ishiyama, T., Okazaki, M., Factor, P., Patterson, G.A., and Schuster, D.P. 2005. *In vivo* molecular imaging characterizes pulmonary gene expression during experimental lung transplantation. *Am. J. Transplant. 5*:1216–1225.

Doubrovin, M., Akhurst, T., Cai, S., Balatoni, J., Alauddin, M., Finn, R., Conti, P., Gelovani-Tjuvajev, J., and Blasberg, R. 2002. Comparison of HSV1-tk PET imaging probes: FIAU and FHBG for early imaging. *J. Nucl. Med. 43*:41P.

Driskell, R.A., and Engelhardt, J.F. 2003. Current status of gene therapy for inherited lung diseases. *Annu. Rev. Physiol. 65*:585–612.

Drumm, M.L., Pope, H.A., Cliff, W.H., Rommens, J.M., Marvin, S.A., Tsui, L.C., Collins, F.S., Frizzell, R.A., and Wilson, J.M. 1990. Correction of the cystic fibrosis defect *in vitro* by retrovirus-mediated gene transfer. *Cell 62*:1227–1233.

Fasbender, A., Lee, J.H., Walters, R.W., Moninger, T.O., Zabner, J., and Welsh, M.J. 1998. Incorporation of adenovirus in calcium phosphate precipitates enhances gene transfer to airway epithelia *in vitro* and *in vivo. J. Clin. Invest. 102*:184–193.

Flotte, T.R., Brantly, M.L., Spencer, L.T., Byrne, B.J., Spencer, C.T., Baker, D.J., and Humphries, M. 2004. Phase I trial of intramuscular injection of a recombinant adeno-associated virus alpha 1-antitrypsin (rAAV2-CB-hAAT) gene vector to AAT-deficient adults. *Hum. Gene. Ther. 15*:93–128.

Frey, P., Townsend, D., Flattet, A., De Gautard, R., Widgren, S., Jeavons, A., Christin, A., Smith, A., Long, A., and Donath, A. 1986. Tomographic imaging of the human thyroid using 124I. *J. Clin. Endocrinol. Metab. 63*:918–927.

Fujiwara, T., Cai, D.W., Georges, R.N., Mukhopadhyay, T., Grimm, E.A., and Roth, J.A. 1994. Therapeutic effect of a retroviral wild-type p53 expression vector in an orthotopic lung cancer model. *J. Natl. Cancer Inst. 86*:1458–1462.

Gahery-Segard, H., Molinier-Frenkel, V., Le Boulaire, C., Saulnier, P., Opolon, P., Lengagne, R., Gautier, E., Le Cesne, A., Zitvogel, L., Venet, A., Schatz, C., Courtney, M., Le Chavalier, T., Tursz, T., Guillet, J.G., and Farace, F. 1999. Phase I trial of recombinant adenovirus gene transfer in lung cancer: longitudinal study of the immune responses to transgene and viral products. *J. Clin. Invest. 100*:2218–2226.

Gambhir, S.S., Bario, J.R., Herschman, H.R., and Phelps, M.E. 1999. Assays for noninvasive imaging of reporter gene expression. *Nucl. Med. Biol. 26*:481–490.

Gambhir, S.S., Barrio, J.R., Phelps, M.E., Iyer, M., Namavari, M., Satyamurthy, N., Wu, L., Green, L.A., Bauer, E., MacLaren, D.C., Nguyen, K., Berk, A.J., Cherry, S.R., and Herschman, H.R. 1999. Imaging adenoviral-directed reporter gene expression in living animals with positron emission tomography. *Proc. Natl. Acad. Sci. USA 96*:2333–2338.

Gambhir, S.S., Bauer, E., Black, M.E., Liang, Q., Kokoris, M.S., Barrio, J.R., Iyer, M., Namavari, M., Phelps, M.E., and Herschman, H.R. 2000. A mutant herpes simplex virus type 1 thymidine kinase reporter gene shows improved sensitivity for imaging reporter gene expression with positron emission tomography. *Proc. Natl. Acad. Sci. USA. 97*:2785–2790.

Gautum, A., Densmore, C.L., and Waldrep, J.C. 2000. Inhibition of experimental lung metastasis by aerosol delivery of PEI-p53 complexes. *Mol. Ther. 2*: 318–323.

Groot-Wassink, T., Aboagye, E.O., Glaser, M., Lemoine, N.R., and Vassaux, G. 2002. Adenovirus biodistribution and noninvasive imaging of gene expression *in vivo* by positron emission tomography using human sodium/iodide symporter as reporter gene. *Hum. Gene Ther. 13*: 1723–1735.

Haberkorn, U., Oberdorfer, F., Gebert, J., Morr, I., Haack, K., Weber, K., Lindauer, M., van Kaick, G., and Schackert, H.K. 1996. Monitoring of gene therapy with cytosine deaminase: *in vitro* studies using ^{3}H-5-fluorocytosine. *J. Nucl. Med. 37*:87–94.

Herschman, R.H. 2003. Molecular imaging: looking at problems, seeing solutions. *Science 302*:605–608.

Herzog, H., Tellmann, L., Qaim, S.M., Spellerberg, S., Schmid, A., and Coenen, H.H. 2002. PET quantitation and imaging of the non-pure positron-emitting iodine isotope 124I. *Appl. Radiat. Isot. 56*:673–679.

Hoffman, E.J., Huang, S.C., Phelps, M.E., and Kuhl, D.E. 1981. Quantitation in positron emission computed tomography: 4. Effect of accidental coincidences. *J. Comput. Assist. Tomogr 5*:391–400.

Ji, L., Fang, B., Yen, N., Fong, K., Minna, J.D., and Roth, J.A. 1999. Induction of apoptosis and inhibition of tumorigenicity and tumor growth by adenovirus vector-mediated fragile histidine triad (FHIT) gene overexpression. *Cancer Res. 59*:3333–3339.

Kay, M.A., Glorioso, J.C., and Naldini, L. 2001. Viral vectors for gene therapy: the art of turning infectious agents into vehicles of therapeutics. *Nat. Med. 7*:33–40.

Kay, M.A., Li, Q., Liu, T.J., Leland, F., Toman, C., Finegold, M., and Woo, S.L. 1992. Hepatic gene therapy: persistent expression of human alpha 1-antitrypsin in mice after direct gene delivery *in vivo. Hum. Gene Ther. 3*:641–647.

Knowles, M.R., Hohneker, K.W., Zhou, Z., Olsen, J.C., Noah, T.L., Hu, P.C., Leigh, M.W., Engelhardt, J.F., Edwards, L.J., Jones, K.R., Grossman, M., Wilson, J.M., Johnson, L.G., and Boucher, R.C. 1995. A controlled study of adenoviral-vector-mediated gene transfer in the nasal epithelium of patients with cystic fibrosis. *N. Engl. J. Med. 333*:823–831.

Kubba, S., Adak, S., Shiller, J., Slovis, B., Coffee, K., Worell, J., Thet, L., Krozely, P., Johnson, D., and Carbone, D. 2000. Phase I trial of Adenovirus p53 in bronchioalveolar cell lung carcinoma (BAC) administered by bronchoalveolar lavage [Abstract]. *Proc. Am. Soc. Clin. Oncol. 19*:487a.

La Perle, K.M., Shen, D., Buckwalter, T.L., Williams, B., Haynam, A., Hinkle, G., Pozderac, R., Capen, C.C., and Jhiang, S.M. 2002. *In vivo* expression and function of the sodium/iodide symporter following gene transfer in the MATLyLu rat model of metastic prostate cancer. *Prostate 50*:170–178.

Mandell, R.B., Mandell, L.Z., and Link, C.J., Jr. 1999. Radioisotope concentrator gene therapy using the sodium/iodide symporter gene. *Cancer Res. 59*:661–668.

Marom, E.M., McAdams, H.P., Erasmus, J.J., Goodman, P.C., Culhane, D.K., Coleman, R.E., Herndon, J.E., and Patz, E.F., Jr. 1999. Staging non-small cell lung cancer with whole-body PET. *Radiology 212*:803–809.

Marsee, D.K., Shen, D.H., MacDonald, L.R., Vadysirisack, D.D., Lin, X., Hinkle, G., Kloos, R.T., and Jhiang, S.M. 2004. Imaging of metastatic pulmonary tumors following NIS gene transfer using single photon emission computed tomography. *Cancer Gene Ther.* 11:121–127.

Morris, J.C., Bergert E.R., Bryant, W.P., and Jensen, C.E. 1997. Binding of immunoglobin G from patients with auto-immune thyroid disease to rat sodium/iodide symporter peptides: evidence for the iodide transporter as an autoantigen. *Thyroid* 7:527–534.

Nakai, H., Storm, T.A., and Kay, M.A. 2000. Increasing the size of rAAV-mediated expression cassettes *in vivo* by inter molecular joining of two complementary vectors. *Nat. Biotechnol.* 18:527–532.

Nemunaitis, J., Swisher, S.G., Timmons, T., Connors, D., Mack, M., Doerksen, L., Weill, D., Wait, J., Lawrence, D.D., Kemp, B.L., Fossella, F., Glisson, B.S., Hong, W.K., Khuri, F.R., Kurie, J.M., Lee, J.J., Lee, J.S., Nguyen, D.M., Nesbitt, J.C., Prez-Soler, R., Pisters, K.M.W., Putnam, J.B., Richli, W.R., Shin, D.M., Walsh, G.L., Merritt, J., and Roth, J. 2000. Adenovirus-mediated p53 gene transfer in sequence with cisplatin to tumors of patients with non-small cell lung cancer. *J. Clin. Oncol.* 18:609–622.

Niu, G., Gaut, A.W., Ponto, L.L., Hichwa, R.D., Madsen, M.T., Graham, M.M., and Domann, F.E. 2004. Multimodality noninvasive imaging of gene transfer using the human sodium iodide symporter. *J. Nucl. Med.* 45:445–449.

Niu, G., Krager, K.J., Graham, M.M., Hichwa, R.D., and Domann, F.E. 2005. Noninvasive radiological imaging of pulmonary gne transfer and expresion using the human sodium iodide transporter. *Eur. J. Nucl. Med. Mol. Imaging.* 32:534–540.

Peebles, D., Gregory, L.G., David, A., Themis, M., Waddington, S.N., Knapton, H.J., Miah, M., Cook, T., Lawrence, L., Nivsarkar, M., Rodeck, C., and Coutelle, C. 2004. Widespread and efficient marker gene expression in the airway epithelia of fetal sheep after minimally invasive tracheal application of recombinant adenovirus in utero. *Gene Ther.* 11:70–78.

Pentlow, K.S., Graham, M.C., Lambrecht, R.M., Daghighian, F., Bacharach, S.L., Bendriem, B., Finn, R.D., Jordan, K., Kalaigian, H., Karp, J.S., Robeson, W.R., and Larson, S.M. 1996. Quantitative imaging of iodine-124 with PET. *J. Nucl. Med.* 37:1557–1562.

Quinton, P.M. 1990. Cystic fibrosis: a disease in electrolyte transport. *FASEB J.* 4:2709–2717.

Rich, D.P., Anderson, M.P., Gregory, R.J., Cheng, S.H., Paul, S., Jefferson, D.M., McCann, J.D., Klinger, K.W., Smith, A.E., and Welsh, M.J. 1990. Expression of cystic fibrosis transmembrane conductance regulator corrects defective chloride channel regulation in cystic fibrosis airway epithelial cells. *Nature* 347:358–363.

Richard, J.C., Factor, P., Welch, L.C., and Schuster, D.P. 2003a. Imaging the spatial distribution of transgene expression in the lungs with positron emission tomography. *Gene Ther.* 10:2074–2080.

Richard, J.C., Janier, M., Decailliot, F., Le Bars, D., Levenne, F., Berthier, V., Lionnet, M., Cinotti, L., Annat, G., and Guérin, C. 2002. Comparison of PET with radioactive microspheres to assess pulmonary blood flow. *J. Nucl. Med.* 43:1063–1071.

Richard, J.C., Zhou, Z., Chen, D.L., Mintun, M.A., Piwnica-Worms, D., Factor, P., Ponde, D.E., and Schuster, D.P. 2004. Quantitation of pulmonary transgene expression with PET imaging. *J. Nucl. Med.* 45:644–654.

Richard, J.C., Zhou, Z., Ponde, D.E., Dence, C.S., Factor, P., Reynolds, P.N., Luker, G.D., Sharma, V., Ferkol, T., Piwnica-Worms, D., and Schuster, D.P. 2003b. Imaging pulmonary gene expression with positron emission tomography. *Am. J. Respir. Crit. Care Med.* 167:1257–1263.

Rommens, J.M., Iannuzzi, M.C., Kerem, B., Drumm, M.L., Melmer, G., Dean, M., Rozmahel, R., Cole, J.L., Kennedy, D., Hidaka, N., Zsiga, M., Buchwald, M., Riordan, J.R., Tsui, L.C., and Collins, F.S. 1989. Identification of the cystic fibrosis gene: chromosome walking and jumping. *Science* 245:1059–1065.

Rosenfeld, M.A., Siegfried, W., Yoshimura, K., Yoneyama, K., Fukayama, M., Stier, L.E., Paakko, P.K., Gilardi, P., Stratford-Perricaudet, L.D.,

Perricaudet, M., Jallat, S., Pavirani, A., Lecocq, J.P., and Crystal, R.G. 1991. Adenovirus-mediated transfer of a recombinant alpha 1-antitrypsin gene to the lung epithelium *in vivo. Science* 252:431–434.

Rosenfeld, M.A., Yoshimura, K., Trapnell, B.C., Yoneyama, K., Rosenthal, E.R., Dalemans, W., Fukayama, M., Bargon, J., Stier, L.E., Stratford-Perricaudet, L.D., Perricaudet, M., Guggino, W.B., Pavirani, A., Lecocq, J.P., and Crystal, R.G. 1992. *In vivo* transfer of the human cystic fibrosis transmembrane conductance regulator gene to the airway epithelium. *Cell* 68:143–155.

Roth, J.A., Nguyen, D., Lawrence, D.D., Kemp, B.L., Carrasco, C.H., Ferson, D.Z., Hong, W.K., Komaki, R., Lee, J.J., Nesbitt, J.C., Pisters, K.M., Putnam, J.B., Schea, R., Shin, D.M., Walsh, G.L., Dolormente, M.M., Han, C.I., Martin, F.D., Yen, N., Xu, K., Stephens, L.C., McDonnell, T.J., Mukhopadhyay, T., and Cai, D. 1996. Retrovirus-mediated wild-type p53 transfers to tumors of patients with lung cancer. *Nat. Med.* 2:985–991.

Ruiz, F.E., Clancy, J.P., Perricone, M.A., Bebok, Z., Hong, J.S., Cheng, S.H., Meeker, D.P., Young, K.R., Schoumacher, R.A., Weatherly, M.R., Wing, L., Morris, J.E., Sindel, L., Rosenberg, M., van Ginkel, F.W., McGhee, J.R., Kelly, D., Lyrene, R.K., and Sorscher, E.J. 2001. A clinical inflammatory syndrome attributable to aerosolized lipid-DNA administration in cystic fibrosis. *Hum. Gene Ther.* 12:751–761.

Saeki, T., Mhashilkar, A., Swanson, X., Zou-Yang, X.H., Sieger, K., Kawabe, S., Branch, C.D., Zumstein, L., Meyn, R.E., Roth, J.A., Chada, S., and Ramesh, R. 2002. Inhibition of human lung cancer growth following adenovirus-mediated mda-7 gene expression *in vivo. Oncogene* 21:4558–4566.

Salgia, R., and Skarin, A.T. 1998. Molecular abnormalities in lung cancer. *J. Clin. Oncol.* 16:1207–1217.

Schuler, M., Herrmann, R., De Greve, J.L., Stewart, A.K., Gatzemeier, U., Stewart, D.J., Laufman, L., Gralla, R., Kuball, J., Buhl, R., Heussel, C.P., Kommoss, F., Perruchoud, A.P., Shepherd, F.A., Fritz, M.A., Horowitz, J.A., Huber, C., and Rochlitz, C. 2001. Adenovirus-mediated wild-type p53 gene transfer in patients receiving chemotherapy for advanced non-small cell lung cancer: results of a multicenter Phase II study. *J. Clin. Oncol.* 19:1750–1758.

Schuler, M., Rochlitz, C., Horowitz, J.A., Schlegel, J., Perruchoud, A.P., Kommoss, F., Bolliger, C.T., Kauczor, H.U., Dalquen, P., Fritz, M.A., Swanson, S., Herrmann, R., and Huber, C. 1998. A Phase I study of adenovirus-mediated wild-type p53 gene transfer in patients with advanced non-small cell lung cancer. *Hum. Gene. Ther.* 9:2075–2082.

Schuster, D.P. 1998. The evaluation of lung function with PET. *Semin. Nucl. Med.* 28:341–351.

Shah, K., Jacobs, A., Breakefield, X.O., and Weissleder, R. 2004. Molecular imaging of gene therapy for cancer. *Gene. Ther.* 11:1175–1187.

Shin, J.H., Chung, J.K., Kang, J.H., Lee, Y.J., Kim, K.I., So, Y., Jeong, J.M., Lee, D.S., and Lee, M.C. 2004. Noninvasive imaging for monitoring of viable cancer cells using a dual-imaging reporter gene. *J. Nucl. Med.* 45:2109–2115.

Song, S., Morgan, M., Ellis, T., Poirier, A., Chesnut, K., Wang, J., Brantly, M., Muzyczka, N., Byrne, B.J., Atkinson, M., and Flotte, T.R. 1998. Sustained secretion of human alpha-1-antitrypsin from murine muscle transduced with adeno-associated virus vectors. *Proc. Natl. Acad. Sci. USA.* 95:14384–14388.

Song, S., Embury, J., Laipis, P.J., Berns, K.I., Crawford, J.M., and Flotte, T.R. 2001. Stable therapeutic serum levels of human alpha-1 antitrypsin (AAT) after portal vein injection of recombinant adeno-associated virus (rAAV) vectors. *Gene Ther.* 8:1299–1306

Spitzweg, C., Dietz, A.B., O'Connor, M.K., Bergert, E.R., Tindall, D.J., Young, C.Y., and Morris, J.C. 2001. *In vivo* sodium iodide symporter gene therapy of prostate cancer. *Gene. Ther.* 8:1524–1531.

Stern, M., Caplen, N.J., Browning, J.E., Griesenbach, U., Sorgi, F., Huang, L., Gruenert, D.C., Marriot, C., Crystal, R.G., Geddes, D.M., and Alton, E.W. 1998. The effect of mucolytic agents on gene transfer across a CF sputum barrier *in vitro. Gene Ther.* 5:91–98.

Stoll, S.M., Sclimenti, C.R., Baba, E.J., Meuse, L., Kay, M.A., and Calos, M.P. 2001. Epstein-Barr virushuman vector provides high-level long-term expression of α1-antitrypsin in mice. *Mol. Ther.* 4:122–129.

Stroobants, S., Verschakelen, J., and Vansteenkiste, J. 2003. Value of FDG-PET in the management of non-small cell lung cancer. *Eur. J. Radiol.* 45:49–59

Swisher, S.G., Roth, J.A., Komaki, R., Gu, J., Lee, J.J., Hicks, M., Ro, J.Y., Hong, W.K., Merritt, J.A., Ahrar, K., Atkinson, N.E., Correa, A.M., Dolormente, M., Dreiling, L., El-Naggar, A.K., Fossella, F., Francisco, R., Glisson, B., Grammer, S., Herbst, R., Huaringa, A., Kemp, B., Khuri, F.R., Kurie, J.M., Liao, Z.X., McDonnell, T.J., Morice, R., Morello, F., Munden, R., Papadimitrakopoulou, V., Pisters, K.M.W., Putnam, J.B., Sarabia, A.J., Shelton, T., Stevens, C., Shin, D.M., Smythe, W.R., Vaporciyan, A.A., Walsh, G.L., and Yin, M. 2003. Induction of p53-regulated genes and tumor regression in lung cancer patients after intratumoral delivery of adenoviral p53 (INGN 201) and radiation therapy. *Clin. Cancer Res.* 9:93–101.

Swisher, S.G., Roth, J.A., Nemunaitis, J., Lawrence, D.D., Kemp, B.L., Carrasco, C.H., Connors, D.G., El-Naggar, A.K., Fossella, F., Glisson, B.S., Hong, W.K., Khuri, F.R., Kurie, J.M., Lee, J.J., Lee, J.S., Mack, M., Merritt, J.A., Nguyen, D.M., Nesbitt, J.C., Perez-Soler, R., Pisters, K.M.W., Putnam, J.B., Richli, W.R., Savin, M., Schrump, D.S., Shin, D.M., Shulkin, A., Walsh, G.L., Wait, J., Weill, D., and Waugh, M.K.A. 1999. Adenovirus-mediated p53 gene transfer in advanced non-small cell lung cancer. *J. Natl. Cancer Inst.* 91:763–771.

Tjuvajev, J.G., Avril, N., Oku, T., Sasajima, T., Miyagawa, T., Joshi, R., Safer, M., Beattie, B., DiResta, G., Daghighian, F., Augensen, F., Koutcher, J., Zweit, J., Humm, J., Larson, S.M., Finn, R., and Blasberg, R. 1998. Imaging herpes virus thymidine kinase gene transfer and expression by positron emission tomography. *Cancer Res.* 58:4333–4341.

Tjuvajev, J.G., Chen, S.H., Joshi, A., Joshi, R., Guo, Z.S., Balatoni, J., Ballon, D., Koutcher, J., Finn, R., Woo, S.L., and Blasberg, R.G. 1999. Imaging adenoviral-mediated herpes virus thymidine kinase gene transfer and expression *in vivo*. *Cancer Res.* 59:5186–5193.

Tjuvajev, J.G., Stockhammer, G., Desai, R., Uehara, H., Gansbacher, B., Watanabe, K., and Blasberg, R. 1995. Imaging gene transfer and expression *in vivo*. *Cancer Res.* 55:6121–6135.

Tovell, D.R., Samuel, J., Mercer, J.R., Misra, H.K., Xu, L., Wiebe, L.I., Tyrrell, D.L., and Knaus, E.E. 1988. The *in vitro* evaluation of nucleoside analogues as probes for use in the non-invasive diagnosis of herpes simplex encephalitis. *Drug. Des. Delivery* 3:213–221.

Wagner, J.A., Messner, A.H., Moran, M.L., Daifuku, R., Kouyama, K., Desch, J.K., Manley, S., Norbash, A.M., Conrad, C.K., Friborg, S., Reynolds, T., Guggino, W.B., Moss, R.B., Carter, B.J., Wine, J.J., Flotte, T.R., and

Gardner, P. 1999. Safety and biological efficacy of an adeno-associated virus vector-cystic fibrosis transmembrane regulator (AAV-CFTR) in the cystic fibrosis maxillary sinus. *Laryngoscope* 109:266–274.

Wagner, J.A., Reynolds, T., Moran, M.L., Moss, R.B., Wine, J.J., Flotte, T.R., and Gardner, P. 1998. Efficient and persistent gene transfer of AAV-CFTR in maxillary sinus. *Lancet* 351:1702–1703.

Wang, G., Williams, G., Xia, H., Hickey, M., Shao, J., Davidson, B.L., and McCray, P.B. 2002. Apical barriers to airway epithelial cell gene transfer with amphotropic retroviral vectors. *Gene Ther.* 9:922–931.

Weber, D.A., Echerman, K.F., Dillman, L.T., and Ryman, J.C. *MIRD: Radionuclide Data and Decay Schemes*. New York: Society of Nuclear Medicine, 1989.

Weissleder, R., Moore, A., Mahmood, U., Bhorade, R., Benveniste, H., Chiocca, E.A., and Basilion, J.P. 2000. *In vivo* magnetic resonance imaging of transgene expression. *Nat. Med.* 6:351–355.

Yang, H.S., Lee, H., Kim, S.J., Lee, W.W., Yang, Y.J., Moon, D.H., and Park, S.W. 2004. Imaging of human sodium-iodide symporter gene expression mediated by recombinant adenovirus in skeletal muscle of living rats. *Eur. J. Nucl. Med. Mol. Imaging* 31:1304–1311.

Zabner, J., Couture, L.A., Gregory, R.J., Graham, S.M., Smith, A.E., and Welsh, M.J. 1993. Adenovirus-mediated gene transfer transiently corrects the chloride transport defect in nasal epithelia of patients with cystic fibrosis. *Cell* 75:207–216.

Zabner, J., Karp, P., Seiler, M., Phillips, S.L., Mitchell, C.J., Saavedra, M., Welsh, M., and Klingelhutz, A.J. 2003. Development of cystic fibrosis and noncystic fibrosis airway cell lines. *Am. J. Physiol. Lung Cell Mol. Physiol.* 284:L844–L854

Zhang, W.W., Fang, X., Mazur, W., French, B.A., Georges, R.N., and Roth, J.A. 1994. High-efficiency gene transfer and high-level expression of wild-type p53 in human lung cancer cells mediated by recombinant adenovirus. *Cancer Gene. Ther.* 1:5–13.

Zhou, H., Pastore, L., and Beaudet, A.L. 2002. Helper-dependent adenoviral vectors. *Methods Enzymol.* 346:177–198.

Zinn, K.R., Chaudhuri, T.R., Krasnykh, V.N., Buchsbaum, D.J., Belousova, N., Grizzle, W.E., Curiel, D.T., and Rogers, B.E. 2002. Gamma camera dual imaging with a somatostatin receptor and thymidine kinase after gene transfer with a bicistronic adenovirus in mice. *Radiology* 223:417–425.

Zuckerman, J.B., Robinson, C.B., McCoy, K.S., Shell, R., Sferra, T.J., Chirmule, N., Magosin, S.A., Propert, K.J., Brown-Parr, E.C., Hughes, J.V., Tazelaar, J., Baker, C., Goldman, M.J., and Wilson, J.M. 1999. A Phase I study of adenovirus-mediated transfer of the human cystic fibrosis transmembrane conductance regulator gene to a lung segment of individuals with cystic fibrosis. *Hum. Gene Ther.* 10:2973–2985.

15

Lung Cancer with Idiopathic Pulmonary Fibrosis: High-resolution Computed Tomography

Kazuma Kishi and Atsuko Kurosaki

Introduction

Idiopathic pulmonary fibrosis (IPF) is one of several idiopathic interstitial pneumonias (American Thoracic Society, 2002). IPF, also known as cryptogenic fibrosing alveolitis (CFA) in the United Kingdom, is a distinctive type of chronic fibrosing interstitial pneumonia of unknown cause limited to the lungs, and usual interstitial pneumonia (UIP) is the histologic pattern that identifies patients with IPF. A radiologic diagnosis of UIP on computed tomography (CT) is based on a bilateral, predominantly basal, predominantly subpleural, reticular pattern, associated with honeycombing and/or traction bronchiectasis.

The association between IPF and lung cancer has been well recognized for many years (Turner-Warwick et al., 1980; Aubry et al., 2002). Several different clinical and pathologic characteristics have been demonstrated in lung cancer cases with IPF; however, high-resolution CT (HRCT) findings are not fully delineated. In this chapter, we address the HRCT features of lung cancer associated with IPF.

Prevalence of Lung Cancer in Idiopathic Pulmonary Fibrosis

The reported frequency of lung cancer in IPF varies widely, from as high as 48.2% to reports of a negative association depending on the type of study (Daniels and Jett, 2005). Autopsy studies report the highest rates of lung cancer in IPF, with prevalence rates of 45.7% and 48.2% in two series reviewing 83 and 70 cases, respectively, supported by the tissue evidence of UIP (Matsushita et al., 1995; Hironaka and Fukayama, 1999). By contrast, two large registry follow-up studies found that 20 of 205 CFA patients and 39 of 890 CFA patients had lung cancer, with prevalence rates of 9.8% and 4.4%, respectively (Turner-Warwick et al., 1980; Hubbard et al., 2000). Contradictory results are based on two death certificate studies (Wells and Mannino, 1996; Harris et al., 1998). They found a lower rate of lung cancer in patients with a death certificate diagnosis of pulmonary fibrosis/CFA than would be expected in the general population or in patients with chronic obstructive pulmonary disease.

Because both are death certificate reviews, their methodology has been criticized for potential underreporting of IPF (Samet, 2000).

Pathogenesis of Lung Cancer in Idiopathic Pulmonary Fibrosis

The mechanisms of developing lung cancer in IPF remain unclear. Persistent inflammation may cause repeated cellular injury and repair of the respiratory epithelium, leading to the atypical or dysplastic epithelial changes. These histologic changes are considered to be part of a precancerous lesion that could potentially progress to invasive carcinoma (Aubry et al., 2002). Recurrent injury and inflammation induced by IPF result in genetic errors that may predispose to the development of lung cancer. Genetic alterations such as microsatellite instability and loss of heterozygosity are frequent in IPF (Vassilakis et al., 2000; Uematsu et al., 2001). Furthermore, point mutations in p53 and ras are found in the epithelial cells of IPF patients, with lung cancer at a higher frequency than in IPF patients without lung cancer (Takahashi et al., 2002).

Clinical Features

The common clinical features of lung cancer associated with IPF show high rates for male patients, smokers, lower lobes, and peripheral type (Mizushima and Kobayashi, 1995). Men account for > 85% of patients, and ~ 90% of patients are smokers or former smokers. Matsushita et al. (1995) reported that the smoking index (mean cigarettes smoked in a day multiplied by the number of years of smoking) in patients with IPF and lung cancer was significantly greater than in lung cancer patients without IPF, and IPF patients without lung cancer. Although cigarette smoking is a common risk factor for both lung cancer and IPF, the incidence of lung cancer is increased in patients with IPF, independent of the effect of cigarette smoking (Hubbard et al., 2000). Regarding lobar localization, occurrence is more frequent in the lower lobe than in the upper lobe. Most tumors are located in peripheral areas of the lung, where fibrosis is predominant (Nagai et al., 1992; Aubry et al., 2002; Kishi et al., 2006).

There has been controversy over the most frequent cell types of the lung cancer associated with IPF. Some authors have reported that squamous cell carcinoma is the most common histologic type in IPF patients (Turner-Warwick et al., 1980; Nagai et al., 1992; Aubry et al., 2002), whereas others showed a predominance of adenocarcinoma (Kawai et al., 1987; Matsushita et al., 1995; Fujita et al., 1999).

Several studies have reported that synchronous multiple lung cancer was found in ~ 15–20% of patients with IPF (Matsushita et al., 1995; Mizushima and Kobayashi, 1995; Sakai et al., 2003). Most of them were double lung cancer, and the incidence of small cell carcinoma was high in the multiple lung cancer associated with IPF (Matsushita et al., 1995; Mizushima and Kobayashi, 1995).

Chest Radiograph

Typical chest radiographic findings of lung cancer developing with IPF have been described as a nodular shadow or a mass and a diffuse reticular infiltrate with honeycombing (Fig. 73a). However, the radiographic detection of lung cancer may be difficult in the background of pulmonary fibrosis. Kawai et al. (1987) reported that chest radiographs demonstrated a nodular shadow in ~ 50% of cases before death. Large opacities suggestive of lung cancer were found at the time of first hospital attendance in 20% of patients (Turner-Warwick et al., 1980).

Computed Tomography and High-resolution Computed Tomography Findings

High-resolution computed tomography now has a central role in the diagnostic approach to nodules as well as diffuse lung diseases. Computed tomography can detect small nodules that are not visible on chest radiographs. Several studies have described CT and HRCT findings of lung cancer associated with IPF or pulmonary fibrosis. Lee et al. (1996) evaluated the CT findings of 32 patients with lung cancer associated with IPF. Scanning techniques used were conventional CT in 24 patients, HRCT in 2 patients, and both conventional CT and HRCT in 6 patients. Computed tomography findings of lung cancer were an ill-defined consolidation-like mass in 53.1% (17/32), nodular in 37.5% (12/32), and undetermined pattern in 9.4% (3/32). Of the 3 patients who showed an undetermined pattern, 1 patient with bronchioloalveolar carcinoma (BAC) demonstrated diffuse air-space consolidation, and 2 patients had massive lymphadenopathy. As a whole, 21 lesions (65.6%) were located at the peripheral portion of the lung and in the lower lobes.

Fujita et al. (1999) conducted a unique study to compare the intensity of lung infiltrates between the side associated with lung cancer and the side without lung cancer. They evaluated the CT findings of 24 lung cancers of 23 patients with pulmonary fibrosis using the intensity scores. One patient revealed synchronous squamous cell carcinomas. All but one of the lung cancers were located at the peripheral subpleural area. The lung masses were located most commonly in the lower lobes (66.7%).

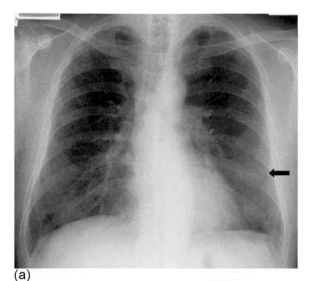

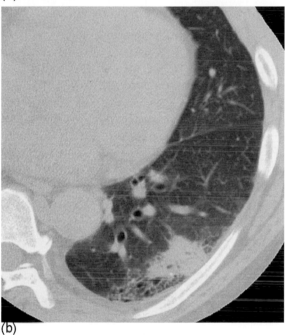

(a)

(b)

Figure 73 An individual with squamous cell carcinoma associated with IPF. (a) A chest radiograph shows an ill-defined nodular lesion on the left lung (arrow) with bilateral ground-glass opacity and a fine reticulation, which are predominant in the lower lung zone. (b) High-resolution computed tomography scan at the level of the left lower lung reveals a lobulated nodule with air bronchogram in the peripheral lung field adjacent to honeycombing. A punctuate calcification was also present in the nodule.

nal lymph node involvement. Therefore, one possibility of the laterality of pulmonary infiltrates was obstruction of the lymphatic vessels by cancer cells. Another possibility was the result of septal thickening of the honeycomb lung adjacent to the lung cancer (Sakai *et al.*, 2003). These authors reviewed the CT scans and pathologic specimens of 57 bronchogenic carcinomas in 47 patients with diffuse pulmonary fibrosis. Most patients were diagnosed as having IPF. Ten cases were synchronous multiple lung cancers. Of the 47 patients, 20 (26 tumors) underwent additional HRCT scans of the lesions. Most tumors (82%) were located in peripheral areas that demonstrated honeycoming. In 50 of the 57 lesions (87.7%), the tumors were round or lobulated with sharp margins. In 7 tumors, the tumor invaded the adjacent honeycomb lung and lacked distinct margins. Intratumor lucency was seen in 13 tumors.

In correlating the specific CT findings and the histology of the 29 resected tumors in 25 patients, those with a sharp margin on CT were discrete masses with intact but compressed adjacent honeycomb lung. Intratumor lucency resulted from dilated bronchi, patent honeycomb cysts, and necrosis. In three cases with indistinct margins, the tumor had spread along the internal surface of the honeycomb cysts, resulting in patent air-spaces within the tumors and a broad area of carcinoma that was carcinoma *in situ* or squamous metaplasia.

Kishi *et al.* (2006) evaluated the HRCT findings of 30 patients with lung cancer associated with IPF. The shape of the lung cancer was a nodular lesion in 93.3% (28/30) of patients and an undetermined pattern in 6.7% (2/30) of

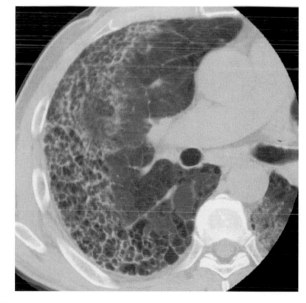

Figure 74 Bronchioloalveolar carcinoma associated with IPF. High-resolution computed tomography scan reveals peripheral honeycombing and diffuse ground-glass opacities resembling acute exacerbation.

In 16 of the 23 patients, it was possible to compare the intensity of lung infiltrates between both sides of the lung. As a result, increased intensity at the side in which the lung cancer developed was demonstrated in 12 of 16 patients (75%). Among these 12 patients, 10 had mediasti-

patients. Of 28 nodular lesions, a well-defined margin was seen in 23 lesions (82.1%) (Fig. 73b). Lobulation and spiculation were found in 24 (85.7%) and 14 (50.0%) lesions, respectively. Air bronchogram was observed in 16 lesions (57.1%). Cavitation was present in 3 lesions of squamous cell carcinoma. Stippled calcification was recognized in 3 lesions. All nodular lesions were located in the peripheral area of the lung, where honeycombing or a reticular pattern indicating fibrosis was recognized.

Of the two patients with an undetermined pattern, one with mucinous bronchioloalveolar carcinoma demonstrated diffuse ground-glass opacities resembling acute exacerbation of IPF (Fig. 74), and the other with adenocarcinoma had extensive pleural thickening mimicking malignant pleural mesothelioma.

In summary, the typical HRCT findings of lung cancer associated with IPF are well-defined nodular lesions with lobulation in peripheral areas of the lung. Because it is difficult to find a small-size lung cancer associated with IPF by chest radiography, routine use (e.g., annual check-up) of CT is recommended for early detection of lung cancer.

References

American Thoracic Society. 2002. American Thoracic Society/European Respiratory Society international multidisciplinary consensus classification of the idiopathic interstitial pneumonias. *Am. J. Respir. Crit. Care Med. 165*:277–304.

Aubry, M-C., Myers, J.L., Douglas, W.W., Tazelaar, H.D., Washington Stephens, T.L., Hartman, T.E., Deschamps, C., and Pankratz, V.S. 2002. Primary pulmonary carcinoma in patients with idiopathic pulmonary fibrosis. *Mayo Clin. Proc. 77*:763–770.

Daniels, C.E., and Jett, J.R. 2005. Does interstitial lung disease predispose to lung cancer? *Curr. Opin. Pulm. Med. 11*:431–437.

Fujita, J., Yamadori, I., Namihira, H., Suemitsu, I., Bandoh, S., Fukunaga, Y., Hojo, S., Ueda, Y., Dobashi, N., Dohmoto, K., and Takahara, J. 1999. Increased intensity of lung infiltrates at the side of lung cancer in patients with lung cancer associated with pulmonary fibrosis. *Lung Cancer 26*:169–174.

Harris, J.M., Cullinan, P., and McDonald, J.C. 1998. Does cryptogenic fibrosing alveolitis carry an increased risk of death from lung cancer? *J. Epidemiol. Community Health 52*:602–603.

Hironaka, M., and Fukayama, M. 1999. Pulmonary fibrosis and lung carcinoma: a comparative study of metaplastic epithelia in honeycombed areas of usual interstitial pneumonia with or without lung carcinoma. *Pathol. Int. 49*:1060–1066.

Hubbard, R., Venn, A., Lewis, S., and Britton, J. 2000. Lung cancer and cryptogenic fibrosing alveolitis. A population-based cohort study. *Am. J. Respir. Crit. Care Med. 161*:5–8.

Kawai, T., Yakumaru, K., Suzuki, M., and Kageyama, K. 1987. Diffuse interstitial pulmonary fibrosis and lung cancer. *Acta Pathol. Jpn. 37*:11–19.

Kishi, K., Homma, S., Kurosaki, A., Motoi, N., and Yoshimura, K. 2006. High-resolution computed tomography findings of lung cancer associated with idiopathic pulmonary fibrosis. *J. Comput. Assist. Tomogr. 30*:95–99.

Lee, H.J., Im, J-G., Ahn, J.M., and Yeon, K.M. 1996. Lung cancer in patients with idiopathic pulmonary fibrosis: CT findings. *J. Comput. Assist. Tomogr. 20*:979–982.

Matsushita, H., Tanaka, S., Saiki, Y., Hara, M., Nakata, K., Tanimura, S., and Banba, J. 1995. Lung cancer associated with usual interstitial pneumonia. *Pathol. Int. 45*:925–932.

Mizushima, Y., and Kobayashi, M. 1995. Clinical characteristics of synchronous multiple lung cancer associated with idiopathic pulmonary fibrosis: a review of Japanese cases. *Chest 108*:1272–1277.

Nagai, A., Chiyotani, A., Nakadate, T., and Kohno, K. 1992. Lung cancer in patients with idiopathic pulmonary fibrosis. Tohoku *J. Exp. Med. 167*:231–237.

Sakai, S., Ono, M., Nishio, T., Kawarada, Y., Nagashima, A., and Toyoshima, S. 2003. Lung cancer associated with diffuse pulmonary fibrosis: CT-pathologic correlation. *J. Thorac. Imag. 18*:67–71.

Samet J.M. 2000. Does idiopathic pulmonary fibrosis increase lung cancer risk? *Am. J. Respir. Crit. Care Med. 161*:1–2.

Takahashi, T., Munakata, M., Ohtsuka, Y., Nishihira, H., Nasuhara, Y., Kamachi-Satoh, A., Dosaka-Akita, H., Homma, Y., and Kawakami, Y. 2002. Expression and alteration of ras and p53 proteins in patients with lung carcinoma accompanied by idiopathic pulmonary fibrosis. *Cancer 95*:624–633.

Turner-Warwick, M., Lebowitz, M., Burrows, B., and Johnston, A. 1980. Cryptogenic fibrosing alveolitis and lung cancer. *Thorax 35*: 496–499.

Uematsu, K., Yoshimura, K., Gemma, A., Mochimaru, H., Hosoya, Y., Kunugi, S., Matsuda, K., Seike, M., Kurimoto, F., Takenaka, K., Koizumi. K., Fukuda, Y., Tanaka, S., Chin, K., Jablons, D.M., and Kudoh, S. 2001. Aberrations in the fragile histidine triad (FHIT) gene in idiopathic pulmonary fibrosis. *Cancer Res. 61*:8527–8533.

Vassilakis, D.A., Sourvinos, G., Spandidos, D.A., Siafakas, N.M., and Bouros, D. 2000. Frequent genetic at the microsatellite level in cytologic sputum samples of patients with idiopathic pulmonary fibrosis. *Am. J. Respir. Crit. Care Med. 162*:1115–1119.

Wells, C., and Mannino, D.M. 1996. Pulmonary fibrosis and lung cancer in the United States: analysis of the multiple cause of death mortality data, 1979 through 1991. *South Med. J. 89*:505–510.

IV

Breast Carcinoma

1

Categorization of Mammographic Density for Breast Cancer: Clinical Significance

Mariko Morishita, Akira Ohtsuru, Ichiro Isomoto, and Shunichi Yamashita

Introduction

Mammography is a powerful X-ray imaging technique for breast cancer diagnosis in routine health checks, as well as for the ongoing examination of breast tumors. In addition to its use as a diagnostic tool, mammography can be used to assess the characteristics of the normal breast. The mammographically dense area is composed of both stromal and epithelial tissue. Several studies have shown that the stroma can provide a microenvironment that promotes tumor growth, invasion, and metastasis as a result of the interaction between the stroma and breast cancer (van Roozendaal *et al.*, 1992; Gache *et al.*, 1998; Noel and Foidart, 1998). Epidemiological surveys have already shown that increased breast density is an independent risk factor for breast cancer (Boyd *et al.*, 1998; Harvey and Bovbjerg, 2004). The level of breast density is due in part to genetic factors. However, breast density is also hormonally responsive and can be influenced by lifestyle factors, such as alcohol intake and diet (Byrne *et al.*, 2000; Maskarinec *et al.*, 2001).

The reasons for the increased incidence of breast cancer throughout the world are currently a subject of intense discussion. Endogenous and exogenous hormonal influences, including hormone replacement therapy (HRT), are thought to be important factors and can affect the breast density of postmenopausal women (Lacey *et al.*, 2002; Brekelmans, 2003; Ghafoor *et al.*, 2003). Breast density has been evaluated as a risk factor. However, the relevant clinical studies have not specifically considered the correlation between breast density and appropriate factors known to be prognostic for tumor growth.

In this chapter, we first describe the mammographic density distribution in breast cancer patients and its relationship to clinicopathological features of tumors and their prognosis. In addition, we review some recent publications regarding tumor-stromal interaction.

Breast Density by Mammography

Wolfe applied a method of classification of breast density and showed the relationship between mammographic parenchymal patterns and the risk of developing breast cancer (Wolfe, 1976). The parenchymal patterns are classified into four categories as follows.

1. *N1 category:* parenchyma composed primarily of fat with, at most, small amounts of dysplasia; no ducts are visible.

2. *P1 category:* parenchyma composed chiefly of fat, with prominent ducts in the anterior portion up to one-fourth of the volume of the breast; a thin band of ducts may also be present, extending into a quadrant.

3. *P2 category:* severe involvement, with a prominent ductal pattern occupying more than one-fourth of the volume of the breast.

4. *DY category:* severe involvement with dysplasia, often obscuring an underlying prominent ductal pattern.

Authors of case-control and cohort studies have found an association between increased breast cancer risk and the Wolfe P2 and DY categories. However, this method depends on visual estimation without practical definition of the percentages of density and is applied inconsistently (Oza and Boyd, 1993).

Attempts to develop a reproducible quantitative method of assessing breast density began in the early 1980s. Visual estimation of the percentage of the breast occupied by breast tissue has been used frequently, with the most common method having six categories, where tissue occupies > 75%, 50–75%, 25–50%, 10–25%, < 10%, or none of the breast (Boyd *et al.*, 1995).

The American College of Radiology (ACR) developed the Breast Imaging-Reporting and Data System (BI-RADS), which is becoming the standard of mammography-reporting terminology and assessment and recommendation categories (American College of Radiology, 1998). The BI-RADS classifies mammographic density into four categories:

1. *BI-RADS category 1:* breast with the density of fat.

2. *BI-RADS category 2:* fatty breast with scattered fibroglandular densities.

3. *BI-RADS category 3:* heterogeneously dense breast.

4. *BI-RADS category 4:* dense breast.

Representative mammography pictures of density categories 1–4 are shown in Figure 75. Category 1 is entirely radiolucent, indicating that the breast is almost completely replaced by fatty tissue (Fig. 75a). In category 2 breasts, there remains some sparse breast gland in an otherwise fatty breast (Fig. 75b). Category 3 refers to a heterogeneously dense breast, whereas category 4 indicates an entirely dense breast (Figs. 75c, 75d). The BI-RADS category system, which is widely used in the United States and allows analysis of large study populations, is based on quantitative assessment, although the categories are again not defined by the percentage of the breast occupied by dense tissue.

Recently, more consistent and quantitative computer-assisted measurements of breast density have been developed. All methods use planimetry in some way. Computer-assisted planimetry was used as early as 1987 (Wolfe *et al.*, 1987). More recently, film mammograms have been digitized, and the area of the breast parenchyma and the total area of the breast are outlined either by use of a mouse or by digital segmentation of the mammogram (Ursin *et al.*, 2001; Byng *et al.*, 1998; Freedman *et al.*, 2001). Future work will focus on the extraction of further texture information and the use of feature selection algorithms in order to provide a more reliable subset of characteristics.

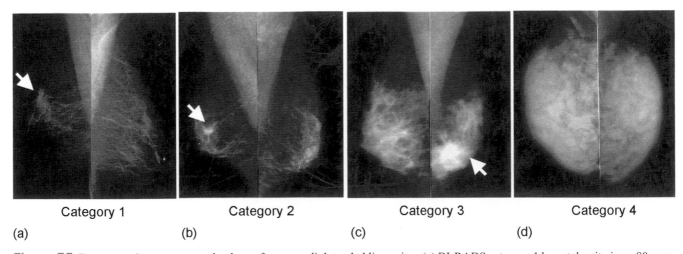

Figure 75 Representative mammographs shown from a mediolateral oblique view. (a) BI-RADS category 1 breast density in an 80-year-old woman. (b) BI-RADS category 2 breast density in a 66-year-old woman. (c) BI-RADS category 3 breast density in a 33-year-old woman. (d) BI-RADS category 4 breast density in a 41-year-old woman. Arrows indicate the locations of the breast tumors. No tumor was visible in panel (d) because of the high breast density.

Clinical Applications of Breast-density Category

Analysis of Patient Characteristics by Breast-density Category

In most cases, the BI-RADS density category is thought to change with patient age. Table 12 displays clinicopathological data by breast-density categories among Japanese breast cancer patients (Morishita et al., 2005). The average age of the patients was 57 years (range: 28–86 years), and the mean ages of the categories were 76, 65, 56, and 45 years, respectively. The breast-density categories were significantly correlated with age, but there was heterogeneity of breast density among patients of the same age group. This is probably due to the individual genetic and hormonal status and/or lifestyle effects. No statistically significant differences across the categories were revealed with the other clinicopathological factors: mean tumor size, axillary lymph-node involvement, steroid receptors (SR; estrogen and progesterone receptor) positivity, and histological tumor grade (Morishita et al., 2005).

Aiello et al. (2005) also reported association between mammographic breast density and breast cancer tumor characteristics. They divided breast density into two groups: fatty (BI-RADS categories 1 and 2) and dense (BI-RADS categories 3 and 4). Tumor size, lymph node status, and lymphatic or vascular invasion were positively associated with breast density, but there was no correlation between breast density and hormone receptor status, histological and nuclear grade, and profiling by immunohistochemistry including Ki-67, p53, p27, cyclin E, Bcl-2, and C-erb-B2 (Aiello et al., 2005).

Steroid Receptor Status and Breast-density Category

All cases in category 1 were SR positive (Table 12). There was a significant correlation between mean tumor size and SR status in categories 2 and 3. The proportion of tumors that were axillary-node positive was also correlated with SR status in categories 3 and 4, and there was a difference in tumor grade by SR status in categories 2 and 4 (Table 12).

Table 12 Tumor Prognostic Factors by BI-RADS Density Category and Tumor SR Status

Factor		Category 1	Category 2	Category 3	Category 4
No. of patients		7	42	72	42
Mean age (y)		76	65	56	45
Mean tumor size (mm)		20.1	27.9	30.2	28.5
Axillary node positivity		1 (14)	16 (38)	28 (39)	17 (37)
Steroid receptor positivity		7 (100)	28 (67)	43 (60)	26 (62)
Tumor grade	1	3 (43)	12 (29)	17 (24)	7 (17)
	2	3 (43)	20 (48)	22 (31)	14 (33)
	3	1 (14)	7 (17)	29 (40)	19 (45)
	unavailable	0 (0)	3 (7.1)	4 (5.6)	2 (4.8)

		Category 1		Category 2		Category 3		Category 4	
		SR(+)	SR(−)	SR(+)	SR(−)	SR(+)	SR( )	SR(+)	SR(−)
No. of patients		7	0	28	14	43	29	26	16
Mean age (y)		76	-	65	65	56	57	45	46
Mean tumor size (mm)		20.1	-	23.4	35.4 *	25.3	37.3 *	22.7	37.9
Axillary node involvement		1	-	9	7	21	7 *	7	10 *
Tumor grade	1	3	-	10	2	13	4	7	0
	2	3	-	14	6 *	12	10	0	4 *
	3	1	-	1	6	16	13	8	11
	unavailable	0	-	3	0	2	2	1	1
NPI score		3.40	-	3.59	4.78 *	4.19	4.38	3.81	5.55 *
Distant metastasis case (%)[#]		0	-	3.8	14	9.8	19	4.3	50

*$P < 0.05$.

#Mean observation period was 38 months.

Comparison of Nottingham Prognostic Index Scores in Breast-density Categories

We calculated the Nottingham Prognostic Index (NPI) score, which is based on tumor size, node involvement, and Scarff–Bloom–Richardson (SBR) grading, by combining several clinicopathological factors in order to clarify the differences between the BI-RADS breast-density categories (Table 12). Although comparison between the categories did not reveal any significant differences in the NPI scores, the sequential correlation tendency between density category and NPI score was borderline significant. There was no significant correlation detected between age and NPI score (Morishita et al., 2005). However, there was an influence of SR status on NPI score: the NPI scores were significantly worse in the SR-negative cases compared with the SR-positive cases in categories 2 and 4 (Table 12).

Patient Prognosis and Breast-density Category

Investigation of patient prognosis after breast surgery excluding deaths from other causes was performed (the mean observation period was 38 months), and there were no significant differences in outcome between the categories. A striking proportion of category 4 SR-negative cases (50%) had distant metastases (compared to 4.3% of the SR positives). The percentages of cases of distant metastasis in the other bands were all equally low. The mortality rate during follow-up of the BI-RADS category 4 SR-negative group was 43% (7/16), compared with 3.6% (1/27) for the category 3 SR-negative group, and there were no fatalities in the other groups (Morishita et al., 2005).

Analytic Considerations

A variety of clinical and analytical data should be considered in the interpretation and use of mammographic breast density as a predictor of disease prognosis. From the small clinical investigation, we have shown a relationship between categories of BI-RADS mammographic density and clinicopathological features in Japanese breast cancer patients (Morishita et al., 2005). The mean age of each density category was the only factor that differed between them: tumor size, lymph node involvement, SR status, histological grade, and NPI score did not differ significantly between the density categories. In the BI-RADS breast-density category 2 and 4 groups, SR status was associated with significant differences in the NPI score. Furthermore, breast-density category 4, SR-negative patients had substantially higher frequencies of distant metastasis and mortality than all other groups.

Breast cancer prognostic factors are measurements available at the time of surgery or diagnosis that are associated with disease-free survival or overall survival. The best-recognized prognostic factors are tumor size, axillary lymph node status, tumor type, standardized pathological grade, and SR status (Bundred, 2001). The evidence concerning the influence of age at the time of diagnosis on the prognosis of patients with primary breast cancer is ambiguous. In many studies, patients < 35 years old with invasive breast cancer were found to have features indicating a poorer prognosis than that of older patients; these included grade 3 histology, lymphatic-vessel invasion, estrogen receptor (ER) negativity, and some genetic mutations (Albain et al., 1994; Nixon et al., 1994; Kroman et al., 2000; Yildirim et al., 2000). The effect of age on clinicopathological features and NPI score has also been examined previously (Kollias et al., 1997). These authors compared patients who were < 35, 35–50, and 51–71 years old, and showed that those who were < 35 years old had a poor prognosis because of a higher proportion of poorly differentiated cancers. By contrast, no difference in prognosis was detected between patients who were 35–50 and 51–71 years old. They, therefore, concluded that age itself had no influence on prognosis.

In the present study, we divided the patients into several age groups (≥ 70, 60–69, 50–59, and ≤ 49 years old) in order to compare the BI-RADS breast-density categories. As expected, NPI score did not significantly differ between these age groups. BI-RADS breast-density category 4 is often seen in younger patients, and in the quoted data in this chapter this category had the lowest mean age (45 years old) (Table 12). The lack of correlation between representative prognostic factors or NPI score and category may therefore be due to the fact that only two of the patients were < 35 years old and in BI-RADS category 3, and only three were in category 4. Thus, the contribution of young patients (< 35 years old) to the results was minimal in these results.

Other investigators have studied differences in mean age and tumor size between the BI-RADS breast-density categories and, in particular, category 1 is correlated with favorable clinicopathological features, such as histological grade 1, ER positivity, and stage I tumors (Roubidoux et al., 2004). All breast cancers in BI-RADS breast-density category 1 were SR-positive from the study. However, there were not significant correlations between other prognostic factors and breast density. At least these results indicate that breast cancers in density category 1 can develop surrounded by tissues that release little, if any, nonsteroid growth factors.

Mammographic density variation is reportedly affected by endogenous and exogenous hormonal influences, such as menstrual, parity, and lactation status; excessive fat; and HRT use (Harvey and Bovbjerg, 2004). Furthermore, estrogen and progesterone are essential for the proliferation and morphogenesis of the normal mammary gland (Hovey et al., 2002; Neville et al., 2002). Taken together, these findings suggest that these steroid hormones might also have a critical role in maintaining the dense breast.

Steroid receptor expression in breast cancer cells is thought to be an important prognostic factor, as well as a marker of tumor differentiation (Bundred, 2001; Dontu et al., 2004; Platet et al., 2004). Many studies have shown that SR positivity is associated with a favorable prognosis and responsiveness to anti-estrogen hormone therapy (Bundred, 2001; Early Breast Cancer Trialists' Collaborative Group, 1998). Many SR-positive and, especially, estrogen receptor-positive tumors proliferate in response to estrogen stimulation (Cordera and Jordan, 2006), whereas the growth of SR-negative, and some SR-positive, tumors tends to be affected by other growth factors, such as insulin growth factor, and by autocrine and/or paracrine mechanisms from the tumor itself or the surrounding stromal cells (Hamelers et al., 2002; Haslam and Woodward, 2003; Seeger et al., 2004; Stull et al., 2004). For this reason, we divided the patients in each breast-density category into two groups according to their SR status, in order to examine differences in tumor characteristics arising in an environment of similar breast density. Comparison between SR-positive and SR-negative cases showed that the latter cases in the same breast-density category had relatively unfavorable features, such as larger tumor size and higher histological grade, although there was no influence of category on axillary lymph node involvement (Table 12).

The NPI score has been shown to constitute a definitive prognostic factor for primary operable breast cancer in the adjuvant setting (Galea et al., 1992; Balslev et al., 1994). In the study by Morishita et al. (2005), the NPI score tended to increase with an increase in the BI-RADS breast-density category, although the effect was not statistically significant. However, there were significant differences in NPI score according to SR status in categories 2 and 4, suggesting that breast density might exert an important microenvironmental influence on SR-negative tumors.

Recent basic studies have given more detailed and promising insights into the interaction between tumor and surrounding stroma cells. Boire et al. (2005) demonstrated a new mechanism of interaction between protease-activated receptor 1 (PAR1) expressed on the tumor and matrix metalloprotease (MMP-1) activity derived from stromal fibroblasts. Protease-activated receptor 1 is both required and sufficient to promote growth and invasion of breast carcinoma cells, and at the same time, MMP-1 can function as a protease agonist of PAR1, leading to the conclusion that MMP-1 in the stromal-tumor microenvironment can alter the character of cancer cells in the tumor microenvironment.

By using a new technique of methylation-specific digital karyotyping, Hu et al. (2005) also added a new aspect by showing that distinct epigenetic alterations occur not only in epithelial and myoepithelial cells but also in stromal fibroblasts from normal breast and breast cancer tissue. This finding suggests that epigenetic changes have a role in establishing the abnormal tumor microenvironment and can contribute to tumor progression. On the other hand, the relevance of adipocyte-derived factors to breast cancer cell survival and growth is also well known, although which specific factors play a key role in this process has only recently been established. Lyengar et al. (2005) showed that collagen VI, which is abundantly expressed in adipocytes, promotes mammary tumor growth-stimulatory and pro-survival effects in part by signaling through the NG/chondroitin sulfate proteoglycan receptor, expressed on the surface of malignant ductal epithelial cells, to sequentially activate Akt and β-catenin and stabilize cyclin D1. Therefore, they concluded that adipocytes play a vital role in defining the extracellular matrix (ECM) microenvironment for normal and tumor-derived ductal epithelial cells and contribute significantly to tumor growth at early stages through secretion and processing of collagen VI.

A schema of the different types of tumor microenvironment and their potential initiation/progression factors is shown in Figure 76. In the state of fatty breast, one of the major tumor growth factors is estrogen for the SR-positive tumor (Fig. 76a), and surrounding adipocytes might provide other factors such as collagen VI, especially for SR-negative tumors (Fig. 76c). In the dense breast environment, the major tumor growth factor is also thought to be estrogen for SR-positive tumors (Fig. 76b), and surrounding stromal fibroblasts contribute to drive tumor initiation/progression through the interaction with SR-negative tumors (Fig. 76d). Since the data presented in Table 12 revealed that patients with dense breast and SR-negative tumors showed extremely poor prognosis, these clinical prognostic data suggest that the tumor microenvironment of abundant stromal fibroblasts could provide a more malignant character to the primary mammary tumor. This also leads to the idea that provision by the tumor microenvironment of various growth factors, cytokines, and/or radicals can contribute not only to direct tumor growth but also to nongenetic chromosomal instability, resulting in acceleration of the malignancy and poor prognosis.

In summary, we analyzed the clinicopathological features of breast cancer patients as a function of mammographic density. It was found that the combined classification of high breast density and SR negativity predicted poor prognosis and significantly elevated frequencies of distant metastasis. Tumor microenvironment measured by mammography has the potential to be a tumor prognostic factor, probably owing to its importance in establishing the tumor microenvironment.

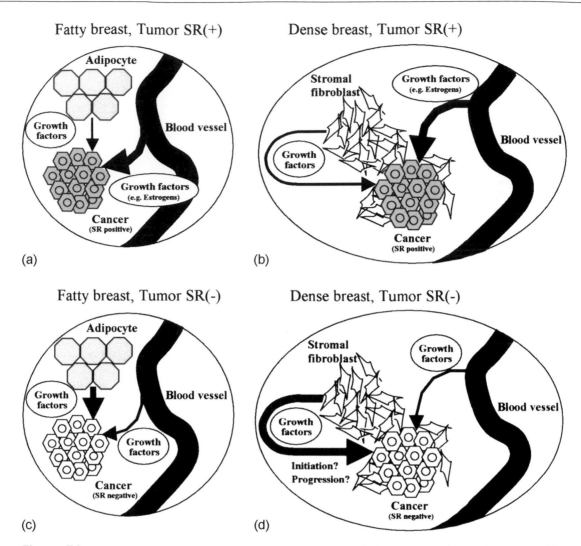

Figure 76 Type of tumor microenvironment. (a) SR-positive tumor in fatty breast background. (b) SR-positive tumor in dense breast background. (c) SR-negative tumor in fatty breast background. (d) SR-negative tumor in dense breast background.

References

Aiello, E.J., Buist, D.S., White, E., and Porter, P.L. 2005. Association between mammographic breast density and breast cancer tumor characteristics. Cancer Epidemiolo. *Biomarkers Prev. 14*:662–668.

Albain, K.S., Allred, D.C., and Clark, G.M. 1994. Breast cancer outcome and predictors of outcome: are there age differentials? *J. Natl. Cancer Inst. Monogr. 16*:35–42.

American College of Radiology. 1998. *Breast Imaging Reporting and Data System (BI-RADS)*, 3rd ed. Reston, VA: American College of Radiology.

Balslev, I., Axelsson, C.K., Zedeler, K., Rasmussen, B.B., Carstensen, B., and Mouridsen, H.T. 1994. The Nottingham Prognostic Index applied to 9,149 patients from the studies of the Danish Breast Cancer Cooperative Group (DBCG). *Breast Cancer Res. Treat. 32*:281–290.

Boire, A., Covic, L., Agarwal, A., Jacques, S., Sherifi, S., and Kuliopulos, A. 2005. PAR1 is a matrix metalloprotease-1 receptor that promotes invasion and tumorigenesis of breast cancer cells. *Cell 120*:303–313.

Boyd, N.F., Byng, J.W., Jong, R.A., Fishell, E.K., Little, L.E., Miller, A.B., Lockwood, G.A., Tritchler, D.L., and Yaffe, M.J. 1995. Quantitative

classification of mammographic densities and breast cancer risk: results from the Canadian National Breast Screening Study. *J. Natl. Cancer Inst. 87*:670–675.

Boyd, N.F., Lockwood, G.A., Byng, J.W., Tritchler, D.L., and Yaffe, M.J. 1998. Mammographic densities and breast cancer risk. *Cancer Epidemiol. Biomark. Prev. 7*:1133–1144.

Brekelmans, C.T. 2003. Risk factors and risk reduction of breast and ovarian cancer. *Curr. Opin. Obstet. Gynecol. 15*:63–68.

Bundred, N.J. 2001. Prognostic and predictive factors in breast cancer. *Cancer Treat. Rev. 27*:137–142.

Byng, J.W., Yaffe, M.J., Jong, R.A., Shumak, R.S., Lockwood, G.A., Tritchler, D.L., and Boyd, N.F. 1998. Analysis of mammographic density and breast cancer risk from digitized mammograms. *Radio-Graphics 18*:1587–1598.

Byrne, C., Colditz, G.A., Willett, W.C., Speizer, F.E., Pollak, M., and Hankinson, S.E. 2000. Plasma insulin-like growth factor (IGF) I, IGF-binding protein 3, and mammographic density. *Cancer Res. 60*:3744–3748.

Cordera, F., and Jordan, V.C. 2006. Steroid receptors and their role in the biology and control of breast growth. *Semin. Oncol. 33*:631–641.

Dontu, G., El-Ashry, D., and Wicha, M.S. 2004. Breast cancer, stem/progenitor cells and the estrogen receptor. *Trends Endocrinol. Metab.* 15:193–197.

Early Breast Cancer Trialists' Collaborative Group. 1998. Tamoxifen for early breast cancer: an overview of the randomised trials. *Lancet* 351:1451–1467.

Freedman, M., San Martin, J., O'Gorman, J., Eckert, S., Lippman, M.E., Lo, S.C., Walls, E.L., and Zeng, J. 2001. Digitized mammography: a clinical trial of postmenopausal women randomly assigned to receive raloxifene, estrogen, or placebo. *J. Natl. Cancer Inst.* 93:51–56.

Gache, C., Berthois, Y., Martin, P.M., and Saez, S. 1998. Positive regulation of normal and tumoral mammary epithelial cell proliferation by fibroblast in coculture. *In Vitro Cell Dev. Biol. Anim.* 34:347–351.

Galea, M.H., Blamey, R.W., Elston, C.E., and Ellis, I.O. 1992. The Nottingham Prognostic Index in primary breast cancer. *Breast Cancer Res. Treat.* 22:207–219.

Ghafoor, A., Jemal, A., Ward, E., Cokkinides, V., Smith, R., and Thun, M. 2003. Trends in breast cancer by race and ethnicity. *CA Cancer J. Clin.* 53:342–355.

Hamelers, I.H., van Schaik, R.F., Sipkema, J., Sussenbach, J.S., and Steenbergh, P.H. 2002. Insulin-like growth factor I triggers nuclear accumulation of cyclin D1 in MCF-7S breast cancer cells. *J. Biol. Chem.* 277:47645–47652.

Harvey, J.A., and Bovbjerg, V.E. 2004. Quantitative assessment of mammographic breast density: relationship with breast cancer risk. *Radiology* 230: 29–41.

Haslam, S.Z., and Woodward, T.L. 2003. Host microenvironment in breast cancer development: epithelial-cell–stromal-cell interactions and steroid hormone action in normal and cancerous mammary gland. *Breast Cancer Res.* 5:208–215.

Hovey, R.C., Trott, J.F., and Vonderhaar, B.K. 2002. Establishing a framework for the functional mammary gland: from endocrinology to morphology. *J. Mammary Gland Biol. Neoplasia* 7:17–38.

Hu, M., Yao, J., Cai, L., Bachman, K.E., van den Brule, F., Velculescu, V., and Polyak, K. 2005. Distinct epigenetic changes in the stromal cells of breast cancers. *Nat. Genet.* 37:899–905.

Kollias, J., Elston, C.W., Ellis, I.O., Robertson, J.F., and Blamey, R.W. 1997. Early-onset breast cancer—histopathological and prognostic considerations. *Br. J. Cancer* 75:1318–1323.

Kroman, N., Jensen, M.B., Wohlfahrt, J., Mouridsen, H.T., Andersen, P.K., and Melbye, M. 2000. Factors influencing the effect of age on prognosis in breast cancer: population based study. *Brit. Med. J.* 320:474–478.

Lacey, J.V. Jr, Devesa, S.S., and Brinton, L.A. 2002. Recent trends in breast cancer incidence and mortality. *Environ. Mol. Mutagen* 39:82–88.

Lyengar, P., Espina, V., Williams, T.W., Lin, Y., Berry, D., Jelicks, L.A., Lee, H., Temple, K., Graves, R., Pollard, J., Chopra, N., Russell, R.G., Sasisekharan, R., Trock, B.J., Lippman, M., Calvert, V.S., Petricoin, E.F., 3rd, Liotta, L., Dadachova, E., Pestell, R.G., Lisanti, M.P.,

Bonaldo, P., and Scherer, P.E. 2005. Adipocyte-derived collagen VI affects early mammary tumor progression *in vivo,* demonstrationg a critical interaction in the tumor/stroma microenvironment. *J. Clin. Inves.* 115:1163–1176.

Maskarinec, G., Meng, L., and Ursin, G. 2001. Ethnic differences in mammographic densities. *Int. J. Epidemiol.* 30:959–965.

Morishita, M., Ohtsuru, A., Hayashi, T., Isomoto, I., Itoyanagi, N., Maeda, S., Honda, S., Yano, H., Uga, T., Nagayasu, T., Kanematsu, T., and Yamashita, S. 2005. Clinical significance of categorisation of mammographic density for breast cancer prognosis. *Int. J. Oncol.* 26:1307–1312.

Neville, M.C., McFadden, T.B., and Forsyth, I. 2002. Hormonal regulation of mammary differentiation and milk secretion. *J. Mammary Gland Biol. Neoplasia* 7:49–66.

Nixon, A.J., Neuberg, D., Hayes, D.F., Gelman, R., Connolly, J.L., Schnitt, S., Abner, A., Recht, A., Vicini, F., and Harris, J.R. 1994. Relationship of patient age to pathologic features of the tumor and prognosis for patients with stage I or II breast cancer. *J. Clin. Oncol.* 12:888–894.

Noel, A., and Foidart, J.M. 1998. The role of stroma in breast carcinoma growth *in vivo. J. Mamm. Gland Biol. Neoplasia* 3:215–225.

Oza, A.M., and Boyd, N.F. 1993. Mammographic parenchymal patterns: a marker of breast cancer risk. *Epidemiol. Rev.* 15:196–208.

Platet, N., Cathiard, A.M., Gleizes, M., and Garcia, M. 2004. Estrogens and their receptors in breast cancer progression: a dual role in cancer proliferation and invasion. *Crit. Rev. Oncol. Hematol.* 51:55–67.

Roubidoux, M.A., Bailey, J.E., Wray, L.A., and Helvie, M.A. 2004. Invasive cancers detected after breast cancer screening yielded a negative result: relationship of mammographic density to tumor prognostic factors. *Radiology* 230:42–48.

Seeger, H., Wallwiener, D., and Mueck, A.O. 2004. Influence of stroma-derived growth factors on the estradiol-stimulated proliferation of human breast cancer cells. *Eur. J. Gynaecol. Oncol.* 25:175–177.

Stull, M.A., Rowzee, A.M., Loladze, A.V., and Wood, T.L. 2004. Growth factor regulation of cell cycle progression in mammary epithelial cells. *J. Mammary Gland Biol. Neoplasia* 9:15–26.

Ursin, G., Parisky, Y.R., Pike, M.C., and Spicer, D.V. 2001. Mammographic density changes during the menstrual cycle. *Cancer Epidemiol. Biomark. Prev.* 10:141–142

van Roozendaal, C.E., van Ooijen, B., Klijn, J.G., Claassen, C., Eggermont, A.M., Henzen-Logmans, S., and Foekens, J.A. 1992. Stromal influences on breast cancer cell growth. *Br. J. Cancer* 65:77–81.

Wolfe, J.N. 1976. Breast patterns as an index of risk for developing breast cancer. *Am. J. Roentgenol.* 126:1130–1137.

Wolfe, J.N., Saftlas, A.F., and Salane, M. 1987. Mammographic parenchymal patterns and quantitative evaluation of mammographic densities: a case-control study. *Am. J. Roentgenol.* 148:1087–1092.

Yildirim, E., Dalgic, T., and Berberoglu, U. 2000. Prognostic significance of young age in breast cancer. *J. Surg. Oncol.* 74:267–272.

2

Breast Tumor Classification and Visualization with Machine-learning Approaches

Tim W. Nattkemper, Andreas Degenhard, and Thorsten Twellmann

Introduction

When diagnosing and monitoring cancer, the evaluation of radiological image data is an important source of objective information. Such evaluation includes the detection and classification of suspicious lesions as well as the classification of tumor tissue under treatment. Identifying and categorizing abnormality relys on a set of consistent criteria based on image features to which the human visual system will respond with particular sensitivity. Many imaging modalities help radiologists reach their decisions on the basis of the appearance of a mass while monitoring dynamic changes. Frequently, this is accomplished by using contrast agents signal changes that can be observed over time.

Magnetic resonance imaging (MRI) is based on the physical principle of nuclear magnetic resonance (NMR). In MRI, contrast agents are transported by the bloodstream as paramagnetic ions chemically bonded in a chelate complex (Heywang-Kobrunner et al., 1997). The contrast agents are mostly injected intravenously, leading to a signal enhancement that can be monitored by a time series of images referred to as the dynamic contrast-enhanced magnetic reso-

nance imaging (DCE-MRI) sequence (Heywang-Kobrunner et al., 1997). This technique is of increasing clinical interest as a diagnostic tool in cancer research (Heywang-Kobrunner et al., 1997), because the ability to differentiate between benign and malignant lesions depends on both enhanced morphology and the rate and amplitude of enhancement related to the vascular characteristics of tissue.

Because the technique permits a combined consideration of morphological and kinetic features, DCE-MRI has become an established image modality with a rich information content for diagnosing and monitoring tumors (Collins and Padhani, 2004), particularly breast tumors. The gadolinium-based paramagnetic contrast agent Gd-DTPA (diethylenetriaminepentaacetic acid) is used to visualize the vascularity of tissue by changing signal intensities in a sequence of n_s magnetic resonance images. Applying high-resolution three-dimensinal (3D) imaging techniques makes it possible to record a volume of signal values (Fig. 77). The resulting image data can be represented as a four-dimensional (4D) data set (three-volume coordinate axes and one-time axis) containing a treasure of information for estimating the state of a tumor. This is important not only in cancer diagnosis but also in treatment obser-

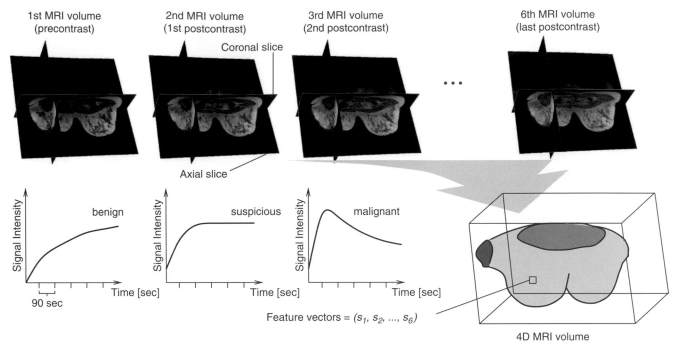

Figure 77 Top: The sequence of six MRI volumes needs to be inspected and evaluated for typical signal time curves. Bottom: The signal curves of the tumors as classified into typical benign, suspicious, and malignant curve shapes. Bottom right: The fused-volume MRI sequence is interpreted as a 4D image.

vation and control. A tissue can be classified by evaluating the short-signal time courses in voxels or ROIs (regions of interest). DCE-MRI is reported to be considerably more sensitive than X-ray mammography (XRM)-based classification (Harms *et al.*, 1993; Heiberg *et al.*, 1996; Heywang-Kobrunner *et al.*, 1997; Nunes *et al.*, 1997; Liu *et al.*, 1998; Knowles *et al.*, 2000; Lucht *et al.*, 2000, 2001), with specificity values reported to be between 37 and 89% (Kuhl *et al.*, 1997). However, analyzing the data—that is, extracting the clinically relevant information from the data set—calls for special software tools that generate meaningful displays from the large data set.

Image Parameters

This chapter will focus on image sequences obtained by using protocols that generate a time series of $n_s = 6$ volumes with scanners equipped with dedicated surface coils enabling simultaneous imaging of both female breasts. The time interval between the single volumes varies between 90 sec and 110 sec depending on the particular protocol. The in-plane resolution of the volume is between 1.4 mm × 1.4 mm and 1.2 mm × 1.2 mm. The slice thickness varies from 2–4 mm. However, the approaches presented here can also be applied to other sequences and will be described on a formal basis independent from the imaging protocol applied.

In clinical practice, the 4D volumes are evaluated by an experienced radiologist. The expert does this by visualizing

the uptake and wash-out characteristics of the tissue. The usual approach is a voxel-wise computing of signal differences in a volume pair. To measure the signal uptake after the bolus injection, the precontrast volume (i.e., a volume acquired before administration of the contrast agent) is subtracted from one of the early postcontrast volumes. The washout can be analyzed by subtracting one of the late postcontrast volumes from a volume measured during the early postcontrast period.

The Contribution of Machine Learning and Artificial Neural Networks

In clinical patient management, there is only limited time for evaluating such DCE-MRI data. Such an evaluation includes the following three visual tasks:

1. Detection of suspicious lesions.

Whereas a primary tumor is detected by sonography and/or X-ray mammography, a full-volume DCE-MRI 3D scan provides an opportunity to search for additional lesions. This is particularly indicated for young women whose breast density reduces the sensitivity of X-ray mammography. Of course, the chances of detecting small lesions are limited by the coarser spatial resolution of DCE-MRI compared with mammography. Machine-learning approaches could contribute to detecting suspicious lesions by screening volumes for signals that are either unusual or similar to those caused by either benign, suspicious, or malignant lesions. This would result in a display of the breast indicating the position and extent of lesions.

2. Classification of signals and lesions.

In practice, the state of the detected lesions is analyzed by computing averaged signals of manually selected ROIs. The placement of such ROIs has to consider the heterogeneity of the tumor tissue in order to prevent any averaging of signals arising from mixtures of benign and malignant tissue—thus causing the characteristics of the averaged signal to become indistinct. Although subtraction images reflect only a small part of the signal curve characteristics, it is still common practice to select ROIs by means of this type of data visualization. Learning algorithms can classify the image data *per voxel* into different classes of tissue states, while taking into account the entire temporal information of the signals associated with individual voxels. In a display of the breast, the tissue class, and thereby the tissue heterogeneity, can be visualized using color. The colors of the lesion voxels and the pattern of color distribution then summarize the complex information of the DCE-MRI signal in one display.

3. Determination of tissue change.

During cancer treatment, tissue change needs to be monitored carefully. The signal features of a tumor need to be visualized (as in the classification scenario) for two different DCE-MRI data sets obtained from two examinations separated by an interval of several weeks. There are two ways to estimate tissue change. First, the classification in (2) is applied for both data sets, and the physician examines both results. In this case, the user forms a mental model of the tumor and uses his or her visual impression to estimate tissue change personally. Second, a less subjective analysis can be performed by direct quantification of tissue change. This can be done by computing and comparing the fractions of different tissue classes in the lesion in different image sequences. This method can support the physician's decision-making process by delivering clear numerical data on the progression or regression of tissues showing specific vascular behaviors.

In patient management, it is necessary to consider extensive data and information from different sources during patient-monitoring and decision-making processes. Because limited time is available, the information hidden in complex DCE-MRI data needs to be compressed. The three evaluation steps comprising (1) detection, (2) classification, and (3) tissue-change determination are used to extract relevant information from the complex image data and transform it to a more appropriate level for patient management.

State-of-the-Art Approaches to Dynamic Contrast-enhanced Magnetic Resonance Visualization

The most common approach when analyzing DCE-MRI data is direct visual inspection. Clinical practice reveals two established strategies of DCE-MRI visualization: subtraction images and the three time points method.

Subtraction Images

For each of the postcontrast images in a data set, one computes a subtraction image using the precontrast image. This results in a set of subtraction images and subtraction volumes depicting high intensities at voxel positions with a strong signal uptake after the bolus injection of the contrast agent. Each volume is inspected for types of signal enhancement indicative of suspicious lesions by browsing through all the two-dimensional image slices of each subtraction volume. This strategy is particularly suited for Task 1: the detection of lesions. Many radiologists consider the signal washout to be important when assessing the malignancy of a lesion (Fig. 77). However, taking this information into account requires the computation of additional subtraction based on one of the late and one of the early postcontrast images. Even if all single visual evaluations can be performed without any problem, the integration of the visual impressions from all subtraction images must be done by the user—a very difficult task requiring long experience in DCE-MRI data analysis.

Three Time Points Method (3TP)

A more sophisticated approach, 3TP (three time points), is based on the idea of simultaneously visualizing both the signal uptake and washout characteristics in one display using a pseudocoloring algorithm (Weinstein *et al.*, 1999). This algorithm maps the temporal signal information associated with each voxel to a color hue and an intensity value. The intensity value is computed from the signal uptake and the color hue from the washout (i.e., the difference between a late and an early postcontrast value). Both mappings are calibrated on the basis of a three-compartment tissue model using the parameters of the applied protocol. The method computes high-contrast pseudocolor displays of tumors that permit an analysis of the uptake/washout characteristics of one case in one display. One limitation of this approach is that only three images can be taken into account owing to the restrictions in the underlying model of the dependencies between local contrast agent concentration and temporal MRI signal. Thus, higher-order signal features cannot be taken into account.

Machine-Learning Alternative

The development of learning algorithms started in the 1980s, when the field of artificial neural networks became very popular among both computational intelligence scientists and signal processing engineers. The new algorithms for adapting a set of simplified information processing units, or neurons, made it possible to learn nontrivial classification functions.

In the 1990s, statistical learning theory paved the way for a new learning paradigm. The entire process of learning a classification problem was translated into an optimization problem, dropping all analogies to learning in biological systems. The new proposed classification architectures, called support vector machines (SVMs), outperformed classic neural network architectures in a large number of studies.

Learning Algorithms

In computer science, learning algorithms have become an alternative to classic deterministic approaches to programming and optimization. Learning algorithms usually work on algebraic representations of the problem domain. Cases, or more generally data, are considered as points in a high-dimensional vector space. Within this space, methods of analytical geometry and algebra are used to measure the similarity of data distances and to build classification algorithms.

In the case of DCE-MRI, the n_s signal values $s(1), \ldots,$ $s(n_s)$ associated with one voxel $\mathbf{p} = (x, y, z)$ are represented by a set of m features $(x_1, \ldots, x_m)$. These feature vectors $\mathbf{x}_\alpha = (x_1, \ldots, x_m)_\alpha$, in turn, are interpreted as points in an m-dimensional space, or $\mathbf{x}_\alpha \in \mathbf{IR}^m$.

Toy Example

The reader who is less familiar with this perspective may consider the following toy example. Assume we have a set of $n_t = 12$ DCE-MRI sequences obtained from 12 patients. In each sequence, a lesion is marked by a region of interest (ROI). From each voxel $\mathbf{p}$ inside the marked lesion, the precontrast signal value $s_\mathbf{p}(1)$, the second postcontrast $s_\mathbf{p}(2)$, and the last postcontrast $s_\mathbf{p}(6)$ are taken as a set of ($m = 3$) features. Hence, the feature vectors are created by the rule ($x_1 = s_\mathbf{p}(1)$, $x_2 = s_\mathbf{p}(2)$, $x_3 = s_\mathbf{p}(6)$). The feature vectors of all voxels from all 12 ROIs can now be plotted in a 3D scatter plot. This results in a display of the signal space spanned by the lesion.

Learning Data Structure

The aim of learning algorithms is to learn structure in data. In medical image data, two kinds of structure are of interest: classification functions and clusters. In the following, we assume that we have a set of n_t DCE-MRI volumes acquired from n_t individuals. All volumes have been recorded using the same parameters (i.e., by applying the same contrast agent and the same imaging protocol). We also assume that one lesion has been detected in each patient, a biopsy has been taken, and it has been given a

histopathological classification. The remainder of this section will give an overview on cluster learning algorithms and different features. The section after that will focus on the problem of tumor classification and monitoring.

Learning Clusters

Displaying the m-dimensional feature space resulting from the data of a set of sequences requires special data-mining algorithms that are able to detect hidden regularities in this space. It is important to know whether tissue types are represented by different signal patterns so that they can be distinguished in DCE-MRI data. Reformulated in machine-learning terms, we have to ask: *Do the different tissue types form distinct clusters in the signal space? And how are these clusters related to tissue categories or tumor development?*

Overview on Clustering Approaches

For the purpose of evaluating the structure of the signal space, unsupervised learning can give new insights on the information in the signal space and the relationships between signal values from different tissue types. In unsupervised learning, the feature vectors $\{\mathbf{x}_\alpha\}$ from one image or set of images are analyzed without considering further information such as the specific classification of tissue. The common principle in clustering is to partition the feature space by grouping similar feature vectors $\mathbf{x}_\alpha$ together, that is, to separate nonsimilar feature vectors into different groups. This partitioning itself is performed by using several different algorithmic approaches. For convenience, the reader who is not used to this mathematical principle of abstraction should consider the case of $m = 3$ outlined in the toy example above. More information on clustering algorithms can be found in Bishop (1997) and Hastie *et al.* (2001).

The different unsupervised approaches can be categorized into four approaches: agglomerate clustering, divisive clustering, vector quantization, and statistical dimension reduction and projection.

Agglomerative Clustering   At the start, each point is regarded as one single cluster, and the procedure is to merge a selected pair of clusters. However, it is first necessary to define a criterion with which to select the particular cluster pairs. All proposed criteria are based on measuring similarities between clusters (i.e., similarities between feature vectors). The two most similar clusters A and B are merged. This step is repeated until all points are merged into a suitable small number of clusters. The following criteria have been proposed for agglomerative clustering:

▲ **Single linkage**: Those two clusters A and B are merged that contain the pair of points showing the strongest similarity.

▲ **Complete linkage**: For all pairs of clusters, the most dissimilar pair of points is computed. Those two clusters A and B are merged that have the most similar of all the dissimilar pairs.

▲ **Average linkage**: Those two clusters A and B are merged that exhibit the minimum average pairwise distance.

▲ **Ward's clustering**: All possible pairs of clusters are evaluated in terms of the inner-cluster variance of the merged outcome cluster.

Divisive Clustering The divisive approach is, in principle, the same as the agglomerative one, but takes the opposite direction. At the beginning, all points are regarded as one large cluster that is then split iteratively into subclusters applying the same criteria as listed above.

Vector Quantization A set of n_c abstract feature vectors $\{\mathbf{u}_k\}$, $k = 1, \ldots n_c$ is used to represent the data structure (i.e., the clusters). The vectors are taken from the same space as the feature vectors $\{\mathbf{x}_\alpha\}$ but are initialized with some more or less randomly set values. They are usually termed prototype vectors (or reference vectors, codebook vectors). Using different algorithms, the prototype vectors $\{\mathbf{u}_k\}$ are moved iteratively into the centers of the data clusters. The most prominent algorithms are the k-means and the neural-gas (Martinetz *et al.*, 1993) algorithms.

A very special form of clustering is achieved by the self-organized map (SOM) (Kohonen, 2000). This links the prototype vectors to nodes in a two-dimensional grid. This grid can be visualized on a screen, and each node is displayed as a graphical symbol with features depending on the values of its prototype vector $\mathbf{u}_k$.

Figure 78 displays the SOM principle. A SOM is applied to the feature vectors extracted from the ROIs of 14 cases. Each voxel inside the ROIs is represented by a feature vector $\mathbf{x}$ containing all the $n_s = 6$ values of the signal time course; that is, $x_1 = s(1)$, $x_2 = s(2)$, . . ., $x_6 = s(6)$ (Fig. 78). The vectors are regarded as clouds of points in a six-dimensional space (Fig. 78, middle). During training, the prototype vectors $\mathbf{u}_k$ are moved iteratively into the cluster centers. After training is finished, each feature vector is associated with both its cluster center and its grid node (Fig. 78, right). The grid is then visualized by evaluating the clusters. Using a SOM, one of the present authors was able to show that the 6D feature space can be projected onto a 2D subspace spanned by the two axes of signal uptake and washout (Nattkemper and Wismueller, 2005). This low-dimensional representation can be used simultaneously to analyze the signal patterns of multiple cases and also for an alternative pseudocoloring.

Statistical Dimension Reduction and Projection To analyze the feature vectors from different ROIs, the m-dimensional feature vectors are projected into a lower dimensional space that is usually IR^2. Several techniques have been proposed for computing this projection. The common idea is to calculate it according to a statistical feature in the data. This strategy is motivated by the idea that the most important information will be preserved in the projection result.

Using the well-known principal component analysis (PCA), we can project signal data into a low-dimensional representation space, retaining the major fraction of variance of the original data. A nonlinear version of PCA, kernel PCA, has also been proposed to account for nonlinear dependencies (Schölkopf *et al.*, 1998). In independent component analysis (ICA), a set of source signals is computed that separates the input signals (i.e., the signal time courses) into statistically independent source signals. A more novel approach is local linear embedding (LLE), in which the mapping depends on local linear approximations and a global fitting of coefficients to account for global nonlinear structures (Varini *et al.*, 2004a).

Various studies have applied clustering and statistical dimension reduction algorithms to DCE-MRI. For example, principal component analysis (PCA) and its nonlinear extensions (kernel PCA) have been examined by Twellmann *et al.* (2004). Temporal kinetic signals of DCE-MRI sequences consisting of two precontrast and five postcontrast images are projected into a low-dimensional representation space and subsequently displayed as gray-value images, referred to as *fusion images* (Fig. 79). As well as reducing the number of images to be analyzed by the human observer, the fusion process also discloses regions exhibiting suspicious signal characteristics that cannot be distinguished from normal tissue in the original images. The potential advantages of analyzing DCE-MRI sequences with ICA (independent component analysis) were investigated by Yoo *et al.* (2002) and Meyer-Bäse *et al.* (2004). Nattkemper and Wismueller (2005), Wismüller *et al.* (2002), and Meyer-Bäse *et al.* (2004) examined the application of vector quantization algorithms for learning prototypes representing data clusters in the signal space. Wismüller *et al.* (2002) subsequently used these prototypes to perform an image segmentation. The authors successfully segmented the lesion mass from surrounding tissue and were able to find subdivisions of lesions in compartments with homogeneous signal characteristics. Chen *et al.* (2004) employed a vector quantization algorithm to segment lesion masses in order to facilitate acquisition of reliable averaged temporal kinetic signals of larger lesion compartments.

Even though unsupervised learning has proved to be of value for DCE-MRI data exploration, its potential is limited in clinical applications and patient management. Clinical patient care is more interested in an exact determination of tissue change. To this end, the tissue needs to be classified to predefined tissue categories representing the degree of

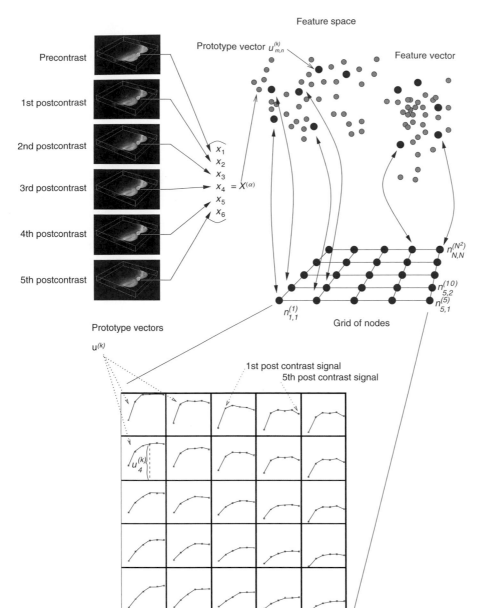

Feature space

Prototype vector $u_{m,n}^{(k)}$

Feature vector

Precontrast

1st postcontrast

2nd postcontrast

3rd postcontrast

4th postcontrast

5th postcontrast

x_1
x_2
x_3
$x_4 = X^{(\alpha)}$
x_5
x_6

$n_{N,N}^{(N^2)}$

$n_{5,2}^{(10)}$

$n_{5,1}^{(5)}$

$n_{1,1}^{(1)}$ Grid of nodes

Prototype vectors

$u^{(k)}$

1st post contrast signal
5th post contrast signal

$u_4^{(k)}$

Figure 78 The feature vectors from a set of DCE-MRI sequences are collected in a training set, and a self-organizing map of $N \times N$ nodes is applied. The construction of the feature vector is displayed upper left. In this application, the signal time curve is taken as feature vector. In the SOM algorithm, the cluster structure of the data is approximated by a set of prototype vectors $\mathbf{u}^k$ that are linked to nodes in a regular $N \times N$ grid (see upper right). After training, the prototype vectors can be visualized in a grid display. They represent the prototypes of the $N \times N$ clusters with changing uptake and washout characteristics.

malignancy. The next section will discuss the problem of feature computation and evaluation.

Human Experts versus Computer Algorithms

DCE-MRI employs both morphological and kinetic features for diagnostic decision making. By using the concept of difference or subtraction, suspicious tissue will usually highlight in the image. This is because tumor tissue has a different temporal rate of contrast agent uptake compared with normal tissue. Based on the bright appearance of the lesion in the difference image, kinetic and morphologic

features are employed in radiological diagnosis to differentiate benign from malignant lesions.

As already pointed out, DCE-MRI is very sensitive for breast cancer diagnosis. However, the specificity is compromised by the observation that not only malignant lesions but also several benign masses take up the contrast agent. Computer-aided diagnosis can increase the objectivity when using DCE-MRI by employing features generated by a computer algorithm. In the following description for breast tumors, the listed features are marked by (†) if they refer to malignancy and marked by (*) if they refer to benign lesions.

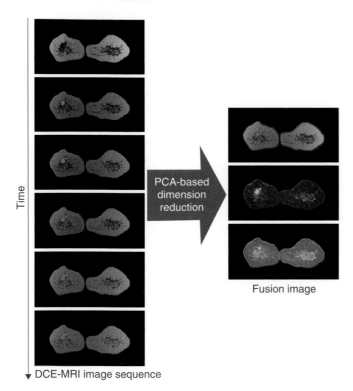

Time

DCE-MRI image sequence

PCA-based dimension reduction

Fusion image

Figure 79 Application of statistical dimension reduction techniques to the DCE-MRI data makes it possible to visualize image sequences through a reduced set of *fusion images*, revealing tissue regions exhibiting suspicious signal characteristics.

Features Determined by a Human Expert

Radiological reports contain observations from the image data presented in a descriptive language based on a set of diagnostic criteria (Abdolmaleki *et al.*, 1997). In clinical practice, the radiologist generally compares the time-intensity plot derived from a selected ROI (i.e., a region ideally contained fully within the suspicious tissue) with one or more reference plots taken from other, clearly distinguishable parts of the image (Heywang-Köbrunner, 1997).

Contrast Agent Uptake Descriptors   The often remarkably different dynamic response of enhanced lesions compared with normal tissue makes it possible to specify descriptive features, also known as contrast agent uptake descriptors (Abdolmaleki *et al.*, 1997; Brown *et al.*, 2000). These descriptors are then designed to represent the temporal behavior of the average value of the selected ROI. The resulting set of features comprises, for example, the *percentage enhancement at maximum signal intensity*, the *maximum-intensity time rate*, or the *duration of monotonic increase in signal intensity*. A high value on the measured percentage enhancement at maximum signal intensity and maximum-intensity time rate, as well as a short duration of

monotonic increase in signal intensity, are viewed as indicators for malignancy.

Topological Patterns   A further set of features can be computed by analyzing the spatial pattern of enhancement, that is, the enhancement homogeneity (Brown *et al.*, 2000). This investigates the degree of heterogeneity in the pattern of a previously defined ROI or the manually or automatically segmented tumor region during contrast agent uptake. As well as distinguishing only homogeneous from heterogeneous enhancement, this also includes more typical enhancing mechanisms such as centrifugal (*) and rim- or ring-like (†) enhancement.

Tumor Morphology   In addition, difference images clearly show the outlines of the enhancing structure, allowing a classification of the tumor morphology on the basis of enhancement (Abdolmaleki *et al.*, 1997; Brown *et al.*, 2000). In this respect, three classes of features can be distinguished: diffusivity of enhancement, type of margins observed, and shape of the mass (Viehweg *et al.*, 2000). The last named is essentially categorized into circularity (*) versus irregularity (†). With respect to the margin, a *lobulated* and well-defined morphology (*) contrasts a *spiculated* or *focal branching* one (†).

Computer-generated Features

The recent development of commercial digital MR imaging systems not only provides a significant improvement in image quality for traditional screening, but also translates into a wealth of information for the analysis and detection of mammographic features by computer. In contrast to descriptive radiological features related to the human visual system, computer-generated features present an alternative approach and can aid decision making (Russ, 1998; Sinha *et al.*, 1997). In general, however, these features mimic the radiological descriptors by reproducing desired information as a set of computed numbers (Gilhuijs and Giger, 1998; Sinha *et al.*, 1997).

Contrast agent uptake descriptors are often single measured values like the *area under the curve* or the *average baseline-to-peak gradient*. These quantities can be accessed directly utilizing a properly defined mathematical prescription consisting of one or more formulas.

Pharmacokinetic Models   As already mentioned, tissue-specific and biologically meaningful parameters can be extracted by fitting the time-series data to some kinetic model derived from physiological considerations. Depending on the modeling approach, specific parameters like vascularity or permeability can be measured.

Various entropy measures can be employed to quantify the observation that malignant lesions take up the contrast agent in a less homogeneous pattern than benign masses.

Shape Descriptors Frequently, standard techniques from the pattern recognition community are applied to compute shape descriptors (Russ, 1998). Different quantities exist here, including compactness or deformation measures (Gilhuijs and Giger, 1998). The diffusivity relates to the sharpness of a margin, measurable as the variance of the gradient crossing the boundary. Malignant lesions are assumed to have less sharp boundaries and are shaped more irregularly. A huge variety of formulas exist to express these observations through numbers (Gilhuijs and Giger, 1998). Features describing the degree of irregularity of the contour of a mass in general depend directly or indirectly on the *effective length* or the *fractal extent* of the border line. However, the measurement error of computer-generated features depends crucially on the accuracy of preprocessing steps such as denoising or segmentation (Gilhuijs and Giger, 1998).

Limitations of Computer-generated Features

Finally, the question arises as to how the features established in the sections Features Determined by a Human Expert and Computer-generated Features are related or can even be mapped. First, this raises the question of how accurately the computational features resemble their descriptive counterparts. For example, if one were to consider the computed slope factor of the contrast agent uptake curve as a possible replacement for the measured duration of monotonic increase, one would expect a superior and much more detailed description because it involves additional information on the curvature. However, interpretation of computed numbers depends crucially on the reliability of an algorithm and the errors in the values obtained. For example, measurement errors, comparable to radiological uncertainty, will weaken the computational analysis, whereas radiologists can balance this by drawing on their long experience.

The Semantic Gap The *semantic gap* is known as the lack of coincidence between the information one can extract from the digital data computationally and the interpretation of the same data by a human expert. Bridging the semantic gap between the simplicity of available visual features and the richness of user or expert semantics is the key issue in building effective content management systems. The latter are of increasing importance in medical data processing as more and more data become available and need to be accessed quickly. The weakness of current systems is that, while the user or expert seeks semantic similarity, the database can only provide similarity on data processing. Therefore, it would be desirable for expert knowledge to be encoded in a language of pixel-based data descriptors. Only this could permit a fast, probably hierarchical access to medical image and information databases.

Extended Feature Spaces and Preprocessing

The high-dimensional data space and the steadily increasing amount of data generated by modern imaging devices has resulted in a high number of degrees of freedom in the unsupervised learning procedure. Therefore, the application of clustering algorithms might benefit crucially from preprocessing the data. Thus, an important problem is to find efficient ways of compressing and encoding the relevant information contained in the data to facilitate its transmission to effective clustering algorithms.

Wavelet Features Wavelets can be used to transform the information in image space into scale space in which the data is decomposed according to the contribution from scales (Mallat, 1989). A variety of measures are available from signal analysis that allow a calculation of the relevant scales and even the relevant information on each scale (Walker, 1999).

Because wavelet transforms provide sparse representations of signals, one of their privileged areas of application has been coding and compression (Walker, 1999). Unlike other applications, however, the compression or reduction of biomedical image data raises delicate issues. Truncation as a preprocessing step must be performed with care in order to preserve all medically relevant information while eventually suppressing irrelevant features. Thus, research in this area must address these specific needs and pay great attention to the issue of validation.

Filtering A wavelet-based scale-space decomposition can be followed by a clustering procedure to construct efficient filtering techniques (Lessmann *et al.*, 2004). Recently, the wavelet-based processing scheme has been extended to investigate the feasibility of automatic tumor detection by receiver operating characteristics (ROCs) analysis utilizing subtraction images in a similar way to the radiological approach discussed earlier (Lessmann *et al.*, 2005). In general, sensitivity is affected by additional contrast enhancement within normal breast parenchyma or within the heart. However, applying the extended wavelet-filtering method has been shown to increase sensitivity for high specificity (Lessmann *et al.*, 2005).

Wavelets Beyond Filtering Depending on the clustering procedure applied, features are generated that capture the characteristics of the image data processed. When the data structure is rather complex, as is generally the case for DCE-MRI data sets, this should impact on the distinction between the different clusters formed in that the feature vectors representing the clusters will contain some overlap. There is a clear need to interpret the computed features because they are data-driven and lack a concrete interpretation. This would require a procedure for mapping the generated feature vectors to meaningful descriptors.

The Semantic Gap

User acceptance and adoption of the currently available automated content annotation systems is greatly impeded by the discontinuity between the simplicity of features or data content descriptions that can currently be computed automatically and the richness of semantics in user queries on medical image data search and retrieval. This problem will become even more complex, because radiological sciences will soon enter the molecular level, including the imaging of cellular processes. Here, descriptive features fail to characterize and differentiate the various patterns observed.

Monitoring Tumor Development

Tumor development is expressed by the changing extent of tissue affected by pathological disorders as well as by the changing structure of the tumors themselves. This is reflected in the fraction and the spatial distribution of malignant, benign, and normal tissue compartments in the heterogeneous tumor tissue. Whereas DCE-MRI provides valuable information for monitoring both aspects of tumor development, integrating and evaluating the much wider range is a challenging task for the radiologist. When performing an initial detection and segmentation of suspicious tissue masses, radiologists need to screen the entire image volume for newly developed lesions or for changes in the extent of previously identified tumors. The challenges arise from the huge numbers of voxels that have to be classified as either normal or suspicious on the basis of their temporal

kinetic signal. Even though the detailed examination of tumor composition requires exploration of only a limited number of temporal kinetic signals, their categorization as indicative for malignant, benign, or normal tissue demands a much more detailed analysis of their signal characteristics. This leads to a substantial demand for computer-aided detection and diagnosis tools that will facilitate analysis of the image data by, for example, providing a visual presentation of an automatic tissue classification based on the exposed temporal kinetic signals.

The task of classifying signal time courses into predefined tissue classes can be translated into a *supervised* learning problem. The basis for learning is a set of input-output pairs $(\mathbf{x}_\alpha, y_\alpha)$ for the to-be-learned classification function. The inputs $\mathbf{x}_\alpha$ are feature vectors representing the time course at a voxel position in a DCE-MRI volume (Fig. 80). The output values y_α represent the different predefined tissue classes.

The Registration Problem In case of considerable patient movement, it has to be taken into account by performing an appropriate registration. This means that the volumes of a single time series have to be spatially aligned. The image registration of contrast-enhanced images can be a nontrivial problem and is itself a subject of many studies. However, because an aligned data set is available in many cases, we shall focus our discussion on such data.

Expert Labels The output values are set to the class label provided by the radiologist or the physician. This can be, for example, $y_\alpha = 0$ for normal tissue when the signal

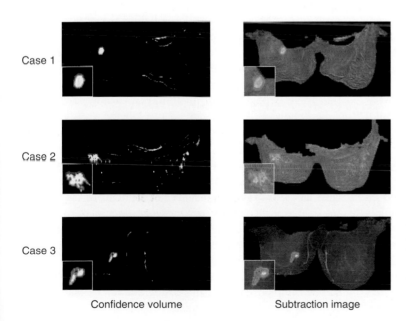

Case 1

Case 2

Case 3

Confidence volume   Subtraction image

Figure 80 Confidence volumes (left column) depicting the local probability of suspicious tissue permit the detection and segmentation of lesions (enclosed by green lines) by examining a single gray-value image per DCE-MRI sequence. In contrast to subtraction images (right column), the confidence volumes depict a lesion with high visual contrast to normal tissue and do not require any further manual tuning of the presentation by the radiologist.

time course $\mathbf{x}_\alpha$ is taken from a nontumor region of normal tissue. If the voxel is located inside a lesion, it can be set to $y_\alpha = 1$ for benign cases and $y_\alpha = 2$ for malignant ones. In this case, the class numbers are arbitrary, and their order does not encode specific information.

Training Sets Different strategies can be used to extract the training set from a set of n_t DCE-MRI volumes (representing n_t patients). The "classic" approach is to select an ROI in each lesion (sometimes the entire lesion can be regarded as an ROI). Then, one representative signal time course is computed for this ROI (e.g., by averaging all signal time courses) and is taken as the feature vector $\mathbf{x}_\alpha$ representing the signal time course characteristics. Hence, in this case, the index α denotes *one* entire lesion. Studies in this field vary mostly in the way the feature vector $\mathbf{x}$ is computed from the ROI.

The straightforward approach of computing average values from a whole-lesion ROI has the disadvantage that tumor tissue shows very heterogeneous time courses due to heterogeneous tissue states inside the lesion. For instance, a malignant lesion may contain flat signals rooting from necrose tumor tissue in the center. In our own studies, we have observed a vast amount of typical benign signals in histopathologically proven malignant tumors (Nattkemper and Wismueller, 2005). As a consequence, a whole-lesion ROI located in such a heterogeneous lesion can be mapped to a feature vector that shows no malignant characteristic. This results in a training set that contains lots of features that are labeled for a certain class but contain components that are totally nontypical for that class. Obviously, it is particularly difficult to learn a correct classification function from such noisy training sets (Nattkemper *et al.*, 2005).

A classification of features determined by radiological experts (see Features Determined by a Human Expert section) has also been tested in an attempt to find feature patterns associated with certain degrees of malignancy (Nattkemper *et al.*, 2005). Different classification architectures were tested: decision trees, *k*-nearest neighbor classifiers, and support vector machines (SVMs). Nonetheless, results clearly showed the limited potential of this approach, since most features showed no clear relation to tumor classes.

In the data-driven approaches, two different strategies have been proposed alongside the strategy of signal averaging. One subset of the proposed features is generated on the basis of observations by radiologists. These features describe the uptake, shape, and morphology, as described earlier. The problem with these values is that they can be heavily corrupted by the averaging process.

The second class of approaches computes texture features of the ROI. To this end, spatial gray-value statistics are computed using, for instance, gray level co-occurrence matrices (GLCMs) (Castellano *et al.*, 2004; Haralick *et al.*, 1973). This is usually motivated by compliance with the broad tradition in medicine of classifying tissue on the basis of its texture morphology. However, although some effort has been put into the computation of texture features, their contribution to improving classification seems limited. One reason may be the low spatial resolution in DCE-MRI that is still too coarse for a sufficiently good texture description.

A more recent setup applies learning algorithms in a different manner. The observed problems caused by tumor heterogeneity have led to a shift from classifying entire lesions to classifying the individual voxel signal time courses. Each voxel is mapped to a tissue class Ω_i (e.g., $\Omega_0 = malignant$, $\Omega_1 = normal$, $\Omega_2 = benign$). The result is displayed in a slice visualization and evaluated by an expert. This expert inspects the topological pattern of tissue classes and classifies the tissue state of the entire lesion.

Another option is to use learning algorithms to classify the lesion on the basis of computed tissue classes and their topological pattern. However, such an application would require a large set of labeled DCE-MRI data (at least 100 cases).

Supervised Learning Algorithms

In situations in which voxels have to be classified into two or more predetermined tissue categories Ω_i on the basis of their temporal signal characteristics, supervised learning algorithms are appropriate tools for data processing and for computing a meaningful and problem-related visualization of the image data. In supervised learning, a classification function

$$C : \mathbf{x} \rightarrow y(\mathbf{x})$$

maps feature vectors $\mathbf{x}$ describing temporal kinetic signals $\mathbf{s}$ to outputs $y(\mathbf{x})$, thereby reflecting the probability that $\mathbf{x}$ belongs to one of the predetermined signal classes Ω_i.

The classification function is learned in a data-driven fashion. Instead of formulating a suitable mathematical function C explicitly, this function is *learned* by an *artificial neural network* (ANN) from a set of labeled *training examples* $\Gamma = \{(\mathbf{x}_\alpha, y_\alpha)\}$, $\alpha = 1, \ldots, N$. To this end, the ANN correlates signal features given by $\mathbf{x}_i$ with user-provided class information reflected by the class labels $\mathbf{y}_i$. After this training process, the adapted ANN is able to predict the unknown class label for examples that are not elements of the training set Γ; that is, it is able to *generalize from seen to unseen data*. Particularly for medical applications, it is important to note that the ability of ANNs to generalize also holds when training examples and unseen signals relate to different groups of patients. Thus, training data for tuning the ANN can be sampled from the increasing amount of patient data stored in digital form in clinical databases. The tuned ANN can subsequently be applied to aid lesion detection and segmentation and quantification of tumor features in new DCE-MRI sequences.

Supervised Detection and Segmentation of Lesions

The need to classify tissue into predetermined classes becomes apparent when tumor masses have to be detected and segmented in order to localize newly grown tumor masses or to monitor changes in the extent of already known tissue disorders. A detailed delineation of the extent and form of affected tissue regions demands a voxel-by-voxel classification of the image data. For each voxel, it is necessary to decide on the basis of the temporal signal characteristics whether the represented tissue belongs to the class of normal tissue (Ω_0) or to the class of suspicious tissue (Ω_1)— that is, the class of tissue exhibiting benign or malignant signal characteristics.

As outlined in the first section, radiologists commonly make this binary decision after an inspection of subtraction images displaying the temporal signal gradient between two images recorded at different points in time (e.g., the gradient between one of the postcontrast images and the precontrast image). However, each subtraction image depicts only a fragment of the available temporal information, and it is frequently insuffcient because the temporal signal courses of benign and malignant tissue exhibit their peak intensity at different points in time. This makes it essential to compute several subtraction images based on different postcontrast images in order to obtain a presentation of all the different tumor compartments. Detection and segmentation of the tumor masses then requires simultaneous examination of the different subtraction images.

Recently, supervised artificial neural networks have been employed for computing more sophisticated visualizations of DCE-MRI data sets, enabling the radiologist to locate regions of suspicious tissue by means of examination of single 3D gray-value images, referred to as *confidence volumes* (Twellmann *et al.*, 2004). In a confidence volume, voxel intensities reflect the probability that the underlying tissue is affected by a pathological disorder. This assessment of the local tissue state is based on evaluation of the entire information provided in the signal **s** by a supervised ANN that has been trained to distinguish a suspicious temporal signal course from a normal one.

Instead of postulating signal models for the two tissue classes, the ANN is trained with the DCE-MRI data of 11 cases in which a radiologist has marked the location of suspicious tissue manually with a cursor after analyzing the data with clinical standard software. Temporal kinetic signals of voxels marked by an expert as lesion voxels constitute the part of the training set Γ that is representative for suspicious tissue. Signals not marked as lesion voxels by the expert serve as examples for normal tissue signals. After tuning the ANN with the set of labeled training examples Γ, the ANN is able to mimic the human expert's binary decision in the assessment of single temporal kinetic signals s. Each feature vector **x** describing a signal **s** is mapped to a confidence value reflecting the posteriori probability $P(\Omega_1|\mathbf{x})$ for a suspicious signal, and this determines the intensity value in the confidence volume.

Figure 79 shows three axial slices of the confidence volumes computed for three unseen DCE-MRI sequences (left column). One of the corresponding subtraction images is presented for each case in the right column. In contrast to the subtraction images displaying the lesions (marked by the green line) with low contrast to normal tissue and thus demanding further tuning of the visual presentation, lesion masses are depicted with high visual contrast to normal tissue in the confidence images. An average *area-under-the-ROC-curve value* (Az) of more than 0.98 shows how accurate the signal classification by the ANN is compared with the radiologist's lesion segmentation.

Supervised Classification of Lesions

Whereas the initial detection and segmentation of lesions can be formulated as a binary classification problem, the subsequent differential diagnosis of tumors requires a more detailed categorization of signal characteristics. Particularly when evaluating malignant lesions in which heterogeneous areas of enhancement are diagnostically important, radiologists have to be provided with visualizations of the image data that take into account the heterogeneity of tumor vascular characteristics (Collins and Padhani, 2004). The visual depiction of heterogeneity may be improved by visualizations in which pseudocolors of voxels reflect a more detailed classification of the local tissue by using a supervised learning algorithm. Such *voxel-mapping* approaches enable radiologists to quantify the amount and the spatial distribution of normal and, in particular, of different types of suspicious tissue in specific tissue regions.

Twellmann *et al.* (2005) extended the binary classification setup employed for the initial detection of suspicious lesions to a multiclass setup, classifying image voxels **p** into the three tissue classes: malignant (Ω_0), normal (Ω_1), and benign (Ω_2). They employed a supervised learning architecture to derive the classification function $C : \mathbf{x}$ $y(\mathbf{x})$ from labeled training data. The training data was sampled from a set of DCE-MRI sequences in which lesions were segmented manually by a radiologist. The entire lesions were labeled either as malignant or benign according to the outcome of a histological examination. The classification function maps each temporal kinetic signal associated with an image voxel to a three-dimensional output vector $y(\mathbf{x})$. The components of $y(\mathbf{x})$ reflect the posteriori probabilities $P(\Omega_0|\mathbf{x})$, $P(\Omega_1|\mathbf{x})$, and $P(\Omega_2|\mathbf{x})$—that is, the probability that the evaluated feature vector $\mathbf{x}$ relates to a temporal kinetic signal $\mathbf{s}$ caused by malignant, normal, or benign tissue, respectively. The trained predictor is applied for a voxel-by-voxel evaluation of unseen image data, supporting radiologists in two aspects of lesion analysis and monitoring.

Quantification of Malignant, Normal, and Benign Tissue Fractions

The fraction of malignant, normal, and benign tumor signals in an ROI (e.g., a whole-lesion ROI) can be quantified by

using the maximum posteriori classification rule to determine the most probable tissue class $\Omega_*(\mathbf{p})$ for each voxel $\mathbf{p}$

$$\Omega_*(p) = \operatorname*{argmax}_{k = 0, \ldots, 2}\left[P(\Omega_k|x_p) \right]$$

This makes it possible to quantify the number of image voxels exposing malignant, normal, and benign signals and to relate them to the total number of voxels, providing an objective assessment of the regression or progression of different tissue classes in the tissue mass in question.

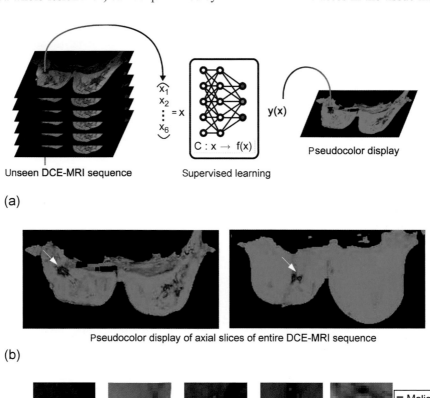

(a)

(b)

(c)

Figure 81 (a) After training with labeled examples, the adapted ANN predicts the probabilities of a malignant, normal, and benign state of the local tissue for each voxel $\mathbf{p}$ of an unseen sequence on the basis of the temporal signal $\mathbf{s_p}$. The outcome $y(\mathbf{x(p)})$ is displayed as a pseudocolor in a new image. (b) The pseudocolor depiction of entire sequences enables radiologists to localize regions affected by malignant (red) or benign (blue) disorders in the normal breast tissue (green) by examining a single image. (c) The ANN-based assessment of lesions tissue coincides broadly with the 3TP-based analysis. However, the data-driven approach based on ANNs does not require radiologists to enunciate their medical expertise in the form of a mathematical model.

Visualization of Tissue Heterogeneity

Applying the trained predictor to the data of ROIs or entire sequences makes it possible to visualize the heterogeneity of the tissue vascularity expressed by spatially varying temporal dynamics of tissue. The local posteriori probabilities $P(\Omega_0|\mathbf{x}_p)$, $P(\Omega_1|\mathbf{x}_p)$, and $P(\Omega_2|\mathbf{x}_p)$ for the three tissue states given by the ANN output $\mathbf{y}(\mathbf{x}_p)$ are depicted as a pseudocolor by mapping the posteriori probability values to the red, green, and blue component of an RGB color. This color is depicted at the position $\mathbf{p}$ in a new three-dimensional pseudocolor image (Fig. 81a). Bright red, green, or blue voxels indicate high probabilities of malignant, normal, or benign tissue states, respectively. Mixtures of the three basic colors suggest less distinct signal characteristics. The pseudocolor display of the image data enables radiologists to perceive the spatial variation of tissue vascularity, providing information for analyzing the lesion architecture and for avoiding any averaging of temporal kinetic signals of different tissue types while analyzing the temporal dynamic of larger areas of lesion tissue.

Figure 81 presents examples for pseudocolor visualizations based on a supervised ANN. The ANN is trained with labeled signal data sampled from 11 sequences and subsequently applied to evaluate an unseen sequence. Figure 81b depicts the pseudocolor visualizations of axial slices of two different DCE-MRI sequences. The lesions (marked by white arrows) are displayed with shadings of red and blue indicating malignant and benign signal characteristics, and can be distinguished clearly from the surrounding normal tissue (green). Figure 81c shows zoomed views of two benign and three malignant lesions. Only lesion voxels marked by an expert are displayed with pseudocolors. In order to illustrate that the ANN-based pseudocoloring provides a reasonable assessment of the lesion voxels, 3TP-based visualizations are depicted below. Comparison of the results of the two techniques (Fig. 81) underlines that the ANN-based approach leads to a similar assessment of the lesion tissue, though trained in a data-driven fashion without any explicit model assumption about the signal.

Summary and Outlook

The main potential of machine-learning applications is the classification of image signals to tissue categories on the basis of a previously collected, labeled set of images with correctly assigned categories. The strength of machine-learning solutions is based on two specific features.

First, the tuning of the classification function is based on labeled training sets. When labeling the training set, the user (a radiologist or a physician) applies his or her primary visual expertise together with the histopathological outcome.

As a result, the algorithm is adapted on the highest possible level of accuracy and expertise because users are applying their original skills.

Second, the machine-learning algorithm has a high degree of noise resistance. This is important, because signals can be corrupted by patient movement, registration errors, or error-prone manual ROI selection. Even though these artifacts can generally not be corrected, they do not significantly interfere with the later training process.

In contrast to learning-based approaches, the model-based approaches for tissue characterization and visualization are based on strict assumptions about the influence of the contrast agent on the tissue signal under a certain imaging protocol. Hence, these approaches have a certain degree of visual transparency, meaning that the user can map the colors to physically defined properties. One disadvantage is that only a small number of image parameters can be considered or the model will become too complex. A higher level of complexity reduces the benefits of easy interpretation through transparency. However, one future trend in medical imaging seems to be an increase in signal dimension (Daisne et al., 2003; Duncan and Ayache, 2000; Loening and Gambhir, 2003; Nattkemper, 2004; Ratib, 2004) and it is questionable whether low-dimensional models like 3TP can be defined on the basis of such data. We believe that it will be worth considering the machine-learning applications presented in this chapter when searching for the hidden regularities in the future image data of medical diagnosis.

Acknowledgments

We thank Oliver Lange and Axel Wismüller for providing the data in our studies. We also want to thank Axel Saalbach for computational support.

References

Abdolmaleki, P., Buadu, L., Naderimansh, H., Murayama, S., Murakami, J., Hashiguchi, N., Yabuuchi, H., and Masuda, K. 1997. Neural network analysis of breast cancer from MRI findings. *Radiat. Med.* 15:283–293.

Bishop, C.M. 1997. *Neural Networks for Pattern Recognition.* Oxford, UK: Oxford University Press.

Brown, J., Buckley, D., Coulthard, A., Dixon, A., Dixon, J., Easton, D., Eeles, R., Evans, D., Gilbert, F., Graves, M., Hayes, C., Jenkins, J., Jones, A., Keevil, S., Leach, M.O., Liney, G.P., Moss, S., Padham, A., Parker, G., L.J., P., Ponder, B., Redpath, T., Sloane, J., Turnbull, L., Walker, L., and Warren, R. 2000. Magnetic resonance imaging screening in women at genetic risk of breast cancer: imaging and analysis protocol for the U.K. multicentre study. U.K. MRI breast screening study advisory group. *Magn. Reson. Imaging 18*:765–776.

Castellano, G., Bonilha, L., Li, L.M., and Cendes, F. 2004. Texture analysis of medical images. *Clin. Radiol. 59*:1061–1069.

Chen, W., Giger, M.L., and Bick, U. 2004. Automated identification of temporal pattern with high initial enhancement in dynamic MR lesions using fuzzy c-means algorithm. In: Fitzpatrick, J.M., and Sonka, M. (Eds.), *Proceedings of SPIE 2004 5370*:607–611.

Collins, D.J., and Padhani, A.R. 2004. Dynamic magnetic resonance imaging of tumor perfusion. *IEEE Eng. Med. Biol.* 23:65–83.

Daisne, J., Sibomana, M., Bol, A., Cosnard, G., Lonneux, M., and Gregoire, V. 2003. Evaluation of a multimodality image (CT, MRI and PET) coregistration procedure on phantom and head and neck cancer patients: accuracy, reproducibility and consistency. *Radiother. Oncol.* 69:237–245.

Duncan, J.S., and Ayache, N. 2000. Medical image analysis: progress over two decades and the challenges ahead. *IEEE Trans. PAMI* 22: 85–105.

Gilhuijs, K., and Giger, M. 1998. Computerized analysis of breast lesions in three dimensions using dynamic magnetic-resonance imaging. *Med. Phys.* 25:1647–1654.

Haralick, R.M., Shanmugam, K., and Dinstein, I. 1973. Textural features for image classification. *IEEE Transactions on Systems, Man and Cybernetics* 3:610–621.

Harms, S., Flamig, D., Hesley, K., Meiches, M., Jensen, R., Evans, W., Savino, D., and Wells, R. 1993. MR imaging of the breast with rotating delivery of excitation off resonance: clinical experience with pathologic correlation. *Radiology* 187:493–501.

Hastie, T., Tibshirani, R., and Friedman, J.H. 2001. *The Elements of Statistical Learning: Data Mining, Inference, and Prediction*. New York: Springer.

Heiberg, E., Perman, W., Herrmann, V., and Janney, C. 1996. Dynamic sequential 3D gadolinium-enhanced MRI of the whole breast. *Magn. Reson. Imaging* 14:337–348.

Heywang-Kobrunner, S.H. 1997. *Contrast-Enhanced MRI of the Breast*. Berlin: Springer.

Heywang, S.H., Hahn, D., Schmidt, H., Krischke, I., Eiermann, W., Bassermann, R., and Lissner, J. 1986. MR imaging of the breast using gadolinium-dtpa. *J. Comp. Ass. Tomography* 10:199–204.

Heywang-Kobrunner, S.H., Viehweg, P., Heinig, A., and Kuchler, C. 1997. Contrast-enhanced MRI of the breast: accuracy, value, controversies, solutions. *Eur. J. Radiol.* 24:94–108.

Knowles, A., Gibbs, P., and Turnbull, L. 2000. Improved classification of breast DCE-MRI using a neural network ensemble. In *Proc. Int. Soc. Mag. Reson. Med. (ISMRM)*, no. 2163.

Kohonen, T. 2000. *Self-Organizing Maps*, Volume 30 of *Series in Information Sciences*. Berlin: Springer.

Kuhl, C., Kreft, B., Bieling, H., Sommer, T., Lutterbey, G., Gieseke, J., and Schild, H. 1997. Dynamic MRI in premenopausal healthy volunteers: normal values of contrast enhancement and cycle phase dependency. *Radiology* 203:137–144.

Lessmann, B., Nattkemper, T., Degenhard, A., Pointon, L., Kessar, P., Khazen, M., and Leach, M.O. 2005. SOM-based wavelet filtering for the exploration of medical images. In *Lecture Notes in Computer Science 3696*: 671–676. Proc. of 15th ICANN. Berlin: Springer.

Lessmann, B., Twellmann, T., Degenhard, A., Nattkemper, T.W., and Leach, M.O. 2004. Wavelet features for improved tumour detection in DCE-MRI. In *Proceedings of Medical Image Understanding and Analysis (MIUA)*, London, UK.

Liu, P.F., Debatin, J.F., Caduff, R.F., Kacl, G., Garzoli, E., and Krestin, G.P. 1998. Improved diagnostic accuracy in dynamic contrast enhanced MRI of the breast by combined quantitative and qualitative analysis. *Br. J. Radiol.* 71:501–509.

Loening, A.M., and Gambhir, S.S. 2003. Amide: a free software tool for multimodality medical image analysis. *Mol. Imaging* 2:131–137.

Lucht, R., Knopp, M., and Brix, G. 2000. Neural network-based classification of signal-time curves obtained from dynamic MR mammographic image series. In *Proc. Intl. Soc. Mag. Reson. Med. (ISMRM)*, Vol. 8.

Lucht, R., Knopp, M., and Brix, G. 2001, January. Classification of signal-time curves from dynamic MR mammography by neural networks. *Magn. Reson. Imaging* 19:51–57.

Mallat, S. 1989. A theory for multiresolution signal decomposition: the wavelet representation. *IEEE Trans. Patt. Anal. Mach. Int.* 11:674–693.

Martinetz, T.M., Berkovich, S.G., and Schulten, K.J. 1993. "Neural-Gas" network for vector quantization and its application to time-series prediction. *IEEE Transactions on Neural Networks* 4:558–569.

Meyer-Bäse, A., Wismüller, A., Lange, O., and Leinsinger, G. 2004. Computer-aided diagnosis in breast MRI based on unsupervised clustering techniques. In *Intelligent Computing: Theory and Applications II. Proceedings of the SPIE, 5421*: 29–37.

Müller, H., Michoux, N., Bandon, D., and Geissbuhler, A. 2004. A review of content-based image retrieval systems in medicine—clinical benefits and future directions. *Int. J. Med. Inform.* 73:1–23.

Nattkemper, T.W. 2004. Multivariate image analysis in biomedicine: a methodological review. *J. Biomed. Inform.* 37:380–391.

Nattkemper, T.W., Arnrich, B., Lichte, O., Timm, W., Degenhard, A., Pointon, L., Hayes, C., Leach, M.O., and U.K. MARIBS Breast Screening Study. 2005. Evaluation of radiological features for breast tumour classification in clinical screening with machine learning methods. *Artif. Intell. Med.* 32:29–139.

Nattkemper, T.W., and Wismueller, A. 2005. Tumour feature analysis with unsupervised machine learning. *Med. Image Anal.* 9:344–351.

Nunes, L., Schnall, M., Siegelman, E., Langlotz, C., Orel, S., Sullivan, D., Muenz, L., Reynolds, C., and Torosian, M. 1997. Diagnostic performance characteristics of architectural features revealed by high spatial-resolution MR imaging of the breast. *AJR Am. J. Roentgenol.* 169:409–415.

Ratib, O. 2004. PET/CT image navigation and communication. *J. Nucl. Med. 45* (Suppl. 1):46–55.

Russ, J.C. 1998. *The Image Processing Handbook*. New York: CRC Press.

Schnall, M.D. 2001. Application of magnetic resonance imaging to early detection of breast cancer. *Breast Cancer Res.* 3:17–21.

Schölkopf, B., Smola, A.J., and Müller, K.R. 1998. Nonlinear component analysis as a kernel eigenvalue problem. *Neural Comput.* 10:1299–1319.

Sinha, S., Lucas-Quesada, F., DeBruhl, N., Sayre, J., Farria, D., Gorczyca, D., and Bassett, L.W. 1997. Multifeature analysis of Gd-enhanced MR images of breast lesions. *J. Magn. Reson. Imag.* 7:1016–1026.

Twellmann, T., Lichte, O., and Nattkemper, T.W. 2005. An adaptive tissue characterization network for model-free visualization of dynamic-contrast enhanced magnetic resonance image data. *IEEE Trans. Med. Imaging* 24:1256–1266.

Twellmann, T., Saalbach, A., Gerstung, O., Leach, M.O., and Nattkemper, T.W. 2004a. Image fusion for dynamic contrast enhanced magnetic resonance imaging. *Biomed. Eng. OnLine* 3:35–37.

Twellmann, T., Saalbach, A., Müller, C., Nattkemper, T.W., and Wismüller, A. 2004b. Detection of suspicious lesions in dynamic contrast-enhanced MRI data. In *Proceedings of EMBC 2004*, 26th Annual Int. Conf. of the IEEE Engineering in Medicine and Biology Society, IEEE Press.

Varini, C., Nattkemper, T.W., Degenhard, A., and Wismüller, A. 2004a, July. Breast MRI data analysis by LLE. In *Proc. of Int. Joint Conf. on Neural Networks*, Budapest, Hungary, pp. 2449–2554.

Varini, C., Nattkemper, T.W., Degenhard, A., and Wismüller, A. 2004b, September. Visualisation of breast tumour DCE-MRI data using LLE. In *Proc. Medical Image Understanding and Analysis (MIUA)*, pp. 97–100.

Viehweg, P., Lampe, D., Buchmann, J., and Heywang-Köbrunner, S. 2000. *In situ* and minimally invasive breast cancer: Morphologic and kinetic features on contrast-enhanced MR imaging. *Magn. Reson. Mat. Phys. Biol. Med.* 11:129–137.

Walker, J. 1999. *A Primer on Wavelets and Their Scientific Applications.* New York: CRC Press.

Weinstein, D., Strano, S., Cohen, P., Fields, S., Gomori, J., and Degani, H. 1999. Breast fibroadenoma: mapping of pathophysiologic features with three-time-point, contrast-enhanced MR imaging–pilot study. *Radiology 210*:233–240.

Wismüller, A., Lange, O., Dersch, D., Leinsinger, G., Hahn, K., Ptz, B., and Auer, D. 2002. Cluster analysis of biomedical image time series. *Int. J. Comput. Vision 46*:103–128.

Yoo, S.S., Choi, B.G., Han, J.Y., and Kim, H.H. 2002. Independent component analysis for the examination of dynamic contrast-enhanced breast magnetic resonance imaging data. *Invest. Radiol. 37*:647–654.

3

Mass Detection Scheme for Digitized Mammography

Bin Zheng

Introduction

Development of computer-aided detection (CAD) schemes for mammography has been attracting extensive research interest for the last two decades. Many independent CAD schemes for the detection of microcalcification clusters and masses have been developed and reported. Currently, commercial CAD products (systems) have been routinely used in a large number of medical institutions around the world to assist radiologists in detecting suspected microcalcification clusters and masses depicted on screening mammograms. In general, CAD systems have successfully helped radiologists to more efficiently search for microcalcification clusters and detect more subtle cancers associated with malignant microcalcifications. However, in the busy clinical practice, radiologists rely less on (or sometimes ignore) the CAD-cued masses, due to the relatively lower performance of current CAD schemes on mass detection. In this chapter, we review and discuss the basic principles (architecture) of CAD schemes for mass detection, the current status of using commercial CAD systems, the technical challenges and issues of further improving CAD performance, and the latest development and research effort in an attempt to search for optimal approaches to use CAD systems in the clinical environ-ments (in particular to increase radiologists' confidence in CAD results of mass detection).

Basic Architecture of Mass Detection Schemes

Computer-aided Detection Schemes Based on a Single Image

Single-image-based CAD schemes are the most commonly reported schemes in mass detection. In these schemes, images acquired from the same patient (e.g., four mammograms acquired from craniocaudal and mediolateral oblique views of the left and right breasts) are processed separately and analyzed. Only the information (or features) extracted and computed from one image is used to identify positive and negative mass regions depicted on images. Although the original mammograms are typically digitized using high-resolution film digitizers (i.e., Lumisys digitizers made by Eastern Kodak Company, Rochester, NY, and Howtek digitizers made by iCAD Inc., in Nashua, NH) with a pixel size of $\leq 50\ \mu m \times 50\ \mu m$, the digitized images used in mass detection schemes are typically subsampled images with the pixel size of approximately $400\ \mu m \times 400\ \mu m$. The mass detection schemes usually involve three processing stages. The first stage applies an image-filtering approach to search for suspected mass regions; the second stage uses a region growth algorithm to define the boundary contour of each initially detected region; and the third stage applies a feature-based machine learning classifier to identify true-positive and false-positive mass regions.

Image Segmentation and Identification of Suspected Mass Locations (Stage One)

The first stage of CAD schemes applies image filtering, subtraction, and segmentation methods to identify the pixels (or areas) associated with suspected mass regions. Because asymmetric patterns in the bilateral images (the same-view mammograms of the left and right breast) are the most visually important symptom used by radiologists in searching for suspected mass regions, the bilateral image subtraction method is a common approach to detect an initially suspicious mass at the early stage of CAD development. Through this approach, bilateral images are registered and subtracted. As a result, symmetric breast tissue patterns of two images are largely erased (deleted) and only asymmetric areas remain in the subtracted image. However, because of morphologic feature differences between two breasts, variation in breast compression, and skewing of the three-dimensional breast relative to the X-ray image projection, it is extremely difficult for mammograms to be placed in registration with sufficient accuracy for subtraction to be performed without generating substantial artifacts. Hence, a high false-positive detection rate of over 50 suspicious areas per image was reported in the early CAD schemes.

Before the early 1990s, great research effort had been made to improve the bilateral image subtraction methods. Instead of image registration through translation and rotation, the more sophisticated nonlinear bilateral subtraction method combined with various feature-extraction techniques was applied to improve image registration and reduce the number of initially suspicious regions (Yin *et al.*, 1994). These transformations require either manually or automated identified landmarks, but there are few appropriate features in mammograms. In addition, it is necessary to adjust the gray scale of the image after image registration so as to optimize the difference image (Mendez *et al.*, 1998). The adjustment methods usually include the variance equalization of pixel-value distribution and global or local histogram equalization. Although local equalization can significantly reduce the magnitude of the difference image, it risks eliminating local difference that may be associated with true-positive masses.

Because of these limitations, bilateral image subtraction methods are no longer used in current CAD schemes. As a result, a number of single-image-based processing methods have been developed and tested in the CAD scheme to search for the initially suspected mass regions. These include but are not limited to (1) a dual Gaussian filtering method using different kernel sizes followed by image subtraction (Zheng *et al.*, 1995), (2) a density-weighted contrast-enhancement (DWCE) segmentation method (Petrick *et al.*, 1996), and (3) a hybrid method combining the detection of radiating patterns of linear spicules and local "bright" pixels (te Brake *et al.*, 1998). The first stage of CAD schemes typically detects a large number of initially suspected mass regions (i.e., 10–30), in particular for the images of dense breasts.

Identification of Mass Boundary Using the Region Growth Algorithm (Stage Two)

The second stage of the CAD schemes uses a region growth and labeling algorithm to identify the boundary contour of each initially detected suspected mass region and compute a set of features to represent the region. A multi-layer topographic region growth algorithm was developed and applied in the second stage of CAD schemes (Zheng *et al.*, 1995). In each initially detected suspected region, the algorithm searches for a pixel with local minimum digital value (I_0) and uses it as a region growth seed. The algorithm also computes the average digital value of all pixels located inside the initially detected suspected region ($\bar{I}_R$). A growth threshold for the first topographic layer is determined as $T_1 = \alpha(\bar{I}_R - I_0)$, where α is a constant (e.g., $\alpha = 0.15$). Hence, the growth threshold is adaptively adjusted based on the digital value distribution of pixels inside each suspected region. After growing the first topographic layer, the algorithm computes following three features: (1) region size of this growth layer ($A = S \times N_{GR}$), which is computed by counting the number of pixels located inside the growth region (N_{GR}) and multiplying the size of each pixel (e.g., $S = 0.16$ mm^2 assuming the pixel size of 400 μm × 400 μm); (2) region circularity $\left(C = \dfrac{N_{GR} \cap N_C}{N_{GR}}\right)$, which is defined as the number of pixels located inside both the growth region (N_{GR}) and the equivalent circle (N_C) divided by the number of pixels inside the growth region alone; and (3) region shape factor $\left(F_8 = \dfrac{P^2}{A}\right)$, where P and A were the perimeter (mm) and size (mm^2) of the growth region, respectively. The algorithm then computes the average digital values of pixels inside the first growth layer ($\bar{I}_1$) and pixels inside the surrounding background of the growth region ($\bar{I}_B$). The background window is defined as 10 mm (e.g., 25 pixels for the subsampled images with pixel size of 400 μm × 400 μm) away from the boundary (contour) of the growth region in all four directions. The growth threshold for the second topographic layer is $T_2 = \alpha(\bar{I}_B - \bar{I}_1)$. After completing the second growth layer, the algorithm computes the same three features as the first growth layer and the change of each pair of two features computed between two growth layers. For example, the size growth ratio was computed as $G = \dfrac{A_2}{A_1}$. A set of classification rules based on these three features and their changes is applied to eliminate false-positive regions and decide whether the region growth continues using a new adaptively determined threshold ($T_i = \alpha(\bar{I}_{Bi} - \bar{I}_i), i = 2, ..., n$) for the next layer. This region growth process is iteratively performed until terminated by the classification rule. Once the final boundary contour of a suspected mass region is defined, a set of morphological, pixel-value distribution, and texture features is computed to represent or describe this mass region.

Since computed image features are used to classify true-positive and false-positive mass regions, accurately detecting boundary contours of suspected mass regions plays an important role in improving CAD performance. Although a multilayer topographic region growth algorithm has been quite successful in defining boundary contour of "typical" suspected mass regions, it has not been proved as an "optimal" or a robust region growth method to segment a fraction of mass regions that either have high margin spiculation or are partially obscured by other dense breast tissue.

To improve accuracy in defining boundary contour of mass regions, a number of other region growth algorithms based on either a Gaussian model or a dynamic contour model have been developed and tested (te Brake and Karssemeijer, 2001). If a suspected mass region is partially obscured by fibroglandular tissue or by "connection" to other low-density areas, it is difficult for a traditional region growth algorithm to detect end points along many radial growth rays. To stop (avoid) the growth region merging into the tissue background, the region can be first multiplied by a Gaussian function centered at the growth seed (x_0, y_0). As a result, each pixel (x, y) has a new digital value, $h(x, y) = f(x, y) N(x, y, x_0, y_0, \sigma^2)$. Using this approach, we see that the distant pixels are typically suppressed and the growth region becomes more circular (and "smooth"); hence, it does not easily "expand" into tissue background. This helps a region growth algorithm find correct end points of a mass region before entering the surrounding normal dense tissue. However, applying this method, it is important for a user to select "optimal" filter width (or σ of Gaussian function) that depends on prior knowledge of the estimated size of a mass region. When partially obscured mass has low circularity, the growth method based on the Gaussian function is likely to fail to find the region boundary.

To solve this problem, a region growth algorithm based on a discrete dynamic contour (snake) model has been tested and applied to segment boundary contours of mass regions. The discrete contour is a deformable curve whose shape is controlled by both internal and external force. The internal force imposes a smoothness constraint on the contour, and the external force determined by the image gradient magnitude moves the vertices to locations in the image with strong gradients, such as the edge of the mass region. The ratio between internal and external force needs to be balanced with weight parameters. In general, this contour model is quite sensitive to the image noise or is easily trapped by the local minimum. The success of this method depends heavily on the selection of an initial contour (start seeds).

A hybrid method that combines both a multilayer topographic region growth algorithm and either a model based on the Gaussian function or a discrete contour model may improve the results in defining mass boundary contours. In this hybrid method, the last successfully generated topographic growth layer plays an important role. Using a Gaussian model-based region growth algorithm, the advantage of this hybrid approach is that neither empirical selection nor complicated prediction of the mass size is required. The initial estimated size is automatically generated by the topographic region growth algorithm. For example, if a suspicious mass passes through the second topographic growth layer (with a size of A_2 in this layer) but fails in the third growth layer, the size of the mass region can be initially estimated as $A_T = A_2 \times G_{1-2}$, where G_{1-2} is the growth ratio between the first and the second growth layers. By using this value as the size of the initial estimated region, the filter width (σ^2 of Gaussian function) can be automatically determined. In the same way, the boundary contour of the final topographic growth layer can also be used as the initial contour of a mass region when the discrete contour model is applied. Because this initial contour is reasonably close to the actual mass boundary, it helps avoid many other local "traps" inside the mass region and speeds up the convergence process of the discrete contour algorithm. Once an initial contour is selected, the discrete contour algorithm can iteratively search for the "optimal" deformed contour until the energy due to the internal and external forces is minimized. Unfortunately, because of the diversity of breast mass regions and their surrounding tissue patterns, neither model (a Gaussian or a discrete contour) has proven to be superior to the other, nor optimal. Accurately defining mass region boundary and computing margin-related features (e.g., level of spiculation) is still a major technical challenge in developing CAD schemes for mass detection.

Classification of Suspected Mass Regions (Stage Three)

The third stage of CAD schemes uses a variety of feature-based machine-learning classifiers to classify true-positive and false-positive mass regions. Many different types of classifiers, such as artificial neural networks (ANNs), Bayesian belief networks (BBNs), decision trees, support vector machines (SVMs), and knowledge-based expert systems, have been developed in an attempt to better classify true-positive and false-positive mass regions. Experimental results from a large number of independent studies indicate that, although a large effort has been made to optimize these classifiers, the performance of these classifiers actually converges to a very similar level. As a result, improvement in CAD performance and robustness might be more dependent on feature selection, database diversity, and size of training samples than on any particular machine-learning paradigm used in developing such schemes.

Computer-aided detection schemes usually extract and compute a large number of image features (including intensity-based, geometrical, morphological, fractal dimension, and texture features). Each feature contains both information (signal) and noise. A large fraction of the computed features are highly correlated (or redundant). The redundant features used in the classifiers make very little contribution to incorporate more information but add noise to the classifiers. Based on statistic theory, too many free parameters result in overfitting in a supervised machine-learning classifier. A curve fitted with too many parameters follows all small details or noise but is very poor for interpolation and extrapolation. Thus, one of the most critical problems in the design and optimization of a machine learning classifier is to select a set of optimal features and an appropriate classifier size for a given application (Bebis *et al.*, 1997). An optimal feature set should meet the following criteria:

1. Large interclass mean distance (discrimination): features in different classes should have significantly different values.

2. Small intraclass variance (reliability): features should have similar values for objects of the same class.

3. Low correlation with other features (independence): features should not be strongly correlated to each other.

4. Insensitivity to extraneous variables (little signal-to-noise ratio dependency): change of other features should not significantly affect the function of this feature in the classifier.

Because many computed features cannot be visually examined or meaningfully explained by human observers, it is impossible to directly or visually select a set of optimal (highly performed) features. Hence, several computerized methods have been applied for feature selection. The goal of these optimization methods is to assemble a small set of features that can generate the best classification performance evaluated by the maximum area under an ROC (receiver operating characteristic) curve (A_z value). The stepwise searching method and genetic algorithm are two popular feature selection methods used in CAD development.

The stepwise method is a well-established statistical approach to search for the features that can enhance the performance of a classifier. The first step search in this method is to define a small number of features as the initial feature model. Many methods may be used to define the initial feature model. For example, one research group reported using a professional statistical program (SPSS for Windows, SPSS, Chicago, IL) to select features (Chan *et al.*, 1995). Since there are only two classes (true-positive and false-

positive mass regions) in this problem, the program calculates the Wilks' lambda values between the two classes when each feature is used individually. The feature that produces the smallest Wilks' lambda is selected into the search model first. Once the initial feature model is selected, the number of features selected in the following steps by the stepwise method is controlled by two parameters, called F-to-enter and F-to-remove. The feature entry step and the feature removal step are alternately performed in the stepwise method. In a feature entry step, each of the features not yet in the model is selected and added into the model one at a time. The Wilks' lambda in each new feature model is then tested based on F statistics. The feature that provides the smallest Wilks' lambda (or the most significant performance improvement) formally enters the next feature model if the F-to-enter value is larger than the F-to-enter threshold. In the feature removal step, a new set of testing is performed to evaluate the performance of the classifier by removing each of the features inside the feature model one at a time. If after removing one feature from the model, there is no significant change in the performance of the classier (or the F-to-remove value is smaller than the F-to-remove threshold), this feature is permanently removed from the feature model. The procedure of stepwise feature selection method is terminated until (1) the F-to-enter and F-to-remove values of all selected features in the model are greater than the predetermined thresholds, and (2) the F-to-enter and F-to-remove values of all unselected features are smaller than the thresholds.

It is clear that the results of feature selection using the stepwise method depend on the optimization of F-to-enter and F-to-remove parameters and predetermined thresholds. The optimal values of F-to-enter and F-to-remove thresholds are often not known in advance. Thus, one has to experiment with these parameters and increase or decrease the number of selected features inside the initial model to obtain the optimal performance.

The genetic algorithm (GA) is another popular method used to select not only optimal features but also the topology of a classifier (i.e., the number of hidden neurons in an artificial neural network). It has been applied to optimize many different types of machine-learning classifiers, including the artificial neural network (ANN), Bayesian belief network (BBN), and distance-weighted k-nearest neighbor (KNN) algorithms. The fundamental principle of GA is based on natural selection (only the strongest survive strategy). The genetic algorithm uses the following five steps to train and optimize a classifier.

1. *Initialization.* The genetic algorithm starts from a population of randomly selected chromosomes, which are represented by binary or gray-level digital strings. Each chromosome consists of a number of genes (bits in the string) and corresponds to a possible solution of the

problem. In a mass detection scheme, a binary coding method is typically used to create chromosomes. Each chromosome has a fixed length of $N + M$ genes. The first N genes represented the extracted image features (where 1 means that the feature is selected and 0 means that the feature is not selected). The following M genes can be used to optimize other parameters of classifiers (i.e., the number of hidden neurons in an ANN, the number of neighbors in a KNN algorithm).

2. *Evaluation.* In this step a fitness function is applied to evaluate the fitness of the entire chromosomes in population. The fitness function or criterion is determined by the specific applications. In the mass detection schemes, the area under the ROC curve (Az) is the most popular fitness criterion. The higher Az value indicates the higher performance of the classifier.

3. *Selection.* This step uses a selection method (e.g., roulette wheel selection, tournament selection, and elite selection) to reward high-fitness chromosomes (with high A_z value) and to eliminate the low fitness ones. Thus, the chromosomes with better fitness levels can expand to take up a larger percentage of the population, while those with poor fitness levels decrease in numbers.

4. *Search.* After population has adjusted itself to take advantage of the higher fitness chromosomes, search operators are used to create a new generation of chromosomes. The two most popular operators are crossover and mutation. The crossover exchanges genes between two chromosomes to produce two offspring in a new generation. The mutation injects random changes to select genes (from 0 to 1 or vice versa in binary-coded chromosomes) to minimize the risk of GA results being trapped inside the local "maximum" during the optimization process.

5. *Termination.* The genetic algorithm continually evolves until one of some terminating conditions is reached. These conditions can be that (1) GA has identified a chromosome that has reached a predetermined fitness value, (2) GA has reached the predetermined maximum generation of evolution, and (3) GA cannot find better chromosomes in the new generations.

The supervised learning methods are commonly used to train and optimize the performance of the classifiers implemented in CAD schemes. Many classifiers can be easily overfitted. Hence, robustness is an important issue in the optimization and evaluation of CAD schemes. Due to the size limitation of the database, CAD schemes are often trained and tested using jackknifing (e.g., leave-one-out), N-fold cross validation, and bootstrapping methods. In general, an increase of the training database size can improve the robustness of CAD schemes. The best way to test CAD performance is to use the cases that have "never been seen" by the schemes during the training process. If a scheme (or a classifier) needs to be re-optimized without using a new

database, the training and testing data sets should be randomly reassigned in an attempt to minimize the training biases. Otherwise, the testing data set will gradually convert into the training samples and the robustness of the scheme will be substantially decreased.

Computer-aided Detection Schemes Based on Multi-image

To interpret screening mammograms and detect suspected masses, radiologists typically examine and compare asymmetrical tissue patterns depicted on bilateral images (the same view of the left and right breast) and matched regions on ipsilateral images CC (cranio caudal) and MLO (mediolateral oblique) views of the same breast), as well as the change of size and contrast of the corresponding regions depicted on images acquired from current and prior examinations. Such a comparison process plays an important role for radiologists to visually detect and diagnose suspected masses. Without combining and incorporating the multi-image-based information into the detection and classification process, current single-image-based CAD schemes limit their capability of achieving high performance. Therefore, a number of research groups have developed and tested multi-image-based CAD schemes. The biggest technical challenge in developing multi-image-based CAD schemes is image registration that includes the detection of the landmarks and identification of the matched regions depicted on different images. Several approaches used in multi-image-based CAD schemes are discussed next.

Image Registration

Current image registration methods used in CAD schemes typically include two steps: global image registration and local region adjustment. Global image registration depends on the automated detection of several image fiducial markers, such as skin line, nipple, and chest wall. For an example, one image registration scheme includes the following computing algorithms.

1. *Skin line detection.* Due to a variety of clinical reasons and in some cases the saturation of a film digitizer at high optical densities, skin lines (skin–air interfaces) are quite fuzzy or inseparable in many images. To compensate such difficulty, the algorithm assumes that a transition curve with the smoothest curvature between breast tissue and air background represents the skin line. An iterative method is applied to search for an optimal threshold that can generate the smoothest transition boundary of tissue–air interface. Since a digitized mammogram typically includes a multimodal gray-level histogram, the algorithm first detects the deepest valley in the low end of the histogram, assuming that the gray level in the valley and the air background peak in the histogram are d_v and d_{air}, respectively. A series of

thresholds is selected as $T_i = T_{i-1} + 16$, $i = 2, 3, \ldots, n$, where the first one is $T_1 = d_v$ and the last one is $T_n \le d_{air} - 16$. At each threshold level, the program tracks the tissue–air interface. Then, the smoothness of all tracked boundaries is compared. The smoothest boundary is defined as the skin line.

2. *Chest wall detection.* To detect the chest wall line depicted on an MLO view image, the image is orientated in such a way that the chest wall always locates in the left side of the image. From the top of the image, the computer program scans horizontally from the left edge of the image to the skin line to find the maximum gradient before reaching the skin line. The program calculates the gray-level change (or gradient) for each pixel. The pixel with the maximum gradient is selected as the candidate for the intercepted pixel between this horizontal scanning line and the chest wall. The program stops scanning when the pixel with maximum gradient is quite close to the left edge of the image (e.g., 10 pixels away). Then, a least squares regression method is applied to fit all the identified maximum gradient pixels into a line, which is defined as the chest wall depicted on the image. Using the least squares regression fitting method, the effect of potential error in search maximum gradient points can be minimized.

3. *Nipple classification.* Similar to the skin line, the visibility of the nipples varies in different images. To automatically detect (or estimate) the locations of nipples in screening mammograms acquired under different machines and X-ray exposure conditions, three hypotheses are used to classify nipples into three categories. (1) A small and obvious protruding area in a smooth skin boundary (interface between breast and background) can indicate the location of a nipple not in profile. (2) If a nipple is in profile, it is likely to be located in an area where the pixel values are not only relatively unchanged but also significantly smaller than that of any other tissue regions near the skin boundary. (3) For a nipple that is invisible in an image, a point in the skin line that has maximum distance to the chest wall is assumed as the location of the nipple.

4. *Nipple detection.* Based on these three assumptions (categories), the computer algorithm uses three steps to detect nipple locations. The first step is a search for a possible protruding area along the skin line. The computer program scans along the skin line. In each scanned point (x_i, y_i), the program selects another point that is 40 tracking pixels (16 mm) away (x_{i+40}, y_{i+40}). A line is drawn between these two points, and the area (e.g., the number of pixels) between this line and the skin line is computed. After scanning the entire skin line, if the maximum size of the areas is larger than a pretrained value, this protruding area is identified as the nipple area. If the nipple location is not detected in the first step, the algorithm continues the second step. By scanning the skin line again, the algorithm computes the gray levels in the tissue side using a square window of 20×20 pixels (8 mm $\times$ 8 mm). Then, the program selects one area with the smallest mean digital value of pixels inside the scanning window and computes (1) the difference between this mean digital value and the digital value at the skin line, and (2) the standard deviation of digital values inside the window. If both the digital value difference and the standard deviation can pass through a set of pretrained thresholds, this area is an identified nipple area. If both steps fail to identify a nipple location, which usually means that the nipple is invisible in the image, the third step follows. The algorithm computes the distance of every pixel in the skin line to the chest wall. The point with the maximum distance is used to represent the location of the nipple. The nipple location is typically used as the origin of coordinates, and the chest wall is used to determine the orientation of the coordinate system. One axis ($\rho = 0$) starts from the nipple location and is perpendicular to the chest wall. Due to the difference in breast compression of mammography, local region adjustment (nonrigid transformation) is typically required after global image registration.

Test of Computer-aided Detection Schemes Based on Multi-image

To match the corresponding suspicious mass regions depicted on two ipsilateral (CC and MLO) view images, one CAD scheme applies and compares two local matching methods, the arc and Cartesian straight line. The study suggests no significant difference ($p > 0.99$) in localization accuracy between the two methods. The study also demonstrates that by using the location measurement and combining CAD results in both CC and MLO views, a CAD scheme based on a simple ipsilateral view eliminates more than half of the false-positive regions while maintaining 90% sensitivity (Chang *et al.*, 1999).

Recently, more studies have focused on the development of CAD schemes that could automatically detect and match mass regions on serial mammograms (acquired from current and prior examinations). One research group developed a three-stage scheme to locate and match two mass regions depicted on the current and prior images. The first stage defines an initial fan-shaped search region based on the global estimation of breast geometry. The second stage refines a search region by warping and alignment (including a two-dimensional affine transformation and nonlinear simplex optimization). The third stage applies correlation measures to iteratively define the "best" matched regions depicted on two images (with the maximum correlation coefficient). Using a database involving 124 pairs of biopsy-proven masses, the computer scheme generated results showing that the average distance between the estimated and the true center of mass regions on the prior mammograms was 4.2 ± 5.7 mm. The 87% of the estimated prior mass regions resulted in an area overlap of at least 50% with true mass locations (Hadjiiski et al., 2001a). The matching accuracy is a key factor for automatic analysis and comparison of interval change of masses in the sequential multi-image-based CAD schemes. To further improve matching accuracy, this research group investigated and compared 11 methods of similarity measures and found that three of them, Pearson's correlation, the cosine coefficient, and Goodman and Kruskal's Gamma coefficient, provided significantly higher accuracy ($p < 0.05$) in matching the corresponding masses on serial mammograms (Filev et al., 2005).

Because to date no method has been proven optimal in detecting and matching the mass regions on sequential images, another research group developed and tested a hybrid algorithm (Timp et al., 2005). After global image registration, two local registration methods based on mass likelihood and gray scale (digital value) correlation were tested. Each method produces a quantitative registration measure. The hybrid registration algorithm then linearly combines two measures to compute a final measure to quantify how well a selected region on the prior image matches the target mass region on the current image. A test involving 389 temporal mass pairs indicated that the hybrid algorithm correctly registered 82% pairs of mass regions, whereas other previously reported methods achieved correct registration in < 72% cases.

Although considerable research effort has been made in the development of multi-image-based CAD schemes to detect and diagnose suspected masses, a number of technical issues and difficulties remain unsolved to date. As a result, multi-image-based CAD schemes have not been used in any commercial CAD systems. Further research is needed for any multi-image-based CAD schemes to be optimally utilized in the clinical environment.

Evaluation and Application of Commercial Computer-aided Detection Systems

For the last several years, commercially-available CAD systems (including Image-Checker from R2 Technology Inc., Los Altos, CA, and Second-Look from iCAD Inc., Nashua, NH) have been routinely used in a large number of medical institutions around the world to assist radiologists in detecting abnormal findings, specifically the regions depicting suspected microcalcification clusters and masses. CAD systems are intended to be used as a "second reader." Specifically, formally approved use of commercial CAD systems calls for radiologists to review and make initial interpretation of complete examination before viewing CAD-cued suspected masses and microcalcification clusters, followed by adjustment (changes) in their initial interpretation as needed. Several studies have suggested that current CAD systems can detect a high percentage of sub-

tle cancers initially missed by radiologists. In a blinded retrospective review, a panel of radiologists found that 67% of 427 cancers were "visible" on the images acquired from "prior" examinations that were diagnosed as negative, but cancers were detected in current examination (1 year later). Among them 115 prior examinations were warranting recalls. Applying a commercial CAD system to process these 115 cases, CAD correctly identified 89 cancers (77%) depicted on these "prior" images with an average one false-positive detection per image (Warren Burhenne et al., 2000). The researchers concluded that using CAD could potentially help radiologists reduce this false-negative rate by 77% without an increase in the recall (false-positive) rate.

To investigate whether such potential can be achieved in clinical practices, many observer performance studies have been conducted in this area. However, no general agreement has yet been reached on the issues of whether the performance of current CAD systems is sufficient for routine clinical use and what is needed in order to realize the full potential benefits of CAD systems in the screening environment. One observer performance study investigated how different CAD cueing performance levels and conditions could affect radiologists' performance in detecting masses and microcalcification clusters (Zheng et al., 2001). In the study, seven radiologists interpreted 120 cases to detect masses and microcalcification clusters five times under five display modes, where the CAD cueing was provided in four modes with a combination of either 90% or 50% sensitivity and either 0.5 or 2 false-positive cues per image. The results of the study clearly indicated that high-performance CAD systems had the potential to significantly improve the detection performance of radiologists ($p < 0.05$). On the other hand, low-performing cueing schemes could actually adversely affect observer performance. This is especially true when a large number of false-positive cues are displayed. The study also found that increasing false-positive cues significantly reduced radiologists' detection sensitivity in noncued areas of the images.

Several large prospective studies have also been performed and reported to assess the potential and actual impact of applying CAD systems in the clinical environment. In one study, two radiologists interpreted 12,860 screening mammograms in a community breast center with and without CAD assistance (Freer and Ulissey, 2001). Applying the CAD system to process these images generated a total of 14,214 marks. One radiologist read and interpreted one screening mammography case without knowing CAD cues, followed immediately by presentation of the CAD results (display of cues). The radiologist could then make changes in his original interpretations. Two sets of detection results before and after viewing CAD cues were recorded for each case. The study results showed that 97.4% of CAD marks were dismissed, and the remaining 2.6% (368) marks were considered actionable by these two radiologists. The study reported that with CAD assistance the two radiologists detected 8 (19.5%) additional cancers (from 41 to 49), with an 18.5% increase in recall rate (from 830 to 986). Among 8 additional detected cancers, 7 were associated with microcalcification clusters and 1 was a mass. The study also found that among 49 finally detected cancers, radiologists initially detected 26 masses and 15 microcalcification clusters and CAD detected 18 masses and 22 microcalcification clusters.

In another recently reported study (the United Kingdom's national breast screening program), 12 radiologists read and interpreted screening mammograms obtained from 6,111 women (Khoo et al., 2005). In the study, readers recorded an initial evaluation, viewed CAD cues, and recorded a final evaluation on each case. In this image database, the CAD system achieved 84% detection sensitivity with a false-positive rate of 1.59 per case (four images). Of the 12 cancers initially missed by the radiologists in the database, 9 of them were detected (correctly cued) by CAD. Eight were associated with masses, and 1 was associated with microcalcifications. However, 7 malignant mass cues were discarded by the radiologists as false-positive cues. In this study, viewing CAD cues increased radiologists' sensitivity by 1.3%. These studies demonstrated that distinguishing between CAD-cued true-positive and false-positive masses was a difficult task for radiologists.

One group of researchers used verified practice and outcome-related databases to compute cancer detection and recall rates for 24 radiologists before and after the introduction of commercial CAD systems into a university-based medical center (Gur et al., 2004a). In the databases, 56,432 screening cases were interpreted "without" CAD, and 59,139 cases were interpreted "with" CAD results. No statistically significant difference was found between two reading environments. The cancer detection rates were 3.49 versus 3.55 per 1000 screening cases ($p = 0.68$), and the recall rates were 11.39% versus 11.40% ($p = 0.96$) "without" and "with" CAD, respectively. The 2% increase in cancer detection rate was attributable to better detection of microcalcifications when using CAD rather than masses.

These and many other studies suggest that in the clinical practice, radiologists usually have high confidence in CAD-cued microcalcifications, but they largely ignore CAD-cued mass regions (D'Orsi, 2001). This is largely because CAD schemes have achieved much higher performance in the detection of microcalcification clusters (e.g., 98% sensitivity reported by several studies), which is often higher than the sensitivity of radiologists, while CAD performance of mass detection is much lower. One prospective study reported that CAD detected 18 out of 27 malignant masses representing 66% sensitivity (Freer and Ulissey, 2001). In another comprehensive evaluation study that included several different categories of images acquired in (1) current and prior examinations, (2) false-

negative cases, (3) recalled but benign cases, and (4) screening normal cases, the mass detection sensitivities of two leading commercial CAD systems were typically found in the range between 65 and 80%, with false-positive mass detection rates between 1 and 1.5 per case (Gur et al., 2004b). In addition, between approximately 40 and 80% of masses (in different categories) were detected by CAD only on one view. The subtle masses detected by CAD schemes only on one view are more likely being discarded by radiologists as false-positive cues.

When using digitized images, the reproducibility of CAD results is another important issue in affecting the overall quality of CAD systems in mass detection. Several research groups have tested the reproducibility of CAD systems by repeatedly scanning a set of film mammograms several times through the digitizer and applying the CAD scheme to these repeatedly digitized images. One study found that although during each of 10 repeated scans (digitization), CAD achieved relatively consistent sensitivity on mass detection (78.4% ± 7.0) and average false-positive marks of 0.66 per image, and 40% of masses (10 out of 25) were not reproducible, which were detected between 1 and 9 times out of the 10 CAD scans and processing (Baker et al., 2004). In addition, masses that are more difficult for the radiologists to detect are also detected less reproducibly by CAD schemes (Taylor et al., 2003).

New Developments in Mass Detection Schemes

Compared to detection of microcalcification clusters, CAD schemes generate more false-positive mass regions with lower detection sensitivity. Unlike CAD-generated cues for microcalcification clusters that can be easily and visually examined and confirmed, visually confirming (or discarding) false-positive mass regions is much more difficult because masses often: (1) have varying characteristics in size, shape, and density; (2) exhibit poor image contrast; (3) have a similar characteristic to the nonuniform high-dense tissue background; and (4) are highly connected to the parenchyma tissue, particularly for spiculated masses.

Several independent studies have demonstrated that even in the laboratory environment using a database with a high percentage of cancer cases (e.g., ≥ 50%), a high-rate of CAD-generated false-positive detections could significantly reduce radiologists' performance in mass detection (Zheng et al., 2001; Alberdi et al., 2004). Owing to the large volume of mammograms performed and the low yield in a screening environment, correctly dismissing a majority of CAD cues (e.g., > 97.4% of CAD marks) (Freer and Ulissey, 2001) without losing true-positive detections is a very difficult task for the radiologists. To date, many radiologists largely ignore CAD-cued mass regions in their clinical practice.

Hence, improving CAD performance on mass detection and increasing radiologists' confidence regarding CAD-cued masses are important issues in current CAD development. A great research effort has been made recently for this purpose.

Improvement of Computer-aided Detection Performance

Several methods have been investigated in an attempt to improve CAD performance, in particular to reduce false-positive detection rate while maintaining the detection sensitivity level. Following are a few examples of increasing CAD sensitivity on the subtle masses that are difficult to be visually detected by the radiologists and reducing the number of false-positive cues that radiologists need to rule out.

Current CAD schemes use an operating threshold of detection scores to cue mass regions with detection scores larger than the threshold and to discard the regions with the scores smaller than threshold. Because the false-positive cues are not uniformly distributed on the cases in a diverse image database, the difficult cases (i.e., the cases with dense breast tissue) typically have more false-positive cues than the easy cases (e.g., the cases dominated by fat tissue). In one study with a limited database of 79 images, 31.7% of images accounted for 63.5% of false-positive cues generated by the CAD scheme (Li et al., 2001). The researchers developed a neural network-based competitive classification strategy in which only the "best" cues were selected from the preclassified suspected regions in each case. The initial results demonstrated that using this strategy could reduce 56% false-positive rates (from 8.36 to 3.72 per image) at the cost of reducing 1% sensitivity.

Recently, another study assessed the performance changes of a CAD scheme by restricting the maximum number of mass regions that were allowed to be cued by CAD as showing positive findings in each case (typically four images) (Zheng et al., 2004a). A database involving 500 cases (2000 images) was used in the study. Among these cases, malignant masses were detected and verified in 300 cases. By setting the threshold of the detection scores at 0.565, the CAD scheme achieved 79% sensitivity with 0.4 false-positive detections per image (or 1.6 per case). By limiting the number of maximum allowed cued regions per case, the false-positive detection rates decreased faster than the true-positive rates (sensitivity). At a maximum of two cues per case, 47% of false-positive cues were eliminated, while 7% of true-positive masses were also missed. To maintain the original detection sensitivity, the operating threshold of detection scores should be reduced in the CAD scheme. In this study, by reducing the threshold to 0.36, the CAD scheme could maintain the same detection sensitivity and eliminate 24.8% false-positive cues. The study demonstrated that limiting the maximum number of cues allowed

per case and adjusting the operating threshold appropriately increased CAD sensitivity in the subset of smaller and subtle masses. In general, this effect is desirable in that it may (1) replace the true-positive cues that are relatively easy to be visually detected with those that are more difficult to be detected, and (2) reduce the number of cues that have to be ruled out by the radiologists.

With increasing compliance following screening recommendations, radiologists could be forced to detect increasingly more subtle abnormalities over time. Current CAD schemes were generally optimized using "current" (latest) images, when the cancers in question were identified by radiologists. Studies have demonstrated that when applying a current CAD scheme to "prior" images depicting visible but more subtle abnormalities, its performance was significantly lower than that when applied to current images. This is due to the substantially different distribution of a large number of features used in the optimization process. Since the supervised machine-learning strategies are commonly used in developing current CAD schemes, the feature or characteristic distribution of the training data set heavily influences the learning process and affects the performance of the CAD schemes on new testing data sets that may or may not be similar to the training data set. In order to improve CAD performance in detecting subtle masses associated with cancers at an "earlier" stage, several studies have demonstrated that current CAD schemes might need to be re-optimized using image databases involving a large number of images that had been originally interpreted as negative but were later proven as malignant during follow-up examinations. For example, one study used an image database involving 140 temporal image pairs (Hadjiiski et al., 2001b). Masses were visible in both "current" and "prior" images. Thirty-five image features were computed to represent each mass region. A stepwise linear discriminant analysis method was applied to optimize CAD schemes. The results indicated that, in order to achieve optimal results for detecting and classifying mass regions depicted on the current and prior images, different features should be selected. The study showed that the CAD scheme trained, using both current and prior images, achieved a test A_z value (area under an ROC curve) of 0.88, while the CAD scheme trained using current image only had a test A_z value of 0.82.

Improvement of Reproducibility of Computer-aided Detection Schemes

The variability of CAD results on the same images during repeated scans can raise a number of important issues regarding the use of CAD systems in the clinical practice. Poor reproducibility can reduce the confidence of radiologists in the CAD results, making them somewhat reluctant to revise the original interpretations without CAD. Questions can also be raised whether radiologists should keep CAD results with the patient's medical records for medico-legal reasons. Although reporting reproducibility may be important to evaluate the overall performance level of a CAD scheme, testing CAD reproducibility is quite a difficult task. Many CAD schemes were developed using publicly available databases of digitized images, and researchers cannot access the original films for re-digitization. Even when using in-house developed databases, re-digitizing images multiple times and evaluating results generated from a large number of repeatedly digitized images is a costly and time-consuming task. As a result, the reproducibility of CAD schemes has rarely been reported.

To solve this problem, one research group investigated and developed a relatively simple method to test the reproducibility of CAD schemes for digitized mammograms without the need for repeated digitization (Zheng *et al.*, 2004b). The method is based on a hypothesis that one of the most important contributing factors to the poor reproducibility of a CAD scheme is the result of small shifts in film positioning. In this method, a computer scheme is applied to generate multiple images (simulating the repeated digitization) by resampling a single originally digitized image multiple times after a series of slight image rotations. By assuming that subsampled images (for mass detection) have a variable matrix size of M columns and N rows, the computer scheme repeatedly rotates each image in increments of $\Delta\alpha = \pm 0.2°$ (i.e., rotated 10 times with $\alpha = \pm0.2°, \pm0.4°, \pm 0.6°, \pm 0.8°$, and $\pm 1.0°$). Two conditions are set up at the scheme: (1) the origin (0, 0) of the coordinate system is established at the left top corner of the image, and (2) a rotation of 0.2 degree shifts the center pixel at the right edge by two pixels in the vertical (y) direction, which represents a maximum linear displacement of 0.8 mm over the whole image. Hence, the rotation center to the right edge of the image is computed as $L = 0.8/\tan (0.2°) = 229$ mm with its coordinate at $(M - L, N / 2)$. For a typical subsampled image with pixel size of 400 μm $\times$ 400 μm, L = 573 (pixels). After each rotation, each pixel $(x, y \mid 0 < x < M, 0 < y < N)$ in the rotated image is shifted to a new location (x', y') as follows:

$$y' = y + (L - x) \times \tan(\alpha)$$

$$x' = x - (\frac{N}{2} - y') \times \tan (\alpha)$$

Because x' and y' are generally not integers, an interpolation (resampling) procedure is performed in which each pixel in the rotated image typically covers partial areas of four pixels in the original image. Hence, the resampled digital value of each pixel after rotation is computed as:

$$I'_{x,y} = \sum_{i = 1}^{2} \sum_{j = 1}^{2} S_{i,j} \times I_{i,j}$$

where the coefficient $S_{i,j}$ is a coverage ratio of the partial area $(0 \leq S_{i,j} \leq 1$, and $S_{1,1} + S_{1,2} + S_{2,1} + S_{2,2} = 1)$. In this manner, the

computer scheme generates a series of images with different pixel-value matrices from each originally digitized image.

Then, a CAD scheme is applied to detect suspicious mass regions depicted on these rotated and resampled images. To evaluate CAD performance and reproducibility, the scheme first searches for all matched regions detected in all images (including the original image and all rotated and resampled images). The matching criterion is that the distance between the centers of any pair of identified regions in corresponding images is smaller than the maximum radial length of the two regions ($d \leq \max (r_k)$) in question. Radial length (r_k) is computed as the length from the region center of gravity (x_c, y_c) to the kth pixel (x_k, y_k) on the boundary (contour) of a detected region. If two regions are considered "unmatched," they are assumed to be different detections by the scheme. Comparing with the "truth" file, reproducibility for both true-positive and false-positive mass detections is computed.

Because the reproducibility of false-positive mass detections is substantially lower than that of true-positive mass detections, averaging detection scores from multiple digitized images cannot only reduce the variations in detection scores (improve reproducibility) but also improve CAD performance. One study digitized each of 92 film mammograms (depicting 44 malignant mass regions) four times (Zheng *et al.*, 2005). Each digitized image was also rotated and resampled four times. A CAD scheme was applied to all images. Researchers compared the reproducibility and detection performance of the CAD scheme using the detecting scores only from the digitized images and the average detection scores of five images (one originally digitized image and four rotated and resampled images). The results indicated a 52% reduction in the variation of detection scores when using average scores. Using average detection scores also substantially improved the overall CAD performance by a 12.6% increase in sensitivity and a 17.3% reduction in the false-positive detection rate.

Interactive Computer-aided Detection Systems

Studies have demonstrated that unless CAD systems had extremely high performance levels and the results could be visually confirmed (e.g., CAD-cued microcalcification clusters), it was quite difficult for radiologists to accept the CAD results without knowing the logic or reasoning for the detection. Current CAD systems cue all detected suspected mass regions if their detection scores are larger than a predetermined (optimized) threshold without explanation of why these regions are identified by the schemes as positive masses. This "black-box" type approach and the relatively low performance are the major factors that reduce radiologists' confidence in CAD-cued mass regions.

To improve radiologists' confidence in and reliance on CAD-cued suspected "masses" and their classification, several interactive CAD schemes have been developed and tested in which "similar" verified regions selected from a reference library are displayed side by side with the suspected mass region. When a suspected region is identified (as positive or negative) and/or classified (as malignant or benign), a set of reference regions "similar" to the one of interest is selected and displayed. These regions are extracted from a reference library of previously processed examinations that includes a large number of verified regions of interest (ROIs) with a wide range of characteristics. An interactive CAD system provides radiologists more information of why the CAD scheme detects a region of interest depicting a positive (or malignant) mass, which includes the detection score (the likelihood of being positive or malignant) and a set of "similar" regions that have been previously verified as positive or negative. Hence, the observers cannot only know the CAD-generated detection score for the cued region (which is not provided in current CAD systems) but also visually compare the CAD-selected "similar" regions. This visual aid explains to the observers why the CAD scheme cues this region as a suspected mass region. The initial studies demonstrated that using such a CAD workstation could help to improve radiologists' performance in both detection sensitivity and specificity (Giger et al., 2002).

Using interactive CAD systems, radiologists can query any suspected mass regions, either originally cued by the CAD scheme or not. When an observer queries a region depicted on the image, the CAD scheme first segments the boundary contour of the queried region using a region growth algorithm. If the CAD-generated boundary contour of the region of interest is not satisfied, the observer can manually draw (edit) the region boundary contour. The CAD scheme computes a set of features to represent the growth region and selects a set of "similar" regions from the reference library. A likelihood score of this queried region being associated with a malignant mass, and a set of reference regions are displayed in the interactive CAD workstation for visual comparison.

The key issue of developing an interactive CAD scheme is to search for the "best" similar reference regions. The similarity index is typically measured and evaluated based on either digital value distribution of pixels or a set of computed image features between a queried region and a reference region. Mutual information (MI) is a popular tool to measure similarity based on digital (pixel) value distributions. Given two compared regions X and Y, the mutual information between two regions is computed as

$$I(X, Y) = \sum_x \sum_y P_{XY}(x, y) \log_2 \frac{P_{XY}(x, y)}{P_X(x) P_Y(y)}$$

where $P_{XY}(x, y)$ is the joint probability density function (PDF) of two regions based on their corresponding pixel values, and $P_X(x)$ and $P_Y(y)$ are the marginal PDFs for region X and Y, respectively. Using MI to measure and evaluate similarity is based on a hypothesis that if two regions are similar, pixels with certain pixel values in one region should correspond to a more clustered distribution of pixel values in another region. The more the two regions are alike, the more information X provides for Y and vice versa. Because MI measures general dependence without making any a priori assumptions, it may be more effective and robust than traditional correlation measures. However, using an efficient method to compute joint and marginal PDFs is important for computing MI. One research group used a histogram approach to estimate PDFs. To improve computation efficiency, after image registration, the PDFs were estimated using a reduced number of 256 equal-size intensity (pixel digital value) bins for the histogram approximation technique. Once a suspected mass region (Q_i) is queried, the CAD scheme computes the MI between the queried region and each of "known" regions stored in the reference library, including both positive (malignant) regions (M) and negative regions (N). By selecting the best k "similar" regions in both positive and negative reference regions, respectively, a decision index (unnormalized detection score) for this queried region is computed as

$$D(Q_i) = \frac{1}{k} \sum_{j=1}^{k} MI(Q_i, M_j) \quad \frac{1}{k} \sum_{j=1}^{k} MI(Q_i, N_j)$$

It is obvious that a queried positive mass region should have a higher score. In one study to test the performance of applying this MI-based method to search for the similar regions and classify suspicious mass regions, a reference library involving 1450 regions was established (Tourassi et al., 2003). Among these reference regions, 809 depict the confirmed masses and 656 were normal. Using a leave-one-out sampling and validation method, the study resulted in the detection performance of $A_z = 0.87 \pm 0.01$.

Many computed image features (including morphological and pixel-value distribution) have also been used in other CAD schemes to measure the similarity between a queried "mass" region and a reference region. For example, a multifeature-based k-nearest neighbor (KNN) algorithm is applied to search for the similar regions in the reference library. Similarity is measured by the distance between a queried region (Q_i) and each of the reference regions (R_j) in a multidimensional (n) feature space.

$$d(Q_i, R_j) = \sqrt{\sum_{r=1}^{n} (f_r(Q_i) - f_r(R_j))^2}$$

The smaller the distance, the higher the degree of the similarity is between any two regions. Based on the sorted list (from the smallest distance to the largest), the first

k-regions in the list are selected as the k-most similar reference regions. A distance weight is defined as:

$$w_i = \frac{1}{d(Q_i, R_j)^2}$$

The probability (detection score) of the queried region being actually positive (or malignant) is computed as:

$$P_{TP} = \frac{\sum_{i=1}^{M} w_i^{TP}}{\sum_{i=1}^{N} w_i^{TP} + \sum_{j=1}^{M} w_j^{FP}}$$

where M is the number of positive mass regions and N is the number of negative regions that were selected in the set of the k "most similar" regions.

In one study, researchers applied a genetic algorithm (GA) with a leave-one-out validation method to define an optimal topology for the KNN algorithm (Zheng *et al.*, 2006). The reference library involved 3000 regions. Among them, 1000 depicted malignant masses and 2000 depicted either benign or CAD-generated false-positive regions. In the study, a binary coding method was applied to create a chromosome used in GA. Each extracted feature corresponds to a gene. Five additional genes were added to determine the optimal number of neighbors used in the KNN. In 3000 training iterations using a leave-one-out optimization method, each region in the reference library was used as a queried region, and the KNN algorithm searched for the k-most similar regions in the remaining 2999 regions. Detection scores of the 3000 regions were then used to compute the A_z value (area under the ROC curve), which was defined as a GA fitness criterion. The GA was terminated when it either converged to the "highest" A_z value (no improvement was accomplished in the next generation) or it reached a predetermined number of maximum generations (e.g., 100). In this study, GA generated an optimal KNN algorithm-based classifier that selected 14 image features from the initial pool of 36 features and included 15 neighbors for comparison ($k = 15$). Using all 3000 test regions in the reference library and the leave-one-out validation method, the A_z value of the optimized KNN classifier was 0.86 ± 0.01.

Despite these encouraging results, the interactive CAD systems achieved limited success in improving radiologists' confidence in CAD-cued mass regions. A large fraction of "the similar regions" selected by the CAD schemes are not as visually similar as one would hope for. There seems to be a substantial difference in similarity measures between the human visual system and the CAD schemes. When radiologists feel that the CAD scheme identified a poor set of reference images for comparison, they will largely ignore this visual aid. This is an important and difficult issue that needs to be solved if interactive CAD schemes are to succeed in the clinical environment. Currently, several approaches to investigate this issue have been proposed and tested, which include the reduction of the interobserver variability in similarity assessment (Muramatsu *et al.*, 2005) and the improvement of visual similarity among the selected reference regions by adding a subjective rating of mass boundary margins (spiculation) into the region selection process (Zheng *et al.*, 2006).

In summary, significant progress in CAD development and evaluation has been made in recent years. Currently, only single-image-based schemes have been implemented and used in the commercial systems. Many new approaches (i.e., multi-image-based schemes and interactive methods) are still under development and test. Commercial CAD systems can detect a high percentage of subtle cancers (e.g., malignant masses) that radiologists initially miss. Many researchers believe that using CAD may help radiologists substantially reduce the false-negative detection rate (Warren Burhenne *et al.*, 2000). However, because of the relatively lower performance level of current single-image-based CAD schemes, including higher false-positive rate, cueing method (cueing the most of both subtle masses and false-positive detections only on one view), and poorer reproducibility, radiologists' confidence in CAD-cued mass regions remains low. As a result, radiologists discard the majority of CAD-cued subtle masses that are missed in their original interpretations (Khoo *et al.*, 2005). Clearly, the full potential benefit of using CAD systems has not been realized in screening mammography. More research and development work is needed for improving CAD performance and searching for the approaches to optimally use CAD systems in the clinical environment.

References

Alberdi, E., Povyakaio, A., Strigini, L., and Ayton, P. 2004. Effects of incorrect computer-aided detection (CAD) output on human decision-making in mammography. *Acad. Radiol.* 11:909–918.

Baker, J.A., Lo, J.Y., Delong, D.M., and Floyd, C.E. 2004. Computer-aided detection in screening mammography: variability in cues. *Radiology* 233:411–417.

Bebis, G., Georgiopoulos, M., and Kasparis, T. 1997. Coupling weight elimination with genetic algorithms to reduce network size and preserve generalization. *Neurocomputing* 17:167–194.

Chan, H.P., Wei, D., Helvie, M.A., Sahiner, B., Adler, D.D., and Goodsitt, M.M. 1995. Computer-aided classification of mammographic masses and normal tissue: linear discriminant analysis in texture space. *Phys. Med. Biol.* 40:857–876.

Chang, Y.H., Good, W.F., Sumkin, J.H., Zheng, B., and Gur, D. 1999. Computeized location of breast lesions from two views: an experimental comparison of two methods. *Invest. Radiol.* 34:585–588.

D'Orsi, C.J., 2001. Computer-aided detection: there is no free lunch. *Radiology* 221:585–586.

Filev, P., Hadjiiski, L., Sahiner, B., Chan, H.P., and Helvie, M.A. 2005. Comparison of similarity measures for the task of template matching of masses on serial mammograms. *Med. Phys.* 32:515–529.

Freer, T.W., and Ulissey, M.J., 2001. Screening mammography with computer-aided detection: prospective study of 12,860 patients in a community breast center. *Radiology 220*:781–786.

Giger, M.L., Huo, Z., Vyborny, C.J., Lan, L., Bonta, I., Horsch, K., Nishikawa, R.M., and Rosenbourgh, I. 2002. Intelligent CAD workstation for breast imaging using similarity to known lesions and multiple visual prompt aids. *Proc SPIE 4684*:768–773.

Gur, D., Sumkin, J.H., Rockette, H.E., Ganott, M., Hakim, C., Hardesty, L.A., Poller, W.P., Dhah, R., and Wallace, L. 2004a. Changes in breast cancer detection and mammography recall rates after the introduction of a computer-aided detection system. *J. Nat. Cancer Inst. 96*:185–190.

Gur, D., Stalder, J.S., Hardesty, L.A., Zheng, B., Sumkin, J.H., Chough, D.M., Shindel, B.E., and Rockette, H.E. 2004b. Computer-aided detection performance in mammographic examination of masses: assessment. *Radiology 223*:418–423.

Hadjiiski, L., Chan, H.P., Sahiner, B., Petrick, N., and Helvie., M.A. 2001a. Automated registration of breast lesions in temporal pairs of mammograms for interval change analysis—local affine transformation for improved localization. *Med. Phys. 28*:1070–1079.

Hadjiiski, L., Sahiner, B., Chan, H.P., Petrick, N., Helvie, M., and Gurcan, M. 2001b. Analysis of temporal changes of mammographic features: computer-aided classification of malignant and benign breast masses. *Med. Phys. 28*:2309–2317.

Khoo, L.A., Taylor, P., and Given-Wilson, R.M. 2005. Computer-aided detection in the United Kingdom national breast screening programme: prospective study. *Radiology 237*:444–449.

Li, L., Zheng, Y., Zhang, L., and Clark, R.A. 2001. False-positive reduction in CAD mass detection using a competitive classification strategy. *Med. Phys. 28*:250–258.

Mendez, A.J., Tahoces, P.G., Lado, M.J., Souto, M., and Vidal, J.J. 1998. Computer-aided diagnosis: automatic detection of malignant masses in digitized mammograms. *Med. Phys. 25*:957–964.

Muramatsu, C., Li, Q., Suzuki, K., Schmidt, R.A., Shiraishi, J., Newstead, G.M., and Doi, K. 2005. Investigation of psychophysical measures for evaluation of similar images for mammographic masses: preliminary results. *Med. Phys. 32*:2295–2304.

Petrick, N., Chan, H.P., Sahiner, B., and Wei, D. 1996. An adaptive density-weighted contrast enhancement filter for mammographic breast mass detection. *IEEE Trans. Med. Imag. 15*:59–67.

Taylor, C.G., Champness, J., Reddy, M., Taylor, P., Potts, H.W., and Given-Wilson, R. 2003. Reproducibility of prompts in computer-aided detection (CAD) of breast cancers. *Clin. Radiol. 58*:733–738.

te Brake, G.M., and Karssemeijer, N. 2001. Segmentation of suspicious densities in digital mammograms. *Med. Phys. 28*:259–266.

te Brake, G.M., Karssemeijer, N., and Hendriks, J.H. 1998. Automated detection of breast carcinomas not detected in a screening program. *Radiology 207*:465–471.

Timp, S., Engeland, S., and Karssemeijer, N. 2005. A regional registration method to find corresponding mass lesions in temporal mammogram pairs. *Med. Phys. 32*:2629–2638.

Tourassi, G.D., Vargas-Voracek, R., Catarious, D.M., and Floyd, C.E. 2003. Computer-assisted detection of mammographic masses: a template matching scheme based on mutual information. *Med. Phys. 30*:2123–2130.

Warren Burhenne, L.J., Wood, S.A., D'Orsi, C.J., Feig, S.A., Kopans, D.B., O'Shaughnessy, K.F., Sickles, E.A., Tabar, L., Vyborny, C.J., and Castellino, R.A. 2000. Potential contribution of computer-aided detection to the sensitivity of screening mammography. *Radiology 215*:554–562.

Yin, F.F., Giger, M.L., Doi, K., Vyborny, C.J., and Schmidt, R.A. 1994. Computerized detection of masses in digital mammograms: automated alignment of breast images and its effect on bilateral-subtraction technique. *Med. Phys. 21*:445–452.

Zheng, B., Chang, Y.H., and Gur, D. 1995. Computerized detection of masses in digitized mammograms using single-image segmentation and a multilayer topographic feature analysis. *Acad. Radiol. 2*:959–966.

Zheng, B., Ganott, M.A., Britton, C.A., Hakim, C.M., Hardesty, L.A., Chang, T.S., and Gur, D. 2001. Soft-copy mammographic readings with different computer-assisted diagnosis cuing environments: preliminary findings. *Radiology 221*:633–640.

Zheng, B., Leader, J.K., Abrams, G., Shindel, B., Catullo, V., Good, W.F., and Gur, D. 2004a. Computer-aided detection schemes: the effect of limiting the number of cued regions in each case. *Am. J. Roentgenol 182*:579–583.

Zheng, B., Gur, D., Good, W.F., and Hardesty, L.A. 2004b. A method to test the reproducibility and to improve performance of computer-aided detection schemes for digitized mammograms. *Med. Phys. 31*:2964–2972.

Zheng, B., Maitz, G.S., Ganott, M.A., Abrams, G., Leader, J.K., and Gur, D. 2005. Performance and reproducibility of a computerized mass detection scheme for digitized mammography using rotated and re-sampled images: an assessment. *Am. J. Roentgenol. 185*:194–198.

Zheng, B., Lu, A., Hardesty, L.A., Sumkin, J.H., Hakim, C.M., Ganott, M.A., and Gur, D. 2006. A method to improve visual similarity of breast masses for an interactive computer-aided diagnosis environment. *Med. Phys. 33*:111–117.

4

Full-field Digital Phase-contrast Mammography

Toyohiko Tanaka, Chika Honda, Satoru Matsuo, and Tomonori Gido

Introduction

Radiologists generally still prefer screen-film (S/F) mammography to full-field digital mammography (FFDM) owing to their demand for high image quality and cost efficiency. However, as a result of advances in image processing, digital mammography is expected to provide higher detectability of suspicious lesions than S/F mammography, especially in radiographically dense breasts (Hasegawa *et al.*, 2003). It is important for screening and diagnostic mammographic examinations to clearly depict tiny calcification grains and ambiguous masses buried in fibroglandular tissues. For this purpose, S/F mammography is designed as a combination of a single-coated intensifying screen and a single-coated emulsion on film in order to obtain as high a resolution as 18–20 cycles/mm with high image sharpness. The spatial resolution of 20 cycles/mm is equivalent to 25 µm pixels in digital magnification mammography (Yip *et al.*, 2001). However, FFDM is hampered by its low spatial resolution due to the pixel size of the acquisition devices such as 40, 50, 54, 70, or 100 µm (Pisano *et al.*, 2005). On the other hand, the spatial resolution of any given imaging system can be improved by geometric magnification mammography (Sickles *et al.*, 1977). In addition, it is reported that geometric magnification mammography has been successfully applied in digital mammography with computed radiography (CR) in order to compensate for its inherent limits in spatial resolution (Funke *et al.*, 1998).

Unfortunately, magnification mammography can lack geometric sharpness depending on the focal spot size of the X-ray tube used and the magnification ratio. However, magnification with a small-focus X-ray tube in appropriate radiography geometry generates an edge effect due to the refraction of X-rays, so that image sharpness is improved by the use of phase contrast in magnification mammography as reported by Honda *et al.* (2002) for S/F mammography. In digital mammography, the magnified image in phase-contrast imaging can be reduced to the original size of the object. It is reported that this technique allows the utilization of phase contrast and magnification in FFDM (Freedman *et al.*, 2003).

Konica Minolta MG, Inc. has developed a full-field digital phase-contrast mammography (PCM) system, which performs phase-contrast imaging with ×1.75 magnification (Tanaka *et al.*, 2005). The magnified PCM images on a CR storage phosphor plate are scanned at a sampling rate of 43.75 µm and then reduced to the original object size with 25 µm pixels for printing on photothermographic film (Gido *et al.*, 2005). The image qualities of the full-field digital PCM system achieved are equivalent to or better than those of conventional S/F mammography. As a result of clinical trials, the full-field digital PCM system exceeded image quality of conventional S/F systems (Tanaka *et al.*, 2005). In this chapter, we briefly discuss the present physical properties and clinical experience of the full-field digital PCM system, in which the phase-contrast technique is utilized.

Historical Background of the Phase-contrast Technique

Phase-contrast techniques provide unique X-ray images owing to the wave nature of X-rays, which has not been utilized in the X-ray imaging field for almost a hundred years. In 1895, Wilhelm Roentgen discovered an unknown ray called an X-ray, and in 1912, Max von Laue demonstrated that an X-ray is an electromagnetic wave with a short wavelength on the order of crystalline lattice distances. After several decades, Somenkov *et al.* (1991) reported improvement in the X-ray image contrast with a monochromatic X-ray employing Cu K α 1 radiation and Si-reflection in phase-contrast imaging of small objects such as a polyethylene cylinder, 0.5 mm in diameter, and a fly. This was the first report on phase-contrast imaging by using X-rays and utilizing their wave property.

In the 1980s, synchrotron radiation facilities were installed in many institutions, and they have provided researchers the opportunities to utilize strong monochromatic parallel-beam X-rays to study phase-contrast techniques and mammography. Arfelli *et al.* (1998) reported a study of a mammographic phantom and of breast tissue with a digital detector, that is, a linear-array silicon detector. In addition, Arfelli *et al.* (2000) reported phase-detection techniques with synchrotron radiation in the X-ray imaging of their mammographic phantom and of *in vitro* breast tissue specimens. Pisano *et al.* (2000) reported diffraction-enhanced imaging of human breast cancer specimens. Takeda *et al.* (2004) carried out a feasibility study of the interferometric X-ray techniques for breast cancer specimens.

On the other hand, Wilkins *et al.* (1996) described phase-contrast imaging with polychromatic X-rays with a high degree of lateral coherence from a micro-focus X-ray tube, showing clear images of small fish. Later, based on the theory of Kotre and Birch (1999), they investigated phase-contrast imaging using a conventional X-ray tube in mammography. In order to obtain high coherence, they placed the object at 2 m or more away from the X-ray tube, and used a conventional S/F mammography system in their experimental study. This imaging configuration was not practical for clinical use because of the extended exposure time. Wu and Liu (2003) investigated theoretical phase-contrast mammography with coherent X-rays from

an X-ray tube with a focal spot of 25 µm, where the distance between the focal spot and the object was 0.65 m. They expected that an X-ray tube with this focal spot size could be developed to emit sufficient intensity of X-rays for medical use. Fitzgerald (2000) extensively reviewed phase-contrast imaging techniques with synchrotron X-rays and micro-focus X-ray tubes, and stated that the phase-contrast technique is useful for mammography in principle. However, he also questioned the clinical use of phase-contrast techniques because of the size of synchrotron radiation facilities and the weakness of X-rays from micro-focus X-ray tubes designed for nondestructive testing.

Ishisaka *et al.* (2000) presented a new method of analyzing the edge effect in phase-contrast imaging with incoherent X-rays from conventional X-ray tubes for medical use. Based on theoretical analysis, Honda *et al.* (2002) succeeded in developing phase-contrast mammography using a conventional molybdenum anode tube with a 100 µm focal spot and conventional high-speed mammography S/F systems with the object placed 1.0 m away from the X-ray tube. Freedman *et al.* (2003) reported improvement of detectability in ACR RMI 156 phantom images by using the phase-contrast technique in digital mammography with a customized mammography unit that had a molybdenum anode tube with a 100 µm focal spot, where the distance between the focal spot and the object was 0.60 m. In these studies, the fundamentals of phase techniques were established for clinical mammography.

Absorption Contrast and Phase Contrast

In order to understand the phase-contrast technique, we need to discuss X-ray imaging using physical optics. An X-ray is an electromagnetic wave as represented, for example, as $A\sin\phi$, which carries information regarding amplitude, A, and phase, ϕ. The X-ray attenuates after passing through an object in air due to a photoelectric effect and Compton scattering, upon which the amplitude decreases from A to B. This change causes differences of image density on a radiogram, and this is referred to as absorption contrast. This principle of X-ray imaging has been known since the discovery of the X-ray. At the same time, after the penetration of the X-ray through the object, a phase shift from ϕ to $(\phi + \Delta)$ takes place in addition to the amplitude change, because the X-ray propagates in the object faster than in the air. The wave after the penetration is represented then as $B\sin(\phi + \Delta)$. The phase shift, Δ, generates refraction and interference. The interference takes place only with coherent waves from unique X-ray sources, such as synchrotron and micro-focus X-ray tubes. However, using conventional medical X-ray tubes, we can utilize refraction as phase shift (Ishisaka *et al.*, 2000). The detection of this phase shift as a difference in

X-ray intensity is defined as phase-contrast imaging (Nugent *et al.*, 1996), whereas the difference in image density due to the phase shift is defined as phase contrast.

Edge Effect Due to Phase Contrast

Electromagnetic waves refract at boundaries of materials with different refractive indexes because phase shift manifests itself after passing through the boundaries. The energy of an X-ray is greater than the resonance energies of the electrons in the atoms of which materials consist, so although the refractive indexes of visible light are greater than unity, the refractive indexes of X-rays are consequently smaller than unity. For example, for glass suspended in the air, the refractive index, n, for visible light is ~ 1.4, whereas that for X-rays of 20 keV is $n = 1 - \delta$ ($\delta = 1.3 \times 10^{-6}$). Thus, although visible light after passing through cylindrical glass is concentrated, X-ray beams expand at a small angle in the range from 10^{-6} to 10^{-7} radian.

In Figure 82, the principle of PCM is illustrated for a cylindrical object suspended in the air, where an X-ray detector is placed away from the object at a distance of R_2. The diverged X-ray flux after passing through the object can be superimposed on the flux, which travels straight through the air. When this happens, the refraction of the X-ray due to the phase shift is detected as a change in X-ray intensity resulting in phase contrast. In actual radiography, absorption contrast is superimposed on the phase contrast at the fringe of the object image, so that we obtain edge enhancement in the X-ray image. This is called the edge effect that results in clear X-ray images. For example, on the right in Figure 82, the fringe of the plastic cylinder with an 8.5 mm diameter is clearly depicted with phase contrast. In addition, the white lines are surrounding the air bubbles in the cylinder due to edge enhancement, whereas they cannot be seen in the contact image with only absorption contrast.

The edge effect or the edge-enhancement effect is well known in the fields of photography and radiography. For example, in photography, the edge effect arises from chemical adjacency effects due to the diffusion of chemical byproducts in monochromatic film development processes. In xeroradiography, the edge effect due to the electrostatic phenomenon was utilized in mammography (Sickles, 2000). Because the edge effect in radiography is useful for detecting abnormalities in medical imaging, it is applied to digital mammography (Higashida *et al.*, 1992) with an unsharp-masking technique. In the edge effect generated by image processing of unsharp-masking, the noise gained before image processing is enhanced and overlapped on the image signals acquired by a digital detector. For this reason, the images with strong edge enhancement resulting from the image processing are often grainy. On the other hand, the edge enhancement due to phase contrast takes place on the detector in image acquisition, and it does not enhance the noises generated in the imaging device. Consequently, the edge effect due to phase contrast does not induce an increase in noise arising from the imaging system used.

Realization of the Phase-contrast Technique in Mammography

In order to perform phase-contrast imaging in mammography, a new mammography unit has been designed, as seen on the left in Figure 82. The distance between a focal spot and an object, R_1, has been first determined to be 0.65 m. This distance is equivalent to that between the focal spot and the object for conventional contact mammography, so that there is no change in the incident angles of X-rays to the object from the conventional contact mammography. Consequently, the structure of the breast in a projected image on the detector is kept consistent from that of the conventional contact mammography.

An X-ray detector should be placed away from an object to obtain phase contrast. The construction of a mammography unit mechanically permits the longest distance from the X-ray source and the detector (source-image distance: SID) to be less than 1.2 m. In addition, our simulation on phase contrast in mammography has revealed that the phase contrast can be obtained most effectively at approximately two times magnification, when the magnified image is demagnified to the original object size. On the other hand, in conventional contact mammography, the size of the cassette used is $8'' \times 10''$. When we employ a magnification ratio of 1.75, we should use a $14'' \times 17''$ cassette for full-field mammography. Then the distance between the object to the detector, R_2, is designed to be 0.49 m in order to obtain $\times 1.75$ magnification with an R_1 of 0.65 m, where the SID is 1.14 m ($0.65 + 0.49 = 1.14$), which is less than 1.2 m. The dedicated mammography unit was designed for phase-contrast imaging with a nominal 100 µm focal spot with no antiscattering X-ray grid. In this new design of a mammography unit, we have attached a plastic cover in order to eliminate shadows of clothing or any item attached to a patient in craniocaudal (CC) view (Fig. 82).

Design of Digital Image Acquisition and Output

The goal of the full-field digital PCM system is to achieve a clinical performance that is equal to or better than that with conventional S/F mammography. Thus, we have designed 25 µm pixels in output for mammographic images because the spatial resolution of S/F mammography is 20 cycles/mm, which is equivalent to a 25 µm pixel in digital mammography. The limit of spatial resolution for the naked

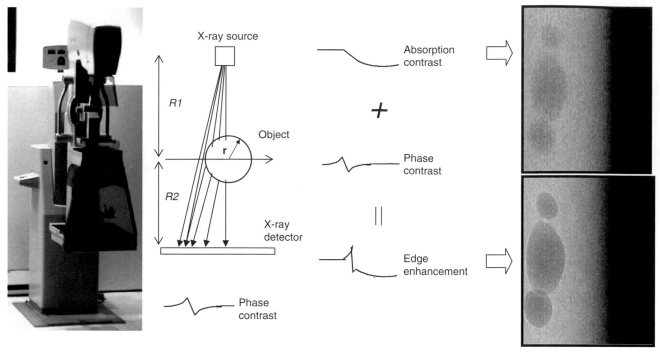

Figure 82 Principle of phase-contrast mammography and a dedicated mammography unit.

human eye is 73 μm when the distance between the eyes and an object is 0.25 m. This suggests that a spatial resolution of 10 cycles/mm is sufficient for reading mammograms with the naked eye because a pixel size of 50 μm at 10 cycles/mm surpasses 6.8 cycles/mm at 73 μm. However, magnifiers are used for reading mammograms, suggesting that more than twice the spatial resolution (i.e., 13.6 cycles/mm) is required in mammography. In order to depict the fringes of calcification grains smoothly, it is necessary to eliminate aliasing noise in the range from 7–14 cycles/mm in digital images, when the X-ray image acquisition system has high enough sharpness such as high values of a modulation transfer function (MTF) in the range of spatial frequency. In other words, to depict smoothly the shape of fine calcification grains with high sharpness, we have designed 25 μm pixels in output digital images (Honda, 2004).

In this system, the acquired image in ×1.75 magnification is reduced in size to the original object for an output hardcopy image (i.e., demagnification). When a phase-contrast magnified image is read at 43.75 μm pitch in a CR unit, then information from each pixel is printed using 25 μm pixels ($43.75 \div 1.75 = 25$) on photothermographic dry film. The film is designed to reach a maximum optical density of 4.0 because the maximum printed density of conventional S/F mammography is greater than 4.0.

Magnification-demagnification Effect in Digital Mammography

Sickles *et al.* (1977) reported the effect of magnification on image quality, particularly resolution, system noise, and contrast in an S/F mammography system. In the full-field digital PCM, an image magnified by 1.75 is reduced in size by the magnification ratio on the printing image. As a result, we can obtain an advantage in image quality such as sharpness and image noise through the magnification-demagnification process in addition to the phase-contrast effect (Ohara *et al.*, 2002).

Sharpness

Shaw *et al.* (2000) have reported that magnification increases sharpness through a rescaling effect that is in competition with geometric unsharpness in digital mammography. The focal spot size of the X-ray source in the mammography unit is as small as 100 μm nominally, and the image sharpness is improved by the rescaling effect, where the improvement of sharpness due to magnification reaches the maximum in ×3 magnification, because geometric unsharpness increases along with the magnification ratio (Ohara *et al.*, 2002).

It is well known that scattered X-rays from an object reduce the image contrast; for this reason, antiscatter X-ray grids are used in conventional contact mammography. However, in the full-field digital PCM system, no antiscatter X-ray grid is used because of the air-gap effect due to the distance between the object and the storage phosphor plate, R_2. Our results of effective scatter fraction measured with the Pb disc method have revealed that the effective scatter ratio at 0.49 m of R_2 is almost equal to that for contact mammography without any grid (Tanaka et al., 2005).

Image Noise

Radiographic mottle due to quantum mottle is increased by magnification in the absence of concurrent increase of X-ray dose to an object because of the reduction in number for the X-ray photon hitting a unit area of the X-ray detector. In addition, it is easily understood that reduction of the magnified image to the original size of the object in printing would gain the increase of the mottle. Equation (1), based on the theoretical study of this effect of noise reduction, is reported by Doi and Imhof (1977).

$$WS_m(u) = WS(u/m)/m^2 \qquad (1)$$

where $WS(u)$ is the wiener spectrum at u cycles/mm of spatial frequency, and m is the magnification factor. The $WS_m(u)$ is related with the image noise. In this equation, the magnification-demagnification process causes the variation of m eventually to equal 1 ($1.75 \times (1/1.75) = 1$). As a result, the image noise in the magnified image is reduced in the image after demagnification. Using this technique, we can perform full-field digital PCM with the equivalent dose of the conventional contact digital mammography with no increase of noise.

If a storage phosphor plate is set apart from an object, then it eliminates scattered X-rays from the object through an air-gap effect, whereas an antiscatter X-ray grid is used in conventional S/F mammography to eliminate the scattered X-rays in order to avoid loss of image contrast. The antiscatter X-ray grid also hinders passing primary X-rays in reaching at an X-ray detector, and this results in increased image noise. Consequently, this air-gap effect gives the full-field digital PCM system an advantage in noise reduction over conventional contact mammography with an antiscatter X-ray grid (Ohara et al., 2002).

Improvement of Image Quality by the Magnification-demagnification Effect

Image quality is demonstrated with MTF for image sharpness, noise power spectrum (NPS) for image noise, and noise equivalent quanta (NEQ) for total image quality at u cycles/mm of spatial frequency, defined as

$$NEQ(u) = MTF^2(u)/NPS(u) \qquad (2)$$

In Figure 83, the experimental results of the image quality for a demagnified image from ×1.75 magnification are shown for (a) presampling MTF and (b) NPS in solid lines compared to conventional contact digital imaging in dotted lines (Gido et al., 2005). The calculated NEQ results with Eq. (2) are also presented in Figure 83c.

The presampling MTF was obtained from the Fourier transition of the line spread function (LSF), measured by a 10 μm-wide slit made of 2 mm-thick tungsten. Measurement of the LSF was conducted at 28 kVp of tube voltage using Mo anode-target and Mo filter. The presampling MTF for digital PCM represented by the solid line is superior to that for contact digital mammography represented by the dotted line in the range of measured spatial frequency through 20 cycles/mm. Note that the Nyquist frequency for the contact digital mammography is 11.4 cycles/mm, because the sampling pitch is 43.75 μm for contact digital mammography in this experiment.

One-dimentional NPSs were measured for images with uniform exposure of X-rays through a 4 cm Lucite filter at 28 kVp of tube voltage and 42 mAs using Mo anode-target and Mo filter. Note that the contact digital mammography was performed with an antiscatter x-ray grid, but digital PCM without it. As a result of the image noise, NPS for the digital PCM represented by the solid line up to 10 cycles/mm is lower than that for digital contact mammography represented by the dotted line.

The NEQs, calculated from values of MTF and NPS with Eq. (2), are shown in Figure 83c. The NEQ for digital PCM is twice or more superior to that of contact digital mammography. The NEQ can also be defined as $NEQ = (S/N)^2_{out}$, where S/N is the signal-to-noise ratio. Based on this consideration, the increase of NEQ for the digital PCM suggests that the digital PCM may provide more image signals or less noise, that is, more information, through the magnification-demagnification effect than the conventional digital mammography with a scatter-X-ray grid.

Improvement of Image Sharpness in Digital Full-field PCM

Qualitative evaluation such as NEQ as we have stated has so far been limited to the image of absorption contrast basis that has been applied to conventional X-ray images. For full-field digital PCM images, improvement of image contrast due to phase contrast should be considered in addition to the assessment stated earlier. Because the phase contrast manifests itself depending on differences between refraction indexes on the boundary of materials, the effects on image

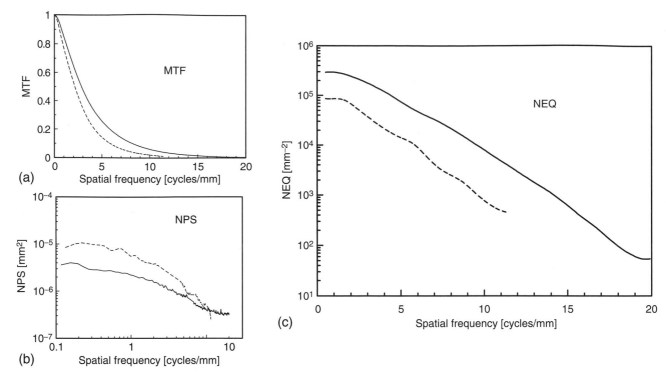

Figure 83 The image quality of demagnified images from magnified images at ×1.75: (a) presampling MTF, (b) NPS, and (c) NEQ.

quality by phase contrast should be evaluated separately based on the refraction indexes from part to part, such as microcalcifications and tumors.

Ohara *et al.* (2002) have conducted theoretical assessment of the edge effect due to phase contrast on image sharpness in addition to the magnification effect. The theoretical consideration suggests that the phase-contrast effect would be more conspicuous in the images of fibroglandular tissue than calcification grains in the breast images. Matsuo *et al.* (2005), in their empirical study of the total image quality evaluation including phase contrast, reported that phase contrast increases image signal as spatial frequency increases.

Clinical Images

We have already conducted mammographic examinations of more than a thousand patients at Shiga University using the full-field PCM system for a period of 2 years, including an initial 6 months for clinical trials, in order to compare the digital mammography system with the conventional S/F mammography system, which had been used in our medical facilities. Here, we describe briefly the results of our clinical trials on clinical images and receiver operating characteristic (ROC) evaluation (Tanaka *et al.*, 2005).

To compare the images of this full-field digital PCM system with a conventional S/F system, we chose for the S/F system a mammography unit, a Senographe DMR (General Electric, USA), with a nominal 0.3 mm focal spot for contact imaging, and employing a UM-MAMMO FINE (screen) / UM-MA HC (film) (Fuji Photo Film, Tokyo). The exposed film was developed with an automatic processor, a Cepros M2 (Fuji Photo Film, Tokyo), with the developing solution temperature at 36°C.

Clinical trials for 38 patients were conducted for a preliminary ROC evaluation of calcification and mass. Mammograms were taken for each patient from a mediolateral (MLO) view for one side of the breast, first with conventional S/F contact mammography and then, on the same day, with the digital PCM system. The X-ray tube voltage and glandular dose for each PCM exposure, with no scattering X-ray grid used, were adjusted to those of the corresponding conventional S/F mammography system, in which an antiscatter X-ray grid was used. The exposure conditions for conventional S/F mammography were determined by an automatic exposure control system in the Senographe DMR mammography unit (General Electric, USA) in advance of digital PCM imaging with the MGU-100B unit (Toshiba, Tokyo). There were 24 microcalcification clusters (12 clusters: category 3; 5 clusters: category 4; 7 clusters: category 5 by BIRADS-interpretation) and 21 masses (9 masses:

category 3; 6 masses: category 4; 6 masses: category 5 by BIRADS-interpretation) in the 38 patients.

For ROC evaluation, three qualified radiologists with expertise in mammography independently read 38 sets of S/F and PCM mammograms arranged in random order. This evaluation of microcalcification and mass was conducted via the continuous convince degree method. From the results, Az values were determined with the ROCKIT computer software application program, obtained from the University of Chicago's Internet homepage. The results of a preliminary ROC evaluation for the full-field digital PCM system were Az = 0.9657 for microcalcification and Az = 0.9651 for masses, both of which were larger than the results for the SF system (Az = 0.8874 for microcalcification and Az = 0.9546 for masses). These results suggest that the clinical breast images in the full-field digital PCM exceed those of the S/F mammography used in this trial.

In treatment of breast cancer, we compared PCM and S/F imaging of a pathologic specimen. The PCM images showed the breast cancer mass more clearly than did the S/F images. In Figure 84, the breast images from a 71-year-old woman are shown in order to compare between (a) the full-field digital PCM and (b) the SF mammography system with the pathological photographs for (c) macroscopic and (d) microscopic images. The pathologic diagnoses were invasive ductal carcinoma, scirrhous type, ly(+), v(+), intraductal spread(−). The phase-contrast image shows the breast cancer mass more clearly than the SF, especially regarding the spiculated structure surrounding the mass.

Clinical Experience

As seen in Figure 82, the construction of the mammography unit for the phase-contrast mammography is different from that of current mammography units. It seems that the bulky attachment under the object holder would hinder positioning of the patient's breasts for mammography in a CC view, although mammography in MLO and lateral views would be conducted in the same manner with the current mammography units regarding positioning. However, in our clinical experience, it has been revealed that a plastic protecting plate against the body of a patient helps the patient to relax by leaning on it in the CC-view position. For patients in a wheel chair, the object holder and the storage phosphor plate holder can be exchanged with a currently used object holder unit for contact mammography.

As for exposure time in the full-field digital PCM system, the average exposure time for 249 exposures (Mo filter: 209; Rh filter: 40) in a month in our medical facilities was 1.29 sec for the average compressed width of 37.2 mm. This average exposure time is shorter than the 1.38 sec for S/F mammography, which was reported as an average exposure time in a 1992 survey in the United States (Conway et al., 1994). In the S/F system, low-intensity reciprocal failure of silver halide materials causes the exposure time to be longer, especially beyond 2 sec for dense breasts (Kimme-Smith et al., 1991), whereas CR obeys the low-intensity reciprocity effect in the exposure time regions of seconds. In addition, the compressed widths of breasts for the Japanese patients would not be so thick as those in the United States (Fujisaki et al., 2002; Young et al., 2005).

Because a 14″ × 17″ plate is used with a sampling ratio at 43.75 μm, the data volume is 128 MB per exposure (i.e., 70 mega-pixels per image acquisition). The transfer of the image data in the full-field digital PCM system requires a period of time depending on the processing speed of the computer used; however, we do not experience any delay in examination. An examination for one patient with four-view images in MLO and CC views takes 7–15 min with the full-field digital PCM system: the four shots for two MLO views and two CC views takes 4 min with the interval between patients being 3 min. As a result, to screen 200 patients, the time necessary would be ~ 2.5 hr. This time is equivalent to that for SF mammography.

For spot compression at ×1.5 magnification, R_1 is 0.43 m and R_2 is 0.71 m, resulting in a magnification ratio of 2.65 in image acquisition. This magnification ratio for mammography is the highest today. Note that the distance of $R_1 + R_2$, SID, is 1.14 m, equal to that for full-field digital PCM. For output images, the image magnified ×2.65 in acquisition is printed on a 1.5-magnified print image with 25 μm pixels (2.65/1.75 = 1.5). Because the improvement of MTF in ×2.65 magnification is greater than that in ×1.75 magnification for full-field mammography, the shape discrimination of calcification grains increases, and smaller size grains are also depicted better than in the full-field mammographic images, although Cowen et al. (1997) reported that the minimum detectable size of microcalcification is 200 μm.

For larger breasts, a configuration of 0.78 m for R_1, the distance between the focal spot and the object, and 0.36 m for R_2, the distance between the object and the storage phosphor plate, could be designed. In this case, the SID is 0.78 m + 0.36 m = 1.14 m, where the magnification ratio is 1.46. The output images reduced by 1.46 times would be depicted with pixels of 30 μm. In this system, we can use softcopy display such as the liquid crystal display (LCD). However, the highest specifications of LCD monitors today are not sufficient for our requirements because the output image is intended to be printed with 25 μm pixels, and our dry film for this image is designed to have a maximum density of 4.0. For monitor diagnosis, computer-aided detection (CAD) would be more accurate using this system than conventional FFDM.

Future Development

In full-field digital PCM, the molybdenum anode X-ray tube used has a 100 μm focal spot, which emits incoherent X-rays. In future, the phase-contrast technique using coherent

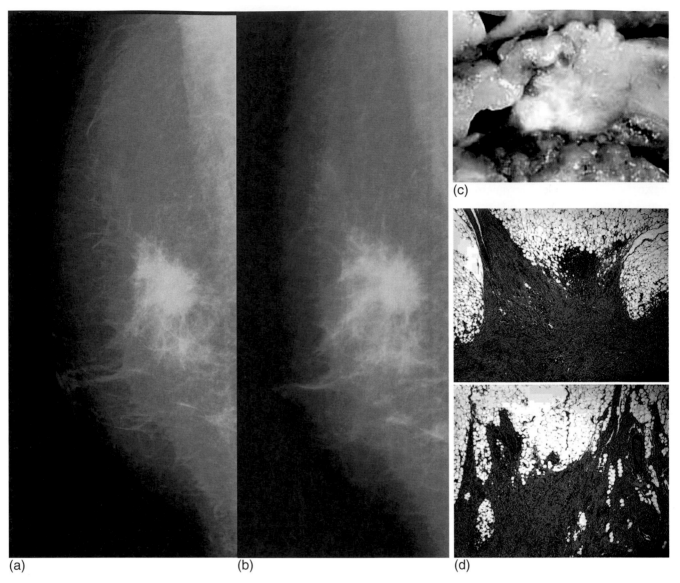

(a)   (b)   (d)

Figure 84 Comparison of clinical images between (a) digital full-field PCM and (b) conventional S/F mammography. Pathological photographs for (c) macroscopic and (d) microscopic specimen.

X-rays will be realized in mammography for widespread clinical use. Wu and Liu (2003) presented a theoretical foundation and design considerations for clinical implementation of X-ray phase-contrast imaging using coherent X-rays. They designed the SID for phase-contrast mammography to be 1 m with an X-ray tube having a 25 μm focal spot, which is not available today for mammography. They also reported the feasibility of a new X-ray tube with a 25 μm focal spot, which emits X-rays strong enough for medical use. They suggested that a small enough pixel size for an acquisition device has to be provided in order to capture the phase contrast when the distance between an object and a detector, R_2, is as short as 0.40 m or less. The progress of technology for X-ray tubes and digital acquisition devices will make phase-

contrast mammography units compact and will further improve image quality.

Phase-contrast techniques have been applied experimentally to general radiography by using a tungsten-anode X-ray tube with a 100 μm focal spot (Donnelly and Price, 2002; Donnelly *et al.*, 2003; Ohara *et al.*, 2002). Gao *et al.* (1999) reported phase-contrast imaging with a tungsten-anode X-ray tube with microfocal spots of 3 μm to 20 μm for images of soft tissue and bone samples. In the past, a tungsten-anode X-ray tube with a 50 μm, 100 μm, or 160 μm focal spot was used for study of magnification radiography with SF systems in skeletal radiology (Genant *et al.*, 1977) and pediatric radiology (Brasch and Gould, 1982). Computed radiography systems are also used for magnification radiology with an X-ray tube having a small focus, such as a 60 µm used in skeletal radiology (Link *et al.*, 1994). It is expected that the phase-contrast technique will be utilized in chest radiology (Shimizu *et al.*, 2000). For phase-contrast general radiology in widespread use, new technology will create X-ray sources, which emit strong enough X-rays from small focal spots for medical use.

A new tabletop synchrotron has been developed by Yamada *et al.* (2002). Sato *et al.* (2004) have developed a new X-ray source, which radiates monochromatic X-rays from cerium targets. The progress of the new technologies on X-ray sources will lead to phase-contrast 3D images and phase imaging such as diffraction-enhanced imaging (Pisano *et al.*, 2000) and interference X-ray imaging (Takeda *et al.*, 2000). These new methods will provide new diagnostic value for patients. New research in the field of phase contrast will continue to be refined and developed in technology, which has the potential to benefit patients on a global scale.

Acknowledgments

The authors thank Dr. Kiyoshi Murata and Mr. Kazutaka Masuda at the Shiga University of Medical Science, Department of Radiology, for their support to this work. They also appreciate the technical aids of Mr. Hisashi Yonekawa, Mr. Sumiya Nagatsuka, Mr. Akira Ishisaka, and Mr. Hiromu Ohara in Konica Minolta Medical & Graphic, Inc.

References

Arfelli,F., Bonvicini, V., Bravin, A.,Cantatore, G., Castelli, E., Palma, L.D., DiMichiel, M., Long, R., Olivo, A., Pani, S., Pontoni, D., Poropat, P., Prest, M., Rashevsky, A., Tronba, G., and Vacchi, A. 1998. Mammography of a phantom and breast tissue with synchrotron radiation and a linear array silicon detector. *Radiology* 208:709–715.

Arfelli, F., Bonvicini, V., Bravin, A., Cantatore, G., Castelli, E., Palma, L.D., DiMichiel, M., Fabriziol, M., Long, R., Menk, R.H., Olivo, A., Pani, S., Pontoni, D., Poropat, P., Prest, M., Rashevsky, A., Ratti, M., Rigon, L., Tronba, G., Vacchi, A., Vallazza, E., and Zanconati, F. 2000. Mammography with synchrotron radiation: phase-detection techniques. *Radiology* 215:286–293.

Brasch, R.C., and Gould, R.G. 1982. Direct magnification radiography of the newborn infant. *Pediat. Radiol. 142*:649–655.

Conway, B.J., Suleiman, O.H., Rueter, F., Antonen, R.G., and Slayton, R.J. 1994. National survey of mammographic facilities in 1985, 1988, and 1992. *Radiology 191*:323–330.

Cowen, A.R., Launders, J.H., Jadav, M., and Brettle, D.S. 1997. Visibility of microcalcifications in computed and screen-film mammography. *Phys. Med. Bio. 42*:1533–1548.

Doi, K., and Imhof, H. 1977. Noise reduction by radiographic magnification. *Radiology 122*:479–487.

Donnelly, E.F., and Price, R.R. 2002. Quantification of the effect of kVp on edge-enhancement index in phase-contrast radiography. *Med. Phys. 29*:999–1002.

Donnelly, E.F., Price, R.R., and Pickens, D.R. 2003. Dual focal-spot imaging for phase extraction in phase-contrast radiology. *Med. Phys. 30*: 2292–2296.

Fitzgerald, R. 2000. Phase-sensitive X-ray imaging. *Physics Today 53*:23–26.

Freedman, M.T., B-Lo, S.C., Honda, C., Makariou, E., Sisney, G., Pien, E., Ohara, H., Ishisaka, A., and Shimada, F. 2003. Phase contrast mammography using molybdenum X-ray: clinical implications in detectability improvement. *Phys. Med. Imag. Proc. SPIE. 5030*:533–540.

Fujisaki, T., Igarashi, I., Takahashi S., Watanabe, K., Nishiyama, K., Abe, S., Saito, H., Fukuda, K., and Matsumoto, M. 2002. Investigation of radiation quality and doses in Japanese routine mammography. *Nippon Acta. Radiologica 62*:436–441 (in Japanese).

Funke, M., Breiter, N., Hermann, K.P., Oestmann, J.W., and Grabbe, E. 1998. Storage phosphor direct magnification mammography in comparison with conventional screen-film mammography-a phantom study. *Brit. J. Radiol. 71*:528–534.

Gao, D., Pogany, A., Stevenson, A.W., Gureyev, T., and Wilkins, S.W., 1999. X-ray phase-contrast imaging study of soft-tissue and bone samples. *Phys. Med. Imag. Proc. SPIE 3659*:346–355.

Genant, H.K., Doi, K., Mall, J.C., and Sickles, E.A. 1977. Direct magnification for skeletal radiology. *Radiology 123*:47–55.

Gido, T., Nagatsuka, S., Amitani, K., Yonekawa, H., Shimoji, M., and Honda, C. 2005. Advanced digital mammography system based on phase contrast technology. *Phys. Med. Imag. Proc. SPIE 5745*:511–518.

Hasegawa, S., Oonuki, K., Nagakubo, J., Kitani, A., Oyama, K., Koizumi, R., and Ohuchi, N. 2003. Breast cancer imaging by mammography: effects of age and breast composition. *J. Jpn. Assoc. Breast Cancer Screen. 12*.101–107 (in Japanese).

Higashida, Y., Moribe, N., Morita, K., Katsuda, N., Hatemura, M., Takada, T., Takahshi, M., and Yamashita, J. 1992. Detection of subtle microcalcifications: comparison of computed radiography and screen-film mammography. *Radiology 183*:483–486.

Honda, C. 2004. Fundamental technical concept of digital phase contrast mammography. *Med. Imag. Information Sci. 21*:230–238 (in Japanese).

Honda, C., Ohara, H., Ishisaka, A., Shimada, F., and Endo, T. 2002. X-ray phase imaging using small focus X ray tubes. *Jpn. J. Med. Phys 22*:21–29 (in Japanese).

Ishisaka, A., Ohara, H., and Honda, C. 2000. A new method of analysis edge effect in phase contrast imaging with incoherent X-ray. *Opt. Rev. 7*:566–572.

Kimme-Smith, C., Bassett, L.W., Gold, R.H., and Chow S. 1991. Increased radiation dose at mammography due to prolonged exposure, delayed processing, and increase film darkening. *Radiology 178*:387–391.

Kotre, C.J., and Birch, I.P. 1999. Phase contrast enhancement of X-ray mammography: a design study. *Phys. Med. Biol. 44*:2853–2866.

Link, T.M., Rummeny, E.J., Lenzen, H., Reuter, I., Roos, N., and Peters, P.E. 1994. Artificial bone erosions: detection with magnification radiography versus conventional high-resolution. *Radiology 192*: 861–864.

Matsuo, S., Katafuchi, T., Toyama, K., Morishita, J., Yamada, K., and Fujita, H. 2005. Empirical evaluation of edge effect due to phase imaging for mammography. *Med. Phys. 32*:2690–2697.

Nugent, K.A., Gureyev, T.E., Cookson, D.F., and Barnea, Z. 1996. Quantitative phase imaging using hard X-rays. *Phys. Rev. Lett. 77*:2961–2964.

Ohara, H., Honda, C., Ishisaka, A., and Shimada, F. 2002. Image quality in digital phase contrast imaging using a tungsten anode X-ray tube with a small focal spot size. *Phys. Med. Imag. Proc. SPIE 4682*:713–723.

Pisano, E.D., Gatsonis, C.A., Hendric, R.E., Tosteson, A.N.A., Fryback, D.G., Bassett, L.W., Baum, J.K., Conant, E.F., Jong, R.A., Rebner, M., and D'Orsi, C.J. 2005. American College of Radiology imaging network digital mammographic imaging screening trials; objectives and methodology. *Radiology 236*:404–412.

Pisano, E.D., Johnston, R.E., Chapman, D., Geradts, J., Iacocca, M.V., Livasy, C.A., Washburn, D.B., Sayers, D.E., Zhong, Z., Kiss, M.Z., and Thomolison, W.C. 2000. Human breast cancer specimens: diffraction-enhanced imaging with histologic correlation—improved conspicuity of lesion detail compared with digital radiography. *Radiology 214*:895–901.

Sato, E., Tanaka, E., Mori, H., Kawai, T., Ichimaru, T., Sato, S., Takayama, K., and Ido, H. 2004. Demonstration of enhanced K-edge angiography using a cerium target X-ray generator. *Med. Phys. 31*:3017–3021.

Shaw, C.C., Liu, X., Lemack, M.S. , Rong, J.X., and Whitman, G.J. 2000. Optimization of MTF and DQE in magnification radiology—theoretical analysis. *Phys. Med. Imag. Proc. SPIE. 3977*:466–475.

Shimizu, K., Ikezoe, J., Ikura, H., Ebara, H., Nagareda, T., Yagi, N., Umetani, K., Yesugi, K., Okada, K., Sugita, A., and Tanaka, M. 2000. Synchrotron radiation microtomography of the lung specimens. *Phys. Med. Imag. Proc. SPIE. 3977*:196–204.

Sickles, E.A. 2000. Breast imaging: from 1965 to the present. *Radiology 215*:1–16.

Sickles, E.A., Doi, K., and Genant, H.K. 1977. Magnification film mammography: image quality and clinical studies. *Radiology 125*:69–76.

Somenkov, V.A., Tkalich, A.K., and Shil'shtein, S.S. 1991. Refraction contrast in X-ray introscopy. *Sov. Phys. Tech. Phys. 36*:1309–1311.

Tanaka, T., Honda, C., Matsuo, S., Noma, K., Oohara, H., Nitta, N., Ota, S., Tsuchiya, K., Sakashita, Y., Yamada, A., Yamasaki, M., Furukawa, A., Takahashi, M., and Murata K. 2005. The first trial of phase contrast imaging for digital full-field mammography using a practical molybdenum X-ray tube. *Invest. Radiol. 40*:385–396.

Takeda, T., Momose, A., Hirano, K., Haraoka, S., Watanabe, T., and Itai, Y. 2000. Human carcinoma: early experience with phase-contrast X-ray CT with synchrotron radiation—comparative specimen study with optical microscopy. *Radiology 214*:298–301.

Takeda, T., Wu, J., Tsuchiya, Y., Yoneyama, A., Lwin, T.T., Aiyoshi, Y., Zeniya, T., Hyodo, K., and Ueno, E. 2004. Interference X-ray imaging of breast cancer specimens at 51 keV X-ray energy. *Jpn. J. Appl. Phys. 43*:5652–5656.

Wilkins, S.W., Guerev, T.E., Gao, D., Pogany, A., and Stevens, A.W. 1996. Phase-contrast imaging using polychromatic hard X-rays. *Nature 384*:335–338.

Wu, X., and Liu, H. 2003. Clinical implementation of X-ray phase-contrast imaging: theoretical foundations and design considerations. *Med. Phys. 30*:2169–2179.

Yamada, H., Hirai, T., Sonoda, Y., Takashige, T., Maki, S., Hasegawa, D., Kuribayasi, M., Hyodo, K., and Matsumoto, M. 2002. X-ray imaging with the novel X-ray source based on the tabletop synchrotron MIRRORCLE. *J. Soc. Photogr. Sci. Tech. Jpn. 65*:452–458. (in Japanese).

Yip, W.M., Pang, S.Y., Yim, W.S., and Kwok, C.S. 2001. ROC curve analysis of lesion detectability on phantoms: comparison of digital spot mammography with conventional spot mammography. *Brit. J. Radiol. 74*:621–628.

Young, K.C., Burch, A., and Oduko, J.M. 2005. Radition dose received in the UK breast screening programme in 2001 and 2002. *Brit. J. Radiol. 78*:207–218.

5

Full-field Digital Mammography versus Film-screen Mammography

Arne Fischmann

Introduction and Historical Perspective

Digital imaging technologies were introduced in the field of general radiology in the early 1980s. The advantages of digital imaging, such as postprocessing, softcopy-reading, three-dimensional imaging, and computer-aided detection (CAD), became obvious within a short time, and, subsequently, many departmental imaging facilities became digital. The only exception was mammography, which was limited to small field detectors for biopsy due to restricted spatial resolution of large detectors or limited size of high-resolution detectors. The latter were in general CCD-chips similar to those used in digital photocameras usually 5 × 5 cm in size. Many of these detectors are still in use for stereotactic biopsy equipment that can either be mounted on a conventional mammography system or installed in dedicated units, such as the Mammotest plus (Fischer-Imaging) or the Stereoguide breast biopsy table (Lorad), for prone breast biopsy.

In 1981, the first computed radiography (CR) system for full-field digital mammography was developed on the basis of photostimulable storage-phosphors. Though improved over CR systems in general radiology, these systems were still limited in spatial resolution, which resulted in inferior image quality compared to standard film-screen mammography (FSM) systems. Only the introduction of double-sided readout and consecutive increase in spatial resolution made this technique feasible for mammography.

The development of new detectors based on a matrix of amorphous silicon with either a scintillator with cesium iodide (CsI) or a detector based on amorphous selenium (aSe) followed in the next years. These detectors have the advantage of direct processing of the images and, therefore, of faster access to image data.

After the first systems were approved by the Food and Drug Administration (FDA) in 2000, several systems became available in the subsequent years, with multiple systems being installed worldwide. Nevertheless, film-screen is by far the most common; in January 2006 there were 832 facilities with 1160 FFDM units, compared to a total of 8872 facilities with 13,640 FDA-accredited units for mammography (www.fda.gov/CDRH/mammography). This means that < 10% of all mammography units were digital. It is expected that FFDM will replace a large number of older film-screen units in the next few years, and that in the near future most mammographic systems will be digital. This evolution is already visible in several European countries and the United States.

Physical Performance of Digital Compared to Film-screen Mammography

Although digital imaging in general radiology gained wide acceptance in a short time, there was resistance against this new technology when mammography came into play. This criticism was based mainly on the limited spatial resolution of digital mammography, which was thought to translate into inferior image quality, caused by incomplete understanding of physical and physiological principles of human perception. The maximal performance of the human eye in the discrimination of structures is spatial resolution of 2–3 line pairs/mm (lp/mm). Therefore, contrast resolution in this frequency band is crucial for visibility, while higher frequency bands are of less importance. In conventional or film-screen mammography, contrast resolution at any frequency is correlated with spatial resolution of high-contrast objects, indicated by lp/mm that can be discriminated with a given system.

On the other hand, spatial resolution of digital mammography systems is limited by pixel size or the so-called Nyquist-Frequency $(2^*\text{pixel-size})^{-1}$, while image quality is dependent on contrast resolution and signal-to-noise ratio (SNR). Therefore, the image quality of digital mammography is independent of maximal spatial resolution (Pisano *et al.*, 2005). We could show that the image quality of a system based on a CsI-scintillator with a spatial resolution of 5 lp/mm was superior to a system based on amorphous selenium with a spatial resolution of 7.5 lp/mm. This superior image quality is translated into better detection of small high-contrast objects of 0.06 mm diameter, which were below the minimal spatial resolution of both systems (Fischmann *et al.*, 2005b).

Although early reports characterizing digital mammography detectors concentrated on line-spread function and modulation transfer function (MTF), there is currently no report without measurements of the detective quantum efficiency (DQE) as an adequate measurement for digital mammography. The definition is: DQE = (SNR out)^2/(SNR in)^2 or, in short, the loss of signal-to-noise ratio through the detector. A perfect imaging system would have a DQE of one (100%) or no loss of signal through the process of detection. Detective quantum efficiency is also dependent on spatial frequency, so all comparison should be done with comparable parameters and publications that compare DQE should be read carefully (Noel and Thibault, 2004). Most film-screen systems have a DQE of ~ 0.3 at 0 lp/mm, the GE Senographe 2000D has a DQE that is almost twice as high, and detectors based on amorphous selenium such as the Lorad Selenia or Siemens Novation or photon-counting systems such as the Sectra Microdose reach even higher values. However, DQE measurements should be interpreted with caution because values are dependent on measurement parameters and cannot be directly translated into image quality. Therefore, the numerous characterizations of detec-

tors and mammography systems cannot replace comparison with trials on phantoms, patients, and screening populations.

Phantom Studies Comparing Full-field Digital Mammography and Film-screen Mammography

Most studies published on digital mammography were performed on phantom images, using different phantoms for digital and conventional mammography. While these studies are a valuable tool in evaluating image quality without unnecessary exposure of patients, the choice of phantom is essential. Owing to increased contrast resolution and availability of wide-contrast levels for digital mammography, observer performance with the American College of Radiology (ACR)-accreditation phantom is constant when digital systems are tested at dose levels used in analog mammography, masking differences in image quality (Huda *et al.*, 2002). The same could be said for the contrast-detail phantom CDMAM 3.2 used in European mammography screening. Current systems for digital mammography at clinical dose are able to detect all gold disks, meaning that possible differences between different systems cannot be detected. Only the development of a special phantom for digital mammography—the CDMAM 3.4—containing smaller and thinner gold disks (Fig. 85) enabled comparison of different FFDM systems.

However, phantoms such as the ACR-phantom or CDMAM 3.2 were used in several studies comparing FFDM to FSM. One example is a study comparing systems based on amorphous selenium and CCD to screen-film, where the digital systems were superior for all dose levels, but differences between digital systems could not be detected (Lee *et al.*, 2003). These results are contrary to a publication where images of an anthropomorphic phantom made with a third-generation CR system and FSM on two different mammography units showed significantly better detection of masses and calcifications for FSM (Kheddache *et al.*, 1999). Therefore, improvements in the later-generation CR systems might explain these contradictory results. This view is supported by recent trials, which showed comparable detection of CR compared to FSM and aSi (Schulz-Wendtland *et al.*, 2004). As spatial resolution of the aSi system is lower than that of CR (5 vs. 9 lp/mm), this again supports the view that spatial resolution is useless in evaluating digital mammography and should be avoided. These results were confirmed in a subsequent trial comparing five different digital units with film-screen: CR and aSi scored higher without significance ($p = 0.058$) compared to FSM, while two systems based on aSe as well as the slot-scanning CCD system showed significantly superior results (Schulz-Wendtland *et al.*, 2004). However, the superiority of aSe systems could not be reproduced in our own trials (Fischmann *et al.*, 2005b).

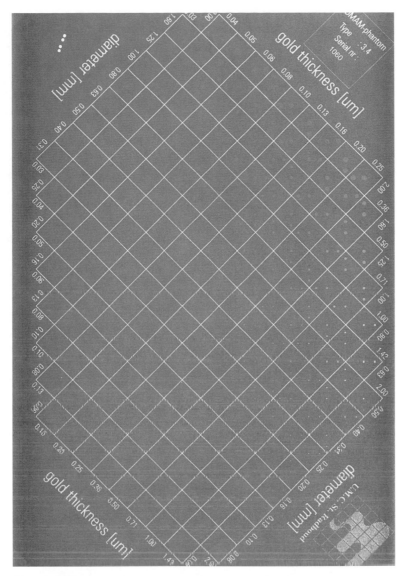

Figure 85 Image of a contrast-detail mammography-phantom type CDMAM 3.4. In a rectangular matrix gold disks of different thickness and diameter are embedded. Image quality can be evaluated by either counting the number of correct observed disks or by calculating an index over the thickness and diameter of the disks that were barely visible.

Simulated Microcalcifications

A detailed comparison of simulated microcalcifications with different systems for digital mammography revealed that detection for calcifications of 125–140 μm was superior with aSi detectors compared to CR and FSM. No differences in the detection of larger or smaller calcifications could be detected, although the aSi detector showed the highest scores (Rong *et al.*, 2002). Similar results were achieved when comparing aSi to CCD and FSM. The detector based on aSi outperformed the other modalities, while FSM was superior to CCD, especially for calcifications from 125–150 μm in size. Interestingly, the addition of anthropomorphic background reversed results, achieving superior imaging quality of CCD over FSM, although only larger calcifications could be detected (Lai *et al.*, 2005). As microcalcifications associated with breast cancer and detected with mammography are generally above 130 μm, the superiority of the aSi system for these calcifications is expected to produce superior detection. This result was supported by our own trials (Fischmann *et al.*, 2005a) as well as other trials presented in the next sections. In the Radiology Department in Tübingen, we also notice the inferior detection of calcifications with CCD detectors in daily clinical practice. It is often difficult to reproduce minute calcifications on a prone biopsy table with a CCD detector, while these calcifications are clearly visible with FFDM units based on aSi and amorphous selenium, which are available in our department. As problems with visualization also occurred for calcifications originally detected on analog film-screen mammography, we could not reproduce the superiority of CCD with an anthropomorphic background. On the other hand, the highest scores were achieved for the RMI-Phantom 152A with the SenoScan, a slot-scanning system based on 4 CCD (Schulz-Wendtland *et al.*, 2004).

When comparing low-contrast objects on FFDM and FSM, results seem to depend on phantom thickness. While measurements with 2 cm polymethyl methacrylate or plexiglass (PMMA) revealed similar results, detection for thicker phantoms is significantly better with digital mammography. This is probably due to better contrast resolution and the possibility of windowing for FFDM. Several trials considered possible dose reduction with FFDM due to higher DQE. A ROC analysis with simulated microcalcifications showed comparable detection and conspicuity of lesions and a possible dose reduction of 37% for normal views. With a magnification of 1.8, a dose reduction of even 51% could be achieved (Obenauer *et al.*, 2003). At 1.8 magnification, contrast-detail resolution of FFDM with 49% standard dose was comparable to FSM, while a dose reduction of 76% significantly reduced image quality of FFDM. It is therefore to be expected that digital mammography might reduce the dose by up to 50% without reduced image quality.

These dose reductions possible with FFDM are not confined to the special surroundings of phantom studies or clinical trials, but are generally found in clinical practice. A survey of different FFDM and FSM systems installed in the United States showed comparable exposure times and glandular dose for thin to intermediate breasts with both modalities, while for thick breasts FFDM reduced exposure time as well as glandular dose with less variability between different mammography units examined. Furthermore, image quality, indicated as signal-to-noise ratio, was superior for digital mammography (Berns *et al.*, 2002). Several systems based on amorphous selenium are on the market, but phantom trials comparing these to FSM are scant, except for the aforementioned trials with the Hologic detector. Other systems with detectors by a different manufacturer (Anrad, Montreal, Canada) are constructed by several manufacturers. Currently, Giotto is the only system on the market, but no publication has compared this system to FSM.

The synopsis of the phantom studies performed up until now is that FFDM is at least equivalent or superior to FSM, with the exception of single-side readout CR systems. As all manufacturers have replaced these by dual-readout systems, equivalence with FFDM can be assumed. Therefore, compar-

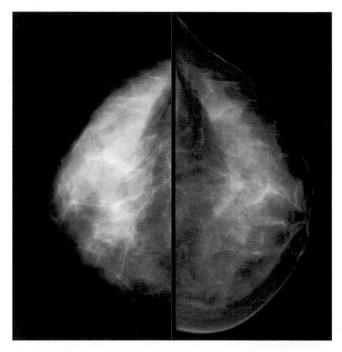

Figure 86 Craniocaudal views of one woman with film-screen mammography (left) and full-field digital mammography (right). Clearly visible are the better depiction of skin and nipple-areola area with digital mammography. Also clearly visible is the difference in the impression of the image, necessitating training in digital mammography.

isons of different mammography systems can be focused. These include new versions of systems that are compared to initial systems, such as the dual-sided CR system, which have proven to be superior to single-side CR. Other trials compared mammography based on amorphous selenium to the CsI-scintillator (Fischmann *et al.*, 2005b) or compared CR to slot-scan, asi, and aSe systems (Schulz-Wendtland *et al.*, 2004). Because these studies are to some extent contradictory, larger clinical trials would be necessary to prove equivalence or show potential differences between the systems.

Clinical or Diagnostic Digital Mammography

An early study on 22 patients with known breast lesions with CR and film-screens showed comparable image quality with increased detectability but inferior characterization of calcifications due to limited spatial resolution. For FDA approval of the GE Senographe 2000D, Hendricks *et al.* (2001) performed digital and conventional mammography with matched dose on 641 women, who presented for diagnostic mammography on several sites. Furthermore, images of 21 women with known breast cancer were included, so that a total of 44 cancers could be evaluated independently by five readers. While the inferiority of FFDM could be refuted with $p = 0.006$, digital mammography found a significant 2% reduction in the recall rate for digital mammograms. Sensitivity was 0.67 for digital and 0.70 for film-screen; corresponding specificity was 0.55 and 0.53, respectively.

In an intraindividual comparison of FFDM and FSM in 200 nonsymptomatic women with automatic exposure control, we did not find any dose reduction with digital mammography. As expected, the image quality of FFDM was superior, especially for dense breasts that were rated to be less dense and better evaluable (Fig. 86). In addition, a higher number of calcifications were detected with FFDM. This again supported the thesis that image quality is to a certain degree dependent on dose and that a superior system can be adjusted either to obtain better image quality at equivalent dose, or alternatively, to lower dose at equivalent image quality.

A study on 34 paired analog and digital examinations of women with known breast lesions showed superior lesion characterization with digital imaging, indicated by Az values of 0.81 versus 0.75, showing a significant difference (Van Woudenberg *et al.*, 2003). In a study on 57 clusters of microcalcification image, quality was assessed to be superior with FFDM in 50% and showed more calcifications in 41% of patients, with equivalence in all the other cases. Sensitivity and specificity were not different (Fischer *et al.*, 2002). Another study on 47 clusters of microcalcifications showed no significant difference between digital and conventional imaging at a comparable dose, although readers

rated digital images to be superior in quality. Nevertheless, inter-reader variability was higher than differences between the systems (Diekmann *et al.*, 2003).

To evaluate rates and causes of disagreement between digital and film-screen, Venta *et al.* (2001) conducted a study on 1147 breasts of 692 women scheduled for diagnostic mammography, who received imaging with both modalities. Assessment of breast density showed complete agreement in 69% ($K = 0.46$), with results being shifted to less dense on digital images. Furthermore, digital images revealed more clustered calcifications (183 vs. 146), more scattered calcifications (329 vs. 254), and more masses (188 vs. 160), but comparable numbers of architectural distortions and focal asymmetries. Comparison of BI-RADS assessment revealed a K-value of 0.20, with the main causes of disagreement being differences in the management approach between two radiologists (52%), additional imaging available for film-screen (34%), and technical differences in only 10%. Based on biopsy recommendation, intermodality disagreement was 4% compared to interobserver disagreement of 15–27%. While detection of calcifications is dependent on the imaging modality and, therefore, superior with FFDM, detection of focal asymmetries and architectural distortions is obscured by irregularities of the breast or "structural noise," rendering differences between detectors less important and emphasizing the importance of good mammographic positioning.

A recent trial on 232 patients with 46 cancers, 88 benign and 98 normal findings, showed Az values (from ROC analysis) of 0.916 with digital and 0.887 with film-screen mammography, which was a nonsignificant difference. Analysis of subgroups (calcifications, masses, malignant, or benign cases) showed no significant difference between the systems (Skaane *et al.*, 2005); however, the number of cases was too small to detect minor differences in this very heterogeneous group.

During the FDA approval of Fischer SenoScan, 676 women were enrolled in several phases: during phase 1, 560 women with an abnormal screening mammogram including 25 cancers were included; during phase 2, an additional 101 cancers from women scheduled for breast biopsy were enrolled; and during phase 3, 25 additional cancers were added. All women included in these phases underwent mammography with both modalities and hardcopy reading of FFDM; images of 247 cases were included in the final reader study due to time constraints. Eight readers evaluated all images of 136 asymptomatic and 111 women with cancer for an ROC analysis. The area under the curve (AUC), were 0.715 for digital and 0.765 for film-screen mammography, sensitivity of 0.66 versus 0.74, and specificity of 0.67 versus 0.60 for digital and film, respectively. None of these differences were statistically significant (Cole *et al.*, 2004). This trial presents several obvious limitations, including complex recruitment procedures as well as exclusion of cases after recruitment without specification of reasons. It was also stated that not all readers were able to read all film-screen examinations because images were removed for patient-care purposes. Considering the differences between the AUC values with a 95% confidence interval of −0.101 to 0.002, only minor changes in reading could have caused significance of the results. Hence, it is concluded that a trial with more power might show the inferiority of the SenoScan system.

Nevertheless, the results from the Digital Mammographic Imaging Screening Trial (DMIST) confirm the equivalence of both modalities independent of the system used for digital mammography. Therefore, it is probable that the Fischer SenoScan is equivalent to film-screen mammography.

The Lorad Digital Breast Imager (LDBI) is based on an array of 12 CCDs and was initially produced by TREX medical (Danbury, CT); its production was stopped after the Lorad Selenia imaging system was introduced. The FDA approval of the LDBI included 200 women, with 48 pathology-proven cancers, evaluated by 12 independent readers. Digital imaging reduced BI-RADS = 0 by 5.5% compared to screen-film as well as decreased false-positive rates, while the ROC analysis showed no significant difference. In general, the LDBI was similar or slightly better than film-screen (www.fda.gov 2002, PMA 010025).

Currently, two different systems with an identical detector based on amorphous selenium are approved by the FDA. While approval for the Lorad Selenia has not been published, the approval of the Siemens Mammomat Novation can be accessed (www.fda.gov 2004, PMA 030010). As both systems are based on the same detector, the clinical data acquired by Hologic were used for the approval of both systems. Physical characteristics and phantom scores were similar for both systems. Currently, results from DMIST are the only published clinical trials on those two systems.

Most phantom studies published on FFDM were performed with the Senographe 2000D; this is even more true for clinical trials or trials concerning patients with known breast disease. Apart from FDA-approval studies and phantom studies, almost all trials published were performed with this system. Although the conclusions drawn can probably be transferred to other systems, this remains unproven. Furthermore, phantom studies indicate that image quality is inferior despite higher DQE for some digital systems. It is therefore highly recommended that further clinical trials on other systems be performed as well as studies comparing the image quality of different systems for FFDM.

Full-field Digital Mammography and Film-screen Mammography in Screening

There is general agreement that screening mammography reduces the rate of death from breast cancer among women of 40 years or older. Meta-analyses of several large trials found a mortality reduction of 15–35% and, furthermore, in most European and North American countries screening mammography has been made available to women ranging in age from 50–70 years. In some cases, the age groups included in screening range from 40–75 years.

After the introduction of digital mammography, efforts were made to transform screening mammography to digital imaging. The logistic and technical advantages of digital mammography rendered it suitable especially for screening: teleradiology and digital reading can facilitate double reading; image archival is simplified; CAD may perhaps even replace double reading or at least reduce false-negative cases, and the potentially unlimited number of copies from each study will ease the transfer of films to reference centers and recall units. However, it is still unclear as to whether differences in image quality and perception from digital mammography would result in diminished detection or increased recall rates. To solve these problems, four large trials on a screening population have been performed, two in the United States and two in the Norwegian breast cancer-screening program in Oslo. The results of these trials are as follows.

The first screening study on digital mammography was performed on 4489 women presenting for screening, which received mammography twice: digital and conventional. As in all trials except DMIST, digital mammography was performed with the Senographe 2000D. Several women presented twice or three times during the enrollment period, so a total of 6736 paired examinations could be evaluated. There was no double reading of each modality, a shortcoming that will be discussed later. Recall and further work-up were performed on the findings of both modalities. A total of 42 cancers were detected, 33 by screen-film and 25 by digital mammography, so that fewer cancers were detected reading digital than film-screen mammography. This difference was not significant and not too surprising considering the small number of cancers detected.

On the other hand, significantly fewer biopsies were based on digital imaging: 38 biopsies on lesions only detected with FFDM versus 87 lesions only with FSM, while 56 biopsies were performed on lesions detected with both modalities, leading to a better PPV for digital imaging. Furthermore, the recall rate for film-screen was higher with 14.9% versus 11.8% for FFDM, which is a significant difference (Lewin et al., 2002). In addition, recall for both modalities was above normal and far above the levels reached in the Oslo trials or in DMIST. It remains to be seen whether these high recall rates are due to legal considerations in a surrounding highly sensitive to missed cancers. Alternatively, they might be due to increased sensitivity in the surrounding of a clinical trial, where every missed cancer would become obvious in a short period. Also, the reduced recall rate could not be verified in other studies. Therefore, it remains unclear whether this lower recall rate for FFDM as well as lower cancer detection was caused by the limited experience of the readers or by differences in the systems. The former statement is supported by the fact that many of the missed cancers were architectural distortions or asymmetric densities. Other cancers were missed, with one of the modalities independent of technology due to differences in positioning or interpretation. It is also possible that lesions generally appeared less suspicious with digital mammography, reducing recall for benign as well as for malignant lesions and, therefore, increasing specificity as most mammographic lesions are benign. It became increasingly obvious that initial hopes for increased detection of cancers as well as fears of excessive false-positive results were not borne out. Once more, the person using the system or interpreting the images was the weakest point in the chain.

Oslo I and II Studies

The so-called Oslo I study compared FFDM (Senographe 2000D) with softcopy reading and FSM in 3683 women aged 50–69 years receiving both imaging modalities (Skaane et al., 2003). A total of 31 cancers (23 digital, 28 conventional, n.s.) were detected, with a slightly higher recall rate (4.6% vs. 3.5%) for FFDM. Still, the recall rate for both modalities was higher in the study group compared to women of the same age group that were screened but not included in the trial (2.6%). Furthermore, cancer detection with FSM for women participating in the study (0.078%) was significantly higher compared to the general screening population (0.0048%), indicating a potential bias. Cancer conspicuity was evaluated by external radiologists and was found to be equal for digital and film-screen mammography. Double reading and consensus conferences were performed independently for both modalities, making this study both complex and expensive. The higher recall rate for digital was explained by the smaller experience of the radiologists with FFDM: a few hundred cases compared to at least 4 years of screening.

As a follow-up to Oslo I, 25,263 women aged 45–69 years were randomized to undergo either FFDM (6997) or FSM (17,911) in the Oslo II study published in 2004 (Skaane et al., 2004). As in Oslo I, both groups received double reading and consensus conference independently. A total of 120 cancers were detected during this study, with cancer detection rates of 0.59% for FFDM and 0.41% with FSM. This result was consistently present in both age groups—45–49 and 50–69 years—with a cancer detection rate for digital versus conventional of 83% versus 54% for older and 27% versus 22% in younger women. This advantage was limited by the significantly higher recall rate for FFDM in older women: 3.8% versus 2.5%, while the difference for younger women was not significant: 3.7% versus 3.0%. This is a major difference from the study by Lewin et al. (2002) as well as from Oslo I, where digital mammography showed lower recall rates. Because

there are no consistent results from different studies on recall rates and because recall did not differ in DMIST, one must suppose that there is no such difference and that results indicating a lower recall rate might be incidental. In conclusion of Oslo II as well as in conclusion of both Norwegian trials, the authors stated that FFDM is equivalent to SFM for breast cancer detection and suited for screening mammography (Skaane et al., 2004).

Digital Mammography Imaging Screening Trial

The largest study so far has been the Digital Mammographic Imaging Screening Trial (DMIST), which included 49,528 asymptomatic women without previous history of breast cancer. These subjects underwent mammography twice: digital and analog mammography in random order. The detectors used were the SenoScan (Fischer Medical), the computed mammography phosphor screen system (Fuji Imaging), the Senographe 2000D (General Electric Medical Systems), the Trex/Hologic System, and the Selenia FFDM System (Hologic). Reading was performed as either hardcopy or softcopy depending on local availabilies. Film-screen mammography units were those being used at the participating sites; imaging parameters, dose, and positioning were matched as closely as possible. A total of 335 cancers were diagnosed in this study. There was no significant difference in cancer detection for digital and film mammography (Pisano et al., 2005). Prespecified subgroup analysis of women 50 years of age or younger, premenopausal women, or women with dense breasts showed a significantly higher AUC for digital mammography, probably owing to the higher contrast resolution of these systems. In addition, radiation dose was lower for digital mammography without compromise in image quality (Pisano et al., 2005).

The limitations of this study were that only one reader evaluated each modality without double reading for each modality, obscuring potential differences between the two modalities through inter-reader variability. Also, it is impossible to deduce differences in quality between systems for digital mammography because no center used multiple systems on one woman. Although differences between the detectors were expected to be found, the power of the study was not sufficient for this purpose. It is to be expected that further results on the cost-effectiveness of digital mammography as well as the quality of life substudy and reader studies comparing soft- and hardcopy display, prevalence, and breast density will produce further insights on the possible advantages of digital mammography.

Now that we have most of the results from the large screening trials on digital mammography as well as numerous results from clinical and phantom trials, it is time to take a step back and consider the facts on digital mammography.

Up to now it has been obvious that digital mammography has not been detecting the majority of cancers in all women, but rather has offered a different method with at least equivalent image quality but advantages in handling and processing. Furthermore, there seem to be advantages in women with dense breast parenchyma, especially those below 50 years of age or under hormone replacement therapy. Consequently, if there is limited availability of digital facilities in one screening unit, these women should be preferably examined on this digital unit to maximize advantages. Further developments that are expected with digital mammography are improved CAD systems and reduction in double reading (see Chapter 4.3).

Financial Considerations of Digital Mammography

Given that to date, the image quality of both modalities is equivalent, further aspects during the transition from conventional to digital need to be considered. One limitation of digital mammography is high initial cost for most digital systems. In general, a full-field digital system including a dedicated workstation and storage system costs ~ €400,000 ($500,000). The only exception to this rule is the phosphor-storage system which can be integrated into an FSM unit already present. Most current mammography systems can easily be adjusted to the requirements of CR mammography without major limitations. Nevertheless, even these units require a dedicated viewing workstation because most workstations for general radiology do not fulfill the minimal requirements for digital mammography. On the other hand, FFDM has financial advantages, such as suspended costs for films and chemistry, reduced spatial requirements for image storage, as well as immediately available previous images with improved workflow for the radiologist. Depending on the number of examinations performed, the additional costs for digital mammography compared to FSM are between $10 and $26 per woman screened, with 5000 to 10,000 tests/year and unit (Ciatto *et al.*, 2006). The DMIST also includes measurements of relative cost-efficiency for digital and film-screen mammography, but results are not yet published.

These calculations are usually based on mammography performed in a setting, where the mammography unit, the radiologist, and eventual second readers are all in the local vicinity. For thinly populated regions, calculations might be different. Here digital mammography can be performed in mobile units mounted on trucks, reducing spatial requirements for chemical processing of FSM. Image transfer to the reading units is possible via Internet, including double reading by specialists in distant regions. Additional costs for transport of films can be avoided, thus reducing the difference in costs for FFDM and FSM. The local situation must

be reviewed as to whether these differences will translate into cost-efficiency for FFDM. The increasing request for digital mammography units will probably lower the prices by a certain amount, making this technology profitable for the general market. Improvement in computing equipment might also reduce time for digital review and retrieval of previous images. The advantages of digital imaging with a dedicated softcopy system are currently under examination in the SCREEN-TRIAL in Europe. The softcopy unit developed for these trials (Mevis breast care, Bremen, Germany) is able to present 100–200 screening cases in 1 hr and is distributed by several manufacturers.

Radiation Dose Considerations

Whereas radiation dose in FSM is given by parameters of the film-screen system, it is highly variable in FFDM. In the FFDM, optimal dose is determined by detector characteristics and image contrast. Generally, image quality is dependent on glandular dose in a broad range limited by minimal radiation level and the saturation level of the detector. It is therefore possible to increase image quality simply by increasing radiation dose. This places a high responsibility on the manufacturers and service technologists, as well as on the individual radiologist, to find a reasonable compromise between image quality and glandular dose. In the first years of digital mammography, this problem was solved by adjusting the digital system to radiation levels of conventional mammography with at least equivalent or even superior image quality, as most phantom and clinical (including our own) studies could show. A recent survey of patient dose for FFDM showed a 28% lower average glandular dose compared to slow film-screen combination, but 37% higher than fast film-screen combinations. Furthermore, dose for digital mammography was ~ 50% lower than dose reference levels for film-screen mammography (Morán *et al.*, 2005). These data are consistent with earlier studies on dose with the Senographe 2000D, which showed a 25% lower dose in a clinical population (Hermann *et al.*, 2002). Dose with different units of the FFDM system was dependent on the calibration of the flat panel, which caused changes in SNR levels as well as in performance of the automatic exposure control (Morán *et al.*, 2005).

Currently, no optimal level has been defined to solve this problem, and different solutions are presented by different or even by individual manufacturers. Some companies offer systems with higher image quality at higher dose, whereas others concentrate on lower but an apparently still diagnostic image quality at minimal dose. At least one manufacturer offers high-quality systems for the U.S. market, where legal requirements support maximal detection rate, and low-dose systems for the European market, where radiation dose considerations are given higher importance. Although image quality and dose are interdependent variables for digital mammography, there is no linear correlation, and presently, major changes in radiation dose only translate into small differences of detection. As "biological noise" or structural irregularities of breast parenchyma are generally by far larger than minimal differences in the signal-to-noise ratio, it has been suggested that the radiation dose be reduced. Recent clinical studies indicate that it is possible to reduce patient dose by 50% without compromising diagnostic quality (Hemdal *et al.*, 2005). It is also possible to reduce dose by omission of the grid with digital mammography, which is possible without reduced image quality (Fischmann *et al.*, 2005b). Because there are several ongoing studies on image quality and radiation dose for digital mammography, it is expected that in the near future different solutions for the image-quality/dose problem will be found.

References

Berns, E.A., Hendrick, R.E., and Cutter, G.R. 2002. Performance of full-field digital mammography to screen-film mammography in clinical practice. *Med. Phys. 29*:830–834.

Ciatto, S., Brancato, B., Baglioni, R., and Turci, M. 2006. A methodology to evaluate differential costs of full-field digital as compared to conventional screen film mammography in a clinical setting. *Eur. J. Radiol. 57*:69–75.

Cole, D., Pisano, E.D., Brown, M., Kuzmiak, C., Braeuning, M.P., Kim, H.H., Jong, R., and Walsh, R. 2004. Diagnostic accuracy of Fischer SenoScan digital mammography versus screen-film mammography in a diagnostic mammography population. *Acad. Radiol. 11*:879–886.

Diekmann, S., Bick, U., von Heyden, H., and Diekmann, F. 2003. Visualization of microcalcifications on mammograms obtained by digital full-field mammography in comparison to conventional film-screen mammography. *RöFo Fortschr. Röntgenstr. 175*:775–779.

Fischer, U., Baum, F., Obenauer, S., Luftner-Nagel, S., von Heyden, D., Vosshenrich and R., and Grabbe, E. 2002. Comparative study in patients with microcalcifications: full-field digital mammography vs. screen-film mammography. *Eur. Radiol. 12*:2679–2683.

Fischmann, A., Siegmann, K.C., Wersebe, A., Claussen, C.D., and Müller-Schimpfle, M. 2005a. Comparison of full-field digital mammography and film-screen mammography: image quality and lesion detection. *Brit. J. Radiol. 78*:312–315.

Fischmann, A., Nykänen, K., Siegmann, K., Wersebe, A., Xydeas, T., Miller, S., and Claussen, C.D. 2005b. Image quality of a prototype full-field digital mammography system based on amorphous selenium. In: Pisano, E. (Ed.), The Proceedings of the 7th, Annual Workshop on Digital Mammography. Chapel Hill: University of North Carolina, Biomedical Research Imaging Center, pp. 011–017.

Hemdal, B., Andersson, I., Grahn, A., Håkansson, M., Ruschin, M., Thilander-Klang, A., Båth, M., Börjesson, S., Medin, J., Tingberg, A., Månsson, L.G., and Mattsson, S. 2005. Can the average glandular dose in routine digital mammography screening be reduced? A pilot study using revised image quality criteria. *Radiat. Prot. Dosimetry 114*:383–388.

Hendricks, R.E., Lewin, J.M., and D'Orsi, C.J. 2001 Non-inferiority study of FFDM in an enriched diagnostic cohort: comparison with screen-film mammography in 625 women. In: Yaffe, M.J. (Ed.), IWDM 2000; 5th, International Workshop on Digital Mammography, *Madison, Wis. Med. Phys.*:475–481.

Hermann, K.P., Obenauer, S., Funke, M., and Grabbe, E.H. 2002. Magnification mammography: a comparison of full-field digital mammography and screen-film mammography for the detection of simulated small masses and microcalcifications. *Eur. Radiol. 12*:2188–2191.

Huda, W., Sajewicz, A.M., Ogden, K.M., Scalzetti, E.M., and Dance, D.R. 2002. How good is the ACR accreditation phantom for assessing image quality in digital mammography? *Acad. Radiol. 9*:764–772.

Kheddache, S., Thilander-Klang, A., Lanhede, B., Månsson, L.G., Bjurstam, N., Ackerholm, P., and Björneld, L. 1999. Storage phosphor and film-screen mammography: performance with different mammographic techniques. *Eur. Radiol. 9*:591–597.

Lai, C.-J., Shaw, C.C., Whitman, G.J., Johnston, D.A., Yang, W.T., Selinko, V., Arribas, E., Dogan, B., and Kappadath, S.C. 2005. Visibility of simulated microcalcifications—a hardcopy based comparison of three mammographic systems. *Med. Phys. 32*:182–194.

Lee, L.D., Yorker, J.G., Jing, Z., and Jeromin, L.S. 2003. Imaging characteristics of a direct conversion full-field digital mammography detector using selenium. In: Peitgen, H.-O (Ed.) *Digital Mammography IWDM 2002*. Heidelberg: Springer, pp. 43–47.

Lewin, J.M., D'Orsi, C.J., Hendrick, R.E., Moss, J.L., Isaacs, P.K., Karellas, A., and Cutter, R.E. 2002. Clinical comparison of full-field digital mammography and screen-film mammography for detection of breast cancer. *Am. J. Roentgenol. 197*:671–677.

Morán, P., Chevalier, M., Ten, J.I., Fernández Soto, J.M., and Vañó, E. 2005. A survey of patient dose and clinical factors in a full field digital mammography system. *Radiat. Prot. Dosimetry 114*:375–379.

Noel, A., and Thibault, F. 2004. Digital detectors for mammography: the technical challenges. *Eur. Radiol. 14*:1990–1998.

Obenauer, S., Hermann, K.P., and Grabbe, E. 2003. Dose reduction in full-field digital mammography: an anthropomorphic breast phantom study. *Br. J. Radiol. 76*:487–482.

Pisano, E.D., Gatsonis, C., Hendrick, E., Yaffe, M., Baum, J.K., Acharyya, S., Conant, E.F., Fajardo, L.L., Basset, L., D'Orsi C., Jong, R., and Rebner, M., for the DMIST Investigators Group 2005. Diagnostic performance of digital versus film mammography for breast-cancer screening. *N. Eng. J. Med. 353*:1773–1783.

Rong, X.J., Shaw, C.C., Johnston, D.A., Lemacks, M.R., Liu, X., Whitman, G.J., Dryden, M.J., Stephens, T.W., Thompson, S.K., Krugh, K.T., and Lai, C-J. 2002. Microcalcification detectability for four mammographic detectors: flat-panel, CCD, CR, and screen/film. *Med. Phys. 29*:2052–2061.

Schulz-Wendtland, R., Hermann, K.-P., Lell, M., Böhner, C., Wenkel, E., Imhoff, K., Schmid, A., Krug, B., and Bautz, W. 2004. Phantom study for the detection of simulated lesions in five different digital and one conventional mammography system. *RöFo Fortschr. Röntgenstr. 176*:1127–1132.

Skaane, P., Balleguier, C., Diekmann, F., Diekmann, S., Piguet, J-C., Young, K., and Niklason, L.T. 2005. Breast lesion detection and classification: comparison of screen-film mammography and full-field digital mammography with soft-copy reading- observer performance study. *Radiology 237*:37–44.

Skaane, P., and Skjennald, A. 2004. Screen-film mammography versus full-field digital mammography with soft copy reading: randomized trial in a population-based screening program—the Oslo II study. *Radiology 232*:197–204.

Skaane, P., Young, K., and Skjennald, A. 2003. Population-based mammography screening: comparison of screen-film and full-field digital mammography with soft copy reading Oslo I study. *Radiology 229*:877–884.

Van Woudenberg, S., Roelofs, T., Hendricks, J., and Karssemejer, N. 2003. Comparison of full-field digital and analog mammograms. In: Peitgen, H.-O (Ed.) *Digital Mammography IWDM 2002*. Heidelberg: Springer, pp. 497–501.

Venta, L.A., Hendrick, R.E., Adler, Y.T., DeLeon P., Mengoni, P.M., Scharl, A.M., Comstock, C.E., Hansen, L., Kay, N., Coveler, A., and Cutter, G. 2001. Rates and causes of disagreement in interpretation of full-field digital mammography and film-screen mammography in a diagnostic setting. *Am. J. Roentgenol. 176*:1241–1248.

6

Use of Contrast-enhanced Magnetic Resonance Imaging for Detecting Invasive Lobular Carcinoma

Carla Boetes and Ritse M. Mann

Introduction

Invasive lobular carcinoma of the breast presents a well-recognized diagnostic problem. The incidence, clinical presentation, and pathological findings for this disease will be discussed briefly. The mammographic and ultrasound findings are shown, and the use of contrast-enhanced magnetic resonance imaging (MRI) in this disease is presented against this background.

Incidence

Invasive lobular carcinoma (ILC) is the second most common form of malignancy of the breast, accounting for 10–15% of all malignancies of the breast. A recent report on its incidence, based on the documentation of 190,458 women with breast cancer in the SEER database (the Surveillance, Epidemiology, and End Results data from the American National Cancer Institute), showed that the proportion of ILC steadily increased from 9.5–15.6% between 1987 and 1999. This rise might be due to the increase in use of combined estrogen and progesterone replacement therapy, also known as complete hormone replacement therapy (CHRT), for prevention of menopausal complaints in this period. It has been shown that CHRT increases the risk of ILC two- to four-fold, while it has only little impact on the risk of the most common form of breast cancer, invasive ductal carcinoma (IDC). Although this knowledge has led to the decreased use of CHRT, its effects on the proportion of ILC are not yet visible (Li *et al.*, 2003).

Presentation

Patients presenting with an ILC are generally older than those presenting with an IDC. Patients often present with a palpable mass or apparent changes of the breast; others may show a mammographic abnormality consisting of a mass or an architectural distortion. However, findings at palpation or mammography can be very subtle, and ILC is easily missed. The tumors are generally slightly larger in patients with ILC than in patients with IDC, and the chance of ILC presenting as a tumor larger than 5 cm is 50% greater than the chance of IDC presenting as such (14% vs. 9%) (Cristofanilli et al., 2005; Arpino et al., 2004). However, axillary lymph nodes are not more often positive.

Invasive lobular carcinomas are characterized by multifocality and multicentricity in the ipsilateral breast and are more often bilateral than other types of invasive breast cancer (15%), although data on the amount of contralateral breast cancer vary widely (Arpino et al., 2004; Sastre-Garau et al., 1996). Metastatic spread of ILC is different from that in IDC. Metastases to lungs, liver, and brain are more common in IDC, while ILC metastases are also found in bone marrow, the leptomeningen, the peritoneal cavity, the retroperitoneal space, and the gynecologic organs (Winston et al., 2000; Lamovec and Bracko, 1991). Tumor characteristics of ILC are generally more benign than those of IDC; receptor status of the estrogen and progesterone receptors is more often positive; while p53, Her-2/Neu, and epidermal growth factor expression are more often normal. However, disease-free and overall survival are not significantly different from IDC and are mostly dependent on tumor stage and age at diagnosis (Arpino et al., 2004).

Pathology

The pathology of ILC may mimic an IDC, showing a firm gray-white mass. However, no visible mass may be present, and although the breast tissue may have a rubbery consistency, there may also be no visible or palpable evidence of malignancy without microscopic examination (Schnitt and Guidi, 2004). Histopathologically, the classic form of ILC is characterized by small round cells that are relatively uniform. The cells are only loosely cohesive due to a typical loss of expression of the adhesion molecule E-cadherin (Simpson et al., 2005). They invade the surrounding stroma in a single-file pattern (Indian filing), resulting in linear strands along the ductuli. Some cause hardly any desmoplastic stromal reaction. These features make them difficult to observe (Rajesh et al., 2003). Other forms of ILC are the solid form, where the tumor cells grow in confluent sheets, and the alveolar form, where tumor cells grow in groups of 20 or more cells connected by delicate fibrovascular tissue. Some other variants have been described, including a tubulolobular variant where small tubules with rounded contours are formed concurrent with the classic appearance; a pleiomorphic variant where the tumor cells are generally larger and show more nuclear variation but grow in the typical (classic) pattern; and a signet ring cell variant with a prominent portion of cells with a signet ring appearance caused by large intracytoplasmatic lumina (Gangane et al., 2002).

A mixed type is described when no single form comprises > 80–85% of the lesion. The classic form is present in 30–77% of patients, most others have either a solid form lesion or a mixed type. Furthermore, low-grade ductal carcinoma in situ (DCIS) is often present around ILC (Schnitt and Guidi, 2004). However, the pathological diagnosis of ILC is difficult and poorly reproducible. Specialized breast pathologists are better able to reproduce such a diagnosis, but even then reproducibility of the diagnosis is no more than moderate with kappa-scores between 0.4 and 0.6. Furthermore, 3% of all breast malignancies have features of both ILC and IDC and cannot readily be assigned to either (Cserni, 1999).

Mammography

The specific growth pattern of ILC hampers the value of mammography. Sensitivity is reported between 50 and 85%. Up to 25% of mammograms are reported to be normal or of low suspicion. The false-negative rate for ILC is higher than for any other type of invasive breast cancer (Molland et al., 2004; Krecke and Gisvold, 1993).

The appearance of ILC on mammograms is variable. Approximately 60% of lesions are reported as a mass, most are spiculated, others are ill-defined, and some are even reported as well-defined masses (Fig. 87). Architectural distortion is seen in 20–30% of cases (Fig. 88), asymmetric density of the breast in 10–15%, and no tumor in 5%. Other findings include skin retraction, nipple retraction, and cutaneous thickening of the skin. In 10% of patients microcalcifications are found, although these are not always associated with the ILC but may occur in surrounding ductal carcinoma in situ (DCIS) or sclerosing adenosis (Evans et al., 2002). A difficulty is that detectable ILC on mammography are of a relatively low radiographic opacity, similar to or even less opaque than normal fibroglandular breast tissue. Many visible lesions are thus reported as benign. Furthermore, ILC is often visible in only one radiologic view, usually craniocaudal (Selinko et al., 2004; Evans et al., 2002; Uchiyama et al., 2001). Size determination with mammography is difficult. The size of > 50% of all ILCs is underestimated; therefore, mammography is unable to accurately measure the extent of ILC (Boetes et al., 2004; Uchiyama et al., 2001).

Ultrasound

Early reports on the use of ultrasound in ILC mention low sensitivity, comparable to the sensitivity of mammography. However, more recently high-resolution ultrasonography has been reported to yield a much higher sensitivity, up to 98% (Selinko et al., 2004).

Invasive lobular carcinomas appear most commonly as hypoechoic masses with or without shadowing in 58% and

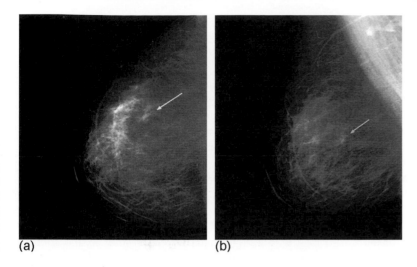

(a)   (b)

Figure 87 (a) Craniocaudal (CC) view of the right breast. A small mass is present at the tip of the arrow. This proved to be an ILC at pathology. (b) Mediolateral oblique (MLO) view of the same lesion.

27%, respectively. Larger ILC may have a distinctive infiltrative pattern, which is harder to recognize, and sometimes only shadowing may be the sole recognizable feature. A significant advantage of imaging with ultrasound is the possibility of easy acquisition of histologic material by an ultrasound-guided biopsy. The correlation of pathological size and size measured with ultrasound is poor, however. Several authors mention Pearson's correlation coefficients between 0.19 and 0.67. The size, as with mammography, is usually underestimated (Kepple *et al.*, 2005; Boetes *et al.*, 2004; Munot *et al.*, 2002; Cawson *et al.*, 2001).

Goal of MRI in the Assessment of Invasive Lobular Carcinoma

Mammography is characterized by its rather poor sensitivity and its inability to accurately measure the size of an ILC. Therefore, it is not an appropriate modality for the assessment of ILC. Ultrasound seems to have a high sensitivity and provides an easy way to obtain histologic confirmation of the diagnosis. However, ultrasound is unable to provide an accurate assessment of tumor size. The typical growth pattern of ILC makes radical surgery difficult, and the secondary mastectomy rate is generally higher than for IDC (Molland *et al.*, 2004). Accurate assessment of tumor size is thus the most important feature that should be provided by MRI. Furthermore, its sensitivity should at least equal that of ultrasound.

Magnetic Resonance Imaging

Magnetic resonance imaging of the breasts is generally performed with a dedicated double breast coil in at least a 1.0 T MRI device. Images are obtained in a three-dimensional (3D) matrix prior to and after the administration of a gadolinium-based contrast agent. This contrast agent is administered intravenously and leaves the plasma through the new formed "leaky" vessels within and surrounding the tumor. Standard pulse sequences are aimed at the visualization

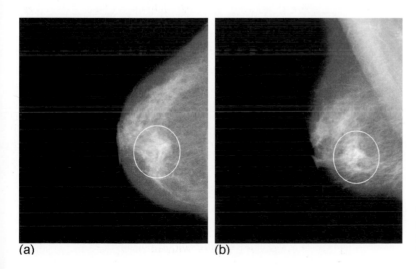

(a)   (b)

Figure 88 (a) CC view of the right breast. (b) MLO view of the same breast. On both views an area of architectural distortion retromammilair (circles) is visible, representing a large ILC. Size determination is very difficult.

of contrast and are therefore T1 weighted. Morphologic features are either examined on high-resolution subtraction images generated from pre- and postcontrast FLASH-series (Fast Low Angle SHot) or on series that are fat-saturated or generated by selective excitation of non–fat-bound protons only, as is the case in the RODEO (ROtating Delivery of Excitation Off-resonance) pulse sequence. Several large prospective studies have shown that MRI has a high sensitivity of around 95% and is more accurate in the determination of disease extent than ultrasound and mammography (Deurloo et al., 2005; Schelfout et al., 2004; Fischer et al., 1999). However, these studies did not explicitly state the value of MRI in cases of ILC. Other authors have, however, presented specific results for ILC in smaller retrospective studies. Direct comparison with ultrasound shows that MRI has a higher sensitivity, although the sensitivity of ultrasound in these studies was not nearly equal to the earlier mentioned results (Boetes et al., 2004; Francis et al., 2001). In 40% of patients MRI is able to show more extensive tumor burden (Weinstein et al., 2001) and Pearson's correlation coefficient with pathologic size has been reported to be between 0.81 and 0.97 (Kepple et al., 2005; Boetes et al., 2004; Munot et al., 2002). Tumor size estimation has been shown to be accurate in 75% of cases, with over- and underestimation occurring in equal proportions.

The pattern of the lesion on MRI correlates well with pathology and is characterized by either a solitary mass with irregular margins in 30% (Fig. 89) or by multiple small enhancing foci (60%) with or without enhancing strands between these foci (Fig. 90), corresponding to clusters of tumor cells connected by single-strand invasion or by normal tissue; sometimes only enhancing strands are visible (Qayyum et al., 2002). The standard practice of imaging both breasts at once reveals contralateral breast tumors in a number of cases.

Dynamic Sequences in Breast Magnetic Resonance Imaging

Standard breast MRI consists mainly of T1-weighted imaging with a high spatial resolution prior to and after an intravenously administered unspecific extracellular gadolinium-based contrast agent. Contrast enhancement is the principal parameter that determines the visibility of any breast lesion in breast MRI. Typically, exchange of contrast from the intravascular space to the extravascular extracellular space and vice versa is much faster in malignant than in benign lesions. This results in different kinetic profiles for different types of lesions, and even distinct differences between the different histological types of breast cancer. A typical malignant pattern shows rapid early enhancement followed by rapid

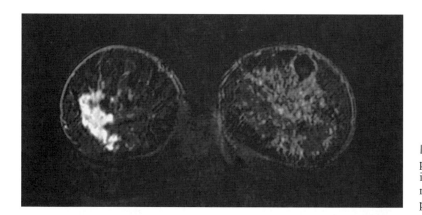

Figure 89 An MRI image, created by subtraction of pre- and postcontrast high-resolution FLASH 3D MRI images, of both breasts in the coronal plane. There are multiple small enhancing foci in the right breast. At pathology this proved to be a large multicentric ILC.

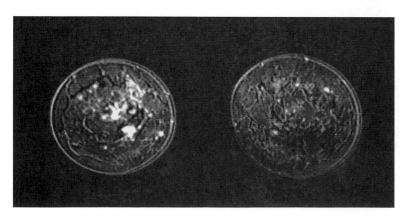

Figure 90 A subtracted MRI image of both breasts in the coronal plane. There is a large mass in the right breast, representing a very large ILC. Note the long spiculae at the superior margins of the tumor suggesting linear spread along the ductuli.

washout of the contrast agent. Some researchers use rapid pulse sequences immediately after administration of the contrast agent to document the outflow of the contrast agent from the tumor vessels. These "dynamic" acquisitions provide information on the contrast-enhancement profile and hence on tumor vasculature. As they document the rather fast transition of contrast agent from the plasma to the extravascular extracellular space, a temporal resolution of only several seconds is needed. Consequently, spatial resolution is diminished in these sequences.

These contrast-enhancement kinetics are dependent mainly on the expression of vascular endothelial growth factor (VEGF), also known as vascular permeability factor (VPF). This factor is a cytokine that stimulates neovascularization by increasing the microvascular permeability for plasma proteins. This is more important than the vascular density within the tumor (Knopp et al., 1999). Quantitative and qualitative analyses of the contrast-enhancement profile

for ILC are, however, less specific as in IDC. Enhancement is often only marginally faster than in healthy tissue, and atypical slow enhancement and absence of washout may occur (Fig. 91). This is probably due to the specific growth pattern of ILC, which does not need extensive neovascularization and thus produces only a few leaky vessels (Yeh et al., 2003; Qayyum et al., 2002; Bazzocchi et al., 2000). Furthermore, expression of VEGF is less extensive in lobular carcinoma than in ductal carcinoma, even though the vascular density is not really different. This might be reflected in the substantial lower and less specific enhancement profiles found for invasive lobular carcinoma (Lee et al., 1998).

False-negative Imaging on Magnetic Resonance Imaging

Even though sensitivity of contrast-enhanced MRI for ILC is high, various authors have reported false-negative

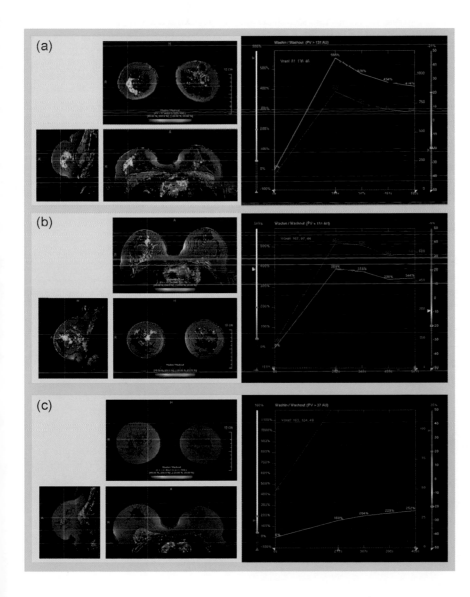

Figure 91 The dynamic profile of three different ILC. The colors within the tumor represent relative enhancement. Relative enhancement increases as colors change from blue through green and yellow to red. Tumor (a) is a very large ILC with a typical malignant enhancement pattern and a dynamic curve showing rapid initial enhancement and subsequent washout of the contrast agent. Tumor (b) shows a less malignant pattern with a plateau phase in the dynamic curve after initial rapid enhancement. Tumor (c) shows no rapid enhancement and no subsequent washout. Although the tumor is readily visible on MRI, it may thus be reported as enhancing breast tissue because the typical malignant pattern is absent.

findings. This might be due to the difficulties described earlier in this chapter regarding enhancement profiles for ILC. There may be no apparent enhancement at all, or enhancement can be attributed to normal glandular tissue, and the absence of a contrast-enhancement profile suggestive of malignancy thus leads to a false benign classification (Munot *et al.*, 2002; Boetes *et al.*, 1997). Especially ILC growing in a single-file pattern without a discrete mass is easily missed. Microscopic disease after excision biopsy may also be missed (Weinstein *et al.*, 2001).

Conclusions

Magnetic resonance imaging has several advantages over mammography and ultrasound in the assessment of ILC. The sensitivity of contrast-enhanced MRI is much higher than the sensitivity of mammography, and although sonography has been shown to have a high sensitivity, comparative studies favor MRI. Magnetic resonance imaging is better capable of accurate assessment of tumor size than any other imaging modality, even though substantial under- or overestimation may still occur. Magnetic resonance imaging often detects a more extensive tumor burden and may detect unsuspected contralateral lesions. However, findings are often subtle, and a typical malignant enhancement profile may be absent; therefore, false-negative findings still do occur. In cases of ILC where breast-conserving therapy is proposed, contrast-enhanced MRI of both breasts appears mandatory.

References

Arpino, G., Bardou, V.J., Clark, G.M., and Elledge, R.M. 2004. Infiltrating lobular carcinoma of the breast: tumor characteristics and clinical outcome. *Breast Cancer Res. 6*:R149–R156.

Bazzocchi, M., Facecchia, I., Zuiani, C., Puglisi, F., Di, L.C., and Smania, S. 2000. [Diagnostic imaging of lobular carcinoma of the breast: mammographic, ultrasonographic and MR findings]. *Radiol. Med. (Torino) 100*:436–443.

Boetes, C., Strijk, S.P., Holland, R., Barentsz, J.O., Van Der Sluis, R.F., and Ruijs, J.H. 1997. False-negative MR imaging of malignant breast tumors. *Eur. Radiol. 7*:1231–1234.

Boetes, C., Veltman, J., van Die, L., Bult, P., Wobbes, T., and Barentsz, J.O. 2004. The role of MRI in invasive lobular carcinoma. *Breast Cancer Res. Treat. 86*:31–37.

Cawson, J.N., Law, E.M., and Kavanagh, A.M. 2001. Invasive lobular carcinoma: sonographic features of cancers detected in a Breast Screen Program. *Australas. Radiol. 45*:25–30.

Cristofanilli, M., Gonzalez-Angulo, A., Sneige, N., Kau, S.W., Broglio, K., Theriault, R.L., Valero, V., Buzdar, A.U., Kuerer, H., Buccholz, T.A., and Hortobagyi, G.N. 2005. Invasive lobular carcinoma classic type: response to primary chemotherapy and survival outcomes. *J. Clin. Oncol. 23*:41–48.

Cserni, G. 1999. Reproducibility of a diagnosis of invasive lobular carcinoma. *J. Surg. Oncol. 70*:217–221.

Deurloo, E.E., Peterse, J.L., Rutgers, E.J., Besnard, A.P., Muller, S.H., and Gilhuijs, K.G. 2005. Additional breast lesions in patients eligible for breast-conserving therapy by MRI: impact on preoperative management and potential benefit of computerised analysis. *Eur. J. Cancer 41*:1393–1401.

Evans, W.P., Warren Burhenne, L.J., Laurie, L., O'Shaughnessy, K.F., and Castellino, R.A. 2002. Invasive lobular carcinoma of the breast: mammographic characteristics and computer-aided detection. *Radiology 225*:182–189.

Fischer, U., Kopka, L., and Grabbe, E. 1999. Breast carcinoma: effect of preoperative contrast-enhanced MR imaging on the therapeutic approach. *Radiology 213*:881–888.

Francis, A., England, D.W., Rowlands, D.C., Wadley, M., Walker, C., and Bradley, S.A. 2001. The diagnosis of invasive lobular breast carcinoma. Does MRI have a role? *Breast 10*:38–40.

Gangane, N., Anshu, Shivkumar, V.B., and Sharma, S. 2002. Pleomorphic lobular carcinoma of the breast. A case report. *Acta Cytol. 46*:909–911.

Kepple, J., Layeeque, R., Klimberg, V.S., Harms, S., Siegel, E., Korourian, S., Gusmano, F., and Henry-Tillman, R.S. 2005. Correlation of magnetic resonance imaging and pathologic size of infiltrating lobular carcinoma of the breast. *Am. J. Surg. 190*:623–627.

Knopp, M.V., Weiss, E., Sinn, H.P., Mattern, J., Junkermann, H., Radeleff, J., Magener, A., Brix, G., Delorme, S., Zuna, I., and van Kaick, G. 1999. Pathophysiologic basis of contrast enhancement in breast tumors. *J. Magn Reson. Imaging 10*:260–266.

Krecke, K.N., and Gisvold, J.J. 1993. Invasive lobular carcinoma of the breast: mammographic findings and extent of disease at diagnosis in 184 patients. *AJR Am. J. Roentgenol. 161*:957–960.

Lamovec, J., and Bracko, M. 1991. Metastatic pattern of infiltrating lobular carcinoma of the breast: an autopsy study. *J. Surg. Oncol. 48*:28–33.

Lee, A.H., Dublin, E.A., Bobrow, L.G., and Poulsom, R. 1998. Invasive lobular and invasive ductal carcinoma of the breast show distinct patterns of vascular endothelial growth factor expression and angiogenesis. *J. Pathol. 185*:394–401.

Li, C.I., Anderson, B.O., Daling, J.R., and Moe, R.E. 2003. Trends in incidence rates of invasive lobular and ductal breast carcinoma. *JAMA 289*:1421–1424.

Molland, J.G., Donnellan, M., Janu, N.C., Carmalt, H.L., Kennedy, C.W., and Gillett, D.J. 2004. Infiltrating lobular carcinoma—a comparison of diagnosis, management and outcome with infiltrating duct carcinoma. *Breast 13*:389–396.

Munot, K., Dall, B., Achuthan, R., Parkin, G., Lane, S., and Horgan, K. 2002. Role of magnetic resonance imaging in the diagnosis and single-stage surgical resection of invasive lobular carcinoma of the breast. *Br. J. Surg. 89*:1296–1301.

Qayyum, A., Birdwell, R.L., Daniel, B.L., Nowels, K.W., Jeffrey, S.S., Agoston, T.A., and Herfkens, R.J. 2002. MR imaging features of infiltrating lobular carcinoma of the breast: histopathologic correlation. *AJR Am. J. Roentgenol. 178*:1227–1232.

Rajesh, L., Dey, P., and Joshi, K. 2003. Fine needle aspiration cytology of lobular breast carcinoma. Comparison with other breast lesions. *Acta Cytol. 47*:177–182.

Sastre-Garau, X., Jouve, M., Asselain, B., Vincent-Salomon, A., Beuzeboc, P., Dorval, T., Durand, J.C., Fourquet, A., and Pouillart, P. 1996. Infiltrating lobular carcinoma of the breast. Clinicopathologic analysis of 975 cases with reference to data on conservative therapy and metastatic patterns. *Cancer 77*:113–120.

Schelfout, K., Van, G.M., Kersschot, E., Colpaert, C., Schelfhout, A.M., Leyman, P., Verslegers, I., Biltjes, I., Van Den, H.J., Gillardin, J.P., Tjalma, W., Van Der Auwera, J.C., Buytaert, P., and De, S.A. 2004. Contrast-enhanced MR imaging of breast lesions and effect on treatment. *Eur. J. Surg. Oncol. 30*:501–507.

Schnitt, S.J., and Guidi, A.J. 2004. *Diseases of the Breast*, 3rd ed. Chapter 34, "Pathology of Invasive Breast Cancer," pp. 541–584.

Selinko, V.L., Middleton, L.P., and Dempsey, P.J. 2004. Role of sonography in diagnosing and staging invasive lobular carcinoma. *J. Clin. Ultrasound 32*:323–332.

Simpson, P.T., Reis-Filho, J.S., Gale, T., and Lakhani, S.R. 2005. Molecular evolution of breast cancer. *J. Pathol. 205*:248–254.

Uchiyama, N., Miyakawa, K., Moriyama, N., and Kumazaki, T. 2001. Radiographic features of invasive lobular carcinoma of the breast. *Radiat. Med. 19*:19–25.

Weinstein, S.P., Orel, S.G., Heller, R., Reynolds, C., Czerniecki, B., Solin, L.J., and Schnall, M. 2001. MR imaging of the breast in patients with invasive lobular carcinoma. *AJR Am. J. Roentgenol. 176*:399–406.

Winston, C.B., Hadar, O., Teitcher, J.B., Caravelli, J.F., Sklarin, N.T., Panicek, D.M., and Liberman, L. 2000. Metastatic lobular carcinoma of the breast: patterns of spread in the chest, abdomen, and pelvis on CT. *AJR Am. J. Roentgenol. 175*:795–800.

Yeh, E.D., Slanetz, P.J., Edmister, W.B., Talele, A., Monticciolo, D., and Kopans, D.B. 2003. Invasive lobular carcinoma: spectrum of enhancement and morphology on magnetic resonance imaging. *Breast J. 9*:13–18.

7

Axillary Lymph Node Status in Breast Cancer: Pinhole Collimator Single–Photon Emission Computed Tomography

Giuseppe Madeddu, Orazio Schillaci, and Angela Spanu

Introduction

One of the most important goals of breast cancer staging is to accurately determine axillary lymph node status because the presence or absence of metastases provides a guide to the most proper therapeutic approach and to defining disease prognosis. With regard to prognosis, when metastases are present, the number of malignant nodes seems to represent the single most important factor in the absence of distant metastases to classify patients as with a better or worse prognosis, the worse prognosis being associated with more than three nodes.

Neither the clinical examination nor the diagnostic imaging procedures commonly employed in identifying primary carcinoma, such as mammography, has proved very accurate in the preoperative prediction of axillary lymph node status. Ultrasonography has a higher sensitivity than mammography,

but it, too, presents limitations, particularly in the identification of nonpalpable axillary involved nodes. Some authors have also indicated magnetic resonance imaging (MRI) as a useful diagnostic tool, but its role seems limited in small metastatic node detection. Thus, at present, axillary lymph node dissection (ALND) represents the procedure of choice to assess axillary lymph node involvement. It is routinely performed in patients with newly ascertained invasive primary breast cancer in spite of its high invasiveness and morbidity. However, the routine use of ALND has been questioned, particularly when carcinomas are at an early stage with a maximum diameter of 10 mm. Its use is questioned still more with a negative axillary clinical examination, because in this case the percentage of lymph node metastases is very low, generally < 20–30%. Therefore, the use of ALND in breast cancer patients should be reassessed, so that this invasive procedure is limited to selected cases.

In order to avoid an unnecessary ALND, both mini-invasive and noninvasive nuclear medicine procedures have recently been proposed as diagnostic tools in the preoperative evaluation of axillary lymph nodes in breast cancer. Among the mini-invasive procedures, radioisotopic lymphatic mapping combined with the mini-invasive radioguided biopsy of the sentinel lymph node (RGBSN) is emerging, particularly in cases with an early-stage carcinoma and clinically negative axillae in which the probability of metastases is very low.

According to the results obtained in studies on this procedure, which has also proved effective in identifying micrometastases in sentinel node (SN), ALND could be recommended or avoided when SN is positive or negative, respectively, for metastases. However, even if in a very low number of cases, the SN may not always be detected or may be false-negative at histology. Moreover, in other cases the SN may be the only metastatic site, and ALND, to which patients should be submitted according to conventional protocol, could be avoided. Thus, in selected patients, other diagnostic tools could be employed in combination with RGBSN, such as noninvasive axillary radioisotopic imaging procedures, which can play a useful role even when RGBSN is not indicated.

Among these procedures, positron emission tomography (PET) with [^{18}F]fluorodeoxyglucose (FDG-PET) has shown high values of sensitivity and specificity in axillary metastasis detection according to the data of some authors (Greco et al., 2001). However, it has not been confirmed by others who have found low sensitivity values, in particular in patients with small and few axillary lymph node metastases (Wahl et al., 2004) and in the early stage (Stages I–II) of the disease (Barranger et al., 2003). At present, the data are still limited and also do not report FDG-PET effectiveness in defining the exact number of lymph nodes.

Scintimammography with different gamma-emitting tumor-seeking agents has also been employed in breast cancer patients in both primary tumor detection and lymph node status evaluation especially the cationic lipophilic radiotracers ^{99m}Tc-sestaMIBI (Tiling et al., 1998) and ^{99m}Tc-tetrofosmin (Spanu et al., 2001a), which present the most favorable physical properties.

The uptake mechanism of cationic lipophilic complexes in malignant cells is not yet well known; however, in vitro studies have demonstrated that the uptake of these two radiotracers is favored by an increased blood flow and capillary permeability and by an elevated metabolic activity of neoplastic cells. This mechanism is strictly dependent on plasma membrane and mitochondrial potentials (Arbab et al., 1996). A mechanism partially related to Na^+/K^+ pump and N^+/H^+ antiport system has also been hypothesized for ^{99m}Tc-tetrofosmin, which predominantly accumulates in the cytosol with a small fraction in the mitochondria, while ^{99m}Tc-sestaMIBI accumulates only in the mitochondria (Arbab et al., 1997).

These cationic lipophilic radiotracers have demonstrated very high sensitivity values in breast cancer staging, in particular when single photon emission computerized tomography (SPECT) is associated with planar acquisition and even more so in nonpalpable and in small number lymph nodes (Schillaci et al., 1997; Spanu et al., 2001a). Neither SPECT nor planar imaging, however, has been able to determine the number of involved nodes, and thus both are unable to provide further important prognostic information, as indicated earlier. This information can also contribute to a more correct selection of patients for adjuvant chemotherapy following surgical treatment of the primary lesion.

Pinhole–SPECT is recognized as having a better spatial resolution than planar and conventional SPECT with a large field of view of parallel-hole collimators, given the more favorable geometric properties of the cone beam collimator. Pinhole–SPECT provides a powerful and widely available tool for the in vivo investigation of regional radioligand distribution in mice and rats. The narrow aperture of the pinhole collimator, combined with short imaging distance and appropriate image reconstruction software, gives a spatial resolution two to three times higher than that of conventional SPECT and comparable to that achieved with dedicated small-animal PET scanners (Bennink et al., 2004). Pinhole–SPECT has also proved its clinical usefulness in the detection of small thyroid nodules (Wanet et al., 1996), in normal and morbid bone and joint studies, and in ankle disease detection (Bahk et al., 1998).

We were the first to employ pinhole–SPECT in patients with malignant and benign tumors. Clinical applications have regarded the identification of neck metastases from differentiated thyroid carcinoma in patients previously submitted to thyroidectomy and radioiodine treatment, as the technique showing a higher per-lesion sensitivity than planar imaging and conventional SPECT with parallel-hole collimators (Spanu et al., 2004a). Pinhole–SPECT was also used in the detection of neck lymph node metastases from Kaposi's sarcoma (Spanu et al., 2003a) and in the identification of hyperfunctioning glands in both primary and secondary hyperparathyroidism, significantly increasing the sensitivity of the conventional planar parathyroid scintigraphy (Spanu et al., 2004b). In these studies we preferred to use tetrofosmin rather than sestaMIBI as the oncologic agent because of its more favorable pharmacokinetics: its faster and greater clearance from the lung and liver allows a higher target to background.

Pinhole–SPECT has also been proposed in breast cancer. The results obtained with two pinhole–SPECT procedures—incomplete (180°) circular orbit SPECT (Tornai et al., 2003) and emission-tuned aperture computed tomography (Fahey et al., 2001)—have appeared particularly encouraging, both of them demonstrating their potential in detecting small-size (< 10 mm) spherical simulated lesions in breast phantom studies. Moreover, after a phantom experimental study (Chiaramida *et al.*, 1998), we were the first to demonstrate the clinical value of pinhole–SPECT in breast cancer patients for axillary lymph node status evaluation, with ^{99m}Tc-tetrofosmin as radiotracer. We have routinely used this procedure for this purpose since 1997, publishing the first data on a large series of patients in 2000 (Spanu *et al.*, 2000).

[99m]Tc-tetrofosmin Pinhole–Single Photon Emission Computed Tomography

Method

Radiolabeling and quality control procedures of ^{99m}Tc-tetrofosmin (Myoview, Amersham Health) are routinely carried out according to the manufacturer's instructions, and a labeling efficiency > 95% is always achieved. In each patient, after obtaining written informed consent, 740 MBq of ^{99m}Tc-tetrofosmin is injected in the controlateral arm of the suspect mammary lesion; when bilateral disease is suspected, the radiotracer is injected in a pedal vein. Axillary pinhole–SPECT images are acquired immediately after conventional planar and 360° SPECT scintimammography with parallel-hole collimators, the former acquisition beginning 10 min after tracer injection.

For axillary pinhole–SPECT a specific software implemented on an Elscint-GE SPX computer is employed. The computer is connected to a circular, high-resolution, single-head gamma camera (SP4HR-Elscint) equipped with a pinhole collimator. Pinhole–SPECT images are acquired over 180° (in clockwise rotation with the step and shoot method) with the patient in a supine anterior position, the arm corresponding to the involved axilla raised over the head. A radioactive [99m]Technetium marker is used to ensure that the axillary region is always at the center of the collimator's field of view. Particular care is given to positioning of the patient to obtain the minimum distance between the center of rotation and the collimator, in all cases within 15 cm. The final position thus obtained determines the starting angle for each individual patient. When the left axilla is studied, the starting angle ranges from −45° to −35°, with an end angle ranging from +135° to +145°. When the right axilla is studied, the starting angle ranges from −145° to −135°, with an end angle ranging from +35° to +45°.

The images are acquired using a matrix size of 128 × 128 and a zoom factor of 2 as fixed by the software acquisition protocol, an angular step of 3°, and an acquisition time per frame/angular step of 30 sec. A pinhole aperture size of 4.45 mm is used because it represents the best compromise between spatial resolution and sensitivity on the basis of both phantom and our clinical studies (Chiaramida *et al.*, 1998; Spanu *et al.*, 2000). Pinhole–SPECT projections are preprocessed by a cone beam algorithm and then processed

by the backprojection filter (BPF) method using a Metz filter (coefficient: 3; FWHM: 14) to obtain 2-pixelwide transaxial slices; the latter are used to reconstruct a final set of 4-pixelwide coronal slices. In patients with unilateral disease, pinhole–SPECT images are acquired only of the axilla corresponding to the involved breast, while in patients with bilateral disease both axillae are studied. In all cases Pinhole–SPECT images are independently evaluated by two experienced nuclear medicine physicians who have had an adequate training to interpret the data.

The site of ^{99m}Tc-tetrofosmin intravenous injection must be revealed to the observers in order to avoid false-positive interpretations of the level of the axilla ipsilateral to the injection site. Particular care is taken to distinguish the axillary region from surrounding muscular structures when evaluating pinhole–SPECT images. In our studies high-quality images have always been obtained in all cases with a clear topographical portrayal of each axilla and important landmarks, as well as an excellent visualization of single or multiple focal areas of increased tracer uptake within axilla. One or more focal area of increased uptake observed in the axilla controlateral to the site injection in respect of the surrounding normal structures is considered as indicating the presence of pathological findings. The scintigraphic results obtained by pinhole–SPECT are always related to the histopathological findings from surgical samples.

Results and Discussion

We performed the first study regarding the use of pinhole–SPECT in breast cancer axillary staging using ^{99m}Tc-tetrofosmin as oncotropic radiotracer (Spanu *et al.*, 2000) in 112 patients with breast lesions, 100 of whom had carcinoma, in comparison with both planar and conventional supine dual-head SPECT scintimammography, which are more frequently used for axillary metastasis detection. The latter, SPECT, was acquired with a body-contouring system, and both procedures always included in the same exam both mammary and axillary regions. The results showed that pinhole–SPECT per-axilla sensitivity and accuracy were markedly higher (100% and 97%) with respect to planar (56.6% and 77%), and also more elevated of SPECT (96.2% and 95%), though only slightly. Specificity was identical for both tomographic procedures (93.6%) but lower than planar (100%).

Other authors (Tiling *et al.*, 1998; Myslivecek *et al.*, 2004) and we also (Spanu *et al.*, 2001a; Schillaci *et al.*, 2002) in other series of patients employed both conventional supine SPECT and planar with cationic lipophilic radiotracers in predicting lymph node status. The results demonstrated sensitivity and accuracy values ranging from 64.7–94.6% and 71.8–92%, respectively, for SPECT and from 41.2–61.3% and 72.9–82.8%, respectively, for planar. These data were partially superimposable with the SPECT and planar data we obtained in the previous study (Spanu *et al.*, 2000), with sensitivity and accuracy values lower than our pinhole–SPECT results, as stated earlier. In addition, pinhole–SPECT distinctly identified a markedly higher number of nodes within axillary cavities than SPECT and far more than planar in our comparative study, also pinpointing their sites. Furthermore, pinhole–SPECT was the only one of the three methods capable of determining the exact number of nodes in many patients with multiple lymph node metastases. Pinhole–SPECT also correctly classified almost 89% of patients as having ≤ 3 or > 3 metastatic nodes for prognostic purposes. Finally, ^{99m}Tc-tetrofosmin proved an excellent oncotropic tracer because it accumulated in all patients with metastases and, except in three cases, had no uptake in patients without metastases.

In a further study we focused on patients with negative axillary clinical examination for whom, as is known, noninvasive diagnostic imaging procedures are the most difficult to stage correctly. We compared pinhole–SPECT with both planar and conventional SPECT also in this group of patients. In the latter study (Spanu *et al.*, 2003b), we enrolled 188 patients with suspected breast cancer on the basis of physical examination and/or mammography and in some cases also of ultrasonography, all of whom were negative for axillary lymph node metastases at clinical examination. At histology, 176 patients had breast cancer, bilaterally in three cases, and in 74 of 179 axillae metastases were ascertained. Pinhole–SPECT was always performed within one week before surgery, and its results were related to the final histological findings from surgical samples obtained by axillary lymph node dissection, which were obtained in all cases.

Pinhole–SPECT significantly improved both the quality and the resolution of the images with respect to conventional SPECT images, with a more distinct visualization of the axillary cavity from the surrounding muscle-skeletal structures. Clearer evidence of the focal areas corresponding to lymph node metastases within the axilla was obtained, with a sharp separation from the surrounding normal tissue activity, in the presence of both one metastatic node and multiple nodes. Thus, pinhole–SPECT, given its higher spatial resolution, also showed in these cases a significantly higher overall sensitivity than SPECT and planar (93.2% vs. 85.1% and 36.5%, respectively; $p < 0.05$ and $p < 0.0005$, respectively) and improved the overall accuracy of SPECT (92.7% vs. 90.5%) and especially of planar (73.2%). Moreover, pinhole–SPECT was more reliable in predicting of the absence metastases when compared with the other two procedures, negative-predictive value (NPV) being 95% for pinhole–SPECT, 90% for SPECT, and 68.9% for planar.

Pinhole–SPECT was false-negative in five patients with one metastatic node each, micrometastatic in 4 of 5 cases

and macrometastatic, but with partial involvement in the remaining case; SPECT and planar were false-negative in these 5 cases and also in 6 and in 42 further cases, respectively. Pinhole–SPECT was false-positive, showing one focal area in 8 cases, while SPECT in 6 and planar in 1 of these, with all 8 cases showing a nonspecific inflammatory reaction for each node at histopathological findings. Except in the above 5 false-negative cases, pinhole–SPECT distinguished single from multiple lymph nodes, correctly detecting 100% of the other cases with single nodes, while SPECT did so in 87.5% and planar in 16.6%.

Pinhole–SPECT also identified a more elevated number of focal areas in patients with > 3 involved nodes in respect of the other procedures. Furthermore, only pinhole–SPECT showed a number of focal areas corresponding to the exact number of involved nodes in 15 of 25 patients with multiple nodes (in 9/12 axillae with 2 nodes, in 4/8 with 3 nodes, and in 2/5 with 4 nodes). In addition, pinhole–SPECT differentiated patients with 3 or fewer metastatic nodes from those with > 3 in 89.8% of cases, thus giving more important prognostic information, in confirmation of our previous results (Spanu *et al.*, 2000). To our knowledge, the latter data, have not been obtained preoperatively by other noninvasive diagnostic imaging procedures, not even by FDG-PET, but only by an invasive method, such as ALND, during operation.

In particular, the markedly high negative-predictive value of ^{99m}Tc-tetrofosmin pinhole–SPECT, if confirmed in a larger series of patients, could suggest a wider use of this method to better select breast cancer patients for ALND. Thus, this invasive procedure could be avoided in unnecessary cases, although the 7% false-negative rate obtained in this study is still too high to permit consideration of pinhole–SPECT as a single diagnostic method for this purpose. The results of these studies seem to indicate that some limitations of pinhole–SPECT imaging of the axilla exist (such as a low sensitivity in identifying small lesions). However, the enhanced spatial resolution of its small aperture associated with an adequate acquisition time per frame/angular step (almost 30 sec/3° in our experience), providing an enhanced contrast and signal-to-noise ratio, outweighs this limitation. Thus, this procedure appears to be an improved method compared to conventional SPECT with parallel collimators. Therefore, in scintimammography, use of pinhole–SPECT should be preferred to conventional SPECT in axillary metastasis detection and even more so to planar, which results in an unreliable method.

In all the aforementioned studies, we used the approximate Feldkamp algorithm and the BPF method for pinhole–SPECT image reconstruction. This is in conformity with the majority of other authors who have employed this tomographic method for both experimental (Tornai *et al.*, 2003) and different clinical purposes (Wanet *et al.*, 1996; Bahk *et al.*, 1998), while an iterative reconstruction of 180°

orbit pinhole–SPECT with ordered subset expectation maximization (PH OS-EM) has been employed by others in phantom studies (Vanhove *et al.*, 2000). The latter procedure obtained a global gain in overall image quality, resolution, and uniformity when compared with BPF reconstruction.

Therefore, the iterative method could also improve pinhole–SPECT *in vivo* performance in small-size axillary lymph node detection and could further reduce false-negative cases, which, however, were very few in our series and were due to a single micrometastasis ($\leq$ 2 mm) in four of five patients or to partial tumor involvement in one patient. It should be emphasized that apart from small size biological tumor factors could also be responsible for a low radiotracer uptake, thus reducing the possibility of visualizing lymph nodes. To date, however, a clinical application of this iterative reconstruction pinhole–SPECT method has not yet been reported.

Nevertheless, it is likely that this iterative method as well as BPF can facilitate micrometastasis identification, the detection limit being probably intrinsic to the Anger camera revelation system. It should be noted that lymph node micrometastasis represents a very important limitation factor for other imaging methods proposed for breast cancer axillary lymph node metastasis detection such as FDG-PET (Keleman *et al.*, 2002; Barranger *et al.*, 2003). Possibly in the future, a pinhole–SPECT system employed with a dedicated small field of view, a very high-resolution gamma camera could further improve the identification of a single micrometastasis. The presence of a single micrometasis should be taken into account, even if its clinical significance — and in particular its impact on the overall survival in breast cancer patients—has not yet been established.

At present, ALND remains the method of choice to identify axillary lymph node micrometastasis. However, RGBSN combined with an appropriate and accurate multisectioning histopathological examination has recently emerged as a mini-invasive approach that offers an alternative to ALND in the axillary staging of selected patients with breast cancer. It also appears to be a very reliable method to detect micrometastasis. In some cases, however, this method gives false-negative results, and, when positive, may not predict the status of the other axillary lymph nodes besides the SN. Moreover, the SN has occasionally not been detected.

On the basis of our encouraging results obtained with ^{99m}Tc-tetrofosmin pinhole–SPECT, we employed this method (Spanu *et al.*, 2001b) comparatively with RGBSN in a large series of T1/T2 patients with breast cancer (*n*-101 cases) without clinical evidence of axillary lymph node metastases or a previous history of excision biopsy. RGBSN evidenced the SN in all cases except 4 with primary tumor > 1.5 cm correctly determined by pinhole–SPECT. In the 97 cases that could be compared, pinhole–SPECT, always per-

formed some days before SN biopsy, obtained a slightly lower overall accuracy with respect to RGBSN (93.8% vs. 94.8%), but a higher negative predictive value (95.2% vs. 92.5%), although the difference was not statistically significant. A case with multiple axillary metastases studied with both pinhole–SPECT, in comparison with SPECT and planar, and RGBSN is illustrated in Figure 92.

Pinhole–SPECT was false-positive in 3 cases and false-negative in 3 further cases (2 of the latter with micrometastasis in the SN), all with one lymph node and correctly diagnosed by RGBSN. [However, the latter was false-negative in 5 further cases and true-positive at pinhole–SPECT.] The combined use of the two procedures achieved 100% accuracy, thus suggesting their complementary use especially in patients with primary carcinomas > 1.5 cm, since in the cases observed in this study, tumor size did not affect pinhole–SPECT performance. However, in smaller carcinomas at an early stage with low risk of lymph node metastases, RGBSN alone achieved 100% accuracy, also identifying micrometastases that represent a diagnostic limitation of pinhole–SPECT. We confirmed these data in a further study on a larger series (Spanu *et al.*, 2003c).

The results obtained suggest that ^{99m}Tc-tetrofosmin pinhole–SPECT is able to predict axillary lymph node status even in the few cases in which RGBSN fails to detect the SN. In the patients in whom RGBSN has identified the SN, the combined use of RGBSN and ^{99m}Tc-tetrofosmin pinhole–SPECT could suggest a management algorithm for selecting patients to be submitted to ALND. This algorithm can be useful even if it should be accurately validated in larger series. We described it in more detail in our previous study (Spanu *et al.*, 2001b), and we also reported it in our later papers (Madeddu and Spanu, 2004; Spanu *et al.*, 2005). In brief, when the two procedures are negative or positive (pinhole–SPECT for more than one metastasis), ALND should be excluded or indicated, respectively. However, ALND might also be indicated when SN biopsy is negative but pinhole–SPECT is positive for more than one metastasis, since this condition has never been associated with false-positive results in our observed cases. On the contrary, ALND could be avoided when biopsy is positive but pinhole–SPECT is negative or positive for only one metastasis, histology excluding further metastatic nodes in this condition.

Conclusions

The pinhole–SPECT procedure with its very high intrinsic spatial resolution seems to have the potential to improve the trade-off between sensitivity and resolution for small-organ or structure imaging, such as the axillary cavity, more than other radioisotopic procedures. Thus, this method could represent one of the noninvasive imaging diagnostic tools of

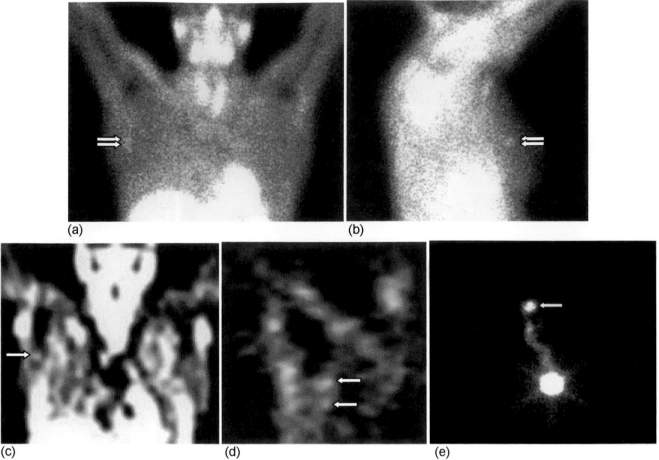

Figure 92 Patient with a Tlc (15 mm) infiltrating ductal carcinoma in the left breast and two metastatic nodes in the ipsilateral axilla. The primary tumor was visible at planar scintigraphy in both anterior (a) and lateral (b) views (double arrows). Planar scintigraphy was false-negative in the right axilla; SPECT (c) was positive for only one lymph node (arrow), while P-SPECT (d) clearly distinguished the two metastatic nodes (arrow). Lymphoscintigraphy (e) detected one sentinel node (arrow), which resulted metastatic. The other metastatic node was ascertained at axillary lymph node dissection.

choice in predicting axillary lymph node status in breast cancer using ^{99m}Tc-tetrofosmin as a suitable oncotropic radiotracer, at least on the basis of our data. The contribution of this procedure is even more significant in nonpalpable and small-size metastatic node detection, and its sensitivity is independent of primary tumor size, with the only important limitation being represented by the detection of micrometastasis.

Pinhole–SPECT can also be used in combination with RGBSN with a significant improvement of accuracy values, thus permitting a more appropriate selection of breast cancer patients in whom ALND could be avoided. The contribution of ^{99m}Tc-tetrofosmin pinhole–SPECT to adding important prognostic information by determining the number of involved nodes in positive patients should be emphasized. This performance has not been reported for other noninvasive imaging procedures including FDG-PET, but only for ALND. On the basis of these considerations,

a wider use of this tomographic procedure is suggested even if in proposing this suggestion we are aware that confirmation in further larger casuistries is necessary before assigning to pinhole–SPECT a definitive role in axillary lymph node staging in breast cancer. On the other hand, axillary ^{99m}Tc-tetrofosmin pinhole–SPECT, such as that which we have utilized, can be performed with a conventional single-head gamma camera equipped with a SPECT system. In our experience, it represents a noninvasive imaging procedure that is simple, easy to perform and read after an adequate training, time saving, and well tolerated. Therefore, it could be routinely applied in any nuclear medicine department.

References

Arbab, A.S., Koizumi, K., Toyama, K., and Araki, T. 1996. Uptake of technetium-99m-tetrofosmin, technetium-99m-MIBI and thallium-201 in tumor cell lines. *J. Nucl. Med. 37*:1551–1556.

Arbab, A.S., Koizumi, K., Toyama, K, Arai, T., and Araki, T. 1997. Ion transport systems in the uptake of 99Tcm-tetrofosmin, 99Tcm-MIBI and 201Tl in a tumor cell line. *Nucl. Med. Commun. 18*:235–240.

Bahk, Y.W., Chung, S.K., Park, Y.H., Kim, S.H., and Lee, H.K. 1998. Pinhole SPECT imaging in normal and morbid ankles. *J. Nucl. Med. 39*:130–139.

Barranger, E., Grahek, D., Antoine, M., Montravers, F., Talbot, J.N., and Uzan, N. 2003. Evaluation of fluorodeoxyglucose positron emission tomography in the detection of axillary lymph node metastases in patients with early-stage breast cancer. *Ann. Surg. Oncol. 10*:622–627.

Bennink, R.J., van Montfrans, C., de Jong, W.J., de Bruin, K., van Deventer, S.J., and de Velde, A.A. 2004. Imaging of intestinal lymphocyte homing by means of pinhole SPECT in a TNBS colitis mouse model. *Eur. J. Nucl. Med. Mol. Imaging. 31*:93–101.

Chiaramida, P., Spanu, A., and Madeddu, G. 1998. 180° Pinhole (P) SPECT and 360° (C) SPECT spatial resolution. An experimental model. *Q. J. Nucl. Med. 42* (Suppl.):11.

Fahey, F.H., Grow, K.L., Webber, R.L., Harkness, B.A., Bayram, E., and Hemler, P.F. 2001. Emission tuned-aperture computed tomography: a novel approach to scintimammography. *J. Nucl. Med. 42*:1121–1127.

Greco, M., Crippa, F. Agresti, R., Seregni, E. Gerali, A, Giovanazzi, R., Micheli, A., Asero, S., Ferrarsi, C., Gennaro, M., Bombardieri, E., and Cascinelli, N. 2001. Axillary lymph node staging in breast cancer by 2-fluoro-2-deoxy-D-glucose-positron emission tomography: clinical evaluation and alternative management. *J. Natl. Cancer Inst. 18*:630–635.

Keleman, P.R., Lowe, V., and Phillips, N. 2002. Positron emission tomography and sentinel lymph node dissection in breast cancer. *Clin. Breast Cancer 3*:73–77.

Madeddu, G., and Spanu, A. 2004. Use of tomographic nuclear medicine procedures, SPECT and pinhole SPECT, with cationic lipophilic radiotracers for the evaluation of axillary lymph node status in breast cancer patients. *Eur. J. Nucl. Med. Mol. Imag. 31* (Suppl. 1):23–34.

Myslivecek, M., Koranda, P., Kaminek, M., Husak, V., Hartlova, M., Duskova, M., and Cwiertka, K. 2004. Technetium-99m-MIBI scintimammography by planar and SPECT imaging in the diagnosis of breast carcinoma and axillary lymph node involvement. *Nucl. Med. Rev. Cent. East. Eur. 7*:151–155.

Schillaci, O., Scopinaro, F., Danieli, R., Tavolaro, R. Cannas, P., Picardi, V., and Colella, A.C. 1997. Technetium-99m sestamibi imaging in the detection of axillary lymph node involvement in patients with breast cancer. *Anticancer Res. 17*:1607–1610.

Schillaci, O., Scopinaro, F., Spanu, A., Donnetti, M., Danieli, R., Di Luzio, E., Madeddu, G., and David, V. 2002. Detection of axillary lymph node metastases in breast cancer with Tc-99m tetrofosmin scintigraphy. *Int. J. Oncol. 20*:483–487.

Spanu, A., Dettori, G., Chiaramida, P., Cottu, P., Falchi, A., Porcu, A., Solinas M.E., Nuvoli, S., and Madeddu, G. 2000. The role of 99mTc-tetrofosmin pinhole-SPECT in breast cancer axillary lymph node staging. *Cancer Biother. Radiopharm. 15*:81–91.

Spanu, A., Dettori, G., Nuvoli, S., Porcu, A., Falchi, A., Cottu, P., Solinas, M.E., Scanu, A.M, Chessa, F., and Madeddu, G. 2001a. 99mTc-

Tetrofosmin SPET in both primary breast cancer and axillary lymph node metastasis detection. *Eur. J. Nucl. Med. 28*:1781–1784.

Spanu, A., Dettori, G., Chessa, F., Porcu, A., Cottu, P., Solinas, P., Falchi, A., Solinas, M.E., Scanu, A.M., Nuvoli, S., and Madeddu, G. 2001b. 99mTc-tetrofosmin pinhole-SPECT (P-SPECT) and radioguided sentinel node (SN) biopsy and in breast cancer axillary lymph node staging. *Cancer Biother. Radiopharm. 16*:501–512.

Spanu, A., Madeddu, G., Cottoni, F., Manca, A., Migaleddu, V., Chessa, F., Masala, M.V., Cossu, A., Falchi, A., Mura, M.S., and Madeddu, G. 2003a. Usefulness of 99mTc-tetrofosmin scintigraphy in different variants of Kaposi's sarcoma. *Oncology 65*:295–305.

Spanu, A., Tanda, F., Dettori, G., Manca, A., Chessa, F., Porcu, A., Falchi, A., Nuvoli, S., and Madeddu, G. 2003b. The role of 99mTc-tetrofosmin pinhole-SPECT in breast cancer nonpalpable axillary lymph node metastasis detection. *Q. J. Nucl. Med. 47*:116–128.

Spanu, A., Dettori, G., Chessa, F., Porcu, A., Stochino, M.B., Cottu, P., Falchi, A., and Madeddu, G. 2003c. Radioguided sentinel node (SN) biopsy vs. 99mTc-tetrofosmin axillary pinhole-SPECT (P-SPECT) in the prediction of breast cancer (BC) axillary lymph node status. *2nd Congress of the World Society of Breast Health, Budapest. 2003. Abstract Book*, 47.

Spanu, A., Solinas, M.E., Migaleddu, V., Falchi, A., Chessa, F., Stochino, M.B., Marongiu, P., Nuvoli, S., and Madeddu, G. 2004a. The role of 99mTc-tetrofosmin neck pinhole (P)-SPECT in the follow up of patients with differentiated thyroid carcinoma (DTC). *Eur. J. Nucl. Med. Mol. Imag. 31* (Suppl. 2):426.

Spanu, A., Falchi, A., Manca, A., Marongiu, P., Cossu, A., Pisu, N., Chessa, F., Nuvoli, S., and Madeddu, G. 2004b. The usefulness of neck pihole SPECT as a complementary tool to planar scintigraphy in primary and secondary hyperparathyroidism. *J. Nucl. Med. 45*:40–48.

Spanu, A., Schillaci, O., and Madeddu, G. 2005. 99mTc labelled cationic lipophilic complexes in malignant and benign tumors: the role of SPECT and pinhole-SPECT in breast cancer, differentiated thyroid carcinoma and hyperparathyroidism. *Q. J. Nucl. Med. Mol. Imag. 49*:145–169.

Tiling, R., Tatsch, K., Sommer, H., Meyer, G., Pechmann, M., Gebauer, K., Munzing, W., Linke, R., Khalkhali, I., and Hahn, K. 1998. Technetium-99m-sestamibi scintimammography for the detection of breast carcinoma: comparison between planar and SPECT imaging. *J. Nucl. Med. 39*:849–856.

Tornai, M.P., Bowsher, J.E. Jaszczak, R.J., Pieper, B.C., Greer, K.L., Hardenberg, P.H., and Coleman, R.E. 2003. Mammotomography with pinhole incomplete circular orbit SPECT. *J. Nucl. Med. 44*:583–593.

Vanhove, C., Defrise, M., Franken, P.R., Everaert, H., Deconinck, F., and Bossuyt, A. 2000. Interest of the ordered subsets expectation maximization (OS-EM) algorithm in pinhole single-photon emission tomography reconstruction: a phantom study. *Eur. J. Nucl. Med. 27*:140–146.

Wahl, R.L., Siegel, B.A., Coleman, R.E., Gatsonis, C.G., and PET Study Group. 2004. Prospective multicenter study of axillary nodal staging by positron emission tomography in breast cancer: a report of the staging breast cancer with PET Study Group. *J. Clin. Oncol. 22*:277–285.

Wanet, P.M., Sand, A., and Abramovici, J. 1996. Physical and clinical evaluation of high-resolution thyroid pinhole tomography. *J. Nucl. Med. 37*:2017–2020.

8

Detection of Small-size Primary Breast Cancer: ^{99m}Tc-tetrofosmin Single Photon Emission Computed Tomography

Angela Spanu, Orazio Schillaci, and Giuseppe Madeddu

Introduction

Breast cancer is the most frequently occurring cancer in women in western countries and is the second leading cause of cancer death. The survival of breast cancer patients is directly related to primary tumor size at the time of diagnosis, with the best prognosis demonstrated for carcinomas < 1 cm without axillary lymph node metastases. The most efficacious possibility of reducing cancer-related deaths is thus represented by an early diagnosis when the tumor is at low risk of metastasis and is at a stage of high curability.

At present, besides clinical examination, X-ray mammography, owing to its high sensitivity, represents the imaging procedure of choice in both breast cancer screening and diagnosis, even in the detection of early-stage *in situ* carcinomas, occult at clinical examination. However, false-negative results can occur, especially in younger women with dense breast. Moreover, in breast cancer diagnosis, mammography presents another important limitation that has low-specificity and positive-predictive value (Kopans, 1992), leading to a

high number of unnecessary biopsies. The addition of high-resolution ultrasound (US) breast imaging can partially complement mammography, improving the diagnosis of breast cancer in dense breast (Kolb *et al.*, 2002). Magnetic resonance imaging (MRI) has proved more sensitive than both mammography and US, but the routine introduction of MRI remains controversial because of its high rate of false-positive results (Bluemke *et al.*, 2004).

To overcome the limitations of mammography, in the last decade there has been a growing interest in the employment of radioisotopic procedures that give a functional imaging of the breasts. At present, scintimammography with the cationic lipophilic radiotracers ^{99m}Tc-sestaMIBI and ^{99m}Tc-tetrofosmin in particular appears to be the radioisotopic imaging procedure of choice in the diagnosis of breast cancer. It can be considered a complementary tool to mammography, contributing to detection of breast cancer in women with indeterminate mammograms, including those with dense breast, and to reducing the number of unnecessary biopsies. Scintimammography, usually performed with the planar acquisition method, in both anterior and lateral views (the latter preferably in a prone position) is, however, limited by its low sensitivity, not superior to 60% according to several casuistries, in the detection of nonpalpable and small-size breast carcinomas < 1 cm in their maximum diameter (Scopinaro *et al.*, 1997). This low-sensitivity value appears unsatisfactory for clinical use and suggests that the procedure should not be used for breast cancer screening. This markedly limits its employment in breast cancer diagnosis, especially in the detection of small-size lesions.

Because of its better contrast resolution and cross-sectional and three-dimensional (3D) images, single photon emission computed tomography (SPECT) is characterized by an intrinsically higher sensitivity than planar imaging. SPECT plays a useful complementary role in certain clinical settings, increasing planar sensitivity, especially in small and deeply located lesions, while representing the procedure of choice in others.

Since its introduction, SPECT has thus presented an indispensable diagnostic tool in the nuclear medicine armamentarium. The recent advances in this technology have generated time-saving devices with two or three detectors and well-designed reconstruction softwares, which have further contributed to its expansion.

During the last years, there has also been a growing interest in employing SPECT in breast cancer diagnosis, mostly as a tool complementary to conventional planar scintimammography. The first SPECT studies were carried out without specific guidelines for the acquisition and processing protocols, which differed significantly from study to study, partially affecting the results and rendering them extremely difficult to compare. These differences include patient positioning (the majority of authors having preferred the supine, others the prone position), a large variability in matrix size (64 × 64 or 128 × 128), angular step (3°, 4°, or 6°), and acquisition timeframe; and the processing protocol adopted. The majority of the authors has preferred the conventional back projection filter (BPF) method, while others prefer the iterative reconstruction.

Despite these technical considerations, the results reported by different authors in detecting primary breast cancer showed that the sensitivity and specificity values of SPECT with the cationic lipophilic radiotracers were satisfactory, in the range of 69–95.8% and 70–91%, respectively. In comparative SPECT studies, ^{99m}Tc-sestaMIBI and ^{99m}Tc-tetrofosmin have given similar results (Obwegeser *et al.*, 1999), suggesting that the two radiotracers can be used without distinction in clinical practice. It has been demonstrated, however, that ^{99m}Tc-tetrofosmin is characterized by more favorable pharmacokinetics, such as faster and greater clearance from the lungs and liver, thus allowing a higher target-to-background ratio (Higley *et al.*, 1993).

When compared with conventional planar scintimammography, SPECT has proved more sensitive in the majority of studies, with a statistically significant difference in most series. However, the increase in sensitivity is generally associated with a decrease in specificity, but not significantly (Spanu *et al.*, 2001). Other authors have also found negative-predictive values of SPECT significantly higher than those of planar scintigraphy, thus suggesting that a negative SPECT study can rule out malignancy with great confidence (Aziz *et al.*, 1999). In contrast, in a few studies, SPECT was used with the patient in a prone position, which has given similar results or has not improved the diagnostic accuracy over planar scintimammography, although the tomographic procedure has proved more useful in determining the extent of the tumor and in precisely localizing the lesion in some cases (Palmedo *et al.* 1996). These results, partially incongruent with those suggesting the superiority of SPECT, may be explained by the high prevalence in certain series of palpable and large carcinomas in which planar scintigraphy tends to be equivalent to SPECT. Technical reasons related to the acquisition or processing methods, such as the high distance between the detector and the breasts, low statistic counts, and the presence of artifacts arising from the heart included in the reconstruction volume, may also be advocated.

With regard to the effectiveness of SPECT in the detection of small-size primary breast carcinomas ≤ 1 cm in size, the specific object of this chapter, only sparse data have been reported in the majority of SPECT studies. However, these studies are generally concordant in assessing that size can affect the sensitivity of SPECT, but less so than planar scan. During the last years, we have investigated the specific performance of SPECT in relation to the size of lesions in a larger series of patients (the most numerous reported in literature in a single study), routinely employing since 1995 supine SPECT in combination with planar scintimammography and using ^{99m}Tc-tetrofosmin rather than ^{99m}Tc-sestaMIBI as radiotracer for the above-mentioned reason.

The Planar and SPECT Scintimammography Method

Written informed consent was obtained from all patients before scintigraphy. All patients received an intravenous injection of 740 MBq of ^{99m}Tc-tetrofosmin in the antecubital vein of the arm contralateral to the mammary lesion. When bilateral disease was suspected, the radiotracer was injected in a pedal vein. Radiolabeling and quality control procedures of the radiotracer were performed according to the manufacturer's instructions. Labeling efficiency was always > 95%. Ten minutes after the injection of the radiotracer, planar imaging was acquired in both supine and lateral views followed by SPECT, using a rectangular, large field of view, dual-head gamma camera (Helix, Elscint, or Millennium VG, General Electric) equipped with low-energy, high-resolution, parallel-hole collimators. A 10% window and a 140 KeV photopeak were selected. Planar images were acquired using a 256 × 256 matrix size, an acquisition time of 600 sec per view, and a zoom factor ranging from 1–1.3, according to the individual patient.

The anterior planar view was obtained with the patient lying in a supine position with the arms raised over the head and with the chest, both breasts and axillae, included in the field of view. Not having a special cushion or table for the prone lateral view, we acquired planar lateral images with the patient in lateral recumbency. The detector was positioned below and as close as possible to the table with the involved breast and corresponding axilla included in the field of view. Single photon emission computed tomography images were acquired over 360° (180° per head) in supine position with the same zoom factor as used for the anterior planar view, with a 64 × 64 matrix size, a 3° angular step, and an acquisition time of 30 sec/frame (total acquisition time: 30 sec). In order to ensure the minimum distance between patient and collimator during rotation, the body-contouring system was always used. After acquisition, SPECT images were normalized to correct for misalignments and radiation time decay and then reconstructed with the backprojection filter method (with a count-optimized Metz filter) without attenuation correction. Sagittal and coronal slices were then obtained from transverse sections.

With two experienced nuclear medicine physicians blinded to the clinical findings, the other diagnostic imaging procedure data and the final histopathological diagnoses independently evaluated planar and SPECT images. However, the observers were informed of the injection site (and thus, indirectly of the affected breast) in order to avoid false-positive interpretations in the axillary region on the same side as the site of injection, possibly caused by radiotracer extravasation. Scintigraphy was considered positive when one or more focal areas of increased uptake with a higher activity compared to the surrounding background was identified in the breast in at least one view (either supine anterior or lateral) for planar imaging and in at least two sequential planes (transverse, coronal, or sagittal) for SPECT. The results were compared with those given by planar imaging; in addition, both were compared with other diagnostic imaging data and subsequently related to the histopathological findings obtained from surgical samples.

Results and Discussion

Our first extensively discussed data (Spanu et al., 2001) refer to 192 consecutive patients with suspected primary breast cancer at mammography and/or clinical examination scheduled to nodular excisional biopsy, which was per-

formed in all cases within one week of scintimammography. With histology, breast cancer was diagnosed in 175 of 192 patients, whereas benign mammary lesions were ascertained in the remaining 17 cases. In total, 212 mammary lesions were excised: 191 were primary carcinomas (171 palpable and 20 nonpalpable) and 21 were benign lesions (18 palpable and 3 nonpalpable).

The lesions were also stratified according to size: 42 carcinomas were ≤ 10 mm, 86 ranged from 11–20 mm, 39 from 21–30 mm, and 24 were > 30 mm; 8 of the benign lesions were ≤ 10 mm and 13 were > 10 mm. In this series SPECT showed a significantly higher overall sensitivity than planar scan (96 vs. 76%). Single photon emission computed tomography sensitivity was significantly higher in both palpable (96.5 vs 75.9%) and nonpalpable carcinomas (90 vs. 45%). With regard to lesion size, SPECT and planar sensitivity were, respectively, 90.5 and 45.2% in carcinomas ≤ 10 mm, 95.3 and 81.4% in carcinomas of 11–20 mm, 95.3 and 81.4% in carcinomas of 21–30 mm, and 100 and 95.8% in carcinomas > 30 mm, with a statistically significant difference in the former three groups, but not in the latter.

Moreover, no statistically significant difference was found in the sensitivity of SPECT in the detection of palpable versus nonpalpable carcinomas (96.5 vs. 90%) or in the detection of the > 10 mm versus the ≤ 10 mm carcinomas (97.3 vs. 90.5%). By contrast, the sensitivity of planar scintimammography differed significantly in these groups (79.5 vs. 45% and 84.6 vs. 45.2%, respectively). Single photon emission computed tomography also proved more sensitive than planar in assessing multifocal/multicentric disease, resulting true-positive in 93.3% of cases, while planar in 66.7%. In addition, the tomographic procedure demonstrated another important advantage: it shows a higher sensitivity (89 vs. 61%) in the detection of carcinomas sited in the internal lower quadrants, which can be obscured in planar images by the high activity present in the liver or in the heart where ^{99m}Tc-tetrofosmin physiologically accumulates, as does the other cationic lipophilic radiotracer ^{99m}Tc-sestaMIBI. Planar scintimammography showed a slightly higher overall specificity than SPECT (86 vs. 76%), but in nonpalpable lesions and in those ≤ 10 mm the two techniques had the same specificity (100%), without any false-positive results. The overall accuracy in differentiating malignant from benign lesions was higher for SPECT with respect to planar scan (93.9 vs. 76.9%).

In the same series, mammography proved more sensitive (97.4% sensitivity value) than SPECT, but at the cost of an extremely lower specificity (30%). Moreover, SPECT was more accurate than the radiologic procedure in the assessment of multifocal/bilateral disease. We concluded that planar scintimammography maintains a role in detecting of palpable carcinomas > 10 mm, whereas SPECT is preferred in nonpalpable and smaller carcinomas, the two categories that are most difficult to detect and to discriminate from benign lesions using conventional diagnostic procedures. Planar scintimammography missed > 50% of these lesions, whereas SPECT missed only 10% and gave no false-positive results.

We confirmed the greater effectiveness of SPECT in small-size carcinoma detection in a second study. This study specifically focused on a larger series of patients with breast lesions ≤ 10 mm in size (Spanu et al., 2002) suspected of malignancy on the basis of clinical and/or mammographic findings. A total of 93 patients were evaluated. In this series too, the scintigraphic data were related to the mammographic data and to histology, which ascertained a primary breast carcinoma in 69 of 93 patients (palpable in 39 cases and nonpalpable in 30 cases), while the remaining 24 patients had benign breast lesions (palpable in 16 cases and nonpalpable in 8 cases). The 69 primary breast carcinomas included 62 invasive ductal, 5 invasive lobular, 1 mucinous, and 1 medullary carcinoma. In this series, the carcinomas were further stratified according to the American Joint Committee on Cancer (AJCC) T pathologic classification as T1a and T1b carcinomas: 14 of the 69 carcinomas were classified as T1a (≤ 5 mm) and 55 as T1b (6–10 mm). The 24 benign mammary lesions included 14 fibrocystic diseases, 6 fibroadenomas, and 4 sclerosing adenosis.

^{99m}Tc-tetrofosmin SPECT scintimammography detected 89.8% of primary breast carcinomas, whereas planar scintimammography 46.4%, and the difference was statistically significant. Single photon emission computed tomography sensitivity was 71.4% in T1a carcinomas and reached 94.5% in T1b tumors. The corresponding sensitivity values for planar were markedly lower (35.7 and 49.1%, respectively), with a statistically significant difference in T1b carcinomas. Single photon emission computed tomography sensitivity was 94.9% in palpable and 83.3% in nonpalpable carcinomas; the corresponding values (48.7 and 43.3%, respectively) demonstrated by planar scintigraphy were significantly lower. Moreover, while SPECT sensitivity was significantly higher in T1b than in T1a carcinomas, no statistical difference in sensitivity was found when palpable carcinomas were compared with nonpalpable ones; planar imaging showed no statistical difference in sensitivity for these groups. In all carcinomas positive both at SPECT and planar, SPECT gave a better localization and characterization of the lesions. SPECT scintimammography was false-negative in 7 carcinomas: 4 T1a nonpalpable carcinomas, including 3 invasive ductal and 1 invasive lobular, and 3 T1b, 2 palpable and 1 nonpalpable, invasive ductal carcinomas. At histology, 5 of 7 carcinomas false-negative at SPECT showed low-grade differentiation, while one had numerous focal areas of necrosis. Planar imaging was false-negative in the above-mentioned 7 cases missed by SPECT as well as in 30 further carcinomas; 5 of these 30 carcinomas were T1a and 25 were T1b; 18 were palpable and 12 were nonpalpable. One T1b carcinoma

negative at planar scintimammography and positive at SPECT is illustrated in Figure 93. Both procedures were true-negative in 23 of 24 benign mammary lesions, showing a 95.8% specificity value.

The only false-positive case observed with both planar and SPECT imaging was a palpable 10 mm fibroadenoma. This lesion was suspicious for malignancy at mammography and had proliferative changes and a high mitotic index at histology. The accuracy in differentiating malignant from benign lesions and the negative-predictive value were significantly higher for SPECT with respect to planar scan (91.4 vs. 59.1% and 76.7 vs. 38.3%, respectively). The positive-predictive value was similar for both procedures (98.4% for SPECT and 97% for planar scintigraphy).

When compared with mammography, SPECT showed a slightly lower sensitivity (89.9 vs. 94%) and a markedly higher specificity (95.8 vs. 36.8%); the accuracy in differentiating malignant from benign lesions was also higher for SPECT (89.9 vs. 81.4%, respectively). Interestingly, the two procedures also resulted complementary, mammography being true-positive in 6 of 7 carcinomas false-negative at SPECT and SPECT in 5 of 5 of the carcinomas negative or indeterminate for dense breast at mammography. In addition, SPECT was true-negative in 16 benign lesions, false positive (11 cases), or indeterminate for dense breast (5 cases) at mammography.

We also emphasize the high quality of SPECT images obtained in this study, while maintaining an acceptable acquisition time for the patient (30 min). All studies were diagnostic and without significant reconstruction artifacts. Furthermore, at SPECT imaging, the breasts always resulted well separated from the thoracic structures, and the carcinomas were clearly visualized as focal areas of increased uptake, overcoming the background activity around the breasts, as a result of a high statistic count rate and a good reconstruction processing, based on an appropriate count-optimized filter. The high statistic count in particular can be explained by a good combination of the parameters selected for acquisition of the images, as well as by employment of the body-contouring system, which ensured minimum distance between the detectors and the breasts during rotation.

More recently, in another study we reviewed 430 patients with suspected breast cancer, including 133 patients with breast lesions ≤ 10 mm in size: a primary breast cancer in 106 cases and a benign lesion in the remaining 27 cases (Spanu et al., 2005a). In this series, too, SPECT confirmed its high sensitivity in the detection of carcinomas of such dimension, with a statistically significant difference when compared with planar scan (88.7 vs. 37.7%) and maintained the same high specificity (96.4%).

Although the data on detection of small-size carcinoma reported by other authors are sparse and with limited series, the study by Mathieu et al. (2005) deserves to be mentioned. These authors employed supine SPECT in comparison with

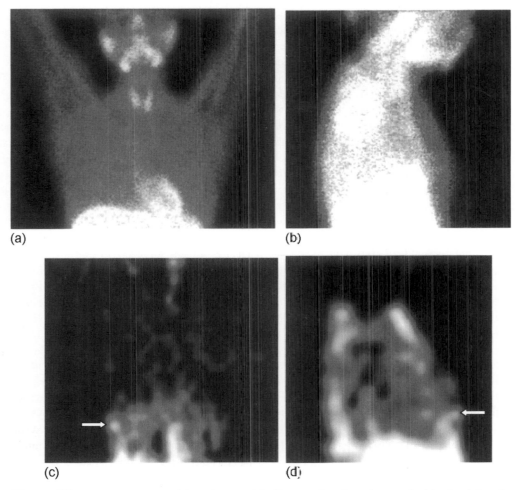

Figure 93 Patient with a palpable T1b (9 mm) infiltrating ductal carcinoma sited in the right sub-areolar region, negative at planar scintimammography in both anterior (a) and lateral (b) views and true-positive (arrow) at SPECT in both coronal (c) and sagittal (d) views.

planar scintimammography, using ^{99m}Tc-sestaMIBI as radiotracer and an iterative algorithm for reconstruction images in a very highly selected series of patients, all with an inconclusive triple diagnosis assessment. This series refers to 104 patients with suspicion of breast cancer, either at initial presentation or after treatment, including 19 patients with one T1b carcinoma at initial presentation, 74% of which resulted true-positive at SPECT, while only 58% at planar. SPECT also demonstrated a higher sensitivity in both T1c (91% sensitivity for SPECT and 65% for planar scan) and T2 carcinomas carcinomas (96% sensitivity for SPECT and 89% for planar scintigraphy), but with a statistically significant difference only in the former. The authors concluded that SPECT imaging is mandatory in clinical practice, permitting a significant increase in sensitivity with respect to planar imaging. They also emphasized the feasibility of the method, indicating that SPECT is not difficult to perform in a supine position (Mathieu *et al.*, 2005).

The results of all the aforementioned studies seem to demonstrate that supine SPECT is a highly sensitive method in detecting small-size (≤ 10 mm) primary breast cancer, such as T1a and T1b carcinomas, whether palpable or non-palpable. The method showed an overall sensitivity of ~ 90%, significantly higher than that obtained with planar scintimammography, which globally missed more than 53% of carcinomas. Thus, the superiority of supine SPECT seems unequivocal.

These results in combination with the high quality of images emphasize the suitability of the acquisition and processing protocol we have selected for the SPECT imaging of the mammary glands and suggest some technical considerations that we have reported in detail in a recent review (Spanu *et al.*, 2005b). In particular, a supine position with the arms raised should be preferred to the prone position, because the latter, prone position, may be affected by constraints imposed by the patient and by the imaging table and gantry.

A good statistic count within the breasts should also be obtained, properly combining the matrix size, angular step, and acquisition timeframe to avoid star artifacts originating from the organs (heart and liver) in which the radiotracer presents a high accumulation and to enable the identification of breast lesions, especially if small in size and deeply located.

In our review, we explained the high statistic counts obtained in our studies with the relatively small matrix size (64 × 64) and the relatively high acquisition time (30 sec associated with an angular step of 3°) as well the employment of the body-contouring system. Moreover, a multidetector camera should preferably be used to save time, although the total acquisition time should be at least 30 min. The filtered backprojection (FBP) method remains the processing method of reference, although it would be interesting to study it in combination with the iterative reconstruction, especially in very small-size breast tumors such as T1a carcinomas. The performance of SPECT obviously depends on both the acquisition and processing protocol and especially on a good combination of the two.

The interpretation of SPECT images is another important point to consider, necessitating a long learning curve to precisely identify and localize the sites of increased uptake in the mammary quadrants.

In our series, the performance of SPECT proved very effective in detecting T1b carcinomas and in encouraging T1a carcinomas. A further improvement of SPECT sensitivity in the latter, T1a carcinomas, is likely to prove very difficult, even with modified acquisition or processing methods, owing to the intrinsic poor spatial resolution of the Anger camera equipped with parallel hole collimators. This goal could probably be partially achieved by equipping the gamma camera with a high-resolution collimator, such as a pinhole. The feasibility of SPECT with a pinhole collimator (pinhole–SPECT) in imaging breast lesions has been successfully tested in phantom studies. Two pinhole–SPECT procedures in particular—incomplete (180°) circular orbit SPECT (ICO-SPECT) and emission tuned-aperture computed tomography (ETACT)—demonstrated in phantom studies their potential in improving the visualization of small-size (< 10 mm) lesions when compared with planar images acquired with the same collimator (Scarfone et al., 1997). However, a 5 mm simulated breast lesion was not detected (Fahey et al., 2001).

Tornai et al. (2003) applied pinhole ICO SPECT *in vivo* to visualize one primary breast carcinoma measuring 3.7 cm at ultrasonography, using ^{99m}Tc-tetrofosmin as radiotracer. The patient was scanned ~ 2 hr after the injection of the radiotracer using a pinhole collimator by both planar and ICO SPECT in the prone position on a specially designed bed. The carcinoma was clearly seen with both acquisition methods, but with a better contrast on tomographic images. The application of this method in human breasts, however, requires further standardization, which is not easy to obtain for anatomical reasons.

The preliminary sensitivity results given by newly developed high-resolution, small field-of-view dedicated breast cameras are also particularly encouraging. These cameras make it possible to obtain planar breast projections corresponding even to those given by mammography. These prototypes have proved capable of increasing the sensitivity of planar scintimammography acquired with a conventional gamma camera, especially in nonpalpable and ≤ 10 mm carcinomas (Scopinaro et al., 1999). Rhodes et al. (2005), using ^{99m}Tc-sestaMIBI as radiotracer, obtained the highest sensitivity (86% in 22 carcinomas ≤ 10 mm globally considered; 75% in T1a and 89% in T1b carcinomas) yet reported with dedicated systems, comparable to those we obtained using supine ^{99m}Tc-tetrofosmin SPECT in a larger series of cases. However, the specificity value of 64% was less satisfactory. The comparison of SPECT with the above newly developed procedures in a large series of small-size primary breast carcinomas could be very useful, and we have a study in progress.

All developments in nuclear medicine techniques are welcome if they lead to an improvement in detecting small-size primary breast carcinoma. In any case, one should always bear in mind that the performance of nuclear medicine imaging is affected not only by physical factors but also by biological ones. Carcinomas characterized by low cellularity and proliferation indices may result false-negative at SPECT, even when large in size. The overexpression in a tumor of P-glycoprotein or of other multidrug resistance proteins may also lead to false-negative results, both ^{99m}Tc-sestaMIBI and ^{99m}Tc-tetrofosmin being substrates of these efflux-pump proteins. On the contrary, lesions small in size, even below the spatial resolution of the detector system, may be evident if characterized by more favorable biological properties that favor radiotracer accumulation.

Conclusions

In conclusion, ^{99m}Tc-tetrofosmin supine SPECT appears to be a highly valuable diagnostic method for detecting small-size primary breast carcinomas ≤ 10 mm in size, markedly improving the sensitivity of planar scintimammography, while maintaining an extremely high specificity value. Single photon emission computed tomography seems to be particularly useful for detecting T1b carcinomas, although the sensitivity value obtained in smaller tumors, such as T1a carcinomas, also appears very interesting. This method is simple to perform in a supine position and can give a simultaneous visualization of the two mammary glands as well as of the two axillary cavities in a relatively short time. It is also easy to interpret following adequate training.

At present, SPECT is available in most nuclear medicine departments. Its wider clinical application to include breast cancer imaging is thus suggested, promoting prospective multicentric trials in a large series of patients with appropriate well-standardized acquisition and processing protocols. Supine SPECT with cationic lipophilic radiotracers has the potential to be considered an important noninvasive diagnostic tool, complementary to mammography also in small-size breast lesion detection. However, further studies are necessary to assess the definitive role that could be assigned to this procedure in the management of patients with breast lesions. The recent development of new dedicated breast cameras also appears interesting, even though their availabity is very limited.

References

Aziz, A., Hashmi, R., Ogawa, Y., and Hayashi, K. 1999. Tc-99m-MIBI scintimammography; SPECT versus planar imaging. *Cancer Biother. Radiopharm. 14*:495–500.

Bluemke, D.A., Gatsonis, C.A., Chen, M.H., DeAngelis, G.A., DeBruhl, N., Harms, S., Heywang-Kobrunner, S.H., Hylton, N., Kuhl, C.K., Lehman, C., Pisano, E.D., Causer, P., Schnitt, S.J., Smazal, S.F., Stelling, C.B., Weatherall, P.T., and Schnall, M.D. 2004. Magnetic resonance imaging of the breast prior to biopsy. *JAMA 292*:2735–2742.

Fahey, F.H., Grow, K.L., Webber, R.L., Harkness, B.A., Bayram, E., and Hemler, P.F. 2001. Emission tuned-aperture computed tomography: a novel approach to scintimammography. *J. Nucl. Med. 42*:1121–1127.

Higley, B., Smith, F.W., Gemmel, H.G., Das Gupta, P., Gvozdanovic, D.V., Graham, D., Hinge, D., Davidson, J., and Lahiri, A. 1993. Technetium-99m-1,2-bis[bis(2-ethoxyethyl) phosphino] ethane: human biodistribution, dosimetry and safety of a new myocardial perfusion imaging agent. *J. Nucl. Med. 34*:30–38.

Kolb, T.M., Lichy, J., and Newhouse, J.H. 2002. Comparison of the performance of screening mammography, physical examination, and breast US and evaluation of factors that influence them: an analysis of 27,852 patients evaluations. *Radiology 225*:165–175.

Kopans, D.B. 1992. The positive predictive value of mammography. *Am. J. Radiol. 158*:521–526.

Mathieu, I., Mazy, S., Willemart, B., Destine, M., Mazy, G., and Lonneux, M. 2005. Inconclusive triple diagnosis in breast cancer imaging: is there a place for scintimammography? *J. Nucl. Med. 46*:1574–1581.

Obwegeser, R., Berghammer, P., Rodrigues, M., Granegger, S., Hohlagschwandtner, M., Kucera, H., Singer, C., Berger, A., Kubista, E., and Sinzinger, H. 1999. A head-to-head comparison between technetium-99m-tetrofosmin and technetium-99m-MIBI scintigraphy to evaluate suspicious breast lesions. *Eur. J. Nucl. Med. 26*:1553–1559.

Palmedo, H., Schomburg, A., Grunwald, F., Mallmann, P., Krebs, D., and Biersack, H.J. 1996. Technetium-99m-MIBI scintimammography for suspicious breast lesions. *J. Nucl. Med. 37*:626–630.

Rhodes, D.J., O'Connor, M.K., Phillips, S.W., Smith, R.L., and Collins, D.A. 2005. Molecular breast imaging: a new technique using technetiumTc scintimammography to detect small tumors of the breast. *Mayo Clin. Proc. 80*:24–30.

Scarfone, C., Jaszczak, R.J., Soo, M.S., Smith, M.F., Greer, K.L., and Coleman, R.E. 1997. Breast tumour imaging using incomplete circular orbit pinhole SPECT: a phantom study. *Nucl. Med. Commun. 18*:1077–1086.

Scopinaro, F., Pani, R., De Vincentis, G., Soluri, A., Pellegrini, R., and Porfiri, L.M. 1999. High-resolution scintimammography improves the accuracy of technetium-99m methoxyisobutylisonitrile scintimammography: use of a new dedicated gamma camera. *Eur. J. Nucl. Med. 26*:1279–1288.

Scopinaro, F., Schillaci, O., Ussof, W., Nordling, K., Capoferro, R., De Vincentis, G., Danieli, R. Ierardi, M., and Picardi, V. 1997. A three center study on the diagnostic accuracy of 99mTc-MIBI scintimammography. *Anticancer Res. 17*:1631–1634.

Spanu, A., Dettori, G., Nuvoli, S., Porcu, A., Falchi, A., Cottu, P., Solinas, M.E., Scanu, A.M., Chessa, F., and Madeddu, G. 2001. 99mTc-Tetrofosmin SPET in both primary breast cancer and axillary lymph node metastasis detection. *Eur. J. Nucl. Med. 28*:1731–1794.

Spanu, A., Schillaci, O., Meloni, G.B., Porcu, A., Cottu, P., Nuvoli, S., Falchi, A., Chessa, F., Solinas, M.E., and Madeddu, G. 2002. The usefulness of 99mTc-tetrofosmin SPECT scintimammography in the detection of small size primary breast carcinomas. *Int. J. Oncol. 21*:831–840.

Spanu A., Schillaci, O., and Madeddu, G. 2005a. 99mTc labelled cationic lipophilic complexes in malignant and benign tumors: the role of SPECT and pinhole-SPET in breast cancer, differentiated thyroid carcinoma and hyperparathyroidism. *Q. J. Nucl. Med. Mol. Imag. 49*: 145–169.

Spanu, A., Dettori, G., Schillaci, O., Chessa, F., Porcu, A., Cottu, P., and Madeddu, G. 2005b. 99mTc-tetrofosmin SPECT: a useful diagnostic tool in breast cancer (BC) detection. *Eur. J. Nucl. Med. Mol. Imag. 32* (Suppl. 1): S 128.

Tornai, M.P., Bowsher, J.E., Jaszczak, R.J., Pieper, B.C., Greer, K.L., Hardenbergh, P.H., and Coleman, R.E. 2003. Mammotomography with pinhole incomplete circular orbit SPECT. *J. Nucl. Med. 44*:583–593.

9

Microcalcification in Breast Lesions: Radiography and Histopathology

Arne Fischmann

Introduction

Marblelike structures in breast tumors were first described by Giovanni Morgagni in the late eighteenth century. This was followed by the discovery of "black spots" detected centrally in carcinoma on a mastectomy specimen radiograph by the Berlin surgeon Albert Salomon in 1913. He interpreted these as "cystic degraded tumor-mass growing in the ducts." Neither researcher recognized the importance of their findings, and these were, for a considerable time, the only description of microcalcifications, except for a case report by R. Finsterbusch and F. Gross in the 1930s on uncommon calcifications in plasma-cell mastitis.

In the 1950s, LeBorgne (1951) established the importance of microcalcifications (MC) detected in 30% of breast disease and named them "clustering of innumerable, punctuate calcifications like fine grains of salt." These microcalcifications were considered pathognomonic for carcinoma. However, it soon became obvious that this correlation was incorrect: a study in 1978 found that calcifications resulted in positive biopsy rates of 10%. In the following years, several publications attempted to improve the classification of calcifications according to shape, number, and distribution (Lanyi, 1986; Le Gal et al., 1976). The final result of this search is the Breast Imaging-Reporting and Data System (BI-RADS™), which was introduced by the American College of Radiology (ACR) in the early 1990s and is currently in its fourth edition (ACR, 2004). However, the importance of LeBorgne's work is visible, as part of his original terminology is still in use.

Histopathology

Breast calcifications are crystals, usually compound of calcium phosphate, often as hydroxyapatite, calcium carbonate, calcium oxalate, or magnesium phosphate. They develop in benign or malignant transformations of breast tissue. Usually, it is not possible to deduce malignancy based on the chemical composition of calcifications, although calcium oxalate is found mainly in benign lesions.

Calcifications are usually localized in the terminal ductulo-lobular unit (Tabar et al., 2001), although exceptions include vascular calcifications in arterial walls. In general,

383

calcifications are deposited in areas where compound lipids are present. It has not yet been concluded as to whether calcifications reflect a degenerative process or are a product of an active living cell. On stained slices, calcifications may be psammoma bodies, granular, laminar, or amorphous. In some instances, they form circular structures referred to as Liesegang's rings within cysts (Tavassoli, 1999).

While most calcifications are easily detectable in hematoxylin-eosin-stained sections, calcium oxalate (or Weddellite) is obscure to standard work-up and only visible in polarized light. As calcium oxalate is found in ~ 6% of breast biopsies for calcifications, and as it is mainly associated with benign disease, detection is mandatory to avoid unnecessary follow-up biopsies.

In plasma-cell mastitis, calcifications are usually deposited around inflamed ducts; Y-shaped and linear distributions are, therefore, often present. As opposed to malignant disease, these calcifications are usually smooth and homogeneous. A small radiolucent area can sometimes be detected in the center correlated to the lumen of the duct, while malignant casting calcifications are centrally dense due to their origin inside the intraductal malignancy. As a result of necrosis in fatty tissue due to either trauma or postoperative changes, lipids are liquefied to form so-called oil cysts. In the periphery of these lipid-containing cysts, fatty acids precipitate as calcium deposits and form the typical thin eggshells.

Radiologic-pathologic correlation of calcifications is difficult because up to 25% of calcifications are lost during embedding and fixation of specimens and, hence, they are undetectable to the pathologist. Especially aqueous fixatives (e.g., paraformaldehyde) have the ability to dissolve calcifications in a few days. Microcalcifications of < 100 μm are, on the other hand, obscure to mammography due to limited spatial resolution. Studies on dedicated specimen-radiography units with sevenfold magnification reveals approximately twice as many calcifications than standard radiography with magnification factors of ~ 2. Fischmann et al. (2004) found that of 851 calcifications, 169 could only be detected in radiography and 103 in histopathology. Only in 30% of all patients could concordance of histological and pathological calcifications be achieved. Therefore, all specimens taken for microcalcifications should undergo specimen radiography. Furthermore, the pathologist should thoroughly search for calcifications marked radiologically.

Detection

Because of their size and distribution, calcifications are impalpable and clinically invisible. Therefore, only lesions with mass effect and associated calcifications are clinically detectable. Presently, mammography is the only imaging modality that can reliably detect and classify MC. Because these are important markers for ductal carcinoma *in situ* and invasive carcinoma, mammography remains in the near future indispensable for breast cancer screening. Large macrocalcifications have been detected with ultrasound (US) for a long time, whereas MC remained obscured due to limited spatial resolution. Although modern US scanners are able to detect smaller calcifications, it is still impossible to classify these according to the underlying pathology. Nevertheless, US is useful in the work-up of microcalcifications by detecting fibrocystic disease underlying lobular or adenomatous calcifications and in performing biopsy (Teh et al., 2000).

The widespread use of screening mammography since the 1970s has led to a substantial increase in diagnosis of calcifications. Approximately 40–50% of mammary carcinomas have calcifications that are mammographically detectable. In particular, intraductal comedocarcinoma and infiltrative carcinoma are frequently associated with calcifications, while lobular carcinomas are rarely calcified. Among the benign lesions, sclerosing adenosis is associated with calcifications with a frequency of 50%. With 58% positive-predictive value in patients treated with breast-conserving therapy, calcifications are a valuable tool in the detection of recurrence (Tavassoli, 1999).

With the development of digital mammography, arguments arose as to whether the increased contrast resolution would facilitate detection of calcifications, or if reduced detail resolution due to pixel size would diminish detection. As a result of both physiological limitations of the human eye and the size of calcifications, it was probable that digital mammography was superior in the detection of small calcifications. Studies on phantoms (Rong et al., 2002) and patients (Fischmann et al., 2005) could show better detection of calcification at equal dose, but large screening studies (Pisano et al., 2005) did not find increased cancer detection with digital mammography. Due to limitations of the Digital Mommographic Imaging Screening Trial (DMIST), the superior classification of microcalcifications with digital mammography from clinical studies (Fischer et al., 2002a) could not be reproduced. Results with modern mammography techniques such as dual-energy subtraction (Lemacks et al., 2002) or tomosynthesis (Lewin et al., 2004) are promising and might increase detection of calcified tumors in screening mammography.

Despite the characteristic appearance of calcium on magnetic resonance imaging (MRI) with low signal on T1w and T2w images, this method has no practical use in the detection or classification of calcifications. This is mainly because of the small size of the single calcification, which is far below the voxel size. Magnetic resonance imaging, therefore, can only detect the underlying disease. Correlation of mammographic suspicious calcifications to contrast-enhancing lesions in breast MRI is, therefore, performed mainly as visual co-registration by the readers.

As this situation is obviously not ideal, ongoing research projects are engaged in computer-aided co-registration of breast structures in mammography and MRI in order to facilitate correlation of lesions in both modalities. Nevertheless, sensitivity for *in situ* carcinoma is ~ 80%, and a negative MRI in case of suspicious MC cannot rule out malignant disease. On the other hand, an increasing number of ductal carcinoma in situ (DCIS) is detected in younger women or women with hormone replacement therapy. For this reason and due to dense and often irregular fibroglandular breast tissue, preoperative MRI can be useful in detecting or ruling out multifocal disease. This is also true for new calcifications after breast-conserving therapy, where it is difficult to distinguish growing liponecrotic calcifications from recurrent breast cancer.

Classification of Breast Calcifications

Mammography is currently the only imaging modality that is able to classify MC with sufficient diagnostic accuracy. Therefore, in all cases of suspected breast disease, mammographic imaging should be performed in at least two planes. These should be preferably in the craniocaudal and mediolateral view; oblique views are of limited use in delineating milk of calcium. Magnified views of calcifications are mandatory with film-screen mammography and should also be performed in at least two directions. Recent studies on digital mammography indicate that digital zoom is sufficient for characterization, while magnification improves only image quality without significant influence on diagnostic accuracy (Fischer *et al.*, 2002b). For the majority of calcifications, this method remains undisputed.

For small powderlike groups of calcifications, and in especially dense breast tissue, further magnification is helpful even with digital mammography. In this case, the additional views often reveal more calcifications than provided by standard views in a wider distribution. Especially in recurrent breast cancer, these additional calcifications are an important marker of tumor distribution changing surgical strategy. At the same time, it is possible that adjacent benign calcifications from adenosis are attributed to the carcinoma leading to larger lumpectomy than necessary. Currently, there are no data on the distribution of minute calcifications in healthy breast tissue or in benign disease.

Hence, the extent to which additional imaging is causing unnecessary therapy and potentially mutilating surgery remains unclear.

Morphology and distribution (especially triangular or linear groups) are the most important parameters when differentiating benign from malignant calcifications. Another independent factor seems to be the size of the group of calcifications. However, classification is less precise for calcifications than that for solid lesions (Hall *et al.*, 1988),

thus necessitating a more aggressive treatment. Analysis of calcifications primarily necessitates the search for additional or underlying solid masses. In many cases, the underlying mass is easier to classify as calcifications improve accuracy of decisions. For example, calcifications in an oval-shaped mass with smooth borders almost certainly identify a fibroadenoma, even if the calcifications do not yet have the typical popcorn shape.

Systematic Classification

While early studies considered calcifications to be pathognomonic for malignancy, it soon became clear that this approach leads to an unnecessary rate of mastectomies and that, therefore, further refinement was desired. As an early precursor of modern systems, Le Gal described five different shapes of calcifications (Le Gal *et al.*, 1976):

Type 1: ring-shaped (centrally radiolucent)
Type 2: punctuate regular
Type 3: dustlike
Type 4: punctuate irregular
Type 5: worm-shaped

These five groups are used in all current classification systems, though with some modifications, especially in terminology (e.g., dustlike calcifications are classified to be indistinct by the BI-RADS lexicon).

While Le Gal *et al.* (1976) concentrated on individual calcifications, Lanyi (1986) was, in the 1970s and 1980s, unremitting in correlating shapes of groups to malignancy. He defined eight different groups:

1. Triangular or trapezoid
2. Squared or rectangular
3. Bottle-shaped
4. Rod-shaped
5. Propeller-, cross- or star-like
6. Rhomboid
7. Linear, branching
8. None of the above

Triangular shapes were often detected in breast carcinomas, which led to the so-called triangular-principle of his work. In the BI-RADS lexicon, Lanyi's triangular calcifications are nowadays called segmental, but the correlation with malignancy can still be maintained. These triangular and linear branching groups are proving useful for classification. Most of the other shapes, however, could not be assigned to any pathology or anatomic structure and so are omitted in the present systems.

In their outstanding teaching atlas of mammography, Tabar *et al.* (2001) classified calcifications according to their histological origin into ductal and lobular, using parameters of shape, size, density, distribution, and number as diagnostic tools. This led to a detailed description of different shapes and distributions of calcifications in the ductuli and the terminal ductulo-lobular unit. In their latest edition, the description is adjusted to include and correlate to the termini used in the BI-RADS lexicon. Thus, this atlas has become a helpful addition, with excellent image material to be used together with BI-RADS for teaching purposes.

Although Tabar *et al.*'s (2001) classification based on the origin is facile for typically benign or malignant calcifications, the exact attribution of indistinct calcifications remains difficult. Here, the teaching atlas goes beyond the BI-RADS lexicon, adding other factors such as size of calcifications or number of calcifications, but still placing emphasis on the limitations of these factors. One helpful tool offered is the analysis of radio-opacity or density of single (intraparticular analysis of density) as well as between different calcifications (inter-particular analysis of density). Although most benign calcifications are of homogeneous density, casting or granular calcifications often have a wide range of densities that are inhomogeneously distributed in calcifications. Therefore, most casting single calcifications can be shown to be aggregates of numerous microscopic particles differentiating in size, length, and shape.

Incidental findings of atypical calcifications are described in the lactating breast with scattered bilateral calcifications with focal clusters; some show branching or linear distribution and granular heterogeneous shape. Despite their appearance of intermediate concern, all these calcifications were of benign origin and decreased during follow-up (Mercado *et al.*, 2002). Similar calcifications were found in association with aprocrine metaplasia (Janzen *et al.*, 2002).

Breast Imaging-Reporting and Data System

Currently, BI-RADS is the most widespread method of describing breast disease. Many western countries have now adapted the terminology and assessment categories, with minor adaptations, into their local or national requirements of breast imaging.

In this section, we present a summary of the descriptors of calcifications in the BI-RADS lexicon.

1. Typically, benign calcifications (Fig. 94) can rule out malignancy with a high probability and are usually pathognomonic. These include skin calcifications; vascular, round, popcornlike (fibroadenoma), large rodlike (plasma-cell mastitis), lucent centered, eggshell, or rim calcifications (lipid necrosis); and milk of calcium, suture, and dystrophic calcifications.

2. Coarse heterogeneous calcifications are of intermediate concern and often represent fibroadenoma, often the aforementioned growing fibroadenoma, fibrosis, or trauma.

3. Amorphous or punctuate calcifications have a diameter of < 0.5 mm, meaning that these are usually

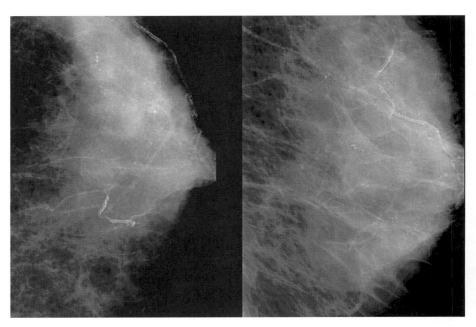

Figure 94 Typically benign calcifications: milk of calcium, round scattered calcifications indicating sclerosing adenosis and vascular calcifications.

indistinct (Fig. 95). Pathologically, they often represent psammom-like structures. As the process of calcification for benign adenomatous lesions is identical to that for low-grade carcinoma of the terminal ductulo-lobular unit, the differentiation of these two entities is almost impossible, with both leading to amorphous calcifications.

4. Fine pleomorphic calcifications are of similar shape but vary in size without casting parts. Consequently, these are more suspicious.

5. Fine-linear or fine-linear branching calcifications are usually associated with comedocarcinoma (Fig. 96).

Groups of calcifications can be classified into diffuse or scattered with a random distribution throughout the breast; regional (> 2 cm³ in an area that does not suggest a ductal distribution); grouped or clustered (< 1 cm³ of volume with at least five calcifications); and linear and segmental, a distribution suggesting involvement of one or multiple ducal units. Segmental and linear distribution is especially associated with malignant disease and might increase suspicion of indistinct or fine pleomorphic calcifications.

Skin calcifications are specified through localization and can either be centrally radiolucent or punctuate, depending on calcification of atheromatous or secretory changes. Usually, they are abundant and distributed over large parts of the breast. Vascular calcifications can be easily detected by their appearance as parallel tracks or tubular structures. As vascular calcifications are easy to diagnose, discussions have arisen on the use of mammography as a screening tool

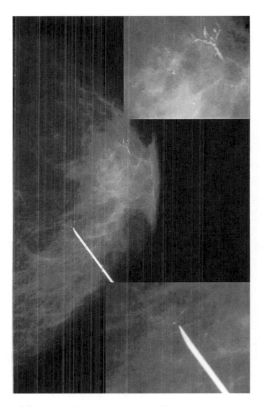

Figure 96 Typically malignant calcifications in a patient with ductal carcinoma *in situ*. The distribution is along the milk ducts with the Y-shaped form of the single calcifications. The needle marked a cluster of amorphous calcifications that represented a second focus of the multifocal malignancy.

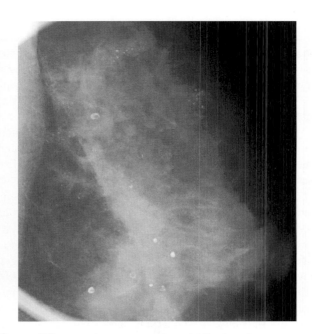

Figure 95 Amorphous or granular calcifications in a segmental distribution. The underlying ductal carcinoma *in situ* is visualized by a faint mass along the ductulo-lobular unit with calcifications in the lobuli.

for cardiac risk. Pecchi *et al*. (2003) showed that the degree of vascular calcifications in breast cancer screening is correlated with cardiovascular atherosclerosis. If excessive vascular calcifications are visible in mammography, further diagnostic steps to detect cardiac or cerebrovascular calcifications should be considered.

Typical calcifications of fibroadenoma are large (> 2–3 mm), popcornlike, and smooth that are found in myxoid degeneration of fibroadenomas. Difficulties can be encountered in the classification of liponecrosis and fibroadenoma with beginning calcifications because these may sometimes mimic casting or pleomorphic calcifications. Due to distribution or correlating sonographic lesions, the underlying pathology can often be suspected. In this case, follow-up after 6 months usually reveals confluent calcifications that can then be correctly identified. In BI-RADS, these calcifications of growing fibroadenoma are named as coarse heterogeneous with a size of > 0.5 mm.

Large (usually > 1 mm diameter) rodlike calcifications can usually be separated from casting calcifications by their smooth external appearance, with often a small central radiolucency also being visible. As previously described, these are typical for periductal or plasma-cell mastitis and are typically benign. Most of these rods are bilateral and are seen in women older than 60 years.

Suture and dystrophic calcifications are a rare entity found in postoperative and irradiated breasts, requiring no further attendance. The latter are similar to eggshells but irregular in shape. The difference between lucent-centered and eggshell calcifications, according to BI-RADS, is the thickness of the calcium deposit being > 1 mm in the lucent-centered and < 1 mm in the eggshall.

Milk of calcium is specified by different appearance in mediolateral (crescent-shaped) and craniocaudal (centrally lucent) views or tea-cup phenomenon. Often, these are clustered in a grapelike group representing the terminal ductulolobular unit. The explanation for this phenomenon is minute calcifications in the ductuli that are dissolved in retained fluid and then sedimentate. If the fluid is reabsorbed, the calcifications aggregate and form round or punctate calcifications of adenomatosis. The difference between round and punctate is generally based on the size of calcifications, with calcifications of > 0.5 mm being called round, and smaller ones punctate. Scattered punctate calcifications are considered benign, whereas clustered punctate calcifications, especially when new or ipsilateral to a known malignancy, require further attendance.

Indistinct or amorphous calcifications are either too small or too hazy to permit a further classification.

Especially for punctate and indistinct calcifications, the shape of the cluster is of special value in differentiating low-grade DCIS from adenomatosis. In a diffuse scattered distribution, they are usually benign, and thus there is no reason for recall in screening mammography. Clustered, regional, segmental, or linear distribution of indistinct calcification might require biopsy, but they are the most difficult kind of calcifications and overlapping between entities is large. The Dutch reference center in Nijmegen advocates a biopsy if three groups of amorphous or indistinct calcifications are present (Holland, pers. com. 2002). We support biopsy especially in linear or segmental distributions with round clusters or bunch of grapelike clusters supporting short-term follow-up.

Most authors state that microcalcifications indicating malignancy are usually < 0.5 mm (Kopans, 1989). A recent computer analysis of 260 cases found that average length of microcalcification is highly correlated to malignancy, with microcalcifications of > 0.41 mm having a 77% association to carcinoma, while smaller calcifications were 71% associated with benign lesions. However, overlap between the two groups was too large to separate these groups, and the difference in length is probably due to fine-linear versus punctate calcifications.

The difference between amorphous and fine pleomorphic calcifications is small, indicating similar probabilities of malignancy. However, fine pleomorphic calcifications are better delineated with a sharper limit and more suspicious appearance. There is also considerable overlap with fine-linear calcifications, especially when these are only sparse in a limited space. Fine-linear calcifications have the highest probability of malignancy. Usually, ductal origin is clearly visible and the irregular shape indicates malignancy, which means that confusion with other entities is generally impossible.

Despite all efforts, diagnostic accuracy for calcifications is lower than for solid lesions. Sensitivity is reported to be as low as 67%. When specificity is low, biopsy should be performed more often: the definite diagnosis of the underlying disease in the case of suspicious calcifications can only be obtained by histology. Because of limited accuracy, the subdivision of BI-RADS 4 into categories 4A to 4C in our experience remains of limited use for calcifications, suggesting additional information that cannot be translated into clinical proceeding, especially as we would suggest stereotactic biopsy for all of these patients regardless of probability of malignancy. In addition, calcifications tend to either underestimate tumor extent due to noncalcified parts of the lesion or to overestimate the extent as benign calcifications are included in the suspicious group. Despite all efforts, the classification of mammography according to histopathology remains difficult in literature as well as in daily practice, and the minimal sensitivity and specificity values of the BI-RADS system are hard to achieve in calcifications.

While classification in most western countries relies on terminology and assessment groups inspired by the BI-RADS, the translation of terminology into assessment categories remains difficult. The differentiation between regional and segmental or between pleomorphic and amorphic is vague. Furthermore, assignment of assessment relies on the reader's experience; several thousand mammographies are required until basic experience can be achieved. To overcome this obstacle especially for residents and persons in initial mammography training, we developed the following matrix system as a basic classification tool (Table 13).

The evaluation of 199 lesions with the matrix system showed for BI-RADS categories 2, 3, 4, and 5 malignancy rates of 0, 5.9, 17.6, and 90.9%, respectively (Müller-Schimpfle et al., 2005). These rates are slightly higher for BI-RADS 3 than is required by the ACR, but are in a comparable range with the publication by Liberman et al. (1998). To compare the matrix to subjective reading, we recently performed a multireader trial with > 200 microcalcifications. This trial showed that readers with mammography experience of > 2 years, translating to at least 1000 carcinomas classified according to BI-RADS, were superior in subjective classification compared to the matrix system in specificity at comparable sensitivity. Furthermore, all residents in our department and most participants in workshops, where we presented the matrix system, described it as a helpful tool in mammography.

Table 13 Microcalcification Matrix System

Through a matrix made from distribution and morphology, the BI-RADS assessment category can be deduced.

Distribution	Morphology		
	Typically Benign (round, rodlike, lucent-centered, vascular, etc.)	Indeterminate (amorphous, coarse, heterogeneous)	Typically Malignant (fine pleomorphic fine-linear/branching)
Diffuse	2	3	4
Clustered	2	3	4
Segmental/regional	3	4	5
Linear/branching	3	5	5

Work-up of Breast Calcifications

Most breast calcifications encountered during screening mammography are singular or scattered calcifications that do not fulfill the minimum criteria of > 5 calcifications in 1 cm². Some authors demand 5 calcifications in 2 cm² and these are therefore to be classified as certainly benign or BI-RADS 2. The second most frequent class is vascular calcifications, followed by plasma-cell mastitis, liponecrotic and fibroadenoma calcifications that are also easily recognized as benign. Cutaneous calcifications can usually be detected in tangential views. One difficulty arises from the artifacts caused by antiperspirants containing particles of aluminum. Especially in axillary calcifications, often of a casting type, additional views should be performed after thorough cleaning of the breast and armpit in order to rule out this possibility. Another artifact rarely encountered in screening mammography is caused by splinters of glass from motor vehicle accidents; these can mimic suspicious calcifications.

Clearly, malignant lesions are easily handled, because they always require biopsy, either by direct mastectomy or breast-conserving therapy or by guided core needle or vacuum-assisted biopsy devices. The extent of preoperative biopsy is usually specified in local or national regulation; for example, the European guidelines demand a > 90% rate of preoperative biopsy.

Difficulties are usually encountered in classifying intermediate calcifications that do not fall into the aforementioned categories, and so further decision is necessary. Although most countries as well as the BI-RADS system advocate short-period (usually 6-month) follow-up, regulations in several countries (e.g., the Norwegian breast cancer screening program) do not permit undecided results leading to 6-month follow-up after thorough work-up. These regulations, therefore, require either bioptic work-up or the decision to consider the lesion to be benign, and the women to return to regular screening. This method clearly reduces the psychological results of follow-up, including uncertainty and anxiety. Especially for microcalcifications, this process seems to be advantageous because follow-up is less precise for microcalcifications compared to solid lesions. Whereas calcified parts remain constant, intraductal carcinoma can progress without detection. A potential disadvantage is a probable increased biopsy rate for indeterminate lesions, which can only be avoided by strict regulation and surveillance of biopsy and recall rates. Differences in reimbursement and costs of follow-up versus final decision are unclear at present.

Biopsy with stereotactic guidance is the preferred diagnostic method for bioptic work-up of suspicious calcifications. This procedure can be performed either in a prone position on a dedicated breast-biopsy table or in an upright position with an extension to standard or digital mammography systems. The latter is less expensive in most of the units. In general, only in large dedicated centers with high biopsy rates can the advantages of prone biopsy weight up higher initial cost. These advantages are lower rates of failure due to patient collapse and better choice of access.

The majority of masses and BI-RADS 5 lesions can be evaluated with core-needle biopsy as malignant diagnosis is easily established. In BI-RADS 4 calcifications, vacuum biopsy is superior to core biopsy as underdiagnosis (e.g., the diagnosis of atypical hyperplasia in a case of ductal carcinoma) is less frequent with vacuum biopsy. Furthermore, vacuum-assisted biopsy makes preoperative management more precise and reduces additional excision biopsy (Siegmann et al., 2003).

Another advantage of the larger volume of vacuum-assisted biopsy has to do with so-called adjacent calcifications that are found in nonmalignant tissue close to malignant lesions. In many specimens from stereotactic biopsy, calcifications are not found in the cores containing calcifications but in cores without any visible calcifications, possibly caused by benign changes of breast parenchyma induced by malignant lesions. Owing to the larger specimens, the underlying malignancy is detected with a higher probability.

Each biopsy for microcalcifications should contain a specimen radiograph to localize the calcifications and to help the pathologist with correct work-up. A postbioptic mammogram is also desired to verify that the correct calcifications are sampled. In larger breasts with dense parenchyma (and sometimes even multiple groups of benign calcifications), it is possible to sample a different group of calcifications than planned, especially with older stereotactic units where contrast resolution is not equivalent to current technology or film-screen mammography. The postbioptic image also serves for further work-up as well as an initial image for further screenings in case of a benign lesion. Although some societies consider a negative biopsy sufficient to rule out malignancy and to return the woman to regular screening, others classify a negative biopsy of a BI-RADS 4 lesion to be BI-RADS 3 and, therefore, to be followed by a further mammography in 6 months. The call for a follow-up is justified by possible sampling errors and risk of underdiagnosis. A positive (or malignant) histology result from stereotactic biopsy should be followed by open excision of remaining malignant tissue to verify complete resection of the malignancy.

Although vacuum-assisted biopsy with an 8G needle can resect benign lesions such as fibroadenoma of up to 3 cm, it is impossible to achieve tumor-free margins with VB, not to mention with core-needle biopsy. From the postbioptic mammogram, further work-up after malignant biopsy can be deduced. In the case of remaining calcifications, localization as described below is facile, and difficulties can only be encountered by mammographically complete biopsy. Here, ultrasound is able to detect a postbioptic hematoma up to 14 days after biopsy, especially after vacuum biopsy (VB), leaving a larger biopsy hole. This hematoma should be localized either by guide-wire placement or with a clip, especially the latter if definitive resection cannot be scheduled in the following days or if a preoperative or neoadjuvant therapy is desired. Placement of a clip should also be considered when complete biopsy of calcifications is to be expected due to small-lesion volume. Subsequently, the clip can already be placed during biopsy either through a guiding sheath or with some systems directly through the biopsy needle. If no calcification or hematoma is visible, localization can be performed according to anatomic landmarks. However, this method is far less precise than the aforementioned methods and requires a larger volume to be excised and thoroughly searched for the biopsy site and possible remaining tumor.

Ultrasound-guided biopsy of calcifications should be performed only if a mammographic and ultrasound-detected lesion can definitively be correlated. However, specimen radiography and postbioptic mammography is obligatory in order to avoid misrepresentation of lesions and to verify biopsy results. The exact localization of the calcifications in the specimen radiographs should be described and communicated to the pathologist to ensure correct work-up, preferably with a copy of the radiograph being given to the pathologist.

If open biopsy is necessary, preoperative localization should be performed. This is also true for breast-conserving therapy of impalpable lesions. Placement of a guide wire is the most commonly used technique for preoperative localization either on a dedicated stereotactic unit or as manual placement with mammographic control images. Other procedures involving injection of coal or dye were equally effective in clinical studies, but bear the disadvantage of more complicated handling during placement and limited visibility in specimen radiography.

Although stereotactic guide-wire localization is easy to use in a precise way in preoperative management, it requires larger initial investments in a dedicated radiography unit. One simple and cost-effective method of guide-wire placement is a dedicated compression plate with multiple holes as described by Goldberg *et al.* (1983): the breast remains fixed after craniocaudal mammography, and the correct hole in the compression plate is searched in the mammogram and a needle is placed through the lesion in the correct position in the x- and y-axis. On a second mammogram in mediolateral view, the correct localization in the z-axis is determined and the needle is retracted accordingly. The disadvantage of this localization is that the access is always from the cranial and, therefore, in an esthetically less desirable position.

Most current equipment for stereotactic biopsy also provides a dedicated tool for needle localization that in most cases permits a preferable access to the lesion. Free-hand localization with mammographic control has the advantage of fewer restrictions in access and potentially best cosmetic results, as the radiologist can define access according to the desired operative procedure. However, the latter should be performed only by experienced specialists to reduce procedural pain and anxiety, as well as radiation dose. Usually with free-hand localization several views are required until the guide wire is placed correctly. A distance from the lesion of < 1 cm is required. A distance of < 5 mm is desired for any localization procedure, especially in smaller breasts.

As with core-needle or vacuum-assisted biopsy, specimens from open biopsy with guide-wire localization should be examined with specimen radiography to verify complete biopsy. These specimens should be marked intraoperatively in at least three directions, and specimen radiography should be performed in two views, enabling the radiologist to communicate the direction of incomplete excision and facilitating additional surgery. In our hospital, specimens are labeled by three strings of different lengths in the anterior, superior, and inferior position. Specimen radiography performed during the procedure is preferable, as direct resection of remaining calcifications is possible and additional surgery can be avoided. It is highly recommended that all histological results from biopsy or surgery be discussed in an interdisciplinary pathologic-surgical-radiological conference. Discrepancies between imaging and pathology require further work-up. Calcifications that were radiologically clearly malignant but showed benign pathology are especially suspicious for insufficient histological processing or biopsy. Incomplete histological processing cannot be excluded, as even small specimens cannot be sliced entirely and are, therefore, only reviewed in representative slices. In these cases, additional specimen radiography is a helpful tool to localize calcifications that were eventually not contained in the initially stained histology sections.

Summary

Calcifications are a common finding in mammography. The most important factor in differentiating malignant from benign lesions is the morphology of individual calcifications. Especially for indistinct calcifications, the shape of the group is an additional factor for further assessment of malignancy. Additional examinations for calcifications are magnified views, while ultrasound and MRI are useful for detecting associated or additional lesions. Assessment of BI-RADS 3 or probably benign calcifications includes systematic control, usually in 6 months. BI-RADS 4 calcifications should generally lead to biopsy, either by core-needle or vacuum-assisted devices. A negative biopsy of BI-RADS 4 calcifications may be downgraded to probably benign. Probably malignant or BI-RADS 5 calcifications require biopsy either as minimal-invasive for preoperative histology or as direct treatment via open surgery. All biopsies should be discussed in interdisciplinary conference. In future developments including digital mammography, tomosynthesis or dual-energy mammography is of clinical value but remains to be proven in further studies.

References

American College of Radiology (ACR). 2004. *ACR BI-RADS-Mammography, in ACR Breast Imaging Reporting and Data System*, 4th ed. *Breast Imaging Atlas*. Reston VA: American College of Radiology.

Finsterbusch, R., and Gross, F. 1934. Kalkablagerungen in den Milch- und Ausfuhrungsgangen beide Brustdrusen. *Rontgen praxis 6*:172.

Fischer, U., Baum, F., Obenauer, S., Luftner-Nagel, S., von Heyden, D., Vosshenrich, R., and Grabbe, E. 2002a. Comparative study in patients with microcalcifications: full-field digital mammography vs. screen-film mammography. *Eur. Radiol. 12*:2679–2683.

Fischer, U., Baum, F., Obenauer, S., Funke, M., Hermann, K.P., and Grabbe. E. 2002b. Digital full field mammography: comparison between radiographic direct magnification and digital monitor zooming. *Radiologe 42*:261–264.

Fischmann, A., Pietsch-Breitfeld, B., Müller-Schimpfle, M., Siegmann, K., Wersebe, A., Rothenberger-Janzen, K., Claussen, C.D., and Janzen, J. 2004. Radiologic-histopathologic correlation of microcalcifications from 11G vacuum biopsy: analysis of 3196 core biopsies. *RöFo Fortschr Röntgenstr. 176*:538–543.

Fischmann, A., Siegmann, K.C., Wersebe, A., Claussen, C.D., and Müller-Schimpfle, M. 2005. Comparison of full-field digital mammography and film-screen mammography: image quality and lesion detection. *Brit. J. Radiol. 78*:312–315.

Goldberg, R.P., Hall, F.M., and Simon, M. 1983. Preoperative localization of nonpalpable breast lesions using a wire marker and perforated mammographic grid. *Radiology 146*:833–835.

Hall, F.M., Storella, J.M., Silverstone, D.Z., and Wyshak, G. 1988. Nonpalpable breast lesions: recommendations for biopsy based on suspicion of carcinoma at mammography. *Radiology 167*:353–358.

Holland, R. Personal communication, 16 April 2002.

Janzen, J., Fischmann, A., and Rothenberger-Janzen, K. 2002. Microcalcifications in association with apocrine metaplasia in fibrocystic breast disease. *Singapore J. Obstet. Gynaecol. 33*:47–49.

Kopans, D.B. 1989. *Breast Imaging*. Philadelphia: J. B. Lippincott Co.

Lanyi, M. 1986. *Diagnostik und Differentialdiagnostik der Mamma-Verkalkungen*. Berlin, Germany: Springer.

LeBorgne, R. 1951. Diagnosis of tumors of the breast by simple roentgenography: calcifications in carcinomas. *Am. J. Roentgenol. 65*:1–11.

Le Gal, M., Durand, J.C., Laurent, M., and Pellier, D. 1976. Management following mammography revealing grouped microcalcifications without palpable tumor. *Nouv. Presse Med. 5*:1623–1627.

Lemacks, M.R., Kappadath, S.C., Shaw, C.C., Liu, X., and Whitman, G.J. 2002. A dual-energy technique for microcalcification imaging in digital mammography—a signal-to-noise analysis. *Med. Phys. 29*:1739–1751.

Lewin, J.M., D'Orsi, C.J., and Hendrick, R.E. 2004. Digital mammography. *Radiol. Clin. North Am. 42*:871–884.

Liberman, L., Abramson, A.F., Squires, F.B., Glassman, J.R., Morris, E.A., and Dershaw, D.D. 1998. The Breast Imaging Reporting and Data System: positive predictive value of mammographic features and final assessment categories. *Am. J. Roentgenol. 171*:35–40.

Mercado, C.L., Koenigsber, T.C., Hamele-Bena, D., and Smith, S.J. 2002. Calcifications associated with lactational changes of the breast: mammographic findings with histologic correlation. *Am. J. Roentgenol. 179*:685–689.

Müller-Schimpfle, M., Wersebe A., Xydeas, T., Fischmann, A., Vogel, U., Fersis, N., Claussen, C.D., and Siegmann, K. 2005. Microcalcifications of the breast: how does radiologic classification correlate with histology? *Acta Radiologica. 46*:774–781.

Pecchi, A., Rossi, R., Coppi, F., Ligabue, G., Modena, M.G., and Romagnoli, R. 2003. Association of breast arterial calcifications detected by mammography and coronary artery calcifications quantified by multislice CT in a population of post-menopausal women. *Radiol. Med. (Torino) 106*:305–312.

Pisano, E.D., Gatsonis, C., Hendrick, E., Yaffe, M., Baum, J.K., Acharyya, S., Conant, E.F., Fajardo, L.L., Basset, L., D'Orsi C., Jong, R., and Rebner, M., for the DMIST Investigators group. 2005. Diagnostic performance of digital versus film mammography for breast-cancer screening. *N. Eng. J. Med. 353*:1773–1783.

Rong, X.J., Shaw, C.C., Johnston, D.A., Lemacks, M.R., Liu, X., Whitman, G.J., Cryden, M.J., Stephens, T.W., Thompson, S.K., Krugh, K.T., and Lai, C.-J. 2002. Microcalcification detectability for four mammographic detectors: flat-panel, CCD, CR, and screen/film. *Med. Phys. 29*:2052–2061.

Salomon, A.L. 1913. Beiträge zur Pathologie und Klinik der Mammakarzinome. *Arch. Klin. Chir. 101*:573–668.

Siegmann, K., Wersebe, A., Fischmann, A., Fersis, N., Vogel, U., Claussen, C.D., and Müller-Schimpfle, M. 2003. Stereotactic vacuum-assisted breast biopsy—success, histologic accuracy, patient acceptance and optimizing the BI-RADS-correlated indication *RöFo Fortschr. Röntgenstr. 175*:99–104.

Tabar, L., Dean, P.B., and Tot, T. 2001. *Teaching Atlas of Mammography*, 3rd ed. Stuttgart, Germany: Thieme Medical Publishers.

Tavassoli, F.A. 1999. *Pathology of the Breast,* 2nd ed. New York: McGraw-Hill.

Teh, W. L., Wilson, A.R., Evans, A.J., Burrell, H., Pinder, S.E., and Ellis, I.O. 2000. Ultrasound guided core biopsy of suspicious mammographic calcifications using high frequency and power Doppler ultrasound. *Clin. Radiol. 55*:390–394.

10

Benign and Malignant Breast Lesions: Doppler Sonography

José Luis del Cura

Introduction

Tumor angiogenesis plays a fundamental role in the growth and extension of malignant tumors, including breast cancer (Gasparini and Harris, 1995). Tumor mass increase and formation of metastasis require the development of new vessels from preexisting ones. To help the recruitment of new vessels, malignant tumors secrete angiogenic factors. Because the process is not physiological and occurs very quickly, the newly formed tumor vessels differ from normal vessels. The new vessels lack muscular layer and follow a tortuous and an anarchic course, with variable calibers and abnormal ramifications. Stenosis, obstructions, coiling, and arteriovenous shunts are also often observed.

Several tumor factors, such as cell-cycle time, duplication time, and tumor volume, will depend on the formation of new vessels. Thus, the characteristics of the tumor vessels will be an indicator of prognosis, and they will also serve as a criterion for distinguishing cancer from benign lesions. The mean vascular density of the tumors has been identified as an important predictive marker for survival in breast cancers and for the likelihood of distal metastasis. Furthermore, mean vascular density has been found to diminish with chemotherapy. Thus, it could be inferred that a technique that allows vascularization assessment could be used to distinguish benign lesions from malignant ones in the breast, and even to predict the prognosis of tumors (Weind et al., 1998). This concept is the basis for the diagnosis of breast tumors by means of techniques such as magnetic resonance imaging (MRI) and Doppler sonography.

Currently, mammography is still the most used imaging technique for diagnosing and monitoring breast carcinoma. Nonetheless, although its use and interpretation are highly standardized, mammography still has important limitations, with a significant number of undetected lesions and indeterminate findings. The technological advances in ultrasound during the past 15 years have made use of this technique widespread in breast diseases. It has become a standard procedure, routinely used in addition to mammography to improve diagnostic effectiveness. Ultrasound has even become the preferred procedure in lesions not visible in mammography, in dense breasts, and in young women. These advances have consisted mainly in equipment resolution improvement, use of high-frequency transducers, incorporation of tissue harmonic imaging, and use of Doppler techniques, including the appearance of new constrast agents.

Doppler Ultrasound Technique in Breast Diseases

Ultrasound examination of the breast is usually performed with high-resolution equipment, using broadband transducers with frequencies of ~ 7.5 MHz. This equipment also enables quality Doppler examinations. The forms of Doppler ultrasound that have been used to explore breasts include spectral Doppler, color Doppler, and power Doppler. The first two, based on mean frequency shift caused by the Doppler effect of circulating blood corpuscles, enable the assessment of vascular flow and its direction. The power Doppler is based on intensity of energy of the Doppler signals. It is essential to use an appropriate technique, because breast lesions tend to be small, with narrow vessels that have a slow flow. The equipment should be adjusted to detect slow flows and to minimize artifacts. It is advisable to lower the filters, use a low-pulse repetition frequency, and adjust the color or spectral Doppler box to the area of interest, making it as small as possible.

Spectral Doppler sonography provides a temporal spectrum of the flow velocities in the vessel studied and allows various flow parameters to be quantified in the tumor. The most important parameters are systolic peak velocity, resistance index, and pulsatility index. The resistance index is calculated as the peak systolic velocity minus the minimum diastolic velocity, divided by the peak systolic velocity. The pulsatility index is calculated as the peak systolic velocity minus the minimum diastolic velocity, divided by mean velocity. Power Doppler ultrasound has the disadvantages of lacking information regarding the direction of flow and yielding more artifacts. However, it is the most sensitive technique for detecting smaller vessels with slower flows, so it is the best technique to study small lesions and, in general, to increase examination sensitivity (Kook *et al.*, 1999). It allows a more detailed survey of intratumor vascularization than conventional color Doppler sonography, and it is able to depict more vessels. This advantage makes it the preferred technique for examining breasts, rather than conventional color Doppler, even in contrast-enhanced examinations.

Enhanced ultrasound significantly increases the ability of Doppler sonography to identify vessels, regardless of the form of Doppler ultrasound used. Contrast agents increase the signal-to-noise ratio, enabling the detection of smaller vessels and those with slower flow, when comparing with unenhanced techniques. Until now, Levovist® has been the contrast agent more widely used, and most of the studies published about contrast-enhanced ultrasound have used this agent. Levovist is a suspension in water of d-galactose microparticles stabilized by adding 0.1% palmitic acid. This contrast produces microbubbles that, when injected, pass into the bloodstream and increase the echoes of the vessels while in transit, increasing Doppler signals. The sonographic contrast action mechanism differs from that of contrasts used in other imaging techniques, such as computed tomography (CT) and magnetic resonance imaging (MRI): sonographic contrast remains in the vessels and enhances them instead of being taken by the lesion. The contrast effect lasts ~ 10 min. and disappears when the bubbles burst. The galactose is excreted physiologically. The product has no contraindications or side effects. However, it does have the disadvantage of raising the cost of the examination.

Breast Doppler Limitations

Currently, there is a high degree of consensus regarding the use and significance of gray-scale ultrasound signs in breast lesions. However, with regard to the contribution of Doppler sonography, consensus does not exist. In recent years, many studies of different qualitative and quantitative diagnosis criteria in Doppler sonographies of breast lesions have been published. The possibility of distinguishing benign lesions from malignant ones has been evaluated, as well as how to predict several prognostic factors, such as axillary involvement and tumor grade. The results obtained have not always been in agreement, and therefore the potential usefulness of Doppler ultrasound in the diagnosis of breast cancer remains undefined. Several factors contribute to these variable outcomes. First, important technological advances have been made in the field of ultrasonography in recent years. As equipment improved, so did Doppler sensitivity in detecting vessels, whether tumoral or normal, reported in the literature. First publications mentioned only flow detection in malignant tumor vessels. Later, flows were detected on unenhanced Doppler in both benign and malignant lesions, making the technique less specific.

Another factor for variability of Doppler examinations is the influence of the hormonal status, particularly the estrogen environment, in the vascularization of breast lesions. Menopause and contraceptives cause significant variations in the characteristics of flow. The resistance and pulsatility indices are higher in the lesions of premenopausal women, whether the lesions are benign or malignant. However, in premenopausal women who take contraceptives, the pulsatility and resistance indices are lower, and the number of vessels detected in benign tumors is significantly higher (Germer *et al.*, 2002).

The changes occurring after administering the contrast, the time it takes for the enhancement to take place, and the duration and degree of enhancement achieved do not depend on lesion vascularization alone. They may also vary depending on factors relating to the patient's own circulation system and the speed at which the contrast was injected (Stuhrmann *et al.*, 2000).

Lesion localization, especially its distance to the transducer, makes the detection of the flow signal vary. Vessels in superficial lesions are easier to be depicted, whereas in deeper lesions ultrasound becomes progressively attenuated, restricting the detection of vessels with less flow. Similarly, it is easier to detect vessels in large lesions because, owing to their size, they have more vessels (Birdwell *et al.*, 1997). Vessels are more frequently detected in palpable lesions, which are usually larger or more superficial than nonpalpable ones. This means that the data obtained in the studies carried out on symptomatic lesions may not be valid for nonpalpable lesions. Regardless of other features, depiction of vessels is easier in malignant lesions because they are more frequently palpable and have a larger average dimension than those in benign lesions (del Cura *et al.*, 2005). Other factors that restrict the usefulness of Doppler ultrasound are lack of standardization of examination procedures, quantitative measurements, and diagnostic criteria. These factors are added to the traditional ultrasonography issue of operator dependency, which often causes high interobserver variability in the results. Therefore, presently, Doppler imaging is scarcely used in sonographic examination and management of breast nodules, despite its availability in most state-of-the-art ultrasound equipment.

On the other hand, despite the high resolution of modern equipment, it is still limited and allows only identification of vascular structures of a certain caliber and flow velocity. This limits the potential usefulness of Doppler ultrasound in the assessment of angiogenesis by correlating vessel detection in sonography with mean vascular density. The studies carried out in this area have not managed to demonstrate a significant correlation between them. On the contrary, they do suggest that mean vascular density and Doppler sonography provide information about different aspects of tumor vascularization: mean vascular density about microvascular network, and Doppler Sonography about vessels with diameters larger than 300 μm (Peters-Engl *et al.*, 1998).

Differentiation of Benign and Malignant Solid Breast Lesions

Tumor Vessel Identification

Vessel detection inside lesions was the first criterion used in Doppler sonography to distinguish between benign and malignant lesions. Flow is detected more frequently in breast cancers than in benign lesions, and the number of vessels depicted is significantly higher in malignancies. These differences are independent of the size of the lesions. Rather, they are due to the nature of tumors (del Cura *et al.*, 2005). Vessel detection in breast lesions may vary widely, depending on equipment sensitivity and examination technique. By using an adequate technique and power Doppler sonography, vessel detection has been described in 36% of benign nodules and 68% of carcinomas.

The problem with this criterion is that, when considered separately, it shows limited sensitivity and specificity in breast cancer diagnosis—68% and 64%, respectively. If a sonographic contrast is employed, the number of vessels detected in benign and malignant lesions increases. This increase occurs mostly in carcinomas, in which flow can be detected in 95% of the tumors after contrast administration, thus increasing sensitivity at the expense of a less important decrease in specificity (Moon *et al.*, 2000). The visualization of four or more vessels in a tumor on enhanced Doppler has been described as indicative of malignancy (Yang *et al.*, 2001). Moreover, the increase in flow observed after administration of contrast can be a diagnostic criterion: a significant increase in flow with new areas of enhancement, when compared to unenhanced studies, suggests malignancy.

Vessel morphology is one of the most useful criteria for distinguishing benign lesions from malignant ones (see Fig. 97). Penetrating vessels—that is, the ones that penetrate from the border of the tumor to its center—are more frequent in malignant tumors, whereas peripheral vessels—those that present around the border of a mass—are more frequent in benign lesions (Raza and Baum, 1997). Branching vessels and those with an irregular course also suggest malignancy. Using penetrating vessels as a criterion for malignancy has shown a sensitivity of 68% and a specificity of 95%. The use of contrasts raises the number of vessels detected, increasing sensitivity and diagnostic confidence.

Quantitative Criteria

The nature of malignant tumor vessels makes blood circulation through these vessels differ from the circulation in normal breasts and benign lesions. Phenomena such as the presence of stenosis and arteriovenous shunts cause a high-speed flow in malignant tumors. This determines the use of peak systolic velocity to differentiate breast lesions. The velocity is significantly higher in malignant lesions. The problem with peak systolic velocity is the necessity of manually tracing the course of the vessel studied to measure it. This limitation makes the measurement frequently inexact and hard to reproduce, because to obtain a correct value, the trajectory of the vessel used to measure peak systolic velocity needs to be identified very precisely. Minor variations in definition of the course of the vessels may cause significant differences in the velocity calculated. The resistance and pulsatility indices are other quantitative criteria based on spectral Doppler sonography. These indices are also higher in malignant lesions due to stenosis and vascular occlusions in tumor vessels that cause an increase in peripheral resistance. Both are ratios, so they are not influenced by the angle between the vessels and the transducer.

The problem posed by quantitative criteria is the overlapping that exists between the values obtained in benign lesions and in malignancies, which restricts its practical usefulness. In general, lesions with a resistance index higher than 0.8 are more likely to be malignant. But it is even more interesting that the observation of absence or

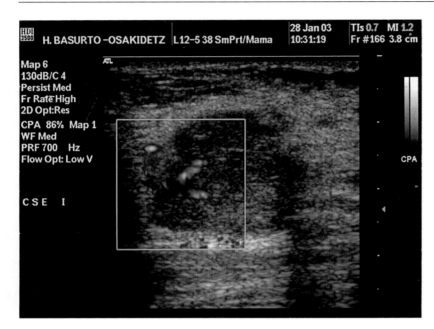

Figure 97 Power Doppler in an infiltrating ductal carcinoma of the breast. Angular, irregular-shaped vessels can be seen penetrating from the border toward the center of the tumor.

inversion of diastolic flow in a lesion (in other words, the existence of a resistance index of 1 or more) is a diagnostic criterion with a very high positive-predictive value (see Fig. 98). Therefore, any lesion in which this sign is observed should be considered highly suspect of malignancy, regardless of any other image or examination finding (del Cura et al., 2005). Despite its very high specificity, this sign has the limitation of being observed in only 15% of malignant tumors.

Semiquantitative Criteria

Earlier we mentioned that the sonographic contrast action mechanism is different from the mechanism of the contrast used in CTs and MRIs. This means that, unlike these techniques, enhanced ultrasound cannot be used to obtain a quantitative assessment of the signal increase caused by the contrast. Some authors have proposed several semiquantitative criteria in Doppler color sonography to differentiate

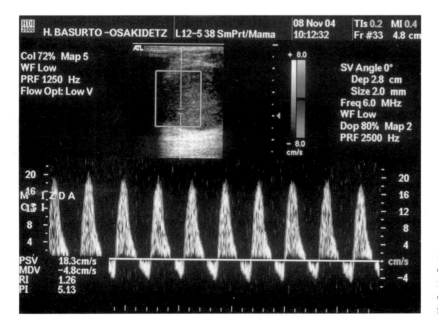

Figure 98 Spectral Doppler in an infiltrating ductal carcinoma of the breast. A high-resistance flow, with flow inversion during the diastole, can be observed, a finding highly suggestive of malignancy.

between malignant and benign breast lesions. These criteria are based on computing the number of colored pixels or vessels in the areas concerned. In some cases, the calculation formulas incorporate a factor that depends on the flow velocity at each point provided by the color Doppler. These criteria have the limitation of being difficult to measure, and they often require special software that is not available in standard equipment. The velocity factor included in the formulas of some criterion is not precise because it does not take into account the course of vessels in relationship to the transducer. Furthermore, the results obtained are poor and sometimes contradictory. Therefore, there is no sufficient scientific evidence to support the use of such methods to diagnose breast cancer, being currently limited to the field of research.

Breast Cancer Prognosis

The higher growth speed observed in more aggressive tumors implies an increased need for blood. Therefore, tumors with a higher histologic grade and a poor prognosis have a more developed vascular network than those that grow more slowly. The mean vascular density is a predictive marker of breast cancer. In the same way, the detection of vessels in Doppler ultrasound may provide information on the speed of tumor growth and metastasic dissemination of the tumor, that is, on its prognosis. It should be remembered, however, that as mentioned earlier, no relationship exists between mean vascular density and the vessels observed in a breast Doppler because the thickness of the vessels identified in each technique differs. The Doppler does not provide information on microvessels.

No relationship has been demonstrated between various parameters of Doppler ultrasound (vessel detection, quantitative criteria) and tumor grade; therefore, the technique is not useful to predict histologic grade in breast neoplasms. Nonetheless, the number of vessels detected by color Doppler ultrasound has proved to be an independent predictive survival factor in women with breast cancer. As a general rule, the higher the vascularization detected in a malignant tumor, the worse the prognosis will be. Although it is a rare finding, identifying more than 10 arteries in a tumor has the same prognostic implications as the presence of axillary metastasic nodes (Watermann et al., 2004).

Assessment of Lymph Node Involvement

The relationship between vascularization of breast tumors studied on Doppler ultrasound and lymph node involvement has focused the interest of scientific literature in recent years. Axillary status is one of the main prognostic indicators in breast cancer, so the information on the existence of axillary metastasis is vital. The technique of sentinel node biopsy has become widely used and has proved effective in detecting the existence or absence of axillary metastases, but it is an expensive technique that depends on the surgeon's experience and may have false-negative results, particularly during the learning period. It would be desirable to count on an alternative noninvasive method for accurate prediction of lymph node involvement. This would enable the identification of patients with a high probability of having axillary metastases in which the sentinel node biopsy technique would not be recommended and who should be sent to axillary lymph node dissection directly. Several Doppler ultrasound criteria have been proposed as potential methods for predicting axillary status before surgery. The criteria are of two types: those that focus on the assessment of primary tumor features to describe signs that allow the prediction of lymph node involvement, and those that assess the vascularization of axillary nodes to deduce whether they are benign or metastatic.

No consensus has been reached on the relationship between vascularization in primary breast cancer and lymph node involvement. Published studies offer contradictory results. In general, there seems to be a slightly higher probability of axillary metastases in tumors in which vessels are detected on power Doppler or color Doppler, but the low specificity of this sign makes it practically useless (Santamaría et al., 2005). Lymph node assessment, however, offers more hopeful perspectives. Conventional ultrasound has important limitations regarding the detection of axillary nodes, particularly when they are benign. The anatomy of the axilla makes its exploration difficult; moreover, lymph nodes are often hard to identify among the axillary fatty content. Classic diagnostic criteria for detection of lymph node metastasis were based on the size of the nodes, thus being of little reliability. However, there are other sonographic signs of malignancy, such as the absence of echogenic fatty hilum or the presence of nodular asymmetric cortical thickening, which are much more useful,. Both criteria have a high positive-prediction value in the larger nodes, but they are not reliable or are difficult to identify in nodes with diameters < 1 cm. For nodes in which ultrasound findings are uncertain, Doppler can be an important ancillary tool for assessment.

Although malignant lymph nodes are generally more vascularized than benign ones, the detection of flow in axillary nodes, as a sole criterion, is of little use for assessing the nature of nodes. Nonetheless, the observation of vessels in the periphery of lymph nodes not connected to hilum vessels in Doppler ultrasound is highly suggestive of malignancy. Used together, the earlier mentioned ultrasonographic criteria for malignancy (absence of fatty hilum, asymmetric cortex thickening) and the existence of peripheral vessels in a Doppler sonography have a high sensitivity and specificity (94 and 82%, respectively) for identifying axillary metastasis (Esen et al., 2005).

Sonographic contrast enables the identification of a significantly higher number of vessels in Doppler ultrasound, particularly in metastatic nodes, including the detection of more peripheral vessels. This increases the sensitivity and diagnostic confidence of Doppler ultrasound in assessing lymph node involvement. Furthermore, malignant nodes enhance more than do corresponding primary breast cancers after contrast injection, whereas benign nodes do not (Yang et al., 2001).

Recurrence versus Scar in Operated Patients

One of the most difficult challenges in breast cancer diagnosis is the differentiation between postoperative changes and tumor recurrence in patients who have had part of the breast resected. The differentiation is generally difficult because the structural distortions that surgery and radiotherapy cause in the breast may present features that are similar to a tumor, both on mammography and sonography. The same criteria used to distinguish a benign lesion from a malignant one on Doppler ultrasound can be used to differentiate between a postoperative scar and a tumor recurrence. Scars are less vascularized than recurrences. Furthermore, their vascularization tends to diminish over time, which makes Doppler ultrasound especially useful in scars that have developed over a longer period of time, because older scars are more difficult to distinguish from tumors on mammography or gray-scale sonography. Detecting vessels, especially when they have malignant features (penetrating, irregular) in a lesion on color or power Doppler, favors the suspicion of recurrence. Injecting a sonographic contrast enhances Doppler sensitivity in such cases. However, it should be remembered that vessels with malignant features can also be detected in some scars, particularly in recent scars and in scars that develop from postoperative hematomas. Also, a small number of recurrences do not show vessels, even after injecting a contrast agent. Therefore, it is still necessary to correlate Doppler findings with other imaging techniques. When in doubt, the nature of the lesion should be confirmed histologically (Stuhrmann et al., 2000).

Treatment Monitoring

Preoperative neoadjuvant chemotherapy of patients with breast cancer has shown similar overall survival when compared to initial surgical treatment, with the advantage that tumor bulk reduction allows a less invasive surgery. Moreover, sometimes there is a full pathological response to chemotherapy, and no evidence of a tumor is found during surgery. This response is an excellent prognostic sign because it indicates a tumor's positive response to chemotherapy. Assessing tumor response is important for planning treatment after chemotherapy. Techniques such as palpation and imaging techniques (mammography, ultrasound, MRI, PET) are used, but the results are often not very sensitive and may even be contradictory. Some studies have used semiquantitative methods to assess the potential usefulness of Doppler to determine tumoral extension after chemotherapy. The response to chemotherapy has been found to cause a reduction in detected vascularity on color Doppler ultrasound in most of the treated tumors (Huber et al., 2000). However, the absence of vascular signals on Doppler ultrasound does not always mean that there has been a full response to chemotherapy. There are cases of residual cancers in which no vessels are detected. This fact limits the practical usefulness of Doppler after neoadjuvant treatment (Roubidoux et al., 2005). The ability of Doppler ultrasound to identify those tumors that will benefit from neoadjuvant treatment and those that will not respond to chemotherapy has also been investigated. This is based on the hypothesis that tumors with a poor vascularization may not receive a sufficient amount of chemotherapy. However, the amount of vascularization, as assessed by Doppler sonography, does not allow prediction of the response to chemotherapy, because the response is similar in highly vascularized tumors and in those that are less vascularized (Roubidoux et al., 2005).

Conclusions

Doppler sonography has long been used for studying breast malignancy and is available in most state-of-the-art equipment used for breast examination. However, its use has not become widespread and is not included in standard breast cancer diagnostic procedures. The American College of Radiology has not included any criteria related to Doppler sonography in their popular Breast Imaging-Reporting and Data System (BI-RADS), which allows standardization of radiological reports of breast examinations. Lack of standardization of the technique, its procedures, and the interpretation of the signs observed, makes its results frequently unreliable and often contradictory, making breast Doppler a highly operator and equipment-dependent technique. Despite the attempt to include objective measures in the interpretation of results, they are still based on subjective criteria, with few exceptions. Finally, the sensitivity and specificity of many of the signs and the techniques used are insufficient. The increase in sensitivity due to technological advances has been achieved at the expense of specificity.

When used appropriately, however, Doppler ultrasound can be a valuable complementary tool in breast examination. Some Doppler sonography signs are highly reliable and can increase the sensitivity and specificity of conventional ultrasound. The detection of penetrating vessels that go through the border of tumor toward the center, or irregular and branched vessels, allows a lesion to be classified as suspicious. Detecting null or inverted diastoles at any vessel of a tumor in spectral Doppler is a highly positive-predictive sign and makes the lesion highly suspicious of malignancy regardless of the appearance that the lesions may have on mammography or on gray-scale sonography. Doppler sonography, therefore, should be a part of routine ultrasound techniques for solid breast lesions, particularly in lesions that ultrasound features suggest are benign, because it is precisely in the presumed benign lesions that the level of diagnostic suspicion is most likely to change.

Doppler sonography should also be a part of routine in the sonographic examination of axillas to detect potential axillary metastases in patients on whom a sentinel node biopsy technique is to be performed. The detection of peripheral vessels that are not connected to the hiliar vessels is a reliable sign of metastatic axillary lymph node infiltration.

The usefulness of contrast agents in breast Doppler sonography raises a number of issues and expectations. The sonographic contrast enables the visualization of a higher number of vessels and increases the technique's sensitivity. In addition, an intensely enhanced lesion after injecting a contrast agent is highly suggestive of malignancy. However, when examining solid breast lesions, the improvement obtained in diagnostic value with contrast-enhanced Doppler over conventional sonography and unenhanced Doppler sonography is relatively limited. Its high cost does not appear to justify routine use. As a general rule, it should only be used in doubtful cases in which conventional sonography and unenhanced Doppler sonography do not provide a sufficiently reliable approach to the nature of a lesion. In any case, continued technological improvements in Doppler sonography open a promising field for potential advances in the use of this technique in breast examination. The new third-generation sonographic contrast agents, of which few results have been published up until now, are part of this new possibility.

References

Birdwell, R.L., Ikeda, D.M., Jeffrey, S.S., and Jeffrey, R.B. 1997 Preliminary experience with power Doppler imaging of solid breast masses. *Am. J. Roentgenol. 169*:703–707.

del Cura, J.L, Elizagaray, E., Zabala, R., Legórburu, A., and Grande, D. 2005. The use of unenhanced Doppler sonography in the evaluation of solid breast lesions. *Am. J. Roentgenol. 184*:1788–1794.

Esen, G., Gurses, B., Yilmaz, M.H., Ilvan, S., Ulus, S., Celik, V., Farahmand, M., and Calay, O.O. 2005. Gray scale and power Doppler US in the preoperative evaluation of axillary metastases in breast cancer patients with no palpable lymph nodes. *Eur. Radiol. 15*:1215–1223.

Gasparini, G., and Harris, A.L. 1995. Clinical importance of the determination of tumor angiogenesis in breast carcinoma: much more than a new prognostic tool. *J. Clin. Oncol. 13*:765–782.

Germer, U., Tetzlaff, A., Geipel, A., Diedrich, K., and Gembruch, U. 2002. Strong impact of estrogen environment on Doppler variables used for differentiation between benign and malignant breast lesions. *Ultrasound Obstet. Gynecol. 19*:380–385.

Huber, S., Medl, M., Helbich, T., Taucher, S., Wagner, T., Rudas, M., Zuna, I., and Delorme, S. 2000. Locally advanced breast carcinoma: computer assisted semiquantitative analysis of color Doppler ultrasonography in the evaluation of tumor response to neoadjuvant chemotherapy (work in progress). *J. Ultrasound Med. 19*:601–607.

Kook, S.H., Park, H.W., Lee, Y.R., Lee, Y.U., Pae, W.K., and Park, Y.L. 1999. Evaluation of solid breast lesions with power Doppler sonography. *J. Clin. Ultrasound 27*:231–237.

Moon, W.K., Im, J., Noh, D., and Han, M.C. 2000. Nonpalbable breast lesions: evaluation with power Doppler US and a microbubble contrast agent—initial experience. *Radiology 217*:240–246.

Peters-Engl, C., Medl, M., Mirau. M., Wanner, C., Bilgi, S., Sevelda, P., and Obermair, A. 1998. Color-coded and spectral Doppler flow in breast carcinomas—relationship with the tumor microvasculature. *Breast Cancer Res. Treat. 47*:83–89.

Raza, S., and Baum, J.K. 1997. Solid breast lesions: evaluation with power Doppler US. *Radiology 203*:164–168.

Roubidoux, M.A., LeCarpentier, G.L., Fowlkes, J.B., Bartz, B., Pai, D., Gordon, S.P., Schott, A.F., Johnson, T.D., and Carson, P.L. 2005. Sonographic evaluation of early-stage breast cancers that undergo neoadjuvant chemotherapy. *J. Ultrasound Med. 24*:885–895.

Santamaría, G., Velasco, M., Farré, X., Vanrell, J.A., Cardesa, A., and Fernández, P.L. 2005. Power Doppler sonography of invasive breast carcinoma: does tumor vascularization contribute to prediction of axillary status? *Radiology 234*:374–380.

Stuhrmann, M., Aronius, R., and Schietzel, M. 2000. Tumor vascularity of breast lesions: potentials and limits of contrast-enhanced Doppler sonography. *Am. J. Roentgenol. 175*:1585–1589.

Watermann, D., Madjar, H., Sauerbrei, W., Hirt, V., Frömpeler, H., and Stickeler, E. 2004. Assessment of breast cancer vascularization by Doppler ultrasound as a prognostic factor of survival. *Oncol. Rep. 11*:905–910.

Weind, K.L, Maier, C.F., Rutt, B.K., and Moussa, M. 1998. Invasive carcinomas and fibroadenomas of the breast: comparison of microvessel distributions—implications for imaging modalities. *Radiology 208*:477–483

Yang, W.T., Metreweli, C., Lam, P.K.W., and Chang, J. 2001. Benign and malignant breast masses and axillary nodes: evaluation with echo-enhanced color power Doppler US. *Radiology 220*:795–802.

11

Response to Neoadjuvant Treatment in Patients with Locally Advanced Breast Cancer: Color-Doppler Ultrasound Contrast Medium (Levovist)

Paolo Vallone

Introduction

Locally advanced breast carcinoma (LABC) (T3-T4) is a pathology that has a poor long-term prognosis. The most common therapeutic protocol includes chemotherapy and/or hormone therapy, sometimes followed by surgery, if necessary. Total remission or a significant partial remission makes it possible to perform surgery on otherwise inoperable patients and reduces the number of radical mastectomies, causing only minor debilitating effects on the patients. Furthermore, the average overall survival improves after neoadjuvant treatment; it is greater than 50% after 3 years (Bonadonna et al., 1991; Smith, 1991). Vascular changes induced by treatment are of particular importance. Some authors have made a correlation between neoplastic vascularization and tumor growth (Folkman, 1990), as well as the rate of metastatic dissemination during surgery (Weinder et al., 1991).

Many malignant tumors are characterized by the development of autonomous blood vessels. This type of vascularization, which is recognized as a distinct sign of malignity (neoangiogenesis), represents a significant indicator of the biological activity of the tumor itself. Direct assessment of not only tumor volume but also vascularization provides precise evidence of the patient's response to the neoadjuvant therapy.

Materials and Methods

From January to December 2002, at the National Tumor Institute of Naples (Italy), 50 patients with LABC were studied. Tumor volume ranged from 8–57 ml, and the patient age ranged from 44–80 years (mean 55 years). All patients underwent the following: complete routine staging including color-Doppler ultrasound (CDU), followed by intravenous injection of contrast medium (Table 14); in all cases we

Table 14 B-mode and Color-doppler Ultrasound before and after Preneoadjuvant Treatment, in Patients with Locally Advanced Breast Carcinoma

Patients	Vol. cc	PDU (vascularization)						Levovist	
		Site	Grade	Poles	IP	IR	SP cm/s	Dose	Increase
1	22	2	2	+	1	0.6	46	300	2
2	18	1	2	+	1.1	0.7	34	300	1
3	33	1	2	+	1.1	0.7	45	300	2
4	67	2	2	+	1.9	0.8	23	300	2
5	12	2	2	+	1.4	0.8	28	300	2
6	15	–	0	0	–	–	–	2 × 300	1
7	20	1	1	+	1.2	0.7	28	300	2
8	32	2	2	+	1.4	0.8	31	300	2
9	18	1	1	+	1.3	0.7	22	2 × 300	1
10	16	1	1	+	1.9	0.9	18	2 × 300	1
11	33	2	2	+	1.7	0.8	33	300	1
12	28	2	2	+	1.6	0.8	27	300	1
13	27	2	2	+	1.4	0.7	35	300	1
14	8	–	0	–	–	–	–	3 × 300	0
15	32	2	2	+	1.2	0.7	21	300	2
16	14	2	1	+	1.0	0.7	24	2 × 300	2
17	12	1	1	+	1.5	0.7	25	2 × 300	1
18	52	2	2	+	1.4	0.8	32	300	2
19	14	1	1	+	1.1	0.6	28	2 × 300	1
20	44	2	2	+	1.2	0.7	20	300	1
21	17	2	1	+	1.3	0.8	27	300	2
22	8	–	0	0	–	–	–	3 × 300	0
23	33	2	1	+	2.2	0.9	20	300	2
24	32	2	1	+	1.3	0.6	25	300	1
25	37	2	1	+	1.3	0.8	30	300	3
26	15	1	1	1	1.9	0.9	23	300	1
27	34	1	1	+	0.9	0.6	45	300	1
28	39	2	0–1	+	1.4	0.8	41	300	3
29	24	1	1	+	1.7	0.8	33	300	1
30	8	2	1	+	1.2	0.7	28	300	1
31	42	2	2	+	1.6	0.8	35	300	2
32	22	1	1	+	1.8	0.7	16	2 × 300	1
33	18	1	1	1	1.6	0.8	24	2 × 300	1
34	32	2	1	+	1.5	0.9	36	300	3
35	10	–	0	–	–	–	–	3 × 300	0
36	45	2	2	+	1.5	0.8	46	300	1
37	42	2	2	+	1.4	0.8	41	300	1
38	35	2	2	+	2.1	0.7	37	300	2
39	58	2	2	+	1.8	0.6	28	300	2
40	16	1	1	1	2.0	0.6	16	2 × 300	1
41	45	2	1	+	1.8	0.8	24	300	3
42	32	2	1	+	1.8	0.9	22	2 × 300	1
43	35	2	2	+	1.8	0.9	36	300	2
44	28	1	1	1	1.2	2.0	27	2 × 300	1
45	26	1	2	+	1.3	1.8	24	300	2
46	24	1	1	1	1.7	1.4	33	2 × 300	1
47	18	1	2	+	1.6	1.5	38	2 × 300	2
48	42	2	2	+	1.8	1.6	29	300	1
49	18	1	2	1	1.7	1.8	32	300	1
50	26	2	2	+	1.2	1.8	27	300	1

performed a core biopsy for microhistological examination of the tumor, using a UNICUT needle (14 G).

The patients were included in one of the two protocols, either hormonal or neoadjuvant chemotherapy. Next, we performed a restaging, using CDU before and after intravenous injection of contrast medium (Table 15). Finally all patients underwent surgery; 60% underwent a radical mastectomy, while the remaining 40% underwent a conservative procedure. Ultrasound was performed using an Aloka 2000 with 7.5 MHz linear probe, before and after continuous intravenous infusion of Levovist (1 ml/s); one to three doses of Levovist were used per patient, at a concentration of 300 mg/ml (Bonadonna *et al.*, 1991). The ultrasound machine was set with pulse repetition frequency (PRF) and filters to detect slow flows. When Levovist is injected into the vein, it caused a reproducible and dose-dependent increase in backscatter; this results in an amplified Doppler signal (~ 10–20 dB).

Morphostructural aspects and volume of the tumor were assessed using B-mode ultrasound; color Doppler was used to assess the site and grade of vascularization (0 = absent, 1 = moderate, 2 = considerable, 3 = marked), and the increases in contrast after injection of Levovist (0 = absent, 1 = moderate, 2 = considerable, 3 = marked) were quantified using an optical-visual scale in relation to the amount of color pixels visualized. The instrumental examinations were always carried out by the same operator. Morphostructural and vascular aspects were correlated with postoperative histopathological findings.

Results

Of the 50 patients, 41 were affected by infiltrating ductal carcinoma and 9 by lobular carcinoma; 70% presented axillary lymph nodes. B-mode ultrasound (Fig. 99a) revealed that all neoplasms had a hypoechogenic and dishomogeneous aspect with irregular margins. In 4 cases (8%), the basic color-Doppler examination did not reveal any vascularization (Fig. 100a); in 18 cases (36%), only peripheral vascularization and in 28 cases (56%) vascularization that was both intra- and perilesional, with the presence of multiple vascular poles in 40 cases (80%). In 4 cases (8%), the grade of vascularization was 0, in 23 (46%) it was 1, and in 23 (46%) it was 2. The pulsatility index (PI) was between 0.9 and 2.2 (mean 1.46), the resistance index (RI) between 0.4 and 0.9 (mean 0.75), and systolic peak (SP) between 16 and 46 cm/s (mean 30.15).

After injection of Levovist, no vascular changes were observed in only 3 cases (6%), while in the remaining 47 cases (94%) an increase in contrast was observed (Fig. 101a), which was classified as grade 1 (52%), grade 2 (34%), or grade 3 (8%). Some patients were examined using CDU-Levovist in the middle of treatment (Fig. 100b). Following

neoadjuvant chemotherapy, the patient presented a partial response, with a deduction in the tumor that was greater than 50% (Fig. 99c) in 44 cases (88%) and less than 50% in 3 cases (6%); a completed response was observed in 2 cases (4%). Only 1 patient (2%) demonstrated a progression of the disease during the course of treatment.

The basic CDU examination (Fig. 99b) that was performed after neochemotherapy revealed the absence of vascularization in 31 out of 50 cases (62%), while in the other 38% of cases the grade vascularization was 1 in 16 cases (32%), 2 in 1 case (2%) and 3 in 2 cases (4%). The PI in these cases was between 1.0 and 1.7 (mean 1.39), the RI between 0.6 and 0.8 (mean 0.67), and the SP between 10 and 45 cm/s (mean 20.5).

After injection of Levovist in 22 cases (44%), nonvascular changes were observed, while in the other 28 cases (56%) we observed an increase in contrast (Fig. 100c), classified as grade 1 (36%) or grade 2 (20%). We did not observe any complications due to intolerance to the Levovist.

The visualization of vascular signals as a parameter of residual tumor biological activity was confirmed by postsurgical histopathological observations. Indeed, the presence of active tumor tissue was revealed in all cases in which vascular signals had been detected echographically (19 cases with basic CDU alone and 28 cases after Levovist). On the other hand, in 6 out of 22 cases, which were negative according to the ultrasound, we observed the absence of active cells with a fibrotic-type residual mass (echo-histological agreement 34/50 = 68%).

Discussion

A preoperative assessment of the response to neoadjuvant chemotherapy in patients with LABC using traditional methods (clinical examination, B-mode ultrasound, and mammography) is mainly based on the morphology and volume of the tumor. A detailed vascular map, which is both qualitative and quantitative, is an important indicator of the biological activity of the residual neoplasm and appears to be necessary for a more accurate assessment of the efficacy of the neoadjuvant treatment (Huber *et al.*, 2000).

Color-Doppler ultrasound is the examination of choice (Cosgrove *et al.*, 1990; Lagalla *et al.*, 1998). Ultrasound is a noninvasive procedure, relatively inexpensive, and easy to perform. In addition, despite technological advances, it is still difficult to evaluate slow blood flows in small vessels present in residual tumor tissue after treatment (Kedar *et al.*, 1994). As such, the use of contrast medium has increased the possibilities of basic color Doppler.

Our results confirmed the efficacy of CDU in assessing patient response to neoadjuvant treatment. The use of Levovist permitted us to identify, prior to neoadjuvant

Table 15 B-mode and Color-doppler Ultrasound before and after Levovist Postneoadjuvant Treatment, in Patients with Locally Advanced Breast Carcinoma

Patients	B-mode Vol. cc	B-mode Grade	CDU Poles	CDU IP	CDU IR	CDU SP cm/s	Levovist Dose	Levovist Increase
1	1.4	0	−	−	−	−	3 × 300	0
2	0.34	0	−	−	−	−	300	1
3	2.8	1	+	1.2	0.7	27	300	2
4	5.1	0	−	−	−	−	300	1
5	−	0	−	−	−	−	3 × 300	0
6	1.1	0	−	−	−	−	3 × 300	0
7	0.46	0	−	−	−	−	300	1
8	3.6	1	+	1.1	0.6	22	300	1
9	1.3	0	−	−	−	−	3 × 300	0
10	0.8	0	−	−	−	−	3 × 300	0
11	1.6	0	−	−	−	−	3 × 300	0
12	2.7	1	1	1.3	0.7	21	300	2
13	3.4	1	+	1.4	0.6	26	300	2
14	3.0	0	−	−	−	−	3 × 300	0
15	10.4	1	1	1.6	0.7	16	300	1
16	0.6	0	−	−	−	−	3 × 300	0
17	1.2	0	−	−	−	−	300	0
18	7.9	1	+	1.2	0.7	10	300	2
19	3.8	1	+	1.6	0.7	10	300	1
20	13.3	2	+	1.0	0.8	22	300	1
21	1.7	1	+	1.5	0.8	21	300	1
22	2.7	0	−	−	−	−	3 × 300	0
23	10.1	1	+	1.5	0.7	25	300	2
24	12.0	1	1	1.2	0.5	18	300	1
25	2.4	0	−	−	−	−	300	2
26	4.8	0	−	−	−	−	3 × 300	0
27	3.8	1	1	1.4	0.6	45	300	1
28	7.1	1	+	1.3	0.6	7.6	300	1
29	16.1	0	−	−	−	−	300	2
30	10.0	1	+	1.2	0.6	22	300	1
31	1.5	0	−	−	−	−	300	1
32	6.4	0	−	−	−	−	3 × 300	0
33	0.8	0	−	−	−	−	3 × 300	0
34	7.0	0	−	−	−	−	3 × 300	0
35	2.0	0	−	−	−	−	3 × 300	0
36	5.0	0	−	−	−	−	3 × 300	0
37	7.4	1	+	1.5	0.7	18	2 × 300	1
38	3.2	0	−	−	−	−	3 × 300	0
39	7.3	1	+	1.6	0.8	16	300	1
40	2.1	0	−	−	−	−	3 × 300	0
41	1.0	0	−	−	−	−	3 × 300	0
42	−	0	−	−	−	−	3 × 300	0
43	4.5	1	+	1;4	0.6	24	300	1
44	0.7	0	−	−	−	−	3 × 300	0
45	2.5	0	−	−	−	−	300	1
46	3.4	1	+	1.6	0.8	18	300	2
47	1.6	0	−	−	−	−	300	2
48	1.9	0	−	−	−	−	300	1
49	2.2	0	−	−	−	−	300	1
50	3.4	1	+	1.6	0.8	23	300	2

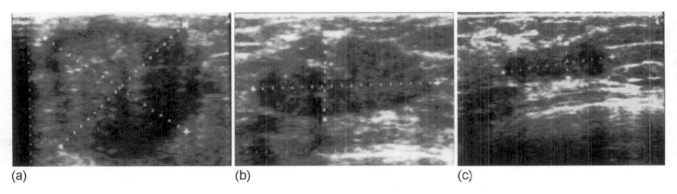

(a) (b) (c)

Figure 99 Carcinoma of the breast: B-mode ultrasound. (a) Base morphology and volume. (b) After two cycles of neoadjuvant chemotherapy: decrease in tumor volume is ~ 50%. (c) After four cycles of chemotherapy: decrease in tumor volume is > 50%.

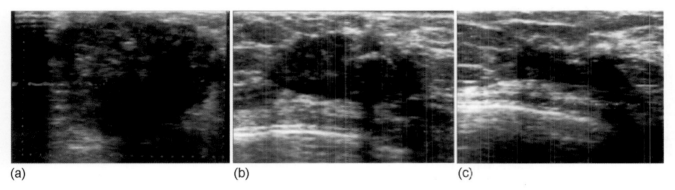

(a) (b) (c)

Figure 100 Carcinoma of the breast: color-Doppler ultrasound. (a) Baseline: grade 1 vascularization. (b) After two cycles of chemotherapy: moderate peripheral vascular signals. (c) After four cycles of chemotherapy: no evidence of vascular spots is present.

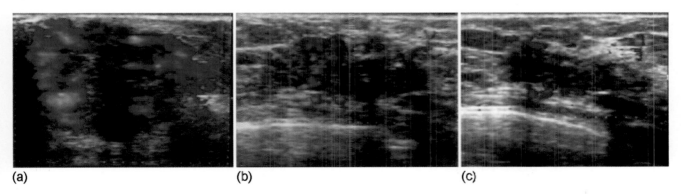

(a) (b) (c)

Figure 101 Carcinoma of the breast: color-Doppler ultrasound after Levovist injection. (a) Grade 3 increase in contrast with rich intralesional and perilesional vascularization. (b) After two cycles of chemotherapy: grade 2 increased compared to baseline. (c) After four cycles of chemotherapy: moderate residual vascularization is evident, but is not evident with basic color Doppler.

treatment, a greater number of vascular signals in almost all of the lesions (94%) with grade 1, 2, and 3 increases in contrast (52%, 34%, and 8% respectively).

Only in 6 cases (12%) were we unable to observe changes after injection of Levovist. This observation was equally evident after neoadjuvant treatment, as we were able to visualize a greater number of vessels in 28 lesions and observe the presence of vascular signals in 9 neoplasms that appeared to be primarily avascular during the basic CDU.

In conclusion, in the carcinoma of the breast, traditional analytic methods are based primarily on monitoring the volume of the tumor during neoadjuvant therapy. Nevertheless, the significance of morphological and volumetric changes cannot be considered without considering the vascular aspect of a lesion as well. Consequently, the value of a quantitative analysis is limited without also incorporating a qualitative assessment. The present study demonstrated that CDU-Levovist is sufficiently accurate in assessing vascular changes in carcinomas of the breast following preoperative therapy. CDU-Levovist makes it possible to obtain a precise qualitative and qualitative vascular map of the tumor, which is a parameter of fundamental importance in establishing the efficacy of the patient's response to neoadjuvant therapy.

It is important to highlight that current, state-of-art ultrasound scanners allow us to objectively and accurately quantify the vascularization degree of the tumor legions. Additionally, recently introduced, so-called second generation contrast media, combined with development of contrast-specific software, have allowed analysis of tumor microvasculature in a real-time, gray-scale modality. Nevertheless, to date, there are now published articles about LABC assessment with contrast-enhanced, gray-scale ultrasound. In the future, development of new contrast media and more sophisticated ultrasound software and scanners will allow for the improvement of using this imaging modality.

References

Bonadonna, G., Valagussa, P., Brambilla, C., Moliterni, A., Zambetti, M., and Ferrarsi, L.S. 1991. Adjuvant and neoadjuvant treatment of breast cancer with chemiotherapy and endocrine therapy. *Semin. Oncol.* 18:515–542.

Cosgrove, D.O., Bamber, J.C., Davey, J.B., McKinna, J.A., and Sinnett, H.D. 1990. Color-Doppler signals from breast tumors. *Radiology* 176:175–180.

Folkman, J. 1990. What is the evidence that tumors are angiogenesis dependent? *J. Natl. Cancer Inst. 82*:4–6.

Huber, S., Medl. M., Helbich, T., Taucher, S., Wagner, T., Rudas, M., Zuna, I., and Delorme. 2000. Locally advanced breast carcinoma: computer assisted semiquantitative analysis of color Doppler ultrasonography in the evaluation of tumor response to neoadjuvant chemotherap (work in progress). *S. J. Ultrasound-Med. 19*(9):601–607.

Kedar, R.P., Congrove, D.O., Smith, I.E., Mansi, J.L., and Bamber, J.C. 1994. Breast carcinoma: measurement of tumor response to primary medical therapy with color-Doppler flow imaging. *Radiology* 190:825–830.

Lagalla, R., Caruso, G., and Finazzo, M. 1998. Monitoring treatment response with color-Doppler and power-Doppler *Eur. J. Radiol. 27* (Suppl. 2):S149–156.

Smith, I.E., 1991. Primary (neoadjuvant) medical therapy. In Powles T.J., and Smith, I.E. (Eds.). *Medical Management of Breast Cancer.* London: Martin Dunitz, pp. 259–265.

Weinder, N., Semple, J.P., Welch, W.R., and Folkman, J. 1991. Tumor angiogenesis and metastasis: correlation in invasive breast carcinoma. *N. Engl. J. Med. 324*:1–8.

12

Magnetic Resonance Spectroscopy of Breast Cancer: Current Techniques and Clinical Applications

Sina Meisamy, Patrick J. Bolan, and Michael Garwood

Introduction

Background

Magnetic resonance spectroscopy (MRS), once used only by chemists for analyzing the structure of chemical compounds, has become a highly valuable tool in medical imaging and research. This technique is used for studying a variety of different disease processes such as metabolic and infectious disorders, cancer, and certain inflammatory and ischemic diseases. The medical application of *in vivo* MRS has historically focused on brain imaging but has recently made great progress in other parts of the body, including breast and prostate. In this chapter we will introduce hydrogen (^{1}H) MRS of the breast, describe current methods and technical issues, and discuss the clinical applications of ^{1}H MRS for diagnosis of breast cancer and therapeutic monitoring.

Early MRS studies of the breast were performed by measuring the resonances from phosphorus atoms (^{31}P). These studies demonstrated that changes in phospholipid metabolism may be used for breast cancer diagnosis and therapeutic monitoring (Leach *et al.*, 1998). However, there has been a growing trend toward measuring ^{1}H atoms instead of ^{31}P atoms. The primary reason for this transition is based on the fact that the larger gyromagnetic ratio and greater abundance of ^{1}H atoms allow for higher spectral sensitivity.

The "Choline Peak"

Numerous *in vivo* (Cecil *et al.*, 2001; Tse *et al.*, 2003) and *ex vivo* (Mackinnon *et al.*, 1997; Gribbestad *et al.*, 1999) ^{1}H MRS studies have demonstrated that neoplastic breast tissue contains elevated levels of choline-containing compounds (tCho), which have methyl protons that resonate at a chemical shift of 3.2 ppm. The primary constituents that make up the tCho resonance are compounds with a trimethylamine moiety backbone [R-$(CH_2)_2$-N$^+$-$(CH_3)_3$] such as free choline, phosphocholine, and glycerophosphocholine. Taurine, glucose, phosphoethanolamine, and myoinositol are other metabolites that have been shown to contribute to the tCho resonance (Sitter *et al.*, 2002). Using high-resolution *ex vivo* ^{1}H spectroscopy, we can visualize these different metabolites as separate peaks in their respective resonance, but with a clinical MR scanner, the resonances from these different compounds are essentially indistinguishable. Therefore, a simplified approach is to consider the resonance at 3.2 ppm to be a single peak, which we herein refer to as the "tCho peak."

Why Is tCho Elevated in Cancer?

The precise mechanism by which neoplastic tissues exhibit elevated levels of tCho is not yet fully understood and still remains an area of active research. However, the current hypothesis is based on the idea that increased levels of phosphocholine, one of the primary metabolites responsible for the tCho signal, is a result of increased synthesis of membranes by rapidly proliferating tissue (Negendank *et al.*, 1992). More recently, researchers have discovered that the augmented levels of tCho in breast cancer are caused by increased choline transport and upregulation of the enzymes choline kinase, protein kinase C, and phospholipase C (Katz-Brull and Degani, 1996; Glunde *et al.*, 2004). Researchers are still working on obtaining a complete picture of the biological processes that lead to elevated levels of tCho in cancer tissue. It is expected that this will help develop newer and more sophisticated anticancer drugs designed to target specific intracellular pathways and molecules.

Even though little is currently known about the molecular relationship between the tCho metabolite profile and its underlying mechanism in cancer cells, many researchers have successfully used MRS to show that the tCho resonance may be used as a noninvasive marker for identifying breast cancer lesions. In addition, some groups have also discovered that the size of the tCho peak can sometimes change in breast tumors that undergo neoadjuvant chemotherapy. However, ^{1}H MRS of the breast is technically challenging because sensitivity can be limiting and spectral artifacts can occur. Therefore, it is important to understand the techniques used in breast ^{1}H MRS and to become familiar with its limitations.

Technique

Tumor Localization

The spectroscopy component of a breast MR exam is performed immediately after the imaging and dynamic portion of the study, while the patient is in the magnet. Once the lesion morphology and time intensity curves have been evaluated, the radiologist can "localize" a volume of interest (VOI) and begin acquiring MRS data. Currently, breast MRS localization is performed by a technique called single-voxel spectroscopy (SVS). The borders of the voxel, which defines the boundaries from where MRS data will be acquired, are delineated using a minimum of three radiofrequency (RF)

pulses, each of which is applied in a different orientation (x, y, and z) of the B_0 gradient. This is typically performed by using point-resolved spatially localized spectroscopy (PRESS) or stimulated-echo acquisition mode (STEAM) pulse sequences. These sequences, however, have imperfect localization and therefore can allow lipid signals from outside the VOI to "leak" in and diminish the spectral quality. Our group has developed an SVS technique that has improved spectral localization using six RF pulses rather than three. This technique has been shown to provide high-quality ^{1}H spectra reproducibly and is known as localization by adiabatic selective refocusing (LASER) (Garwood and DelaBarre, 2001). With LASER, the total time required to collect the spectrum for each voxel is ~ 8 min. This includes 2 min for shimming, power calibrations, and acquisition of a water spectrum, followed by ~ 6 min of signal averaging. Figure 102 shows a representative example of a single-voxel ^{1}H spectrum from a patient with invasive ductal carcinoma.

With respect to tumor localization, others have evaluated the use of magnetic resonance spectroscopic imaging (MRSI) as an alternative to SVS (Jacobs *et al.*, 2004). Unlike SVS, MRSI is performed by using a grid to acquire multiple spectra rather than a single spectrum. The primary advantage of MRSI over SVS is the ability to obtain information about the spatial distribution of metabolites. This can be very helpful in many cases such as those with multiple lesions, heterogeneous lesions, or lesions whose morphologic shape prevent adequate placement of a single voxel. Even more valuable is the potential of MRSI to serve as a safety net by labeling neoplastic lesions that were never identified by the radiologist. However, performing breast MRSI is even more technically challenging than SVS, especially when considering quantification.

Technical Issues

Currently, most of the MRS applications that are available have been tested and designed for brain imaging. For this reason, many researchers and clinicians who are performing breast MRS are forced to use a commercial spectroscopy software package that was developed and optimized mainly for brain studies. Although this method is currently considered a tolerable one at 1.5T, it can pose serious problems because, unlike the brain, the breast has a heterogeneous distribution of fat and glandular cells that produce large lipid signals that can give rise to contaminant peaks around 3.2 ppm (near the tCho peak). To remedy this problem, our group has developed a technique called echo-time averaging, which can suppress these artifactual peaks (also known as lipid sidebands) (Bolan *et al.*, 2002). In addition to lipid sidebands, breast adipose tissue can pose yet another challenge for *in vivo* ^{1}H MRS. When the radiologist is planning the voxel, any adipose tissue that is inadvertently included within the voxel will create a partial volume effect, which will

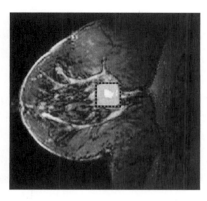
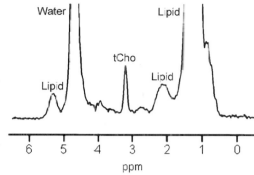

Figure 102 Example of a localized breast ^{1}H MR spectrum of a patient with invasive ductal carcinoma acquired at high field (4 Tesla). The sagittal high-spatial-resolution 3D fat-suppressed fast low-angle shot (FLASH) MR image of the breast (TR/TE = 13.5/4.1) was obtained 7 min after gadopentetate dimeglumine injection. The box surrounding the lesion depicts the MR spectroscopic voxel. The spectrum on the right shows the resonances obtained from this lesion.

reduce the effective volume for spectroscopy. Furthermore, excess adipose tissue within the voxel can also limit the ability to optimize (or "shim") the homogeneity of the magnetic field, which in turn leads to broad resonances and reduced signal-to-noise ratio (SNR). Breast lesion size and shape are two other factors that present challenges to breast ^{1}H MRS. It has been shown that tCho levels in voxels < 1 cm^3 may be difficult to measure accurately (Bolan *et al.*, 2003). This finding is primarily a limitation of the SNR. However, given that SNR increases at least linearly with magnetic field (Vaughan *et al.*, 2001), the ability to detect tCho in diffuse or small lesions can be improved by using a higher magnetic field strength such as a 3 Tesla MR system. Therefore, given the above technical issues and limitations surrounding breast MRS, it is imperative that voxels be planned in such a way that maximizes coverage of the lesion while minimizing the inclusion of adipose tissue.

Respiratory Artifact

Respiratory motion has been shown to reduce the quality of breast MRS. This form of susceptibility artifact does occur in brain imaging but is much more noticeable in the breast because of its close anatomical proximity to the lungs. Respiratory motion artifact reduces the quality of breast MRS by causing B$_0$ field distortions that lead to "shot-to-shot" frequency shifts. If this correction is not performed, the frequency shifts will decrease spectral resolution and increase the peak fitting errors. This becomes most problematic in cases where MRS measurements are being acquired from small lesions with low SNR. Our group has recently formulated a method that corrects for this artifact by retrospectively shifting each spectrum prior to averaging (Bolan *et al.*, 2004). Based on data obtained from our study, when

this artifact was not corrected, quantification errors increased by an average of 28%.

Quantification

Previous breast MRS studies performed at 1.5 T were based on the assumption that tCho is detectable only in malignant breast tissue and that the presence of a tCho peak with an SNR ≥ 2 is considered positive for malignancy. This semiquantitative method assumes that the sensitivity of measuring tCho is invariable from one patient to the next. However, studies performed by our group have shown that this assumption can be violated because tCho measurements are primarily limited by SNR, and several factors affecting SNR can vary between patients by a factor of 100 (Bolan *et al.*, 2003). Factors affecting SNR in tCho measurements include voxel size, partial volume of adipose tissue in the voxel, distance between the voxel and the receiver coil, and variable coil loading. Therefore, a method to quantify *in vivo* breast ^{1}H MRS measurements had to be designed that corrects for these factors and compensates for differences in sensitivity. However, the above-mentioned factors are not the only basis that makes quantification of breast MRS a necessity. For example, quantification is also required when performing *in vivo* breast MRS at high field because the increased sensitivity allows for the detection of tCho in benign lesions as well as normal fibroglandular breast tissue (Bolan *et al.*, 2003). Also, we will see later that therapeutic monitoring or any kind of longitudinal study requires implementation of a method to quantify spectra.

A variety of techniques have been designed to quantify MRS, each of which is based on referencing and spectral fitting. Our group has developed a method to quantify *in vivo* breast MRS by using the intravoxel water resonance as an internal reference (Bolan *et al.*, 2003). Some groups have proposed quantifying breast MRS by using an external standard for reference (Bakken *et al.*, 2001). However, using an internal reference is much more robust because it automatically compensates for a variety of factors such as voxel size and coil efficiency. In addition, using an internal reference also compensates for the partial-volume effect from adipose tissue, which allows calculations to be made in a molal concentration (mmol/kg of water) for water-soluble metabolites such as tCho. It should be noted that both the internal and external referencing methods require correction for differences in relaxation rates, which at times may be difficult to measure in certain patients. With respect to the spectral fitting technique, our group utilizes the time domain frequency domain (TDFD) approach, which allows for a flexible lineshape by using a time-domain model (Slotboom *et al.*, 1998). This method was chosen because it has superb frequency selection properties, such that the residuals are evaluated and minimized over a specific region of the frequency domain. This becomes very crucial in cases where fitting is being performed on weak signals that are surrounded by multiple large-signal peaks (i.e., small tCho peak surrounded by large lipid peaks). Based on our patient series thus far, malignant breast lesions typically present with a tCho concentration ([tCho]) that is greater than or equal to 1.00 mmol/kg water. Using this cut-off point, we find that the sensitivity and specificity of breast ^{1}H MRS are 72% and 83%, respectively (Meisamy *et al.*, 2005).

How Reliable Is the tCho Measurement?

The last component of the quantification method involves calculating the error estimation of the tCho measurement. Various methods have been designed for estimating the error in quantitative MRS. The most common method, which is the method currently used by our group, is based on the Cramer-Rao minimum variance bound (Cavassila *et al.*, 2001). Estimating the measurement error in quantitative breast MRS is clinically important and, as we will see later, can become crucial in cases where the radiologist is assessing a lesion with indeterminate characteristics. Not all malignant breast lesions exhibit high levels of tCho. This may be due to certain physiologic or technical factors such as large adipose tissue content, low cellularity, or low SNR. In either case, the MRS measurement may be indeterminate, and the error assessment must make this known to the radiologist. Thus, the primary purpose of the error estimation is to provide the radiologist with an indication of whether the low or absent tCho signal is a true reflection of the low [tCho] or if it is the result of the earlier mentioned limiting factors. The errors are prohibitively large in two cases: (1) the error estimation is calculated to be greater than the lesion [tCho]; or (2) a lesion shows no quantifiable tCho and the calculated error estimation is greater than the cut-off [tCho] indicative of malignancy (i.e., ≥ 1.00 mmol/kg water). Either case will indicate

to the radiologist that the MRS measurement is indeterminate and should not be used as part of the image interpretation or decision for patient management.

Clinical Applications

Diagnosis

Contrast-enhanced MRI offers nearly 100% sensitivity for detecting breast cancer, yet variable specificity (30–90%) (Orel and Schnall, 2001). This may be partly because not all malignant breast lesions enhance or have a specific time-signal intensity curve pattern (Schnall *et al.*, 2006). Furthermore, experience has shown that a small percentage of breast malignancies, seen on MRI, have benign morphologic features that are impossible to differentiate from invasive malignancies. For these reasons, researchers have studied the clinical application of ^{1}H MRS for distinguishing benign from malignant breast lesions prior to biopsy. The first article on this topic was published in 1998 and proposed that tCho may be used as a noninvasive marker for detecting breast malignancies (Roebuck *et al.*, 1998). Later, this hypothesis was further verified by numerous studies at various institutions that showed similar results. Interestingly, a combined analysis of the first five publications on this subject showed that using only tCho for the detection of breast malignancy had an overall sensitivity of 83% and an overall specificity of 85% (Katz-Brull *et al.*, 2002). However, these earlier studies did not consider lesion morphology or time-signal intensity curves.

More recently, two studies have aimed at evaluating whether the addition of breast ^{1}H MRS can improve the specificity of a standard breast MRI examination. One group of researchers included a single-voxel ^{1}H MRS measurement with a single-slice T_2^*-weighted perfusion measurement to a traditional breast MRI examination (Huang *et al.*, 2004). Their results showed that the addition of ^{1}H MRS increased the specificity of the breast MR exam from 62.5–87.5%, and with the addition of the perfusion measurement, the specificity increased even further to 100%. Our group recently conducted an observer performance study at high field (4 Tesla) by using four experienced breast radiologists' to retrospectively evaluate 55 breast MR examinations (Meisamy *et al.*, 2005). The results from this study showed that adding quantitative ^{1}H MRS to the standard breast MRI exam not only resulted in the radiologists' ability to detect a higher number of malignant lesions (higher sensitivity) but also facilitated their ability to distinguish benign from malignant lesions (higher specificity). We also found that the addition of quantitative ^{1}H MRS resulted in higher reader accuracy and interobserver agreement regarding patient management when compared with the values achieved by using MR imaging alone.

Sample Diagnostic Cases

Figures 103 and 104 are example cases taken from our observer performance study. The case shown in Figure 103 is from a patient with invasive ductal carcinoma in which ^{1}H MRS helped the radiologists correctly diagnose a breast lesion with equivocal MR imaging characteristics. The high-spatial-resolution 3D MR images in Figure 104 show an area with minimal regional enhancement with a time-signal intensity curve that was described by all four readers as slow enhancement during the initial phase, followed by plateau enhancement during the delayed phase. After the radiologists evaluated the lesion morphology and time-signal intensity curve, three of the four readers recommended a 6-month follow-up and one of the readers recommended a biopsy. For the second interpretation, the radiologists were provided with ^{1}H MRS measurements, which showed a [tCho] that indicated a malignant lesion ([tCho] = 1.78 mmol/kg water ± 0.56). At that point, the the readers who recommended a 6-month follow-up changed their decision and recommended a biopsy and the fourth reader kept the recommendation of biopsy. The case shown in Figure 104 is from a patient with invasive ductal carcinoma, which demonstrates the impor-

tance of quantification when assessing breast MRS measurements. The high-spatial-resolution 3D MR images from Figure 103 show a lesion measuring only 0.39 cm^3 with a time-signal intensity curve that was described by all four readers as fast enhancement during the initial phase, followed by an enhancement plateau during the delayed phase. After the radiologists evaluated the lesion morphology and time-signal intensity curve, all four readers recommended biopsy. Even after the readers were provided with the ^{1}H MRS measurements, which demonstrated no quantifiable amount of tCho ([tCho] = 0 mmol/kg = 1.73), the readers did not change their decision to biopsy. Had the spectrum been assessed qualitatively, the radiologist may have been misled by the absent tCho signal and thus they may have changed their decision to not biopsy. However, when assessed quantitatively, the large error from the tCho measurement provided the radiologist with an important piece of information in that the spectrum acquired from this lesion was in fact not reliable. The inability to detect tCho in this lesion most likely resulted from its small voxel size and hence low SNR ratio. As discussed previously, voxels < 1 cm^3 yield large errors in tCho measurements.

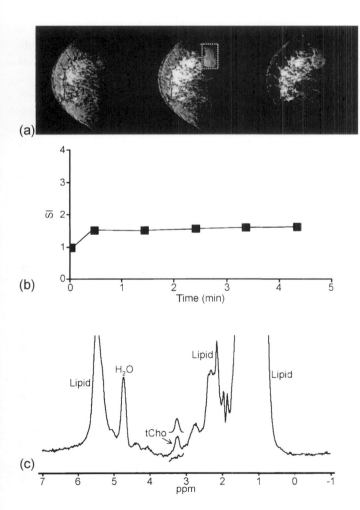

(a)

(b)

(c)

Figure 103 Breast MR imaging case acquired at high field (4 Tesla) in which ^{1}H MR spectroscopic findings led to altered treatment recommendations. (a) Sagittal high-spatial-resolution 3D fat-suppressed fast low-angle shot (FLASH) MR images of the breast (TR/TE = 13.5/4.1) obtained before (left) and 7 min after (middle) gadopentetate dimeglumine injection and with subtraction (right) show an 8.3 cm^3 lesion. The box surrounding the lesion depicts the MR spectroscopic voxel. (b) Time-signal intensity (SI) curve measured from the lesion depicted in (a). All four readers described this curve as showing slow enhancement during the initial phase, followed by plateau enhancement during the delayed phase. After evaluating the morphologic features and time-signal intensity curve of the lesion, three of the four readers did not recommend biopsy; rather, they recommended a 6-month follow-up examination. The fourth reader recommended biopsy. (c) ^{1}H MR spectra measured from the lesion. The spectral peaks of mobile lipid, water, and tCho are labeled. The lines above and below the tCho peak represent the fitted tCho peak and the residual of the fit, respectively. The [tCho] measured from this lesion was 1.78 mmol/kg ± 0.56. When the tCho measurement was presented to the readers in the second interpretation, three of the four readers changed their decision and recommended biopsy; the fourth reader kept the recommendation of biopsy. This patient received a diagnosis of invasive ductal carcinoma. (Figure previously printed in *Radiology*, Meisamy *et al.*, 2005.)

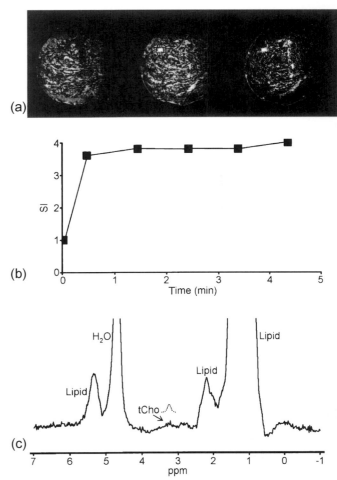

Figure 104 Breast MR imaging case in which the measured tCho did not lead to a change in the recommended lesion management. (a) Sagittal high-spatial-resolution 3D fat-suppressed fast low-angle shot (FLASH) MR images of the breast (TR/TE = 13.5/4.1) obtained before (left) and 7 min after (middle) gadopentetate dimeglumine injection and with subtraction (right) show a 0.39 cm³ lesion. The box surrounding the lesion depicts the MR spectroscopic voxel. (b) Time-signal intensity (SI) curve measured from the lesion depicted in (a). All four readers described this curve as showing fast enhancement during the initial phase, followed by an enhancement plateau during the delayed phase. After evaluating the morphologic features and time-signal intensity curve of the lesion, all four readers recommended biopsy. (c) ¹H MR spectra measured from the lesion. The spectral peaks of mobile lipids, water, and tCho are labeled. The line above the tCho peak represents the minimal [tCho] that is detectable with use of the quantification procedure. The [tCho] measured from this lesion was 0 mmol/kg ± 1.73. When the tCho measurement was presented to the four readers in the second interpretation, none of them changed the recommendation of biopsy. This patient received a diagnosis of invasive ductal carcinoma. (Figure previously printed in *Radiology*, Meisamy *et al.*, 2005.)

Therapeutic Monitoring with Early Feedback

Neoadjuvant chemotherapy (NCT) has become the standard of care for the treatment of patients with locally advanced breast cancer. Such chemotherapy prior to breast surgery does not offer any survival benefits over postoperative chemotherapy, yet patients who undergo NCT are more likely to undergo breast-conserving surgery. Furthermore, using NCT permits *in vivo* monitoring of tumor response.

Currently, no standardized criteria have been delineated that can individually detect early response to NCT. Current clinical applications such as physical exam, ultrasound, and mammography have been shown to be only moderately useful for predicting residual pathologic tumor size after NCT (Herrada *et al.*, 1997; Chagpar *et al.*, 2006). With respect to treatment monitoring, studies have shown that changes in lesion size measured on MRI, neoplastic phenotype, dynamic contrast enhancement (DCE), and extraction flow product correlate with clinical response (Esserman *et al.*, 2001; Choyke *et al.*, 2003; Delille *et al.*, 2003). However, changes in lesion size or DCE measured with MRI are not detected until at least 6 weeks following NCT (Rieber *et al.*, 2002). With MRS, one can measure changes in the tumor's

metabolic activity, which is known to occur far in advance from changes in tumor size and DCE. There were two early reports on the use of ¹H MRS as a potential means of monitoring tumor response in breast cancer patients undergoing NCT (Kvistad *et al.*, 1999; Jagannathan *et al.*, 2001). They both found that in a majority of patients who underwent NCT for breast cancer, the tCho signal either disappeared or decreased.

More recently, our group discovered that changes in the [tCho] could serve as an early noninvasive marker for predicting response to NCT (Meisamy *et al.*, 2004). We found that patients who were nonresponders had a [tCho] measured within 24 hr after the first dose of NCT that was greater than or equal to the pretreatment baseline measurement, and those who were objective responders had a [tCho] measured within 24 hr after the first dose of NCT that was less than the baseline measurement. The results from our study also showed that changes in [tCho] within 24 hr after the first dose of NCT were significantly different ($p = 0.007$) between objective responders and nonresponders, as defined by the RECIST (Response Evaluation Criteria in Solid Tumors) classification system. In addition, we also found that changes in [tCho] between baseline and 24 hr after the first dose of

NCT showed significant positive correlation ($R = 0.79$, $p = 0.001$) with changes in lesion size measured after the final dose of NCT. Other researchers have investigated other components of the MRS data as a means of predicting response to NCT. Recently, one study found that a small or absent decrease in water to fat ratios, measured by MRS after the second dose of NCT, accurately predicted the final size of tumor volumes in 50% of the patients who were classified as nonresponders (Manton *et al.*, 2006).

Sample Therapeutic Monitoring Cases

Figures 105 and 106 show example cases taken from our NCT study. The case shown in Figure 105 is from a known breast cancer patient whose tCho measurements predicted an objective response to four cycles of Adriamycin and Cyclophosphamide (AC). The [tCho] from baseline to within 24 hr after the first dose of NCT decreased by 20%, yet the tumor size did not change. At that point, it was predicted that the patient would have an objective response. Sixty-four days later, after the fourth dose of NCT, the tumor size decreased by 58% which, according to the RECIST classification system, was compatible with an objective response. Since this patient was found to have palpable lymphadenopathy, she was continued on a second NCT regimen with paclitaxel prior to surgery. It is of interest to note that 24 hr after the second dose of paclitaxel, the [tCho] dramatically increased by 355% whereas the lesion size did not change. Although not shown in Figure 105, the patient's tumor size did not change even after the fourth dose of paclitaxel, whereas the [tCho] increased by an additional 17%. Based on the change in [tCho] after initiation of paclitaxel, it is possible that ^{1}H

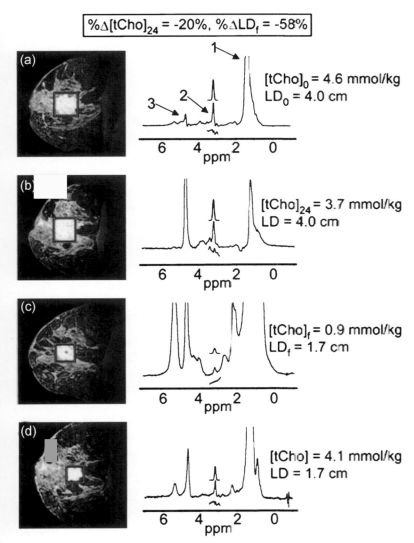

Figure 105 Sagittal three-dimensional gadolinium-enhanced fat-suppressed fast low-angle shot (FLASH) (TR/TE = 13.5/4.1). MR images (left) and corresponding spectra (right) of the right breast in a 43-year-old objective responder with invasive ductal carcinoma and positive lymph nodes. On MR images, boxes surrounding enhancing lesions depict spectroscopy voxels. The labeled spectral peaks in (a) arise from lipid (1), tCho (2), and water (3). On all spectra, the lines above and below the tCho peak represent the fitted tCho peak and the residual of the fit, respectively. Data were obtained 2 days prior to starting AC (a), within 24 hr after the first dose of AC (b), after the fourth dose of AC (c), and after the second dose of paclitaxel (d). Change in [tCho] by 24 hr ($\%\Delta[tCho]_{24}$) was −20%, which predicts an objective response to AC. Change in tumor size (LD) after the fourth dose ($\%\Delta LD_f$) was −58%, which is compatible with an objective response. LD_0 and $[tCho]_0$ = parameters measured at baseline, $[tCho]_{24}$ = tCho concentration measured at 24 hr after first dose, LD_f and $[tCho]_f$ = parameters measured after fourth dose. (Figure previously printed in *Radiology*, Meisamy *et al.*, 2004.)

$$\%\Delta[\text{tCho}]_{24} = 50\%, \ \%\Delta\text{LD}_f = -7\%$$

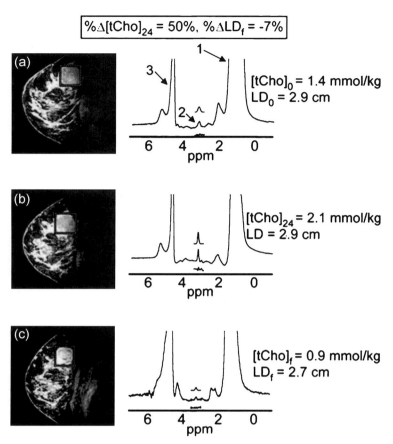

(a)

$$[\text{tCho}]_0 = 1.4 \ \text{mmol/kg}$$
$$\text{LD}_0 = 2.9 \ \text{cm}$$

(b)

$$[\text{tCho}]_{24} = 2.1 \ \text{mmol/kg}$$
$$\text{LD} = 2.9 \ \text{cm}$$

(c)

$$[\text{tCho}]_f = 0.9 \ \text{mmol/kg}$$
$$\text{LD}_f = 2.7 \ \text{cm}$$

Figure 106 Sagittal three-dimensional gadolinium-enhanced fat-suppressed fast low-angle shot (FLASH) (TR/TE = 13.5/4.1). MR images (left) and corresponding spectra (right) of the right breast in a 42-year-old nonresponder with invasive ductal carcinoma. On MR images, boxes surrounding enhancing lesions depict spectroscopy voxels. The labeled spectral peaks in (a) arise from lipid (1), tCho (2), and water (3). On all spectra, the lines above and below the tCho peak represent fitted tCho peak and residual of the fit, respectively. Data were obtained 1 day prior to starting AC (a), within 24 hr after the first dose (b), and after the fourth dose (c). The $\%\Delta[\text{tCho}]_{24}$ was 50%, which predicts no response to AC. The $\%\Delta\text{LD}_f$ was −7%, which is compatible with a nonresponder to AC. See Figure 104 for an explanation of abbreviations. (Figure previously printed in *Radiology*, Meisamy *et al.*, 2004).

MRS may potentially be used to predict clinical response between various agents. However, further work is needed to evaluate this hypothesis. The example case shown in Figure 106 is taken from a known breast cancer patient whose tCho measurements predicted no response to four cycles of AC. The [tCho] from baseline to within 24 hr after the first dose of NCT increased by 50% and the tumor size did not change. At that point, it was predicted that the patient would be a nonresponder. Sixty-five days later, after the fourth dose of NCT, the tumor size decreased by only 7%, which, according to the RECIST classification system, was compatible with a nonresponder.

In vivo ^{1}H MRS is also helping expedite testing of experimental chemotherapeutic agents in research laboratories. A recent study on an experimental anticancer drug (PX-478) found that the tCho resonance, measured in tumor-inoculated mice, significantly decreased within 12 and 24 hr after treatment (Jordan *et al.*, 2005). Similar to our findings discussed earlier, this study also suggested that changes in tCho resonance, measured with *in vivo* ^{1}H MRS, can serve as a noninvasive biomarker for assessment of drug response.

In conclusion, as demonstrated in this chapter, breast ^{1}H MRS has emerged as a promising tool for improving the diagnosis and clinical management of breast cancer. In addition, the newly developed techniques discussed in this chapter show that breast ^{1}H MRS can be a feasible and reliable complement to the standard breast MR examination. However, larger multicenter trial studies are still needed before the tCho biomarker can be widely accepted.

Acknowledgment

We would like to acknowledge the following grants from the NIH (grants CA92004, RR08079, and RR00400) and the DOD Breast Cancer Research program (DAMD 17-01-1-0331).

References

Bakken, I.J., Gribbestad, I.S., Singstad, T.E., and Kvistad, K.A. 2001. External standard method for the *in vivo* quantification of choline-containing compounds in breast tumors by proton MR spectroscopy at 1.5 Tesla. *Magn. Reson. Med. 46*:189–192.

Bolan, P.J., DelaBarre, L., Baker, E.H., Merkle, H., Everson, L.I., Yee D., and Garwood, M. 2002. Eliminating spurious lipid sidebands in ^{1}H MRS of breast lesions. *Magn. Reson. Med. 48*:215–222.

Bolan, P.J., Henry, P.G., Baker, E.H., Meisamy, S., and Garwood, M. 2004. Measurement and correction of respiration-induced B_0 variations in breast ^{1}H MRS at 4 Tesla. *Magn. Reson. Med. 52*:1239–1245.

Bolan, P.J., Meisamy, S., Baker, E.H., Lin, J., Emory, T., Nelson, M., Everson, L.I., Yee, D., and Garwood, M. 2003. *In vivo* quantification of choline compounds in the breast with ^{1}H MR spectroscopy. *Magn. Reson. Med. 50*:1134–1143.

Cavassila, S., Deval, S., Huegen, C., van Ormondt, D., and Graveron-Demilly, D. 2001. Cramer-Rao bounds: an evaluation tool for quantitation. *NMR Biomed. 14*:278–283.

Cecil, K.M., Schnall, M.D., Siegelman, E.S., and Lenkinski, R.E. 2001. The evaluation of human breast lesions with magnetic resonance imaging and proton magnetic resonance spectroscopy. *Breast Cancer Res. Treat. 68*:45–54.

Chagpar, A.B., Middleton, L.P., Sahin, A.A., Dempsey, P., Buzdar, A.U., Mirza, A.N., Ames, F.C., Babiera, G.V., Feig, B.W., Hunt, K.K., Kuerer, H.M., Meric-Bernstam, F., Ross, M.I., and Singletary, S.E. 2006. Accuracy of physical examination, ultrasonography, and mammography in predicting residual pathologic tumor size in patients treated with neoadjuvant chemotherapy. *Ann. Surg. 243*:257–264.

Choyke, P.L., Dwyer, A.J., and Knopp, M.V. 2003. Functional tumor imaging with dynamic contrast-enhanced magnetic resonance imaging *J. Magn. Reson. Imaging. 17*:509–520.

Delille, J., Slanetz, P.J., Yeh, E.D., Halpern, E.F., Kopans, D.B., and Garrido, L. 2003. Invasive ductal breast carcinoma response to neoadjuvant chemotherapy: noninvasive monitoring with functional MR imaging-pilot study. *Radiology 228*:63–69.

Esserman, L., Kaplan, E., Partridge, S., Tripathy, D., Rugo, H., Park, J., Hwang, S., Kuerer, H., Sudilovsky, D., Lu, Y., and Hylton, N. 2001. MRI phenotype is associated with response to doxorubicin and cyclophosphamide neoadjuvant chemotherapy in stage III breast cancer. *Ann. Surg. Oncol. 8*:549–559.

Garwood, M., and DelaBarre, L. 2001. The return of the frequency sweep: designing adiabatic pulses for contemporary NMR. *J. Magn. Reson. 153*:155–177.

Glunde, K., Jie, C., and Bhujwalla, Z.M. 2004. Molecular causes of the aberrant choline phospholipid metabolism in breast cancer. *Cancer Res. 15*:4270–4276.

Gribbestad, I.S., Sitter, B., Lundgren, S., Krane, J., and Axelson, D. 1999. Metabolic composition in breast tumors examined by proton nuclear resonance spectroscopy. *Anticancer Res. 19*:1737–1746.

Herrada, J., Iyer, R.B., Atkinson, E.N., Sneige, N., Buzdar, A.U., and Hortobagyi, G.N. 1997. Relative value of physical examination, mammography, and breast sonography in evaluating the size of the primary tumor and regional lymph node metastases in women receiving neoadjuvant chemotherapy for locally advanced breast carcinoma. *Clin. Cancer Res. 3*:1565–1569.

Huang, W., Fisher, P.R., Dulaimy, K., Tudorica, L.A., O'Hea, B., and Button, T.M. 2004. Detection of breast malignancy: diagnostic MR protocol for improved specificity. *Radiology 232*:585–591.

Jacobs, M.A., Barker, P.B., Bottomley, P.A., Bhujwalla, Z., and Bluemke, D.A. 2004. Proton magnetic resonance spectroscopic imaging of human breast cancer: a preliminary study. *J. Magn. Reson. Imaging 19*:68–75.

Jagannathan, N.R., Kumar, M., Seenu, V., Coshic, O., Dwivedi, S.N., Julka, P.K., Srivastava, A., and Rath, G.K. 2001. Evaluation of total choline from in-vivo volume localized proton MR spectroscopy and its response to neoadjuvant chemotherapy in locally advanced breast cancer. *Br. J. Cancer 84*:1016–1022.

Jordan, B.F., Black, K., Robey, I.F., Runquist, M., Powis, G., and Gillies, R.J. 2005. Metabolite changes in HT-29 xenograft tumors following HIF-1alpha inhibition with PX-478 as studied by MR spectroscopy *in vivo* and *ex vivo*. *NMR Biomed. 18*:430–439.

Katz-Brull, R., and Degani H. 1996. Kinetics of choline transport and phosphorylation in human breast cancer cells: NMR application of the zero trans method. *Anticancer Res. 16*:1375–1380.

Katz-Brull, R., Lavin, P.T., and Lenkinski, R.E. 2002. Clinical utility of proton magnetic resonance spectroscopy in characterizing breast lesions. *J. Natl. Cancer Inst. 94*:1197–1203.

Kvistad, K.A., Bakken, I.J., Gribbestad, I.S., Ehrnholm, B., Lundgren, S., Fjosne, H.E., and Haraldseth, O. 1999. Characterization of neoplastic and normal human breast tissues with *in vivo* (1)H MR spectroscopy. *J. Magn. Reson. Imaging 10*:159–164.

Leach, M.O., Verrill, M., Glaholm, J., Smith, T.A., Collins, D.J., Payne, G.S., Sharp, J.C., Ronen, S.M., McCready, V.R., Powles, T.J., and Smith, I.E. 1998. Measurements of human breast cancer using magnetic resonance spectroscopy: a review of clinical measurements and a report of localized 31P measurements of response to treatment. *NMR Biomed. 11*:314–340.

Mackinnon, W.B., Barry, P.A., Malycha, P.L., Gillett, D.J., Russell, P., Lean, C.L., Doran, S.T., Barraclough, B.H., Bilous, M., and Mountford, C.E. 1997. Fine-needle biopsy specimens of benign breast lesions distinguished from invasive cancer *ex vivo* with proton MR spectroscopy. *Radiology 204*:661–666.

Manton, D.J., Chaturvedi, A., Hubbard, A., Lind, M.J., Lowry, M., Maraveyas, A., Pickles, M.D., Tozer, D.J., and Turnbull, L.W. 2006. Neoadjuvant chemotherapy in breast cancer: early response prediction with quantitative MR imaging and spectroscopy. *Br. J. Cancer 13*:427–435.

Meisamy, S., Bolan, P.J., Baker, E.H., Bliss, R.L., Gulbahce, E., Everson, L.I., Nelson, M.T., Emory, T.H., Tuttle, T.M., Yee, D., and Garwood, M. 2004. Neoadjuvant chemotherapy of locally advanced breast cancer: predicting response with *in vivo* (1)H MR spectroscopy—a pilot study at 4 T. *Radiology 233*:424–431.

Meisamy, S., Bolan, P.J., Baker, E.H., Pollema, M.G., Le, C.T., Kelcz, F., Lechner, M.C., Luikens, B.A., Carlson, R.A., Brandt, K.R., Amrami, K.K., Nelson, M.T., Everson, L.I., Emory, T.H., Tuttle, T.M., Yee, D., and Garwood, M. 2005. Adding *in vivo* quantitative ^{1}H MR spectroscopy to improve diagnostic accuracy of breast MR imaging: preliminary results of observer performance study at 4.0 T. *Radiology 236*:465–475.

Negendark, W.G., Brown, T.R., Evelhoch, J.L., Griffiths, J.R., Liotta, L.A., Margulis, A.R., Morrisett, J.D., Ross, B.D., and Shtern, F. 1992. Proceedings of a National Cancer Institute workshop: MR spectroscopy and tumor cell biology. *Radiology 185*:875–883.

Orel, S.G., and Schnall, M.D. 2001. MR imaging of the breast for the detection, diagnosis, and staging of breast cancer. *Radiology 220*:13–30.

Rieber, A., Brambs, H.J., Gabelmann, A., Heilmann, V., Kreienberg, R., and Kühn, T. 2002. Breast MRI for monitoring response of primary breast cancer to neo-adjuvant chemotherapy. *Eur. Radiol. 12*:1711–1719.

Roebuck, J.R., Cecil, K.M., Schnall, M.D., and Lenkinski, R.E. 1998. Human breast lesions: characterization with proton MR spectroscopy. *Radiology 209*:269–275.

Schnall, M.D., Blume, J., Bluemke, D.A., DeAngelis, G.A., DeBruhl, N., Harms S., Heywang-Kobrunner, S.H., Hylton, N., Kuhl, C.K., Pisano, E.D., Causer, P., Schnitt, S.J., Thickman, D., Stelling, C.B., Weatherall, P.T., Lehman, C., and Gatsonis, C.A. 2006. Diagnostic architectural and dynamic features at breast MR imaging: multicenter study. *Radiology 238*:42–53.

Sitter, B., Sonnewald, U., Spraul, M., Fjosne, H.E., and Gribbestad, I.S. 2002. High-resolution magic angle spinning MRS of breast cancer tissue. *NMR Biomed. 15*:327–337.

Slotboom, J., Boesch, C., and Kreis, R. 1998. Versatile frequency domain fitting using time domain models and prior knowledge. *Magn. Reson. Med. 39*:899–911.

Tse, G.M., Cheung, H.S., Pang, L.M., Chu, W.C., Law, B.K., Kung, F.Y., and Yeung, D.K. 2003. Characterization of lesions of the breast with proton MR spectroscopy: comparison of carcinomas, benign lesions, and phyllodes tumors. *Am. J. Roentgenol. 181*:1267–1272

Vaughan, J.T., Garwood, M., Collins, C.M., Liu, W., DelaBarre, L., Adriany, G., Andersen, P., Merkle, H., Goebel, R., Smith, M.B., and Ugurbil, K. 2001. 7T vs. 4T: RF power, homogeneity, and signal-to-noise comparison in head images. *Magn. Reson. Med. 46*:24–30.

13

Breast Scintigraphy

Orazio Schillaci, Angela Spanu, and Giuseppe Madeddu

Introduction

Carcinoma of the breast is the most common malignancy in women. It represents 26% of all invasive tumors in the female population in the United States, and it accounts for 15% of all female cancer deaths. For 2007, the American Cancer Society estimated 178,480 new cases and 40,460 female deaths from this disease (Jemal *et al.*, 2007). Trends in incidence and mortality show that there has been a small but steady annual increase in breast cancer incidence during the last 30 years, whereas the mortality rate has declined steadily since the beginning of the 1990s (Jemal *et al.*, 2007). This benefit is attributed to earlier detection of breast cancer by mammographic screening.

Mammography is currently the best imaging modality for early identification of breast cancer: its findings are based on anatomic changes in the breast, and it is the method of choice in screening asymptomatic women (Tabar *et al.*, 2001). This technique has some limitations including reduced sensitivity because not all breast carcinomas are evident on mammograms, especially in dense or dysplastic breasts (Birdwell *et al.*, 2001). Moreover, its specificity and positive-predictive value are low. The main limitation is that it cannot always differentiate benign lesions from malignant ones. This is especially the case in women with dense breasts, in those who have architectural distortion of their breasts following radiation therapy or

surgery, or in those with breast implants (Birdwell *et al.*, 2001). Therefore, lesions detected by mammography are frequently submitted to biopsy, and the final outcome is that many women without cancer are biopsied. The drawbacks of mammography have led to the development of complementary modalities for breast cancer imaging, including breast scintigraphy or, as it is often called, scintimammography.

Breast Scintigraphy

Planar Method

Breast scintigraphy or scintimammography is a nuclear medicine imaging technique that uses radionuclides to image malignant breast tumors; it requires the administration of a single photon-emitting radiotracer to the patient and a gamma-camera for imaging. The ideal radiopharmaceuticals for scintimammography should have high and specific tumor uptake and minimal activity within the normal breast (Schillaci and Buscombe, 2004a). The most widely currently used are Tc-99m sestamibi and Tc-99m tetrofosmin, two small cationic complexes of technetium originally introduced and routinely used for myocardial perfusion imaging, and then proposed as tumor-seeking agents. Tc-99m sestamibi uptake and its retention in cancer cells depend on several factors, such as regional blood flow (Scopinaro *et al.*, 1994), plasma and mithocondrial membrane potential, angiogenesis, and tissue metabolism, with ~ 90% of tracer activity concentrating in the mitochondria. Similar mechanisms have also been suggested for Tc-99m tetrofosmin. However, it has been demonstrated that in tumor cell lines tetrofosmin uptake depends on both cell membrane and mitochondria potentials, predominantly accumulating in the cytosol with only a small fraction inside the mitochondria (Arbab *et al.*, 1996). Moreover, both of these radiopharmaceuticals are transport substrates for the P-glycoprotein (Pgp), a M_r 170,000 plasma membrane protein encoded by the multidrug resistance gene (MDRI) and for the MDR-related protein (MRP_1), which function as energy-dependent efflux pumps for many drugs (Hendrikse *et al.*, 1999).

The radiopharmaceutical agent (dose: ~ 740 MBq) is injected into the antecubital vein of the opposite arm to the known breast abnormality, or in a dorsalis pedis vein when both breasts have lesions, to avoid artifactual axillary lymph node uptake from the extravasated radiotracer via lymphatic drainage. After the injection, patients are usually imaged in the prone position, which provides improved separation of the breast tissue from the myocardium and the liver, which always show a high uptake that may mask overlying breast activity. The prone position also allows evaluation of deep breast tissue from the myocardium, the liver, and the thoracic wall, and provides natural landmarks of breast contours, which are very important for lesion localization. Two lateral views (left and right) in the prone position are usually performed with a special cushion enabling the examined breast to hang freely through an opening very close to the collimator of the gamma-camera. An anterior chest image with the patient in a supine position is also acquired for better localization of primary tumors in the inner quadrants to visualize the axillary and possibly internal mammary lymph node involvement (Schillaci and Buscombe, 2004a). Moreover, the anterior image is acquired in all patients in whom planar imaging is followed by supine single photon emission computed tomography (SPECT) acquisition.

Planar Results

Khalkhali *et al.* (1994) reported the first series with a relatively large number of patients on the clinical application of planar scintimammography. In this study performed on 59 patients, with abnormal mammogram and physical examination, and thus warranting biopsy or fine-needle citology of the breast, sensitivity of scintimammography was 95.8%, specificity 86.8%, positive-predictive value 82.1%, and negative-predictive value 97.1%. On the basis of these results, the authors concluded that scintimammography is very sensitive and able to improve the specificity of mammography, and thus potentially useful to reduce the high rates of negative biopsies. Since this first study, numerous other studies have been published on the clinical usefulness of scintimammography. A review including 2009 patients from 20 studies has yielded the following results in evaluating the diagnostic accuracy of breast scintigraphy in patients with mammary lesions (total number = 2304 lesions): sensitivity 85%, specificity 89%, negative-predictive value 84%, positive-predictive value 89%, and global accuracy 86% (Taillefer, 1999). The aggregated overall summary estimates of a recent meta-analysis selecting 64 unique studies (Liberman *et al.*, 2003) with data on 5340 patients and including 5354 breast lesions were: sensitivity 85.2%, specificity 86.6%, negative-predictive value 81.8%, positive-predictive value 88.2%, and accuracy 85.9%. It is worth noting that 80% of the studies yielded sensitivity and specificity values > 80%, and nearly half of them values > 90%. Moreover, in more than 5660 cases reported until now, the sensitivity and specificity of Tc-99m sestamibi scintimammography in detecting primary breast cancer were 83.8% and 86.4%, respectively (Taillefer, 2005).

The results of all these studies, however, refer to primary carcinomas independently of their size, while it is of the utmost importance to highlight that the sensitivity of scintimammography is strictly dependent on the size of lesions. A multicenter study on 420 patients with 449 breast lesions (21% were benign) reported sensitivity values of 26%, 56%, 95%, and 97% for T1a, T1b, T1c, and T2 breast cancers, respectively (Scopinaro *et al.*, 1997). In particular, sensitivity was significantly different between malignant lesions ≤ 1 cm (46.5%) and those > 1 cm (96%). Similar results regarding sensitivity are obtained when breast lesions are divided into palpable and nonpalpable, the nonpalpable always showing a lower sensitivity (Scopinaro *et al.*, 1997), as confirmed by the review of Liberman *et al.* (2003), who reported sensitivity of 87.8% for patients with palpable breast mass and 66.8% for patients without a palpable lesion. Moreover, the results of a North American multicenter trial involving 673 patients in 42 institutions (Khalkhali *et al.*, 2000) indicate a sensitivity for breast cancer detection of 87% and 61% for palpable and nonpalpable lesions, respectively. However, besides size, biological tumor factors, which may affect the net radiotracer uptake in cancer, have also to be taken into account in lesion *in vivo* visualization by scintigraphy.

Single Photon Emission Computed Tomography Method

The capacity of scintimammography to detect small tumors is critical for its clinical development and acceptance because the other breast-imaging modalities are increasingly used for the early identification of small suspicious lesions. The acquisition of tomographic images through SPECT with its better contrast resolution can play a role in increasing the sensitivity of planar scintimammography. Until now, discordant results have been reported in the studies comparing tomographic and planar imaging in primary breast cancer diagnosis, in part owing to the inhomogeneous selection of cases and to the different technical protocols employed in the various casuistries. In SPECT acquisition, it is important to emphasize that good quality imaging can be obtained only with the patient in the supine position and the arms up because SPECT with patients in a prone position is limited by the geometric constraints of the patient, imaging table, and gantry (Schillaci and Buscombe, 2004a).

Single Photon Emission Computed Tomography Results

In a group of 63 patients with 66 mammographically suspicious breast lesions, we observed a sensitivity of 92.9% for supine SPECT and 85.7% for planar imaging, whereas accuracy was 91% and 88%, respectively. Moreover, supine SPECT yielded a significantly higher sensitivity than planar images in particular in both ≤ 1 cm and nonpalpable breast cancers, without any decrease in specificity (Schillaci et al., 1997), thus indicating that SPECT acquisition is mandatory if scintimammography is performed for imaging small lesions.

Spanu et al. (2001) studied a very large series of patients with breast lesions: 175 with carcinoma and 17 with benign nodules for a total of 191 primary carcinomas and 21 benign lesions. The carcinomas were stratified according to both palpability and size (0–10 mm, 11–20 mm, 21–30 mm, and > 30 mm). A significantly higher sensitivity for SPECT with respect to planar was obtained in each group except for lesions > 30 mm, with the greatest difference in nonpalpable and in ≤ 10 mm carcinomas. SPECT detected almost 90% of these carcinomas, while planar only 50%. Planar showed an overall specificity that was slightly higher than that of SPECT (86% vs. 76%), but not in nonpalpable lesions or in those ≤ 10 mm, in which the two techniques had the same specificity (100%). A case observed in this study is illustrated in Figure 107.

In another study, Spanu et al. (2002) focused on a larger selected series of patients (96 cases) with single breast lesions ≤ 10 mm, further stratified into T1a (≤ 5 mm) and T1b (6–10 mm), obtaining sensitivity values of 71.4% and 94.5%, respectively, for SPECT and of 35.7% and 49.1%, respectively, for planar, with the same high-specificity values (95.8%) for both procedures. In both aforementioned studies, high-quality images were obtained in all cases by employing a dual-head gamma-camera equipped with high-resolution collimators, with body-contouring system and maintaining an acceptable acquisition time (30 min) and by using a count-optimized filter in the filtered backprojection (FBP) processing method.

Although SPECT imaging has better contrast resolution than planar, it can sometimes be difficult to obtain a clear and accurate definition of the sites of focal areas of radiopharmaceutical uptake, in particular in tumors with low tracer uptake, and breast contours can be less evident. However, an adequate training of nuclear medicine physicians can facilitate the interpretation of SPECT data, and this training can be obtained in nuclear medicine centers where a SPECT system is available. More recently, the co-registration of SPECT images with those obtained through radiological imaging, allowing a precise correlation of functional and anatomical data on the same image, seems to give additional information.

The commercial availability of a hybrid gamma-camera/CT scanner, which is able to provide, in addition to scintigraphic data, cross-sectional X-ray transmission images, has facilitated the fusion of anatomical maps and SPECT images. The first clinical applications of this new technology are very encouraging; in breast imaging, SPECT/CT correlative data have proven particularly useful in the most difficult cases, facilitating the interpretation of SPECT findings with a more accurate anatomical assessment of sites of abnormal activity (Schillaci et al., 2004b; Schillaci et al., 2005).

Dedicated Imaging Systems

Methods

The best option for improving planar scintimammography sensitivity in small tumor detection seems to be the use of dedicated gamma-cameras specifically built for breast imaging. Until now, scintimammography has usually been performed with conventional gamma-cameras that are large, bulky, and not specifically designed for imaging the breast, making them less than optimal for this application. The use of small field-of-view high-resolution cameras allows both greater flexibility in patient positioning (improving breast imaging by limiting the field of view and reducing image contamination from other organs, i.e., liver and heart) and breast compression, with an important increase in the target-to-background ratio (Schillaci et al., 2005). The detector can be placed directly against the chest, and a mild compression is possible, reducing breast thickness and improving camera sensitivity. Moreover, by design, the dedicated cameras are also able to provide better intrinsic and extrinsic spatial resolution than standard ones, with an enhancement in contrast resolution of small lesions (Brem et al., 2002). These cameras can be easily attached to an adapter to fit on most upright mammography machines, replacing the radiographic Bucky. Another important advantage is the possibility of acquiring scintigraphic scans with the same mammographic views (craniocaudal and lateral

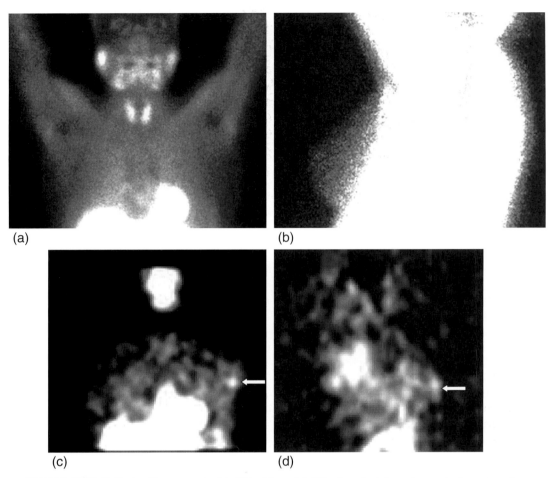

Figure 107 Patient with a non palpable T1b (10 mm) infiltrating ductal carcinoma in the external upper quadrant of the left breast, negative at planar scintimammography in both anterior (a) and lateral (b) views and positive (arrow) at SPECT in both coronal (c) and sagittal (d) views.

oblique), thereby simplifying the comparison of the two kinds of images.

Results

The results obtained in a limited number of studies indicate a better sensitivity of high-resolution cameras when compared to conventional, large field-of-view cameras, especially in detecting small breast cancers. Scopinaro *et al.* (1999) performed a comparison between a 10 cm FOV detector and a standard Anger-camera in a group of 53 patients. Breast cancers were found in 31 patients, 16 sized ≤ 1 cm and 15 sized > 1 cm. Both the high-resolution camera and the conventional one detected 14 out of 15 cancers > 1 cm. In the group of small tumors, the difference between the sensitivity of the two detectors was significant: 81% for the dedicated camera and only 50% for the Anger camera. Brem *et al.* (2002) evaluated 50 patients with 58 lesions with a size ranging from 0.3–6.0 cm: 41 (71%) were nonpalpable and 30 (52%) benign. The sensitivity of Tc-99m sestamibi scintimammography was 64% for the standard gamma-camera and 79% for the dedicated one. The sensitivity for tumors ≤ 1 was 67% for the dedicated camera and 47% for the standard one. The sensitivity for nonpalpable cancer was 72% and 56%, respectively.

The results of a study specifically designed to evaluate the usefulness of a dedicated breast camera as a screening modality have been recently reported: Coover *et al.* (2004) demonstrated cancer that was otherwise undetectable by conventional methods in three out of five scintimammography positive cases; only one of the three carcinomas identified by the dedicated gamma-camera was also detected by a standard camera. Rhodes *et al.* (2005) have recently used a cadmium zinc thelluride (CZT) dedicated camera in a group of 40 women. The detector head was composed of an array of 80 × 80 CZT elements (each of 2.5 × 2.5 mm), with a

FOV of 20 × 20 cm. Scintimammography with Tc-99m sestamibi detected 33 out of 36 malignant lesions (sensitivity 92%) in 26 patients. In particular, the sensitivity was 86% in tumors < 1 cm and 100% in larger ones, and 75% in T1a and 89% in T1b tumors. Their data indicate that high-resolution imaging permits the visualization of smaller and deeper breast cancers, overcoming the main limitations of conventional scintimammography. At present, however, dedicated imaging systems exist in only a few nuclear medicine centers, unlike planar and SPECT conventional systems. Moreover, comparative studies between dedicated gamma-camera and SPECT images in a large series of small-size breast carcinomas would be advisable.

Clinical Indications of Breast Scintigraphy or Scintimammography

In any patient with doubtful results at mammography or other diagnostic tests, scintimammography might further improve sensitivity and, in particular, specificity values (Schillaci and Scopinaro, 1999). Moreover, because radiopharmaceutical uptake is independent of breast density, scintimammography can be very useful in the population subgroup with mammographically dense breasts.

Furthermore, scintimammography is particularly indicated in patients with doubtful microcalcifications or parenchyma distortions at mammography, as well as in the presence of scar tissue in the breast following biopsy, surgery, radiation, or chemotherapy. In these situations mammographic interpretations are often difficult in evaluating breasts that have been submitted to these therapeutic procedures. On the contrary, scintimammography is not affected by post-therapeutic tissue morphological changes.

In addition, scintimammography may play a role in identifying multifocal-multicentric breast cancer and in detecting the possible primary breast tumor in patients with metastatic axillary lymph node involvement. The detection of multicentric lesions is of the utmost importance because it can alter the surgical management of the patient (i.e., total mastectomy instead of quadrantectomy). It has been reported that scintimammography is able to assess the presence of multifocal-multicentric disease, as well as bilateral breast cancers, with a higher sensitivity when compared to mammography/ultrasound (Cwikla et al., 2001; Spanu et al., 2001). However, because of the limited data available in this specific application, together with the low sensitivity of conventional planar scintimammography in visualizing small additional malignant lesions, this potential indication deserves further studies in larger series with a dedicated imaging system and/or with SPECT. Moreover, the high performance of MRI in this field should be taken into account.

In patients with axillary lymph node metastases due to adenocarcinoma, but with negative mammography and

ultrasound, scintimammography may also be useful for detecting the possible primary tumor in the breast. Nevertheless, data are still insufficient for recommending a routine clinical use of breast scintigraphy for this application. Finally, scintimammography can be useful for both monitoring and predicting response to neoadjuvant chemotherapy in patients with locally advanced breast cancer. In one study protocol including two scintigraphies before and after neoadjuvant chemotherapy, scintimammography with a cationic lipophilic radiotracer proved accurate in predicting the presence or absence of tumor after treatment, and useful for the *in vivo* detection of intrinsic and acquired resistant cancers, a very important factor for planning the best therapeutic strategy (Mezi et al., 2003).

It is concluded that mammography remains the first imaging modality for the early detection of breast cancer. However, scintimammography can be useful when mammography is nondiagnostic and in certain specific conditions in which additional information is required in order to reach a more accurate diagnosis with noninvasive methods. Scintimammography can also increase specificity and thus potentially reduce the number of unnecessary biopsies. Single photon emission computed tomography and dedicated gamma-camera systems are suggested rather than conventional planar imaging in small-size breast carcinoma detection.

References

Arbab, A.S., Koizumi, K., Toyama, K., and Araki, T. 1996. Uptake of technetium-99m-tetrofosmin, technetium-99m-MIBI and thallium-201 in tumor cell lines. *J. Nucl. Med.* 37:1551–1556.

Birdwell, R.L., Ikeda, D.M., O'Shaughnessy, K.F., and Sickles, E.A. 2001. Mammographic characteristics of 115 missed cancers later detected with screening mammography and the potential utility of computer-aided detection. *Radiology* 219:192–202.

Brem, R.F., Schoonjans, J.M., Kieper, D.A., Majewski, S., Goodman, S., and Civelek, C. 2002. High-resolution scintimammography: a pilot study. *J. Nucl. Med.* 43:909–915.

Coover, L.R., Caravaglia, G., and Kuhn, P. 2004. Scintimammography with dedicated breast camera detects and localizes occult carcinoma. *J. Nucl. Med.* 45:553–558.

Cwikla, J.B., Buscombe, J.R., Holloway, B., Parbhoo, S.P., Davidson, T., McDermott, N., and Hilson, A.J. 2001. Can scintimammography with ⁹⁹ᵐTc-MIBI identify multifocal and multicentric primary breast cancer? *Nucl. Med. Commun.* 22:1287–1293.

Hendrikse, N.H., Franssen, E.J., van der Graaf, W.T., Vaalburg, W., and de Vries, E.G. 1999. Visualization of multidrug resistance *in vivo*. *Eur. J. Nucl. Med.* 26:283–293.

Jemal, A., Siegel, R., Ward, E., Murray, T., Xu, J., and Thun, M.J. 2007. Cancer statistics, 2007. CA *Cancer J. Clin.* 57:43–66.

Khalkhali, I., Mena, I., Jouanne, E., Diggles, L., Venegas, R., Block, J., Alle, K., and Klein, S. 1994. Prone scintimammography in patients with suspicion of carcinoma of the breast. *J. Am. Coll. Surg.* 178:491–497.

Khalkhali, I., Villanueva-Meyer, J., Edell, S.L., Connolly J.L., Schnitt S.J., Baum J.K., Houlihan, M.J., Jenkins R.M., and Haber S.B. 2000. Diagnostic accuracy of 99mTc-sestamibi breast imaging: multicenter trial results. *J. Nucl. Med.* 41:1973–1979.

Liberman, M., Sampalis, F., Mulder, D.S., and Sampalis, J.S. 2003. Breast cancer diagnosis by scintimammography: a meta-analysis and review of the literature. *Breast Cancer Res. Treat. 80*:115–126.

Mezi, S., Primi, F., Capoccetti, F., Scopinaro, F., Modesti, M., and Schillaci, O. 2003. *In vivo* detection of resistance to anthracycline based neoadjuvant chemotherapy in locally advanced and inflammatory breast cancer with technetium-99m sestamibi scintimammography. *Int. J. Oncol. 22*:1233–1240.

Rhodes, D.J., O'Connor, M.K., Phillips, S.W., Smith, R.L., and Collins, D.A. 2005. Molecular breast imaging: a new technique using technetiumTc scintimammography to detect small tumors of the breast. *Mayo Clin. Proc. 80*:24–30.

Schillaci, O. 2005. Hybrid SPECT/CT: a new era for SPECT imaging? *Eur. J. Nucl. Med. Mol. Imag. 32*:521–524.

Schillaci, O., and Buscombe, J.R. 2004a. Breast scintigraphy today: indications and limitations. *Eur. J. Nucl. Med. Mol. Imaging 31* (Suppl. 1): S35–S45.

Schillaci, O., Cossu, E., Buonuomo, O., Granai, A.V., Pistolere, C.A., Danieli, R., and Simonetti, G. 2005. Dedicated breast camera: is it the best option for scintimammography? *J. Nucl. Med. 46*:550.

Schillaci, O., Danieli, R., Manni, C., and Simonetti, G. 2004b. Is SPECT/CT with a hybrid camera useful to improve scintigraphic imaging interpretation? *Nucl. Med. Commun. 25*:705–710.

Schillaci, O., and Scopinaro, F. 1999. Tc-99m sestamibi scintimammography: where is it now? *Cancer Biother. Radiopharm. 14*:417–422.

Schillaci, O., Scopinaro, F., Danieli, R., Tavolato, R., Picardi, V., Cannas, P., and Centi Colella, A. 1997. 99Tcm-sestamibi scintimammography in patients with suspicious breast lesions: comparison of SPET and planar images in the detection of primary tumours and axillary lymph node involvement. *Nucl. Med. Commun. 18*:839–845.

Scopinaro, F., Pani, R., De Vincentis, G., Soluri, A., Pellegrini, R., and Porfiri, L.M. 1999. High-resolution scintimammography improves the accuracy of technetium-99m methoxyisobutylisonitrile scintimammography: use of a new dedicated gamma camera. *Eur. J. Nucl. Med. 26*:1279–1288.

Scopinaro, F., Schillaci, O., Scarpini, M., Mingazzini, P.L., Di Macio, L., Banci, M., Danieli, R., Zerilli, M., Limiti, M.R., and Centi Colella, A. 1994. Technetium-99m sestamibi: an indicator of breast cancer invasiveness. *Eur. J. Nucl. Med. 21*:984–987.

Scopinaro, F., Schillaci, O., Ussof, W., Nordling K, Capoferro, R., De Vincentis, G., Danieli, R., Ierardi, M., Picardi, V., Tavolato, R., and Centi Colella, A. 1997. A three center sudy on the diagnostic accuracy of Tc-99m MIBI scintimammography. *Anticancer Res. 17*: 1631–1634.

Spanu, A., Dettori, G., Nuvoli, S., Porcu, A., Falchi, A., Cottu, P., Solinas, M.E., Scanu, A.M, Chessa, F., and Madeddu, G. 2001. ^{99m}Tc-Tetrofosmin SPET in both primary breast cancer and axillary lymph node metastasis detection. *Eur. J. Nucl. Med. 28*:1781–1784.

Spanu, A., Schillaci, O., Meloni, G.B., Porcu, A., Cottu, P., Nuvoli, S., Falchi, A., Chessa, F., Solinas, M.E., and Madeddu, G. 2002. The usefulness of 99mTc-tetrofosmin SPECT scintimammography in the detection of small size primary breast carcinomas. *Int. J. Oncol. 21*:831–840.

Tabar, L.K., Vitak, B., Chen, H.H.T., Yen, M.F., Duffy, S.W., and Smith, R.A. 2001. Beyond randomized controlled trials: organized mammographic screening substantially reduces breast cancer mortality. *Cancer 91*:1724–1731.

Taillefer, R. 1999. The role of Tc-99m sestamibi and other conventional radiopharmaceuticals in breast cancer diagnosis. *Semin. Nucl. Med. 24*:16–40.

Taillefer, R. 2005. Clinical applications of ^{99m}Tc-sestamibi scintimammography. *Semin. Nucl. Med. 35*:100–115.

14

Primary Breast Cancer: False-negative and False-positive Bone Scintigraphy

Hatice Mirac Binnaz Demirkan and Hatice Durak

Introduction

Breast cancer is the most common cancer and the second leading cause of cancer death among women in developed countries (Jemal *et al.*, 2004). The skeleton is the most frequent site of metastatic spread by hematogenous dissemination of cancer cells. The existence of a link between Batson's system and the Azygos and the Hemi-Azygos systems and the Internal Mammary system, via thoracic wall veins (i.e., lateral and intercostal veins), could explain the spread from breast cancer to bone. A particular site of dissemination is the sternum: its involvement results from local dissemination from a pathological internal mammary chain (O'Sullivan and Cook, 2002). Bones are often involved in 5–10% of breast cancer patients in the first stages; but in advanced disease, it may be detected in 65–75% of cases at autopsy. This explains the interest in bone scintigraphy in initial staging and the follow-up of breast cancer. Because of its high sensitivity and the ability to image the entire skeleton, bone scan is widely employed (Maffioli *et al.*, 2004). This chapter illustrates the procedures and technical aspects of bone scintigraphy, clinical application in breast cancer,

pitfalls and problems encountered and their solutions, and potential advances.

Search Strategy and Selection Criteria

Data for this review were obtained by searching the PubMed database using a combination of the following terms: bone metastases, breast cancer, bone scintigraphy, bisphosphonates, Paget disease of bone, osteoporosis, bone remodeling, bone marrow micrometastases, computurized tomography, magnetic resonance imaging, FDG-positron emission tomography, drug interactions, radiopharmaceuticals, procedures guidelines, tumor imaging, and nuclear medicine. References were restricted to journal articles published in English up to February 2006.

Procedures and Technical Aspects of Bone Scan

Bone scintigraphy (BS) is one of the most well-established methods in nuclear medicine and has provided valuable data in the assessment and management of neoplastic diseases since being first described in the early 1960s. Since the introduction of Technetium 99m (Tc 99m)-labeled phosphates and phosphonates in 1971, bone scintigraphy has become one of the most commonly performed procedures in nuclear medicine clinics (Subramanian and McAfee, 1971). In all scintigraphic procedures, labeled radiopharmaceuticals are used to visualize a metabolic process in the body. Diagnostic radiopharmaceuticals are molecules labeled with gamma or positron-emitting radionuclides. After they are administered into the body, the biochemical pathway that the labeled radiopharmaceutical enters into can be imaged using a gamma-camera or positron emission tomography (PET). Thus, the scintigraphic images reflect the biochemical status of the organs under study, which most of the time appears before the anatomical changes.

In the case of bone scintigraphy, the visualized metabolic process is the osteoblastic activity or, in other words, the bone remodeling. The metastatic bone lesions are visualized indirectly due to the bone formation aggravated by the tumor invasion at the metastatic sites.

The most commonly used radiopharmaceuticals for bone scintigraphy are Tc 99m-labeled diphosphonates, such as Tc 99m methylene diphosphonate (MDP), Tc 99m hydroxyl methylene diphosphonate (HDP), and Tc 99m hexamethylene diphosphonate (HMDP) (Bombardieri et al., 2006). These compounds are commercially available in vials in lyophilized form and are prepared by adding the required amount of Tc 99m pertechnetate (Bombardieri et al., 2006). For an adult patient, 740–1110 MBq (20–30 mCi) activity is administered by intravenous injection. For children, 9–11 MBq/kg (250–300 μCi/kg) activity is recommended, with a minimum of 20–40 MBq (0.05–1.0 mCi) (Donohoe et al., 2006). Bone is the critical organ, receiving most of the radiation dose, 0.063 mGy/MBq (0.23 rad/mCi) (Bombardieri et al., 2006; Donohoe et al., 2006).

Adverse reactions are quite rare and mild, reported as 1 in 800 for methylene diphosphonate. Reactions to diphosphonate are erythema, nausea, vomiting, and malaise, with an onset usually of 2–3 hr after injection (Sampson, 1993). The blood clearance is high, and only 3% of the administered activity remains in the blood 3 hr after injection. Maximum bone accumulation may be reached as early as 1 hr after injection (Bombardieri et al., 2006). The biological half-life of phosphonates is 26 hr. With normal renal function, > 30% of the unbound complex goes through glomerular filtration within 1 hr. Elimination via the intestines is insignificant (Bombardieri et al., 2006). Patients are instructed to drink one or more liters of water (4–8 glasses) between the injection time and the imaging time, if not contraindicated well-hydrated and instructed to. Patients are asked to void frequently between the time of injection and the time imaging, as well as immediately prior to the scan (Bombardieri et al., 2006; Donohoe et al., 2006).

After intravenous injection, Tc 99m-labeled phosphonates reach the bones by blood flow. The amount of blood flow reaching the bones is crucial where alterations in blood flow influence the amount of compound extracted from the blood by the bones. Decreased delivery of the radiopharmaceutical reduces the accumulation in the bones, whereas increased vascularity is associated with increased concentration in the extracellular fluid by passive diffusion, resulting in increased accumulation (Murray, 1994). The inhibition of sympathetic control causes a significant diffuse augmented uptake as seen in reflex sympathetic dystrophy syndrome (Ozturk et al., 2004; Ryan and Fogelman, 1996; Murray, 1994).

The extraction efficiency of the bones is another important factor. The bones extraction capacity depends on capillary permeability, opening of microvascular pathways, acid-base balance, parathyroid hormone levels, and other hormonal factors (Bombardieri et al., 2006; Murray, 1994). When phosphonates reach the bone extracellular fluid, the contact of the radiopharmaceutical with reactive bone formation is facilitated (Murray, 1994). Phosphonates concentrate in the mineral phase of bone, nearly two-thirds in hydroxyapatite crystals and one-third in calcium phosphate (Bombardieri et al., 2006). The P-C-P (phosphorus-carbon-phosphorus) portion of the bisphosphonate structure has high affinity for hydroxyapatite, and they rapidly localize on the bone surface (Hirabayashi and Fujisaki, 2003). In addition, there may be some binding with the collagen moiety (Murray, 1994). Approximately 50% of the injected activity accumulates in the skeleton (Bombardieri et al., 2006).

A good-quality bone scan (Fig. 108) should show a good contrast between bone and soft tissue, but this does not always mean a good lesion detection. The lesion-to-bone ratio should be high for an improved lesion visualization (Murray, 1994). The osteoblastic response in the bone appears as increased uptake, generally indicating a repairing process and new bone formation. Lesions that do not create an osteoblastic reaction usually do not show any significant accumulation. In some lesions, such as metastatic lesions, bone destruction and osteoclastic activity are accompanied by reactive new bone formation. This condition results with increased focal accumulation of the radiopharmaceutical. So, a sclerotic or a lytic lesion in direct radiography may be visualized as focal areas of increased uptake in bone scintigraphy, if the lesion has induced a reactive osteoblastic response. The focal areas of uptake are influenced by the amount of metabolic response, and sometimes the uptake is far higher and greater than the actual lesion size (Murray, 1994). This mechanism of uptake may be induced by many other bone pathologies other than metastasis, which makes bone scintigraphy a highly sensitive but a less specific diagnostic test for the diagnosis of bone metastasis.

Routine static images are usually obtained 2–5 hr after injection. Whole-body scintigraphy in both anterior and

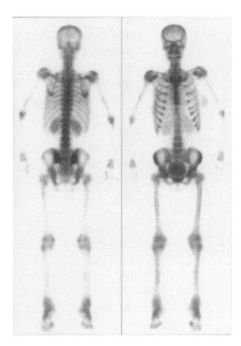

Figure 108 Normal whole-body bone scan (left: posterior; right: anterior).

posterior projections may be preferred in order to visualize the complete skeleton. A low-energy high-resolution or an ultrahigh-resolution collimator should be used (Bombardieri et al., 2006; Donohoe et al., 2006; Hirabayashi and Fujisaki, 2003). Minimizing the patient to collimator distance is very important for improving the image quality (Brown et al., 1993). The scanning speed should be adjusted so that a whole-body image contains > 1.5 million counts in 256 × 1024 or greater matrix. Complementary spot images may be obtained in order to better visualize the areas with suspicious lesions. Imaging with pinhole collimator may be necessary for small lesions (Bombardieri et al., 2006; Donohoe et al., 2006; Brown et al., 1993). It is optional to obtain a 128 × 128 or a 256 × 256 matrix and > 200,000 counts. In some patients, especially in children, spot views may be preferred to whole-body imaging (Bombardieri et al., 2006; Donohoe et al., 2006). When spot views are used, care should be taken not to skip any regions of the skeleton. The first spot view, usually the posterior projection of the chest, is acquired for ~ 500,000 to 1 million counts. Other remaining spot views are then acquired for the same time as the first view.

In some patients single photon emission computed tomography (SPECT) may be performed to delineate and locate the lesions more precisely, especially in patients with low back pain. Hips, temporomandibular joints, skull, and knees are more clearly evaluated with a good-quality bone SPECT (Brown et al., 1993). Bone SPECT helps in the differential diagnosis of malignant and benign spinal lesions

and the detection of metastatic cancer in the spine (Sarikaya et al., 2001). The optimal acquisition parameters for bone SPECT are reported as 360° circular orbit, 60–120 steps, 64 × 64 or greater matrix, and 10–40 sec/stop (Bombardieri et al., 2006; Donohoe et al., 2006; Hirabayashi and Fujisaki, 2003; Brown et al., 1993). In a study conducted by the National Cancer Institute of Milan, 118 patients with bone metastases from different tumor types, bone SPECT has been shown to improve the predictive value of bone scintigraphy (Savelli et al., 2001).

In some patients, the blood flow of a lesion needs to be evaluated, so three-phase bone scintigraphy is indicated. The camera is positioned over the region of interest, and the dynamic acquisition is started with the tracer injection. Flow images are acquired in a 64 × 64 or greater matrix at 1–3 sec/frame. Dynamic flow phase is immediately followed by blood pool (tissue phase) images and completed within 10 min of tracer injection. Images of the area of interest are acquired in a 128 × 128 or greater matrix for ~ 3–5 min with a count density of ~ 300,000 counts/image (Bombardieri et al., 2006; Donohoe et al., 2006).

A good-quality bone scan is very likely to provide an early diagnosis of metastasis before any morphological abnormality is seen. Whole-body imaging is easily performed with gamma camera, permitting the survey of the whole skeleton (Soderlund, 1996). Care should be taken to prepare the radiopharmaceutical according to the manufacturer's recommendations. Many national and international organizations have published guidelines for bone scintigraphy, so for optimal results, imaging processing and display parameters should be designed according to the published guidelines. Interpretation of the images needs expertise because many benign lesions also produce an osteoblastic response that appears as focally increased uptake. There may be many metabolic, hormonal, or therapy-induced conditions inducing focal or diffuse bone uptake. Carcinoma polyarthritis, hypertrophic pulmonary osteoarthropathy, flare phenomenon (which is the enhanced uptake in metastases due to hormonal treatment, chemotherapy, or radiotherapy), colony-stimulating factor causing increased uptake in axial skeleton and juxta-articular areas, osteonecrosis due to radiotherapy, or corticosteroid use may all interfere with correct evaluation (Stokkel et al., 1993). Primary benign or malignant bone tumors, sports medicine injuries, fractures, stress injuries, skeletal trauma, osteomyelitis, postprosthesis infection and loosening, septic arthritis, avascular necrosis, reflex sympathetic dystrophy syndrome, enthesopathies and biomechanical stress lesions, inflammatory arthropathies, Paget disease, other metabolic bone diseases, costochondritis, and miscellaneous bone conditions also create increased uptake in the bones (Fig. 109) (Love et al., 2003; Scharf and Zhao, 1999; Ryan and Fogelman, 1996).

The bone scan images should be evaluated with history, clinical findings, laboratory tests, and findings of other

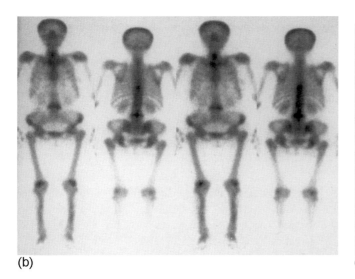

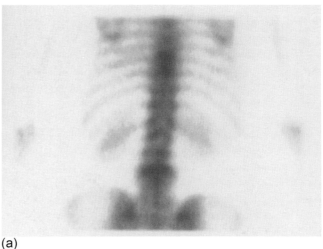

(b)   (a)

Figure 109 Anterior and posterior whole-body bone scan (a), and spot posterior spine image (b) of a patient with breast cancer show focal increased uptake at thoracic 8, 9, and lumber third vertebrae due to Paget's disease of bone, which are falsely interpreted as bone metastases.

imaging modalities of the patient for a more correct interpretation. If the scan is not typical for multiple metastases, correlation with radiological methods is necessary for symptomatic or scintigraphy positive areas to increase the specificity of the scintigraphy and to avoid false-negative results (Soderlund, 1996). Extraosseous radiopharmaceutical uptake is not surprising; soft tissue, organs, and some tumors are reported to show Tc 99m MDP uptake (Ryan and Fogelman, 1996; Murray, 1994). Image artifacts caused by the instruments and processing and contamination artifacts should be well recognized (Forstrom *et al.*, 1996). The most common contamination artifact is the spilling of radioactive urine on the patient during voiding; the patient should be warned to avoid such a possibility (Howarth *et al.*, 1996).

The typical appearance of multiple metastases is focal areas of increased uptake or, in other words, hot spots (Fig. 110) (Love *et al.*, 2003; Van der Wall, 1994). Multiple foci of varying size and intensity distributed asymmetrically through the axial skeleton in a patient with a known primary tumor are very likely to represent a metastatic bone involvement. Ten percent of the lesions can be seen in the skull or appendicular skeleton. Sometimes the metastatic involvement becomes so disseminated that an appearance of superscan can be seen, which manifests as intense uptake through the skeleton with faint or absent renal accumulation (Van der Wall, 1994). Adverse reactions to the radiopharmaceuticals are rare. For bone-seeking radiopharmaceuticals, one adverse reaction per 800 injections has been reported. Erythema, nausea, vomiting, and malaise are quoted as the typical diphosphonate reactions, which usually start 2–3 hr after injection (Sampson, 1993).

Causes influencing the quality of the scan other than radiopharmaceutical kit preparation and technical accuracy are drug interactions. Many drugs interfere with the radiopharmaceutical activity, resulting in reduced bone uptake, masking the lesions, or increasing bone, soft tissue, and other organ accumulation such as in the liver and kidneys.

Administration of diphosphonates (such as etidronate, pamidronate, or alendronate) has shown to interfere with the uptake of Tc 99m MDP (Fig. 111) (Demirkan *et al.*, 2005; Murphy *et al.*, 1997), though contradictory reports are also present in the literature (Roudier *et al.*, 2003). These drugs are used in the treatment of Paget's disease of the bone, hypercalcemia associated with malignancy, and osteoporosis. Their structures are similar to MDP, so competition occurs between the drug and the radiopharmaceutical for the skeletal binding sites. The result is decreased skeletal visualization and high soft tissue uptake. Murphy *et al.* (1997) defined loss of bone definition below the midthigh and indistinguishable rib uptake and absence of uptake on metastatic bone lesions when patients were imaged during etidronate treatment. Thus, imaging the patients either before or 2–4 weeks after etidronate treatment is recommended.

Aluminium-containing drugs, such as antacids and nifedipine, are reported to reduce the bone uptake, and cortisone treatment has been shown to reduce the sensitivity of bone scintigraphy in bone trauma (Hesslewood and Leung, 1994; Sampson, 1993). Iron preparations are reported to cause high blood pool, liver, and renal activity, as well as focal accumulation of the bone-seeking radiopharmaceutical at the sites of intramuscular iron injections. Some chemotherapeutic agents such as cyclophosphamide, vincristine, and doxorubicin may cause increased renal retention (Park *et al.*, 1997; Hesslewood and Leung, 1994), while methotrexate

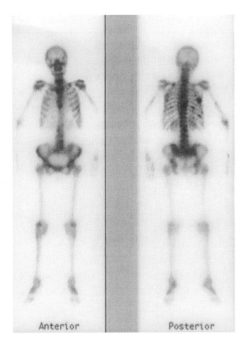

Figure 110 Anterior and posterior whole-body bone scan of a patient with breast cancer revealing multiple sites of focal increased osteoblastic activity in the spine, ribs, scapulae, and pelvis due to disseminated bone metastases.

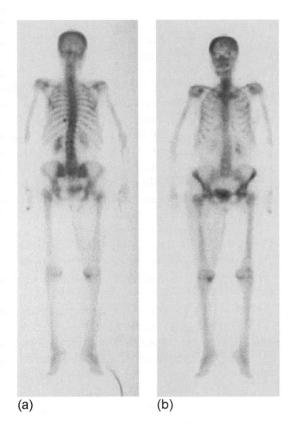

(a)   (b)

Figure 111 Posterior (a) and anterior (b) whole-body bone scan of a patient with breast cancer falsely interpreted as normal bone scan under the alendronate treatment for osteoporosis, although extensive bone metastases were revealed by CT and MRI scans.

sodium and cyclophosphamide are reported to cause increased bone uptake of the Tc 99m dicarboxypropane diphosphonic acid in rats (Jankovic and Djokic, 2005).

It should be kept in mind that bone scintigraphy visualizes the metabolic reaction of the bones, reflecting increased bone turnover and osteoblastic activity (Hamaoka et al., 2004). False-negative lesions of bone scintigraphy when compared to MRI or ^{18}F-2-deoxy-2-fluoro-D-glucose-PET are mostly bone marrow metastases (Abe et al., 2005; Aydin et al., 2005; Altehoefer et al., 2001). Bone scintigraphy is a good whole-body screening method rather than a diagnostic test and can be complementary for characterizing the lesion types if used in correlation with other imaging modalities (Hamaoka et al., 2004).

Clinical Applications in Breast Cancer

Bone scintigraphy (BS) has been recommended for stage II (T3 or N1), stage III, and stage IV diseases, while it has limited value for lower tumor stages or routine follow-up because of the adoption of an almost minimalist policy using evidence-based guidelines and worldwide economic restrictions (Maffioli et al., 2004; O'Sullivan and Cook, 2002; Koizumi et al., 2001). Yet some important studies clearly demonstrated that intensive follow-up does not improve the survival of patients with early breast cancer. Cost/benefit ratios and, more generally, cost-effectiveness analyses are becoming standard tools for physicians. These factors have to be taken into account when assessing whether a patient should undergo a medical procedure. From a psychological point of view, it is obvious that when a patient undergoes a diagnostic test (e.g., bone scan), she has a fear of the outcome, but the reassurance provided by a negative test has a great impact on the patient's quality of life. Problems can arise when the examination yields a doubtful finding. This can generate anxiety and fear until a further test clarifies the condition. In order to minimize the patient's anxiety and reduce the number (and costs) of additional tests, the nuclear physician should in some cases be more definitive in diagnosing the abnormality detected on the bone scan.

Finally, in most cases the same clinician will require that further examinations be undertaken in order to assist in resolving doubtful cases (Palli et al., 1999). The American Society of Clinical Oncology (ASCO) guidelines, which aim to determine an effective, evidence-based, postoperative surveillance strategy for the detection and treatment of recurrent breast cancer, state that "the data are insufficient to suggest the routine use of bone scans" (Smith et al., 1999). However, it is reported that, at least in Europe, these guidelines are not being completely implemented. In fact, except in cases of very early breast cancer, bone scan is routinely used in clinical practice for staging and surveillance (Maffioli *et al.*, 2004). Although guidelines may emphasize the relevance of economic problems, the reader needs to bear in mind that a number of parameters can vary from country to country. Thus, the social impact of implementing a procedure in the work-up of a patient may differ and should be calculated for each individual region in light of each patient's individual circumstances. Compliance with the guidelines does reduce the risk of overperscription (Loprinzi *et al.*, 2003; Myers *et al.*, 2001).

Osteolytic bone lesions are quite common in breast cancer, owing to the predominant activity of the osteoclasts. Bone remodeling occurs in discrete areas, termed *basic multicellular units*, and it is estimated that the entire adult skeleton is remodeled every 2–4 years (Jilka, 2003; Parfitt, 2002). The coupling of osteoblastic bone deposition and osteoclastic bone resorption in discrete skeletal foci called bone-remodeling units is responsible for continuous turnover of the skeleton. At the "front" of a remodeling site, activated osteoclasts, which are derived from mononuclear precursor cells in the monocyte-macrophage lineage, attach to the mineral surface and seal off a roughly circular area by means of a ringlike attachment of the cell's plasma membrane. Hydrolytic enzymes are secreted by the overlying osteoclast into the newly created extracellular compartment, and secreted proteins lead to acidification, dissolution of mineral, and release of collagen breakdown products, which pass into the circulation. Behind this resorptive front, osteoblasts that are derived from mesenchymal stem cells migrate into the demineralized pocket and lay down new bone matrix, composed primarily of type I collagen. The new matrix is then mineralized, and the pocket created by osteoclastic resorption is thus refilled with bone. In the healthy adult skeleton, the remodeling sequence is balanced, but in the presence of metastasis, osteoclast-osteoblast coupling is disrupted, and there is increased resorptive activity with release of collagen breakdown products. Thus, in osteolytic metastases, it appears that dysfunctional osteoclast-osteoblast units are the mediators of bone destruction, not the tumor cells themselves (Roodman, 2004; Rodan, 2003).

Currently, the median survival for women with bone metastases from breast cancer is 3 years and the 5-year survival rate is 20% (Gralow, 2001). Two-thirds of patients with bone metastases subsequently develop a skeletal-related event (SRE) (Table 16) (Coleman, 1997). Controlling bone metastases is therefore an essential part of management. Over the years many new therapies have been approved that can palliate metastatic breast cancer. With the introduction of these new drugs, postrecurrence treatment has changed, and the survival and quality of life of these patients have improved.

In addition to conventional therapies, such as chemotherapy, endocrine therapy, radiation therapy, and analgesics, bisphosphonates are now an essential component of the

Table 16 Skeletal-related Events (SREs)

Vertebral fractures
Pathological nonvertebral fractures
Spinal cord compression
Surgery for bone complications
 (fractures or impending fractures)
Radiotherapy for bone complications
 (uncontrolled bone pain/impending fracture)
Hypercalcemia

treatment of bone metastases (Coleman, 2004). Osteoporosis, hypercalcemia of malignancy, and Paget's disease of the bone are the other indications of bisphosphonate use. The most commonly used bisphosphonates in oncology—clodronate, pamidronate, zoledronic acid, and ibandronate—are potent inhibitors of osteoclast-induced bone resorption by inhibiting the attachment of osteoclasts to bone matrix and enhancing programmed cell death. They have all been shown to significantly reduce SREs in patients with bone metastases from breast cancer (Table 17) (Coleman, 2004; Lipton, 2003). Moreover, the effect of bisphosphonates is hypothesized to be additive to the effect of antineoplastic drugs as adjuvant treatment. But because of the inconsistent, evolving data found in studies with bisphosphonate therapy to prevent osseous metastases, its use is not recommended in the adjuvant setting.

Among these agents, intravenous pamidronate and zoledronic acid have demonstrated the greatest clinical benefit based on conservative end points. Subsequently, zoledronic acid, a new-generation bisphosphonate, was shown to significantly reduce the risk of skeletal complications by an additional 20% compared with pamidronate, and it is the only modern bisphosphonate to be directly compared with an active agent. It has become the new standard of care for the treatment of bone metastases in breast cancer patients (Berenson, 2005; Rosen *et al.*, 2003). The ASCO 2003 update on the role of bisphosphonates and bone health issues in women with breast cancer states that lytic or mixed lytic/blastic disease on plain radiographs or abnormal bone scan with normal radiographs, but CT (computerized tomography) or MRI (magnetic resonance imaging) scan showing bone destruction, are indications for starting bisphosphonates. The guideline suggests that once initiated, intravenous bisphosphonates should be continued until evidence of substantial decline in a patient's general performance status. There is no evidence addressing the consequences of stopping bisphosphonates after one or more adverse skeletal events (Hillner *et al.*, 2003).

Osteoporosis, which constitutes a major public health problem, is a common skeletal disorder characterized by compromised bone strength predisposing to an increased susceptibility to fracture. Because of the high prevalence rates of both low bone mass and breast carcinoma in women,

Table 17 Bisphosphonates for Treatment of Patients with Bone Metastases from Breast Cancer[a]

Bisphosphonate	Relative Potency	Dose, mg	Schedule	Infusion Time
First Generation				
Clodronate	1	1600–3200	Daily	Oral
Second Generation				
Pamidronate	20	90	Every 3–4 wks	2 hr
Ibandronate	857	6	Every 3–4 wks	1–2 hr
		50	Daily	Oral
New Generation				
Zoledronic acid	16,700	4	Every 3–4 wks	15 min

[a]Data from Lipton (2003).

these two diseases commonly coexist in the same individual (Van Poznak and Sauter, 2005). Oncology professionals, especially medical oncologists, need to pay careful attention to the routine and regular assessment of these women's bone health. Osteoporosis is defined as a disease of increased skeletal fragility accompanied by low bone mineral density (a *T* score for bone mineral density below −2.5) and microarchitectural deterioration by the World Health Organization. Bone strength reflects the integration of bone density and bone quality. Bone mineral density (BMD) measurement is a widely accepted quantitative method for diagnosing osteoporosis and predicting fracture risk (Kanis, 2002; Cummings *et al.*, 2002). The bisphosphonates, which are often considered a first-line therapy for postmenopausal osteoporosis, are the most widely prescribed antiresorptive agents.

ASCO guidelines recommend the screening of breast cancer patients for osteoporosis risk. The high-risk group (all women age > 65 years, postmenopausal women of any age receiving aromatase inhibitors, premenopausal women with therapy-associated premature menopause, and all women age 60–64 years with special circumstances) is recommended to be screened for BMD by DEXA (Dual Energy X-Ray Absorptiometry bone scan) (Hillner *et al.*, 2003). If osteoporosis is proven, physicians have to begin therapy. In the body, systemic bisphosphonate is bound preferentially to active bone sites, whereas the drug enters the osteoclasts. So, its effect should be at least partly determined by its amount at the skeleton and its pharmacokinetic. Since low bone mass due to their age or anticancer treatments and bone metastases also mostly coexist in postmenopausal breast cancer patients, some biphosphonates are also retained in osteoporotic bone sites besides metastatic regions (Hoffman *et al.*, 2001; Fleisch, 1998). For an effective treatment, it can be suggested that appropiate dose adjustment of bisphosphonates may be needed to be done according to the bone mineral density. But more studies of this subject are needed.

Paget's disease of bone (PD) is the second most common bone disease after osteoporosis. The disease is characterized by focal regions of highly exaggerated bone remodeling. Three pathologic phases have been described: the lytic phase (incipient-active) in which osteoclasts predominate; the mixed phase (active) in which osteoblasts cause repair superimposed on the resorption; and the blastic phase (late-inactive) in which osteoblasts predominate. Frequent sites of involvement include the skull (25–65%), spine (30–75%), pelvis (30–75%), and proximal long bones (25–30%). Monostotic disease (10–35%) is more often seen in the axial skeleton, although any site can be the sole region of involvement. Polyostotic disease (65–90%) is more frequent than monostotic disease, tends to have right-sided predominance, and usually involves lower extremities. Bone scintigraphy typically demonstrates marked increased uptake of radionuclide in all phases of Paget's disease (Roodman and Windle, 2005; Smith *et al.*, 2002).

Osseous metastases, osteoporosis, Paget's disease of bone, bisphosphonate treatment, multidrug use due to comorbidity, and association of one or more of these conditions in breast cancer patients should be considered during screening and follow-up of those patients with bone scan.

Pitfalls of Bone Scan Encountered in Breast Cancer Patients and Their Solutions with Potential Advances

99mTc-MDP (Technetium 99m–methylene diphosphonate) is a short-lived tracer used for radionuclide studies of bone. Its protein binding is subtantial, varying with time and by individual. Like bisphosphonates, 99mTc-MDP is cleared from the plasma by uptake into bone and elimination via renal excretion. Plasma protein binding of 99mTc-MDP is starting from ~ 25–30% immediately after tracer administration and increasing to 45–55% at 4 hr and 60–70% at 24 hr. Free 99mTc-MDP, but not the bound fraction, is available for clearance to the skeleton (Moore *et al.*, 2003; Blake *et al.*, 2002; Hoffman *et al.*, 2001). Biphosphonate treatment (e.g., alendronate, zoledronate) during radionuclide bone assessment could cause competition between the drug and the tracer by blocking the entrapment and accumulation of the tracer in bone and causing a false-negative scintigraphy (Demirkan *et al.*, 2005).

This phenomen may be the result not only of decreased skeletal uptake of free 99mTc-MDP due to bisphosphonate treatment, but also of increased plasma protein binding and renal clearance. Further bioavailability determination and quantitative studies are needed to quantify the bone uptake and clearance of biphosphonates and 99mTc-labeled phosphates and phosphonates. There have been many developments in imaging techniques and radiopharmaceuticals over the years, allowing more reliable detection of metastatic spread to bone. So, another possible explanation for the false-negative bone scan could be that bone metastases, indicating hematogenous tumor spread, are detected earlier by CT, MRI, PET, or PET/CT, which are sensitive to bone marrow abnormalities and yield information on tumor extent (Demirkan *et al.*, 2005; Buscombe *et al.*, 2004).

Clinicians act on results of BS to determine the best form of primary treatment. Although bone scans are sensitive, they are relatively nonspecific. It is important to determine whether an abnormal scan is due to metastases or benign conditions such as osteoarthritis, Paget's disease of bone, or previous trauma. The false-positive rate varies from 1.6 to as high as 22%, while the false-negative rate varies from 0.96 to 13% (Buscombe *et al.*, 2004; Yip and Paramsothy, 1999).

Magnetic resonance imaging could be more effective in the early diagnosis of distant metastases because (1) metastases often originate from red marrow, (2) MRI can visualize cortical involvement, and (3) scintigraphic evidence depends on the cortical infiltration, which happens after the localization of disease in the red marrow. Magnetic resonance imaging may differentiate metastases from osteoporotic fracture and other causes of increased bone metabolism such as degenerative disease. During an evaluation of the comparative impact of MRI and BS in bone metastases of breast cancer, in 13% of patients with bone metastases BS was false-negative. In 49% of patients with spinal metastases MRI showed more extensive disease. It is beyond doubt that, owing to restricted availability, high cost, and the long duration of a whole-body study, MRI cannot be routinely used as a standard procedure for screening or follow-up (Altehoefer *et al.*, 2001).

[18]F-2-deoxy-2-fluoro-D-glucose-PET (FDG-PET) is able to explore the whole body and has great potential for tumor staging with a single procedure because it can detect not only the primary tumor and nodal metastases but also skeletal and visceral metastases (Eubank *et al.*, 2002). FDG-PET and BS have been shown to be complementary in the detection of skeletal metastases. FDG-PET is more sensitive than BS for detection of lytic metastases or lesions predominantly involving bone marrow, accounting for cases that are positive at FDG-PET and negative at BS, but BS is more sensitive than FDG-PET for detection of osteoblastic metastases (Abe *et al.*, 2005; Nakai *et al.*, 2005).

Another tracer, ^{18}F-fluoride, has been used in PET to assess skeletal metastases. The rationale behind this approach is that the accumulation of ^{18}F in bone is ~ twofold higher than that of 99mTc-MDP and its blood clearance is faster, resulting in a high bone-to-background ratio. Studies in which ^{18}F was compared with 99mTc-MDP demonstrated on a lesion-by-lesion basis that the area under the receiver operating characteristic (ROC) curve was 0.99 for PET and 0.74 for BS. A problem is that ^{18}F-PET detects a larger number of benign lesions than conventional scintigraphy, thus, increasing the rate of false-positive results. Another problem is that ^{18}F-fluoride is a specific radiopharmaceutical for the skeleton, whereas FDG visualizes not only skeletal lesions, but also soft tissue and visceral lesions. This means that there is a single radiopharmaceutical and a single diagnostic modality for the complete evaluation of all possible sites of breast cancer spread (Buscombe *et al.*, 2004). Data from the literature suggest that in breast cancer patients FDG-PET allows complete tumor staging with a single whole-body examination, potentially leading to the diagnosis of a significant number of metastases, which would have been missed or not correctly diagnosed by CT, US, MRI, or BS alone (Van der Hoeven *et al.*, 2004; Dose *et al.*, 2002).

The follow-up of operable breast cancer patients is the subject of continuous discussion. The current tendency is to reduce instrumental examinations during follow-up to a minimum in asymptomatic patients. It is, therefore, important to identify categories of patients with a high risk of recurrence in whom an adequate program of diagnostic check-ups would be acceptable. Not only asymptomatic but also patients with the presence of clinical symptoms or with a progressive increase in biochemical marker (CA15-3) may benefit from whole-body FDG-PET, which could limit the number of diagnostic tests. In a recent study from Japan, the sensitivity, specificity, and accuracy of [18]FDG-PET were found to be 84%, 99%, and 95%, respectively. Although the results were comparable with BS, the combination of [18]FDG-PET and BS has improved the sensitivity (98%) and accuracy (97%) of detection. So, [18]FDG-PET should play a complementary role in detecting bone metastasis with BS (Abe *et al.*, 2005). There has been a longstanding interest in fused images of anatomical information, such as that provided with CT with biological information obtainable with PET. By combining multiple diagnostic studies within a single examination, significant logistic advantages can be expected if the combined PET/CT examination is to replace separate PET and CT scans, thus resulting in significantly accelerated diagnostics (Bockisch *et al.*, 2004).

In breast cancer, several large prospective studies have documented that the presence of disseminated tumor cells (DTC) at primary diagnosis predicts the subsequent occurrence of overt metastases in bone and other organs (Wiedswang *et al.*, 2003; Gebauer *et al.*, 2001; Braun *et al.*, 2000). Advances in the development of immunhisto-chemical (e.g., cytokeratins, epithelial mucin antigen) and molecular/assays (e.g., comparative genomic hybridiza-tion, mutational analysis, fluorescence in situ hybridization, reverse transcriptase–polymerase chain reaction, gene expression profiling with microarrays) now enable specific detection of DTC even at the single-cell stage (Woelfle *et al.*, 2003; Kraus *et al.*, 2003; Lugo *et al.*, 2003; Rama-swamy *et al.*, 2003). In total, 20–40% of patients with various epithelial tumors (e.g., carcinomas of the breast, prostate, colon, or lung) harbor occult metastatic cells in their bone marrow (BM), even in the absence of any lymph node metastases (stage N0) and clinical signs of overt dis-tant metastases (stage M0) (Lugo *et al.*, 2003). Bone marrow, therefore, not only is a frequent site of overt metas-tases in some tumor types (in particular breast and prostate cancer) but also seems to be a homing organ for DTC derived from various primary sites.

These DTCs can reside in a state of dormancy for many years before they either grow out into overt skeleton metas-tases (e.g., breast cancer) or recirculate from BM into other organs such as liver or lung, where better growth conditions might exist (e.g., colon cancer). Bone marrow, which might be an important reservoir for metastatic epithelial tumor cells, therefore, is a useful indicator organ to screen for the presence of early disseminated cancer cells (Gray, 2003; Fidler, 2003). Several recent prospective studies, including a total of > 3000 breast cancer patients, demonstrated that the presence of micrometastatic cells in BM has a strong prog-nostic impact on patient survival. In this context, lack of lymph node (micro-) metastases is not an indicator for the absence of DTC in BM. This supports the view that screen-ing of both compartments can provide independent prognos-tic information during the initial staging work for breast cancer–like lymphomas (Wiedswang *et al.*, 2003; Gerber *et al.*, 2001; Braun *et al.*, 2001; Mansi *et al.*, 1999).

References

Abe, K., Sasaki, M., Kuwabara, Y., Koga, H., Baba, S., Hayashi, K., Takahashi, N., and Honda, H. 2005. Comparison of 18FDG-PET with 99mTc-HMDP scintigraphy for the detection of bone metastases in patients with breast cancer. *Ann. Nucl. Med.* 19:573–579.

Altehoefer, C., Ghanem, N., Högerle, S., Moser, E., and Langer, M. 2001. Comparative detectability of bone metastases and impact on therapy of magnetic resonance imaging and bone scintigraphy in patients with breast cancer. *Eur. J. Radiol.* 40:16–23.

Aydin, A., Yu, J.Q., Zhuang, H., and Alavi, A. 2005. Detection of bone marrow metastases by FDG-PET and missed by bone scintigraphy in widespread melanoma. *Clin. Nucl. Med.* 30:606–607.

Berenson, J.R. 2005. Recommendations for zoledronic acid treatment of patients with bone metastases. *The Oncologist* 10:52–62.

Blake, G.M., Park-Holohan, S.J., and Fogelman, I. 2002. Quantitative stud-ies of bone in postmenopausal women using (18) F-fluoride and (98 m) Tc-methylene diphosphonate. *J. Nucl. Med.* 43:338–345.

Bockisch, A., Beyer, T., Antoch, G., Freudenberg, L.S., Kuhl, H., Debatin, J.F., and Muller, S.P. 2004. Positron emission tomography/computed tomography imaging protocols, artifacts, and pitfalls. *Mol. Imaging Biol.* 6:188–199.

Bombardieri, E., Aktolun, C., Baum, R.P., Bishof-Delaloye, A., Buscombe, J., Chatal J.F., Maffioli L., Moncayo R., Mortelmans L., and Reske SN. Bone scintigraphy—procedures guidelines for tumour imaging. (Retrieved January 21, 2006) http://www.eanm.org/scientific_info/ guidelines/gl_onco_bone.pdf.

Braun, S., Cevatli, B.S., Assemi, C., Janni, W., Kentenich, C.R., Schindlbeck, C., Rjosk, D., and Hepp, F. 2001. Comparative analysis of micrometastasis to the bone marrow and lymph nodes of node-negative breast cancer patients receiving no adjuvant chemotherapy. *J. Clin. Oncol.* 19:1468–1475.

Braun, S., Pantel, K., Muller, P., Janni, W., Hepp, F., Kentenich, C.R., Gastroph S., Wischnik, A., Dimpfl, T., Kindermann, G., Riethmuller, G., and Schlimok, G. 2000. Cytokeratin-positive cells in the bone mar-row and survival of patients with stage I, II, or III breast cancer. *N. Engl. J. Med.* 342:525–533.

Brown, M.L., O'Connor, M.K., Hung, J.C., and Hayostek, R.J. 1993. Technical aspects of bone scintigraphy. *Radiol. Clin. North. Am.* 31:721–730.

Buscombe, J.R., Holloway, B., Roche, N., and Bombardieri, E. 2004. Position of nuclear medicine modalities in the diagnostic work-up of breast cancer. *Q. J. Nucl. Med. Mol. Imaging.* 48:109–118.

Choy, D., Murray, I.P., and Hoschl, R. 1981. The effect of iron on the biodis-tribution of bone scanning agents in humans. *Radiology* 140:197–202.

Coleman, R.E. 1997. Skeletal complications of malignancy. *Cancer* 80:1588–1594.

Coleman, R.E. 2004. The role of bisphosphonates in breast cancer. *The Breast.* 13:S19–S28.

Cote, R.J., Rosen, P.P., Lesser, M.L., Old, L.J., and Osborne, M.P. 1991. Prediction of early relapse in patients with operable breast cancer by detection of occult bone marrow micrometastases. *J. Clin. Oncol.* 9:1749–1756.

Cummings, S.R., Bates, D., and Black, D.M. 2002. Clinical use of bone densitometry: scientific review. *J. Am. Med. Assoc.* 288:1889–1897.

Demirkan, B., Baskan, Z., Alacacioglu, A., Gorken, I.B., Bekis, R., Ada, E., Osma, E., and Alakavuklar, M. 2005. False negative bone scintigraphy in a patient with primary breast cancer: a possible transient phenome-non of bisphosphonate (alendronate) treatment. *Tumori.* 91:77–80.

Donohoe, K.J., Brown, M.L., Collier, B.D., Carretta, R.F., Henkin, R.E., O'Mara, R.E., and Royal, H.D. Society of Nuclear Medicine Procedure Guideline for Bone Scintigraphy. Society of Nuclear Medicine Procedure Guidelines Manual. (Retrieved January 21, 2006) http:// interactive.snm.org/docs/pg_ch34_0403.pdf.

Dose, J., Bleckmann, C., Bachmann, S., Bohuslavizki, K.H., Berger, J., Jenicke, L., Habermann, C.R., and Janicke, F. 2002. Comparison of flu-orodeoxyglucose positron emission tomography and "conventional diagnostic procedures" for the detection of distant metastases in breast cancer patients. *Nucl. Med. Commun.* 23:857–864.

Eubank, W.B., Mankoff, D.A., Vesselle, H.J., Eary, J.F., Schubert, E.K., Dunnwald, L.K., Lindsley, S.K., Gralow, J.R., Austin-Seymour, M.M., Ellis, G.K., and Livingston, R.B. 2002. Detection of locoregional and distant recurrences in breast cancer patients by using FDG-PET. *RadioGraphics* 22:5–17.

Fidler, I.J. 2003. The pathogenesis of cancer metastasis: The "seed and soil" hypothesis revisited. *Nat. Rev. Cancer* 3:453–458.

Fleisch, H. 1998. Biphosphonates: mechanisms of action. *Endocr. Rev.* 19:80–100.

Forstrom, L.A., Dunn, W.L., O'Connor, M.K., Decklever, T.D., Hardyman, T.J., and Howarth, D.M. 1996. Technical pitfalls in image acquisition, processing, and display. *Semin. Nucl. Med. 26*:278–294.

Gebauer, G., Fehm, T., Merkle, E., Beck., E.P., Lang, N., and Jager, W. 2001. Epithelial cells in bone marrow of breast cancer patients at time of primary surgery: clinical outcome during long term follow-up. *J. Clin. Oncol. 19*:3669–3674.

Gerber, B., Krause, A., Muller, H., Richter, D., Reimer, T., Makovitzky, J., Herrnring, C., Jeschke, U., Kundt, G., and Friese, K. 2001. Simultaneous immunohistochemical detection of tumor cells in lymph nodes and bone marrow aspirates in breast cancer and its correlation with other prognostic factors. *J. Clin. Oncol. 19*:960–971.

Gralow, J.R. 2001. The role of bisphosphonates as adjuvant therapy for breast cancer. *Curr. Oncol. Rep. 3*:506–515.

Gray, J.W. 2003. Evidence emerges for early metastasis and parallel evolution of primary and metastatic tumors. *Cancer. Cell 4*:4–6.

Hamaoka, T., Madewell, J.E., Podoloff, D.A., Hortobagyi, G.N., and Ueno, N.T. 2004. Bone imaging in metastatic breast cancer. *J. Clin. Oncol. 22*:2942–2953.

Hesslewood, S., and Leung, E. 1994. Drug interactions with radiopharmaceuticals. *Eur. J. Nucl. Med. 21*:348–356.

Hillner, B.E., Ingle J.N., Chlebowski, R.T., Gralow, J., Yee, G.C., Janjan, N.A., Cauley, J.A., Blumenstein, B.A., Albain, K.S., Lipton, A., and Brown, S. 2003. American Society of Clinical Oncology 2003 update on the role of bisphosphonates and bone health issues in women with breast cancer. *J. Clin. Oncol. 21*:4042–4057.

Hirabayashi, H., and Fujisaki, J. 2003. Bone-specific drug delivery systems: approaches via chemical modification of bone-seeking agents. *Clin. Pharmacokinet. 42*:1319–1330.

Hoffman, A., Stepensky, D., Ezra, A., Van Gelder, J.M., and Golomb, G. 2001. Mode of administration-dependent pharmacokinetics of bisphosphonates and bioavailability determination. *Int. J. Pharm. 220*:1–11.

Howarth, D.M., Forstrom, L.A., O'Connor, M.K., Thomas, P.A., and Cardew, A.P. 1996. Patient-related pitfalls and artifacts in nuclear medicine imaging. *Semin. Nucl. Med. 26*:295–307.

Jankovic, D.L., and Djokic, D.D. 2005. Alteration of the organ uptake of the 99mTc radiopharmaceuticals, 99mTc-DPD, 99mTc-DMSA, 99mTc-tin colloid and 99mTc-MAA, induced by the applied cytotoxic drugs methotrexate sodium and cyclophosphamide. *Nucl. Med. Communications 26*:415–419.

Jemal, A., Tiwari, R.C., Murray, T., Ghafoor, A., Samuels, A., Ward, E., Feuer, E.J., and Thun, M.J. 2004. Cancer statistics, 2004. *CA Cancer J. Clin. 54*:8–29.

Jilka, R.L. 2003. Biology of the basic multicellular unit and the pathophysiology of osteoporosis. *Med. Pediatr. Oncol. 41*:182–185.

Kanis, J.A. 2002. Diagnosis of osteoporosis and assessment of fracture risk. *Lancet 359*:1929–1936.

Koizumi, M., Yoshimoto, M., Kasumi, F., and Ogata, E. 2001. What do breast cancer patients benefit from staging bone scintigraphy? *Jpn. J. Clin. Oncol. 31*:263–269.

Kraus, J., Pantel, K., Pinkel, D., Albertson, D.G., and Speicher, M.R. 2003. High-resolution genomic profiling of occult micrometastatic tumor cells. *Genes Chromosomes Cancer 36*:159–166.

Lipton, A. 2003. Management of metasatic bone disease and hypercalcemia of malignancy. *Am. J. Cancer. 2*:427–438.

Loprinzi, C.C., Hayes, D., and Smith, T. 2003. Doc, shouldn't we be getting some tests? *J. Clin. Oncol. 21*:108s–111s.

Love, C., Din, A.S., Tomas, M.B., Kalapparambath, T.P., and Palestro, C.J. 2003. Radionuclide bone imaging: an illustrative review. *Radiographics 23*:341–358.

Lugo, T.G., Braun, S., Cote, R.J., Pantel, K., and Rusch, V. 2003. Detection and measurement of occult disease for the prognosis of solid tumors. *J. Clin. Oncol. 21*:2609–2615.

Maffioli, L., Florimonte, L., Pagani, L., Butti, I., and Roca, I. 2004. Current role of bone scan with phosphonates in the follow-up of breast cancer. *Eur. J. Nucl. Med. Mol. Imag. 31* (Suppl. 1):S143–S148.

Mansi, J.L., Gogas, H., Bliss, J.M., Gazet, J.C., Berger, U., and Coombes, R.C. 1999. Outcome of primary-breast-cancer patients with micrometastases: a long-term follow-up study. *Lancet 354*:197–2002.

Martin, T.J., and Moseley, J.M. 2000. Mechanisms in the skeletal complications of breast cancer. *Endocr. Relat. Cancer 7*:271–284.

Moore, A.E., Hain, S.F., Blake, G. M., and Fogelman, I. 2003. Validation of ultrafiltration as a method of measuring free 99mTc-MDP. *J. Nucl. Med. 44*:891–897.

Morgan-Parkes, J.H. 1995. Metastases: mechanism, pathways, and cases. *Am. J. Radiol. 164*:1075–1082.

Murphy, K.J., Line, B.R., and Malfetano, J. 1997. Etidronate therapy decreases the sensitivity of bone scanning with methylene diphosphonate labelled with technetium-99m. *Can. Assoc. Radiol. J. 48*:190–202.

Murray, I.P.C. 1994. Bone Scintigraphy: the procedure and interpretation. In Murray, I.P.C., Ell, P.J., and Strauss, H.W. (Eds.), *Nuclear Medicine in Clinical Diagnosis and Treatment. Volume 2.* New York: Churchill Livingstone, pp. 909–934.

Myers, R.E., Johnston, M., Pritchard, K., Levine, M., Oliver, T., and the Breast Cancer Disease Site Group of the Cancer Care Ontario Practice Guidelines Initiative. 2001. Baseline staging tests in primary breast cancer: a practice guideline. *Can. Med. Assoc. J. 164*:1439–1444.

Nakai, T., Okuyama, C., Kubota, T., Yamada, K., Ushijima, Y., Taniike, K., Suzuki, T., and Nishimura, T. 2005. Pitfalls of FDG-PET for the diagnosis of osteoblastic bone metastases in patients with breast cancer. *Eur. J. Nucl. Med. Mol. Imag. 32*:1253–1258.

O'Sullivan, J.M., and Cook, G.J. 2002. A review of the efficacy of bone scanning in prostate and breast cancer. *Q. J. Nucl. Med. 46*:152–159.

Ozturk, E., Mohur, H., Arslan, N., Entok, E., Tan, K., and Ozguven, M.A. 2004. Quantitative three-phase bone scintigraphy in the evaluation of intravenous regional blockade treatment in patients with stage-I reflex sympathetic dystrophy of upper extremity. *Ann. Nucl. Med. 18*:653–658.

Palli, D., Russo, A., Saieva, C., Ciatto, S., Roselli del Turco, M., Distante, V., and Pacini, P. 1999. Intensive vs. clinical follow-up after treatment of primary breast cancer: 10-year update of a randomized trial. National Research Council Project on Breast Cancer Follow-up. *J. Am. Med. Assoc. 281*:1586–1592.

Parfitt, A.M. 2002. Targeted and nontargeted bone remodeling: relationship to basic multicellular unit origination and progression. *Bone 30*:5–7.

Park, C.H., Kim, H.S., Shin, H.Y., and Kim, H.C. 1997. Hepatic uptake of Tc-99m MDP on bone scintigraphy from intravenous iron therapy (Blutal). *Clin. Nucl. Med. 22*:762–764.

Pecherstorfer, M., Schilling, T., Janisch, S., Woloszczuk, W., Baumgartner, G., Ziegler, R., and Ogris, E. 1993. Effect of clodronate treatment on bone scintigraphy in metastatic breast cancer. *J. Nucl. Med. 34*:1039–1044.

Ramaswamy, S., Ross, K.N., Lander, E.S., and Golub, T.R. 2003. A molecular signature of metastasis in primary solid tumors. *Nat. Genet. 33*:1–6.

Rodan, G. 2003. The development and function of the skeleton and bone metastases. *Cancer 97*:726–732.

Roodman, G. 2004. Mechanisms of bone metastases. *N. Engl. J. Med. 350*:1655–1664.

Roodman, G.D., and Windle, J.J. 2005. Paget disease of bone. *J. Clin. Invest. 115*:200–208.

Rosen, L.S., Gordon, D., Kaminski, M., Howell, A., Belch, A., Mackey, J., Apffelstaedt, J.,Hussein, M.A., Coleman, R.E., Reitsma, D.J., Chen, B.L., and Seaman J.J. 2003. Long-term efficacy and safety of zoledronic acid compared with pamidronate disodium in the treatment of skeletal complications in patients with advanced multiple myeloma or breast carcinoma. A randomized, double-blind, multicenter, comparative trial. *Cancer 98*:1735–1744.

Rostom, A.Y., Powe J., Kandil, A., Ezzat, A., Bakheet, S., El-Khwsky, F., El-Hussainy, G., Sorbris, R., and Sjoklint, O. 1999. Positron emission tomography in breast cancer: a clinicopathological correlation of results. *Br. J. Radiol. 72*:1064–1068.

Roudier, M.P., Vesselle, H., True, L.D., Higano, C.S., Ott, S.M., King, S.H., and Vessella, R.L. 2003. Bone histology at autopsy and matched bone scintigraphy findings in patients with hormone refractory prostate cancer: the effect of bisphosphonate therapy on bone scintigraphy results. *Clin. Exp. Metastasis 20*:171–180.

Ryan, P.J., and Fogelman, I. 1996. Isotope imaging. *Basillieres. Clin. Rheumatol. 10*:589–613.

Sampson, C.B. 1993. Adverse reactions and drug interactions with radiopharmaceuticals. *Drug Saf. 8*:280–294.

Sarikaya, I., Sarikaya, A., and Holder, L.E. 2001. The role of single photon emission computed tomography in bone imaging. *Semin. Nucl. Med. 31*:3–16.

Savelli, G., Maffioli, L., Maccauro, M., De Deckere, E., and Bombardieri, E. 2001. Bone scintigraphy and the added value of SPECT (single photon emission tomography) in detecting skeletal lesions. *Q. J. Nucl. Med. 45*:27–37.

Scharf, S., and Zhao, Q.H. 1999. Radionuclide bone scanning in routine clinical practice. *Prim. Care. Pract. 3*:521–528.

Smith, S.E., Murphey, M.D., Motamedi, K., Mulligan M.E., Resnik, C.S., and Gannon, F.H. 2002. Radiologic spectrum of Paget disease of bone and its complications with pathologic correlation. *RadioGraphics 22*:1191–1216.

Smith, T.J., Davidson, N.E., Schapira, D.V., Grunfeld, E., Muss, H.B., Vogel V.G. 3rd, and Somerfield, M.R. for the ASCO Breast Cancer Surveillance Expert Panel. 1999. American Society of Clinical Oncology 1998 update of recommended breast cancer surveillance guidelines. *J. Clin. Oncol. 17*:1080–1082.

Soderlund, V. 1996. Radiological diagnosis of skeletal metastases. *Eur. Radiol. 6*:587–595.

Stokkel, M.F., Valdes Olmos, R.A., Hoefnagel, C.A., and Richel, D.J. 1993. Tumor and therapy associated abnormal changes on bone scintigraphy: old and new phenomena. *Clin. Nucl. Med. 18*:821–828.

Subramanian, G., and McAfee, J.G. 1971. A new complex of 99mTc for skeletal imaging. *Radiology 99*:192–196.

Van der Hoeven, J.J.M., Krak, N.C., Hoekstra, O.S., Comans, E.F.I., Boom, R.P.A., Van Geldere, D., Meijer, S., Van der Wall, E., Buter, J., Pinedo, H.M., Teule, G.J.J., and Lammertsma, A.A. 2004. ^{18}F-2-fluoro-2-deoxy-D-glucose positron emission tomography in staging of locally advanced breast cancer. *J. Clin. Oncol. 22*:1253–1259.

Van der Wall, H. 1994. The evaluation of malignancy: metastatic bone disease. In: Murray, I.P.C., Ell, P.J., and Strauss, H.W. (Eds.), *Nuclear Medicine in Clinical Diagnosis and Treatment, Volume 2*. New York: Churchill Livingstone, pp. 949–962.

Van Poznak, C. 2001. How are biphosphonates used today in breast cancer clinical practice? *Semin. Oncol. 28*:69–74.

Van Poznak, C., and Sauter, N.P. 2005. Clinical management of osteoporosis in women with a history of breast carcinoma. *Cancer 104*:443–456.

Wiedswang, G., Borgen, E., Karesen, R., Kvalheim, G., Nesland, J.M., Qvist, H., Schlichting, E., Sauer, T., Janbu, J., Harbitz, T., and Naume, B. 2003. Detection of isolated tumor cells in bone marrow is an independent prognostic factor in breast cancer. *J. Clin. Oncol. 21*:3469–3478.

Woelfle, U., Cloos, J., Sauter, G., Riethdorf, L., Janicke, F., van Diest, P., Brakenhoff, R., and Pantel, K. 2003. Molecular signature associated with bone marrow micrometastasis in human breast cancer. *Cancer Res. 63*:5679–5684.

Yip, C.H., and Paramsothy, M. 1999. Value of routine 99mTc-MDP bone scintigraphy in the detection of occult skeletal metastases in women with primary breast cancer. *The Breast 8*:267–269.

15

Improved Sensitivity and Specificity of Breast Cancer Thermography

E.Y-K. Ng

Introduction

Breast cancer is one of the most common malignancies among women in the world today (Dixon, 1999). It constitutes 18% of cancer among women, and this figure is increasing every year. In Singapore, the figure stands at 20% (Breast Cancer Foundation, 2005). Each year, more than 1000 women are diagnosed with breast cancer, and the age group with the highest incidence is the fifties to sixties category.

Breast cancer is caused by the epithelial cells, which are found along the terminal duct lobular unit (Dixon, 1999). In general, breast cancer can be categorized as invasive and noninvasive based on the cells' characteristic pattern. The abnormal growth of epithelial cells can lead to the forming of tumors (Hirshaut and Pressman, 1996). If the growth is restricted, the cancer is classified as benign. If the growth is rapid or has the ability to progress to other regions of the body, it is classified as malignant or carcinoma. More often than not, the malignant breast is firm and irregular in shape. However, confirmation can only be made after mammography and clinical examination (biopsy).

It is difficult to detect breast cancer. Currently, many methods are available for breast cancer detection. Figure 112 summarizes these methods based on the underlying principles of utilizing wave theory, heat energy, audio/magnetic field, and electric properties; many of these methods are discussed in this volume. At present, mammography is the gold standard for breast cancer detection. The other methods mainly play a complementing role by providing additional critical data for breasts unsuitable or difficult to be analyzed using X-rays (Fok et al., 2003).

Thermography is a noninvasive screening method that is economical and quick and does not inflict any pain on the patient. It is a relatively straightforward imaging method that detects the variation of temperature on the surface of human skin. Thermography is widely used in the medical arena. However, thermograms alone are not sufficient for the medical practitioner to make a diagnosis since thermography is a test of physiology and is complementary to the other techniques. Hence, thermography, when used adjunctively with other laboratory and outcome assessment tools such as anatomical techniques including mammography, ultrasound, and computed tomography (CT) scanning, may contribute to the best possible evaluation of breast health. In this chapter, an integrative biostatistical method comprising regression, artificial neural network (ANN), and receiver operating characteristic (ROC) approaches are utilized to analyze the thermograms.

Thermography application in breast cancer is a procedure that applies a thermal imager to detect and record the heat pattern of the breast surface. The underlying principle of this approach is that if there is a tumor beneath the breast, it will yearn for more nutrients to supplement its growth (with angiogenesis). As a result of this increase in metabolism rate, the tumor site's temperature will tend to increase through heat transfer progresses from the tumor in all directions; thus, the effect of tumor is not to be seen as a precise localized hot spot, but as a diffused area on the breast surface (Ng and Sudharsan, 2001a).

Breast thermography holds great potential in detecting an early breast lump. In fact, it has been reported that it can detect breast cancer 10 years earlier than the traditional golden method—mammography (Dixon, 1999). Although research and investigation of thermography application in breast cancer has been ongoing for the last three decades, the results have not been consistent thus far. Clinical examination has an accuracy rate of ~ 70% (Breast Cancer Foundation, 2005). It is desired that thermography equals or exceeds this accuracy rate with high sensitivity and specificity. However, due to inconsistencies in diagnosis from breast cancer thermograms, it is not yet used in Singapore as an adjunct tool to mammography.

The present work thus seeks to achieve an improved and reliable level of consistency in the use of breast cancer thermography by virtue of a novel and unique approach encompassing biostatistical method and ANN. More specifically for ANN, the use of RBFN (Battelle, 2005) will be the main focus. Neural network (NN) is a pattern recognition program that has the ability to predict the outcome based on the various inputs fed into the program. Hence, it will possess the ability to predict whether the breast is healthy or cancerous. Linear regression is incorporated to increase the accuracy of the results by selecting only useful and relevant inputs to predict the outcome.

Image Analysis Tools

Thermography

Infrared thermography was originally developed for military purposes. In recent times, its applications have been extended to engineering applications and medical imaging. Infrared thermography makes use of a thermal imager to detect the infrared radiation and to measure the heat pattern of the object surface or human skin (Ng, 2005; Ng and Kaw, 2005; Ng and Kee, 2005). It is passive in nature—that is, it will not emit any harmful radiation or subject the patient to any risk. Hence, it is considered to be a physiological test compared to anatomical tests such as CT imaging or X-rays. In addition, it is a noncontact screening process, making it a hygienic procedure. Other advantages of thermography include high portability and real-time imaging, which made it possible for the data to be recorded in a computer for processing.

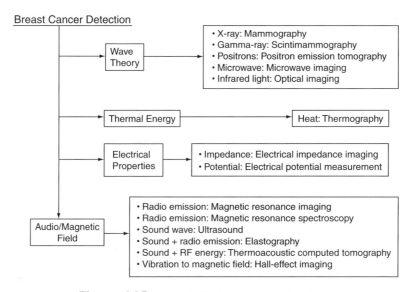

Figure 112 Methods for breast cancer detection.

The imager converts the thermal energy to electrical signals in order to display the temperature profile of the subject. The temperature is an illustration of many colors, each of which indicates a particular temperature. The thermal imager should be used in an indoor environment where external factors such as ambient temperature, humidity, and electrical sources can be controlled. The target distance between the imager and the target is usually between 0.5 m and 6 m. The selected distance should optimize the resolution of the color display of heat patterns. The ideal ambient condition is between 20°C and 25°C. As for relative humidity, it should range from 40–60%.

Artificial Neural Networks

Artificial neural networks (ANNs) are a group of techniques for numerical learning (Hopgood, 2000). They consist of many nonlinear computational elements called neurons. These neurons, also known as network nodes, are linked to one another. Through this weighted interconnection, they form the main architect of the network. It is not realistic to expect ANN to emulate biological NN, which is responsible for the behavior of humans and animals. However, ANN has the capability to assist us in some tasks. This includes nonlinear estimation, classification, clustering, and content-addressable memory.

Two or more inputs are connected to a node in an ANN (Hopgood, 2000). Each of them has a weighted linkage attached to it. Based on the input values, a node has the ability to perform simple calculation. Both inputs and outputs are real numbers or integers between −1 and 1. All the input data have to be normalized before being fed into the program. The output from one individual node can either be inputted into another node or be part of the NN's overall output. Each node performs its calculation and function independently from the rest of the nodes. The only association between the nodes is that the output from a node might be the input for another node. This type of architect is also known as a parallel structure, which allows for the exploration of numerous hypotheses. In addition, this parallel architect also allows the NN to make full use of conventional personal computers.

ANN's best advantage is that the tolerance of failure of an individual node or neuron is relatively high. This includes the weighted interconnection, because it might be erroneous too. The weights can be obtained through use of a trained algorithm and through iteration and adjustments. The eventual transfer function is obtained with regard to the desired output. When given a set of inconsistent or incomplete data, ANN is able to give an approximate answer rather than a wrong answer. The performance of the network will undergo a gradual degradation should there be any failures from individual nodes in the network. This is very useful in the medical arena because it is often difficult to run a comprehensive test. The

disadvantage of using ANN is that it does not have the capability to predict and forecast accurately beyond the range of previously trained data. In other words, the predicted outcome is based on the available set of data.

Backpropagation

Backpropagation (back-prop) or BP is a general-purpose network paradigm (Hopgood, 2000). It is a very popular algorithm used for system modeling, classification, and forecasting. The algorithm is based on a gradient-proportional descent function, and it is characterized as continuous and differentiable. One appropriate candidate is the sigmoid function, $f_t(a)$ (a is an activation), and its derivative is given by

$$f_t'(a) = f_t(a) (1 - f_t(a))$$

The BP algorithm can be further improved by the optimization of weights. A good example is the delta rule, which has the ability to deduce the errors of the weights (hidden layers), δ_{Aj},

$$\delta_{Aj} = f_t'(y_{Aj}) \Sigma \delta_{Bi} w_{Bij}$$

This describes δ for neuron j in layer A, which connects to i neurons in layer B, where $B = A + 1$; y_{Ai} = output from node i on level A; δ_{Bi} = error term associated with node i on level B; w_{Bij} = weight between node i on level B and node j on level B-2. In general, the gradient-proportional descent method has its limitations in that it is not effective in the minimum's region. To remedy this weakness, a momentum term is usually added into the parent algorithm. This will result in the weights being dependent on previous alterations in weights. The value of momentum lies between the range of 0 and 1. The value 0 implies no momentum, while the value 1 signifies the maximum that can be allocated. Many researches have used the value 0 for the first few training cycles and progressively increase to 0.9.

Backpropagation is a nonlinear regression technique. In other words, it does not assume that the relationship between the input variables and output is linear. The network's main objective is to minimize the global error. It has the ability to analyze very diverse, complicated data, and thus it can mimic multidimensioned functions, which are very useful especially in medical and engineering applications. However, for complicated or noisy data, BP will require hundreds of thousands of training cycles, and the central processing unit times will take a substantial amount of time to process. It is hard to predetermine the number of hidden layers and the amount of processing elements for each hidden layer. Often, the trial and error approach has to be used. Mathematically, BP learning has the danger of getting trapped in a local minimum, and this will prevent it from minimizing the global error.

Each layer of processing elements (PEs) is interconnected to the next layer. Each layer's PEs, however, is not connected to one another. As the name of this type of NN suggests, information is propagated or sent back to and from the network during the learning phase. This will ensure that the connection weights are set in such a way that the desired outcome will be produced. In many BP networks (BPNs), the hyperbolic tangent function is used as the transfer function. The output of this differentiable function ranges from −1 to 1, which is the scale that many users prefer the outcome to be. In many engineering and medical classification applications, a three-layered BP is deemed to be enough to solve complex problems.

Radial Basis Function Network

Radial basis function network (RBFN) is a kind of feedforward and unsupervised learning paradigm. A simple RBFN consists of three separate layers—input, hidden, and output. The first part of the training cycle involves the clustering of input neurons. Mathematically, the clustering is done using the Dynamic K-Means algorithm (Hopgood, 2000). At the end of the clustering process, the radius of the Gaussian functions at the middle of the clusters will be equivalent to the distance between the two nearest cluster centers. During the training, the RBFN will be required to accomplish two tasks: (1) to determine the middle of each hypersphere and (2) to obtain its radius. The first task is done by allocating the weights of the PEs. This can be achieved by using an unsupervised clustering algorithm. It is important to note that the output neuron in the prototypical layer of a RBFN is in a function of the Euclidean distance. This distance measures from the input neuron to the weighted neuron. The unsupervised learning phase in the hidden layer of RBFN is followed by another different supervised learning phase. It is at this stage that the output neurons will be trained to associate each individual cluster with their own distinct shapes and sizes.

Normally, the training speed of RBFN is faster than the backpropogation networks (BPN). It has the ability to detect data that are not within the norm, and thus it can make better decisions during classification problems. Nevertheless, the first few thousand training cycles of RBFN are unsupervised, and as a result, important information could be overlooked. The bounded transfer function may hinder the network's ability to solve regression problems, and when compared to BP, it is not as effective in providing a compact distributed function. The input and output neurons of RBFN and perceptron are alike (Hopgood, 2000). The major difference lies in the hidden neuron. In most cases, it is governed by the Gaussian function. This is different from other processing neurons that produce an output based on the weighted sum of the inputs. On the contrary, the input neurons of the RBFN are not involved in the processing of information. Their sole function is to input the given data to the receiving nodes. Using a lin-

ear transfer function, these receiving nodes will decide the weights to be allocated to each PE that follows. They are governed by the following transfer functions:

$$y_i = f_r(r_i) \qquad r_i = \sqrt{\sum_{j=1}^{n}(x_j - w_{ij})^2}$$

where w_{ij} represents the amount of weights allocated to the inputs of the neuron i, and f_r represents the Gaussian function, which is the preferred choice of most researchers.

$$f_r(r_i) = exp(-r_i^2/2\sigma_i^2)$$

where σ_i represents the standard deviation of the Gaussian distribution. Every neuron at each hidden layer will have its own unique σ_i value.

Biostatistical Methods

Linear Regression Analysis

Regression analysis, also known as least squares regression, is a statistical technique used to determine the unique curve or line that "best fits" all the data points. The underlying principle is to minimize the square of the distance of each data point to the line itself. In regression analysis, there are two variables: dependent and independent. The dependent variable is to be estimated or predicted. The most important result obtained in the analysis is the R-squared, or coefficient of determination. R-squared is an indication of how tightly or sparsely clustered the data points are, and it is a value that lies between 0 and 1. In other words, it is a measure of the correlation between the two variables. Correlation refers to the predictability of the change in the dependent variable given a change in the independent variable.

Linear regression (LR) is using a straight line to fit the data points. It is a simple, yet effective, way to obtain the correlation between two variables. However, a few assumptions are made in using LR. First, a linear relationship is assumed between the two variables, which might not always be the case. Second, the dependent variable is assumed to be normally distributed, with the same variance with its corresponding value of independent variable. Mathematically, the LR model is given by $Y = Ax + B$.

Receiver Operating Characteristic Curves

Receiver operating characteristic (ROC) curves are used to assess the diagnostic performance of a medical test to discriminate unhealthy cases from healthy cases (ROC, 2005). Very often a medical test, perfect separation between unhealthy and healthy cases, is not possible if we are to discriminate them based on a threshold value. To illustrate this phenomenon, let's call the threshold value γ in which the majority of those without the disease will be correctly diagnosed as healthy (TN). Similarly, the majority of those with the disease will be correctly diagnosed as unhealthy (TP). However, there will also be one group of diseased patients wrongly diagnosed as healthy (FN) and one group of healthy patients wrongly diagnosed as unhealthy (FP). Table 18 summarizes all the possibilities—TN, TP, FN, and FP and their respective algebraic representation. With that, four important criterions can be defined—sensitivity, specificity, positive-predictive Value (PPV), and negative-predictive Value (NPV)—and they are commonly used in ROC analysis to assess the credibility of the test. The mathematical formulas are summarized as follows:

▲ Sensitivity: the probability that the test is positive in the unhealthy population = $a/(a + b)$.
▲ Specificity: the probability that the test is negative in the healthy population = $d/(c + d)$.
▲ PPV: given a positive forecast, the probability that it is correct = sensitivity/$(1 - $ specificity$)$.
▲ NPV: given a negative forecast, the probability that it is correct = $(1 - $ sensitivity$)$/specificity.

In the ROC curves analysis result, both sensitivity and specificity are displayed for all criteria. This allows the user to choose the optimum criterion, which ought to have high values for both sensitivity and specificity. The value of sensitivity is inversely proportional to that of specificity. This principle can be easily illustrated by the threshold value γ. A low γ will ensure that those with the disease will be detected. But this will also cause those without the disease to be classified as diseased. On the other hand, a high γ will allow correctly categorizing the healthy group but will miss out on the diseased group.

An example of the user-iterative ROC curve can be found in ROC (2005). The vertical axis is the sensitivity, while the horizontal axis shows the $(100 - $ specificity$)$. Once again this reinforces the trade-off between sensitivity and specificity. The area under the ROC curve is important information obtained in the analysis. The value lies from 0.5–1. A value of 0.5 implies that the test cannot discriminate the unhealthy from healthy group, whereas a value of 1 implies that the test can distinguish the two groups perfectly.

Data Acqusition

Data collection for breast cancer was done from the Department of Radiology Diagnostics, Singapore General hospital (Ng *et al.*, 2001, 2002). Ninety patients for breast thermography were chosen at random. It was ensured that patients were within the recommended period of the 5th to 12th and 21st day after the onset of the menstrual cycle for breast thermography since the vascularization is at a basal level with least engorgement of blood vessels (Ng *et al.*, 2001). The accuracy of thermography in women whose thermal images are taken in a suitable period is higher (80%) than the total population of patients (73%). For the analysis here, thermograms of 8 patients are excluded for analysis as 7 of the patients have a history of mastectomy and 1 patient has a highly distorted breast on one side. Thus, thermograms of 82 patients were used in the analysis: 30 asymptomatic patients (age = 51 ± 8), 48 patients (age = 46 ± 10) with benign breast abnormality on either side of the breasts, and 4 patients (age = 45 ± 5) with cancer on either side of the breasts.

The thermal imager used was Avio TVS-2000 MkII ST (Fok *et al.*, 2002). It possesses a wide range of capabilities, including image enhancement, freeze-frame mode, automated tracking of the heat pattern, and recording. The venue was an indoor environment where the room temperature was between 20°C and 22°C (within ± 0.1°C) and the humidity was about 60% ± 5%. Heat sources such as sunlight or other electrical appliances were reduced to a minimum due to its effect on the ambient temperature. Prior to the screening, the patients were instructed to abstain from alcohol, cigarettes, and any form of drugs that affect the body's biological system and thereby result in a change in body temperature. In addition, the patient's breast surface had to be free from powder or ointments.

Procedures for Thermal Imaging

The patients are required to abstain from any physical activities for 20 minutes before the start of thermal screening (Ng *et al.*, 2001). The rationale is to reduce the body's metabolism rate so as to allow the overall body temperature to stabilize. During the thermographic examination, the patients are required to take off their top clothing, and their hands will be positioned behind their heads. During the imaging, three thermograms are taken: one frontal image and two lateral images. Each image is then improvised digitally to enhance the resolution.

Figures 113a and b show two typical examples of thermograms—healthy and cancerous. In the healthy state, an

Table 18 Basic Mathematical Formulas for ROC Analysis

Test Result	Disease Present	Number n	Disease Absent	Number m	Total
Positive	True-positive (TP)	a	False-positive (FP)	c	$a + c$
Negative	False-negative (FN)	b	True-negative (TN)	d	$b + d$
Total		$a + b$		$c + d$	

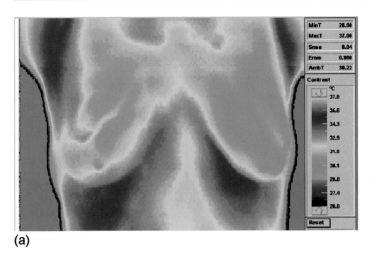
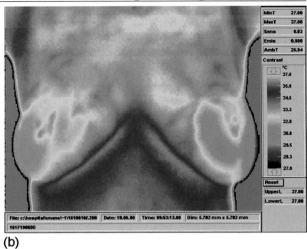

(a)   **(b)**

Figure 113 Typical normal and abnormal thermograms. (a) Symmetrical thermogram of a healthy 50-year-old female. (b) Cancerous thermogram of a 50-year-old (invasive lobular Ca grade I in left lateral breast).

individual's breasts are generally symmetrical, although they may differ slightly in size, and the breast outline as portrayed by the thermogram is smooth and regular with a convex contour. The thermal image and quantitative levels of heat are usually similar but never identical in the two breasts. The background on which the breast is superimposed is cool compared with the upper thorax and the inframammary fold, and the nipple itself stands out as a round cool spot, frequently outlined by a faintly warmer zone. The distinguishing hallmark of abnormality (cancerous), however, is vascular discrepancy encompassing

(a) Differences in the number and calibre of veins and their quantitative thermal measurements.

(b) Focal areas of heat whether periareolar or localized elsewhere and not necessarily identified with a particular vein.

(c) Diffuse increase in background temperature occasionally associated with breast enlargement. An additional abnormality is the loss of the regular convex contour of the breast outline referred to as the "edge sign."

Temperature data are extracted from the breast thermograms. The thermograms consist of many colored pixels, each representing a temperature. From the thermograms alone, it is possible for an experienced medical practitioner to diagnose abnormalities such as a cyst. After every pixel's temperature is compiled, biostatistical technique can be used to analyze them, such as determining the mean, median, and modal temperature of the breast region.

Designed Integrated Approach

The proposed advanced approach is a multipronged method comprising LR, RBFN, and ROC analysis. It is a novel and powerful integrated technique that can be used to analyze complicated and large measured data.

Step 1: Linear Regression

Linear regression (LR) reflects the correlation between the variables and the actual health status (healthy or cancerous) of the subject, which is decided by mammography and biopsy. Hence, LR is used to decide if a particular variable should be used for inputs in the Train File. In other words, a variable will be used as input in the NN if and only if it has a strong correlation with the outcome (health status of the patient).

The following data were compiled and collected from each subject (Ng *et al.*, 2001, 2002):

Temperature Data from Thermograms
Mean temperature of left breast; mean temperature of right breast; median temperature of left breast; median temperature of right breast; modal temperature of left breast; and modal temperature of right breast.

Biodata from questionnaire
Age of patient; family history of breast cancer; hormone replacement therapy; age of menarche patient; presence of palpable lump; previous breast surgery/biopsy; presence of nipple discharge; pain in the breast; menopause at the age of > 50 years, and first child at the age of > 30 years old.

Step 2: ANN RBFN/BPN

Based on the various inputs fed into the network, RBFN is trained to produce the desired outcome, which is either positive (1) for cancer and benign cases or negative (0) for healthy cases. Different combinations of Learn Rule, Transfer Rule, and Options will be tested under RBFN's wide umbrella. When this is done, the RBFN algorithm will possess the ability to predict the outcome when there are new input variables. The advantages of using RBFN include fast training, superior classification, and decision-making abilities as compared to other networks such as BP. For this breast cancer study, conventional BP training (Ng et al., 2002) and testing are also included, and the results are compared with those of RBFN.

Step 3: ROC Analysis

Next, ROC is used to evaluate the accuracy, sensitivity, and specificity of the outcome of RBFN Test files to check whether the RBFN is well built.

Flowchart of the Proposed Method

The software needed for all the processes include:

▲ Image J: to view thermograms from thermal imager and to extract temperature data.
▲ MS Excel Statistical Toolbox: to normalize raw temperature data and to perform statistical analysis (e g., mean, median, standard deviation).
▲ MedCal: to determine the correlation of each variable with the output (health status).
▲ NeuralWorks Pro II: to provide training and testing of data; also, to build an algorithm for the data.
▲ MedCal: to evaluate the effectiveness of the computed method.

Figure 114 presents the entire process in a flowchart, including the steps prior to advanced integrated technique (AIT). Either RBFN or BPN can be used for the AIT.

Results and Discussion

Summarized Results for Step 1: Linear Regression

The percentage of score is obtained from the number of correct predictions over the total number of cases made by the ANN. Various combinations of Learn Rule, Transfer Rule, and Options were tested. In general, the coefficient of determination is low for both temperature-related data and biodata since the relationship between the variables and the health status (healthy or unhealthy) is not obvious. However, it is noted that the coefficient of determination for the temperature-related data is generally higher than that of the biodata. This reinforces the fact that thermography can be used as an adjunct tool as there suggests a strong correlation between the surface temperature of the breast and the health status of the patient. The variables with the highest and

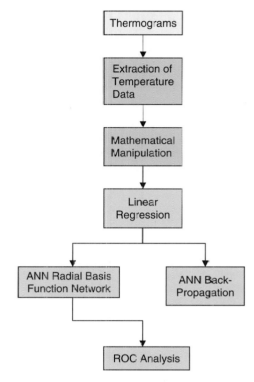

Figure 114 Flowchart of advanced integrated technique for BC thermograms. green boxes: pre-processing, blue boxes: advanced technique.

lowest coefficients of determination are the modal temperature of the right breast and the "First child at more than 30 years old" criterion.

Selected Results for Step 2: ANN RBFN/BPN

The highest level of accuracy attained by both ANN RBFN and BP is 80.95%. However, as many as 37 RBFN have achieved 80.95%, whereas only 9 BPNs manage to achieve this score. Hence, the RBFN is superior and credible to BP in the prediction of the breast cancer, since data are rather complicated and large with 10 input variables (selected from LR).

Selected Results (with Area > 0.85) for Step 3: ROC Analysis

The accuracy rate in Step 2 is only based on the number of correct predictions. However, it does not take into account the percentage of correct predictions out of the positive cases or the percentage of correct predictions out of the negative cases. Hence there is a need for ROC analysis on the selected RBFN with a high accuracy rate, in order to further verify its effectiveness.

Evaluating the RBFN with ROC (Tables 19 and 20) shows that the NN model is well built. The area under the

Table 19 Selected Results for ROC Analysis for RBFN SLP, with Selected and Various Combination of Learn Rule and Transfer Rule (with Various Options Tested)

Learn Rule	Transfer Rule	Option	Area under Curve	Sensitivity	Specificity
Delta Bar Delta	TanH	—	0.888	75	94.1
Delta Bar Delta	TanH	Connect Prior	0.888	75	94.1
Delta Bar Delta	TanH	Connect Bias	0.899	81.2	94.1

Table 20 Selected Results for ROC Analysis for ANN RBFN MLP, with Selected Combination of Learn Rule, Transfer Rule, and Options Were Tested with Different Hidden Neurons

Learn Rule	Transfer Rule	Option	Hidden Neuron	Area under Curve	Sensitivity	Specificity
Delta Rule	DNNA	Connect Prior	1	0.866	78.1	94.1
QuickProp	Linear	Connect Bias	1	0.869	75	82.4
QuickProp	Linear	Connect Bias	3	0.858	75	88.2
QuickProp	Linear	Connect Bias	4	0.869	78.1	82.4
Delta Bar Delta	TanH	Connect Prior	0	0.899	81.2	84.1
Delta Bar Delta	TanH	Connect Prior	2	0.877	84.4	76.5
Delta Bar Delta	TanH	Connect Prior	4	0.914	65.6	100
Delta Bar Delta	TanH	Connect Prior	5	0.89	81.2	88.2

curve for most RBFNs is higher than 0.85. These RBFNs also achieved high sensitivities (> 75%) and high specificities (about 90%). This indicates that the overall diagnostic performance is competitive with that of mammography. The best performing RBFN is a multilayered perceptron (MLP) with Delta Bar Delta as the Learn Rule, TanH as the Transfer Rule, and Connect Prior as the selected Option. The number of hidden neurons is 5, and the ROC area achieves 0.89. Although its area under the curve is not the highest, it possesses very high sensitivity (81.2%) and high specificity (88.2%). In brief, the NNs with Delta Bar Delta as the Learn Rule outperformed other Learn Rules.

Ultimately, the proposed AIT analysis should be integrated for the clinical application. Figure 115 outlines a possible *clinical protocol* with the use of multimodality approaches, including numerical simulation (Ng and Sudharsan, 2000, 2001b, 2004), patients' biodata, laser Doppler perfusion imaging (LDPI) (Seifalian *et al.*, 1995), and the data interpretation of thermograms. To overcome a manual data analysis that is highly inefficient and prone to human errors, a computer-aided tool to assist the specialist in the analysis of thermograms is desirable. Figure 116 further shows the framework of the proposed system for analysis of the thermograms. The system relies on the current thermogram approach along with a database, which supplies clinical test results, patient records, physiological data, historical thermo-images, genetic information, and so on, to automatically analyze the likelihood of breast cancer development. The analysis module has been discussed earlier in "Designed Integration Approach." Besides the analysis, the system should allow for 3D reconstruction of suspected tumor development based on the heat patterns associated with the thermograms. This visualization module would enable the specialist to have a good picture of the location, size, and topology of the suspected tumor. The information would be useful, as further tests could then be conducted to search for tumor development in that area. One possible visualization module is documented as the Thermal Texture Mapping (TTM) algorithm by Qi *et al.* (2002) and Wang and Wang (2003).

Conclusions and Future Trends

Thermography measurements allow changes in tumor angiogenesis to be evaluated and may assist in dynamic monitoring of therapy. Its application is like an unpolished gemstone, waiting for us to unleash its full potential. The major potential problems are the perception and acceptance of clinical and insurance colleagues to the entire concept of thermographic evaluation. Through the use of ANN and biostatistical methods, progress is made in thermography application with regard to achieving a higher level of consistency. This is made possible with the introduction of the novel AIT in thermogram analysis. In all, the chapter investigated the diagnostic potential of infrared thermography in light of recent technological advances, utilizing existing advanced thermographic equipment and applying the multidisciplinary experience in data analysis to breast care. Importantly, the work does not seek to replace or provide an alternative

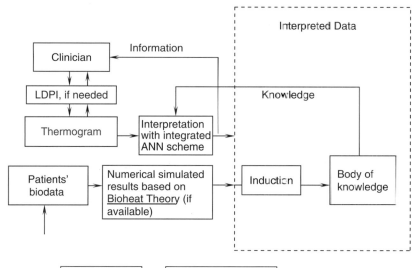

Figure 115 Outline of a clinical protocol with the use of multimodality environment, including simulation, patients' biodata, LDPI, and interpretation of thermograms.

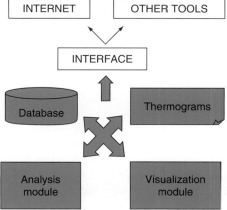

Figure 116 Framework for the computer-aided analysis of thermograms.

to existing mainstays of screening (clinical examination and mammography), but to provide a scientifically valid adjunct to existing breast cancer services.

The next research focus on breast cancer detection using thermography would be the attempt to pinpoint the actual location and size of the tumor (such as TTM) if the diagnosis is positive. This would definitely be a gigantic step forward in the fight against breast cancer.

Acknowledgments

The author would like to thank Mr. Kee E.C. for his help in data preparation; also, Drs. Ng F.C. and Sim L.S.J., Senior Consultant Radiologists, Department of Diagnostic Radiology of the Singapore General Hospital, Singapore, on their interpretation of postsurgery on vessel phenomenon/complications.

References

Battelle: http://www.battelle.org/pipetechnology/ (accessed November 10, 2005).

Breast Cancer Foundation: http://www.bcf.org.sg/ (accessed November 14, 2005).

Dixon, M. 1999. *ABC of Breast Diseases*. Edinburgh, Scotland: BMJ Publishing Group.

Fok, S.C., Ng, E.Y-K., and Tai, K. 2002. Early detection and visualization of breast tumor with thermogram and neural network. *J. Mech. Med. Biol.* 2:185–196.

Fok, S.C., Ng, E.Y-K., and Thimm, G.L. 2003. Developing case-based reasoning for discovery of breast cancer. *J. Mech. Med. Biol.* 3:231–246.

Hirshaut, Y, and Pressman, P. 1996. *Breast Cancer. The Complete Guide.* New York: Bantam Trade Paperback.

Hopgood, A. 2000. *Intelligent Systems for Engineers and Scientists.* Boca Raton, FL: CRC Press.

Ng, E.Y-K. 2005. Is thermal scanner losing its bite in mass screening of fever due to SARS? *Med. Phys.* 32:93–97.

Ng, E.Y-K., and Kaw, G.J.L. 2005. IR scanners as fever monitoring devices: physics physiology and clinical accuracy. In: Diakides, N. (Ed.), *Biomedical Engineering Handbook.* Boca Raton: CRC Press, pp. 1–20.

Ng, E.Y-K., and Kee E.C. 2007. Fever mass screening tool for infectious diseases outbreak: integrated artificial intelligence with bio-statistical approach in thermogram analysis. In: Diakides, N., and Bronzino, J.D. (Eds.), *Biomedical Engineering Handbook, Infrared Imaging Spin-off edition.* Boca Raton, Fl: CRC Press, pp. 16–19.

Ng, E.Y-K., and Fok, S.C. 2003. A framework for early discovery of breast tumor using thermography with artificial neural network. *Breast J.* 9:341–343.

Ng, E.Y-K., Fok, S.C., Peh, Y.C., Ng, F.C., and Sim, L.S.J. 2002. Computerized detection of breast cancer with artificial intelligence and thermograms. *Int. J. Eng. Med. Tech.* 26:152–157.

Ng, E.Y-K., and Sudharsan, N.M. 2000. Can numerical simulation adjunct to thermography be an early detection tool? *J. Thermol. Int.* (formerly *Eur. J Med. Therm.*) 10:119–127.

Ng, E.Y-K., and Sudharsan, N.M. 2001a. Effect of blood flow, tumour and cold stress in a female breast: a novel time-accurate computer simulation. *Int. J. Eng. Med.* 215:393–404.

Ng, E.Y-K, and Sudharsan, N.M. 2001b. Numerical computation as a tool to aid thermographic interpretation. *Int. J. Eng. Med. Tech.* 25:53–60.

Ng, E.Y-K, and Sudharsan, N.M. 2004. Numerical modelling in conjunction with thermography as an adjunct tool for breast tumour detection, *BMC Cancer, Medline J.* 4:1–26.

Ng, E.Y-K, Ung, L.N., Ng, F.C., and Sim, L.S.J. 2001. Statistical analysis of healthy and malignant breast thermography. *Int. J. Eng. Med. Tech.* 25:253–263.

Qi, H., Liu, Z.Q., and Wang, C. 2002. Breast cancer identification through shape analysis in thermal texture maps. Annual Int. Conference of the IEEE Engineering in Medicine Biological. Procedure 2: 1129–1130.

Receiver Operating Characteristics (ROC): http://www.medcalc.be/. (accessed December 13, 2005).

Seifalian, AM., Chaloupka, K., and Parbhoo, S.P. 1995. Laser Doppler perfusion imaging—a new technique for measuring breast skin blood flow. *Int. J. Microcirc. 15*:125–130.

Wang B., and Wang, Y.L. 2003. TTM analysis of 221 cases of microcirculation depression. Proc. of the 1st Conference of Thermal Texture Mapping (TTM) Technology in Medicine and Engineering, Houston, TX, pp. 105–107.

16

Optical Mammography

Sergio Fantini and Paola Taroni

Introduction

White-light transillumination was introduced into medicine in the early 1800s, but only in the late 1920s was it applied for breast imaging and visualization of breast lesions. In a darkened room, the breast was illuminated with powerful white light, and the transmitted image was observed directly by eye on the other side of the pendant breast, looking for shadows that might reveal the presence of a pathologic condition. The procedure was simple and inexpensive, and it proved useful to identify hematomas and liquid cysts. However, it did not reach wide applicability, because it did not enable discrimination between malignant and benign solid lesions, and some physicians had problems with image interpretation. In the 1980s, new systematic attempts were made to apply optical techniques to breast imaging, and led to a method of transillumination called diaphanography, or lightscanning. A tungsten filament lamp, filtered to select red and near-infrared (NIR) light, was used for illumination, typically through a fiber-optic handheld illuminator applied to the breast, and the breast shadow was recorded on an infrared-sensitive film or with a video camera connected to a black-and-white monitor and videotape recorder. In a diaphanography examination, light was diffused throughout the breast and randomly scattered. Opaque lesions formed shadows on the surface of the breast that acted as a screen. The deeper the lesion, the greater the

distance from the screen, and the less the contrast. This implies an inherent limitation of the technique, since small lesions will only appear with high contrast if they are not too far from the surface. Typically, high absorption, showing as a dark shadow, was regarded as the most specific sign of abnormality, but also asymmetry between the two breasts and abnormal vasculature (again visualized as dark shadows) were considered in the interpretation of the images.

Initial studies, performed on a small series of selected patients, suggested the possibility of successfully detecting solid breast tumors (Ohlsson *et al.*, 1984). Those positive outcomes fostered the research on diaphanography: commercial instruments became available and systematic trials were carried out, including blind studies performed prospectively in a screening population to compare diaphanography with X-ray mammography. Contrasting results were obtained. In some cases, optical methods compared favorably with X-rays (Wallberg *et al.*, 1985), while other studies were negative about the diagnostic potential of diaphanography (Monsees *et al.*, 1987). The latter studies highlighted a high number of "equivocal" or false-positive findings, which could only partially be retrospectively explained with technical limitations such as limited field of view, insufficient illumination level, or inadequate positioning of the light source. The sensitivity was also significantly lower than with mammography (about 60–70% versus 90–98%), especially when small lesions were considered. Moreover, previous knowledge of the lesion location seemed to be a critical factor, leading to much better results in retrospective analysis of data that were originally acquired and first analyzed prospectively (e.g., an increase in sensitivity from 76 –94% was achieved by Bartrum and Crow, 1984).

Despite the mixed results, even the researchers who would not recommend diaphanography as a screening method stressed its advantages in terms of noninvasiveness, good patient acceptance, quick and easy examination, and cost-effective instrumentation. They also suggested potential adjunct roles for diaphanography, like improving the positive yield of biopsies among patients recommended for surgery or following up equivocal lesions at frequent intervals. However, the negative results in the diagnostic discrimination of solid breast tumors discouraged further development of diaphanography.

A first reason for the failures of optical methods in their early application to the detection and classification of breast lesions is the insufficient training and experience of the investigators performing the examination and, more importantly, interpreting the images. But a second and more conceptual limitation of diaphanography was associated with the instrumentation used and the lack of a physical model to quantitatively describe light propagation in breast tissue. As a result, diaphanography could not address crucial challenges in optical imaging associated with the diffusive nature of light propagation in breast tissue, the strong attenuation of light, and the sensitivity to breast boundaries. Furthermore, diaphanography did not take full advantage of the diagnostic value of the spectral information that can be obtained with optical mammography. More recently, starting in the 1990s, technical improvements on both instrumental aspects and theoretical modeling have suggested possibilities to better exploit the optical properties of healthy and diseased breast for diagnostic purposes, and have opened new opportunities to the application of optical methods for breast cancer detection.

Sources of Intrinsic Optical Contrast in Breast Tissue

A practical problem associated with optical techniques is the strong light attenuation in biological tissues, which prevents whole-body examinations. Even in more favorable cases, dealing with a few centimeters of soft tissue, as in the case of breast imaging, the attenuation of ultraviolet or blue/green light is too strong for any practical application. As a result, red and near-infrared (NIR) light is always used in practice for its relatively low attenuation. The red-NIR spectral region, say from 600–1000 nm, is sometimes called the optical diagnostic window. As described above, diaphanography relied on the direct visualization of a shadow resulting from a localized increase in light attenuation. Two optical

phenomena contribute to the attenuation of light: absorption and scattering. Absorption annihilates the incoming photons, thus reducing the intensity of the transmitted light. Scattering events are essentially changes in the direction of propagation of photons. Even though photons continue to propagate in the medium, their change in direction contributes to the attenuation of the light transmitted through tissue along the original direction of propagation. To fully understand the diagnostic potential of optical data, one needs to consider the origin of both absorption and scattering phenomena.

Different substances typically absorb at different wavelengths. Thus, at least in principle, the measurement of the absorption spectrum of a medium allows for identification of its constituents. Moreover, the higher the amount of a specific constituent, the stronger its relative contribution to the overall absorption of tissue. Thus, the absorption properties of tissue also provide information on the concentration of its various constituents. Early studies have shown that hemoglobin contributes markedly to breast absorption at red and NIR wavelengths. This can have important diagnostic implications, since areas of high vascularization can be readily identified. Moreover, the two forms of hemoglobin (oxy-hemoglobin and deoxy-hemoglobin) have different absorption spectra. Therefore, changes in oxygen saturation levels can also be optically measured. Water and lipids show characteristic strong absorption peaks in the NIR. The balance of water and lipids in breast tissue depends on factors like age and hormonal status. Furthermore, the water and lipid content of pathologic lesions is likely to differ from that of healthy tissue, making its assessment potentially useful for diagnostic purposes. From an experimental point of view, to derive information on tissue composition, measurements must be performed at several wavelengths. In fact, the absorption coefficient μ_a of tissue (which is related to the absorption probability per unit distance) at any wavelength λ is due to the superposition of the contributions from its constituents, where each contribution is given by the product of the specific absorption ε and the concentration c of the constituent. In an equation:

$$\mu_a(\lambda) = \sum_i \varepsilon_i(\lambda) c_i \qquad (1)$$

In a simplified description of breast tissue, we can consider four main constituents absorbing significantly in the red and NIR, namely, oxy-hemoglobin (HbO_2), deoxy-hemoglobin (Hb), water, and lipids. Their specific absorption spectra are known from the literature. Thus, by measuring the absorption of breast tissue at a minimum of four different wavelengths, one can estimate the concentration of each constituent.

Scattering is essentially due to the presence of interfaces, or refractive index discontinuities, at a microscopic level. For visible and NIR light, scattering is believed to originate mostly from cell nuclei and subcellular organelles. The scattering properties are affected by both the size and density of these scattering centers. So their assessment can provide information on the microscopic structure of tissue and, in particular, on the local density of cellular nuclei and organelles. The wavelength dependence of the transport scattering coefficient μ_s' can be expressed as follows using a simple empirical approximation to Mie theory:

$$\mu_s'(\lambda) = a\lambda^{-b} \tag{2}$$

This description is in agreement with the experimental finding that the transport-scattering coefficient decreases progressively upon increasing wavelength, with no characteristic peaks. The scattering amplitude a provides information on the density of the scattering centers (higher values of a correspond to denser tissues), while the scattering power b is related to their size (smaller scattering centers lead to steeper slopes).

Because both absorption and scattering contribute to light attenuation, a direct visualization of the transmitted light as performed in diaphanography cannot enable the discrimination between the two optical phenomena, which have independent origins and can even compensate each other. For example, a highly vascularized region of low-density tissue combines a higher absorption associated with a higher blood concentration and a lower scattering associated with a lower density. These two effects tend to balance each other, leading to overall low contrast, if any. Hence, the possibility of discriminating between absorption and scattering is beneficial and, at least in principle, could increase the sensitivity of optical techniques. Providing more information on the nature of the detected abnormality could also aid its identification, thus affecting positively even the specificity to cancer detection. The assessment of the optical properties (i.e., absorption and scattering coefficients) of tissue *in vivo* in a clinical environment has become technically feasible in the last two decades. This implies the potential to derive noninvasively information on tissue composition and structure that can be profitably used for diagnostic purposes. Consequently, the biomedical community has shown a renewed interest in optical mammography, that is, optical imaging for the detection and characterization of breast lesions.

Principles of Optical Mammography

Breast imaging approaches can be classified into those based on a direct projection of optical data and those based on solution of the inverse imaging problem of tomography. The latter approaches yield a more rigorous spatial reconstruction of the breast optical properties, but are more complex in terms of data acquisition, analysis, and interpretation. Consequently, instruments for breast optical tomography have been developed only recently, and only initial studies on human subjects have been reported. By contrast, larger human studies based on projection imaging have been performed over the last 20 years.

For projection imaging, the breast is typically positioned between plane parallel plates, similar to what is done in conventional X-ray mammography, but with a much milder degree of compression so as not to cause discomfort to the patient even when the full examination requires several minutes to be completed. As already in the case of diaphanography, the light illuminates one side of the breast and the transmitted light is collected on the opposite side. However, discrete wavelengths are typically used, not broadband light. This is done to take advantage of the different absorption properties of tissue constituents at different wavelengths and derive their concentrations, as described earlier. Moreover, the breast is not fully illuminated. The light is generally coupled to an optical fiber that provides a relatively small (1–2 mm) illumination spot on the breast, and the light transmitted through the breast is collected with another optical fiber on the opposite side of the breast. Images are built by raster-scanning the two fibers in tandem over the compressed breast and collecting data every 1–2 mm. This measurement step determines the pixel size of the images. This may seem to set an undesired technical limit to the spatial resolution of optical images but this is not the case. In fact, at red and NIR wavelengths, the attenuation in breast tissue is dominated by scattering. Contrary to what occurs with X-rays, which mostly propagate straight through tissue, optical photons undergo hundreds of scattering events per centimeter traveled within tissue. So, a collimated light beam injected into tissue rapidly broadens upon propagation. As a consequence, the shadow cast by an optical inhomogeneity (e.g., a region of altered vascularization) is expected to appear bigger than the real size of the inhomogeneity, and the effect is more marked for deeper inhomogeneities. This sets a physical limitation to the spatial resolution that can be achieved by imaging at optical wavelengths.

The spatial resolution of optical mammography cannot be easily quantified because it depends on several factors (optical properties of the inhomogeneity and surrounding tissue, optical contrast, depth of the inhomogeneity). However, as a rule of thumb, we could say that an accurate estimate of the size is possible typically around 1 cm and above. Smaller objects, down to a few millimeters, can still be detected, provided that their optical contrast is large enough, but their image will be significantly bigger than their real size, thus preventing any accurate estimate of their dimensions. Consequently, optical imaging cannot compete with X-ray mammography on the ground of morphologic information. In particular, it will not be possible to visualize small calcifications that are a key element for diagnosing malignant lesions in X-ray images. However, optical data can provide different pieces of information. Specifically, functional information is available through the assessment of total hemoglobin content and oxygen saturation, and potentially of other constituents and related roles.

For breast optical tomography, the geometry of illumination and collection is more complex than for projection imaging, with a number of possible arrangements. For a parallel-plate geometry (similar to the case of projection imaging), more illumination and collection points are used at the planes of illumination and collection, whereas for a circular geometry, arrays of illumination and collection optical fibers are arranged around the pendulous breast. The larger number of illumination and collection points yields data that is suitable for tomographic image reconstruction of the breast optical properties.

Optical measurements are noninvasive, safe, and do not cause any discomfort to the patient. The optical instrumentation is relatively simple and cost-effective, as compared to other diagnostic imaging equipment that is routinely in a clinical setting. Consequently, optical imaging could be effectively applied even as a complementary technique. Moreover, promising preliminary results have recently been achieved in monitoring neoadjuvant chemotherapy, where repeated measurements can be performed with no risk for the patient, thus potentially allowing the optimized development of individual therapeutic protocols.

Continuous-wave Approaches: Dynamic Measurements and Spectral Information

While continuous-wave approaches to optical mammography are conceptually similar to the diaphanography techniques of the 1980s, a number of technical advances in the data collection and data analysis have been introduced since the 1990s to generate much richer functional information with respect to the transillumination images of diaphanography. The term *continuous-wave* indicates that the light source emission is constant with time. On the one hand, this fact limits the information content of the measured data to the overall attenuation (or optical density) contributed by the combination of absorption and scattering events within breast tissue. On the other hand, continuous-wave methods are technologically straightforward, provide a high signal-to-noise ratio, and ideally lend themselves to real-time dynamic measurements and to spectral measurements over a broad, continuous spectrum.

The real-time measurement of dynamic processes within the breast has the potential to provide novel functional information that was not previously sensed by other diagnostic imaging modalities. For example, the oscillatory hemodynamics associated with arterial pulsation or respiration may reflect the local vascular impedance and compliance, which can in turn be affected by cancerous modifications. Furthermore, the dynamic features of the return to equilibrium in response to a mechanical perturbation (for example, a transient application of pressure to the breast) may reveal spatial patterns that can be associated with the presence of cancerous lesions.

We have already discussed how multiwavelength data can provide functional and metabolic information, which represents the greatest promise of optical mammography for diagnostic imaging and distinguishes it from X-ray mammography and ultrasonography of the breast. The measurement of a continuous optical spectrum, as readily afforded by continuous-wave methods, is a most effective way to identify the relative concentrations of the various absorbing species in breast tissue. Within the spectral region considered in optical mammography, which is typically within the 600–1000 nm wavelength range, the absorption of deoxy-hemoglobin decreases with wavelength, with the exception of a peak at ~ 758 nm; the absorption of oxy-hemoglobin shows a broad valley with a minimum at ~ 692 nm and a broad peak with a maximum at ~ 924 nm; the absorption of water shows a relatively strong peak at ~ 975 nm; and the absorption of lipids shows a peak at ~ 924 nm. The scattering spectrum of breast tissue is featureless and decreases with wavelength (see Eq. (2)). It has a wavelength power dependence that is typically in the range $\lambda^{-0.4}$ to $\lambda^{-1.5}$ (Shah et al., 2004), which is a weaker wavelength dependence than the Rayleigh limit (~ λ^{-4}) for particles that are much smaller than the wavelength. Even though single-wavelength, continuous-wave measurements are generally not able to discriminate absorption from scattering contributions, full spectral data can accomplish such a discrimination by taking advantage of the featureless scattering spectrum of breast tissue. It is well-established that cancer is associated with a higher concentration of hemoglobin in breast tissue (Fantini et al., 1998; Tromberg et al., 2000; Grosenick et al., 2003; Dehghani et al., 2003), while it is still unclear whether hemoglobin saturation provides a reliable intrinsic source of contrast for cancer. With regard to water and lipids, case studies have indicated that cancer, relative to healthy breast tissue, typically has a higher water concentration (Tromberg et al., 2000; Jakubowski et al., 2004) and a lower lipids content (Jakubowski et al., 2004).

Time-resolved Approaches

Continuous-wave measurements of the attenuation of light transmitted through the breast do not allow one to fully exploit the diagnostic potential of optical mammography that is associated with separate measurements of absorption and scattering properties of breast tissue. Time-dependent methods, where the light source emission is not constant with time and the optical detection is time-resolved, afford the measurement of absorption and scattering features of breast tissue. Time-resolved approaches are implemented in the time domain or in the frequency domain. These two implementations differ in the instrumentation used, but the data collected in the time domain and frequency domain are mathematically related by a temporal Fourier transformation.

In time-domain measurements, a short light pulse (~ 100 ps duration) is injected into the tissue. Scattering and absorption events occurring during propagation through tissue cause attenuation, delay, and broadening of the injected pulse. From a qualitative point of view, we can say that the scattering essentially delays the detected pulse, as each scattering event changes the direction of photon propagation. Thus, photons move along "zigzag" trajectories that are much longer than the distance between the injection and the detection points. So photon detection is delayed: the higher the scattering, the longer the delay. The absorption determines how steep the temporal tail of the detected pulse is. We see the effects of the absorption mostly at long times, on the pulse tail, because the longer the photons stay in the medium, the higher the probability they undergo an absorption event. Thus, strong absorption means that many photons are removed from the temporal tail of the pulse and its slope becomes steeper. This holds qualitatively, but to get a quantitative estimate of the absorption and scattering properties, we need a suitable theoretical model of light propagation that correlates the shape and delay of the transmitted pulse to the absorption and scattering properties of breast tissue. Generally, the diffusion approximation to the radiative transport theory (Patterson et al., 1989) is used, as it provides a simple analytical solution that can be readily applied for the interpretation of clinical data. In its simplest derivation, the model holds only for a homogeneous medium. So it does not allow the estimate of local values of the optical properties. It can only provide average values measured over the light path between the injection and the collection point. More realistic theories that describe the heterogeneity of breast tissue are just starting to be developed, and models that take into account at least the presence of a localized inhomogeneity (i.e., a pathologic lesion) have been applied only recently, and not routinely yet, for the interpretation of patient data (Torricelli et al., 2003; Grosenick et al., 2005).

The idea of optical diagnosis of breast lesions relies on optical contrast, on the different optical properties of lesion and surrounding tissue. This clearly violates the hypothesis of homogeneous medium. Moreover, the healthy breast tissue itself is markedly heterogeneous. Even under such conditions, the diffusion approximation still provides effective information on the scattering properties. On the contrary, absorption images are of difficult use because the inadequacy of the model limits the contrast and spatial resolution, thus hindering the detection of lesions, especially small ones. However, the availability of the time distribution of the transmitted photons can have direct applications, as the scattering and absorption events modify the pulse shape in a different way and at different times. Consequently, a convenient temporal selection of the detected photons can yield information on the optical properties. In particular, early-arriving photons, on the raising edge of the pulse, are mostly (even though not only) affected by the scattering properties. Conversely, late-arriving photons, on the tail of the detected pulse, are mostly sensitive to the absorption properties. Thus, if only "late" photons are selected from the transmitted pulse at each measurement position during the breast scan, special intensity images—so called delayed gated images—can be built. The intensity in each "pixel" is related to the absorption properties in the corresponding position: high transmitted intensity indicates low absorption. The method is clearly not quantitative, as the absorption coefficient is not assessed. However, it is relatively simple, not requiring data analysis based on a theoretical model, and allows one to compare two locations and determine where the absorption is higher or lower and whether the difference is small or large. Such a procedure is commonly used to detect spatial changes in the absorption properties and to identify areas of abnormal absorption, whereas diffusion theory is used to translate time-domain data into scattering images.

Time-domain optical mammography has been applied in clinical trials (Grosenick et al., 2005; Taroni et al., 2005) according to the measurement scheme where the slightly compressed breast is raster-scanned in a transmission geometry, and time-domain data are collected at every measurement position. Spectral information is obtained by injecting picosecond pulses at different wavelengths and collecting independently each of the transmitted pulses. As already described, time-gated intensity images are routinely used to track absorption changes as a function of position. If the imaging wavelengths are chosen so that each wavelength isolates the contribution of a specific constituent, namely, a single constituent absorbs at each wavelength, then each image can be used to investigate the spatial distribution of a different constituent. In practice, it is usually difficult to find wavelengths where just one constituent absorbs. So, it is not possible to isolate single constituents, but still, with proper choice of the imaging wavelengths, each of them can dominate a different image. In particular, Politecnico di Milano (Milan, Italy) has developed a prototype that operates at 4–7 wavelengths between 637 and 985 nm (Taroni et al., 2005). Using that instrument, it was shown that images at wavelengths shorter than 685 nm are dominated by deoxy-hemoglobin, at 916 nm by lipids, and at 975 nm by water, while 785 nm enhances the sensitivity to oxy-hemoglobin. An example is shown in Figure 117, which displays images acquired at 683, 785, 916, and 975 nm from a 55-year-old patient with a 1.8 cm invasive ductal carcinoma in her left breast. The cancer is detected because of its marked vascularization that causes strong absorption at the two shorter wavelengths (Figs. 117a and 117b). Moreover, the uniform and dark appearance of the image at 916 nm (Fig. 117c) indicates a high lipid content, revealing the adipose nature of the breast, in agreement with what derived from the X-ray mammogram (Fig. 117e). Further details on image interpretation are reported in the next section on the interpretation of optical mammograms.

The frequency-domain approach is based on modulating the intensity of the light source (at a frequency f that is typically in the order of 100 MHz) and performing phase-sensitive detection of the modulated optical signal. One can fully describe the modulated optical signal using three parameters, namely, the average intensity (DC intensity), the amplitude (AC amplitude), and the phase (Φ) of the intensity oscillations. These three parameters can provide sufficient information to characterize both the absorption and scattering properties of tissue. In particular, the phase measurement is directly associated with the time-delay experienced by the probing intensity-wave in tissue, which is also directly measured in the time-domain approach. Frequency-domain instruments have been developed for direct projection imaging of the breast (Fantini *et al.*, 2005), as well as optical tomographic imaging of the breast (Dehghani *et al.*, 2003). A frequency-domain prototype originally developed by Siemens AG, Erlangen, Germany (Götz *et al.*, 1998), has produced a clinical data set of optical mammograms in a planar transmission geometry. Figure 118 shows a picture of this prototype (panel (a)) and two representative optical mammograms (panels (b) and (c)) obtained on a 53-year-old patient with a 3 cm invasive ductal carcinoma in her left breast. The optical mammograms shown in Figure 118 (panels (b) and (c)) are direct projection images of the slightly compressed left breast taken in a craniocaudal (cc) projection. The total time required to scan the breast is 2–3 min. The frequency-domain optical data have been initially processed with an algorithm designed to enhance tumor contrast by minimizing the effects of the breast geometry on the optical data. Then a spatial second derivative algorithm is applied to enhance the spatial information content of the image and the visualization of blood vessels (Fig. 118b). Finally, data at four wavelengths (690, 750, 788, and 856 nm) are combined to yield a measure of tissue saturation (StO2) that results in an oxygenation image (Fig. 118c) The optical mammograms in Figure 118 illustrate the potential of this imaging technique to detect angiogenic signatures and oxygenation data associated with breast tumors. In particular, the invasive ductal carcinoma of Figure 118 (indicated by the arrow in panels (b) and (c)) is associated with a high density of blood vessels (Fig. 118b) and low levels of oxygenation (Fig. 118c).

Interpretation of Optical Mammograms

A number of structures of the healthy breast can be identified in optical images. First, blood vessels and highly vascularized regions, like those surrounding the lactiferous ducts in the nipple area, are detected due to the strong hemoglobin absorption at short wavelengths (below 800 nm). The mammary gland is characterized by marked absorption at 975 nm, possibly due to its high water content as compared to the surrounding more adipose breast tissue, even though a contribution could also come from collagen, which shows significant absorption at 975 nm. Moreover, lipids have a characteristic absorption peak around 924 nm. Thus, optical images collected around this wavelength of 924 nm provide

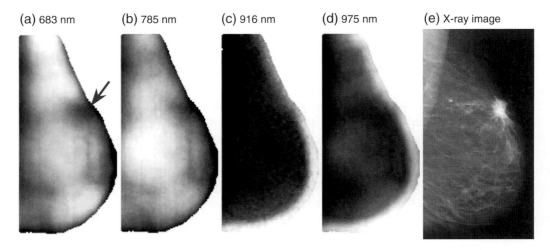

(a) 683 nm   (b) 785 nm   (c) 916 nm   (d) 975 nm   (e) X-ray image

Figure 117 Late-gated intensity images at 683 nm (a), 785 nm (b), 916 nm (c) and 975 nm (d), and X-ray mammogram (e) of the left breast (oblique view) of a 55-year-old patient bearing an invasive ductal carcinoma (max. diameter = 1.8 cm), indicated by the red arrow. The images were acquired with the time-resolved multiwavelength optical mammograph developed by Politecnico di Milano (Milan, Italy).

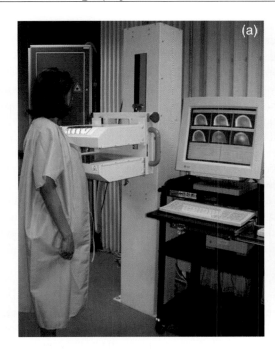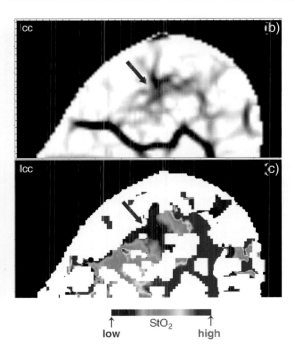

Figure 118 (a) Photograph of a prototype for frequency-domain (70 MHz) optical mammography developed by Siemens AG, Medical Engineering (Erlangen, Germany). The slightly compressed breast is optically scanned to obtain 2D projection images at four wavelengths (690, 750, 788, and 856 nm). The scanning time is about 2 min per image. (b) Second-derivative image at 690 nm, and (c) oxygenation image from data at all four wavelengths of the left (l) breast in craniocaudal (cc) view of a 53-year-old patient affected by invasive ductal carcinoma (indicated by the arrow in panels (b) and (c)). Cancer size is 3 cm.

information on lipid distribution, and regions that appear dark in those images, due to strong absorption, show good correspondence with areas that are relatively transparent to X-rays. The scattering images of healthy breasts are generally uniform, except for the mammary gland that in some cases reveals slightly lower scattering, especially at the longest wavelengths.

Concerning breast lesions, cancers are usually identified through the detection of associated neovascularization. Thus, they appear as strongly absorbing areas at short wavelengths. The contrast is often higher at 637–685 nm than at 780–785 nm, suggesting the presence of deoxygenated blood. However, low oxygenation has not yet been proven to be a reliable index for identifying malignant lesions. These qualitative observations are based on the visual inspection of images acquired at different wavelengths. However, they are confirmed by the quantitative estimate of blood content and oxygenation in the lesion and surrounding tissue. Such estimates are obtained applying "inhomogeneous" models of breast tissue that account for the presence of a localized lesion and allow one to quantify its optical properties and estimate its composition. These models indicate that the blood content in the tumor area is typically two to five times higher than in the surrounding

tissue. However, such values are often exceeded. On the contrary, no systematic and reliable findings have yet been reported for the oxygenation level of breast cancer with respect to the oxygenation level of healthy tissue or benign breast lesions.

At least in principle, the scattering images could provide diagnostically useful information. Actually, the neoplastic transformation affects the entire tissue architecture, altering cell density and nuclear volume, degrading the extracellular matrix upon invasion, and forming a complex network of new blood vessels. All these processes may modify the scattering properties of tissues, especially when poorly differentiated invasive lesions develop. Experimentally, changes in scattering are often observed, especially when the lesion involves a significant volume. The detection and identification of cysts are performed relying on their liquid nature that leads to low scattering. Cysts may be of several types, containing a clear fluid, a turbid liquid, or even big floating particles. The effect of these various structures on the slope of the scattering spectrum is different. Consequently, the higher or lower contrast at different wavelengths provides information on the nature of the cyst. For example, a marked increase in contrast at long wavelengths, corresponding to a very steep scattering spectrum, will suggest the presence of a clear fluid. In some cases, specific absorption features may also identify fluid-filled cysts, as a result of their relatively high water concentration, low lipid concentration, and high concentration of deoxygenated blood.

The detection of fibroadenomas is often more challenging for optical mammography. When identified, fibroadenomas are generally characterized by high absorption around 975 nm and sometimes even in the red spectral region (637–685 nm). The marked absorption at 975 nm is likely related to their high water content, but it might also be due, at least in part, to collagen. In the early stages of a clinical test of the only time-domain instrument currently featuring wavelengths longer than 900 nm (Taroni *et al.*, 2005), data acquisition was characterized by limited signal levels at wavelengths longer than 900 nm. Such technical limitation has likely reduced the potential for detecting fibroadenomas in that clinical test. However, instrumental upgrade is ongoing and is expected to improve the diagnostic potential for fibroadenomas in the near future.

Optical images are routinely acquired in two views: craniocaudal and either mediolateral (90°) or oblique (45°). If localization of the lesion is required in both views, the detection rate (sensitivity) of optical mammography is around 80% for both cancers and cysts. As already observed, the detection of fibroadenomas is often difficult, with only 39% of fibroadenomas identified in both views. If the criterion for positive classification is relaxed to lesion localization in just one view, the sensitivity increases significantly, reaching 92–96% for cancers and 90% for cysts.

On average, the optical contrast of detected cancers increases progressively with their size, but no such correlation with size has been reported for other lesion types. This different trend observed for malignant and benign lesions is likely due to the fact that cancers are detected in intensity images at short wavelengths, where all blood-reach structures, not only the neovascularization associated with tumor development, are highlighted and can hamper the detection of the lesion. Thus, a bigger lesion size can significantly increase the optical contrast. On the contrary, cysts are detected in scattering images, which are rather uniform for healthy breasts. Thus, no main dependence on the lesion size and generally higher contrast can be expected. Demographic parameters, such as age and body mass index (BMI), seem to have no influence on lesion detection or on the contrast for detected lesions, either malignant or benign ones of any kind.

Prospects of Optical Mammography

X-ray mammography, the current gold standard for breast cancer screening, is less effective in women younger than 50, who have radiographically dense breasts, leading to high rates of both false-negative and false-positive cases. By contrast, no clear dependence on age was observed for optical imaging in terms of cancer detection rate or image contrast. The possible effect of breast density was also investigated. Five mammographic parenchymal patterns were identified following Tabàr's classification (Gram *et al.*, 1997). Dense breasts (type IV and V) were compared with adipose breasts, more transparent to X-rays (type II and III). The detection rate is a few percent points higher for dense breasts, with slightly lower average value of the detection contrast. However, the difference between the two categories is not significant (Taroni *et al.*, 2005). Because most findings are based on retrospective studies, involving patients with lesions previously identified in X-ray mammograms, there is no information on how optical mammography performs in patients with false-negative X-ray results. Consequently, the results might be somehow biased. Nevertheless, there are strong indications that optical imaging is not negatively affected by the radiological density of breast. For this reason, optical mammography may find a niche of clinical applicability in a population of younger women, where X-ray mammography is not applicable or suffers from degraded performance.

It is also envisioned that optical mammography may effectively complement X-ray mammography. First, the information about hemodynamic, oxygenation, and water/lipids composition provided by optical mammography is complementary to the fine structural information provided by X-ray mammography. The combination of these complementary pieces of diagnostic information has the potential to result in a more effective breast imaging modality than X-ray mammography alone. Second, the optical scattering spectrum correlates with breast density. Consequently, a prescreening optical mammogram may identify breasts at high risk for cancer and indicate cases in which X-ray exposure should be avoided if conventional mammography is not expected to be effective.

There is also promise in the combination of ultrasound and optical techniques. On the one hand, it is possible to complement the information content of optical mammography and ultrasound imaging, similarly to the way that it is envisioned to complement the information content of optical and X-ray mammography. On the other hand, ultrasound and light may be combined into a truly hybrid diagnostic tool either by using a focused ultrasound beam to label or tag optical photons that have traveled through the ultrasound focal volume (ultrasonic tagging of light), or by using photoacoustics to generate high-frequency pressure waves (ultrasound) as a result of localized areas of increased optical absorbance. It has also been suggested that the rich spatial information provided by magnetic resonance imaging (MRI) can provide crucial *a priori* information for the implementation of optical tomographic reconstruction algorithms. In this sense, MRI and optical mammography can also be combined into a hybrid imaging tool.

Another potential clinical role of optical mammography is in the area of monitoring the effectiveness of therapeutic procedures and performing post-treatment follow-up. This potential results from features of optical mammography such as its safety, lack of discomfort, noninvasiveness, real-time capability, implementation in portable instrumental units, and cost-effectiveness.

While there is an obvious emphasis on basing optical mammography on the intrinsic optical contrast provided by the human breast (mostly from oxy-hemoglobin, deoxy-hemoglobin, water, and lipids), research efforts are also being aimed at assessing the potential offered by extrinsic optical contrast agents. Although extrinsic contrast agents introduce the need for an intravenous injection, thus making the procedure invasive, they can offer unprecedented opportunities and possibly lead to a more powerful approach (in terms of both sensitivity and specificity) to the optical detection of breast cancer. For example, it is possible to detect specific enzyme activity by using auto-quenched near-infrared fluorescence probes, whose fluorescence emission is restored in the presence of enzymes that are overexpressed by tumors. Other approaches use near-infrared fluorescent dyes (typically, the clinically approved indocyanine green or structurally related dyes) that exhibit preferential accumulation in cancerous tissue. Some animal studies have shown promising results of cancer detection based on extrinsic contrast agents, but more research is needed to fully appreciate the potential of this approach in humans.

In conclusion, optical mammography is a potentially powerful imaging modality for the human breast, which provides diagnostic information that is not available from other current imaging tools. Optical mammography may play an important clinical role as a stand-alone technique for breast cancer detection (especially in younger women) or for follow-up to treatment, and can effectively complement other diagnostic imaging modalities such as X-ray mammography, ultrasound imaging, and magnetic resonance imaging. The potential of combining multiple imaging tools for the detection and diagnosis of cancer is enormous, and optical methods can play an important role in developing such synergistic combinations of imaging tools.

Acknowledgments

We acknowledge support from the National Institutes of Health (Grant CA95885) and the National Science Foundation (Award BES-93840).

References

Bartrum, R.J., and Crow, H.C. 1984. Transillumination lightscanning to diagnose breast cancer: a feasibility study. *AJR* 142:409–414.

Dehghani, H., Pogue, B.W., Poplack, S.P., and Paulsen, K.D. 2003. Multiwavelength three-dimensional near-infrared tomography of the breast: initial simulation, phantom, and clinical results. *Appl. Opt.* 42:135–145.

Fantini, S., Heffer, E.L., Pera, V.E., Sassaroli, A., and Liu, N. 2005. Spatial and spectral information in optical mammography. *Technol. Cancer Res. Treat.* 4:471–482.

Fantini, S., Walker, S.A., Franceschini, M.A., Kaschke, M., Schlag, P.M., and Moesta, K.T. 1998. Assessment of the size, position, and optical properties of breast tumors *in vivo* by non-invasive optical methods. *Appl. Opt.* 37:1982–1989.

Götz, L., Heywang-Köbrunner, S.H., Schütz, O., and Siebold, H. 1998. Optische mammographie an präoperativen patientinnen. *Akt. Radiol.* 8:31–33.

Gram, I.T., Funkhouser, E., and Tabar, L. 1997. The Tabàr classification of mammographic parenchymal patterns. *Eur. J. Radiol.* 24:131–136.

Grosenick, D., Moesta, K.T., Wabnitz, H., Mücke, J., Stroszczynski, C., Macdonald, R., Schlag, P.M., and Rinneberg, H. 2003. Time-domain optical mammography: initial clinical results on detection and characterization of breast tumors. *Appl. Opt.* 42:3170–3186.

Grosenick, D., Wabnitz, H., Moesta, K.T., Mucke, J, Schlag, P.M., and Rinneberg, H. 2005. Time-domain optical mammography (part II): optical properties and tissue parameters of 87 carcinomas. *Phys. Med. Biol.* 50:2451–2468.

Jakubowski, D.B., Cerussi, A.E., Bevilacqua, F., Shah, N., Hsiang, D., Butler, J., and Tromberg, B.J. 2004. Monitoring neoadjuvant chemotherapy in breast cancer using quantitative diffuse optical spectroscopy a case study. *J. Biomed. Opt.* 9:230–238.

Monsees, B., Destouet, J.M., and Totty, W.G. 1987. Light scanning versus mammography in breast cancer detection. *Radiology* 163:463–465.

Ohlsson, B., Gundersén, J., and Nilsson, D.M. 1984. Diaphanography: a method for evaluation of the female breast. *World J. Surg.* 4:701–707.

Patterson, M.S., Chance, B., and Wilson, B.C. 1989. Time-resolved reflectance and transmittance for the noninvasive measurement of tissue optical properties. *Appl. Opt.* 28:2331–2336.

Shah, N., Cerussi, A.E., Jakubowski, D., Hsiang, D., Butler, J., and Tromberg, B.J. 2004. Spatial variations in optical and physiological properties of healthy breast tissue. *J. Biomed. Opt.* 9:534–540.

Taroni, P., Torricelli, A., Spinelli, L., Pifferi, A., Arpaia, F., Danesini, G., and Cubeddu, R. 2005. Time-resolved optical mammography between 637 and 985 nm: clinical study on the detection and identification of breast lesions. *Phys. Med. Biol.* 50:2469–2488.

Torricelli, A, Spinelli, L., Pifferi, A., Taroni, P., and Cubeddu, R. 2003. Use of a nonlinear perturbation approach for *in vivo* breast lesion characterization by multi-wavelength time-resolved optical mammography. *Opt. Expr.* 11:853–867.

Tromberg, B.J., Shah, N., Lanning, R., Cerussi, A., Espinoza, J., Pham, T., Svaasand, L., and Butler, J. 2000. Non-invasive *in vivo* characterization of breast tumors using photon migration spectroscopy. *Neoplasia* 2:26–40.

Wallberg, H, Alveryd, A., Bergvall, U., Nasiell, K., Sundelin, P., and Troell, S. 1985. Diaphanography in breast carcinoma: correlation with clinical examination, mammography, cytology and histology. *Acta Radiol. Diagn.* 26:33–44.

17

Digital Mammography

John M. Lewin

Introduction

Traditional film mammography has been shown to reduce mortality from breast cancer in large randomized trials. Consequently, mammography is the only technique recommended and widely used for screening. Because mammographic sensitivity is far from perfect, research and development efforts aimed at improving early detection of breast cancer continue. At a workshop sponsored by the National Cancer Institute in 1991, an expert panel reviewed all the potential breast cancer screening technologies used They concluded that, of all the technologies presented, full-field (i.e., whole breast) digital mammography held the greatest promise to improve breast cancer detection (Shtern, 1992). Since that time, several companies have developed digital mammography systems. At the time of writing, five companies—General Electric (GE), Fischer Imaging, Hologic, Fuji, and Siemens—have adopted commercial digital mammography systems with Food and Drug Administration approval for use in the United States. At least three other companies—Sectra, Planmed, and Kodak— have devices in clinical use in other countries.

Technical Advantages of Digital Mammography

Unlike traditional screen-film mammography, where the image is captured, displayed, and stored on film, digital mammography decouples these three tasks. The image is captured by the digital detector but is displayed on a monitor or film, usually after mathematical processing by a computer. By separating these tasks, the typical trade-off between dynamic range, the overall range of shades of gray that can be imaged by the system, and contrast resolution, the ability to distinguish between small differences in shades of gray, can be avoided, and both can be optimized (Suryanarayanan et al. 2002; Cooper et al., 2003). The result is a detector with both higher contrast resolution and equal or better dynamic range than the combination of film and phosphorescent screen used in screen-film mammography. In addition, because the digital detector captures more of the incoming X-ray photons than the screen-film combination, most digital systems produce images with lower noise than a typical screen-film system operating at an equivalent radiation dose. The fraction of X-ray photons captured by a system (digital or film) and converted to image information is summarized in a value termed detective quantum efficiency (DQE). The DQE of an imaging system can be used to compare systems in terms of image noise for a given dose. This measure is a function of spatial frequency, making comparison more complicated, because one system may be better than its competitor at one frequency but worse at another and because systems with different pixel sizes are operating at different maximum spatial frequencies.

One parameter in which screen-film mammography is superior to digital is spatial resolution. Digital mammography systems have pixel sizes between 50 and 100 microns, limiting the maximum spatial resolution for high-contrast objects to between 5 and 10 line-pairs/mm, versus ~ 12–15 for screen-film. This should at least, theoretically, enable screen-film mammography to resolve smaller objects as long as they are of high contrast. In practice, however, objects tend to have lower contrast as they get smaller, so contrast resolution and spatial resolution are inevitably intertwined, with both playing critical roles in the detection of small objects.

Technologies Used for Digital Mammography

Digital mammography detectors currently in use can be classified as flat-panel full-field capture devices, computed radiography, or scanning systems. Flat-panel detectors use either amorphous silicon or amorphous selenium as their substrate. The GE system uses an amorphous silicon flat-panel detector, directly coupled to a cesium-iodide (CsI) crystal. The CsI crystal converts incoming X-ray photons to light photons, which are then detected by the amorphous silicon detector. The pixel size on this system is 100 microns, giving it the lowest high-contrast spatial resolution of all the systems, a limitation compensated for by excellent noise characteristics and a large dynamic range. The Hologic and Siemens systems use a flat-panel detector made from amorphous selenium. This detector directly converts the incoming X-ray photons to an electrical voltage. Pixel size with this system is 70 microns.

Fuji and Kodak have systems based on computed radiography technology. This technology, widely used for general radiography, is adapted to mammography by improving both spatial resolution and detection efficiency. In computed radiography, a phosphorescent plate is exposed in a standard film mammography unit and then carried to a readout device, where a laser excites the phosphor, releasing the latent image. Efficiency is improved for mammography by reading out both sides of the phosphor (as opposed to one side for general radiology), while spatial resolution is improved by decreasing the spot size of the laser. This results in a pixel size of 50 microns. The Fischer system is a slot-scanning system in which a collimated X-ray beam (fan beam) sweeps across the breast to expose a moving detector. This system uses multiples of charge-coupled device (CCD) chips coupled to a CsI crystal to create the digital image. The use of a CCD array allows for very high spatial resolution, 50 microns in standard mode and 25 microns in high-resolution mode, but limits dynamic range, since CCDs are relatively easily saturated. The scanning mode increases the image time, but use of the slot beam decreases X-ray scatter, eliminating the need for an antiscatter grid and thus reducing patient dose to levels below that needed for flat-panel or computed radiography systems. An even lower dose is

achieved by the Sectra system. This system uses multiple, tightly collimated, slit fan beams to scan the breast simultaneously. The device uses a photon-counting detector composed of silicon strips in an edge-on geometry. This detector has extremely high detective efficiency, resulting in the ability to generate a mammogram using much less radiation than that needed for other detector types.

Clinical Advantages of Digital Mammography

Four large trials comparing the diagnostic performance of digital mammography to that of screen-film for screening have been completed to date. Of these, the first three, the Colorado–Massachusetts trial (Lewin *et al.*, 2002), the Oslo I trial (Skaane *et al.*, 2003), and the Oslo II trial (Skaane and Skjennald, 2004), showed no significant difference in cancer detection sensitivity. The ACRIN D-MIST trial also showed no significant difference for the entire cohort but did show statistically significant advantages for digital mammography for the overlapping subgroups of women with dense breasts, women < 50 years old, and premenopausal women (Pisano *et al.*, 2005). This result is the first clinical validation of what has always been expected of this new technology, given its superior technical abilities, which should allow it to "see" through dense breast tissue better than screen-film.

Aside from this benefit in sensitivity for some groups of women, digital mammography has numerous operational benefits as compared to screen-film. These include both operational advantages and true advantages in diagnostic ability or confidence. Though not easily measured, these advantages positively impact the patient as well as her physicians. Like other digital modalities, digital mammography allows for digital storage and transmission of each study, eliminating lost films and eventually eliminating the need for a film library. Images can be sent electronically to several treating physicians simultaneously, or given to the patient, without any loss of quality. This is an important operational change. At present, however, images cannot generally be sent electronically between institutions due to incompatible systems and security firewalls, so centers still must make copies when patients change to another location and request their prior studies. Because the cost of creating a CD is about one-tenth that of printing four films for a screening study, some centers are creating image CDs, as is common practice with computed tomography (CT) and magnetic resonance imaging (MRI). The Mammography Quality Standards Act, however, requires that printed films be made available upon request for comparison to a later study done elsewhere.

Also important to mammography is the elimination of film artifacts, such as dust and the structured noise caused by film processing. Digital modality also reduces the variability in contrast, density, dose, and exposure time associated with film emulsion and processing. The film processor, in particular, is a major source of variability and requires daily quality assurance to monitor and correct changes. Insufficient contrast is not an issue with digital images, and, because the detector is sealed, there are no dust artifacts.

From a patient's perspective, the most important advantage of a digital system is speed. This is especially true in diagnostic studies during which the radiologist reviews each image or pair of images prior to deciding the next step in the work-up. By not having to wait for films to develop, the length of a diagnostic mammography examination is shortened. Wire localizations are affected more dramatically because for this procedure the patient must stay in compression while the last image taken is processed and then reviewed by the radiologist. By reducing the time between exposure and image display to 10–20 sec (from ~ 3.0 min for film processing, including carrying the films to and from the processor), both total procedure time and patient discomfort are markedly decreased. Decreasing examination time has positive effects for the physician, technologist, and radiology administrator as well, the last benefiting from the efficiencies in terms of technologist and room costs because examinations can be scheduled more frequently on a digital unit. In most cases, these savings do not completely offset the increased price of the digital equipment, but, along with decreased film and film library costs, they do provide partial justification for choosing digital mammography.

Digital mammography is ideally suited for microcalcification characterization with magnification or high-resolution views due to the low image noise, the ability to handle a wide variety of exposure settings, and the ability to magnify the image on the screen. Small clinical studies have shown a benefit to digital systems in terms of depicting and characterizing microcalcifications (Hermann *et al.*, 2002; Fischer *et al.*, 2002; Rong *et al.*, 2002; Diekmann *et al.*, 2003). Low noise is the most important feature for this task because image noise is what most hides subtle calcifications, making them difficult to assess. Wide dynamic range and high contrast allow the image to be acquired at a higher voltage but a shorter exposure time, reducing patient motion. One disadvantage of digital mammography is that it has lower spatial resolution, countered by the use of geometric magnification (accomplished by moving the breast away from the detector), which enlarges the projected image of the calcifications above the lower limit of even the lowest spatial resolution digital system. An important aspect realized early in the dissemination of the technology is that magnification of the image on the monitor alone is no substitute for true geometric magnification or imaging in a special high-resolution mode. While it had been hoped that digital mammography would eliminate the need to recall

patients for special magnification views, this has not turned out to be the case.

Because of the large dynamic range, digital mammography is ideal for imaging implants. A single exposure can be displayed to optimally show the details of the implant itself, at the expense of the surrounding breast tissue, or, by adjusting the window and level settings, one can optimally show the breast tissue at the expense of seeing into or through the implant. For the same reason, large dynamic range, digital mammography is also ideal for imaging the skin and tissues immediately deep to the skin. This tissue is typically blackened on a well-exposed film mammogram, requiring a special hot light to partially recover the information. As shown above, processing of the digital image allows the skin to be routinely evaluated without any extra effort, such as additional windowing or leveling. Although the skin is usually of no importance, it can be thickened in some disease processes, including inflammatory carcinoma. In addition, special views are sometimes performed to localize indeterminate calcifications to the skin in order to prove that they are benign. Digital mammography is ideal for these skin views both because it shows the skin and because the multistep process used to obtain these views is similar to that used for wire localization and is accomplished much faster using a digital system.

Advanced Applications of Digital Mammography

Although digital mammography used in the same manner as film mammography will not revolutionize breast cancer detection and diagnosis, certain advanced applications made possible by the use of digital images have shown promise in early studies.

Tomosynthesis

Tomosynthesis involves the mathematical processing of a set of planar images to create tomographic slices, similar in appearance to the old linear tomograms. The concept and mathematical formulation of tomosynthesis originated years ago, but application of this process to the breast has only become feasible with the advent of digital mammography (Niklason *et al.*, 1997). For breast tomosynthesis, multiple mammographic images are acquired at various angles. Each image is acquired with a fraction of the typical X-ray dose of a mammogram, so that the total radiation dose is equivalent to that of a standard mammogram. The detector can remain stationary or move under the breast; the breast must stay immobile. There is currently much excitement about this technique. Because overlapping tissue can both hide a cancer and simulate one, it is a source of both false-positive and false-negative mammograms. It is hoped that tomosynthesis will be able to greatly reduce these errors.

Several companies, including GE, Hologic, Siemens, and Planmed, are actively pursuing tomosynthesis.

Contrast-enhanced Digital Mammography

Another way proposed to improve the cancer detection ability of mammography is to use a contrast agent. It is well known from years of experience with MRI that cancers will be enhanced with intravenous contrast. Contrast-enhanced CT of the breast has also been demonstrated with good results. Compared to CT, however, the sensitivity of mammography to iodinated contrast agent is small, so some sort of subtraction must be used to eliminate unenhanced normal tissue and thereby better show enhancing cancers. Two subtraction methods have been tried: temporal subtraction, in which a post-contrast image is subtracted from a precontrast image, and dual-energy subtraction, in which a low-energy image and a high-energy image, both obtained after iodine injection, are mathematically combined in such a way that unenhanced breast tissue is eliminated but iodine is seen clearly. This is possible because of the different X-ray absorption spectra of iodine versus breast tissue. Both temporal subtraction (Jong et al., 2003, Diekmann et al., 2005) and dual-energy subtraction (Lewin et al., 2003) have been tested in small pilot studies on volunteers with suspicious mammographic or palpable lesions. Both techniques performed well in demonstrating cancers, with few false-positives. One advantage of the dual-energy technique is that the breast does not have to remain immobilized during and after the contrast agent injection. This allows multiple projections to be obtained versus the single projection possible with temporal subtraction. Several chapters in this volume discuss in detail the use of contrast-enhancing agents in diagnosing cancer.

In conclusion, the use of digital mammography continues to increase as more companies enter the market and more users convert in order to gain the operational benefits of digital mammography. In addition, there is now evidence that digital modality performs better than screen-film mammography for screening in some groups of women. Meanwhile, because it is a relatively new technology, improvements and refinements are being introduced at a much faster pace than those of screen-film. Perhaps, most importantly, the use of digital systems is paving the way for new methods of cancer detection, such as tomosynthesis and contrast-enhanced mammography.

References

Cooper, V.N., Oshiro, T., Cagnon, C.H., Bassett, L.W., McLeod-Stockmann, T.M., and Bezrukiy, N.V. 2003. Evaluation of detector dynamic range in the X-ray exposure domain in mammography: a comparison between film-screen and flat panel detector systems. *Med. Phys. 30*:2614–2621.

Diekmann, S., Bick, U., von Heyden, H., and Diekmann, F. 2003. Visualization of microcalcifications on mammographies obtained by digital full-field mammography in comparison to conventional film-screen mammography. *Rofo. 175*:775–779.

Diekmann, F., Diekmann, S., Taupitz, M., Bick, U., Winzer, K.J., Huttner, C., Muller, S., Jeunehomme, F., and Hamm, B. 2005. Use of iodine-based contrast media in digital full-field mammography—initial experience. *Rofo. 175*:342–345.

Fischer, U., Baum, F., Obenauer, S., Luftner-Nagel, S., von Heyden, D., Vosshenrich, R., and Grabbe, E. 2002. Comparative study in patients with microcalcifications: full-field digital mammography vs. screen-film mammography. *Eur. Radiol. 12*:2679–2683.

Hermann, K.P., Obenauer, S., Funke, M., and Grabbe, E.H. 2002. Magnification mammography: a comparison of full-field digital mammography and screen-film mammography for the detection of simulated small masses and microcalcifications. Eur *Radiology. 12*:2188–2191.

Jong, R.A., Yaffe, M.J., Skarpathiotakis, M., Shumak, R.S., Danjoux, N.M., Gunesekara, A., and Plewes, D.B. 2003. Contrast-enhanced digital mammography: initial clinical experience. *Radiology 228*:842–850.

Lewin, J.M., D'Orsi, C.J., Hendrick, R.E., Moss, L.J., Isaacs, P.K., Karellas, A., and Cutter, G.R. 2002. Clinical comparison of full-field digital mammography to screen-film mammography for breast cancer detection. *Am. J. Roentgenol. 179*:671–677.

Lewin, J.M., Isaacs, P.K., Vance, V., and Larke, F.J. 2003. Dual-energy contrast-enhanced digital subtraction mammography—feasibility. *Radiology 229*:261–268.

Niklason, L.T., Christian, B.T., Niklason, L.E., Kopans, D.B., Castleberry, D.E., Opsahl-Ong, B.H., Landberg, C.E., Slanetz, P.J., Giardino, A.A., Moore, R., Albagli, D., DeJule, M.C., Fitzgerald, P.F., Fobare, D.F., Giambattista, B.W., Kwasnick, R.F., Liu, J., Lubowski, S.J., Possin, G.E., Richotte, J.F., Wei, C.Y., and Wirth, R.F. 1997. Digital tomosynthesis in breast imaging. *Radiology 205*:399–406.

Pisano, E.D. Gatsonis, C., Hendrick, E., Yaffe, M., Baum, J.K., Acharyya, S., Conant, E.F., Fajardo, L.L., Bassett, L., D'Orsi, C., Jong, R., and Rebner, M. Digital Mammographic Imaging Screening Trial (DMIST) Investigators Group. 2005. Diagnostic performance of digital versus film mammography for breast-cancer screening. *N. Engl. J. Med. 353*:1773–1783.

Rong, X.J., Shaw, C.C., Johnston, D.A., Lemacks, M.R., Liu, X., Whitman, G.J., Dryden, M.J., Stephens, T.W., Thompson, S.K., Krugh, K.T., and Lai, C.J. 2002. Microcalcification detectability for four mammographic detectors: flat-panel, CCD, CR, and screen/film. *Med. Phys. 29*:2052–2061.

Shtern, F. 1992. Digital mammography and related technologies: a perspective from the National Cancer Institute. *Radiology 183*:629–630.

Skaane, P., and Skjennald, A. 2004. Screen-film mammography versus full-field digital mammography with soft-copy reading: randomized trial in a population-based screening program—the Oslo II Study. *Radiology 232*:197–204.

Skaane, P., Young, K., and Skjennald, A. 2003. Population-based mammography screening: comparison of screen-film and full-field digital mammography with soft-copy reading—Oslo I Study. *Radiology 229*:877–884.

Suryanarayanan, S., Karellas, A., Vedantham, S., Ved, H., Baker, S.P., and D'Orsi, C.J. 2002. Flat-panel digital mammography system: contrast-detail comparison between screen-film radiographs and hard-copy images. *Radiology 225*:801–807.

18

Screening for Breast Cancer in Women with a Familial or Genetic Predisposition: Magnetic Resonance Imaging versus Mammography

Mieke Kriege, Cecile T.M. Brekelmans, and Jan G.M. Klijn

Introduction

Breast cancer is the most common cancer in females. Worldwide over 1 million new cases are diagnosed each year, and 375,000 women die from breast cancer. The incidence varies over the world and is highest in the United States, followed by Europe.

A strong family history of breast and ovarian cancer combined with young ages at diagnosis of affected family members are high risk factors for breast cancer. To date, two genes have been identified with a high-penetrance susceptibility to breast cancer: BRCA1 and BRCA2. Recently, a low-penetrance susceptibility gene was also identified: Chek2. Apart from these genes, mutations of the high-penetrance susceptibility cancer genes TP53, PTEN, and STK11/LKB1 are associated with breast cancer. The other cases of familial breast cancer are possibly caused by

multiple low-penetrance genes, environmental factors, or a combination of both (King et al., 2003). The prevalence of BRCA1/2 mutations is estimated as 0.23% in the general Caucasian population and 2–3% in breast cancer patients. This percentage increases in younger age groups with breast cancer.

Mutations in the BRCA1 and BRCA2 genes are associated with an early onset of breast cancer: for BRCA1 gene mutation carriers, the cumulative breast cancer risk is 20% by age 40, 50% by age 50, and 87% by age 70; for BRCA2, these statistics are 12% by age 40, 28% by age 50, and 84% by age 70 (Ford et al., 1995). Population-based studies found slightly lower percentages: a cumulative lifetime risk of 65% for BRCA1 and 45% for BRCA2 at age 70 (Antoniou et al., 2003). If no gene mutation is detected or no gene mutation analysis is performed, models exist that can estimate the cumulative lifetime risk and age-specific risk of breast cancer based on family history, such as the Claus or BRCAPRO model.

Current risk-reducing strategies in BRCA1/2 mutation carriers include prophylactic mastectomy, oophorectomy, or both, and chemoprevention. Bilateral prophylactic mastectomy is associated with a > 90% breast cancer risk reduction (~ 100%). In the studies by Meijers-Heijboer *et al.* (2001) and Hartmann *et al.* (2001), no breast cancer cases were detected in the group with preventive mastectomy, whereas in the study of Rebbeck *et al.* (2004) two breast cancer cases were found after prophylactic mastectomy in 102 women. One patient had metastatic disease, and the other patient developed breast cancer after subcutaneous, not total, mastectomy (Rebbeck *et al.*, 2004). Studies regarding prophylactic oophorectomy report a risk reduction of ~ 50% for breast cancer (Kauff *et al.*, 2002). Currently, chemoprevention is under investigation. A meta-analysis of the tamoxifen prevention trials showed a 38% (95% confidence interval [CI] 28–46) overall reduction in breast cancer incidence (Cuzick *et al.*, 2003). However, in women with estrogen receptor (ER)-negative tumors, there was no reduction in the incidence of breast cancer, while the reduction of ER-positive tumors was 48% (95% CI 36–58). As tumors in BRCA1 mutation carriers often are ER-negative, it is anticipated that tamoxifen is not an effective chemopreventive agent in these women. However, Narod *et al.* (2000) showed a significant reduction (~ 50%) of the risk of contralateral breast cancer by tamoxifen treatment in BRCA1 mutation carriers affected with primary breast cancer. The side effects of chemoprevention with tamoxifen include an increase in the incidence of endometrium cancer (relative risk [RR] 2.4 [95% CI 1.5–4.0]) and venous thromboembolic events (RR 1.9 [95% CI 1.4–2.6]).

For mutation carriers for whom preventive mastectomy is not acceptable, screening is another option, aiming at reduction of breast cancer mortality, possibly combined with prophylactic oophorectomy and chemoprevention. In women without a proven BRCA1/2 mutation, but with a high cumulative lifetime risk of breast cancer due to a family history, these risk-reducing strategies are less frequently offered, and screening is therefore the main option for reducing breast cancer mortality.

Especially after identification of the BRCA1 and BRCA2 genes in the mid-1990s, demand for breast cancer screening in women with a high familial risk increased. Mammography screening can reduce breast cancer mortality 20–25% and is cost-effective in women 50–70 years of age with an average risk of breast cancer. Also in women 40–50 years of age with an average risk, mortality reduction was found, but less than in women 50–70 years old, and cost-effectiveness was questionable. Also, when mammographic screening for those under 50 years was limited to high-risk women, the efficacy was never proven. In the screening study of Brekelmans *et al.* (2001) the detection rate of breast cancer in women

with a familial or genetic predisposition ranged from three invasive cancers in 1000 women-years in women with a 15–30% cumulative lifetime risk to 33 invasive cancers per 1000 women-years in BRCA1/2 mutation carriers. In women with a 15–30% cumulative lifetime risk this is ~ 3 times more than in women of the same age in the general population, and for BRCA1/2 carriers, even 24 times more frequently.

The sensitivity of screening in high-risk women varied from 50–91% between different studies, and the percentage tumors with positive lymph nodes from 10–45%. Especially in BRCA1/2 carriers, the sensitivity of mammography was low: Brekelmans *et al.* (2001) found a sensitivity of 56% (5/9) in this subgroup of high-risk women compared to 81% in high-risk women without a BRCA1/2 mutation. Possible reasons include a high tumor growth rate, as well as atypical mammographic and specific histopathologic characteristics (such as prominent pushing margins) in BRCA1/2 mutation carriers as compared with controls of the same age (Tilanus-Linthorst *et al.*, 2002). Only the study by Kollias *et al.* (1998) compared the tumor characteristics of a screened high-risk group and age-matched symptomatic controls and found no major differences with respect to tumor size, nodal status, and histological grade.

Mammography is the only screening method for breast cancer that is extensively evaluated and widely used. Because of the low sensitivity of mammography in BRCA1/2 mutation carriers and young premenopausal women with high breast density, especially for these groups there is an interest in better imaging methods. Ultrasound is evaluated in young women with dense breasts and women with a genetic risk for breast cancer. It may be more sensitive, but less specific, than mammography (Warner *et al.*, 2004).

In a diagnostic setting, MRI is a sensitive breast imaging modality, especially in detecting multicentric disease. The reported specificity is variable, ranging from 37–100%. Benign fibroadenomas and fibrocystic disease often cause false-positive MRI results. The reported sensitivity of MRI for ductal carcinoma *in situ* (DCIS) in a diagnostic setting is variable and ranges from 40–100%. Low-grade DCIS in particular is more frequently associated with slow uptake of contrast and no washout, mimicking of a benign lesion, or no contrast uptake (Neubauer *et al.*, 2003). Although there are few data about the potential of MRI to detect invasive lobular carcinoma (ILC) in a diagnostic setting, the sensitivity of MRI in detecting this carcinoma might be slightly lower than that for detecting invasive ductal carcinoma (IDC). Nevertheless, most lesions are visible on MRI (Boetes *et al.*, 2004). However, ILC and medullary carcinoma can demonstrate slow uptake of contrast and no washout or no contrast uptake at all, producing a false-negative examination (Neubauer *et al.*, 2003). At the same time, a recent study found that MRI was more sensitive than mammography for IDC, ILC, and DCIS (Berg *et al.*, 2004). In another study, tumors missed by MRI were characterized by a diffuse growth pattern and small size (< 5 mm) (Teifke *et al.*, 2002). These difficulties make it uncertain whether results in the diagnostic setting can be translated to screening programs for high-risk women.

Magnetic Resonance Imaging Screening Studies

In the late 1990s, breast cancer screening studies including MRI were being set up in women with a genetic susceptibility. Initial results of small pilot studies were published (Podo *et al.*, 2002; Stoutjesdijk *et al.*, 2001), followed by the first results of four large prospective trials (Kriege *et al.*, 2004; Kuhl *et al.*, 2005; Leach *et al.*, 2005; Warner *et al.*, 2004). Some of the pilot studies included retrospective data, while in others an MRI was performed in selected cases (i.e., dense breast tissue) after an occult mammography. The four large studies all had a cross-sectional design; that is, each woman was screened by both mammography and MRI. The aim of all the studies was to compare the sensitivity, specificity, and other screening parameters of MRI with those of mammography. In only one study were characteristics of detected tumors compared with those of age-matched control groups (Kriege *et al.*, 2004).

Results

All small pilot studies described a very high sensitivity (100%) of MRI, while sensitivity of mammography was < 50% in all studies. However, in most studies the specificity of MRI was lower than that of mammography. The number of detected cancers in these studies was small and varied between 3 and 14, so no conclusions about the effectiveness of screening expressed in early diagnosis can be drawn from these studies.

Beginning in 2004, larger prospective series were published (Table 21). The multicenter Dutch MRISC study had 1909 participants, including 358 mutation carriers, with a median follow-up time of 2.9 years (Kriege *et al.*, 2004). A total of 51 tumors were diagnosed, of which 50 were breast cancers. The overall sensitivity was 40% for mammography and 71% for MRI. For invasive cancers this was 33% and 80%, respectively. Mammography detected five and MRI one of six cases of DCIS. The overall specificity was 95% for mammography and 90% for MRI. To date, this is the only MRI screening study that has compared the tumor characteristics of screen-detected breast cancers with those of age-matched symptomatic controls. The characteristics of the tumors detected within the screening program group were significantly more favorable when compared with both control groups with respect to tumor size, nodal status, and grade of tumor differentiation. In the study group, 43% of the tumors were 1 cm or smaller; in the control groups this percentage was only 14% and 13%, respectively. The node positivity rate was 21% in the screened group compared with 52 and 56% in the control groups.

Another study, a Canadian study by Warner *et al.* (2004), included 236 BRCA1/2 mutation carriers, and 22 breast cancers were found with six cases of DCIS. The sensitivity was 36% for mammography and 77% for MRI. Of six cases of DCIS, three were visible on mammography and four on MRI. The specificity (based on biopsy rate) was 99.8% for mammography and 95.4% for MRI. The tumor characteristics of detected cancers were favorable: 56% (9/16) of the invasive cancers were 1 cm or smaller, all invasive cancers were 2 cm or smaller, and 87% (13/15) had negative lymph nodes.

In the recently published multicenter British MARIBS study (Leach *et al.*, 2005), 649 evaluable women were recruited, including 120 (19%) BRCA1/2 gene mutation carriers. During a follow-up period between 2 and 7 years, 35 breast cancers (including six cases of DCIS) were detected in a total follow-up time of 1861 years. The sensitivity was 40% for mammography and 77% for MRI; the specificity was 93% for mammography and 81% for MRI. Of the six cases of DCIS, five were visible on mammography and two on MRI. Of the invasive tumors, 38% (11/29)

Table 21 Study Characteristics and Results of MRI Screening Studies

Study	Women	Mutation Carriers	Scans	Tumors Detected	Sensitivity		Specificity		Positive Bioptic Rate	Tumor Stage Invasive Cancers	
					MRI	Mammo-graphy	MRI	Mammo-graphy		≤ 1 cm	No
	N	N(%)	N	N	%	%	%	%	%	%	%
Kriege *et al.* (2004)	1909	358 (19)	4169	50 (6x DCIS)	71	40	90	95	60	43	79
Warner *et al.* (2004)	236	236 (100)	457	22 (6x DCIS)	77	36	95	99.7	39	56	87
Leach *et al.* (2005)	649	120 (18)	1881	35 (6x DCIS)	77	40	81	93	56	38	81
Kuhl *et al.* (2005)	529	43 (8)	1452	43 (3x DCIS)	91	33	97	97	?	38	79

were 1 cm or smaller and 69% (20/29) were 2 cm or smaller; 81% (21/26) had negative lymph nodes.

Kuhl et al. (2005) detected 43 breast cancers (including nine cases of DCIS) in 41 patients in a single-center study in which 529 women participated. Both women with a family and personal history of breast cancer were included. The sensitivity of MRI was 91% and that of mammography 33%. Specificity was 97% for both MRI and mammography. However, this was a single-center study performed in a highly experienced center. Furthermore, ultrasound was performed every 6 months. Tumor characteristics were presented for the 31 women without a personal history of breast cancer; of these 31 breast cancers, 7 were DCIS (23%). Nine of the 24 invasive cancers (38%) were ≤ 1 cm, 22 (92%) were ≤ 2 cm, and 19 (79%) had negative lymph nodes.

The participants of these MRI screening studies differ with respect to hereditary risk, age, and inclusion of women with previous breast cancer. Nevertheless, all studies found a higher sensitivity for MRI than for mammography. Specificity of MRI varied between 81 and 97%, which might be partly explained by the different definition of this parameter: while some studies defined a test as false-positive only when a biopsy with a negative result was performed, others defined tests as false-positive after a certain Breast Imaging-Reporting and Data System (BI-RADS) score of the imaging. Except for the study of Kuhl et al. (2005), all studies found a lower specificity for MRI than for mammography. The positive biopsy rate varied between 39 and 60% in the different studies.

There is no randomized study comparing the tumor characteristics of the screened and nonscreened group in women with a genetic risk for breast cancer, and it is not expected that this type of study will ever be performed in this setting. The percentage of cases of DCIS in the prospective studies varied from 12 to 27%, which is relatively high, but in agreement with other screening trials. However, detection of DCIS plays only a small role in reducing breast cancer mortality (Duffy et al., 2003). The tumor stage of the detected tumors was favorable; the percentage of node-negative tumors varied from 79–87%, and 38–56% of the invasive tumors were 1 cm or smaller.

Among some of the women participating in the MRISC study, the influence of screening on the quality of life and psychological consequences is being investigated. No important negative effects of screening on the short-term quality of life and general psychological distress have been found (Rijnsburger et al., 2004; van Dooren et al., 2005). However, some subgroups of women appeared to be more vulnerable to psychological distress: younger women (< 40 years) excessively examining their own breasts (i.e., at least once a week), women overestimating their own risk of developing breast cancer, and women who were closely involved in the breast cancer process of their sister (van Dooren et al., 2005). Women who were recalled to additional tests experienced increased anxiety, but not more than women without a hereditary risk (Watson et al., 2005). The long-term effects and the effect of a false-positive result have to be studied further.

Discussion

MRI screening presents some pitfalls compared to mammography screening; breast MRI screening is a costly and time-consuming imaging method, requiring very experienced radiologists. Further drawbacks of the MRI include its relative low specificity compared to mammography and the high number of probably benign findings. The possible anxiety and costs caused by these false-positive results have to be taken into account in the decision making about screening. Improvement of the MRI's specificity is very important in order to reduce unneeded additional investigations and additional costs. Neither the technique nor the criteria for interpretation are standardized, and MRI is not feasible in patients with pacemakers, aneurysm clips, or with severe claustrophia and the availability is limited.

An MRI-guided biopsy is sometimes needed to obtain a histologic diagnosis of nonpalpable lesions detected by MRI and not visible on mammography or ultrasound, a technique that is not available in every center. Different systems exist for MRI-guided biopsy techniques (core and vacuum biopsy), but all have limitations. With many systems only the lateral side of the breast can be accessed, and this is often not the shortest way to the lesion. Another limitation is the inability to verify lesion removal in many cases.

Another disadvantage is spontaneous hormone-induced enhancement of the glandular tissue. To decrease this problem, the MRI should be performed in the second week of the menstrual cycle to avoid false-positive results. An intravenous contrast medium is needed and allergic reactions to gadolinium contrast agents may occur, but serious allergic reactions are extremely rare.

An advantage of MRI in comparison with mammography is the absence of radiation risk. Studies in atomic bomb survivors and women exposed to radiation for treatment reasons, such as women treated for Hodgkin disease, found that radiation is a risk factor for breast cancer. In screening for breast cancer, radiation doses are much lower than in atomic bomb survivors or women treated for Hodgkin disease, and it is still unknown how serious the problem of radiation-induced tumors is. Model studies, however, suggest that it is a minor problem in women 50–75 years of age with an average risk. Another advantage of MRI is the lower level of reported pain (Rijnsburger et al., 2004). Investigation to improve evaluation of MRI is important, especially to improve specificity. As yet, there is no study investigating mortality reduction because of short follow-up. Predictions of mortality reduction and cost-effectiveness analyses with a computer simulation model (MISCAN) are currently being performed for the Dutch MRISC study.

In addition to breast cancer mortality reduction and financial costs, other important questions include a subgroup analysis for different hereditary risk and age groups. It is important to offer MRI, an expensive and time-consuming method, only to women for whom MRI has a high additional value. These subgroup analyses are also being performed in the Dutch MRISC study.

In conclusion, MRI is a much more sensitive method than mammography and can detect invasive tumors that are occult on mammography. However, inconsistent results of MRI's sensitivity in detecting DCIS have been reported. Most studies found a lower specificity for MRI than for mammography. Screening programs including MRI are able to detect breast cancer in a favorable stage in high-risk women. Because of the high incidence of breast cancer and a relative low sensitivity of mammography in BRCA1/2 mutation carriers, currently we recommended screening with MRI in this group. In other risk groups a longer follow-up is needed. Screening of this group of women by MRI should only be performed in a research setting.

References

Antoniou, A., Pharoah, P.D., Narod, S., Risch, H.A., Eyfjord, J.E., Hopper, J.L., Loman, N., Olsson, H., Johannsson, O., Borg, A., Pasini, B., Radice, P., Manoukian, S., Eccles, D.M., Tang, N., Olah, E., Anton-Culver, H., Warner, E., Lubinski, J., Gronwald, J., Gorski, B., Tulinius, H., Thorlacius, S., Eerola, H., Nevanlinna, H., Syrjakoski, K., Kallioniemi, O.P., Thompson, D., Evans, C., Peto, J., Lalloo, F. Evans, D.G., and Easton, D.F. 2003. Average risks of breast and ovarian cancer associated with BRCA1 or BRCA2 mutations detected in case series unselected for family history: a combined analysis of 22 studies. *Am. J. Hum. Genet.* 72:1117–1130.

Berg, W.A., Gutierrez, L., NessAiver, M.S., Carter, W.B., Bhargavan, M., Lewis, R.S., and Ioffe, O.B. 2004. Diagnostic accuracy of mammography, clinical examination, US, and MR imaging in preoperative assessment of breast cancer. *Radiology* 233:830–849.

Boetes, C., Veltman, J., van Die, L., Bult, P., Wobbes, T., and Barentsz, J.O. 2004. The role of MRI in invasive lobular carcinoma. *Breast Cancer Res. Treat.* 86:31–37.

Brekelmans, C.T., Seynaeve, C., Bartels, C.C., Tilanus-Linthorst, M.M., Meijers-Heijboer, E.J., Crepin, C.M., van Geel, A.A., Menke, M., Verhoog, L.C., van den, O.A., Obdeijn, I.M., and Klijn, J.G. 2001. Effectiveness of breast cancer surveillance in BRCA1/2 gene mutation carriers and women with high familial risk. *J. Clin. Oncol.* 19:924–930.

Cuzick, J., Powles, T., Veronesi, U., Forbes, J., Edwards, R., Ashley, S., and Boyle, P. 2003. Overview of the main outcomes in breast-cancer prevention trials. *Lancet* 361:296–300.

Duffy, S.W., Tabar, L., Vitak, B., Day, N.E., Smith, R.A., Chen, H.H., and Yen, M.F. 2003. The relative contributions of screen-detected in situ and invasive breast carcinomas in reducing mortality from the disease. *Eur. J. Cancer* 39:1755–1760.

Ford, D., Easton, D.F., and Peto, J. 1995. Estimates of the gene frequency of BRCA1 and its contribution to breast and ovarian cancer incidence. *Am. J. Hum. Genet.* 57:1457–1462.

Hartmann, L.C., Sellers, T.A., Schaid, D.J., Frank, T.S., Soderberg, C.L., Sitta, D.L., Frost, M.H., Grant, C.S., Donohue, J.H., Woods, J.E., McDonnell, S.K., Vockley, C.W., Deffenbaugh, A., Couch, F.J., and Jenkins, R.B. 2001. Efficacy of bilateral prophylactic mastectomy in BRCA1 and BRCA2 gene mutation carriers. *J. Natl. Cancer Inst.* 93:1633–1637.

Kauff, N.D., Satagopan, J.M., Robson, M.E., Scheuer, L., Hensley, M., Hudis, C.A., Ellis, N.A., Boyd, J., Borgen, P.I., Barakat, R.R., Norton, L., Castiel, M., Nafa, K., and Offit, K. 2002. Risk-reducing salpingo-oophorectomy in women with a BRCA1 or BRCA2 mutation. *N. Engl. J. Med.* 346:1609–1615.

King, M.C., Marks, J.H., and Mandell, J.B. 2003. Breast and ovarian cancer risks due to inherited mutations in BRCA1 and BRCA2. *Science* 302:643–646.

Kollias, J., Sibbering, D.M., Blamey, R.W., Holland, P.A.M., Obuszko, Z., Wilson, A.R.M., Evans, A.J., Ellis, I.O., and Elston, C.W. 1998. Screening women aged less than 50 years with a family history of breast cancer. *Eur. J. Cancer* 34:878–883.

Kriege, M., Brekelmans, C.T., Boetes, C., Besnard, P.E., Zonderland, H.M., Obdeijn, I.M., Manoliu, R.A., Kok, T., Peterse, H., Tilanus-Linthorst, M.M., Muller, S.H., Meijer, S., Oosterwijk, J.C., Beex, L.V., Tollenaar, R.A., De Koning, H.J., Rutgers, E.J., and Klijn, J.G. 2004. Efficacy of MRI and mammography for breast-cancer screening in women with a familial or genetic predisposition. *N. Engl. J. Med.* 351:427–437.

Kuhl, C.K., Schrading, S., Leutner, C.C., Morakkabati-Spitz, N., Wardelmann, E., Fimmers, R., Kuhn, W., and Schild, H.H. 2005. Mammography, breast ultrasound, and magnetic resonance imaging for surveillance of women at high familial risk for breast cancer. *J. Clin. Oncol.* 23:8469–8476.

Leach, M.O., Boggis, C.R., Dixon, A.K., Easton, D.F., Eeles, R.A., Evans, D.G., Gilbert, F.J., Griebsch, I., Hoff, R.J., Kessar, P., Lakhani, S.R., Moss, S.M., Nerurkar, A., Padhani, A.R., Pointon, L.J., Thompson, D., and Warren, R.M. 2005. Screening with magnetic resonance imaging and mammography of a U.K. population at high familial risk of breast cancer: a prospective multicentre cohort study (MARIBS). *Lancet* 365:1769–1778.

Meijers-Heijboer, H., van Geel, A.N., van Putten, W.L.J., Henzen-Logmans, S.C., Seynaeve, C., Menke-Pluymers, M.B.E., Bartels, C.C.M., Verhoog, L.C., van den Ouweland, A.M.W., Niermeijer, M.F., Brekelmans, C.T.M., and Klijn, J.G.M. 2001. Breast cancer after prophylactic bilateral mastectomy in women with a BRCA1 or BRCA2 mutation. *N. Eng. J. Med.* 345:159–164.

Narod, S.A., Brunet, J.S., Ghadirian, P., Robson, M., Heimdal, K., Neuhausen, S.L., Stoppa-Lyonnet, D., Lerman, C., Pasini, B., de los, R.P., Weber, B., and Lynch, H. 2000. Tamoxifen and risk of contralateral breast cancer in BRCA1 and BRCA2 mutation carriers: a case-control study. Hereditary Breast Cancer Clinical Study Group. *Lancet* 356:1876–1881.

Neubauer, H., Li, M., Kuehne-Heid, R., Schneider, A., and Kaiser, W.A. 2003. High grade and non-high grade ductal carcinoma in situ on dynamic MR mammography: characteristic findings for signal increase and morphological pattern of enhancement. *Br. J. Radiol.* 76:3–12.

Podo, F., Sardanelli, F., Canese, R., D'Agnolo, G., Natali, P.G., Crecco, M., Grandinetti, M.L., Musumeci, R., Trecate, G., Bergonzi, S., De Simone, T., Costa, C., Pasini, B., Manuokian, S., Spatti, G.B., Vergnaghi, D., Morassut, S., Boiocchi, M., Dolcetti, R., Viel, A., De Giacomi, C., Veronesi, A., Coran, F., Silingardi, V., Turchetti, D., Cortesi, L., De Santis, M., Federico, M., Romagnoli, R., Ferrari, S., Bevilacqua, G., Bartolozzi, C., Caligo, M.A., Cilotti, A., Marini, C., Cirillo, S., Marra, V., Martincich, L., Contegiacomo, A., Pensabene, M., Capuano, I., Burgazzi, G.B., Petrillo, A., Bonomo, L., Carriero, A., Mariani-Costantini, R., Battista, P., Cama, A., Palca, G., Di Maggio, C., D'Andrea, E., Bazzocchi, M., Francescutti, G.E., Zuiani, C., Londero, V., Zumui, I., Gustavino, C., Centurioni, M.G., Iozzelli, A., Panizza, P., and Del Maschio, A. 2002. The Italian multi-centre project on evaluation of MRI and other imaging modalities in early detection of breast cancer in subjects at high genetic risk. *J. Exp. Clin. Cancer Res. 21*:115–124.

Rebbeck, T.R., Friebel, T., Lynch, H.T., Neuhausen, S.L., van't Veer, L., Garber, J.E., Evans, G.R., Narod, S.A., Isaacs, C., Matloff, E., Daly, M.B., Olopade, O.I., and Weber, B.L. 2004. Bilateral prophylactic mastectomy reduces breast cancer risk in BRCA1 and BRCA2 mutation carriers: the PROSE Study Group. *J. Clin. Oncol. 22*:1055–1062.

Rijnsburger, A.J., Essink-Bot, M.L., van Dooren, S., Borsboom, G.J., Seynaeve, C., Bartels, C.C., Klijn, J.G., Tibben, A., and De Koning, H.J. 2004. Impact of screening for breast cancer in high-risk women on health-related quality of life. *Br. J. Cancer 91*:69–76.

Stoutjesdijk, M.J., Boetes, C., Jager, G.J., Beex, L., Bult, P., Hendriks, J.H., Laheij, R.J., Massuger, L., van Die, L.E., Wobbes, T., and Barentsz, J.O. 2001. Magnetic resonance imaging and mammography in women with a hereditary risk of breast cancer. *J. Natl. Cancer Inst. 93*:1095–1102.

Teifke, A., Hlawatsch, A., Beier, T., Werner, V.T., Schadmand, S., Schmidt, M., Lehr, H.A., and Thelen, M. 2002. Undetected malignancies of the breast: dynamic contrast-enhanced MR imaging at 1.0 T. *Radiology 224*:881–888.

Tilanus-Linthorst, M., Verhoog, L., Obdeijn, I.M., Bartels, K., Menke-Pluymers, M., Eggermont, A., Klijn, J., Meijers-Heijboer, H., van der Kwast, T., and Brekelmans, C. 2002. A BRCA1/2 mutation, high breast density and prominent pushing margins of a tumor independently contribute to a frequent false-negative mammography. *Int. J. Cancer 102*:91–95.

van Dooren, S., Seynaeve, C., Rijnsburger, A.J., Duivenvoorden, H.J., Essink-Bot, M.L., Tilanus-Linthorst, M.M., Klijn, J.G., De Koning, H.J., and Tibben, A. 2005. Exploring the course of psychological distress around two successive control visits in women at hereditary risk of breast cancer. *Eur. J. Cancer 41*:1416–1425.

Warner, E., Plewes, D.B., Hill, K.A., Causer, P.A., Zubovits, J.T., Jong, R.A., Cutrara, M.R., DeBoer, G., Yaffe, M.J., Messner, S.J., Meschino, W.S., Piron, C.A., and Narod, S.A. 2004. Surveillance of BRCA1 and BRCA2 mutation carriers with magnetic resonance imaging, ultrasound, mammography, and clinical breast examination. *JAMA 292*:1317–1325.

Watson, E.K., Henderson, B.J., Brett, J., Bankhead, C., and Austoker, J. 2005. The psychological impact of mammographic screening on women with a family history of breast cancer—a systematic review. *Psychooncology 14*:939–948.

19

Mammographic Screening: Impact on Survival

James S. Michaelson

Introduction

Although randomized trials have shown that cancer screening saves lives, they provide little insight into the magnitude of the reduction in death rates that has been achieved, or could be achieved, by screening. The reason is that trials may not reach all individuals in the intervention group, may not have achieved the optimal screening interval, or may contain some patients in the nontreatment arm who have chosen to have screening outside of the trial. Furthermore, the impact of screening during actual practice may be lower than its impact during a trial, where special attention may be paid to compliance. Data on the actual survival of patients who are using screening are also very difficult to collect, and thus may frequently be unavailable. Because at least 15 years are required before one can reach an accurate measure of lethality, our knowledge of the impact of screening on survival during the most recent two decades must invariably be incomplete. To be able to estimate the actual effect of screening on the cancer death rate, we have developed two methods: (1) a Computer Simulation Model of Cancer Growth and Detection, which can estimate the impact of various patterns of screening use, particularly various screening

intervals, on the cancer death rate; and (2) the SizeOnly Equation, which can estimate the risk of cancer death from data on tumor size. Here I will review both the application of these methods to breast carcinoma and the general medical implications of these findings; readers interested in the mathematical details and data can find them in the references.

Why Screening Works

Screening is believed to reduce cancer death by bringing cancers to medical attention at smaller, and thus more survivable, sizes. Surprisingly, although this idea has been appreciated for almost a century, there had not been a rigorous explanation for why this results in a lower level of cancer death, nor had there been a mathematical way to capture the relationship between tumor size and risk of cancer death. We found that a starting point could be made from an appreciation that the main cause of death for many cancers, including breast carcinoma, is the spread of cancer cells. If one or more cancer cells has spread to the periphery before the cancer has been removed by surgery and/or radiation therapy, then cancer will remain in the body and may give rise to distant, lethal, metastatic disease (Michaelson, 1999; Michaelson *et al.*, 1999, 2002a, 2005). Consider *p* as the probability, for every cell in a tumor of *N* cells, that a cell will leave the mass and give rise to distant metastatic disease. We have been able to develop simple mathematical expressions for estimating the value of this probability of spread from data on the survival rates of patients with tumors of various sizes (Michaelson, 1999; Michaelson *et al.*, 1999, 2002a, 2005). Surprisingly, these calculations found that for breast carcinoma, renal cell carcinoma, and melanoma, the value of *p* changes as tumors increase in size, *N*, such that the relationship between *p* and *N* is well fit to a power function:

$$p = aN^b \qquad (1)$$

where $b \sim -2/3$ for all three cancers that we have examined, while the value of *a* is characteristic of each malignancy. A very likely explanation for why the value of *p* declines with tumor size in a way that is captured by Eq. (1) is that as tumors get bigger there are more cells to "push aside" before a cancer cell can get out. In fact, we have been able to demonstrate this mathematically (Michaelson *et al.*, 2005).

As we might imagine, given that *p* is the probability that a cell will leave a mass of cancer and give rise to distant metastatic disease, and *N* is the number of cells in that mass of cancer, the overall chance that the mass has given rise to such a lethal event, *L*, is roughly *p* times *N*. (For more precise methods of calculating the value of *L*, see Michaelson, 1999 and Michaelson *et al.*, 2002a, 2005.) Because Eq. (1) allows us to estimate the value of *p*, and because breast carcinomas have been found to grow exponentially over the sizes that they are seen clinically, this made it possible to develop a computer simulation that could recapitulate the simultaneous day-to-day increase in tumor cell number, *N*, and lethality, *L* (Michaelson *et al.*, 1999, 2001). The results of this computer program revealed an essential and unobvious feature of breast cancer biology: the chance of lethal metastatic disease does not increase gradually over time, but changes dramatically over a relatively short period. For example, the simulation results suggest that while 92% of breast carcinomas of 7 mm are curable by local excision, by 1½ years, when the tumors have reached 18 mm, only 75% will still be curable, and in an additional 1½ years, having reached 47 mm, only 33% are curable. This result of the simulation provides a likely explanation for why mammographic screening works: the rate of breast carcinoma growth, the probability of breast cancer spread, and the mammographic detectability of breast cancers, all have such fortuitous values that mammography is capable of finding tumors just before the point in time when there is an explosive increase in the fraction of cancers incurable by local treatment.

Cancers Become More Lethal as They Increase in Size

We could also use Eq. (1) to derive an expression, which we have called the SizeOnly Equation (Michaelson *et al.*, 2002a), for relating tumor size, *D*, to the risk of cancer death, *L*:

$$L = 1 - e^{-QD^Z} \qquad (2)$$

where *e* is the exponential constant, $Z = 1.33$ and $Q = 0.0062$. The SizeOnly Equation has proven to be remarkably good at predicting the risk of death for a considerable number of populations of breast carcinoma patients, as well as for subpopulations of patients whose tumors were detected at screening or detected on clinical grounds (Michaelson *et al.*, 2002a, 2003d). As we shall see, this has proved to be a very useful tool for gauging the impact of screening from data on the sizes of the tumors found in women who used or did not use screening.

Present and Future Life-saving Impact of Screening

Only two studies (the Health Insurance Plan of New York (HIP) and Swedish Two-County Trials) have had the statistical power to detect the survival difference between women who are screened and women who are not screened. No trials have as yet compared different usages of screening, such as different screening intervals. Nor are there likely ever to be such trials, which would be prohibitively expensive to carry out and would not yield results for decades. Thus, we must rely on other methods to estimate the impact of such various usages of screening, such as various screening intervals, on the reduction in breast cancer death. One such approach is computer simulation (Michaelson *et al.*, 1999, 2000, 2001; Blanchard *et al.*, 2004, 2006). We saw the core of our simulation in the "Why Screening Works" section. However, before this simulation could be used to make useful predictions of the consequences of various usages of screening, the simulation had to be provided with accurate information on cancer growth, detection, and lethality. To do so, we developed new mathematical methods and collected new data (Michaelson *et al.*, 2001) for estimating the sizes at which cancers become detectable at screening and on clinical grounds (Michaelson *et al.*, 2003b), the growth rate of breast carcinoma (Michaelson *et al.*, 2003c), and the probability of the spread of cancer cells (Michaelson *et al.*, 2002a, 2005). The computer simulation was also provided with data from the U.S. Census and SEER (Surveillance Epidemiology and End Results) national cancer data repositories on the age-associated incidence of breast carcinoma (Ries *et al.*, 2000), U.S. life expectancy by age, and age structure of the U.S. population (Anonymous, 1998). Our simulation also incorporated information on the costs of screening: both the direct costs associated with screening itself and the indirect costs, such as those that might arise from such expenses as biopsies used to rule out cancer in women with spurious signs that appear at screening. These cost data were used in the simulation to calculate the cost/benefit values of screening.

These simulation results revealed that high levels of breast cancer survival should be achievable if women are screened with sufficient frequency. For example, the simulation results indicated that for women age 65, breast cancer survival of > 90% should be achievable if the women utilize screening at least once a year, and even higher survival levels with more frequent screening. In contrast, screening every 3 years would appear to achieve a more modest survival of ~ 85%, while screening every 5 years yields a survival rate of ~ 75%.

The simulation method made it possible to estimate the expected survival rates for both individuals and the population as a whole (estimated for all women, including those not using screening). The simulation was also made to provide values for the benefit of screening, in terms of "Cancer Free Years of Life Saved" per mammogram or per woman. For example, the simulation results revealed that the American Cancer Society recommendation of yearly screening from age 40 should be a highly effective strategy, yielding a populationwide 88% chance of breast cancer survival, which corresponds to ~ 66% populationwide reduction in death. Less intensive patterns of screening yielded lower levels of breast cancer survival. In contrast, the simulation revealed that the pattern of screening used in the United Kingdom of one mammogram every 36 months from age 50 to age 70 is capable of achieving only a 12% populationwide reduction in death.

Surprisingly, the simulation revealed that there might even be additional benefit from screening as frequently as twice a year from age 30. Such a strategy would appear capable of achieving a populationwide 91% chance of breast cancer survival, which corresponds to a 74% reduction in death, in comparison to women who do not use screening. Using a well-established mathematical technique, the equimarginal method (Samuelson and Nordhaus, 1998; Friedman 1990), it could be seen that this was also an efficient usage of screening, reaching the maximal reduction in death that was practical. Little additional benefit was to be derived by screening more frequently than twice a year, or screening women younger than 30. At the other end of the age spectrum, the simulation found no upper age limit where women no longer receive benefit. The simulation also revealed that a biannual screening strategy for women age 30 and older is cost-effective, with a cost of ~ $10,000 for each cancer-free year of life saved. Most medical procedures are far more expensive for the years of life that they save (Tengs et al., 1995). Organ transplant operations may have costs that reach into the millions of dollars per year of life saved. Indeed, these calculations tell us that even at this most intensive usage of screening, mammography remains, after immunization, one of our cheapest ways of saving lives.

Life-saving Potential of Screening

What do these simulation results mean in terms of the number of breast cancer deaths that might be prevented by the optimized use of screening? Reaching the predicted survival level of 88% by prompt annual attendance screening from age 40, as predicted by the simulation, would mean an enormous reduction in breast cancer death because the current level of breast cancer survival is believed to be ~ 55–70%. Because there are more than 40,000 breast cancer deaths in the United States per year, this translates into tens of thousands of lives saved. The simulation also indicates that twice-yearly screening from age 30 might reach populationwide breast cancer survival levels of 91%. Because more than 200,000 women are found to have breast carcinoma in the United States each year, this translates into more than 5000 extra lives saved. The simulation results also indicate that women's widespread failure to follow the current guidelines probably leads to much higher breast cancer death rates, perhaps two or three times higher, than might be expected among the few women now using screening regularly.

Tumor Size and Survival

A second source of information on the benefit of screening could be found in data on the sizes of the cancers seen among women who used, or did not use, screening (Michaelson et al., 2001, 2002a, 2002b, 2003d). As we have noted, when such size data are available, the SizeOnly Equation provides a way to estimate the survival of such women. A particularly informative data set included information on 810 breast carcinoma patients treated at Massachusetts General Hospital in the 1990s (Michaelson et al., 2001). Of the 810 cancers, 204 were found clinically in women who had never used screening, 427 were found by mammography at screening, and 179 were found as palpable masses in women who had had a previous negative screening mammogram (Michaelson et al., 2001). The 179 cancers detected on clinical ground in women who had had at least one previous negative mammogram were especially informative because the time since the previous negative mammogram was known for each patient; 68 of the cancers were found within a year of the previous mammogram, and 111 were found more than a year afterward. By using tumor growth data to backcalculate the likely size of each of these 111 cancers, it could be seen that virtually all would have been too small to have been detected at screening at the time of the previous mammogram. Thus, almost all of these 111 clinically detected cancers appeared at larger, and thus more lethal, size because the women had failed to come back on time for their annual mammograms. Calculations made with the Size Only Equation with data on the size of the cancers seen in the 179 women who had never used screening indicated that these women could expect a 25% breast carcinoma death rate. Similar calculations made with data on the size of the cancers seen in the 492 women with cancers found either at screening (427) or clinically within a year of the mammogram (68) indicated that these women could expect a 16% breast carcinoma death rate (Michaelson *et al.*, 2002a, 2003d). In other words, women who attend screening regularly can expect to reduce their risk of breast cancer death by a third, to 16%. Note that the breast carcinoma survival rate estimated from size data for women who attend screening regularly, at 84%, agrees closely to the value of 88% estimated by the simulation model stated earlier.

These size data also gave evidence of how the effectiveness of screening is reduced when women do not return on time for their annual mammograms (Michaelson *et al.*, 2001, 2002a, 2003d). This can be seen by considering the mixture of breast carcinomas found at screening together with the palpable cancers found within various periods of time after the negative mammogram. Such data reveal that the greater the delay in return to screening, the greater will be the average size of the cancers in the population. Very few palpable cancers are seen within the first 6 months after a negative mammogram, but they begin to appear in considerable numbers from ~ 6 months onward (Michaelson *et al.*, 2003b, d). By ~ 9 months, these palpable cancers appear at a regular rate that continues for many years. We call the period soon after a mammogram, when few palpable cancers are seen, the "protective shadow of mammography." The data on the time course of the appearance of palpable cancers and the impact of these cancers on the average size of the cancers in the screening populations revealed that this protective shadow lasts only ~ 6 to ~ 9 months (Michaelson *et al.*, 2003b, c). These size data could also be translated into expectations of survival for women who used screening at various intervals by the Size Only Equation. These data suggest that once one year has passed there is no grace period for return to screening; any amount of delay in return will increase tumor size and thus the level of breast carcinoma death.

These size data also provided a means for estimating the impact of patient age and the density of the breast on the efficiency of detection (Michaelson *et al.*, 2003d). This could be accomplished by examining the mixture of cancers found at screening and after screening as palpable masses. These studies revealed that although the cancers tended to be larger for women with denser breasts and younger women than for women with less dense breasts and older women, women of all ages and density groups who used screening had smaller, and thus less lethal, cancers than women who did not use screening. This appeared to be the case even for women as young as

30 years. Thus, these calculations indicate that although screening is likely to be less effective in women in their thirties than in older women, screening should still be expected to lead to a reduction in tumor size and lethality in this age group.

False-positives

Screening is not without its negative consequences, particularly "false-positives" that lead to biopsies and other medical interventions among women who subsequently are not found to have cancer. One widely quoted study, by Elmore *et al.* (1998), "estimate[d] that among women who do not have breast cancer, 18.6% will undergo a biopsy after 10 mammograms." Remarkably, no women in this study had undergone 10 mammograms. It has long been appreciated that if a radiologist does not have access to a patient's previous mammogram, the patient is at much higher risk for a false-positive outcome. Thus, populationwide estimates of the rate of false-positives, such as those provided by Elmore *et al.* (1998), are fraught with the confounding influence of the intermittent use of screening on the overall false-positive rate. Fortunately, the very large number of women who had undergone screening at Massachusetts General Hospital made it possible to examine the false-positive rate among women who had chosen to use screening with various degrees of regularity (Michaelson, in press). These studies revealed that while the overall populationwide rate of false-positives leading to biopsies and false-positive mammography assessments were similar to those found in other studies, such as those reported by Elmore *et al.* (1998), much of the burden of these false-positive events was borne by women who used screening intermittently. For example, 2.9% of the women who had a screening examination in 1996 and received five mammograms over the next 5 years had false-positives leading to biopsies, while 4.6% of women who utilized only three mammograms over the 5-year period had a biopsy not revealing cancer. Among women who used screening regularly, the risk of having a biopsy not revealing cancer declined to 0.25% per year after several years of screening, a value that is lower than the risk of these events among women not using screening. These findings are reassuring, for they indicate that increased use of screening should not be expected to lead to increased rates of false-positives. In fact, the somewhat counterintuitive, but encouraging, lesson from these data is that prompt attendance to mammographic screening actually leads to a reduced occurrence of false-positive mammographic results and unnecessary biopsies (Michaelson, in press).

How is Screening Actually Used?

The database of women who attended screening at Massachusetts General Hospital over the past two decades provided an unusually rich source of information on the patterns of screening, and indeed, made possible detailed analysis of the patterns of screening use to be carried out to date (Blanchard *et al.*, 2004; Colbert *et al.*, 2004). These studies revealed that most women begin screening close to their fortieth birthday, as recommended, but that prompt return after that is rare: very few return on time for their subsequent mammograms, and many never return. It is this failure of women to return promptly for their annual mammograms that is the critical failure-point in our ability to use screening to its maximal life-sparing potential.

That most women begin screening near the age recommended could be seen from data on women screened at Massachusetts General Hospital between 2000 and 2002. The median age of the women attending the first mammograms was 40.4; 60% of women began screening by age 40, and almost 90% by age 50 (Colbert *et al.*, 2004). These data agree with national results. For example, the Behavioral Risk Factor Surveillance Survey found that by 1997, 85% of women over the age of 40 report having had at least one mammogram.

These favorable findings indicating that most women are beginning screening near their fortieth birthdays were not seen among specific subpopulations of women, particularly women of lower socioeconomic status. Thus, while the median age of first mammogram for women in the population as a whole was 40.4 years, the median age of first mammogram was 41.0 for African-American women, 40.3 for Hispanic women, 41.2 for women without a primary care physician, 46.6 for women without private health insurance, 49.3 for women who did not speak English, and 55.3 for women who both lacked private health insurance and spoke a language other than English (Colbert *et al.*, 2004).

As noted, most women begin screening near their fortieth birthday as recommended, but most do not come back on time, or do not come back at all. This failure has very negative health consequence. For example, of the women who had a negative mammogram at Massachusetts General Hospital in 1992, only 6% had nine more screening mammograms during the next 10 years, whereas 40% of these women had fewer than five mammograms over the decade and 18% never returned. The median number of mammograms used during the decade was five, and computer simulation analysis suggests that this degrades the life-sparing benefit of screening by ~ 50% (Blanchard *et al.*, 2004). These findings agree with those made in other populations. For example, Ulcickas-Yood *et al.* (1999) reported that only 16% of the women who had a mammogram between 1983 and 1993 at the largest HMO in Michigan took advantage of all five mammograms during the 5-year period following the index mammogram. Sabogal *et al.* (2001), using 1992–1998 California Medicare data, found that only 30% of non-HMO women age 65 and older who utilized screening did so regularly without missing screening more than 2 years in a row. Phillips *et al.* (1998), using several sources of data, found

that while 70% of women age 50–74 have had at least one mammogram, only 16% have utilized annual screening.

Women from traditionally underserved socioeconomic, racial, and ethnic groups; women without insurance; and women who did not speak English were found to be less likely to return on time, although these associations were small in magnitude. Women attending their first mammogram or those who had not previously returned promptly for screening were also less likely to return on time. Women age 55–65 had higher levels of usage than younger or older women. However, none of the subpopulations of women sorted by age, race, ethnicity, zip code, income, previous screening use, or medical history approached either of the extremes of widespread failure to return or prompt annual screening over extended periods of time. Although women who returned on time for the last exam were more likely to return on time for their next exam, even this correlation was not particularly strong. These data indicate that it is not the characteristics of the woman, but the performance of the system, that is the main reason many women do not return on time. Issues such as the shortage of screening facilities, and thus long waiting time for making appointments, no doubt contribute to the problem. Practical difficulties in reminding and tracking women would seem to be the likely reasons many women failed to return on time for their mammograms.

As might be expected, women with a prior breast cancer had a higher degree of screening use than the population as a whole (Blanchard *et al.*, 2004). However, even among this group, considerable numbers of women did not return on time, or indeed at all, for screening. Since annual screening in this group serves the double purpose of detecting local recurrence and detecting second breast cancers, for which these women are at higher risk, the fact that utilization is far from ideal in this population is of considerable concern.

Women obtaining their first mammogram were found to be at particularly high risk for not returning. Indeed, one in four of such women will never return for a subsequent mammogram (Blanchard *et al.*, 2004).

Present Status of Breast Cancer Screening

Both the computer simulation studies and the studies of the sizes of the breast carcinomas found among women who use screening indicate that very high breast carcinoma survival rates (~ 90%) should be achievable by screening if women follow the American Cancer Society's recommendation of prompt annual screening from age 40. However, studies of the actual utilization of screening suggest that we fail to achieve most of the life-saving benefit of screening principally because most women fail to return on time, or not at all, for their annual mammograms.

The simulation studies indicate that even higher survival rates might be achieved by twice yearly screening from age 30. Such a screening strategy would also appear to be highly cost-effective. Screening more frequently than twice a year, or in women younger than age 30, appears to yield very little additional benefit. The transition from once yearly screening from age 40, to twice yearly screening from age 30, would appear to have the potential to save as many as 5000 lives each year in the United States. It is striking that ~ 5% of the invasive cancers occur in women younger than 40 (Ries *et al.*, 2000). Furthermore, since younger women have many more years of productive life, the potential loss in years of life is even greater, with ~ 10% of the potential years of life that could be lost to breast cancer being located in women younger than 40. Analysis of the sizes of the cancer seen in the screening populations of women of various ages (Michaelson *et al.*, 2003d) also suggests that screening should be effective in this group, in terms of bringing cancer to medical attention at smaller, and thus less lethal, sizes. As regards the benefit of screening twice a year, it is relevant that the "protective shadow of mammography"—the time period after a negative mammogram when the rate of appearance of larger palpable masses is reduced—only lasts ~ 6 months. This provides empirical support for the computer simulation results that indicated that we could expect benefit by reducing the screening interval to as frequently as twice a year. Finally, the data do not support concerns that increasing the frequency of screening will increase the occurrence of false-positive events; this suggests that women who attend screening regularly actually have a lower level of such negative events (Michaelson, in press).

One in four women will attend their first mammogram and never return again (Blanchard *et al.*, 2004). Clearly, the screening experience itself is discouraging many women from receiving the benefit of screening. One year after the single screening experience, the cancers found among these women will be just as large and just as lethal as the cancers found among women who never use screening. The simulation data suggest that this will double the risk of cancer death among these women.

Many women find mammography unpleasant and stressful. One impressive study found that if women are provided with a button that controls the compression of the mammography paddle, they perceive less pain associated with screening (Kornguth *et al.*, 1993) Although this study was published 14 years ago, such a device has yet to become available commercially. It seems plausible that greater attention to the human-factors aspect of screening would appear to have a considerable impact on the reduction in breast cancer death by increasing the percentage of women who return.

Among the women who do return, 2, 3, or more years elapse between visits (Blanchard *et al.*, 2004). Both the simulation studies, and the studies carried out on the time course of the appearance of larger palpable cancers after a negative mammogram, show that there is no grace period for return to screening. Once a year has passed, women are back in the same group as women who never use screening, in terms of the regular rate of appearance of larger, palpable masses (Michaelson *et al.*, 2003b). Our computer simulation studies indicate that this alone probably reduces the life-sparing potential of screening by 50%. A number of studies have shown that as many as 40% of women who make appointments for screening examinations will forget to show up (McCoy *et al.*, 1991; Margolis *et al.*, 1993). It seems plausible that greater attention to tracking and reminding women, so as to help them make and then attend their annual screening examinations, could lead to considerable reductions in breast cancer death.

Analysis of the actual screening usage pattern among women who use screening at Massachusetts General Hospital (Blanchard *et al.*, 2004; Colbert *et al.*, 2004) and elsewhere (Ulcickas-Yood *et al.*, 1999; Sabogal *et al.*, 2001; Phillips *et al.*, 1998) indicates that the greatest reduction in breast cancer death should be achievable simply by finding ways to encourage women to return on time for their annual screening mammograms. Most women, perhaps as many as 85%, begin to go to screening close to their fortieth birthdays. Thus, little is to be gained from populationwide efforts to encourage entry into the screening process. Instead, public health efforts should be focused on those subpopulations of women at highest risk for not using screening: women without private insurance, women without a primary care physician, and women who do not speak English. Most of our efforts to reduce breast cancer death should be concentrated on finding ways to encourage women who have already come for a mammogram to return *promptly* and *repeatedly*.

References

Anonymous. 1998. *Statistical Abstracts of the United States 1998.* Washington, DC: U.S. Government Printing Office.

Beckett, J.R., Kotre, C.J., and Michaelson, J.S. 2003. Analysis of benefit: risk ratio and mortality reduction for the U.K. Breast Screening Programme. *Br. J. Radiol.* 76:309–320.

Blanchard, K., Colbert, J., Kopans, D., Moore, R., Halpern, E., Hughes, K., Tanabe, K., Smith, B., and Michaelson, J. 2006. The risk of false positive screening mammograms, as a function of screening usage. *Radiology* 240:335–342.

Blanchard, K., Weissman, J., Moy, B., Puri, D., Kopans, D., Kaine, E., Moore, R., Halpern, E., Hughes, K., Tanabe, K., Smith, B., and Michaelson, J. 2004. Mammographic screening: patterns of use and estimated impact on breast carcinoma survival. *Cancer* 101:495–507.

Cady, B., and Michaelson, J.S. 2001. The life-sparing potential of mammographic screening. *Cancer* 91:1699–1703.

Colbert, J., Bigby, J.A., Smith, D., Moore, R., Rafferty, E., Georgian-Smith, D., D'Alessandro, H.A., Yeh, E., Kopans, D.B., Halpern, E., Hughes, K., Smith, B.L., Tanabe, K.K., and Michaelson, J. 2004. The age at which women begin mammographic screening. *Cancer* 101:1850–1859.

del Carmen, M.G., Hughes, K.S., Halpern, E., Rafferty, E., Kopans, D., Parisky, Y.R., Sardi, A., Esserman, L., Rust, S., and Michaelson, J. 2003. Racial differences in mammographic breast density. *Cancer* 98:590–596.

Elmore, J.G., Barton, M.B., Moceri, V.M., Polk, S., Arena, P.J., and Fletcher, S.W. 1998. Ten-year risk of false positive screening mammograms and clinical breast examinations. *N. Engl. J. Med.* 338:1089–1096.

Friedman, D.D. 1990. *Price Theory: An Intermediate Text.* Cincinnati, OH: South-Western Publishing Co.

Jones, J.L., Hughes, K.S., Kopans, D.B., Moore, R.H., Howard-McNatt, M., Hughes, S.S., Lee, N.Y., Roche, C.A., Siegel, N., Gadd, M.A., Smith, B.L., and Michaelson, J.S. 2005. Evaluation of hereditary risk in a mammography population. *Clin. Breast Cancer* 6:38–44.

Kornguth, P.J., Rimer, B.K., Conaway. M.R., and Sullivan, D.C. 1993. Impact of patient controlled compression on the mammography experience. *Radiology* 186:99–102.

Margolis, K.L., Lurie, N., McGovern, P.G., and Slater, J.S. 1993. Predictors of failure to attend scheduled mammography appointments at a public teaching hospital. *J. Gen. Intern. Med.* 8:602–605.

McCoy, C.B., Nielsen, B.B., Chitwood, D.D., Zavertnik, J.J., and Khoury, E.L. 1991. Increasing the cancer screening of the medically underserved in south Florida. *Cancer* 67:1808–1813.

Michaelson, J.S. 1999. The table of molecular discreteness in normal and cancerous growth. *Anticancer Res.* 19:4853–4867.

Michaelson, J.S. 2001. Using information on breast cancer growth, spread, and detectability to find the best ways to use screening to reduce breast cancer death. *J. Women's Imag.* 3:54–57.

Michaelson, J., Halpern, E., and Kopans, D. 1999. A computer simulation method for estimating the optimal intervals for breast cancer screening. *Radiology* 212:551–560.

Michaelson, J.S. Optimal lifelong breast cancer screening strategies determined by a computer simulation model of invasive breast cancer growth and spread. In press.

Michaelson, J.S., Kopans, D.B., and Cady, B. 2000. The breast cancer screening interval is important. *Cancer* 88:1282–1284.

Michaelson, J.S. Satija, S., Moore, R., Weber, G., Garland, G., and Kopans, D.B. 2001. Observations on invasive breast cancers diagnosed in a service screening and diagnostic breast imaging program. *J. Women's Imag.* 3:99–104.

Michaelson, J.S., Wyatt, J., Weber, G., Moore, R., Kopans, D.B., and Hughes, K. 2002a. The prediction of breast cancer survival from tumor size. *Cancer* 95:713–723.

Michaelson, J.S., Satija, S., Moore, R., Weber, G., Garland, A., Phuri, D., and Kopans, D.B. 2002b. The pattern of breast cancer screening utilization and its consequences. *Cancer* 94:37–43.

Michaelson, J.S., Satija, S., Kopans, D.B., Moore, R.A., Silverstein, M., Comegno, A., Hughes, K., Taghian, A., Powell, S., and Smith, B. 2003a. Gauging the impact of breast cancer screening, in terms of tumor size and death rate. *Cancer* 98:2133–2143.

Michaelson, J.S., Satija, S., Moore, R., Weber, G., Garland, A., Kopans, D.B., and Hughes, K. 2003b. Estimates of the sizes at which breast cancers become detectable on mammographic and or clinical grounds. *J. Women's Imag.* 5:11–20.

Michaelson, J.S., Satija, S., Moore, R., Weber, G., Garland, A., and Kopans, D.B. 2003c. Estimates of the breast cancer growth rate and sojourn time from screening database Information. *J. Women's Imag.* 5:3–10.

Michaelson, J.S., Silverstein, M., Sgroi, D., Cheongsiatmoy, J.A., Taghian, A., Powell, H. K., Cogmegno, A., Tanabe, K., and Smith, B.A. 2003d. The effect of tumor size and nodal status on the lethality of breast cancer. *Cancer* 98 2133–2143.

Michaelson. J.S., Cheongsiatmoy, J.A., Dewey, F., Silverstein, M., Sgroi, D., Smith, B., and Tanabe, K.K. 2005. The spread of human cancer cells occurs with probabilities indicative of a nongenetic mechanism. *J. Cancer* 93:1244–1249.

Phillips, K.A., Kelikowse, K., Baker, L.C., Chang, S.W. and Brown, M.L. 1998. Factors associated with women's adherence to mammography screening guidelines. *HSR: Health Sci. Res.* 33:29–53.

Ries, L.A.G., Eisner, M.P., Kosary, C.L., Hankey, B.F., Miller, B.A., Clegg, L., and Edwards, B.K. (Eds). 2000. *Seer Cancer Statistics Review, 1973–1997.* Bethesda, MD: National Cancer Institute.

Sabogal, F., Merrill, S.S., and Packel, L. 2001. Mammography rescreening among older California women. *Health Care Financ Rev.* 22:63–75.

Samuelson, P.A., and Nordhaus. W. D. 1998. *Economics.* 16th edition. New York: McGraw-Hill.

Tengs, T.O., Adams, M.E., Pliskin, J.S., Safran, D.G., Siegel, J.E., Weinstein, M.C., and Graham, J.D. 1995. Five hundred life-saving interventions and their cost-effectiveness. *Risk Analysis* 15:369–390.

Ulcickas-Yood, M., McCarthy, B.D., Lee, N.C., Jacobsen, G., and Johnson, C.C. 1999. Patterns and characteristics of repeat mammography among women 50 years and older. *Cancer Epidemiol. Biomark. Prev.* 8:595–599.

20

False-positive Mammography Examinations

Pamela S. Ganschow and Joann G. Elmore

Introduction

It would be wonderful if screening and diagnostic mammography were always completely accurate in diagnosing breast cancer. A perfect "gold standard" test would have 100% sensitivity, which means that all individuals with breast cancer would be correctly identified by a positive test. It would also be 100% specific in that all of the individuals who do not have breast cancer would be told that their test is negative. Unfortunately, the perfect clinical test is difficult to achieve. This chapter reviews the topic of false-positive test results in mammography. We begin by defining false-positive examinations and describing false-positive rates in the United States and abroad. We then review factors that have been associated with false-positive examinations, including characteristics of the patients, the radiologists, and the health-care system. We also examine the impact of false-positive examinations on patients and health systems, and we end with suggestions of ways to reduce the false-positive rate while maintaining high sensitivity.

Definitions

Defining a false-positive test result in mammography might seem simple, but it is not. First, we need to establish definitions of "false," "positive," and "mammography examination." In general terms, a false-positive examination is a positive examination in an individual with no breast cancer. A woman can have either a positive or a negative interpretation, and she can have either a breast cancer diagnosed or no cancer, resulting in four possible outcomes, one of which is a false-positive outcome. A simplified 2×2 table is often used when discussing these definitions (Fig. 119). A false-positive rate is usually defined as the total number of individuals without breast cancer who have a positive test (in the shaded cell labeled "FP") divided by the total number of individuals with no breast cancer. The false-positive rate is therefore equal to $(1 - \text{specificity})$.

A "positive" examination is generally considered one in which an abnormality is noted and additional evaluation is recommended. However, there is variability in what is assumed to be "positive" and how such findings are documented. Over the years, clinicians have used many different methods to report their interpretation of mammography examinations. Assessment and recommendations have ranged from quantitative scores to qualitative descriptive terms.

In the United States, the American College of Radiology (ACR) took the lead in coordinating efforts to address the quagmire of variability in mammography reporting. The ACR worked in cooperation with the National Cancer

		Breast Cancer Status		
		Positive (+)	Negative (−)	
Mammogram Assessment	Positive (+)	TP	FP	Total Tests +
	Negative (−)	FN	TN	Total Tests −
		Total Individuals with Cancer	Total Individuals without Cancer	Total Tests

Legend	Calculation of Performance Measures
TP = True Positive	Recall = (Total Test +)/(Total Tests)
FP = False Positive	Sensitivity = TP/(TP + FN)
FN = False Negative	Specificity = TN/(FP + TN)
TN = True Negative	Positive Predictive Value = TP/(TP + FP)
	Negative Predictive Value = TN/(FN + TN)
	False-Positive Rate = FP/(FP + TN)
	False-Positive Rate = 1− Specificity

Figure 119 Indices for calculating accuracy in mammography practice, with the false-positive cell shaded.

Institute, the Centers for Disease Control and Prevention, the Food and Drug Administration, the American Medical Association, the American College of Surgeons, and the College of American Pathologists. The result of this collaboration was the publication of the Breast Imaging-Reporting and Data System (BI-RADS™) breast-imaging atlas. This atlas, initially published in 1992, has been regularly updated in an effort to standardize mammography reporting and reduce confusion in breast imaging interpretations. Even with the use of BI-RADS assessment and recommendation categories, however, a single definition of a "positive" examination has remained elusive. A positive screening examination is defined in the BI-RADS atlas as "one for which a recall is initiated (BI-RADS category 0) or one that requires a tissue diagnosis (BI-RADS categories 4 and 5)" (ACR, 2003). The BI-RADS atlas adds a cautionary note that their definition differs from the definition used in the final U.S. Mammography Quality Standards Act (MQSA). The MQSA limits the definition of a "positive" examination to those that recommend tissue diagnosis (MQSA, 1992).

The definitions of positive results agreed upon by the United States Breast Cancer Surveillance Consortium (BCSC) are also slightly different from those stated in the BI-RADS atlas (Carney *et al.*, 2003). The BCSC is a rich resource for studies designed to assess the delivery and quality of breast cancer screening and related patient outcomes in the United States (Ballard-Barbash *et al.*, 1997). The BSCS investigators consider a positive examination to be one that has a BI-RADS assessment of category 0, 4, 5, or any category 3 with a recommendation for immediate additional evaluation or biopsy (Carney *et al.*, 2003).

Now that a positive examination has been defined (in a few different ways), the definition of a "false"-positive examination is needed. In general terms, if no breast cancer is diagnosed in a specific period of time, then the mammogram is considered a "false"-positive examination. However, some studies report the cancer outcomes based on cancers diagnosed within 1 year, and other studies report outcomes diagnosed within a 2-year period.

The three definitions of a false-positive (FP) examination defined in the BI-RADS atlas are as follows (ACR, 2003):

1. FP_1: No known tissue diagnosis of cancer within 1 year of a positive screening examination (BI-RADS™ category 0, 4, or 5).

2. FP_2: No known tissue diagnosis of cancer within 1 year after recommendation for biopsy or surgical consultation on the basis of a positive examination (BI-RADS category 4 or 5).

3. FP_3: Benign tissue diagnosis of cancer within 1 year after recommendation for biopsy on the basis of a positive examination (BI-RADS™ category 4 or 5).

Finally, the type of "examination" needs to be defined. Mammography examinations are usually categorized as "screening" or "diagnostic," depending on whether or not a woman has current symptoms or signs of breast cancer. Although data on these two categories are usually analyzed separately, false-positive rates will vary depending on whether screening examinations are considered exclusively if diagnostic examinations are also included.

Current Estimates of False-positive Rates in the United States and International Guidelines

Given the varying ways in which a false-positive examination can be defined, it is not surprising that the published reports of false-positive rates have also varied widely, ranging from < 1% to > 10% (IARC, 2002). In general, the false-positive rate of screening mammography in the United States is estimated to be approximately 10% and is currently twice the rate seen in other countries (Elmore et al., 2003; Smith-Bindman et al., 2003). In addition, the U.S. false-positive rate has been noted to increase over time (Elmore et al., 1998). Concern over medical malpractice liability might be one potential reason for this increase.

Interestingly, the suggested goals for the recall rate differ substantially between countries. While the recall rate is not exactly the same as a false-positive rate (Fig. 119), given the low prevalence of breast cancer, the majority of women recalled will have a false-positive examination. In the United States, the goal for an acceptable recall rate is < 10% (Basset et al., 1994). The 2006 recommendations from the Commission of the European Communities describes both acceptable levels and desirable levels for recall rates, and also differentiates goals for the initial screen and subsequent regular screening examinations. The acceptable recall rates in European countries is < 7% for initial screening and < 5% for subsequent regular screening; the desirable recall rates are < 5% for initial screening and < 3% for subsequent regular screening (European Commission, 2006).

One would expect that a higher false-positive rate, similar to that in the United States, would translate into an increase in the number of breast cancers detected compared with countries that have lower recall rates. Importantly, however, the lower false-positive rates seen in some European countries do not appear to occur at the expense of an increase in the false-negative rate (or lower sensitivity). Simply stated, the United Kingdom and other countries seem to have been able to limit false-positive rates to half of the current U.S. rates without missing more cancers (Elmore et al., 2003; Smith-Bindman et al., 2003). In fact, studies performed in both the United States and Europe suggest that cancer detection rates level off at a recall rate of 4–5%, with a disproportionate increase in false-positive rates associated with higher recall rates (Otten et al., 2005; Yankaskas et al., 2001). A more detailed discussion of the differences in false-positive rates between the United States and other countries can be found near the end of this chapter.

Cumulative False-positive Rates

Although a false-positive rate of < 10% may not seem problematic, concern has been raised regarding the cumulative risk for a false-positive mammogram. This cumulative risk is estimated to be much higher than 10% because breast cancer screening for most individuals is recommended regularly for many years. For example, if a woman begins screening at age 40 and has a mammogram every year, she will have received 30 screening mammography examina-

tions by the time she turns 70 years of age. Even if she begins screening at age 50 and has examinations every 2 years, she will have had 10 examinations by the time she is 70. Discussed below are several studies that have shown that a woman's chance of having a false-positive mammography examination increases as she undergoes more examinations.

One retrospective cohort study provided estimates of the cumulative false-positive rate of screening mammography in the United States (Elmore et al., 1998). This study, performed in a health plan in the Northeast, involved a medical review of 2400 women over a 10-year period. A median of four mammograms were obtained per woman over the decade study period. False-positive examinations were noted in 6.5% of the screening mammography examinations. Of the women who were screened, approximately 24% had at least one false-positive mammogram. Bayesian modeling techniques were used to provide estimates of a woman's risk after repeated screenings up to 10 screening examinations (Fig. 120). After 10 screening examinations, it was estimated that approximately one-half of the women would have experienced at least one false-positive examination (the estimated cumulative false-positive rate was 49.1%; 95% confidence interval [CI] 40.3–64%).

With the false-positive rate varying among countries, cumulative false-positive rates reported in the literature have also varied. In the screening mammography program in British Columbia, Canada, the cumulative risk for a false-positive mammogram after 10 screens was estimated to be 35% for women 50–59 years of age (Olivotto et al., 1998). A study in Norway found a considerably lower cumulative risk for false-positive examinations (20%) over two decades for biennial screening (Hofvind et al., 2004). To better assess the potential impact of these varying false-positive rates on estimates of the cumulative false-positive risk, the International Agency for Research on Cancer (IARC) has estimated the cumulative risk of ever having a false-positive result under various conditions: varying false-positive rates at the first screen and the subsequent screens and varying frequencies of screening during a 20-year program of screening (every 3 years, every 2 years, or annually) (IARC, 2002). If the false-positive rate is 2% at the first screen and 1% at subsequent screens, the cumulative false-positive risk is estimated to be 10% after 10 screening examinations.

Predictors of False-positive Mammograms

Much effort has been spent investigating potential reasons for false-positive examinations. Predictors of false-positive mammograms have been noted at the level of the patient, the radiologist, and the system. Patient factors associated with a higher likelihood of a false-positive mammogram include younger age of the patient, a higher breast

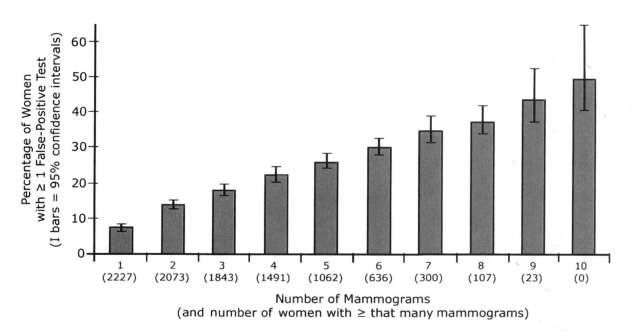

Figure 120 Estimated 10-year risk of having at least one false-positive screening mammogram according to the total number of screening mammograms performed. *Source:* Reprinted from Elmore *et al.*, 1998.

density on mammogram, and the use of hormone therapy at the time of mammogram. In addition, patient history (e.g., prior breast biopsies) and whether this is the first or subsequent mammogram are also important predictors. Factors associated with an increased risk of false-positive mammograms at the level of the radiologist include the younger age of the radiologist or having fewer years of clinical experience. The impact of the radiologists' annual volume of interpretation is not clear. System factors that have been associated with higher false-positive rates include whether or not prior mammograms were available for comparison. All of these factors are discussed in greater detail below and can potentially affect the accuracy of mammography, either individually or collectively, thereby increasing the rate of false-positive mammograms.

Patient Factors

Patient age

Randomized trials and cohort studies have shown that the false-positive rate of mammography is higher in young women as compared to older women (NIH Consensus Development Panel, 1997). A number of large cohort studies have examined the effects of age and other risk factors on the accuracy of mammography in detail. One of these studies, conducted in the United States by the BCSC and involving over 300,000 women, demonstrated that after controlling for breast density and hormone therapy, the false-

positive rate of mammography was much higher in younger women than in older women (9% vs. 6%, respectively) (Carney *et al.*, 2003). Another study performed by the BCSC examining the impact of family history on the accuracy of mammography showed that accuracy was influenced primarily by age, not by family history. This finding raises the question of whether there is enough evidence to support the recommendation for beginning screening mammography at a younger age for women with a family history of breast cancer (NCI, 2004).

The higher false-positive rate of mammography in a younger population, in combination with the lower incidence rates of breast cancer, greatly increases the chance of having a false-positive mammogram among women undergoing mammography in their forties. This concept is demonstrated in a figure from a review article on mammography published in the *New England Journal of Medicine* (Fletcher and Elmore, 2003). Figure 121 shows estimates of the number of false-positive mammograms among 1000 women in three different age groups who undergo annual mammography in the United States for 10 years. For example, among 1000 women in their forties, ~ 560 women, or more than half, may experience a false-positive mammogram over a decade, while only 15 women will be diagnosed with breast cancer. The number of women in their sixties with false-positive examinations ($N = 360$) is markedly lower, with more women diagnosed with breast cancer ($N = 37$).

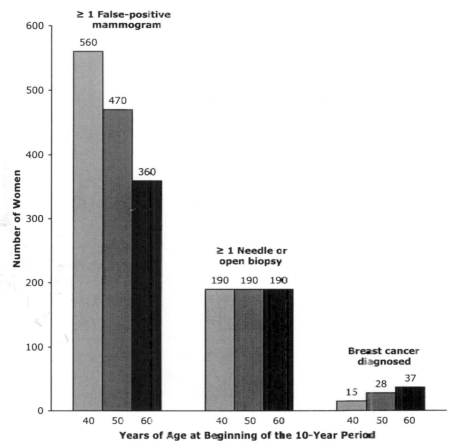

Figure 121 Chances of false-positive mammograms, need for biopsies, and development of breast cancer among 1000 women who undergo annual mammography for 10 years. *Source:* Reprinted from Fletcher and Elmore, 2003.

Breast Density

The normal process of breast senescence involves fatty replacement of denser glandular tissue as a woman ages. While studies have shown that younger age is independently associated with lower accuracy of mammography, it appears to be due, in part, to the degree of breast density, and vice versa. The BCSC cohort study examining over 300,000 women reported that the false-positive rate was lower in older women with fatty breasts and higher in younger women with dense breasts (Carney *et al.*, 2003). Another study demonstrated a dramatic reduction in the accuracy of mammography among women with extremely dense breasts compared to women with almost complete fatty replacement of their breasts (sensitivity 30 vs. 80%, respectively) (Mandelson *et al.*, 2000). In this study, after controlling for age and hormone therapy use, the false-positive rate of mammography was still much higher for women with dense breasts compared with women with predominately fatty breasts (11 vs. 3%, respectively).

Studies also suggest that breast density varies throughout different phases of the menstrual cycle and with use of hormone therapy (discussed later in this chapter). Stopping hormone therapy 10–30 days before repeat mammograms for postmenopausal women and obtaining mammograms during the first two weeks of the menstrual cycle for premenopausal women may decrease breast density and reduce mammographic abnormalities (Harvey *et al.*, 1997; White *et al.*, 1998). Studies are currently underway to more definitely assess the accuracy of mammography by phase of menstrual cycle (NCI, 2004).

Hormone Therapy

Hormone therapy has been associated with an increased false-positive rate of mammography. Hormone therapy may affect the false-positive rate by inhibiting the normal involution of the glandular breast tissue, leading to an increase and/or persistence in breast density, instead of the age-related decrease in density that is normally expected (Rutter *et al.*, 2001). Data from the large BCSC cohort study suggest that the use of hormone therapy increases breast density, thereby affecting the improvement in mammography accuracy typically seen with increasing age (Carney *et al.*, 2003). Data from a substudy of the Women's Health Initiative (WHI), a randomized trial, showed that mammogram density increased by 6% at year one in women receiving combination therapy with estrogen and progesterone, as compared to a 0.91% decrease in the placebo group (McTiernan *et al.*, 2005). In this same study, there was a fourfold increase in the risk of having an abnormal mammogram among women receiving estrogen and progesterone compared to the placebo group. These data are consistent with a study by Banks *et al.* (2004) in which hormone therapy was associated with an increased false-positive rate. Current evidence does not indicate whether the association of hormone therapy and the accuracy of mammography is limited primarily to therapy with estrogen and progesterone in combination as opposed to therapy with estrogen alone. However, the data from Banks *et al.* (2004) suggest that either type of preparation may be associated with an increase in the false-positive rate.

Impact of the Clinical/Patient History

A clinical history is usually obtained from each woman before her mammogram, and, in addition to information on her age and use of hormone therapy, information is often obtained on race/ethnicity, body mass index (BMI), history of previous breast biopsies, family history of breast cancer, and any breast symptoms. Several studies examining women from a variety of ethnic backgrounds have found no association between race or ethnicity and false-positive rates of mammography (Gill and Yankaskas, 2004; Kerlikowske *et al.*, 2005). However, the effect of BMI on the accuracy of mammography remains controversial. In one study, overweight and obese women were more likely to have false-positive mammography examinations, even after controlling for age and breast density, compared with normal or underweight women (Elmore *et al.*, 2004). The changes in false-positive rates occurred with no change in the sensitivity of mammography. The authors hypothesized that larger breast size in obese women may increase the search area, and that obese women may have a thicker volume of breast tissue that might lead to decreased image quality. These findings contrast with those seen in a substudy of the U.K. Million Women Study, in which the false-positive rates were slightly higher and sensitivity was much lower for women with a lower BMI (< 25 vs. ≥ 25) (Banks *et al.*, 2004).

The clinical history of multiple previous breast biopsies, on the other hand, appears to be a strong patient predictor of false-positive mammography examinations (Christiansen *et al.*, 2000). The relative risk (RR) increase after one biopsy is RR 1.56 (95% CI 1.03–2.38) and increases to RR 3.42 (95% CI 1.55–5.92) after three biopsies. Family history of breast cancer has also been associated with an increased risk for false-positive examinations (RR 1.24, 95% CI 1.01–1.47) (Christiansen *et al.*, 2000). Less well studied but also possibly associated with an increase in the risk for false-positive mammography examinations is the presence of breast symptoms (Elmore *et al.*, 1997). Although the presence of suspicious symptoms would usually lead to a diagnostic examination, there are some women who report mild symptoms and/or chronic abnormalities at the time of the screening examination.

While age, hormone use, and breast density all appear to impact Fp rates through changes in the appearance of the mammogram image, it is less clear how factors such as family history and breast symptoms impact the risk of false-positive mammograms. A study by Elmore *et al.* (1997) found that even though overall diagnostic accuracy was not altered by the clinical history, an alerting history such as the presence of a breast symptom or a family history of breast cancer (or both) increased the number of recommended work-ups among patients without cancer. These findings suggest that some patient characteristics may emerge as risk factors because of radiologists' reactions to clinical information rather than to changes in mammogram appearance.

First Mammogram versus Subsequent Mammogram and Intervals between Examinations

The risk of having a false-positive mammogram is twice as high for women undergoing their first mammogram, compared with women who have subsequent screening mammograms at regular intervals (RR 2.0) (Elmore *et al.*, 2002). During this first screening examination, the radiologist is unable to compare findings with prior films to assure stability over time; thus, a higher false-positive rate is expected. Longer periods between mammograms (> 18 months vs. ≤ 18 months) have also been associated with a moderate increase in the risk of a false-positive mammogram (RR 1.37, 95% CI 1.07–1.68) (Elmore *et al.*, 2002).

Radiologist Factors

Experience

The impact that radiologists' interpretive volume and years of experience have on accuracy has recently received a great deal of attention, though conflicting findings make a consensus difficult. Radiologist-level characteristics that have been associated with false-positive mammograms are the radiologist's age and time since graduation from medical school. Radiologists who are younger (in their thirties and forties) and who are more recently trained (5–15 years since graduation from medical school) have been noted to have false-positive examination rates that are two to four times higher than those of their older counterparts (Elmore *et al.*, 2002). One study found that less experienced radiologists had higher sensitivity for breast cancer, but at the expense of higher false-positive rates (Barlow *et al.*, 2004). Investigators suggest that the association between age and years of clinical experience with false-positive rates may reflect a more recent trend in training that emphasizes sensitivity over specificity (Elmore *et al.*, 2002). Studies conflict as to whether the false-positive rate is associated with radiologists' "experience" as gained through interpreting a high volume of mammograms annually (Barlow *et al.*, 2004; Beam *et al.*, 2003; Esserman *et al.*, 2002; Smith-Bindman *et al.*, 2005). A higher annual volume may be associated with varying accuracy due to different thresholds for calling a mammogram positive (Barlow *et al.*, 2004).

Variability of False-positive Rates

Several studies have identified variability in interpretive performance of radiologists, even in controlled cancer-enriched laboratory conditions where efforts to perform accurately would likely be heightened (Beam *et al.*, 1996; Elmore *et al.*, 1994; Kerlikowske *et al.*, 1998). The use of BI-RADS assessment and recommendation categories in the United States has facilitated our ability to monitor breast cancer screening performance in the community. Unfortunately, even with the use of a standardized BI-RADS taxonomy, the variability in interpretation among clinicians persists (Kerlikowske *et al.*, 1998). In one study of variability, Elmore *et al.* (1994) found perfect agreement in only 7% of 150 diagnostic and screening mammograms interpreted by 10 radiologists. The false-positive rate in this study, which included diagnostic examinations and a deliberate enrichment of cancer cases, ranged from 11–65%. The study also found significant variability in follow-up recommendations, with a range of 3–20% among radiologists for biopsy recommendations among women without cancer.

The majority of studies on variability in mammography interpretive practice have focused on clinical features of the films as the sole sources of variation. This does not fully explain decision making in a complex practice environment. As discussed previously, one study evaluating the impact of knowing a patient's clinical history on mammography interpretations found that diagnostic suspicion may be altered when clinical history is known (Elmore *et al.*, 1997). Findings from enriched test sets and "test" situations of mammography interpretation may not reflect accuracy in the real clinical setting (Rutter and Taplin, 2000). Data from the community setting are thus helpful in examining the phenomena of variability. However, while many variables can be held constant in a "test" situation, marked heterogeneity is present in the real world of screening mammography.

One study that evaluates variability in U.S. community practice shows the importance of adjusting for heterogeneity among patients when comparing levels of accuracy among radiologists (Elmore *et al.*, 2002). In this study using community data, wide variation in false-positive rates was noted among 24 radiologists, with a range of 2.6–15.9%. The initial appearance of wide variability, however, is reduced by half after differences in patient characteristics are considered (e.g., some radiologists might interpret more films from younger women with dense breasts). Additional adjustments for radiologists' characteristics reduced the variability even further (Elmore *et al.*, 2002). Although this statistical adjustment reduces some of the apparent variability among radiologists, it is important to point out that wide variability still remains.

Facility and System Factors

Two Views versus One View in Screening Mammography

It is standard practice in the United States and in many international breast cancer screening programs to include two views of each breast (mediolateral-oblique and craniocaudal views) during a screening mammography examination, especially on the prevalence screen. Several studies, done primarily in the United Kingdom, have shown that two-view mammography is associated with higher cancer detection rates than one-view mammography (oblique view only), although at the expense of a higher biopsy rate and greater program costs (Bryan *et al.*, 1995; Warren *et al.*, 1996).

Double Reading and Computer-aided Detection

Most international screening programs use double reading to interpret screening mammography examinations (IARC, 2002). However, a double reading can be performed in a number of different ways. For example, the two readers might interpret the films together or independently. In addition, an examination might be called "positive" if either one of the two readers notes an abnormality, a consensus of the two readers might be required for any disagreements in interpretation, or a third reader might be brought in to decide the final assessment in the case of disagreements. The false-positive rate therefore may vary depending on the technique used for the double reading. In the Netherlands, all films are read independently at a central location by two experienced radiologists, and consensus is required before an examination is called "positive." It has been suggested that this double-reading technique may contribute to the low referral rate in that country (IARC, 2002).

Computer-aided detection (CAD) software is currently used as a second reading device in several mammography (and other radiology) programs throughout the United States, and is designed to improve the detection rates (sensitivity) of mammography while minimizing the use of a limited resource of mammographers. However, the effect of CAD application on the actual interpretation of screening mammography remains ambiguous. A large study of 43 mammography facilities (a total of 429, 345 mammograms) found that the use of CAD is significantly associated with reduced accuracy in the interpretation of screening mammograms, and that the increased biopsy rate associated with

CAD does not clearly improve the detection of invasive breast cancer (Fenton et al., 2007). An increase in the recall rate has been noted in four other studies (Birdwell et al., 2005; Cupples et al., 2005; Freer and Ulissey, 2001; Khoo et al., 2005). An increased false-positive rate may follow logically from the design of current CAD software as CAD programs can insert 1.5–4 marks on the average screening mammogram, the overwhelming majority of which are false-positives (Castellino, 2002). One prior study of CAD reported no increase in recall rate after CAD implementation within a large academic practice (Gur et al., 2004), although the proportion of women receiving first-time screening was higher (40%) at the beginning of the 3-year study period than at the end (30%). Because first-time screens have higher recall rates, the imbalance of this patient characteristic during the study period could have obscured increases in the measurement of the false-positive rate after CAD implementation (Elmore and Carney, 2004).

Prior Mammograms for Comparison and Interval between Screening Examinations

The unavailability of prior mammograms for comparison is associated with a 1.5- to 2.0-fold increase in the rate of false-positive mammography examinations (Christiansen et al., 2000; Elmore et al., 2002). The length of time between prior films may also be relevant, with the comparison of 1-year prior films providing lower false-positive rates than a comparison of films taken 2 years apart (Sumkin et al., 2003). This relationship has been noted in the United States, where the screening system is not centrally organized as it is in many other countries (see Blanchard et al., 2006, and Chapter 4.19 in this volume).

A woman may have a prior mammogram abnormality that has been investigated in the past and/or that is stable on serial imaging and deemed benign until (for whatever reason, not uncommonly insurance changes, change/loss of job, etc.) she changes facilities where her mammograms are performed. Then, without prior mammograms for comparison at this new facility, the mammogram abnormality is considered "new," and recommendations for additional work-up and/or close follow-up may be made unless prior films can be obtained.

Predicting the Cumulative Risk of False-positive Mammograms

Earlier we described how the cumulative risk of a false-positive examination varies dramatically based on the false-positive rate. The cumulative risk of a false-positive mammogram over time also varies substantially among women and radiologists. One study evaluated the relative impact across the continuum of a variety of risk factors known or thought to be associated with false-positive mammograms (Christiansen et al., 2000). The risk of experiencing a false-positive examination after nine screening mammograms was estimated to be as low as 5% for women with low-risk characteristics or as high as 100% for women with multiple high-risk factors. The radiologists' tendency to call a mammogram "positive" had the largest estimated effect on a woman's risk of experiencing a false-positive mammogram (Christiansen et al., 2000). Most factors increase the likelihood of a woman having a false-positive mammogram only moderately with risk ratios under 2.0. Because women have multiple mammograms over time, even modest risk ratios substantially increase the cumulative risk of a woman experiencing at least one false-positive mammogram during her lifetime. Given the magnitude of the cumulative risk for a false-positive mammogram, the potentially harmful outcomes from false-positive examinations cannot be ignored, especially as they relate to anxiety and morbidity on the individual level and cost to society on the public health level. These potential harms associated with false-positive mammograms are discussed in greater detail in the following section.

Significance of False-positive Mammography Examination

Impact on Patients

Women who have a false-positive mammography examination may have heightened anxiety and breast cancer worry, which appears to persist for several months after the examination. Almost half (47%) of the women with a high-suspicion abnormal mammogram have reported mammography-related anxiety and breast cancer worry (Barton et al., 2004; Lerman et al., 1991). In one study, these psychological effects led to an impairment in mood in 26% of women, while 17% reported a compromised ability to engage in daily activities (Lerman et al., 1991). Heightened levels of anxiety, while decreasing after the initial abnormal mammogram, appear to persist for at least 3 months (Barton et al., 2004; Lerman et al., 1991) and may last as long as 18 months (Gram et al., 1990). These psychological effects may occur despite additional evaluation excluding the diagnosis of breast cancer (Lerman et al., 1991).

One study in the United States by Barton et al. (2004) evaluated interventions to decrease a woman's anxiety after an abnormal mammogram. This study, involving a controlled trial of ~ 8500 women, found that the immediate reading of screening mammograms was more effective than an educational intervention (or no intervention) in decreasing short-term anxiety levels among women with false-positive mammograms. Instead of the traditional "batch" reading of screening mammograms typically performed after women have left the mammography facility, this intervention involved the immediate interpretation of mammograms. This allowed for additional diagnostic imaging by mammography or ultrasound if needed during the same appointment and resolution of many of the abnormal findings before the woman left the mammography facility. This intervention was considered effective because the majority of women who received immediate readings did not know they had an abnormal mammogram and because additional work-up (except for biopsies) was performed immediately. Interestingly, the educational intervention, which included both a videotape presentation and written materials containing detailed information about breast cancer risk and coping skills, did not help to reduce women's anxiety levels (Barton et al., 2004).

Fortunately, the psychological distress of a false-positive examination does not appear to impact subsequent adherence to future screening; instead, it may actually be associated with an increase in health-care utilization. Several studies have demonstrated no negative impact on future screening behaviors (Burman et al., 1999; Lampic et al., 2003; Lerman et al., 1991; Pinckney et al., 2003). In fact, women with false-positive mammograms, especially those without previous mammography, may be more likely to return for the next scheduled screening (Burman et al., 1999). In one study, subsequent ambulatory visits for breast- and nonbreast-related concerns increased for the 12 months after a false-positive mammogram (Barton et al., 2001).

Several studies investigating women's attitudes regarding false-positive mammography have found that the majority of women are highly tolerant of false-positive examinations (Ganott et al., 2006; Schwartz et al., 2000). In one study involving ~ 500 women, more than half of the participants (63%) considered 500 or more false-positives per one life saved as reasonable, and almost 40% reported that they would tolerate up to 10,000 or more (Schwartz et al., 2000). In a larger, more recent study involving 1500 women, more than 80% reported that they would have preferred the inconvenience of and anxiety associated with a higher recall rate if it resulted in the possibility of earlier breast cancer detection (Ganott et al., 2006). In addition to the anxiety potentially associated with false-positive examinations, the additional downstream evaluations can lead to discomfort and morbidity. As noted in Figure 120, after 10 years of annual screening of 1000 women, 190 would undergo a breast biopsy with less than 40 cancers diagnosed. Therefore, ~ 150 women out of 1000 would undergo breast biopsies due to false-positive interpretations (Fletcher et al., 2003). If skin infections set in after a biopsy, the morbidity can be increased (Elmore et al., 1998).

Impact on the Health-care System

From a public health perspective, the financial cost of evaluating all of the patients with false-positive results can be substantial. In one U.S. study using Medicare and health plan reimbursement allowances, the cost of all downstream appointments and testing after false-positive examinations was approximately one-third the cost of performing the initial screening (Elmore et al., 1998). From these data, it has been estimated that for every $1 million spent on screening in the United States, an additional $330,000 will be spent on the work-up of false-positive tests.

A rough estimation can be made of the financial impact of false-positive examinations in the United States. If the ~ 54 million American women who are currently between the ages of 40 and 69 (U.S. Census Bureau, 2006) were to receive breast cancer screening annually for 10 years, ~ 26.5 million (49%) of those women would have at least one false-positive mammogram (Elmore et al., 1998). If, for simplicity, we assume that screening costs $100 per exam, the total cost of screening examinations would be $54 billion (54 million women × $100 per examination × 10 examinations over a decade). Using data from Elmore et al. (1998) showing that the cost of the diagnostic work-up is approximately one-third that of screening, we can then calculate the overall cost of false-positive examinations in the United States: the diagnostic work-up for all of the false-positive examinations during this decade would be an additional $18 billion ($54 billion × 0.33), bringing the total cost of mammography for this population of women from $54 billion to $72 billion.

Recall Rates in the United States versus Other Countries

As briefly discussed earlier, striking differences have been suggested between false-positive rates for screening examinations in the United States and those in other countries. Many studies have only limited ability to follow up on the cancer outcomes for all women; thus, false-positive rates are less commonly reported than overall recall rates. However, since the majority of women who are recalled for additional examinations do not have breast cancer, the recall rate is similar to the false-positive rate.

A meta-analysis of 32 published reports noted higher recall rates in North American countries than in other countries (Elmore et al., 2003). Most of the data representing North America was from the United States. Overall, the weighted mean percentage of screening mammograms judged abnormal was significantly higher in North American programs compared to elsewhere (8.4 vs. 5.6%, respectively). Some data were available in these published studies to adjust for possible differences in the women screened or the years the mammograms were obtained. The recall rate was still 2–4% higher in the North American screening programs than it was in programs from other countries, after adjusting for covariates such as percentage of women who were less than 50 years of age and the calendar year in which the mammogram was obtained. As the recall rate went up, the positive predictive value (PPV) of a biopsy obtained as a result of the mammogram decreased.

The higher recall rate would have been acceptable if the yield of cancers diagnosed increased appreciably. However, while the frequency and number of ductal carcinoma *in situ* cases diagnosed increased, no significant difference was noted in invasive cancers.

A second report compared screening mammography performance between the United States and the United Kingdom among women 50 years or older and noted similar findings (Smith-Bindman *et al.*, 2003). In this study, three large mammography registries or screening programs were compared. The two programs from the United States were the Breast Cancer Surveillance Consortium (BCSC, $N = 978,591$) and the National Breast and Cervical Cancer Early Detection Program (NBCCEDP, $N = 613,388$), and the program from the United Kingdom was the National Health Service Breast Screening Program (NHSBSP, $N = 3.94$ million). The recall rates in the United States were twice as high as those in the United Kingdom. When reviewing data by different age groups, the authors found that, among women age 50–54 years, the recall rates were 14.4% in the BCSC and 12.5% in the NBCCEDP versus only 7.6% in the NHSBSP. The negative open surgical biopsy rates were also noted to be twice as high in the U.S. than in the U.K. settings, yet cancer detection rates remained similar; cancer detection rates per 1000 mammogram screening examinations were 5.8, 5.9, and 6.3, in the BCSC, NBCCEDP, and NHSBSP, respectively (Smith-Bindman *et al.*, 2003).

These studies are consistent in suggesting that the U.S. programs interpret a higher percentage of mammograms as abnormal without evident benefit in the yield of cancers detected per 1000 screens, although an increase in ductal carcinoma *in situ* detection was noted in one study (Elmore *et al.*, 2003). There are numerous potential reasons for this variability in recall rate, including differences in the patients, the radiologists, and the health-care systems. The potential reasons are discussed by Elmore *et al.* (2003) and are listed in Table 22.

Characteristics of the population screened might explain part of the difference in recall rates between the United States and United Kingdom. For example, women less than 50 years of age, who have a higher recall rate than women over 50, are actively screened in the United States but are not normally screened in other countries. National regulations are another possible explanation for international variation in recall rates. For example, there are different regulations regarding the minimum volume of mammography examinations required of a radiologist: $> \sim 500$ per year in the United States and > 5000 in the United Kingdom (Smith-Bindman *et al.*, 2003). In addition, the screening systems are centralized in many countries, as opposed to the diverse and spread-out systems of the United States. Features of the mammography examination might explain part of this international variation and include the use of one versus two views and the use of single versus double readings for the examinations.

Finally, malpractice concerns in the United States have been cited as an additional potential reason for its higher recall rate. Failure to detect breast cancer has been one of the most frequent medical legal allegations against physicians

Table 22 **Possible Explanations for the Variability Noted among Published Studies of Screening Mammography**

Characteristics of the population screened
- Age (e.g., percentage of women < 50 years)
- Initial versus subsequent screening examination
- Presence of risk factors for breast cancer
- Presence of breast symptoms
- Self-referral versus physician referral

Features of the mammography examination
- Equipment type and year
- One or two views of each breast
- Single or double readings
- Technician training

Features of physicians interpreting the mammogram
- Experience of the physician
- Level of personal comfort with ambiguity
- Individual thresholds to label film as abnormal

Features of the health-care system
- Malpractice concerns
- Financial incentives
- Private versus academic/public programs
- Different stated goals for the percentage of mammograms judged abnormal and positive predictive value
- Quality control and auditing procedures
- Variability of definitions used to calculate outcomes

Source: Reprinted from Elmore *et al.*, 2003.

for many years (Physician Insurers Association of America, 2002). Because the majority of women recalled after a screening mammogram do not have breast cancer, North American screening programs are attempting to more efficiently deal with false-positive rates, as well as to lower the recall rate without reducing the cancer detection rate.

Efforts to Reduce False-positive Mammograms and to Better Deal with Expected False-positive Screenings

False-positive examinations can never be completely eliminated. If the threshold for interpreting a mammography examination is reduced too much, breast cancers might also be missed. Still, efforts to decrease false-positive rates in the United States may help to minimize the psychological and financial burdens associated with the large number of false-positive mammograms each year. However, reducing the false-positive rate is not a simple task (Sickles, 2000).

As discussed earlier, other countries such as the United Kingdom and Norway have recall rates that are almost half of the rates seen in U.S. screening programs without any apparent reduction in the ability to detect breast cancers (i.e., not at the expense of mammography sensitivity) (Elmore et al., 2003; Fletcher and Elmore, 2005; Smith-Bindman et al., 2003). Although some of the factors that impact recall rates and false-positive rates are not modifiable, such as patient age and family history, some factors can be considered or modified.

Having prior films, especially more recent films (< 18 months), available for comparison during imaging might lower the rate of false-positive examinations. This may be facilitated by encouraging women to receive serial screenings at the same mammography facility or, when that is not possible, by encouraging women to bring prior films from another facility to the new screening site. With more widespread use of digital mammography and electronic storage of films, the often burdensome task of accessing and transferring prior films from one mammography site to another should become easier for patients and mammography facilities alike.

As discussed in the section on breast density, another way to possibly decrease false-positive mammography examinations may be to schedule them in the first two weeks of a woman's menstrual cycle for premenopausal women or consider temporary cessation (10–30 days) of hormones for postmenopausal women using hormone therapy. This may be more difficult at public facilities and resource-poor institutions where access to any appointment for a screening mammogram can be difficult, even without the added burden of timing the examination with a woman's menstrual cycle. Similarly, for postmenopausal women using hormone therapy, temporary cessation of hormones 10–30 days prior to

the examination may be beneficial in reducing false-positive examinations. However, this suggestion may be problematic for women who are taking hormone therapy to manage severe menopausal symptoms, for patients may not tolerate a return of their menopausal symptoms following hormone therapy cessation. In addition, mammography image quality can be maximized and false-positive rates potentially decreased by using appropriately trained technicians, ensuring proper breast positioning, and educating women about the importance of standing completely still during compression.

Finally, it is important to educate women undergoing screening about false-positive examinations so that they are prepared for the possibility of a false-positive result. With the high levels of media attention that breast cancer receives (Lerner, 2001), many women are anxious regarding their own risk of breast cancer and may have unrealistic expectations about the accuracy of technology such as mammography. Immediate reading and work-up of image-detected abnormalities may lead to a decrease in the anxiety experienced by women with false-positive mammograms (Barton et al., 2004). Education may also help alter women's expectations of mammography screening.

In conclusion, screening mammography examinations are simply a screening tool; they are designed to identify women with abnormalities that suggest the need for further testing. There is an obvious trade-off between interpreting too many mammography examinations as positive and missing an abnormality that will lead to a diagnosis of breast cancer. Although the threshold for missing a cancer varies among patients, radiologists, and health-care systems, data from European mammography programs suggest that decreasing the false-positive rates in the United States might be achieved with minimal or no change in the cancer detection rates. By pursuing more effective approaches to mammogram interpretation and breast imaging management nationwide, we may be able to reduce false-positive rates (Sickles, 2000). The cost and resources needed to accomplish this in the United States need to be further examined. A perfect balance is difficult to achieve.

Acknowledgment

Dr. Elmore is supported by National Cancer Institute grant K05CA104699.

References

American College of Radiology (ACR). 2003. *Breast Imaging Reporting and Data System Atlas (BI-RADS™)*. 4th ed. Reston, VA.
Ballard-Barbash, R., Taplin, S.H., Yankaskas, B.C., Ernster, V.L., Rosenberg, R.D., Carney, P.A., Barlow, W.E., Geller, B.M., Kerlikowske, K., Edwards, B.K., Lynch, C.F., Urban, N., Chrvala, C.A., Key, C.R., Poplack, S.P., Worden, J.K., and Kessler, L.G. 1997. Breast cancer surveillance

consortium: a national mammography and screening outcomes database. *Am. J. Roentgenol. 169*:1001–1008.

Banks, E., Reeves, G., Beral, V., Bull, D., Crossley, B., Simmonds, M., Hilton, E., Bailey, S., Barrett, N., Briers, P., English, R., Jackson, A., Kutt, E., Lavelle, J., Rockall, L., Wallis, M.G., Wilson, M., and Patnick, J. 2004. Impact of use of hormone replacement therapy on false positive recall in the NHS breast screening programme: results from the million women study. *BMJ 328*:1291–1292.

Barlow, W.E., Chi, C., Carney, P.A., Taplin, S.H., D'Orsi, C., Cutter, G., Hendrick, R.E., and Elmore, J.G. 2004. Accuracy of screening mammography interpretation by characteristics of radiologists. *J. Natl. Cancer Inst. 96*:1840–1850.

Barton, M.B., Moore, S., Polk, S., Shtatland, E., Elmore, J.G., and Fletcher, S.W. 2001. Increased patient concern after false-positive mammograms: clinician documentation and subsequent ambulatory visits. *J. Gen. Intern. Med. 16*:150–156.

Barton, M.B., Morley, D.S., Moore, S., Allen, J.D., Kleinman, K.P., Emmons, K.M., and Fletcher, S.W. 2004. Decreasing women's anxieties after abnormal mammograms: a controlled trial. *J. Natl. Cancer Inst. 96*:529–538.

Bassett, L.W., Hendrick, R.E., Bassford, T.L., Butler, P.F., Carter, D.C., and DeBor, J.D. 1994. *Quality Determinants of Mammography: Clinical Practice Guideline No. 13*. Rockville, MD: Agency for Health Care Policy and Research Publication No. 95-0632.

Beam, C.A., Conant, E.F., and Sickles, E.A. 2003. Association of volume and volume-independent factors with accuracy in screening mammogram interpretation. *J. Natl. Cancer Inst. 95*:282–290.

Beam, C.A., Layde, P.M, and Sullivan, D.C. 1996. Variability in the interpretation of screening mammograms by US radiologists: findings from a national sample. *Arch. Intern. Med. 156*:209–213.

Birdwell, R.L., Bandodkar, P., and Ikeda, D.M. 2005. Computer-aided detection with screening mammography in a university hospital setting. *Radiology 236*:451–457.

Blanchard, K., Colbert, J.A., Kopans, D.B., Moore, R., Halpern, E.F., Hughes, K.S., Smith, B.L., Tanabe, K.K., and Michaelson, J.S. 2006. Long-term risk of false-positive screening results and subsequent biopsy as a function of mammography use. *Radiology 240*:335–342.

Bryan, S., Brown, J., and Warren, R. 1995. Mammography screening: an incremental cost effectiveness analysis of two view versus one view procedures in London. *J. Epidemiol. Community Health 49*:70–78.

Burman, M.L., Taplin, S.H., Herta, D.F., and Elmore, J.G. 1999. Effect of false positive mammograms on interval breast cancer screening in an HMO. *Ann. Intern. Med. 131*:1–6.

Carney, P.A., Dietrich, A.J., Freeman, D.H., and Mott, L.A. 1993. The periodic health examination provided to asymptomatic women: an assessment using standardized patients. *Ann. Intern. Med. 119*:129–135.

Carney, P.A., Miglioretti, D.L., Yankaskas, B.C., Kerlikowske, K., Rosenberg, R., Rutter, C., Geller, B.M., Abraham, L.A., Dignan, M., Cutter, G., and Ballard-Barbash, R. 2003. Individual and combined effects of breast density, age, and hormone replacement therapy use on the accuracy of screening mammography. *Ann. Intern. Med. 138*:168–175.

Castellino, R.A. 2002. Computer-aided detection in oncologic imaging: screening mammography as a case study. *Cancer J. 8*:93–99.

Christiansen, C.L., Wang, F., Barton, M.B., Kreuter, W., Elmore, J.G., Gelfand, A.E., and Fletcher, S.W. 2000. Predicting the cumulative risk of false-positive mammograms. *J. Natl. Cancer Inst. 92*:1657–1666.

Cupples, T.E., Cunningham, J.E., and Reynolds, J.C. 2005. Impact of computer-aided detection in a regional screening mammography program. *Am. J. Roentgenol. 185*:944–950.

Elmore, J.G., Wells, C.K., Lee, C.H., Howard, D.H., and Feinstein, A.R. 1994. Variability in radiologists' interpretations of mammograms. *N. Engl. J. Med. 331*:1493–1499.

Elmore, J.G., Wells, C.K., and Howard, D.H. 1997. The impact of clinical history on mammographic interpretations. *JAMA 277*:49–52.

Elmore, J.G., Barton, M.B., Moceri, V.M., Polk, S., Arena, P.J., and Fletcher, S.W. 1998. Ten-year risk of false positive screening mammograms and clinical breast examinations. *N. Engl. J. Med. 338*:1089–1096.

Elmore, J.G., Miglioretti, D.L., Reisch, L.M., Barton, M. B., Kreuter, W., Christiansen, C.L., and Fletcher, S.W. 2002. Screening mammograms by community radiologists: variability in false-positive rates. *J. Natl. Cancer Inst. 94.18*:1373–1380.

Elmore, J.G., Nakano, C.Y., Koepsell, T.D., Desnick, L.M., D'Orsi, C.J., and Ransohoff, D.F. 2003. International variation in screening mammography interpretations in community-based programs. *J. Natl. Cancer Inst. 95*:1384–1393.

Elmore, J.G., and Carney, P.A. 2004. Computer-aided detection of breast cancer: has promise outstripped performance? (Editorial) *J. Natl. Cancer. Inst. 96*:162–163.

Elmore, J.G., Carney, P.A., Abraham, L.A., Barlow, W.E., Egger, J.R., Fosse, J.S., Cutter, G.R., Hendrick, R.E., D'Orsi C.J., Paliwal, P., and Taplin, S.H. 2004. The association between obesity and screening mammography accuracy. *Arch. Intern. Med. 164*:1140–1147.

Elmore, J.G., Carney, P.A., Taplin, S., D'Orsi, C.J., Cutter, G., and Hendrick, E. 2005. Does litigation influence medical practice? The influence of community radiologists, medical malpractice perceptions and experience on screening mammography. *Radiology 236*:37–46.

Esserman, L., Cowley, H., Eberle, C., Kirkpatrick, A., Chang, S., Berbaum, K., and Gale, A. 2002. Improving the accuracy of mammography: volume and outcome relationships. *J. Natl. Cancer Inst. 94*:369–375.

European Commission. 2006. *European guidelines for quality assurance in mammography*. Luxembourg: Office for Official Publications of the European Communities.

Fenton, J.J., Taplin, S.H., Carney, P.H., Abraham, L.A., Sickles, E.A., D'Orsi, C., Berns, E.A., Cutter, G., Hendrick, R.E., Barlow, W.E., and Elmore, J.G. 2007. Influence of computer-aided detection on performance of screening mammography. *N. Eng. J. Med. 356*:1399–1409.

Fletcher, S.W., and Elmore, J.G. 2003. Mammographic screening for breast cancer. *N. Engl. J. Med. 348*:1672–1680.

Fletcher, S.W., and Elmore, J.G. 2005. False-positive mammograms—can the USA learn from Europe? *Lancet 365*: 7–8.

Freer, T.W., and Ulissey, M.J. 2001. Screening mammography with computer-aided detection: prospective study of 12,860 patients in a community breast center. *Radiology 220*:781–786.

Ganott, M.A., Sumkin, J.H., King, J.L., Klym, A.H., Catullo, V.J., Cohen, C.S., and Gur, D. 2006. Screening mammography: do women prefer a higher recall rate given the possibility of earlier detection of cancer? *Radiology 238*:793–800.

Gill, K.S., and Yankaskas, B.C. 2004. Screening mammography performance and cancer detection among black women and white women in community practice. *Cancer 100*:139–148.

Gram, I.T., Lund, E., and Slenker, S.E. 1990. Quality of life following a false positive mammogram. *Br. J. Cancer 62*:1018–1022.

Gur, D., Sumkin, J.H., Rockette, H.E., Ganott, M., Hakim, C., Hardesty, L., Poller, W.R., Shah, R., and Wallace, L. 2004. Changes in breast cancer detection and mammography recall rates after the introduction of a computer-aided detection system. *J. Natl. Cancer Inst. 96*:185–190.

Harvey, J.A., Pinkerton, J.V., and Herman, C.R. 1997. Short-term cessation of hormone replacement therapy and improvement of mammographic specificity. *J. Natl. Cancer Inst. 89*:1623–1625.

Hofvind, S., Thoresen, S., and Tretli, S. 2004. The cumulative risk of a false-positive recall in the Norwegian Breast Cancer Screening Program. *Cancer 101*:1501–1507.

Humphrey, L.L., Helfand, M., Chan, B.K., and Woolf, S.H. 2002. Breast cancer screening: a summary of the evidence for the U.S. Preventive Services Task Force. *Ann. Intern. Med. 137*:347–360.

International Agency for Research on Cancer (IARC). 2002. *Breast Cancer Screening*. IARC Handbooks of Cancer Prevention Vol.7. Lyon, France: IARC Press.

Kerlikowske, K., Creasman, J., Leung, J.W., Smith-Bindman, R., and Ernster, V.L. 2005. Differences in screening mammography outcomes among White, Chinese, and Filipino women. *Arch. Intern. Med.* 165:1862–1868.

Kerlikowske, K., Grady, D., Barclay, J., Frankel, S. D., Ominsky, S. H., Sickles, E. A., and Ernster, V. 1998. Variability and accuracy in mammographic interpretation using the American College of Radiology Breast Imaging Reporting and Data System. *J. Natl. Cancer Inst.* 90:1801–1809.

Khoo, L.A., Taylor, P., and Given-Wilson, R.M. 2005. Computer-aided detection in the United Kingdom National Breast Screening Programme: prospective study. *Radiology 237:*444–449

Lampic, C., E. Thurfjell, and P. O. Sjoden. 2003. The influence of a false-positive mammogram on a woman's subsequent behaviour for detecting breast cancer. *Eur. J. Cancer 39*:1730–1737.

Lerman, C., Trock, B., Rimer, B.K., Boyce, A., Jepson, C., and Engstrom, P.F. 1991. Psychological and behavioral implications of abnormal mammograms. *Ann. Intern. Med. 114*:657–661.

Lerner, B.H. 2001. *The Breast Cancer Wars.* New York: Oxford University Press.

Mammography Quality Standards Act (MQSA) of 1992. 42 U.S.C. § 263b(f).

Mandelson, M.T., Oestreicher, N., Porter, P.P., White, D., Finder, C.A., Taplin, S.H., and White, E. 2000. Breast density as a predictor of mammographic detection: comparison of interval vs. screen-detected cancers. *J. Natl. Cancer Inst. 92*: 1081–1087.

McTiernan, A., Martin, C.F., Peck, J.D., Aragaki, A.K., Chlebowski, R.T., Pisano, E.D., Wang, C.Y., Brunner, R.L., Johnson, K.C., Manson, J.E., Lewis, C.E., Kotchen, J.M., and Hulka, B.S. 2005. Estrogen-plus-progestin use and mammographic density in postmenopausal women: women's health initiative randomized trial. *J. Natl. Cancer Inst.* 97:1366–1376.

National Cancer Institute (NCI). 2004. *Breast Cancer Surveillance Consortium: Evaluating Screening Performance in Practice.* NIH Publication No. 04-5490. Bethesda, MD: National Cancer Institute, National Institutes of Health, U.S. Department of Health and Human Services.

NIH Consensus Statement Online. 1997 Jan 21–23. Breast Cancer Screening for Women Ages 40–49. [cited 2007, March 29]; 15(1):1–35. http://consensus.nih.gov/1997/1997BreastCancerScreening103html.htm.

Olivotto, I.A., Kan, L., and Coldman, A.J. 1998. False positive rate of screening mammography. *N. Engl. J. Med. 338*:1089–1096

Otten, J.D., Karssemeijer, N., Hendriks, J.H., Groenewoud, J.H., Fracheboud, J., Verbeek, A.L., de Koning, H.J., and Holland, R. 2005. Effect of recall rate on earlier screen detection of breast cancers based on the Dutch performance indicators. *J. Natl. Cancer Inst.* 97:748–754.

Physician Insurers Association of America. 2002. *Breast Cancer Study: Third Edition.* Washington, DC: Physician Insurers Association of America.

Pinckney, R.G., Geller, B.M., Burman, M., and Littenberg, B. 2003. Effect of false-positive mammograms on return for subsequent screening mammography. *Am. J. Med. 114*:120–125.

Rutter, C.M., Mandelson, M.T., Laya, M.B., and Taplin, S. 2001. Changes in breast density associated with initiation, discontinuation, and continuing use of hormone replacement therapy. *JAMA 285*:171–176.

Rutter, C.M., and Taplin, S.H. 2000. Assessing mammographers' accuracy: a comparison of clinical and test performance. *J. Clin. Epidemiol.* 53:443–450.

Schwartz, L.M., Woloshin, S., Sox, H.C., Fischhoff, B., and Welch, H.G. 2000. US women's attitudes to false positive mammography results and detection of ductal carcinoma in situ: cross sectional survey. *BMJ 320*:1635–1640.

Sickles, E.A. 2000. Successful methods to reduce false-positive mammography interpretations. (Review). *Radiol. Clin. North Am.* 38:693–700.

Smith-Bindman, R., Chu, P.W., Miglioretti, D. L., Sickles, E.A., Blanks, R., Ballard-Barbash, R., Bobo, J. K., Lee, N. C., Wallis, M. G., Patnick, J., and Kerlikowske, K. 2003. Comparison of screening mammography in the United States and the United Kingdom. *JAMA 290*:2129–2137.

Smith-Bindman, R., Chu, P., Miglioretti, D.L., Quale, C., Rosenberg, R.D., Cutter, G., Geller, B., Bacchetti, P., Sickles, E.A., and Kerlikowske, K. 2005. Physician predictors of mammographic accuracy. *J. Natl. Cancer Inst.* 97:358–367.

Sumkin, J.H., Holbert, B.L., Herrmann, J.S., Hakim, C.A., Ganott, M.A., Poller, W.R., Shah, R., Hardesty, L.A., and Gur, D. 2003. Optimal reference mammography: a comparison of mammograms obtained 1 and 2 years before the present examination. *Am. J. Roentgenol.* 180:343–346.

Taplin, S.H., Rutter, C.M., Finder, C., Mandelson, M.T., Houn, F., and White, E. 2002. Screening mammography: clinical image quality and the risk of interval breast cancer. *Am. J. Roentgenol. 178*:797–803.

U.S. Census Bureau. National sex and age population estimates for 2005. Accessed June 8, 2006. Available: http://www.census.gov/popest/national/asrh/NC-EST2005-sa.html.

Warren, R.M., Duffy, S.W., and Bashir, S. 1996. The value of the second view in screening mammography. *Br. J. Radiol. 69*:105–108.

White, E., Velentgas, P., Mandelson, M.T., Lehman, C.D., Elmore, J.G., Porter, P., Yasui, Y., and Taplin, S.H. 1998. Variation in mammographic breast density by time in menstrual cycle among women aged 40–49 years. *J. Natl. Cancer Inst. 90*:906–910.

Yankaskas, B.C., Cleveland, R.J., Schell, M.J., and Kozar, R. 2001. Association of recall rates with sensitivity and positive predictive values of screening mammography. *Am. J. Roentgenol. 177*:543–549.

21

Breast Dose in Thoracic Computed Tomography

Eric N.C. Milne

Introduction

According to the National Center for Health Statistics, the United States has the highest utilization rate of computed tomography (CT) in the world. In the year 2004–2005, a total of 172.5 CT scans were performed for every 1000 persons. With a female population of > 150 million, this translates to ~ 26 million CT scans per year in female patients. It is difficult to determine accurately how many of these 26 million scans in female patients will involve irradiation to the breasts (which are particularly mutagenic) because, in addition to thoracic scans, the breasts will often be partially or wholly within the radiation field of an upper abdominal scan. With this in mind our best estimate of the percentage of all CT scans that will cause irradiation to the breast is 35%, resulting in 9 million scans per year. Of these, ~ 3 million will be in children under the age of 15 (Brenner *et al.*, 2001).

The fact that CT has proven to be one of the greatest diagnostic advances in the history of medicine renders the ground very infertile to any suggestion that CT can have significant costs to the patient in terms of inducing cancer. I would like to emphasize at the outset that overall the benefits of CT carried out for good clinical indications far outweigh its "risks." However, there is a large subgroup of female patients (9 million per year, as above), in whom the cost in terms of induced breast cancer may prove to outweigh the benefits and who deserve special consideration before a CT scan is performed.

Despite our century-old knowledge of the hazards of ionizing radiation, many highly reputable clinics and organizations still fail to warn patients regarding these dangers. To quote the American College of Radiology web site, "information for patients," "the effective radiation dose from chest CT is ~ 8.0 mSv which is about the same as an average person receives from background radiation in three years." According to the Mayo Clinic, "during the CT exam you're briefly exposed to radiation. But doctors and other scientists believe that CT scans provide enough valuable information to outweigh the risks." On the other hand, an increasing number of articles give dire warnings regarding the risks of CT. For example, "2,500 cancer deaths can be predicted from one year of 200 mAs pediatric CT exams" (Brenner *et al.* 2001) and, "Imaging X-rays Cause Cancer: A Call to Action for Caregivers and Patients" (Semelka, 2005).

These views diverge so widely because in general both radiologists and referring clinicians underestimate the level of irradiation given by a CT scan (Lee *et al.*, 2004). In particular, radiologists, having been brought up on the thesis that the absorbed dose to an organ must always be less than the measured skin dose (which is correct for standard non-CT radiology), find it difficult to conceive that this may not be true for computed tomography. In particular, it is untrue for CT of the female thorax in which the absorbed dose to the breast may be larger than the skin dose. As a simple test of the reader's personal concept of how much radiation the breast receives during CT of the thorax, I would like to pose the following question: which receives the greater amount of radiation, the bowel during a barium enema or the breast during a thoracic CT? The answer is that the amount of radiation absorbed by the bowel during a barium enema and that absorbed by the breast during a CT scan of the thorax is the same (de Gonzales and Darbys, 2004). A second less cogent reason for the general feeling that the radiation dose for CT is not very high may be the adoption of the Grey and Sievert units to quantify radiation. 1.0 G seems a lot better than 100 Rads, and 1.0 mSv considerably better on paper than 100 mRem, even though they are the same! For that reason, the absorbed breast doses reported in this chapter are quoted in Rads.

Many articles concerning the dose given to the lungs by CT of the thorax do not even mention the much greater dose absorbed by the breast (Brenner, 2004). Having had some experience in my days as a chest physician of fluoroscoping patients weekly to establish whether their pneumothoraces (immobilization treatment for tuberculosis) required "refilling," and with the subsequent discovery that a high percentage of the female patients developed breast cancer on the fluoroscoped side (Miller *et al.*, 1989), I became concerned 14 years ago that the dose to the breast given by thoracic CT might be approaching these carcinogenic levels.

Considering that mathematical simulations and phantom studies were not sufficiently realistic to duplicate the working world of CT, where patient size, chest wall thickness, and breast size and consistency differ greatly, and scans may be repeated several times, both precontrast and postcontrast, for follow-up or for better visualization. Furthermore, a chest scan and an abdominal scan may be done on the same patient, providing a double dose of irradiation to the breast. We began therefore to measure the dose to the breast directly, instead of predicting it mathematically or simulating it with a plastic phantom, by using thermoluminescent dosimeters applied to the patient during thoracic and upper abdominal scans (Milne *et al.*, 1992).

Methodology

We examined 58 consecutive female patients, age 20–75 years, undergoing chest CT. There were no other selection criteria. Thermoluminescent dosimeters (TLD chips) were placed on the lateral, medial, superior, and inferior aspects of both breasts, and individual TLDs were placed on each nipple and on the chest wall at the xiphi-sternal junction (11 TLDs per patient in all) (Fig. 122). The TLDs remained in place during the entire examination, including any repeat examinations that were requested and any abdominal study

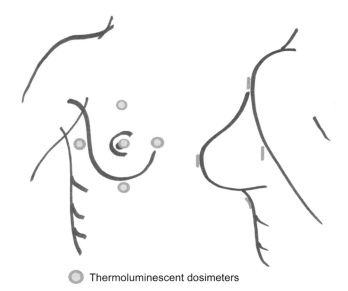

○ Thermoluminescent dosimeters

Figure 122 Demonstrating the location of the thermoluminescent dosimeters. Five on each breast and one between the breasts on the chest wall.

that was being done at the same time. The patients' age and chest circumference were measured, and brassiere cup size was recorded as a measure of breast volume for later correlation with the dose findings. The number of CT "cuts" per examination was also noted.

Results

The results were quite surprising. In 56 of the 58 patients we found that the skin dose at the nipple was lower (range 0.5–12.3; mean 4.4 R) than the chest wall dose (range 4.0–22.0 R; mean 6.3 R). Because the nipple is closer to the X-ray source, it would have been anticipated that the nipple dose would be higher than the chest wall dose. The reason for this apparent discrepancy is important to an understanding of why the breast-absorbed dose is higher than the skin dose in CT and will be explained in the Discussion section. The total number of "cuts" per patient varied from 15 to 80 (mean 33). The calculated breast absorbed dose (BAD) ranged from 1.8–19.6 R, equivalent to 360–1500 chest X-rays, or 6–65 mammograms (the largest dose was from the patient who had 80 CT cuts performed). The mean breast absorbed dose was 7.1 R.

The relationship between the breast-absorbed dose and thickness (of the compressed breast) is well documented for mammography (Boone *et al.*, 2005). We have shown a similar excellent linear relationship for CT of the thorax between the breast-absorbed dose and the volume of the noncompressed breast (determined by the patients' bra cup size), ranging from 32A to 42D (Fig. 123). A very large

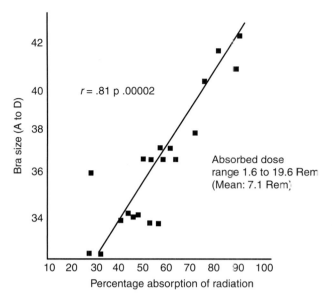

Figure 123 Graph of breast volume (measured by bra cup size) versus percent of radiation absorbed by the breast.

r = .81 p .00002

Absorbed dose range 1.6 to 19.6 Rem (Mean: 7.1 Rem)

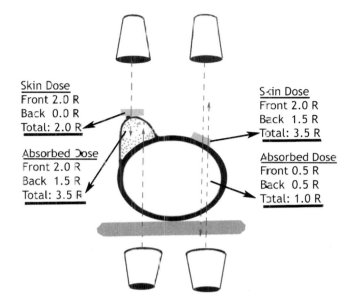

Skin Dose
Front 2.0 R
Back 0.0 R
Total: 2.0 R

Absorbed Dose
Front 2.0 R
Back 1.5 R
Total: 3.5 R

Skin Dose
Front 2.0 R
Back 1.5 R
Total: 3.5 R

Absorbed Dose
Front 0.5 R
Back 0.5 R
Total: 1.0 R

(▬ = Thermoluminescent Dosimeters)

Figure 124 On the patient's left side there is no breast. The dose to the TLD from the front is 2.0 R. The dose to the TLD from the back is 1.5 R (2.0 minus 0.5 R absorbed in the lungs and mediastinum). Skin dose 3.5 R and lung absorbed dose 0.5 R. On the right side, the dose to the TLD from the front is 2.0 R. (In fact, because the nipple is closer to the X-ray source, the dose from the front would be higher; the increase is omitted for simplicity.) Dose to the TLD from the back is close to zero because the radiation is absorbed in the breast; therefore skin dose equals only 2.0 R. Most of the radiation from the front, (2.0 R) and the back (1.5 R) will be absorbed in the breast, giving skin dose 2.0 R, breast absorbed dose 3.5 R, and lung absorbed dose 0.5 R.

dense breast can absorb virtually all of the radiation falling on the breast from both the front and the back (Boone *et al.*, 2005).

Our figures seemed very high at the time this study was carried out, and editors were unwilling to accept our result. However, our findings have been confirmed much more recently by other investigators, quoting CT surface doses of 2–6 R (Semelka, 2006; Parker *et al.*, 2005) and by the International Commission on Radiological Protection, which quotes the "effective" dose in CT as being equivalent to 400 chest X-rays (Biological Effects and Ionizing Radiation (BEIR) VII, 2006). It should be noted that the quoted numbers refer to single CT exams, whereas in our series, carried out in the working clinical environment, many patients had multiple scans, reflected in our higher breast absorbed doses.

Discussion

Why did the TLD on the nipple record a skin dose lower than the skin dose for the bare chest wall? The reason is that the TLD on the bare chest wall receives direct irradiation from the front (e.g., 2.0 R) plus irradiation from the back, which has passed through the thorax and has been only partially absorbed by the lungs and mediastinum (e.g., 0.5 R). This leaves 1.5 R to impinge on the TLD, giving a total skin dose of 3.5 R for an absorbed dose to the lung of 0.5 R. The nipple TLD must have received a slightly higher dose from the front because it is closer to the anode than the bare skin surface. However, it received much less radiation from the back because most of the radiation that had traversed the lung with a small amount of absorption (0.5 R) was then

absorbed by the breast (varying with its size and density), leaving, for example, 2.0 R for the nipple skin dose and 3.5 R for the breast-absorbed dose (Fig. 124). This study was carried out on single-slice CT apparatus, and it might be thought that newer machines with much faster scan times and multiple slices would have reduced these doses. In fact, the opposite is true. In practice, the average radiation dose per CT scan has actually increased (10–30%) with the use of spiral CT and ultrafast multislice scanners (IRCP, 2005), so the problem is increasing, not diminishing.

Cancer Risks

Without doubt, large doses of ionizing radiation to the breast, as in the previously mentioned study on the carcinogenic effects of chest fluoroscopy, can cause breast cancer. There remains considerable debate, however, particularly in the field of mammography, as to whether "low-dose" ionizing radiation can also cause breast cancer. The ICRP defines low dose as, "from near-zero to 1.0 R." However, low dose is not what we are concerned about in this chapter. We have shown in our study that the amount of radiation absorbed by the breast during CT of the thorax far exceeds the boundaries of low-dose irradiation. The mean absorbed dose in our study was 7.1 R (the average dose for a biplane mammogram being 300 mR). Therefore, the dose to the breast from a thoracic CT is equivalent to having a mammogram every year for 23 years or having 1500 chest films. The highest dose in our series was 19.6 R, equivalent to the cumulative dose of 65 mammograms.

Using estimates, based on linear extrapolation, as recommended by the International Committee on Radiation Protection, we find that 1.0 R of low-dose irradiation would cause 100 cancers per 100,000 patients (BEIR VII, 2006). The mean dose of 7.1 R given to the breasts by thoracic CT would be expected to cause 710 additional cancers in 100,000 patients. Among the huge number of CTs now being carried out in women in the United States (26 million per year), 9 million involve thoracic or upper abdominal studies. Using the same extrapolation, we see that these CTs would result in an increase of 63,900 breast cancers. However, the risk of cancer induction appears to diminish rapidly with increasing age. Even at the high-radiation levels we have measured, it is highly unlikely that this number of cancers would result. If 25% of all the patients receiving thoracic CT were under the age of 40, we would expect an increase in breast cancers of ~ 16,000.

In the majority of cases of thoracic CT (those with good clinical indications), the benefit will almost always exceed the cost. However, the risks for young females are clearly higher than is commonly believed. In such cases we would suggest that CT should be performed only for specific clinical indications and where an alternative nonionizing imaging methodology is not available. It should never be used as a screening procedure in young females, for example, as has been proposed for following cystic fibrosis patients (with biannual CTs) for detecting and quantifying coronary artery calcification (which gives the highest recorded doses to the breast; de Gonzales and Darbys, 2004) and for screening for lung cancer.

Many recent articles discuss CT for screening for lung cancer using "low-dose" techniques, that is, reducing the radiation level to the point at which noise begins to obtrude on the signal but attempting to maintain sufficient signal to make the scan of diagnostic value. This does appear to be possible but is now a moot point because evidence has emerged that CT screening for lung cancer has the same efficacy as plain chest X-ray screening (i.e., virtually none). Neither technique has any appreciable effect on the mortality rate from lung cancer (Swensen et al., 2005).

Computed Tomography of the Breast

In view of the preceding data, it would appear strange that several centers are now working on the development of CT of the breast, but the doses required to image the breast satisfactorily with direct CT are much lower than those obtained during CT of the thorax—first, because the breast is imaged coronally and the beam does not need to traverse the chest, and second, because a lower kV can be used. At 80 kV excellent images can be obtained in an average sized (14 cm diameter) breast, at a dose of 400 mR for a fatty breast and 700 mR for a dense breast, which is close to the range presently used for X-ray mammography (Boone et al., 2005). As breast size and density increase, the dose required to produce a satisfactory image also increases, and for a 17.0 cm diameter dense breast the dose can be as high as 1.4 R. Even at this level the dose to the breast with this direct CT examination is less than the dose obtained indirectly during thoracic CT. The removal of overlapping anatomic detail, which obscures pathology on conventional mammography, permits small lesions to be detected with greater assurance and can be expected to yield a sensitivity greater than that obtainable by mammography.

Reducing Radiation Dose

Attempting to reduce the amount of radiation down to a level at which noise is just acceptable is one way to reduce dose to the patient and appears successful in the laboratory environment, but it is debatable whether radiologists would accept this in nonuniversity clinical practice. If any type of imaging examination needs to be carried out to answer a clinical problem, the resultant image needs to be consistently of top quality, which necessitates having sufficient quanta striking the recording medium. Radiologists are extremely sensitive to image quality and are unwilling to accept any degradation of their images. Indeed, the trend in CT has been to progressively increase the amount of radiation used to provide (to the radiologists' eye) the highest quality of images (ICRP, 2005). In an attempt to stop this 10–30% escalation of dose, many equipment manufacturers now have built-in methods of tailoring the dose to the patient's weight and size. Methods of automatically modulating mA as the CT beam traverses tissue of different linear absorption and thickness are under development. If this type of automation is not available on any particular scanner, the simpler but still effective traditional way of reducing radiation dose while maintaining image quality is to employ well-trained technologists who are educated not only concerning the hazards of CT radiation and the need to reduce radiation dose but also on what constitutes excellent image quality The experience and skill of a good technologist allow assessment of a patient's build with considerable accuracy and calculation for any particular machine (all of which have their own foibles) that exposure will give excellent diagnostic images at the lowest dose for that patient.

Imaging without Using Ionizing Radiation

Even the low-dose levels given by mammography are unacceptable in certain classes of patients, including younger patients, those with high-risk factors, familial or genetic, for developing breast cancer, and those with dense breasts where the sensitivity of mammography is very low. In addition to the problems posed by radiation, any imaging technique that provides solely morphologic information (including mammography and ultrasound) is rarely able to distinguish malignant tissue from benign tissue *in vivo*. This is well illustrated by the fact that a pathologist, when presented with a 350 times magnified image of the tissue from the breast (giving a resolution that is unachievable in the intact patient) still cannot usually make a diagnosis without staining the tissue. To differentiate between benign and malignant tissues and thereby reduce the very large number of negative biopsies presently being carried out (80% in the United States and 60% in Europe), it is necessary to add functional information to the image. Without functional data, imaging can only act as a triage procedure to decide whether the patient should be biopsied or followed up, and not as a true diagnostic procedure.

Optical Imaging

In order both to provide the functional information that is essential for making an *in vivo* diagnosis of malignancy or benignancy and to eliminate the need for ionizing radiation, many centers have been developing optical methods of imaging the breast. Various techniques used (several of which are described in detail elsewhere in this volume), rely either on imaging the obligatory increase in vascularity (angiogenesis) and associated increase in hemoglobin levels that occur within or around malignant tumors, or in mapping and quantifying the ratio between oxy- and deoxyhemoglobin concentrations in normal versus malignant tissue.

The principal problem that has held back *in vivo* optical imaging of patients is the difficulty of reconstructing an image formed by light. In contrast to X-ray images, there is a huge amount of scattering. This problem is less significant in very small experimental animals, and many optical imaging systems for this purpose (of particular value for assessing the effects of new pharmaceuticals and for following biological processes at the molecular level) are already available commercially. There has been some reluctance, however, to push forward these systems into human imaging largely because of predictions that only poor spatial resolution could be obtained, which would make optical imaging *in vivo* of little value. This may have been an error in thinking, because unlike mammography, there is much less need for high spatial resolution when one is imaging angiogenesis. This is because the volume of angiogenesis is always much larger than the volume of the tumor tissue (Milne, 1967). For example, a malignant lesion that is only 3–4 mm in size on mammography is usually seen as a 3–6 cm area of angiogenesis on optical imaging.

Several of these optical imaging devices are quite far-advanced and are being tested on intact human patients, but only one device, computed tomographic laser mammography (CTLM), has been put into commercial production. Using an 808 nM laser beam and conventional X-ray CT geometry, but with the gantry lying horizontally under the tabletop, the CTLM system replaces the X-ray tube with a laser diode that rotates around the breast (descending from base to nipple) to produce 1–4 mm thick sections of the uncompressed breast, reconstructed in coronal, axial, and sagittal planes, plus a translucent animated 3D picture of the normal and abnormal distribution of blood within the breast. To date, more than 7000 cases have been examined using this CTLM system. The system is proving to be of particular value in the young and dense breast and for following the effects of treatment of breast cancer (Helbich, 2004; Milne, 2006). A fuller description of CTLM is outside the scope of this present chapter and will be reserved to a later date.

Ultrasound

Ultrasound has been utilized quite extensively in the dense breast and in high-risk populations. In the fatty breast it adds nothing to conventional mammography, but in the dense breast it has the potential to discover many small lesions. In dense breasts it can achieve a higher level of sensitivity than mammography. Unfortunately, standard ultrasound provides quite limited functional information. As a result, a large percentage of the discovered "lesions" turn out to be false-positives, giving a very low specificity and leading to an unacceptably high level of negative biopsies.

Magnetic Resonance Imaging

Magnetic resonance imaging (MRI) would seem to answer simultaneously both of the problems of ionizing radiation and functional information. There is no argument that its spatial resolution is good (though much less so than mammography) and the contrast resolution is excellent, but the technique is very costly in terms not only of dollars but of patient time, and of the physician's and technologist's time and expertise needed to acquire the images and to interpret them. In addition to these factors, injection of contrast media is necessary, and one of the most important areas in breast diagnosis, the display and differential diagnosis of microcalcification, is poorly served by MRI (Bluemke *et al.*, 2004). Presently, therefore, the MRI appears to be a valuable problem-solving tool for individually selected cases but does not answer the need for screening patients.

In conclusion it would seem that the topic of radiation dose in CT, which was very unpopular a few years ago, has reappeared on the scene with considerable vigor. The present concern about the carcinogenic effects of radiation has been driven mainly by far-seeing and critical radiologists

(Semelka, 2005; Brenner and Hall, 2004), but has now reached the referring clinicians who have rightly become concerned about ordering procedures that have the potential to cause cancer in their patients. The prospect of litigation on the behalf of patients who have developed breast cancer after having had thoracic CT is very real. Because of the high prevalence of breast cancer and the very high utilization levels of CT, a large number of women who have had one or more thoracic and/or abdominal CTs will undoubtedly develop breast cancer, and some of them may attribute this (rightly or wrongly) to the radiation received. At that time both clinicians and radiologists will need to be very well educated concerning the doses of radiation given by CT and their carcinogenic risk. They will also have to be able to demonstrate that any thoracic or upper abdominal CT carried out in a younger woman was done for very good clinical indications and, at optimum radiation levels, that no suitable alternative nonionizing imaging technique was available.

References

Bleumke, D., Gatsonis, C.A., Cheu, M.H., DeAngelis, G.A., DeBruhl, N., and Harms, S. 2004. Magnetic resonance images of the breast prior to biopsy *J. Am. Med. Assoc. 292*:2735–2742.

Boone, J.M., Kwan, L.C., Seibert, J.A., Shah, N., and Lindfors, K.K. December 2005. Technique factors and their relationship to radiation dose in pendant geometry breast. *Med. Phys. 32*:3767–3776.

Brenner, D.J. 2004. Radiation risks potentially associated with low-dose CT screening of adult smokers for lung cancer. *Radiology 231*:440–445.

Brenner, D.J., and Elliston, C.D. 2005. Radiation risks of body CT: what to tell our patients and other questions. *Radiology 234*:968–970.

Brenner, D., Elliston, C., Hall, E., and Berdon, W. 2001. Estimated risks of radiation-induced fatal cancer from pediatric CT. *Am. J. Roentgenol. 176*:289–296.

Brenner, D.J., and Hall, E.J., 2004. Risk of cancer from diagnostic X-rays. *Lancet 363*:2192–2193.

de Gonzalez, A., and Darbys, S. 2004. Risk of cancer from diagnostic X-rays: estimates for the U.K. and 145 other countries. *Lancet 363*:345–351.

Health Risks from Exposure to Low Levels of Ionizing Radiation. 2006. BEIR VII. Phase 2. Publ. Nat. San Diego, CA: Academic Press.

Helbich, T. 2004. Computed tomography laser mammography. *Eur. Hosp. 13*:4–5.

ICRP. 2005. *Managing Patient Dose in Computed Tomography.* IRCP Publ. 87.

Lee, C.I., Harms, A.H., Monico, E.P., Brink, J.A., and Forman, H.P. 2004. Diagnostic CT scans: assessment of patient, physician and radiologist awareness of radiation dose and possible risks. *Radiology 231*:393–398.

Miller, A.B., Howe, G.R., and Sherman, G.J. 1989. Mortality from breast cancer after irradiation during fluoroscopic examinations in patients being treated for tuberculosis *N. Eng. J. Med. 321*:1285–1289.

Milne, E.N.C. 1967. The circulation of primary bronchogenic carcinoma and pulmonary metastases. *Am. J. Roentgenol. 100*:603–610.

Milne, E.N.C., Cyrlak, D., Rabbani, B., and Roeck, W. 1992. Absorbed dose to the breast during thoracic CT. Syllabus: Fleischner Society Annual Meeting.

Parker, M.S., Hui, F.K., Camacho, M.A., Chung, J.K., Broga, D.W., and Sethi, N.N. 2005. Female breast radiation exposure during CT pulmonary angiography. *Am. J. Roentgenol. 185*:1228–1233.

Semelka, R.S. 2005. Radiation risk from CT scans: a call for patient-focused imaging. Medscape (http://www.medscape.com/viewarticle/ 496297).

Semelka, R.S. 2006. Imaging X-rays cause cancer: a call to action for caregivers and patients. Medscape (http://www.medscape.com/viewprogram/ 5063 pnt).

Swensen, S.J., Jerr, J.R., Hartman, T.E., Midthun, D.E., Mandrekar, S.J., and Hillman, S.L., Sykes, A.M., Augenbaugh, G.L., Bungum, A.O., and Allen, K.L. 2005. CT screening for lung cancer: five-year prospective experience. *Radiology 235*:259–265.

22

Absorbed Dose Measurement
in Mammography

Marianne C. Aznar and Bengt Å. Hemdal

Introduction

Mammography refers to the X-ray examination of the human breast and has been used for several decades in the diagnosis of breast disease and for mass screening of asymptomatic women. As with any X-ray examination, there is a small risk of carcinogenesis involved with irradiating the breast tissue. The benefit-to-risk ratio of this procedure is most important in the screening situation and depends on several factors, including the age of the subject. For example, it has been estimated that when saving the lives of 20 women (age 40–50 years) through yearly mammography screening, one lethal cancer may be induced by the radiation (Andersson and Janzon, 1997). This issue has been generating a lot of debate and has resulted in many different screening protocols worldwide.

Regardless of the approach chosen, it is crucial that the radiation doses delivered in mammography be monitored. A screening examination will almost always include two exposures of each breast, in mediolateral oblique (MLO) and possibly in craniocaudal (CC) projections. Suboptimal imaging techniques can lead to poor image quality and/or a higher absorbed dose than needed. Both would result in an unnecessary radiation risk for the patient, and low image quality carries the additional risk of a tumor being undetected. This realization has led to the development of quality assurance protocols in an attempt to standardize the imaging techniques and other procedures involved. Such protocols provide guidelines for image quality and absorbed dose in mammography (ACR, 1999; EC, 2006) or concentrate on dosimetry (EC, 1996).

The image quality must enable the detection of clusters of microcalcifications (each with an extension of only 0.1–0.5 mm) and of tumors with low contrast compared to their surroundings, even in dense and large breasts. However, the radiation dose delivered to the breast must not be higher than needed for this purpose. In general terms, this is expressed by the ALARA principle: the radiation dose should be *As Low As Reasonably Achievable*. In other words, the relationship between dose and image quality must be optimized. Guidance is given in multiple national and international documents, also for digital radiology, for instance, by the International Commission on Radiological Protection (ICRP, 2004). In order to perform the necessary quality control and optimization work, tools are needed for estimation of both image quality and dose.

This chapter briefly describes methods to estimate the relevant radiation dose values that are necessary for quality control and optimization, and that also can be used as a basis for risk estimates and comparison between different mammography techniques and groups of women. The characteristics of various types of dosimeters that could be used for the necessary measurements will be discussed. Screen-film mammography (SFM) is still the dominant technique worldwide, but digital mammography (DM) is increasingly being used, as can be seen in other chapters of this volume. Fortunately, dose measurements can at large be performed in the same way with SFM and DM. Instead, both techniques present pitfalls, which hopefully the readers of this chapter will avoid.

Estimation of Absorbed Dose to the Breast

The dose received by an individual woman during a mammographic examination is influenced by many parameters, such as the breast anatomy as well as the technical equipment and the examination technique. The parameters related to the mammography unit include the choice of anode (also called target), filter, and voltage (also called potential, in kV or kVp) across the X-ray tube. As a consequence, the X-ray tubes produce photons with different spectra of energies, the maximum energy being determined by the tube voltage and the distribution of energies mainly by the anode and filter used. The energy spectrum will influence the quality of the mammogram as well as the dose delivered to the breast.

The patient's specific anatomy will also have an influence. Larger and/or more glandular breasts, for instance, generally receive a higher dose owing to the necessary increase in exposure. On most modern mammography units, the selection of anode, filter, and tube voltage can be changed so that higher and then more penetrating photon energies are produced in these cases. This can actually be beneficial for both image quality and dose. Some modern mammography units are able to automatically adjust the selection of anode, filter, and tube voltage based on the thickness and even the glandularity of the breast. In DM, higher photon energies are generally used compared with SFM, which is one reason for the potential dose reduction in DM.

Geometrical factors also have to be considered. In most other medical X-ray imaging, dose measurements are performed with the center of the X-ray field of view along the central axis of the X-ray beam. In mammography, however, the X-ray tube usually is oriented with the focus closer towards the patient. This orientation deliberately causes a more or less pronounced "heel effect"; that is, the intensity of the X-rays basically decreases, while their mean energy increases toward the stand. Hence, the point of measurement has to be defined carefully.

Concepts and Quantities Used

Strictly speaking, the basic dosimetric quantity in mammography is either the *kerma* (*k*inetic *e*nergy *r*eleased per unit *ma*ss) or the *exposure* (charge of ions produced in air per unit mass). Exposure is the oldest quantity and can be measured directly with an ionization chamber. A translation can be performed between these basic quantities, exposure being essentially the ionization equivalent of air kerma, but contrary to exposure, kerma can also be expressed in other matter than air. Kerma relates conceptually to the *energy imparted* to matter as opposed to absorbed dose, which refers to the *energy absorbed* within a finite volume of matter (air or tissue). In conditions where electron equilibrium is achieved, absorbed dose and kerma are equal. These conditions are easily achieved in diagnostic radiology (where the photon energies are low compared to radiation therapy), and dosimeters are sometimes described as measuring "absorbed dose in air" instead of "air kerma" for simplification purposes.

The concepts and quantities used in mammography can differ between dose protocols, which describe how dose measurements should be performed. The names and units used in this chapter follow mainly the International Commission on Radiation Units and Measurements (ICRU, 2005). However, none of the dose protocols referred to here (EC, 1996, 2006; ACR, 1999) follows the ICRU exactly, and even contradictions exist. The objective is to clarify some of these in the following.

Incident Air Kerma, K, and Incident Exposure, X

For both kerma and exposure, the word "incident" stands for a point where X-rays "enter" the breast of the patient. These are the basic quantities in mammography and most other medical X-ray imaging. In mammography they are measured (directly or indirectly) under the compression paddle without backscatter (cf. "Skin Dose, D_{skin}") at a distance, usually 4 cm (ACR, 1999) to 6 cm (EC, 1996), from the edge of the breast support and laterally centered.

Incident Air Kerma, K
Incident air kerma, K, is usually expressed as the kinetic energy of charged particles in air per unit mass with the unit gray (1 Gy = 1 J/kg). In some dose protocols (EC, 1996, 2006), K is called ESAK (*e*ntrance *s*urface *a*ir *k*erma) as opposed to the ICRU terminology, where "entrance surface" means that backscatter is in fact included.

Incident Exposure, X
Incident exposure, X, is always expressed as the charge of ions in air of one sign produced per mass unit. The recommended unit is C/kg, but in the dose protocol that prescribes this quantity (ACR, 1999), the non-SI unit roentgen is used (1 R = 2.58×10^{-4} C/kg). It is also stated that air kerma (Gy) is given by the exposure (R) times 0.00873 Gy/R.

Skin Dose, D_{skin}
Skin dose, D_{skin}—short for absorbed dose to the skin—is the energy absorbed to the skin of the breast per unit mass with the unit gray (1 Gy = 1 J/kg). Skin dose is more generally

called ESD (entrance surface dose) and includes backscatter of photons that have interacted in the breast or phantom. The D_{skin} value can be derived from K by multiplication with the *backscatter factor*, which is relatively low for the X-ray beams used in mammography (about 1.10 with only a few percents of variation). Alternatively, D_{skin} can of course be measured with a dosimeter on the skin and K calculated from that value (see the discussion on direct *in vivo* measurements in "Dosimeters for Direct *in vivo* Measurements").

X-ray Tube Output, Y

Incident air kerma or skin dose can be measured indirectly by measurement of the *X-ray tube output*, Y, that is, the incident air kerma at a certain calibration distance from the focal spot divided by the tube loading used. The tube loading (in mAs) is the product of the tube current (in mA) and the exposure time (in seconds). The X-ray tube output is measured without the patient present (EC, 1996). For each exposure of a breast or a phantom simulating a breast, the anode/filter, tube voltage, tube loading, and compressed breast thickness must be registered. The incident air kerma (in mGy) for that exposure is then simply given by the product of the tube output (in mGy/mAs) and the tube loading (in mAs) for that exposure with a correction for the difference in heights above the breast support for the dosimeter and the entrance surface of the breast, respectively. This height for the breast is equivalent to the compressed breast thickness (cf. "Compressed Breast Thickness, T"). Note that this is valid only if the same beam quality (i.e., the same anode/filter and tube voltage) was used on both occasions. If multiple beam qualities are used, the tube output has to be measured for each one of them. A corresponding X-ray tube output (R/mAs) based on exposure (R) can be used (ACR, 1999) in a similar way as described here.

Half-value Layer (HVL)

The *half-value layer* (HVL) of the X-ray beam is expressed as the thickness (mm) of a material, in mammography aluminum of high purity, which is needed to reduce the air kerma or exposure to half of its value without the material. The measurement must be performed with a dosimeter in so-called good geometry; that is, the aluminum has to be at the same distance from or closer to the focus of the X-ray tube than to the dosimeter, and a diaphragm also has to be present close to the aluminum in order to avoid scattered radiation to the dosimeter. Each dose protocol prescribes a distance, usually 4–6 cm, from the edge of the breast support to the center of the dosimeter, as for the measurement of tube output. For the protocols mentioned here, the compression paddle has to be present and can be used as a support for the aluminum foils during the measurement.

The HVL is a relatively simple way of characterizing the beam quality, which more accurately could be described as the energy distribution of the X-rays (i.e., the X-ray spectrum). In the American dose protocol (ACR, 1999), the conversion factor c_G (cf. "Estimation of AGD") is given as a function of both the HVL and the X-ray tube voltage. Modern mammography units have a very stable, almost constant tube voltage, but for older units a correction factor can be used depending on the tube voltage waveform. It is then relevant to use the unit kVp, which stands for peak kV, since the tube voltage could vary substantially during the exposure.

Compressed Breast Thickness, T

Firm compression of the breast in mammography is necessary to achieve both a high image quality and a low dose. Therefore, this condition is always assumed in the dose protocols. The *compressed breast thickness*, T, is an important parameter for patient dose measurements, primarily because the conversion factors c_G and c (cf. "Estimation of AGD") depend on T. Also, if output measurements are used (cf. "X-ray Tube Output, Y"), T has to be known in order to calculate air kerma, K. To be accurate, T should be estimated at the same distance from the edge of the breast support as for tube output and HVL (cf. "Half-value Layer [HVL]"). The average error should be less than 5 mm (EC, 1996). Modern mammography units usually have a scale indicating the height of the compression paddle over the breast support. At least this scale has to be calibrated as a function of compressed breast thickness. Unfortunately, the result of the calibration depends on the size and form of the object as well as on the compression force used. In order to improve the accuracy to an average error of about 2 mm, another method has been proposed (Burch and Law, 1995) involving radio-opaque markers attached to the compression paddle. The distance between the markers as seen on the image will increase as the compressed breast thickness increases. This relationship can generally be determined independently of the object or the compression force (possible flexion of the compression paddle in the direction perpendicular to the anode-cathode axis will, however, influence the accuracy of the measurement). It has been demonstrated (Aznar *et al.*, 2005) that the reading of the mammograms is not disturbed if markers with small sizes and characteristic shape are used (in this case lead bullets with a diameter of 1–2 mm) (Fig. 125).

Breast Glandularity

The *breast glandularity*—the fraction of glandular tissue in the breast—is expressed in relative terms (0–100%) under the assumption that the breast only contains glandular and adipose tissue. Different strategies have been used to determine breast glandularity. It can be estimated from the mammogram by experienced radiologists or, if digital images are used, with a computer program. Alternatively, the tube loading (mAs) value provided by the automatic exposure control can be used if a calibration is performed at the same beam quality with phantoms representing various breast glandularities.

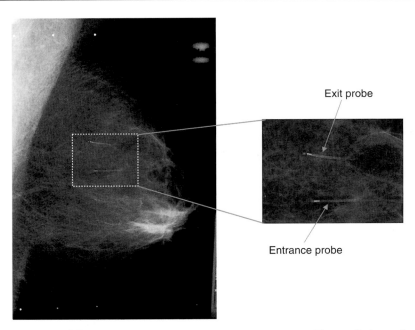

Figure 125 Impact of the OSL probes on a mammogram. The small size and characteristic shape of the probes (their crystals appear as white objects close to the end of each probe) indicate only minimal interference with the reading of the mammogram. Images of lead markers attached to the compression paddle are also seen, close to the short sides of the image, three on each side. One pair of lead markers and the OSL probes are located approximately 6 cm from the edge of the breast support. Reprinted from Aznar *et al.*, 2005 with permission from the *British Journal of Radiology.*

An estimation of breast glandularity is necessary if the conversion factor *c* (cf. "Estimation of AGD") is used.

Average Glandular Dose (AGD)

Average glandular dose (AGD) and *mean glandular dose* (MGD) are commonly used names instead of the more correct (ICRU, 2005) *average absorbed dose in glandular tissue*, D_G, which refers to the energy absorbed in the glandular tissue of the breast per unit mass and is expressed in gray (1 Gy = 1 J/kg). The risk from radiation in mammography is currently best described by the AGD.

From Measurement to Dose Estimate

It is important to differentiate between measurements performed directly (cf. "Skin Dose, D_{skin}") on the breast of a patient (or a phantom simulating a breast) and "indirect" measurements (cf. "X-ray Tube Output, Y") performed in the absence of a patient/phantom.

Both phantom and patient measurements play an important role. Phantom measurements can be used to assess the performance of the mammography system. They are also efficient tools for optimization and comparison between mammography units and centers as well as with diagnostic reference levels (DRL) (cf. "Dose Limits and Diagnostic Reference Levels") based on standard breasts. A phantom (e.g., 4.5 cm thick) used to simulate a standard breast (e.g.,

5 cm thick with 50% glandularity) must be exposed using the same anode, filter, and tube voltage as the corresponding breast would have been (not as a 4.5 cm breast in this case). The dose value presented must always refer to the standard breast; for example, AGD for a phantom makes no sense, since a phantom does not contain glandular tissue. Not only the thickness and material of the phantom, but also the area and form should be considered and is prescribed in the dose protocols. For digital mammography it can be very important that phantoms used for extensive dose estimates cover the whole imaging detector in order to avoid so-called ghost images—that is, remaining signals from previous exposures that can adversely affect a mammogram.

Patient measurements are generally used to investigate the dose level and variation for a population of women at a specific center. This is needed, since studies have shown that such measurements resulted in higher absorbed dose estimates than measurements performed on standard phantoms (Young *et al.*, 2005). DRLs can also be provided for patients, and in that case appropriate patient doses have to be determined. For proper risk estimates, AGD for relevant groups of patients should be estimated. In order to achieve the best clinical relevance of optimization, both image quality and AGD should be determined for patients.

Consequently, AGD is the most useful dose parameter in mammography and should be estimated in the first place. If that is not possible, skin dose or even incident air kerma or exposure can be useful for regular control and other limited quality controls.

Estimation of Skin Dose

A dosimeter on a breast or on a phantom simulating the breast gives the skin dose, D_{skin}, directly, provided the measurement is performed correctly (i.e., as prescribed by a dose protocol). The incident air kerma or the incident exposure estimated from X-ray tube output measurements can easily be converted to skin dose.

Estimation of AGD

The average glandular dose (AGD) is calculated from the incident air kerma, K, or incident exposure, X, by multiplication with one or several conversion factors provided by the dose protocol used. As has been described, K or X can be determined by direct or indirect measurements. If skin dose is measured, K or X can easily be estimated (cf. the inverse of "Estimation of Skin Dose").

Unfortunately, there is no universal standard procedure to determine AGD, and dose protocols can differ significantly. The most commonly used protocols will be briefly described in this section.

The American dose protocol (ACR, 1999) is the most widespread. It uses a standard breast 4.2 cm thick with 50% glandular breast tissue that is simulated with the so-called accreditation phantom. The phantom contains structures, and the image of the phantom is used to evaluate image quality. In contrast, the European protocol (EC, 1996) uses a standard breast that is 5 cm thick with 50% glandular breast tissue and is simulated with a 4.5 cm thick polymethyl methacrylate (PMMA, also called Plexiglas) phantom. This phantom has no structures, which is in accordance with most other phantoms used for breast dosimetry.

The incident exposure (ACR, 1999) or the incident air kerma (EC, 1996) to the surface of the phantom is measured and converted to AGD using protocol-specific conversion factors. The measurement is conveniently performed without the phantom. However, as long as the dosimeter is positioned outside the region in which the exposure settings are determined for the mammography unit in question, it can be performed with the phantom in place, as is usually done according to the American dose protocol (ACR, 1999). These different ways to measure are at large equivalent, since the X-ray tube output varies very little in the direction perpendicular to the anode-cathode axis. Instead, a pitfall could be the preexposure used in most modern mammography systems for the selection of exposure settings. The dose from this preexposure must be included in the dose value estimated, which might be complicated by the use of different X-ray beam quality for the preexposure and main exposure.

In the European protocol, conversion factors intended to be used to calculate AGD for a group of patients are provided for breasts of thicknesses 3–8 cm and a glandularity (cf. "Breast Glandularity") of 50%. A more recent dose protocol is provided in the European Guidelines (EC, 2006) and includes conversion factors for AGD estimates for breasts of various thicknesses (2.1–10.3 cm) and glandular contents (97-3%) simulated with PMMA phantoms of thickness 2–8 cm. As an example, a 4.5 cm thick PMMA phantom stimulates a 5.3 cm thick breast with glandular content 29%. Also, conversion factors for patients in standard age groups 40–49 and 50–64 are presented (i.e., the assumed glandularity for patients in these age groups is considered).

A summary and comparison of conversion factors published by various authors has been presented recently (ICRU, 2005). Notably, the conversion factors used by Dance et al. (2000) are mentioned—sometimes among other factors—in the major protocols (EC, 1996, 2006; ACR, 1999). Those conversion factors are of three types: c_G, c, and s. The first one, c_G, converts air kerma to AGD for a breast with an assumed glandular content of 50%. To be precise, the glandular content in all the breast models used for calculation of these conversion factors is 50% only in the central region of the breast with superficial layers of 4–5 mm of adipose tissue. However, this is usually not considered when breast glandularity is estimated clinically and is in fact a discrepancy in AGD estimates, though small in practice. In order to calculate AGD for breasts with other glandular content than 50%, c factors are given. The factor s corrects for the anode/filter combination used.

Conversion factors c_G are given as a function of the half-value layer and the compressed breast thickness. Consequently these parameters have to be known in order to calculate AGD. Naturally, the breast glandularity must be known if c factors are used.

Dose Limits and Diagnostic Reference Levels

The traditional method of dose constraint is the use of dose limits—prescribed levels of a dose parameter, usually AGD, either to a standard breast or to groups of patients. If the dose value in a specific case exceeds the dose limit, the dose constraint is not fulfilled. There is then a risk to be satisfied with a fulfillment of this dose constraint, without performing any optimization, which could have resulted in an even lower dose. The further risk is that the dose limit is selected to be so high that only obviously suboptimal systems fail to fulfill the dose constraint.

Examples of dose limits are found in both the United States and Europe. AGD for the standard breast 4.2 cm thick must not exceed 3 mGy in screen-film mammography, according to the American dose protocol (ACR, 1999). The European protocol (EC, 1996) states a limiting value of 2.8 mGy and 3.6 mGy at net optical density 1.2 and 1.6, respectively, for the AGD to the 5 cm thick standard breast.

Another concept is the use of diagnostic reference levels (DRLs). Optimization is first performed. If the dose value in the actual case is above the DRL, use of the mammography system can be permitted and even desirable, provided that the diagnostic accuracy is higher than for a system with a dose value below the DRL. On the other hand, if the dose value lies below the DRL, further optimization may still be required.

The use of dose protocols has allowed national comparative studies, which have shown a wide range of variation of absorbed doses among mammography facilities (Gentry and DeWerd, 1996; Kruger and Schueler, 2001; Young *et al.*, 2005). As a result, DRLs have been defined as the AGD to a specific group of patients. For example, a recent British study (Young *et al.*, 2005) confirmed the appropriateness of a national DRL of 3.5 mGy for 5.5 cm thick breasts. In Sweden, there is a legal need to perform patient AGD measurements in mammography screening for at least 20 patients with a mean compressed breast thickness 5.0 ± 0.5 cm each year (SSI, 2002), according to the European dose protocol (EC, 1996), with a DRL of only 1.3 mGy.

Obviously there is no universal standard for dose constraints. This can be partly explained by the lack of a standard for the way breast dosimetry should be performed. Another explanation is the lack of consensus regarding image quality demands. Nevertheless it is important to estimate AGD levels as well as image quality in an accurate way and to document the procedures in a way that may permit comparisons between various centers and regions.

Dosimeters for Indirect Measurements

Dosimeters must fulfill specific requirements as specified in the dose protocols. As an example (EC, 1996), they should have a dynamic range covering at least 0.5–100 mGy, a total uncertainty of less than ± 10%, including the energy response for the beam qualities involved, and a precision better than ± 5%. Because of these requirements and other factors such as user-friendliness and cost, to date only ionization chambers and semiconductors are in widespread use for indirect measurements and HVL determinations in mammography.

Ionization Chambers

Ionization chambers are the "gold standard" dosimetry system in medical imaging. In principle, electrodes collect the ion pairs created in a small volume of air. The measurements have to be corrected for variation in atmospheric pressure and temperature. A recent study (DeWerd *et al.*, 2002) evaluated 10 commercially-available ionization chambers and found that they varied considerably in volume (0.2–15 cm³), composition, and geometry, but also that their energy response allowed AGD determination within about 2% for a wide range of beam qualities. Ionization chambers are fragile and not very user-friendly, but provide valuable tools for accurate dosimetry.

Semiconductors

Semiconductor systems are commercially available for use in mammography. When the detector is exposed to radiation, electron-hole pairs are created and a current is generated, proportional to the radiation absorbed by the detector. Semiconductor systems are sturdy, sensitive, and relatively easy to handle. However, they are energy-dependent and must be calibrated with care (Witzani *et al.*, 2004). It has been demonstrated (Hemdal *et al.*, 2005) that noninvasive measurements of HVL can be performed accurately on a Sectra MicroDose Mammography (MDM) system even without a diaphragm with a sensitive and well-collimated detector with simultaneous correction for the energy dependence. The Sectra system is a scanning multislit digital mammography system with a multislit precollimator device that precludes HVL measurements according to the usual geometry prescribed in the dose protocols.

Dosimeters for Direct *in vivo* Measurements

When dosimeters are used for direct *in vivo* measurements—when they are present during the acquisition of a patient image—it is fundamental that the reading of the mammogram is not disturbed by the dosimeters. This can be achieved in different ways.

The first strategy is to use dosimeters small enough that relevant structures (e.g., clusters of microcalcifications or small tumors) are not obscured and with shapes that are characteristic enough not to be confused with such structures or any other structure of the breast. Even if these conditions are fulfilled, the radiologist has to get used to and accept the view of dosimeters in the mammogram.

The second strategy is to use thin dosimeters (in direction from focus to image detector) with a similar atomic composition as normal breast tissue making them basically invisible. Nevertheless, the radiologist must be informed about their presence, because they may still be seen in patients with thin, adipose breasts, or with the use of a low X-ray beam energy, an image detector with high-contrast resolution, or an extreme window setting in digital mammography. Such dosimeters are likely to have a relatively large surface in order to be sensitive enough, which increases the risk of interfering with the reading of the mammogram.

A third strategy is simply to locate the dosimeter away from the breast. The obvious disadvantage is that it is no longer a "true" *in vivo* measurement close to the position prescribed by the dose protocol used. It is then necessary to correct for differences in the X-ray beam (e.g., the heel effect), focus-detector distance (due to, e.g., the X-ray beam geometry and the flexure of the compression paddle), and scatter conditions.

Direct measurements can also be made with the dosimeter on the surface of a phantom simulating a breast. The reading of images will not be disturbed unless the phantom contains structures that are used to evaluate image quality. Such *in vitro* measurements are quite likely comparatively easy to handle, as for the accreditation phantom; therefore this section will focus on *in vivo* measurements.

Thermoluminescence Detectors

The use of thermoluminescence detectors (TLDs) has been widespread in all areas of medical X-ray imaging for the last 30 years. TL detectors are small crystals which, when exposed to radiation, actually store the radiation energy at defects inside the crystal structure. After the exposure is completed, the stored energy is released at high temperatures in the form of low-energy photons (such as visible light). These photons can be detected, and the resulting signal is proportional to the absorbed dose in the TL detector. TLDs are found in various solid shapes, or they can be reduced to a powder and used in small bags. TLDs are most commonly made of lithium fluoride (LiF) which has a relatively stable energy response across the range of energies used in mammography. One limitation of TLDs is a phenomenon called "fading": if stored at room temperature after a measurement and before reading, the detectors are going to "lose" some of their signal as a function of time, and the absorbed dose estimate will be inaccurate. This and other pitfalls can at least partly be overcome through a complicated procedure with oven storage under specified conditions both before and after the reading of the dosimeters. TLDs are consequently relatively cumbersome to use, but they have the interesting advantage of enabling mail-based services and surveys.

Dose protocols can mention special requirements for TLDs. As an example (EC, 1996), they should have an accuracy and precision better than ± 10%. This protocol also provides procedures for sending TLDs to a central laboratory for calibration and reading. This procedure has been used for interinstitute comparison studies such as the one reported by Gentry and DeWerd (1996), which involved 4400 women in 170 mammography facilities across the United States. This is one of many studies using small TLDs with a characteristic shape that are visible in the mammogram.

Another strategy mentioned earlier is placing on the patient's breast TLDs that are made thin enough not to degrade the diagnostic quality of the mammogram (they should not mask a lesion or be mistaken for one). This strategy has been used by Warren-Forward and Duggan (2004).

The results are promising but yet not quite convincing that the reading of mammograms cannot be affected under any circumstances.

Novel *in vivo* Techniques

In the past several years, some dosimetry systems other than TLDs have shown a potential for *in vivo* measurements in mammography. They differ mainly from TLDs in terms of increased sensitivity and potential user-friendliness. Although those novel techniques are not yet evaluated and commercially available in the way TLDs are, and hence, are far from widespread in clinical application, a short review is given here.

MOSFET

Metal oxide semiconductor field effect transistor (MOSFET) detectors are a relatively new and exciting development in the world of dosimetry in medical imaging. In essence, a MOSFET is a silicon semiconductor to which a voltage is applied. Exposure to radiation will result in a shift in voltage, directly proportional to the amount of energy absorbed in the detector. Although their clinical use is at present limited, MOSFETs have gained popularity because of their small size and the possibility of immediate readout. Recent articles (Dong *et al.*, 2002; Benevides and Hintenlang, 2006) report on their application in mammography, and the results are encouraging in terms of sensitivity. However, in the direct measurements used, the detectors were not positioned on the breast surface (Dong *et al.*, 2002) or phantom surface (Benevides and Hintenlang, 2006), but substantially further from the edge of the breast support than the prescribed distance (4–6 cm) in various dose protocols. This was done in order to minimize the degradation of the diagnostic quality of the image. Consequently, no "true" *in vivo* measurements are made, and the impact of this discrepancy on dose results should be evaluated further. Other potential pitfalls of MOSFET detectors include their limited lifetime (they have to be replaced regularly as they accumulate dose) and their energy dependence.

Optically Stimulated Luminescence (OSL)

The phenomenon of optically stimulated luminescence (OSL) is closely linked to thermoluminescence, but uses light as a stimulation source instead of heat. As a result, one can read absorbed doses from a small crystal linked to an optical fiber immediately after exposure. A prototype instrument using aluminum oxide (Al_2O_3:C) crystals was shown to be well suited to *in vivo* measurements in mammography (Aznar *et al.*, 2005). A small cylindrical crystal (diameter 0.48 mm and length 2 mm) was coupled to the end of a 1mm diameter optical fiber cable. Because of their small size and characteristic shape, these probes can be placed on the breast surface during the examination, with minimal interference with the reading of the mammogram. This new technique was tested over a range of clinically relevant X-ray energies, which showed that the high sensitivity of the crystals permitted the measurement of both entrance and exit doses (i.e., also doses on the inferior surface of the breast). Figure 124 shows an example of *in vivo* measurements where the presence of the small and characteristic probes did not disturb the reading of the mammograms. Both probes were positioned about 6 cm from the edge of the breast support, one attached underneath the compression paddle and the other on the breast support. One potential drawback of this system is the significant energy dependence owing to the fact that the crystals are not tissue equivalent. However, the first results indicate a potential for use in routine quality control and *in vivo* dose measurements in mammography.

Summary and Conclusions

In mammography, optimization is necessary to achieve sufficient image quality while delivering the lowest possible absorbed dose to the glandular tissue of the breast. In this chapter, tools for the determination of absorbed dose have been described in the context of indirect measurements involving X-ray tube output measurement and direct measurements performed during image acquisition.

Several dose protocols have been published to ensure that all measurements are performed with care, are reproducible, and can be compared among institutions. The results of measurements can be presented as kerma or skin dose, but the average absorbed dose in glandular tissue (AGD) is the most relevant quantity as it more accurately relates to the risk of radiation-induced carcinogenesis.

Phantoms can be used to simulate standard breasts and provide information about the performance of the mammography system. They are also efficient tools for optimization and comparison between mammography units and centers as well as with diagnostic reference levels (DRLs) based on standard breasts.

DRLs for patients are also used requiring the determination of appropriate patient doses. For proper risk estimates, AGD for relevant groups of patients should be evaluated. This is done most accurately if conversion factors given as a function of breast glandularity are used. In order to achieve the best clinical relevance of optimization, both image quality and AGD should be determined by patient measurements.

The standard dosimeter for indirect measurements is the ionization chamber, but other dosimeters that fulfill the requirements of the dose protocol used can also be considered if calibrated properly. There are already commercially-available semiconductors that, if correctly used, are at least as sensitive, stable, and accurate as ionization chambers while being much less fragile.

The standard dosimeter for direct *in vivo* measurements is the TLD. But despite its positive features, it is generally recognized that a dosimeter with higher precision and immediate readout is preferred for *in vivo* mammography measurements. Hence, novel *in vivo* techniques such as MOSFET and OSL might be alternatives for measuring the absorbed dose directly to individual women without compromising the reading of the mammograms.

References

American College of Radiology (ACR). 1999. *Mammography Quality Control Manual.* Reston, VA.

Andersson, I., and Janzon, L. 1997. Reduced breast cancer mortality in women under age 50: updated results from the Malmö Mammographic Screening Program. *J. Natl. Cancer Inst. Monogr. 22*:63–67.

Aznar, M.C., Hemdal, B., Medin, J., Marckmann, C.J., Andersen, C.E., Bøtter-Jensen, L., Andersson, I., and Mattsson, S. 2005. *In vivo* absorbed dose measurements in mammography using a new real-time luminescence technique. *Br. J. Radiol. 78*:328–334.

Benevides, L.A., and Hintenlang, D.E. 2006. Characterization of metal oxide semiconductor field effect transistor dosimeters for application in clinical mammography. *Med. Phys. 33*:514–520.

Burch, A., and Law, J. 1995. A method for estimating compressed breast thickness during mammography. *Br. J. Radiol. 68*:394–399.

Dance, D.R., Skinner, C.L., Young, K.C., Becket, J.R., and Kotre, C.J. 2000. Additional factors for the estimation of mean glandular breast dose using the UK mammography dosimetry protocol. *Phys. Med. Biol. 45*:3225–3240.

DeWerd, L.A., Micka, J.A., Laird, R.W., Pearson, D.W., O'Brien, M., and Lamperti, P. 2002. The effect of spectra on calibration and measurement with mammographic ionization chambers. *Med. Phys. 29*: 2649–2654.

Dong, S.L., Chu, T.C., Lee, J.S., Lan, G.Y., Wu, T.H., Yeh, Y.H., and Hwang, J.J. 2002. Estimation of mean glandular dose from monitoring breast entrance skin air kerma using a high sensitivity metal oxide semiconductor field effect transistor (MOSFET) dosimeter system in mammography. *Appl. Radiat. Isot. 57*:791–799.

European Commission (EC). 1996. *European Protocol on Dosimetry in Mammography.* Report EUR 16263. Luxembourg: Office for Official Publications of the European Communities.

European Commission (EC). 2006. *European Guidelines for Quality Assurance in Breast Cancer Screening and Diagnosis,* 4th ed. Luxembourg: Office for Official Publications of the European Communities.

Gentry, J.R., and DeWerd, L.A. 1996. TLD measurements of *in vivo* mammographic exposures and the calculated mean glandular dose across the United States. *Med. Phys. 23*:899–903.

Hemdal, B., Herrnsdorf, L., Andersson, I., Bengtsson, G., Heddson, B., and Olsson, M. 2005. Average glandular dose in routine mammography screening using a Sectra MicroDose Mammography unit. *Radiat. Prot. Dosimetry 114*:436–443.

International Commission on Radiological Protection (ICRP). 2004. *Managing Patient Dose in Digital Radiology.* ICRP Publication 93. Oxford, UK: Elsevier.

International Commission on Radiation Units and Measurements (ICRU). 2005. *Patient Dosimetry for X-rays Used in Medical Imaging.* ICRU Report 74. Oxford, UK: University Press.

Kruger, R.L., and Schueler, B.A. 2001. A survey of clinical factors and patient dose in mammography. *Med. Phys. 28*:1449–1454.

Swedish Radiation Protection Authority (SSI). 2002. Regulations and General Advice on Diagnostic Standard Doses and Reference Levels within Medical X-ray Diagnostics, SSI FS 2002:2. Stockholm, in Swedish. (Unofficial translation to English available at http://www.ssi.se/forfattning/eng_forfattlista.html.)

Warren-Forward, H.M., and Duggan, L. 2004. Towards *in vivo* TLD dosimetry in mammography. *Br. J. Radiol.* 77:426–432.

Witzani, J., Bjerke, H., Bochud, F., Csete, I., Denoziere, M., de Vries, W., Ennow, K., Grindborg, J.E., Hourdakis, C., Kosunen, A., Kramer, H.M.,

Pernicka, F., and Sander, T. 2004. Calibration of dosemeters used in mammography with different X-ray qualities: EUROMET Project no. 526. *Radiat. Prot. Dosimetry 108*:33–45.

Young, K.C., Burch, A., and Oduko, J.M. 2005. Radiation doses received in the UK breast screening program in 2001 and 2002. *Br. J. Radiol. 78*:207–218.

23

Metastatic Choriocarcinoma to the Breast: Mammography and Color Doppler Ultrasound

Naveen Kalra and Vijaynadh Ojili

Introduction

Metastases to the breast from extramammary tumors are rare, with a reported incidence of 0.5–6.6% in various autopsy series (Bartella *et al.*, 2003) and 0.4–2.6% in various clinical series (Silverman *et al.*, 1987). Choriocarcinoma is a rare extramammary site for the origin of breast metastases. Differentiating a metastatic lesion from a primary breast lesion assumes importance because the prognosis and the treatment of the two is entirely different. The major challenge for a radiologist is to differentiate metastatic malignancy from primary breast lesions. It has been stated that there is a close correlation between the size of the metastatic lesion on palpation and on mammography owing to the absence of desmoplastic reaction around the lesion. On the other hand, most primary scirrhous breast carcinomas have associated desmoplastic reaction and are apparently larger on palpation than on mammography. Imaging the breast with mammography and color Doppler may aid in the diagnosis of metastases and circumvent the need for a biopsy.

Most patients have a known diagnosis of primary carcinoma at the time of presentation with breast metastases. Breast metastases usually occur years after the diagnosis of the primary cancer. They appear on an average two years after the discovery of the primary tumor. In some studies, however, breast metastasis is the first manifestation of malignancy in up to 25% of patients.

The origin of metastases to the breast has been reported in the literature from numerous primary sites. Several studies have attempted to classify the sites of origin in the order of their frequency. The most common primary site is the contralateral breast, but this site is specifically excluded from most series that also exclude hematological malignancies such as leukemia and lymphoma. In the most recent review of the literature that excluded only the contralateral breast malignancy as the primary site, the most common primary tumor sources for breast metastases in order of decreasing frequency are lymphomas, melanomas, rhabdomyosarcomas, and lung and ovarian tumors (Vizcaino *et al.*, 2001). Rhabdomyosarcoma is the most frequent site of origin of breast metastases in adolescent females. In men, carcinoma of the prostate is the most common extramammary site for breast metastases, and those on estrogen therapy are more predisposed to this.

Seventy percent of patients with breast metastases are less than 50 years of age; these patients are younger than those with primary breast carcinoma. It has been hypothesized that the breasts of younger patients have a better blood supply, which enhances the possibility of tumor metastases. However, a recent study reported a mean age of 57.4 years for breast metastases, which is not less than that for primary tumors (David *et al.*, 2002).

Clinically, most metastases present as rapidly growing, painless, palpable, firm breast masses that are usually mobile. Approximately 50% of the breast metastases are found to be superficial, and they are in the plane between skin and breast. Metastatic lesions do not result in thickening or retraction of the skin and nipple. Furthermore, their superficial extraductal location precludes nipple discharge. Axillary lymph node involvement is frequently encountered. Diffuse skin involvement or associated subcutaneous nodules can occur, especially with melanoma metastases. Metastases to the breast may present as a solitary lesion (85%), as multiple lesions (11%), or as diffuse involvement (4%) (Demirkazik *et al.*, 1996). They show slight predominance in the left breast. The most common site of involvement is the upper outer quadrant. The incidence of bilateral lesions may be as high as 50% (DiBonito *et al.*, 1991).

The usual sites of metastases of choriocarcinoma are lungs and vagina followed by brain, liver, and kidney. Breast metastases from choriocarcinoma are extremely rare. There are no clinical signs that help us to distinguish these lesions from other metastases. The patient is usually a diagnosed case of choriocarcinoma who presents with a breast nodule with rising human chorionic gonadotropin (HCG) levels on follow-up. Radiologically, it is difficult to distinguish metastases from primary breast lesions because there are no specific imaging characteristics for metastasis to the breast. However, a correlation of mammographic findings with those on color Doppler sonography helps differentiate metastases from primary breast lesions (Kalra *et al.*, 2005).

Mammography

The most common mammographic appearance of metastases is of one or more well-circumscribed masses located in the upper and outer quadrant of the breast. Typically, there is no spiculation, architectural distortion, skin thickening, or other signs of surrounding desmoplastic reaction, which characterize the majority of primary breast carcinomas. However, the mammographic findings of breast metastases can be variable and range from normal to a pattern of diffuse skin thickening that simulates inflammatory carcinoma. The mammographic appearances of metastatic hematologic malignancies can vary from discrete to ill-defined masses that may be obscured by benign proliferative breast changes.

Pleomorphic microcalcifications are one of the characteristic features of primary breast cancer on mammography. The presence of recognizable calcifications in a mass on a mammogram virtually excludes metastatic disease to the breast. The only exception is metastatic medullary thyroid cancer and metastatic carcinoma of the ovary in which Psammoma bodies–related calcification may be seen. As mentioned earlier, metastases to the breast do not cause a surrounding desmoplastic reaction in the adjacent normal breast. Typically, then, there is a close correlation between the palpable size of the mass on clinical examination and its size on mammography. This contrasts with primary carcinoma of the breast in which the mammographic abnormality is smaller than the palpable mass.

Benign lesions such as cysts or fibroadenomas can have an appearance similar to breast metastases on mammography. Because the margins of a metastasis are slightly more irregular than those of a fibroadenoma or cyst, magnification and spot compression images may be useful. Medullary, mucinous, and papillary carcinomas of the breast should also be included in the differential diagnosis of breast metastases on mammography. It is in this context that the role of gray-scale and color Doppler sonography needs to be evaluated. Choriocarcinoma metastases present as well-defined masses on mammography with no spiculations, calcifications, or architectural distortion (Fig. 126a, b). These metastases have been reported in the lower inner quadrant of the breast (Kalra *et al.*, 2005).

Ultrasonography and Color Doppler

Ultrasonography reliably distinguishes cystic from solid breast lesions. A variety of sonography findings have been reported in patients with breast metastases. Characteristic lesions are rounded or oval with a low echogenicity and a well-defined posterior wall. They have macrolobulations and are aligned with their long axis parallel to the skin (Francois *et al.*, 2005). Multiple lesions in the breast maintain the same sonographic characteristics. Since there is considerable overlap in the gray-scale sonographic appearance of breast metastasis and benign breast lesions such as fibroadenomas, color Doppler sonography can be used to assess the vascularity of the lesion for better characterization.

Malignant tumors stimulate the growth of blood vessels (neovascularization) by releasing angiogenesis factor. Because color Doppler sonography can detect neovascularization, it has the potential to distinguish benign from malignant lesions. Color Doppler findings in patients with breast metastases have been described only rarely in the literature. These findings are similar to those described earlier for differentiating between malignant and benign masses. These include the presence of both central and peripheral neovascularity, penetrating vessels, and certain spectral patterns

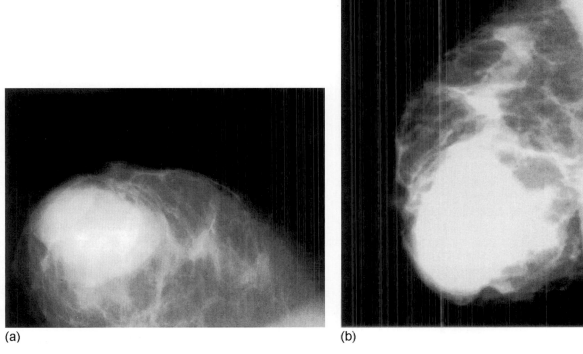

(a)   (b)

Figure 126 Mammograms of the right breast craniocaudal (a) and oblique (b) views in a woman who had been diagnosed to have choriocarcinoma after a molar pregnancy 1 year back. There is a large, well-defined lobulated high-density mass in the lower inner quadrant. No microcalcification or architectural distortion is seen.
Source: Reprinted with permission from the *American Journal of Roentgenology.*

such as demonstration of venous signal or a combination of venous, high-impedance pulsatile and turbulent signals. A penetrating vessel is a continuous vascular signal from outside the lesion to inside it (Lee *et al.*, 2002). The spectral pattern shows the signal pattern of blood flow. Venous signal is denoted by continuous flow with little variation in velocity. High-impedance flow is denoted by a high peak in systole with a sharp dropoff and little or no flow in diastole. Turbulent flow is high-velocity flow with marked spectral broadening (McNicholas *et al.*, 1993).

Choriocarcinoma metastases are seen as well-defined low-echogenicity lesions on ultrasound (Fig. 127). They have been reported to have peripheral and central neovascularity and numerous penetrating vessels on color Doppler (Fig. 128). The spectral trace of these vessels shows a high-impedance flow with a sudden drop in systolic flow and no flow in diastole (Fig. 129) (Kalra *et al.*, 2005). If the color Doppler sonographic features suggest a malignant lesion, then that lesion being primary breast carcinoma can be excluded by correlation with mammographic findings. Primary breast carcinomas on mammography appear as spiculated mass lesions with surrounding architectural distortion. They show pleomorphic microcalcifications, skin thickening, and nipple retraction. However, well-defined

primary breast cancers such as mucinous and medullary carcinomas may be indistinguishable from breast metastases on both mammography and sonography.

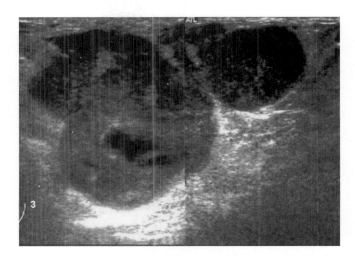

Figure 127 Gray-scale ultrasound of a choriocarcinoma metastasis shows a well-defined, lobulated, hypoechoic solid mass. No posterior acoustic shadowing or calcification is seen within the mass.
Source: Reprinted with permission from the *American Journal of Roentgenology.*

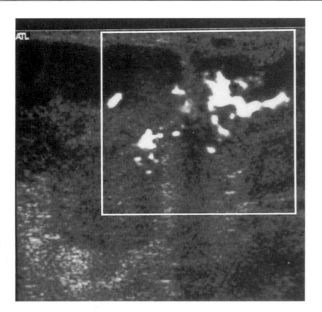

Figure 128 Power Doppler sonogram of the same mass lesion shows numerous intralesional penetrating blood vessels.
Source: Reprinted with permission from the *American Journal of Roentgenology*.

The use of magnetic resonance imaging in evaluating breast metastases has been reported in isolated cases due to the rarity of the condition. Magnetic resonance imaging may be useful in young patients who have dense breast parenchyma and in metastatic melanoma, which has high signal on T1-weighted images and low signal on T2-weighted images suggesting a melanin-containing tumor (Ho *et al.*, 2000).

Tissue Diagnosis

Fine-needle aspiration cytology or biopsy diagnosis of metastatic malignancy of the breast may be needed in order to avoid unnecessary mastectomy and to ensure appropriate chemotherapy. The differentiation of primary breast malignancy from metastatic disease is based on both cytological and architectural findings. The metastatic lesions are typically well demarcated and show a pattern of displacing or engulfing mammary ducts rather than arising from them. Associated *in situ*, atypical hyperplastic lesions are not seen in surrounding breast parenchyma. The cytopathology of metastatic malignancies of the breast includes features that are not usually seen in primary breast carcinomas, such as clear cytoplasm, intracytoplasmic pigment, undifferentiated small cells, and malignant cells of hematopoetic origin. Ancillary studies performed on the fine-needle aspiration material, including special stains and immunocytochemistry, may contribute to a definitive diagnosis in most cases. Choriocarcinoma metastases show both mononuclear and multinuclear cells with abundant cytoplasm and prominent nucleoli (Kumar *et al.*, 1991). The cytoplasm shows positivity for HCG antibody.

Metastases to the breast carry a poor prognosis. As these metastases indicate wide dissemination of the disease, prolonged survival after diagnosis is unusual. The 1-year survival rate is < 20% in most reported series (Bartella *et al.*, 2003). Prolonged survival occurs in malignancies that have an indolent clinical course, such as carcinoid tumors, and in malignancies that can be effectively treated with chemotherapy, such as lymphoma, ovarian carcinoma, and choriocarci-

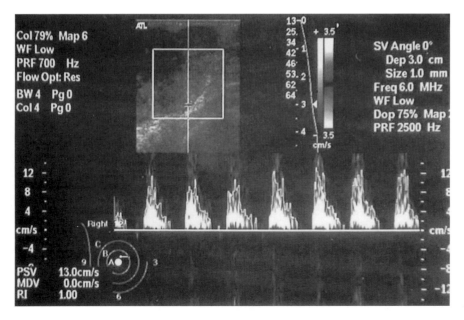

Figure 129 Spectral trace of the neovessels shows high-impedance flow as indicated by absent flow in diastole.
Source: Reprinted with permission from the *American Journal of Roentgenology*.

noma. The results of chemotherapy for choricarcinoma may result in up to 100% cure or remission except in those patients who have high-risk disease.

In conclusion, metastatic choriocarcinoma to the breast is distinctly unusual and has been described in isolated case reports. The data on the imaging findings is scant. There are no specific clinical features for the diagnosis of choriocarcinoma metastases to the breast. Similarly, there is no single radiological investigation that can reliably diagnose these metastases. Evaluating mammography together with color Doppler sonography may aid in the prebiopsy diagnosis of choriocarcinoma metastases in a patient with known primary disease. A correct diagnosis of metastases to the breast is needed to avoid unnecessary surgery and to provide appropriate systemic chemotherapy.

References

Bartella, L., Kaye, N., Perry, N.M., Malhotra, A., Evans, D., Ryan, D., Wells, C., and Vinnicombe, S.J. 2003. Metastases to the breast revisited: radiological—histopathological correlation. *Clin. Radiol.* 58:524–531.

David, O., Gattuso, P., Razan, W., Moroz, K., and Dhurandhar, N. 2002. Unusual cases of metastases to the breast: a report of 17 cases diagnosed by fine needle aspiration. *Acta. Cytol. 46*:377–385.

Demirkazik, F.B., Baskan, O., Aydingoz, U., Tacal, T., and Firat, P. 1996. Case report: squamous cell carcinoma of the skin metastasizing to the breast-imaging findings. *Er. J. Radiol. 69*:678–680.

DiBonito, L., Luchi, M., Giarelli, L., Falconieri, G., and Viehl, P. 1991. Metastatic tumors to the female breast: an autopsy study of 12 cases. *Pathol. Res. Pract. 187*:432–436.

Francois, C., Rangachari, B., and Bova, D. 2005. Mammography and sonography of pathologically proven adrenal cortical carcinoma metastatic to the breast. *Am. J. Roentgenol. 184*:1279–1281.

Ho, L.W., Wong, K.P., Chan, J.H., Chow, L.W., Leung, E.Y., and Leong, L. 2000. MR appearance of metastatic melanoma in the breast. *Clin. Radiol. 55*:572–573.

Kalra, N., Ojili, V., Gulati, M., Prasad, G.R., Vaiphei, K., and Suri, S. 2005. Metastatic choriocarcinoma to the breast: appearance on mammography and Doppler sonography. *Am. J. Roentgenol. 184*:S53–S55.

Kumar, P.V., Esfahani, F.N., and Salimi, A. 1991. Choriocarcinoma metastatic to the breast diagnosed by fine needle aspiration. *Acta Cytol. 35*:239–242.

Lee, S.W., Choi, H.Y., Baek, S.Y., and Lim, S.M. 2002. Role of color and power Doppler imaging in differentiating between malignant and benign solid breast masses. *J. Clin. Ultrasound. 30*:459–464.

McNicholas, M.M., Mercer, P.M., Miller, J.C., McDermott, E.W., O'Higgins, N.J., and MacErlean, D.P. 1993. Color Doppler sonography in the evaluation of palpable breast masses. *Am. J. Roengenol. 161*:765–771.

Silverman, J.F., Feldman, P.S., Covell, J.L., and Frable, W.J. 1987. Fine needle aspiration cytology of neoplasms metastatic to the breast. *Acta Cytol. 31*:291–300.

Vizcaino, I., Torregrosa, A., Higueras, V., Morote, V., Cremades, A., Torres, V., Olmos, S., and Molins, C. 2001. Metastasis to the breast from extramammary malignancies: a report of four cases and a review of literature. *Eur. Radiol. 11*:1659–1665.

24

Detection and Characterization of Breast Lesions: Color-coded Signal Intensity Curve Software for Magnetic Resonance–based Breast Imaging

Federica Pediconi, Fiorella Altomari, Luigi Carotenuto, Simona Padula, Carlo Catalano, and Roberto Passariello

Introduction

Breast cancer is the second leading cause of cancer death among women in the developed world. Unlike many other forms of cancer, awareness among women of the risks associated with breast cancer is high and derives from many sources, including health education programs promoting screening, extensive media coverage, and first-hand knowledge from friends or relatives with the diagnosis. Although early breast cancer detection is achievable with mammography, limitations in sensitivity and specificity remain for this technique and sonography.

Breast magnetic resonance imaging (MRI) is creating a revolution in breast diagnosis and intervention. In the last decade, breast MRI has evolved from being an investigational technique to a clinically valuable tool for breast cancer detection and diagnosis. Because of its high sensitivity and effectiveness in dense breast tissue, MRI can be a valuable addition to the diagnostic work-up of a patient with a breast abnormality or biopsy-proven cancer.

The major limitation of breast MRI is the low-to-moderate specificity, which in combination with high sensitivity can lead to unnecessary biopsy, patient anxiety, and cost. Nonetheless, consistent findings are emerging that show contrast-enhanced MRI to be effective for early detection of cancer in high-risk women, and superior to mammography for identifying and demonstrating the extent of diffuse and multifocal breast cancer.

There are a number of clinical indications for which breast MRI is believed to add value to the conventional clinical and diagnostic work-up, including (1) evaluation of patients with axillary carcinoma and negative mammographic and clinical findings (Cup Syndrome), (2) evaluation of women with questionable mammographic findings and previous breast surgery to distinguish postsurgical scars from recurrent carcinoma, and (3) staging of the extent of a cancer diagnosed by percutaneous needle biopsy. Other indications are less well accepted and are being evaluated largely in the research setting (Kuhl, 2002).

The sensitivity of MRI to breast carcinoma and its high staging accuracy, has led to the emerging role of MRI in breast cancer screening for women identified to be at high risk and to the current use for assessing tumor response to neoadjuvant chemotherapy. Numerous studies have suggested that MR imaging may be a useful clinical tool for the diagnosis, staging, and management of breast cancer.

Contrast-enhanced magnetic resonance imaging (MRI) of the breast was first performed in the late 1980s in women with biopsy of proven carcinomas. Heywang and colleagues (1989) demonstrated that breast carcinomas showed significant enhancement following the administration of contrast material. However, further investigations showed that not only do malignant lesions enhance, but benign lesions can also show a similar degree of enhancement. Thresholds for significant enhancement were used with normalized units of enhancement to attempt to differentiate more reliably between the different tissues; however, overlap existed (Heywang et al., 1989). Therefore, multiple differing attempts were made to develop more defining characteristics to distinguish benign from malignant processes.

The earliest MRI studies of the breast were performed with a T1-weighted spin echo sequence before and after intravenous administration of contrast agent, with an imaging time of at least 5 min and a slice thickness of 5 mm. With the development of fast T1-weighted gradient echo pulse sequences, imaging of the breast in thinner contiguous sections along with dynamic MRI of the whole breast became feasible. With this new technique, it became possible to repeat the same image at short time intervals, and therefore characterize an enhancement lesion over a shorter interval time. Through use of these data, a signal intensity curve can be generated, and, thereby, the curve rate and velocity can be analyzed.

Studies with faster imaging technique demonstrated that the initial phase of rapid contrast uptake during the first 2 to 3 min contained valuable information to distinguish between benign and malignant tissue. Subsequently, a flurry of studies was performed using these new techniques. Using a two-dimensional (2D) gradient echo sequence, Kaiser and Zeitler (1989) found that malignancies showed a sudden increase in signal intensity of 100% within the first 2 min. Gradual, mild contrast uptake was seen in benign tissue. Other investigators made several different attempts by varying the imaging parameters and changing the enhancement criteria to improve specificity.

Boets et al. (1994) used a turbo FLASH subtraction technique and classified lesions as suspicious if they enhanced within 11.5 sec after the aorta enhancement. Gilles et al. (1994) used a T1-weighted spin echo sequence with subtraction imaging and obtained an acquisition time of 47 sec. They classified any enhancement concomitant with early normal vascular enhancement as a positive finding for malignancy, obtaining a sensitivity of 95% and a specificity of 53%. However, the validity of all these criteria was questioned by later investigators, who found even higher signal intensities in benign lesions such as fibroadenomas (Orel et al., 1994). These investigators found that while cancers tend to enhance faster than benign lesions, there is still a clear overlap in enhancement rates of benign and malignant lesions.

A more recent study (Kuhl et al., 1999) analyzed not only the enhancement pattern of a lesion in its early phase, but also the intermediate and late phases. By using a 2D dynamic technique, it qualitatively analyzed the shape of the time-signal intensity curve of suspicious lesions over time and described three different curves: type I (steady) curve responds to a straight or slightly curved enhancement pattern with the enhancement progressively increasing over time; type II (plateau) curve levels off after the initial sharp level of enhancement; and type III (washout) curve has a drop in signal intensity after the initial upstroke, indicating washout of contrast. These curves were generated on only focal masslike lesions that appeared morphologically suspicious and showed a signal intensity increase of > 60% on the first postcontrast images. A region of interest was placed in the area of the most rapid strongest enhancement and was quantified by the change in signal intensity before and after the injection of gadopentetate dimeglumine (Gd-DTPA). Type I curve was rated indicative of a benign lesion, type II was suggestive of malignancy, and type III was indicative of malignant lesion. Using these curves, they achieved a sensitivity of 91% and a specificity of 83%. Although this technique appears promising, the dynamic approach to differentiating benign and malignant lesions has not been fully corroborated by other studies. Some investigators have found no significant difference in the enhancement characteristics between benign and malignant lesions.

In a study of 74 lesions, Harms et al. (1993) showed significant overlap between malignant and benign lesions such as fibroadenomas, sclerosing adenosis, and proliferative fibrocystic change, and they suggest that analyzing a lesion morphologic characteristic may help to improve the specificity of MRI. Similar to its use in mammography and ultrasound, border characteristics such as well defined or spiculated may be a useful adjunct to enhancement features. Subsequently, Orel et al. (1994) evaluated both the morphologic and enhancement characteristics of suspicious breast lesions. They used a fat-saturated spoiled gradient echo sequence to acquire high-resolution images along with temporal information. Their data confirmed some of the previous studies that signal intensities and enhancement characteristics overlapped between benign and malignant lesions, particularly fibroadenomas.

Although carcinomas had a tendency toward more rapid enhancement and washout, there was still a significant overlap with enhancement patterns of fibroadenomas. In the morphologic analysis of lesions, they discovered that architectural features were helpful in differentiating between benign and malignant lesions. Carcinomas exhibited irregular borders and rim enhancement, while fibroadenomas often had lobulated borders, with nonenhancing internal septations.

In a study of 192 patients, Nunes et al. (1997) exclusively analyzed architectural features to develop a tree-shaped interpretation model to distinguish benign from malignant lesions. Masses with irregular borders and rim enhancement were associated with carcinoma, while masses with lobulated borders and internal septations were associated with fibroadenomas. Nonmass enhancement was also described, which included ductal and regional enhancement. Ductal enhancement correlated with ductal carcinoma in situ, while regional enhancement was not particularly predictive of either benign or malignant disease. Currently, both dynamic and morphologic data are seen as helpful in the assessment of breast MRI lesions.

Computer-aided Diagnosis: Features and Applications

Over the last decade or so, many investigators have carried out basic studies and clinical applications toward the development of modern computerized schemes for detection and characterization of lesions in radiological images, based on computer vision and artificial intelligence. These methods and techniques are generally called computer-aided diagnosis (CAD) schemes. Computer-aided diagnosis, by definition, is the use of computerized image analysis in interpreting an image. Computer-aided diagnosis assist the interpreter by drawing attention to areas in the image that might have been overlooked and by giving probability of malignancy estimates for regions of interest. It provides a second opinion. A radiologist's ability to interpret images can be improved significantly by using CAD.

The development of CAD reached a new phase in June 1998 when the Food and Drug Administration (FDA) approved the first commercial unit of detection of breast lesion in mammograms for marketing and sale for clinical use. Various observer performance studies have documented the benefit of radiologists using a computer aid in their interpretation process. Computerized enhancement, analysis and visualization of three-dimensional medical images have touched both diagnostic radiology (e.g.,

enhanced interpretation) and radiation therapy (e.g., treatment planning). Similar strides have been made with breast imaging with the aims of increased patient management.

As low-dose, spiral CT becomes routine, CT images may potentially be used for the screening of disease such as lung cancer, utilizing computerized detection of pulmonary nodules in CT images of the thorax. Image segmentation and visualization techniques are being investigated as a means of viewing representations of cardiac and abdominal structures such as in virtual colonoscopy. Vascular imaging based on either biplane, computer tomography angiography (CTA), and intravascular ultrasound will benefit greatly from developed computerized methods for fusion and visualization. The efficient and effective use of CAD will depend on well-implemented picture archiving and communication system (PACS), which will transport images, patient data, and CAD results to required sites within and about medical centers.

Computer-aided diagnosis research in breast imaging is now including digital mammography, ultrasound, and magnetic resonance imaging to improve breast cancer detection and characterization. A variety of reasons may explain why breast cancer is missed on mammograms. Some breast cancers simply are not seen on mammograms and may remain hidden by dense tissue until a lump is felt. Other cancers are difficult to see because they blend into the background of fibroglandular tissue and are overlooked at screening. On retrospective evaluation, these cancers are occasionally detected; however, they might be missed a second time. Still other cancers are located in areas difficult to visualize (e.g., subtle calcifications on the burned-out edge of the image). Occasionally, cancers are missed for no other reason than momentary distraction or inattention of the screening radiologist. Therefore, computers and CAD were applied to help detect breast cancer at an earlier stage.

In the research community, Winsberg et al. (1967) may have been the first investigators to report on CAD. They considered the difficulty of viewing large volumes of screening mammograms before such screening was accepted in the clinical community. Other researchers, such as Ackerman and Gose, slowly expanded the concept of CAD. These investigators used a computer to extract four properties of lesions including (1) calcification, (2) speculation, (3) roughness, and (4) shape. Eventually, these properties and many others came to be termed features in the CAD community, and the number of features gradually but substantially expanded. Image processing (IP) was a natural tool for this purpose because it provides a set of techniques that enhances features of interest and de-enhances others with the application of many types of filters. Image processing also has techniques for quantifying visual features and for providing metrics to measure geometric, topologic, or other characteristics that describe images.

Segmentation, which means separating an image into regions of similar attributes, is a typical early-stage operation under active study in IP. A common example of segmentation is the automated detection of the skin line of the breast and the separation of the image into the breast area, the directly exposed (background) area, and perhaps the unexposed (film edge) area. This particular segmentation method limits the area of the mammogram that the computer analyzes to save computation time. It also may avoid the mistake of producing marks on objects that are not in the breast. In the early detection of breast cancer, segmentation is used to determine the boundary of masses, separating the image into regions inside and outside of the mass. This is a particularly challenging type of segmentation because of the proximity of masses to normal parenchymal tissue. Once a lesion is segmented, the computer algorithm uses IP techniques to measure the pertinent features and to describe features such as the border of the lesion. Feature extraction, or the quantification of relevant features, is the portion of the computer code that is most important for good performance. A neural network is the term for a computer code that helps make a decision based on the value of certain features (color or size). The required neural network collects input (e.g., the color of the object) and provides output. Although CAD determines a great contribution to the interpretation process, the radiologist makes final diagnosis.

The role of CAD in sonography is not the same as that in screening mammography. It is used to provide the second opinion for interpreting a sonographically detected tumor and to improve diagnostic confidence. Image texture analysis plays an important role in the CAD ultrasound system.

Many researchers also demonstrated that texture analysis with co-occurrence matrix parameters could help improve the differentiation of benign from malignant breast lesions. Three-dimensional images were acquired using a conventional ultrasound machine and a mechanical transducer-guiding system developed at University of Michigan Health System (UMHS). Then, they were analyzed by a CAD system and read by breast radiologists. After the physicians scored each mass on a scale of the likelihood of malignancy, they were shown the score that the CAD system assigned to the same mass, based on algorithms for mass shape, shadowing, and border characteristics. Many results show that a CAD system improved the ability of highly experienced radiologists to tell cancerous tumors from benign growths on ultrasound breast scans. Such scans are often performed after a suspicious finding on a screening mammogram, to help determine if a biopsy is needed.

Computer-aided Detection for Breast Magnetic Resonance Imaging

Magnetic resonance imaging is the most sensitive technique currently available for imaging primary or recurrent breast cancer. It has been shown to be extraordinarily useful for predicting disease extent, multifocal or multicentric lesions, differentiating scar from recurrent cancer, identifying primary cancer in young high-risk patients, and evaluating tumor response to neoadjuvant chemotherapy. Unfortunately, MR imaging does not provide adequate discrimination of tumoral tissue without use of a contrast agent because of the similarity of relaxing times between normal and pathological tissues. On the grounds that the standard clinical procedure includes the injection of a contrast agent containing gadolinium compounds, which produces a higher enhancement of pathologic tissue.

Several studies have clearly demonstrated the importance of contrast-enhancement MRI for breast lesion detection and characterization. The pathophysiological basis of lesion contrast enhancement in breast MRI has not yet been fully elucidated, but some fundamental facts are known that should help us understand the technique's specific strengths and weaknesses in terms of lesion detection and differential diagnosis. It is a well-established fact that malignant lesions release angiogenic factors (e.g., vascular endothelial growth factor (VEGF)) that induce sprouting and growth of preexisting capillaries, as well as the *de novo* formation of new vessels (Knopp *et al.*, 1999). As revealed by histological and electronmicroscopic studies, these capillaries exhibit a pathologic vessel wall architecture with leaky endothelial linings. Thus, the effect of angiogenic activity is twofold: there is an increased vascularity (vessel density) leading to a focally increased inflow of contrast material, plus an increased vessel permeability leading to an accelerated extravasation of contrast material at the site of the tumor. Because the regular capillary architecture is only poorly reconstructed, arteriovenous shunts are another hallmark of tumor-induced angiogenesis, leading to perfusion shortcuts. To date, however, it is unclear what exactly determines the degree of contrast material enhancement seen on the MR image. Lesion enhancement is determined by a variety of contributing factors, including vessel permeability, but also contrast material diffusion rates, composition of the interstitial tumor matrix, and baseline and postcontrast tissue T1 relaxation times. At the same time, the problem is that hypervascularity (or lack thereof) is not pathognomonic for malignant or benign lesions. Almost all benign neoplastic lesions, and many benign nonneoplastic states, go along with a significant hypervascularity or hyperemia. Accordingly, contrast enhancement, or even strong and rapid contrast enhancement, is not a feature that is reserved for malignant lesions. However, a low vascular density can also be found in some malignant changes, even though a "nonenhancing" invasive breast cancer is rather rare.

In terms of characterization, contrast-enhanced MR mammography (CE-MRM) can help to distinguish between benign and malignant lesions on the basis of lesion morphology and in most cases through the profile of the curve for lesion signal intensity enhancement over time (SI/T curve). Typically, MR images acquired with high spatial resolution are needed for accurate assessment of the morphologic characteristics and internal architecture of lesions, while fast imaging protocols with high temporal resolution are needed to evaluate the enhancement kinetics of lesions. Although marked differences exist between the enhancement slopes of benign and malignant lesions (malignant lesions tend to enhance earlier and more strongly than benign lesions), the conventional approach to the assessment of lesion-enhancement kinetics is based on the washout patterns of lesions. Specifically, the SI/T curves have been classified into four types according to their washout pattern, from type Ia and Ib, which are indicative of benign lesions, through type II (borderline) and type III, which are indicative of a malignant lesion (Kuhl, 2002).

Unfortunately, the postprocessing evaluation of breast MRI exams, including images subtraction and analysis of the kinetic behavior of the contrast agent inside the lesion, is typically very time consuming (~ 15 min per lesion) and is largely operator dependent. For these reasons we have developed a new software, fully digital imaging and communication in medicine (DICOM) compatible, for a qualitative, semiautomatic, and nonparametric analysis of the kinetics of the contrast agent. The software loads a complete magnetic resonance mammography (MRM) study and automatically performs the necessary postprocessing operation such as images subtraction, contrast adjustment, and eventually image realignment. It also provides the capability of a semiautomatic, signal intensity versus time (SI/T) analysis, in order to produce a single false color map (FCM) for each scan plane, according to the signal enhancement and the qualitative kinetic behavior of the contrast agent in each pixel of the plane. Thus, each pixel color conveys information about the washout behavior of the contrast agent, according to four typical SI/T trends. The intensity of each hue is further modulated with information related to the earliness of the maximum enhancement peak.

Contrast agent behavior is described in literature, following one of two different approaches:

1. A quantitative/parametric approach, based on an analytic description of a pharmacokinetic model of the contrast agent evolution.
2. A qualitative approach, based on a well-known classification and identification of typical SI/T behavior, statistically related to different degrees of tumor malignancies.

Both approaches are typically justified by a bicompartmental pharmacokinetic model (usually the Tofts and Kermode models), such as the one described in the following chart.

The differential equations fully describing the model are:

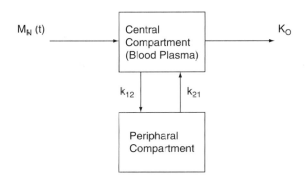

$$V_c \frac{\partial C_c}{\partial t} = k_{21} V_p C_p - (k_{12} + k_{out}) V_p C_p + M_{in} \qquad (1)$$

$$V_p \frac{\partial C_p}{\partial t} = k_{12} V_c C_c - k_{21} V_p C_p$$

where V_c, V_p, C_c, and C_p represent volumes and contrast agent concentrations of the central and peripheral compartment, respectively.

In the case of an ideal contrast agent bolus, the concentration $C(t)$ can be described as follows:

$$C(t) = \frac{A}{a-b}(e^{-bt} - e^{-at}) \qquad (2)$$

where A, a, and b are constants related to the compartmental parameters k_{12}, k_{21}. These parameters are generally related to physiopathological parameters such as microvascular permeability, reference area, and extracellular volume fraction.

Of course, the SI/T curve clearly resembles the $C(t)$ curve, filtered by the instrument (the MR scanner) sensibility to the local contrast agent concentration.

The parametric approach directly uses the expression in Eq. (2) (or similar expressions derived from more complex models), in order to find the optimal values for parameters A, a, and b that best fit the SI/T curve measured in the single pixel. This kind of approach permits a quantization of the physiologic parameters related to the evolution of the mean of contrast in intra- and extracellular space. Nevertheless, it requires detailed knowledge of the infusion technique (dose, velocity, initial concentration, etc.) and larger computation time in order to fit the double exponential expression in Eq. (2) with the measured data. In addition, the range of significant excursions for parameters A, a, and b is not well defined yet, yielding to a difficult implementation of an exhaustive color-mapping scheme.

Finally, correct usage of this parametric approach should require sufficient temporal resolution in order to correctly fit the nonlinear expression in Eq. (2). This high temporal resolution could be guaranteed only with a low spatial resolution, preventing morphological valuation of the lesion, which is surely one of the most important features for a correct lesion characterization.

On the other hand, the qualitative approach we have used in our software reduces the complexity of the previous method, using a subset of linear approximation of Eq. (2) that has been proven to correlate well with four different degrees of lesion malignancy. This simplification clearly reduces dependence on the contrast agent injection technique and is still feasible with a relatively low temporal resolution (in the order of a few minutes between each complete volume scan), so that a simultaneous morphological characterization of the lesion could also be performed. The necessary drawback of this approach is related to the lack of a quantitative representation of the (physiological) parameters of the model that could only be qualitatively deduced by SI/T behavior.

The four typical linear SI/T behaviors that have been proven to correlate with four different types of lesions are presented in Figure 130.

Characterization Algorithm

Each image in each postcontrast sequence is previously realigned (registered), if necessary, to the corresponding image in the precontrast (basal) sequence in order to reduce movement artifacts. (The registration algorithm will be described in the next section.) The realigned images are then automatically subtracted in order to enhance only the difference related to the contrast agent. A significant enhancement threshold is defined on the basis of a global breast volume enhancement evaluation in order to automatically identify all suspected lesions that could also be seen by the human eye.

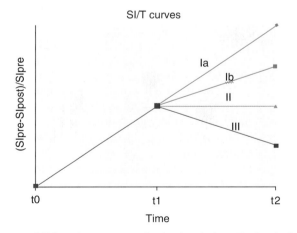

Figure 130 SI/T types: Ia, Ib= benign lesion, II= borderline, III= malignant lesion.

From all pixels in each scan plane, corresponding to an over-threshold enhancement within the first 2 minutes postcontrast injection, an SI/T curve is extracted and partitioned in a three-linear piecewise segmentation. According to the combination of angular coefficients of each of the three linear segments, the SI/T curve is associated with one (or none) of the four trends in Figure 129. Thus, a different color is associated with the pixel according to the hue scheme (Fig. 131).

Red hues are associated with malignant lesions, while green ones are associated with benign ones. Each hue is further modulated according to the earliness of the maximum enhancement peak, according to four different degrees. Earlier peaks exhibit darker intensities; later ones exhibit brighter intensities (Fig. 132).

According to this type of visualization and associating increasing values to each pixel according to increasing degrees of malignancy, the single-plane FCMs can easily be assembled in a powerful false color 3D rendered image, applying the maximum intensity projection (MIP) algorithm and thus producing a FCM MIP for each postcontrast sequence.

Registration Algorithm

Correct analysis of the SI/T curve requires perfect alignment for each pixel in each homologous scan plane, precontrast, and postcontrast injection. Slight patient movement during the exam can cause one or more misalignments, generating false enhanced areas and thereby canceling real contrast-agent-related enhancements. In order to reduce these movement artifacts, many registration algorithms have been proposed. This algorithm typically tends to minimize the differences between homologous images, presuming that only slight differences between them could be justified by the presence of the contrast agent. Let M be the image acquired before the contrast injection and F the same image acquired postcontrast injection. Correct registration of the two images should provide an opportune transformation $T(F)$ on F, so that the transformed image $M'=T(F)$ is the one minimizing a particular goal function related to the distance from the basal image M.

The various registration techniques differ from one another in three different aspects:

1. The particular type of transformation $T(\bullet)$ that "moves" the postcontrast image F.

2. The "distance function" used as the goal function (least square, cross-correlation, mutual information, etc.).

3. The optimization algorithm used to find the optimal transformation that minimizes the goal function in a relatively short time.

Our software makes use of a classical linear "rigid body" transformation (i.e., a subclass of affine transformation), consisting of a rigid (i.e., not deforming) roto-translation of F.

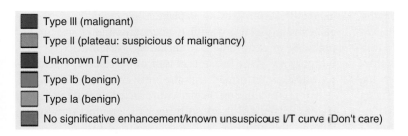

- Type III (malignant)
- Type II (plateau: suspicious of malignancy)
- Unknonwn I/T curve
- Type Ib (benign)
- Type Ia (benign)
- No significative enhancement/known unsuspicous I/T curve (Don't care)

Figure 131 Degree of malignancy/color scheme.

Let x, y be the coordinates of a generic point P in F and x', y' the coordinates of the same point after application of the $T(\bullet)$ transformation:

$$\begin{pmatrix} x' \\ y' \\ 1 \end{pmatrix} = \begin{pmatrix} \cos(\vartheta) & \sin(\vartheta) & tx \\ -\sin(\vartheta) & \cos(\vartheta) & ty \\ 0 & 0 & 1 \end{pmatrix} \begin{pmatrix} x \\ y \\ 1 \end{pmatrix}$$

where (tx, ty, θ) are the three parameters uniquely identifying the current transformation.

We have chosen as the goal function the minimum square error function defined as

$$MMSE = \frac{\sum_{i,j} (F(i,j) - M(i,j))^2}{N^2}$$

with N = total number of pixels in each image.

In order to find the optimal transformation (i.e., the optimal (tx, ty, θ) parameters) minimizing the MMSE function, we have used a Downhill-Simplex algorithm, which, though not having a demonstrated polynomial convergence, presents the great advantage of a simple numerical implementation because of the total absence of derivatives.

To assess the efficacy of the new software for detecting and characterizing MR breast lesions, in our study we preoperatively evaluated 36 consecutive women with suspected breast cancer based on mammographic and sonographic examinations on CE-MRM. Images were analyzed with the new software package and separately with a standard display method. Statistical comparison was performed of the confidence for lesion detection and characterization between the two methods and of the diagnostic accuracy for characterization compared with histopathologic findings. A total of 68 lesions were detected at final diagnosis in the 36 evaluated patients, and all were assessed histologically. Fifty-four lesions were confirmed to be malignant and comprised 19 invasive ductal carcinomas (IDCs), 22 invasive lobular carcinomas (ILCs), 8 ductal carcinomas *in situ* (DCIS), 1 lobular carcinoma *in situ* (LCIS), 1 mucinous carcinoma (MC), 2 intraductal papillomas with DCIS foci, and 1 radial scar with DCIS foci. The remaining 14 lesions were all confirmed to be benign and comprised 1 radial scar, 8 fibroadenomas, 3 lobular hyperplasia, and 2 fibrocystic diseases. All 68 (100%) lesions were detected with both methods, and good correlation with histopathologic specimens was obtained. Confidence for both detection and characterization was significantly ($P = 0.025$) better with the color-coded method, although no difference ($P = 0.05$) between the methods was noted in terms of the sensitivity, specificity, and overall accuracy for lesion characterization. Excellent agreement between the two methods was noted for determination of both lesion size (kappa = 0.77) and SI/T curves (kappa = 0.85) (Pediconi *et al.*, 2005).

Conclusions

Mammography is currently the best imaging technique for the early detection and diagnosis of breast cancer. Although numerous advances and improvements in mammography in the past decades have greatly improved image quality, the technique is not without shortcomings that limit its sensitivity and specificity. Multiple areas of research have therefore been sought not only to improve film/screen mammography, but also to consider entirely new techniques in the study of breast cancer.

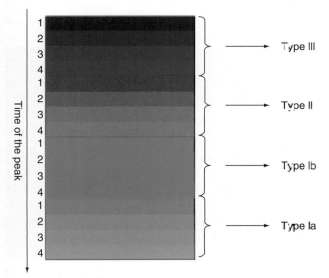

Figure 132 Modulation of the hue according to the earliness of the maximum enhancement peak.

CE-MRM has emerged as a viable complementary examination to mammography and sonography, with sensitivities of up to 90% reported. A major limitation of CE-MRM as a routine diagnostic tool is a comparatively low specificity for characterizing detected lesions. In large part this is due to the difficulty in distinguishing benign from malignant lesions on the basis of morphologic features alone (Liu *et al.*, 1998).

Two principal strategies have evolved to improve specificity: rapid dynamic imaging after gadolinium enhancement and high spatial resolution imaging. CE-MRI, besides detecting morphological information of the lesion, can provide dynamic information on the contrast medium behavior related to the intensity of MRI signal over time.

Generally, the signal intensity-time curves of malignant lesions have characteristic washout profiles, whereas the curves of benign lesions more frequently have a steady increase or plateau profile (Kuhl and Schild, 2000). Unfortunately, postprocessing evaluation, involving image subtraction and analysis of the kinetic behavior of lesions, is usually very time consuming and operator-dependent. Previous studies have used pharmacokinetic modeling to condense dynamic MR imaging data into color maps of quantitative parameters of gadolinium enhancement that are overlaid onto corresponding gray-scale anatomic images. However, the overlaying of these maps may obscure morphologic features of enhancing lesions, which may be important predictors of the risk of malignancy.

In looking to overcome this potential drawback, we have developed a DICOM-compatible software package, which is able to automatically load a complete breast MRI study and to automatically register and subtract images before creating FCMs for each scan plane. The approach compares the qualitative parameters of the SI/T evolution of each point with the standard lesion curve types (different hues for each type) and modulates the color brightness with the earliness of the enhancement peak. This qualitative approach to lesion classification greatly reduces the computational effort and the time required for examination analysis even on an individual PC-based system. We tested the efficacy of the new software for the detection and characterization of MR breast lesions, evaluating 36 patients with suspected breast cancer based on mammographic and sonographic examinations, and we correlated the results with histopathological specimens. Images were analyzed both with a standard display method and separately with the new software using a user-friendly interactive mechanism.

All detected lesions were characterized by using morphological patterns and qualitative observation of the SI/T curve extracted from a region of interest (ROI) manually selected inside the lesion. During the assessment, the observers also had the opportunity to place up to three regions of interest (ROIs) per lesion for the purpose of generating signal intensity-time curves. The radiologists were asked to enter the results into a database specifying global lesion detection (lesion detection score of 0, 1, or 2 to each lesion detected, with 0 = uncertain, 1 = possibly or probably present, and 2 = definitely present), lesion size, morphology of the lesion, and type of SI/T curve. The analysis was also performed with the software.

For this evaluation, readers were asked to characterize all lesions detected with the software by using the same parameters of the previous one. Evaluations of lesions obtained with and without software analysis were compared and correlated to histopathological result. Our study of 36 consecutive patients with 68 lesions reveals excellent agreement between the color-coded method with the traditional display method in terms of the sensitivity, specificity, and overall accuracy for lesion detection and characterization. It also suggests greater confidence on the part of the blinded readers for lesion detection and characterization with the color-coded method. Our results revealed excellent correlation between the FCM and the ROI-based SI/T curve analysis, with the advantage of a fast, semiautomatic, and largely operator-independent procedure in the case of the color-coded approach (Fig. 133).

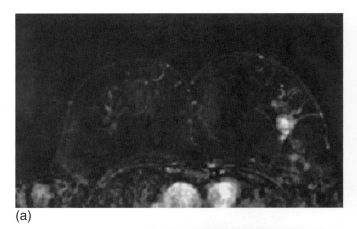
(a)

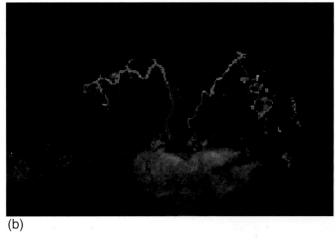
(b)

Figure 133 CE-MRM. Suspected lesion in the left breast (a), and MIP reconstruction of the corresponding lesion (b).

The semiautomatic pixel-by-pixel analysis of the ROI makes the procedure largely operator independent and can help radiologists distinguish different patterns inside a single lesion, perhaps allowing reduction of false-negative evaluation. At least, the qualitative approach to the lesion classification greatly reduces the computational effort and the time required for an exam analysis even on single-PC-based architecture. The excellent performance obtained with the color-coded method may be considered of value both for the detection and preliminary staging of breast cancer, and for the evaluation of lesions before surgery. In this regard, improved visualization of the degree of tumor infiltration would be advantageous for presurgical planning.

At present a major limitation of CE-MRM as a routine diagnostic tool is a comparatively low specificity for the characterization of detected lesions. Our study demonstrated an excellent correlation between lesion characterization with the traditional display method and the intensity-modulated color-coded display method. We found disagreement in only two lesions. Therefore, regarding lesion margins and enhancement morphology, no significant differences were discernible between the two methods. The intensity-modulated color-coded display method efficiently condenses high-spatial-resolution and rapid dynamic breast MR imaging data into a single comprehensive image that is a visually intuitive tool for characterizing the morphology, pharmacokinetics, and extent of breast lesions. Currently, most breast MR examinations comprise a review of enhanced subtracted anatomic images without co-registered pharmacokinetic data, followed by a selective assessment of any suspicious focal regions with ROI analysis of the SI/T dynamic data (Jacobs *et al.*, 2003). The possibility of shifting from a retrospective interrogation strategy to a prospective review may facilitate rapid assessment of both dynamic and morphologic features throughout the entire breast. The possible benefits of a combined rapid dynamic and high-spatial-resolution breast MR imaging include a more practical examination for breast screening due to the more rapid interpretation achievable, and the potential to more readily highlight unsuspected abnormally enhancing foci with suspicious pharmacokinetic features. Although the study is limited in that only 36 patients and 68 histologically proven lesions were evaluated, the comparison of the two methods provides a clear indication of the utility of the intensity-modulated color-coded display method.

The software is fully reliable and easy to use, and can be considered a good CAD system. Computer-aided detection algorithms have allowed radiologists to regain efficiency while maintaining optimized acquisition techniques. The primary reason for this quick adoption of CAD for breast MR is that the CAD software enables readers to increase their efficiency while potentially improving their overall accuracy. The full benefits of CAD for breast MR are realized when the interpreting radiologist has a thorough understanding of the algorithms used and of CAD's limitations.

References

Ackerman, L.V., and Gose, E.E. 1972. Breast lesion classification by computer and xeroradiograph. *Cancer 30*:1025–1035

Boets, C., Barentsz, J.O., and Mus, R.D. 1994. MR characterization of suspicious breast lesions with a gadolinium-enhanced turbo FLASH subtraction technique. *Radiology 193*:777–781.

Essermann, L., Hylton, N., and Yassa, L. 1999. Utility of magnetic resonance imaging in the management of breast cancer: evidence for improved preoperative staging. *J. Clin. Oncol. 17*:110–119.

Gilles, R., Guinebretiere, J.M., and Lucidarme, C. 1994. Nonpalpable breast tumors: diagnosis with contrast-enhanced subtraction dynamic MR imaging. *Radiology 191*:625–631.

Harms, S., Flaming, D.P., and Hesley, K.L. 1993. MR imaging of the breast with rotating delivery of excitation off resonance: clinical experience with pathologic correlation. *Radiology 187*:493–501.

Hayton, P., Brady, M., Tarassenko, L., and Moore, N. 1997. Analysis of dynamic MR breast images using a model of contrast enhancement. *Med. Image Anal. 1*:207–224.

Heywang, S.H., Wolf, A., and Pruss, E. 1989. MR imaging of the breast using Gd-DTPA: use and limitations. *Radiology 71*:95–103.

Jacobs, M.A., Barker, P.B., and Bluemke, D.A. 2003. Benign and malignant breast lesions: diagnosis with multiparametric MR imaging. *Radiology 229*:225–232.

Kaiser, W.A., and Zeitler, E. 1989. MR imaging of the breast: fast imaging sequences with and without Gd-DTPA-preliminary observations. *Radiology 170*:681–686.

Knopp, M.V., Weiss, E., and Sinn, H.P. 1999. Pathophysiologic basis of contrast enhancement in breast tumors. *J. Magn. Res. Imaging 10*:260–266.

Kuhl, C.K. 2002. High-risk screening: multi-modality surveillance of women at high risk for breast cancer (proven or suspected carries of a breast cancer susceptibility gene). *J. Exp. Clin. Cancer Res. 21*:103–106.

Kuhl, C.K., Mielcareck, P., and Klaschik, S. 1999. Dynamic breast MR imaging: are signal intensity time course data useful for differential diagnosis of enhancing lesions? *Radiology 211*:10–110.

Kuhl, C.K., and Schild, H.H. 2000. Dynamic image interpretation of MRI of the breast. *J. Magn. Reson. Imag. 12*:965–974.

Liu, P.F., Debatin, J.F., and Caduff, R.F. 1998. Improved diagnostic accuracy in dynamic contrast-enhanced MRI of the breast by combined quantitative and qualitative analysis. *Br. J. Radiol. 71*:501–509.

Nunes, L.W., Schnall, M.D., and Orel, S.G. 1997. Breast MR imaging: interpretation model. *Radiology 202*:833–841.

Orel, S.G. 1999. Differentiating benign from malignant enhancing lesions identified at MR imaging of the breast: are time-signal intensity curves an accurate predictor? *Radiology 211*:5–7.

Orel, S.G., Schanall, M.D., and Rivolsi, V.A. 1994. Suspicious breast lesions: MR imaging with radiologic-pathologic correlation. *Radiology 190*:485–493.

Pediconi, F., Catalano, C., Vencitti, F., Ercolani, M., Carotenuto, L., Padula, S., Moriconi, E., Roselli, A., Giacomelli, L., Kirchin, M.A., and Passariello, R. 2005. Color-coded automated signal intensity curves for detection and characterization of breast lesions: preliminary evaluation of a new software package for integrated magenetic resonance-based breast imaging. *Invest. Radiol. 40*:448–457.

Winsberg, F., Elkin, M., and Macy, J. 1967. Detection of radiographic abnormalities in mammograms by means of optical scanning and computer analysis. *Radiology 89*:211–215.

25

Detection of Breast Malignancy: Different Magnetic Resonance Imaging Modalities

Wei Huang and Luminita A. Tudorica

Introduction

Breast disease is the second leading cause of cancer death among women and is a significant health-care problem in the United States. More than 211,000 American women were projected to receive a diagnosis of breast cancer in the year 2005, and ~ 40,000 of them would die of the disease (Jemal *et al.*, 2005). Imaging plays a crucial role in all aspects of breast cancer care, including early detection through screening, diagnosis, image-guided biopsy, treatment planning, and follow-up (Schnall, 2003).

Major Breast Imaging Modalities

Conventional X-ray mammography is at present the diagnostic mainstay. However, even when performed optimally, mammographic sensitivity is between 69 and 90% (Mandelson *et al.*, 2000). Although some mammographically occult lesions may be palpable, others will defy detection. Tumors may be missed because of poor technique or observer error, or because of the size and nature of the lesions relative to the surrounding breast tissue, which may obscure them. Decreased sensitivity is a particular problem in the dense breast, following surgery or radiotherapy, adjacent to implants or in the younger population. Mammographic specificity may be sacrificed to improve sensitivity. The reported specificity of mammography ranges from 10–64% (Fischer *et al.*, 1999). Up to 75% of mammographically demonstrated indeterminate or suspicious masses are found to be benign at biopsy (Rankin, 2000).

Ultrasound (US) is an excellent method for assessing palpable abnormalities, differentiating between cystic and solid lesions, and classifying solid masses (Berg and Gilbreath, 2000). It also allows accurate needle placement for biopsy. However, US is highly operator dependent with demonstrated intra- and interobserver variability in the interpretation of breast sonograms (Baker and Soo, 2002). Consequently, US may be more sensitive but less specific than mammography (Warner *et al.*, 2001). Ultrasound used as a mammography supplement detects additional cancers, but it also increases the false-positive rate. Furthermore, US is time consuming, and it is difficult to ensure that the entire breast has been imaged.

Excellent soft tissue resolution and contrast, combined with tomography to avoid tissue image overlap, and the lack of ionizing radiation, makes magnetic resonance imaging (MRI) an attractive imaging modality. Over the past two decades, tremendous advances have been made in breast MRI. Initial experience with only T_1- or T_2-weighted acquisitions was disappointing (Heywang et al., 1987). However, the introduction of gadolinium (Gd) contrast reagent (CR) use in breast MRI (Kaiser and Zeitler, 1989) has enabled substantial improvement in this technique. In particular, the so-called dynamic contrast-enhanced (DCE) method, which monitors CR passage through mammary tissue following a bolus intravenous injection, has become an integral part of many routine breast MRI examination protocols. The popularity of this approach stems from many reports that even the qualitative MRI signal intensity time courses from regions of interest (ROIs) exhibit reproducible patterns that appear capable of discriminating benign from malignant lesions, and even different types of malignancies (Daniel et al., 1998; Knopp et al., 1999; Kuhl et al., 1999). Although results to date have varied significantly, the sensitivity of breast T_1-weighted DCE-MRI for detection of malignancy has consistently been reported to be excellent (88–100%) (Kuhl et al., 1999; Huang et al., 2004). However, the specificity of this method has been rather more variable and unsatisfactory. Review of the literature yields a wide specificity range, from 37–97% (Kuhl et al., 1999; Huang et al., 2004). The patterns of contrast enhancement, for example, of benign fibroadenomas may overlap with those of malignant lesions (Brinck et al., 1997).

Breast Dynamic Contrast-enhanced Magnetic Resonance Imaging

Breast DCE-MRI involves acquisition of a time series of T_1-weighted images, during which an intravenous CR bolus injection is made. The images are usually acquired with gradient echo pulse sequence, such as spoiled gradient-recalled acquisition at steady state (GRASS) sequence, with short echo time TE (< 8 msec), short repetition time TR (< 20 msec), and low flip angle (10–30°). The rationale is that most tumors have increased and/or "leaky" (increased capillary permeability) vascularization reflective of angiogenesis (Folkman, 1995). Malignant tissues usually show more rapid and larger increases in DCE-MRI signal intensities, followed by faster washout, as compared to benign tissues. Besides lesion morphology, evaluation of CR pharmacokinetics from the signal intensity time course is the major aspect of image interpretation. The three general approaches (Buadu et al., 1996; Daniel et al., 1998) to analysis of the enhancement curve have been (1) subjective (qualitative) assessment of the overall shape, (2) empirical quantitative characterization, and (3) more sophisticated analytical pharmacokinetic modeling.

Subjective Assessment

Subjective (qualitative) visual assessment of the enhancement time course has often been used for clinical diagnosis. The most commonly identified enhancement patterns are characterized as (1) persistent (signal intensity rises persistently), (2) plateau (signal intensity remains fairly constant after reaching maximum), and (3) washout (signal intensity immediately decreases after reaching maximum) (Knopp et al., 1999). It has been generally found that benign lesions tend to exhibit the persistent pattern and that malignant lesions tend to exhibit the washout pattern, whereas the plateau pattern can be seen in both benign and malignant lesions.

Figure 134 shows the results of a DCE-MRI study from a typical clinical diagnostic breast MRI protocol. One precontrast and three postcontrast volumetric sagittal image sets were acquired for DCE-MRI with ~ 2 min temporal resolution. Panels a, b, c, and d demonstrate results for a malignant invasive ductal carcinoma (IDC) lesion and a lesion with benign fibrocystic changes, respectively. For the signal intensity-time course plots, the horizontal axis represents the DCE-MRI data set number as sequentially acquired (over time): 0—precontrast, 1—first postcontrast, 2—second postcontrast, 3—third postcontrast; and the vertical axis represents the percentage of signal intensity change, $[S - S_0]/ S_0 \times 100\%$, where S_0 is the precontrast signal intensity. The plots were obtained from the ROIs drawn within the contrast-enhanced lesions. The IDC lesion exhibits the typical washout pattern (Fig. 134b), while the benign lesion shows the persistent pattern (Fig. 134d). The color-coding scheme defines the washout pattern as red (Fig. 134a), the plateau pattern as green (not shown here), and the persistent pattern as blue (Fig. 134c). One study (Liberman et al., 2002) shows that for such a DCE breast MRI protocol, the positive predictive value for the washout pattern is only 33%, while the other two patterns (persistent and plateau) exhibit a 24% positive predictive value, demonstrating significant overlap of malignant and benign breast disease when the diagnosis is based on the shape of signal time course curves.

The advantage of the qualitative approach is the simple visual inspection of enhancement kinetics. However, there are major drawbacks, such as: (1) the restriction to one (or a small number) of substantially sized ROI per tumor: the pharmacokinetic assessment necessary to evaluate the crucial tumor fine-scale heterogeneity is not humanly possible; (2) the requirement of subjective selection of the ROI for signal time-course analysis leads to significant curve shape variability; and (3) the overlap of enhancement patterns for

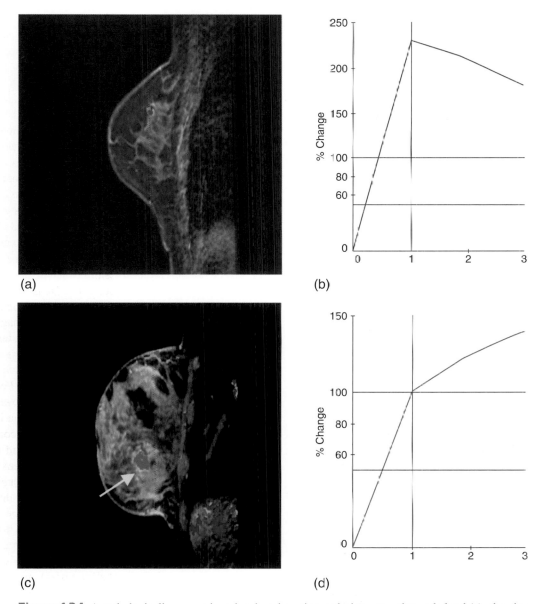

Figure 1 34 A pathologically proven invasive ductal carcinoma lesion was color-coded red (a), showing a washout DCE-MRI pattern (b). A pathologically proven benign lesion with fibrocystic changes (arrow) was color-coded blue (c), showing a persistent DCE-MRI pattern (d).

benign and malignant lesions leads to low specificity, even with good sensitivity.

Empirical Quantitative Characterization

Various investigators (Buadu *et al.*, 1996; Daniel *et al.*, 1998) have employed empirical characterizations of the enhancement curves to differentiate benign and malignant lesions. These metrics include uptake slope, initial area under the curve (IAUC), signal intensity increase, arrival time, time-to-peak, and washout slope. Although these measurements do not yield quantities directly related to physiologic parameters, they do provide clinically useful results. Importantly, two separate studies (Buckley *et al.*, 1997; Buadu *et al.*, 1996) found excellent correlation between the uptake slope and the mean microvessel density (MVD). The latter was obtained from histologic analysis. These independent results strongly confirm the basic rationale for DCE-MRI.

A major advantage of this approach is that the empirical quantities can be mapped on a pixel-by-pixel basis. Another advantage is that no arterial input function (AIF) is required for data analysis (see below). However, due to large variations in pulse sequences, acquisition parameters, and data postprocessing schemes employed at different imaging sites, the optimal empirical parameter threshold values considered suggestive of malignancy have varied widely. As a result, the specificity in breast cancer diagnosis using this approach is also limited.

Besides insufficient specificity, the main problem common to both the qualitative and empirical methods is considerable difficulty with study reproducibility. There are differences of pulse sequences, acquisition parameters, slice numbers, number of acquisition volumes, temporal and spatial resolutions, CR, CR dose, hardware and software available, clinical indications, results desired, and personal experience. Because of the variable results, no single standardized and generally accepted technique has emerged for DCE breast MRI examinations. This has made it difficult to obtain meaningful comparisons between different cancer types and between data from different imaging sites (Rankin, 2000; Orel and Schnall, 2001).

Analytical Pharmacokinetic Modeling

The preceding comments point to the need for determining fundamental pharmacokinetic/physiologic parameters. Thus, researchers have increasingly performed analytical pharmacokinetic modeling of CR uptake and washout in breast lesions (Knopp et al., 1999; Tofts et al., 1995; Su et al., 2003). The pharmacokinetic parameters most often determined are variants of K^{trans}, a pseudo-first-order rate constant for CR transfer between blood plasma and extracellular extravascular space [EES] (extravasation), which includes blood flow (perfusion) and vessel permeability contributions; and v_e, a measure of the EES volume fraction. The advantage of the analytical approach is that the parameters are, in principle, pharmacophysiological properties fully independent of MRI pulse sequences, data acquisition parameters, CR dosage, scanner field strength, available hardware and software, and so on. Thus, they should be comparable between different studies at different sites and could establish a basis for standardization of the breast DCE-MRI method.

A unique aspect of pharmacokinetic modeling is the requirement for an AIF. The absolute accuracy of K^{trans} and v_e depends on the AIF accuracy (Yankeelov et al., 2003). Determination of AIF accuracy, particularly for the breast, has been considered extensively (reviewed in Port et al., 2001). Recently, significant progress has been reported for the reference tissue method (Yang et al., 2004) for estimating AIF.

There are two major classes of DCE-MRI pharmacokinetic models. The first is the heterogeneous class, in which CR and (some nuclear magnetic resonance [NMR] properties of) water are not assumed to be always uniformly distributed within each compartment entered (reviewed in Moran and Prato, 2004; Buckley, 2002). The second is the homogeneous class, in which such assumptions are made ("well-mixed"). Although the first class may be the more realistic, the second class is the more practical, because, for a given number of compartments, it has fewer potentially variable parameters: it is more parsimonious. The heterogeneous class always requires compartmental geometry parameters that are not necessary in the homogeneous class.

In the early 1990s, several analytical models (Larsson et al., 1990; Tofts and Kermode, 1991; Brix et al., 1991) of the latter homogeneous class were introduced to derive pharmacokinetic parameters from DCE-MRI data. These models all share the assumptions of the fast exchange of all mobile (NMR visible) protons within the tissue (see below) (Tofts, 1997). Although the extracted pharmacokinetic parameters are different for different versions (e.g., $k^{PS\rho}$ in Tofts and Kermode, 1991; k_{ep} in Larsson et al., 1990; and k_{21} in Brix et al., 1991), they are related by simple equations (Tofts, 1997). Tofts et al. (1999) later published a review article to standardize the quantities and symbols of the pharmacokinetic parameters. By far the most widely applied of the DCE-MRI analytical models is the so-called Tofts Model, or the Standard Model (SM) (Yankeelov et al., 2003, 2005; Li et al., 2005).

The relationship between the ^1H_2O longitudinal relaxation rate constant, $R_1(\equiv 1/T_1)$ and the CR concentration, [CR], is the basis for analytical modeling of DCE-MRI data. In the SM, the investigators appropriately chose the Kety (or Kety-Schmidt) pharmacokinetic rate law from nuclear medicine. The two generations of the SM (Tofts, 1997; Tofts and Kermode, 1991; Larsson et al., 1990) share the constraint to the linear dependence of the ^1H_2O R_1 on [CR] that is characteristic of homogeneous aqueous CR solutions. (Another SM version introduced by Brix et al. [1991] had the even further simplified assumption of a linear relationship between MRI signal and [CR].) However, its use for tissue ^1H_2O is equivalent to assuming that the cytolemmal CR barriers do not exist or that the equilibrium transcytolemmal water exchange effectively is infinitely fast compared to the NMR shutter speed for this process—the fast-exchange limit [FXL]—during the CR bolus passage through the lesion (Yankeelov et al., 2003; Landis et al., 1999, 2000).

In DCE-MRI, the CR plays the role of the nuclear medicine tracer. However, in nuclear medicine, the tracer is detected and quantified directly; it is also the signal molecule. In MRI, the CR is detected indirectly, via its effect on ^1H_2O: thus, water is the signal molecule. This has two major consequences: (1) the CR and water are never equally distributed in tissue (in some SM versions, this is taken into account), and (2) water equilibrium exchange between compartments where CR is unequally distributed has the pertinent kinetics to have very significant effects. The SM (defined as employing $R_1 = r_1[CR] + R_{10}$, where r_1 is the CR relaxivity and R_{10} is the relaxation rate constant in the absence of CR) has neglected the latter, ignoring the effects of shutter-speed variation during CR bolus passage through the lesion.

In the last few years, Springer et al. (1999) have shown that, though fast, equilibrium transcytolemmal water exchange is not infinitely so (Landis et al., 1999). The incorrect assumption that it is infinitely fast can lead to significant errors in estimating instantaneous [CR] values (Landis et al., 2000), and thus K^{trans} and v_e values (Yankeelov et al., 2003). These errors are greater at the lower magnetic field strengths (Springer and Rooney, 2001) used in clinical examinations, such as 1.5T, because of the greater [CR] values required. These investigators have recently introduced the Shutter-Speed Model (SSM) to incorporate the effects of finite equilibrium intercompartmental water exchange into analytical modeling of DCE-MRI data (Yankeelov et al., 2003). It is effectively a two-compartment model, but the compartments are the major water loci, the interstitial and intracellular spaces. The SSM allows the equilibrium transcytolemmal water exchange system to transiently exit the FXL and enter the fast-exchange regime [FXR] (i.e., R_1 and [CR] are nonlinearly related [Landis et al., 1999]) during the CR bolus passage (Yankeelov et al., 2003; Landis et al., 2000). Using this approach, it was found through theoretical simulation, experimental animal tumor model (Yankeelov et al., 2003; Zhou et al., 2004), and human pathology (Yankeelov et al., 2005; Li et al., 2005) data analyses, that the SM produces systematic errors; that is, the pharmacokinetic parameters (K^{trans} and v_e) can be underestimated by factors of two to three, after even a rather standard bolus CR injection. Furthermore, the SSM explains much of the CR dose dependence in literature reports. Exactly as predicted by theory (Yankeelov et al., 2003), the rodent tumor studies showed that the SM K^{trans} and v_e values decreased with increasing CR dose (Zhou et al., 2004). The SSM analyses of the same data effectively completely eliminated this CR dose dependence (Zhou et al., 2004). Of course, such fundamental pharmacokinetic parameters as K^{trans} and v_e should not be CR dose dependent.

A preliminary study (Li et al., 2005) shows that the more accurate K^{trans} values estimated by the SSM analyses provide much better discrimination of benign and malignant breast lesions than those returned by the SM analysis. Figure 135 displays pixel-by-pixel parametric maps of K^{trans} for patients with pathologically proven benign fibroadenoma (FA) and malignant IDC, respectively. Whole-slice maps from SM fittings are given in panels a and b and from SSM fittings in panels c and d (they have common color scales). Thus, one can compare the SM K^{trans} (Figs. 135a, b) maps of FA and IDC with the SSM K^{trans} (Figs. 135c, d) maps. In general, most breast tissue appears as dark blue: there is very little CR

uptake in normal glandular tissue and fatty tissues (Yankeelov et al., 2005). As expected from the above, the greatest distinction between the FA and IDC lesions is seen in the SSM K^{trans} maps, Figs. 135c and d, respectively. Careful inspection of Fig. 135d would reveal subtumor heterogeneity, with a dendritic "hot region" having K^{trans} values ≥ 0.8: the K^{trans} value for a tiny sublesion ROI in the IDC is 1.2 $(min)^{-1}$ (Fig. 135d).

In summary, even though DCE breast MRI has been shown to have higher sensitivity in the detection of breast malignancy compared to X-ray mammography and US, its specificity is limited and essentially no better than those of the other two imaging modalities. This results in unnecessary (or benign) biopsies, and undesirable complications and consequences in patient care. Furthermore, there is currently no standard protocol for DCE breast MRI in terms of either image acquisition or data interpretation. It is difficult to compare results from study to study and from site to site. This lack of standardization also contributes to the low specificity, as false-positive diagnoses are preferable to false-negative diagnoses (missing cancer). Hence, there is great need for standardizing DCE breast MRI and improving its specificity. The authors believe that the SSM analytical modeling of DCE breast MRI data will offer this solution. This method provides quantitative high-resolution mapping of fundamental pathophysiologic quantities in breast lesions, such as tumor perfusion and/or microvascular permeability, and EES volume fraction. These quantities are fully independent of MRI pulse sequences, data acquisition parameters, CR, CR dosage, scanner field strength, and various commercial hardware and software and thus can be compared between different studies at different sites. Not limited to just cancer diagnosis, DCE-MRI can be used for follow-up monitoring of breast cancer response to surgical intervention or chemotherapy (Knopp et al., 2003). The SSM approach for DCE breast MRI may prove to be very valuable in monitoring the effects of cancer treatment with anti-angiogenic drugs, as the K^{trans} value is directly related to tumor vessel permeability and/or blood flow. By taking into account transcytolemmal water exchange kinetics, the SSM analysis provides a more accurate determination of this pharmacokinetic parameter than the SM analysis.

Breast ^{1}H Magnetic Resonance Spectroscopy

Recently, investigators have explored the possibility of using ^{1}H magnetic resonance spectroscopy (MRS) to improve specificity in breast cancer diagnosis. In vivo ^{1}H MRS measurement provides the biochemical characteristics of investigated tissue. This technique has been approved by the U.S. Food and Drug Administration for imaging the brain and prostate and is widely used in clinical settings.

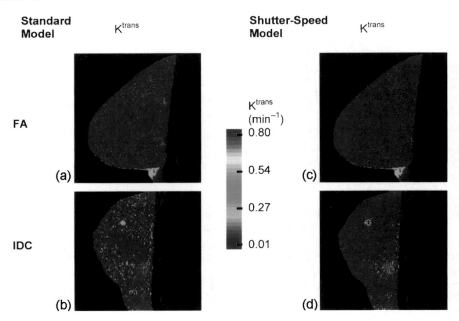

Figure 135 The results of DCE-MRI parametric mappings are shown. Standard model K^{trans} maps of the entire image slices containing a fibroadenoma (FA) and an invasive ductal carcinoma (IDC) are presented in panels (a) and (b), respectively, while shutter-speed model K^{trans} maps of the same slices are seen in panels (c) and (d). The parametric color scale for these maps is given.
Source: Adapted from Li *et al.* (2005).

In cancer studies, the typical diagnostic value of ¹H MRS is based on the detection of elevated resonance signal of choline-containing compounds (Cho), which serves as the marker of active tumor (Negendank, 1992). Several *ex vivo* MRS studies have shown elevated Cho, such as phosphocholine and glycerophosphocholine (Aboagye and Bhujwalla, 1999) in cancerous human mammary cells. Multiple *in vivo* ¹H MRS studies aimed at improving the discrimination of benign and malignant breast lesions have been done at several centers (Huang *et al.*, 2004; Bolan *et al.*, 2003; Bartella *et al.*, 2006). In addition to being used for breast cancer diagnosis, *in vivo* proton MRS has also been used to monitor breast cancer response to chemotherapies (Meisamy *et al.*, 2004).

Most breast ¹H MRS studies are single-voxel measurements using the point-resolved spectroscopy sequence (PRESS) or its variations. Typical acquisition parameters are TE ≥135 ms to reduce lipid signal, TR = 1.5 – 3.0 sec. The number of scan averages is usually between 128 and 256, resulting in net data acquisition time from 3.2–12.8 min. An extra 5–10 min is needed for prescan setup of MRS voxel shimming and water suppression. In cases of multiple suspicious lesions in one breast, multivoxel MR spectroscopic imaging (MRSI) is the preferable technique, with the capability of measuring multiregional metabolite levels in data acquisition time comparable to that of a single-voxel study.

However, because of the difficulty in obtaining good shimming in a relatively large breast region within a reasonable time frame, only a few studies have shown breast MRSI data with acceptable quality (Jacobs *et al.*, 2004; Hu *et al.*, 2005). For discrimination of benign and malignant breast lesions, except for a study (Bolan *et al.*, 2003) quantifying Cho concentration using water signal as the internal reference, the majority of the ¹H MRS studies are based on detection or nondetection of the Cho peak, or its signal-to-noise (S/N) ratio (Huang *et al.*, 2004). Figure 136 shows such examples. Figure 136a shows the placement of the MRS voxel encompassing the suspicious contrast-enhanced lesion. Figure 136b demonstrates a magnified spectrum from a pathologically proven malignant tumor, showing apparent detection of the Cho peak, while Figure 136c depicts a magnified spectrum from a pathologically proven benign lesion, showing nondetection of Cho peak. There was only noise-level signal at the Cho resonance frequency (3.23 ppm). Previous *in vivo* breast ¹H MRS studies (Huang *et al.*, 2004; Bartella *et al.*, 2006) have reported sensitivity of 70–100% and specificity of 67–100% in breast cancer detection. The latter shows improvement compared to the specificity of conventional breast MRI protocols that include DCE-MRI.

There are several drawbacks for breast ¹H MRS. Prior contrast-enhanced MRI is usually required for lesion localization and MRS voxel placement. The accumulation of CR in the

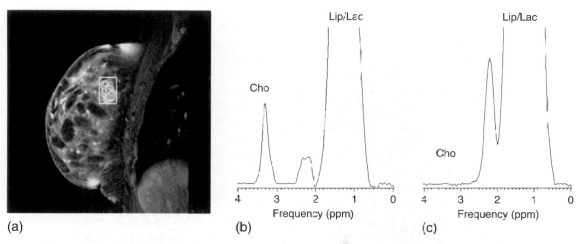

Figure 136 (a) The rectangular box encompassing the enhanced lesion demonstrates the placement of MRS voxel for the single-voxel ^{1}H MRS study. (b) Magnified proton spectrum acquired from the region of a pathologically proven malignant breast tumor. The spectrum was collected with a PRESS sequence (2000/135). An apparent Cho (choline-containing compounds) peak was detected at 3.23 ppm. (c) The same type of magnified proton spectrum as in panel (b), collected from the region of a pathologically proven benign lesion. No Cho peak was detected with only noise-level signal at 3.23 ppm.
Lip: lipid, Lac: lactate.

lesion can affect MRS quality. In addition, the scanning time (including prescan adjustment time) is relatively long (10–25 min) and the spatial resolution is poor. Fine tumor heterogeneity cannot be assayed. It is difficult to achieve sufficient simultaneous suppression of water and lipid resonances, causing difficulties in quantitative Cho concentration determination. Thus, the majority of ^{1}H MRS studies are non- cr semiquantitative. Furthermore, because of the difficulty in detecting weak Cho signal from a small lesion in reasonable scanning duration under a clinical setting, the sensitivity of ^{1}H MRS detection breast malignancy drops greatly when the lesion size is < 2 cm in diameter (Katz-Brull et al., 2002). With expected improvement in the S/N ratio, higher field (such as 3T) MR scanners may allow ^{1}H MRS to investigate smaller lesions and still maintain high sensitivity.

It appears that breast ^{1}H MRS could be best used as a supplement to MRI for diagnosis of relatively large breast lesions. One study (Bartella et al., 2006) shows that with addition of MRS information to MRI results, benign biopsy could have been spared in 58% of the studied population, and none of the cancers would have been missed.

Breast T_2^*-Weighted Perfusion Magnetic Resonance Imaging

Another MRI technique that has been recently investigated with the goal of improving specificity is dynamic susceptibility contrast (DSC) T_2^*-weighted perfusion MRI (Huang et al., 2004; Kvistad et al., 2000). The rationale for this method is that malignant tumors have higher

vascularity than benign tumors. The data acquisition and processing of breast T_2^*-weighted perfusion MRI are similar to those of the well-established brain perfusion MRI method. A T_2^*-weighted gradient echo pulse sequence is usually employed to acquire images over a time course during which a bolus intravenous injection of CR is carried out. For data processing, the standard area-under-curve algorithm (Rosen et al., 1990) is used to construct the relative blood volume map of the corresponding breast image slice. The MR signal versus time curve can be converted to ΔR_2^* versus time curve based on the relationship: $\Delta R_2^* = \Delta(1/T_2^*) = -(1/TE)\ln(S_t/S_0)$. where S_t is the signal intensity at time t and S_0 is the baseline signal intensity prior to contrast injection. The ΔR_2^* versus time curve can then be analyzed using gamma-variate fit and the area under the curve is computed, which is proportional to blood volume.

Figure 137 shows the relative blood-volume maps of an image slice containing a pathologically proven malignant lesion (panel a) and another image slice containing a pathologically proven benign lesion (panel b). Compared to the normal breast tissue area, hyperintensity was observed in the lesion area on the map (Fig. 137a), revealing high vascularity of the malignant tumor. For the benign lesion (location indicated by the arrow, Fig. 137b) where contrast enhancement was seen in the DCE-MRI study, there was no obvious enhancement on the map compared to normal breast tissue area. While DCE-MRI provides assessment of the combined effects of blood flow and vessel permeability, DSC perfusion MRI measures relative

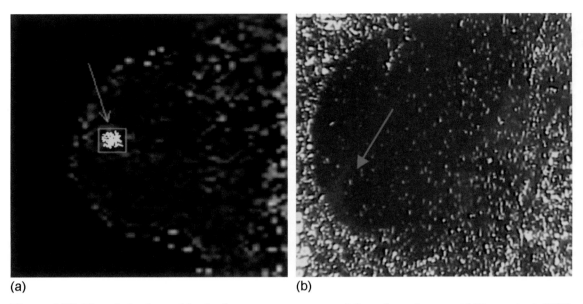

(a)   (b)

Figure137 The relative breast blood-volume maps reconstructed from dynamic susceptibility contrast (DSC) T_2^*-weighted perfusion MRI studies. (a) Compared to normal breast tissue areas, hyperintensity was observed in the region of a pathologically proven malignant tumor (arrow). (b) No enhancement was seen in a pathologically proven benign lesion (arrow), even though contrast enhancement was observed in the same lesion in the T_1-weighted DCE-MRI study.

blood volume, thus offering additional information for diagnostic purpose. One study (Huang *et al.*, 2004) shows that by adding perfusion MRI to a protocol of DCE-MRI and ^{1}H MRS, the specificity of breast cancer diagnosis can be improved to 100%.

The T_2^*-weighted MR signal change of DSC perfusion MRI is a negative one, however, and much more difficult to analytically interpret compared with T_1-weighted DCE-MRI. The quantitative tumor blood-volume determination is rendered inaccurate by the fundamentally underdetermined nature of the underlying physical phenomenon (Springer *et al.*, 1999) and the unknown extent of CR extravasation. Therefore, in clinical practice DSC perfusion MRI is typically constrained to be a qualitative or semiquantitative technique. Furthermore, the addition of DSC perfusion MRI to a routine MRI diagnostic protocol that includes DCE-MRI will require the patient to undergo two CR bolus injections. The two injections have to be separated by a sufficient time period to allow the washout of the first CR dose and not affect the intended purpose of the second CR injection.

In conclusion, DCE breast MRI has become an integral part of a diagnostic breast MRI exam protocol. With excellent sensitivity, its specificity needs to be improved to reduce unnecessary (benign) biopsies. Furthermore, DCE-MRI needs to be standardized in data acquisition and interpretation to allow for study reproducibility and study comparison between different imaging sites. The SSM DCE breast MRI approach may offer such solutions. By providing

information in lesion biochemical composition and blood volume, respectively, ^{1}H MRS and DSC perfusion MRI are useful adjuncts to a regular MRI protocol for improving accuracy in breast cancer diagnosis. It is conceivable that combination of the two or all three MR techniques (Huang *et al.*, 2004) provides better diagnostic outcome than using one method alone.

References

Aboagye, E.O., and Bhujwalla, Z.M. 1999. Malignant transformation alters membrane choline phospholipid metabolism of human mammary epithelial cells. *Cancer Res. 59:*80–84.

Baker, J.A., and Soo, M.S. 2002. Breast US: assessment of technical quality and image interpretation. *Radiology 223:*229–238.

Bartella, L., Morris, E.A., Dershaw, D.D., Liberman, L., Thakur, S.B., Moskowitz, C., Guido, J., and Huang, W. 2006. Proton MR spectroscopy with choline peak as malignancy marker improves positive predictive value for breast cancer diagnosis: preliminary study. *Radiology 239:*686–692.

Berg, W.A., and Gilbreath, P.L. 2000. Multicentric and multifocal cancer: whole-breast US in preoperative evaluation. *Radiology 214:*59–66.

Bolan, P.J., Meisamy, S., Baker, E.H., Lin, J., Emory, T., Nelson, M., Everson, L.I., Yee, D., and Garwood, M. 2003. *In vivo* quantification of choline compounds in the breast with 1H MR spectroscopy. *Magn. Reson. Med. 50:*1134–1143.

Brinck, U., Fischer, U., Korabiowska, M., Jutrowski, M., Schauer, A., and Grabbe, E. 1997. The variability of fibroadenoma in contrast-enhanced dynamic MR mammography. *Am. J. Roentgenol. 168:*1331–1334.

Brix, G., Semmler, W., Port, R., Schad, L.R., Layer, G., and Lorenz, W.J. 1991. Pharmacokinetic parameters in CNS Gd-DTPA enhanced MR imaging. *J. Comput. Assist. Tomogr. 15:*621–628.

Buadu, L.D., Murakami, J., Murayama, S., Hashiguchi, N., Sakai, S., Masuda, K., Toyoshima, S., Kuroki, S., and Ohno, S. 1996 Breast lesions: correlation of the contrast medium enhancement patterns on MR images with histopathologic findings and tumor angiogenesis. *Radiology 200*:639–649.

Buckley, D.L. 2002. Uncertainty in the analysis of tracer kinetics using dynamic contrast-enhanced T_1-weighted MRI. *Magn. Reson. Med. 47*:601–606.

Buckley, D.L., Drew, P.J., Mussurakis, S., Monson, J.R., and Horsman, A. 1997. Microvessel density of invasive breast cancer assessed by dynamic Gd-DTPA enhanced MRI. *J. Magn. Reson. Imaging 7*:461–464.

Daniel, B.L., Yen, Y.F., Glover, G.H., Ikeda, D.M., Birdwell, R.L., Sawyer-Glover, A.M., Black, J.W., Plevritis, S.K., Jeffrey, S.S., and Herfkens, R.J. 1998. Breast disease: dynamic spiral MR imaging. *Radiology 209*:499–509.

Fischer, U., Kopka, L., and Grabbe, E. 1999. Breast carcinomas: effect of preoperative contrast-enhanced MR imaging on the therapeutic approach. *Radiology 213*:881–888.

Folkman, J. 1995. Angiogenesis in cancer, vascular, rheumatoid and other disease. *Nat. Med. 1*:27–31.

Heywang, S.H., Bassermann, R., Fenzl, G., Nathrath, W., Hahn, D., Beck, R., Krischke, I., and Eiermann, W. 1987. MRI of the breast—histopathologic correlation. *Eur. J. Radiol. 7*:175–182.

Hu, J., Vartanian, S.A., Xuan, Y., Latif, Z., and Soulen, R.L. 2005. An improved 1H magnetic resonance spectroscopic imaging technique for human breast: preliminary results. *Magn. Reson. Imaging 23*:571–576.

Huang, W., Fisher, P.R., Dulaimy, K., Tudorica, L.A., O'Hea, B., and Button, T.M. 2004. Detection of breast malignancy: diagnostic MR protocol for improved specificity. *Radiology 232*:585–591.

Jacobs, M.A., Barker, P.B., Bottomley, P.A., Bhujwalla, Z., and Bluemke, D.A. 2004. Proton magnetic resonance spectroscopic imaging of human breast cancer: a preliminary study. *J. Magn. Reson. Imaging 19*:68–75.

Jemal, A., Murray, T., Ward, E., Samuels, A., Tiwari, R.C., Ghafoor, A., Feuer, E.J., and Thun, M.J. 2005. Cancer Statistics, 2005. *CA Cancer J. Clin. 55*:10–30.

Kaiser, W.A., and Zeitler, E. 1989. MR imaging of the breast: fast imaging sequences with and without Gd-DTPA. Preliminary observations. *Radiology 170*:681–686.

Katz-Brull, R., Lavin, P.T., and Lenkinski, R.E. 2002. Clinical utility of proton magnetic resonance spectroscopy in characterizing breast lesions. *J. Natl. Cancer Inst. 94*:1197–1203.

Knopp, M.V., von Tengg-Kobligk, H., and Choyke, P.L. 2003. Functional magnetic resonance imaging in oncology for diagnosis and therapy monitoring. *Mol. Cancer Ther. 2*:419–426.

Knopp, M.V., Weiss, E., Sinn, H.P., Mattern, J., Junkermann, H., Radeleff, J., Magener, A., Brix, G., Delorme, S., Zuna, I., and van Kaick, G. 1999. Pathophysiologic basis of contrast enhancement in breast tumors. *J. Magn. Reson. Imaging 10*:260–266.

Kuhl, C.K., Mielcareck, P., Klaschik, S., Leutner, C., Wardelmann, E., Gieseke, J., and Schild, H.H. 1999. Dynamic breast MR imaging: are signal intensity time course data useful for differential diagnosis of enhancing lesions? *Radiology 211*:101–110.

Kvistad, K.A., Rydland, J., Vainio, J., Smethurst, H.B., Lundgren, S., Fjosne, H.E., and Haraldseth, O. 2000. Breast lesions: evaluation with dynamic contrast-enhanced T1-weighted MR imaging and with T2*-weighted first-pass perfusion MR imaging. *Radiology 216*:545–553.

Landis, C.S., Li, X., Telang, F.W., Coderre, J.A., Micca, P.L., Rooney, W.D., Latour, L.L., Vetek, G., Palyka, I., and Springer, C.S. 2000. Determination of MRI contrast agent concentration time course *in vivo* following bolus injection: effect of equilibrium transcytolemmal water exchange. *Magn. Reson. Med. 44*:563–574.

Landis, C.S., Li, X., Telang, F.W., Molina, P.E., Palyka, I., Vetek, G., and Springer, C.S. 1999. Equilibrium transcytolemmal water exchange kinetics in skeletal muscle *in vivo*. *Magn. Reson. Med. 42*:467–478.

Larsson, H.B.W., Stubgaard, M., Frederiksen, J.L., Jensen, M., Henriksen, O., and Paulson, O.B. 1990. Quantitation of blood-brain barrier defect by magnetic resonance imaging and gadolinium-DTPA in patients with multiple sclerosis and brain tumors. *Magn. Reson. Med. 16*:117–131.

Li, X., Huang, W., Yankeelov, T.E., Tudorica, A., Rooney, W.D., and Springer, C.S. 2005. Shutter-speed analysis of contrast reagent bolus-tracking data: preliminary observations in benign and malignant breast disease. *Magn. Reson. Med. 53*:724–729.

Liberman, L., Morris, E.A., Lee, M.J.Y., Kaplan, J.E., LaTrenta, L.R., Menell, J.H., Abramson, A.F., Dashnaw, S.M., Ballon, D.J., and Dershaw, D.D. 2002. Breast lesions detected on MR imaging: features and positive predictive value. *Am. J. Roentgenol. 179*:171–178.

Mandelson, M.T., Oestreicher, N., Porter, P.L., White, D., Finder, C.A., Taplin, S.H., and White, E. 2000. Breast density as a predictor of mammographic detection: comparison of interval- and screen-detected cancers. *J. Natl. Cancer Inst. 92*:1081–1087.

Meisamy, S., Bolan, P.J., Baker, E.H., Bliss, R.L., Gulbahce, E., Everson, L.I., Nelson, M.T., Emory, T.H., Tuttle, T.M., Yee, D., and Garwood, M. 2004. Neoadjuvant chemotherapy of locally advanced breast cancer: predicting response with *in vivo* (1)H MR spectroscopy—a pilot study at 4 T. *Radiology 233*:424–431.

Moran, G.R., and Prato, F.S. 2004. Modeling 1H exchange: an estimate of the error introduced in MRI by assuming the fast exchange limit in bolus tracking. *Magn. Reson. Med. 51*:816–827.

Negendank, W. 1992. Studies of human tumors by MRS: a review. *NMR Biomed. 5*:303–324.

Orel, S.G., and Schnall, M.D. 2001. MR imaging of the breast for the detection, diagnosis, and staging of the breast cancer. *Radiology 220*:13–30.

Port, R.E., Knopp, M.V., and Brix, G. 2001. Dynamic contrast-enhanced MRI using Gd-DTPA: interindividual variability of the arterial input function and consequences for the assessment of kinetics in tumors. *Magn. Reson. Med. 45*:1030–1038.

Rankin, S.C. 2000. MRI of the breast. *British J. Radiol. 73*:806–818.

Rosen, B.R., Belliveau, J.W., Vevea, J.M., and Brady, T.J. 1990. Perfusion imaging with NMR contrast agents. *Magn. Reson. Med. 14*:249–265.

Schnall, M.D. 2003. Breast MR imaging. *Radiol. Clin. N. Am. 41*:43–50.

Springer, C.S., Patlak, C.S., Palyka, I., and Huang, W. 1999. Principles of susceptibility contrast-based functional MRI: the sign of the functional MRI response. In: Moonen C.T.W., and Bandettini P.A. (Eds.), *Functional MRI*. Berlin: Springer-Verlag, Chapter 9 pp. 91–102.

Springer, C.S., and Rooney, W.D. 2001. B_0-dependence of the CR-determined exchange regime for equilibrium transcytolemmal water transport: implications for bolus-tracking studies. *Proc. Intl. Soc. Magn. Reson. Med.* 2241.

Su, M.Y., Cheung, Y.C., Fruehauf, J.P., Yu, H., Nalcioglu, O., Mechetner, E., Kyshtoobayeva, A., Chen, S., Hsueh, S., McLaren, C.E., and Wan, Y. 2003. Correlation of dynamic contrast enhancement MRI parameters with microvessel density and VEGF for assessment of angiogenesis in breast cancer. *J. Magn. Reson. Imaging 18*:467–477

Tofts, P.S. 1997. Modeling tracer kinetics in dynamic Gd-DTPA MR imaging. *J. Magn. Reson. Imaging 7*:91–101.

Tofts, P.S., Berkowitz, B., and Schnall, M.D. 1995. Quantitative analysis of dynamic Gd-DTPA enhancement in breast tumors using a permeability model. *Magn. Reson. Med. 33*:564–568.

Tofts, P.S., Brix, G., Buckley, D.L., Evelhoch, J.L., Henderson, E., Knopp, M.V., Larsson, H.B., Lee, T.Y., Mayr, N.A., Parker, G.J., Port, R.E., Taylor, J., and Weisskoff, R.M. 1999. Estimating kinetic parameters from dynamic contrast-enhanced T1-weighted MRI of a diffusable tracer: standardized quantities and symbols. *J. Magn. Reson. Imaging 10*:223–232.

Tofts, P.S., and Kermode, A.G. 1991. Measurement of the blood-brain barrier permeability and leakage space using dynamic MR imaging: 1. Fundamental concepts. *Magn. Reson. Med. 17*:357–367.

Warner, E., Plewes, D.B., Shumak, R.S., Catzavelos, G.C., Di Prospero, L.S., Yaffe, M.J., Goel, V., Ramsay, E., Chart, P.L., Cole, D.E., Taylor,

G.A., Cutrara, M., Samuels, T.H., Murphy, J.P., Murphy, J.M., and Narod, S.A. 2001. Comparison of breast magnetic resonance imaging, mammography, and ultrasound for surveillance of women at high risk for hereditary breast cancer. *J. Clin. Oncol. 19*:3524–3531.

Yang, C., Karczmar, G.S., Medved, M., and Stadler, W.M. 2004. Estimating the arterial input function using two reference tissue in dynamic contrast-enhanced MRI studies: fundamental concepts and simulations. *Magn. Reson. Med. 52*:1110–1117.

Yankeelov, T.E., Rooney, W.D., Huang, W., Dyke, J.P., Li, X., Tudorica, A., Lee, J.-H., Koutcher, J.A., and Springer, C.S. 2005. Evidence for shutter-speed variation in CR bolus-tracking studies of human pathology. *NMR Biomed. 18:*173–185.

Yankeelov, T.E., Rooney, W.D., Li, X., and Springer, C.S. 2003. Variation of the relaxographic "Shutter-Speed" for transcytolemmal water exchange affects the CR bolus-tracking curve shape. *Magn. Reson. Med. 50:*1151–1169.

Zhou, R., Pickup, S., Yankeelov, T.E., Springer, C.S., and Glickson, J.D. 2004. Simultaneous measurement of arterial input function and tumor pharmacokinetics in mice by dynamic contrast enhanced imaging: effects of transcytolemmal water exchange. *Magn. Reson. Med. 52:*248–257.

26

Breast Lesions: Computerized Analysis of Magnetic Resonance Imaging

Kenneth G.A. Gilhuijs

Introduction

The standard imaging techniques for detecting and characterizing breast lesions are mammography and ultrasonography. Complementing each other, mammography visualizes differences in tissue density, while ultrasonography shows differences between solid masses and cysts. Although the combination of these techniques is effective in the majority of cases, problems occur in breasts with dense fibroglandular tissue that may obscure the presence of malignant disease. This problem is of particular concern in premenopausal women who have relatively more dense fibroglandular breast tissue than postmenopausal women.

Magnetic resonance imaging (MRI) was introduced to breast imaging in the early 1980s in ongoing efforts to improve both the detection of breast cancer and the specificity in distinguishing between benign and malignant lesions. This modality shows contrast between tissues that have different proton (hydrogen) density. A magnetic field aligns the spins of the protons, and a radiofrequency field produced by a dedicated breast coil feeds energy into this system by changing the angle of the spins. When the

radiofrequency field stops, the spins are realigned with the magnetic field, releasing the absorbed energy, which is translated into a gray-value intensity image. The chemical composition of water and fatty tissue contains more protons than that of fibroglandular tissue, resulting in excellent tissue contrast at MRI. Because MRI provides tomographic information along cross sections through tissue rather than projections of anatomy like mammography, it was anticipated that MRI would result in superior definition of tumors in dense fibroglandular breasts. Unfortunately, tumors and fibroglandular tissue were found to be much alike in terms of proton composition, which made MRI suffer from the same major limitation as mammography. Combined with considerable overlap in the proton composition between benign and malignant lesions, the application of MRI resulted in far from obvious advantage over conventional breast imaging to increase neither the sensitivity nor the specificity of breast cancer detection.

It was not until the introduction of the MRI-specific contrast agent in 1986 that MRI became a useful adjunct to mammography (Heywang et al., 1986). The paramagnetic contrast agent (typically gadolinium-diethyltriaminepentaacetic acid:

Gd-DTPA) changes the way the energy is released from neighboring protons, resulting in local enhancement of the signal. Applied intravenously, the agent accumulates faster in tumor tissue than in normal tissue due to distinct differences between the neovascular system in tumors and normal vascularity. Contrast-enhanced (CE) MRI visualizes the uptake of the contrast agent over time using repetitive measurements. What makes the technique so attractive, despite its relatively high cost and less widespread availability compared to mammography and ultrasonography, is that it visualizes the functional behavior of tumors. This complementary information allows detection of tumors that are occult at conventional imaging even in the absence of dense fibroglandular tissue (Kelcz and Santyr, 1995; Heywang-Kobrunner *et al.*, 1997). Two decades after its introduction, the reported sensitivities of CE MRI to detect invasive breast cancer are ~ 100%. The number of CE MRI examinations of the breast has increased considerably in care centers throughout the world, and even at present new subgroups of patients are identified who benefit from this technique.

Nonetheless, CE MRI is not without limitations. The first problem is its limited ability to visualize early-stage cancer (ductal carcinoma *in situ*: DCIS). The reported sensitivity for DCIS varies widely and may be as low as 45%. A likely explanation is the subtle functional behavior of the early-stage cancers resulting in insufficient accumulation of the contrast agent within. A second problem is the limited ability to discriminate between benign and malignant lesions. The reported specificity varies and may be as low as 37%. This lack of specificity is caused primarily by the overlap in functional behavior between several types of benign lesions (e.g., fibroadenoma, adenosis) and malignant tumors. Standardized guidelines for the reporting of lesions in CE MRI have recently been formulated (Reston, 2003), but despite these guidelines, variations in image interpretation continue to exist within and between readers (Stoutjesdijk *et al.*, 2005). Further contributing to these variations is the lack of standardization in MRI technique. Numerous acquisition protocols exist to excite the protons and to accumulate the results, yielding images with different appearance that emphasize different aspects of tumors. It is not likely that standardization of techniques will be realized in the near future because new developments in MRI technology increase rather than decrease the variety of protocols. As a result, it is likely that variations in reading will also increase. In addition, as MRI technology continues to advance, the amount of data to be analyzed continues to increase. In the past decade, the number of images resulting from a single examination increased from several hundred at most up to > 1000 at present. Moreover, interpretation of CE MRI more often involves comparing multiple examinations of the same patients over time.

The advances in MRI technology have triggered new developments in breast-MRI workstations. Tools for computer-supported image analysis are rapidly becoming essential to keep the workload within acceptable limits and to aid in interpreting the increasingly complex signals that underlie these images. Automated preprocessing is required to emphasize relevant information within and across examinations. The ultimate goal of computerized analysis is to improve the sensitivity, specificity, and consistency of image interpretation within and between readers while reducing the workload to analyze large data sets.

Mechanisms of Functional Imaging Using Magnetic Resonance Imaging

To fully appreciate the rationale for computerized analysis of CE MRI, it is important to review the basic mechanisms of contrast uptake in the breast and the limitations of CE MRI to visualize this uptake. Magnetic resonance imaging acquisition protocols are defined by the sequences at which the radiofrequency field in the coil is turned on and off. Typical techniques are T_1-weighted spin-echo and gradient-echo imaging. The patient is examined in prone orientation using a dedicated bilateral breast coil. A series of precontrast images is made, followed by intravenous application of the contrast agent at a well-defined rate by means of a power injector. Because the membranes of blood vessels formed by the tumor are more permeable than normal blood-vessel membranes, the contrast agent leaks faster into spaces between tumor cells than into other regions. As a result, within 2 min after its application, the contrast agent is more prominently present in viable tumor tissue than in other tissues. The amount of accumulated contrast agent at any time depends on the available space accessible to the agent (extravascular and extracellular space, or leakage space, denoted as V_e), the rate at which the agent permeates from the blood vessels to this free space (transfer constant, denoted as K^{trans}), and the rate at which the agent flows back into the vessels (denoted as k_{ep}) (Tofts *et al.*, 1999). Typically, after 2 min, the backflow from the tumor to the vascular system overtakes the inflow. After 10 min, the concentration in surrounding tissues exceeds the concentration in the tumor. As a result, it becomes more difficult to differentiate between breast tumors and fibroglandular tissue. Typically, two or more series of postcontrast images are acquired within this 10-min window in order to visualize the behavior of the contrast uptake in the different tissues. The time between the postcontrast series varies between several seconds and several minutes, depending on the acquisition protocol used.

The reason for this variation is that CE MRI imposes a trade-off between temporal and spatial resolution. Typically, detailed images of morphology of the uptake cannot be made in rapid succession, nor can rapidly acquired images visualize detailed morphology. Magnetic resonance imaging acquisition protocols that favor high spatial resolution often depict one breast at a time, yielding voxel sizes smaller than 1 mm at a temporal resolution of several minutes. Relatively high sensitivity for DCIS is typically reported with these protocols. Protocols favoring high temporal resolution yield a large number of series of both breasts at intervals of < 2 sec and at spatial resolution of several millimeters up to 1 cm. A typical trade-off involves multiple series of both breasts at 60- to 90-sec intervals and at 1 mm spatial resolution. Analysis of the morphology of enhancement is likely to yield superior ability to discriminate between benign and malignant lesions at high spatial resolution than at low spatial resolution. On the other hand, analysis of the kinetics of the enhancement is likely to yield more accurate results at high temporal resolution. Recent developments in MRI-acquisition protocols aim to close the gap between these extremes. An example is key-hole imaging where complex patterns in the breast are sampled more often than less complex homogeneous regions. As a result, images can be acquired more rapidly without significant loss of information in regions that contain morphology of interest.

Current developments in contrast agents and MRI technique will continue to increase the amount and complexity of the information offered to the radiologists. Despite what is often described, CE MRI does not visualize the actual presence of the contrast agent, but rather its effect on the protons in surrounding tissue and water. Different MRI-specific agents may have different ability to change the release of energy from nearby protons. New developments are macromolecular contrast agents that remain in the neovascular system rather than diffuse to extravascular and extracellular space. Macromolecular contrast agents result in different uptake kinetics than conventional contrast agents, which may need to be taken into consideration in the image analysis. First studies on the use of these agents report superior potential to predict histologic tumor grade compared with the use of conventional contrast agents (Daldrup-Link and Brasch, 2003).

Another MRI technique to visualize evidence of functional processes related to malignant disease is MRI spectroscopy. This technique allows assessment of the molecular composition of breast tissues. Elevated choline levels are typically associated with the presence of malignant disease and may be picked up by MRI spectroscopy. One of the main challenges, however, is to identify the relatively weak signal associated with choline in the presence of the much stronger signal produced by fat. In addition, the technique is currently limited by its relatively poor spatial resolution (~ 1 cm). These problems will be reduced with the higher magnetic field strength of the new generation MRI units (3 Tesla vs. 1.5 Tesla). Nonetheless, even at 1.5 Tesla field strength, MRI spectroscopy has already shown promising

potential to complement CE MRI in the discrimination between benign and malignant lesions (Jacobs et al., 2005) and in the assessment of response of breast tumors to neoadjuvant chemotherapy. Wide accessibility to these techniques will, however, depend critically on tools that effectively visualize relevant information from these complex signals.

Interpretation of Contrast-enhanced Magnetic Resonance Imaging

The presence of malignant disease is correlated with several temporal and morphological characteristics of the contrast enhancement at CE MRI. To visualize the enhancement, subtraction images are typically computed. The precontrast series subtracted from the postcontrast series (wash-in series) emphasize the early kinetics caused by transfer of the contrast agent from the neovascular system to the leakage space. The first postcontrast series subtracted from the last postcontrast series (washout series) visualize early departure of the contrast agent from the leakage space back to the neovascular system (also referred to as late kinetics) (Fig. 138). In addition to fast early kinetics and pronounced late kinetics, internal enhancement such as heterogeneous uptake is often indicative of malignant disease. On the other hand, inhomogeneity caused by dark nonenhancing separations may be more indicative of benign disease (fibroadenoma). In addition to these aspects, the shape of the enhancing lesion, the appearance of its margins, and the distribution of possible multifocal patterns of enhancement are also taken into account (Reston, 2003). Typically used methods to visualize the contrast enhancement are multiplanar reformatting (MPR) and maximum-intensity projection (MIP). In the first approach, cross sections through the CE MRI volume are calculated at various depths and directions—typically the directions along the three orthogonal axes—and are updated interactively in response to actions of the operator. In the second approach, the wash-in volume is projected in one or multiple directions. The resulting MIP views allow superior assessment of multifocal enhancement patterns that extend into multiple cross sections (Fig. 138).

Although the examinations are typically performed by radiologists experienced in the reading of breast MRI, moderate to poor agreement between observers has been reported in particular for morphological characteristics such as internal enhancement and lesion shape. The Breast Imaging-Reporting and Data System (BI-RADS™) lexicon for standardized reporting of CE MRI examinations of the breast has not completely resolved this issue (Stoutjesdijk et al., 2005).

To improve the consistency of temporal and morphological analysis, several computerized methods have been developed to describe the uptake and morphology of the contrast agent in a quantitative fashion. These methods are typically

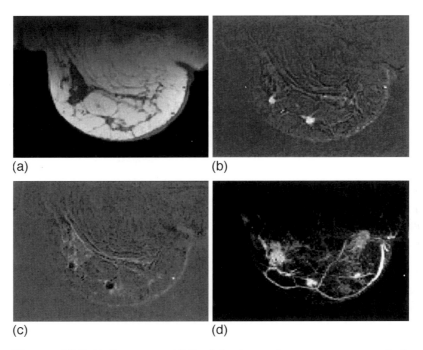

Figure 138 (a) Precontrast MRI of the breast (transverse cross section). (b) Wash-in image showing increased uptake (white) after 90 sec in two ductal cancers. (c) Washout image showing lack of enhancement (dark) in the tumors after 450 sec. (d) MIP view showing extensive multifocal disease throughout the same breast.

aimed at computer-aided detection (distinguishing between normal tissue and suspicious tissue), computer-aided diagnosis (distinguishing between benign and malignant lesions), and computer-aided monitoring of treatment (distinguishing between tumors that respond favorably to neoadjuvant chemotherapy and those that do not). To understand the strengths and limitations of computerized analysis, we will focus in more detail on their five main components: reduction of motion artifacts, computerized extraction of temporal features, lesion segmentation, computerized extraction of morphological features, and classification of extracted features into two or more mutually exclusive groups.

Reduction of Motion Artifacts

Movement and deformation of the breasts during CE MRI often occur to a varying extent and are typically caused by involuntary contraction of the pectoral muscle in response to the power injection of the contrast agent. The motions result in misalignment of tissues between contrast series, affecting both temporal and morphological characteristics of observed contrast enhancement and leading to reduced ability to detect and characterize breast lesions. Advanced breast-MRI workstations typically offer methods to automatically realign the breast tissue between pre- and postcontrast series. These methods identify corresponding

structures in the precontrast and postcontrast images. This is a particularly daunting task because the most relevant structures change appearance with the presence of the contrast agent. Examples of applied techniques are mutual-information analysis, optical-flow analysis, and finite-element analysis. Once corresponding structures have been identified, one set of images is deformed (warped) to align these structures with those in the target images. Deformation of the tissue between corresponding structures is estimated by linear or elastic interpolation. Methods for the reduction of motion artifacts currently have two major limitations: accuracy and computation time.

The accuracy of motion reduction depends on three factors. The first is whether the identification and recovery of corresponding structures are done on a slice-by-slice basis (two-dimensional [2D]) or volumetrically (three-dimensional [3D]). At present, methods based on 2D analysis have largely been abandoned because they do not allow recovery of tissue shifts perpendicular to the plane of the images. The second factor that determines the accuracy of motion reduction is whether rigid-body alignment or nonrigid alignment is employed. Rigid-body alignment uses only translations and rotations to realign corresponding structures in pre- and postcontrast images. It corrects for displacements and rotations of the breast, but is not capable of handling deformations. Nonrigid registration attempts to align all corresponding structures, but is typically constrained by biomechanical models of breast tissue in order to obtain physically meaningful alignment estimates. Although non-rigid registration has the potential to reduce motion artifacts more accurately than rigid-body registration (Bruckner *et al.*, 2000), a disadvantage is its more extensive computation time.

Typically, a trade-off exists between computation time and complexity of the deformations that can be properly handled. Motion reduction in full 3D using nonrigid alignment may currently take over half an hour to compute. Although methods for motion reduction have demonstrated potential to provide clinically relevant information in the presence of large motion artifacts (Fig. 139), they often result in residual misalignment. The validation of the methods is challenging because the "true alignment" is unknown. Alternatively, consistency checks such as preservation of breast volume before and after registration have been used, as well as methods that provide confidence bounds on feasible tissue shifts based on the biomechanical properties of various tissues. In clinical practice, it is advisable to always inspect the original nonsubtracted images in addition to the motion-reduced views, especially near the boundaries of lesions. The process of interpolation in the motion reduction tends to smooth the images and change the appearance of the boundaries. Moreover, the impact of the constraints imposed by elastic models has not been investigated in detail near (rigid) tumors.

Computerized Extraction of Temporal Features

The image-intensity values at CE MRI are not directly related to a single physical property, such as electron density in computed tomography (CT), but rather to a number of complex interacting mechanisms that underlie the uptake of the contrast agent. In addition, the intensity values are dependent on the acquisition protocol and may also be affected by postprocessing steps applied by the MRI units. The main challenge in the interpretation of CE MRI is, therefore, to summarize relevant information on voxel-intensity values over time in quantitative descriptions that are more or less consistent across patients, across units from different manufacturers, and across hospitals. Two approaches are typically pursued: empirical analysis (describing the observed changes in image intensity over time) and pharmacokinetic analysis (describing the underlying properties of contrast uptake derived from the observed changes in image intensity). Often, a summary statistic is produced for subgroups of voxels in a manually or automatically established region of interest (ROI).

Methods for empirical analysis of CE MRI are either feature-based or curve-based. Either way, normalization of the signal-intensity values is essential to compare data from different examinations. In other words, calculated features must be in physical units such as seconds or percentages rather than in absolute numbers that represent local image intensity. Feature-based methods extract properties from the time-signal intensity curve such as initial enhancement (relative wash-in), late enhancement (relative washout), time to peak enhancement, and signal-enhancing ratio (SER). An often applied approach to visualize this information is to produce a color overlay image in which the color at each voxel location indicates the value of the feature at that location.

Yet another feature-based approach to describe the kinetics of the contrast uptake is to categorize the time-signal intensity curve into three basic shapes defined by three points in time—for example, at 0 sec, 2 min, and 6 min after contrast injection. Type I curves show increase of signal between the second and third time point, type II curves

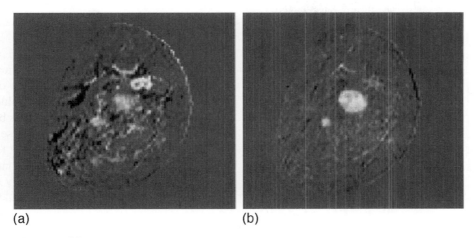

(a)   (b)

Figure 139 Example of nonrigid motion reduction in CE MRI of the breast. (a) Original wash-in image (coronal cross section) giving rise to uncertainty in tumor extent. (b) Motion-reduced wash-in image providing superior assessment of tumor extent.

show stable signal, and type III curves show decreasing signal. Type I curves are suggestive of benign disease, and type III curves are indicative of malignant disease (Kuhl *et al.*, 1999). Again, color overlays have been used to visualize the curve type at each voxel location (Fig. 140). An example of such processing was reported by Kelcz and colleagues (Kelcz *et al.*, 2002). Curve-based methods for temporal feature extraction employ all time points along the time-signal intensity curve to assess the probability of malignancy at each voxel location. If images from additional acquisition protocols are available (e.g., from T_2-weighted imaging), these may also be used. Using multivariate analysis, we can identify combinations of signal intensity values that discriminate between different tissue types such as glandular tissue, benign lesions, and malignant lesions.

Although the empirical methods described here show consistent results across different studies, they rely solely on observed changes in image intensity over time and do not address the underlying causes for these changes. It is conceivable that different combinations of contributing factors can result in comparable observations of enhancement. As a result, subtle differences in tissue properties, expressed by differences in the underlying mechanisms of contrast uptake, may remain unnoticed. Pharmacokinetic analysis uses theoretical models to estimate these underlying components from the time-signal intensity curve (Tofts *et al.*, 1999). The typically employed two-compartment models describe the transfer of contrast agent between the blood vessels (first compartment) and the extravascular extracellular space (second compartment). The permeability of the membrane between these compartments is typically denoted as k_{31}. The transfer constant K^{trans} from the first compartment to the second is proportional to the product of the surface area of the blood vessels and the permeability of the membrane. By fitting the theoretical models to the time-signal intensity values, estimates are obtained

for the pharmacokinetic parameters. Again, the value of these parameters is often visualized at each voxel location by color overlays.

Promising results have been reported to distinguish between benign and malignant lesions using K^{trans} or k_{31}, and even to detect DCIS using k_{ep} (Mariano *et al.*, 2005). Pitfalls of pharmacokinetic analysis, however, are the difficulty of comparing parameters between studies and the typically long computation times. Although effort has been made to standardize the terminology and symbols used in pharmacokinetic analysis, large differences still exist between the methods implemented in various studies. Moreover, patient motion, differences in acquisition protocols, differences in accuracy of the radiofrequency field produced by various breast coils, and differences in postprocessing applied by different MRI systems continue to complicate the generalization of findings from single studies to the general population. Although pharmacokinetic parameters were shown to have potential benefit over conventional empirical methods, a consensus on this issue has not yet been reached.

Computerized Extraction of the Region of Interest

Although color overlay of temporal features of contrast uptake is a useful tool for quickly identifying suspicious regions, it is not uncommon for breast lesions to show benign-type as well as malignant-type signal intensity curves. Proceeding from characterization of voxels to characterization of lesions, average time-signal intensity curves are typically calculated for the voxels within a manually indicated ROI. This ROI is usually limited to the most enhancing parts of the lesion, because the contrast uptake in these regions has been shown to be more indicative of malignancy than the overall uptake. Manual selection of the

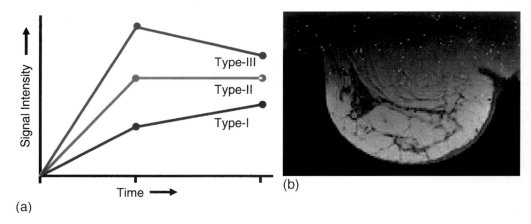

(a)

(b)

Figure 140 (a) The shape of the time-signal intensity curve is correlated with presence of malignant disease, ranging from type I (blue) to type III (red). (b) Corresponding color overlay, showing the curve type at each pixel location.

ROI has, however, two disadvantages. First, variations in selection of the ROI within and between radiologists lead to variations in interpretation, which, in turn, contribute to variations in the specificity to distinguish between benign and malignant lesions. Second, manual selection of ROIs is time consuming, especially when multifocal enhancement occurs.

Several investigators developed tools for computerized selection of the ROI. Important considerations of these tools are the required level of interaction from the operator, the robustness of results, and the question of whether the tools operate on single slices or volumetrically in 3D. Methods often first segment the entire lesion using a manually or automatically chosen criterion that separates lesion voxels from background voxels. Typical criteria are thresholds in feature images computed from precontrast and postcontrast image intensity values. After segmentation of the lesion, the most enhancing subregions are established from a manually or automatically chosen threshold of initial enhancement. Some methods require selection of a rectangular box around the lesion to start the procedure; others require selection of a single seed point within the lesion; and yet other methods do not require any user interaction at all. Variations in selection of the ROI will decrease with increasing levels of automation. However, as the level of automation increases, demands on the accuracy and robustness of the methods also increase. One of the challenges at present is to automatically differentiate between normal major blood vessels in the breast and enhancing blood vessels related to the tumor; it is generally undesirable to include the major normal blood vessels in the ROI.

Yet another potential application of computerized extraction of the ROI is to improve the sensitivity of assessing response to neoadjuvant chemotherapy. The main challenge in this application is to detect subtle differences in enhancement between different CE MRI examinations of the same patient prior to and during neoadjuvant chemotherapy. When variations in selection of the ROI are reduced, incomplete response to therapy may be anticipated sooner, offering the possibility for early changes to the treatment plan. An example of a computer-extracted feature to evaluate response to therapy is the change of volume of initial enhancement above a predefined threshold (Partridge *et al.*, 2005).

Computerized Extraction of Morphological Features

When CE MRI of the breast was first introduced, its interpretation was primarily focused on either temporal or morphological features. The current consensus is that these features complement each other and need to be combined to improve the detection and characterization of breast lesions (Reston, 2003). Interestingly, developments in computerized analysis of CE MRI appear to progress through the same process. At present, the majority of commercially-available breast-MRI workstations provide computerized support for the analysis of temporal features only. Nonetheless, morphological features in particular are prone to variations in interpretation between radiologists, which may likely be reduced by computerized analysis.

Although more challenging than computerized extraction of temporal features, several research groups have developed methods to automatically quantify the morphology of contrast enhancement. These methods are boundary-based, pattern-based, or shape-based, and are typically preceded by (semi-)automatic segmentation of suspicious enhancement. Mathematical descriptions of morphology may mimic the observations of radiologists, albeit aimed at increased reproducibility, or target subtle spatial properties that are more difficult to assess by eye.

Boundary-based methods analyze the border between the enhancing lesion and the background. Examples of boundary features are the fractal dimension of the border, the mean image intensity gradient (sharpness) of the border, and the variation in the sharpness. The fractal dimension describes the complexity of the border ranging from smooth (small fractal dimension) to spiculated (large fractal dimension); spiculated borders are more indicative of underlying malignant disease than smooth borders.

Pattern-based methods describe the pattern of enhancement in and around the lesion. Examples of these methods are texture analysis and radial-gradient analysis. Each of these analyses quantifies the inhomogeneity of the enhancement. Complex inhomogeneous enhancement is more often associated with malignant disease than smooth homogeneous enhancement. Shape-based methods describe the form of the enhancing regions. The features are typically calculated from the binary segmented lesion. Examples of shape features are compactness, irregularity, and circularity. Decreasing compactness, increasing irregularity, and decreasing circularity are associated with increasing likelihood of malignant disease.

Important considerations for computerized extraction of morphology are the voxel resolution and lesion size. Evidently, when lesions approach the size of one voxel, descriptive values of morphology converge regardless of the nature of the underlying disease. In fact, the efficacy of morphological assessment breaks down before the one-voxel limit is reached due to the partial-volume effect (smoothing of image structures in voxels that cover enhancing as well as nonenhancing tissue). The partial-volume effect also limits the efficacy of temporal features to characterize small lesions. Although minimum requirements on lesion size in relation to spatial resolution have not yet been reported for computer-extracted descriptors of morphology, it is evident that morphological feature extraction imposes higher demands on spatial resolution, especially for small lesions (< 1 cm), than temporal feature extraction.

Another important consideration for computerized extraction of morphology is whether the procedures operate on a slice-by-slice basis (2D) or volumetrically (3D). Evidence suggests that the efficacy of many morphological features improves when the enhancing areas are analyzed in 3D rather than in 2D.

Computerized Classification of Features of Enhancement

The process of categorizing enhancing lesions in two or more mutually exclusive groups based on the value of one or multiple features of enhancement is referred to as classification. Examples of classification are differentiation between benign and malignant disease, and differentiation between favorable and nonfavorable responders to neoadjuvant chemotherapy. In conventional clinical setting, the classification is performed on the basis of manually assessed characteristics of enhancement using an *a priori* set of guidelines. Several guidelines based on decision trees and scoring systems have been reported, and more recently, the BI-RADS lexicon for CE MRI of the breast.

Attempts to further improve the objectivity and consistency of the interpretation of enhancement lead to methods for computerized classification. Automated classifiers are either based on manually rated features expressed in terms of scores or categories, automatically extracted features quantified on a continuous scale, or on a combination of both. The classification is either performed at voxel level, categorizing tissue at each voxel location using, for example, color overlays, or at lesion level, providing overall assessment of connected regions of enhancement. Methods for computerized classification of features typically consist of four consecutive steps: construction of a training set, model fitting, selection of an operating point, and model validation.

Construction of the training set is one of the most crucial aspects of building a classifier. The training set must contain cases with known outcome from both categories and must be representative for the population to which the classifier will be applied. The larger the training set, the smaller the risk of inadequate generalization to new cases, provided that no bias was introduced in the selection of these cases. Unfortunately, bias is unavoidable in any training set of finite size, and a choice must be made between various selection strategies to minimize the effect of the bias. The first option is to select cases that constitute a broad range of possible observations in both categories. The disadvantage is that rare events may be overrepresented at the cost of performance in more common cases. The second option is consecutive inclusion of cases encountered in the clinic without any prior selection. This approach results in a representative set of observations, but may not generalize to hospitals where different patient populations occur. It has been reported, for instance, that the boundaries of enhancing tumors in women screened due to high lifetime risk of breast cancer have sharper transitions to surrounding breast tissue than the boundaries of sporadic breast tumors. The third option is consecutive inclusion using multiple training sets, one for each anticipated population. Although this approach may minimize selection bias, rarely are sufficient numbers of cases available to allow such stratification into subgroups. One may argue which approach is most effective, but unfortunately the employed inclusion criteria for the training set are rarely reported in sufficient detail to assess how well a computerized classification is likely to generalize across hospitals.

The second step in computerized classification of features of enhancement is to combine the values of multiple features in such a way that the likelihood of a positive test result (e.g., malignant disease or unfavorable response) is obtained, ranging between 0% and 100%. To combine the features, a mathematical model is employed, whose parameters determine how the feature values are merged. Typically used models, sorted in order of increasing complexity, are logistic regression and linear discriminant analysis, followed by quadratic discriminant analysis and artificial neural networks. The parameter values in these models are established by fitting the model to the cases in the training set in such a way that the fraction of misclassified lesions (false-positives and false-negatives) is minimal. How well the classifier succeeds in this task depends on the number of parameters in the model: the larger this number, the smaller the errors in the training set. The downside, however, is decreasing accuracy of the model for new cases (i.e., poor generalization). The main challenge, therefore, is to select a model with no more than the minimum number of parameters that is required to achieve optimum performance for new cases. Because the number of parameters increases with the number of included features and the complexity of the model, methods for automated classification involving small numbers of training cases (< 100), combined with relatively large numbers of features (> 4), and complex classification models (e.g., artificial neural networks) must be carefully considered.

The third step in computerized classification is the selection of a threshold value in the likelihood of a positive test result, often referred to as the operating point. A straightforward approach commonly applied in current breast-MRI workstations is to use the color red for enhancing areas that are considered to be at risk of malignancy (e.g., areas with initial enhancement values larger than 100%, or areas with type III enhancement curves). When a classifier is used to combine multiple features, likelihood values larger than or equal to the operating point are considered to be a positive test result. Different operating points yield different trade-offs between the true-positive and false-positive fraction. Large true-positive fractions (high sensitivity for a positive test result) are typically associated with large false-positive fractions (many biopsies on benign lesions or changes of therapy in patients who respond favorably to therapy). The optimum trade-off usually follows from a cost-benefit analysis. A commonly applied tool to select the operating point is receiver-operating characteristic (ROC) analysis. The ROC curve shows the relationship between the true-positive and false-positive fraction for operating points ranging between 0% and 100%. The area under the ROC curve, denoted as A_z, is an overall measure of the performance of the classifier to discriminate between the two groups. A_z ranges between 0.5 (chance performance) and 1.0 (perfect performance).

The fourth step in computerized classification is validation of the model. Ranging from low to high chance of overestimating the performance, three approaches exist: prospective testing, cross validation, and consistency analysis. In the first approach, the model is tested on new cases Cross-validation techniques split the available set of cases in an independent training set and a test set. This process may be repeated multiple times. Examples are leave-one-out validation, N-fold cross validation, and jack-knifing. Consistency analysis involves training and testing the model on the same set of cases without an independent test set. Results obtained from consistency analysis are likely to considerably overestimate the performance of the classifier, unless the number of cases is very large.

Current Status and Future Role of Computerized Analysis of Breast Magnetic Resonance Imaging

Many methods for computerized analysis of CE MRI of the breast have been reported, and their number is steadily growing. Comparison of these methods is, however, challenging. The first problem is that most studies involve relatively small numbers of lesions and that the inclusion criteria are often missing. A second problem is the lack of uniformity in reporting of the results. Some report A_z values, while others give a true-positive and false-positive fraction at a fixed operating point (that varies between studies). In addition, some investigators use prospective testing, others use cross validation, and yet others consistency analysis. Despite these limitations, prospectively tested computerized techniques that do not require manual rating or classification of uptake features appear to result in comparable or improved performance of expert radiologists to distinguish between benign and malignant enhancement (Kelcz et al., 2002; Vergnaghi et al., 2001; Deurloo et al., 2005). Promising results on computerized assessment of the time-signal intensity curves and tumor size have also been reported to complement the reading of radiologists in the assessment of response of breast tumors to neoadjuvant chemotherapy (Partridge et al., 2005; Demartini et al., 2005).

Nonetheless, current commercially-available computer-aided diagnosis (CAD) workstations for CE MRI of the breast are still in their infancy compared to their CAD counterparts available for mammography and chest radiography. Although they adhere to the general concept of CAD as a diagnosis performed by a human operator supported by an automated computer analysis that offers advice, current CAD stations for CE MRI of the breast typically lack automated analysis of morphology, multivariate classification, and automated selection of the operating point. On the other hand, the rationale for the development of these CAD tools cannot simply be extrapolated from screening mammography to CE MRI. A major difference is that CE MRI is typically applied to different populations of women than the general breast cancer screening population. Because of its relatively high cost and less widespread availability, CE MRI is employed as a problem-solving tool in symptomatic women, or as a screening tool in asymptomatic premenopausal women at high lifetime risk of developing breast cancer. This discrepancy puts high demands on CAD tools to fulfill their classical role of reducing the number of biopsies on benign lesions, reducing workload, and reducing the number of missed cancers (Gilhuijs et al., 2002).

The underlying question as to whether CAD for MRI will reduce the number of biopsies on benign lesions is how often CAD is expected to replace biopsies. This is a challenging task because the current indications for CE MRI of the breast involve high *a priori* risk of having or developing breast cancer. The threshold to biopsy is therefore low, requiring very high accuracy of alternative methods to exclude presence of malignant disease. On the other hand, obtaining histopathological proof of MRI-detected lesions can be a daunting task. Lesions detected at MRI are not always visible at ultrasonography, not even when directed by MRI findings. Although MRI-compatible biopsy devices are currently commercially available, they are not widely accessible to centers that perform CE MRI of the breast. Moreover, the efficacy of these devices may be limited for lesions < 1 cm. The management of patients with suspicious breast lesions detected at CE MRI would be greatly facilitated if benign lesions can be identified with high certainty to allow a shift from biopsy to follow-up imaging in at least a subgroup of women. Although follow-up imaging resolves the diagnosis of many benign lesions that are caused by hormone-related enhancement, it also increases the risk of delaying cancer diagnosis, which is particularly undesirable for the (rapidly growing) breast cancers in women with a genetic predisposition for the disease. Because the prevalence of benign findings in these women is up to 30 times higher than the prevalence of malignant disease, techniques that improve the specificity of reading without cost to the sensitivity are highly desirable.

Related to the issue of biopsies on benign lesions is the question of whether CAD will reduce the workload of analyzing large data sets. The underlying question here is twofold: can we safely dismiss regions not indicated by the computer, and how many false-positive regions are indicated that require additional work-up? Currently, there is insufficient data to answer these questions in detail. Although the majority of breast-MRI workstations provide tools to highlight regions that exhibit suspicious temporal behavior of enhancement, the threshold (operating point) that defines the separation between benign and malignant disease is often still manually chosen. If the risk of underestimation is real, the nonhighlighted regions may need to be examined as well, canceling any possible benefit of reduced workload. Conversely, if false-positive highlights occur, the follow-up on these regions may considerably increase the overall workload. Fibroadenomas and some benign fibrocystic changes are known to mimic the behavior of malignant tumors by exhibiting type III uptake curves with rapid initial enhancement and early washout. Ultimately, the efficacy of computerized detection of malignant lesions is expected to improve with the addition of computerized assessment of the morphology of enhancement. Nonetheless, current breast-MRI workstations are already showing promising results to reduce the laborious process of manually obtaining uptake curves from multiple regions.

CAD's potential role in reducing the number of missed cancers is related to the underlying question of how many cancers are initially overlooked at CE MRI of the breast, but visible on retrospect. The risk of overlooking invasive ductal breast cancer is expected to be small, because of the very high sensitivity of CE MRI for these cancers. When early signs of invasive breast cancer are visible on previous CE MRI examinations of the same breast, they are often attributed to possibly benign causes. Here, the main challenge of CAD is to improve the characterization of BI-RADS 3 (possibly benign) lesions in order to minimize delayed diagnosis of malignant disease (e.g., in women screened due to high lifetime risk of breast cancer). Another potential role of CAD is to enhance subtle signs of lobular and intraductal cancers, which are often inadequately visualized by CE MRI, even at retrospective analysis. Here, new MRI technique and pharmacokinetic modeling may prove useful to augment the enhancement of these types of lesions.

In conclusion, although results from various studies demonstrate a potential benefit of computerized analysis for CE MRI of the breast, prospective validation of these findings in larger studies is required. These validations should include assessment of the impact of these tools on the efficacy of the radiologists and the variations between radiologists, not only within institutes, but also across institutes and various MRI techniques.

References

American College of Radiology (ACR). 2003. BI-RADS Breast Imaging Reporting and Data System Atlas: BI-RADS Atlas. 4th ed. Reston, VA. American College of Radiology.

Bruckner, T., Lucht, R., and Brix, G. 2000. Comparison of rigid and elastic matching of dynamic magnetic resonance mammographic images by mutual information. *Med. Phys. 27*:2456–2461.

Daldrup-Link, H.E., and Brasch, R.C. 2003. Macromolecular contrast agents for MR mammography: current status. *Eur. Radiol. 13*:354–365.

Demartini, W.B., Lehman, C.D., Peacock, S., and Russell, M.T. 2005. Computer-aided detection applied to breast MRI: assessment of CAD-generated enhancement and tumor sizes in breast cancers before and after neoadjuvant chemotherapy. *Acad. Radiol. 12*:806–814.

Deurloo, E.E., Muller, S.H., Peterse, J.L., Besnard, A.P., and Gilhuijs, K.G. 2005. Clinically and mammographically occult breast lesions on MR images: potential effect of computerized assessment on clinical reading. *Radiology 234*:693–701.

Gilhuijs, K.G., Deurloo, E.E., Muller, S.H., Peterse, J.L., and Schultze Kool, L.J. 2002. Breast MR imaging in women at increased lifetime risk of breast cancer: clinical system for computerized assessment of breast lesions initial results. *Radiology 225*:907–916.

Heywang, S.H., Hahn, D., Schmidt, H., Krischke, I., Eiermann, W., Bassermann, R., and Lissner, J. 1986. MR imaging of the breast using gadolinium-DTPA. *J. Comput. Assist. Tomogr. 10*:199–204.

Heywang-Köbrunner, S.H., Viehweg, P., Heinig, A., and Kuchler, C. 1997. Contrast-enhanced MRI of the breast: accuracy, value, controversies, solutions. *Eur. J. Radiol. 24*:94–108.

Jacobs, M.A., Barker, P.B., Argani, P., Ouwerkerk, R., Bhujwalla, Z.M., and Bluemke, D.A. 2005. Combined dynamic contrast enhanced breast MR and proton spectroscopic imaging: a feasibility study. *J. Magn. Reson. Imaging 21*:23–28.

Kelcz, F., Furman-Haran, E., Grobgeld, D., and Degani, H. 2002. Clinical testing of high-spatial-resolution parametric contrast-enhanced MR imaging of the breast. *AJR Am. J. Roentgenol. 179*:1485–1492.

Kelcz, F., and Santyr, G. 1995. Gadolinium-enhanced breast MRI. *Crit. Rev. Diagn. Imaging 36*:287–338.

Kuhl, C.K., Mielcareck, P., Klaschik, S., Leutner, C., Wardelmann, E., Gieseke, J., and Schild, H.H. 1999. Dynamic breast MR imaging: are signal intensity time course data useful for differential diagnosis of enhancing lesions? *Radiology 211*:101–110.

Mariano, M.N., van den Bosch, M.A., Daniel, B.L., Nowels, K.W., Birdwell, R.L., Fong, K.J., Desmond, P.S., Plevritis, S., Stables, L.A., Zakhour, M., Herfkens, R.J., and Ikeda, D.M. 2005. Contrast-enhanced MRI of ductal carcinoma *in situ*: characteristics of a new intensity-modulated parametric mapping technique correlated with histopathologic findings. *J. Magn. Reson. Imaging 22*:520–526.

Partridge, S.C., Gibbs, J.E., Lu, Y., Esserman, L.J., Tripathy, D., Wolverton, D.S., Rugo, H.S., Hwang, E.S., Ewing, C.A., and Hylton, N.M. 2005. MRI measurements of breast tumor volume predict response to neoadjuvant chemotherapy and recurrence-free survival. *AJR Am. J. Roentgenol. 184*:1774–1781.

Stoutjesdijk, M.J., Futterer, J.J., Boetes, C., van Die, L.E., Jager, G., and Barentsz, J.O. 2005. Variability in the description of morphologic and contrast enhancement characteristics of breast lesions on magnetic resonance imaging. *Invest. Radiol. 40*:355–362.

Tofts, P.S., Brix, G., Buckley, D.L., Evelhoch, J.L., Henderson, E., Knopp, M.V., Larsson, H.B., Lee, T.Y., Mayr, N.A., Parker, G.J., Port, R.E., Taylor, J., and Weisskoff, R.M. 1999. Estimating kinetic parameters from dynamic contrast-enhanced T(1)-weighted MRI of a diffusable tracer: standardized quantities and symbols. *J. Magn. Reson. Imaging 10*:223–232.

Vergnaghi, D., Monti, A., Setti, E., and Musumeci, R. 2001. A use of a neural network to evaluate contrast enhancement curves in breast magnetic resonance images. *J. Digit. Imaging 14*:58–59.

Optical Imaging Techniques
for Breast Cancer

Alexander Wall and Christoph Bremer

Introduction

Optical imaging, particularly in combination with contrast-enhancing fluorophors, is an appealing concept for breast cancer imaging because nonionizing radiation (permitting continuous or repeated exposure) is applied and fluorophors emitting in the near-infrared range can sufficiently penetrate the breast tissue even from deeper tissue section. Moreover, in the near-infrared range the tissue autofluorescence is virtually zero; therefore, even picomolar amounts of tracers can be sensitively detected with novel imaging devices. In this chapter, we present a brief perspective on optical imaging techniques for breast cancer and discuss current advances in this field. Specific emphasis is placed on fluorophor design for molecular contrast agents.

Generally, medical optical imaging encompasses many imaging technologies that use light from the ultraviolet to the infrared region to investigate tissue optical characteristics. The interaction of light with tissue offers unique contrast mechanisms for imaging: scattering, absorption, and fluorescence of intrinsic and extrinsic tissue elements reveal information on structure, physiology, biochemistry, and molecular function (Ntziachristos and Chance, 2001). Because of the important information that light reveals,

optical imaging has found many applications for *in vivo* tissue measurements for surface structures (skin, endoscopic procedures, etc.), but also for noninvasive investigation of the internal function of large organs such as the breast (Ntziachristos *et al.*, 2000).

There are, however, fundamental differences between optical imaging of surface structures and of large organs (e.g., human breast tissue). Light photons that travel inside the tissue do not travel straight paths as do X-ray photons, but rather they diffuse and follow randomly (Yodh and Chance, 1995). The reason for that is the interaction between photons and tissue in the form of absorption, scattering, and reflection, which results in two major problems. First, optical imaging, a high-resolution technique for surface imaging, becomes an imaging method with millimeter-scale resolution when probing large organs (Ntziachristos and Chance, 2001). Second, only very small amounts of light are transmitted through the breast tissue. This places special demands on both detector and fluorophor technology. Most optical breast imagers operate in the near-infrared range between 700 and 900 nm, also called the "diagnostic window" (Weissleder, 2001). In this "diagnostic window," tissues exhibit low absorbance by oxy- and deoxy-hemoglobin and thus photons can traverse the tissue several centimeters.

Intrinsic optical properties of the breast tissue can offer meaningful information. By imaging of scattering, for example, structural characteristics and the concentrations of organelles can be visualized. More importantly, imaging of the absorption coefficient at appropriately selected wavelengths can quantify the concentrations of water and oxy-hemoglobin and deoxy-hemoglobin and may thus serve as a surrogate marker for tumor angiogenesis.

Tomographic Imaging

In the recent decade, a breakthrough in breast optical imaging was made by the introduction of highly sensitive CCD cameras, improved laser, and light emmiting diodes (LED) technology as well as the development of algorithms for three-dimensional (3D) image reconstruction (optical tomography). Optical tomography is based on delivering light in the near-infrared range (NIR), typically a laser beam through optical fibers, to several locations on the surface and measuring transmitted and/or backreflected intensities to yield 3D quantitative tomographic images of the internal optical properties of organs (Hielscher et al., 2002). Besides diffuse optical tomography (DOT), various other names are commonly used, such as photon migration tomography (PMT), medical optical tomography (MOT), fluorescence-mediated tomography (FMT), computed tomography–laser mammography (CTLM), or simply optical tomography (OT) (Hielscher et al., 2002).

The image-reconstruction problem in DOT suffers from the strong scattering of near-infrared photons in biological tissue. Unlike in X-ray-based imaging, the probing light does not propagate on a straight line from the source to the detector. Therefore, standard backprojection algorithms, as employed in X-ray-based computerized tomography (CT), have limited applicability, and more complex image-reconstruction algorithms must be employed for image reconstruction (Hielscher et al., 2002). Generally, DOT uses a theoretic model (typically, a numeric or analytic solution of the diffusion equation) to describe the propagation of photons into diffuse media and to predict the measurements of the experimental arrangement (forward problem). Then, inversion methods, which are based on this forward model, reconstruct the optical properties of the breast under investigation by operating on a set of light measurements that are taken through this tissue (Hawrysz and Sevick-Muraca, 2000).

In comparison with 2D techniques, optical tomography offers superior quantification accuracy and can yield quantified 3D determination of absorption, scattering, vascularization, oxygenation, and contrast agent uptake in either fluorescence or absorption mode (Ntziachristos and Chance, 2001).

Similar to 2D operating devices, three types of tomographic systems are used for clinical and experimental applications: steady-state (continuous) domain (CW) systems, as well as time-domain (TD) and frequency-domain systems (Hebden et al., 1997; Hielscher et al., 2002). Recently, a computed tomographic laser light-based CW scanner for the breast (computed tomography–laser mammography; CTLM Scanner Model 1020, Imaging Diagnostic Systems, Inc., Ft. Lauderdale, FL), was applied for optical mammography in patients presenting with various breast lesions (Floery et al., 2005). In this study, 100 patients with 105 known breast lesions were examined. The histopathology of all lesions was confirmed using biopsy or surgery. Using CTLM, increased absorption of laser light, related to increased vascularization causing elevated tissue hemoglobin concentration, was observed significantly more often in malignant than in benign breast tumors (70% vs. 32.7%, $P = 0.028$). Moreover, Grades II (71.4%) and III (76.5%) cancers more often showed increased absorption than Grade I cancers (50.0%). However, a variety of benign lesions (fibroadenomas, papillomas) and high-risk lesions (i.e., atypical ductal hyperplasia) also demonstrated increased absorption with CTLM. This is not surprising because an overlap in the microvessel density between benign and malignant lesions has been reported in the literature. Moreover, some of the examined carcinomas (15/50, 30.0%) did not show at all increased absorption with CTLM. Thus, these lesions would have been missed using CTLM alone (Floery et al., 2005).

The authors therefore conclude that diffuse optical tomography may be a useful diagnostic adjunct to X-ray mammography, providing supplemental tissue information. However, as a "stand-alone-modality," nonenhanced diffuse optical tomography is limited with respect to sensitivity and specificity depending on tumor histology (Floery et al., 2005). These findings are in line with several observations made in the 1980s and 1990s. Whereas few studies in the 1980s reported favorable results for optical mammography compared with X-ray mammography (Ntziachristos and Chance, 2001), other studies discouraged the use of nonenhanced optical mammography mainly due to poor sensitivity and specificity. Because of this limited sensitivity and specificity, optical mammography based on intrinsing tissue contrast cannot compete with X-ray mammography. However, experimental data suggest that both the sensitivity and specificity can be improved dramatically by application of optic contrast agents.

Nonspecific Contrast Agents (Perfusion-type Contrast Agents)

Optical imaging using fluorescent contrast agents is a newly emerging imaging tool for breast cancer imaging. Similar to CT or MRI, injected fluorophores can be detected in vivo such as Gd-chelates in MRI or iodinated molecules in CT. Fluorochromes for optical imaging should ideally fulfill the following criteria in order to be suitable for *in vivo* diagnostic imaging:

1. Excitation and emission maxima in the NIR range (650–950 nm).
2. High extinction coefficient.
3. High quantum yield (i.e., strong fluorescence signal).
4. No photo-bleaching.
5. No photosensitizing effects.
6. Hydrophilicity.
7. Low toxicity.

Cyanine dyes (CDs) represent one of the most prominent classes of optical contrast agents with ideal properties for *in vivo* imaging. The chromophore's optical properties are adjustable, offering high-extinction coefficients with the desired absorption and emission ranges throughout the visible to the near-infrared range (Bornhop *et al.*, 2001). At the same time, CD can be conjugated to specific ligands, which can impart molecular specificity within the optical probes (see below). Generally, fluorophores can be divided into three subtypes depending on their contrasting mechanism: (1) nonspecific perfusion agents, (2) target-specific fluorochromes, and (3) so-called smart or protease-sensing probes (Fig. 141).

Indocyanine green (ICG, cardiogreen) is one example of a nonspecific fluorochrome that has been used clinically for many years for hepatic function testing, cardiac physiology tests, and so on (Schad *et al.*, 1977). Experimental studies have proven the feasibility of tumor detection using nonspecific CD fluorochromes (Reynolds *et al.*, 1999; Licha *et al.*, 2000). Moreover, in a clinical study on patients with various breast lesions, ICG proved to be an efficient tool for lesion detection when compared with Gd-enhanced MRI. In this study, lesions seen by Gd-DTPA-enhanced MRI could also be detected by ICG-enhanced diffuse optical tomography (DOT), thus providing the first proof of concept that contrast-enhanced DOT can be utilized for detection of breast neoplasms (Ntziachristos *et al.*, 2000). Because ICG rapidly binds to albumin, ICG-enhanced DOT simply shows perfusion/accumulation effects with no additional molecular information on the tissue. Parameters such as fast uptake by liver tissue, a small quantum yield for fluorescence, and low stability in watery solutions make ICG less suitable as an optical contrast agent for detection of breast tumors (Riefke *et al.*, 1997).

NIR 96010 (Indotricarbocyanine-diglucamine, Schering, Germany) is one example of recently synthesized derivatives of ICG with improved photophysical and pharmacological characteristics (Riefke *et al.*, 1997; Licha *et al.*, 2000; Boehm *et al.*, 2001). In comparison to indocyanine green (ICG), NIR96010 shows low-plasma protein binding of only 10 (ICG, 95–100%) and a low-partition coefficient of 0.005 (ICG = ∞). The fluorescence maximum in plasma

(790 nm) is in a shorter spectral range compared to ICG (829 nm). Fluorescence quantum yield in plasma is 7.6% (ICG, 3–5%). Experimental data suggest that this perfusion-type optical tracer is capable of discriminating between different degrees of tumor vascularity (Wall *et al.*, 2005). In xenograft models of human breast cancer using cell lines with different degree of angiogenesis, the amount of fluorochrome retention closely correlated with surrogate markers of angiogenesis such as the microvessel density or the expression level of the vascular endothelial growth factor (VEGF). It is thus conceivable that optical imaging with perfusion-type agent may be capable of extracting relevant biological information (e.g., degree of tumor vascularization). However, compared to CT or MRI, optical imaging, especially in the NIR, is several orders of magnitude more sensitive to detect very low concentrations of fluorophores. Therefore, optical methods can potentially be exploited for detecting molecular signatures of breast cancer noninvasively. In this context, the recent development of optical

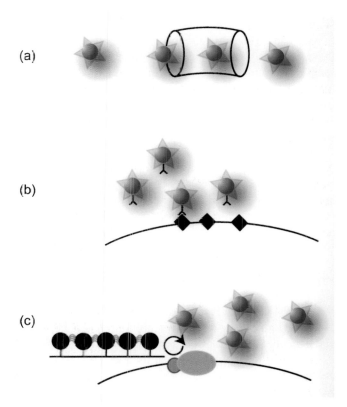

(a)

(b)

(c)

Figure 141 Design of different optical contrast agents. Perfusion-type contrast agents (a) demonstrate simple perfusion and/or permeability properties of the tissue. Targeted probes (b) bind via specific ligands to receptors on cell surface. "Smart" probes (c) are activated by an enzymatic conversion from their native state (little to no signal) to become brightly fluorescent after enzymatic cleavage. *Source:* Reprinted with permission from Bremer *et al.*, Optical based molecular imaging: contrast agents and potential medical applications. *European Radiology* 2003, 13:231–243. Copyright Springer Verlag.

probes with molecular specificity holds great promise to significantly enhance the diagnostic sensitivity and specificity for breast cancer.

Fluorochromes with Molecular Specificity

Fluorochromes with molecular specificity can be divided into smart and target-specific optical probes (Fig. 141).

Smart Probes

One way to impart molecular specificity into optical contrast agents is to design activatable probes, which undergo conformational changes after interaction with a specific enzyme ("smart" probes) (Weissleder et al., 1999). The first generation of this type of molecular contrast agent consisted of a long circulating macromolecular carrier molecule (polylysine backbone) shielded by multiple methoxypolythylene-glycol sidechains (PLL-MPEG). Multiple cyanine dyes (e.g., Cy 5.5) are loaded onto this carrier molecule, resulting in a fluorescence signal quench due to fluorescence resonance energy transfer (FRET) among the fluorochromes (Weissleder et al., 1999). Thus, in the native state background, fluorescence is minimal, whereas after enzymatic cleavage the fluorochromes are released and a strong fluorescence signal can be detected. This first generation of protease-sensing optical probes is activated by lysosomal cystein or serin proteases such as cathepsin-B (Weissleder et al., 1999). Using this probe, subcutaneously implanted tumors even in the submillimeter range could easily be visualized in nude mice 24 hr after intravenous injection of the probe.

Remarkably, multiple enzymes, especially proteases, are involved in a whole variety of pathologies ranging from carcinogenesis to immune diseases (Edwards and Murphy, 1998). In breast cancer specifically, high-expression levels of thiol proteases such as cathepsin-B have been linked to highly aggressive tumors and poor clinical outcome (Barbi et al., 1994). Thus, protease-sensing optical probes can be envisioned to serve as a sensitive tool for early cancer detection or cancer screening tests, respectively. The selectivity of this smart optical probe can be tailored to other enzymes by insertion of specific peptide stalks between the carrier and the fluorochromes. Using this approach, smart optical probes have been developed for targeting, for example, matrix-metalloproteinase-2, thrombin, or caspases (Bremer et al., 2001; Tung et al., 2002; Messerli et al., 2004). Based on experimental and clinical data, smart enzyme-sensing probes should not only allow for early tumor detection but also for lesion differentiation because highly invasive cancers have been linked to a higher load of proteases (McCarthy et al., 1999). In fact, experimental studies revealed a stronger degree of probe activation (and thus a higher fluorescence signal) in highly aggressive breast cancers expressing higher levels of cathepsin-B compared to well-differentiated breast tumors with lower loads of the target enzyme (Bremer et al., 2002). Besides tumor xenografts, efficient probe activation, and therefore tumor delineation, could be observed in a spontaneous breast cancer model that more closely resembles the clinical scenario of carcinogenesis (Fig. 142) (Bremer et al., 2005). Smart probes have also been applied for monitoring anticancer therapies that aim at enzyme inhibition of tumor-associated proteases (Bremer et al., 2001). These studies suggested that sufficient enzyme inhibiton could be monitored early and noninvasively.

While the "smart probes" outlined here are interesting agents for an *in vivo* molecular imaging, the probe design is rather complex and the molecules are large (~ 500 kD). Therefore, these probes exhibit a prolonged blood half-life, which is not desirable for clinical applications.

Thus, in recent years, new smart probes with reduced molecular size have been developed, which are currently under investigation for *in vivo* applications. These probes consist of an NIR fluorochrome linked to a "acceptor" molecule (e.g., second fluorochrome or quencher) through an enzyme-sensitive linker. The quenching-dequenching mechanism is basically identical to the probes outlined earlier. Different low-molecular-weight smart probes have been designed for visualization of caspase-activity (Bullok and Piwnica-Worms, 2005) and protein-kinase A-activity (Law et al., 2005). While the *in vitro* results of these small protease-sensing probes are promising, *in vivo* applications may be more difficult since rapid clearance of the probes may counteract sufficient probe accumulation at the target of interest.

Targeted Probes

Besides smart enzyme-sensing probes, a whole variety of other target-specific fluorochromes have been described that allow the sensitive detection of cellular proteins such as cell receptors, which are involved in breast cancer development (e.g., epidermal growth factor [EGF] receptor) (Ke et al., 2003). These probes consist of a fluorophore with excitation and emission in the NIR (e.g., Cy 5.5) and an affinity ligand such as an antibody, antibody fragment, a peptide, or a nonpeptidic lead substance (Becker et al., 2001). In the initial phase, after probe application, unbound targeted fluorochromes may spoil the signal-to-noise ratio because they contribute to the background signal. However, after washout of the unbound probes, the SNR increases substantially.

By conjugation of flourescein and carbocyanine dyes to various peptide motifs, tumor-associated cell receptors such as the somatostatin, bombesin, vasointestinal peptide, folate, and the EGF receptor could be selectively targeted in experimental studies (Achilefu et al., 2002; Moon et al., 2003). Molecular markers of endothelial proliferation such as the $\alpha v \beta_3$ integrin could be successfully visualized by linking

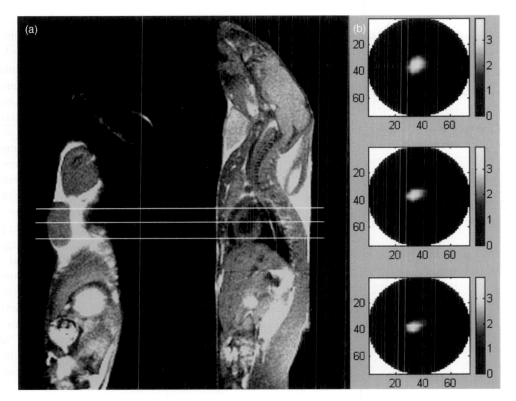

Figure 142 Application of a cathepsin-sensing probe for *in vivo* tumor detection. Detection of spontaneous mammary cancer in mice using fluorescence mediated tomography (FMT). The mouse was injected with a cathepsin-sensing optical probe. FMT images were obtained at the cross sections illustrated in the corresponding sagittal MR images (a). The injected optical probe accumulated strongly in the tumor and yielded strong fluorescence signal that could be detected in the corresponding axial FMT slice (b).
Source: Reprinted with permission from: Bremer *et al.*, Optical imaging of spontaneous breast tumors using protease sensing "smart" optical probes, *Invest. Radiology* 2005, *40*:321–327, copyright: Lippincott Williams & Wilkins.

RGD peptide sequences to cyanine dyes (Gurfinkel *et al.*, 2005). Moreover, the conjugation of a cyanin dye to more complex proteins such as Annexin V allowed fluorescent-labeling apoptotic or necrotic cells and tissues, facilitating an early assessment of treatment response to chemotherapy (Ntziachristos *et al.* 2004). Antibody-based targeted fluorochromes have been designed to visualize L-selectin expression as an early proinflammatory marker (Licha *et al.*, 2005). A bisphosphonate-NIR dye-conjugate showed distinct binding affinity to hydroxylapatite (HA), allowing for the detection of osteoblastic activity *in vivo* (Zaheer *et al.*, 2001). More recently, small nonpeptidic molecules (i.e., drug precursors) could be labeled with cyanine dyes through polyethylene glycol (PEG) linkers and still preserve the binding affinity of the ligands (Hoeltke *et al.*, 2005). Furthermore, the affinity of the single-ligand steric compostition can greatly influence the target affinity of a probe. In this respect multimeric affinity ligands may substantially increase the ability of the optical probe to bind to the target structure.

Generally, potential clincal indications for the application of targeted optical probes arise from the specific molecular structure that can be displayed by the fluorochrome. Imaging of certain tumor cell receptors (e.g., EGF receptor) may potentially allow disease detection at an early stage, to stratify tumor types for novel molecular-targeted therapies, and to assess treatment effects by targeting cellular apoptosis substantially earlier as compared to conventional response markers (e.g., tumor size regression).

Multimodality Probes

Because optical imaging is a highly sensitive imaging method but limited with respect to anatomical resolution, a combination, for example, of an optical and an MRI tracer is an interesting concept not only for colocalization but also for combining, for example, preoperative MR imaging with intraoperative optical imaging (Kircher *et al.*, 2003). Josephson *et al.* (2002) described a hybrid iron-oxide-based nanoparticle that was conjugated with a fluorochrome (Cy 5.5). The surface of the nanoparticles was covered with aminated, cross-linked dextran that allowed covalent binding of Cy 5.5 via protease-sensitive (or protease-resistant) peptides. This probe could successfully be applied for imaging lymph nodes in a mouse model by both MRI as well as near-infrared fluorescence reflectance imaging (FRI). Modification of this probe sensitive to Annexin V (marker of apoptosis) was presented by Schellenberger *et al.* (2004).

Outlook

The impact of developing molecular optical imaging, and in particular molecular DOT, of the breast is potentially enormous. First, selected molecular activity can be achieved with high sensitivity because background fluorescence is quenched. Second, cancers can be detected at their molecular onset before anatomic changes become apparent. Therefore, therapies can be initiated at a very early stage, which is the single most important strategy in achieving high survival rates. Third, specific cancer parameters such as growth kinetics, angiogenesis growth factors, tumor cell markers, and genetic alterations could be studied without perturbing the tumor environment. Finally, this additional information could aid in the development of novel targeted drugs and therapies, and could allow assessment of their efficacy at the molecular level.

The importance of this imaging strategy is further amplified by considering that photon technology can detect single photons, so that it can resolve fluorescent molecules at nanomolar to picomolar concentrations, and requires instrumentation that is of relatively low cost and that uses nonionizing radiation.

References

Achilefu, S., Jimenez, H.N., Dorshow, R.B., Bugaj, J.E., Webb, E.G., Wilhelm, R.R., Rajagopalan, R., Johler, J., and Erion, J.L. 2002. Synthesis, *in vitro* receptor binding, and *in vivo* evaluation of fluorescein and carbocyanine peptide-based optical contrast agents. *J. Med. Chem.* 45:2003–2015.

Barbi, G.P., Margallo, E., Margiocco, M., Paganuzzi, M., Marroni, P., Costanzi, B., Gatteschi, B., Tanara, G., Spina, B., and Nicolo, G. 1994. Evaluation of cathepsin D as prognostic predictor in breast cancer. *Oncology* 51:329–333.

Becker, A., Hessenius, C., Licha, K., Ebert, B., Sukowski, U., Semmler, W., Wiedenmann, B., and Grotzinger, C. 2001. Receptor-targeted optical imaging of tumors with near-infrared fluorescent ligands. *Nat. Biotechnol.* 19:327–331.

Boehm, T., Hochmuth, A., Malich, A., Reichenbach, J.R., Fleck, M., and Kaiser, W.A. 2001. Contrast-enhanced near-infrared laser mammography with a prototype breast scanner. *Invest. Radiol.* 36:573–581.

Bornhop, D.J., Contag, C.H., Licha, K., and Murphy, C.J. 2001. Advance in contrast agents, reporters, and detection. *J. Biomed. Opt.* 6:106–110.

Bremer, C., Tung, C., Bogdanov, A., Jr., and Weissleder, R. 2002. Imaging of differential protease expression in breast cancers for detection of aggressive tumor phenotypes. *Radiology* 222:814–818.

Bremer, C., Ntziachristos, V., Weitkamp, B., Theilmeier, G., Heindel, W., and Weissleder, R. 2005. Optical imaging of spontaneous breast tumors using protease sensing "smart" optical probes. *Invest. Radiol.* 40:321–327.

Bremer, C., Tung, C., and Weissleder, R. 2001. *In vivo* molecular target assessment of matrix metalloproteinase inhibition. *Nat. Med.* 7:743–748.

Bullok, K., and Piwnica-Worms, D. 2005. Synthesis and characterization of a small, membrane-permeant, caspase-activatable far-red fluorescent peptide for imaging apoptosis. *J. Med. Chem.* 48:5404–5407.

Edwards, D.R., and Murphy, G. 1998. Cancer: proteases—invasion and more. *Nature* 394:527–528.

Floery, D., Helbich, T.H., Riedl, C.C., Jaromi, S., Weber, M., Leodolter, S., and Fuchsjaeger, M.H. 2005. Characterization of benign and malignant breast lesions with computed tomography laser mammography (CTLM): initial experience. *Invest. Radiol.* 40:328–335.

Gurfinkel, M., Ke, S., Wang, W., Li, C., and Sevick-Muraca, E.M. 2005. Quantifying molecular specificity of alphavbeta 3 integrin-targeted optical contrast agents with dynamic optical imaging. *J. Biomed. Opt.* 10:034019.

Hawrysz, D.J., and Sevick-Muraca, E. 2000. Developments toward diagnostic breast cancer imaging using near-infrared optical measurements and fluorescent contrast agents. *Neoplasia* 2:388–417.

Hebden, J.C., Arridge, S.R., and Delpy, D.T. 1997. Optical imaging in medicine. 1. Experimental techniques. *Phys. Med. Biol.* 42:825–840.

Hielscher, A.H., Bluestone, A.Y., Abdoulaev, G.S., Klose, A.D., Lasker, J., Stewart, M., Netz, U., and Beuthan, J. 2002. Near-infrared diffuse optical tomography. *Dis. Markers* 18:313–37.

Hoeltke, C., von Wallbrunn, A., Heindel, W., Schaefers, M., and Bremer, C. 2005. A fluorescent photoprobe for the imaging of endothelin receptors: 4th annual meeting of the Society for Molecular Imaging, September 7–10, 2005, Köln. *Molecular Imaging* 4:234.

Josephson, L., Kircher, M.F., Mahmood, U., Tang, Y., and Weissleder, R. 2002. Near-infrared fluorescent nanoparticles as combined MR/optical imaging probes. *Bioconjug. Chem.* 13:554–560.

Ke, S., Wen, X., Gurfinkel, M., Charnsangavej, C., Wallace, S., Sevick-Muraca, E.M., and Li, C. 2003. Near-infrared optical imaging of epidermal growth factor receptor in breast cancer xenografts. *Cancer Res.* 63:7870–7875.

Kircher, M.F., Mahmood, U., King, R.S., Weissleder, R., and Josephson, L. 2003. A multimodal nanoparticle for preoperative magnetic resonance imaging and intraoperative optical brain tumor delineation. *Cancer Res.* 63:8122–8125.

Law, B., Weissleder, R., and Tung, C.H. 2005. Mechanism-based fluorescent reporter for protein kinase A detection. *Chembiochem.* 6:1361–1367.

Licha, K., Debus, N., Emig-Vollmer, S., Hofmann, B., Hasbach, M., Stibenz, D., Sydow, S., Schirner, M., Ebert, B., Petzelt, D., Buhrer, C., Semmler, W., and Tauber, R. 2005. Optical molecular imaging of lymph nodes using a targeted vascular contrast agent. *J. Biomed. Opt.* 10:041205.

Licha, K., Riefke, B., Ntziachristos, V., Becker, A., Chance, B., and Semmler, W. 2000. Hydrophilic cyanine dyes as contrast agents for near-infrared tumor imaging: synthesis, photophysical properties and spectroscopic *in vivo* characterization. *Photochem. Photobiol.* 72:392–398.

McCarthy, K., Maguire, T., McGreal, G., McDermott, E., O'Higgins, N., and Duffy, M.J. 1999. High levels of tissue inhibitor of metalloproteinase-1 predict poor outcome in patients with breast cancer. *Int. J. Cancer* 84:44–48.

Messerli, S.M., Prabhakar, S., Tang, Y., Shah, K., Cortes, M.L., Murthy, V., Weissleder, R., Breakefield, X.O., and Tung, C.H. 2004. A novel method for imaging apoptosis using a caspase-1 near-infrared fluorescent probe. *Neoplasia* 6:95–105.

Moon, W.K., Lin, Y., O'Loughlin, T., Tang, Y., Kim, D.E., Weissleder, R., and Tung, C.H. 2003. Enhanced tumor detection using a folate receptor-targeted near-infrared fluorochrome conjugate. *Bioconjug. Chem.* 14:539–545.

Ntziachristos, V., and Chance, B. 2001. Probing physiology and molecular function using optical imaging: applications to breast cancer. *Breast Cancer Res.* 3:41–46.

Ntziachristos, V., Schellenberger, E.A., Ripoll, J., Yessayan, D., Graves, E., Bogdanov, A. Jr., Josephson, L., and Weissleder, R. 2004. Visualization of antitumor treatment by means of fluorescence molecular tomography with an annexin V-Cy5.5 conjugate. *PNAS 101*:12294–12299.

Ntziachristos, V., Yodh, A.G., Schnall, M., and Chance, B. 2000. Concurrent MRI and diffuse optical tomography of breast after indocyanine green enhancement. *PNAS 97*:2767–2772.

Reynolds, J.S., Troy, T.L., Mayer, R.H., Thompson, A.B., Waters, D.J., Cornell, K.K., Snyder, P.W., and Sevick-Muraca, E.M. 1999. Imaging of spontaneous canine mammary tumors using fluorescent contrast agents. *Photochem. Photobiol. 70*:87–94.

Riefke, B., Licha, K., and Semmler, W. 1997. Contrast media for optical mammography. In German. *Radiologe 37*:749–755.

Schad, H., Brechtelsbauer, H., and Kramer, K. 1977. Studies on the suitability of a cyanine dye (Viher-Test) for indicator dilution technique and its application to the measurement of pulmonary artery and aortic flow. *Pflugers Arch. 370*:139–144.

Schellenberger, E.A., Sosnovik, D., Weissleder, R., and Josephson, L. 2004. Magneto/optical Annexin V, a multimodal protein. *Bioconjug. Chem. 15*:1062–1067.

Tung, C.H., Gerszten, R., Jaffer, F.A., and Weissleder, R. 2002. A novel near-infrared fluorescence sensor for detection of thrombin activation in blood. *Chembiochem. 3*:207–211.

Wall, A., Licha, K., Schirner, M., von Wallbrunn, A. Heindel, W., and Bremer, C. 2005. Differentiation of angiogenetic tumor burden in human breast cancer xenografts using a new perfusion type optical tracer. *Radiology Supplement*, 91st RSNA Meeting Program, 213.

Weissleder, R. 2001. A clearer vision for *in vivo* imaging. *Nat. Biotechnol. 19*:316–317.

Weissleder, R., Ting, C.H., Mahmood U., and Bogdanov, A. Jr., 1999. *In vivo* imaging of tumors with protease-activated near-infrared fluorescent probes. *Nat. Biotechnol. 17*:375–378.

Yodh, A.G., and Chance, B. 1995. Spectroscopy and imaging with diffusing light. *Physics Today 48*:34–40.

Zaheer, A., Lenkinski, R.E., Mahmood, A., Jones, A.G., Cantley, L.C., and Frangion, J.V. 2001. *In vivo* near-infrared fluorescence imaging of osteoblastic activity. *Nat. Biotechnol. 19*:1148–1154.

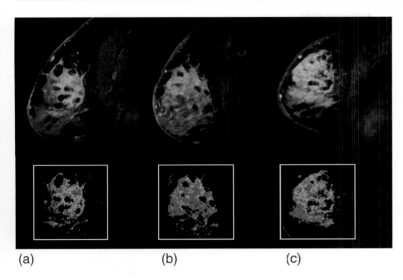

(a)　　　　　　(b)　　　　　　(c)

Figure 145 A 41-year-old female subject with invasive ductal carcinoma, Grade III, studied with MRI while undergoing neoadjuvant chemotherapy treatment. Shown are a representative sagittal contrast-enhanced MRI image (top) with corresponding SER map of the segmented tumor region (bottom): before initiation of chemotherapy (a), after one cycle of chemotherapy (b), and following completion of four cycles of chemotherapy (c). This woman presented with a large initial MRI tumor volume of 71 cm³ (6.2 cm diameter) and exhibited increasing tumor volume throughout treatment (28% overall increase). The vascular parameters of her tumor did not reflect a good treatment response: PE did not change significantly with treatment, SER increased 23% by the end of treatment, and although the washout volume (red) showed a promising 24% drop after one cycle, it subsequently increased to 65% above the initial value by the end of treatment. On pathology at the time of surgery, 8 cm of residual disease and nine involved lymph nodes were identified. This subject experienced disease recurrence 8 months after surgery.

changes in tumor size, and correspond to tumor regression at the end of treatment. For the subject in Figure 143, who was judged to have a good clinical response, early decreases were observed in SER, PE, and washout volume (regions indicated in red) of the tumor after one cycle of chemotherapy. In contrast, the subject in Figure 144 who was a poor responder demonstrated little change in these vascular parameters with treatment. The MRI signal intensity-time curve demonstrates the absorption and clearance of the contrast agent, and represents vascular permeability and blood perfusion in the tissue. Rapid uptake and washout of contrast are recognized characteristics of invasive malignancies, as a result of tumor angiogenic properties. Effective therapy leads to a decrease in tumor vascular permeability, which is reflected by a decrease in the rate of enhancement (uptake). A subsequent regression of vessel density in this region may cause a decrease in the overall signal intensity. Further therapeutic effects in the tissue tend to result in accumulation or reduced washout of contrast from the interstitial space, represented by a persistent pattern of enhancement. These functional vascular changes in the tumor are therefore reflected in the signal intensity-time curve, which transforms from a rapid rise and washout to a slow rise and persistent enhancement (Knopp et al., 2003).

Conclusions

Effective measures of treatment response are essential in order to tailor regimens and maximize patient benefit from neoadjuvant chemotherapy. Clinical assessment of tumor regression during treatment is currently the most common method of ascertaining response and treatment effectiveness. However, clinical examinations lack precision and are somewhat subjective. The size of residual disease in the breast measured by pathology is also a valuable predictor for patient survival, but this end-point measure can only be obtained at the time of surgery and cannot be used to improve treatment planning. Alternatively, contrast-enhanced MRI has demonstrated promising value for monitoring treatment and may predict outcome to therapy earlier and more accurately than other methods.

In clinical practice, lesion size is typically measured by longest diameter on palpation, imaging, and pathology assessments. Moreover, current RECIST recommendations to assess treatment response in solid tumors are based on the change in the longest diameter of the tumor (Therasse et al., 2000). However, one-dimensional characterization of tumor response may be less sensitive than 3D volumetric measurements. Functional tumor-volume measures, calculated as described by applying enhancement criteria on a pixel-by-pixel basis to segment tumors in three-dimensional MRI data sets, more accurately capture the extent of irregularly shaped tumors, multifocality, and diffuse shrinkage of lesions during treatment. Indeed, tumor volume measured on MRI using this technique shows stronger association with the length of recurrence-free survival than one-dimensional tumor diameter measured by MRI, clinical examination, or pathology, and is potentially more predictive of treatment outcome.

Contrast-enhanced MRI also allows characterization of the vascular properties of breast tumors, and changes in tumor vascularity with treatment may precede detectable changes in tumor size. This is particularly important for evaluating new antiangiogenic treatments that work to control cancer by preventing the formation of neovasculature, but may not cause substantial reductions in tumor size. Consequently, monitoring the functional effects of these drugs requires imaging-based

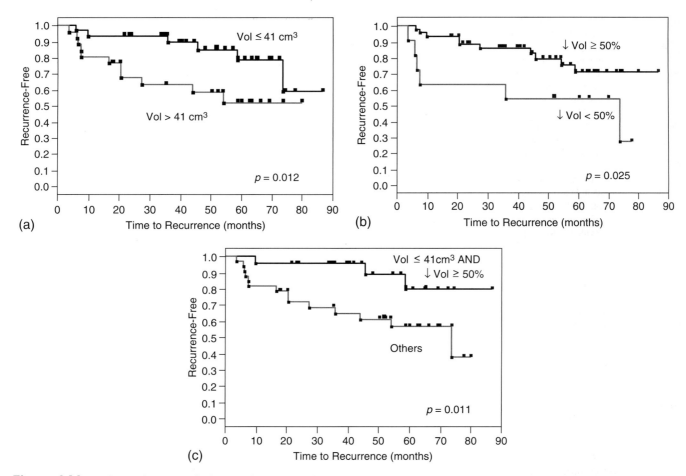

Figure 146 Kaplan-Meier curves for length of recurrence-free survival (RFS). Multivariate Cox analysis showed initial MRI tumor volume ($p = 0.005$) and final change in MRI tumor volume ($p = 0.003$) to be the most significant independent predictors. A significant difference in RFS ($p = 0.012$, Wilcoxon test) was observed when subjects were divided based on initial MRI tumor volume (mean = 41 cm³) (a). The 3-year RFS rate was 90% for patients with smaller tumor volumes of 41 cm³ or less ($n = 32$), compared with 60% for those with larger tumors ($n = 26$). When subjects were divided based on the amount of tumor shrinkage they experienced during treatment, there were also significant differences in RFS ($p = 0.025$, Wilcoxon test) (b). The 3-year RFS rate was 85% for patients with 50% or greater reduction in tumor volume ($n = 47$), compared with 55% for those with less than 50% tumor shrinkage ($n = 11$) during treatment. Combining those two significant factors, longer RFS was observed in the group of patients with initial tumor volumes less than 41 cm³ and at least 50% reduction in tumor volume during treatment (92% 3-year RFS, $n = 25$), when compared with all other patients (65% 3-year RFS, $n = 33$), $p = 0.011$, Wilcoxon test (c).

approaches such as contrast-enhanced MRI that reflect changes in tissue vascularity (Padhani, 2002).

In summary, contrast-enhanced MRI can be a valuable tool for monitoring neoadjuvant treatment response by providing sensitive detection of alterations in both tumor size and vascular function for early indication of treatment efficacy. Based on results from initial clinical studies, it is clear that MRI will play an important role in future treatment monitoring and in evaluation of new cancer therapeutics. Larger multi-institutional trials are currently underway to further refine recommendations for using MRI to assess treatment response.

Acknowledgments

The author thanks Dr. Nola Hylton at the University of California, San Francisco, and her research staff for providing MRI illustrations and figures.

References

Abraham, D.C., Jones, R.C., Jones, S.E., Cheek, J.H., Peters, G.N., Knox, S.M., Grant, M.D., Hampe, D.W., Savino, D.A., and Harms, S.E. 1996. Evaluation of neoadjuvant chemotherapeutic response of locally advanced breast cancer by magnetic resonance imaging. *Cancer* 78:91–100.

Boetes, C., Mus, R.D., Holland, R., Barentsz, J.O., Strijk, S.P., Wobbes, T., Hendriks, J.H., and Ruys, S.H. 1995. Breast tumors: comparative accuracy of MR imaging relative to mammography and US for demonstrating extent. *Radiology 197*:743–747.

Bonadonna, G., Valagussa, P., Brambilla, C., Ferrari, L., Moliterni, A., Terenziani, M., and Zambetti, M. 1998. Primary chemotherapy in operable breast cancer: eight-year experience at the Milan Cancer Institute. *J. Clin. Oncol. 16*:93–100.

Degani, H., Gusis, V., Weinstein, D., Fields, S., and Stranc, S. 1997. Mapping pathophysiological features of breast tumors by MRI at high spatial resolution. *Nat. Med. 3*:780–782.

Esserman, L., Hylton, N., George, T., and Weidner, N. 1999a. Contrast-enhanced magnetic resonance imaging to assess tumor histopathology and angiogenesis in breast carcinoma. *Breast J. 5*:13–21.

Esserman, L., Hylton, N., Yassa, L., Barclay, J., Frankel, S., and Sickles, E. 1999b. Utility of magnetic resonance imaging in the management of breast cancer: evidence for improved preoperative staging. *J. Clin. Oncol. 17*:110–119.

Fisher, B., Bryant, J., Wolmark, N., Mamounas, E., Brown, A., Fisher, E.R., Wickerham, D.L., Begovic, M., DeCillis, A., Robidoux, A., Margolese, R.G., Cruz, A.B., Jr., Hoehn, J.L., Lees, A.W., Dimitrov, N.V., and Bear, H.D. 1998. Effect of preoperative chemotherapy on the outcome of women with operable breast cancer. *J. Clin. Oncol. 16*:2672–2685.

Hayes, C., Padhani, A.R., and Leach, M.O. 2002. Assessing changes in tumour vascular function using dynamic contrast-enhanced magnetic resonance imaging. *NMR Biomed. 15*:154–163.

Hylton, N.M. 2001. Vascularity assessment of breast lesions with gadolinium-enhanced MR imaging. *Magn. Reson. Imaging Clin. N. Am. 9*:321–332, vi.

James, K., Eisenhauer, E., Christian, M., Terenziani, M., Vena, D., Muldal, A., and Therasse, P. 1999. Measuring response in solid tumors: unidimensional versus bidimensional measurement. *J. Natl. Cancer Inst. 91*:523–528.

Knopp, M.V., Brix, G., Junkermann, H.J., and Sinn, H.P. 1994. MR mammography with pharmacokinetic mapping for monitoring of breast cancer treatment during neoadjuvant therapy. *Magn. Reson. Imaging Clin. N. Am. 2*:633–658.

Knopp, M.V., Giesel, F.L., Marcos, H., von Tengg-Kobligk, H., and Choyke, P. 2001. Dynamic contrast-enhanced magnetic resonance imaging in oncology. *Top. Magn. Reson. Imaging 12*:301–308.

Knopp, M.V., von Tengg-Kobligk, H., and Choyke, P.L. 2003. Functional magnetic resonance imaging in oncology for diagnosis and therapy monitoring. *Mol. Cancer Ther. 2*:419–426.

Kuerer, H.M., Newman, L.A., Smith, T.L., Ames, F.C., Hunt, K.K., Dhingra, K., Theriault, R.L., Singh, G., Binkley, S.M., Sneige, N., Buchholz, T.A., Ross, M.I., McNeese, M.D., Buzdar, A.U., Hortobagyi, G.N., and Singletary, S.E. 1999. Clinical course of breast cancer patients with complete pathologic primary tumor and axillary lymph node response to doxorubicin-based neoadjuvant chemotherapy. *J. Clin. Oncol. 17*:460–469.

Kuhl, C.K., Mielcareck, P., Klaschik, S., Leutner, C., Wardelmann, E., Gieseke, J., and Schild, H.H. 1999. Dynamic breast MR imaging: are signal intensity time course data useful for differential diagnosis of enhancing lesions? *Radiology 211*:101–110.

Malur, S., Wurdinger, S., Moritz, A., Michels, W., and Schneider, A. 2001. Comparison of written reports of mammography, sonography and magnetic resonance mammography for preoperative evaluation of breast lesions, with special emphasis on magnetic resonance mammography. *Breast Cancer Res. 3*:55–60.

Padhani, A.R. 2002. Dynamic contrast-enhanced MRI in clinical oncology: current status and future directions. *J. Magn. Reson. Imaging 16*:407–422.

Partridge, S.C., Gibbs, J.E., Lu, Y., Esserman, L.J., Sudilovsky, D., and Hylton, N.M. 2002. Accuracy of MR imaging for revealing residual breast cancer in patients who have undergone neoadjuvant chemotherapy. *AJR Am. J. Roentgenol. 179*:1193–1199.

Partridge, S.C., Gibbs, J.E., Lu, Y., Esserman, L.J., Tripathy, D., Wolverton, D.S., Rugo, H.S., Hwang, E.S., Ewing, C.A., and Hylton, N.M. 2005. MRI measurements of breast tumor volume predict response to neoadjuvant chemotherapy and recurrence-free survival. *AJR Am. J. Roentgenol. 184*:1774–1781.

Therasse, P., Arbuck, S.G., Eisenhauer, E.A., Wanders, J., Kaplan, R.S., Rubinstein, L., Verweij, J., Van Glabbeke, M., van Oosterom, A.T., Christian, M.C., and Gwyther, S.G. 2000. New guidelines to evaluate the response to treatment in solid tumors. European Organization for Research and Treatment of Cancer, National Cancer Institute of the United States, National Cancer Institute of Canada. *J. Natl. Cancer Inst. 92*:205–216

Wasser, K., Klein, S.K., Fink, C., Junkermann, H., Sinn, H.P., Zuna, I., Knopp, M.V., and Delorme, S. 2003. Evaluation of neoadjuvant chemotherapeutic response of breast cancer using dynamic MRI with high temporal resolution. *Eur. Radiol. 13*:80–87.

Weatherall, P.T., Evans, G.F., Metzger, G.J., Saborrian, M.H., and Leitch, A.M. 2001. MRI vs. histologic measurement of breast cancer following chemotherapy: comparison with X-ray mammography and palpation. *J. Magn. Reson. Imaging 13*:868–875.

Yeh, E., Slanetz, P., Kopans, D.B., Rafferty, E., Georgian-Smith, D., Moy, L., Halpern, E., Moore, R., Kuter, I., and Taghian, A. 2005. Prospective comparison of mammography, sonography, and MRI in patients undergoing neoadjuvant chemotherapy for palpable breast cancer. *AJR Am. J. Roentgenol. 184*:868–877.

29

Defining Advanced Breast Cancer: ^{18}F-fluorodeoxyglucose-Positron Emission Tomography

William B. Eubank

Introduction

Breast cancer is the most common nonskin cancer and the second leading cause of cancer death in women (Greenlee *et al.*, 2000). There are 40,000 women per year dying of breast cancer in the United States, and most breast cancer victims die of progressive metastatic disease (Greenlee *et al.*, 2000). Because optimal treatment of patients with recurrent breast cancer depends on knowing the true extent of disease, accurate staging of these patients is an important public health problem. This is a useful application of ^{18}F-fluorodeoxyglucose-positron emission tomography (FDG-PET) when it is performed to complement conventional imaging (CI) such as computed tomography (CT), magnetic resonance imaging (MRI), and bone scintigraphy. The additional metabolic information provided by FDG-PET increases the accuracy of detecting recurrent or metastatic lesions (Isasi *et al.*, 2005). This is particularly true in the evaluation of anatomic regions that have been previously treated by surgery or radiation (Eubank *et al.*,

1998) where the discrimination between post-treatment scar and recurrent tumor can be problematic. ^{18}F-fluorodeoxyglucose-positron emission tomography can also significantly impact the choice of treatment, especially in patients with more advanced or recurrent disease (Yap *et al.*, 2001; Eubank *et al.*, 2004).

The recognition that breast cancer is a systemic disease, even in its early stages, led to the current approach to treatment, which combines local measures such as surgery and radiotherapy with systemic treatment (Hortobagyi, 2000). For most clinical trial studies, local failure is defined as any recurrence of tumor in the ipsilateral chest wall or mastectomy scar; regional failure is defined as any recurrence of tumor in the ipsilateral supraclavicular, infraclavicular, axillary, or internal mammary (IM) nodes; and recurrence of tumor in any other site is considered as distant failure (Taghian *et al.*, 2004). In general, systemic therapy is used at almost all disease stages; however, isolated locoregional disease or single sites of metastatic recurrence are also treated with surgery and radiation therapy (Schwaibold *et al.*, 1991; Probstfeld and O'Connell, 1989).

An equally important clinical need is monitoring systemic therapy to assess the success or failure of a particular form of systemic treatment. Many solid tumors respond poorly to systemic therapy; however, breast cancer is one of the more chemotherapy-sensitive solid tumors (Hortobagyi, 2000). Women with locally advanced or metastatic breast cancer can have prolonged remissions (Machiavelli *et al.*, 1998; Feldman *et al.*, 1986). Those who have failed first-line chemotherapy still have a number of reasonable choices for second-line therapy with substantial response rates (Hortobagyi, 2000). In addition, a number of other systemic options are available besides cytotoxic chemotherapy, including hormonal and other biologically targeted therapies (Honig and Swain, 1993; Wakeling *et al.*, 2001). However, the ability to predict and evaluate systemic therapy response in these patients is limited. Because we currently rely on changes in tumor size to assess response, it takes several weeks to months to evaluate efficacy (Husband, 1996; Tannock and Hill, 1992). For therapies that are potentially cytostatic, such as hormonal therapy, it can be impossible to discern tumor response from slow disease progression when relying on anatomically based measures of response. This is an area where biochemical imaging using PET offers significant advantages and where PET is likely to play a clinically important role. The potential of FDG-PET to provide more accurate and earlier detection of breast cancer recurrences as well as a means of improving assessment of treatment response will hopefully translate into more effective treatment strategies and better health outcomes for these patients in the future.

Positron Emission Tomography Principles

Positron Emission Tomography Instrumentation

Positron-electron annihilation after positron emission leads to two opposing 511 keV photons. Positron emission tomographs are designed to detect "coincident" photon pairs along all possible projection lines through the body to reconstruct quantitative maps of tracer concentration. Tomographs primarily collect annihilation photon counts from the patient (emission scans); however, they also use transmission or attenuation scanning to correct for the body's absorption of photon pairs. Traditionally, the transmission scan was performed using a 511 keV source rotating around the patient to perform attenuation correction. More recently, tomographs containing both a positron tomography and an X-ray computed tomography (CT) device have come into widespread use and have replaced older PET-only devices in

many centers (Alessio *et al.*, 2004). Because the lower energy X-ray photons have different photon attenuation than the higher energy annihilation photons from PET, the use of X-ray CT scanning for attenuation correction can introduce bias into the quantitative data produced by PET. However, in clinical practice, the combination of functional information from PET and detailed anatomic information from CT more than offsets this potential pitfall. Increasingly sophisticated algorithms for extrapolating 511 keV attenuation correction information from CT data appear to be able to provide reliable quantitative information, perhaps even in the presence of the iodinated intravenous contrast often used for CT (Alessio *et al.*, 2004). Commercially-available dedicated PET and PET/CT tomographs achieve high sensitivity to annihilation photon pairs using a ring of detectors and blocks of small crystals surrounding the patient. The practical spatial resolution utilizing current instrumentation for body imaging is 5–10 mm, with torso scanning times as low as 20–30 min per patient.

Fluorodeoxyglucose (FDG)

The positron emitter most commonly used in routine clinical applications is F-18 (in the form of [F-18]-FDG). With a nearly 2-hour half-life, FDG can be produced in regional tracer production facilities and shipped to facilities that are within a 1- to 2-hour flight of the production facility. Fluorodeoxyglucose is transported into cells and phosphorylated in parallel to glucose; however, unlike glucose, it is not a substrate for enzymatic reactions beyond phosphorylation. Furthermore, it is not readily dephosphorylated in most tissues, including tumors, and the phosphorylated compound cannot cross cell membranes. Therefore, phosphorylated FDG is "metabolically trapped" in the cell as FDG-6P.

The rate of FDG uptake and trapping is a quantitative indicator of glucose metabolism. Static measures of FDG uptake normalized to the injected dose, frequently referred to as the standard uptake value (SUV), provide an approximate indicator that correlates with FDG metabolism (Spence *et al.*, 1998): $SUV = A/(ID/BW)$, where A is the tissue tracer content (mCi/g), ID is injected dose (mCi), and BW is patient weight (kg). Although less precise than kinetic determinations, SUV is conveniently implemented in a routine clinical setting.

Early studies by Warburg (Warburg, 1956) established that glucose metabolism is elevated in tumors in comparison to normal tissues. The observation that FDG accumulates in most untreated tumors led to the concept that increased FDG uptake reflects increased glucose metabolism in tumors. While this is undoubtedly an important cause of uptake in tumors, some recent work (Spence *et al.*, 1998) has suggested that the handling of FDG relative to glucose is different in tumors versus normal tissue in a way that may increase the prominence of FDG uptake in tumors. Ongoing studies seek to elucidate the nature of FDG uptake in tumors and will provide further insights into the biologic significance of increased FDG uptake in tumors.

Axillary Node Staging

The status of axillary lymph nodes carries important prognostic information and is considered key in choosing breast cancer treatment. Early studies of FDG-PET for breast cancer staging focused on detecting axillary nodal metastases. They showed sensitivities largely of 85–90% and higher (reviewed in Eubank et al., 2004), generating enthusiasm for FDG-PET as a noninvasive method for axillary staging and eliminating the need for axillary dissection. However, these early studies included a population with more advanced primary tumors than found in typical screened populations. More recent series with a larger fraction of T1 primary tumors show much lower sensitivity for axillary nodal metastases, as low as 50% (Wahl et al., 2004). Another study indicated that the sensitivity of FDG-PET for axillary nodal metastases depended upon both the axillary tumor burden and the FDG uptake in the primary tumor (van der Hoeven et al., 2002). Therefore, the limited sensitivity of FDG-PET for axillary nodal disease in early-stage breast cancer (versus more advanced disease) is likely a function of the lower burden of disease in the axilla, even when metastases exist, and perhaps an indicator that smaller, and possibly less aggressive, lesions have lower FDG uptake. In practice, most centers continue to use surgical staging of the axilla, but limit morbidity through the use of sentinel node lymphatic mapping.

Recent studies comparing preoperative FDG-PET with pathologic results from sentinel lymph node (SLN) biopsy in patients with early-stage breast cancer show sensitivity in a range of 20–50% (reviewed in Eubank et al., 2004), with false-negative FDG-PET occurring predominantly in small-sized (10 mm or less) metastatic sentinel nodes (Barranger et al., 2003). Although recent data do not support the routine use of FDG-PET for axillary staging of early breast cancer, FDG-PET may be complementary to SLN mapping and other standard axillary procedures in patients with more advanced tumors and/or equivocally palpable axillary nodes. One concern in more advanced disease, especially with palpable axillary nodes, is that a SLN "packed" with a large volume of disease may not be visualized at mapping because lymph flow is diverted around it, resulting in a potential false-negative examination (Wagner et al., 1999). A clearly positive FDG-PET in selected patients with a high risk of nodal metastases carries high positive predictive value and may identify patients with evidence of nodal metastases. This could indicate the need for standard axillary nodal dissection or other diagnostic and therapeutic approaches, rather than SLN biopsy. This algorithm for evaluating patients at high risk for axillary metastases may be practical and cost-effective, as suggested by other investigators (Schirrmeister et al., 2001; Wojcik et al., 1997).

Detection of Locoregional and Distant Recurrences

18F-fluorodeoxyglucose-positron emission tomography can contribute in significant ways to the clinical management of patients with suspected locoregional or distant recurrences. Because FDG-PET provides functional information, it is often complementary to conventional staging methods such as physical examination, cross-sectional imaging (CT or MR), and bone scintigraphy, which rely more on changes in morphology to detect disease recurrence. This is particularly true in the evaluation of anatomic regions that have been previously treated by surgery or radiation (Eubank et al., 1998) where the discrimination between post-treatment scar and recurrent tumor can be problematic. Because of its high sensitivity in the detection of metabolically active tissue, FDG-PET can help define the extent of disease when conventional imaging (CI) is equivocal or negative and recurrence is suspected. Earlier recognition of recurrent disease will hopefully provide more effective treatment options and improve survival in this group of patients.

Locoregional Recurrences

Recurrence in the breast, skin of the breast, axillary nodes, chest wall, and supraclavicular nodes are the most common sites of first locoregional recurrence after primary surgical resection (Kamby et al., 1987). In recent years, the shift toward breast-conserving surgery and local radiation therapy for early breast cancer has heightened concern over locoregional recurrence (Schmolling et al., 1997). The incidence of locoregional recurrence after breast conservation treatment ranges from 5–22% (Huston and Simmons, 2005). Independent risk factors associated with locoregional recurrence in this group of patients include positive margins at surgical resection, tumors with extensive intraductal component, high-grade ductal carcinoma in situ (DCIS), patient age under 40 years, and absence of radiation after breast conservation therapy (Huston and Simmons, 2005; Leborgne et al., 1995).

Among patients treated with mastectomy, axillary node dissection, and adjuvant chemotherapy, the most common sites of locoregional recurrence are the chest wall (68% of locoregional recurrences) and supraclavicular nodes (41% of locoregional recurrences) (Katz et al., 2000). Recurrent disease at both of these sites is associated with poor prognosis in terms of survival after recurrence (Aristei *et al.*, 2000). Factors that predict an increased risk of chest wall or supraclavicular node recurrence include four or more positive axillary nodes, tumor size $\geq$ 4 cm, and extranodal extension of $\geq$ 2 mm (Katz *et al.*, 2000). Supraclavicular node recurrence is technically considered stage IV disease and is generally viewed as a harbinger of more widely disseminated disease. However, patients with supraclavicular node involvement as the sole site of disseminated disease may benefit from aggressive local radiotherapy.

One clinical situation in which FDG-PET has been shown to be helpful is in the evaluation of previously treated patients with symptoms of brachial plexopathy. This debilitating condition can be secondary to tumor recurrence in the axilla or chest wall or due to scarring of tissue neighboring the brachial plexus from previous surgery or radiation. Since the signs and symptoms of locoregional recurrence often overlap with the side effects of treatment (Iyer *et al.*, 1996; Bagley *et al.*, 1978) and patients with tumor recurrence may benefit from surgical resection (Toi *et al.*, 1997; Hathaway *et al.*, 1994), it is important to distinguish one from the other. One study (Hathaway *et al.*, 1998) showed the value of combining the functional information of FDG-PET and the anatomic information from dedicated MR imaging to decide whether patients would benefit from further surgery. Other studies (Ahmad *et al.*, 1999) have confirmed these early findings.

Intrathoracic Lymphatic Recurrences

Lymphatic drainage to the internal mammory (IM) nodal chain is an important pathway of spread of disease both at the time of initial diagnosis and after primary treatment of breast cancer. Data from sentinel node lymphoscintigraphy series in patients with early breast cancer at our institution reveals that overall prevalence of drainage to the IM nodes is 17% (Byrd *et al.*, 2001). This is a similar prevalence to that shown in early extended radical mastectomy series where metastasis to IM nodes occurred in close to one in five women with operable (stage II–III) breast cancer (Veronesi *et al.*, 1985; Lacour *et al.*, 1983).

Metastasis to IM nodes can occur from tumor located anywhere in the breast; however, in our series, IM drainage was significantly less frequent in tumors located in the upper outer quadrant (10%) compared with the other three quadrants and subareolar portion of the breast (17–29%) (Byrd *et al.*, 2001). Metastasis to the IM and axillary nodes usually occurs synchronously, but infrequently (4–6% incidence) it may be isolated to the IM chain (Noguchi *et al.*, 1993). The prognosis of patients with IM and axillary nodal metastasis is significantly worse compared with patients with only axillary node disease (Cody and Urban, 1995), suggesting that IM nodal chain is a conduit for more widespread dissemination of disease.

The importance of IM nodal detection and treatment remains controversial (Freedman *et al.*, 2000). Unlike axillary nodes, IM nodes are not routinely biopsied as part of an individual patient's staging work-up and their status is generally unknown. There has been reluctance to biopsy IM nodes because (1) early radiotherapy trials (before the era of routine adjuvant chemotherapy) failed to demonstrate a clear benefit in survival with IM chain radiation and (2) there is a relatively high complication risk (pneumothorax and bleeding) associated with IM nodal sampling. A recent large, prospective, randomized radiotherapy trial has shown a benefit in systemic relapse-free and overall survival from aggressive regional nodal irradiation (including IM field) following lumpectomy or mastectomy, even in patients with fairly limited spread to the axilla (Ragaz *et al.*, 2005). These data suggest that eradication of residual locoregional metastasis (including IM nodal disease) has a strong systemic effect.

Fluorodeoxyglucose uptake in the IM nodal chain has been anectodally reported in some of the studies that have focused on detection of primary tumor or axillary staging (Schirrmeister *et al.*, 2001; Greco *et al.*, 2001). In one study of 85 patients who underwent FDG-PET before axillary node dissection, 12 (14%) had uptake in the IM region, but there was no histologic confirmation of these nodes (Schirrmeister *et al.*, 2001). Our experience with imaging patients with locally advanced breast cancer shows that the prevalence of IM-FDG uptake can be as high as 25% and that the presence of IM-FDG uptake predicts treatment failure patterns of disease consistent with IM nodal involvement and progression (Bellon *et al.*, 2004). An example of FDG uptake in the IM nodal chain is shown in Figure 147. A recent study (Tran *et al.*, 2005) showed that the likelihood of extra-axial lymph node findings on FDG-PET was affected by the position of the primary tumor (medial versus lateral). The presence of extra-axial nodal uptake on FDG-PET, combined with medial tumor location, indicated a high risk of subsequent disease progression. A preliminary study (Bernstein *et al.*, 2000) showed the feasibility of detecting IM nodal metastases in early-stage patients using FDG-PET. This system may prove to be an ideal method of noninvasively staging this important nodal region and may aid in selecting patients who would potentially benefit most from directed IM nodal radiotherapy. However, further work needs to be done to confirm FDG-PET findings with histopathology.

Neoplastic spread to mediastinal nodes is also common in patients with advanced disease and as a site of recurrence in patients who have undergone axillary node dissection and radiation (Fig. 148). As is true of IM nodes, mediastinal nodes are rarely sampled in breast cancer patients. Computed tomography, the conventional method of staging these nodes, relies on size criteria to determine the presence or absence of disease. This method has been proven significantly less accurate than FDG-PET in patients with non-small cell lung cancer where histologic analysis is used as

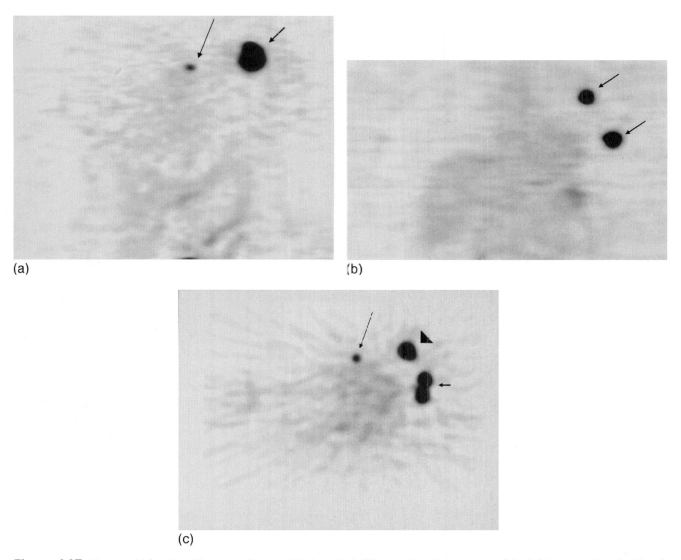

(a)

(b)

(c)

Figure 147 52-year-old female with newly diagnosed high-grade infiltrating ductal carcinoma of the left breast with palpable left axillary lymph nodes. (a) Anterior coronal FDG-PET image shows hypermetabolic focus in the upper medial left breast (short arrow; maximum SUV = 17.8) consistent with the primary tumor and a hypermetabolic focus in the left internal mammary region (long arrow; maximum SUV = 7.2) consistent with internal mammary nodal metastasis. (b) A more posterior coronal image shows intense uptake in several left axillary nodes (arrows; maximum SUV = 22.8). (c) An axial image shows uptake in the primary tumor (arrowhead), axillary nodes (short arrow), and left internal mammary node (long arrow).

the gold standard (Vansteenkiste *et al.*, 1998; Scott *et al.*, 1996). In our retrospective series of 73 patients with recurrent or metastatic breast cancer who underwent both FDG-PET and chest CT (Eubank *et al.*, 2001), FDG uptake in mediastinal or IM nodes was two times more prevalent than suspiciously enlarged nodes by CT. This suggests that PET is a much more sensitive technique for detecting nodal disease. In the subset of patients with confirmation, the sensitivity of FDG-PET was significantly higher (85%) than CT (50%), with nearly the same level of specificity (90% for PET and 83% for CT). Ten of 33 (30%) patients suspected of having only locoregional recurrence by CT and clinical

examination had mediastinal or IM-FDG uptake; risk factors associated with mediastinal or IM-FDG uptake in these patients were recurrent chest wall invasion and three or more positive axillary nodes.

Distant Metastases

The detection of skeletal metastases, the most common site of distant disease in breast cancer, poses additional challenges. Nearly 70% of patients with advanced disease have bone metastasis (Coleman and Rubens, 1987). Breast cancer is one of a several tumor types that can give rise to

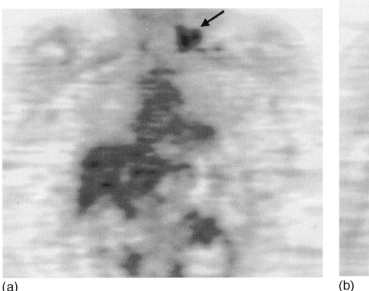
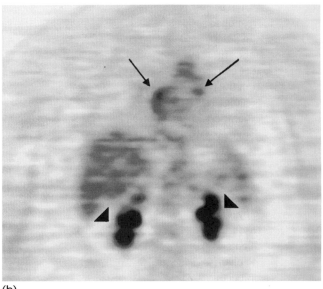

(a)   (b)

Figure 148 62-year-old female with a history of an ER/PR positive, HER-2/neu negative infiltrating lobular carcinoma of the right breast. She underwent a lumpectomy and limited axillary node staging with one positive sentinel node but no further axillary node sampling. She did receive adjuvant chemotherapy, radiation, and 2 years of anti-estrogen therapy. Recent tumor markers were elevated. (a) A coronal FDG-PET image shows uptake in the left supraclavicular region (arrow; maximum SUV = 4.4) consistent with nodal metastases. (b) A more posterior coronal image shows uptake in the mediastinum (arrows; maximum SUV = 3.3) suspicious for nodal metastases in this region. Arrowheads indicate activity in the renal collecting systems.

metastases that are either osteolytic or osteoblastic. ^{18}F-fluorodeoxyglucose-positron emission tomography appears to be complementary to bone scintigraphy, which is the standard procedure used for surveying the skeleton for metastases. One study showed that FDG-PET detected the osteolytic metastases, often missed by bone scintigraphy, while FDG-PET often missed osteoblastic metastases, for which bone scintigraphy has high sensitivity (Cook *et al.*, 1998). Subsequent studies have confirmed these findings (Nakai *et al.*, 2005). Based on these results, the combination of bone scintigraphy and CT remains the standard for distant metastasis staging in most centers, with FDG-PET used to help clarify staging in the cases of difficult or equivocal conventional staging. Evolving data suggest that [F-18]-fluoride-PET may improve bone metastasis detection compared to bone scintigraphy (Schirrmeister *et al.*, 1999) and may play a role in breast cancer bone metastasis staging in the future. In addition, combined FDG-PET/CT may provide "one-stop shopping" for staging, as CT is reasonably sensitive for the osteoblastic lesions missed by FDG-PET and vice versa. Ongoing studies are evaluating the performance of PET/CT for breast cancer staging (Zangheri *et al.*, 2004).

Several studies have shown that whole-body FDG-PET is a highly accurate method compared with CI for restaging patients who have previously undergone primary treatment for breast cancer. For detecting distant metastases, the reported sensitivity and specificity of FDG-PET range from

80–97% and 75–94%, respectively, on a per patient basis (Isasi *et al.*, 2005). One advantage of FDG-PET is that large areas can be surveyed; it is therefore able to evaluate sites of recurrent disease that can be extensive and separated by large anatomic distances. Several investigations have shown the added benefit of FDG-PET to CI in asymptomatic patients with elevated tumor marker serum levels and negative or equivocal CI (Lonneux *et al.*, 2000; Liu *et al.*, 2002; Suarez *et al.*, 2002). Although the ability of FDG-PET to detect skeletal metastases compared to bone scan has been met with mixed success in these series, as previously discussed, FDG-PET has been proven significantly more accurate in the evaluation of nodal disease and equal or superior to CI for visceral metastases (Moon *et al.*, 1998; Gallowitsch *et al.*, 2003; Eubank *et al.*, 2001). A retrospective study of 61 patients (Vranjesevic *et al.*, 2002) showed that FDG-PET is significantly more accurate than a combination of CI studies to predict the outcome (disease-free survival) of patients being reevaluated after primary treatment of breast cancer. These data support the practice at our institution of using FDG-PET as a complement to CI studies for the evaluation of patients with suspected recurrent disease.

Response to Therapy

Breast cancer was one of the first diseases to which FDG-PET was applied to evaluate response to systemic therapy (reviewed in Biersack et al., 2004). Serial FDG-PET as a measure of response has been most extensively applied to the neoadjuvant or presurgical chemotherapy of larger primary tumors. In this setting, FDG-PET estimates of response were compared to histopathologic assessment of response from the post-therapy surgical specimen. Because biochemical changes precede anatomic changes in tumor response to systemic treatment, it is logical to surmise that measuring changes in glucose metabolism might provide an early and more accurate assessment of response than changes in tumor size, the current gold standard.

[18]F-fluorodeoxyglucose-positron emission tomography is also useful for monitoring the response of metastatic breast cancer to therapy. One study (Gennari et al., 2000) showed an average decrease in lesion SUV of 72% following chemotherapy among patients who showed clinical response to treatment compared to no change in nonresponders. The responders also showed an appreciable drop in lesion FDG uptake following the first course of chemotherapy. These results have recently been confirmed by a larger series showing a decline in FDG uptake of 28% after the first cycle and 46% after the second cycle in patients with metastatic breast cancer responding to chemotherapy (Dose Schwarz et al., 2005).

A novel FDG-PET application was used to measure early response of breast cancer to tamoxifen in patients with locally advanced and metastatic breast cancer (Mortimer et al., 2001). In this study the authors reported an increase in FDG uptake 7–10 days after institution of tamoxifen, termed *metabolic flare*, and postulated that this effect was associated with the early agonist response to tamoxifen and predicted subsequent tumor shrinkage. These data show a clear *in vivo* correlation between early post-treatment estrogen receptor (ER) agonist effect and increase in glucose utilization by tumor cells. This is a good example of how PET can be used to characterize tumors *in vivo*, evaluate response to treatment, and provide important prognostic information.

An application of increasing clinical importance is monitoring the response of bone metastases to treatment. Evaluating response to treatment in patients with bone-dominant metastases using conventional imaging, including bone scintigraphy and MR, can be problematic. These methods detect reactive changes in bone adjacent to tumor that may not be a true representation of pathologic response (Nielsen et al., 1991). In a retrospective study, we evaluated the response of skeletal metastases to therapy using serial FDG-PET (Stafford et al., 2002) and found a strong correlation between the quantitative change in FDG-SUV and overall clinical assessment of response (combination of conventional imaging, tumor markers, and clinical examination) and change in tumor marker, CA 27.29. An example of therapeutic response to systemic therapy in a patient with metastatic bone disease is shown in Figure 149. Preliminary analysis has also shown that the change in FDG uptake in serial scans of treated bone-dominant breast cancer predicts

time-to-progression and the likelihood of skeletal-related events (Tam et al., 2005). ^{18}F-fluorodeoxyglucose-positron emission tomography is likely to play an important role in monitoring bone metastasis response to therapy, given the difficulty of evaluating response in patients with bone-dominant breast cancer, the widening array of therapies for this patient group, and the prolonged survival that can be obtained with successful treatment.

Impact of FDG-PET on Patient Management

A distinguishing feature of the management of breast cancer is the need for a multidisciplinary approach for optimal local and regional control of the disease. Many strategies and agents offer patients with advanced breast cancer a therapeutic benefit, including surgery, radiation, chemotherapy, and hormonal therapy. Choosing the most appropriate therapy depends primarily on accurately defining the extent of disease. As of October 2002, PET has been approved for payment by the Center for Medicare and Medicaid Services in the United States for staging or restaging of patients with recurrent or metastatic disease, especially when conventional staging studies are equivocal. Despite this approval, there is relatively little insight into which patients with recurrent or metastatic breast cancer benefit the most from FDG-PET. A previous investigation (Yap et al., 2001) showed that FDG-PET had a major impact on the management of patients with breast cancer who underwent restaging for local or distant recurrences. FDG-PET changed the clinical stage in 36% of patients and induced changes in therapy in 58% of the patients. In our retrospective study of 125 patients with advanced breast cancer undergoing CI and FDG-PET for staging (Eubank et al., 2004), the extent of disease was changed in 67% (increased in 43% and decreased in 24%) of patients and the therapeutic plan was altered in 32% of patients based on FDG-PET findings. This system altered therapy most frequently in two subgroups of patients: (1) patients with suspected or proven locoregional recurrence, under consideration for aggressive local therapy (FDG-PET altered treatment in 44%) and (2) patients with known metastases being evaluated for response to therapy (FDG-PET altered treatment in 33%). These results indicate that FDG-PET should not be used as the sole restaging tool in patients with recurrent or metastatic disease but to answer specific questions that will likely impact their management. Future prospective trials using oncologist-directed questionnaires will help to further define the role and provide data for the cost-benefit analysis of FDG-PET in staging patients with advanced breast cancer.

A few preliminary studies have shown promise for FDG-PET/CT in the evaluation of patients with recurrent or

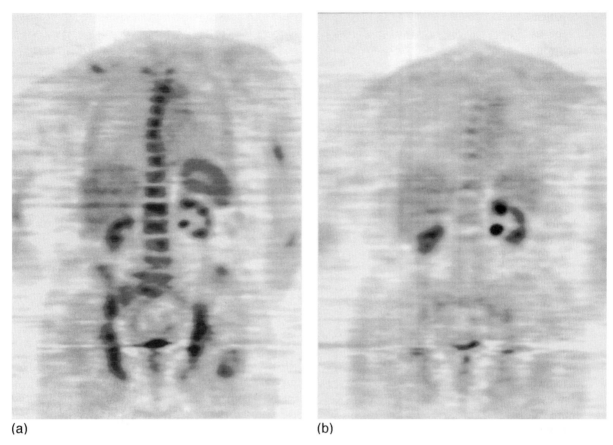

(a)   (b)

Figure 149 67-year-old female with recently diagnosed ER/PR positive infiltrating ductal carcioma of the left breast and widespread skeletal metastases. (a) Coronal FDG-PET image before treatment shows multiple foci of increased activity in the spine, ribs, pelvis, bilateral humeri, and femora consistent with widespread skeletal metastases with a predominant uptake pattern in the medullary space. (b) Coronal image after completing five months of aromatase inhibitor therapy shows complete resolution of hypermetabolic foci consistent with excellent response to therapy.

metastatic breast cancer (Pelosi *et al.*, 2004; Buck *et al.*, 2003). The accurate registration of metabolic and morphologic images from these integrated systems has in turn improved the accuracy in localizing FDG uptake; this advance can potentially lead to an improvement in specificity (reduction in false-positive results) and staging of disease compared to the separate evaluation of CT and PET data. PET/CT also shows promise in radiation treatment planning by modifying target tissue volumes selected for irradiation based on the combination of morphologic and metabolic data (Ciernik *et al.*, 2003). The impact of PET/CT on patient outcomes will need to be addressed in future studies.

Beyond FDG: Future Applications of PET to Breast Cancer

Although FDG continues to play an increasingly important role in the clinical management of breast cancer, it is likely that other PET radiopharmaceuticals will also play a

role in the management of breast cancer in the near future. Energy metabolism is associated not only with tumor growth, but also with a variety of other biological processes, including inflammation and tissue repair in response to damage. As breast cancer treatment becomes more targeted and individualized to a particular patient and her tumor's biologic characteristics, more specific PET radiopharmaceuticals will help guide treatment selection. Positron emission tomography can help guide treatment selection by (1) quantifying the therapeutic target, (2) identifying resistance factors, and (3) measuring early response to therapy. A number of positron-emitting radiopharmaceuticals measuring a number of aspects of breast cancer biology have been developed and tested in patients. We will focus on one of these newer tracers in the following section.

Estrogen Receptor Imaging

One of the most promising applications of using PET with a tracer other than FDG has been in the *in vivo* assessment of

ER status of a patient's tumor burden. This is an example of how PET can be used to quantify the therapeutic target, namely, the ER, which is present in the majority of patients with breast cancer. Estrogen receptor expression in breast cancer is an indicator of prognosis and predicts the likelihood of responding to anti-estrogen therapy. The assessment of ER expression in primary breast cancer by *in vitro* assay of biopsy material, most typically by immunohistochemistry, is part of the standard care of breast cancer patients and weighs heavily in the choice of therapy. A variety of agents has been tested for PET-ER imaging (Katzenellenbogen *et al.*, 1997). The close analog of estradiol, the labeled estrogen. [F-18]- 16 α-[fluoroestradiol] (FES), has shown the most promise in quantifying the functional ER status of breast cancer, either in the primary tumor or in metastatic lesions (Katzenellenbogen *et al.*, 1997). Studies have shown that the quantitative level of FES uptake in primary tumors correlates with the level of ER expression measured by *in vitro* assay by radioligand binding (Mintun *et al.*, 1988) and in preliminary data by immunohistochemistry in our laboratory. [F-18]-16 α-[fluoroestradiol]-PET provides sufficient image quality to image metastatic lesions with high sensitivity in patients with ER positive tumors (Dehdashti *et al.*, 1995) at an acceptable radiation dose to the patient (Mankoff, *et al.*, 2001).

An important use of FES-PET will be to image and characterize the entire volume of disease in an individual patient, especially in patients with recurrent or metastatic breast cancer, where tissue sampling at all sites is not feasible. Studies using FES-PET have shown heterogeneous FES uptake within the same tumor and between metastatic lesions (Dehdashti *et al.*, 1995). This type of comprehensive evaluation of functional ER status of the entire disease burden in patients will likely give important information about prognosis and help guide treatment selection. Positron emission tomography–estrogen receptor imaging can be used, in analogy to assay of ER in biopsy specimens, to predict the likelihood of response to hormonal therapy and thereby guide appropriate selection of patients for this type of treatment. The study by Dehdashti *et al.* (1999) showed that a higher level of FES uptake in advanced tumors predicts a greater chance of response to tamoxifen. Serial FES-PET can also assess the functional response to hormonal therapy, or ER blockade in the case of tamoxifen, in the primary tumor or metastasis (Mortimer *et al.*, 2001). High degrees of ER blockade in the primary tumor (~ 50% decrease in SUV from baseline) portended a good response to therapy. Preliminary results in our center show similar results for patients with recurrent or metastatic breast treated with a variety of hormonal agents (Linden *et al.*, 2004). These exciting preliminary results show the potential of PET-ER imaging to help guide appropriate, individualized breast cancer treatment and point the way for future studies and clinical use.

In conclusion, one of the major strengths of FDG-PET in breast cancer imaging is in the evaluation of patients with suspected locoregional recurrence or distant metastasis. In general, FDG-PET is more sensitive than conventional imaging for the detection of recurrent disease. Because of its ability to more accurately stage patients with advanced breast cancer, FDG-PET has a significant impact on choice of treatment and management in this patient group. Future applications of PET will likely involve other tracers, in addition to FDG, to better characterize tumor biology and more effectively measure response to therapy.

References

Ahmad, A., Barrington, S., Maisey, M., and Rubens, R.D. 1999. Use of positron emission tomography in evaluation of brachial plexopathy in breast cancer patients. *Br. J. Cancer 79*:478–482

Alessio, A.M., Kinahan, P.E., Cheng, P.M., Vesselle, H., and Karp, J.S. 2004 PET/CT scanner instrumentation, challenges, and solutions. *Radiol. Clin. North Am. 42*:1017–1032.

Aristei, C., Marsella, A.R., Chionne, F., Panizza, B.M., Marafioti, L., Mosconi, A.M., Cherubini, R., and Colozza, M. 2000. Regional node failure in patients with four or more positive lymph nodes submitted to conservative surgery followed by radiotherapy to the breast. *Am. J. Clin. Oncol. 23*:217–221.

Bagley, F.H., Walsh, J.W., Cady, B., Salzman, F.A., Oberfield, R.A., and Pazianos, A.G. 1978. Carcinomatous versus radiation-induced brachial plexus neuropathy in breast cancer. *Cancer 41*:2154–2157.

Barranger, E., Grahek, D., Antoine, M., Montravers, F., Talbot, J.N., and Uzan, S. 2003. Evaluation of fluorodeoxyglucose positron emission tomography in the detection of axillary lymph node metastases in patients with early-stage breast cancer. *Ann. Surg. Oncol. 10*:622–627.

Bellon, J.R., Livingston, R.B., Eubank, W.B., Gralow, J.R., Ellis, G.K., Dunnwald, L.K., and Mankoff, D.A. 2004. Evaluation of the internal mammary (IM) lymph nodes by FDG-PET in locally advanced breast cancer (LABC). *Am. J. Clin. Oncol. 27*:407–410.

Bernstein, V., Jones, A., Mankoff, D.A., and Davis, N. 2000. Assessment of internal mammary lymph nodes by fluorodeoxyglucose positron emission (FDG PET) in medial hemisphere breast cancer. *J. Nucl. Med. 41*:289P (abstract).

Biersack, H.J., Bender, H., and Palmedo, H. 2004. FDG-PET in monitoring therapy of breast cancer. *Eur. J. Nucl. Med. Mol. Imaging. 31* (Suppl. 1):S112–117.

Buck, A.K., Wahl, A., Eicher, U., Blumstein, N., Schirrmeister, H., Helms, G., Glatting, G., Neumaier, B., and Reske, S.N. 2003. Combined morphological and functional imaging with FDG-PET/CT for restaging breast cancer—impact on patient management. *J. Nucl. Med. 44*:78P (abstract).

Byrd, D.R., Dunnwald, L.K., Mankoff, D.A., Anderson, B.O., Moe, R.E., Yeung, R.S., Schubert, E.K., and Eary, J.F. 2001. Internal mammary lymph node drainage patterns with breast cancer documented by breast lymphoscintigraphy. *J. Surg. Oncol. 8*:234–240.

Ciernik, F., Dizendorf, E., Baumert, B.G., Reiner, B., Burger, C., Davis, J.B., Lutolf, U.M., Steinert, H.C., and Von Schulthess, G.K. 2003. Radiation treatment planning with an integrated positron emission and computer tomography (PET/CT): a feasibility study. *Int. J. Radiat. Oncol. Biol. Phys. 57*:853–863.

Cody, H.S., 3rd, and Urban, J.A. 1995. Internal mammary node status: a major prognosticator in axillary node-negative breast cancer. *Ann. Surg. Oncol. 2*:32–37.

Coleman, R.E., and Rubens, R.D. 1987. The clinical course of bone metastases from breast cancer. *Br. J. Cancer 55*:61–66.

Cook, G.J., Houston, S., Rubens, R., Maisey, M.N., and Fogelman, I. 1998. Detection of bone metastases in breast cancer by 18FDG PET: differing metabolic activity in osteoblastic and osteolytic lesions. *J. Clin. Oncol. 16*:3375–3379.

Dehdashti, F., Flanagan, F.L., Mortimer, J.E., Katzenellenbogen, J.A., Welch, M.J., and Siegel, B.A. 1999. Positron emission tomographic assessment of "metabolic flare" to predict response of metastatic breast cancer to antiestrogen therapy. *Eur. J. Nucl. Med. 26*:51–56.

Dehdashti, F., Mortimer, J.E., Siegel, B.A., Griffeth, L.K., Bonasera, T.J., Fusselman, M.J., Detert, D.D., Cutler, P.D., Katzenellenbogen, J.A., and Welch, M.J. 1995. Positron tomographic assessment of estrogen receptors in breast cancer: a comparison with FDG-PET and in vitro receptor assays. *J. Nucl. Med. 36*:1766–1774.

Dose Schwarz, J., Bader, M., Jenicke, L., Hemminger, G., Janicke, F., and Avril, N. 2005. Early prediction of response to chemotherapy in metastatic breast cancer using sequential 18F-FDG PET. *J. Nucl. Med. 46*:1144–1150.

Eubank, W.B., and Mankoff, D.A. 2004. Current and future uses of positron emission tomography in breast cancer imaging. *Semin. Nucl. Med. 34*:224–240.

Eubank, W.B., Mankoff, D.A., Bhattacharya, M., Gralow, J., Linden, H., Ellis, G., Lindsley, S., Austin-Seymour, M., and Livingston, R. 2004. Impact of [F-18]-Fluorodeoxyglucose PET on defining the extent of disease and management of patients with recurrent or metastatic breast cancer. *AJR 83*:479–486.

Eubank, W.B., Mankoff, D.A., Schmiedl, U.P., Winter, T.C., 3rd, Fisher, E.R., Olshen, A.B., Graham, M.M., and Eary, J.F. 1998. Imaging of oncologic patients: benefit of combined CT and [F-18]-fluorodeoxyglucose positron emission tomogoraphy scan interpretation in the diagnosis of malignancy. *AJR 171*:1103–1110.

Eubank, W.B., Mankoff, D.A., Takasugi, J., Vesselle, H., Eary, J.F., Shanley, T.J., Gralow, J.R., Charlop, A., Ellis, G.K., Lindsley, K.L., Austin-Seymour, M.M., Funkhouser, C.P., and Livingston, R.B. 2001. 18Fluorodeoxyglucose positron emission tomography to detect mediastinal or internal mammary metastases in breast cancer. *J. Clin. Oncol. 19*:3516–3523.

Feldman, L.D., Hortobagyi, G.N., Buzdar, A.U., Ames, F.C., and Blumenschein, G.R. 1986. Pathological assessment of response to induction chemotherapy in breast cancer. *Cancer Res. 46*:2578–2581.

Freedman, G.M., Fowble, B.L., Nicolaou, N., Sigurdson, E.R., Torosian, M.H., Boraas, M.C., and Hoffman, J.P. 2000. Should internal mammary lymph nodes in breast cancer be a target for the radiation oncologist? *Int. J. Radiat. Oncol. Biol. Phys. 46*:805–814.

Gallowitsch, H.J., Kresnik, E., Gasser, J., Kumnig, G., Igerc, I., Mikosch, P., and Lind, P. 2003. [18F]- fluorodeoxyglucose positron-emission tomography in the diagnosis of tumor recurrence and metastases in the follow-up of patients with breast carcinoma: a comparison to conventional imaging. *Invest. Radiol. 38*:250–256.

Gennari, A., Donati, S., Salvadori, B., Giorgetti, A., Salvadori, P.A., Sorace, O., Puccini, G., Pisani, P., Poli, M., Dani, D., Landucci, E., Mariani, G., and Conte, P.F. 2000. Role of 2-[18F]-fluorodeoxyglucose (FDG) positron emission tomography (PET) in the early assessment of response to chemotherapy in metastatic breast cancer patients. *Clin. Breast Cancer 1*:156–161; discussion 162–163.

Greco, M., Crippa, F., Agresti, R., Seregni, E., Gerali, A., Giovanazzi, R., Micheli, A., Asero, S., Ferraris, C., Gennaro, M., Bombardieri, E., and Cascinelli, N. 2001. Axillary lymph node staging in breast cancer by 2-fluoro-2-deoxy-D- glucose-positron emission tomography: clinical evaluation and alternative management. *J. Natl. Cancer Inst. 93*:630–635.

Greenlee, R.T., Murray, T., Bolden, S., and Wingo, P.A. 2000. Cancer Statistics, 2000. *CA Cancer J. Clin. 50*:7–33.

Hathaway, C.L., Rand, R.P., Moe, R., and Marchioro, T. 1994. Salvage surgery for locally advanced and locally recurrent breast cancer. *Arch. Surg. 129*:582–587.

Hathaway, P.B., Mankoff, D.A., Maravilla, K.R., Austin-Seymour, M.M., Ellis, G.K., Gralow, J.R., Cortese, A.A., Hayes, C.E., and Moe, R.E. 1998. The value of combined FDG-PET and magnetic resonance imaging in the evaluation of suspected recurrent local-regional breast cancer: preliminary experience. *Radiology 210*:807–814.

Honig, S.H., and Swain, S.M. 1993. Hormonal manipulation in the adjuvant treatment of breast cancer. In DeVita, V.T., Hellman, S., and Rosenberg, S.A. (Eds.), *Important Advances in Oncology*. Philadelphia: J.B. Lippincott, pp. 103–123.

Hortobagyi, G.N. 2000. Developments in chemotherapy of breast cancer. *Cancer 88*:3073–3079.

Husband, J.E. 1996. Monitoring tumor response. *Eur. Radiol. 6*:775–785.

Huston, T.L., and Simmons, R.M. 2005. Locally recurrent breast cancer after conservation therapy. *Am. J. Surg. 189*:229–235.

Isasi, C.R., Moadel, R.M., and Blaufox, M.D. 2005. A meta-analysis of FDG-PET for the evaluation of breast cancer recurrence and metastases. *Breast Cancer Res. Treat. 90*:105–112.

Iyer, R.B., Fenstermacher, M.J., and Libshitz, H.I. 1996. MR imaging of the treated brachial plexus. *AJR 167*:225–229.

Kamby, C., Rose, C., Ejlertsen, B., Andersen, J., Birkler, N.E., Rytter, L., Andersen, K.W., and Zedeler, K. 1987. Stage and pattern of metastases in patients with breast cancer. *Eur. J. Cancer Clin. Oncol. 23*:1925–1934.

Katz, A., Strom, E.A., Buchholz, T.A., Thames, H.D., Smith, C.D., Jhingran, A., Hortobagyi, G., Buzdar, A.U., Theriault, R., Singletary, S.E., and McNeese, M.D. 2000. Locoregional recurrence patterns after mastectomy and doxorubicin-based chemotherapy: implications for postoperative irradiation. *J. Clin. Oncol. 18*:2817–2827.

Katzenellenbogen, J.A., Welch, M.J., and Dehdashti, F. 1997. The development of estrogen and progestin radiopharmaceuticals for imaging breast cancer. *Anticancer Res. 17*:1573–1576.

Lacour, J., Le M., Caceres, E., Koszarowski, T., Veronesi, U., and Hill, C. 1983. Radical mastectomy versus radical mastectomy plus internal mammary dissection: ten year results of an international cooperative trial in breast cancer. *Cancer 51*:1941–1943.

Leborgne, F., Leborgne, J.H., Ortega, B., Doldan, R., and Zubizarreta, E. 1995. Breast conservation treatment of early stage breast cancer: patterns of failure. *Int. J. Radiat. Oncol. Biol. Phys. 31*:765–775.

Linden, H.M., Stekhova, S., Link, J.M., Gralow, J.R., Livingston, R.B., Ellis, G.K, Peterson, L.M., Schubert, E.K., Petra, P., Krohn, K.A., and Mankoff, D.A. 2004. HER2 expression and uptake of 18F-Fluoroestradiol (FES) predict response of breast cancer to hormonal therapy. *J. Nucl. Med. 45*:85P (abstract).

Liu, C.S., Shen, Y.Y., Lin, C.C., Yen, R.F., and Kao, C.H. 2002. Clinical impact of [18F]FDG-PET in patients with suspected recurrent breast cancer based on asymptomatically elevated tumor marker serum levels: a preliminary report. *Jpn. J. Clin. Oncol. 32*:244–247.

Lonneux, M., Borbath, I., Berliere, M., Kirkove, C., and Pauwels, S. 2000. The place of whole-body FDG PET for the diagnosis of distant recurrence of breast cancer. *Clin. Positron Imaging 3*:45–49.

Machiavelli, M.R., Romero, A.O., Pérez, J.E., Lacava, J.A., Dominguez, M.E., Rodriguez, R., Barbieri, M.R., Romero-Acuna, L.A., Romero-Acuna, J.M., Langhi, M.J., Amato, S., Ortiz, E.H., Vallejo, C.T., and Leone, B.A. 1998. Prognostic significance of pathological response of primary tumor and metastatic axillary lymph nodes after neoadjuvant chemotherapy for locally advanced breast carcinoma. *Cancer J. Sci. Am. 4*:125–131.

Mankoff, D.A., Peterson, L.M., Tewson, T.J., Link, J.M., Gralow, J.R., Graham, M.M., and Krohn, K.A. 2001. [18F]fluoroestradiol radiation dosimetry in human PET studies. *J. Nucl. Med. 42*:679–684.

Mintun, M.A., Welch, M.J., Siegel, B.A., Mathias, C.J., Brodack, J.W., McGuire, A.H., and Katzenellenbogen, J.A. 1988. Breast cancer: PET imaging of estrogen receptors. *Radiology 169*:45–48.

Moon, D.H., Maddahi, J., Silverman, D.H.S., Glaspy, J.A., Phelps, M.E., and Hoh, C.K. 1998. Accuracy of whole-body [fluorine-18]-FDG PET for the detection of recurrent or metastatic breast carcinoma. *J. Nucl. Med. 39*:431–435.

Mortimer, J.E., Dehdashti, F., Siegel, B.A., Trinkaus, K., Katzenellenbogen, J.A., and Welch, M.J. 2001. Metabolic flare: indicator of hormone responsiveness in advanced breast cancer. *J. Clin. Oncol.* 19:2797–2803.

Nakai, T., Okuyama, C., Kubota, T., Yamada, K., Ushijima, Y., Taniike, K., Suzuki, T., and Nishimura, T. 2005. Pitfalls of FDG-PET for the diagnosis of osteoblastic bone metastases in patients with breast cancer. *Eur. J. Nucl. Med. Mol. Imaging* 32:1253–1258.

Nielsen, O.S., Munro, A.J., and Tannock, I.F. 1991. Bone metastases: pathophysiology and management policy. *J. Clin. Oncol.* 9:509–524.

Noguchi, M., Ohta, N., Thomas, M., Kitagawa, H., and Miyazaki, I. 1993. Risk of internal mammary lymph node metastases and its prognostic value in breast cancer patients. *J. Surg. Oncol.* 52:26–30.

Pelosi, E., Messa, C., Sironi, S., Picchio, M., Landoni, C., Bettinardi, V., Gianolli, L., Del Maschio, A., Gilardi, M.C., and Fazio, F. 2004. Value of integrated PET/CT for lesion localisation in cancer patients: a comparative study. *Eur. J. Nucl. Med. Mol. Imaging* 31:932–939.

Probstfeld, M.R., and O'Connell, T.X. 1989. Treatment of locally recurrent breast carcinoma. *Arch. Surg.* 124:1127–1129, discussion 1130.

Ragaz, J., Olivotto, I.A., Spinelli, J.J., Phillips, N., Jackson, S.M., Wilson, K.S., Knowling, M.A., Coppin, C.M., Weir, L., Gelmon, K. Le, N., Durand, R., Coldman, A.J., and Manji, M. 2005. Locoregional radiation therapy in patients with high-risk breast cancer receiving adjuvant chemotherapy: 20-year results of the British Columbia randomized trial. *J. Natl. Cancer Inst.* 97:116–125.

Schirrmeister, H., Guhlmann, A., Kotzerke, J., Santjohanser, C., Kuhn, T., Kreienberg, R., Messer, P., Nussle, K., Elsner, K., Glatting, G., Trager, H., Neumaier, B., Diederichs, C., and Reske, S.N. 1999. Early detection and accurate description of extent of metastatic bone disease in breast cancer with fluoride ion and positron emission tomography. *J. Clin. Oncol.* 17:2381–2389.

Schirrmeister, H., Kuhn, T., Guhlmann, A., Santjohanser, C., Horster, T., Nussle, K., Koretz, K., Glatting, G., Rieber, A., Kreienberg, R., Buck, A.C., and Reske, S.N. 2001. [F-18]-2-deoxy-2-fluoro-D-glucose PET in the preoperative staging of breast cancer: comparison with the standard staging procedures. *Eur. J. Nucl. Med.* 28:351–358.

Schmolling, J., Maus, B., Rezek, D., Fimmers, R., Holler, T., Schuller, H., and Krebs, D. 1997. Breast preservation versus mastectomy—recurrence and survival rates of primary breast cancer patients treated at the UFK Bonn. *Eur. J. Gynaecol. Oncol.* 18:29–33.

Schwaibold, F., Fowble, B.L., Solin, L.J., Schultz, D.J., and Goodman, R.L. 1991. The results of radiation therapy for isolated local regional recurrence after mastectomy. *Int. J. Radiat. Oncol. Biol. Phys.* 21:299–310.

Scott, W.J., Gobar, L.S., Terry, J.D., Dewan, N.A., and Sunderland, J.J. 1996. Mediastinal lymph node staging of non-small-cell lung cancer: a prospective comparison of computed tomography and positron emission tomography. *J. Thorac. Cardiovasc. Surg.* 111:642–648.

Spence, A.M., Muzi, M., Graham, M.M., O'Sullivan, F., Krohn, K.A., Link, J.M., Lewellen, T.K., Lewellen, B., Freeman, S.D., Berger, M.S., and Ojemann, G.A. 1998. Glucose metabolism in human malignant gliomas measured quantitatively with PET, 1-[C-11]-glucose and FDG: analysis of the FDG lumped constant. *J. Nucl. Med.* 39:440–448.

Stafford, S.E., Gralow, J.R., Schubert, E.K., Rinn, K.J., Dunnwald, L.K., Livingston, R.B., and Mankoff, D.A. 2002. Use of serial FDG PET to measure the response of bone-dominant breast cancer to therapy. *Acad. Radiol.* 9:913–921.

Suarez, M., Perez-Castejon, M.J., Jimenez, A., Domper, M., Ruiz, G., Montz, R., and Carreras, J.L. 2002. Early diagnosis of recurrent breast cancer with FDG-PET in patients with progressive elevation of serum tumor markers. *Q. J. Nucl. Med.* 46:113–121.

Taghian, A., Jeong, J., Mamounas, E., Anderson, S., Bryant, J., Deutsch, M., and Wolmark, N. 2004. Patterns of locoregional failure in patients with

operable breast cancer treated by mastectomy and adjuvant chemotherapy with or without tamoxifen and without radiotherapy: results from five National Surgical Adjuvant Breast and Bowel Project randomized clinical trials. *J. Clin. Oncol.* 22:4247–4254.

Tam, S.L., Gralow, J.R., Livingston, R.B., Linden, H.M., Ellis, G.K., Schubert, E.K., Dunnwald, L.K., and Mankoff, D.A. 2005. Serial FDG-PET to monitor treatment of bone-dominant metastatic breast cancer predicts time to progression (TTP). *J. Clin. Oncol.* 28:36s (abstract).

Tannock. I.F., and Hill, R.P. 1992. *The Basic Science of Oncology.* New York: McGraw-Hill.

Toi, M., Tanaka, S., Bando, M., Hayashi, K., and Tominaga, T. 1997. Outcome of surgical resection for chest wall recurrence in breast cancer patients. *J. Surg. Oncol.* 37:853–863.

Tran, A., Pio, B.S., Khatibi, B., Czernin, J., Phelps, M.E., and Silverman, D.H. 2005. 18F-FDG PET for staging breast cancer in patients with inner-quadrant versus outer-quadrant tumors: comparison with long-term clinical outcome. *J. Nucl. Med.* 46:1455–1459.

van der Hoeven, J.J., Hoekstra, O.S., Comans, E.F., Pijpers, R., Boom, R.P., van Geldere, D., Meijer, S., Lammertsma, A.A., and Teule, G.J. 2002. Determinants of diagnostic performance of [F-18]fluorodeoxyglucose positron emission tomography for axillary staging in breast cancer. *Ann. Surg.* 236:619–624.

Vansteenkiste, J.F., Stroobants, S.G., De Leyn, P.R., Dupont, P.J., Bogaert, J., Maes, A., Deneffe, G.J., Nackaerts, K.L., Verschakelen, J.A., Lerut, T.E., Mortelmans, L.A., and Demedts, M.G. 1998. Lymph node staging in non-small-cell lung cancer with FDG-PET scan: a prospective study on 690 lymph node stations from 68 patients. *J. Clin. Oncol.* 16:2142–2149.

Veronesi, U., Cascinelli, N., Greco, M., Bufalino, R., Morabito, A., Galluzzo, D., Conti, R., De Lellis, R., Delle, Donne V., and Piotti, P. 1985. Prognosis of breast cancer patients after mastectomy and dissection of internal mammary nodes. *Ann. Surg.* 202:702–707.

Vranjesevic, D., Filmont, J.E., Meta, J., Silverman, D.H., Phelps, M.E., Rao, J. Valk, P.E., and Czernin, J. 2002. Whole-body 18F-FDG PET and conventional imaging for predicting outcome in previously treated breast cancer patients. *J. Nucl. Med.* 43:325–329.

Wagner, J.D., Schauwecker, D., Davidson, D., Coleman, J.J, 3rd, Saxman, S., Hutchins, G., Love, C., and Hayes, J.T. 1999. Prospective study of fluorodeoxyglucose-positron emission tomography imaging of lymph node basins in melanoma patients undergoing sentinel node biopsy. *J. Clin. Oncol.* 17:1508–1515.

Wahl, R.L., Siegel, B.A., Coleman, R.E., and Gatsonis, C.G. 2004. Prospective multicenter study of axillary nodal staging by positron emission tomography in breast cancer: a report of the staging breast cancer with PET Study Group. *J. Clin. Oncol.* 22:277–285.

Wakeling, A.E., Nicholson, R.I., and Gee, J.M. 2001. Prospects for combining hormonal and nonhormonal growth factor inhibition. *Clin. Cancer Res.* 7:4350s–4355s, discussion 4411s–4412s.

Warburg, O. 1956. On the origin of cancer cell. *Science.* 123:309–314.

Wojcik, C., Yahonda, A., McFarlane, D., and Wahl, R.L. 1997. Economic analysis of sentinel-node surgery versus preoperative FDG-PET in the management of patients with intermediate thickness cutaneous melanomas. *J. Nucl. Med.* 38:34P (abstract).

Yap, C.S., Seltzer, M.A., Schiepers, C., Gambhir, S.S., Rao, J., Phelps, M.E., Valk, P.E., and Czernin, J. 2001. Impact of whole-body 18FDG PET on staging and managing patients with breast cancer: the referring physician's perspective. *J. Nucl. Med.* 42:1334–1337.

Zangheri, B., Messa, C., Picchio, M., Gianolli, L., Landoni, C., and Fazio, F. 2004. PET-CT and breast cancer. *Eur. J. Nucl. Med. Mol. Imaging* 31:S135–142.

Leiomyoma of the Breast Parenchyma: Mammographic, Sonographic, and Histopathologic Features

Aysin Pourbagher and M. Ali Pourbagher

Introduction

Leiomyomas, benign smooth muscle tumors, usually occur in the uterus or gastrointestinal tract, less frequently in skin and soft tissue, and rarely in the breast (Kotsuma et al., 2001). Leiomyoma of the breast parenchyma proper, first described by L.W. Strong in 1913, is a rare neoplasm (Diaz-Arias et al., 1989). There are two distinct tumor sites in the breast. The most frequent site is found in the region extending from the nipple to the subareolar location (Kaufman and Hirsch, 1996). Leiomyomas occurring in the breast parenchyma are rare except for those involving the subareolar region (Son et al., 1998). Leiomyomas can occur in both women and men (Velasco et al., 1995). Most breast leiomyomas occur in the right breast and more frequently in the outer quadrant (Sidoni et al., 1999). These tumors occur predominantly in late middle-aged women and can be clinically difficult to discern from carcinoma (Diaz-Arias et al., 1989). Leiomyosarcoma of the breast is one of the rarest primary breast tumors, with fewer than 20 cases reported in the literature. This rare primary breast tumor appears to be the most common in women in their sixth decade. The tumors may be deep within the breast parenchyma or superficial in association with the nipple-areolar complex. The usual mammographic finding is a dense, circumscribed, noninvasive lesion (Lee et al., 2004).

The histogenesis of leiomyomas is still controversial (Nazario et al., 1995), with various theories being offered as to the origin of intraparenchymal leiomyomas. Diaz-Arias et al. (1989) proposed five sources: teratoid origin with extreme overgrowth of the myomatous elements, embryologically displaced smooth muscle from the nipple, angiomatous smooth muscle, a multipotent mesenchymal cell, and myoepithelial cells.

Immunohistochemical and ultrastructural studies of a leiomyoma of the breast supported a smooth muscle rather than a myoepithelial origin for these lesions (Diaz-Arias et al., 1989). The frequent occurrence of these tumors near the nipple may be related to the abundance of smooth muscle cells around the nipple and areola (Kaufman and Hirsch, 1996). The smooth muscle cells of the blood vessels of the breast are believed to be the most likely primary source of parenchymal leiomyomas (Heyer *et al.*, 2006). Leiomyoma of the breast has no known relationship with epithelial tumors or carcinomas (Son *et al.*, 1998).

The leiomyoma is usually a firm, freely movable, painless, and palpable mass upon physical examination. The lesion often appears as a clearly demarcated and slow-growing one, and tumor size varies from 0.5 cm to as large as 13.8 cm. The duration of the mass is 1 month to 26 years, and the range in age is from 34–69 years in reported cases (Kotsuma *et al.*, 2001). None of the cases developed a recurrence. Two of them were an incidental finding on routine mammography. Neither nipple discharge nor retraction has been recorded, even in the centrally located tumors. No lymphadenopathy has been noted. The usual presentation of parenchymal leiomyomas is a painless breast mass noted by a patient on self-examination or by an examining physician. As opposed to parenchymal leiomyomas, areolar leiomyomas are more frequently tender (Kaufman and Hirsch, 1996). A family history or genetic linkage has not been reported (Son *et al.*, 1998). It is impossible to distinguish benign from malignant lesions by physical examination alone. Only a few reports have described the radiologic features of intraparenchymal leiomyomas. In comparison with other lesions commonly found in the breast, the leiomyoma had no specific mammographic and sonographic findings.

Mammographic Appearance

Only nine mammographic reports on leiomyoma exist. In 1989, xeromammographic findings were reported for the first time (Diaz-Arias *et al.*, 1989). It was a circumscribed soft tissue mass without calcification, skin change, nipple retraction, or architectural distortion. Mammographic and sonographic findings were reported for the first time (Son *et al.*, 1998). Mammographically, the leiomyomas are usually sharply circumscribed, well-defined lesions that are round, ovoid, or smoothly lobulated similar to fibroadenoma. Smooth margins and the mammographic lack of radial extensions are the primary findings in intraparenchymal leiomyomas. But the margins may also be ill-defined (Tamir *et al.*, 1995). Most lesions are homogeneous and have a moderate to high density. Leiomyomas do not calcify in either location.

Sonographic Appearance

Only six sonographic reports on leiomyoma exist. Some of them note sonography findings of a well-circumscribed solid mass similar to fibroadenoma. Sonographically, intraparenchymal leiomyomas appear as well marginated and slow-growing, hypoechoic lesions with a homogeneous echotexture, sometimes with embedded semicystic components. One report described a case of breast leiomyoma that has isoechoic or slightly hyperechoic appearance on sonography (Son *et al.*, 1998). There can be a lack of posterior shadowing, or posterior enhancement. Ill-defined margins may be found (Tamir *et al.*, 1995). In addition, one report described posterior acoustic shadowing, highly echogenic halo, and central tumor vessel on Doppler imaging (Heyer *et al.*, 2006).

We reported a 47-year-old woman without any complaint admitted to our hospital for routine screening mammography. Physical examination revealed no palpable mass in either breast. On mammography, well-circumscribed dense mass was located in parenchyma of the medial half of her left breast (Fig. 150). The mass showed no calcification, and no other lesions were detected in either breast. No axillary lymphadenopathy was noted on the mammogram. On sonography, the mass appeared as a well-defined solid oval and hypoechoic compared with the breast parenchyma, all of which suggested fibroadenoma. There was no acoustic enhancement or shadowing (Fig. 151). Although the patient's radiologic findings throughout 2 years of follow-up suggested that the lesion was a fibroadenoma, unexplained significant expansion over the course of 1 year made it necessary to excise the mass *en bloc*. Histologic examination of the surgical specimen identified the mass as a leiomyoma (Pourbagher *et al.*, 2005).

Histopathology is necessary for definitive diagnosis. Histologic descriptions have been variable in length and content, but neither cytologic atypia, hypercellularity, high mitotic rate, nor necrosis has been reported. The reported number of leiomyomas of the breast are not sufficient to allow differentiation from fibroadenomas in the absence of histopathology (Kotsuma *et al.*, 2001). The common histologic features of leiomyomas of the breast are identical to those observed in leiomyomas at other sites: groups of interfacing bundles of spindle-shaped cells with blunt-ended nuclei and eosinophilic cytoplasm (Kaufman and Hirsch, 1996). Further confirmation can be carried out by positive periodic acid-Schiff (PAS) stain (which is digested by diastase), silver impregnation, and immunohistochemical antibody staining against vimentin, desmin, and smooth muscle actin, which verify that the tumor arises purely from smooth muscle (Kotsuma *et al.*, 2001). If the differentiation between a leiomyoma and a highly differentiated leiomyosarcoma proves difficult on frozen sections, paraffin-embedding is required in order to distinguish these entities (Heyer *et al.*, 2006).

The histopathologic differential diagnoses for leiomyoma of the breast include adenoleiomyoma, cystosarcoma phyllodes, fibroadenoma with prominent smooth muscle,

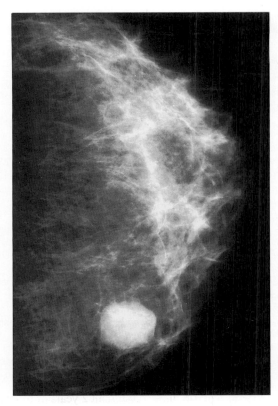

Figure 150 Craniocaudal mammogram shows well-circumscribed dense mass in parenchyma of inner quadrant of left breast.

Figure 151 On sonography, mass appears oval, hypoechoic, and well defined, all of which suggest fibroadenoma.

fibromatosis, benign spindle cell tumor of the breast, fibrous histiocytoma, myoepithelioma, myoid hamartoma, and leiomyosarcoma. The first three lesions can be ruled out by thorough sectioning because they contain epithelial/ductal structures. Fibromatosis, benign spindle cell tumor of the breast, fibrous histiocytomas, and myoepitheliomas are lesions composed of varying mixtures of fibroblasts, myofibroblasts, undifferentiated mesenchymal cells, and myoepithelial cells rather than purely smooth muscle cells (Diaz-Arias et al., 1989). In myoid hamartoma, benign glandular elements are seen scattered between the spindle cell bundles, and evidence of sclerosing adenosis is usually seen at the edge of the lesion. In leiomyosarcoma, the lesion is more cellular, and cellular pleomorphism, mitotic activity, and sometimes necrosis are present (Chaudhary and Shousha, 2004).

Perhaps the most important differential diagnosis is leiomyosarcoma of the breast (Diaz-Arias et al., 1989). It is particularly important to differentiate these two neoplasms because of the risk of local recurrence or distant spread with leiomyosarcoma (Pourbagher et al., 2005). In most cases of breast leiomyosarcoma reported to date, neither axillary lymph node involvement nor metastatic lesion was present at the time of diagnosis. However, the possibility of spread

must be monitored long term because there is potential for recurrence or distant metastasis later; metastases can even be found more than 10 years after excision of a breast leiomyosarcoma. A long period of disease-free survival is no guarantee of a cure. Recurrence tends to be local or to occur via hematogenous spread (Munitiz et al., 2004).

The distinction between leiomyoma and leiomyosarcoma is not always an easy one. The distinguishing features of sarcomatous change can usually be adequately identified on histopathologic examination (Kaufman and Hirsch, 1996). Histologically, leiomyosarcomas feature prominent cytologic atypia, with 2–16 mitoses per 10 high-power fields, atypical mitoses, vascular invasion, and necrosis (Pourbagher et al., 2005). Investigation of mitosis, necrosis, and vascular invasion should help to exclude malignancy from the differential diagnosis (Kotsuma et al., 2001). In estimating the prognosis of leiomyomas, the mitotic index must be evaluated together with evidence of necrosis and hypercellularity. Mitotic rates of 1–3 × 10 HPF are considered a gray area. These leiomyomas may be prone to recurrence or degeneration to a malignancy (Boscaino et al., 1994).

Also of potential interest is the effect of the anti-estrogen agent tamoxifen on the growth of breast leiomyomas.

Tamoxifen has been shown to cause a sudden, rapid increase in the growth rate of uterine leiomyomas (Son *et al.*, 1998). Recent reports of tamoxifen-related complications in uterine leiomyomas may have implications for these tumors when they occur in the breast (Dilts *et al.*, 1992). One study indicated that this drug promotes formation of parenchymal leiomyomas of the breast and causes these masses to enlarge (Son *et al.*, 1998). The effect of the anti-estrogen agent, tamoxifen, on the growth and differentiation of breast leiomyomas is unknown (Kaufman and Hirsch, 1996). The recent increased use of tamoxifen in women with breast cancer should alert clinicians and radiologists to look for a possible increased incidence of breast leiomyoma.

In our patients, after the initial follow-up mammography and sonography evaluation, the patient started taking sibutramine and orlistat as treatment for obesity. Two years after the mammography and sonography, examination showed that the mass had significantly enlarged (Pourbagher *et al.*, 2005). No study has yet investigated the effects of either of these agents on leiomyoma in the breast or any other part of the body. Further investigation will be required to obtain reliable information about the effects of anti-obesity drugs on leiomyoma.

The usual management of breast leiomyoma is complete excision with wide margins. In rare cases, relapse has been reported when total excision was not accomplished at the time of first operation (Boscaino *et al.*, 1994).

References

Boscaino, A., Ferrara, G., Orabona, P., Donofrio, V., Staibano, S., and De Rosa, G. 1994. smooth muscle tumors of the breast: clinicopathologic features of two cases. *Tumori 80*:241–245.

Chaudhary, K.S., and Shousha, S. 2004. Leiomyoma of the nipple, and normal subareolar muscle fibres, are oestrogen and progesterone receptor positive. *Histopathology 44*:626–628.

Diaz-Arias, A.A., Hurt, M.A., Loy, T.S., Seeger, R.M., and Bickel, J.T. 1989. Leiomyoma of the breast. *Hum. Pathol. 20*:396–399.

Dilts, P.V., Hopkins, M.P., Chang, A.E., and Cody, R.L. 1992. Rapid growth of leiomyoma in a patient receiving tamoxifen. *Am. J. Obstet. Gynecol. 166*:167–168.

Heyer, H., Ohlinger, R., Schimming, A., Schwesinger, G., and Grunwald, S. 2006. Parenchymal leiomyoma of the breast-clinical, sonographic, mammographic and histological features. *Ultraschall. Med. 27*:55–58.

Kaufman, H.L., and Hirsch, E.F. 1996. Leiomyoma of the breast. *J. Surg. Oncol. 62*:62–64.

Kotsuma, Y., Wakasa, K., Yayoi, E., Kishibuchi, M., Watatani, M., and Sakamoto, G. 2001. A case of leiomyoma of the breast. *Breast Cancer 8*:166–169.

Lee, J., Li, S., Torbenson, M., Liu, Q.Z., Lind, S., Mulvihill, J.J., Bane, B., and Wang, J. 2004. Leiomyosarcoma of the breast: a pathologic and comparative genomic hybridization study of two cases. *Cancer Genet. Cytogenet. 149*:53–57.

Munitiz, V., Rios, A., Canovas, J., Ferri, B., Sola, J., Canovas, P., Illana, J., and Parrilla, P. 2004. Primitive leiomyosarcoma of the breast: case report and review of the literature. *The Breast 13*:72–76.

Nazario, A.C., Tanaka, C.I., de Lima G.R., Gebrim, L.H., and Kemp, C. 1995. Leiomyoma of the breast: a case report. *Rev. Paul. Med. 113*:992–994.

Pourbagher, A., Pourbagher, M.A., Bal, N., Oguzkurt, L., and Ezer, A. 2005. Leiomyoma of the breast parenchyma. *Am. J. Roentgenol. 185*:1595–1597.

Sidoni, A., Lüthy, M., Bellezza, G., Consiglio, M.A., and Bucciarelli, E. 1999. Leiomyoma of the breast: case report and review of the literature. *Breast 8*:289–290.

Son, E.J., Oh, K.K., Kim, E.K., Son, H.J., Jung, W.H., and Lee, H.D. 1998. Leiomyoma of the breast in a 50-year-old woman receiving tamoxifen. *Am. J. Roentgenol. 171*:1684–1686.

Tamir, G., Yampolsky, I., and Sandbank, J. 1995. Parenchymal leiomyoma of the breast: report of a case and clinicopathological review. *Eur. J. Surg. Oncol. 21*:88–89.

Velasco, M., Ubeda, B., Autonell, F., and Serra, C. 1995. Leiomyoma of the male areola infiltrating the breast tissue. *Am. J. Roentgenol. 164*:511–512.

31

Detection of Breast Cancer: Dynamic Infrared Imaging

Terry M. Button

Introduction: Infrared and Its Detection

Infrared radiation (IR) is electromagnetic radiation between visible light (0.75 μm) and microwave radiation (1000 μm). Infrared is further broken down into near IR (0.75–1.4 μm and commonly used in fiber-optic telecommunication), short-wavelength IR (1.4–3 μm with strong water absorption), mid-wavelength IR (3–8 μm), long-wavelength (8–15 μm), and far-infrared (15–1000 μm). All objects with a temperature above absolute zero emit infrared radiation from their surface. Infrared imaging is the passive formation of an image using this radiation. Infrared spectral radiance (I) is related to the temperature and is given by Planck's law of black body radiation:

$$I(v, T) = \frac{2hv^3}{c^2} \frac{1}{e^{\frac{hv}{kT}} - 1} \qquad (1)$$

where v is the frequency, T is the temperature (Kelvin), h is Planck's constant, c is the speed of light, and k is Boltzmann's constant. The relationship between the wavelength for peak intensity (λ_{max}) and temperature is given by the Wien formula:

$$\lambda_{max} = 2897 \, (\mu m)/T(^\circ K) \qquad (2)$$

At physiologic temperature, the peak intensity is at approximately 9.5 μm. In this range, the emissivity of skin is approximately 1, and there are no significant absorption bands (carbon dioxide is at 4.2 μm). For biomedical applications of passive infrared imaging, long wavelength is, therefore, ideal.

A simple device used in early applications of noncontact infrared was a thermopile. When a circuit is made of two dissimilar metals, and the junction between the two wires are held at different temperatures, an electromotive force is produced. The effect is multiplied by using a number of junctions in series in the circuit. Imaging devices in the 1970s employed a scanning mirror and a single sensor. The sensor was either a photodetector or a thermal detector.

Photodetector systems often employed Ge:Au or InSb sensors cooled by liquid nitrogen. These systems allowed the acquisition of a single frame typically in less than a minute. Resolution was the order of 0.1°C. Often these detectors were sensitive to 2–6 μm infrared, and filters were often employed to eliminate photons in the range of 2–3 μm. This is important because the emissivity of skin falls to 0.7 and irradiation from skin at physiologic temperature is near zero in this range so that only noise would be collected. A thermistor or thermopile bolometer operates as a thermal detector.

Thermistor-based devices are often sensitive in the range of 2–16 μm and operate on the basis that detector resistance changes with temperature, resulting in bias current change.

More recent infrared imagers employ infrared optics and arrays of sensors that were devised for military applications. This allows for rapid image acquisition. Microbolometer arrays typically employ a layer of vanadium oxide mounted on silicon. Infrared incident on the detector layer changes its resistance, which is read out by measuring the resulting change in bias current. Microbolometers do not require cooling but have the disadvantage of low sensitivity if used for rapid image formation.

Quantum well infrared photodetectors (QWIPs) employ layers of semiconductor materials such as gallium arsenide. These photodetectors operate by photoexcitation of electrons between ground and excited-state subbands of quantum wells that are artificially fabricated alternately using thin layers of two different, high-bandgap semiconductor materials. The structure is designed so that the photoexcited electrons can escape from the potential wells and be collected as photocurrent. These sensors are designed to be sensitive to a narrow but useful infrared band (9–12 μm). While cooling the detector is a requirement, very high sensitivity (0.01°C) and frame rates (400 fps) are possible.

History of Infrared for Breast Cancer Detection

Using a handheld thermopile (Lloyd-Williams and Handley, 1961), 54 of 57 of breast cancers were reported detectable by IR methods in the early 1960s. Soon after, IR imaging was introduced to the United States using a Barnes thermograph that required several minutes to produce a single IR image (Gershen-Cohen et al., 1965). The very encouraging results published using IR for breast cancer detection prompted its inclusion in the Breast Cancer Detection Demonstration Project (BCDDP). The BCDDP compared the static infrared imaging available at that time to X-ray mammography for breast cancer detection. Unfortunately, this was undertaken as a competitive process and did not examine the potential complementary aspects of an anatomical imaging technology, X-ray mammography, and a physiologic imaging technology, infrared, as is currently done in say PET-CT.

Several BCDDP sites reported poor sensitivity and reliability of IR imaging (Moskowitz et al., 1976; Threatt et al., 1980), and their findings were key aspects of the BCDDP report and recommendations. While the results from these sites are inconsistent with numerous previous multicenter trials, the consensus was that it was difficult to make a proper interpretation with IR because criteria did not seem objective. One investigator (Threatt et al., 1980) even suggested computer recognition capabilities. Currently, this

capability of computer-aided diagnosis (CAD) is routinely employed in X-ray mammography and is under investigation for breast magnetic resonance (Jambawalikar, 2005). In the application to magnetic resonance texture feature, extraction and machine-learning tools have been developed (Jambawalikar, 2005) that may be applicable to CAD for infrared breast cancer detection.

The BCDDP found that static infrared imaging was "not a suitable substitute for mammography in routine screening" (Anonymous, 1979). This finding had the unintended effect of essentially eliminating infrared as a clinical option in breast cancer detection. This was clearly not the intent in asmuch as the National Institutes of Health consensus for Breast Cancer Screening, as a result of the BCDDP, recommended placing greater emphasis on research with noninvasive techniques, such as thermography, ultrasound, and biologic markers (Anonymous, 1978).

Infrared research since the BCDDP has been sparse but noteworthy. A multi-imaging strategy was used, and the advantages of infrared were shown to include enhanced diagnostic accuracy, reduced unnecessary surgery, and improved prognostic ability (Isard, 1984). Gamagami (1996) used static infrared to study angiogenesis and reported that hypervascularity and hyperthermia could be shown in 86% of nonpalpable breast cancers. He also reported that in 15% of these cases, static infrared imaging helped detect cancers that were not detected by mammography. Keyserlingk et al. (2000) conducted infrared analysis of 100 successive cases of breast cancer and found that the detection rate of mammography improved from 83 to 93% with the addition of infrared. He also noted that the average tumor size for those cases undetected by mammography was 1.66 cm, while those undetected by IR imaging averaged 1.28 cm (Keyserlingk et al., 2000). He suggested that tumor-induced infrared patterns depend on early vascular and metabolic changes possibly induced by regional nitric oxide diffusion rather than strictly on tumor size (Keyserlingk et al., 2000).

Research conducted after the BCDDP has also demonstrated the potential of infrared for early breast cancer detection. Gautherie et al. (1987) investigated the thermal and vascular disorders associated with early stages of breast malignancy for more than 25,000 asymptomatic (59%) and symptomatic (41%) women. All patients underwent infrared, mammographic, and physical examination. Two hundred and four (21.3%) of the 958 patients who, on their first visit, had an abnormal infrared imaging study but no findings at physical examination or mammography, were found to develop cancer within the next 3 years.

Despite newly available infrared technology, primarily from military research and development, as well as compelling statistics of > 70,000 documented cases showing the contribution of infrared to detect small breast cancers, few legitimate imaging researchers have shown an interest in this area and few noteworthy publications have been made

(Keyserlingk *et al.*, 2000). This is surprising in view of the current consensus regarding the importance of vascular-related events associated with tumor initiation and growth that finally provide a plausible explanation for the IR findings associated with the early development of smaller tumors (Keyserlingk *et al.*, 2000).

It is most likely that infrared is not currently accepted by mainstream medical imagers because of some of its supporters. A simple Internet search reveals many claims and statements from individuals that appear to lack credibility. This may be the result of infrared not being regulated as X-ray-based imaging technologies are. A potential model that may eventually help overcome this credibility issue is an accreditation process similar to that in place by the American College of Radiology (ACR) for magnetic resonance (MR). Magnetic resonance is not heavily regulated; however, it is very difficult to get insurance reimbursement for MR services without this accreditation. Accreditation also becomes a symbol of competence in the field. It is imperative that these issues be addressed for IR.

Dynamic Infrared Imaging

Quantum well infrared photodetectors (QWIPs) constructed as a focal plane array were developed as part of the Star Wars initiative (Gunapala *et al.*, 1997) and are now available for biomedical research. Detectors available for this purpose contain an aGaAs QWIP, are configured as a 256×256 scanned focal plane array, and allow for rapid imaging (Breiter *et al.*, 2000). This new generation of infrared sensor has improved thermal ($\sim 0.01°C$) and spatial

(1 mm) resolution and can image dynamically (400 frames per sec). Arbar (1994) first proposed image collection and processing strategies employing these new rapid sensors that have collectively come to be called dynamic infrared imaging (DIRI).

High sensitivity and frame rate allow dynamic interrogation of biological systems. Figure 152 demonstrates the process of collecting successive infrared frames followed by fast Fourier transform (FFT). The result is the spectrum of infrared flux oscillation frequencies. Potential frequencies that may be useful for image formation include vasomotion and cardiogenic. Vasomotion is spontaneous rhythmic variations in blood flow due to oscillation in blood vessel radius and has a frequency of the order of 0.1 Hz. Cardiogenic is the result of the heart supplying warm fresh blood to the surface with each beat (1 Hz). Other frequencies may be useful (Anbar *et al.*, 1999, 2001), and other methods of image analysis are possible (Ecker *et al.*, 2002).

An important question centers on the ability to detect such small infrared flux modulations. This question can be answered by modeling the signal-to-noise ratio (SNR). For this purpose, the following parameters were used and are representative of available DIRI hardware:

FOV—field of view (200 mm)
$n \times n$ — image matrix surface area elements (128×128 after averaging)
ε — emissivity of the skin (0.98)
A — lens aperture diameter (5.4 cm)
r — distance from surface area element to detector (1 m)
f — fraction of infrared photons collected that are incident on active detector (0.75)

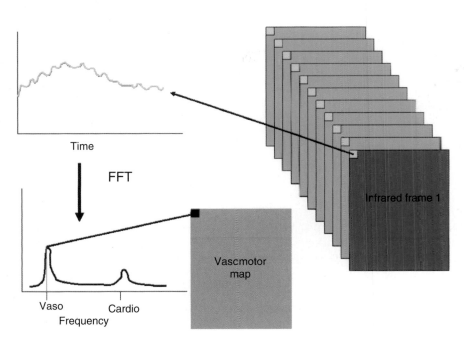

Time

FFT

Vasomotor map

Vaso Cardio

Frequency

Infrared frame 1

Figure 152 Dynamic infrared image formation at the vasomotor frequency.

i — intrinsic detector efficiency (10%)

τ — detector integration time (0.01 sec)

N — infrared photons emitted per unit area by a black body at temperature T (from Planck's Law, Eq. (1), above)

The number of photons detected D from a unit area element in the integration time τ is given by

$$D = N \,[FOV/n]^2 \,\varepsilon \, f \, i \, \tau \, [A^2/(16\pi r^2)] \qquad (3)$$

The noise in the number of detected events at each element follows Poisson counting statistics, and the standard deviation is $D^{1/2}$. Assuming that the cardiogenic oscillation is a sine wave having an amplitude of 0.005°C about physiologic temperature and a frequency of 1 Hz, an FFT can be applied to the theoretical signal function (sine wave with counting noise), and in this way the impact of acquisition time, the role of apodizing functions, and zero filling may be examined. Using a Hanning filter, zero filling, and a 20.48 acquisition time at 100 fps provided the cardiogenic frequency peak shown in Figure 153. The resulting signal-to-noise ratio (SNR; peak value divided by noise adjacent to the peak) was approximately 13. In principle, cardiogenic oscillations can be detected with the proper detector, geometry, and acquisition parameters.

Clinical and experimental studies have demonstrated that peripheral microcirculation is influenced by multiple mechanisms, including respiration (Eriksen and Lossius, 1995), heart rate, vasomotion (Kvernmo et al., 1998), movement (Riechelmann and Krause, 1994), and temperature. Laser Doppler flowmetry (LDF) has been used as a reliable method to monitor the hemodynamic changes of microcirculation clinically and experimentally. The spectra derived from LDF and measured simultaneously by dynamic infrared for human subjects have been studied (Button et al., 2004).

Dynamic infrared images of the upper extremities (forearms and hands) of six volunteers (total 24 testing areas) were acquired under normal indoor temperature (18°C). The extremity under study was stabilized on a table. Simultaneously, a contact laser probe from a PF5010 laser Doppler system (Perimed AB, Jarfalla, Sweden) was placed in the DIRI field of view, and the local hemodynamic parameter (flux) was monitored and recorded. An audio metronome was used to regulate the subject's respiratory rate (0.3 Hz) during acquisition.

Fast Fourier transformation was performed on both DIR image data and flux curves from laser Doppler. The frequency spectra of adjacent test points were then compared. The laser Doppler test probe and DIRI surface area element had roughly the same area. Multiple oscillation frequencies were demonstrated on both DIRI and laser Doppler spectra as shown in Figure 154. Both spectra demonstrated a low-frequency complex (< 0.2 Hz) assumed to be vasomotion, a respiratory frequency (0.3 Hz) peak, and a cardiogenic frequency (1.2–1.5 Hz) peak. In some test areas (17/24, 70.8% for LDF and 7/24, 29.2% for DIRI) one or more high-frequency, low-intensity components were also noted. These components were at two times and three times the corresponding cardiogenic frequency.

The locations of LDF and DIRI spectral peaks correlated very well for all subjects. Each subject demonstrated three primary frequency peaks that are apparent on both the LDF, and DIRI spectra corresponding to the vasomotor, respiratory and cardiac frequencies. The peak height at each of these frequencies was noted to vary between subjects and with different sample areas for the same subject. High-frequency components that were observed are exact multiples

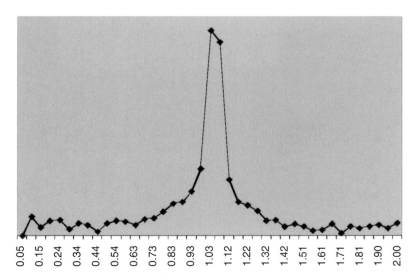

Figure 153 Theoretical cardiogenic DIRI spectra from a surface area element.

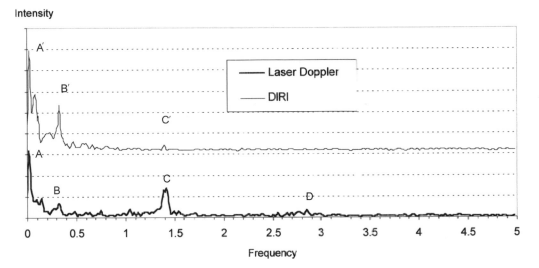

Figure 154 Comparison of laser Doppler and dynamic infrared derived spectra.

of the corresponding heart rate and are, therefore, harmonics of the cardiogenic signal. These results demonstrate that DIRI is able to detect skin temperature variations due to physiologic hemodynamics.

The work described above employed a single-channel LDF device. This probe only allows assessment of perfusion of a single small location at a time and also requires direct contact for accurate measurement. Laser Doppler imaging devices have recently become available that allow for analysis of a large area (Wardell *et al.*, 1993). Unfortunately, the SNR associated with these devices is compromised compared to contact probes. In addition, LDF imaging requires data collection time of several minutes and provides essentially a single-time-point blood velocity map. Complete analysis of vasomotor and cardiogenic perfusion components can be added to this acquisition but only for a few points at an additional time investment. The advantage of DIRI over laser Doppler techniques is that such analysis can be rapidly accomplished for all picture elements in a large field of view.

Mechanism for Breast Cancer Detection with Dynamic Infrared

Breast cancer detection by infrared can be described using two simple models: primitive tumor vasculature or tumor-generated nitric oxide (NO). In the first model, angiogenesis leads to the formation of new vessels associated with malignancy that lack smooth muscle fibers rendering them unreceptive to control by vasoconstrictors (Andrade *et al.*, 1992). As a result of these primitive vessels, greater blood flow and larger modulations such as cardiogenic are in

evidence. Note that in this model the increased flow and modulations are limited to the tumor only. This signal must somehow reach the surface for detection.

An alternative model is based on the finding that, unlike other types of cancer, tumors of the breast are greatly influenced by steroid hormones. For example, progesterone activates inducible nitric oxide synthase (iNOS). Patients with iNOS positive breast carcinomas have a significantly worse overall survival rate versus those with negative stains (Loibl *et al.*, 2005). Because NO is a vasodilator, increased perfusion results. While this signal may be associated with the tumor itself, it will also be associated with any tissue into which NO has diffused. This mechanism also predicts larger infrared flux modulations from, for example, the cardiac cycle (cardiogenic DIRI signal).

Preliminary published reports using an animal model of malignancy (Button *et al.*, 2004) support the NO model. In that study, athymic mice were implanted with ~ 2×10^6 A431 human epithelial cancer cells subcutaneously on the inner left thigh. Mice were anesthetized using ketamine and xylazine prior to dynamic infrared imaging on days 7, 8, and 11, post-tumor initiation. On day 8, three mice were imaged 1 hr after i.p. injection of 50 mg/kg of N(G)-nitro-L-arginine methyl ester (LNAME). This agent blocks vasodilation by NO. In addition, a control mouse was topically (tail) treated with 2% nitroglycerine ointment and imaged. Nitroglycerine produces NO. Mice were imaged periodically until they woke.

Image patterns characteristic of malignancy were found to be removed by LNAME. Introducing NO (nitroglycerine) into a normal mouse mimicked the infrared modulation patterns of the femoral artery of a tumor-bearing mouse. These findings support the hypothesis that DIRI allows for transdermal cancer detection due to malignancy-generated NO and subsequent vasodilation. Finally, evidence is available from a different animal model that suggests that thermal convection from the tumor is probably not the source of high-infrared intensities (Xie *et al.*, 2004). This finding also favors the NO model over the primitive tumor vasculature model.

A key disadvantage of static infrared-based breast cancer detection is that other pathologies can result in false-positives. For example, simple infection can mimic the response, increased infrared flux, which is being sought to detect malignancy. In addition, unstable room conditions may cause similar errors. This is not the case for dynamic infrared-based detection. These oscillations cannot easily be mimicked by environmental effects and require vasodilation, which is a better indication of malignancy.

Applications of Dynamic Infrared Imaging

A variety of potential applications of DIRI have started to appear in the literature. Anbar *et al.* (1999, 2001) examined the potential applications of high-frequency components of infrared flux modulations in an effort to detect breast cancer. Janicek *et al.* (2002) studied infrared modulations to monitor tumor response to therapy. Dynamic infrared imaging has been employed in neurosurgery to define tumor margins, map brain function, detect seizure foci, and examine treatment of vascular malformations (Ecker *et al.*, 2002). Recently, Garbade *et al.* (2006) investigated the potential of dynamic infrared imaging for noninvasive analysis of graft flow estimation in beating heart surgery. Other applications under study include perforator identification for skin grafts, melanoma detection, and wound healing.

Breast cancer detection studies have been initiated (Button *et al.*, 2004) and are now continuing using the BioscanIR (Advanced Biophotonics, Bohemia, NY) dynamic infrared imaging system. This QWIP-based system is cooled (by compressor) to 58°K to reduce noise. The camera is calibrated at each start-up. The current system includes a small black body mounted directly on the camera, which can be placed over the germanium lens for calibration. In this process, a uniformity correction map and calibration information are stored. During data acquisition, the uniformity-corrected infrared flux images are outputted to a grabber board and stored. Acquisitions shown here are 51.2 sec acquisitions at 10 frames per sec, but current studies are breath-hold 20.48 sec acquisitions at 100 frames per sec, producing 2048 total IR images. Because the integration time in the early 10 fps studies was only 20 msec, the new shorter breath-hold data collection protocol greatly improves SNR and provides much fewer motion artifacts.

The first processing step is to improve SNR by averaging the data to a 128 square matrix. This is accomplished by taking the average value for each 2×2 in the 256×256 matrix to represent a single element of the 128×128 matrix. While this degrades resolution by a factor of 2, SNR is improved by a factor of 2. Next, the infrared intensity versus time function at each pixel is first-order baseline corrected, followed by the application of a Hanning filter and FFT. The resulting spectroscopic data can be displayed as a magnitude image of the intensity at a single frequency (peak image) or an image of the intensity integrated over a frequency range (integration image). The spatial resolution of the processed images is ~ 0.8 mm. Dynamic infrared images are formed at the cardiogenic and vasomotor frequencies for review. A pulser is employed to determine the cardiogenic frequency, and the largest spectral peak above background between 0.024 and 0.147 is taken to be the vasomotor frequency.

Twenty-nine patients scheduled for breast biopsy were recruited into the initial study and imaged the same day prior to the biopsy (Button *et al.*, 2004). For each breast to be biopsied, a single lateral view of the hanging breast was acquired with the patient lying prone. The current acquisition protocol includes gravity-based breast stabilization (weight of the breast onto a friction plate) to minimize motion and provides two views of each breast. The cardogenic and vasomotor DIR images of the 29 patients previously studied were incorporated into a PowerPoint (Microsoft Corp., Redmond, WA) presentation and were reviewed by two mammographers. Training images (Fig. 155) were included to demonstrate image features associated with malignancy. Detection of breast cancer was based on high and irregular signal or extremely high intensity surrounding superficial vessels on vasomotor or cardiogenic DIRI. Each study was scored: 1—not suspicious; 2—moderately suspicious; and 3—highly suspicious.

Cardiogenic images are particularly clear and easy to understand: NO reduces vasoconstriction leaving arterioles with a larger radius, resulting in a larger cardiogenic signal as previously described. Sample benign and malignant breast cardiogenic dynamic infrared images are compared in Figure 156. The average score of all malignant breasts was 2.13 ± 0.58. For breasts with invasive malignancy the average score was 2.33 ± 0.41. For breasts with benign processes, the average score was 1.64 ± 0.67. Employing a score of 1 as negative, while scores of 2 or 3 are considered positive, the overall sensitivity as scored by both mammographers (invasive cancer and DCIS) using DIRI was 81%, while the overall specificity was 54%. Sensitivity and specificity for invasive cancers was 92% and 53%, respectively (Button *et al.*, 2004).

These breast imaging results demonstrate that vasomotor and cardiogenic dynamic infrared images can be derived with current technology. It appears that both image sets may be useful in assessing disease state as shown here for breast cancer. The observed sensitivity and specificity of DIRI for

Normal breast

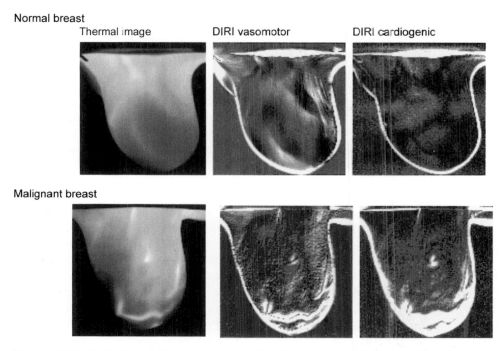

Thermal image DIRI vasomotor DIRI cardiogenic

Malignant breast

Figure 155 Single-frame infrared, vasomotor, and cardiogenic dynamic infrared images of a patient diagnosed with advanced breast cancer. Display window settings are identical for normal and malignant.

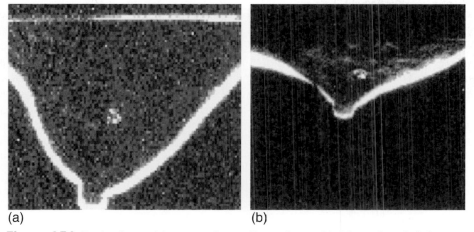

(a) (b)

Figure 156 Benign breast (a) compared to malignant breast (b) (11 mm invasive) for cardiogenic DIRI. In each case a marker was placed on the breast to focus the system, and it appears as an artifact.

cancer detection with a single breast view for a small population are an improvement over conventional mammography. For example, the patient population being studied was composed of patients with positive X-ray mammography findings. Under such conditions, only one in four biopsies is positive. Potential improvements are obvious and have been implemented: breath-hold acquisition, a simultaneous view of both breast to use symmetry, multiple views of both breasts, stabilization, and motion rejection and correction software.

Pitfalls of Dynamic Infrared Imaging

Motion is known to degrade medical image quality, especially when images separated in time are processed. Respiratory and cardiac-related motion particularly degrades image quality in breast DIRI. One essential aspect for minimizing motion is acquisition geometry and protocol. Breath-hold acquisitions are now being conducted and markedly reduce motion-related artifacts. In addition, a stable geometry will reduce motion. A gravity-based breast stabilization system has been designed for this purpose. This system allows the weight of the breast to rest on a friction plate at 45 degrees to vertical, greatly stabilizing the breast. Use of this hardware is currently under study.

Another consideration is image registration or motion correction. Rigid-body motion correction can readily be accomplished. For example, we have applied such correction (translation, rotation, and shear) to dynamic infrared image data in Figure 157 (Jambawalikar *et al.*, 2002). As can be seen, the cardiogenic DIRI has significantly more vascular information and improved SNR as a result. Unfortunately, such correction does not generally work for DIRI because the hanging breast is not a rigid body. Nonrigid correction is currently being examined.

Conclusion: The Future of Infrared and Dynamic Infrared Imaging for Breast Cancer Detection

The goal in screening is an imaging technology that does not employ ionizing radiation and has very high sensitivity and specificity for breast cancer. Dynamic and static infrared imaging may provide this ideal. Clearly, in the time of the BCDDP this was not the case. Sensors were slow, and had poor sensitivity and stability. Research funded by the military has removed these difficulties and provided dynamic capabilities.

Infrared imaging does not employ ionizing radiation, so its use for screening can begin at an earlier age and at greater frequency without risk. Infrared appears to be equally effective even in dense breasts (Keyserlingk *et al.*, 2000), allowing effective breast cancer detection for premenopausal women. While X-ray mammography has been shown to be effective in reducing breast cancer mortality in women ages 50–69, it is less clear for women in their forties (Anonymous, 1997). The limitations of X-ray mammography for women in their forties include poor sensitivity, 55% for film screen and 70% for digital (Pisano *et al.*, 2005), and a false-positive rate of 3 in 4 in general and as high as 7 in 8 for a dense breast (Anonymous, 1997), leading to unnecessary (in retrospect) breast biopsies. Finally, earlier screening is associated with increased risk of radiation-induced malignancy (Berrington de Gonzalez and Reeves, 2005). These facts are particularly distressing given that breast cancer is the single leading cause of death for women ages 40–49 in the United States (Anonymous, 1997), and breast cancer genesis may often be premenopausal (Simpson *et al.*, 1988).

It is important to have some perspective on the magnitude of this failure of X-ray mammography for premenopausal women. In the United States in 2005, ~ 60,000 new breast cancers were detected in women < 50 years (Anonymous, 2006). Of these, 46,000 were invasive and 14,000 were *in situ* cancers (Anonymous, 2006). Based on the fact that 92% of mammographic systems in this country are film-screen and 8% are digital X-ray mammography systems, ~ 26,280 cancers are currently *missed* in women younger than 50 years of age. In addition, X-ray mammography results in ~ 290,000 unnecessary (in retrospect) breast biopsies each year in this population. Unnecessary biopsies have complications, including hemorrhage, abscess, pain, missed lesions, and complications of general anasthesia for patients unsuitable for local anasthesia. In addition to complications, invasive breast procedures can leave scars that make subsequent mammographic interpretation difficult in nearly half of patients. These false-positive results lead to tremendous unnecessary anxiety and expense. While stereotactic and ultrasound-guided breast biopsy have reduced the morbidity of definitive tissue

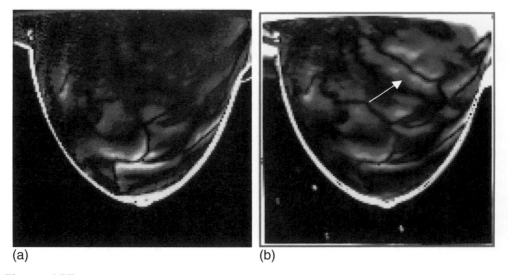

(a)   (b)

Figure 157 Comparison of uncorrected (a) to motion corrected (b) breast cardiogenic DIRI.

diagnosis, they may lead to prolonged and additional mammographic surveillance and subsequent anxieties. There is clearly room for improvement.

Early infrared technology was generally reported as having a sensitivity of 85% and a specificity of 87% (Jones, 1983). With this early technology, 17,280 of those missed cancers could be detected each year! On this basis, it is difficult to understand why infrared plays no role in screening. It is simpler to understand, however, if you look at the situation from the clinician's perspective. If a patient has a positive X-ray mammogram, the lesion in question is visible and can be sampled (biopsied). In the case of infrared, which is physiologically based and not anatomical, a positive finding could be very distressing, particularly if the X-ray mammogram was negative. The clinician is stuck with knowledge that there may be a malignancy in the breast, but cannot visualize it or sample it. It is my conjecture that for this reason early infrared was not acceptable clinically.

For infrared or dynamic infrared imaging to be useful, an equally sensitive anatomic imaging technology is required. Fortunately, dynamic contrast-enhanced (DCE) magnetic resonance (MR) is developing into that role (Huang *et al.*, 2004). Simple DCE-MR is in routine application but has some limitations. As a screening tool, it is not practical; it is too expensive and too complex. It has near ~ 100% sensitivity and ~ 70% specificity, which, with the addition of magnetic resonance spectroscopy (MRS), improves to 87% (Huang *et al.*, 2004). Even for ductal carcinoma *in situ* (DCIS), DCE-MR can improve the surgical planning and provides a sensitivity of 86% and a positive predictive value of 84% (Chung *et al.*, 2005).

A change in the breast cancer screening paradigm for premenopausal women from digital X-ray mammographic screening followed by biopsy for positive findings to infrared followed by DCE-MR and MRS may have a significant impact. In women under 50 years of age, there would be 9000 fewer missed cancers per year and 210,000 fewer unnecessary (in retrospect) breast biopsies per year at a "cost" of 900,000 DCE-MR studies per year. These projections may improve significantly with computer-aided IR breast cancer detection and dynamic infrared capabilities as discussed here.

Implementation of a new imaging technology often requires considerable time for adequate training of support and physician staff. Existing imaging technologists, such as radiologic, ultrasound, or nuclear medicine technologists, could easily be trained to competently acquire image data using a QWIP IR imaging system. Computer-aided detection (CAD) would probably be an important aspect of clinical integration. Such work applied to static and dynamic infrared imaging is underway (Jambawalikar, 2005).

Acknowledgments

The author acknowledges support from Advanced Biophotonics, Inc., and the New York State Center for Biotechnology. The author holds a financial interest in Advanced Biophotonics, Inc. The author acknowledges useful discussions and advice from Dr. Haifang Li and the technical assistance of Veronica Geronimo-Aghera in preparation of this chapter

References

Anonymous. 1978. National Institutes of Health/National Cancer Institute consensus development meeting on breast cancer screening: issues and recommendations. *J. Natl. Cancer Inst. 60*:1519–1521.

Anonymous. 1979. Report of the working group to review the national cancer institute breast cancer detection demonstration projects. *J. Natl. Cancer Inst. 62*:641–709.

Anonymous. 1997. National Institutes of Health Consensus Development Conference Statement: Breast Cancer Screening for Women Ages 40–49, January 21–23, 1997. National Institutes of Health Consensus Development Panel. *J. Natl. Cancer Inst. 89*: 1015–1026.

Anonymous. 2006. *Breast Cancer Facts and Figures 2005–2006.* Atlanta, GA: American Cancer Society.

Anbar, M. 1994. Hyperthermia of the cancerous breast: analysis of mechanism. *Cancer Lett. 84*:23–29.

Anbar, M., Brown, C., and Milescu, L. 1999. Objective identification of cancerous breasts by dynamic area telethermometry (DAT). *Thermology Int. 9* 127–133.

Anbar, M., Milescu, L., Naumov, A., Brown, C., Button, T., Carty, C., and Dulaimy, K. 2001. Detection of cancerous breasts by dynamic area telethermometry. *IEEE Eng. Med. Biol. Mag. 20*:80–91.

Andrade, S.P., Bakhle, Y.S., Hart, I., and Piper, P.J. 1992. Effects of tumour cells on angiogenesis and vasoconstrictor responses in sponge implants in mice. *Br. J. Cancer. 66*:821–825.

Berrington de Gonzalez, A., and Reeves, G. 2005. Mammographic screening before age 50 years in the U.K.: comparison of the radiation risks with the mortality benefits. *Br. J. Cancer 93*:590–596.

Breiter, R., Cabanski, W., Ziegler, J., Walther, M., and Schneider, H. 2000. High-performance focal plane array modules for research and development. *SPIE AeroSense 4020*:257–266.

Button, T.M., Li, H., Fisher, P., Rosenblatt, R., Dulaimy, K., Li, S., O'Hea, B., Salvitti, M., Geronimo, V., Geronimo, C., Jambawalikar, S., Carvelli, P., and Weiss, R. 2004. Dynamic infrared imaging for the detection of malignancy. *Phys. Med. Biol. 49*:3105–3116.

Chung, A., Sacuaf, R., Scharre, K., and Phillips, E. 2005. The impact of MRI on the treatment of DCIS. *Am. Surg. 71*:705–710

Dewhirst, M.W. 1998. Concepts of oxygen transport at the microcirculatory level. *Semin. Radiat. Oncol. 8*:143–150.

Ecker, R.D., Goerss, S.J., Meyer, F.B., Cohen-Gadol, A.A., Britton, J.W., and Levine, J.A. 2002. Vision of the future: initial experience with intraoperative real-time high-resolution dynamic infrared imaging. *J. Neurosurg. 97*:1460–1471.

Eriksen, M., and Lossius, K. 1995. A causal relationship between fluctuations in thermoregulatory skin perfusion and respiratory movements in man. *J. Auton. Nerv. Syst. 53*: 223–229.

Gamagami, P. 1996. Indirect signs of breast cancer: Angiogenesis study. *Atlas of Mammography*, Cambridge, MA, Blackwell Science. pp. 231–258.

Garbade, J., Ullmann, C., Hollenstein, M., Barten, M.J., Jacobs, S., Dhein, S., Walther, T., Gummert, J.F., Falk, V., and Mohr, F.W. 2006. Modeling of temperature mapping for quantitative dynamic infrared coronary angiography for intraoperative graft patency control. *J. Thorac. Cardiovasc. Surg. 131*:1344–1351.

Gautherie, M., Haehnel, P., Walter, J.P., and Keith, L.G. 1987. Thermovascular changes associated with *in situ* and minimal breast cancers: results of an ongoing prospective study after four years. *J. Reprod. Med. 32*:833–842.

Gershen-Cohen, J., Haberman, J., and Brueschke, E.E. 1965. Medical thermography: a summary of current status. *Radiol. Clin. North Am. 3*:403–431.

Gunapala, S.D., Liu, J.K., Park, J.S., Sundaram, M., Shott, C.A., Hoelter, T., Lin, T.L., Massie, S.T., Maker, P.D., Muller, R.E., and Sarusi, G. 1997. 9.mu.m cutoff 256 × 256 GaAs/AlxGa.sub.1-x As quantum well infrared photodetector focal plane array camera. *IEEE Trans. Electron Devices 44*:51–57.

Huang, W., Fisher, P.R., Dulaimy, K., Tudorica, L.A., O'Hea, B., and Button, T.M. 2004. Detection of breast malignancy: diagnostic MR protocol for improved specificity. *Radiology 232*:585–591.

Isard, H.J. 1984. Other imaging techniques. *Cancer 53* (Suppl.):658–664.

Jambawalikar, S. 2005. Dissertation: application of texture analysis to dynamic contrast enhanced breast magnetic resonanc imaging. Stony Brook University, Stony Brook, NY.

Jambawalikar, S., Li, H., and Button, T. 2002. Automatic registration of breast images for dynamic infrared imaging (#185). First Annual Meeting of the Society of Molecular Imaging. *Mol. Imag. 1*:211.

Janicek, M.J., Merriam, P., Potter, A., Silberman, S., Dimitrijevic, S., Fauci, M., and Demetri, G.D. 2002. Imaging responses to Imatinib mesylate (Gleevec, STI571) in gastrointestinal stromal tumors (GIST): vascular perfusion patterns with Doppler ultrasound (DUS) and dynamic infrared imaging (DIRI). *ASCO* abstract 333.

Jones C.H. 1983. Thermography of the female breast. In: Parsons, C.A. (Ed.), *Diagnosis of Breast Disease*. Baltimore, MD: University Park Press, pp. 214–234.

Keyserlingk, J.R., Ahlgren, P.D., Yu, E., Belliveau, N., and Yassa, M. 2000. Functional infrared imaging of the breast. *J. IEEE Eng. Med. Biol. 19*:30–41.

Kvernmo, H.D., Stefanovska, A., Bracic, M., Kirkeboen, K.A., and Kvernebo, K. 1998. Spectral analysis of the laser Doppler perfusion signal in human skin before and after exercise. *Microvasc. Res. 56*:173–182.

Lloyd-Williams, K., and Handley, R. 1961. Infrared thermometry in the diagnosis of breast cancer. *Lancet 2*:1378–1381.

Loibl, S., Buck, A., Strank, C., von Minckwitz, G., Roller, M., Sinn, H.P., Schini-Kerth, V., Solbach, C., Strebhardt, K., and Kaufmann, M. 2005. The role of early expression of inducible nitric oxide synthase in human breast cancer. *Eur. J. Cancer. 41*:265–271.

Moskowitz, M., Milbrath, J., Gartside, P., Zermeno, A., and Mandel, D. 1976. Lack of efficacy of thermography as a screening tool for minimal and stage I breast cancer. *N. Engl. J. Med. 295*:249–252.

Pisano, E.D., Gatsonis, C., Hendrick, E., Yaffe, M., Baum, J.K., Acharyya, S., Conant, E.F., Fajardo, L.L., Bassett, L., D'Orsi, C., Jong, R., and Rebner, M., Digital Mammographic Imaging Screening Trial (DMIST) Investigators Group. 2005. Diagnostic performance of digital versus film mammography for breast-cancer screening. *N. Engl. J. Med. 353*:1773–1783.

Riechelmann, H., and Krause, W. 1994. Autonomic regulation of nasal vessels during changes in body position. *Eur. Arch. Otorhinolaryngol. 251*:210–213.

Simpson, H.W., Candlish, W., Pauson, A.W., McArdle, C.S., Griffiths, K., and Small, R.G. 1988. Genesis of breast cancer is in the premenopause. *Lancet. 2(8602)*:74–76.

Terborg, C., Gora, F., Weiller, C., and Rother, J. 2000. Reduced vasomotor reactivity in cerebral microangiopathy: a study with near-infrared spectroscopy and transcranial Doppler sonography. *Stroke 31*:924–929.

Threatt, B., Norbeck, J.M., Ullman, N.S., Kummer, R., and Roselle, P.F. 1980. Thermography and breast cancer: an analysis of a blind reading. *Ann. NY Acad. Sci. 335*:501–527.

Tooke, J.E., and Williams, S.A. 1987. Capillary blood pressure. *Adv. Exp. Med. Biol. 220*:209–214.

Wardell, K., Jakobsson, A., and Nilsson, G.E. 1993. Laser Doppler perfusion imaging by dynamic light scattering. *IEEE Trans. Biomed. Eng. 40*:309–316.

Xie, W., McCahon, P., Jakobsen, K., and Parish, C. 2004. Evaluation of the ability of digital infrared imaging to detect vascular changes in experimental animal tumours. *Int. J. Cancer 108*:790–794.

32

Phyllodes Breast Tumors: Magnetic Resonance Imaging

Aimée B. Herzog and Susanne Wurdinger

Introduction

Phyllodes tumors are rare fibroepithelial tumors of the female breast, accounting for 2–3% of fibroepithelial lesions (Liberman et al., 1996) and 0.3–1% of all breast neoplasms (Rowell et al., 1993). The fundamental characteristic of this special kind of intracanalicular fibroadenoma is the epithelial and stromal component similar to benign fibroadenomas, but the stromal component is neoplastic. Areas of necrosis, hemorrhage, or cystic spaces develop due to fast enlargement. Usually, phyllodes tumors become conspicuous as a result of their rapid growth. They are divided into benign, borderline, or malignant subtypes. Both benign and malignant phyllodes tumors have a tendency to recur if not completely excised (Parker and Harries, 2001). Wide local excision margin or mastectomy, if necessary, is currently the preferred surgical treatment for all subtypes of this neoplasia (Kapiris et al., 2001). One-third of this neoplasia is able to spread; most of the metastases are found in the lungs, and one-third spreads into the bones (Contarini et al., 1982). Phyllodes tumors arise mostly in women aged between 35 and 55 years (Reinfuss et al., 1996).

On mammography, phyllodes tumor manifests as a well-defined round or oval mass. Plaquelike, coarse calcifications are rarely seen (Page and Williams, 1991). A circumscribed, inhomogeneous, solid-appearing mass that can contain cystic spaces with well-defined margins is the most common manifestation on ultrasound (Chao et al., 2002). A typical phyllodes tumor has smooth walls, is oval or lobulated, may have intramural cysts, and exhibits internal echoes. Particularly in small phyllodes tumors, sonography and mammography do not allow distinction of fibroadenomas or breast cancer and a phyllodes breast tumor.

Diagnostic tools for breast imaging were extended by magnetic resonance imaging (MRI), and analysis of lesions could be refined. The value of dynamic breast MRI for detection of multifocal and bilateral breast cancer has been proven (Fischer et al., 2004). Despite its limited specificity, dynamic breast MRI helps to differentiate between benign and malignant breast lesions. Description of evaluated characteristics may lead to better specification of phyllodes breast tumors on MR imaging. However, the number of phyllodes breast tumors reported on MRI is limited. For specification there is given an overview of morphological signs of phyllodes breast on dynamic MRI evaluated in studies.

581

Magnetic Resonance Imaging

Morphology

The first MRI case report on the benign phyllodes tumor was by Grebe et al. (1992). The study described one 9 cm lesion. The benign tumor had well-defined margins and appeared visibly lobulated. In other MRI case studies, benign phyllodes tumors were described as oval or multilobulated, solid masses with septations (Farria et al., 1996; Ogawa et al., 1997). Eight patients with benign phyllodes tumors that were more than 3 cm in size were imaged as spotted, inhomogeneous tumors with cystic components or septations inside (Kinoshita et al., 2004). In 2002 Cheung et al. (2002) reported a recurrent phyllodes tumor of borderline malignancy that measured 14 cm. It showed leaflike projections, which can be interpreted as heterogeneous inner structure with septations. In the fat-suppressed technique, it showed solid and cystic areas. Further information on dynamic breast MRI was described in a study with 24 phyllodes breast tumors (Wurdinger et al., 2005). Most phyllodes breast tumors had smooth margins (87.5%, $n = 21$), were round or lobulated (100%, $n = 24$), and had a heterogeneous internal structure (70.8%, $n = 17$); 11 lesions (45.8%) showed nonenhancing septations.

Tumor size of benign phyllodes breast tumors ranged from 9–50 mm (average size 23 mm). The malignant tumor was 75 mm. Five tumors were smaller than 10 mm, 11 tumors were between 10 and 20 mm, and 8 tumors were larger than 20 mm. The tumors showed different morphological features depending on the size. Therefore, tumors were distributed in:

Category (1) tumors $\leq$ 10 mm (5 lesions)
Category (2) tumors > 10 mm, $\leq$ 20 mm (11 lesions)
Category (3) tumors > 20 mm, $\leq$ 30 mm (4 lesions)
Category (4) tumors > 30 mm (4 lesions)

Typical signs like lobulated shape, internal septations, and heterogeneous structure developed when the tumor reached a certain size. Tumors up to 10 mm in size were circumscribed in 60%, structured heterogeneous in 60%, and lobulated in 20%. In category (4) > 30 mm, all lesions were circumscribed and heterogeneous and lobulated in 75%. So, the phyllodes tumors > 30 mm had the most typical shape. We saw a primary phyllodes tumor (20 mm) that had nonenhancing septations. But its benign recurrence (15 mm) did not have any septation. Forty percent of the lesions < 10 mm had septations; septations were shown in 45.5% of the lesions between 10 mm and 20 mm and 75% between 20 mm and 30 mm, respectively. However, 50% of the lesions > 30 mm had no septations. Tumors, also phyllodes tumors, reveal regressive changes after reaching a certain size. Phyllodes tumors often grow very fast (Feder et al., 1999). Rapid tumor growth causes an insufficient vascular situation and may result in focal necrosis with heterogeneous appearance.

Therefore, the typical phyllodes tumor is characterized by smooth margins, a round or lobulated shape in combination with heterogeneous internal structure. If a benign imposing neoplasia shows heterogeneous structure and septations, a phyllodes tumor should be considered. However, fibroadenoma is also possible. It often shows septations (Orel et al., 1994). In patients under 50 years with a family history of breast cancer, the presence of a medullary carcinoma should be considered if a well-circumscribed lesion is seen (Kuhl, 1998).

Signal Intensity

Phyllodes breast tumors are hypointense or isointense lesions compared with adjacent breast tissue on T1-weighted images (Grebe et al., 1992; Farria et al., 1996). On T2-weighted images, one-third had a hyperintense signal. Two-thirds (16 phyllodes breast tumors out of 24) were hypointense, and three of these showed bright cystic areas inside the tumor (Wurdinger et al., 2005). Several analyses have shown that imaged tumors had low signal intensity compared with fibroadenoma (Kinoshita et al., 2004) or have shown inhomogeneous signal intensity, including a hypointense area and internal septation in the solid portion on T2-weighted images (Kitamura et al., 1999). Other authors also found heterogeneous, partly high signal intensity (Grebe et al., 1992; Buadu et al., 1996; Farria et al., 1996). Usually, phyllodes tumors have low signal intensity in native T1-weighted scans and in a high number in T2-weighted images.

Using the fat-suppressed technique, Cheung et al. (2002) reported hemorrhage and cystic spaces as signs of regression in the solid mass, which were confirmed by histopathology, and they suggested that these spaces can be pathognomonic for phyllodes tumors. In T1- and T2-weighted sequences, phyllodes tumors showed inhomogeneous signal intensity for the presence of cystic areas with internal septation and hemorrhage (Kitamura et al., 1999; Kinoshita et al., 2004; Franceschini et al., 2005). Hence, cysts and hemorrhage are supposed to be an effect of rapid growth and size, with regressive changes occurring in large tumors. Therefore, these appearances can be assessed as typical, but not clearly as pathognomonic for phyllodes tumors. Intratumoral cysts were occasionally observed in other well-defined lesions, such as medullary carcinoma or fibroadenoma (Harper et al., 1983).

Grebe et al. (1992) described a large benign phyllodes tumor showing a border of high signal intensity with high share of water near the chest wall. This could be interpreted as perifocal edema. A surrounding edema could be described in 16.7% of the tumors in T2-weighted images in our study. These perifocal edemas were found around tumors > 20 mm. The displacing effect of the tumor is exceptionally strong while its phase of growing is ongoing. Lymphatic vessels are pressed and sealed. Phyllodes tumors can measure 3–5 cm at time of diagnosis (Bick, 2000). It is caused by sudden rapid growth, which mostly leads to the clinical manifestation (Feder et al., 1999).

Contrast-enhancement Characteristics

Benign, borderline, or malignant phyllodes tumors were observed (Grebe et al., 1992; Buadu et al., 1996; Cheung et al., 2002) and had a heterogeneous enhancement of gadolinium-diethylenetriaminepentaacetate (DTPA). Buadu et al. (1996) reported one malignant phyllodes tumor with a plateau-phenomenon on dynamic breast MRI. Several authors demonstrated rapidly and markedly enhancing multilobulated lesions (Farria et al., 1996; Ogawa et al., 1997; Franceschini et al., 2005). Farria et al. (1996) described an increase of signal intensity between 50 and 75% during the first minute after gadolinium injection. Some authors described gradually and rapidly enhancing types of time-signal intensity curves (Kitamura et al., 1999; Kinoshita et al., 2004).

In our study (Wurdinger et al., 2005), three types of contrast medium enhancement were differentiated: the type 1 enhancement pattern indicated benign lesions, and types 2 and 3 patterns were suggestive of malignancy. In the type 1 pattern, two-thirds of phyllodes breast tumors were classified as lesions with slow initial contrast enhancement and a persistent delayed phase. The type 2 contrast enhancement pattern, with fast initial and plateau phases, was seen in 12% of phyllodes tumors. Twenty percent were classified as the type 3 pattern of contrast enhancement, with a fast initial phase and a washout phenomenon. The malignant tumor belonged to those with inhomogeneous enhancement and washout phenomenon. The phyllodes tumors with suspicious courses of contrast enhancement had a fast signal increase of contrast medium comparable to the descriptions of Farria et al. (1996). The study of Kitamura et al. (1999) showed cases of the type 1 with histopathological apparent hemorrhage, necrosis, cystic dilatation, and a less densely stromal component than the rapid type.

Buadu et al. (1996) proposed a different contrast-enhancement pattern for benign and malignant phyllodes tumor. They described two benign phyllodes tumors that had an amount of microvessels comparable to those in malignant tumors in the same report. Among the malignant lesions the highest mean microvessel density was seen in the malignant phyllodes tumor. Compared to the other benign neoplasia, the two cystadenomas showed a high mean signal increase; compared to the malignant lesions, the cystosarcoma had the highest mean signal increase after application of the contrast medium. However, a differentiation of benign and malignant phyllodes tumors by contrast-enhancement pattern could not be found in our study with one malignant and seven benign phyllodes tumors with suspicious contrast enhancement. This review confirms that the vast majority of phyllodes tumors enhance the contrast medium in an inhomogeneous way. Slow and fast dynamic patterns are described and may reflect changes in tumor nutrition. The findings suggest that phyllodes tumors more often reveal suspicious signal courses such as rapid increase and following plateau or washout phenomenon than other benign breast lesions.

The phyllodes tumor that shows a continuous signal increase can be misdiagnosed as another benign lesion; that is, as fibroadenomas, which have up to an 80% monophasic course of contrast enhancement (Heywang-Köbrunner and Beck, 1996; Kuhl et al., 1999). Furthermore, up to 2% of breast carcinomas have a slow contrast enhancement, in particular lobular and medullary carcinomas (Wurdinger et al., 2001).

In conclusion, the results suggest that the phyllodes tumor described by breast MRI is a well-circumscribed lobulated lesion with septations or inhomogeneous internal structure. A typical sign is the presence of cystic areas that can contain hemorrhage. Mostly, the solid mass has low signal intensity in the native T1-weighted images and in the T2-weighted images. The contrast enhancement is inhomogeneous and mainly continuous, but more frequently than in other benign lesions; the tumor shows suspicious contrast enhancement. An edema surrounding the tumor may be observed. Small and large phyllodes tumors can have different structures. The large tumors more often show sharp margins and a lobulated shape; they are inhomogeneous before the injection of contrast medium, and they have more frequent septations. The most important differentiations of well-defined phyllodes tumor are the fibroadenoma and circumscribed breast carcinoma similar to medullary or papillary cancer.

Guidelines

The preoperative diagnosis of phyllodes tumor of the breast is still a challenge for gynecologists and radiologists. Dynamic breast MRI is a useful investigation tool for diagnosis before using interventional methods. The images in slices can contribute to a completely defined tumor expansion and clearly defined margins, even near the chest wall. Therefore, the extension of the operative intervention can be exactly determined preoperatively for an optimal surgical management. Our study may help lead to a better understanding of the MR imaging characteristics of the phyllodes tumor and describes the criteria in which it should be considered. The outlined features are summarized in Table 23. Guidelines are shown in order to define the important marks and to improve the preoperative diagnosis.

Table 23 Guidelines: Features of Typical Phyllodes Tumor in Breast MRI

▲ Patient > 35 years
▲ Sudden growth
▲ Solid mass
▲ Well-circumscribed margins
▲ Round or lobulated shape
▲ Heterogeneous internal structure with septations, hemorrhage, cystic areas
▲ Low signal intensity in T1- and T2-weighted images
▲ Surrounding, perifocal edema
▲ Inhomogeneous enhancement of the contrast medium
▲ Continuous or rapid enhancement of contrast agent

References

Bick, U. 2000. Typische und ungewöhnliche Befunde in der MR-Mammographie. *Rofo Fortschr. Geb. Röntgenstr. Neuen Bildgeb. Verfahr.* 172:415–428.

Buadu, L.D., Murakami, J., Murayama, S., Hashiguchi, N., Sakai, S., Masuda, K., Toyoshima, S., Kuroki, S., and Ohno, S. 1996. Breast lesions: correlation of contrast medium enhancement patterns on MR images with histopathologic findings and tumor angiogenesis. *Radiology* 200:639–649.

Chao, T.C., Lo, Y.F., Chen, S.C., and Chen, M.F. 2002. Sonographic features of pyllodes tumors of the breast. *Ultrasound Obstet. Gynecol.* 20:64–71.

Cheung, H.S., Tse, G.M.K., and Ma, T.K.F. 2002. "Leafy" pattern in phyllodes tumour of the breast: MRI-pathologic correlation. *Clin. Radiol.* 57:230–236.

Contarini, O., Urdaneta, L.F., Hagan, W., and Stephenson, S.E., Jr. 1982. Cystosarkoma phylloides of the breast: a new therapeutic proposal. *Am. Surg.* 48:157–166.

Farria, D.M., Gorczyca, D.P., Barsky, S.H., Sinha, S., and Basset, L.W. 1996. Benign phyllodes tumor of the breast: MR imaging features. *AJR Am. J. Roentgenol.* 167:187–189.

Feder, J.M., de Paredes, E.S., Hogge, J.P., and Wilken, J.J. 1999. Unusual breast lesions: radiologic-pathologic correlation. *Radiographics* 19:11–26.

Fischer, U., Zachariae, O., Baum, F., von Heyden, D., Funke, M., and Liersch, T. 2004. The influence of preoperative MRI of the breasts on recurrence rate in patients with breast cancer. *Eur. Radiol.* 14:1725–1731.

Franceschini, G., D'Ugo, D., Masetti, R., Palumbo, F, D'Alba, P.F., Mule, A., Costantini, M., Belli, P, and Picciocchi, A. 2005. Surgical treatment and MRI in phyllodes tumors of the breast: our experience and review of the literature. *Ann. Ital. Chir.* 76:127–140.

Grebe, T., Wilhelm, K., Brunier, A., and Mitze, M. 1992. MR-Tomographie des Cystosarkoma phylloides. Ein Fallbeispiel. *Acta Radiol.* 2:376–378.

Harper, A.P., Kelly-Fry, E., Noe, J.S., Bies, J.R., and Jackson, V.P. 1983. Ultrasound in the evaluation of solid breast masses. *Radiology* 146:731–736.

Heywang-Köbrunner, S.H., and Beck, R. 1996. *Contrast-enhanced MRI of the Breast*, 2nd ed. Berlin, Germany: Springer-Verlag, pp. 59–117.

Kapiris, I., Nasiri, N., A'Hern, R., Healy, V., and Gui, G.P.H. 2001. Outcome and predictive factors of local recurrence and distant metastases following primary surgical treatment of high-grade malignant phyllodes tumours of the breast. *Eur. J. Surg. Oncol.* 27:723–730.

Kinoshita, T., Fukutomi, T., and Kubochi, K. 2004. Magnetic resonance imaging of benign phyllodes tumors of the breast. *Breast J.* 10:232–236.

Kitamura, K., Makino, S., Yoshimura, H., Jyouko, T., Yoshioka, H., and Hanai, J. 1999. Phyllodes tumor of the breast with cystic portion: MR imaging with histopathological correlation. *Nippon Igaku Hoshasen Gakkai Zasshi* 59:239–244.

Kuhl, C.K. 1998. MRI of the breast. *Advances in MRI Contrast* 5:56–69.

Kuhl, C.K., Mielcarek, P., Klaschik, S., Leutner, C., Wardelmann, E., Gieseke, J., and Schild, H.H. 1999. Dynamic breast MR imaging: are signal intensity time course data useful for differential diagnosis of enhancing lesions? *Radiology* 211:101–110.

Liberman, L., Bonaccio, E., Hamele-Bena, D., Abramson, A.F., Cohen, M.A., and Dershaw, D.D. 1996. Benign and malignant phyllodes tumors: mammographic and sonographic findings. *Radiology* 198:121–124.

Ogawa, Y., Nishioka, A., Tsuboi, N., Yoshida, D., Inomata, T., Yoshida, S., Moriki, T., and Toki, T. 1997. Dynamic MR appearance of benign phyllodes tumor of the breast in a 20-year-old woman. *Radiat. Med.* 15:247–250.

Orel, S.G., Schnall, M.D., LiVolsi, V.A., and Troupin, R.H. 1994. Suspicious breast lesions: MR imaging with radiologic-pathologic correlation. *Radiology* 190:485–493.

Page, J.E., and Williams, J.E. 1991. The radiologic features of phyllodes tumor of the breast with clinico-pathological correlation. *Clin. Radiol.* 44:8–12.

Parker, S.J., and Harries, S.A. 2001. Phyllodes tumours. *Postgrad. Med. J.* 77:428–435.

Reinfuss, M., Mitus, J., Duda, K., Stelmach, A., Rys, J., and Smolak, K. 1996. The treatment and prognosis of patients with phyllodes tumor of the breast: an analysis of 170 cases. *Cancer* 77:910–916.

Rowell, M.D., Perry, R.R., Hsiu, J.G., and Barranco, S.C. 1993. Phyllodes tumors. *Am. J. Surg.* 165:376–379.

Wurdinger, S., Herzog, A.B., Fischer, D.R., Marx, C., Raabe, G., Schneider, A., and Kaiser, W.A. 2005. Differentiation of Phyllodes breast tumors from fibroadenomas on MRI. *AJR Am. J. Roentgenol.* 185:1–5.

Wurdinger, S., Kamprath, S., Eschrich, D., Schneider, A., and Kaiser, W.A. 2001. False-negative findings of malignant breast lesions on preoperative magnetic resonance mammography. *Breast* 10:131–139.

Index

B